HANDBOOK OF PLANT CELL CULTURE,
Volume 1

Techniques for Propagation and Breeding

KT-447-976

Editors

David A. Evans
DNA Plant Technology Corporation

William R. Sharp
DNA Plant Technology Corporation

Philip V. Ammirato
Barnard College, Columbia University
DNA Plant Technology Corporation

Yasuyuki Yamada
Kyoto University

Macmillan Publishing Co.
A Division of Macmillan, Inc.
NEW YORK

Collier Macmillan Publishers
LONDON

Copyright © 1983 by Macmillan Publishing Co.
A Division of Macmillan, Inc.

Macmillan Publishing Co.
A Division of Macmillan, Inc.
866 Third Avenue, New York, N. Y. 10022

Collier Macmillan Canada, Inc.

Printed in the United States of America

printing number
1 2 3 4 5 6 7 8 9 10

Library of Congress Catalog Card Number: 82-73774

ISBN 0-02-949230-0

Library of Congress Cataloging in Publication Data

This treatise is dedicated to our mentors:

Elton F. Paddock (DAE)

James E. Gunckel (WRS)

F. C. Steward (PVA)

Azuma Okuda (YY)

£55·0

HANDBOOK OF
PLANT CELL CULTURE

In Preparation

Volume 2, Crop Species
Editors: William R. Sharp, David A. Evans, Philip V. Ammirato, Yasuyuki
Yamada

Volume 3, Crop Species
Editors: Philip V. Ammirato, David A. Evans, William R. Sharp, Yasuyuki
Yamada

Contents

Section II. Molecular Techniques

PART C: MODIFICATIONS AND APPLICATIONS

Preface

The past few years have witnessed a dramatic increase in our ability to manipulate and study plant cells in culture. The variety of techniques that collectively comprise "plant cell culture" have permitted investigations at many levels--molecular, cellular, and organismal--and have been applied to a range of disciplines--biochemistry, genetics, physiology, anatomy, and cell biology. Some of the most important tools in modern genetics have emerged from work with cultured plant cells. These have not only permitted a better understanding of Mendelian genetics and cytoplasmic inheritance, but have also resulted in new and unique organisms that could not have been derived from sexual crosses being developed from anther culture and protoplast fusion products.

The growing realization of the potentialities of plant cell culture for plant propagation and breeding has itself provided a substantial impetus for research. Commercial plant propagation using shoot tip cultures is now being implemented, new breeding lines have been selected using protoplast fusion, and new varieties have been developed via somaclonal variation and from plants derived from cultured anthers. Furthermore, the increasing competence to manipulate plant DNA and genes has led to the appreciation that cell culture will be necessary to recover modified plants. All in all, it is now readily apparent that cell culture is the keystone to progress in plant biotechnology. The application and development of current techniques is opening the door to a second green revolution.

Although many of the recent developments in this field have been chronicled in the proceedings of scientific meetings and the publications in refereed journals, no unified summary of this exciting field is available. To remedy this, we have tried to bring together in this treatise much of the information that until now has been scattered. Three aspects of recent research serve as focal points. First, as many developments have been reported in recent years, there is a need for a critical review of the literature. Each chapter contains a discussion of key contributions for the topic, including summary tables detailing relevant information. Second, in previous reviews, little emphasis has been placed on the actual techniques employed. This series, in contrast, focuses on each technique, supplying the reader with the detailed protocols that serve as the foundation for current research. Finally, recognizing the tremendous potential for this field in fostering new agricultural developments, emphasis has also been placed on the applications of cell culture to crop improvement.

It was recognized that in a review such as this no uniform conclusions could be made on the value of each technique for crop improvement. Some cell culture techniques, such as shoot tip propagation, can easily be applied to a wide range of crop species, but other techniques, such as protoplast regeneration, are limited to a small number of closely related species. Different plants or plant groups present special problems for propagation or improvement, and cell culture techniques have been directed toward solving these problems, with varying degrees of success. To present such a diverse assortment of topics, we have divided the first three volumes into a book on techniques and applications and two books on crop species. In this first volume, Techniques for Propagation and Breeding, we have included chapters on both basic and specialized techniques, the latter including genetic and molecular procedures. There are also a number of chapters on modifications of cell culture techniques for specific goals and on agricultural applications to which these techniques are being directed. Volumes 2 and 3 will present strategies for the use of the tools for improvement of specific crop species.

It is our hope that this multi-volume treatise will be useful to all corporate and academic scientists and students interested in keeping abreast of this breakthrough area of biotechnology. In addition to specific protocols, key references to each topic have been highlighted to direct the reader to other useful publications. This is especially important, because it is virtually impossible in any compendium to include all relevant information. It is especially difficult to do so in a rapidly developing field where new areas of interest develop in the time between planning and publication. Recognizing this, it is our plan to remedy any oversight and to report emerging techniques in future volumes in this series. It is our objective to develop and maintain a current, comprehensive, and useful summary of research in the exciting area of plant cell culture.

ACKNOWLEDGMENTS

We are most grateful for the continued support of our publication editors at Macmillan: Sarah Greene and Frances Tindall. In addition, we wish to thank Janis Bravo for administrative assistance in collecting and editing manuscripts and corresponding with contributors. We would also like to thank our typists, Susan Dale and Karen Selover, for their excellent talents. Finally, our sincere appreciation goes to the many colleagues who have aided us in one way or another in bringing this treatise to completion.

CONTRIBUTORS

M. Alves de Lima	Department of Plant Pathology, Ohio State University, Columbus, OH
P.V. Ammirato	Department of Biology, Barnard College, Columbia University, New York, NY and DNA Plant Technology Corporation, Cinnaminson, NJ
D. Aviv	Department of Plant Genetics, Weizmann Institute of Science, Rehovot, Israel
Y.P.S. Bajaj	Tissue Culture Laboratory, Punjab Agricultural University, Ludhiana, India
J.E. Bravo	DNA Plant Technology Corporation, Cinnaminson, NJ
M.P. Bridgen	Department of Horticulture, Virginia Polytechnic Institute and State University, Blacksburg, VA
H.A. Collin	Department of Botany, University of Liverpool, Liverpool, United Kingdom
D. Cress	Cell Culture and Nitrogen Fixation Laboratory, USDA, Beltsville, MD
O.J. Crocomo	Department of Biochemistry, University of Sao Paulo, Piracicaba, Brazil
J. Dyck	Prairie Regional Laboratory, National Research Council, Saskatoon, Canada
D.A. Evans	DNA Plant Technology Corporation, Cinnaminson, NJ
B. Finkel	Western Regional Research Center, USDA, Berkeley, CA
M.S. Fitter	Department of Biochemistry, State University of New York, Stony Brook, NY
C.E. Flick	DNA Plant Technology Corporation, Cinnaminson, NJ
Y. Fujita	Research Center, Mitsui Petrochemical Industries, Yamaguchi, Japan
E. Galun	Department of Plant Genetics, Weizmann Institute of Science, Rehovot, Israel
Yu.Yu. Gleba	Institute of Botany, Academy of Sciences of the Ukrain. SSR, Kiev, USSR
C.Y. Hu	Biology Department, William Paterson College of New Jersey, Wayne, NJ
K. Hughes	Department of Botany, University of Tennessee, Knoxville, TN
H.-J. Jacobsen	Institute for Genetics, University of Bonn, Federal Republic of Germany
A.D. Krikorian	Department of Biochemistry, State University of New York, Stony Brook, NY
S.D. Kung	Department of Biological Sciences, University of Maryland - Baltimore County, Catonsville, MD
P.O. Larsen	Department of Plant Pathology, Ohio State University, Columbus, Ohio
C. McDaniel	Department of Biology, Rensselaer Polytechnic Institute, Troy, NY

M. Martin — Department of Plant Pathology, Ohio State University, Columbus, OH

B. Matthews — Cell Culture and Nitrogen Fixation Laboratory, USDA, Beltsville, MD

D.P. Maxwell — Department of Plant Pathology, University of Wisconsin, Madison, WI

H. Medina-Filho — Department of Genetics, Agronomic Institute, Campinas, Brazil

S.A. Miller — DNA Plant Technology Corporation, Cinnaminson, NJ

W. Nitzsche — Max-Planck-Institute for Plant Breeding, Koln, Federal Republic of Germany

N. Ochoa-Alejo — Center for Biological Research, La Paz, Mexico

S.A. O'Connor — Department of Biochemistry, State University of New York, Stony Brook, NY

K. Ohyama — Department of Agricultural Chemistry, University of Kyoto, Kyoto, Japan

V. Raghavan — Department of Botany, Ohio State University, Columbus, OH

B. Reisch — Department of Pomology and Viticulture, NY State Agricultural Experiment Station, Geneva, NY

L.H. Rhoades — Department of Plant Pathology, Ohio State University, Columbus, OH

R.M. Riedel — Department of Plant Pathology, Ohio State University, Columbus, OH

A.P. Ruschel — CENA, University of Sao Paulo, Piracicaba, Brazil

F. Sato — Department of Agricultural Chemistry, Kyoto University, Kyoto, Japan

W.R. Sharp — DNA Plant Technology Corporation, Cinnaminson, NJ

G.L. Staby — Department of Horticulture, Ohio State University, Columbus, OH

F.C. Steward — 1612 Inglewood Dr., Charlottesville, VA

M. Tal — Department of Biology, Ben-Gurion University of the Negev, Beer Sheva, Israel

S.D. Tanksley — Department of Horticulture, New Mexico State University, Las Cruces, NM

J. Ulrich — Western Regional Research Center, USDA, Berkeley, CA

P.B. Vose — International Atomic Energy Agency, Vienna, Austria

P.J. Wang — Institute of Botany, Academia Sinica, Taipei, Taiwan

M. Watts — Department of Botany, University of Liverpool, Liverpool, England

L. Wetter — Prairie Regional Laboratory, National Research Council, Saskatoon, Canada

Y. Yamada — Department of Agricultural Chemistry, University of Kyoto, Kyoto, Japan

Reprinted courtesy of *Omni Magazine* © 1983.

CHAPTER 1
Reflections on Aseptic Culture

F.C. Steward

For me the interest in, and initial contact with, the problems of tissue and cell cultures arose out of a long period (1929-1940) of research upon the recrudescence of active growth and metabolism that occurs in thin slices of potato tuber (only a few cells thick) when they survive in well-aerated water or very dilute single-salt solutions. This directed attention not only to the well-known reactivation of cell division as it occurs in moist air at a potato surface, but to the prior events in the cells. These events centered around (a) respiration, enhanced but limited by access to oxygen and maintained by carbohydrate from starch hydrolysis, (b) protein synthesis from endogenous supplies of nonprotein nitrogen compounds, and (c) the ability of the cells to harness this metabolism and available energy to the uptake of water and accumulation of inorganic salts which occurred concomitantly. Thus the fully mature cells of potato tubers remained, even after a long period of quiescence in the storage organ, capable when appropriately aroused of many of the activities associated with growing cells. All this occurred independently of buds on the tuber which normally give rise to shoots when they grow.

It was a new turn in this ongoing research (after the hiatus of World War II) that led specifically into the main topic here in question. A search was made for experimental material that could be manipulated aseptically in solution and could be maintained either in a state of relative quiescence, on the one hand, or contrasted with the same clonal material prompted to grow rapidly by cell proliferation, on the other. This aim was best satisfied by the use of explants drawn from the storage root of the carrot. These explants, sufficiently minute (less than 3 mg) to be obtained far enough from vascular cambium (ca. 2-3

1

mm), and remote from the vascular supplies to lateral roots, contained only mature *phloem parenchyma* cells which were ideal for the purpose. Their endogenous content of growth promoting stimuli were unable to induce more than a minimal swelling by intake of water, even under the most favorable conditions of oxygen supply and temperature. However, when the tissue explants had free access to a normally complete heterotrophic nutrient medium, containing essential salts, sugars, and vitamins, and to appropriate stimuli to their growth and cell division, they could grow apace. The effective stimuli that released this rapid though unorganized growth were those contained in the natural liquid endosperm of the coconut or in a solution or extract from certain morphologically similar sources. By the use of this system, under standardized conditions especially devised for its experimental control, the many contrasts between the normally quiescent carrot cells and their rapidly growing counterparts were described, and the causal agents in the ambient media that aroused all this activity were investigated (cf. Steward, 1968 and references there cited).

The following general points should be stressed here. Although many surviving mature cells of angiosperms retain an innate capacity to grow, it was realized early that the release of this capacity (i.e., growth induction) and its full expression was not to be attributed solely to any single agent (certainly not to the then ubiquitous auxin, IAA, alone). On the contrary, it results from the interactions of numerous constituents of the ambient media that need to be present acceptably to the plant material in question. Moreover, this "acceptability" must take account of inherent differences that arise from the morphology at the source, from the variety within the species, from the environment during its cultivation, and especially of its state of development. Above all, these factors vary greatly from one specific source to another. All this bred an early scepticism concerning the merit of the highly touted specific "tissue culture" media often identified with the names of their respective advocates. But nature easily overrides whatever limitations or virtues are implied by these numerous specific recipes; nature recognizes no proprietary rights to the governing principles of growth, nor should we.

But the early excursion into the use of aseptically cultured carrot phloem explants to study their growth induction and metabolism would not alone have led to an invitation to write this introduction were it not for another unexpected development. This came when it was realized that the rapidly proliferating small carrot explants, growing at their best rates in slowly rotated special flasks, will spontaneously release free and viable cells to the ambient medium. These cells, which bear little visible resemblance to the cells of the original carrot explants, were first seen to survive, then to divide and, later, to give rise in subcultures to cell populations in which small clusters of cells could either proliferate randomly or grow in an organized way. This organized growth could produce true somatic facsimiles of carrot embryos from which normal carrot plants could be grown (cf. Steward et al., 1961 and references there cited).

Thus a second landmark in this continuing odyssey was reached. The carrot cells, originally developed in situ to fulfill their functions as parenchyma of a secondary phloem of the carrot root, are not only able to fulfill the various metabolic attributes of growing cells as they proliferate, but they may also, when isolated as free cells, develop in an organized way and produce embryos.   The embryos so formed are morphogenetically similar to, and as fully competent as, zygotic embryos.   Hence the totipotentiality of the original cells of the carrot was conceived.   It is this characteristic inherent in plant cells that underlies the uses to which they may be adapted in culture and that is the subject matter of this volume.

Again, however, this potentiality is not to be lightly regarded as released by any one single exogenous factor or condition, for a multiplicity of interacting factors and conditions enter into the successful realization of the somatic embryogenesis, even from carrot cells. Moreover, a few isolated embryo-like forms do not constitute a morphogenetically competent somatic embryonic culture.   When this competence is properly achieved, and expressed, literally thousands of embryos can be developed from aliquots of a given source by spreading cells on semisolid media in small dishes (cf. Steward and Krikorian, 1979).   This initially surprising outcome has been routinely accomplished for carrot innumerable times, repeated with various other umbellifers, accomplished with comparable ease for some plants, but with greater difficulty for others.   And, even yet, some plants, despite careful investigation, are still recalcitrant in this respect (cf. Krikorian, 1982 and references there cited)!

Therefore it should not be lightly assumed that by using any new source, the objectives of true somatic embryogenesis will be accomplished casually or without due and often prolonged preparation.   There is no universally applicable formula, although many guiding principles may be, and have been enunciated.

The events recounted above pose obvious philosophical problems. How do cells in situ, e.g., those in the environment of the carrot root, create such specific morphology and biochemistry whereas other cells, e.g., of the carrot leaf in situ, have equally prescribed but very different characteristics?   Both sets of organ-specific characteristics, whether morphological or biochemical, are essentially inherited.   But when cultures are made from isolated cells, their innate genetic characteristics, which are expressed in situ in such different ways, are often submerged in culture.   Moreover, the highly organ-specific characteristics of plants, which are expressed during normal ontogeny, may be highly sensitive to external environmental controls, as by day and night length or day and night temperatures and their interactions.   This is well illustrated by recent studies, of which one on the potato plant is a case in point (Steward et al., 1981).   Even proliferating carrot explants, free of developing organs and lacking root- and shoot-growing points, may respond significantly to similar environmental stimuli.   All this raises profound questions of the controls over the expression of genetic information during development and its persistence in cells to be reexpressed in clonal somatic embryogenesis.   Insofar as tissue or cell cultures are motivated by recapitulating characteristics seen in the developed plant in isolated unorganized cultures, this may be (and usually is) difficult to achieve without also recreating the appropriate morphological setting which does happen when somatic embryogenesis is induced.   To "turn-on" at will one set of inherited characteristics without the cooperation of others with which they are normally asso- ciated still presents a challenge (cf. Krikorian and Steward, 1969; Steward and Krikorian, 1975, 1979).

From time to time during the course of the events briefly recounted above, their significance has been summarized.   This was done with respect to the all important first steps of growth induction, whether measured in terms of tissue mass or cell numbers or size (Steward and Shantz, 1954); with respect to the causative agents and their inter- actions and to the metabolic responses they induce (Steward et al., 1964); to the visible signs of renewed activity in the cells (Steward et al., 1968); and to comparative studies on materials other than carrot (Steward et al., 1970).   Latterly, the implications of the work on the lines indicated has been placed in its historical setting and also in relation to events of the current scene (Krikorian, 1975; Steward and Krikorian, 1975, 1979).

After all this, therefore, the temptation is to say here: "We now have little to add and less to take away!" Nevertheless, and in fulfill- ment of the charge to write a brief introduction to this volume, the plan is to touch upon selected aspects of the current status of the aseptic culture of isolated segments, explants, cells, or protoplasts from plants.

This whole subject is ripe for review.   It should be seen for its implications for concepts of development and morphogenesis, on the one hand, and with regards to its prospects for agriculture and industrial practices, on the other.   This overview is timely, but seen against the pressing urge for genetic engineering and freewheeling claims for "clonal cultivations," it even becomes imperative.   Aspects that seem to merit attention here are developed below, albeit with the minimum of detail.

NOMENCLATURE: SOURCES AND IDENTITY OF CULTURES

The art of growing, aseptically and heterotrophically, isolated plant parts as explants on appropriate media has become loosely identified as plant tissue culture.   The culture of surviving parts or segments of vascular plants first became familiar under conditions adopted in the necessarily aseptic culture of fungi and bacteria.   As such, it extended the nutritional methods of microbiology to higher plants.   Root tip cul- ture, whether in liquid or on semisolid media, soon raised the problems of the potentially unlimited growth of their tips.   But now the terms "tissue" and "cell culture" embrace procedures and practices which may be applied to plant materials from sources that may cover the entire morphological range.   This range may be from immature seeds, embryos, or buds to segments of the axis that include, preformed, all the compo- nent parts of shoots or roots, to isolates that are so reduced that they may, or purport to, exclude everything but their respective apical meristems, to developing organs or explants from developed organs, to cells, or even to surviving subcellular organelles. The sources of the different cultures should reflect their nomenclature, the techniques used, and the objectives to be achieved; it will also bear upon the order of difficulty their cultivation will encounter.

Two examples of loose terminology and fuzzy thinking, though embedded in time-honored usage, should be mentioned. How often have the phrases "plant tissues" cultured "in vitro" recurred? This is often so even when the plant materials in question are not strictly tissues nor the "in vitro" conditions in any true sense to be contrasted with "in vivo" as these terms are, or were, classically used to distinguish syntheses or processes that do not, or do, require the essential organization that supports life. In this sense, root tip cultures, e.g., of tomato or tobacco, in a minimal liquid medium are at least organ cultures (which comprise all the essentials of the radicle of the embryo) and everything that happens to them thereafter in their potentially unlimited subculture is certainly in vivo, even though the culture medium is encased in a glass (or plastic) vessel! That other favored example of longevity in culture, namely proliferated carrot root cultures seemingly indefinitely continued by their subculture, is equally inappropriately described as "in vitro." The feature common to both of these cultural practices was, and is, that they exposed isolated segments from angiosperms to culture medium which, being aseptic, could contain both inorganic and organic nutrients. The common problem that both systems encountered at the outset was to obtain viable starting materials that were aseptic and to maintain the cultures free of subsequent contamination (the devices and rules for these procedures, though in principle relatively simple, are dealt with and commented upon throughout this volume).

NUTRITION AND THE CAPACITY TO GROW

Thus the arts and cultural practices here in question must furnish to minute starting materials from higher plants the exogenous equivalents of the nutrient requirements they require in situ, whether these would normally derive from photosynthesis in leaves with light or from roots or (perhaps more subtly) from their interactions. But the exogenously supplied nutrients, however elaborately concocted, depend for their success on endogenous cells or centers that are still able to grow, or if this capacity lies dormant, able to be reactivated.

Even embryos, although endowed with all the potentialities to become plants, may have their ability to grow variously restricted by the onset of dormancy. This may occur even when embryos contain all the nutrients required for subsequent growth within cotyledons or swollen parts of their axis, or when they are available from surrounding endosperm via absorbing organs (i.e., situations illustrated by the case of coconut, of legumes, and of buckwheat). Hence dormancy or quiescence once established, totally or partially, may need to be overcome.

GROWTH INDUCTION: THOUGHTS ON PRIOR ORGANIZATION

This need for growth induction is a widespread feature to be faced in this field of aseptic cultivation of plant parts. One needs to know whether a seeming inability of explants to grow is inherent at the site from which cultures were being sought, due, for example, to the accumulation of endogenous growth inhibitors, or it merely reflects the need for a stimulus from adjacent centers (e.g., buds), or some deficiency (or excess) in the culture medium. Any of these situations can occur. However, the general experience is that growth induction occurs more spontaneously in cells of dicotyledonous plants, even in their parenchyma produced secondarily from cambia, than in other plants which lack vascular cambia or secondary growth. This rule-of-thumb generalization, however, should not imply that the great array of monocotyledonous plants, with their agricultural importance, are outside the range of these cultural practices; rather, the means to induce their growth may be more elusive and require more subtle combinations of stimuli to activate rapid proliferations. Also, be it said that to induce growth of explants from different organs may encounter special difficulties or even (as in the often successful use of floral parts, including anthers) some advantages.

However, many proliferated cultures, which are appropriately called tissue cultures, ultimately have a cambial origin, for they may emanate from centers in situ which still contain active or reactivated vascular or cork cambia or from derivative cells which, though quiescent, may be reactivated to divide (e.g., as in many wound-healing phenomena).

By contrast, cultures of angiosperm shoot or root tips should be designated as organized cultures of shoot or root tips. They are in fact analogous to short vegetative cuttings! Here, however, the question arises as to how successfully may *small* explanted tips respond in an organized way solely to exogenously supplied nutrients. Even today only a relatively few angiosperms have easily yielded continuously viable root tip cultures. The successes no doubt trace to the fact that even within a very short distance from its tip, a root, as explanted, has all the essentials and primordia for its continued organized growth. Even when excised very close to the tip, procambial strands which are able to generate deep-seated cells that produce the lateral organs (roots), are present and both the main axis and the laterals contain (in dicotyledons) putative vascular cambia. The situation in shoot tips, with their superficial growth of lateral organs and contrasted arrangement and development of vascular bundles, is very different.

An apical shoot segment that excludes all leaf primordia and their recognizable initials is so small that the remaining central dome of meristem, unlike the comparable situation in a root tip (with its more conspicuous quiescent center), seems to lack essential basipetal organization or stimuli for its viable organized growth. Therefore successful root tip cultures, though by no means universal (even in dicotyledons), are certainly more frequent than are organized shoot tip cultures. But both cases (i.e., root tip and shoot tip cultures) should be regarded properly as cultures that are fully organized from the outset and are strictly analogous to small cuttings. As such, their ongoing culture, strictly aseptic micropropagation, occurs when their requirements for continued growth can be met exogenously.

Morphologically speaking, true "tissue" cultures would seem to be more specific and their demands more prescribed. Having been laid down initially to fulfill specific, though often heterogenous functions (e.g., as in pith, cortex, xylem, and phloem), their continued activity in situ is subject to endogenous controls that rarely persist in culture and, when they do proliferate in culture, the cells only retain in part the features they had in situ. The larger the original explant, the slower the proliferated growth, and the longer it persists in subculture, the more such cultures seem to acquire and express distinctive features and limitations of their own. These features, however, be it said, may cease to have much relevance to the understanding of the plant of their origin for the cultures may lose, or at least no longer express, a morphological or biochemical totipotency. By contrast, the sooner free cells detach from proliferated cultures, the more rapid their growth and multiplication, the more chance they have of being brought into fully competent, totipotent cell cultures. The provision of exogenous growth promoting stimuli and the elimination or suppression of endogenous inhibitors are potent causal factors here.

CELL CULTURES: TOTIPOTENCY

It is here that the relevance of cell cultures with recognizable totipotency emerges, and it is conveniently illustrated by the case of carrot-root phloem cultures. First, small explants are obtained quite specifically from the secondary phloem of the storage root; these are induced to grow rapidly in solutions that both induce and maintain their growth by rapid proliferation. In this state the metabolism and form of the cells are quite different from their counterparts in situ. Second, the definitive point is that, when freed from the surface of the proliferating explants, they can grow rapidly and independently as true cell cultures in liquid suspensions of minute cells, and the small aggregates they produce, when filtered, will continue to grow, under the appropriate culture conditions, as true somatic facsimiles of carrot zygotic embryos in situ. These cultures should be properly identified as morphogenetically competent, totipotent, cells. But it should be recognized that the steps to realize this goal include in sequence, (1) aseptic isolation of small tissue explants, (2) the induction of their rapid proliferated growth under specific conditions, (3) the spontaneous release from the explants of free cells which continue as viable autonomous units, and (4) in so doing they express morphological and biochemical features traceable to, and reexpressed from, the carrot zygote from which all development began.

Lastly, it should be noted that those carrot cells in culture represent true clonal growth of the carrot plant from which they originated. They retain the full inheritance and totipotency of their nuclei and the full ability of their cytoplasm to foster its expression. In this respect clonal cell propagation in plants is to be distinguished from "cloning," as it is often referred to in animals where during development the cells, although they may have retained the competence of their nuclei, have lost the competence that the cytoplasm of the fertilized egg possessed to foster its expression.

Needless to say, and flowing from the discussion of nomenclature, the ultimate aim of aseptic plant culture should be to achieve, where possible, full control over all the stages outlined above and, finally, to achieve fully competent totipotent cells in culture from which all their morphogenetic-biochemical responses may be recapitulated and investigated.

The fact should be faced that the potentialities that have been amply demonstrated for carrot—shown to be feasible for several other umbelliferae (Steward, Kent, and Mapes, 1966) and with greater or lesser difficulty extended to other plants (Steward and Krikorian, 1979; Krikorian, 1982)—are nevertheless far more difficult to release generally than might be expected or acknowledged. Does this mean that some hitherto unrecognized general principle governing the release of genetic information awaits discovery and application? Or is it that the combinations of stimuli and nutrients to be applied in synergistic and sequential combinations, coupled with the combinations of environmental variables that may affect even normal development (Steward, 1976) are so subtle that their chance discovery in the more recalcitrant situations is a too remote possibility? The suggestion here is that each recalcitrant situation requires systematic and persistent investigation moving from each partial success until a final unique program is achieved.

PROTOPLASTS

However, the above is concerned with cells. There also remain the problems inherent in the use of isolated protoplasts and cellular organelles.

Surviving freely suspended plant protoplasts had been known and investigated for their osmotic properties as early as the 1930s, and for many years attempts were being made to observe their growth and regeneration. Since protoplasts of angiosperms are so universally bounded by walls (though interconnected by plasmodesmata, except in special situations where, as at syngamy, fusions are essential), it is not surprising that their independent survival and development presented difficult and protracted problems. When cultures of viable freely suspended totipotent carrot cells were first being routinely obtained and observed (as in the 1960s), and when Cocking was perfecting his early techniques for the use of free protoplasts, it was an attractive idea to bring both techniques together. If protoplasts could be prepared from suspensions of demonstrably morphogenetically competent cells, should they not more easily, like fertilized eggs, develop into embryos than protoplasts prepared de novo from such mature cells as those of leaves? With the cooperation of Dr. Cocking, his then-purified enzyme preparations were used and the directions for isolating the protoplasts adopted. Very good, uniform, and seemingly "rugged" surviving carrot protoplasts were obtained, but from them somatic embryogenesis did not ensue, not even remotely approaching that of the cell preparations from which they came (Steward et al., 1975). This is but another example of the oft-encountered experience that cultural practices may encounter unpredictable consequences (or even opportunities) when barriers developed during normal development are disturbed. Nevertheless, it is interesting to note that the successful development of carrot embryos from carrot protoplasts has in fact, subsequently, been achieved (Krikorian and Steward, 1979 and unpublished). This is not an isolated example or experience, but may be taken here to illustrate the important point that cellular totipotency is at best fragile, easily disrupted by adverse conditions and may require much painstaking observational work and trials in its restoration (one can properly speak here about the "state of the art," for art it is).

Because free, viable, surviving protoplasts lend themselves to so much manipulation and because modern genetics and molecular biology require their controlled use, all this is now an important area of investigation. Cell and tissue culture genetics now has a familiar sound. However, it should be stressed that its uses (though real) are as limited in practice as are the success with which genetically modified protoplasts can be regenerated into plants.

Finally one may ask, "But what of isolated organelles—nuclei, plastids, mitochondria?" The great riddle of cellular organization is to know how so many discrete viable self-duplicating organelles, within their respective plant protoplasms, perform their specific functions so smoothly and effectively. Since so much is now known about the enzymology and intermediary biochemistry of the vital processes, why is it so difficult to achieve and maintain these events in noncellular preparations? Or even, for that matter, why is it that such limited success has been achieved in causing aseptic proliferated cultures to rival their counterparts in situ in the production of distinctive metabolites or special storage products?

If the chapters to come not only appreciate the successes of aseptic cultural practices but also recognize the limitations which they may encounter when routinely applied to various plant sources and to preparations from various morphological levels, their value will be commendable.

REFERENCES

Krikorian, A.D. 1975. Excerpts from the history of plant physiology and development. In: Historical and Current Aspects of Plant Physiology: A Symposium Honoring F.C. Steward (P. Davies, ed.) pp. 9-97. Cornell University, College of Agricultural and Life Sciences, Ithaca, New York.

_____ 1982. Cloning higher plants from aseptically cultured tissues and cells. Biol. Rev. 57:151-218.

_____ and Steward, F.C. 1969. Biochemical differentiation: The biosynthetic potentialities of growing and quiescent tissue. In: Plant Physiology, A Treatise (F.C. Steward, ed.) Vol. VB, pp. 227-326. Academic Press, New York.

_____ and Steward, F.C. 1979. Is gravity a morphological determinant in plants at the cellular level? In: COSPAR, Life Sciences and Space Research 17 (R. Holmquist, ed.) pp. 271-284. Pergamon Press, Oxford and New York.

Steward, F.C. 1968. Growth and Organization in Plants. Addison-Wesley, Reading, Massachusetts.

_____ 1976. Multiple interactions between factors that control cells and development. In: Perspectives in Environmental Biology, Vol. 2 (N. Sunderland, ed.) pp. 9-23. Pergamon Press, Oxford and New York.

_____ and Krikorian, A.D. 1975. The culturing of higher plant cells: Its status, problems and potentialities. In: Form, Structure and Function in Plants. Prof. B.M. Johri Commemoration Volume. University of Delhi, India (H.Y. Mohan Ram, J.J. Shah, and C.K. Shah, eds.) pp. 144-170. Sarita Prakashan, Meerut.

_____ and Krikorian, A.D. 1979. Problems and potentialities of cultured plant cells in retrospect and prospect. In: Plant Cell and Tissue Culture: Principles and Applications (W.R. Sharp, P.O. Larsen, E.F. Paddock, V. Raghavan, eds.) pp. 221-262. Ohio State University Press, Columbus.

_____ and Shantz, E.M. 1954. The growth of carrot tissue explants and its relation to the growth factors in coconut milk. Annee Biologique 30:399-415.

_____ with Shantz, E.M., Pollard, J.K., Mapes, M.O., and Mitra, J. 1961. Growth induction in explanted cells and tissues: Metabolic and morphogenetic manifestations. In: 19th Annual Growth Symposium: Synthesis of Molecular and Cellular Structure (D. Rudnick, ed.) pp. 193-246. Ronald Press, New York.

_____ with Shantz, E.M., Mapes, M.O., Kent, A.E., and Holsten, R.D. 1964. Growth-promoting substances from the environment of the embryo. In: Regulateurs Naturels de la Croissance Vegetale, pp. 45-58. Editions du Centre National de la Recherche Scientifique, Paris.

_____, Kent, A.E., and Mapes, M.O. 1966. The culture of free plant cells and its significance for embryology and morphogenesis. In: Current Topics in Developmental Biology (A. Monroy and A.A. Moscona, eds.) Vol. I, pp. 113-154. Academic Press, New York.

_____, Israel, H.W., and Mapes, M.O. 1968. Growth regulating substances: Their roles observed at different levels of cellular organization. In: Biochemistry and Physiology of Plant Growth Substances (F. Wightman and G. Setterfield, eds.) pp. 875-892. Runge Press, Ottawa.

_____, Ammirato, P.V., and Mapes, M.O. 1970. Growth and development of totipotent cells: Some problems, procedures and perspectives. Ann. Bot. 34:761-788.

_____, Israel, H.W., Mott, R.L., Wilson, H.J., and Krikorian, A.D. 1975. Observations on the growth and morphogenesis in cultured cells of carrot (Daucus carota L.). Phil. Trans. Roy. Soc. B 273:33-53.

_____, Moreno, U., and Roca, W.M. 1981. Growth, form and composition of potato plants as affected by environment. Ann. Bot. (Supplement 2) 48:1-45.

PART A
BASIC TECHNIQUES OF PLANT CELL CULTURE

CHAPTER 2

Organogenesis

C.E. Flick, D.A. Evans, and *W.R. Sharp*

The success of many in vitro techniques in higher plants depends on the success of plant regeneration. Indeed, the application of some techniques, e.g., in vitro mutant isolation and protoplast fusion, to cell cultures is limited in many crop species because of the inability to regenerate plants. Only a few crop species such as tobacco and carrot can be fully exploited in vitro because of their ease in regeneration.

It has become increasingly clear in recent years that through plant regeneration from in vitro cultures a vast resevoir of genetic variability is available. As many crop species have been extensively inbred, genetic variability has been reduced by widespread cultivation of a limited number of cultivars. In vitro culture may generate a new pool of genetic variability useful for development of new crop varieties (see Chapter 25).

In this chapter we have attempted to review the current status of in vitro organogenesis. Through the use of tables with information regarding conditions useful for regeneration from various species we hope to outline the general procedures that should be taken for development of plant regeneration techniques from recalcitrant crop species. Species have been grouped by families, as species belonging to the same family exhibit similar requirements for regeneration.

LITERATURE REVIEW

A large number of species has been regenerated from in vitro cultures. Those families which undergo plant regeneration via embryo-

genesis have been omitted, as embryogenesis has been discussed elsewhere (see Chapter 3). It is evident that although plants of many crop
species can be regenerated from cell culture, some of the more important crops, notably cereals and legumes, have lagged behind in development of these techniques. We have attempted to emphasize plants of
agricultural interest.

Solanaceae

Solanaceous species have been used as model systems of in vitro
studies. Totipotency was first demonstrated with *Nicotiana tabacum* by
regeneration of mature plants from single cells (Vasil and Hildebrandt,
1965). The first successful production of haploid plants by the in vitro
culture of excised anthers was achieved by using *Datura innoxia* (Guha
and Maheshwari, 1964). Plant regeneration from isolated protoplasts
was accomplished first with *N. tabacum* (Takebe et al., 1971), and the
first somatic hybrid was obtained from two *Nicotiania* species, *N. glauca*
and *N. langsdorfii* (Carlson et al., 1972).

TOBACCO AND OTHER NICOTIANA SPECIES. Varieties of cultivated
N. tabacum are easy to manipulate in vitro. The effects of auxins and
cytokinins on explants and tissue cultures are quite specific and reproducible. Culture medium MS, the nutrient solution most often used for
in vitro cultivation of plant species, was formulated as a result of
growth experiments with *N. tabacum*. Callus and suspension cultures
have been initiated from leaf or stem explants of *N. tabacum* and many
Nicotiana spp. using MS medium with the addition of 4.5 µM 2,4-D and
2 g/liter casein hydrolysate (Flick and Evans, Vol. 2, this series).
Callus can be initiated on MS medium with other hormone concentrations, e.g., 11.4 µM IAA and 2.3 µM KIN (Murashige and Skoog, 1962),
but an auxin is always necessary. Callus can be maintained on MS or
a similar medium with 2,4-D; casein hydrolysate is unnecessary. Suspension cultures also can be readily obtained on B5 or MS media with
2.3-4.5 µM 2,4-D. Shoot regeneration from callus and suspension culture can be obtained for most *Nicotiana* species by the removal of
2,4-D and the addition of a cytokinin, e.g., subculture to a solid MS
medium with 5 µM 6BA. For *N. tabacum*, numerous hormone combinations have been successfully utilized for shoot regeneration (e.g., Tran
Thanh Van and Trinh, 1978; Nitsch et al., 1967). It is also possible to
induce shoot regeneration by replacing 2,4-D with combinations of
auxins and cytokinins, e.g., 11.4 µM IAA and 9.3 µM KIN (Murashige and
Nakano, 1967) or 22.8 µM IAA and 46.5-66.5 µM KIN (Sacristan and
Melchers, 1969). Shoot formation from callus has been reported for 19
Nicotiana spp. (Table 1). Shoots can be obtained from *N. tabacum*
callus in approximately 3 weeks and can be multiplied using shoot cul-
ture on MS medium with 5 µM 6BA. Roots can be induced on MS,
Hoagland, or White's medium with no hormones (Bourgin et al., 1979;
Murashige and Nakano, 1967; Nagata and Takebe, 1971) or on one-half
strength MS medium with 25-75 µM 3-aminopyridine (Phillips and

Table 1. Shoot Formation from Cultured Explants of *Nicotiana* Species on MS Medium

SPECIES	EXPLANT	GROWTH REGULATORS FOR SHOOT FORMATION (μM)	REFERENCE
N. acuminata	Stem	10 2iP, 1 IAA	Helgeson, 1979
N. africana	Leaf	5 6BA	Evans, unpublished
N. alata	Floral branches	10 KIN, 1 IBA	Tran Thanh Van & Trinh, 1978
N. debneyi	Leaf	5 6BA	Evans, unpublished
N. glauca	Leaf	5 6BA	Evans, unpublished
N. glutinosa	Leaf	5 6BA	Evans, unpublished
N. goodspeedii	Stem	10 2iP, 1 IAA	Helgeson, 1979
N. longiflora	Stem	4.7 KIN, 0.06 IAA	Ahuja and Hagen, 1966
N. megalosiphon	Stem	10 2iP, 1 IAA	Helgeson, 1979
N. nesophila	Leaf	5 6BA	Evans, unpublished
N. otophora	Floral branches	10 KIN, 1 IBA	Tran Thanh Van & Trinh, 1978
N. plumbaginifolia	Leaf	5 KIN, 1 6BA	Evans, unpublished
	Floral branches	10 KIN, 1 IBA	Tran Thanh Van & Trinh, 1978
N. repanda	Leaf	5 6BA	Evans, unpublished
N. rustica	Shoot	45.7 IAA, 11.9 KIN	Walkey & Woolfitt, 1968
N. stocktonii	Leaf	5 6BA	Evans, unpublished
N. suaveolens	Leaf	5 6BA	Evans, unpublished
N. sylvestris	Stem	10 2iP, 1 IAA	Helgeson, 1979
	Leaf	5 6BA	Flick, unpublished
		17.1 IAA, 0.9 KIN	Ogura & Tsuji, 1977
N. tabacum	Cell culture	1 6BA	Gamborg et al., 1979

Table 1. Cont.

SPECIES	EXPLANT	GROWTH REGULATORS FOR FOR SHOOT FORMATION (µM)	REFERENCE
N. tomentosiformis	Floral branches	10 KIN, 1 IBA	Tran Thanh Van & Trinh, 1978
SEXUAL HYBRIDS			
N. debneyi x *N. tabacum*	Stem	4.7 KIN, 0.06 IAA	Ahuja & Hagen, 1966
N. tabacum x *N. glauca*	Leaf	5 6BA	Evans, unpublished
N. tabacum x *N. sylvestris*	Leaf	5 6BA	Evans, unpublished
N. glauca x *N. tabacum*	Leaf	5 6BA	Evans unpublished

Collins, 1979; Evans et al., 1980).   Roots appear rapidly, and it is likely that in some cases root primordia are already present.

Some sexual hybrids between *Nicotiana* spp. produce tumors (Kostoff, 1943).   It has been shown that tumorous cells of *N. glauca* x *N. langsdorfii* are hormone autotrophic when grown in vitro (Ahuja and Hagen, 1966).   Auxin autotrophy is not limited to tumors, as cells of a number of *Nicotiana* spp. are also auxin autotrophic, including *N. debneyi* (Smith and Mastrangelo-Hough, 1979); *N. repanda* (Evans, unpublished); and *N. knightiana* (Maliga et al., 1977).

POTATO AND RELATED SPECIES.   Dihaploid lines of *Solanum tuberosum* are also amenable to tissue culture studies (Binding et al., 1978). Tuber, shoot tip, hypocotyl, leaf, and stem explants have been used to initiate callus with morphogenetic potential.   Callus can be initiated from tuber explants on modified MS medium (Lam, 1975) with 2.3 µM IAA, 1 µM GA, and 3.7 µM KIN, or from shoots or stems on MS medium with 9.1 µM 2,4-D (Wang and Huang, 1975).   Callus formation from wild *Solanum* spp. requires a high auxin:cytokinin ratio.   NAA and 2,4-D generally have been utilized, although IAA and 6BA also have been used successfully (Table 2).   Shoot induction has been observed in stem or shoot-derived callus when 4.7-46.5 µM KIN has been substituted for 2,4-D.   Root formation occurs when young shoots are transferred to MS medium with 2.2 µM 6BA and 0.3 µM GA or to MS medium without hormones.

A similar protocol can be followed for *S. xanthocarpum* (Rao and Narayanaswamy, 1968), *S. sisymbriifolium* (Fassuliotis, 1975), *S. dulcamara, S. nigrum* (Zenkteler, 1972) *S. khasianum* (Bhatt et al., 1979), and chlorophyll-deficient *S. chacoense* (Gamborg, personal communication). A modified MS or White (1963) medium with 2.3-27.1 µM 2,4-D can be used to initiate or maintain callus of each species.   For plant regeneration 2,4-D is removed or the concentration reduced from 4.5 to 0.5 µM 2,4-D for *Solanum xanthocarpum*.   In most cases a cytokinin is also added to shoot regeneration medium (Table 2).   Root induction occurs for *S. sisymibriifolium* following transfer to Nitsch medium with 7.4 µM IAA.   *Solanum melongena* (eggplant) hypocotyl sections do not respond to 2,4-D (Kamat and Rao, 1978); however, callus can be initiated by using 5 µM NOA or 43 µM NAA (Matsuoka and Hinata, 1979).   Shoots were obtained when callus was subcultured onto MS medium with 2.3 µM ZEA or 4.7 µM KIN, or when NAA was replaced with 1 µM 6BA; MS medium with no hormones was used to obtain root formation.

TOMATO AND RELATED SPECIES.   MS medium generally has been used for tissue culture studies with species of *Lycopersicon*, particularly the cultivated tomato, *L. esculentum* (Padmanabhan et al., 1974). Callus can be induced from most tomato explants, but leaf or hypocotyl sections are most often used (Table 3).   The auxin 2,4-D has been used only sparingly in regeneration studies.   Combinations of IAA and NAA with KIN and 6BA have been used to initiate callus proliferation. Phenolic oxidation in leaf and hypocotyl sections can be overcome

Table 2. Plant Regeneration of *Solanum* Species

SPECIES	GROWTH REGULATORS		SHOOT MEDIUM	EXPLANT	REFERENCE
	CALLUS (µM)	SHOOT FORMATION (µM)			
S. dulcamara	Not reported	4.7 KIN, 5.7 IAA	MS	Leaf	Zenkteler, 1972
S. khasianum	Not reported	10 6BA, 0-10 IAA	MS	Leaf	Bhatt et al., 1979
S. laciniatum	10 IAA, 0.1 ZEA	10 6BA, 0.1-10 IBA	MS	Leaf	Davies & Dale, 1979
S. melongena (eggplant)	4.2 NAA	1 6BA	MS	Hypocotyl	Matsuoka & Hinata, 1979
S. melongena	5 NOA	4.7 KIN, 5.7 IAA	MS	Hypocotyl	Kamat & Rao, 1978
S. melongena	13.5 2,4-D	0.49 2iP, 0.1 ABA	MS	Pith	Fassuliotis et al., 1981
S. nigrum	10 IAA, 1-10 6BA	10 IAA, 10 6BA	MS	Leaf	Bhatt et al., 1979
S. nigrum	Not reported	4.7 KIN, 5.7 IAA	MS	Leaf	Zenkteler, 1972
S. sisymbriifolium	2.26 2,4-D, 17.1 IAA, 49.2 2iP	0.03-1.7 IAA, 73.8 2iP	MS	Stem pith	Fassuliotis, 1975
S. tuberosum (potato)	2.28 IAA, 1.04 GA, 3.73 KIN	1.8 6BA	MS	Tuber	Lam, 1975
S. tuberosum	9.05 2,4-D	4.7-46.5 KIN	MS	Shoot, stem	Wang & Huang, 1975
S. xanthocarpum	27.15 2,4-D	0.09 2,4-D	White or MS	Shoot	Rao & Narayana-swamy, 1968

Table 3. Plant Regeneration of Tomato and Related Species

SPECIES	GROWTH REGULATORS		SHOOT MEDIUM	EXPLANT	REFERENCE
	CALLUS (μM)	SHOOT FORMATION (μM)			
L. esculentum 'Apidice,' 'Porphyre'	0.5–2.5 IAA, 1–20 6BA	10 6BA, 0.5 NAA	MS	Leaf	Ohki et al., 1978
L. esculentum 'Bizon'	22.8–34.3 IAA, 9.3–18.6 KIN	91.3 IAA, 18.6 KIN	MS	Stem, leaf, cotyledon	Vnuchkova, 1978
L. esculentum 'Karnatak'	2.8 IAA	9 6BA, 2.8 IAA	MS	Hypocotyl, cotyledon	Gunay & Rao, 1980
L. esculentum hybrid Pol × Pusa, Pol. cultivar	5 2iP, 2 IAA	1 ZEA, 1 IAA	MS	Leaf	Dhruva et al., 1978
L. esculentum 'Rheinlands Rhum'	2.3–27.9 KIN, 2.7–32.2 NAA	8.9 6BA, 1.1 IAA	MS	Leaf	Tal et al., 1977
L. esculentum 'Rutgers'	11.4–22.8 IAA 9.3–18.6 KIN	22.8 IAA, 18.6 KIN	MS	Leaf	Padmanabhan et al., 1974
L. esculentum 'Rutgers,' 'EP-7'	4.5 2,4-D	5 6BA	MS	Hypocotyl, leaf, shoot	Flick & Evans, unpublished
L. esculentum 'Starfire'	1 6BA, 1 NAA	1–10 6BA, 0.1 IAA	MS	Leaf	Kartha et al., 1976
L. esculentum 'VFNT Cherry'	10.7 NAA, 4.4 6BA	9.1 ZEA, 0–2.9 IAA	MS	Leaf	Meredith, 1979
L. peruvianum	Not reported	198 ADE	White	Root	Norton & Boll, 1954

19

Table 3. Cont.

| SPECIES | GROWTH REGULATORS | | SHOOT MEDIUM | EXPLANT | REFERENCE |
	CALLUS (μM)	SHOOT FORMATION (μM)			
L. peruvianum	0-27.9 KIN, 10.7-32.2 NAA	8.9 6BA, 0-11.4 IAA, 27.9 KIN	MS	Leaf	Tal et al., 1977
Solanum pennellii	27.9 KIN, 2.9 NAA	8.9 6BA, 0-0.1 IAA	MS	Leaf	Tal et al., 1977

20

either by placing cultures in the dark until callus forms or by adding 500 mg/liter polyvinylpyrrolidone (PVP) to the culture medium (Flick and Evans, unpublished). Shoot formation can occur at a low frequency on callus induction medium or preferably on media with a higher cytokinin: auxin ratio or with cytokinin alone (Kartha et al., 1976). Roots can be obtained on MS medium with 10.7 μM NAA or 11.8 μM IAA (Tal et al., 1977). Coleman and Greyson (1977) have shown that GA_3 at 0.01-100 μM enhances root formation. Similar techniques for plant regeneration have been applied successfully to two wild *Lycopersicon* species (Table 3). Tomato does not seem to be as amenable to tissue culture techniques as do other solanaceous species (Herman and Haas, 1978), since callus cultures lose the ability to undergo shoot morphogenesis when subcultured. Plant regeneration has not yet been routinely obtained from long-term suspension cultures (cf., Meredith, 1978). Normal plantlets could be obtained from 20% of tomato callus lines after 4 months in culture (Meredith, 1979), whereas only abnormal plantlets incapable of root formation could be obtained after 17 months. Shoot regeneration from callus has been obtained by using 6BA and IAA but not KIN and NAA (Zenkteler, 1972). The cytokinin:auxin ratio appears to be more important for the control of tomato shoot formation than do the specific hormones utilized.

PETUNIA AND DATURA SPECIES. *Petunia* spp. can be manipulated easily in vitro. However, most of the information about these species is concerned with the feasibility of plant regeneration from protoplasts. All *Petunia* spp. have been cultured on MS medium (Table 4). Stem or leaf explants have been used for plant regeneration, but roots (Colijn et al., 1979) also have been cultured on combinations of 0-107.4 μM NAA and 0.4-35.2 μM 6BA, although 2,4-D may be used alone (Sangwan and Harada, 1976) or in combination with a low concentration (0.9 μM) of 6BA (Rao et al., 1973a). Shoot regeneration has been achieved in all species by reducing the concentration of auxin while simultaneously increasing the concentration of cytokinin. This has been accomplished in a few cases by substituting a weak auxin for a strong auxin, such as 22.8 μM IAA for 22.6 μM 2,4-D (Frearson et al., 1973). The four species of *Petunia* examined can be regenerated in vitro (Table 4).

The genus *Datura* has not been studied in great detail, although *D. innoxia* represents ideal material for studies of anther culture (Guha and Maheshwari, 1964), plant regeneration (Engvild, 1973), and somatic hybridization (Schieder, 1978). Callus of *D. innoxia* can be induced and maintained on MS medium with 1 μM 2,4-D and plants can be regenerated from callus with 1 μM 6BA. Roots can be obtained on growth regulator-free MS medium or on Nitsch medium with 1 μM IAA (Sopory and Maheshwari, 1976). *Datura metel* and *D. meteloides* have been regenerated in vitro on B5 medium with 2.2 μM 6BA and 8 μM NAA, and shoot formation has occurred when NAA was eliminated and 6BA was increased to 4.4 μM. Shoots have been obtained in *D. innoxia* from cell suspension cultures (Hiraoka and Tabata, 1974), stem segments (Engvild, 1973), and leaf sections (Evans and Gamborg, unpublished).

Table 4. Plant Regeneration of *Petunia* and *Datura* Species

| SPECIES | GROWTH REGULATORS | | SHOOT MEDIUM | EXPLANT | REFERENCE |
	CALLUS (µM)	SHOOT FORMATION (µM)			
Datura innoxia	1 2,4-D or 10 NAA	1 6BA	MS	Stem	Engvild, 1973
D. innoxia	1 2,4-D	100 KIN	MS	Suspension	Hiraoka & Tabata, 1974
D. innoxia (haploid)	15% CW	9.3 KIN, 9.9 ADE	Nitsch	Stem, leaf	Sopory & Maheshwari, 1976
P. hybrida 'Blue Dansy,' 'Comanche,' 'Gypsy'	22.6 2,4-D	11.9 KIN	MS	Leaf	Frearson et al., 1973
P. hybrida 'Celestial' x 'Blue Dedder'	0–107.4 NAA, 0.4–35.2 6BA	0.3–2.7 NAA, 1.1–8.9 6BA	MS	Root	Colijn et al., 1979
P. hybrida 'Cascade'	0.9 6BA				
P. inflata	4.5 2,4-D	2.3 6BA	MS	Leaf	Hayward & Power, 1975
P. parodii	4.5 2,4-D	2.3 6BA	MS	Leaf	Hayward & Power, 1975
P. pendula	Not reported	8.9 6BA, 5.4 NAA	MS	Floral peduncle	Pelletier & Delise, 1969

OTHER SOLANACEOUS SPECIES. Plants have been regenerated via organogenesis from eight other solanaceous species (Table 5). In each case callus could be induced with an auxin:cytokinin ratio >1, and shoots regenerated with auxin: cytokinin ratio <1. White's, MS, or Poirier-Hamon et al. (1974) basal medium was used in each case, and root formation was obtained on hormone-free basal medium or medium with auxin (NAA or IAA). Nearly all possible explants have been used (Table 5) for induction of organogenesis among these solanaceous species. In addition to the species listed in Table 5, three species of *Scopalia*, *S. carniolica*, *S. lurida*, and *S. physaloides* (Wernicke and Kohlenbach, 1975), have been regenerated from anther cultures. *Physalis alkekengi* (Zenkteler, 1972) could not be regenerated in vitro.

A total of 42 solanaceous species have been regenerated in vitro. Organogenesis is the primary mode of plant regeneration in this family. MS medium has been used for 39 of the 42 species cultured in vitro, although B5 medium may be equally useful. Callus can be induced in most species by using 4.5 μM 2,4-D, and 5 μM 6BA is sufficient for plant regeneration in most species. In those species in which this treatment has not been successful, a combination of auxins and cytokinins generally has been used. Among solanaceous species, if the auxin:cytokinin ratio is >1, callus formation is enhanced; if <1, shoot formation is obtained. This general rule applies to 38 of the 42 species tested (Tables 1-5). The only exceptions are *S. dulcamara*, *S. nigrum*, and *Scopalia parviflora*, and in each case IAA, known to degrade in vitro, and also not as effective an auxin as 2,4-D or NAA, was used as the auxin in combination with a cytokinin at a lower concentration for shoot induction.

Cruciferae

The Cruciferae include many important agricultural crops (e.g., broccoli, cabbage, brussels sprouts, rapeseed, kale, Chinese kale, cauliflower, and horseradish) that have been regenerated in vitro. With the exception of horseradish, all of these species belong to the genus *Brassica*. Another cruciferous species, *Arabidopsis thaliana*, has been valuable as a genetic tool (Redei, 1975) and is also amenable to tissue culture and plant regeneration. The emphasis on research in the Cruciferae has been on the application of tissue culture to crop improvement and vegetative propagation.

Callus can be induced routinely and maintained from many explants. Shoots can be induced from callus, usually by increasing the cytokinin: auxin ratio. The potential for organogenesis in callus cultures of cruciferous species decreases rapidly with time, particularly after 6-8 months in vitro (Negrutiu and Jacobs, 1978b). Organogenic potential of *Arabidopsis* cultures can be increased by manipulation of the culture medium.

At least eight species of Cruciferae have been regenerated in vitro (Tables 6 and 7). In *Brassica oleracea* (cauliflower) and *Arabidopsis*, cellular response to hormones in the culture medium varies among cultivars or geographical races. MS medium is used exclusively for

Table 5. Plant Regeneration of Miscellaneous Solanaceous Species

| | GROWTH REGULATORS | | SHOOT | | |
SPECIES	CALLUS (μM)	SHOOT FORMATION (μM)	MEDIUM	EXPLANT	REFERENCE
Browallia viscosa	10.7 NAA, 2.2 6BA	4.6 ZEA	MS	Leaf	Power & Berry, 1979
B. viscosa	4.5 2,4-D	25 2iP	MS	Leaf	Welsh & Sink, 1981
Capsicum annuum (pepper)	4.5 2,4-D	4.4-8.9 6BA, 0-5.7 IAA	MS	Cotyledon, hypocotyl	Gunay & Rao, 1978
Hyoscyamus niger (henbane)	0.5 2,4-D, 0.5 KIN	None	MS	Seedlings	Dhoot & Henshaw, 1977
Physalis minima	4.5 2,4-D	4.4 6BA	White	Stem, leaf	Bapat & Rao, 1977
Physalis peruviana	Not reported	18.6 KIN, 11.4 IAA	MS	Leaf	Zenkteler, 1972
Salpiglossis sinuata	0.5 KIN, 0.5-5.4 NAA	9.3 KIN 0.5 NAA	MS	Leaf, flower petal	Lee et al., 1977
Scopalia parviflora	10 2,4-D	10 IAA, 0.1-10 KIN	MS	Stem	Tabata et al., 1972

regeneration of cruciferous species (Tables 6 and 7). The source of the explant used to induce callus can be important in determining the organogenic response of a culture.

BRASSICA. Agriculturally important *Brassica* species include *B. oleracea* (cauliflower, brussels sprouts, cabbage, kale), *B. napus* (rape), and *B. alboglabra* (Chinese kale). Callus cultures of *B. oleracea* are usually initiated in the presence of both an auxin and cytokinin (Table 6). Although 2.3 μM KIN is most often used (Clare and Collin, 1974; Lustinec and Horak, 1970; Horak et al., 1975), concentrations as low as 0.5 μM (Bajaj and Nietsch, 1975) and as high as 14 μM KIN (Baroncelli et al., 1973) have been used successfully. Numerous hormones can be used to satisfy the auxin requirement for callus initiation and maintenance, i.e., 2,4-D, IAA, and NAA, at concentrations ranging from 0.9 μM for 2,4-D to 11.4 μM for IAA (Table 6). Callus induction of kale and Chinese kale has been promoted by 10.7 μM NAA and 2.2 μM 6BA (Horak et al., 1975; Zee and Hui, 1977). *Brassica napus* is the only *Brassica* species that requires no cytokinin for callus growth, and it has been cultured successfully on 2.3-4.5 μM 2,4-D (Stringham, 1979; Kartha et al., 1974). Growth responses to 5 μM 6BA in *B. oleracea* (cauliflower) vary among different genotypes (Baroncelli et al., 1973).

Plants can be readily regenerated from callus cultures of *Brassica*. Normally a cytokinin, e.g., 2.3-93 μM KIN, must be included in the regeneration medium (Table 6). Auxin concentration in the regeneration medium can be quite variable. In some cases no auxin is required (Clare and Collin, 1974), or a high concentration of a less active auxin, e.g., 5.7-11.4 μM IAA for red cabbage (Bajaj and Nitsch, 1975), may be added to the cytokinin to achieve plant regeneration. *Brassica alboglabra* shoot regeneration has been obtained only in the presence of both 5.4-21.4 μM NAA and 2.3-4.7 μM KIN (Zee and Hui, 1977; Zee et al., 1978). Shoot regeneration of *B. napus* is induced by the presence of 5 μM 6BA (Stringham, 1979; Thomas et al., 1976; Kartha et al., 1974). In contrast to regeneration from most callus cultures, 0.1-2.8 μM GA_3 is necessary for plant regeneration. In the absence of GA_3 green callus is induced, but no shoots are formed (Kartha et al., 1974).

ARABIDOPSIS. Callus growth of *Arabidopsis thaliana* is enhanced by addition of a cytokinin and an auxin to the culture medium as in *Brassica* species. Callus can be initiated on modified B5 medium (Negrutiu et al., 1975) with 10 μM 2,4-D and 0.25 μM KIN and maintained on 5 μM 2,4-D and 0.24 μM KIN (Negrutiu et al., 1978b). Callus also has been initiated and maintained on 43 μM NAA and 0.25 μM KIN (Negrutiu et al., 1975). The source of the explant has been shown to effect subsequent plant regeneration from callus (Negrutiu et al., 1978b). Callus initiated from anthers has the highest organogenic potential, as well as retaining the capacity to regenerate for as long as 18 months. Seed-, stem-, and leaf-derived callus retain the capacity to regenerate for only 6-8 months.

Table 6. Plant Regeneration of *Brassica* Species

SPECIES	GROWTH REGULATORS		SHOOT MEDIUM	EXPLANT	REFERENCE
	CALLUS (µM)	SHOOT FORMATION (µM)			
B. aboglabra (chinese kale)	10.7 NAA, 2.2 BA	21.4 NAA, 2.3 KIN	MS	Cotyledon	Zee & Hui, 1977
B. juncea (leaf-mustard cabbage)	Not reported	2.7 NAA, 8.9 6BA	MS	Hypocotyl cotyledon	Hui & Zee, 1978
B. juncea	5.4 NAA, 0.89 6BA	8.9 6BA, 5.4 NAA	MS	Cotyledon	George & Rao, 1980
B. juncea	4.5 2,4-D	5 6BA, 0.1–0.4 GA	MS	Leaf	Stringham, 1979
B. napus (haploid rape)	5.7–11.4 IAA, 4.7 KIN	46–51 IAA, 14–28 KIN	MS	Leaf	Johnson & Mitchell, 1978
B. oleracea (broccoli)	Not reported	51–57 IAA, 37–42 KIN	MS	Leaf rib	Johnson & Mitchell, 1978
B. oleracea (broccoli)	Not reported	51–57 IAA, 14–93 KIN	MS	Stem	Johnson & Mitchell, 1978
B. oleracea (broccoli)	11.4 IAA, 2.3 KIN, 4.9 IBA	None	MS	Petiole, stem	Clare & Collin, 1974
B. oleracea (brussels sprouts)	0.9 2,4-D, 14 KIN	23.2 KIN	MS	Leaf vein, petal	Baroncelli et al., 1973
B. oleracea (cauliflower)	10.7 NAA, 2.3 KIN	2.4 KIN	MS	Stem, pith	Horak et al., 1975
B. oleracea (kale)	4.5 2,4-D, 0.5 KIN	11.4 IAA, 9.3 KIN	MS	Seeds, cotyledon	Bajaj & Nietsch, 1975
B. oleracea (red cabbage)		5.7 IAA, 2.3 KIN	MS	Hypocotyl	

Table 7. Regeneration of Cruciferous Species

SPECIES	GROWTH REGULATORS		SHOOT MEDIUM	EXPLANT	REFERENCE
	CALLUS (μM)	SHOOT FORMATION (μM)			
Amoracia lapathi-folia (horse-radish)	Not reported	5.4 NAA, 0.5–2.3 KIN	MS	Leaf	Meyer & Milbrath, 1977
Arabidopsis thaliana	4.5 2,4-D, 0.2 KIN	0.2 IAA, 4.7 KIN	B5 or PG	Seed, leaf, stem, anther	Negrutiu et al., 1975
Crambe maritima	11.4 IAA, 3.7 KIN, 10% CW	11.4 IAA, 3.7 KIN, 10% CW	MS	Root	Bowes, 1976
Iberis amara (candytuft)	1–10 2,4-D	10 KIN	MS	Leaf, stem	Mudgal et al., 1981
Lobularia maritima	10 IAA, 20 2,4-D	600 ADE, 5 6BA, 0.1–1 IAA	MS	Internodal stem	Khanna & Chopra, 1977
Sinapis alba	4.5 2,4-D	0.9 2,4-D, 9.3 KIN	MS	Cotyledon, hypoco-tyl, root	Bajaj & Bopp, 1972
Sisymbrium irio	5.7 IAA, 2.3 KIN	2.9 IAA, 14–23 KIN	MS	Stem	Pareek & Chandra, 1978a

Regeneration of shoots from *Arabidopsis* callus cultures can be obtained by the removal or reduction of the auxin concentration in the medium with simultaneous increase in the cytokinin concentration. A low concentration of auxin is sometimes retained in the shoot regeneration medium, e.g., 1 μM KIN and 0.1 μM GA_3 (Negrutiu et al., 1978b). GA_3 is sometimes added to shoot regeneration medium, e.g., 10 μM 6BA and 1 μM GA_3 (Negrutiu et al., 1978a), although no growth requirement for GA_3 has been demonstrated.

In vitro morphogenesis of *Arabidopsis thaliana* can be influenced by the following: (1) selection of a geographic race of high regenerative capacity (Negrutiu, 1976; Negrutiu et al., 1975); (2) subculturing callus at 4-week rather than 8-week intervals (Negrutiu and Jacobs, 1978b); (3) growth of callus in low light (Negrutiu and Jacobs, 1978a); (4) filter sterilization of culture medium (Negrutiu and Jacobs, 1978a); (5) replacement of NH_4^+ by glutamine as the nitrogen source (Negrutiu and Jacobs, 1978a); (6) treatment of cultures for 3-6 days at 4 C (Negrutiu and Jacobs, 1978b); (7) removal of all auxins from the culture medium prior to transfer to regeneration medium (Negrutiu and Jacobs, 1978b); and (8) a 20-day passage of young cultures in glucose-free medium or transfer of old cultures to culture medium containing 6% glucose as high glucose concentrations inhibit morphogenesis in young cultures (Negrutiu and Jacobs, 1978b). A progressive decline in shoot regeneration accompanies increasing age of callus. The origin of explants also may affect the longevity of the regenerative ability of a callus culture (Negrutiu et al., 1978b). Chromosome instabilities, e.g., increases in ploidy observed with age of callus (Negrutiu et al., 1975), may contribute to a decline in organogenesis with callus age.

OTHER CRUCIFEROUS SPECIES. Six other cruciferous species have been regenerated in vitro (Table 7). Generally, with the exception of *Amoracia lapathiofolia*, callus has been initiated and maintained on MS medium in the presence of an auxin:cytokinin ratio >1. Cytokinin is not always included in callus medium. Regeneration is induced by reduction of the auxin concentration in the medium with an increase in the cytokinin concentration. Shoots were regenerated directly from leaf pieces of *A. lapathiofolia* with 5.4 μM NAA and 0.5-2.3 μM KIN (Meyer and Milbrath, 1977).

Although cruciferous species are quite amenable to growth and plant regeneration in vitro, genotype has considerable influence on the success of callus culture and plant regeneration. Hence hormone concentration for callus and shoot culture can vary considerably. Although the potential for regeneration from callus culture can be lost within 6 months of initial culture, specific modifications of the culture technique can prolong regenerative viability. In vitro systems are not being utilized for improvement of cruciferous crops (Anderson et al., 1977).

Leguminosae

Plant regeneration has been quite difficult among the legumes. Forage legumes, e.g., clovers, are more amenable to in vitro plant regeneration than are seed legumes (Phillips and Collins, Vol. 2 of this
series). Twenty-five legume species have been regenerated in vitro,
but in most cases regeneration is at low frequency or limited by the
source of explants.

At least eight different media have been used successfully to obtain
plant regeneration (Table 8). MS medium is the medium most often
used for the regeneration of legumes. Based on the various media
used, it is likely that the media traditionally used for plant regeneration in other plant families may not be appropriate for legumes.

Unlike most plant families, for legume species few generalizations can
be made with regard to the role of plant hormones in regeneration.
The auxins most often used for callus induction were 2,4-D, NAA, IAA,
2,4,5-T, and picloram (Table 8). A cytokinin, either 6BA or KIN, has
been used for callus formation in each of the species except *Acacia
koa* (Skolmen and Mapes, 1976) and *Trigonella* spp. (Sen and Gupta,
1979), in which coconut water was used, and *Alhagi camelorum* and
Lotus corniculatus in which 2,4-D was used alone. In most legume
species, the auxin:cytokinin ratio is high for callus initiation, but a
number of exceptions are evident (Table 8). Most legume species require higher concentrations of cytokinins than do other plant families.

Most seed legumes have a higher propensity for root formation than
for shoot formation. For most species, the frequency of root initiation
is quite high despite the concentrations of auxins and cytokinins. Only
root initiation was observed in attempts to obtain plant regeneration
for *Psophocarpus tetragonolobus* (winged bean) (Bottino et al., 1979),
Glycine max (soybean) (Evans et al., 1976), and *Phaseolus vulgaris*
(French bean) (Haddon and Northcote, 1976). Root formation occurs
prior to shoot regeneration with *Stylosanthes hamata* (Scowcroft and
Adamson, 1976). Forage legumes will form roots but at a lower frequency than seed legumes. Difficulty in root formation has been
reported in both *Trifolium pratense* (Phillips and Collins, 1979) and
Lathyrus sativus (Mukhopadhyay and Bhojwani, 1978).

Plant regeneration has been difficult to obtain for seed legumes. In
general, the concentration of auxin is reduced or the concentration of
cytokinin is increased. The hormone concentrations successfully used
for callus initiation in seed legumes were usually highly specific (Table
8). In some cases unique additives were necessary. Bean seed extract
was required for regeneration of *Phaseolus vulgaris* (Crocomo et al.,
1976), while 5 kRad of gamma irradiation was used to obtain plant
regeneration in *Cajanus cajan* (Shama Rao and Narayanaswamy, 1975).
Wild soybeans can be regenerated, but cultivated soybean cannot yet be
routinely regenerated in vitro (Kameya and Widholm, 1981). Shoot
primordia have, however, been obtained from soybean hypocotyls (Kimball and Bingham, 1973; Oswald et al., 1977).

The hormone requirements are much less specific for forage legumes,
e.g., no hormones are required for plant regeneration in alfalfa (Walker
et al., 1979), whereas a wide range of concentrations of various hor-

Table 8. Regeneration of Legume Species

SPECIES	GROWTH REGULATORS		SHOOT MEDIUM	EXPLANT	REFERENCE
	CALLUS (µM)	SHOOT FORMATION (µM)			
Acacia koa (koa)	1% CW, 11.3 2,4-D	22 6BA	MS	Root sucker tip	Skolmen & Mapes, 1976
Albizzia lebbeck (indian walnut)	None	10.7 NAA	MS	Hypocotyl	Ghayal & Maheshwari, 1981
Alhagi camelorum (Persian manna)	2.9 IAA or 2.25 2,4-D	4.4 6BA	B5	Root, hypocotyl, leaf	Bharal & Rashid, 1981
Arachis hypogaea (peanut)	4.5 2,4-D, 9.3 KIN	22 IAA	MS	Immature embryo	Bajaj et al., 1981
Cajanus cajan T21 (pigeon pea)	Not reported	4.7 KIN, 0.06 IAA, 400 mg/liter CH	White	Hypocotyl	Sama Rao & Narayanswamy, 1975
Ceratonia siliqua (carob-tree)	8.9 6BA, 5.4 NAA	9.3 KIN	MS	Cotyledon, hypocotyl	Martins-Loucao & Rodriguez-Barrueco, 1981
Crotalaria burhia	1.1 2,4-D 1.3 NAA, 1.2 KIN	0.5-1.3 NAA, 2.2 6BA	MS	Stem, leaf	Raj Bhansali et al., 1978a
Crotalaria juncea	26.9 NAA	23.3-46.5 KIN	MS	Stem, leaf	Ramawat et al., 1977
Crotalaria medicagenia	2.3-14.3 2,4-D	2.2-8.9 6BA	MS	Stem, leaf	Raj Bhansali et al., 1978b
Glycine canescens	Not reported	22 6BA, 0.54 NAA	MS	Hypocotyl	Kameya & Widholm, 1981

Table 8. Cont.

SPECIES	GROWTH REGULATORS CALLUS (µM)	SHOOT FORMATION (µM)	SHOOT MEDIUM	EXPLANT	REFERENCE
Glycine tomentella	Not reported	22 6BA, 0.54 NAA	MS	Hypocotyl	Kameya & Widholm, 1981
Indigofera enneaphylla	2.3 2,4-D, 4.7 6BA	2.9 IAA, 4.7 6BA	B5	Cotyledon, hypocotyl	Bharal & Rashid, 1979a
Lathyrus sativus 'LSD-6'	2.3 2,4-D, 4.7 6BA	2.9 IAA, 4.7 6BA	B5	Shoot apex	Mukhopadhyay & Bhojwani, 1978
Lotus corniculatus (birdsfoottrefoil)	4.5 2,4-D	0.44 6BA	B5	Stem	Swanson & Tomes, 1980
Medicago sativa (alfalfa)	9 2,4-D, 9.3 KIN, 10.7 NAA	None, 2 g/liter YE	Blaydes	Hypocotyl	Bingham et al., 1975
Phaseolus vulgaris 'Bico de Ouro', (pinto bean)	18.6 KIN, 5.7 IAA, 1/4 bean seed/ml	11.4 IAA, 5.4 NAA, 0.9 KIN, 1/4 bean seed/ml	67-V	Leaf	Crocomo et al., 1976
Pisum sativum (pea)	10.7 NAA, 4.7 6BA	22.2 6BA, 1.1 IAA	MS	Epicotyl	Malmberg, 1979
Psophocarpus tetragonolobus (winged bean)	5.4–26.9 NAA, 0.5 KIN	None	Not reported	Hypocotyl, cotyledon	Venketeswaran & Huhtinen, 1978
Stylosanthes guyamensis (pencil flower)	9.3 KIN, 9.0 2,4-D	89 6BA, 5.4–10.8 NAA	MS	Hypocotyl, root, stem	Meijer & Broughton, 1981

31

Table 8. Cont.

SPECIES	GROWTH REGULATORS		SHOOT MEDIUM	EXPLANT	REFERENCE
	CALLUS (µM)	SHOOT FORMATION (µM)			
Stylosanthes hamata	9 2,4-D, 0.2 KIN	14 KIN	SH	Radicle, cotyledon	Scowcroft & Adamson, 1976
Trifolium alexandrinum (Berseem clover)	7 KIN, 5.4 NAA	2.3 KIN, 2.7 NAA	MS	Hypocotyl, suspension culture	Mokhtarzadeh & Constantin, 1978
Trifolium incarnatum (crimson clover)	10 2,4-D, 11 NAA, 10 KIN	11 NAA, 15 ADE	B5, SH	Hypocotyl	Horvath Beach & Smith, 1979
Trifolium pratense (red clover)	0.5 6BA, 0.25 picloram	0.05-44.4 6BA, 0.03 picloram	PC-L2	Cotyledon, meristem	Phillips & Collins, 1979
Trifolium repens (Ladino clover)	19.6 2,4,5-T, 0.5 KIN	2.3 2,4-D, 0.5 KIN	BM	Seed	Oswald et al., 1977
Trigonella corniculata	2.7 NAA, 15% CW	2.7 NAA, 15% CW	MS	Leaf	Sen & Gupta, 1979
Trigonella foenumgracum	2.7 NAA, 15% CW	2.7 NAA, 15% CW	MS	Leaf	Sen & Gupta, 1979

mones have resulted in plant regeneration in red clover (see Table 4 in Phillips and Collins, 1979). Although meristematic tissue has been used to regenerate plants among legumes, plant regeneration also can be initiated from nonmeristematic tissue of a number of species. Plants have been obtained from callus derived from hypocotyl, ovary, cotyledon, meristem, leaf, radicle, seed, shoot apex, and cell suspension cultures (Table 8); however, explant source can greatly affect the frequency of plant regeneration. In red clover, the frequency of plant regeneration was 1% from cotyledons and 30-80% from meristematic tissues (Phillips and Collins, 1979).

Initial regeneration frequencies of 12% were obtained from hypocotyl explants of alfalfa (Bingham et al., 1975), but this frequency was increased by selecting for plant regeneration. After two cycles of selection, the frequency of regeneration was increased in one genetic line of alfalfa from 12 to 67%. In addition, intervarietal differences have been observed in plant regeneration. Of 14 genetic lines of *Pisum sativum* tested for regeneration, only 6 could be regenerated after 2 months in vitro (Malmberg, 1979). Only 2 of 6 lines could be regenerated after 6 months in vitro. Intervarietal differences in plant regeneration also were reported for five cultivars of red clover (Phillips and Collins, 1979) and nine cultivars of alfalfa (Bingham et al., 1975). The report of distinct phenotypic variation among lines and the relative ease with which selection can increase the regeneration frequency suggest that in legumes, the ability to regenerate plants is inherited. Selection for regeneration has not yet been tested in other plant families. Malmberg (1979) has suggested that screening a large number of genetic lines may be useful in attempts to achieve plant regeneration.

Shoots can be rooted very rapidly from seed legumes. In nearly all cases, roots were obtained in medium with an auxin; 1 µM IBA (*Acacia*), 0.6-26.9 µM NAA (*Crotalaria*), and 0.6 µM IAA (*Stylosanthes*) have been used. In most cases when a cytokinin is present in the medium (KIN or 6BA), the auxin:cytokinin ratio has been greater than 10, e.g., 5.7 µM IAA:0.4 µM 6BA (berseem clover); 0.5 µM NAA:0.05 µM KIN (*Psophocarpus*); and 10.7 µM NAA:0.9 µM 6BA (*Indigofera*). In *Lathyrus*, *Medicago*, *Trifolium pratense*, and other forage legumes, root formation has been more difficult to induce. Phillips and Collins (1979) developed an effective rooting medium for red clover which used reduced minerals and 3-aminopyridine.

Rapidly growing cell suspension cultures capable of plant regeneration have been established from only a few legume species. Suspension cultures of alfalfa were initiated in Blaydes medium with 9 µM 2,4-D and 9.3 µM KIN (McCoy and Bingham, 1977). After 3 weeks in suspension culture, cells plated onto basal medium produced plants from 96% of the colonies. Regeneration capacity was reduced to 9% after 3 months in culture (McCoy and Bingham, 1977). Suspension cultures of berseem clover capable of plant regeneration could be induced in MS medium with 10.7 µM NAA and 0.95-1.0 µM 2iP (Mokhtarzadeh and Constantin, 1978). Plating efficiency of the suspension was 4%, with resulting colonies undergoing plant regeneration when placed on shoot inducing medium (Table 8).

Compositae

Few species of compositae have been regenerated from callus cultures. Only *Cichorium endiva* (endive) (Vasil and Hildebrandt, 1966a), *Lactuca sativa* (lettuce) (Doerschug and Miller, 1967), *Cynara scolymus* (globe artichoke) (Devos et al., 1979), and *Stevia rabaudiana* (Yang and Chang, 1979) have been of agricultural interest.

Tissue culture propagation of *Chrysanthemum morifolium* (chrysanthemum) has been studied extensively (Table 9). *Brachycome dichromosomatica* and *Crepis capillaris* have been used for cytogenetic studies of callus culture because of their low chromosome number.

Twelve species of *Compositae* have been regenerated from in vitro callus culture (Table 9). Callus cultures are usually initiated and maintained in culture medium with an auxin:cytokinin ratio >1 (Table 9). Although 0.5-53.7 µM NAA is most commonly used, 3-27 µM 2,4-D, 22.8-28.5 µM IAA, and 4.9 µM IBA have been used (Table 9). In most instances the auxin is combined with 0.57-4.7 µM KIN (Table 9). Only in *Chrysanthemum* has callus initiation and maintenance in the presence of an auxin:cytokinin ratio <1 been reported; however, callus also may be cultured with 10.7 µM NAA alone (Bush et al., 1976; Earle and Langhans, 1974a,b).

Plant regeneration is induced by reducing the auxin concentration and/or increasing the concentration of cytokinin. Kinetin is most effective in regeneration media at concentrations from 0.19 to 19 µM. Only in *Stevia rabaudiana* (Yang and Chang, 1979) is 6BA used, at 4.4-8.8 µM. Regeneration of *Chrysanthemum* callus in enhanced by 29 µM GA_3. Callus cultures of composite species do not readily lose the ability to regenerate. Regeneration from 3- to 5-year-old *Crepis capillaris* callus (Husemann and Reinert, 1976), and 14-month-old *B. dichromosamatica* callus (Gould, 1979) has been reported.

Brachycome dichromosomatica has been maintained as a diploid in callus culture for 14 months (Gould, 1979). Plants regenerated from this callus were also diploid. The chromosome stability of haploid and diploid *Crepis capillaris* was examined over a 1-year period in callus culture (Sacristan, 1971). Cultures that were initially haploid produced more polyploid than diploid cultures. Chromosome rearrangements and aneuploidy were more common in polyploids as well. Regeneration from these callus cultures was not reported.

Miscellaneous Dicots

Plants from 27 additional dicotyledonous families including 45 additional species have been regenerated in vitro (Table 10).

Several species of Scrophulariaceae have been regenerated. *Torenia fournieri* has been studied extensively (e.g., Bajaj, 1972). Shoots were initiated in numerous hormone treatments for both leaf (Bajaj, 1972) and internode (Kamada and Harada, 1979) explants. Five different cytokinins could be used to initiate shoot formation in *T. fournieri*. In most cases 4.4 µM 6BA, 4.6 µM ZEA, or 7.3 µM 4-phenylurea was sufficient to initiate shoot formation (Kamada and Harada, 1979), whereas

Table 9. Regeneration of Compositae Species

| SPECIES | GROWTH REGULATORS | | SHOOT MEDIUM | EXPLANT | REFERENCE |
	CALLUS (μM)	SHOOT FORMATION (μM)			
Brachycome di-chromosomatica	2.7 NAA, 2.3 KIN	21.5 NAA, 9.3-14 KIN	Miller B	Leaf	Gould, 1978, 1979
Centaurea cyanus (bachelor's buttons)	3 2,4-D	Not reported	Bonner	Stem	Torrey, 1975
Chrysanthemum morifolium (chrysanthemum)	46.5 KIN, 5.4 NAA	9.3 KIN, 0.11 NAA, 29 GA	LS or MS	Shoot tips, petals	Bush et al., 1976; Earle & Langhans, 1974a,b
Cichorium endiva (endive)	27 2,4-D	0.19 KIN	MS or Hilde-brandt D	Embryo	Vasil & Hilde-brandt, 1966a
Crepis capillaris (hawk's beard)	4.9 IBA, 4.7 KIN	18.6 KIN	White	Hypocotyl	Jayakar, 1971
Cynara scolymus (globe arti-choke)	1 ZEA, 1 2,4-D	1 6BA, 0.1 2,4-D	MS	Cotyledon, petiole	Devos et al., 1979
Gazania splendens	4.5 2,4-D, 2.3 KIN	11.4 IAA, 2.3 KIN	MS	Leaf	Landova & Landa, 1974
Lactuca sativa (lettuce)	28.5 IAA, 2.3 KIN	2.3-4.6 KIN	O	Seedling	Doerschug & Miller, 1967
Parthenium hys-terophorus	10 2,4-D, 1 KIN	5 6BA, 10 IAA	MS	Stem	Subramanian & Subba Rao, 1980
Pterotheca falconeri	11.4 IAA, 15% CW	11.4 IAA	MS	Roots, hy-pocotyl, stem, leaf, petiole, cotyledon	Mehra & Mehra, 1971

Table 9. Cont.

| SPECIES | GROWTH REGULATORS | | SHOOT MEDIUM | EXPLANT | REFERENCE |
	CALLUS (µM)	SHOOT FORMATION (µM)			
Stevia rabaudiana	53.7 NAA, 9.3 KIN	4.4-8.8 IAA	MS	Seed, leaflets	Yang & Chang, 1979
Taraxacum officinale (dandelion)	11.4 IAA	1.1 NAA, 2.8 KIN	White	Secondary roots	Bowes, 1970

Table 10. Regeneration in Miscellaneous Dicotyledonous Species

SPECIES	GROWTH REGULATORS CALLUS (μM)	GROWTH REGULATORS SHOOT FORMATION (μM)	SHOOT MEDIUM	EXPLANT	REFERENCE
ARALIACEAE					
Hedera helix (English ivy)	21.8 NAA, 9.3 KIN	5.4 NAA, 2.3 KIN, 200 mg/l CH	MS	Stem	Banks, 1979
BEGONIACEAE					
Begonia x *chiemantha* 'Astrid'	Not reported	0.5 NAA, 2.2–4.4 6BA	White	Petiole	Fonnesbech, 1974
Begonia x *hiemalis* (Rieger begonia)	Not reported	4.6 KIN, 5.4 NAA	MS	Leaf	Takayama & Misawa, 1981
CACTACEAE					
Mammillaria woodsii (barrel cactus)	Not reported	9.8 IBA, 9.3 KIN	MS	Stem	Kolar et al., 1976
CHENOPODIACEAE					
Beta vulgaris (sugar beet)	5.7 IAA, 0.5 KIN	4.7 KIN, 0.5 GA	PBO	Leaf	deGreef & Jacobs, 1979
CONVOLVULACEAE					
Convolvulus arvensis	0.2 2,4–D, 15% CW	4.7 KIN, 15% CW	LS	Stem	Hill, 1967
Ipomoea batatus (sweet potato)	5.4 NAA	3.8 ABA, 0.1 KIN, 0.2 2,4–D	White	Root tuber	Yamaguchi & Nakajima, 1974
CORYLACEAE					
Betula pendula (birch)	11 IAA, 23 KIN, 1 g/l CH	11 IAA, 30 mg/l ADE, 23 KIN	ABM	Flower cluster (catkin)	Srivastava & Steinhauer, 1981

37

Table 10. Cont.

SPECIES	GROWTH REGULATORS CALLUS (µM)	SHOOT FORMATION (µM)	SHOOT MEDIUM	EXPLANT	REFERENCE
CRASSULACEAE					
Crassula argenta (Jade tree)	100 2iP	0.1 6BA	MS	Leaf	Paterson & Rost, 1981
Sedum telephium (stonecrop)	4.5 2,4-D	44.4 6BA	B5	Immature leaf	Brandao & Selema, 1977
EBENACEAE					
Diospyros kaki (Japanese persimmon)	16.1 NAA, 0.5 KIN	5.4 NAA, 0.5-4.7 KIN	MS	Immature embryo	Yokoyama & Takeuchi, 1976
EUPHORBIACEAE					
Euphorbia pulcherrima (poinsettia)	20 NAA, 2 KIN	10-50 2iP, 0.5 NAA	MS	Internode, petiole	deLanghe et al., 1974
Manihot esculenta (cassava)	0.4 6BA, 1.1 NAA, 0.9 GA	0.4 6BA, 1.1 NAA	MS	Stem	Tilquin, 1979
Putranjiva roxburghii	12 IAA, 24 KIN	12 IAA, 24 KIN	White	Endosperm	Srivastava, 1973
GERANIACEAE					
Pelargonium spp. (geranium)	5.4 NAA	0.5 NAA	MS	Stem, petiole, root	Skirvin & Janick, 1976
HAMAMELIDACEAE					
Liquidambar styraciflua (sweet gum)	Not reported	3.5-7.0 6BA, 0.54-5.4 NAA	Blaydes	Hypocotyl	Sommer & Brown, 1980
HYDROPHYLLACEAE					
Phlox drummondii	4.5 2,4-D, 10% CW	4.5 IAA, 10% CW	White	Flower bud	Konar & Konar, 1966

Table 10. Cont.

SPECIES	GROWTH REGULATORS		SHOOT MEDIUM	EXPLANT	REFERENCE
	CALLUS (μM)	SHOOT FORMATION (μM)			
LABIATAE					
Perilla frutescens	0.45 2,4-D	4.4 6BA, 0.54 NAA	MS	Leaf	Tanimoto & Harada, 1980
LINACEAE					
Linum usitatissium (flax)	0.45 2,4-D	10 6BA	MS	Hypocotyl	Gamborg & Shyluk, 1976
MORACEAE					
Broussonetia kazi-noki (paper mulberry)	Not reported	0.5 6BA	MS	Hypocotyl	Ohyama & Oka, 1978
OXALIDACEAE					
Averrhoa carambola (carambola tree)	0.9 2,4-D, 4.9–9.8 2iP	2.25 2,4-D, 9.8–49.2 2iP	MS	Leaf	Litz & Conover, 1980
PASSIFLORACEAE					
Passiflora suberosa (passion-flower)	4.4 6BA, 5.4 NAA	4.4 6BA, 5.4 NAA	MS	Leaf	Scorza & Janick, 1976
PIPERACEAE					
Peperomia "Red Ripple"	11.6–46.5 KIN 26.9 NAA	11.6–23.2 KIN 2.7 NAA	SH	Leaf	Henry, 1978
Peperomia viridis	Not reported	9.3 KIN, 5.4 NAA	MS	Leaf	Hui & Zee, 1981
POLYGONACEAE					
Fagopyrum esculentum (buckwheat)	22.6–45.2 2,4-D	3 g/liter YE	MS	Cotyledon, hypocotyl	Yamane, 1974

Table 10. Cont.

SPECIES	GROWTH REGULATORS		SHOOT MEDIUM	EXPLANT	REFERENCE
	CALLUS (μM)	SHOOT FORMATION (μM)			
F. esculentum (buckwheat)	4.7 KIN, 4.5 2,4-D	10 6BA, 1 IAA	B5	Cotyledon	Srejovic & Neskovic, 1981
PORTULACACEAE *Mesembryanthemum floribundum* (mid-day flower)	11.4 IAA, 20% CW	None	MS	Root, stem hypocotyl	Mehra & Mehra, 1972
PRIMULACEAE *Anagallis arvensis* (pimpernel)	4.5 2,4-D, 0.5 KIN	2.9 IAA, 4.9 2iP, 3 g/liter CH	MS	Stem, leaf hypocotyl	Bajaj & Mader, 1974
Cyclamen persicum	53.7 NAA, 11.6 KIN	None	White	Corm	Loewengerg, 1969
RANUNCULACEAE *Consolida orientalis*	9.1–22.6 2,4-D, 10% CW	4.5 2,4-D, 10% CW	White	Flower bud	Nataraja, 1971
Coptis japonica (goldthread)	4.5 2,4-D, 0.5 KIN	None	MS	Petiole	Syono & Furuya, 1972b
ROSACEAE *Malus pumila* (apple)	22 NAA, 9.3 KIN, 15% CW, 2.9 GA	13.5 2,4-D, 13.8 KIN	White	Cotyledon	Mehra & Sachdeva, 1979
Prunus amygdalis (almond)	26.9 NAA, 10% CW	26.9 NAA, 2.3–4.7 KIN	MS	Leaf, cotyledon, embryo	Mehra & Mehra, 1974
Rosa hybrida (rose)	Not reported	8.9 6BA, 0.5 NAA	MS	Shoot tip	Skirvin & Chu, 1979

Table 10. Cont.

| SPECIES | GROWTH REGULATORS | | SHOOT MEDIUM | EXPLANT | REFERENCE |
	CALLUS (µM)	SHOOT FORMATION (µM)			
ROSACEAE					
Rubus spp. (blackberry)	Not reported	0.4 6BA, 0.3 GA, 4.9 IBA	MS	Shoot tip	Broome & Zimmerman, 1978
RUTACEAE					
Citrus limettoides	1.1 2,4-D	500 mg/l ME, 2.2 6BA, 1.2 KIN, 1.0 NAA	MS	Stem	Raj Bhansali & Arya, 1979
C. madurensis	0.4-4.4 6BA, 0.4 2,4-D	None	MS	Stem	Grinblat, 1972
SALIACACEAE					
Populus tremuloides (2n, quaking aspen)	0.9-9.1 2,4-D	0.4-1.3 6BA	Wolter & Skoog	Stem	Wolter, 1968
P. tremuloides (3n, quaking aspen)	0.2 2,4-D, 4.7 KIN	0.7 6BA	Wolter & Skoog	Root, stem	Winton, 1968, 1970
Populus heterophylla (black cottonwood)	5.6 IAA, 0.9 KIN	0.7 6BA	Wolters	Flower	Bawa & Stettler, 1972
SANTALACEAE					
Santalum acuminata	Not reported	8.8 6BA	MS	Stem	Barlass et al., 1980
S. lanceolatum	Not reported	5.4 NAA	MS	Stem	Barlass et al., 1980

Table 10. Cont.

SPECIES	GROWTH REGULATORS		SHOOT MEDIUM	EXPLANT	REFERENCE
	CALLUS (µM)	SHOOT FORMATION (µM)			
SCROPHULARIACEAE					
Digitalis purpurea (foxglove)	4.5 2,4-D, 0.5 KIN	0.6 IAA, 4.7 KIN	MS	Seedling	Hirotani & Furuya, 1977
Limnophila chinensis	2.3 KIN	4.7–9.3 KIN	MS	Stem	Sangwan et al., 1976
Mazus pumilus	0.9 KIN	0.9–2.3 KIN, 1 IAA	MS	Floral internodes	Raste & Ganapathy, 1970
Torenia fournieri	4.5 2,4-D	5.7 IAA, 9.3 KIN	MS	Leaf	Bajaj, 1972
Verbascum thapsus	Not reported	None	MS	Stem	Caruso, 1971
WINTERACEAE					
Drimys winteri	0.54–54 NAA, 0.47–4.6 KIN	0.54 NAA, 4.6 KIN	ABM	Stem	Jordan & Cortes, 1981

numerous combinations of cytokinin and auxin also have been successful (Bajaj, 1972). Shoot formation was induced from internode segments when cultured in 0.5 µM NAA with either 3.25 µM SD8339, a cytokinin that has been used with *N. tabacum* (Nitsch et al., 1967), or 4.7 µM KIN. Organogenesis of *Torenia* also could be regulated by application of amino acids; shoot formation was observed in cultures with glutamic acid and aspartic acid (Kamada and Harada, 1979). Organogenesis was also obtained in cultures of *Verbascum thapsus*, *Digitalis purpurea*, *Limnophila chinensis*, and *Mazus pumilus*, but few hormone concentrations were investigated for any of these species (Table 10).

Two species of *Ranunculaceae* undergo organogenesis. Plants have been obtained from *Consolida orientalis* and *Coptis japonica* (Table 10). In *Consolida*, organogenesis was obtained in White medium with CW and IAA. In vitro regeneration of *Delphinium brunonianum* and *Clematic fouriana* has been unsuccessful (Nataraja, 1971).

Many of the other dicotyledonous species examined are economically important. Among species capable of plant regeneration, MS medium is most often used (24 species), although White medium has been used for the culture of eight species in eight different families. Five additional media were used for plant regeneration. Medium variation has not been studied extensively or demonstrated to be family-specific for plant regeneration. In Convolvulaceae and Primulaceae, one species of each family was regenerated in White medium and one species in MS medium (Table 10).

In many species shoot regeneration via organogenesis was achieved by subculturing onto medium with a higher cytokinin:auxin ratio. In some species the cytokinin:auxin ratio was held constant to achieve regeneration. Up to 13 different explants have been used to achieve organogenesis. A maximum of 11-50 shoots per leaf explant and 20-40 shoots per shoot tip were reported for *Peperomia* spp. and *Rubus* spp., respectively (Henry, 1978; Broome and Zimmerman, 1978).

Few angiosperm trees of economic importance have been regenerated in vitro from cell cultures. In most cases, an explant from juvenile tissue is used, e.g., embryo, hypocotyl, cotyledon, root sprout, stem segments, flower bud primordia. In only two instances has mature tissue been used. Wolter (1968) regenerated plants from 7-year-old callus initiated from the cambial region of the stem of diploid *Populus tremuloides* (quaking aspen). Mehra and Mehra (1974) regenerated diploid plants from *Prunus amygdalus* (almond) when stem and leaf sections were used as explants.

Graminaceae

In vitro plant regeneration can be induced in species of numerous families of dicotyledons, but monocotyledonous species have been more recalcitrant. This is unfortunate, as graminaceous species represent one of the most important sources of nutrition. Because of the large number of agriculturally important species in this family, most investigators using graminaceous species for studies on plant regeneration have restricted themselves to the cultivated crops. Hence comparisons of regenerative capacity between cultivated crops and wild relatives have not been possible.

Species within the two subfamilies, Poacoideae (grasses) and Panicoideae (cereals), have been cultivated in vitro with limited success. Meristematic tissues are generally used to initiate callus cultures capable of plant regeneration. Thus it has been suggested that regenerated plants have been derived solely from preorganized structures (Thomas and Wernicke, 1978). Nonetheless, there has been evidence that plants regenerated from tissue culture may be useful in crop improvement in both corn (Gengenbach et al., 1977) and sugarcane (Heinz et al., 1977).

POACOIDEAE. In vitro propagation may prove to be particularly useful among the grasses. These species are often propagated vegetatively in nature and may be both polyploid and uniquely variant in chromosome number (Brown, 1972). Polyploidy and chromosome instability exists in *Saccharum* spp. (2n = 10X = 80 - 120) permitting genotypes to withstand changes in chromosome number. These instabilities may be useful for the initiation of genetic variation in vegetatively propagated grasses. Chromosomal variation has been observed in callus-derived plants of both sugarcane (Heinz and Mee, 1969) and Italian ryegrass (Ahloowalia, 1975), and those variants have been used successfully in breeding programs to select for disease resistance (Heinz et al., 1977) and other agronomic traits (Ahloowalia, 1976).

Sugarcane is certainly the most malleable of the graminaceous species examined in vitro. Although the immature inflorescence is the most successfully cultured explant, various explants may be used for plant regeneration, including apical meristems, young leaves, and pith parenchyma (Liu and Chen, 1976). As plants have been regenerated from long-term callus cultures in sugarcane, it appears that plants may arise from completely unorganized tissue (Nadar and Heinz, 1977). Some evidence exists that plants obtained in vitro may originate from single cells via somatic embryogenesis (Nadar et al., 1978). MS medium, when supplemented with 2-13.6 μM 2,4-D, has been used for callus initiation. Plants could be obtained when the 2,4-D was removed from the medium. Cytokinin (9.3 μM KIN) has been present in some regeneration media used for sugarcane (e.g., Heinz et al., 1977), but it may be unnecessary (Liu et al., 1972). Shoots, when obtained in sugarcane, reportedly have been difficult to root (Heinz et al., 1977); however, young shoots, when separated from growing callus and placed on hormoneless medium or medium with IAA, form roots in a high percentage of cultures.

MS medium has been used to culture other grasses except *Dactylis glomerata* (SH medium). For *D. glomerata*, growth in SH medium, a medium devised specifically for monocots, was greater than on either B5 or MS medium (Conger and Carabia, 1978).

Explant source may limit callus initiation and proliferation. In most cases young organized tissue such as immature inflorescences, embryos, caryopses, peduncle, or mesocotyl has been used to initiate callus. High 2,4-D or 2,4-D with KIN was used to initiate callus formation from these complex explants (Table 11).   Removal or reduction of 2,4-D results in shoot formation in most grass species (Table 11).   In addition, plants can be obtained from callus culture of some species that are over 1 year old (Table 11).   The frequency of plant regeneration has not been reported for most grass species; nonetheless, plant regeneration in the reported species varied from 3.1% for *Festuca* (Lowe and Conger, 1979) to 66% for *Phragmites* (Sangwan and Gorenflot, 1975). Large numbers of shoots per explant have been reported for ryegrass (20–50 plantlets per culture; Ahloowalia, 1975) and Indian grass (5–20 plantlets per culture; Chen et al., 1979).   Albinos have been recovered in tissue cultures of some grass species, e.g., sugarcane (Evans and Crocomo, unpublished), following plant regeneration from callus.   This phenotypic variation may reflect an underlying variation in chromosome number.

Root formation has been initiated when shoots were transferred to fresh medium with no hormones (Kasperbauer et al., 1979; Chen et al., 1979), with reduced mineral concentration (Conger and Carabia, 1978; Ahloowalia, 1975; Lowe and Conger, 1979), or with a high auxin:cytokinin ratio (Sangwan and Gorenflot, 1975).

PANICOIDEAE (CEREALS).   Emphasis has been placed exclusively on agriculturally important species in studies of plant regeneration in the cereals (Table 12).   Plant regeneration has been reported to occur in most cultivated cereal species, but probably involved existing meristematic centers (Rice et al., 1978; King et al., 1978).   Extensive research has been carried out with *Zea mays* (corn), *Triticum aestivum* (wheat), and *Oryza sativa* (rice).   Corn is certainly the most recalcitrant of these species, as plants have been obtained only from organized explants.

In *Zea mays*, callus can be initiated from a number of explants, but the ability to regenerate plants has been limited (Green, 1977).   Immature embryos have been used most often for initiation of callus capable of plant regeneration (e.g., Green and Phillips, 1975).   MS medium has been used most frequently.   While 2,4-D has been the sole hormone used for callus initiation, the concentration has been varied from 2.3 µM (Rice et al., 1978) to 67.8 µM (Harms et al., 1976).   Callus proliferation can be enhanced by the addition of 21.5 µM NAA and 0.24 µM 2iP to 4.5 µM 2,4-D (Green and Phillips, 1975).   The scutellum of the immature embryo must be oriented upwards when immature embryos are cultured (Green, 1977), and downwards when young seedlings are cultured (Harms et al., 1976) for maximum callus induction.   Plant regeneration has been obtained by removing the 2,4-D (Green and Phillips, 1975; Harms et al., 1976; Freeling et al., 1976; Rice et al., 1978).   King et al. (1978) have suggested that reports of plant regeneration from cultured corn explants and most other cereals represent repression of shoot primordia during callus initiation followed by de-repression of preexisting primordia when the 2,4-D is removed.   Nonetheless, the capacity for plant regeneration has been maintained following subculture of scutellar-derived callus for 19–20 months (Green and Phillips, 1975; Freeling et al., 1976).   At least 16 *Zea mays* culti-

Table 11. Plant Regeneration among Species of the Grasses

SPECIES	GROWTH REGULATORS		SHOOT MEDIUM	EXPLANT	REFERENCE
	CALLUS (µM)	SHOOT FORMATION (µM)			
Andropogon gerar-dii (big blue-stem)	2.3 2,4-D, 0.9 KIN	<4.5 2,4-D	MS	Infloresc-ence	Chen et al., 1977
Dactylis glomerata (orchard grass)	6.8 2,4-D, 1.0 KIN	4.5 2,4-D, 1.0 KIN	SH	Caryopses (explant-down)	Conger & Carabia, 1978
Festuca arundi-nacea (tall fescue)	40.7 2,4-D	2.3 2,4-D	MS	Embryo	Lowe & Conger, 1979
Lolium multiflorum x Lolium perenne (ryegrass)	31.7 IAA, 1.0 KIN, 6.8 2,4-D	18.6 IAA, 0.5 KIN, 3.4 2,4-D	MS	Plumule, embryo	Ahloowalia, 1975
Phragmites commu-nis (Indian grass)	4.5 2,4-D	None	MS	Stem	Sangwan & Gorenflot, 1975
Saccharum spp. (sugarcane)	13.6 2,4-D	0 or 26.9 NAA	MS	Leaf, in-floresc-ence	Heinz & Mee, Nadar & Heinz, 1977
Sorghastrum nutans (Indian grass)	22.6 2,4-D, 0.9 KIN	<4.5 2,4-D	MS	Infloresc-ence	Chen et al., 1979
HYBRIDS					
Lolium multiflorum x Festuca arun-dinacea	9-18 2,4-D	1.1 2,4-D	MS	Stem, ped-uncle	Kasperbauer et al., 1979
Lolium multiflorum	Not reported	9.0 2,4-D, 0.88 6BA	MS	Immature embryo	Dale, 1980

Table 12. Plant Regeneration in Cereals

SPECIES	GROWTH REGULATORS		SHOOT MEDIUM	EXPLANT	REFERENCE
	CALLUS (μM)	SHOOT FORMATION (μM)			
Avena sativa (oats) 12 cultivars	2.3–13.6 2,4-D	None	B5	Immature embryo	Cummings et al., 1976
A. sativa 'Tiger'	9.1 2,4-D, 10.7 CPA	30 NAA, 10 IAA, 1.5 6BA	SH	Hypocotyl	Lorz et al., 1976
Eleusine coracana (finger millet)	45.2 2,4-D, 10–15% CW	1.1 NAA or 1.1 IAA	MS	Mesocotyl	Rangan, 1976
Hordeum vulgare 'Himalaya' (barley)	10 IAA, 1.5 2,4-D, 1.5 2iP	None	MS	Apical meristem	Cheng & Smith, 1975
H. vulgare 'Akka'	4.5 2,4-D	None	B5	Immature embryo	Dale & Deambrogio, 1979
Oryza sativa 'Kyote Ashi' (rice)	10 2,4-D	None	MS	Root	Nishi et al., 1968
O. sativa CI8970-S	9–27.1 2,4-D	0–9.8 2iP	MS	Roots, leaves	Henke et al., 1978
O. sativa 'Krasnodarskii'	9–18.1 2,4-D	79.9 IAA, 9.3 KIN	MS	Immature embryo	Davoyan & Smetanin, 1979
Panicum antidotle	9.0 2,4-D	None	MS	Inflorescence	Bajaj et al., 1981
Panicum maximum (guinea grass)	9.0 2,4-D	None	MS	Inflorescence	Bajaj et al., 1981
Panicum miliaceum (millet)	9.0 2,4-D	None	MS	Embryo	Bajaj et al., 1981

47

Table 12. Cont.

SPECIES	GROWTH REGULATORS CALLUS (µM)	GROWTH REGULATORS SHOOT FORMATION (µM)	SHOOT MEDIUM	EXPLANT	REFERENCE
Panicum miliaceum (common millet)	45.2 2,4-D	None	MS	Mesocotyl	Rangan, 1974
Paspalum scrobiculatum (Koda millet)	45.2 2,4-D	1.1 NAA	MS	Mesocotyl	Rangan, 1976
Pennisetum typhoideum (bulrush millet)	45.2 2,4-D	1.1 IAA	MS	Mesocotyl	Rangan, 1976
Sorghum bicolor 'N. Dakota' (sorghum)	22.6-67.8 2,4-D	26.9 NAA	MS	Seedling	Masteller & Holden, 1970
S. bicolor	9.0 2,4-D, 0.47 KIN	0.47 KIN	MS	Leaf	Wernicke & Bretten, 1980
S. bicolor line X4004	5 2,4-D, 10 ZEA	0.5 IAA	MS	Immature embryo	Gamborg et al., 1977
Triticum aestivum 'Tobari 66' (wheat)	10 Dicamba	1 IAA, 0.1 6BA	B5, T	Rachis	Dudits et al., 1975
Triticum aestivum 'Mengavi'	22.6 2,4-D, 5.4 CPA	23.3 KIN, 5.4 NAA	Basal	Embryo	Chin & Scott, 1977
T. aestivum 'Chinese Spring'	4.5-9 2,4-D	4.6 ZEA, 5.7 IAA	T,B5,MS	Rachis, embryo, seed	Gosch-Wackerle et al., 1979
T. durum	27 NAA	27 NAA	Smith	Mesocotyl	Bennici & D'Amato, 1978
T. longissiumum wild species	0.1 2,4-D	5.7 IAA, 4.6 ZEA	MS	Immature embryo	Gosch-Wackerle et al., 1979
Zea mays (corn)	9 2,4-D	None	MS	Immature embryo	Green & Phillips, 1975

Table 12. Cont.

SPECIES	GROWTH REGULATORS CALLUS (μM)	GROWTH REGULATORS SHOOT FORMATION (μM)	SHOOT MEDIUM	EXPLANT	REFERENCE
Zea mays 'Prior,' 'Inrakorn' lines M9473, M0003	67.8 2,4-D	None	MS	Mesocotyl	Harms et al., 1976
HYBRIDS					
Triticale (Triticum x Secale)	27.1 2,4-D	None	MS	Immature embryo	Sharma et al., 1978

vars have been tested for in vitro regenerative capacity with varied results. Only 3 of these 16 genetic lines have resulted in consistent plant regeneration: lines A188 and A188 X R-navajo (Green, 1977) and cultivar "Pior" (Harms et al., 1976). Green (1977) has concluded that regeneration in corn is genotype-dependent.

Cultivated wheat, *Triticum aestivum*, has been investigated in greater detail than has corn. Of 15 cultivars examined in six different laboratories, plant regeneration has occurred in only 5 cultivars. Callus proliferation has been classified extensively for both wild and cultivated species of *Triticum*. Gosch-Wackerle et al. (1979) initiated callus from the rachis and embryos of seven species of *Triticum*, including diploid (*T. monococcum, T. longissimum, T. speltoides,* and *T. tauschii*, 2n = 14), tetraploid (*T. timephevii* and *T. turgidum*, 2n = 28), and hexaploid (*T. aesticum*, 2n = 42) species. T-medium (after Dudits et al., 1975) with 4.5 µM 2,4-D was used to initiate callus from embryos. *Triticum tauschii* formed the least amount of callus on these media. Results from other laboratories collaborate the ability to initiate callus from these seven species (Shimada et al., 1969; Prokhorov et al., 1974). In addition, callus has been induced from an aneuploid series of ditelosomics (Gosch-Wackerle et al., 1979). Only one of seven lines, ditelo 1A, has a callus growth rate less than that of normal wheat. Plant regeneration has been obtained with three *Triticum* species (Table 12). For *T. aestivum*, the cultivar "Chinese Spring" has been used widely, but regeneration also has been achieved with cultivars "Salmon," "Maris Ranger," "Mengavi," and "Tobari 66" (Table 12). Plant regeneration has been observed from callus derived from different explants including rachises, shoots, seeds, and embryos (Table 12). The greatest frequency of regeneration (45-68%) was obtained from embryo-derived callus that had been initiated on MS medium with 9 µM 2,4-D and was subcultured after the first or second transfer on medium with 4.6 µM ZEA and 5.7 µM IAA (Gosch-Wackerle et al., 1979). The callus rapidly lost its organogenic capability.

Rice, *Oryza sativa*, is perhaps the easiest cereal species to regenerate in vitro. Callus can be obtained from numerous young explants of rice using MS medium with 9-45.2 µM 2,4-D (Davoyan and Smetanin, 1979). Shoot regeneration has been obtained from callus derived from seeds (Nishi et al., 1973), immature and mature embryos, root tips, scutellum, plumule, stem, and panicle (Davoyan and Smetanin, 1979). In each case shoots can be obtained following removal of 2,4-D from the medium. Although a cytokinin or auxin is unnecessary for shoot formation, it can be enhanced by addition of 0.3-19.7 µM 2iP (Henke et al., 1978) or 79.9 µM IAA and 9.3 µM KIN (Davoyan and Smetanin, 1979). The frequency of plant regeneration may reach 100%, e.g., 4-month-old callus cultures derived from young root and leaf explants. The age of rice callus is inversely proportional to its regenerative capacity.

Oats, barley, and some other cereals can also be regenerated (Table 12). In each species the medium for callus proliferation includes 2,4-D. In no case has 2,4-D been included in the regeneration medium. Young explants which almost certainly have contained organized shoot apices have been used for callus initiation. In *Avena sativa* regenerative capacity can be retained in callus cultures maintained for 12-18 months, but in most species the ability to regenerate is lost after only a few subcultures (1-4 months). In *Avena* (oats), genotype represents an important factor in plant regeneration. Of 24 genotypes initiated in vitro (Cummings et al., 1976) 9 could not be regenerated, while an additional 5 had a very low frequency of plant regeneration. Callus was initiated from seeds on MS medium with 22 μM 2,4-D (Carter et al., 1967) or SH medium with 9.1 μM 2,4-D and 10.7 μM CPA (Lorz et al., 1976) or from immature embryos on B5 medium with 2.4-13.6 μM 2,4-D. The 2,4-D must be removed for plant regeneration in each case. Although hormoneless medium is sufficient for plant regeneration from each explant, Lorz et al. (1976) have used a combination of three hormones to achieve regeneration (Table 12).

Hordeum vulgare (barley) plants have been obtained from callus derived from shoot apices (Cheng and Smith, 1975; Koblitz and Saalbach, 1976), mature embryos (Kartel and Maneshina, 1978), and immature embryos (Dale and Deambrogio, 1979). Plants have been obtained from embryo-derived callus of four out of seven barley cultivars. Greater shoot regeneration was obtained for immature embryos on B5 medium (50%) than on MS medium (19%) when callus was transferred from medium with 2,4-D to hormoneless medium (Dale and Deambrogio, 1979). Plants of *Sorghum bicolor* have been regenerated from callus derived from seedling shoots (Masteller and Holden, 1970), immature embryos (Gamborg et al., 1977), and mature embryos (Thomas et al., 1977). Although organized tissue was used, multiple shoot formation was obtained. MS is better than B5 medium for plant regeneration of sorghum (Gamborg et al., 1977). Callus can be induced with 5-67.8 μM 2.4-D (Table 12). Plants can be regenerated if subcultured on regeneration medium with 1-2 months. The frequency of regeneration for immature embryos is 20-50%. Plant regeneration of *Sorghum* is auxin-dependent. Shoots could be obtained from immature embryos in the presence of 2,4-D, but only if a high concentration of cytokinin was also present, e.g., 10-50 μM ZEA. Maximum shoot and plantlet production was obtained when callus was subcultured on medium with IAA (Gamborg et al., 1977), NAA (Masteller and Holden, 1970), or with a combination of 6BA, NAA, and GA_3 (Thomas et al., 1977).

Four species of millets have been regenerated from mesocotyl tissue on MS medium in vitro (Rangan, 1974, 1976). Callus was induced with 4.5 μM 2,4-D with or without 10-15% CW. Regeneration was obtained when the 2,4-D was eliminated (*P. miliaceum*) or replaced with IAA or NAA. The frequency of regeneration for all four species has ranged from 36-80% 3-4 months after callus initiation.

Secale cereale (rye) is one major cereal species that has not been tested for in vitro regeneration, although anther cultures of rye have been regenerated (Thomas et al., 1975). Nonetheless, hybrids of wheat and rye (Triticale AD-20) have been regenerated in vitro and behave similarly to wheat (Prokhorov et al., 1974).

Liliaceae

Several agriculturally important species of Liliaceae have been regenerated from callus culture, i.e., *Allium cepa* (onion), *Allium sativum*
(garlic), and *Allium porum* (leek). Other horticulturally important Liliaceae also regenerate in vitro (Table 13).

At least 21 species of Liliaceae have been regenerated from callus
(Table 13). A successful regeneration protocol has been applied for
several *Haworthia* species (Kaul and Sbaharwal, 1972). Callus cultures
usually have been initiated and maintained on MS or LS medium. Explant source does not affect the success of culture as dormant tissues,
e.g., bulbs, and metabolically active tissues, e.g., stem segments, leaf,
seed, petal, and inflorescence segments, have been successfully cultured
(Table 13). Liliaceae callus is usually initiated and maintained in the
presence of high concentrations of auxin, i.e., 11.4–45.6 μM IAA, 0.64–
43 μM NAA, or 0.54–9.1 μM 2,4-D. In most instances, a cytokinin,
usually 1–10 μM KIN, is also included in the culture medium. *Lilium
longiflorum* callus can be initiated in the absence of phytohormones
(Sheridan, 1968).

Reduction of auxin concentration or increase of cytokinin concentration induces the regeneration of shoots from callus cultures of Liliaceae
species (Table 13). Shoot regeneration is often enhanced by culturing
in the dark (Kato, 1978; Stimart and Ascher, 1978; Havranek and
Novak, 1973). Chromosome instabilities have been associated with in
vitro culture of liliaceous species, although these do not occur if plants
are directly regenerated from explants (Reuther, 1978; Hussey, 1976).
Increasing chromosome number is correlated with length of time in
callus culture in *Allium sativum* (garlic) (Havranek and Novak, 1973),
Ornithogalum thyrsoides (Hussey, 1976), and probably *Allium cepa*
(onion) (Davey et al., 1974; Hussey, 1976; Fridborg, 1971). Shoot
regeneration from onion callus greater than 1 year old is rare (Fridborg, 1971). Sheridan (1975), however, has maintained diploid callus of
Lilium longiflorum for over 6 years with no change in chromosome
number or regenerative ability. Plants regenerated from this callus are
diploid (Sheridan, 1975).

Miscellaneous Monocots

Four additional monocot species, representing four plant families,
have been regenerated in vitro (Table 14). All these monocot species
require auxin to initiate callus proliferation with reduction of auxin
necessary to obtain shoot development. All but *Dioscorea deltoidea*
require cytokinin in the callus medium, while all but *Anthurium andraeanum* require reduced auxin concentrations in the shoot induction
medium. In most species young embryonic tissue has been used, but
plant regeneration has also been successful from mature leaf explants
(Pierik, 1976). These monocot species may be maintained as callus for
extended time periods prior to shoot formation. Callus of *D. deltoidea*
has been subcultured for at least 12 months and retains the capability
for plant regeneration (Grewal and Atal, 1976). *Agave* species and

Table 13. Regeneration of Liliaceae Species

SPECIES	GROWTH REGULATORS		SHOOT MEDIUM	EXPLANT	REFERENCE
	CALLUS (μM)	SHOOT FORMATION (μM)			
Allium cepa (onion)	5 2,4-D	5 2iP	B5	Bulb	Fridborg, 1971
A. porrum (leek)	Not reported	20–29 2iP, 5.4–11 NAA	BDS	Leaf base	Dunstan & Short, 1979
A. sativum (garlic)	10 CPA, 2 2,4-D, 0.5 KIN	10 IAA, 10 KIN	AZ	Stem, bulb, leaf	Abo El-Nil, 1977
A. sativum (garlic)	9.5 2,4-D, 11.7 IAA, 9.3 KIN	11.4 IAA, 93 KIN	MS	Leaf	Havranek & Novak, 1973
Aloe pretoriensis	Not reported	0.91 2,4-D, 4.7 KIN	LS	Seed	Groenewald et al., 1975
Cordyline terminalis (ti)	5.4 NAA	4.4 6BA	MS	Stem	Kunisaki, 1975
Dracaena marginata	2.3 2,4-D	4.6 KIN	MS	Stem & node	Chua et al., 1981
Haworthia spp.	0.91–9.1 2,4-D, 0.93–9.3 KIN	4.7 KIN	MS	Inflorescence, gynoecia	Kaul & Sabharwal, 1972
H. planifolia 'Setulifera'	4.7 KIN, 0.9 2,4-D	18.6 KIN	LS	Leaf	Wessels et al., 1976
Helonopsis orientalis	1 NAA, 1 6BA	1 6BA	MS	Stem, leaf	Kato, 1975
Hemerocallis (daylily)	4.5 2,4-D, 4.7 KIN	4.7 KIN	MS	Petal	Heuser & Apps, 1976

53

Table 13. Cont.

SPECIES	GROWTH REGULATORS		SHOOT MEDIUM	EXPLANT	REFERENCE
	CALLUS (µM)	SHOOT FORMATION (µM)			
Hyacinthus hybrids	2.7–43 NAA, 0.5–9.1 2,4-D	<45.7 IAA, <2.7 NAA, <0.1 2,4-D	MS	Bulb, leaf inflorescence, stem, ovary	Hussey, 1976
H. orientalis (hyacinth)	0.44–1.3 6BA, 0.54–1.6 NAA	13 6BA, 1.6 NAA	MS	Flower bud	Kim et al., 1981
Lilium auratum *L. speciosum*	0.5 NAA	5.4 NAA, 4.4 6BA	MS	Peduncle, bulbscale, petal	Takayama & Misawa, 1979
L. auratum	Not reported	0.44 6BA	MS	Leaf	Niimi & Onozawa, 1979
L. japonicum	Not reported	0.44 6BA	MS	Leaf	Niimi & Onozawa, 1979
L. longiflorum (Easter lily)	11.4 IAA	0.2 NAA	LS	Bulb, stem, apex	Stimart & Ascher, 1978; Sheridan, 1968
L. rubellum	Not reported	5.4 NAA	MS	Leaf	Niimi & Onozawa, 1979
Muscari botryoides (grape hyacinth)	11.4–45.6 IAA, 0.69–43 NAA 0.54–99.1 2,4-D	<11.4 IAA, <0.14 2,4-D	MS	Bulb, leaf inflorescence, stem, ovary	Hussey, 1975

Table 13. Cont.

| | GROWTH REGULATORS | | | | |
SPECIES	CALLUS (μM)	SHOOT FORMATION (μM)	SHOOT MEDIUM	EXPLANT	REFERENCE
Narcissus (Lord Nelson)	Not reported	44 6BA, 5.4 NAA	Seabrook et al., 1976	Leaf, base, or scape	Seabrook & Cumming, 1982
Ornithogalum thyrsoides	11.4–45.6 IAA, 0.6–43 NAA, 2.3–9.1 2,4-D	None or 0.2– 0.6 NAA	MS	Stem, leaf, sepal, bulb	Hussey, 1976
Scilla sibirica	0.69–43 NAA, 0.54–9.1 2,4-D	<11.4 IAA, <0.64 NAA	MS	Bulb, leaf, infloresc- ence, stem ovary	Hussey, 1975

Table 14. Plant Regeneration in Miscellaneous Monocotyledonous Species

SPECIES	GROWTH REGULATORS		SHOOT MEDIUM	EXPLANT	REFERENCE
	CALLUS (μM)	SHOOT FORMATION (μM)			
AGAVACEAE					
Agave	4.5 2,4-D, 23.2 KIN	0.9 2,4-D, 4.7 KIN	LS	Seed	Groenewald et al., 1977
ARACEAE					
Anthurium andraeanum	0.4 2,4-D, 3.2 PBA	3.2 PBA	MS	Leaf	Pierik, 1976
BROMELIACEAE					
Ananas sativus (pineapple)	28.1 NAA, 29.7 IBA, 9.8 KIN	9.7 NAA, 9.8 IBA, 9.8 KIN	MS	Axillary bud	Mathews et al., 1976
A. sativus	Not reported	9.7 NAA, 9.8 IBA, 9.7 6BA	MS	Leaf	Mathews & Rangan, 1979
DIOSCOREACEAE					
Dioscorea deltoidea (yam)	4.5 2,4-D	1.2 IBA, 2.2 6BA	MS	Hypocotyl	Grewal & Atal, 1976
D. deltoidea	5.4-26.9 NAA, 10% CW	10% CW	MS	Tuber	Mascarenhas et al., 1976

Anthurium andraeanum also have been maintained in vitro for nearly a year prior to plant regeneration.

Although MS medium has been used for each species, specific nutrient requirements also may exist for these monocot species. Growth of *A. andraeanum* is enhanced if the NH_4NO_3 concentration is reduced from 20.6 to 2.57 mM (Pierik, 1976), while proliferation of *Ananas sativus* is enhanced by the addition of 1.23 mM NaH_2PO_4 to MS medium (Mathews et al., 1976). These variations suggest that common media formulations often used for dicots may be insufficient for monocots.

Gymnosperm Tree Species

Plants have been regenerated from 14 gymnosperm tree species (Winton, 1978; Thorpe and Biondi, Vol. 2 of this series). Both organogenesis and embryogenesis occur. Most research has been concerned with vegetative propagation of forest trees. The chief limiting factor for regeneration of gymnosperm trees is usually explant source, and it is necessary to use juvenile tissue, e.g., embryos or seedlings, for in vitro propagation (Table 15). The use of juvenile tissue precludes the selective propagation of a commercially desirable mature tree. Regeneration of tree species has been reviewed (Bonga, 1977; Winton, 1978; Sommer and Brown, 1979). Although Winton's review includes an extensive list of tree species regenerated in vitro, it has been suggested that in some cases, regeneration merely represents derepression of preexisting shoot primordia.

Achievements in regeneration of plants from gymnosperms are similar to those with angiosperm trees. In several instances, regeneration has been achieved from mature needles, e.g., *Pseudotsuga menziesii* (Winton and Verhagen, 1977), *Pinus sylvestris* (Scotch pine) (Borman and Jansson, 1980), and *Picea abies* (Norway spruce) (Jansson and Borman, 1980) and lateral branch shoot tips (Coleman and Thorpe, 1977), as well as from juvenile tissues (Table 15). Most gymnosperms can be regenerated only from juvenile tissue such as embryos and seedlings (Table 15). In many instances only regeneration directly from the explant with no intervening callus growth has been reported, e.g., *Picea glauca* (Campbell and Durzan, 1975); *Pinus banksiana* (Campbell and Durzan, 1975); and *Pinus pinaster* (David and David, 1977).

Various culture media have been used to regenerate gymnosperms. Because regeneration has been achieved from Douglas fir using several different culture media (Table 15), the composition of the culture medium may not be crucial to success. Shoots usually can be regenerated in the presence of 0.5-50 μM 6BA in the culture medium. A low concentration of auxin is included in the regeneration medium. Cheng (1976) demonstrated that as little as 0.5-5 μM NAA enhances shoot regeneration in Douglas fir. Regeneration can occur when the cytokinin:auxin ratio is <1. Douglas fir has been regenerated on medium with 4.5 μM 2,4-D or on medium with 24.7 μM NOA with 0.4 μM 6BA (Winton and Verhagen, 1977). Harvey and Grasham (1969) reported callus culture of 12 conifer species, but apparently were not able to regenerate plants.

Table 15. Regeneration of Pinaceae Species (Gymnosperm)

SPECIES	GROWTH REGULATORS		SHOOT MEDIUM	EXPLANT	REFERENCE
	CALLUS (µM)	SHOOT FORMATION (µM)			
Biota orientalis	5.5 NAA, 10 IAA or IBA or 5 2,4-D	2.2 6BA, 5 2iP	Lin & Stabe LP (Von Arnold and Eriksson, 1977)	Cotyledon, embryo	Thomas et al., 1977; Von Arnold & Eriksson, 1978
Cupressus sempervirens	Not reported	2.2 6BA	Lin & Staba	Cotyledon	M.J. Thomas et al., 1977
C. arizonica	Not reported	2.2 6BA	Lin & Staba	Cotyledon	M.J. Thomas et al., 1977
C. macrocarpa	Not reported	2.2 6BA	Lin & Staba	Cotyledon	M.J. Thomas et al., 1977
Picea abies (Norway spruce)	5 6BA, 5 NAA	5 6BA, 0.5 NAA	Sommer et al., 1975	Needles	Jansson & Bornman, 1980
P. glauca (white spruce)	Not reported	0-0.1 NAA, 10 6BA	Campbell & Durzan	Hypocotyl	Campbell & Durzan, 1975
Pinus banksiana (gray pine)	Not reported	0-0.1 NAA, 10 6BA	Campbell & Durzan	Hypocotyl	Campbell & Durzan, 1975
P. pinaster	Not reported	10 6BA	Campbell & Durzan	Cotyledon	David & David, 1977
P. radiata	Not reported	22 6BA	SH	Embryo	Horgan & Aitken, 1981
P. strobus (white pine)	0.05-10 2,4-D, NAA, or IBA	0.5-25 IBA, 0.44-4.44 6BA	MS	Embryo	Minocha, 1980

Table 15. Cont.

SPECIES	GROWTH REGULATORS		SHOOT MEDIUM	EXPLANT	REFERENCE
	CALLUS (μM)	SHOOT FORMATION (μM)			
P. sylvestris (scotch pine)	30 IBA, 5 6BA	20 6BA	Sommer et al., 1975	Needle, shoot	Bornman & Jansson, 1980
Pseudotsuga menziesii (Douglas fir)	4.5 2,4-D	45 2,4-D	Winton & Verhagen	Stem	Winton & Verhagen, 1977
P. menziesii (Douglas fir)	0.4 6BA (24.7 NOA)	0.4 6BA (24.7 NOA)	Winton & Verhagen	Needle	Winton & Verhagen, 1977
P. menziesii (Douglas fir)	5 IAA, IBA, 6BA, 2iP	0.5-1 6BA	Modified MS	Embryo	Cheng, 1975
P. menziesii (Douglas fir)	15-30 NAA, 5 6BA, IBA	5 6BA, 0.5-5 NAA	Modified MS	Cotyledon	Cheng, 1975; Cheng & Voqui, 1977
Thuja plicata (western red cedar)	Not reported	1 6BA, 0.1 NAA	MS	Juvenile tissue: cotyledon, stem, shoot tip	Coleman & Thorpe, 1977
T. plicata	Not reported	50 6BA, 0.1 NAA	MS	Mature tissue: branch, shoot tip	Coleman & Thorpe, 1977
T. occidentalis (white cedar)	Not reported	4.4 6BA	Lin & Staba	Cotyledon	M.J. Thomas et al., 1977

Transfer of in vitro regenerated shoots of gymnosperm tree species to soil is quite often unsuccessful, as sometimes there is no viable vascular connection between roots and shoots (Bonga, 1977). Excised conifer shoots are very difficult to root (Winton, 1978). In vitro regeneration of gymnosperm tree species will be economically feasible only if high frequency regeneration from tissue of mature trees and rooting of shoots can be achieved.

PROTOCOLS

Plant Growth

The age and physiological state of the parent plant may contribute to the success of organogenesis in cell cultures. Of course the explant source will dictate the age of the plant used. Leaf explants can be taken from young preflowering plants, e.g., tobacco; however, Zea mays cultures are most often initiated from immature embryos. Nevertheless, it is imperative that the plant be as young and healthy as possible. The transistion from plant to culture is stressful, and old or senescent plants may not survive the transition. The importance of the age of the plant is most obvious in initiation of cultures from tree species, where usually callus can only be initiated from juvenile tissue, and not explants from mature trees.

The success in initiation of cultures may depend on whether the parent plant is greenhouse or field grown. It is easier to control the development of a plant in the greenhouse, where water and temperature can be monitored and controlled. In addition, control of diseases and general cleanliness may enable easier establishment of contamination-free cultures from plants in the greenhouse. Obviously, for some species, greenhouse growth may not be practical, e.g., large trees or other perennials.

The season of the year can affect callus initiation from explants, especially when the donor plant is field grown. Seasonal variations in the concentration of endogenous auxins have been observed (Wodzicki, 1978) and also have been reported for establishment of potato meristem (Mellor and Stace-Smith, 1969) and conifer (Harvey and Grasham, 1969) cultures. For these species, spring and summer were found to be optimum seasons for starting cultures. Developmental stage and physiological state of the plant at the time of culture must be considered, because such factors as dormancy of the cambium or lateral buds, induction of flowering, etc., may affect the response and/or the success of initiating cultures.

Explants

Establishment of callus growth with subsequent organogenesis or embryogenesis has been obtained from many species of plants cultured in vitro. Most viable plant cells can be induced to undergo mitosis in vitro. Callus can be established from many explants. Plant regeneration has been successfully accomplished from numerous explants (Table 15). For any given species or variety, a particular explant may be necessary for successful plant regeneration; e.g., embryonic tissue is required for certain cereals. Explants consisting of shoot tips or isolated meristems, which contain mitotically active cells, have been especially successful for callus initiation and subsequent plantlet regeneration (Murashige, 1974, 1979). Explants from both mature and immature organs can be regenerated directly or can be induced to proliferate and form callus on the appropriate culture medium prior to plant regeneration. Size and shape of the explant may be critical. The increased cell number present in explants of greater mass increases the probability of obtaining a viable culture. Yeoman (1970) has shown that differences in the critical size of carrot root and artichoke tuber explants reflect cell size. In each case the optimum explant contains 20–25,000 cells.

Culture Medium

The essential components of plant cell culture medium have been summarized (Gamborg et al., 1976). MS medium (Murashige and Skoog, 1962) contains a high concentration of nitrogen as ammonium (20 mM), while White's medium (White, 1963) is ammonium-free. As high ammonium concentrations reduce cell growth in some plant systems (Gamborg and Shyluk, 1970), it may be difficult to transfer cells from White's (W), B5 (Gamborg et al., 1968), or SH (Schenk and Hildebrandt, 1972) medium onto MS medium. White's medium is a low salt medium, while the other three media are classified as high salt media. SH medium resembles B5 medium, with some salts present in slightly higher concentrations. LS medium (Linsmaier and Skoog, 1965) has also been used for plant regeneration. The macro- and micronutrients of LS and MS media are identical. As differences between these two media are found only in concentrations of vitamins and organic supplements, we refer only to MS medium. Other modifications to the organic additives of MS medium have been reported (Gamborg and Wetter, 1975). The organic supplements required in plant culture media include a carbon source and vitamins. Sucrose is used as a carbon source but may be substituted with glucose. Other sugars are used less often. Vitamins most often added to culture medium include inositol, nicotinic acid, pyridoxine, thiamine, calcium pantothenate, and biotin. Thiamine is required for plant growth, while the remainder enhance growth in some systems. Other vitamins have been used in plant systems. Vitamin E has been shown to regulate cell aggregation (Oswald et al., 1977), while vitamin D may enhance root formation (Buchala and Schmid, 1979). Various plant extracts or undefined additives are sometimes added to the culture medium to increase cell growth.

Maintenance of Cultures

The maintenance of cultures can determine whether a culture retains its organogenic potential. The most important factor in maintaining morphogenic potential is the maintenance of chromosome stability.

Variation of chromosome number of plants in long-term cell suspension cultures has been well documented (Kao et al., 1970). Chromosomal variation in cell suspension and callus cultures has been reviewed extensively (Krikorian et al., Ch. 16). It has been suggested that the progressive increase in variation of cultured cells is proportional to a progressive loss in organ-forming capacity (Torrey, 1967). Murashige and Nakano (1967) have shown that shoot-forming capacity of aneuploid *Nicotiana tabacum* is severely reduced, but Sacristan and Melchers (1969) were able to regenerate numerous aneuploids of *N. tabacum* with relative ease. Unfortunately, the chromosome instability of cells in suspension culture has been compared with cells of regenerated plants in only a few cases (D'Amato, 1977). The chromosome variability of regenerated plants is always less than in the callus from which plants were derived; however, despite the occurrence of a wide range of chromosome numbers in callus culture, only diploid plants have been regenerated from *Daucus carota* (Mitra et al. 1960), *Oryza sativa* (Nishi et al., 1968), *Prunus amygdalus* (Mehra and Mehra, 1974), and *Triticum aestivum* (Shimada et al., 1969). These species range in somatic chromosome number from $2n = 2X = 16$ for *P. amygdalus* to $2n = 6X = 42$ for *T. aestivum*. Consequently, under these plant regeneration conditions, diploids are selectively favored for these species. Orton (1980) has compared chromosome number of suspension and callus cultures of *Hordeum vulgare*, *H. jubatum*, and their interspecific hybrid with plants regenerated from these cultures. In each case it is evident that no polyploid and a greatly reduced number of aneuploid callus cells are capable of plant regeneration. Identical chromosome number is insufficient to conclude genetic stability. Only in long-term cultures of *Lilium longiflorum* have regenerated plants been shown to have the normal karyotype (Sheridan, 1975). Polyploid plants have been recovered from a number of cultures (D'Amato, 1977), including *Nicotania tabacum* (Kasperbauer and Collins, 1972), *Lilium longiflorum* (Sheridan, 1975), and haploid *Pelargonium* (Bennici, 1974). Polyploid changes are quite common in plants regenerated from anthers whether regenerated directly or via callus intermediate (D'Amato, 1977).

A wide range of aneuploid plants have been recovered from tissue cultures of numerous species. Reduction in chromosome number has been observed in plants regenerated from callus cultures of triploid ($2n = 3X = 21$) ryegrass hybrids (Ahloowalia, 1976), while a wide range of aneuploids having additions and reductions in chromosome number were obtained in sugarcane (Liu and Chen, 1976) and *N. tabacum* (Sacristan and Melchers, 1969). Each of these chromosomal variants is associated with phenotypic variation, including agriculturally useful traits such as disease resistance (Krishnamurthi and Tlaskal, 1974). Variability associated with aneuploidy, if not accompanied by a concomitant depression of yield, is particularly valuable in vegetatively propagated agricultural crops. Aneuploidy and morphological variability have been observed in plants regenerated from protoplasts (Matern et al., 1978) and in most somatic hybrid plants (e.g., Melchers and Sacristan, 1977).

Chromosome number mosaicism of regenerated plants has been reported in *Nicotiana* (Ogura, 1975), *Hordeum* (Orton, 1980), *Triticum durum* (Bennici and D'Amato, 1978), *Saccharum* (Liu and Chen, 1976), and *Lycopersicon peruvianum* (Sree Ramulu et al., 1976). The common occurrence of chromosome number mosaicism in regenerated plants suggests that plantlets originate from two or more initial cells (Bennici and D'Amato, 1978) or that new chromosome variability is generated in vivo after plant regeneration (Orton, 1980). Somatic mosaicism also has been reported in somatic hybrid plants derived from protoplast fusion where plants are presumably derived from single cells (Maliga et al., 1978). There have been suggestions that the chromosome number mosaicism of regenerated plantlets is reduced during the subsequent development of regenerated plants to mature plants. Nonetheless, this mosaicism may be established as periclinal or mericlinal "chimeras" (Sree Ramulu et al., 1976) or transmitted to subsequent generations (Ogura, 1976). The maintenance of chromosome mosaicism in vivo in regenerated plants may be under genetic control (Ogura, 1978).

Evans and Gamborg (1982) have suggested that the frequency of subculture can effect the chromosome stability of cell cultures. In order to maintain chromosome stability, cultures are subcultured frequently, e.g., every 3-4 days for tobacco. With frequent subculturing, aneuploid cells do not accumulate in the cell culture. Cells are subcultured in late exponential growth and hence never enter a stationary phase of growth. Therefore cells do not lag when subcultured. Using frequent subculture, a suspension culture of *N. tabacum* has been maintained for more than 5 years with the normal chromosome number of $2n = 48$ and the ability to regenerate normal plants.

Two modes of cell culture are generally used: (1) the cultivation of clusters of cells on a solid medium (e.g., agar, gelatin, filter paper, or millipore filters) and (2) the cultivation of cell suspensions in liquid medium. It is best to initiate new cultures in solid medium, as some essential nutrients may readily leak from small explants placed in large volumes of liquid medium. A suspension cell culture is usually initiated by placing friable callus into liquid culture medium. The suspension usually consists of free cells and aggregates of 2-100 cells. The frequency of subculture of each should increase when cultures are established to achieve optimal growth rate and genetic stability (Evans and Gamborg, 1982).

Plant Regeneration

Growth regulator concentrations in the culture medium are critical to the control of growth and morphogenesis, as first indicated by Skoog and Miller (1957). Generally, a high concentration of auxin and a low concentration of cytokinin in the medium promotes abundant cell proliferation with the formation of callus. Often, 2,4-D is used alone to initiate callus. On the other hand, low auxin and high cytokinin concentrations in the medium result in the induction of shoot morphogenesis. Auxin alone or with a very low concentration of cytokinin is important in the induction of root primordia.

Organogenesis can be induced from either cell suspension or callus cultures. As in most instances, cells must be transferred from high auxin to high cytokinin concentration culture medium. The efficiency of organogenesis can be effected if a residue of auxin is transferred with the cells into high cytokinin concentration regeneration medium. Shoots are almost always regenerated on solid medium. One need only transfer cells from callus medium to regeneration medium and continue normal subculturing on regeneration medium. Regeneration of shoots from suspension cultures is usually more efficient if cells are first subcultured 1-2 times in suspension culture medium with plant growth regulators for regeneration, i.e., high cytokinin. After subculture in regeneration medium, the auxin level is reduced and shoot regeneration on solid regeneration medium is more efficient.

Tobacco has been used as a model system for in vitro studies on regeneration, since the classical studies of Skoog and Miller (1957). It is one of the few systems for which some basic correlates for organogenesis have been elucidated (Thorpe, 1980).

The following is the protocol for organogenesis using tobacco leaf explants:

1. Plants can be grown under normal greenhouse conditions but must be used prior to flowering. The youngest fully expanded leaves are collected from tobacco plants either 1-2 hours after sunrise or from dark-pretreated tobacco plants.
2. Leaves are washed with mild detergent in water and rinsed first with 70% EtOH, then distilled water.
3. Leaves are sterilized in 20% commercial hypochlorite solution for 20 minutes. It is necessary to expose all leaf surfaces to the sterilizing solution. This is often done by placing leaves in a small beaker on a gyrotary shaker.
4. Leaves are rinsed in sterile distilled water three times and air dried for approximately 5 minutes.
5. Leaves are cut into 1 x 1 cm sections and transferred to culture medium.
6. Leaf explants should be placed onto agar-based MS medium with 1-5 μM 6BA for shoot formation. Shoots should appear in 3-4 weeks.
7. Alternatively, if callus formation is desired, leaf explants should be placed onto MS medium with 4.5 μM 2,4-D and 1-2 g/liter of casein hydrolysate. Callus formation should be visible in 4 weeks.
8. Cultures should be placed in the light (ca. 1000 lux) with 12-24 hours of daylight each day. Both shoot and callus formation can be obtained when cultures are grown at 22-28 C., i.e., ambient temperature should be adequate in most cases.
9. When young shoots appear, they should be transplanted to rooting medium. Root formation can be observed in either hormone-free MS medium or in half-strength MS medium.

10. Rooted plants can be transplanted to Jiffy 7 pots. Humidity must be high for the first few days following transplantation to Jiffy 7 pots.

In addition to direct plant regeneration, presented in the above protocol, it is possible to obtain either callus-mediated plant regeneration or plant regeneration from cell suspension cultures in tobacco. If callus growth is initiated and maintained in medium with 2,4-D, the 2,4-D must be replaced with a cytokinin to achieve shoot formation. In this manner plant regeneration has been obtained from long-term suspension cultures of tobacco (Gamborg et al., 1979). Numerous combinations of growth regulating auxin and cytokinins have been varied by researchers attempting to achieve plant regeneration in tobacco.

This general protocol for in vitro manipulation of *Nicotiana* species can be adapted for use in developing an organogenic cell culture system for many crop species. The variables most likely to be changed are: (1) donor plant growth, (2) explant source, (3) culture medium, (4) plant growth regulator concentrations for callus initiation, (5) growth regulator concentrations for callus maintenance, and (6) culture conditions for plant regeneration. Most importantly, these are not independent variables; it is essential that each interact correctly with all others.

FUTURE PROSPECTS

Few crop species have been recently added to the list of species undergoing organogenesis from in vitro cultures. At this time regeneration from long-term cultures is the most pressing problem. Although direct organogenesis from explants may be possible, in many instances regeneration from long-term cultures is impossible. Future emphasis should be placed on design of systems for regeneration of plants from long-term cultures. Whereas in the past a great deal of emphasis has been placed on investigation of plant growth regulator concentrations required for callus growth and organogenesis, little work has concentrated on other aspects of plant cell culture.

The ability of a cell culture to regenerate plants is probably related to more than the growth regulator composition of the culture medium. All those factors which have been discussed in this review should be considered in development of protocols for plant regeneration. Probably one of the most important factors in attaining plant regeneration is the method of maintenance of callus and suspension cultures. As discussed earlier in this review, aneuploidy can severely restrict regeneration potential. Hence future research should emphasize establishment and maintenance of chromosomally stable cell cultures. This in turn may enable expansion of the number of crop species that can be regenerated from cell culture.

With the development of protocols for the regeneration of plants from in vitro cell cultures a number of other cellular and molecular techniques will be available for use in plant improvement. Genetic engineering of higher plants will depend on the ability to regenerate plants from single cells, i.e., protoplasts. Protoplasts are useful in many respects, e.g., isolation of mutants, feeding of liposome encapsulated DNA, and somatic hybridization via protoplast fusion, but all these techniques depend on the ability to regenerate plants from long-term cell or protoplast cultures.

A new source of genetic variability is available in plants regenerated from cell culture. This somaclonal variation is a useful source of variability only if plants can be efficiently regenerated from cell cultures. This variability is not only present in plants directly regenerated from cell culture, but is also obvious in plants regenerated from protoplast fusion products (e.g., Evans et al., 1982). In this instance somatic hybrids demonstrate greater variability than sexual hybrids, even between sexually compatible species. Somaclonal variability can only be exploited for those crops that regenerate from cell culture.

The uniqueness of cellular and molecular genetics in higher plants as compared to animal cells is dependent upon the potential for regeneration of plants from cell cultures. Without consistent regeneration of important crop species, agricultural applications of plant cell culture will be limited.

KEY REFERENCES

Evans, D.A., Sharp, W.R., and Flick, C.E. 1981. Growth and behavior of cell cultures: Embryogenesis and organogenesis. In: Plant Tissue Culture: Methods and Applications in Agriculture (T.A. Thorpe, ed.) pp. 45–113. Academic Press, New York.

D'Amato, F. 1977. Cytogenetics of differentiation in tissue and cell cultures. In: Plant Cell, Tissue, and Organ Culture (J. Reinert and Y.P.S. Bajaj, eds.) pp. 343–357. Springer–Verlag, New York.

Murashige, T. 1974. Plant propagation through tissue cultures. Ann. Rev. Plant Physiol. 25:135–166.

Thorpe, T.A. 1980. Organogenesis in vitro: Structural, physiological and biochemical aspects. Int. Rev. Cytol. Suppl. 11A:71–112.

REFERENCES

Abo El-Nil, M.M. 1977. Organogenesis and embryogenesis in callus cultures of garlic (*Allium sativum* L.). Plant Sci. Lett. 9:259–264.

Ahloowalia, B.S. 1975. Regeneration of ryegrass plants in tissue culture. Crop Sci. 15:449–452.

Ahloowalia, B.S. 1976. Chromosomal changes in parasexually produced ryegrass. In: Current Chromosome Research (K. Jones and P.E. Brandham, eds.) pp. 115–122. Elsevier/North–Holland, Amsterdam.

Ahuja, M.R. and Hagen, G.L. 1966. Morphogenesis in *Nicotiana debneyi-tabacum*, *N. longiflora* and their tumor-forming hybrid derivatives. Dev. Biol. 13:408–423.

Anderson, W.C., Meagher, G.W., and Nelson, A.G. 1977. Cost of propagating broccoli plants through tissue culture. HortScience 12:543-544.

Bajaj, Y.P.S. 1972. Effect of some growth regulators on bud formation by excised leaves of *Torenia fournieri*. Z. Pflanzenphysiol. 66:284-287.

_____ and Bopp, M. 1972. Growth and organ formation in *Sinapis alba* tissue culture. Z. Pflanzenphysiol. 66:378-381.

_____ and Mader, M. 1974. Growth and morphogenesis in tissue cultures of *Angallis arvensis*. Physiol. Plant. 32:43-48.

_____ and Nietsch, P. 1975. In vitro propagation of red cabbage (*Brassica oleraceae* L. var. capitata). J. Exp. Bot. 26:883-890.

_____, Kumar, P., Labana, K.S., and Singh, M.M. 1981. Regeneration of plants from seedling explants and callus cultures of *Arachis hypogaea* L. Ind. J. Exp. Biol. 19:1026-1029.

_____, Sidhu, B.S., and Dubey, V.K. 1981. Regeneration of genetically diverse plants from tissue cultures of forage grass—*Panicum* spp. Euphytica 30:135-140.

Banks, M.S. 1979. Plant regeneration from callus from two growth phases of English ivy, *Hedera helix* L. Z. Pflanzenphysiol. 92:349-353.

Barlass, M., Grant, W.J.R., and Skene, K.G.M. 1980. Shoot regeneration in vitro from native Australian fruit-bearing trees—Quandong and plum bush. Aust. J. Bot. 28:405-409.

Baroncelli, S., Buiatti, M., and Bennici, A. 1973. Genetics of growth and differentiation in vitro of *Brassica oleracea* var. botrytis. Z. Pflanzenzuchtg. 70:99-107.

Bennici, A. 1974. Cytological analysis of roots, shoots, and plants regenerated from suspension and solid in vitro cultures of haploid *Pelargonium*. Z. Pflanzenzuchtg. 72:199-205.

_____ and D'Amato, F. 1978. In vitro regeneration of Durum wheat plants. 1. Chromosome numbers of regenerated plantlets. Z. Planzenzuchtg. 81:305-311.

Bharal, S. and Rashid, A. 1979. Regeneration of plants from tissue cultures of the legume, *Indigofera enneaphylla* Linn. Z. Pflanzenphysiol. 92:443-447.

Bhatt, P.N., Bhatt, D.P., and Sussex, I.M. 1979. Organ regeneration from leaf desks of *Solanum nigrum*, *S. dulcamara*, and *S. khasianum*. Z. Pflanzenphysiol. 95:355-362.

Binding, H., Nehls, R., Schieder, O., Sopory, S.K., and Wenzel, G. 1978. Regeneration of mesophyll protoplasts isolated from dihaploid clones of *Solanum tuberosum*. Physiol. Plant. 43:52-54.

Bingham, E.T., Hurley, L.V., Kaatz, D.M., and Saunders, J.W. 1975. Breeding alfalfa which regenerates from callus tissue in culture. Crop Sci. 15:719-721.

Bonga, J.M. 1977. Applications of tissue culture in forestry. In: Plant Cell, Tissue, and Organ Culture (J. Reinert and Y.P.S. Bajaj, eds.) pp. 93-107. Springer-Verlag, New York.

Bornman, C.H. and Jansson, E. 1980. Organogenesis in cultured *Pinus sylvestris* tissue. Z. Pflanzenphysiol. 96:1-6.

Bottino, P.J., Maire, C.E., and Goff, L.M. 1979. Tissue culture and organogenesis in the winged bean. Can. J. Bot. 57:1773-1776.

Bourgin, J.P., Chupeau, Y., and Missonier, C. 1979. Plant regeneration from mesophyll protoplasts of several *Nicotiana* species. Physiol. Plant. 45:288-292.

Bowes, B.G. 1970. Preliminary observations on organogenesis in *Taraxacum officinale* tissue cultures. Protoplasma 71:197-202.

_____ 1976. In vitro morphogenesis of *Crambe maritima* L. Protoplasma 89:185-188.

Brandao, I. and Salema, R. 1977. Callus and plantlets development from cultured leaf explants of *Sedum telephium* L. Z. Pflanzenphysiol. 85:1-8.

Broome, O.C. and Zimmerman, R.H. 1978. In vitro propagation of blackberry. HortScience 13:151-153.

Brown, W.V. 1972. Textbook of cytogenetics. C.V. Mosby, St. Louis.

Buchala, A.J. and Schmid, A. 1979. Vitamin D and its analogues, a new class of plant growth substances affecting rhizogenesis. Nature 280: 230-236.

Bush, S.R., Earle, E.D., and Langhans, R.W. 1976. Plantlets from petal segments, petal epidermis, and shoot tips of the periclinal chimera, *Chrysanthemum morifolium* "Indianapolis." Am. J. Bot. 63:729-737.

Campbell, R.A. and Durzan, D.J. 1975. Induction of multiple buds and needles in tissue cultures of *Picea glauca.* Can. J. Bot. 53:1652-1657.

Carlson, P.S., Smith, H.H., and Dearing, R.D. 1972. Parasexual interspecific plant hybridization. Proc. Nat. Acad. Sci. USA 69:2292-2294.

Carter, O. Yamada, Y., and Takahashi, E. 1967. Tissue culture of oats. Nature 214:1029-1030.

Caruso, J.L. 1971. Bud formation in excised stem segments of *Verbascum thapsus.* Am. J. Bot. 58:429-431.

Chen, C.H., Stenberg, N.E., and Ross, J.G. 1977. Clonal preparation of big bluestem by tissue culture. Crop Sci. 17:847-850.

Chen, C.H., Lo, P.F., and Ross, J.G. 1979. Regeneration of plantlets from callus cultures of Indian grass. Crop Sci. 19:117-118.

Cheng, T.-Y. 1975. Adventitious bud formation in culture of Douglas-fir (*Pseudotsuga menziesii* Mirb. Franco). Plant Sci. Lett. 5:97-102.

_____ 1976. Vegetative propagation of western hemlock (*Tsuga heterophylla*) through tissue culture. Plant Cell Physiol. 17:1347-1350.

_____ and Smith, H.H. 1975. Organogenesis from callus culture of *Hordeum vulgare.* Planta 123:307-310.

_____ and Voqui, T.H. 1977. Regeneration of Douglas-fir plantlets through tissue culture. Science 198:306-307.

Chin, J.C. and Scott, K.J. 1977. Studies on the formation of roots and shoots in wheat callus cultures. Ann. Bot. 41:473-481.

Chua, B.U., Kunisaki, J.T., and Sagawa, Y. 1981. In vitro propagation of *Draceana marginata* "Tricolor." HortScience 16:494.

Clare, M.V. and Collin, H.A. 1974. The production of plantlets from tissue cultures of brussels sprout (*Brassica oleracea* L. var gemmifera D.C.). Ann. Bot. 38:1067-1076.

Coleman, W.K. and Greyson, R.I. 1977. Promotion of root initiation by gibberellic acid in leaf discs of tomato (*Lycopersicon esculentum*) cultured in vitro. New Phytol. 78:47-54.

Colijn, C.M., Kool, A.J., and Nijkamp, H.J.J. 1979. Induction of root and shoot formation from root meristems of *Petunia hybrida.* Protoplasma 99:335-340.

Conger, B.V. and Carabia, J.V. 1978. Callus induction and plantlet regeneration in orchardgrass. Crop Sci. 18:157–159.

Crocomo, O.J., Sharp, W.R., and Peters, J.E. 1976. Plantlet morphogenesis and the control of callus growth and root induction of *Phaseolus vulgaris* with the addition of a bean seed extract. Z. Pflanzenphysiol. 78:456–460.

Cummings, D.P., Green, C.E., and Stuthman, D.D. 1976. Callus induction and regeneration in oats. Crop Sci. 16:465–470.

Dale, P.J. 1980. Embryoids from cultured immature embryos of *Lolium multiflorum*. Z. Pflanzenphysiol. 100:73–77.

_____ and Deambrogio, E. 1979. A comparison of culture induction and plant regeneration from different explants of *Hordeum vulgare*. Z. Pflanzenphysiol. 94:65–67.

Davey, M.R., Mackenzie, I.A., Freeman, G.G., and Short, K.C. 1974. Studies of some aspects of the growth, fine structure and flavour production of onion tissue grown in vitro. Plant Sci. Lett. 3:113–120.

David, A. and David, H. 1977. Manifestations de diverses potentialiters organogenes d'organes ou de fragments d'organes de Pin maritime (*Pinus pinaster* Sol.) en cultive in vitro. C.R. Acad. Sci. Paris 284:627–630.

Davies, M.E. and Dale, M.M. 1979. Factors affecting in vitro shoot regeneration on leaf discs of *Solanum lanciniatum* Ait. Z. Pflanzenphysiol. 92:51–60.

Davoyan, E.I. and Smetanin, A.P. 1979. Callus production and regeneration of rice plants. Fiziologiya Rastenii 26:323–329.

DeGreef, W. and Jacobs, M. 1979. In vitro culture of the sugarbeet: description of a cell line with high regeneration capacity. Plant Sci. Lett. 17:55–61.

DeLanghe, E., Debergh, P., and Van Rijk, R. 1974. In vitro culture as a method for vegetative propagation of *Euphorbia pulcherrima*. Z. Pflanzenphysiol. 71:271–274.

Devos, P., DeLanghe, E., and DeBruijne, E. 1979. Influence of 2,4-D on the propagation of *Cynara scolymus* L. in vitro. Meded. Fak. Landbouwwet. Gent. 27:829–836.

Dhoot, G.K. and Henshaw, G.G. 1977. Organization and alkaloid production in tissue cultures of *Hyoscyamus niger*. Ann. Bot. 41:943–949.

Dhurva, B., Ramakrishnan, T., and Vaidyanathan, C.S. 1978. Regeneration of hybrid tomato plants from leaf callus. Curr. Sci. 47:458–460.

Doerschug, M.R. and Miller, C.O. 1967. Chemical control of adventitious organ formation in *Lactuca sativa* explants. Am. J. Bot. 54:410–413.

Dudits, D., Nemet, G., and Haydu, Z. 1975. Study of callus growth and organ formation in wheat (*Triticum aestivum*) tissue cultures. Can. J. Bot. 53:957–963.

Dunstan, D.I. and Short, K.C. 1979. Shoot production from cultured *Allium porrum* tissues. Sci. Hort. 11:37–43.

Earle, E.D. and Langhans, R.W. 1974a. Propagation of *Chrysanthemum* in vitro. I. Multiple plantlets from shoot tips and the establishment of tissue cultures. J. Am. Soc. Hort. Sci. 99:128–132.

_____ 1974b. Propagation of *Chrysanthemum* in vitro. II. Production, growth and flowering of plantlets from tissue cultures. J. Am. So. Hort. Sci. 99:352–358.

Engvild, K.C. 1973.   Shoot differentiation in callus cultures of *Datura innoxia*. Physiol. Plant. 28:155-159.

Evans, D.A. and Gamborg, O.L. 1982.   Chromosome stability of plant cell suspension cultures. Plant Cell Rep. 1:104-107.

Evans, D.A., Sharp, W.R., and Paddock, E.F. 1976.   Variation in callus proliferation and root morphogenesis in leaf tissue cultures of *Glycine max* strain T219. Phytomorphology 26:379-384.

Evans, D.A., Wetter, L.R., and Gamborg, O.L. 1980.   Somatic hybrid plants of *Nicotiana glauca* and *Nicotiana tabacum* obtained by protoplast fusion. Physiol. Plant. 48:225-230.

Evans, D.A., Flick, C.E., Kut, S.A., and Reed, S.M. 1982.   Comparison of *Nicotiana tabacum* and *Nicotiana nesophila* hybrids produced by ovule culture and protoplast fusion. Theor. Appl. Genet. 62:193-198.

Fassuliotis, G. 1975.   Regeneration of whole plants from isolated stem parenchyma cells of *Solanum sisymbriifolium*. J. Am. Soc. Hort. Sci. 100:636-638.

_____, Nelson, B.V., and Bhatt, D.P. 1981.   Organogenesis in tissue culture of *Solanum melongena* cv. Florida Market. Plant Sci. Lett. 22:119-125.

Fonnesbech, M. 1974.   The influence of NAA, BA and temperature on shoot and root development from *Begonia* x *cheimantha* petiole segments grown in vitro. Plant Physiol. 32:49-54.

Frearson, E.M., Power, J.B., and Cocking, E.C. 1973.   The isolation, culture, and regeneration of *Petunia* protoplasts. Dev. Biol. 33:130-137.

Freeling, M., Woodman, J.C., and Cheng, D.S.K. 1976.   Developmental potentials of maize tissue cultures. Maydica 21:97-112.

Fridborg, G. 1971.   Growth and organogenesis in tissue cultures of *Allium cepa* var. proliferum. Physiol. Plant. 25:436-440.

Gamborg, O.L. and Shyluk, J.P. 1970.   The culture of plant cells with ammonium salts as the sole nitrogen source. Plant Physiol. 45:598-600.

_____ 1976.   Tissue culture, protoplasts, and morphogenesis in flax. Bot. Gaz. 137:301-306.

Gamborg, O.L. and Wetter, L.R. 1975.   Plant tissue culture methods. National Research Council, Saskatoon.

Gamborg, O.L., Miller, R.A., and Ojima, K. 1968.   Plant cell cultures. I. Nutrient requirements of suspension cultures of soybean root cells. Exp. Cell Res. 50:151-158.

Gamborg, O.L., Shyluk, J.P., Brar, D.S., and Constabel, F. 1977.   Morphogenesis and plant regeneration from callus of immature embryos of sorghum. Plant Sci. Lett. 10:67-74.

Gamborg, O.L., Shyluk, J.P., Fowke, L.C., Wetter, L.R., and Evans, D.A. 1979.   Plant regeneration from protoplasts and cell cultures of *N. tabacum* sulfur mutant (Su/Su). Z. Pflanzenphysiol. 95:255-264.

George, L. and Rao, P.S. 1980.   In vitro regeneration of mustard plants (*Brassica juncea* var RAI-5) on cotyledon explants from non-irradiated, irradiated and mutagen-treated seed. Ann. Bot. 46:107-112.

Gharyal, P.K. and Maheshwari, S.C. 1981.   In vitro differentiation of somatic embryoids in a leguminous tree—*Albizzia lebbeck* L. Naturwissen. 68:379-380.

Gosch-Wackerle, G., Avivi, L., and Galun, E. 1979. Induction, culture and differentiation of callus from immature rachises, seeds and embryos of *Triticum*. Z. Pflanzenphysiol. 91:267-278.

Gould, A.R. 1978. Diverse pathways of morphogenesis in tissue cultures of the composite *Brachycome lineariloba* (2n=4). Protoplasma 97:125-135.

_____ 1979. Chromosomal and phenotypic stability during regeneration of whole plants from tissue cultures of *Brachycome dichromosomatica* (2n=4). Aust. J. Bot. 27:117-121.

Green, C.E. 1977. Prospects for crop improvement in the field of cell culture. HortScience 12:131-134.

_____ and Phillips, R.L. 1975. Plant regeneration from tissue cultures of maize. Crop Sci. 15:417-421.

Grewal, S. and Atal, C.K. 1976. Plantlet formation in callus cultures of *Dioscorea deltoidea* Wall. Ind. J. Exp. Biol. 14:352-353.

Grinblat, U. 1972. Differentiation of *Citrus* stem in vitro. J. Am. Soc. Hort. Sci. 97:599-603.

Groenewald, E.G., Koeleman, A., and Wessels, D.C.J. 1976. Callus formation and plant regeneration from seed tissue of *Aloe pretoriensis* Pole Evans. Z. Pflanzenphysiol. 75:270-272.

Guha, S. and Maheshwari, S.C. 1964. In vitro production of embryos from anthers of *Datura*. Nature 204:497.

Gunay, A.L. and Rao, P.S. 1978. In vitro plant regeneration from hypocotyl and cotyledon explants of red pepper (*Capsicum*). Plant Sci. Lett. 11:365-372.

_____ 1980. In vitro propagation of hybrid tomato plants (*Lycopersicon esculentum* L.) using hypocotyl and cotyledon explants. Ann. Bot. 45:205-207.

Haddon, L. and Northcote, D.H. 1976. The influence of gebberellic acid and abscisic acid on cell and tissue differentiation of bean callus. J. Cell Sci. 20:47-55.

Harms, C.T., Lorz, H., and Potrykus, I. 1976. Regeneration of plantlets from callus cultures of *Zea mays* L. Z. Pflanzenzuchtg. 77:347-351.

Harvey, A.E. and Grasham, J.L. 1969. Procedures and medium for obtaining tissue cultures of 12 conifer species. Can. J. Bot. 47:547-549.

Havranek, P. and Novak, F.J. 1973. The bud formation in the callus cultures of *Allium sativum* L. Z. Pflanzenphysiol. 68:308-318.

Hayward, C. and Power, J.B. 1975. Plant production from leaf protoplasts of *Petunia parodii*. Plant Sci. Lett. 4:407-410.

Heinz, D.J., Krishnamurthi, M., Nickell, L.G., and Maretzki, A. 1977. Cell tissue and organ culture in sugarcane improvement. In: Plant Cell, Tissue and Organ Culture. (J. Reinert and Y.P.S. Bajaj eds.) pp. 3-17. Springer-Verlag, New York.

Heinz, D.J. and Mee, G.W.P. 1969. Plant differentiation from callus tissue of *Saccharum* species. Crop Sci. 9:346-348.

Helgeson, J.P. 1979. Tissue and cell-suspension culture. In: *Nicotiana*: Procedures for Experimental Use (R.D. Durbin, ed.) pp. 52-59. USDA, Washington, D.C.

Henke, R.R. Mansur, M.A., and Constantin, M.J. 1978. Organogenesis and plantlet formation from organ- and seedling-derived calli of rice (*Oryza sativa*). Physiol. Plant. 44:11-14.

Henry, R.J. 1978. In vitro propagation of *Peperomia* "Red Ripple" from leaf discs. HortScience 13:150–151.

Herman, E.B. and Haas, G.J. 1978. Shoot formation in tissue cultures of *Lycopersicon esculentum* (Mill.). Z. Pflanzenphysiol. 89:467–470.

Heuser, C.W. and Apps, D.A. 1976. In vitro plantlet formation from flower petal explants of *Hemerocallis* cv. Chipper Cherry. Can. J. Bot. 54:616–618.

Hill, G.P. 1967. Morphogenesis in stem callus of *Convolvulus arvensis*. Ann. Bot. 31:437–446.

Hiraoka, N. and Tabata, M. 1974. Alkaloid production by plants regenerated from cultured cells of *Datura innoxia*. Phytochemistry 13:1671–1675.

Hirotani, M. and Furuya, T. 1977. Restoration of cardenolide synthesis in redifferentiated shoots from callus cultures of *Digitalis purpurea*. Phytochemistry 16:610–611.

Horak, J., Lustinec, J., Mesicek, J., Kaminek, M., and Polaco-Kova, D. 1975. Regeneration of diploid and polyploid plants from the stem pith explants of diploid marrow stem kale (*Brassica oleraceae* L.). Ann. Bot. 39:571–577.

Horvath Beach, K. and Smith, R.R. 1979. Plant regeneration from callus of red and crimson clover. Plant Sci. Lett. 16:231–237.

Hui, L.H. and Zee, S.-Y. 1978. In vitro plant formation from hypocotyls and cotyledons of leaf-mustard cabbage (*Brassica juncea* Coss). Z. Pflanzenphysiol. 89:77–80.

_____ 1981. In vitro propagation of *Peperomia viridis* using medium supplemented with ginseng powder. HortScience 16:86–87.

Husemann, W. and Reinert, J. 1976. Regulation of growth and morphogenesis in cell cultures of *Crepis capillaris* by light and phytohormones. Protoplasma 90:353–367.

Hussey, G. 1975. Totipotency in tissue explants and callus of some members of the *Liliaceae*, *Iridaceae*, and *Amaryllidaceae*. J. Exp. Bot. 26:253–262.

_____ 1976. Plantlet regeneration from callus and parent tissue in *Ornithogalum thrysoides*. J. Exp. Bot. 27:375–382.

Jansson, E. and Bornman, C.H. 1980. In vitro phyllomorphic regeneration of shoot buds and shoots in *Picea abies*. Physiol. Plant. 49:105–111.

Jayakar, M. 1971. In vitro flowering of *Crepis capillaris*. Phytomorphology 20:410–412.

Johnson, B.B. and Mitchell, E.D. 1978. In vitro propagation of broccoli from stem, leaf, and leaf rib explants. HortScience 13:246–247.

Jordan, M. and Cortes, I. 1981. Shoot organogenesis in tissue culture of *Drimys winteri*. Plant Sci. Lett. 23:177–180.

Kamada, H. and Harada, H. 1979. Influence of several growth regulators and amino acids on in vitro organogenesis of *Torenia fournieri*. J. Exp. Bot. 30:27–36.

Kamat, M.G. and Rao, P.S. 1978. Vegetative multiplication of eggplants (*Solanum melongena*) using tissue culture techniques. Plant Sci. Lett. 13:57–65.

Kameya, T. and Widholm, J. 1981. Plant regeneration from hypocotyl sections of *Glycine* species. Plant Sci. Lett. 21:289–294.

Kao, K.N., Miller, R.A., Gamborg, O.L., and Harvey, B.L. 1970. Variations in chromosome number and structure in plant cells grown in suspension cultures. Can. J. Genet. Cytol. 12:297-301.

Kartel, N.A. and Maneshina, T.V. 1978. Regeneration of barley plants in callus-tissue cultures. Soviet Plant Physiol. 25:223-226.

Kartha, K.K., Michayluk, M.R., Kao, K.N., Gamborg, O.L., and Constabel, F. 1974. Callus formation and plant regeneration from mesophyll protoplasts of rape plants (*Brassica napus* L. cv. Zephyr). Plant Sci. Lett. 3:265-271.

Kartha, K.K., Gamborg, O.L., Shyluk, J.P., and Constabel, F. 1976. Morphogenetic investigations on in vitro leaf culture of tomato (*Lycopersicon esculentum* (Mill.) cv. Starfire) and high frequency plant regeneration. Z. Pflanzenphysiol. 77:292-301.

Kasperbauer, M.J., Buckner, R.C., and Bush, L.P. 1979. Tissue culture of annual rygrass x tall fescue F_1 hybrids: Callus establishment and plant regeneration. Crop Sci. 19:457-460.

Kasperbauer, M.J. and Collins, G.B. 1972. Reconstitution of diploids from anther-derived haploids in tobacco. Crop Sci. 12:98-101.

Kato, Y. 1975. Adventitious bud formation in etiolated stem segments and leaf callus of *Heloniopsis orientalis* (Liliaceae). Z. Pflanzenphysiol. 75:211-216.

———— 1978. Induction of adventitious buds on undetached leaves, excised leaves and leaf fragments of *Heloniopsis orientalis* (Liliaceae). Physiol. Plant. 42:39-44.

Kaul, K. and Sabharwal, P.S. 1972. Morphogenetic studies on *Haworthia*: establishment of tissue culture and control of differentiation. Am. J. Bot. 59:377-385.

Khanna, R. and Chopra, R.N. 1977. Regulation of shoot-bud and root formation from stem explants of *Lobularia maritima*. Phytomorphology 27:267-274.

Kim, Y.J., Hasegawa, P.M., and Bressan, R.A. 1981. In vitro propagation of hyacinth. HortScience 16:645-647.

Kimball, S.L. and Bingham, E.T. 1973. Adventitious bud development of soybean hypocotyl sections in culture. Crop Sci. 13:758-760.

King, P.J., Potrykus, I., and Thomas, E. 1978. In vitro genetics of cereal: Problems and perspectives. Physiol. Veg. 16:381-399.

Koblitz, H. and Saalbach, G. 1976. Callus cultures from apical meristems of barley (*Hordeum vulgare*). Biochem. Physiol. Pflanzen. 170:97-102.

Kolar, Z., Bartek, J., and Vyskot, B. 1976. Vegetative propagation of the cactus *Mamillaria woodsii* Craig through tissue cultures. Experiential 32:668-669.

Konar, R.N. and Konar, A. 1966. Plantlet and flower formation in callus culture from *Phlox drummondii* Hook. Phytomorphology 16:379-382.

Kostoff, D. 1943. Cytogenetics of the genus *Nicotiana*. States Printing House, Sofia, Bulgaria.

Krishnamurthi, M. and Tlaskal, J. 1974. Fiji disease resistant *Saccharum officinarum* var. Pindar sub-clones from tissue cultures. Proc. Int. Soc. Sugar Cane Technol. 15:130-137.

Kunisaki, J.T. 1975. In vitro propagation of *Cordyline terminalis* Kunth. HortScience 10:601-602.

Lam, S.L. 1975. Shoot formation in potato tuber discs in tissue culture. Am. Pot. J. 52:103-106.

Landova, B. and Landa, Z. 1974. Organogenesis in callus culture of *Gazania splendens* Moore induced on new medium. Experientia 30:832-834.

Lee, C.W., Skirvin, R.M., Soltero, A.I., and Janick, J. 1977. Tissue culture of *Salpiglossis sinuata* L. from leaf discs. HortScience 12:547-549.

Linsmaeier, E.M. and Skoog, F. 1965. Organic growth factor requirements of tobacco tissue cultures. Physiol. Plant. 18:100-127.

Litz, R.E. and Conover, R.A. 1980. Partial organogenesis in tissue cultures of *Averrhoa carambola*. HortScience 15:735.

Liu, M.C. and Chen, W.H. 1976. Tissue and cell culture as aids to sugarcane breeding. I. Creation of genetic variation through callus culture. Euphytica 25:393-403.

Liu, M.C., Huang, Y.J., and Shih, S.C. 1972. The in vitro production of plants from several tissues of *Saccharum* species. J. Agric. Assoc. China 77:52-58.

Loewenberg, J.R. 1969. Cyclamen callus culture. Can. J. Bot. 47:2065-2067.

Lorz, H., Harms, C.T., and Potrykus, I. 1976. Regeneration of plants from callus in *Avena sativa* L. Z. Pflanzenzuchtg. 77:257-259.

Lowe, K.W. and Conger, B.V. 1979. Root and shoot formation from callus cultures of tall fescue. Crop Sci. 19:397-400.

Lustinec, J. and Horak, J. 1970. Induced regeneration of plants in tissue cultures of *Brassica oleracea*. Experientia 26:919-920.

Maliga, P., Kiss, Z.R., Nagy, A.H., and Lazar, G. 1978. Genetic instability in somatic hybrids of *Nicotiana tabacum* and *Nicotiana knightiana*. Molec. Gen. Genet. 163:145-151.

Maliga, P., Lazar, G., Joo, F., Nagy, A.H., and Menczel, L. 1977. Restoration of morphogenic potential in *Nicotiana* by somatic hybridization. Molec. Gen. Genet. 157:291-296.

Malmberg, R.L. 1979. Regeneration of whole plants from callus cultures of diverse genetic lines of *Pisum sativum* L. Planta 146:243-244.

Martins-Loucao, M.A. and Rodriguez-Barrueco, C. 1981. Establishment of proliferating callus from roots, cotyledons and hypocotyls of carob (*Ceratonia siliqua* L.) seedlings. Z. Pflanzenphysiol. 103:297-303.

Mascarenhas, A.F., Hendre, R.R., Nadgir, A.L., Ghugale, D.D., Godbole, D.A., Prabhu, R.A., and Jagannathan, V. 1976. Development of plantlets from cultured tissue of *Dioscorea deltoidea* Wall. Ind. J. Exp. Bot. 14:604-606.

Masteller, V.J. and Holden, D.J. 1970. The growth of and organ formation from callus tissue of sorghum. Plant Physiol. 45:362-364.

Matern, U., Strobel, G., and Shepard, J. 1978. Reaction to phytotoxins in a potato population derived from mesophyll protoplasts. Proc. Nat. Acad. Sci. 75:4935-4939.

Mathews, V.H. and Rangan, T.S. 1979. Multiple plantlets in lateral bud leaf explant in vitro cultures of pineapple. Sci. Hort. 11:319-328.

_____, and Narayanaswamy, S. 1976. Micro-propagation of *Ananas sativus* in vitro. Z. Pflanzenphysiol. 79:450-454.

Matsuoka, H. and Hinata, K. 1979. NAA-induced organogenesis and embryogenesis in hypocotyl callus of *Solanum melongena* L. J. Exp. Bot. 30:363-370.

McCoy, T.J. and Bingham, E.T. 1977. Regeneration of diploid alfalfa plants from cells grown in suspension culture. Plant Sci. Lett. 10:59-66.

Mehra, A. and Mehra, P.N. 1972. Differentiation in callus cultures of *Mesembryanthemum floribundum*. Phytomorphology 22:171-176.

_____ 1974. Organogenesis and plantlet formation in vitro in almond. Bot. Gaz. 135:61-73.

Mehra, P.N. and Sachdeva, S. 1979. Callus cultures and organogenesis in apple. Phytomorphology 29:310-324.

Meijer, E.G.M. and Broughton, W.J. 1981. Regeneration of whole plants from hypocotyl-, root-, and leaf-derived tissue cultures of the pasture legume *Stylosanthes guyanensis*. Physiol. Plant. 52:280-284.

Melchers, G. and Sacristan, M.D. 1977. Somatic hybridization of plants by fusion of protoplasts. II. The chromosome numbers of somatic hybrid plants of four different fusion experiements. In: La culture des tissus et des cellules des vegetaux (R. Gautheret, ed.) pp. 169-177. Masson et Cie, Paris.

Mellor, F.C. and Stace-Smith, R. 1969. Development of excised potato buds in nutrient culture. Can. J. Bot. 47:1617-1621.

Meredith, C.P. 1978. Response of cultured tomato cells to aluminum. Plant Sci. Lett. 12:17-24.

_____ 1979. Shoot development in established callus cultures of cultivated tomato (*Lycopersicon esculentum* Mill.) Z. Pflanzenphysiol. 95:405-411.

Meyer, M.M. and Milbrath, G.M. 1977. In vitro propagation of horseradish with leaf pieces. HortScience 12:544-545.

Minocha, S.C. 1980. Callus and adventitious shoot formation in excised embryos of white pine (*Pinus strobus*). Can. J. Bot. 58:366-370.

Mitra, J., Mapes, M.O., and Steward, F.C. 1960. Growth and organized development of cultured cells. IV. The behavior of the nucleus. Am. J. Bot. 47:357-368.

Mokhtarzadeh, A. and Constantin, M.J. 1978. Plant regeneration from hypocotyl and anther-derived callus of berseem clover. Crop Sci. 18:567-572.

Mudgal, A.K., Goel, S., Gupta, S.C., and Chopra, R.N. 1981. Regeneration of *Iberis amara* plants from in vitro cultured leaf and stem explants. Z. Pflanzenphysiol. 101:179-182.

Murashige, T. 1979. Principles of rapid propagation. In: Propagation of Higher Plants Through Tissue Culture (K.W. Hughes, R. Henke, and M. Constantin, eds.). USDOE, Washington, D.C.

_____ and Nakano, R. 1967. Chromosome complement as a determinant of the morphogenetic potential of tobacco cells. Am. J. Bot. 54:963-970.

_____ and Skoog, F. 1962. A revised medium for rapid growth and bioassays with tobacco tissue cultures. Physiol. Plant. 15:473-497.

Nadar, H.M. and Heinz, D.J. 1977. Root and shoot development from sugarcane callus tissue. Crop Sci. 17:814-816.

Nadar, H.M. Soepraptopo, S., Heinz, D.J., and Ladd, S.L. 1978. Fine structure of sugarcane (*Saccharum* spp.) callus and the role of auxin in embryogenesis. Crop Sci. 18:210–216.

Nagata, T. and Takebe, I. 1971. Plating of isolated tobacco mesophyll protoplasts on agar medium. Planta 99:12–20.

Nakano, H., Tashiro, T., and Maeda, E. 1975. Plant differentiation in callus tissue induced from immature endosperm of *Oryza sativa* L. Z. Pflanzenphysiol. 76:444–449.

Nataraja, K. 1971. Morphogenetic variations in callus cultures derived from floral buds and anthers of some members of *Ranunculaceae*. Phytomorphology 21:290–296.

Negrutiu, I. 1976. In vitro morphogenesis in *Arabidopsis thaliana*. Arabidopsis Inform. Serv. 13:181–187.

———— and Jacobs, M. 1978a. Factors which enhance in vitro morphogenesis of *Arabidopsis thaliana*. Z. Pflanzenphysiol. 90:423–430.

———— and Jacobs, M. 1978b. Restoration of morphogenetic capacity in long-term callus culture of *Arabidopsis thaliana*. Z. Pflanzenphysiol. 90:431–441.

————, Beeftink, F., and Jacobs, M. 1975. *Arabidopsis thaliana* as a model system in somatic cell genetics. I. Cell and tissue culture. Plant Sci. Lett. 5:293–304.

————, Jacobs, M., and Cachita, D. 1978a. Some factors controlling in vitro morphogenesis of *Arabidopsis thaliana*. Z. Pflanzenphysiol. 86:113–124.

————, Jacobs, M., and DeGreef, W. 1978b. In vitro morphogenesis of *Arabidopsis thaliana*: The origin of the explant. Z. Pflanzenphysiol. 90:363–372.

Niimi, Y. and Onozawa, T. 1979. In vitro bulblet formation from leaf segments of lilies, especially *Lilium rubellum* Baker. Sci. Hort. 11:379–389.

Nishi, T., Yamada, Y., and Takahashi, E. 1968. Organ redifferentiation and plant restoration in rice callus. Nature 219:508–509.

———— 1973. The role of auxins in differentiation of rice tissues cultured in vitro. Bot. Mag. (Tokyo) 86:183–188.

Nitsch, J.P., Nitsch, C., Rossini, L.M.E., and Bui Dang Ha, D. 1967. The role of adenine in bud differentiation. Phytomorphology 17:446–453.

Norton, J.P. and Boll, W.G. 1954. Callus and shoot formation from tomato roots in vitro. Science 119:220–221.

Ogura, H. 1975. The effects of a morphactin, chloroflurenol, on organ redifferentiation from calluses cultured in vitro. Bot. Mag. (Tokyo) 88:1–8.

———— 1976. The cytological chimeras in original regenerates from tobacco tissue cultures and in their offsprings. Jap. J. Genet. 51:161–174.

———— 1978. Genetic control of chromosomal chimerism found in a regenerate from tobacco callus. Jap. J. Genet. 53:77–90.

———— and Tsuji, S. 1977. Differential responses of *Nicotiana tabacum* L. and its putative progenitors to de- and redifferentiation. Z. Pflanzenphysiol. 83:419–426.

Ohki, S., Bigot, C., and Mousseau, J. 1978. Analysis of shoot-forming capacity in vitro in two lines of tomato (*Lycopersicon esculentum* Mill) and their hybrids. Plant Cell Physiol. 19:27–42.

Ohyama, K. and Oka, S. 1978. Bud and root formation in hypocotyl segments of *Broussonetia kazinoki* Siev. in vitro. In: Proceedings 4th International Congress Plant Tissue and Cell Culture, Aug. 20-25, 1978, p. 33. Univ. of Calgary Press, Calgary.

Orton, T.J. 1980. Chromosomal variation in tissue cultures and regenerated plants of *Hordeum*. Theor. Appl. Genet. 56:101-112.

Oswald, T.H., Smith, A.E., and Phillips, D.V. 1977. Callus and plant regeneration from cell cultures of Ladino clover and soybean. Physiol. Plant. 38:129-134.

Padmanabhan, V., Paddock, E.F., and Sharp, W.R. 1974. Plantlet formation from *Lycopersicon esculentum* leaf callus. Can. J. Bot. 52:1429-1432.

Pareek, L.K. and Chandra, N. 1978. Differentiation of shoot buds in vitro in tissue cultures of *Sisymbrium irio* L. J. Exp. Bot. 29:239-244.

Paterson, K.E. and Rost, T.L. 1981. Callus formation and organogenesis from cultured leaf segments of *Crassula argentea*: Cytokinin-induced developmental pattern changes. Am. J. Bot. 68:965-972.

Pelletier, G. and Delise, B. 1969. Sur la faculte de regeneration des plantes entieres par culture in vitro du pedoncule floral de *Petunia pendula*. Ann. Amelior. Plantes 19:353-355.

Phillips, G.C. and Collins, G.B. 1979. In vitro tissue culture of selected legumes and plant regeneration from callus cultures of red clover. Crop Sci. 19:59-64.

Pierik, R.L.M. 1976. *Anthurium andraeanum* plantlets produced from callus tissues cultivated in vitro. Physiol. Plant. 37:80-82.

Poirier-Hamon, S., Rao, P.S., and Harada, H. 1974. Culture of mesophyll protoplasts and stem segments of *Antirrhinum majus* (snapdragon) growth and organization of embryoids. J. Exp. Bot. 25:752-758.

Power, J.B. and Berry, S.F. 1979. Plant regeneration from protoplasts of *Browallia viscosa*. Z. Pflanzenphysiol. 94:469-471.

Prokhorov, M.N., Chernova, L.K., and Filin-Koldakov, B.V. 1974. Growing wheat tissues in culture and the regeneration of an entire plant. Doklady Akad. Nauk 214:472-475.

Raj Bhansali, R. and Arya, H.C. 1979. Organogenesis in *Citrus limettoides* (sweet lime) callus culture. Phytomorphology 29:97-100.

Raj Bhansali, R., Kumar, A., and Arya, H.C. 1978. Morphogenesis in somatic callus culture of *Crotolaria* spp. Proceedings 4th International Congress Plant Tissue and Cell Culture, Aug. 20-25, 1978 p. 34. Univ. of Calgary Press, Calgary.

Rangan, T.S. 1974. Morphogenic investigations on tissue cultures of *Panicum miliaceum*. Z. Pflanzenphysiol. 72:456-459.

———— 1976. Growth and plantlet regeneration in tissue cultures of some Indian millets: *Paspalum scrobienlatum* L., *Eleusine coracana* Gaertn and *Pennisetum typhoideum* Pers. Z. Pflanzenphysiol. 78:208-216.

Rao, P.S., Handro, W., and Harada, H. 1973. Hormonal control of differentiation of shoots, roots and embryos in leaf and stem cultures of *Petunia inflata* and *Petunia hybrida*. Physiol. Plant. 28:458-463.

Rao, P.S. and Narayanaswamy, S. 1968. Induced morphogenesis in tissue cultures of *Solanum xanthocarpum*. Planta 81:372-375.

Raste, A.P. and Ganapathy, P.S. 1970. In vitro behavior of inflorescence segments of *Mazus pumilus*. Phytomorphology 20:367–374.

Redei, G.P. 1975. *Arabidopsis* as a genetic tool. Ann. Rev. Genet. 9:111–127.

Reuther, G. 1978. Cloning of female and male asparagus strains by tissue culture. Gartenbauwissenschaft 43:1–10.

Rice, T.B., Reid, R.K., and Gordon, P.N. 1978. Morphogenesis in field crops. In: Propagation of Higher Plants Through Tissue Culture. (K. Hughes, R. Henke, and M.Constantin, eds.) pp. 262–277. USDOE, Washington, D.C.

Sacristan, M.D. 1971. Karotypic changes in callus cultures from haploid and diploid plants of *Crepis capillaris* L. Wallr. Chromosoma 33:273–283.

———— and Melchers, G. 1969. The karyotypic analysis of plants regenerated from tumorous and other callus cultures of tobacco. Molec. Gen. Genet. 105:317–333.

Sangwan, R.S. and Gorenflot, R. 1975. In vitro culture of *Phragmites* tissues. Callus formation, organ differentiation and cell suspension culture. Z. Pflanzenphysiol. 75:256–259.

Sangwan, R.S. and Harada, H. 1975. Chemical regulation of callus growth, organogenesis, plant regeneration, and somatic embryogenesis in *Antirrhinum majus* tissue and cell cultures. J. Exp. Bot. 26:868–881.

Sangwan, R.S., Norreel, B., and Harada, H. 1976. Effects of kinetin and gibberellin A_3 on callus growth and organ formation in *Limnophila chinensis* tissue culture. Biol. Plant. 18:126–131.

Schenk, R.U. and Hildebrandt, A.C. 1972. Medium and techniques for induction and growth of monocotyledonous and dicotyledonous plant cell cultures. Can. J. Bot. 50:199–204.

Schieder, O. 1978. Somatic hybrids of *Datura innoxia* Mill. + *Datura discolor* Bernh. and of *Datura innoxia* Mill + *Datura stramonium* L var. tatula L. Molec. Gen. Genet. 162:113–119.

Scorza, R. and Janick, J. 1976. Tissue culture in *Passiflora*. In: Proceedings 24th Congress American Society Horticultural Science, Tropical Region (Mayaguez, Puerto Rico) pp. 179–183. Am. Soc. Hort. Sci., Alexandria, Virginia.

Scowcroft, W.R. and Adamson, J.A. 1976. Organogenesis from callus cultures of the legume, *Stylosanthes hamata*. Plant Sci. Lett. 7:39–42.

Seabrook, J.E.A. and Cumming, B.G. 1982. In vitro morphogenesis and growth of *Narcissus* in response to temperature. Sci. Hort. 16:185–190.

Sen, B. and Gupta, S. 1979. Differentiation in callus cultures of leaf of two species of *Trigonella*. Physiol. Plant. 45:425–428.

Shama Rao, H.K. and Narayanaswamy, S. 1975. Effect of gamma irradiation on cell proliferation and regeneration in explanted tissue of pigeon pea, *Cajanus cajan* (L.) Mills P. Radiat. Bot. 15:301–305.

Sharma, G.C., Bello, L.L., and Sapra, V.T. 1978. Callus induction and regeneration of plantlets from immature embryos in hexaploid triticale (x *Tricosecale* Wittmack). In: Propagation of Higher Plants Through Tissue Culture (K. Hughes, R. Henke, and M. Constantin, eds.) p. 258. USDOE, Washington, D.C.

Sheridan, W.F. 1968. Tissue culture of the monocot *Lilium*. Planta 82: 189-192.

_____ 1975. Plant regeneration and chromosome stability in tissue cultures. In: Genetic Manipulations With Plant Materials (L. Ledoux, ed.) pp. 263-295. Plenum Press, New York.

Shimada, T., Saskuma, T., and Tsunewaki, K. 1969. In vitro culture of wheat tissues. I. Callus formation, organ redifferentiation and single cell culture. Can. J. Genet. Cytol. 11:294-304.

Skirvin, R.M. and Chu, M.C. 1979. In vitro propagation of "Forever Yours" rose. HortScience 14:608-610.

Skirvin, R.M. and Janick, J. 1976. Tissue culture-induced variation in scented *Pelargonium* spp. J. Am. Soc. Hort. Sci. 101:281-290.

Skolmen, R.G. and Mapes, M.O. 1976. *Acacia koa* Gray plantlets from somatic callus tissue. J. Hered. 67:114-115.

Skoog, F. and Miller, C.O. 1957. Chemical regulation of growth and organ formation in plant tissues cultivated in vitro. In: Biological Action of Growth Substances. Symp. Soc. Exp. Biol. 11:118-131.

Smith, H.H. and Mastrangelo-Hough, I.A. 1979. Genetic variability available through cell fusion. In: Plant Cell and Tissue Culture: Principles and Applications (W.R. Sharp, P.O. Larsen, E.F. Paddock, and V. Raghaven, eds.) pp. 265-285. Ohio State Univ. Press, Columbus.

Sommer, H.E. and Brown, C.L. 1979. Applications of tissue cultures to forest tree improvement. In: Plant Cell and Tissue Culture: Principles and Applications (W.R. Sharp, P.O. Larsen, E.F. Paddock, and V. Raghaven, eds.) pp. 461-492. Ohio State Univ. Press, Columbus.

_____ 1980. Embryogenesis in tissue cultures of sweetgum. For. Sci. 26:257-260.

Sopory, S.K. and Maheshwari, S.C. 1976. Morphogenetic potentialities of haploid and diploid vegetative parts of *Datura innoxia*. Z. Pflanzenphysiol. 77:274-277.

Sree Ramulu, K., Devreux, M., Ancora, G., and Laneri, U. 1976. Chimerism in *Lycopersicon peruvianum* plants regenerated from in vitro cultures of anthers and stem internodes. Z. Pflanzenzuchtg. 76:299-319.

Srejovic, V. and Neskovic, M. 1981. Regeneration of plants from cotyledon fragments of buckwheat (*Fagopyrum esculentum* Moench.). Z. Pflanzenphysiol. 104:37-42.

Srivastava, P.S. 1973. Formation of triploid "plantlets" in endosperm cultures of *Putranjiva* x *roxburghii*. Z. Pflanzenphysiol. 69:270-273.

_____ and Steinhauer, A. 1981. Regeneration of birch plants from catkin tissue cultures. Plant Sci. Lett. 22:379-386.

Stimart, D.P. and Ascher, P.D. 1978. Tissue culture of bulb scale sections for asexual propagation of *Lilium longiflorum* Thunb. J. Am. Soc. Hort. Sci. 103:182-184.

Stringham, G.R. 1979. Regeneration in leaf-callus of haploid rapeseed (*Brassica napus* L.). Z. Pflanzenphysiol. 92:459-462.

Swanson, E.B. and Tomes, D.T. 1980. Plant regeneration from cell cultures of *Lotus corniculatus* and the selection and characterization of 2,4-D tolerant cell lines. Can. J. Bot. 58:1205-1209.

Syono, K. and Furuya, T. 1972. The differentiation of *Coptis* plants in vitro from callus cultures. Experientia 28:236.

Tabata, M., Yamaoto, H., Hiraoka, N., and Konoshima, M. 1972. Organization and alkaloid production in tissue cultures of *Scopalia parviflora.* Phytochemistry 11:949–955.

Takayama, S. and Misawa, M. 1979. Differentiation in *Lilium* bulb scales grown in vitro. Effect of various cultural conditions. Physiol. Plant. 46:184–190.

―――― 1981. Mass propagation of *Begonia* x *hiemalis* plantlets by shake culture. Plant Cell Physiol. 22:461–467.

Takebe, I., Labib, G., and Melchers, G. 1971. Regeneration of whole plants from isolated mesophyll protoplasts of tobacco. Naturwissen. 58:318–320.

Tal, M., Dehan, K., and Heikin, H. 1977. Morphogenetic potential of cultured leaf sections of cultivated and wild species of tomato. Ann. Bot. 41:937–941.

Tanimoto, S. and Harada, H. 1980. Hormonal control of morphogenesis in leaf explants of *Perilla frutescens* Britton var. crispa *Decaisne* f. *virida-crispa* Makino. Ann. Bot. 45:321–327.

Thomas, E. and Wernicke, W. 1978. Morphogenesis in herbaceous crop plants. In: Frontiers of Plant Tissue Culture (T.A. Thorpe, ed.) pp. 402–410. Univ. of Calgary Press, Calgary.

Thomas, E., Hoffmann, F., and Wenzel, G. 1975. Haploid plantlets from microspores of rye. Z. Pflanzenzuchtg. 75:106–113.

Thomas, E., Hoffmann, F., Potrykus, I., and Wenzel, G. 1976. Protoplast regeneration and stem embryogenesis of haploid androgenetic rape. Molec. Gen. Genet. 145:245–247.

Thomas, E., King, P.J., and Potrykus, I. 1977. Shoot and embryo-like structure formation from cultured tissues of *Sorghum bicolor.* Naturwissen. 64:587.

Thomas, M.J., Duhoux, E., and Vazart, J. 1977. In vitro organ initiation in tissue cultures of *Biota orientalis* and other species of the *Cupressaceae.* Plant Sci. Lett. 8:395–400.

Tilquin, J.P. 1979. Plant regeneration from stem callus of cassava. Can. J. Bot. 57:1761–1763.

Torrey, J.G. 1967. Morphogenesis in relation to chromosomal constitution in long-term plant tissue cultures. Physiol. Plant. 20:265–275.

―――― 1975. Tracheary element formation from single isolated cells in culture. Physiol. Plant. 35:158–165.

Tran Thanh Van, K. and Trinh, H. 1978. Morphogenesis in thin cell layers: Concept, methodology and results. In: Frontiers of Plant Tissue Culture (T.A. Thorpe, ed.) pp. 37–48. Univ. of Calgary Press, Calgary.

Vasil, I.K. and Hildebrandt, A.C. 1966a. Variation of morphogenetic behavior in plant tissue cultures. I. *Cichorium endiva.* Am. J. Bot. 53:860–869.

Vnuchkova, V.A. 1978. Development of a method for obtaining regenerate tomato plants under tissue-culture conditions. Soviet Plant Physiol. 24:884–889.

Von Arnold, S. and Erikkson, T. 1978. Induction of adventitious buds on embryos of Norway spruce grown in vitro. Physiol. Plant. 44:283–287.

Walker, K.A., Wendeln, M.L., and Jaworski, E.G. 1979. Organogenesis in callus tissue of *Medicago sativa.* The temporal separation of induction processes from differentiation processes. Plant Sci. Lett. 16:23–30.

Walkey, D.G.A. and Woolfitt, J.M.G. 1968. Clonal multiplication of *Nicotiana rustica* L. from shoot meristems in culture. Nature 220:1346–1347.

Wang, P.-J. and Huang, L.-C. 1975. Callus cultures from potato tissue and the exclusion of potato virus X from plants regenerated from stem tips. Can. J. Bot. 53:2565–2567.

Welsh, K.J. and Sink, K.C. 1981. Morphogenetic responses of *Browalia* leaf sections and callus. Ann. Bot. 48:583–590.

Wernicke, W. and Kohlenbach, H.W. 1975. Antherenkulturen bei *Scopolia*. Z. Pflanzenphysiol. 77:89–93.

Wessels, D.C.J., Groenewald, E.G., and Koeleman, A. 1976. Callus formtion and subsequent shoot and root development from leaf tissue of *Haworthia planifolia* var. *setulifera* v. Poilln. Z. Pflanzenphysiol. 78:141–145.

White, P.R. 1963. A Handbook of Plant Tissue Culture. Jacques Cottell Press, Lancaster, Pennsylvania.

Winton, L. 1968. Plantlets from aspen tissue culture. Science 160:1234–1235.

_____ 1970. Shoot and tree production from aspen tissue cultures. Am. J. Bot. 57:904–909.

_____ 1978. Morphogenesis in clonal propagation of woody plants. In: Frontiers of Plant Tissue Culture (T.A. Thorpe, ed.) pp. 419–426. Univ. of Calgary Press, Calgary.

_____ and Verhagen, S.A. 1977. Shoots from Douglas-fir cultures. Can. J. Bot. 55:1246–1250.

Wodzicki, T.J. 1978. Seasonal variation of auxin in stem cambial region of *Pinus silvestris*. Acta Soc. Bot. Polon. 47:225–231.

Wolter, K.E. 1968. Root and shoot initiation in aspen callus culture. Nature 219:509–510.

Yamaguchi, T. and Nakajima, T. 1974. Hormonal regulation of organ formation in cultured tissue derived from root tuber of sweet potato. In: Plant Growth Substances (S. Tamura, ed.) pp. 1121–1127. Hirokawa-Shoten, Tokyo.

Yamane, Y. 1974. Induced differentiation of buckwheat plants from subcultured calluses in vitro. Jap. J. Genet. 49:139–146.

Yang, Y.W. and Chang, W.C. 1979. In vitro plant regeneration from leaf explants of *Stevia rebaudiana* Bertoni. Z. Pflanzenphysiol. 93:337–343.

Yeoman, M.M. 1970. Early development in callus cultures. Int. Rev. Cytol. 29:383–409.

Yokoyama, T. and Takeuchi, M. 1976. Organ and plantlet formation from callus in Japanese persimmon (*Diospyros kaki*). Phytomorphology 26:273–275.

Zee, S.Y. and Hui, L.H. 1977. In vitro plant regeneration from hypocotyl and cotyledons of Chinese kale (*Brassica alboglabra* Bailey). Z. Pflanzenphysiol. 82:440–445.

Zee, S.Y., Wu, S.C., and Yue, S.B. 1978. Morphogenesis of the hypocotyl explants of Chinese kale. Z. Pflanzenphysiol. 90:155–163.

Zenkteler, M. 1972. In vitro formation of plants from leaves of several species of the Solanaceae family. Biochem Physiol. Pflanzen 163:509–512.

CHAPTER 3
Embryogenesis

P.V. Ammirato

The initiation and development of embryos from somatic tissues in plant culture, rather than the maturation of excised zygotic embryos, was first recognized by Steward (1958) and Reinert (1958, 1959) in cultures of *Daucus carota* (carrot) tissue derived from the storage taproot. Since 1958, investigations of somatic embryogenesis in carrot cultures, both of the cultivated and wild varieties, have been widespread (cf., Tisserat et al., 1979, pp. 33-35) and the numbers of other species that have yielded somatic embryos in culture, either directly on explanted material or after tissue proliferation, have increased. Tisserat et al. (1979) reported somatic embryogenesis in 32 families, 81 genera, and 132 species.

In addition to the development of somatic embryos from sporophytic cells, embryos have been fostered from generative cells, such as in the classic work of Guha and Maheshwari (1964) with *Datura innoxia* microspores and Nitsch (1969) with *Nicotiana tabacum* microspores. Triploid embryos have also been observed in endosperm cultures of *Santalum album* (Lakshmi Sita et al., 1980).

Adventive or asexual embryogenesis is the development of embryos from cells that are not the product of gametic fusion. It is a well known natural phenomenon. Cells of the nucellus or inner integument may develop into embryos; members of the Rutaceae and especially *Citrus* species are perhaps best known for nucellar-derived embryos, often in addition to the expected zygotic embryo (Esan, 1973). Cells within the embryo sac proper, such as the synergid or antipodal, may also develop into embryos bearing the reduced or gametic chromosome number. The proembryo, embryo, or its suspensor may also give rise to multiple embryos. In addition, there are examples of embryos arising naturally from endospermal cells, as in the case of *Brachiaria setigera* (Muniyamma, 1977). More unusual, embryos have formed within anthers, as with *Narcissus biflorus* (Koul and Karihaloo, 1977). For an excellent review of the natural occurrence of asexual embryogenesis, see Tisserat et al. (1979).

Naturally occurring (in vivo) adventive or asexual embryogenesis is known to occur in many species; however, its occurrence is generally restricted to intra-ovular tissue. What is particularly striking about embryogenesis in plant cultures is the development of embryos from somatic cells (epidermis on hypocotyls, vascular parenchyma in petioles, or storage parenchyma in secondary root phloem) in addition to their formation from unfertilized gametic cells and tissues typically associated with in vivo asexual embryogenesis (e.g., nucellus).

Although the list of species for which somatic embryogenesis has been reported is long, the number of clear-cut examples is somewhat smaller. Somatic embryos should closely resemble their zygotic counterparts with the appropriate root, shoot, and cotyledonary organs. Extraneous proliferations should be absent. They should be capable of growth into plants. Finally, there should be no vascular connection with the mother tissue as determined by histological sectioning. According to Haccius (1978), "the most distinctive characteristic of an embryo is its anatomically discrete (closed) radicular end." Unfortunately, for too many reports, convincing documentation of the formation of well-developed true embryos capable of growth into plants is lacking.

Fortunately, somatic embryogenesis has been sufficiently observed in enough species and families to demonstrate that it is not a phenomenon restricted to just a few taxa. The possibility exists that cells from any plant, given the appropriate stimuli and conditions, could be fostered to embark on the embryogenic pathway. At the present time, our understanding of these stimuli and conditions is fragmentary. The potentialities of somatic embryogenesis, however, encourage periodic summaries of the information at hand.

This chapter will look at some of the species for which somatic embryogenesis has been reported and the conditions under which it was fostered, paying attention to the variables that seem particularly important for success in both initiating embryo growth and manipulating their development. It will restrict itself almost entirely to somatic embryogenesis from sporophytic tissues, referring to androgenic embryos occasionally. For a more substantial discussion of androgenesis, the reader is directed to the section on haploid plants (see Chapter 6). The biochemical events associated with somatic embryogenesis, a key area of research, are discussed in Chapter 20. This review will not discuss the culture of excised zygotic embryos, although it is an interesting and important area of investigation. There are many excellent reviews that can be consulted (Raghaven, 1976; Norstog, 1979; Yeung et al., 1981).

LITERATURE REVIEW

Brief History

The earliest work on somatic embryogenesis was conducted on carrot cultures, and a substantial number of more recent investigations have utilized this plant. In a way, it has become the proverbial "model system," and much of our limited understanding of somatic embryogeny has come from work with this plant—especially during the first decade of research.

The earliest successes were achieved in media supplemented with coconut milk or water (Steward et al., 1958) and attention was focused on the role of the complex naturally occurring liquid endosperms that normally bathe zygotic embryos in nourishing young somatic embryos (Steward, 1963). Subsequent investigations showed that both the induction of embryogenic growth and the promotion of maturation in carrot cultures could be achieved with totally defined media lacking CW (Reinert, 1959; Kato and Takeuchi, 1963). However, investigations of liquid endosperms yielded valuable information on growth promoting systems (Steward and Shantz, 1959) and CW continues to be extrememly useful both in embryo induction and maturation (e.g., Litz and Conover, 1980, 1982; Pence et al., 1979; Thomas et al., 1977; Vasil and Vasil, 1980, 1981a,b).

During this early period of research, it was demonstrated that (1) the presence of an auxin or auxin-like substance was critical for embryo initiation and the lowering of the auxin concentration or its complete absence fostered maturation (Halperin and Wetherell, 1964; Halperin, 1966; Steward et al., 1967) and (2) reduced nitrogen was important for both initiation (Halperin and Wetherell, 1965; Halperin, 1966) and maturation (Ammirato and Steward, 1971). Thus the basic protocol calling for a primary medium with an auxin source and a secondary medium devoid of growth regulators, both containing a substantial supply of reduced nitrogen, was established.

At first, it appeared that the explanted tissue on the primary medium underwent cellular dedifferentiation to produce a mass of unorganized cells and cell clusters; transfer of these cells to a secondary medium prompted the initiation of embryonic development (Steward et al., 1964). Later observations (Halperin, 1966; Halperin and Jensen, 1967) suggested that embryo initiation probably occurred during the primary culture and that the presence of auxin in the medium prevented their maturation. Further subcultures served to maintain an embryogenic population; embryos would mature when the auxin was removed. In many cultures this is apparently what is occurring. Thus the auxin-plus and auxin-minus cultures are not comparable to "callus" and "embryo" populations or "pre-induction" and "post-induction" cultures but are in reality small globular proembryos in the former and maturing embryonic stages in the latter.

Sharp et al. (1980) describe two routes to somatic embryogenesis. The first is direct embryogenesis where embryos initiate directly from tissue in the absence of callus proliferation. This occurs through "pre-embryogenic determined cells" (PEDC) where the cells are already committed to embryonic development and need only to be released. The second is indirect embryogenesis where some cell proliferation is required. This occurs in differentiated, non-embryogenic cells or "induced embryogenic determined cells" (IEDC). The nature of the induction of embryogenic cells and the question of "competence" (cf., Street, 1979) remain elusive and a worthy subject of further investigation.

The factors involved in maturation of somatic embryos, once cells are fostered to follow that pathway, also are poorly understood. However, it is clear from a number of studies, including one on caraway (Ammirato, 1974, 1977) that somatic embryo development is extremely plastic and subject to cultural and environmental variables. Under certain conditions, it is possible to have embryogenic cells but fail to provide the proper conditions to foster their normal maturation. Therefore we must be concerned not only with those species that have yielded somatic embryos in culture but also the variables and conditions that have regulated their initiation and development.

Embryogenic Taxa

An impressive number of species of the Umbelliferae, the carrot or parsley family, have produced somatic embryos in culture (Table 1). Most of the species follow the basic protocol: growth initiation on a medium with an auxin (2,4-D, NAA, IAA) and somatic embryo maturation when the cells are removed to a medium free of growth regulators. There are a number of cases where a cytokinin, such as kinetin, proved useful in either the establishment of the cell lines or in promoting somatic embryo maturation. In general, cytokinins are not needed.

Table 2 lists Solanaceous species that have generated somatic embryos in culture. *Nicotiana* species, and especially *N. tabacum* (tobacco), have been the "model system" for studies of organogenesis in cultures (see Chapter 2). Interestingly, members of this genus, or of the entire family, have not readily produced somatic embryos from cultured somatic tissues. Until recently, the only cases were those of Haccius and Lakshmanan (1965) where *N. tabacum* 'Samsun' was grown under conditions of high light intensity (10,000–15,000 lux) and a second report using dark grown petioles (Prabhudesai and Narayanaswamy, 1973). With members of the Solanaceae, it has proven easier to foster embryos from microspores using anther culture, as in the classic work on *N. tabacum* (Nitsch, 1969). More recently, embryos have been observed in cultures derived from protoplasts of *N. tabacum* (Lorz et al., 1977), *Lycopersicon peruvianum* (Zapata and Sink, 1981), and *Nicotiana sylvestris* (Facciotti and Pilet, 1979).

In addition to members of the Umbelliferae and Solanaceae, a range of dicotyledonous families have produced somatic embryos in culture (Table 3). The legumes (Leguminosae), so important agronomically, have proven especially difficult to grow in culture. There have been some notable recent successes. At the moment, the forage legumes such as *Medicago sativa* (alfalfa: Walker et al., 1979; Kao and Michayluk, 1981) and *Trifolium pratense* (red clover: Collins and Phillips, 1980) appear

Table 1. Somatic Embryogenesis in the Umbelliferae

	GROWTH REGULATORS				
SPECIES	1° MEDIUM (μM)	2° MEDIUM (μM)	MEDIUM	EXPLANT	REFERENCE
Ammi major	28.5 IAA	11.4 IAA	MS	Hypocotyl	Grewal et al., 1976
Anethum graveo-lens (dill)	10.7 NAA	None	MS	Embryo	Steward et al., 1970
A. graveolens (dill)	2.3 2,4-D, 2.3 KIN	None	W	Inflor.	Sehgal, 1978
Apium graveolens (celery)	2.2 2,4-D, 2.7 KIN	2.7 KIN	MS	Petiole	Williams & Collin, 1976; Zee & Wu, 1979
Carum carvi (caraway)	10.7 NAA	None	MS	Petiole	Ammirato, 1974
Conium maculatum (poison hemlock)	10.7 NAA	None	MS	Embryo	Steward et al., 1970
Coriandrum sativum (coriander)	10.7 NAA	None	MS	Embryo	Steward et al., 1970
C. sativum (coriander)	2.3 2,4-D	None	MS	Petiole	Zee, 1981
Daucus carota (carrot)	4.5 2,4-D	None	MS	Stor. root	Halperin & Wetherell, 1964
Foeniculum vulgare (fennel)	27.6 2,4-D, ±1 KIN	None	N	Stem	Maheshwari & Gupta, 1965
Petroselium hor-tense (parsley)	27 2,4-D or 0.5 NAA	None	H-C	Petiole	Vasil & Hilde-brandt, 1966
Pimpinella anisum (anise)	5 NAA	None	MS	Hypocotyl	Huber et al., 1978
Sium suave (water parsnip)	10.7 NAA	None	MS	Embryo	Ammirato & Steward, 1971

Table 2. Somatic Embryogenesis in the Solanaceae

SPECIES	GROWTH REGULATORS 1° MEDIUM (µM)	2° MEDIUM (µM)	MEDIUM	EXPLANT	REFERENCE
Atropa belladonna	10.7 NAA, 2.3 KIN	2.7 NAA, 12.4 NOA	W	Root	Thomas & Street, 1970
A. belladonna	18.5 KIN, 5.4 CPA	0.5-0.9 KIN	MS	Leaf protoplasts	Lorz & Potrykus, 1979
A. belladonna	11.4 NAA, 0.5 KIN	0.5 KIN	MS	Cultured protoplasts	Gosch et al., 1975
Nicotiana sylvestris	0.9 2,4-D, 3.2 NAA, 3.7 KIN	0.02 KIN	MS	Haploid leaf protoplasts	Facciotti & Pilet, 1979
N. tabacum (tobacco)	6 CPA, 2.5 KIN, or 10 Dicamba, 50 BA	None	MS	Leaf protoplasts	Lorz et al., 1977
Lycopersicon peruvianum	5.4 NAA, 4.5 KIN	None	MS	Leaf protoplasts	Zapata & Sink, 1981
Petunia hybrida 'Rose du Ceil'	4.5 2,4-D, 0.9 BA	0.5 ZEA	MS	Stem, leaf	Rao et al., 1973
Solanum melongena (eggplant)	4.3 NAA, 1 BA	1 BA	MS	Hypocotyl	Matsuoka & Hinata, 1979
S. melongena (eggplant)	5 NOA	4.7 KIN	MS	Hypocotyl	Kamat & Rao, 1978

Table 3. Somatic Embryogenesis in Dicotyledonous Species (Except Umbelliferae and Solanaceae)

SPECIES	GROWTH REGULATORS 1° MEDIUM (μM)	2° MEDIUM (μM)	MEDIUM	EXPLANT	REFERENCE
AQUIFOLIACEAE					
Ilex aquifolium (holly)	Not reported	None	LS	Cotyledon	Hu & Sussex, 1971
I. aquifolium (holly)	None	None	MS	Embryo	Hu et al., 1978
ARALIACEAE					
Hedera helix	21.5 NAA, 9.3 KIN	5.4 NAA, 2.3 KIN, act. charcoal	MS,W	Adult stem	Banks, 1979
Panax ginseng (ginseng)	2.2 2,4-D, 0.8 KIN	0.4 2,4-D	MS	Pith	Chang & Hsing, 1980
ASCLEPIACEAE					
Pergularia minor	9.1 2,4-D, 10% CW	0.6 IAA	MS	Stem	Prabhudesai & Narayana-swamy, 1972
Tylophora indica	4.5 2,4-D, 5.2 BTOA	0-0.5 2,4-D	W	Stem	Rao & Narayana-swamy, 1972
BETULACEAE					
Corylus avellana (hazel)	4.5 2,4-D	None	MS	Embryo	Radojevic et al., 1975
BRASSICACEAE					
Brassica oleraceae (cauliflower)	5.7 IAA, 2.3 KIN	0.05-0.5 NAA, 2.3 KIN	MS	Leaf	Pareek & Chandra, 1978
CARICACEAE					
Carica papaya (papaya)	1 NAA, 10 2iP	0.1 NAA, 0.01 BA	W	Petiole	de Bruijne et al., 1974
C. papaya (papaya)	20% CW	None	W	Ovule	Litz & Conover, 1982

Table 3. Cont.

SPECIES	GROWTH REGULATORS 1° MEDIUM (µM)	2° MEDIUM (µM)	MEDIUM	EXPLANT	REFERENCE
CARICACEAE					
Carica stipulata	1 NAA, 2 BA	1 NAA, 2 BA, 1% charcoal	MS	Peduncle	Litz & Conover, 1982
CRUCIFERAE					
Cheiranthus cheiri	4.5 2,4-D	0.5 2,4-D	MS	Seedling	Khanna & Staba, 1970
CUCURBITACEAE					
Cucurbita pepo (pumpkin)	4.9 IBA	1.4 2,4-D, 10% water-melon sap	MS	Cotyledon, hypocotyl	Jelaska, 1972, 1974
HAMAMELIDACEAE					
Liquidambar styraciflua (sweet gum)	5.3 NAA, 8.8 BA	None	Blaydes	Hypocotyl	Sommer & Brown, 1980
LEGUMINOSAE					
Albizzia lebbeck (East Indian Walnut)	None	None	B5	Hypocotyl	Gharyl & Maheshwari, 1981
Medicago sativa (alfalfa)	9 2,4-D, 1.2 KIN	0.3 NAA, 2.3 BA	MS	Leaf proto-plasts	Dos Santos et al., 1980
M. sativa (alfalfa)	4.5 2,4-D, 0.5 ZEA riboside	0.5-4.5 2,4-D or 0.3-5.4 NAA, 0.3-4.7 BA, 0.5-9.3 ZEA riboside	MS	Shoot tips	Kao & Michayluk, 1981

Table 3. Cont.

SPECIES	GROWTH REGULATORS		MEDIUM	EXPLANT	REFERENCE
	1° MEDIUM (µM)	2° MEDIUM (µM)			
Trifolium pratense (red clover)	0.25 PIC, 0.44 BA	0.045 2,4-D, 14.8 ADE	SL	Seedling	Phillips & Collins, 1980
LORANTHACEAE *Nuytsia floribunda*	24.6 IBA, 23.2 KIN, 2 g/l CH	24.6 IBA, 23.2 KIN, 2 g/l CH	W	Embryo	Nag & Johri, 1969
MALVACEAE *Gossypium klotzschianum* (wild cotton)	0.5 2,4-D	11.4 IAA, 4.7 KIN	MS	Hypocotyl	Price & Smith, 1979
PAPAVERACEAE *Eschscholzia californica*	500 mg/l CH		N	Placenta	Kavathekar & Ganapathy, 1973
Macleaya cordata (plume poppy)	5.0 2,4-D, 5.0 KIN	None	B	Leaf	Kohlenbach, 1978
Papaver somniferum (opium poppy)	9.0 2,4-D, 1.2 KIN	None	MS	Hypocotyl	Nessler, 1982
RANUNCULACEAE *Nigella damascena* (fennel flower)	9.1 2,4-D, 10% CW	9.1 2,4-D	MS	Flower meristem, bud, pedicel	Raman & Greyson, 1974
N. sativa	2.7 NAA, 15% CW	2.9 IAA, 500 mg/l CH	MS	Leaf, root, stem	Banerjee & Gupta, 1976
Ranunculus sceleratus	5.7 NAA, 10% CW	5.7 NAA, 10% CW	W	Stem	Konar et al., 1972

Table 3. Cont.

SPECIES	GROWTH REGULATORS 1° MEDIUM (μM)	2° MEDIUM (μM)	MEDIUM	EXPLANT	REFERENCE
RANUNCULACEAE					
R. sceleratus (cursed crowfoot)	4.4 BA, 16.1 NAA	None	MS	Mesophyll proto-plasts	Dorion et al., 1975
Thalictrum urbaini (meadow rue)	10.2 IAA, 9.3 KIN	4.7 KIN, 5.7 IAA, 2.9 GA$_3$	MS	Petiole	Yang & Chang, 1980
RUBIACEAE					
Coffea arabica 'Mundo Novo' (coffee)	2 KIN, 2 2,4-D	2.5 KIN, 0.5 NAA	MS	Leaf	Sondahl & Sharp, 1977
C. canephora	2 KIN, 10 2,4-D	2.5 KIN, 0.5 NAA	MS	Leaf	Sondahl & Sharp, 1979
C. canephora	0.5 KIN, 0.5 2,4-D	0.5 KIN, 0.5 2,4-D	MS	Shoot	Staritsky, 1970
C. congensis	2 KIN, 4 2,4-D	2.5 KIN, 0.5 NAA	MS	Leaf	Sondahl & Sharp, 1979
C. dewevrei	10 KIN, 6 2,4-D	2.5 KIN, 0.5 NAA	MS	Leaf	Sondahl & Sharp, 1979
RUTACEAE					
Citrus sinensis 'Shamouti' (orange)	0.5 KIN	1 g/l ME	MS	Ovule	Kochba & Spiegel-Roy, 1973
C. sinensis	None	None	MT	Protoplasts	Vardi et al., 1975
SANTALACEAE					
Santalum album (sandalwood)	9.1 2,4-D, 23.2 KIN	None	W	Embryo	Rao, 1965
S. album (sandalwood)	4.5 2,4-D, 0.9–2.3 KIN	1.5–5.8 GA$_3$	MS, W	Stem	Lakshmi Sita et al., 1979

Table 3. Cont.

SPECIES	GROWTH REGULATORS 1° MEDIUM (μM)	2° MEDIUM (μM)	MEDIUM	EXPLANT	REFERENCE
SCROPHULARIACEAE					
Antirrhinum majus 'Kymosy blanc'	4.5 2,4-D	1.2 NOA, 10% CW	MS	Leaf, stem	Sangwan & Harada, 1975
Paulownia tomentosa	4.5 2,4-D		MS	Ovule	Radojevic, 1979
P. tomentosa	5.7 IAA, 0.5-5 KIN	None	MS	Embryo	Radojevic, 1979
STERCULIACEAE					
Theobroma cacao (cacao)	10% CW, 8 NAA, IAA, or 2,4-D	None	MS	Immature embryo	Pence et al., 1979, 1980
VITACEAE					
Vitis spp. (grapes)	4.5 2,4-D, 0.4 BA	10.7 NAA, 0.4 BA	MS	Flower, leaf	Krul & Worley, 1977
Vitis vinifera	5 2,4-D, or 5 NOA, 1-5 BA	10 NOA, 1 BA	N	Ovule	Srinivasan & Mullins, 1980

more amenable to in vitro culture than seed legumes.   As with the Solanaceae, a number of recent sucesses have been with protoplast-derived cultures (Kao and Michayluk, 1980; Dos Santos et al., 1980). Also of interest is the increasing number of woody plants that have responded in culture, e.g., *Citrus* (Kochba and Spiegel-Roy, 1973), *Corylus* (Radojevic et al., 1975), *Coffea* (Sondahl and Sharp, 1977), *Liquidambar* (Sommer and Brown, 1980), *Albizzia* (Gharyal and Mahesh-wari, 1981) and *Paulownia* (Radojevic, 1979).   To date, there are a number of dicotyledonous families where many species have been regenerated via organogenesis but where somatic embryogenesis has not been observed, e.g., Compositae and Gesneriaceae.

The monocots in general (Table 4) and the Gramineae (grasses and grains) in particular have proven especially difficult to grow in culture and regenerate somatic embryos (Table 5).   Members of the Liliaceae appear to grow readily and regenerate plants via organogenesis (see Chapter 2) and somatic embryogenesis (Table 4).   In the past it has proven easier to grow androgenic embryos from cereal anther cultures than to do so from somatic tissue, e.g., *Oryza sativa* cv. indica (Guha et al., 1970), *Triticale* (Sun et al., 1973), *Triticum aestivum* (Picard and Buyer, 1975) and *Secale cereale* (Thomas and Wenzel, 1975; Thomas et al., 1975).   Recently there have been successes with the millets, *Penni-stum americanum* (Vasil and Vasil, 1980, 1981a), *P. americanum* x *pur-pureum* (Vasil and Vasil, 1981b), *P. pupureum* (Haydu and Vasil, 1981), and *Panicum maximum* (Lu and Vasil, 1981, 1982).   In addition, somatic embryos have been observed in sorghum, *Sorghum bicolor* (Wernicke and Brettell, 1980; Brettell et al., 1980) rice, *Oryza sativa* (Wernicke et al., 1981) wheat, *Triticum aestivum* (Osias-Akins and Vasil, 1982) and corn, *Zea mays* (Lu et al., 1982).   Of particular interest are the embryogenic cultures that have been developed from vegetative rather than embry-onic tissue: immature leaves and inflorescences seem particularly amen-able.

There are few examples of somatic embryogenesis ouside the angio-sperms: *Biota orientalis* (Konar and Oberoi, 1965) and *Pinus palustris* (Sommer et al., 1975) among the gymnosperms and *Zamia integrifolia* (Norstog and Rhamstine, 1967) among the cycads.

Tables 1-5 list specific information pertaining to the culture regimes for each species.   Somatic embryo development generally follows the transfer of cells or callus to media lacking auxin, with reduced levels of the same auxin, or with similar or reduced levels of a weaker auxin. However, with a number of species, and in particular for the grasses, (e.g., Vasil and Vasil, 1980, 1981a,b; Ozias-Akin and Vasil, 1982), soma-tic embryo initiation and maturation occurs on the primary medium. Transfer to a secondary medium is needed for their growth into plants. From the tables, it is evident that each species has been grown with a protocol specifically (and often empirically) developed for it.   The next section will discuss some of the variables that affect somatic embryo-genesis and maturation in culture with a view to developing some basic concepts that may prove useful in either promoting somatic embryo-genesis in species that have so far resisted or in controlling somatic embryo maturation in those species where they can be grown.

Table 4. Somatic Embryogenesis in Monocotyledonous Species (Except Gramineae)

SPECIES	GROWTH REGULATORS 1° MEDIUM (μM)	GROWTH REGULATORS 2° MEDIUM (μM)	MEDIUM	EXPLANT	REFERENCE
ARACACEAE (PALMAE)					
Elaeis guineensis (oil palm)	4.5 2,4-D, 2.3 KIN	5.7 IAA	HE	Embryo	Rabechault et al., 1970
Phoenix dactylifera (date palm)	452 2,4-D, 4.9 2iP	None	MS	Ovule	Reynolds & Murashige, 1979
P. dactylifera (date palm)	135 2,4-D, 13.5 2iP, act. charcoal	None	MS	Axillary bud	Tisserat & De Mason, 1980
DIOSCOREACEAE					
Dioscorea floribunda	4.5 2,4-D	None or 0.1 ABA (500 mg/l glutamine)	MS	Embryo	Ammirato, 1978a
IRIDACEAE					
Iris spp.	4.5 2,4-D, 0.4 BA	0.6 IAA	MS	Shoot apex	Reuther, 1977a,b
LILIACEAE					
Allium sativum (garlic)	10 CPA, 2 2,4-D, 0.5 KIN	10 IAA, 20 KIN	AZ	Stem	Abo El-Nil, 1977
Asparagus officinalis (asparagus)	5.4 NAA, 4.7 KIN	0.5–5.7 IAA, 0.44–17.7 BA	LS, MS	Shoot, cladode	Reuther, 1977a,b
Hemerocallis spp. 'Autumn Blaze' (daylily)	9 2,4-D, 10% CW	None	MS, W, SH	Ovary	Krikorian & Kann, 1981
ZINGIBERACEAE					
Zingiber officinale (ginger)	Not reported	4.4 BA	MS	Rhizome	Hosoki & Sagawa, 1977

Table 5. Somatic Embryogenesis in the Gramineae

SPECIES	GROWTH REGULATORS		MEDIUM	EXPLANT	REFERENCE
	1° MEDIUM (μM)	2° MEDIUM (μM)			
PANICOIDEAE (CEREALS)					
Oryza sativa (rice)	4.7 2,4-D	None	MS	Immature leaf	Wernicke et al., 1981
Panicum maximum	11.3–45 2,4-D	None, or 0.9 2,4-D	MS	Immature embryo, immature inflor.	Lu & Vasil, 1982
P. maximum	11.3–45 2,4-D, 5–15% CW	2.9 GA$_3$	MS	Leaf	Lu & Vasil, 1981
Pennisetum americanum (pearl millet)	11.3 2,4-D, 5% CW	None	LS	Immature embryo and proto- plasts	Vasil & Vasil, 1980
P. americanum (pearl millet)	11.3 2,4-D, 5% CW	0.04–0.08 ABA	MS	Immature embryo suspension	Vasil & Vasil, 1981a
P. americanum (pearl millet)	11.3 2,4-D, ± 5% CW	None, or 0.04 ABA	MS	Immature inflor.	Vasil & Vasil, 1981b
P. americanum x P. purpureum	11.3 2,4-D	None, or 0.04 ABA	MS	Immature inflor.	Vasil & Vasil, 1981b
P. purpureum	2.3 2,4-D, 5.4 NAA, 2.3 BA, 5% CW	None	MS	Leaf	Haydu & Vasil, 1981
Sorghum bicolor	11.6–23.4 2,4-D	None	MS	Immature embryo	Thomas et al., 1977
S. bicolor	11.6–23.4 2,4-D, 10% CW	None	MS	Mature embryo	Thomas et al., 1977

Table 5. Cont.

SPECIES	GROWTH REGULATORS 1° MEDIUM (µM)	2° MEDIUM (µM)	MEDIUM	EXPLANT	REFERENCE
S. bicolor	4.5–13.5 2,4-D	None, or 0.5 KIN	MS	Leaf	Wernicke & Brettell, 1980
S. bicolor	0.9–4.5 2,4-D, 0.3–4.7 BA 0.5–4.7 KIN	None	MS	Immature inflor.	Brettell et al., 1980
Triticum aestivum (wheat)	9.0 2,4-D	None	MS	Immature embryo, young in-florescence	Ozias-Akins & Vasil, 1982
Zea mays (maize)	2.3 2,4-D (12% sucrose)	2.9 GA$_3$	MS	Immature embryo	Lu et al., 1982
POACOIDEAE (GRASSES)					
Bromus inermus (bromegrass)	4.5 2,4-D	None	B5	Mesocotyl	Gamborg et al., 1970
Dactylis glomerata (orchardgrass)	20 2,4-D	1 2,4-D	LS	Mature embryo	McDaniel et al., 1982
Lolium multiflorum (ryegrass)	9 2,4-D, 0.9 BA	None	MS	Immature embryo	Dale, 1980

Key Variables

EXPLANTS. For *D. carota*, the most "cultured" of all plant species, just about any part of the plant body taken at any time of development has successfully produced somatic embryos in culture: excised embryos (Steward et al., 1964), hypocotyl (Fujimura and Komamine, 1979a), young roots (Smith and Street, 1974), taproots (Steward et al., 1958), petioles (Halperin, 1966), peduncle (Halperin and Wetherell, 1964), and protoplasts (Kameya and Uchimiya, 1972). This is apparently the case for many other taxa.

For some species, however, only certain regions of the plant body may respond in culture. This is apparently the case for many monocots, and especially for members of the Graminaceae. For these plants, regions of actively dividing cells seem to respond most readily. Immature embryos have been extremely useful, both in the grasses *Bromus* (Gamborg et al., 1970) and *Lolium* (Dale, 1980), the cereals *Pennisetum* (Vasil and Vasil, 1980), *Panicum* (Lu and Vasil, 1982), *Triticum* (Ozias-Akins and Vasil, 1982), and *Zea* (Lu et al., 1982). In many of these cultures, the orientation of the excised embryo on the primary medium is critical: the embryonic root-shoot axis must be in contact with the medium—the "face down" position (Vasil and Vasil, 1981b). The scutellum therefore sits on top and its proliferations generate the embryogenic tissue. It appears to be a particularly responsive tissue. When wheat plants were sprayed with 2,4-D shortly after flowering, seeds developed with multiple shoots (Ferguson et al., 1979): histological studies traced them to adventive embryos that had formed on the scutellum of zygotic embryos.

Embryogenic culture in the cereals have also been derived from young vegetative material, e.g., the basal 5-8 cm of the youngest 3-4 leaves in millet (Haydu and Vasil, 1981) or sorghum (Wernicke and Brettell, 1980). Immature inflorescences have also proven suitable in the millets (Vasil and Vasil, 1981a,b) and sorghum (Brettell et al., 1980).

Floral or reproductive tissue, in general, has proven to be an excellent source of embryogenic material, e.g., ovules (*Carica papaya*: Litz and Conover, 1982; *Citrus* species: Kochba and Button, 1974; *Paulownia tomentosa*: Radojevic, 1979; *Vitis vinifera*: Srinivasan and Mullins, 1980) and peduncles (*Carica stipulata*: Litz and Conover, 1980). Immature ovary tissue is preferable in daylily (*Hemerocallis*: Krikorian and Kann, 1981).

In the case of English ivy, *Hedera helix*, which progresses through a clear-cut juvenile-adult phase transition, only cultures initiated from adult material produced somatic embryos (Banks, 1979). In summary, embryonic, meristematic, and reproductive tissues appear to have a propensity for embryogenic growth.

The physiological state of the plant from which the explant is taken is also extremely important, as is the season during which material is removed (see Chapter 2).

With somatic embryos reared from haploid microspores, diploid, triploid, and tetraploid somatic cells and from triploid endosperm tissue, the underlying ploidy level clearly does not appear to affect the ability

of the cells to be reprogrammed to undergo somatic embryogenesis.
The ploidy level of established cultures, however, may change and
affect somatic embryogenesis.

Within certain species, different genotypes have shown varying potentialities for somatic embryogenesis, e.g., in *Medicago sativa* (Kao and
Michayluk, 1980), *Trifolium praetense* (Keyes et al., 1980), and *Daucus
carota* (Steward et al., 1975).   Three cultivars of *Zea mays* produced
cultures capable of shoot organogenesis (Green and Phillips, 1975;
Green, 1977) and only one cultivar, A188, readily produced somatic
embryos (Green, personal communication).   Recently Lu et al. (1982)
reported somatic embryogenesis from all 12 genotypes of *Zea mays*
studied, but the frequency of response varied considerably from one to
another.

CULTURE MEDIUM.   Somatic embryos have been grown on a range of
media from the relatively dilute White's medium (White, 1963), to the
more concentrated formulations of Gamborg et al. (1968), Schenk and
Hildebrandt (1972), and Murashige and Skoog (1962).   The B5, SH, and
MS media are all classified as high salt media, and MS, in particular,
has about ten times the salt concentration of White's medium.   In their
survey of somatic embryogenesis in crop plants, Evans et al. (1981)
noted that 70% of the explants were cultured on a MS medium or a
modification of MS.   The choice of medium can affect embryogenic cultures.   Mouras and Lutz (1980), while reaffirming the benefits of
"strong" mineral solutions for the growth of embryos, emphasized that
"weak" salt solutions appear to preserve the potentiality in embryogenic suspensions.   This has also been observed in the author's laboratory.

A key element of the MS medium is the presence of high levels of
nitrogen in the form of ammonium nitrate.   The requirement for nitrogen, in the case of carrot somatic embryos at least, can be satisfied
by high concentrations of inorganic nitrogen in the form of nitrate
(Reinert, 1967).   However, the benefits of reduced nitrogen, in addition
to nitrate, for both embryo initiation and maturation seem well established.   White's medium, being particularly low in nitrate nitrogen and
without reduced nitrogen, needs to be supplemented to support continued somatic embryogeny.   For somatic embryogenesis, the optimal
concentration and form of the nitrogen supplies appears to be dependent on the auxin concentration (cf., Sharp et al., 1980, pp. 273-275).

The source of reduced nitrogen can be in the form of complex
addenda such as CW (Steward and Shantz, 1959; Tulecke et al., 1961)
or CH (Ammirato and Steward, 1971), a mixture of amino acids (Kato
and Takeuchi, 1966), a single amino acid such as L-glutamine or L-
alanine (Wetherell and Dougall, 1976), or the presence of ammonium ion
(Halperin, 1966; Ammirato and Steward, 1971).   In all these cases there
is also nitrate in the medium.   There are cases, however, where ammonium ion can serve as the sole nitrogen source (Dougall and Verma,
1978).   Of all the amino acids, L-glutamine seems to play a special
role.   In a comparison of amino acids added singly, glutamine most
readily promoted carrot somatic embryos (Kamada and Harada, 1979b).

It has proven beneficial, even when added to media already containing reduced nitrogen in the form of ammonium ion, e.g., *Gossypium* (Price and Smith, 1979), *Carica* (Litz and Conover, 1980), and *Dioscorea* (Ammirato, 1978a).

Of the other salts present in all culture media, potassium ion has been shown to be essential for somatic embryogenesis (Reinert et al., 1967). The effects of suboptimal concentrations are especially evident if nitrogen levels are low.

Another feature of the MS medium and other media important to somatic embryogenesis is the presence of chelated iron, often in the form of iron EDTA. In embryogenic anther cultures, iron EDTA was superior to other iron chelates and to nonchelated iron salts (Heberle-Bors, 1980). In the absence of iron, embryo development fails to pass from the globular to the heart-shaped stage. This was observed in androgenic embryos of *Nicotiana tabacum* (Nitsch, 1969) and *Atropa belladonna* (Heberle-Bors, 1980) and in somatic embryos of *Daucus carota* (Havranek and Vagera, 1979). In the latter case the effect was used to "stage" embryo development: globular embryos were produced and "stored" in iron-free medium; embryo maturation proceeded when iron was restored.

GROWTH REGULATORS: Auxins. With some species, e.g., *Citrus sinensis* (Vardi et al., 1975) or *Ilex aquifolium* (Hu et al., 1978), somatic embryos have arisen without the presence of exogenous plant regulators. These are most likely examples of predetermined embryogenic cells. For cultures derived from differentiated tissues, growth regulators in the medium and especially auxins or auxin in combination with cytokinin appear essential for the onset of growth and the induction of embryogenesis (Fujimura and Komamine, 1980a).

Some species readily produce embryogenic cultures with a range of different auxins. Embryogenic carrot cultures, for example, have been produced with NAA (Ammirato and Steward, 1971), 2,4-D (Halperin, 1966), and IAA (Sussex and Frei, 1968). With other species, the choice is more limited, e.g., in millet cultures, as in the case of other cereals and grasses, only 2,4-D was effective (Vasil and Vasil, 1980, 1981a,b). Of all the auxin or auxin-like plant growth regulators, 2,4-D has proven extremely useful, being used in 57.1% of successful embryogenic cultures (Evans et al., 1981). More uncommon auxins also have been successfully employed, for example, 2-benzothiazole acetic acid (BTOA) in *Tylophora indica* (Rao and Narayanaswamy, 1972), para-chlorophenoxyacetic acid (pCPA) in *Allium sativa* (Abo El-Nil, 1977), and picloram in *Trifolium pratense* (Collins and Phillips, 1980).

As seen in Tables 1-5, the auxin for the primary and secondary media may be the same or different; one auxin or several may be used in the same medium. For many cultures, a change in auxin type or concentration is a necessary prelude to somatic embryo development.

Effective concentration ranges are 0.5-27.6 µM for 2,4-D and 0.5-10.7 µM for NAA. Unusually high levels have been successful in certain species, e.g., 452 µM 2,4-D for *Phoenix dactylifera* (Reynolds and Murashige, 1979). In *Medicago* (Kao and Michayluk, 1981), the optimal concentration of growth regulators for somatic embryogenesis for each genotype was different and specific. Unfortunately, growth regulator levels must be determined empirically.

Cytokinins. As seen in Tables 1-5, cytokinins have been important in a number of species. Evans et al. (1981) reported that cytokinins were used in the primary medium for 65.4% of the crop species and 21.4% of the noncrop species. For the secondary culture, cytokinins were used in 34.6% of the crop species and 14.1% of the noncrop species. The effective concentration range for KIN is 0.5-5.0 µM. Cytokinins are important in fostering somatic embryo maturation (Fujimura and Komamine, 1980a) and especially cotyledon development (Ammirato and Steward, 1971). The requirement for a cytokinin may be specific: ZEA but not BA or KIN promoted somatic embryogenesis in carrot cultures (Fujimura and Komamine, 1975). Cytokinins are sometimes required for growth of embryos into plantlets (Kavathekar et al., 1978). In addition, they are especially useful in low-density cultures.

Gibberellins. Gibberellins are rarely incorporated in primary culture media or used for maintenance of the line. However, they have proved useful in a number of cases in fostering embryo maturation or in stimulating the rooting and subsequent growth of plants, e.g., in *Santalum album* (Lakshmi Sita et al., 1979), *Panicum maximum* (Lu and Vasil, 1981), *Citrus sinensis* (Kochba et al., 1974, and *Zea mays* (Lu et al., 1982). In combination with ZEA and ABA, GA_3 promoted more normal development of caraway somatic embryos (Ammirato, 1977). Kavathekar et al. (1978) overcame somatic embryo dormancy with GA_3 in *Eschescholzia californica.*

Complex and Sequential Treatments. There are numerous examples where combinations of plant growth regulators were important in initiating and maintaining growth and in fostering somatic embryo development. In *Carum carvi* normal development could be manipulated by altering the relative levels of ZEA, GA_3, and ABA (Ammirato, 1977).

Equally important is the sequence of media and, especially, the growth regulators (Steward et al., 1967). For many species one media is used for the initial callusing and for maintenance of the callus or suspension culture; a second medium is used for somatic embryo maturation; and a third to allow their growth into plants. For an increasing number of cases, however, a more complex sequence of media changes is needed. In *Panax ginseng* (Chang and Hsing, 1980) different media were needed for initiation of primary cultures, maintenance of cultures, initiation of somatic embryo development, proliferation of embryos, and stimulation of maturation. In *Hemerocallis* (Krikorian and Kann, 1981) a gradual sequence of changes involving growth regulators, coconut water, and different salt solutions were used to generate somatic embryos and plantlets. An elaborate sequence of media may be necessary in those cultures where somatic embryogenesis is now lacking.

GROWTH INHIBITORS: Abscisic Acid. The role of inhibitors in somatic embryogenesis and especially ABA, a naturally occurring growth regulator, has been slowly emerging. When added at noninhibitory levels (usually 0.1-1 µM) to *Carum* cultures, ABA permitted embryo maturation to proceed but (1) inhibited abnormal proliferations, including initiation of accessory/adventitious embryos and (2) repressed precocious germination (Ammirato, 1973, 1974). Normal dicotyledonous embryos, strikingly similar to zygotic embryos, were grown. Since then, the ability of ABA to inhibit embryo initiation has been observed in *Daucus carota* (Kamada and Harada, 1979a; Tisserat and Murashige, 1977). The normalizing effect of ABA on embryo maturation has also been observed with *Daucus carota* (Kamada and Harada, 1981) and *Pennistum americanum* (Vasil and Vasil, 1981a). ABA promoted somatic embryo development in *Citrus* ovular callus (Kochba et al., 1978). Figure 1 illustrates the growth of *Carum* and *Daucus* somatic embryos and the effects of ABA in controlling maturation.

Working with anther cultures, Nitsch and Nitsch (1969) observed that ABA did not prevent embryo development from *Nicotiana tabacum* microspores but did inhibit their germination. Imamura and Harada (1980b) demonstrated an increase in yield after soaking anthers for 3 days in ABA prior to explanting.

These responses to ABA have also been observed in cultures of excised zygotic embryos. Ihle and Dure (1972) reported an inhibition of precocious germination in excised barley embryos; Choinski et al. (1981) saw a similar response with excised cotton embryos. Excised immature barley embryos completed normal embryonic development free of callusing on media with ABA (Norstog and Blume, 1974; Umbeck and Norstog, 1979).

Anti-auxins. 5-hydroxynitrobenzylbromide and 7-azaindole promoted somatic embryo development in citrus cultures (Kochba and Spiegel-Roy, 1977). Fujimura and Komamine (1979b) reported that the anti-auxins 2,4,6-trichlorophenoxyacetic acid and p-chlorophenoisobutyric acid inhibited somatic embryogenesis but pointed out that their effects were at the early stages of embryogeny, when young somatic embryos are also sensitive to auxin in the medium.

8-azaguanine, an inhibitor of cytokinin activity, has been shown to promote somatic embryo development in citrus cultures (Kochba and Speigel-Roy, 1977). When added to *Carum carvi* somatic embryos, 8-AG permitted somatic embryo maturation but prevented the formation of accessory embryos and callus, an effect similar to ABA (Ammirato, 1978b).

Growth promoters, even if not added to culture media, are found in embryogenic cultures (Carr and Reid, 1968). ABA, antiauxins, and other growth inhibitors may serve to promote somatic embryo maturation by countering the effects of growth promoters. Their addition to culture media may permit somatic embryo maturation to proceed under conditions when it normally would not occur.

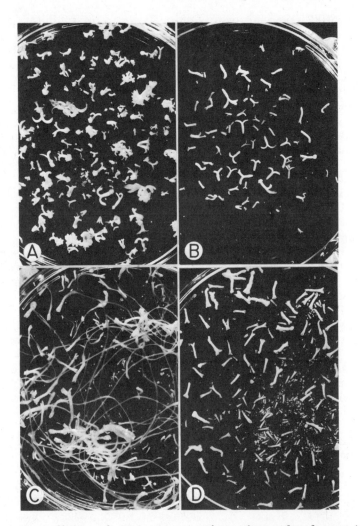

Figure 1.   The effects of ABA on somatic embryo development in um-
bellifer cultures (all x 2).   (1a) *Carum carvi* (caraway) somatic embryos
grown on MS basal medium in complete darkness in tumble tubes.
Many embryos have proliferated to produce additional accessory embryos or callus.   (1b) *Carum* somatic embryos grown on MS medium
with 0.1 µM ABA.   Most are normal dicotyledonous embryos free of
extraneous proliferations.   (1c) *Daucus carota* (carrot) somatic embryos
grown on MS medium in complete darkness in tumble tubes.   Radicles
and some cotyledons have elongated extensively.   *Carum* (a) and *Daucus*
show the range of embryonic structures found in different members of
the Umbelliferae.   (1d) *Daucus* somatic embryos grown on MS with 0.1
µM ABA.   The radicle has remained short; hypocotyls are relatively free
of proliferations.   Cotyledons are not as well developed as those on
ABA-treated *Carum* somatic embryos (b).

ACTIVATED CHARCOAL. The addition of activated charcoal to the medium has proven useful for somatic embryo development in many cultures, including date palm, *Phoenix dactylifera* (Tisserat and De Mason, 1980) and *Carica stipulata* (Litz and Conover, 1980). It has aided *Daucus carota* somatic embryo development, even in cultures that failed to produce somatic embryo when auxin is eliminated (Fridborg and Eriksson, 1975; Drew, 1979a). In *Hedera helix* cultures, callus from mature stems gave rise to embryos only on medium containing activated charcoal; if omitted, only undifferentiated callus growth occurred (Banks, 1979). Activated charcoal has been particularly useful in anther culture (see Chapter 6).

Analysis of its effects show media with charcoal had substantially lower levels of phenylacetic and p–OH benzoic acids, compounds that inhibited somatic embryogenesis when added to cultures (Fridborg et al., 1978). Activated charcoal also has been shown to absorb 5-hydroxymethylfurfural, an inhibitor formed by sucrose degradation during autoclaving, as well as substantial amounts of auxins and cytokinins (Weatherhead et al., 1978). Thus it may absorb inhibitors that would prevent growth as well as reduce the level of growth promoters that would cause continued proliferation. On the other hand, activated charcoal will absorb certain iron chelates, namely FeEGTA and FeEDDHA, thereby preventing the transition from globular to heart-shaped embryos (Heberle-Bors, 1980); it does not appear to absorb FeEDTA.

CARBOHYDRATES. Sucrose appears to be the most effective reduced carbon source for somatic embryogenesis, although many other mono- and disaccharides can be successfully employed (Verma and Dougall, 1977). In certain cases, sugars other than sucrose have been used to advantage. High levels of glucose (6-10%) produced optimal somatic embryo development in one carrot line (Homes, 1967). Galactose stimulated somatic embryo development in citrus ovular callus cultures (Kochba et al., 1978). Changes in carbohydrates may prove beneficial; the growth of habituated, nonembryogenic ovular callus from the Shamouti orange for 4 weeks on agar without sucrose followed by transfer to media with sucrose stimulated somatic embryogenesis (Kochba and Button, 1974).

Raising the sucrose concentration in the primary medium to 12% benefited the formation of embryogenic callus from the scutellum of immature embryos of *Zea mays* (Lu et al., 1982). By increasing the sucrose concentration in *Daucus carota* and *Sium suave* somatic embryo cultures, precocious germination was prevented (Ammirato and Steward, 1971); a similar response was seen by adding hexitols such as inositol and sorbitol. In these cases the additional sucrose and the addition of hexitols served to raise the osmotic concentration of the medium. Unfortunately, embryos grown under these conditions showed substantially higher levels of accessory/adventitious embryos along their axes. By adding inositol while simultaneously lowering the sucrose concentration (and maintaining the cultures in darkness), *Daucus carota* somatic embryos matured free of extraneous proliferations and without germinating precociously (Steward et al., 1975).

Soluble starch added to *Medicago sativa* cultures permitted somatic embryo growth and prevented their dedifferentiation into callus; it also led to the aging of single cells and loose cell clusters (Kao and Michayluk, 1981). Reduction of the vitamin level to one-tenth normal concentrations had a similar effect, leading to the suggestion that starch absorbed molecules onto its surface, thereby reducing the availability of hormones and vitamins for callusing.

ENVIRONMENTAL CONDITIONS: Light. Somatic embryogenesis has occurred under a variety of light/dark regimes. In one case, *Nicotiana tabacum* (Haccius and Lakshmanan, 1965), high light intensities were required for somatic embryony. In other cases, *Daucus* (Ammirato and Steward, 1971) and *Carum* (Ammirato, 1974), embryo maturation proceeded more normally in complete darkness. Newcomb and Wetherell (1970) found that short exposures to far-red illumination enhanced somatic embryogenesis.

Temperature. There has been little work on the effects of temperature, and especially cold treatments, on somatic embryogenesis, although such treatments have proven valuable in inducing androgenic embryos (Nitsch, 1974; Nitsch and Norreel, 1973). In *Citrus* nucellus-derived cultures, embryogenic potential was reduced as the temperature was lowered from 27 to 12 C (Esan, 1973). Cold treatment of *Eschscholzia* somatic embryos partially overcame innate dormancy (Kavathekar et al., 1977).

CULTURE VESSELS AND THE STATE OF THE MEDIUM. The original reports that somatic embryogenesis will proceed on callus cultures grown on semisolid medium (Reinert, 1958, 1959) and in liquid suspension cultures (Steward et al., 1958) illustrate the range of conditions under which somatic embryos can be grown. Somatic embryos have been raised on or imbedded within semisolid medium in flasks and jars with loose-fitting plugs or caps that allow for substantial gas exchange or with tight-fitting caps sealed with parafilm or tape that severely restricts flow. Somatic embryos have developed in liquid media contained in slowly rotating nipple flasks (Steward and Shantz, 1956) and tumble tubes (Steward et al., 1952), in Erlenmeyer flasks on rotary shakers (Halperin and Wetherell, 1964), and in test tubes rotating around their longitudinal axes (Halperin, 1966). They have also formed on filter paper bridges perched above liquid media (Smith, 1973; McWilliam et al., 1974; Drew, 1979a,b).

The general conclusion might be that the culture vessel, the position of the embryos (floating or submerged), and the physical state of the medium (semisolid or liquid, poorly or well-aerated) have little effect. However, work with caraway somatic embryos has shown that the method of culture does affect maturation. Figure 2 shows comparable populations grown in the same amount and type of culture media but in three different vessels. Somatic embryos grown in Erlenmeyer flasks on a rotary shaker, where there is considerable agitation and aeration, are

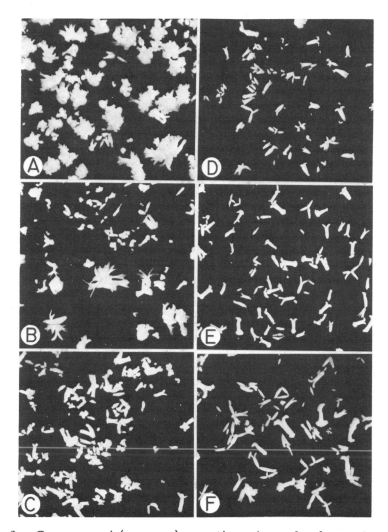

Figure 2. *Carum carvi* (caraway) somatic embryo development as affected by the culture vessel. Growth was in MS in complete darkness.
Identical volumes were used in each vessel and aliquots of the same
suspension were inoculated into each (all x 2.6). (2a) Somatic embryo
development in Erlenmeyer flask placed on a rotary shaker. Almost
every embryo proliferated producing additional embryos and callus. (2b)
Growth in tumble tubes on a klinostat. Although many embryos have
produced accessory embryos, cotyledonary development is normal.
There are also some single embryos. (2c) Growth in test tubes rotated
on their long axis. The extent of proliferations on the embryos is less
than seen in flasks (a) and tumble tubes (b), but cotyledons developed
poorly. Comparable populations grown with 1 μM ABA in the medium:
(2d) Flasks. (2e) Tumble tubes. (2f) Test tubes. Most embryos developed normally.

considerably more abnormal (Fig. 2a) than those grown in tumble tubes which provide gentle agitation and aeration (Fig. 2b). Those grown in test tubes where they are continually submerged show still fewer abnormal proliferations, but cotyledons develop poorly (Fig. 2c). However, with ABA in the medium, normal embryos matured in all vessels—flasks (Fig. 2d), tumble tubes (Fig. 2e), and test tubes (Fig. 2f).

GASES. There have been only a few studies of the effects of gases within the culture vessel or dissolved in the medium. Kessel and Carr (1972) demonstrated that dissolved oxygen levels below a critical level favored embryogenesis, while higher levels promoted callus and root development. (More recent work (Kessel et al., 1977) traced the effect of increased adenosine levels in the low oxygen cultures.)

Tisserat and Murashige (1977) reported that ethanol and ethylene inhibited somatic embryo development in *Citrus* and *Daucus* cultures; Kochba et al. (1978) reported a stimulation in *Citrus* cultures by ethephan, which liberates ethylene. Dunwell (1979) found that a small atmospheric volume in the culture vessel repressed somatic embryo development in tobacco microspore cultures and traced the effect in part to ethylene. Imamura and Harada (1980a) showed that a pretreatment of tobacco anthers with reduced atmospheric pressure attained by aspiration increased the level of pollen embryogenesis. Rajasekhar et al. (1971) reported evidence of an unidentified volatile inhibitor of cell division that accumulated in suspension cultures under conditions of low aeration. Clearly, the availability, uptake, evolution, and dispersion of various gases can affect somatic embryogenesis. Surprisingly little attention has been devoted to this area.

DENSITY. Halperin (1967) reported a direct relation between the density of embryogenic cells in the suspension and the degree of embryo maturation. This response to density has been noted many times in cell cultures (Street, 1977). Growth of somatic embryos below the "minimum effective density" can be fostered by the use of conditioned medium (Hari, 1980; Warren and Fowler, 1981). Media conditioned by *Hordeum vulgare* anthers or ovaries substantially increased the yield of androgenic embryos in anther cultures (Xu et al., 1981). Suspension cultures can sometimes be grown below this "minimum effective density" if growth regulators and amino acids are added (cf., Street, 1977 pp. 61-102). Recent work has shown that when *Carum carvi* somatic embryos are grown at low density, their maturation can be promoted and controlled by the application of ZEA and ABA in proper concentrations (Fig. 3). The effect of low density becomes especially critical if attempts to synchronize populations by sieving and centrifugation severely reduce the population density (Huber et al., 1978).

SYNCHRONY. Populations of somatic embryos typically show a wide range of sizes and stages of development because (1) at the time of transfer from the maintenance medium to the medium that will allow

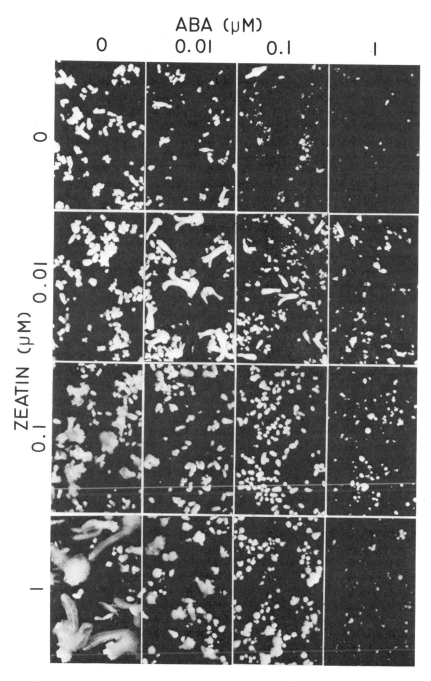

Figure 3.   The interactions of ABA and ZEA on the development of
Carum somatic embryos from small proembryonic units (<74 μm) inoculated at low density.   Growth was in continuous light.   Embryos failed
to mature in unsupplemented MS, and both ABA and ZEA were required
to promote growth and foster normal embryo maturation.   A small but
representative portion of each crop is shown (x 4.5).

development, there is a range of proembryonic cell clusters from those with just a few cells to those substantially larger and (2) somatic embryogenesis is to some extent repetitive (this varies from species to species) so that new embryogenic centers may arise, either from cell clusters or maturing embryos. Having all the embryos pass through each stage simultaneously would be very valuable for both theoretical studies (biochemical analysis, e.g.) and for practical applications, such as mechanical planting or artificial seed production. Attempts at synchrony must address the problems of uniformity of inoculum and preventing adventive/accessory embryogenesis.

The most common method for attaining some degree of uniformity, at least in terms of the starting population, is by means of sieving at the time of transfer to the secondary medium. A graded series of stainless steel mesh sieves (Halperin, 1966; Ammirato, 1974) or nylon mesh (Fujimura and Komamine, 1975) have proved adequate. Passing the suspension through glass beads has also been effective (Warren and Fowler, 1978). Sieving followed by centrifugation in 16% Ficoll solution containing 2% sucrose has isolated a population of *Daucus* cell aggregates from 3 to 10 cells each; these develop synchronously when moved to growth regulator-free medium (Fujimura and Kmomine, 1979a, 1980b).

For controlling "repetitive embryogenesis," ABA has proven effective in *Carum carvi* (Ammirato, 1974) and *Pennisetum americanum* (Vasil and Vasil, 1981a).

GENETIC STABILITY. Changes in the ploidy level have been noted in many cultures leading to mixed populations of polyploids and aneuploids, and plants regenerated from such cultures often show a range of chromosome complements (Sunderland, 1977; D'Amato, 1978). The embryogenic capacity of cultures has been seen to decrease and disappear during progressive subculturing (Syono, 1965), and this loss of potential has been traced, at least in certain cases, to the change in chromosome complement where aneuploids gradually replace diploid cells (Smith and Street, 1974).

This loss of potential may not necessarily be permanent. In habituated *Citrus sinensis* cultures, embryogenesis was restored by eliminating sucrose or aging the tissue (Kochba and Button, 1974). By changing the sequence of growth regulators with each subculture, a nonembryogenic line of *Daucus carota* gradually regained embryogenic capacity (Chandra, 1981). Whether this was due to a selective enrichment of a small number of embryogenic cells remaining in the suspension, or the reinduction of cells that were epigenetically changed was not determined. However, there is evidence that chromosomal, genetic and epigenetic changes occur as cells are cultured.

Although some *Daucus* suspensions showed a range of chromosomal abnormalities, the regenerated plants were almost entirely diploid (Mitra et al., 1960; Mok et al., 1976). At that time it was thought that only cells with an unaltered chromosome complement could develop into somatic embryos and plants. However, in studies of somatic embryogenesis in long-term *Daucus carota* cultures, embryos and plants could be grown, but they were often sterile (Sussex and Frei, 1968).

Changes in the chromosome karyotype rather than complement have been seen in somatic embryos and plants, e.g., *Hemerocallis* (Krikorian et al., 1981). In *Bromus inermis* cultures (Gamborg et al., 1970), the resulting somatic embryos were albino. In *Lolium multiflorum* (Dale, 1980), more than 50% of the regenerated plants were albino.

The maintenance of chromosomal and genetic integrity is essential if the goal of somatic embryogenesis is clonal reproduction. There are a number of studies demonstrating that frequent subculturing can effectively minimize the extent of chromosomal changes in cell cultures (Bayliss, 1977; Sunderland, 1977; Evans and Gamborg, 1982). Krikorian (1982) reported that the phenotype of *Daucus carota* plants raised from somatic embryos was normal, provided the suspensions from which they are derived were maintained for relatively short periods, i.e., less than a year. Thus the frequent establishment of fresh cultures from plant materials and careful attention to the subculture regime may help maintain genetic and chromosomal stability. On the other hand, somaclonal variation that has arisen spontaneously appears to have tremendous potential for producing novel and useful varieties (Larkin and Skowcroft, 1981), as has been the case in *Solanum tuberosum* protoplast-derived plants (Shepard et al., 1980). For more complete discussions, see Chapters 16 and 25.

PROTOCOL

The protocol for *Daucus carota* represents the classic example of somatic embryogenesis in culture:

1. Place a 0.5-1 cm petiole explant or a 0.5 cm^3 storage root explant on MS agar medium + 4.5 µM 2,4-D using aseptic technique.
2. After 4 weeks growth in light or darkness, there should be sufficient callus growth.
3. Subculture 2.0-3.0 g of callus into 50 ml of MS liquid medium + 4.5 µM 2,4-D contained in a 250 ml Erlenmeyer flask or a culture flask (Steward and Shantz, 1956). Erlenmeyers are placed on a gyrotary shaker and maintained at 25 C and agitated at 125-160 rpm; culture flasks are placed on a klinostat and rotated at 1 rpm. The presence or absence of light is not critical at this stage.
4. Erlenmeyer flasks are subcultured at 14-18 days; culture flasks at 21-28 days. Rapidly proliferating cultures can be placed on a more rapid subculture schedule.
5. Aseptically aspirate and/or decant the stale medium from step 4 and resuspend the cells onto culture medium devoid of 2,4-D for initiation of embryo development.
6. For a more uniform embryo population, pass the proembryo suspension through a series of stainless steel mesh sieves, starting at 200 µM and proceeding to finer sieves. For carrot, the 74 µM sieve produces a fairly dense suspension of single and small multiple clumps. Sieving with the 43 µM sieve produces a very fine suspension of solely single clumps but at low density.

7. Transfer the washed and sieved suspension to 50 ml of MS basal medium in 250 ml Erlenmeyer flasks or 10 ml of the medium in the culture tubes of Steward et al. (1952). For more normal embryo development and to inhibit precocious germination, especially root elongation, 0.1-1 μM ABA can be added to the culture medium at the time of inoculation. ABA can be dissolved slowly in distilled water at ambient temperature in complete darkness; it should be filter-sterilized and added to autoclaved and cooled medium.

8. For more normal development, cultures should be grown in complete darkness. Embryos should appear in about 8 days and reach mature size in 10-15 days.

9. Somatic embryos can be placed out on agar medium devoid of 2,4-D for plantlet development.

10. Plantlets are transferred to Jiffy pots or vermiculite for subsequent development.

FUTURE PROSPECTS

The ability to raise somatic embryos in cell cultures creates a number of opportunities not available when plants are regenerated via organogenesis. One distinct advantage is that somatic embryos are bipolar structures bearing both root and shoot apices. In one step, both meristems necessary for complete plant growth have been initiated. In organogenesis, root and shoot development are often mutually exclusive, and a sequence of media changes is necessary to generate an entire plant. Since cell or tissue transfers are expensive in terms of material and personnel time and increase the chance for contamination, somatic embryogenesis offers an attractive alternate for plant regeneration.

In addition, embryogenic cultures optimally can produce large numbers of embryos per culture flask, many more than the multiple shoots generated adventitiously via organogenesis. And, when grown in liquid medium as suspension cultures, embryos usually separate from each other and float freely in the medium. Thus they do not have to be manually separated. Large numbers can be moved easily to the appropriate vessels for growth into plants. Suspension cultures of somatic embryos therefore are amenable to mechanical handling. Fluid drilling (Currah et al., 1974; Gray, 1981) has been suggested as one way to deliver large quantities of somatic embryo from culture flask to field (Evans et al., 1981).

Embryos are natural organs of perennation, many of which typically become dormant. If dormancy could be induced in somatic embryos, the possibility arises that they could be incorporated into artificial seeds either by coating or encapsulation (Durzan, 1980; Murashige, 1980). These artificial units could then be handled like normal seeds and stored, shipped, planted, and so forth. The difficulties in achieving this are impressive: synchronous or staged populations of somatic embryos of the proper genotype need to be produced, the capability of inducing, maintaining and then breaking dormancy established, and the means to nourish the young "germling" while not encouraging the growth of microorganisms, such as those that lead to damping-off in zygotic seedlings, must be found. Still, it is an intriguing and potentially valuable prospect.

Because of their innate properties as embryos, somatic embryos may prove useful for long-term storage, such as in germplasm banks. Induced dormancy, artificial seeds, cold storage, dry storage, or cryogenic preservation may play roles in this endeavor.

The biosynthetic potentiality of cells and tissues in culture has long been of interest. Generally speaking, actively dividing cells in callus or suspension culture do not produce the chemicals found in the mature, nondividing differentiated cells. However, in recent years selection techniques have isolated actively dividing cell cultures that do synthesize certain metabolites (see Chapter 23). Somatic embryogenesis may be another alternative to the production of important chemicals. For example, Al-Abta et al. (1979) demonstrated that flavor compounds in celery tissues were absent in undifferentiated callus but were produced by somatic embryos, especially at the most advanced stages. The secondary product composition of mature embryos and plantlets were comparable to that of intact plants (see Chapter 24). Somatic embryos in culture may be another way to achieve secondary product synthesis in culture.

Finally, although the list of species that have been reported to generate somatic embryos is impressive, it is relatively short in comparison to those regenerated via organogenesis (see Chapter 2) and shorter still in comparison to the list of plants of agronomic and horticultural interest. One major goal of future work is the continuation of efforts in fostering somatic embryogenesis in these still recalcitrant species. In addition, even in those plants where somatic embryos have been cultured, there is variation in response from variety to variety, from clone to clone, and from experiment to experiment. For practical applications to occur, basic research must continue alongside. We must be able to routinely generate somatic embryos and manipulate their growth. For this, a better understanding of the controlling factors and underlying mechanisms is necessary.

It is surprising that, in the 25 years since somatic embryos were first observed, there has not been a greater effort to use this remarkable tool for basic research. The striking phenomenon of differentiated or otherwise committed somatic cells embarking on an embryogenic pathway provides a superb opportunity for exploring some of the fundamental questions of growth, differentiation, and development. The competence of cells to retain and express totipotency, the nature of the events that mark the transition to embryonic growth, and the factors, endogenous and exogenous, that mediate the orderly sequence of changes wherein a small ball of cells develops into a complex embryonic plant can, and should, be answered by continued investigations in this area.

KEY REFERENCES

Kohlenbach, H.W. 1978. Comparative somatic embryogenesis. In: Frontiers of Plant Tissue Culture (T.A. Thorpe, ed.) pp. 59-66. Univ. Calgary Press, Calgary.

Raghavan, V. 1976. Experimental embryogenesis in vascular plants. Academic Press, New York.

Sharp, W.R., Sondahl, M.R., Caldas, L.S., and Maraffa, S.B. 1980. The physiology of in vitro asexual embryogenesis. Hortic. Rev. 2:268-310.

Street, H.E. 1979. Embryogenesis and chemically induced organogenesis. In: Plant Tissue and Cell Culture: Principles and Applications. (W.R. Sharp, P.O. Larsen, E.F. Paddock, and V. Raghavan, eds.) pp. 127-153. Ohio State Univ. Press, Columbus.

Tisserat, B., Esan, B.B., and Murashige, T. 1979. Somatic embryogenesis in angiosperms. Hortic. Rev. 1:1-78.

REFERENCES

Abo El-Nil, N.M. 1977. Organogenesis and embryogenesis in callus cultures of garlic (*Allium sativum* L.). Plant Sci. Lett. 9:259-264.
Al-Abta, S., Galpin, I.J., and Collin, H.A. 1979. Flavour compounds in tissue cultures of celery. Plant Sci. Lett. 16:129-134.
Ammirato, P.V. 1973. Some effects of abscisic acid on the development of embryos from caraway cells in suspension culture. Am. J. Bot. 60(suppl):22-23.
_____ 1974. The effects of abscisic acid on the development of somatic embryos from cells of caraway (*Carum carvi* L.). Bot. Gaz. 135:328-337.
_____ 1977. Hormonal control of somatic embryo development from cultured cells of caraway: Interactions of abscisic acid, zeatin and gibberellic acid. Plant Physiol. 59:579-586.
_____ 1978a. Somatic embryogenesis and plantlet development in suspension cultures of the medicinal yam, *Dioscorea floribunda*. Am. J. Bot. 65(suppl.):89.
_____ 1978b. The effects of 8-azaguanine on the development of somatic embryos from cultured caraway cells. Plant Physiol. 61(suppl.):46.
_____ and Steward, F.C. 1971. Some effects of the environment on the development of embryos from cultured free cells. Bot. Gaz. 132:149-158.
Banerjee, S. and Gupta, S. 1976. Embryogenesis and differentiation in *Nigella sativa* leaf callus in vitro. Physiol. Plant. 38:115-120.
Banks, M.S. 1979. Plant regeneration from callus of two growth phases English ivy, *Hedera helix* L. Z. Pflanzenphysiol. 92:349-353.
Bayliss, M.W. 1977. Factors affecting the frequency of tetraploid cells in a predominantly diploid suspension of *Daucus carota*. Protoplasma 92:109-115.

Brettell, R.I.S., Wernicke, W., and Thomas, E. 1980. Embryogenesis from cultured immature inflorescence of *Sorghum bicolor*. Protoplasma 104:141–148.

Carr, D.J. and Reid, D.M. 1968. The physiological significance of the synthesis of hormones in roots and of their export to the shoot system. In: Biochemistry and Physiology of Plant Growth Substances (F. Wightman and G. Setterfield, eds.) pp. 1169–1185. Runge Press, Ottawa.

Chandra, N. 1981. Clonal variation in morphogenetic response of cultured cells of carrot, *Daucus carota* L. Indian J. Exp. Biol. 19:876–877.

Chang, W–C. and Hsing, Y–I. 1980. Plant regeneration through somatic embryogenesis in root-derived callus of ginseng (*Panax ginseng* C. A. Meyer). Theor. Appl. Genet. 57:133–136.

Choinski, J.S., Jr., Trelease, R.N., and Doman, D.C. 1981. Control of enzyme activities in cotton cotyledons during maturation and germination. III. In vitro embryo development in the presence of abscisic acid. Planta 152:428–435.

Collins, G.B. and Phillips, G.C. 1980. In vitro tissue culture and plant regeneration in *Trifolium pratense* L. In: Proceedings USA–NSF and France–CNRS. Seminar on Plant Regeneration (in press).

Currah, I.E., Gray, D., and Thomas, T.H. 1974. The sowing of germinated vegetable seeds using a fluid drill. Ann. Appl. Biol. 76:311–318.

Dale, P.J. 1980. Embryoids from cultured immature embryos of *Lolium multiflorum*. Z. Pflanzenphysiol. 100:73–77.

D'Amato, F. 1978. Chromosome number variation in cultured cells and regenerated plants. In: Frontiers of Plant Tissue Culture (T.A. Thorpe, ed.) pp. 287–295. Univ. Calgary Press, Calgary.

de Bruijne, E., de Langhe, E., and Van Rijk, R. 1974. Action of hormones and embryoid formation in callus cultures of *Carica papaya*. Meded Fac. Landbouwwett. Rijksuniv. Gent. 39:637–645.

Dorion, N., Chupeau, Y., and Bourgin, J.P. 1975. Isolation, culture and regeneration into plants of *Ranunculus sceleratus* L. leaf protoplasts. Plant Sci. Lett. 5:325–331.

Dos Santos, A.V.P., Outka, D.E., Cocking, E.C., and Davey, M.R. 1980. Organogenesis and somatic embryogenesis in tissues derived from leaf protoplasts and leaf explants of *Medicago sativa*. Z. Pflanzenphysiol. 99:261–270.

Dougall, D.K. and Verma, D.C. 1978. Growth and embryo formation in wild-carrot suspension cultures with ammonium ion as the sole nitrogen source. In Vitro 14:180–182.

Drew, R.L.K. 1979a. Effect of activated charcoal on embryogenesis and regeneration of plantlets from suspension cultures of carrot *Daucus carota* L.). Ann. Bot. 44:387–389.

———— 1979b. The development of carrot (*Daucus carota* L.) embryoids (derived from cell suspension cultures) into plantlets on a sugar-free basal medium (Note). Hortic. Res. 19:79–84.

Dunwell, J.M. 1979. Anther culture in *Nicotiana tabacum*: The role of the culture vessel atmosphere in pollen embryo induction and growth. J. Exp. Bot. 30:419–428.

Durzan, D.J. 1980. Progress and promise in forest genetics. In: Paper Science and Technology—The Cutting Edge, pp. 31-60. The Institute of Paper Chemistry, Appleton, Wisconsin.

Esan, E.B. 1973. A detailed study of adventive embryogenesis in the Rutaceae. Ph.D. dissertation. Univ. California, Riverside.

Evans, D.A. and Gamborg, O.L. 1982. Chromosome stability of cell suspension cultures of *Nicotiana* species. Plant Cell Reports 1:104-107.

Evans, D.A., Sharp, W.R., and Flick, C.E. 1981. Growth and behavior of cell cultures: Embryogenesis and organogenesis. In: Plant Tissue Culture: Methods and Applications in Agriculture (T.A. Thorpe, ed.) pp. 45-113. Academic Press, New York.

Facciotti, D. and Pilet, P-E. 1979. Plants and embryoids from haploid *Nicotiana sylvestris* protoplasts. Plant Sci. Lett. 15:1-6.

Ferguson, J.D., McEwan, J.M., and Card, K.A. 1979. Hormonally induced polyembryos in wheat. Physiol. Plant. 45:470-474.

Fridborg, G. and Erikkson, T. 1975. Effects of activated charcoal on growth and morphogenesis in cell cultures. Physiol. Plant. 34:306-308.

Fridborg, G., Pedersen, M., Landstron, L-E., and Eriksson, T. 1978. The effect of activated charcoal on tissue cultures: Adsorption of metabolites inhibiting morphogenesis. Physiol. Plant. 43:104-106.

Fujimura T. and Komamine, A. 1975. Effects of various growth regulators on the embryogenesis in a carrot cell suspension culture. Plant Sci. Lett. 5:359-364.

_____ 1979a. Synchronization of somatic embryogenesis in a carrot cell suspension culture. Plant Physiol. 64:162-164.

_____ 1979b. Involvement of endogenous auxin in somatic embryogenesis in a carrot cell suspension culture. Z. Pflanzenphysiol. 95:13-19.

_____ 1980a. Mode of action of 2,4-D and zeatin on somatic embryogenesis in a carrot cell suspension culture. Z. Pflanzenphysiol. 99:1-8.

_____ 1980b. The serial observation of embryogenesis in a carrot cell suspension culture. New Phytol. 86:213-218.

Gamborg, O.L., Miller, R.A., and Ojima, K. 1968. Plant cell cultures. I. Nutrient requirements of suspension cultures of soybean root cells. Exp. Cell Res. 50:151-158.

Gamborg, O.L., Constabel, F., and Miller, R.A. 1970. Embryogenesis and production of albino plants from cell cultures of *Bromus inermis*. Planta 95:355-358.

Gharyal, P.K. and Maheshwari, S.C. 1981. In vitro differentiation of somatic embryoids in a leguminous tree, *Albizzia lebbeck* L. (Short comm.) Naturwissenschaften 68:379-380.

Gosch, G., Bajaj, Y.P.S., and Reinert, J. 1975. Isolation, culture, and induction of embryogenesis in protoplasts from cell suspensions of *Atropa belladonna*. Protoplasma 86:405-410.

Gray, D. 1981. Fluid drilling of vegetable seeds. Hortic. Rev. 1:1-27.

Green, C.E. 1977. Prospects for crop improvement in the field of cell culture. HortScience 12:131-134.

_____ and Phillips, R.L. 1975. Plant regeneration from tissue cultures of maize. Crop Sci. 15:417-421.

Grewal, S., Sachdeva, U., and Atal, C.K. 1976. Regeneration of plants by embryogenesis from hypocotyl cultures of *Ammi majus* L. Indian J. Exp. Biol. 14:716-717.

Guha, S. and Maheshwari, S.C. 1964. In vitro production of embryos from anthers of *Datura*. Nature 204:497.

Guha, S., Iyer, R.D., Gupta, N., and Swaminathan, M.S. 1970. Totipotency of gametic cells and production of haploids in rice. Curr. Sci. 39:174-176.

Haccius, B. 1978. Question of unicellular origin of non-zygotic embryos in callus cultures. Phytomorphology 28:74-81.

_____ and Lakshmanan, K.K. 1965. Adventiv-embryonen aus *Nicotiana* Kallus, der bei hohen Lichtintensitaten kultiviert wurde. Planta 65:102-104.

Halperin, W. 1966. Alternative morphogenetic events in cell suspensions. Am. J. Bot. 53:443-453.

_____ 1967. Population density effects in embryogenesis in carrot cell cultures. Exp. Cell Res. 48:170-173.

_____ and Jensen, W.A. 1967. Ultrastructural changes during growth and embryogenesis in carrot cell cultures. J. Ultrastruct. Res. 18:428-443.

_____ and Wetherell, D.R. 1964. Adventive embryony in tissue cultures of the wild carrot, *Daucus carota*. Am. J. Bot. 51:274-283.

_____ and Wetherell, D.R. 1965. Ammonium requirement for embryogenesis in vitro. Nature 205:519-520.

Hari, V. 1980. Effect of cell density changes and conditioned media on carrot cell embryogenesis. Z. Pflanzenphysiol. 96:227-231.

Havranek, P. and Vagera, J. 1979. Regulation of in vitro androgenesis in tobacco through iron-free media. Biol. Plant. 21:412-417.

Haydu, Z. and Vasil, I.K. 1981. Somatic embryogenesis and plant regeneration from leaf tissues and anthers of *Pennisetum purpureum* Schum. Theor. Appl. Genet. 59:269-274.

Heberle-Bors, E. 1980. Interaction of activated charcoal and iron chelates in anther cultures of *Nicotiana* and *Atropa belladonna*. Z. Pflanzenphysiol. 99:339-347.

Homes, J.L.A. 1967. Influence de la concentration en glucose sur la development et la différenciation d'embryons formés dans des tissus de carrotte cultivés in vitro. In: Les Cultures de Tissus de Plantes. pp. 49-60. Coll. Nat. CNRS, Paris.

Hosoki, T. and Sagawa, Y. 1977. Clonal propagation of ginger (*Zingiber officinale* Roscoe) through tissue culture. HortScience 12:451-452.

Hu, C.Y. and Sussex, I.M. 1971. In vitro development of embryoids on cotyledons of *Ilex aquifolium*. Phytomorphology 21:103-107.

Hu, C.Y., Ochs, J.D., and Mancini, F.M. 1978. Further observations on *Ilex* embryoid production. Z. Pflanzenphysiol. 89:41-49.

Huber, J., Constabel, F., and Gamborg, O.L. 1978. A cell counting procedure applied to embryogenesis in cell suspensions |of anise (*Pimpinella anisum* L.). Plant Sci. Lett. 12:209-315.

Ihle, J.N. and Dure, L.S., III. 1972. The developmental biochemistry of cotton seed embryogenesis and germination. III. Regulation of the biosynthesis of enzymes utilized in germination. J. Biol. Chem. 247:5048-5055.

Imamura, J. and Harada, H. 1980a.   Stimulatory effects of reduced atmospheric pressure on pollen embryogenesis.   Naturwissenschaften 67:357–358.

_____ 1980b.   Effects of abscisic acid and water stress on the embryo and plantlet formation in anther cultures of *Nicotiana tabacum* cv. Samsun. Z. Pflanzenphysiol. 100:285–289.

Jelaska, S. 1972.   Embryoid formation by fragments of cotyledons and hypocotyls in *Cucurbita pepo*. Planta 103:278–280.

_____ 1974.   Embryogenesis and organogenesis in pumpkin explants. Physiol. Plant. 31:257–261.

Kamada, H. and Harada, H. 1979a.   Studies on the organogenesis in carrot tissue cultures.   I. Effects of growth regulators on somatic embryogenesis and root formation. Z. Pflanzenphysiol. 91:255–266.

_____ 1979b.   Studies on the organogenesis in carrot tissue cultures. II. Effects of amino acids and inorganic nitrogenous compounds on somatic embryogenesis. Z. Pflanzenphysiol. 91:453–463.

_____ 1981.   Changes in the endogenous levels and effects of abscisic acid during somatic embryogenesis of *Daucus carota* L.   Plant Cell Physiol. 22:1423–1429.

Kamat, M.G. and Rao, P.S. 1978.   Vegetative multiplication of eggplants (*Solanum melongena*) using tissue culture techniques.   Plant Sci. Lett. 13:57–65.

Kameya, Y. and Uchimiya, H. 1972.   Embryoids derived from isolated protoplasts of carrot.   Planta 103:356–360.

Kao, K.N. and Michayluk, M.R. 1980.   Plant regeneration from mesophyll protoplasts of alfalfa. Z. Pflanzenphysiol. 96:135–141.

_____ 1981.   Embryoid formation in alfalfa cell suspensions from different plants. In Vitro 17:645–648.

Kato, H. and Takeuchi, M. 1963.   Morphogenesis in vitro starting from single cells of carrot root. Plant Cell Physiol. 4:243–245.

_____ 1966.   Embryogenesis from the epidermal cells of carrot hypocotyl. Sci. Pap. College Gen. Educ. Univ. Tokyo 16:245–254.

Kavathekar, A.K. and Ganapathy, P.S. 1973.   Embryoid differentiation in *Eschscholzia californica*. Curr. Sci. 42:671–673.

_____ and Johri, B.M. 1977.   Chilling induces development of embryoids into plantlets in *Eschscholzia*. Z. Pflanzenphysiol. 81:358–363.

_____ and Johri, B.M. 1978.   In vitro responses of embryoids of *Eschscholzia californica*. Biol. Plant. 20:98–106.

Kessel, R.H.J. and Carr, A.H. 1972.   The effect of dissolved oxygen concentration on growth and differentiation of carrot (*Daucus carota*) tissue. J. Exp. Bot. 23:996–1007.

Kessel, R.H.J., Goodwin, C., and Philip, J. 1977.   The relationship between dissolved oxygen concentration, ATP and embryogenesis in carrot (*Daucus carota*) tissue cultures. Plant Sci. Lett. 10:265–274.

Keyes, G.J., Collins, G.B., and Taylor, N.L. 1980.   Genetic variation in tissue cultures of red clover. Theor. Appl. Genet. 58:265–271.

Khanna, P. and Staba, J. 1970.   In vitro physiology and morphogenesis of *Cheiranthus cheiri* var. *Cloth of Gold* and *C. cheiri* var. *Goliath*. Bot. Gaz. 131:1–5.

Kochba, J. and Button, J. 1974.   The stimulation of embryogenesis and embryoid development in habituated ovular callus from the 'Shamouti' orange (*Citrus sinensis*) as affected by tissue age and sucrose concentration. Z. Pflanzenphysiol. 73:415-421.

Kochba, J. and Spiegel-Roy, P. 1973. Effect of culture media on embryoid formation from ovular callus of 'Shamouti' orange (*Citrus sinensis*). Z. Pflanzenphysiol. 69:156-162.

Kochba, J. and Spiegel-Roy, P. 1977. The effects of auxin, cytokinins and inhibitors on embryogensis in habituated ovular callus of the 'Shamouti' orange (*Citrus sinensis*). Z. Pflanzenphysiol. 81:283-288.

Kochba, J., Button, J., Spiegel-Roy, P., Bornman, C.H., and Kochba, M. 1974. Stimulation of rooting of *Citrus* embryoids by gibberellic acid and adenine sulphate. Ann.Bot. 38:795-802.

Kochba, J., Spiegel-Roy, P., Neumann, H., and Saad, S. 1978. Stimulation of embryogenesis in *Citrus* ovular callus by ABA, ethephon, CCC and alar and its suppression by GA. Z. Pflanzenphysiol. 89:427-432.

Konar, R.N. and Oberoi, Y.P. 1965. In vitro development of embryoids on the cotyledons of *Biota orientalis*. Phytomorphology 15:137-140.

Konar, R.N., Thomas, E., and Street, H.E. 1972. The diversity of morphogenesis in suspension cultures of *Atropa belladonna* L. Ann. Bot. 36:249-258.

Koul, A.K. and Karihaloo, J.L. 1977. In vivo embryoids from anthers of *Narcissus biflorus* Curt. Euphytica 26:97-102.

Krikorian, A.D. 1982. Cloning higher plants from aseptically cultured tissues and cells. Biol. Rev. 57:151-218.

———— and Kann, R.P. 1981. Plantlet production from morphogenetically competent cell suspensions of daylily. Ann. Bot. 47:679-686.

————, Staicu, S., and Kann, R.P. 1981. Karyotype analysis of a daylily clone reared from aseptically cultured tissues. Ann. Bot. 47:121-131.

Krul, W.R. and Worley, J.F. 1977. Formation of adventitious embryos in callus cultures of 'Seyval,' a French hybrid grape. J. Am. Soc. Hort. Sci. 102:360-363.

Lakshmi Sita, G., Raghava Ram, N.V., and Vaidyanathan, C.S. 1979. Differentiation of embryoids and plantlets from shoot callus of sandalwood. Plant Sci. Lett. 15:265-270.

———— 1980. Triploid plants from endosperm cultures of sandalwood by experimental embryogenesis. Plant Sci. Lett. 20:63-69.

Larkin, P.J. and Scowcroft, J.M. 1981. Somaclonal variation—a novel source of variability from cell cultures for plant improvement. Theor. Appl. Genet. 60:197-214.

Litz, R.E. and Conover, R.A. 1980. Somatic embryogenesis in cell cultures of *Carica stipulata* (Report). Hortic. Sci. 15:733-734.

———— 1982. In vitro somatic embryogenesis and plant regeneration from *Carica papaya* L. ovular callus. Plant Sci. Lett. 26:153-158.

Lorz, H. and Potrykus, I. 1979. Regeneration of plants from mesophyll protoplasts of *Atropa belladonna*. Experientia 35:313-314.

Lorz, H., Potrykus, I., and Thomas, E. 1977. Somatic embryogenesis from tobacco protoplasts. Naturwissenschaften 64:439-440.

Lu, C. and Vasil, I.K. 1981. Somatic embryogenesis and plant regeneration from leaf tissues of *Panicum maximum* Jacq. Theor. Appl. Genet. 59:275-280.

_____ 1982. Somatic embryogenesis and plant regeneration in tissue cultures of *Panicum maximum* Jacq. Am. J. Bot. 69:77–81.

_____ and Ozias-Akins, P. 1982. Somatic embryogenesis in *Zea mays* L. Theor. Appl. Genet. 62:109–112.

Maheshwari, S.C. and Gupta, G.R.P. 1965. Production of embryoids in vitro from stem cells of *Foeniculum vulgare*. Planta 67:384–386.

McWilliam, A.A., Smith, S.M., and Street, H.E. 1974. The origin and development of embryoids in suspension cultures of carrot (*Daucus carota*). Ann. Bot. 38:243–250.

Matsuoka, H. and Hinata, K. 1979. NAA-induced organogenesis and embryogenesis in hypocotyl callus of *Solanum melongena* L. J. Exp. Bot. 30:363–370.

McDaniel, J.K., Conger, B.V., and Graham, E.T. 1982. A histological study of tissue proliferation, embryogenesis and organogenesis from tissue cultures of *Dactylis glomerata*. Protoplasma 110:121–128.

Mitra, J., Mapes, M.O., and Steward, F.C. 1960. Growth and organized development of cultured cells. IV. The behavior of the nucleus. Am. J. Bot. 47:357–368.

Mok, M.C., Gabelman, W.H., and Skoog, F. 1976. Carotenoid synthesis in tissue cultures of *Daucus carota* L. J. Am. Soc. Hortic. Sci. 101:442–449.

Mouras, A. and Lutz, A. 1980. Induction, répression et conservation des propriétés embryogénétiques des cultures de tissus de Carotte sauvage. Bull. Soc. Bot. Fr. 127:93–98.

Muniyamma, A. 1977. Triploid embryos from endosperm in vivo (Short comm.). Ann. Bot. 41:1077–1079.

Murashige, T. 1980. Plant growth substances in commercial uses of tissue culture. In: Plant Growth Substances (F. Skoog, ed.) pp. 426–434. Springer-Verlag, New York.

_____ and Skoog, F. 1962. A revised medium for rapid growth and bioassays with tobacco tissue cultures. Physiol. Plant. 15:473–497.

Nag, K.K. and Johri, B.M. 1969. Organogenesis and chromosomal constitution in embryo callus of *Nuytsia floribunda*. Phytomorphology 19:405–408.

Nessler, C.L. 1982. Somatic embryogenesis in the opium poppy, *Papaver somniferum*. Physiol. Plant. 55:453–458.

Newcomb, W. and Wetherell, D.F. 1970. The effects of 2,4,6-trichlorophenoxyacetic acid on embryogenesis in wild carrot tissue culture. Bot. Gaz. 131:242–245.

Nitsch, J.P. 1969. Experimental androgenesis in Nicotiana. Phytomorphology 19:389–404.

_____ 1974. Pollen culture—A new technique for mass production of haploid and homozygous plants. In: Haploids in Higher Plants (K.J. Kasha, ed.) pp. 123–135. Univ. Guelph Press, Ontario.

_____ and Nitsch, C. 1969. Haploid plants from pollen grains. Science 163:85–87.

_____ and Norreel, B. 1973. Effet d'un choc thermique sur le pouvoir embryogène du pollen de *Datura innoxia* cultivé dans l'anthère ou isolé de l'anthère. C. R. Acad. Sci. 276:303–306.

Norstog, K. 1979. Embryo culture as a tool in the study of comparative and developmental morphology. In: Plant Tissue Culture: Principles and Applications (W.R. Sharp, P.O. Larsen, E.F. Paddock, and V. Raghavan, eds.) pp. 179-202. Ohio Stat Univ. Press, Columbus.

_____ and Blume, D. 1974. Abscisic acid promotion of development of excised immature barley embryos. In: Haploids in Higher Plants: Advances and Potential (K.J. Kasha, ed.). Proceedings First International Symposium, Univ. Guelph, Ontario.

_____ and Rhamstine, E. 1967. Isolation and culture of haploid and diploid cycad tissues. Phytomorphology 17:374-381.

Ozias-Akins, P. and Vasil, I.K. 1982. Plant regeneration from cultured immature embryos and inflorescences of Triticum aestivum (wheat): Evidence for somatic embryogenesis. Protoplasma 110:95-105.

Pareek, L.K. and Chandra, N. 1978. Somatic embryogenesis in leaf callus from cauliflower (Brassica oleracea var. Botrytis). Plant Sci. Lett. 11:311-316.

Pence, V.C., Hasegawa, P.M., and Janick, J. 1979. Asexual embryogenesis in Theobroma cacao L. J. Am. Soc. Hortic. Sci. 104:145-148.

_____ 1980. Initiation and development of asexual embryos of Theobroma cacao L. in vitro. Z. Pflanzenphysiol. 98:1-14.

Phillips, G.C. and Collins, G.B. 1980. Somatic embryogenesis from cell suspension cultures of red clover. Crop Sci. 20:323-326.

Picard, E. and Buyer, J. 1975. Nouveaux resultats concernant la culture d'anthères in vitro de blé tendre (Triticum aestivum L.). Effet d'un choc thermique et de la position de l'anthère dans l'epi. C. R. Acad. Sci. 281:127-130.

Prabhudesai, V.R. and Narayanaswamy, S. 1973. Differentiation of cytokinin-induced shoot buds and embryoids on excised petioles of Nicotiana tabacum. Phytomorphology 23:133-137.

_____ 1974. Organogenesis in tissue cultures of certain Asclepiads. Z. Pflanzenphysiol. 71:181-185.

Price, H.J. and Smith, R.H. 1979. Somatic embryogenesis in suspension cultures of Gossypium klotzschainum Anders. (Short comm.) Planta 145:305-307.

Rabechault, H., Ahee, J., and Guenin, G. 1970. Colonies cellulaires et formes embryoides obtenues in vitro à partir de cultures d'embryons de Palmier à huile (Elaeis guineensis Jacq. var. dura Becc.). C. R. Acad. Sci. Ser. D 270:3067-3070.

Radojevic, L. 1979. Somatic embryos and plantlets from callus cultures of Paulownia tomentosa Steud. Z. Pflanzenphysiol. 91:57-62.

_____, Vujicic, R., and Nesovic, M. 1975. Embryogenesis in tissue culture of Corylus avellana L. Z. Pflanzenphysiol. 77:33-41.

Rajasekhar, W., Edwards, M., Wilson, S.B., and Street H.E. 1971. Studies on the growth in culture of plant cells. V. The influence of shaking rate on the growth of suspension cultures. J. Exp. Bot. 22:107-117.

Raman, K. and Greyson, R.J. 1974. In vitro induction of embryoids in tissue cultures of Nigella damascena. Can. J. Bot. 52:1988-1989.

Rao, P.S. 1965. In vitro induction of embryonal proliferation in Santalum album L. Phytomorphology 15:175-179.

_____ and Narayanaswami, S. 1972. Morphogenetic investigations in callus cultures of Tylophora indica. Physiol. Plant. 27:271-276.

_____, Handro, W., and Harada, H. 1973. Hormonal control of differentiation of shoots, roots and embryos in leaf and stem cultures of *Petunia inflata* and *Petunia hybrida*. Physiol. Plant. 28:458–463.

Reinert, J. 1958. Morphogenese und ihre Kontrolle an Gewebekulturen aus Carotten. Naturwissenschaften 45:344–345.

_____ 1959. Uber die Kontrolle der Morphogenese und die Induktion von Advientiveembryonen an Gewebekulturen aus Karotten. Planta 58:318–333.

_____, Tazawa M., and Semenoff, S. 1967. Nitrogen compounds as factors of embryogenesis in vitro. Nature 216:1215–1216.

Reuther, G. 1977a. Adventitious organ formation and somatic embryogenesis in callus cultures of *Asparagus* and *Iris* and its possible application. Acta Hortic. 78:217–224.

_____ 1977b. Embryoide differenzierungsmuster im Kalus der Gattungen *Iris* und *Asparagus*. Ber. Dtsch. Bot. Ges. 90:417–437.

Reynolds, J.F. and Murashige, T. 1979. Asexual embryogenesis in callus cultures of palms. In Vitro 15:383–387.

Sangwan, R.S. and Harada, H. 1975. Chemical regulation of callus growth, organogenesis, plant regeneration, and somatic embryogenesis in *Antirrhinum majus* tissue and cell cultures. J. Exp. Bot. 26:868–881.

Schenk, R.U. and Hildebrandt, A.C. 1972. Medium and techniques for induction of growth of monocotyledonous and dicotyledonous plant cell cultures. Can. J. Bot. 50:166–204.

Sehgal, C.B. 1968. In vitro development of neomorphs in *Anethum graveolens* L. Phytomorphology 18:509–514.

_____ 1978. Differentiation of shoot bud and embryoids from inflorescence of *Anethum graveolens* in cultures. Phytomorphology 28:291–297.

Shepard, J.F., Bidney, D., and Shanin, E. 1980. Potato protoplasts in crop improvement. Science 208:17–24.

Smith, S.M. 1973. Embryogenesis in tissue cultures of the domestic carrot, *Daucus carota* L. Ph.D. Thesis. University of Leicester.

_____ and Street, H.E. 1974. The decline of embryogenic potential as callus and suspension cultures of carrot (*Daucus carota* L.) are serially subcultured. Ann. Bot. 38:223–241.

Sommer, H.E. and Brown, C.L. 1980. Embryogenesis in tissue cultures of sweetgum (Note). For. Sci. 26:257–260.

Sommer, H.E., Brown, C.L., and Kormanik, P.P. 1975. Differentiation of plantlets in long-leaf pine (*Pinus palustris* Mill.) tissue cultured in vitro. Bot. Gaz. 135:196–200.

Sondahl, M.R. and Sharp, W.R. 1977. High frequency induction of somatic embryos in cultured leaf explants of *Coffea arabica* L. Z. Pflanzenphysiol. 81:395–408.

_____ 1979. Research in *Coffea* and applications of tissue culture methods. In: Plant Cell and Tissue Culture: Priniciples and Applications (W.R. Sharp, P.O. Larsen, E.F. Paddock, and V. Raghavan, eds.) pp. 527–584. Ohio State Univ. Press, Columbus.

Srinivasan, C. and Mullins, M.G. 1980. High-frequency somatic embryo production from unfertilized ovules of grapes. Sci. Hortic. 13:245–252.

Staritsky, G. 1970. Embryoid formation in callus cultures of coffee. Acta Bot. Neerl. 19:509–514.

Steward, F.C. 1963. Carrots and coconuts: Some investigations on growth. In: Plant Tissue and Organ Culture—A Symposium (P. Maheshwari and N.S. Ranga Swamy, eds.) pp. 178–197. International Society of Plant Morphologists, Delhi.

Steward, F.C. and Shantz, E.M. 1956. The chemical induction of growth in plant tissue cultures. In: The Chemistry and Mode of Action of Plant Growth Substances (R.L. Wain and F. Wightman, eds.) pp. 165–187. Academic Press, New York.

———— 1959. The chemical regulation of growth: Some substances and extracts which induce growth and morphogenesis. Annu. Rev. Plant Physiol. 10:379–404.

Steward, F.C., Caplin, S.M., and Millar, F.K. 1952. Investigations on growth and metabolism of plant cells. I. New techniques for the investigation of metabolism, nutrition and growth in undifferentiated cells. Ann. Bot. 16:58–77.

Steward, F.C., Mapes, M.O., and Mears, K. 1958. Growth and organized development of cultured cells. II. Organization in cultures grown from freely suspended cells. Am. J. Bot. 45:705–708.

Steward, F.C., Mapes, M.O., Kent, A.E., and Holsten, R.D. 1964. Growth and development of cultured plant cells. Science 143:20–27.

Steward, F.C., Kent, A.E., and Mapes, M.O. 1967. Growth and organization in cultured cells: Sequential and synergistic effects of growth regulating substances. Ann. N. Y. Acad. Sci. 144:326–334.

Steward, F.C., Ammirato, P.V., and Mapes, M.O. 1970. Growth and development of totipotent cells: Some problems, procedures and perspectives. Ann. Bot. 34:761–787.

Steward, F.C., Israel, H.W., Mott, R.L., Wilson, H.J., and Krikorian, A.D. 1975. Observations on growth and morphogenesis in cultured cells of carrot (*Daucus carota* L.). Philos. Trans. R. Soc. London Ser. B 273:33–53.

Street, H.E. 1977. Plant Tissue and Cell Culture, 2nd ed. Univ. California Press, Berkeley.

Sun, C.W., Wang, C.C., and Chu, Z.C. 1973. Cytological studies on the androgenesis of *Triticale*. Acta Bot. Sinica 15:163–173.

Sunderland, N. 1977. Nuclear cytology. In: Plant Tissue and Cell Culture (H.E. Street, ed.) pp. 177–205. Univ. California Press, Berkeley.

Sussex, I.M. and Frei, K.A. 1968. Embryoid development in long term tissue cultures of carrot. Phytomorphology 18:339–349.

Syono, K. 1965. Changes in organ formating capacity of carrot root callus during subcultures. Plant Cell Physiol. 6:403–419.

Thomas, E. and Street, H.E. 1970. Organogenesis in cell suspension cultures of *Atropa belladonna* L. and *Atropa belladonna* cv. lutea Doll. Ann. Bot. 34:657–669.

Thomas, E. and Wenzel, G. 1975. Embryogenesis from microspores of rye. Naturwissenschaften 62:40–41.

Thomas, E., Hoffman, F., and Wenzel, G. 1975. Haploid plantlets from microspores of rye. Z. Pflanzenphysiol. 75:106–113.

Thomas, E., King, P.J., and Potrykus, I. 1977. Shoot and embryo-like structures from cultured tissues of *Sorghum bicolor*. Naturwissenschaften 64:587.

Tisserat, B. and De Mason, D.A. 1980. A histological study of development of adventive embryos in organ cultures of *Phoenix dactylifera* L. Ann. Bot. 46:465–472.

Tisserat, B. and Murashige, T. 1977. Repression of asexual embryogenesis in vitro by some plant growth regulators. In Vitro 13:799–805.

Tulecke, W., Weinstein, L.H., Rutner, A., and Laurencot, H.J., Jr. 1961. The biochemical composition of coconut water (coconut milk) as related to its use in plant tissue culture. Contr. Boyce Thompson Inst. 21:115–128.

Umbeck, P.F. and Norstog, K. 1979. Effects of abscisic acid and ammonium ions on morphogenesis of cultured barley embryos. Bull. Torrey Bot. Club 106:110–116.

Vardi, A., Spiegel-Roy, P., and Galun, E. 1975. *Citrus* cell culture: Isolation of protoplasts, plating densities, effect of mutagens and generation of embryos. Plant Sci. Lett. 4:231–236.

Vasil, I.K. and Hildebrandt, A.C. 1966. Variations of morphogenetic behavior in plant tissue cultures. II. *Petroselinum hortense*. Am. J. Bot. 53:869–874.

Vasil, V. and Vasil, I.K. 1980. Isolation and culture of cereal protoplasts. II. Embryogenesis and plantlet formation from protoplasts of *Pennisetum americanum*. Theor. Appl. Genet. 56:97–100.

_____ 1981a. Somatic embryogenesis and plant regeneration from suspension cultures of pearl millet (*Pennisetum americanum*). Ann. Bot. 47:669–678.

_____ 1981b. Somatic embryogenesis and plant regeneration from tissue cultures of *Pennisetum americanum* and *P. americanum* x *P. purpureum* hybrid. Am. J. Bot. 68:864–872.

Verma, D.C. and Dougall, D.K. 1977. Influence of carbohydrates on quantitative aspects of growth and embryo formation in wild carrot suspension cultures. Plant Physiol. 59:81–85.

Walker, K.A., Wendeln, M.L., and Jaworski, E.G. 1979. Organogenesis in callus cultures of *Medicago sativa*. The temporal separation of induction processes from differentiation processes. Plant Sci. Lett. 16:23–30.

Warren, G.S. and Fowler, M.W. 1978. Cell number and cell doubling times during development of carrot embryoids in suspension culture. Experientia 34:356.

_____ 1981. Physiological interactions during the initial stages of embryogenesis in cultures of *Daucus carota* L. New Phytol. 87:481–486.

Weatherhead, M.A., Burdon, J., and Henshaw, G.G. 1978. Some effects of activated charcoal as an additive to plant tissue culture media. Z. Pflanzenphysiol. 89:141–147.

Wernicke, W. and Brettell, R. 1980. Somatic embryogenesis from *Sorghum bicolor* leaves (Letter). Nature 287:138–139.

_____, Wakizuka, T., and Potrykus, I. 1981. Adventitious embryoid and root formation from rice leaves (Short comm.). Z. Pflanzenphysiol. 103:361–366.

Wetherell, D.F. and Dougall, D.K. 1976. Sources of nitrogen supporting growth and embryogenesis in cultured wild carrot tissue. Physiol. Plant. 37:97–103.

White, P.R. 1963.   A Handbook of Plant and Animal Tissue Culture. Jaques Cattel Press, Lancaster, Pennsylvania.

Williams, L. and Collin, H.A. 1976.   Embryogenesis and plantlet formation in tissue cultures of celery.   Ann. Bot. 40:325-332.

Xu, Z.H., Huang, B., and Sunderland, N. 1981.   Culture of barley anthers in conditioned media.   J. Exp. Bot. 32:767-778.

Yang, Y.W. and Chang, W.C. 1980.   Embryoid formation and subsequent plantlet regeneration from callus cultures of *Thalictrum urbaini* Hayata. (Ranunculaceae).   Z. Pflanzenphysiol. 97:19-24.

Yeung, E.C., Thorpe, T.A., and Jensen, C.J. 1981.   In vitro fertilization and embryo culture.   In: Plant Tissue Culture: Methods and Applications in Agriculture (T.A. Thorpe, ed.) pp. 253-271.   Academic Press, New York.

Zapata, F.J. and Sink, K.C. 1981.   Somatic embryogenesis from *Lycopersicon peruviantum* leaf mesophyll protoplasts.   Theor. Appl. Genet. 59:265-268.

Zee, S.Y. 1981.   Studies on adventive embryo formation in the petiole explants of coriander (*Coriandrum sativum*).   Protoplasma 107:21-26.

_____ and Wu, S.C. 1979.   Embryogenesis in the petiole explants of Chinese celery.   Z. Pflanzenphysiol. 93:325-335.

CHAPTER 4
Protoplast Isolation and Culture

D.A. Evans and *J.E. Bravo*

With the appropriate enzyme treatment it is possible to isolate protoplasts from virtually any plant species or any type of plant tissue.
However, the ability to isolate protoplasts capable of sustained cell
division with subsequent callus or plant regeneration is limited to a
small, albeit ever increasing, list of plant species. It is the purpose of
this chapter to summarize the reports of successful protoplast isolation,
with emphasis on subsequent plant regeneration, and to present generalized protocols for the isolation and culture of protoplasts that should
optimize the morphogenetic potential of isolated protoplasts.

The term protoplast refers to all the components of a plant cell
excluding the cell wall. Protoplasts were first isolated using mechanical methods. In most cases, the yield of protoplasts was small, and
only extremely large and highly vacuolated cells could be used for isolation. In addition, protoplast yields were not uniform using mechanical
isolation. The use of cell wall degrading enzymes was soon recognized
as the preferred method to release large numbers of uniform plant
protoplasts. Under appropriate conditions, in a limited number of plant
species, these protoplasts have been successfully cultured to resynthe-
size a cell wall and undergo cell division. While protoplasts were
hailed as a unique method to work with isolated single cells, totipotency was not demonstrated until 1971, when protoplasts of tobacco
were regenerated into intact plants (Takebe et al., 1971). A large gap
between expectation and performance plagued protoplast research
during the mid-1970s, especially following the successful production of
the first interspecific somatic hybrid plants produced by protoplast
fusion (Carlson et al., 1972). Innumerable review articles touted the

potential application of protoplast isolation, culture, and fusion as means to develop new, unique crop varieties.   Without exception, this promise of new crop varieties has remained unfulfilled.   However, during the 1980s it is more likely that new crop varieties will be developed.   As we are only now beginning to realize the unique genetic features of fused protoplasts (see Chapters 8 and 9), it is not surprising that this new technology has not been fully exploited.   The primary limitation of application of protoplast techniques to crop improvement remains the inability to routinely regenerate plants from isolated protoplasts.

As will be described in detail, protoplasts do not merely represent "wall-less cells" as originally envisaged.   Treatment of plant cells with crude enzyme preparations that contain many undefined contaminating compounds undoubtably has a deleterious effect on cell viability.   Also, the high osmotic conditions of early culture and the accumulation of metabolic products in protoplast culture media may also adversely affect cell viability.   These deleterious effects on protoplast cultures may be reflected in abnormal wall resynthesis, delayed first division, accumulation of phenolic compounds, and eventual cell death.   It is indeed true that a far greater number of plant species are capable of somatic organogenesis and embryogenesis from cultured explants (see Chapters 2 and 3) than are capable of regeneration of plants from protoplasts.   The nutrient and environmental conditions for successful plant regeneration seem crucial to the success of protoplast regeneration experiments; however, a wide range of treatments and growth conditions will permit regeneration from cultured explants.   Most reports of plant regeneration from cultured protoplasts have not examined many of these presumably crucial parameters in detail.   Some authors emphasize different key factors to the success of protoplast regeneration experiments and ignore factors which had previously been considered crucial to success.   These essential variables include types and concentrations of enzymes used for isolation, environmental conditions during isolation, growth conditions of the donor plant, or components of the protoplast culture medium.   Most of these factors are discussed in this chapter, but it should be realized that no systematic investigation of all variables has been completed for any species or group of species.   It is impossible to identify the one or two key factors that limit regeneration.   Hence, protoplast isolation and culture is still empirical, and as such appears to be more an artistic method than a scientific method.

ENZYMES USED FOR PROTOPLAST ISOLATION

While the plant cell wall does not represent a physiological barrier, it is a formidable mechanical boundary in most higher plants.   The cell wall offers support, prevents the outer membranes from bursting when osmotic pressures are altered, and prohibits entry of infecting microorganisms.   As the cell wall also interferes with most cellular genetic modifications, a great deal of effort has been directed toward isolation of plant protoplasts.   The three primary components of the plant cell

wall have been identified as cellulose, hemicellulose, and pectic substances. Cellulose ranges from 25% of the dry weight of oat coleoptile
cell walls to 50% of the walls of young cotton hairs (Setterfield and
Bayley, 1961). Cellulose is a linear polymer of D-glucose with a β-1,4
linkage. The molecular weight of cellulose may vary between 50,000
and 2,500,000 d in different species. When hydrolysed with cellulase,
this component of the cell wall produces cellobiose and upon complete
hydrolysis, glucose. Hemicelluloses are less well defined and appear to
be mixed polymers of glucose, galactose, mannose, arabinose, and xylose
(Selby, 1973). Most of these carbohydrates are linked with both β-1,4
and β-1,3 linkages. The average cell wall is composed of 53% hemicelluloses. Pectic substances comprise about 5% of the primary cell
wall and are important in the structure of the cell wall and in establishing connections between plant cells. Pectin is a polymer of methyl-
D-galacturonate connected with α-1,4 linkages with molecular weights
of 25,000 to 360,000 d (Northcote, 1958). In addition, cell walls
contain extensin, a complex glycoprotein that is attached to the cellulose fibrils, and many undefined lipids.

Cocking (1960) first used enzymes to release plant protoplasts by
applying an extract of hydrolytic enzymes to isolate protoplasts from
tomato root tips. Since that time many enzyme formulations have been
used to isolate protoplasts, and the most frequently used hydrolytic
enzymes are now commercially available. Most enzymes have been isolated from microorganisms, and these isolates have activity for different
components in the plant cell wall. Hence, the effective concentration
and ratio of each enzyme may need to be altered when attempting to
release protoplasts from a new species or tissue.

A source of cellulase and hemicellulase is crucial for release of
protoplasts. Cellulase (Onozuka) R10 is partially purified from the mold
of *Trichoderma reesei* (formerly *T. viride*; Tomita et al., 1968). The
preparation commercially available from Kinki Yakult Co. (Japan) or
Calbiochem (US) contains both cellulase and hemicellulase activity.
This enzyme has been used most frequently for protoplast isolation
(Ruesink, 1980). Cellulases were discovered by identifying microorganisms that were capable of growth on cellulose as the sole carbon
source. The cellulases appear to be complexes with multiple distinct
activities. Two components that have been separated using column
chromatography are necessary for hydrolysis of cellulose (Selby, 1973).
The $C1$ component is essentially a chain-separating enzyme, while the
Cx component is responsible for subsequent depolymerization to glucose.
The two components act synergistically to digest cellulose.

Additional hemicellulase may also be necessary for recalcitrant
tissues. The most frequently used hemicellulase has been Rhozyme
HP150 (Table 1). Pectinases are also used in most protoplast isolation
procedures. The most frequently used pectinase is Macerozyme (Macerase), which was derived from the fungus *Rhizopus*. When used alone
pectinase has been effective in cell separation from leaf explants (Usui
and Takebe, 1969). Some reports have stated that pectinase has been
used alone to release protoplasts from fruit tissue (e.g., berries of
Solanum nigrum; Raj and Herr, 1970). However, it should be noted that
fruit tissue produces hydrolyzing enzymes during the ripening process.

More frequently, pectinase is combined with cellulase to result in a one-step isolation of plant protoplasts. A more potent pectinase, Pectolyase Y23, has been suggested as an alternative enzyme that releases tobacco leaf protoplasts in 25 min when combined with Cellulysin (Nagata and Ishii, 1979). Pectolyase has high activity of pectin lyase and polygalacturonase.

Table 1. Source of Commercial Enzymes for Protoplast Isolation

ENZYME	COMMERCIAL SOURCE	ORGANISM
Cellulases		
Driselase	Kyowa Hakko Kogyo Co. (Japan) Plenum Scientific, Hackensack, NJ.	a Basidomycete
Cellulysin (Onozuka R10)	Calbiochem, San Diego, CA.	*Trichoderma reesei* (formerly *T. viride*)
Cellulase	Sigma Chemical, St. Louis, MO.	*Aspergillus niger*
Hemicellulases		
Rhozyme HP150	Corning Glass, Corning, NY.	*A. niger*
Hemicellulase	Sigma Chemical, St. Louis, MO.	*A. niger*
Pectinases		
Macerase (Macerozyme)	Calbiochem, San Diego, CA.	*Rhizopus* spp.
Pectinol AC	Corning Glass, Corning, NY.	*A. niger*
Pectolyase Y23	Kikkoman Shoyu Co., Ltd. (Japan)	*A. japonicus*
Pectinase	Sigma Chemical, St. Louis, MO.	*A. niger*
PATE (pectic-acid-acetyl transferase)	Hoechst (Germany)	

Other enzymes have been used to isolate protoplasts. In most cases these enzymes are used in combination with the enzymes in Table 1 and are used to treat a tissue that does not release protoplasts easily. For example, helicase, derived from snails, has been used in combination with macerase and cellulase to release protoplasts from potato tubers (Lorenzini, 1973). Colonase, a stronger pectinase, has been used in combination with Cellulysin to release protoplasts from rice callus (Wakasa, 1973). Glusulase, frequently used to produce yeast protoplasts, has been used with Cellulysin to release barley aleurone protoplasts (Taiz and Jones, 1971). Zymolase, a crude enzyme preparation from *Arthrobacter luteus*, has been effective in releasing protoplasts from pollen tetrads (Wakasa, 1973). These infrequently used enzymes are not normally necessary to release leaf mesophyll or cell suspension cultured protoplasts.

Most commercial enzyme preparations contain toxic substances and unnecessary impurities. In many cases these include contamination with ribonucleases, proteases, lipases and various other enzymes, phenolics, and salts. Impurities have resulted in inconsistency between batches of enzymes. Such inconsistency has been a particular problem for batches of Driselase (Eriksson, 1977). Schenk and Hildebrandt (1969) purified commercial cellulases to eliminate toxic compounds in an attempt to increase yield and viability of protoplasts. Very highly purified cellulases were less effective in release of protoplasts than partially purified preparations, suggesting that some additional contaminating enzymes are necessary for efficient cell wall digestion. However, partially purified preparations had improved yield and viability of protoplasts from 19 species when compared to untreated commercial enzyme preparations (Schenk and Hildebrandt, 1969). Other authors have reported improved yields and viability by using desalted enzyme preparations. The protocol used in our laboratory to desalt commercial enzyme preparations is summarized in Table 2. The desalting is best carried out on a column, as dialysis tubing may be degraded by the enzymes. Desalted enzymes facilitate the ease of protoplast isolation (Kao et al., 1971; Vasil et al., 1975) and improve protoplast viability and plating efficiency (Patnaik et al., 1981).

Table 2. Protocol for Desalting Commercial Enzyme Preparations.

---

1. Dissolve 10.0 g of commercial enzyme in 100 ml distilled H_2O. Stir on slowest speed of magnetic stirrer for 1 hr at room temperature.
2. Centrifuge (Sorvall S-34 rotor) 8000-9000 rpm (8000 x g) for 15 min to pellet insoluble materials. Collect supernatent and load on to a Biogel P6 column (4.5 cm x 48.5 cm) equilibrated with distilled H_2O.
3. Allow enzyme to run into column, followed by 20 ml distilled H_2O. When this enters column, place 50 ml of distilled H_2O on column and continuously drip distilled H_2O from a reservoir to elute the enzyme.
4. Enzyme will consist of light and dark brown areas on the column. As soon as the first light brown fractions are collected, begin testing for protein as follows:
    (a) Mix 1 drop of fraction with 2 drops of 10% TCA (4 C).
    (b) Foaming and cloudiness indicates protein is present.
    (c) Test every 5 fractions until negative reaction is obtained.
5. Combine all protein containing fractions in glass petri plates freeze, and lyophilize for 15-16 hr. Lyophylized enzyme should be stored at -20 C until needed for protoplast isolation.
6. This procedure can be used for all commercial enzymes, although slight differences in the TCA reactions are observed with commercially available preparations. Macerase generally gives a weak reaction, whereas cellulase and driselase have an intense reaction when mixed with TCA.

PROCEDURES FOR PROTOPLAST ISOLATION

Osmotic Conditions

Protoplasts released directly into standard cell culture medium will burst. Hence the pressure that is mechanically supported by the plant cell wall must be replaced with an appropriate osmotic pressure. The osmotic pressure between cell interior and exterior must be balanced or transfer of cells to a plasmolyzing solution will induce stress on the plant cell. It has been observed that when osmotic pressure is too high, metabolism and growth are impaired. This can be monitored by reduced uptake of amino acids across the plasma membrane (Ruesink, 1978) and reduced cell wall regeneration (Pearce and Cocking, 1973). The synthesis of new cell wall material is influenced by both the type and concentration of osmoticum. Hence protoplasts should be just slightly plasmolyzed during isolation. Commonly used protoplast isolation solutions are outlined in Table 3.

Table 3. Sample Protoplast Isolation Solutions

	TOBACCO LEAF	TOBACCO SUSPENSION[a]	POTATO LEAF[b]
Enzymes			
Cellulysin	0.5%	0.5%	2%
Pectinase	0.25%	0.25%	1%
Macerase	0.25%	0.25%	
Rhozyme	0.25%	0.25%	1%
Osmoticum			
Glucose	0.38 M	0.34 M	0.7 M
Supplemental Chemicals			
$CaCl_2 \cdot 2H_2O$	4 mM	3 mM	6 mM
MES	–	1.5 mM	3 mM
$NaH_2PO_4 \cdot 2H_2O$	–	–	0.7 mM
Time of Incubation	5 hr	6 hr	4-16 hr

[a]Flick and Evans, in press.
[b]Gamborg et al., 1981.

Osmotic pressure is manipulated by adding various sugars or sugar alcohols to the isolation and culture solutions used for protoplasts. Mannitol, sorbitol, glucose, and sucrose have all been frequently used. Mannitol and sorbitol, separately or in combination, have been used most often, with mannitol preferred for isolation of leaf mesophyll protoplasts. Glucose has been used successfully as an alternative to these hexitols (Kao and Michayluk, 1974) for cultured cells. Mineral salts, particularly KCl and $CaCl_2$ (Horine and Ruesink, 1972), have also been used, but there has never been good evidence that they are preferable to mannitol or sorbitol.

Concentrations of 0.23-0.90 M mannitol have been successfully used during isolation of protoplasts. The effective osmotic concentration will depend on the leaf cell osmotic pressures at the time of isolation. Endogenous cell osmotic pressures are markedly influenced by environmental conditions (Shepard and Totten, 1975), and can be manipulated by dark pretreatment of plants, use of young leaf tissues, etc.

Upon transfer of protoplasts to culture medium, it may be appropriate to use metabolically active osmotic stabilizers (e.g., glucose and sucrose) along with metabolically inert osmotic stabilizers, such as mannitol. Active substances will be gradually metabolized by the protoplasts during early growth and cell wall regeneration, resulting in a gradual reduction in the osmoticum. This eliminates sudden changes in osmotic potential when regenerated cells are transferred to a nutrient media for regeneration (Eriksson, 1977).

Donor Tissue

While leaf tissue has been used most frequently for protoplast isolation, protoplasts have been isolated from many other plant tissues (Table 4). Usually types and concentrations of enzymes are similar for most of the tissues used. However, certain tissues, such as aleurone, root tips, or pollen cells, require much higher concentrations of enzymes or longer incubation times to release protoplasts. For example, using concentrations of enzymes similar to those used for leaf tissue, protoplasts of pollen tetrads were not released for 30 hr (Redenbaugh et al., 1980).

When protoplasts are isolated from tissue of the intact plant, the growth conditions of the plant are crucial to the success of protoplast culture. Watts et al. (1974) demonstrated that conditioning of plants is critical for successful culturing of leaf protoplasts of *Nicotiana tabacum*. The plants must be undergoing rapid growth with ideal conditions of 22 C, 15-hr daylight of 10,000-20,000 lux and fed weekly with high nitrogen fertilizer. During summer it was important to reduce the light intensity. These authors observed that the age of the plant was also critical to success of protoplast isolation. Plants older or younger than 40-60 days were unsatisfactory for protoplast isolation. These authors suggested that poor protoplast release was related to senescence when the plants flowered. In addition, striking differences in protoplast release were observed when three varieties of tobacco were compared.

Similarly, Shepard and Totten (1977) maintained that specific plant growth conditions were necessary for sustained growth and development of protoplast cultures of *Solanum tuberosum*. Potato plants were raised from potato tubers in environmentally controlled growth rooms prior to protoplast isolation. Plants were maintained in high light intensity (15,000 lux) with a 12-hr light period until 4-10 days before protoplast isolation. At this time plants were transferred to a room with lower light intensity (7,000 lux) with a 6-hr light period. The plants were maintained at 24 C and 70-75% relative humidity throughout and were fertilized daily with a dilute fertilizer solution (1 g/liter of 20-20-20 fertilizer). Variation from this procedure reduced survival of the potato protoplasts.

Table 4. Isolation of Protoplasts from Various Plant Tissues

TISSUE	PLANT	ENZYMES	OSMOTICUM	CONDITIONS	REFERENCE
Aleurone	*Hordeum*	Cellulase 7.0% then Glusulase 50.0%	Sorbitol 0.7M	35 C, 3 hr	Taiz & Jones, 1971
Coleoptile	*Avena*	Cellulase 2.0%	Mannitol 0.5M	Room temp., 1.5 hr	Ruesink & Thimann, 1965
Crown gall	Parthenocissus	Cellulase 2.0%, Macerozyme .01%	Mannitol 0.65M	90 C, 16 hr	Scowcroft et al., 1973
Epidermis	*Nicotiana*	Cellulase 1.5%, Macerozyme 0.5%, Rhozyme 1.0%	Mannitol 0.65M	20 C, 5 hr	Davey et al., 1973
Fibers	*Gossypium*	Cellulase, Macerase	Not reported	Room temp., 3 hr	Beasley et al., 1974
Fruit placenta	*Lycopersicon*	Pectinol R10 20.0%	Sucrose 0.58M	27 C, 2 hr	Gregory & Cocking, 1965
Guard cell	*Allium*	Cellulase 4.0%	Mannitol 0.23M	6 hr	Zeiger & Helper, 1976
Petal	*Nemesia*	Cellulase 0.5%, Pectinase 2.0%	Mannitol 0.6M	22 C	Hess & Endress, 1973
Pollen tetrads	*Datura*	Cellulase 5.0%, Helicase, 2.0%	Sucrose 0.29M	28 C, 6 hr	Rajasekhar, 1973
Pollen mother cells, microspores, pollen tetrad	*Ulmus*	Cellulase R10 0.25% Rhozyme 0.25%, Pectinase 0.125%	Glucose 0.68M	25 C, 20 hr	Redenbaugh et al., 1980

Table 4. Cont.

TISSUE	PLANT	ENZYMES	OSMOTICUM	CONDITIONS	REFERENCE
Root	*Phaseolus*	Rhozyme 2.0%, Meicelase 4.0%, Macerozyme 0.3%	Mannitol 0.71M	25 C, 16 hr	Xu et al., 1981
Root nodule	*Glycine*	Cellulase 1.5%, Macerozyme 0.5%, Rhozyme 1.0%	Mannitol 0.65M	20 C, 5 hr	Davey et al., 1973
Shoot apex	*Pisum*	Driselase 1.0%, Cellulase 1.0%, Pectinase 0.05%, Hemicellulase 0.5%	Sorbitol 0.25M Mannitol 0.25M	26 C, 4.5 hr	Gamborg et al., 1975
Tubers	*Solanum*	Macerozyme 2.0%, Cellulase 5.0%, Helicase 1.5% then Cellulase 5%	Sucrose 0.44M	29 C, 1 hr	Lorenzini, 1973

Kao and Michayluk (1980) established optimum conditions for release of protoplasts from alfalfa. Leaves were taken from young plants maintained in a growth chamber at 21 C with 12-hr illumination and fertilized weekly. Illumination of 12,000 lux was preferred to higher intensities. Leaves were detached and prepared for isolation after the 12-hr dark period. Protoplasts were released from four successive leaves to ascertain the effect of leaf age on protoplast viability. A longer incubation time was required to release protoplasts from older leaves. For most individual plants used, protoplasts from the youngest alfalfa leaves had the highest plating efficiency.

Preisolation growth conditions of plants were also critical for release of viable protoplasts from eggplant. Plants were grown at 7500 lux with 16-hr illumination at 26 C. Only young plants, with four or less leaves were used for mesophyll protoplast isolation (Bhatt and Fassuliotis, 1981). Similarly, young plants were required to release protoplasts from rice leaves (Deka and Sen, 1976).

In cultivated crops in which plant regeneration has been difficult from protoplasts, e.g., tomato, particular attention has been directed to the growth of donor plants to establish conditions suitable for release of viable protoplasts. Shepard (1981) developed a procedure similar to pretreatment of potato plants, to optimize release of protoplasts of tomato. Plants were maintained in high light, 12-hr photo period, at 24 C and 70-75% relative humidity, conditions that were identical to those used for the initiation of potato plants. Seven days prior to protoplast isolation, tomato plants were transferred to 30 C under 16-hr daylight with only 500 lux for optimum protoplast release. Cassells and Barlass (1978) also emphasized that tomato plants must be grown under low-light intensity for optimum protoplast release. They routinely used calcium nitrate fertilization. Tal and Watts (1979) observed that optimum protoplast release was obtained when tomato plants were grown in 82% relative humidity and maintained at low temperatures.

Preincubation of tissue in vitro has been used in combination with growth of donor plants under optimum conditions to increase the release and viability of protoplasts. Kao and Michayluk (1980) found that protoplast yield from alfalfa was enhanced considerably if leaflets, with the lower epidermis removed, were placed onto culture medium prior to exposure to digestive enzymes. The leaflets were incubated for 36-48 hr in dark in a modified cell culture medium enriched with glucose, xylose, sodium pyruvate, citric acid, malic acid, fumaric acid, and the growth regulators 2,4-D and ZEA-riboside (Kao and Michayluk, 1980). Preincubation resulted in a significantly higher plating efficiency for all the plants tested.

Because growth condition of the donor plant is crucial for successful release of viable protoplasts, it is not surprising that some researchers have used in vitro shoot cultures to generate sufficient leaf material for large-scale protoplast isolation. Binding (1974, 1976) used shoot cultures as a source of petunia and tobacco protoplasts, respectively. Schieder (1977) has used shoot cultures of *Datura* species for routine isolation of protoplasts that are capable of subsequent plant regeneration. In order to use these protoplasts in somatic hybridization, both chlorophyll deficient and normal green shoots have been maintained on B5 medium (Gamborg et al., 1968), supplemented with 4.4 μM BA. The shoots of chlorophyll deficient *Datura innoxia* were grown in 500 lux, while wild type *Datura* spp. were cultured in 3000 lux. All cultures were maintained with 14-hr-day length at 15-18 C. Negrutiu and Mousseau (1980) established shoot cultures of *Nicotiana sylvestris* for protoplast isolation. Maximum protoplast yield was obtained using shoot cultures maintained on MS medium with 0.01 μM NAA and .03 M sucrose. The optimum light intensity for protoplast release was ascertained as 2500 lux when 1000-12,000 lux was monitored. Protoplasts released from shoot cultures are already axenic and hence do not have to be exposed to sodium hypochlorite treatment prior to protoplast release.

As with protoplast isolation from the intact plant, protoplast release from cell suspension and callus cultures is also dependent on growth conditions of the donor cells. Wallin et al. (1977) described successful pretreatment of cell suspension cultures of *Haplopappus gracilis* to increase protoplast yield. Optimum growth conditions were obtained by subculturing cells every 2-3 days and by shaking cells at 60 rpm at 28 C. Three days after subculture to new media, protoplast yield declined. A number of medium components influenced protoplast yield. Increased auxin concentrations, particularly NAA, enhanced yield. The type and concentration of sugar was also important. If cells were subcultured to medium with reduced concentration of sucrose or with glucose substituted for sucrose one day before isolation, protoplast yield increased. In addition, protoplast yield was increased when cysteine, methionine, and 2-mercaptoethanol were added to the culture medium. The compounds were not synergistic in the improvement of protoplast yield. Wallin et al. (1977) suggest that these pretreatments change the chemical composition of the plant cell wall, so that less enzyme is needed for protoplast release, thereby reducing toxic enzyme effects.

Fukunaga and King (1978) examined the effects of nitrogen sources on release of protoplasts from many crop species. They tested nitrate, ammonium, glutamine, and arginine as sole nitrogen sources. Cell growth was optimum in potassium nitrate medium, and poorest in ammonium sulfate medium. Protoplast release, however, was optimum in ammonium sulfate medium for alfalfa, flax, rice, soybean, tobacco, and wheat. Glutamine medium was equally good for protoplast release from soybean cell cultures, while arginine medium was equally good for rice. It was suggested that cells grown in nitrate were resistant to enzyme activity.

Isolation of protoplasts from cultured cells has both advantages and disadvantages when compared to isolation from the intact plant. Cultured cells are already free of contamination. As cultured cells are already conditioned to growth in vitro, it may not be difficult to induce cell division when protoplasts are cultured. As was reported for *Haplopappus*, culture conditions can be readily manipulated to enhance protoplast release. On the other hand, cultured cells are frequently aneuploid (see Chapter 16), which may impair subsequent plant regeneration.

In order to minimize the detrimental effect of enzymes, leaf tissue is often plasmolysed prior to enzyme treatment (Cocking, 1972). This treatment has been shown to enhance recovery of viable protoplasts. Protoplasts are more stable if allowed to equilibrate in osmoticum for 1 hr before enzyme is added. Preplasmolysis markedly decreases the leakage of electrolytes that would otherwise occur during protoplast isolation (Cocking, 1972), prevents uptake of exogenous enzymes into the cytoplasm during isolation, and lessens the osmotic shock (Ruesink, 1980).

Incubation Methods

As the goal of protoplast release is to remove the cell wall without impairing cell viability, the isolation conditions are extremely important for success. Many factors have been identified that influence the quantity or quality of protoplast release, including pH, light intensity, temperature, time of incubation, concentration and type of osmoticum, and use of protecting chemicals. As discussed above, commercial enzyme preparations contain many hydrolysing enzymes that are capable of damaging the plant cells. The damaging effect of enzymes can be minimized by reducing the length of enzyme treatment. It is therefore not surprising that variations in the isolation conditions have influenced yield and viability of protoplasts. If the temperature is raised, the duration of enzyme treatment may be reduced. On the other hand, if the temperature is lowered to 10 C, the incubation time must be prolonged (Butenko, 1979). Many procedures have used multiple temperatures for successful isolation. Though temperature has been identified as a critical variable, no consensus procedures have been adopted. For example, Potrykus et al. (1977) recommended that corn internode sections be exposed to enzymes for 16 hr at 12 C and then for 6 hr at 32 C for optimum release. On the other hand, for the same tissue, Vasil and Vasil (1980) recommended incubation for 1 hr at room temperature, followed by 19 hr at 14 C.

While the effect of light on isolation of protoplasts has not been discussed in detail, protoplasts are usually isolated in the dark (Gill et al., 1981), or in low-light intensity (Chellappan et al., 1980). The pH of the enzyme isolation solution has been varied considerably, usually between 5.4-6.2. It has been suggested that higher pH, 6.0-7.0, is most favorable to release mesophyll protoplasts of *Phaseolus* (Pelcher et al., 1974). However, a lower pH, 5.8, has been used to release mesophyll protoplasts of *Glycine* (Schwenk et al., 1981), a closely related seed legume. Similar variation has been reported for shaking versus stationary enzyme treatment during isolation. In general, gentle shaking is used for release of protoplasts from cell suspension cultures, while stationary culture is usually used to isolate mesophyll protoplasts (Kao, 1975). Hence the environmental conditions of protoplast isolation are usually quite critical, but the methods used to ascertain these optimum conditions have been mostly empirical.

Protoplasts are isolated by treating tissues with a mixture of wall degrading enzymes in solution with other chemicals. These include a mixture of salts, or in some cases, a complete culture medium, a buffer, and an osmotic stabilizer. Calcium chloride has been added at

elevated concentrations, e.g., 6.0 mM (Gamborg, 1977) to increase membrane stability of the isolated protoplasts. Magnesium chloride has also
had a positive effect on release of stable protoplasts. When the
enzymes are dissolved directly into the protoplast culture medium (Kao
and Michayluk, 1980), the shock to protoplasts that could otherwise
occur by transfer from isolation solution to culture medium is reduced.
The enzyme isolation solution also may include a phosphate or MES [2-
(N-morpholino)ethanesulfonic acid] buffer to minimize the shift to acidic
pH that may occur during protoplast isolation (Gamborg, 1976). Also,
cells that are broken during protoplast isolation release hydrolytic
enzymes that damage the remaining protoplasts. Compounds such as
0.5% potassium dextran sulfate (Passiatore and Sink, 1981), which interfere with contaminating proteins present in crude enzyme preparations,
have been added to protoplast isolation solutions to minimize protoplast
damage.

Purification Steps

Following enzyme treatment, a mixture of undigested cells, components of broken or burst cells, and protoplasts is obtained. The mixture should be partially purified to eliminate broken and undigested
cells. A number of techniques have been used, with varying degrees of
success.

The most frequently used purification technique is filtration-centrifugation. The protoplast mixture is passed through a filter, generally 40-
100 μm pore size, to retain undigested cells, cell clumps, and vascular
tissue. The separated protoplasts and cell fragments are recovered,
then centrifuged. The speed of centrifugation, usually 100xg, should be
sufficient to precipitate protoplasts while cell fragments continue to
float. The fragments are decanted, and the remaining protoplasts are
resuspended in isolation solution. The centrifugation and decantation
step is repeated to recover a purified protoplast preparation that is resuspended and cultured in protoplast medium. Filtration has the advantage, when compared to other techniques of protoplast purification, of
using the same osmotic solution during the entire purification procedure. On the other hand, protoplasts are susceptible to breakage during passage through filters and therefore filters should not be used with
delicate protoplast preparations. This method is useful for isolation of
cell culture protoplasts and most leaf mesophyll protoplasts.

For mesophyll protoplasts that are extremely delicate, filtration
results in excess cell breakage. Consequently, for a number of species,
flotation has been used to purify mesophyll protoplasts (Gamborg et al.,
1981). As protoplasts have a relatively low density when compared to
organelles or wall containing fragments, many types of gradients have
been used that are sufficient to float protoplasts and allow debris to
sediment. A concentrated solution of sucrose or sorbitol may be combined with the enzyme-protoplast mixture and, by using the appropriate
speed of centrifugation, protoplasts are recovered at the top of the
centrifuge tube. Concentration of sucrose varies between 0.3 M
sucrose (Shepard and Totten, 1977) to 0.6 M sucrose (Day et al., 1981).

For *Solanum* and *Nicotiana* spp. (Shepard and Totten, 1977, and Wilson et al., 1980, respectively), 350xg for 3-10 minutes is preferred. For tomato, reduced centrifugation speed (40-80xg) is necessary (Zapata et al., 1981). Sorbitol has been used in place of sucrose for *Arabidopsis* (Somerville et al., 1981) and maize (Day et al., 1981). In most cases flotation procedures have not been examined extensively. Generally, reports have included only a single concentration of sucrose at a single speed of centrifugation. Also, this method has been combined with a filtration step prior to centrifugation for less delicate protoplasts (Day et al., 1981; Passiatore and Sink, 1981) or a gel filtration step after centrifugation (Halim and Pearce, 1980). The sucrose flotation method results in less protoplast breakage, but may, in some cases, damage protoplasts due to osmotic shock. The most critical variables for protoplast isolation using this method are the concentration of sucrose or sorbitol and the speed of centrifugation, as both affect the stability of fragile protoplasts.

Numerous other gradients have been suggested to aid in protoplast isolation. Two step Ficoll (polysucrose) gradients with 6% on top and 9% below, dissolved in MS medium with 7% sorbitol, have been used to purify *Daucus carota* suspension culture protoplasts following centrifugation for 5 min at 150xg (Gosch et al., 1975). Following centrifugation, the debris—chloroplasts, vascular elements, and walled cells—sedimented to the bottom of the centrifuge tube while protoplasts were at the top. Ficoll gradients have also been used to purify a variety of other protoplasts (Larkin, 1976; Watts et al., 1974).

A number of discontinuous gradients have been used to purify protoplasts. In general, the protoplasts are mixed in enzyme isolation solution with a high molecular weight substance. The intact protoplasts are recovered in a layer between two phases. Wernicke et al. (1979) layered a protoplast mixture onto 20% Percoll with 0.25 M mannitol and 0.1 M calcium chloride. After centrifugation at 200xg, the protoplasts of *Hyoscyamus muticus* were recovered above the Percoll. As these high molecular weight substances, such as Ficoll and Percoll, are osmotically inert, their use may be preferable to use of sucrose flotation. By using these gradients, the same osmotic strength is maintained throughout the isolation procedure. Kanai and Edwards (1973) used a dextran-polyethylene glycol two phase system to recover intact protoplasts of many species. This method should be carefully examined, as PEG has the secondary effect of agglutinating plant protoplasts. Similarly, protoplasts of *Lolium* have been collected on top of Sepharose J539 beads (Keller and Stone, 1978). Collection of protoplasts at the phase separation can also be achieved using mixtures of sucrose and sorbitol. Barley protoplasts have been collected at phase-border between 0.5 M sucrose and 0.14 M sucrose mixed with 0.36 M sorbitol.

Sample Protocols

The protocol used in our laboratory for isolation of mesophyll protoplasts of *Nicotiana* spp. is summarized in Table 5. A number of factors

138   Basic Techniques of Plant Cell Culture

listed in the protocol may be critical for successful protoplast release. (1) The method of sterilization may be crucial for isolation of viable protoplasts. As with initiation of explants for callus formation, extended sterilization treatment could reduce the viability of resultant cultures. Sterilization treatment should be minimized for leaf tissue to be used for protoplast culture. Success has been reported in the use of in vitro shoot cultures as donor material for leaf protoplast isolation. Such shoot cultures do not require sterilization (Schieder, 1977). (2) As epidermal cells are highly resistant to enzyme treatment, removal of the epidermis is desirable prior to exposure to the enzyme mixture. Lower leaf epidermis can be easily removed if turgor is reduced in leaf cells; therefore, most authors incubate donor plants in the dark for 12 to 24 hr before isolation of protoplasts. Alternatively, lower epidermis may be removed by rubbing the surface with an abrasive brush (Shepard and Totten, 1977). (3) In most cases, the protoplast-enzyme mixture is incubated in the dark or in low light intensity. Usually mesophyll protoplasts are isolated at room temperature (23-28 C) with slow shaking (30-50 rpm) or are isolated in stationary culture with occasional stirring.

Table 5. Isolation of Mesophyll Protoplasts from *Nicotiana* Species

1. The plant to be used for protoplast isolation is placed in the dark for 24 hr. Plants used prior to flowering have the most consistent protoplast yields.
2. Sterilize the youngest, fully expanded leaves in 8% Clorox for 10 min, then rinse three times with sterile distilled water.
3. Allow sterilized leaves to air dry under sterile conditions.
4. Peel lower epidermis from leaf with fine dissecting forceps (Ladd Instrument Corp.). Cut the peeled leaf into sections and transfer ca., 0.5 g leaf tissue into 2 ml of enzyme solution in a 60 x 15 mm petri dish.
5. Place leaf, peeled side down, into the following enzyme mixture (contained in a sterile 60 x 15 mm petri dish):
   Cellulase   0.5%
   Macerase   0.5%
   Driselase   0.125%
   Dissolved in KM medium 8P
   (see Chapter 7)
   pH = 5.6-5.8.
6. Incubate peeled leaf in enzyme solution in the dark, shaking occasionally until protoplasts are released (4-5 hr).
7. The digested protoplast mixture contains protoplasts, cell fragments (including chloroplasts), and undigested leaf material. This mixture can then be purified.

The procedure routinely used in our laboratory for isolation, culture, and plant regeneration of protoplasts from suspension cultures of *Nicotiana* spp. is summarized in Table 6. A number of key variables for the success of this protocol should be considered. (1) The isolation of protoplasts from suspension cultures is often more difficult than isolation from leaf tissue, due to the increased cell wall thickness of cells grown in high sucrose concentrations. This problem can be minimized

Table 6.   Isolation and Regeneration of Protoplasts from *Nicotiana* spp.
            Suspension Cultures.

---

1. In a sterile 60 x 15 mm petri dish, combine 2.0 ml of a rapidly
    growing suspension culture with 2.0 ml of the following enzyme
    mixture:

    Cellulase   2.0%
    Macerase   1.0%
    Driselase   1.0%
    Dissolved in EIS[1]
    pH = 5.6–5.8

    We maintain cell suspensions by subculturing every 3–4 days in MS
    medium containing 4.5 μM 2,4-D.

2. Incubate in the dark while shaking (30–50 rpm) for 5 hr. At this
    time, preparation will still contain some undigested cells.

3. Pass the protoplast-enzyme mixture through a stainless steel mesh,
    with 67 μm pore size, to separate protoplasts from undigested
    cells.

4. Wash enzyme from protoplasts by centrifuging for 4 min at 40 x g
    in a sterile conical test tube.

5. Decant and discard supernatent with sterile pasteur pipette.
    Gently resuspend the pelleted protoplasts in 4.0 ml of Kao and
    Michayluk (1975) 8p medium and repeat centrifugation and decan-
    tation steps.

6. Resuspend washed protoplasts in Medium 8p (2.0 ml added to 0.5
    ml protoplast pellet).

7. Place resuspended protoplasts in small droplets in a 60 x 15 mm
    sterile petri dish. Seal plates with a double layer of Parafilm to
    avoid contamination and dehydration of protoplasts.   Store
    protoplasts under diffuse light in plastic boxes.

8. First divisions of protoplasts should occur within 2–5 days.

9. Add fresh Medium 8p to droplets every 1–2 weeks until rapidly
    proliferating callus is visible.

10. Shoot regeneration from callus is accomplished by adding MS
    medium containing 5.0 μM 6BA to the liquid protoplast-derived
    suspension at 2 week intervals.   Shoots should regenerate within
    6 weeks.

11. Regenerated shoots are rooted by transferring to half strength MS
    agar medium containing 0.11 μm 3–aminopyridine.

12. When an extensive root system has formed, the regenerated shoots
    are transferred to peat pellets under high humidity for further
    growth.

13. Gradually decrease the humidity and increase the light intensity
    before transferring regenerated plants to greenhouse conditions.

---

[1]Enzyme isolation solution (EIS) contains 0.7 M glucose, 3.0 mM MES,
6.0 mM $CaCl_2 \cdot 2H_2O$, and 0.7 mM $NaH_2PO_4 \cdot H_2O$.

by using only very rapidly growing cultures in the log phase of growth.
Older cultures contain a higher percentage of senescing cells with
thick cell walls that take much longer to digest than cell walls of

rapidly growing young cells. (2) The filtration-centrifugation system for protoplast purification can be efficiently applied to cell suspension culture derived protoplasts.   Cell suspensions usually contain large aggregates. When single protoplasts are isolated, they become physically separated from the aggregates, while the undigested cells remain in aggregates which do not pass through the filter.   As mentioned above, repeated centrifugations are necessary to eliminate any remaining enzyme from the protoplast preparation.   (3) Cultured protoplasts rapidly deplete the glucose in the culture medium and hence, glucose concentration may be critical.   This depletion, however, is directly proportional to the reduced requirements for high osmotic solutions when the cell wall is resynthesized.   (4) Throughout the callus and subsequent shoot regeneration procedure, it is important that high humidity is maintained and that light intensity is slowly increased.

CULTURE PROCEDURES

Medium Composition

Detailed descriptions of the components of protoplast culture media have been published (Gamborg, 1977; Eriksson, 1977).   As nutritional requirements of cultured plant cells and protoplasts are very similar, it is not surprising that protoplast media are usually modifications of frequently used cell culture media. The cell culture media most commonly used as a basis for protoplast media are B5 medium (cf., Kao and Michayluk, 1975) and MS medium (cf., Nagata and Takebe, 1971). Alterations in these and other cell culture media have been suggested for optimum growth of protoplasts.

It has been proposed that concentrations of iron, zinc, and ammonium in the standard cell culture medium may be too high for some protoplasts (von Arnold and Eriksson, 1977).   Ammonium, in particular, has been found detrimental to protoplast survival, and media have been devised for many species such as potato (Upadhya, 1975), tomato (Zapata et al., 1981), and tobacco (Caboche, 1980) that are devoid of ammonium.   Calcium concentration, on the other hand, should be increased 2-4 times over the concentrations normally used for cell cultures (Eriksson, 1977).   Increased calcium concentration may be important for membrane stability.

While glucose may be the preferred carbon source for most protoplasts (Gamborg, 1977), other carbon sources, including sucrose, may be preferred or necessary for some species such as brome grass (Michayluk and Kao, 1975). Uchimiya and Murashige (1976) have shown that tobacco protoplasts grow equally well on sucrose, cellobiose, or glucose. Most protoplast media contain a mixture of carbon sources.   For tomato, sucrose and glucose are mixed in a 2:1 ratio (Zapata et al., 1981). In many systems, addition of a secondary carbohydrate may be beneficial.   Kao and Michayluk (1975) found addition of ribose to be beneficial. In some cases the preferred carbon source is also the preferred osmoticum (Kao and Michayluk, 1974, for glucose). On the other hand, in some cases a nonmetabolizible osmoticum may be necessary.

For example, for pea mesophyll protoplasts, only mannitol and sorbitol could be used as osmoticum (von Arnold and Eriksson, 1977).

Numerous organic nutrients have been added to protoplast culture media. In most cases, vitamin requirements are the same for plant cells and protoplasts. Von Arnold and Eriksson (1977) suggested addition of folic acid to stimulate division in cultures of pea mesophyll protoplasts. Kao and Michayluk (1974) have suggested addition of several vitamins, organic acids, sugars, sugar alcohols, and undefined nutrients such as casamino acids and CW for culture of protoplasts in very low densities. In most cases, though, many of these factors are unnecessary for culture of protoplasts. For example, while CH or casamino acids have been added to many different protoplast cultures (cf., Constabel, 1975), Uchimiya and Murashige (1976) eliminated CH from their final culture medium, as no benefit could be attributed from its use.

Types and concentrations of growth regulators are the media components that have been varied most frequently. This is not surprising as subtle changes in growth regulators have been shown to have dramatic effects on cultured cells. 2,4-D is the growth regulator most commonly used in protoplast media; however, in some species, other regulators are preferred. For tobacco protoplasts, Uchimiya and Murashige (1976) observed a higher rate of cell reproduction in cultures with NAA than in cultures with 2,4-D or IAA. Also, in tobacco, cytokinin is apparently unnecessary to induce cell division in cultured protoplasts. Von Arnold and Eriksson (1977) reported the requirement for both an auxin (2,4-D) and a cytokinin (2iP) to induce cell division in pea mesophyll protoplasts.

Density of Culture

Published procedures suggest that protoplasts should be cultured at a density of 5000 to 100,000 cells/ml. However, additional interest has been generated in the culture of single isolated protoplasts. Culture of isolated protoplasts would be particularly useful for gene modification experiments such as protoplast fusion or selection of single-cell clones following mutagenesis. Kao and Michayluk (1975) completed the most extensive studies on nutritional requirements of cultured cells and protoplasts at different densities using Vicia hajastana (vetch). Vetch cells were cultured at nine different densities between 2 and 5600 cells/ml. A relatively simple B5 medium could support growth for >250 cells/ml, but additional components were needed to achieve growth at lower densities. Only cell medium 8, which contained B5 with 2,4-D, ZEA, NAA, and a number of organic acids, supplemental sugars and undefined growth additives could support growth at <10 cells/ml. Similarly, for protoplasts cultured in droplet culture, only medium 8p could support growth below 10 protoplasts per ml.

Kao (1977) modified this technique to culture isolated heterokaryocytes following protoplast fusion. Gleba (1978) has used microisolation to culture individual haploid protoplasts of N. tabacum. Protoplasts were placed in extremely small droplets, <2 µl, so that the effective protoplast concentration remained high. Gleba (1978) found that culture of protoplasts at high density for up to 3 days before microisolation resulted in a higher plating efficiency.

Raveh and Galun (1975) used irradiated feeder cells as an aid to culture protoplasts of tobacco at low density. Using a feeder layer, they were able to successfully culture 5-50 protoplasts per ml. This method therefore permits an evaluation of protoplast-derived clones.

Culture Techniques

Several methods have been described for the culture of protoplasts. These include droplet culture, agar culture, coculturing, feeder layers, and hanging droplets.

The liquid droplet method was developed by Kao et al. (1971) and involves suspending protoplasts in culture media and pipetting 0.1-0.2 ml droplets into 60 x 15 mm plastic petri dishes. Five to seven drops can be cultured per plate. The plates are then sealed with parafilm for incubation (Constabel, 1975). This method is convenient for subsequent microscopic examination of protoplast development using an inverted microscope. Fresh media is easily added directly to the developing suspension at 5-7 day intervals. In many cases, however, the cultured protoplasts clump together at the center of the droplet.

In agar culture, protoplasts are either allowed to regenerate cell walls in liquid culture before they are mixed with 0.6% agar (Coutts and Wood, 1977) or are placed in an agar culture media as soon as they are isolated (Cella and Galun, 1980). One method successfully applied to tobacco protoplasts is to mix protoplasts with an equal volume of medium prepared in agar maintained at 45 C (Nagata and Takebe, 1971). Small aliquots of the protoplast-agar mixture are then poured into plates. Using this method, protoplasts remain in a fixed position so that protoplast clumping is avoided and separate clones can be monitored.

The agar culture technique has been coupled with the concept of feeder protoplasts to increase the plating efficiency of some recalcitrant protoplasts. Raveh et al. (1973) isolated *N. tabacum* protoplasts, which were X-irradiated to block cell division, and then mixed the irradiated protoplasts with a low density of normal, viable *N. tabacum* protoplasts. The mixture of viable and inviable protoplasts was then poured with 0.6% agar into petri dishes. Alternatively, a feeder layer could be used. Feeder cells, X-irradiated and incapable of division, are mixed with 0.6% agar, and then plated. Viable protoplasts are then mixed with 0.6% agar and plated in a layer on top of the X-irradiated feeder cells (Cella and Galun, 1980).

Another protoplast culture technique that has been reported to improve the plating efficiency of some protoplast species is coculturing. A reliable, fast growing protoplast preparation, e.g., *N. tabacum* suspension culture, is mixed in varying ratios with protoplasts of a more recalcitrant species. The fast growing protoplasts presumably provide the other species with growth factors and undefined diffusable chemicals which aid in regeneration of a cell wall and cell division (Evans,

1979). Similarly, Menczel et al. (1978) isolated individual protoplast-derived cells using a micropipette and transferred these to nurse cultures of albino cells. Colonies derived from the single-transferred protoplast could be visually distinguished from albino colonies by their green color. Menczel et al. (1978) have used this method to recover separate hybrid clones derived following protoplast fusion.

The hanging droplet method (Potrykus et al., 1976) allows culture of fewer protoplasts per droplet than the conventional droplet technique. This facilitates analysis of small groups of protoplasts and is accomplished by culturing protoplasts in inverted liquid droplets of 0.25-0.50 µl. A more appropriate method for culturing single protoplasts is the microisolation method used by Kao (1977) and Gleba (1978). In this method, single protoplasts are mechanically isolated and placed into separate culture wells. These individual protoplasts can then be monitored for extended periods of time.

Protoplast Viability Tests

The most frequently used staining methods for assessing protoplast viability are fluoroscein diacetate (FDA), phenosafranine (Widholm, 1972) and Calcofluor White (Nagata and Takebe, 1970; Galbraith, 1981). As FDA accumulates in the cell membrane, viable, intact protoplasts fluoresce yellow-green within 5 min. FDA dissolved in 5.0 mg/ml acetone is added to the protoplast culture at 0.01% final concentration. The chlorophyll from broken protoplasts fluoresces red. Therefore, the percentage of viable protoplasts in a preparation can be easily calculated. Protoplasts treated with FDA must, however, be examined between 5-15 min after staining as FDA becomes disassociated from the membrane after 15 min.

Phenosafranine, also used at a final concentration of 0.01%, is specific for dead protoplasts. As soon as the stain is mixed with a protoplast preparation, the inviable protoplasts stain red. Viable cells remain unstained by phenosafranine, even after being in contact with the stain for 30 min.

Calcofluor White (CFW), perhaps the most commonly used stain to ascertain protoplast viability, can detect the onset of cell wall regeneration. CFW binds to the β-linked glucosides in the newly synthesized cell wall. Optimum staining is achieved when 0.1 ml of protoplasts is mixed with 5.0 µl of a 0.1% v/v solution of CFW. Cell wall synthesis is observed by a ring of fluorescence around the plasma membrane. Cell division plates can also be stained using CFW.

Other methods which have been used to determine protoplast viability include observations of cyclosis (Pelcher et al., 1974), exclusion of Evans blue dye (Kanai and Edwards, 1973), variation of protoplast size with osmotic changes (Kanai and Edwards, 1973), oxygen uptake studies (Taiz and Jones, 1971), and photosynthetic activity (Kanai and Edwards, 1973).

Protoplast viability tests are important tools, especially when protoplast preparations are intended for biochemical studies. However, the true test of protoplast viability is the ability for protoplasts to undergo continued mitoses, regenerate callus, and ultimately regenerate plants.

Results: Wall Formation, Cell Division, and Callus Formation

It has traditionally been accepted that cell wall regeneration is a prerequisite for nuclear and cell division (Schilde-Rentschler, 1977). However, in at least one instance, nuclear division has been observed without cell wall regeneration (Meyer and Herth, 1978). The process of cell wall regeneration has been followed through studies of cultured tobacco protoplasts using electron microscopy (Nagata and Yamaki, 1973; Burgess and Fleming, 1974; Fowke and Gamborg, 1980). These studies suggest that newly isolated protoplast cell membranes contain protruding microtubules that function in the orientation of newly synthesized cellulose microfibrils. These cellulose microfibrils, which are synthesized after a lag period of 2 days in tobacco, are transported across the cell membrane where, initially, many fibrils are lost in the surrounding medium. Asamizu et al. (1977) reported that the cellulose fibrils that are synthesized first are shorter than those synthesized later. Some of these newly synthesized fibrils, however, become oriented around the microtubules. This random cellulose matrix eventually becomes thicker and the fibrils become oriented parallel to the plasmalemma, resulting in a continuous cell wall.

The rate and regularity of cell wall regeneration depend on the plant species and the state of differentiation of the donor cell used for protoplast isolation. Protoplasts from leaf mesophyll cells of *Nicotiana*, *Petunia*, *Datura*, and *Brassica* all form new cell walls very quickly. Within 24 hr spherical protoplasts become oval as viewed in the light microscope, and the cell wall can be detected using CFW stain (see above). On the other hand, leaf protoplasts of cereals and legumes may require about 4 days for cell wall regeneration. There have been isolated reports where mesophyll protoplasts were unable to form a cell wall (Giles, 1972).

Incomplete cell wall resynthesis, termed protoplast budding, has been observed in a number of species. It has been proposed that budding occurs when pectin is not incorporated into the new cell wall (Hanke and Northcote, 1974) and is the result of weakened areas in the newly synthesized cell wall (Fowke and Gamborg, 1980).

In addition to genetic factors controlling cell wall synthesis, composition of the culture medium is also important for regeneration of the cell wall. For example, concentrations of sucrose in excess of 0.3 M and of sorbitol in excess of 0.5 M inhibit cell wall formation (Shepard and Totten, 1977). In recalcitrant species, such as oat, a number of chemicals have been added that are important for wall resynthesis and cell division (Kaur-Sawhney and Galston, in press). There is evidence that, in some species, certain growth regulators are required for resynthesis of a cell wall (Uchimiya and Murashige, 1976). However, conventional growth regulators such as 2,4-D and ZEA are apparently more important in regulation of cell division than wall regeneration (Takeuchi and Komamine, 1978).

As first division may be dependent on wall formation, division has been observed in fewer species than wall formation. Many factors are important in initiation of cell division including genotype of the donor plant, culture medium, environmental culture conditions, and condition of the donor tissue used for protoplast isolation. For example, proto-plasts isolated from cell cultures containing rapidly growing cells often undergo first cell divisions sooner than leaf mesophyll protoplasts (Vasil, 1976). Many other factors affecting the condition of donor tissue influ-ence the ability of cultured protoplasts to undergo cell division, includ-ing dark pretreatment, age of plant, fertilization of plants, and applica-tion of pesticides to donor plants.

In most cases plating efficiency of cultured protoplasts is quite low. For some species capable of plant regeneration from protoplasts, initial plating efficiencies of <1% have been reported (Banks and Evans, 1976). However, this may indicate that researchers have not attempted to optimize plating efficiencies, as reports of 1% to 90% have appeared in the literature. While Banks and Evans (1976) reported 1% plating effici-ency for *Nicotiana sylvestris* leaf mesophyll protoplasts, Nagy and Maliga (1976), using modified procedures, reported 60-90% plating efficiency for this same species.

As the resynthesized cell wall is reportedly different than the cell wall of intact plants, incomplete cytokinesis has been frequently ob-served during first division resulting in spontaneous fusion and produc-tion of multinucleate protoplasts (Brar et al., 1979). In most cases, these multinucleate protoplasts will not undergo continued growth. However, polyploid plants have frequently been observed among plants regenerated from protoplasts, and it is likely that some multinucleate protoplasts resulting from nuclear fusion are capable of continued growth and differentiation.

In some cases a lag phase is observed before the first division. A lag phase of up to 25 days was reported for cotton protoplasts (Bhoj-wani et al., 1977). However, in most cases first cell division usually occurs 2-7 days after healthy protoplasts are cultured. Subsequent divisions occur more rapidly resulting in multicellular clumps. Fresh culture medium should be added every 1-2 weeks when protoplasts are growing rapidly. Visible callus formation usually occurs in the initial medium used to culture the protoplasts in 2-3 weeks. When the cells begin to grow rapidly, the osmoticum can be slowly reduced when new culture medium is added. After callus is formed, shoot regeneration medium can be substituted for protoplast medium. Usually, callus is transferred from liquid droplet culture to solid agar culture for shoot regeneration and subsequent plantlet regeneration.

PLANT REGENERATION

Plant regeneration from protoplasts has often been viewed as a phe-nomenon restricted to the Solanaceae. Indeed, 38 solanaceous species have been regenerated from cultured protoplasts (Table 7). These include 17 *Nicotiana* species, 6 *Petunia* species, and 6 *Solanum* species. While most of these species have no economic value, some economically important crops (tobacco, potato, eggplant, and pepper) and ornamentals (*Petunia, Salpiglossis, Browallia,* and *Nicotiana*) have been regenerated from protoplasts. Unfortunately, even among the Solanaceae, where most effort on protoplast regeneration has been directed, only one cul-

Table 7. Procedures for Plant Regeneration from Protoplasts of Solanaceous Species

SPECIES	ISOLATION MIXTURE	PROTOPLAST CULTURE MEDIUM (µM)[1]	PROTOPLAST CULTURE CONDITIONS[2]	REGENERATION MEDIUM (µM)	REFERENCE
Atropa belladonna (suspension)	Cellulase R10 1.0%, Macerozyme 0.5%, Sorbitol 0.6 M	MS, KIN 0.93, NAA 11.0	28 C, dark	MS, KIN 0.93	Gosch et al., 1975
A. belladonna (mesophyll)	Macerozyme R10 0.5%, Cellulase R10 0.5%, Hemicellulase 0.2%, Mannitol 0.4 M	NT, Mannitol 0.4 M, KIN 18.6, CPA 5.4	12 C, dark for 1 day, then 24 C 800 lux (12:12)	MS, IAA 22.8, KIN 11.6	Lorz & Potrykus, 1978
A. belladonna (mesophyll)	Cellulase R10 0.5%, Macerozyme 0.1%, Mannitol 0.5 M	MS, NAA 11.0, KIN 0.47	27 C, dark	MS, NAA 11.0	Bajaj et al., 1978
Browallia viscosa (suspension)	Rhozyme (HP) 2%, Meicelase 4%, Macerozyme 0.3%	8p, 2,4-D 0.90, ZEA 2.3, NAA 5.4	27 C, 700 lux	MS, ZEA 4.6	Power & Berry, 1979
Capsicum annuum (mesophyll)	Cellulase R10 2.0%, Macerozyme 0.4%, Mannitol 0.5M	DPD, 2,4-D 4.5, NAA 5.4, 6BA 4.4	25 C, dark	MS, IAA 22.8, KIN 11.9	Saxena et al., 1981
Datura innoxia Haploid and Diploid (mesophyll)	(Preplasmolysis, Mannitol 0.3 M) Macerozyme 1.0% Cellulase SS 3.0% Mannitol 0.5 M	DPD, NAA 2.7, 6BA 2.2, ZEA 1.8	25 C, 5000 lux	B5, 6BA 2.2	Schieder, 1975
D. metel Haploid and Diploid, D. metaloides (mesophyll)	(Preplasmolysis, Mannitol 0.3 M) Macerozyme 1.0%, Cellulase 3.0%, Mannitol 0.5 M	V47, 6BA 2.2, NAA 8.0	28 C, 1500 lux	MS, 6BA 2.2	Schieder, 1977

Table 7. Cont.

SPECIES	ISOLATION MIXTURE	PROTOPLAST CULTURE MEDIUM (µM)[1]	PROTOPLAST CULTURE CONDITIONS[2]	REGENERATION MEDIUM (µM)	REFERENCE
Hyoscyamus muticus (mesophyll)	Macerozyme R10 0.5%, Cellulase R10 0.5%, Driselase 0.2%, Mannitol 0.5 M, CaCl 0.2 M	NT, KIN 18.6, CPA 3.8	24 C, 800 lux (24:0)	MS, KIN 4.6, NAA 0.54	Lorz et al., 1979
H. muticus Haploid (mesophyll)	Macerozyme R10 0.5%, Cellulase R10 0.25%, Mannitol 0.25 M	NT, 6BA 4.4, pCPA 21.5	17 C for 12 hr, 25 C, 2 wks, dark	MS, 6BA 1.1, NAA 0.32	Wernicke et al., 1979
Lycopersicon peruvianum (mesophyll)	(Preplasmolysis, Mannitol 0.5 M) Macerozyme R10 0.5%, Cellulase R10 1.0%, Mannitol 0.5 M	Z, 2,4-D 4.5, NAA 2.7, 6BA 2.2	29 C, dark	MS, 6BA 10.0, IAA 1.0	Muhlbach, 1980
Nicotiana acuminata *N. alata* *N. glauca* *N. langsdorffii* *N. longiflora* *N. otophora* *N. paniculata* (mesophyll)	Macerozyme R10 0.2%, Cellulase R10 1.0%, Driselase 0.5%, Mannitol 0.44 M	T₀, NAA 16.0, 6BA 4.4	25 C, 2 days dark, then 400–700 lux (16:8)	Various	Bourgin et al., 1979
N. alata	Driselase 2.5%, Mannitol or Sorbitol 0.23–0.26 M	MS, NAA 11.0, 6BA 2.2	25 C, 18–22 µEm²s⁻¹	MS, ZEA 4.56	Passiatore & Sink, 1981

147

Table 7. Cont.

SPECIES	ISOLATION MIXTURE	PROTOPLAST CULTURE MEDIUM (µM)[1]	PROTOPLAST CULTURE CONDITIONS[2]	REGENERATION MEDIUM (µM)	REFERENCE
N. forgetiana × N. sanderae (mesophyll)	Driselase 1.0%, Macerozyme R10 1.0%, Cellulase R10 1.0%, Mannitol 0.23 M	MS, NAA 16.0, 6BA 4.4	25 C, 18–22 µEm⁻²s⁻¹	MS, ZEA 4.56	Passiatore & Sink, 1981
Nicotiana alata × N. sanderae (flower petal)	Cellulase 1.0%, Macerase 0.5%, Driselase 0.25%, Glucose 0.38 M	8p, 2,4-D 0.90, 6BA 2.2, NAA 5.4	25 C, diffuse light	MS, 6BA 5.0	Flick & Evans, unpublished
N. alata Haploid (mesophyll)	Macerozyme R10 0.02%, Cellulase R10 0.1%, Driselase 0.05%, Mannitol 0.44 M	Author's, 6BA 4.4, NAA 16.0	24–27 C, 800 lux (16:8)	MS, 6BA 4.4, IAA 2.9	Bourgin & Missonier, 1978
N. debneyi (mesophyll and suspension)	Cellulysin 2.0%, Macerase 0.5%, Driselase 0.5%, Mannitol 0.38 M	Author's, KIN 9.3, pCPA 10.7	23 C, dark	MS, 6BA 4.4, IAA 2.9	Scowcroft & Larkin, 1980
N. debneyi (mesophyll)	Cellulase 0.8%, Macerozyme 0.2%, Xylanase 0.2%, Mannitol 0.4 M	NT, NAA 16.0, 6BA 4.4		LS, KIN 4.6, IAA 5.7	Piven, 1981
N. langsdorfii N. nesophila N. repanda N. stocktonii (mesophyll)	Cellulase R10 1.0%, Pectinase 0.5%, Hemicellulase 0.5%, Glucose 0.38 M	8p, 2,4-D 0.9, ZEA 2.3, NAA 5.4	25 C, diffuse light	MS, 6BA 5.0	Evans, 1979

Table 7. Cont.

SPECIES	ISOLATION MIXTURE	PROTOPLAST CULTURE MEDIUM (μM)[1]	PROTOPLAST CULTURE CONDITIONS[2]	REGENERATION MEDIUM (μM)	REFERENCE
N. otophora N. sylvestris N. tabacum N. sylvestris x N. otophora N. tabacum x N. otophora (mesophyll)	Meicelase P 5.0%, Macerozyme 0.5%, Mannitol 0.49 M	F5, NAA 8.0, 6BA 4.4, 2,4-D 2.3	25 C, 0-700 lux	MS, KIN 11.6, NAA 21.5	Banks & Evans, 1976
Nicotiana plumbaginifolia (mesophyll)	Cellulase R10 2.0%, Macerozyme 0.6%, Mannitol 0.5 M	Ohyama, 2,4-D 4.5, 6BA 4.4	25 C, 48 hr dark then, 2500 lux (16:8)	MS, IAA 22.8, KIN 11.9	Gill et al., 1978
N. rustica (mesophyll)	Cellulase R10 3%, Mannitol 0.5 M	Ohyama, 2,4-D 4.5, 6BA 4.4	25 C, dark for 2 days, then (8:16)	MS, IAA 22.8, KIN 11.9	Gill et al., 1979
N. sylvestris (mesophyll)	(Preplasmolysis, Sucrose 0.4 M) Cellulase R10 2.0%, Macerozyme 0.5%, Sucrose 0.4 M	Kao--#3, 2,4-D 0.45, 6BA 0.88, NAA 5.4	28 C, 300 lux for 3 days, then 1500 lux	LS, 6BA 4.4	Nagy & Maliga, 1976
N. sylvestris Haploid (mesophyll)	Cellulase R10 0.3%, Driselase 0.03%, Pectinol 0.15%, Sorbitol 0.34 M	MS, 2,4-D 0.9, NAA 3.2, KIN 3.7	23 C, dark	MS, KIN 0.23	Facciotti & Pilet, 1979

Table 7. Cont.

SPECIES	ISOLATION MIXTURE	PROTOPLAST CULTURE MEDIUM (μM)[1]	PROTOPLAST CULTURE CONDITIONS[2]	REGENERATION MEDIUM (μM)	REFERENCE
N. tabacum (mesophyll)	Cellulase 2.0%, Mannitol 0.7 M	MS, NAA 16.0, 6BA 4.4	28 C, 2800 lux	B_3, IAA 22.8, KIN 11.9	Nagata & Takebe, 1971
N. tabacum (epidermis)	Meicelase 4.0%, Macerozyme 0.4%	NT, NAA 16.0, 6BA 4.4	25 C, 2000 lux	B_3, IAA 22.8, KIN 11.9	Davey et al., 1974
N. tabacum Haploid (mesophyll)	Macerozyme 0.5%, Mannitol 0.6 M, then, Cellulase 2.0%, Mannitol 0.7 M	Author's, 2,4-D 4.5, 6BA 4.4	28 C, 500 lux (18:6)	MS, ZEA 0.46	Nitsch & Ohyama, 1971
N. tabacum (suspension)	Cellulase R10 1.0%, Pectinase 0.5%, Rhozyme 0.5%, Glucose 0.35 M	Author's, NAA 1.0, 2,4-D 1.0, ZEA 1.0	24–26 C, diffuse light	MS, 6BA 5.0	Gamborg et al., 1979
N. tabacum (mesophyll)	Cellulase 5.0%, Macerozyme 2.5%, Mannitol 0.4–0.6 M	NT, NAA 16.0, 6BA 4.4	Not reported	MS, IAA 22.8, KIN 11.9	Vasil & Vasil, 1974
Petunia hybrida (stem callus)	Cellulase 3.0%, Macerozyme 3.0%, Mannitol 0.4 M	NT, NAA 2.7, 6BA 0.88	Not reported	MS, NAA 0.54, 6BA 14–56.7	Vasil & Vasil, 1974
Petunia axillaris P. hybrida x P. parodii (mesophyll)	Meicelase P 2.0%, Macerozyme 0.2%, Mannitol 0.71 M	MS, 2,4-D 4.5, 6BA 2.2	25 C, 1000 lux	MS, NAA 0.27, 6BA 2.2	Power et al., 1976

Table 7. Cont.

SPECIES	ISOLATION MIXTURE	PROTOPLAST CULTURE MEDIUM (μM)[1]	PROTOPLAST CULTURE CONDITIONS[2]	REGENERATION MEDIUM (μM)	REFERENCE
P. inflata	Meicelase P 3.0%, Macerozyme 0.3%, Mannitol 0.71 M	MS, NAA 11.0, 6BA 2.2	25 C, 1000 lux	MS, KIN 2.3	Power et al., 1976
P. violaceae (mesophyll)		NAA 27.0, 6BA 4.4	25 C, 1000 lux	MS, ZEA 4.6	Power et al., 1976
P. hybrida (mesophyll)	(Preplasmolysis, Mannitol 0.4 M) Pectinase 2.0% then, Cellulase 2.0%, Mannitol 0.4 M	Author's, 2,4-D 6.3, 6BA 1.8	28 C, 700 lux (16:8)	NT, NAA 11.0, 6BA 2.3	Durand et al., 1973
P. hybrida (mesophyll)	(Preplasmolysis, Mannitol 0.71 M) Cellulase P1500 1.2% or Macerozyme 0.4%, Mannitol 0.71 M	Author's, NAA 8.0, 6BA 4.4	22 C, 700 lux	MS, KIN 11.9, IAA 22.8	Frearson et al., 1973
P. hybrida Haploid and Diploid (mesophyll)	(Preplasmolysis, Mannitol 0.3–0.4 M) then, Cellulase 3.0%, Macerozyme 2.0%, Mannitol 0.4–0.6 M	Author's, NAA 6.0, 6BA 2.0	27 C, 50–300 lux	NT, 6BA 5.0, NAA 2.0	Binding, 1974
P. parodii (mesophyll)	Meicelase 3.0%, Macerozyme 0.3%, Mannitol 0.7 M	F5, NAA 8.0, 6BA 4.4	25 C, 1000 lux	MS, IAA 8.6	Hayward & Power, 1975

151

Table 7. Cont.

SPECIES	ISOLATION MIXTURE	PROTOPLAST CULTURE MEDIUM (μM)[1]	PROTOPLAST CULTURE CONDITIONS[2]	REGENERATION MEDIUM (μM)	REFERENCE
P. parodii (mesophyll)	Meicelase 5.0%, Macerozyme 0.5%, Mannitol 0.71 M	MS or 8p, NAA 11.0, 6BA 2.2	25 C, dark	MS, 6BA 4.4, IAA 11.0	Patnaik et al., 1981
P. parviflora (mesophyll)	(Preplasmolysis, Mannitol 0.49 M) Rohament P 1.0%, Mannitol 0.71 M then, Meicelase P 4.0%, Macerozyme 0.4%, Mannitol 0.71 M	MS or 8p, NAA 11.0, 6BA 2.2	25 C, 1000 lux	MS, ZEA 4.6	Sink & Power, 1977
Salpiglossis sinuata (callus)	Driselase 1.0%, Pectinase 0.75%, Cellulase R10 2.0%, Mannitol 0.44 M	MS, IAA 2.9, 2,4-D 4.4, 6BA 2.2	27 C, 20 $\mu Em^{-2}s^{-1}$	MS, 2iP 4.9	Boyes & Sink, 1981; Boyes et al., 1980
Solanum dulcamara (mesophyll)	Macerozyme 0.3%, Meicelase 1.5%, Mannitol 0.5 M	V-KM, 6BA 2.5, 6BA 2.5, 2,4-D 6.0, NAA 5.7	Not reported	NT or B5, 6BA 5.0, NAA 2.0, Folic Acid 1.0	Binding & Nehls, 1977
S. luteum (shoot tip)	Macerozyme 0.3%, Meicelase 1.5%, Mannitol 0.5 M	V-KM, 6BA 2.5, NAA 5.0, 2,4-D 0.5	26 C, 1000–2000 lux	V-KM (various hormones)	Binding et al., 1981
S. melongena (mesophyll)	(Preplasmolysis, Mannitol 0.7 M) Cellulase R10 1.0%, Macerozyme 0.2%, Mannitol 0.7 M	8p, 2,4-D 0.9, 6BA 4.4, NAA 5.4	25 C, dark	MS, IAA 1.0, 2iP 20.0	Bhatt & Fassuliotis, 1981

Table 7. Cont.

SPECIES	ISOLATION MIXTURE	PROTOPLAST CULTURE MEDIUM (µM)[1]	PROTOPLAST CULTURE CONDITIONS[2]	REGENERATION MEDIUM (µM)	REFERENCE
S. melongena (mesophyll)	Cellulase R10 1.5%, Macerozyme 0.5%, Mannitol 0.5 M	Author's, 2,4-D 2.3, 6BA 4.4, NAA 5.4	23 C, dark for 2 wks, then 3000 lux	MS, 6BA 4.4	Saxena et al., 1981
S. nigrum (mesophyll)	(Preplasmolysis, Mannitol 0.5 M) Macerozyme 0.3%, Meicelase 1.5%, Mannitol 0.25 M, Sorbitol 0.25 M	V-KM, 8p (modified), 6BA 2.5, 2,4-D 5.0, NAA 5.0	Dark, 10 days, then 800 lux	B5, 6BA 2.5	Nehls, 1978
S. phureja (callus and suspension)	Cellulase SS 2.5%, Cellulase II 1.0%, Driselase 0.5%, Sorbitol 0.5 M	B5, 2,4-D 4.5	25 C, 1000 lux (12:12)	B5, ZEA 9.1	Schumann et al., 1980
S. phureja x S. chacoense f. gibberulosum (mesophyll)	Cellulysin 0.4%, Mannitol 0.3-0.4 M	Nitsch, 2,4-D 4.5, 6BA 4.4	Room temp, dark	Upadyha, 1975 ZEA 0.91, NAA 11.0	Grun & Chu, 1978
S. tuberosum (mesophyll)	(Preplasmolysis, Sucrose 0.35 M) then, Macerozyme R10 0.1%, Cellulase R10 0.5%, Mannitol 0.35 M	Lam, NAA 11.0, 6BA 2.2	24 C, 500 lux	Lam, IAA 0.57, ZEA 2.3	Shepard & Totten, 1977

Table 7. Cont.

SPECIES	ISOLATION MIXTURE	PROTOPLAST CULTURE MEDIUM (μM)[1]	PROTOPLAST CULTURE CONDITIONS[2]	REGENERATION MEDIUM (μM)	REFERENCE
S. tuberosum British cultivars (mesophyll)	(Preplasmolysis, Sucrose 0.35 M) then, Macerozyme R10 0.1%, Cellulase R10 0.5%	Lam, NAA 11.0, 6BA 2.2	24 C, 500 lux	Lam, ABA 0.71	Gunn & Shepard, 1981
S. tuberosum Dihaploid (mesophyll)	Macerozyme R10 1.0%, Meicelase 15–20%, Mannitol 0.5 M	8p, 6BA or ZEA 2.5, NAA 5.0, 2,4-D 5.0	25–30 C, continuous light	MS, KIN 15.0, IAA 5.0	Binding et al., 1978
S. tuberosum Tetraploid	Meicelase 1.5%, Pectolyase 0.1%, Mannitol 0.5 M	Author's, 2,4-D 4.5–23.0, ZEA 2.3	25 C, 400 lux for 5 days, then 16 hr 1000 lux light	MS, 6BA 0.44	Thomas, 1981

[1]Culture media for Table 7, 8, 9:

BSH	Schenk & Hildebrandt, 1972	Kao, 1977	Kao, 1977
B3	Sacristan & Melchers, 1969	Lam	Lam, 1975
		LS	Linsmaier & Skoog, 1965
B5	Gamborg et al., 1968	MS	Murashige & Skoog, 1962
DPD	Durand et al., 1973	Nitsch	Nitsch & Ohyama, 1971
F5	Frearson et al., 1973	Nitsch & Nitsch	Nitsch, J.P. & Nitsch, C., 1969
Green	Green, 1977	NT	Nagata & Takebe, 1971
Kao #3	Kao et al., 1974		

Ohyama	Ohyama & Nitsch, 1971
8p, 6p	Kao & Michayluk, 1975
T0	Bourgin et al., 1976
Upadyha	Upadyha, 1975
V-KM	Binding & Nehls, 1977
V47	Binding, 1974
Z	Zapata et al., 1977

[2]Culture conditions: Light-dark cycle expressed (hours light:hours dark), e.g., (15:9) = 15 hr light and 9 hr dark.

tivar of an important crop, tomato, has been regenerated from protoplasts.

Protoplasts isolated from callus, cell suspension, leaf shoot tip, and flower petal have all been regenerated. Most of the methods for protoplast regeneration vary between species and donor tissue (Table 7). Hence it is difficult to make generalizations for this group of plant species. Most authors use either a MS-based (including NT medium) or B5-based (including Kao and Michayluk 8p medium) culture medium to initiate growth of protoplasts (Table 7). Although hormone concentrations are varied between varieties and species, an auxin and a cytokinin are almost always used in the initial culture medium. In nearly every case the auxin concentration is higher than the cytokinin concentration in the initial culture medium. This auxin requirement is in striking contrast to the composition of the regeneration medium where at least 18 species have been regenerated in the absence of auxin, using only a cytokinin (e.g., ZEA, 6BA, or 2iP).

Protoplasts of many of the solanaceous species are cultured in the dark for a short time prior to transfer to the light. In most cases protoplasts are cultured in diffuse light, 400-1500 lux. While most protoplasts are cultured at room temperature, elevated temperatures (28-29 C) have been recommended for some species (e.g., Gosch et al., 1975; Muhlbach, 1980).

In recent years, the list of nonsolanaceous species capable of plant regeneration from protoplasts has been steadily expanding (Table 8). This list includes 28 species of both monocots and dicots. Among these 28 species are a number of economically important crops including carrot, endive, cassava, alfalfa, millet, clover, rapeseed, asparagus, cabbage, and citrus. Plant regeneration from protoplasts is still somewhat limited to certain plant families. For example, this list includes many cruciferous and umbelliferous species. However, it is encouraging that recently regeneration was successfully reported among legumes (*Medicago*) and cereals (*Pennisetum* and *Panicum*), two families traditionally considered to be impossible to regenerate.

For most reports of plant regeneration, detailed experiments comparing media composition and environmental conditions were not completed. Hence it is impossible to identify generalized procedures for plant regeneration among these plants. Quite a few authors have developed their own culture medium after having limited success with traditional media. This suggests that composition of culture media may be more critical for cultured protoplasts than for cultured explants. In most cases room temperature has been used during protoplast culture, suggesting that temperature is not critical. However, optimum light intensity varies dramatically, from complete darkness to 80,000 lux, between different plant species. The plant regeneration medium used also varies between species, and is usually based on conditions necessary to regenerate plants from callus of the particular species in use. For example, carrot callus undergoes embryogenesis in hormoneless B5 or MS medium, while *Panicum* cell suspension cultures regenerate with 1-2 μM 2,4-D in MS medium. Hence the strategy for plant regeneration has been to recover rapidly growing callus from protoplasts and to transfer callus to a species specific regeneration medium.

FUTURE PROSPECTS

As solanaceous species were first regenerated from protoplasts, it is not surprising that this plant family has been used extensively in cellular genetic studies. Tobacco, tomato, *Datura, Petunia,* and potato have all been used for somatic hybridization (see Chapter 7), mutant isolation (see Chapter 10), and for gene transfer experiments (see Chapter 14). As the list of species capable of regeneration now includes many nonsolanaceous crops (Table 8), it is anticipated that cellular genetic experiments will be extended during the next few years to other economically important crops. Recent successes with alfalfa (Kao and Michayluk, 1980) and millet (Vasil and Vasil, 1980) are encouraging. Despite these successes, the legumes and cereals remain difficult to culture. Many researchers have directed effort to culturing protoplasts of these families with very limited success (cf., Kaur-Sawhney and Galston, in press). Promising results have been made consisting of sustained cell divisions or root organogenesis from protoplasts of cereals such as corn, wheat, and rice; and legumes such as soybean, bean, vetch, and cowpea (Table 9). Crops in other important plant families have also produced callus or roots from cultured protoplasts (Table 9) suggesting that protoplast regeneration might soon be extended to an even wider range of crop plants.

Table 8. Procedures for Plant Regeneration from Protoplasts of non-Solanaceous Species

SPECIES	ISOLATION MIXTURE	PROTOPLAST CULTURE MEDIUM (µM)[1]	PROTOPLAST CULTURE CONDITIONS[2]	REGENERATION MEDIUM (µM)	REFERENCE
Antirrhinum majus (mesophyll)	Macerozyme 0.2%, Cellulase 2.0%, Mannitol 0.7 M	Author's, 2,4-D 4.5, 6BA 2.2	25 C, 2000 lux	Author's, 2,4-D 4.5, 6BA 2.2	Poirier–Hamon et al., 1974
Arabidopsis thaliana (suspension)	Cellulase R10 1.0%, Macerozyme R10 0.25%, Driselase 0.5%, Mannitol 0.4 M or Sucrose 0.4 M	B5 with 0.4 M Glucose, 2,4-D 4.5, 6BA 0.67	25 C, dark 4 days, then 800–1000 lux	B5, IAA 5.7, 6BA 0.44	Xuan & Menczel, 1979
Asparagus officinalis (cladodes)	Pretreat: ground in 0.7 M Mannitol Macerozyme 1.0%, Cellulase SS 3.0%, Mannitol 0.9 M	Author's, 6BA 4.4, NAA 5.4	25 C, dark	Author's, NAA 5.4, 6BA 17.7	Bui-Dang-Ha & Mackenzie, 1973
Brassica napus (mesophyll)	Cellulase P1500 0.5%, Hemicellulase 0.5%, Sorbitol 0.25% then Driselase 0.5%, Hemicellulase 0.5%, Mannitol 0.25 M	B5, 6BA 1.0, NAA 10.0	26 C, 100 lux	MS, 6BA 5.0, GA 0.1	Kartha et al., 1974
B. napus Haploid (mesophyll)	PATE 0.1%, Cellulase R10 0.5%, Mannitol 0.5 M	Nitsch & Nitsch, NAA 2.7, 2,4-D 1.13, 6BA 0.44	25 C, dark 2 days, then, 2000 lux (8:16)	MS (hormone free)	Thomas et al., 1976

Table 8. Cont.

SPECIES	ISOLATION MIXTURE	PROTOPLAST CULTURE MEDIUM (μM)[1]	PROTOPLAST CULTURE CONDITIONS[2]	REGENERATION MEDIUM (μM)	REFERENCE
B. napus B. oleracea Cabbage (roots)	(Preplasmolysis, Mannitol 0.21 M) Rhozyme 2.0%, Meicelase 4.0%, Macerozyme 0.3%, Mannitol 0.71 M	8p/8 2:1, 2,4-D 0.9, ZEA 2.3, NAA 5.4	23 C, 14 days dark, then, 700 lux	MS, NAA 11.0, 6BA 4.4	Xu et al., 1982
B. napus Haploid (mesophyll)	(Preplasmolysis, Mannitol 0.55 M) PATE 0.03%, Cellulase R10 0.17% Mannitol 0.55 M	Author's, NAA 2.7, 2,4-D 1.13, 6BA 0.44	25 C, dark	NT, IAA 1.1, ZEA 4.6	Kohlenbach et al., 1982
B. napus Haploid (stem embryo)	PATE 0.2%, Cellulase 0.3%, Driselase 0.3%, Mannitol 0.65 M	Nitsch (1969), 2,4-D 4.5, ZEA 2.3	25 C, dark	MS (various hormones)	Kohlenbach et al., 1982
Bromus inermis (suspension)	Rhozyme 2.0%, Pectinase 1.0%, Cellulase P5000 2.0%, Sorbitol 0.35 M, Mannitol 0.35 M	B5, 2,4-D 2.3	24–26 C	B5 (hormone free)	Kao et al., 1973
Citrus sinensis (ovule callus)	Pectinase 1.0%, Sucrose 0.14 M then, Cellulase 1.0%, Mannitol 0.28 M, Sorbitol 0.28 M, (repeated again)	Author's, Mannitol 0.23 M, Sorbitol 0.23 M, NAA 5.4, 6BA 4.4	Continuous light	Author's, X-ray irrad. (near lethal dose)	Vardi et al., 1975

Table 8. Cont.

SPECIES	ISOLATION MIXTURE	PROTOPLAST CULTURE MEDIUM (µM)[1]	PROTOPLAST CULTURE CONDITIONS[2]	REGENERATION MEDIUM (µM)	REFERENCE
Cichorium endiva C. intybus Gaillardia grandiflora Helianthus annuus Nigella arvensis Senecio jacobea S. silvaticus S. viscosus (shoot tip or mesophyll)	Macerozyme 0.3%, Meicelase 1.5%, Mannitol 0.5 M	V-KM, 6BA 2.5, NAA 5.0, 2,4-D 0.5	26 C, 1000–2000 lux	V-KM (various hormones)	Binding et al., 1981
Daucus carota (suspension)	Cellulase 1.5%, Sorbitol 0.6 M	MS (hormones not reported)	28 C, dark	MS (hormone free)	Gosch et al., 1975
D. carota (root)	Pectinase 0.1%, Cellulase 5.0%, Mannitol 0.8 M	Author's, NAA 0.54	25 C, 3000 lux	Author's, KIN 4.6, NAA 0.54	Kameya & Uchimiya, 1972
D. carota (callus)	Cellulase 1.0%, Hemicellulase or Pectinase 0.5%, Mannitol 0.57 M	B5, 2,4-D 0.45	23 C, 800–1000 lux	B5 (hormone free)	Grambow et al., 1972
D. carota (suspension–proembryo aggregates)	Onozuka P1500 4.0%, Driselase 1.0%, Pectinase 1.0%, Rhozyme 1.0%, Sorbitol 0.35 M, Mannitol 0.55 M	8p, NAA 1, ZEA 0.5	26 C	B5 (hormone free)	Dudits et al., 1976

Table 8. Cont.

SPECIES	ISOLATION MIXTURE	PROTOPLAST CULTURE MEDIUM (μM)[1]	PROTOPLAST CULTURE CONDITIONS[2]	REGENERATION MEDIUM (μM)	REFERENCE
Digitalis lanata (mesophyll)	Cellulase R10 1.0%, Macerozyme R10 0.2%, Mannitol 0.55 M	Author's, NAA 8.0, 6BA 2.7, 2,4,5–T 1.0	26–27 C, 1000–2000 lux	MS, IAA 1.1	Li, 1981
D. purpurea (suspension)	(Preplasmolysis, Mannitol 0.3 M) Cellulase 2.0%, Mannitol 0.3 M	MS (conditioned medium) KIN 30.0	<45 C	Not reported	Diettrich et al., 1980
Hemerocallis spp. (suspension)	Cellulysin 1.0%, Macerozyme 0.5%, Sorbitol 0.3 M, Mannitol 0.3 M	MS, KIN 2.3	24 C, 14 hr light	BSH, 2,4–D 2.3, pCPA 10.7, KIN 0.47	Fitter & Krikorian, 1981
Manihot esculenta (mesophyll)	(Preplasmolysis: CaCl$_2$ 1 mM, NH$_4$NO$_3$ 1 mM, NAA 5.4 μM, 6BA 22.0 μM) Macerozyme R10 0.5%, Cellulase R10 1.0%, Sucrose 0.3 M	Author's, 6BA 2.2, NAA 5.4, 2,4–D 0.9	24 C, 4000 lux	MS, 6BA 0.2, IAA 1.4, GA .04	Shahin & Shepard, 1980
Medicago sativa (mesophyll)	Pectinase 0.25%, Cellulase R10 0.5%, Hemicellulase 0.5%, Sorbitol 0.25 M, Mannitol 0.25 M	Author's, ZEA 0.46	23–25 C, 10–20 lux (8:16)	Author's, (hormone free)	Kao & Mich-ayluk, 1980

Table 8. Cont.

SPECIES	ISOLATION MIXTURE	PROTOPLAST CULTURE MEDIUM (µM)[1]	PROTOPLAST CULTURE CONDITIONS[2]	REGENERATION MEDIUM (µM)	REFERENCE
Medicago sativa (mesophyll)	(Preplasmolysis, Distilled H_2O) Cellulase R10 1.0%, Macerozyme 0.5%, Rhozyme HP150 1.0%, Sorbitol 0.3 M, Mannitol 0.3 M	Author's, ZEA 2.3, NAA 5.4, 2,4-D 0.9	25 C, dark for 2–3 days, then, 1800–2300 lux (16:8) 1800–2300 lux	Author's (various)	Johnson et al., 1981
Nemesia strumosa (mesophyll)	Pectinase 2.0% then, Cellulase 2.0%, Mannitol 0.45 M	Nitsch, 2,4-D 4.5, 6BA 4.4	27 C, 80,000 lux, (16:8)	NT, NAA 2.7, 6BA 2.2	Hess & Leipoldt, 1979
Paricum maximum (suspension)	0.75–1.5 ml susp. mixed with: Cellulase R10 2.0%, Pectolyase 0.15%, Driselase 0.25%, Rhozyme 0.5%, Mannitol 0.15 M, Sorbitol 0.15 M	8p, 2,4-D 0.9, ZEA 2.3	27 C, dark 3 days, then (16:8) diffuse light	MS, 2,4-D 1.1–2.3	Lu et al., 1981
Pelargonium spp. (mesophyll)	Pectinase 0.1%, Cellulase 5.0%, Mannitol 0.8 M	MS, IAA 5.7, 6BA 4.4	25 C, 3000 lux	MS (hormone free)	Kameya, 1975
Pennisetum americanum (suspension)	Cellulysin 2.0%, Macerozyme 1.0%, Driselase 0.5%, Rhozyme 0.5%, Sorbitol 0.25 M, Mannitol 0.25 M	8p, 2,4-D 4.5, 6BA 4.4	27 C, (8:16)	8p (hormone free)	Vasil & Vasil, 1980

Table 8. Cont.

SPECIES	ISOLATION MIXTURE	PROTOPLAST CULTURE MEDIUM (µM)[1]	PROTOPLAST CULTURE CONDITIONS[2]	REGENERATION MEDIUM (µM)	REFERENCE
Ranunculus sceleratus (mesophyll)	Cellulase R10 0.1%, Driselase 0.05%, Macerozyme R10 0.02%, Mannitol 0.55–0.60 M	Author's, NAA 16.0, 6BA 4.4	25–28 C, dark 2 days, then (18:6) light	MS half strength, NAA 16.0, 6BA 4.4	Dorion et al., 1975
Senecio vulgaris (mesophyll, callus)	Macerozyme 0.3%, Meicelase 1.5%, Mannitol 0.5 M	V-KM, 6BA 2.5, NAA 5.0, 2,4-D 0.5	28 C, 1000 lux	NT, 6BA 2.5, NAA 10.0	Binding & Nehls, 1980
Trifolium repens (suspension)	Cellulysin 2.0%, Cellulase 2.0%, Driselase 1.0%, Mannitol 0.25 M, Sorbitol 0.25 M	B5, KIN 5.0, 2,4-D 25.0	27 C, dark	MS, 2iP 2.0, IAA 0.5	Gresshoff, 1980

[1]See footnote in Table 7
[2]See footnote in Table 7

Table 9. Recovery of Callus from Protoplast Cultures of Economically Important Plant Species

SPECIES	SOURCE OF PROTOPLASTS	PROTOPLAST CULTURE MEDIUM[1]	PROTOPLAST HORMONES (μM)	CULTURE CONDITIONS[2]	RESULTS	REFERENCE
Brassica campestris (rapeseed)	Mesophyll (8 wk. old plants)	NT	NAA 4.6, 6BA 4.4	Dark for 10 days then 3500 lux (15:9)	Callus	Schenk & Hoffmann, 1979
B. rapa (turnip)	Mesophyll (4–6 wk. old plants)	Author's	2,4-D 0.9, ZEA 2.3, NAA 5.4	1000 lux, 24 C	Callus & roots	Ulrich et al., 1980
Coffea arabica (coffee)	Callus (4–5 week)	Author's	NAA 2.5, KIN 10.0	Diffuse light, 26 C, (12:12)	Callus	Sondahl et al., 1980
Glycine max (soybean)	Pods (<1 cm x 4 cm)	Kao (1977)	2,4-D 0.4, ZEA 2.3, NAA 5.4	27 C, dark	Callus	Zieg & Outka, 1980
Ipomea batatas (sweet potato)	Petiole (3–4 wk. old plants) Mesophyll	MS half strength	NAA 2.7, 6BA 2.2 / 2,4-D 4.5, KIN 2.3	28 C, 1000 lux (24:0) / 28 C, 2000 lux, (16:8)	Callus	Bidney & Shepard, 1980
Oryza sativa (rice)	Mesophyll (30–60 day plants)	Author's	2,4-D 6.3	26 C, 2000 lux	Callus & roots	Deka & Sen, 1976
Pinus pinaster (pine)	Cotyledons	MS	Pretreat 6 days with 2,4-D 13.6, NAA 32.2, 6BA 8.8,	25 C, dark	Callus	David & David, 1979

163

Table 9. Cont.

SPECIES	SOURCE OF PROTOPLASTS	PROTOPLAST CULTURE MEDIUM[1]	PROTOPLAST HORMONES (μM)	CULTURE CONDI- TIONS[2]	RESULTS	REFERENCE
Phaseolus vulgaris (bean)	Leaf	Modified B5	2,4-D 4.5, KIN 9.2	23-25 C	Callus	Pelcher et al., 1974
Pisum sativum (pea)	Shoot tips (3 day seedlings)	B5	2,4-D 0.45, Kinetin 9.3	23 C, 150 lux, (16:8)	Callus	Gamborg et al., 1975
Rosa spp. cv. Soraya (rose)	Floral bud callus (subculture monthly, used 8 days fr. sub.)	MS	Not reported	25-30 C, 1000 lux	Callus	Krishnemurthy et al., 1979
Saccharum officinarum (sugarcane)	Shoot tips	8p	2,4-D 0.9, ZEA 2.3, NAA 5.4	22-24 C, dark	Callus	Evans et al., 1981
Trifolium arvense (clover)	Suspension (1 day)	Modified B5	2,4-D 9.0	27 C, dark	Callus	White & Bhojwani, 1981
Vicia faba (bean)	Suspension (14 day)	6p	IAA 16.0, 2,4-D 0.9, 6BA 2.2	25 C, dark	Callus	Roper, 1981
V. hajastana (vetch)	Suspension (1-2 days fr. sub- cult.)	8p	2,4-D 0.9, ZEA 2.3, NAA 5.4	24 C, 45 lux, (10:14)	Callus	Kao & Mich- ayluk, 1975
Vigna sinensis (cowpea)	Mesophyll (8-12 day old plants)	NT with low Mannitol	NAA 16.0, 6BA 4.4	22-25 C, low light	Callus	Jha & Roy, 1979

164

Table 9. Cont.

SPECIES	SOURCE OF PROTOPLASTS	PROTOPLAST CULTURE MEDIUM[1]	PROTOPLAST HORMONES (μM)	CULTURE CONDI-TIONS[2]	RESULTS	REFERENCE
Zea mays (maize)	Mesocotyl susp. (sub. 7 days) used 4–8 days	Green (1977)	2,4-D 1.1	Not dis-cussed	Callus	Chourey & Zurawshi, 1981

[1]See footnote in Table 7
[2]See footnote in Table 7

KEY REFERENCES

Caboche, M. 1980.   Nutritional requirements of protoplast-derived haploid tobacco cells grown at low cell densities in liquid medium. Planta 149:7-18.

Eriksson, T. 1977.   Technical advances in protoplast isolation and cultivation.   In: Plant Tissue Culture and Its Biotechnical Application (W. Barz, E. Reinhard, and M.H. Zenk, eds.) pp. 313-322.   Springer-Verlag, Berlin.

Gamborg, O.L., Shyluk, J.P., and Shahin, E.A. 1981.   Isolation, fusion, and culture of plant protoplasts.   In: Plant Tissue Culture: Methods and Applications in Agriculture (T.A. Thorpe, ed.) pp. 115-154.   Academic Press, New York

Kao, K.N. and Michayluk, M.R. 1975.   Nutritional requirements for growth of *Vicia hajastana* cells and protoplasts at a very low population density in liquid media.   Planta 126:105-110.

REFERENCES

Asamizu, T., Tanaka, K., Takebe, I., and Nishi, A. 1977.   Change in molecular size of cellulose during regeneration of cell wall on carrot protoplasts. Physiol. Plant. 40:215-218.

Bajaj, Y.P.S., Gosch, G., Grobler, A., Ottma, M., and Webber, A. 1978. Production of polyploid and aneuploid plants from anthers and mesophyll protoplast of *Atropa belladonna* and *Nicotiana tabacum*. Ind. J. Exp. Biol. 16:947-953.

Banks, M.S. and Evans, P.K. 1976.   A comparison of the isolation and culture of mesophyll protoplasts from several *Nicotiana* species and their hybrids. Plant Sci. Lett. 7:409-416.

Beasley, C.A., Ting, I.P., Linkins, A.E., Birnbaum, E.H., and Delmar, D.P. 1974.   Cotton ovule culture: A review of progress and a preview of potential.   In: Tissue Culture and Plant Science (H.E. Street, ed.) pp. 169-192.   Academic Press, London.

Bhatt, D.P. and Fassuliotis, G. 1981.   Plant regeneration from mesophyll protoplasts of eggplant. Z. Pflanzenphysiol. 104:81-89.

Bhojwani, S.S., Evans, P.K., and Cocking, E.C. 1977.   Protoplast technology in relation to crop plants: Progress and problems.   Euphytica 26:343-360.

Bidney, D.L. and Shepard, J.F. 1980.   Colony development from sweet potato petiole protoplasts and mesophyll cells.   Plant Sci. Lett. 18:335-342.

Binding, H. 1974.   Cell cluster formation by leaf protoplasts from axenic cultures of haploid *Petunia hybrida* L.   Plant Sci. Lett. 2:185-188.

_____ 1976.   Somatic hybridization experiments in *Solanaceous* species.   Molec. Gen. Genet. 144:171-175.

_____ and Nehls, R. 1977.   Regeneration of isolated protoplasts to plants in *Solanum dulcamara* L.   Z. Pflanzenphysiol. 85:279-280.

_____ and Nehls, R. 1980. Protoplast regeneration to plants in *Senecio vulgaris* L. Z. Pflanzenphysiol. 99:183-185.

_____, Nehls, R., Schieder, O., Sopory, S.K., and Wenzel, G. 1978. Regeneration of mesophyll protoplasts isolated from dihaploid clones of *Solanum tuberosum*. Physiol. Plant. 43:52-54.

_____, Nehls, R., Kock, R. Finger, J., and Mordhorst, G. 1981. Comparative studies on protoplast regeneration in herbaceous species of the dicotyledonaceae class. Z. Pflanzenphysiol. 101:119-130.

Bourgin, J.P. and Missonier, C. 1978. Culture of haploid mesophyll protoplasts from *Nicotiana alata*. Z. Pflanzenphysiol. 87:55-64.

_____, and Chupeau, Y. 1976. Culture de protoplastes de mesophylle de *Nicotiana sylvestris* Spegazzini et Comes haploide et diploide. C. R. Acad. Sci. Paris 282:1853-1856.

Bourgin, J.P., Chupeau, Y., and Missonier, C. 1979. Plant regeneration from mesophyll protoplasts of several *Nicotiana* species. Physiol. Plant. 45:288-292.

Boyes, C.J. and Sink, K.C. 1981. Regeneration of plants from callus-derived protoplasts of *Salpiglossis*. J. Am. Soc. Hort. Sci. 106:42-46.

Boyes, C.J., Zapata, F.J., and Sink, K.C. 1980. Isolation, culture and regeneration to plants of callus protoplasts of *Salpiglossis sinuata* L. Z. Pflanzenphysiol. 99:471-474.

Brar, D.S., Rambold, S. Gamborg, O., and Constabel, F. 1979. Tissue culture of corn and sorghum. Z. Pflanzenphysiol. 95:377-388.

Bui-Dang-Ha, K. and Mackenzie, I.A. 1973. The division of protoplasts from *Asparagus officinalis* L. and their growth and differentiation. Protoplasma 78:215-221.

Burgess, J. and Fleming, E.N. 1974. Ultrastructural observations of cell wall regeneration around isolated tobacco protoplasts. J. Cell Sci. 14:439-449.

Butenko, R.G. 1979. Cultivation of isolated protoplasts and hybridization of somatic plant cells. Int. Rev. Cytol. 59:323-373.

Carlson, P.S., Smith, H.H., and Dearing, R.D. 1972. Parasexual interspecific plant hybridization. Proc. Nat. Acad. Sci. USA 69:2292-2294.

Cassells, A.C. and Barlass, M. 1978. A method for the isolation of stable mesophyll protoplasts from tomato leaves throughout the year under standard conditions. Physiol. Plant. 42:236-242.

Cella, R. and Galun, E. 1980. Utilization of irradiated carrot cell suspensions as feeder layer for cultured *Nicotiana* cells and protoplasts. Plant Sci. Lett. 19:243-252.

Chellappan, K.P., Seeni, S., and Gnanam, A. 1980. The isolation, culture, and organ differentiation from mesophyll protoplasts of *Mollugo nudicaulis Lam*, a C_4 species. Experientia 36:60-61.

Chourey, P.S. and Zurawshi, D.B. 1981. Callus formation from protoplasts of a maize cell culture. Theor. Appl. Genet. 59:341-344.

Cocking, E.C. 1960. A method for the isolation of plant protoplasts and vacuoles. Nature 187:962-963.

_____ 1972. Plant cell protoplasts—isolation and development. Ann. Rev. Plant Physiol. 23:29-50.

Constabel, F. 1975. Isolation and culture of plant protoplasts. In: Plant Tissue Culture Methods (O.L. Gamborg and L.R. Wetter, eds.) pp. 11-21. Nat. Res. Council of Canada, Saskatoon.

Coutts, R.H.A. and Wood, K.R. 1977. Improved isolation and culture methods for cucumber mesophyll protoplasts. Plant Sci. Lett. 9:45–51.

Davey, M.R., Cocking, E.C., and Bush, E. 1973. Isolation of legume root nodule protoplasts. Nature 244:460–461.

Davey, M.R., Frearson, E.M., Withers, L.A., and Power, J.B. 1974. Observations on the morphology, ultrastructure and regeneration of tobacco leaf epidermal protoplasts. Plant Sci. Lett. 2:23–27.

David, A. and David, H. 1979. Isolation and callus formation from cotyledon protoplasts of pine (*Pinus pinaster*). Z. Pflanzenphysiol. 94:173–177.

Day, D.A., Jenkins, C.L.D., and Hatch, M.D. 1981. Isolation and properties of functional mesophyll protoplasts and chloroplasts from *Zea mays*. Aust. J. Plant Physiol. 8:21–29.

Deka, P.C. and Sen, S.K. 1976. Differentiation in calli originated from isolated protoplasts of rice (*Oryza sativa* L.) through plating technique. Molec. Gen. Genet. 145:239–243.

Diettrich, B., Neumann, D., and Luckner, M. 1980. Protoplast-derived clones from cell cultures of *Digitalis purpurea*. Planta Medica 38:375–382.

Dorion, N., Chupeau, Y., and Bourgin, J.P. 1975. Isolation, culture and regeneration into plants of *Ranunculus sceleratus* L. leaf protoplasts. Plant Sci. Lett. 5:325–331.

Dudits, D., Kao, K.N., Constabel, F., and Gamborg, O.L. 1976. Embryogenesis and formation of tetraploid and hexaploid plants from carrot protoplasts. Can. J. Bot 54:1063–1067.

Durand, J., Potrykus, I., and Donn, G. 1973. Plantes issues de protoplastes de Petunia. Z. Pflanzenphysiol. 69:26–34.

Evans, D.A. 1979. Chromosome stability in plants regenerated from mesophyll protoplasts of *Nicotiana* species. Z. Pflanzenphysiol. 95:459–463.

_____, Crocomo, O.J., and DeCarvalho, M.T.V. 1980. Protoplast isolation and subsequent callus regeneration in sugarcane. Z. Pflanzenphysiol. 98:355–358.

Facciotti, D. and Pilet, P.E. 1979. Plants and embryoids from haploid *Nicotiana sylvestris* protoplasts. Plant Sci. Lett. 15:1–6.

Fowke, L.C. and Gamborg, O.L. 1980. Applications of protoplasts to the study of plant cells. Int. Rev. Cytol. 68:9–51.

Fitter, M.S. and Krikorian, A.D. 1981. Recovery of totipotent cells and plantlet production from daylily protoplasts. Ann. Bot 48:591–597.

Frearson, E.M., Power, J.B., and Cocking, E.C. 1973. The isolation, culture and regeneration of Petunia leaf protoplasts. Dev. Biol. 33:130–137.

Fukunaga, Y. and King, J. 1978. Effects of different nitrogen sources in culture media on protoplast release from plant cell suspension culture. Plant Sci. Lett. 11:241–250.

Galbraith, D.W. 1981. Microfluorimetric quantitation of cellulose biosynthesis by plant protoplasts using Calcofluor White. Physiol. Plant. 53:111–116.

Gamborg, O.L. 1976. Plant protoplast isolation, culture, and fusion. In: Cell Genetics in Higher Plants (D. Dudits, G.L. Farkas, P. Maliga, eds.) pp.107–128. Akademiai Kiado, Budapest.

_____ 1977.   Culture media for plant protoplasts.   In: Handbook of Nutrition and Food (M. Rechcigl, ed.) pp. 415–422.   CRC Press, Cleveland.

_____, Miller, R.A., and Ojima, K. 1968.   Nutrient requirements of suspension cultures of soybean root cells.   Exp. Cell Res. 50:151–158.

_____, Shyluk, J.P., and Kartha, K.K. 1975.   Factors affecting the isolation and callus formation in protoplasts from the shoot apices of *Pisum sativum* L.   Plant Sci. Lett. 4:285–292.

_____, Shyluk, J.P., Fowke, L.C., Wetter, L.R., and Evans, D.A. 1979.   Plant regeneration from protoplasts and cell cultures of *Nicotiana tabacum sulfur* (Su/Su).   Z. Pflanzenphysiol. 95:255–264.

Giles, K.L. 1972.   An interspecific aggregate cell capable of cell wall regeneration.   Plant Cell Physiol. 13:207–210.

Gill, R., Rashid, A., and Maheshwari, S.C. 1978.   Regeneration of plants from mesophyll protoplasts of *Nicotiana plumbaginifolia* Viv.   Protoplasma 96:375–379.

_____ 1979.   Isolation of mesophyll protoplasts of *Nicotiana rustica* and their regeneration into plants flowering in vitro.   Physiol. Plant. 47:7–10.

_____ 1981.   Dark requirement of cell regeneration and colony formation by mesophyll protoplasts of *Nicotiana plumbaginifolia* Viv.   Protoplasma 106:351–354.

Gleba, Yu.Yu. 1978.   Microdroplet culture.   Naturwissen. 65:158–159.

Gosch, G., Bajaj, Y.P.S., and Reinert, J. 1975.   Isolation, culture, and induction of embryogenesis in protoplasts from cell-suspension of *Atropa belladonna*.   Protoplasma 86:405–410.

Grambow, H.J., Kao, K.N., Miller, R.A., and Gamborg, O.L. 1972.   Cell division and plant development from protoplasts of carrot cell suspension cultures.   Planta 103:348–355.

Green, C.E. 1977.   Prospects for crop improvement in the field of cell culture.   Hort. Sci. 12:131–134.

Gregory, D.W. and Cocking, E.C. 1965.   The large scale isolation of protoplasts from immature tomato fruit.   J. Cell. Biol. 24:143–146.

Gresshoff, P.M. 1980.   In vitro culture of white clover: Callus, suspension, protoplast culture, and plant regeneration.   Bot. Gaz. 141:157–164.

Grun, P. and Chu, L-J. 1978.   Development of plants from protoplasts of *Solanum* (Solanaceae).   Am. J. Bot. 65:538–543.

Gunn, R.E. and Shepard, J.F. 1981.   Regeneration of plants from mesophyll-derived protoplasts of British potato (*Solanum tuberosum* L.) cultivars.   Plant Sci. Lett. 22:97–101.

Halim, H. and Pearce, R.S. 1980.   An electrophoresis method for bulk manipulation of isolated protoplasts from higher plants.   Biochem. Physiol. Pflanzen. 175:123–129.

Hanke, D.E. and Northcote, D.H. 1974.   Cell wall formation by soybean callus protoplasts.   J. Cell Sci. 14:29–50.

Hayward, C. and Power, J.B. 1975.   Plant production from leaf protoplasts of *Petunia parodii*.   Plant Sci. Lett. 4:407–410.

Hess, D. and Endress, R. 1973.   Anthocyansythese in isolierten protoplasten von *Nemesia strumosa* var. Feuerkonig.   Z. Pflanzenphysiol. 68:441–449.

Hess, D. and Leipoldt, G. 1979. Regeneration of shoots and roots from isolated mesophyll protoplasts of *Nemesia strumosa*. Biochem. Physiol. Pflanzen. 174:411–417.

Horine, R.K. and Ruesink, A.W. 1972. Cell wall regeneration around protoplasts isolated from *Convolvulus* tissue culture. Plant Physiol. 50:438–445.

Jha, T.B. and Roy, S.C. 1979. Rhizogenesis from *Nigella sativa* protoplasts. Protoplasma 101:139–142.

Johnson, L.B., Stuteville, D.L., Higgins, R.K., and Skinner, D.Z. 1981. Regeneration of alfalfa plants from protoplasts of selected regenerated S clones. Plant Sci. Lett. 20:297–304.

Kameya, T. 1975. Culture of protoplasts from chimeral plant tissue of nature. Jap. J. Genet. 50:417–420.

_____ and Uchimiya, H. 1972. Embryoids derived from isolated protoplasts of carrot. Planta 103:356–360.

Kanai, R. and Edwards, G.E. 1973. Separation of mesophyll protoplasts and bundle sheath cells from maize leaves for photosynthetic studies. Plant Physiol. 51:1133–1137.

Kao, K.N. 1975. A method for fusion of plant protoplasts with polyethylene glycol. In: Plant Tissue Culture Methods (O.L. Gamborg and L.R. Wetter, eds.) pp. 22–27. Nat. Res. Council of Canada, Saskatoon.

_____ 1977. Chromosomal behavior in somatic hybrids of soybean—*Nicotiana glauca*. Molec. Gen. Genet. 150:225–230.

_____ and Michayluk, M.R. 1974. A method for high-frequency intergenetic fusion of plant protoplasts. Planta 115:355–367.

_____ and Michayluk, M.R. 1980. Plant regeneration from mesophyll protoplasts of alfalfa. Z. Pflanzenphysiol. 96:135–141.

_____, Gamborg, O.L., Miller, R.A., and Keller, W.A. 1971. Cell division in cells regenerated from protoplasts of soybean and *Haplopappus gracilis*. Nature 232:124.

_____, Gamborg, O.L., Michayluk, M.R., Keller, W.A., and Miller, R.A. 1973. The effects of sugars and inorganic salts on cell regeneration and sustained division in plant protoplasts. Protoplastes et Fusion de Cellules Somatique Vegetales Colloques Internation. C.N.R.S. 212:207–213.

_____, Constabel, F., Michayluk, M.R., and Gamborg, O.L. 1974. Plant protoplast fusion and growth of intergeneric hybrid cells. Planta 120:215–227.

Kartha, K.K., Michayluk, M.R., Kao, K.N., Gamborg, O.L., and Constabel, F. 1974. Callus formation and plant regeneration from mesophyll protoplasts of rape plants (*Brassica napus* cv. *Zephyr*). Plant Sci. Lett. 3:265–271.

Kaur-Sawhney, R. and Galston, A.W. 1983. Oats. In: Handbook of Plant and Cell Culture Volume 2 (W.R. Sharp, D.A. Evans, P.V. Ammirato, and Y. Yamada, eds.). MacMillan, New York (in press).

Keller, F. and Stone, B.A. 1978. Preparation of *Lolium* protoplasts and their purification using an anti-galactan-sepharose conjugate. Z. Pflanzenphysiol. 87:167–172.

Kohlenbach, H.W., Wenzel, G., and Hoffman, F. 1982. Regeneration of *Brassica napus* plantlets in cultures from isolated protoplasts of haploid stem embryos as compared with leaf protoplasts. Z. Pflanzenphysiol. 105:131–142.

Krishnamurthy, K.V., Hendre, R.R., Godbole, D.A., Kulkarni, V.M., Mascarenhas, A.F., and Jagannathan, V. 1979. Isolation and regeneration of rose bud callus protoplasts (*Rosa* sp. cv. *Soraya*). Plant Sci. Lett. 15:135–137.

Lam, S.L. 1975. Shoot formation in potato tuber discs in tissue culture. Am. Pot. J. 52:103–106.

Larkin, P.J. 1976. Purification and viability determinations of plant protoplasts. Planta 128:213–216.

Li, X.H. 1981. Plantlet regeneration from mesophyll protoplasts of *Digitalis lanata* Ehrh. Theor. Appl. Genet. 60:345–347.

Linsmaier, E. and Skoog, F. 1965. Organic growth factor requirements of tobacco tissue culture. Physiol. Plant. 18:100–127.

Lorenzini, M. 1973. Obtention de protoplastes de tubercule de pomme de terre. Compt. Rend. Acad. Sci., Paris D 276:1839–1842.

Lorz, H. and Potrykus, I. 1978. Investigations on the transfer of isolated nuclei into plant protoplasts. Theor. Appl. Genet. 53:251–256.

Lorz, H., Wernicke, W., and Potrykus, I. 1979. Culture and plant regeneration of *Hyoscyamus* protoplasts. Planta Medica 36:21–29.

Lu, C.Y., Vasil, V., and Vasil, I.K. 1981. Isolation and culture of protoplasts of *Panicum maximum* Jacq. (Guinea Grass): Somatic embryogenesis and plantlet formation. Z. Pflanzenphysiol. 104:311–318.

Menczel, L., Lazar, G., and Maliga, P. 1978. Isolation of somatic hybrids by cloning *Nicotiana* heterokaryons in nurse cultures. Planta 143:29–32.

Meyer, Y. and Herth, W. 1978. Chemical inhibition of cell wall formation and cytokinesis, but not of nuclear division in protoplasts of *Nicotiana tabacum* L. cultivated in vitro. Planta 142:253–262.

Michayluk, M.R. and Kao, K.N. 1975. A comparative study of sugars and sugar alcohols on cell regeneration and sustained cell division in plant protoplasts. Z. Pflanzenphysiol. 75:181–185.

Muhlbach, H.P. 1980. Different regeneration potentials of mesophyll protoplasts from cultivated and a wild species of tomato. Planta 148:89–96.

Nagata, T. and Ishii, S. 1979. A rapid method for isolation of mesophyll protoplasts. Can. J. Bot. 57:1820–1823.

Nagata, T. and Takebe, I. 1970. Cell wall regeneration and cell division in isolated tobacco mesophyll protoplasts. Planta 92:301–308.

_____ 1971. Plating of isolated tobacco mesophyll protoplasts on agar medium. Planta 99:12–20.

Nagata, T. and Yamaki, T. 1973. Electron microscopy of isolated tobacco mesophyll protoplasts cultured in vitro. Z. Pflanzenphysiol. 70:452–459.

Nagy, J.I. and Maliga, P. 1976. Callus induction and plant regeneration from mesophyll protoplasts of *Nicotiana sylvestris*. Z. Pflanzenphysiol. 78:453–455.

Negrutiu, I. and Mousseau, J. 1980. Protoplast culture from in vitro grown plants of *Nicotiana sylvestris* Spegg. and Comes. Z. Pflanzenphysiol. 100:373–376.

Nehls, R. 1978. Isolation and regeneration of protoplasts from *Solanum nigrum* L. Plant Sci. Lett. 12:183-187.

Nitsch, J.P. and Nitsch, C. 1969. Haploid plants from pollen grains. Science 163:85-87.

Nitsch, J.P. and Ohyama, K. 1971. Obtention de plantes a partir de protoplastes haploides cultives in vitro. C.R. Acad. Sci. Paris 273:801-804.

Northcote, D.H. 1958. The cell walls of higher plants: Their composition, structure and growth. Biol. Revs. Cambridge Phil. Soc. 33:53-102

Ohyama, K. and Nitsch, J.P. 1972. Flowering haploid plants obtained from protoplasts of tobacco leaves. Plant Cell Physiol. 13:229-236.

Passiatore, J.E. and Sink, K.C. 1981. Plant regeneration from leaf mesophyll protoplasts of selected ornamental *Nicotiana* species. J. Am. Soc. Hort. Sci. 106:799-803.

Patnaik, G., Wilson, D., and Cocking, E.C. 1981. Importance of enzyme purification for increased plating efficiency and plant regeneration from single protoplasts of *Petunia parodii*. Z. Pflanzenphysiol. 102:199-205.

Pearce, R.S. and Cocking, E.C. 1973. Behavior in culture of isolated protoplasts from "Paul's Scarlet Rose" suspension culture cells. Protoplasma 77:165-180.

Pelcher, L.E., Gamborg, O.L., and Kao, K.N. 1974. Bean mesophyll protoplasts: Production, culture and callus formation. Plant Sci. Lett. 4:107-111.

Piven, N.M. 1981. Regeneration of whole plants from isolated leaf mesophyll protoplasts of *Nicotiana debneyi Domin*. Tsitol. Genetika 15:35-39.

Poirier-Hamon, S., Rao, P.S., and Harada, H. 1974. Culture of mesophyll protoplasts and stem segments of *Antirrhinum majus* (snapdragon): Growth and organization of embryoids. J. Exp. Bot. 25:752-758.

Potrykus, I., Harms, C.T., and Lorz, H. 1976. Problems in culturing cereal protoplasts. In: Cell Genetics in Higher Plants (D. Dudits, G.L. Farakus, and P. Maliga, eds.) pp. 129-140. Akademiai Kiado, Budapest.

_____, and Thomas, E. 1977. Callus formation from stem protoplasts of corn (*Zea mays* L.). Molec. Gen. Genet. 156:347-350.

Power, J.B. and Berry, S.F. 1979. Plant regeneration from protoplasts of *Browallia viscosa*. Z. Pflanzenphysiol. 94:469-471.

Power, J.B., Frearson, E.M., George, D., Evans, P.K., Berry, S.F., Hayward, C., and Cocking, E.C. 1976. The isolation, culture and regeneration of leaf protoplasts in the genus *Petunia*. Plant Sci. Lett. 7:51-55.

Raj, B. and Herr, J.M., Jr. 1970. The isolation of protoplasts from placental cells of *Solanum nigrum* L. Protoplasma 69:291-300.

Rajasekhar, E.W. 1973. Nuclear divisions in protoplasts isolated from pollen tetrads of *Datura metel*. Nature 246:223-224.

Raveh, D. and Galun, E. 1975. Rapid regeneration of plants from tobacco protoplasts plated at low densities. Z. Pflanzenphysiol. 76:76-79.

Raveh, D., Huberman, E., and Galun, E. 1973. In vitro culture of tobacco protoplasts: Use of feeder techniques to support division of cells plated at low densities. In Vitro 9:216-222.

Redenbaugh, M.K., Westfall, R.D., and Karnosky, D.F. 1980. Protoplast isolation from *Ulmus americana* L. pollen mother cells, tetrads and microspores. Can. J. For. Res. 10:284-289.

Roper, W. 1981. Callus formation from protoplasts derived from cell suspension culture of *Vicia faba* L. Z. Pflanzenphysiol. 101:75-78.

Reusink, A.W. 1978. Leucine uptake and incorporation by *Convolvulus* tissue culture cells and protoplasts under severe osmotic stress. Physiol. Plant. 44:48-56.

_____ 1980. Protoplasts of plant cells. Methods Enzmology 69:69-84.

_____ and Thimann, K.V. 1965. Protoplasts from the Avena coleoptile. Proc. Nat. Acad. Sci. USA 54:56-64.

Sacristan, M.D. and Melchers, G. 1969. The caryological analysis of plants regenerated from tumorous and other callus cultures of tobacco. Molec. Gen. Genet. 105:317-333.

Saxena, P.K., Gill, R., Rashid, A., and Maheshwari, S.C. 1981. Plantlet formation from isolated protoplasts of *Solanum melongena* L. Protoplasma 106:355-359.

Schenck, H.R. and Hoffmann, F. 1979. Callus and root regeneration from mesophyll protoplasts of basic *Brassica* species: *B. campestris, B. oleracea, B. nigra*. Z. Pflanzenphysiol. 82:354-360.

Schenk, R.U. and Hildebrandt, A.C. 1969. Production of protoplasts from plant cells in liquid culture using purified commercial cellulases. Crop Sci. 9:629-631.

_____ 1972. Medium and techniques for induction and growth of monocotyledonous and dicotyledonous plant cell cultures. Can. J. Bot. 50:199-204.

Schieder, O. 1975. Regeneration of haploid and diploid *Datura innoxia* Mill. mesophyll protoplasts to plants. Z. Pflanzenphysiol. 76:462-466.

_____ 1977. Attempts in regeneration of mesophyll protoplasts of haploid and diploid wild type lines and those of chlorophyll-deficient strains from different Solanaceae. Z. Pflanzenphysiol. 84:275-281.

Schilde-Rentschler, L. 1977. Role of the cell wall in the ability of tobacco protoplasts to form callus. Planta 135:177-181.

Schumann, U., Koblitz, H., and Opatrny, Z. 1980. Plant recovery from long-term callus cultures and from suspension culture derived protoplasts of *Solanum phureja*. Biochem. Physiol. Pflanzen. 175:670-675.

Schwenk, F.W., Pearson, C.A., and Roth, M.R. 1981. Soybean mesophyll protoplasts. Plant Sci. Lett. 23:153-155.

Scowcroft, W.R. and Larkin, P.J. 1980. Isolation, culture and plant regeneration from protoplasts of *Nicotiana debneyi*. Aust. J. Plant Physiol. 7:635-644.

Scowcroft, W.R., Davey, M.R., and Power, J.B. 1973. Crown gall protoplasts—isolation, culture and ultrastructure. Plant Sci. Lett. 1:451-456.

Selby, K. 1973. Components of cell wall degrading enzymes with particular reference to the cellulases. Protoplastes et Fusion de Cellules Somatiqaues Vegetales Colloques Internation. C.N.R.S. 212:33-40.

Setterfield, G. and Bayley, S.T. 1961. Structure and physiology of cell walls. Annu. Rev. Plant Physiol. 12:35–62.

Shahin, E.A. and Shepard, J.F. 1980. Cassava mesophyll protoplasts: Isolation, proliferation, and shoot formation. Plant Sci. Lett. 17:459–465.

Shepard, J.F. 1980. Protoplasts as sources of disease resistance in plants. Annu. Rev. Plant Physiol. 19:145–166.

_____ and Totten, R.E. 1975. Isolation and regeneration of tobacco mesophyll cell protoplasts under low osmotic condition. Plant Physiol. 55:689–694.

_____ and Totten, R.E. 1977. Mesophyll cell protoplasts of potato: Isolation, proliferation, and plant regeneration. Plant Physiol. 60:313–316.

Sink, K.C. and Power, J.B. 1977. The isolation, culture and regeneration of leaf protoplasts of *Petunia parviflora* Juss. Plant Sci. Lett. 10:335–340.

Somerville, C.R., Somerville, S.C., and Ogren, W.L. 1981. Isolation of photosynthetically active protoplasts and chloroplasts from *Arabidopsis thaliana*. Plant Sci. Lett. 21:89–96.

Sondahl, M.R., Chapman, M.S., and Sharp, W.R. 1980. Protoplast liberation, cell wall reconstitution, and callus proliferation in *Coffea arabica* L. callus tissues. Turrialba 30:161–165.

Taiz, L. and Jones, R.L. 1971. The isolation of barley-aleurone protoplasts. Planta 101:95–100.

Takebe, I., Labib, G., and Melchers, G. 1971. Regeneration of whole plants from isolated mesophyll protoplasts of tobacco. Naturwissen. 58:318–320.

Takeuchi, Y. and Komamine, A. 1978. Composition of the cell wall formed by protoplasts isolated from cell suspension cultures of *Vinca rosea*. Planta 140:227–237.

Tal, M. and Watts, J.W. 1979. Plant growth conditions and yield of viable protoplasts isolated from leaves of *Lycopersicon esculentum* and *L. peruvianum*. Z. Pflanzenphysiol. 92:207–214.

Thomas, E. 1981. Plant regeneration from shoot culture–derived protoplasts of tetraploid potato (*Solanum tuberosum* cv. Maris Bard). Plant Sci. Lett. 23:81–88.

_____, Hoffmann, F., Potrykus, I., and Wenzel, F. 1976. Protoplast regeneration and stem embryogenesis of haploid androgenetic rape. Molec. Gen. Genet. 145:245–247.

Tomita, Y., Suzuki, H., and Nisizawa, K. 1968. Chromatographic patterns of cellulase components of *Trichoderma viride* grown on the synthetic and natural media. J. Ferment. Technol. 46:701–710.

Uchimiya, H. and Murashige, T. 1976. Influence of the nutrient medium on the recovery of dividing cells from tobacco protoplasts. Plant Physiol. 57:424–429.

Ulrich, T.H., Chowdhury, J.B., and Widholm, J.M. 1980. Callus and root formation from mesophyll protoplasts of *Brassica rapa*. Plant Sci. Lett. 19:347–354.

Upadhya, M.D. 1975. Isolation and culture of mesophyll protoplasts of potato (*Solanum tuberosum* L.). Pot. Res. 18:438–445.

Usui, H. and Takebe, I. 1969. Division and growth of single mesophyll cells isolated enzymatically from tobacco leaves. Devel. Growth, Differen. 11:143-151.

Vardi, A., Spiegel-Roy, P., and Galun, E. 1975. Citrus cell culture: Isolation of protoplasts, plating densities, effect of mutagens and regeneration of embryos. Plant Sci. Lett. 4:231-236.

Vasil, I.K. 1976. The progress, problems and prospects of plant protoplast research. Adv. Agron. 28:119-160.

_____, Vasil, V., Sutton, W.D., and Giles, K.L. 1975. Protoplasts as tools from the genetic modification of plants. In: Proceedings International Symposium Yeast and Other Protoplasts, p. 82. University of Nottingham.

Vasil, V. and Vasil, I.K. 1974. Regeneration of tobacco and petunia plants from protoplasts and culture of corn protoplasts. In Vitro 10:83-96.

_____ 1980. Isolation and culture of cereal protoplasts. Part 2: Embryogenesis and plantlet formation from protoplasts of Pennisetum americanum. Theor. Appl. Genet. 56:97-99.

Von Arnold, S. and Eriksson, T. 1977. A revised medium for growth of pea mesophyll protoplasts. Physiol. Plant. 39:257-260.

Wakasa, K. 1973. Isolation of protoplasts from various plant organs. Jap. J. Genet. 48:279-289.

Wallin, A., Glimelius, K., and Eriksson, T. 1977. Pretreatment of cell suspensions as a method to increase the protoplast yield of Haplopappus gracilus. Physiol. Plant. 40:307-311.

Watts, J.M., Motoyoshi, F., and King, J.M. 1974. Problems associated with the production of stable protoplasts of cells of tobacco mesophyll. Ann. Bot. 38:667-671.

Wernicke, W., Lorz, H., and Thomas, E. 1979. Plant regeneration from leaf protoplasts of haploid Hyoscyamus muticus L. produced via anther culture. Plant Sci. Lett. 15:239-249.

White, D.W.R. and Bhojwani, S.S. 1981. Callus formation from Trifolium arvense protoplast-derived cells plated at low densities. Z. Pflanzenphysiol. 102:257-261.

Widholm, J.M. 1972. The use of fluoroscein diacetate and phenosafranine for determining viability of cultured plant cells. Stain Tech. 47:189-194.

Wilson, H.M., Styer, D.J., Conrad, P.L., Durbin, R.D., and Helgeson, J.P. 1980. Isolation of sterile protoplasts from unsterilized leaves. Plant Sci. Lett. 18:151-154.

Xu, Z.H., Davey, M.R., and Cocking, E.C. 1981. Isolation and sustained division of Phaseolus aureus (mung bean) root protoplasts. Z. Pflanzenphysiol. 104:289-298.

_____ 1982. Plant regeneration from root protoplasts of Brassica. Plant Sci. Lett. 24:117-121.

Xuan, L.T. and Menczel, L. 1980. Improved protoplast culture and plant regeneration from protoplast-derived callus in Arabidopsis thaliana. Z. Pflanzenphysiol. 96:77-80.

Zapata, F.J., Evans, P.K., Power, J.B., and Cocking, E.C. 1977. Effect of temperature on the division of leaf protoplasts of Lycopersicon esculentum and Lycopersicon peruvianum. Plant Sci. Lett. 8:119-124.

Zapata, F.J., Sink, K.C., and Cocking, E.C. 1981.　Callus formation from leaf mesophyll protoplast of three *Lycopersicon* species:　*L. esculentum,* c.v. *Walter, L. pimpinellifolium* and *L. hirsutum* f. *glabratum.* Plant Sci. Lett. 23:41–46.

Zieg, R.G. and Outka, D.A. 1980.　The isolation, culture and callus formation of soybean pod protoplasts.　Plant Sci. Lett. 18:105–114.

Zieger, E. and Hepler, P.K. 1976.　Production of guard cell protoplasts from onion and tobacco.　Plant Physiol. 58:492–498.

CHAPTER 5
Meristem, Shoot Tip, and Bud Cultures

C.Y. Hu and *P.J. Wang*

This chapter describes in vitro techniques for culturing stem meristems, shoot tips, and apical buds. The explant of meristem culture may either be the apical dome (the apical meristem) or, more frequently, the apical dome plus a few subjacent leaf primodia (the subapical meristematic region). The shoot tip explant is usually taken from the tender tip of the growing shoot measuring 2 cm or less in length. Bud cultures are initiated from either terminal or axillary buds, usually with the stem segment attached, using either a growing or dormant shoot.

APPLICATIONS (Review of Literature)

Micropropagation

Following the successful rapid multiplication of orchids by shoot meristem culture (Morel, 1965) there has been an increasing interest in recent years in the application of tissue culture techniques as an alternative means of asexual propagation of economically important plants. The size of the propagule in culture is so minute, that the in vitro asexual propagation technique has been referred to as "micropropagation". Shoot tip and bud cultures are preferred over meristem culture in micropropagation when viral elimination is not part of the objective. Use of larger explants is desirable as they are easier to dissect and have much higher survival and growth rates than smaller explants.

ADVANTAGES. In vitro micropropagation techniques are now often preferred to conventional practices of asexual propagation in several greenhouse species because of the following potential advantages: (1) Only a small amount of plant tissue is needed as the initial explant for regeneration of millions of clonal plants in one year. In comparision, it would take years to propagate an equal number of plants by conventional methods. (2) Many plant species are highly resistant to conventional bulk propagation practices. In vitro micropropagation provides a possible alternative method for these species. (3) The in vitro technique provides a method for speedy international exchange of plant materials. If handled properly, the sterile status of the culture eliminates the danger of disease introduction. Thus, the period of quarantine is reduced or unnecessary. (4) The in vitro stocks can be quickly proliferated at any time of the year (Boxus et al., 1977; Huang and Millikan, 1980) after shipping or storage, while propagation with conventional practices is highly season-dependent.

There are three possible routes available for in vitro propagule multiplication: (a) enhanced release of axillary buds; (b) production of adventitious shoots through organogenesis; and (c) somatic embryogenesis. Each method of propagation has its merits and shortcomings. We will discuss the advantages of the first method.

The merit of using axillary bud proliferation from meristem, shoot tip, or bud cultures as a means of regeneration is that the incipient shoot has already been differentiated in vivo. Thus, to establish a complete plant, only elongation and root differentiation are required. In vitro organogenesis and embryogenesis, on the other hand, must undergo developmental changes which usually involve the formation of callus with subsequent reorganization into plantlets. This has not been easy to achieve in most plants. The induction of axillary bud proliferation seems to be applicable in many cases; e.g., carnation (Roest and Bokelmann, 1981) and soybean (Evans, 1981; Kartha et al., 1981b), where methods of organogenesis and embryogenesis fail.

If an intermediary callus has been involved, as in the case in most regeneration via organogenesis and embryogenesis, the frequency of genetic changes is increased; especially in the form of polyploidization and aneuploidization resulting from mitotic abnormalities (Bayliss, 1973; Mahlderg et al., 1975), as has been observed in many plant species (Edallo et al., 1981; Novak, 1981; Lester and Berbee, 1977). Plants derived from meristem, shoot tip, and bud cultures are generally phenotypically homogenous, thereby indicating genetic stability. For example, McCown and Amos (1979), who had employed shoot tip culture for birch micropropagation, observed only one visually abnormal shoot even after hundreds of thousands of shoots were regenerated. The vast majority of all of the plantlets regenerated from meristem cultures have been found to remain in the diploid state (Ancora et al., 1981; Hasegawa et al., 1973; Murashige et al., 1974).

Although the rate of plantlet multiplication by means of organogenesis and embryogenesis is astonishing, their regeneration capacity usually diminishes rapidly after a number of subcultures, and eventually this morphogenic potential is completely lost (Kehr and Schaeffer, 1976; Yie and Liaw, 1977). The initial multiplication rate for axillary bud proliferation, on the other hand, is rather slow.   The rate, nevertheless, increases during the first few subcultures and eventually reaches a steady plateau during subsequent subculture cycles (see "Proliferation Rate" subsection).   Through such a geometrical progression, the production of millions of plants from a single explant in a single year can be conservatively estimated.   This multiplication rate is quite feasible for commercial production of many species.   Moreover, once a stock multiple-shoot culture is established, it can continuously serve as the source-propagule, instead of having to restart from fresh explant cultures periodically.

BRIEF HISTORY.   Georges Morel was the pioneer in applying shoot tip culture as a clonal multiplication tool.   After his success in cloning the orchid, *Cymbidium*, in vitro clonal multiplication gained momentum in the 1970s.   Among numerous contributors in the field, Murashige should be credited as the dominant figure in the establishment of micropropagation techniques.   He developed the concept of developmental stages (Murashige, 1974):   Stage I: explant establishment; Stage II: multiplication of the propagule; and Stage III: rooting and hardening for planting into soil.   This concept stimulates the awareness that a single medium usually is not sufficient for in vitro plant multiplication and regeneration.   Transferring the propagules through a series of specially designed chemical and physical environments at each developmental stage holds the key to success.

The greatest success using this technique has been achieved in herbaceous horticultural species.   This success is partially due to the weak apical dominance and strong root regenerating capacities of many herbaceous plants, and partially due to the financial support from the greenhouse and nursery industries.   Many of those crops are currently propagated commercially using in vitro procedures.   Examples of successful herbaceous ornamental species propagated through in vitro axillary bud proliferation include *Anigozanthos, Anthurium, Cephalotus, Chrysanthemum, Dianthus, Fuchsia, Gerbera, Gladiolus, Gloxinia, Gypsophila, Hosta, Phlox,* and *Saxifragra;* the herbaceous vegetables include *Allium, Arachis, Asparagus, Beta, Brassica, Cicer, Cynara, Phaseolus, Rheum,* and *Fragaria;* and the agronomic and other herbaceous genera *Capsella, Glycine, Lolium, Stellaria, Stevia, Vigna,* and *Zea* (see Table 1 for references).

Compared to herbaceous plants, the micropropagation of woody species has lagged far behind.   The greatest difficulty is experienced at Stage III, the root induction, especially when explants are taken from mature trees.   Difficulty during Stage I, when the primary culture is established, is also frequently encountered.   This is partially due to the existence of large quantities of polyphenolic compounds in the tissue of many woody species (see "Polyphenol" subsection) and partially due to the difficulty of breaking the physiological quiescent state of the axillary buds.   Compared to Stages I and III, the induction of multiple shoots, Stage II is not as difficult.

A breakthrough was made by Jones in 1976 in the rooting of in vitro apple shoots when he supplemented the culture medium with phlor

	EXPLANT	STAGE I (ESTABLISHMENT)	STAGE II (AXILLARY BUD PROLIFERATION)	STAGE III (ROOTING)	REFERENCE
...sativum ...rlic)	Meristem dome 0.2 mm	MS[1]		MS + 1 NAA[2]	Wang & Huang, 1947
A. sativum (garlic)	Bud tip 5–8 mm	B5 + 0.5 2iP + 0.1 NAA	Same	B5 + 0.01 2iP + 0.2 NAA	Bhojwani, 1980b
Anigozanthos (5 spp.) (kangaroo paws)	Apical and axillary buds of rhizome	1/2 LS + 0.45 BA + 0.47 NAA	1/2 LS + 0.1 IBA	1/2 LS liq + 0.3 IBA + polyurethane foam	McComb & Newton, 1981
Anthurium andreanum	Bud 2 mm or less	MS + 15% CW rotating drum	MS + 0.2 BA 3.1fold		Kunisake, 1980
Araucaria cunninghamii (hoop pine)	Seedling nodal segment 4 mm	BS (med. level salt, sucrose & vit)		Same + 1.2 IBA sporadically	Haines & de Fossard, 1977
Arachis hypogaea (peanut)	Meristem 0.4–0.5 mm	MSB + 0.02 BA + 1.9 NAA 75% into whole plant	MSB + 0.23 BA + 1.9 NAA		Kartha et al., 1981b
Atriplex canescens	Seedling shoot tip 4 mm	MS + 0.01–0.1 KIN .01 + 0.1 IAA + 0.5 GA	Same 168fold/2 mo	MS + 2 IAA	Wochok & Sluis, 1980
Beta vulgaris (sugarbeet)	In vitro bud	GEW + 0.5 BA + 0.01 GA	Same 20–25fold	GE + 5 IBA 95%	Atanassov, 1980
Betula alleghaniensis	Seedling tip and nodal 5–10 mm	MS + 5 ZEA + 80 ADE sulfate	Same 10–15fold	MS + 1 IBA	Minocha, 1980
B. papyrifera (paper birch)	Seedling tip and nodal 5–10 mm	MS + 5 ZEA + 80 ADE sulfate	Same 10–15fold	MS + 1 IBA	Minocha, 1980

SPECIES	EXPLANT	STAGE I (ESTABLISHMENT)	STAGE II (AXILLARY BUD PROLIFERATION)	STAGE III (ROOTING)	REFERENCE
B. platyphylla (Asiatic white birch)	Seedling tip and nodal segment	GD + 1.4 BA, 10% sucrose	Same 20-30fold/3-4 wk	1:1 peat to perlite 100%	McCown & Amos, 1979
Brassica campestris (Chinese cabbage)	Nodal segment 2-4 mm	1/3 MS macro + 1 BA		1/3 MS + 0.1 BA + 1 NAA	Kuo & Tsay, 1977
B. oleracea var. botrytis (cauliflower)	Meristem				see Grout, 1975
B. oleracea var. capitata (white cabbage)	Meristem 0.5-2 mm dia.	MS liq + 2.6 KIN + 8 IAA filter-paper-bridge	MS + 13 KIN 3.8fold/2 wk	MS liq filter-paper-bridge	Walket et al., 1980
Capsella bursa-pastoris (shepherd's purse)	Meristem 0.3-0.8 mm	LS liq + 2.6 KIN + 8 IAA	LS liq + 10 KIN	LS liq	Walkey & Cooper, 1976
Carica papaya (papaya)	Mature plant shoot tip 1-1.5 mm	MS + 0.45 NA + 0.2 NAA	Same 7fold/3 wk	MS + 1 NAA	Litz & Conover, 1978
Citrus x Poncirus (carrizo citrange)	Seedling shoot tip 5 mm	Knops macro + MS micro	Same + BA + 100 ADE sulfate 3fold	MT + 1 NAA	Kitto & Young, 1981
Caphalotus follicularis (Australian pitcher plant)	Shoot tip	1/2 LS + 1 BA + 0.1 IBA	1/2 LS + 0.1 BA + 1 NAA 5-10fold/3-4 mo	Same	Adams et al., 1979
Chrysanthemum morifolium	Shoot tip 0.5 mm	MS + 2 KIN + 0.02 NAA	Same	MS	Earle & Langhans, 1974a
Cicer arietinum (chick pea)	Meristem 0.4-0.5 mm	MSB + 0.02-0.2 BA	Same 10fold	MS + 0.2 IBA	Kartha et al., 1981b

Table 1. Cont.

SPECIES	EXPLANT	STAGE I (ESTABLISHMENT)	STAGE II (AXILLARY BUD PROLIF- ERATION)	STAGE III (ROOTING)	REFERENCE
Citrullus lanatus (watermelon)	Seedling shoot tip 1–3 mm	LS + 1 KIN + 0.05 IAA 4.5fold/5 wk	SL + 2 KIN + 0.05 IAA 10.3fold/5 wk	LS + 2 IAA	Barnes, 1979
Coffea arabica (coffee)	Seedling shoot tip meristem 0.3 mm	MSB + 1–2 BA (or ZEA) + 0.2 NAA	Same	1/2 MS + 0.2 IBA–sucrose	Kartha et al., 1981a
Cryptomeria japonica	3–yr–old tree shoot tip	MS	3–5fold; shoot elongation: + 1 NAA	1/3 MS + 1 NAA 87%	Wang, 1978
Cynara scolymus (globe artichoke)	Shoot tip and nodal segment 1.5–2 cm	MS + 10 KIN	MS + 5 KIN 4.5fold/3 wk	1/2 MS + 2 NAA + 10 ascorbic acid 84%	Ancora et al., 1981
Dactylis glomerata (cocksfoot)	Meristem 0.15 mm	MS + 0.2 KIN + 0.01 2,4-D		Same	Dale, 1977b
Dianthus caryophyllus (carnation)	Shoot tip and nodal segment 5 mm	MS + 1 BA + 0.01 IAA	Same 4.8– 20.4fold	MS + 0.1 IAA	Roest & Bokelmann, 1981
Eucalyptus (10 spp.)	Seedling shoot	HB + 0.34 BA + 0.32 KIN + 1 IBA	Same >3fold/ 3 wk	HB + 1 IBA 5 spp.	Hartney & Baker, 1980
E. citriodora	20–yr–old tree shoot segment 10 mm	MS liq + 0.2 KIN + 0.3 BA rotary shaker	Same + agar 10–15fold	WH to MS	Gupta et al., 1981
E. grandis	Seedling nodal segment 5 mm	NN formula K 49% bud grow		Same 33%	deFossard et al., 1974

SPECIES	EXPLANT	STAGE I (ESTABLISHMENT)	STAGE II (AXILLARY BUD PROLIFERATION)	STAGE III (ROOTING)	REFERENCE
Festuca (3 spp.) (fescue)	Meristem <0.2 mm	MS + 0.2 KIN + 0.01 2,4-D		Same	Dale, 1977b
Fragaria spp. (strawberry)	Young plantlets	Knop + MS micro + 1 IBA	Same + 1 BA 15–20fold/ 2 mo	MS + 0.2 BA + 0.2 NAA	Kartha et al., 1980
Fragaria x ananassa (strawberry)	Runner tip meristem	MS + 0.2 BA + 0.2 IBA + 0.035 GA	MS + 2.3 BA	MS + 0.2 BA + 0.2 NAA	Kartha et al., 1980
F. virginiana x F. chiloensis	Runner tip	MS + 1 BA + 0.1 GA	Same	MS + 5 IBA + 1000 PG[3]	James, 1979
Gerbera jamesonii	Shoot tip				see Murashige, 1974
Gladiolus sp.	Axillary bud 2–3 mm	MS + 2 KIN + 0.1 NAA	Same 5–8fold/ 2 wk	MS + 0.5 NAA + 0.3% AC[3]	Ziv, 1979
Gloxinia sp.	Shoot tip				see Murashige, 1974
Glycine max (soybean)	Bud	MSB + 0.23 BA + 0.02 IBA + 100 ADE	Same	MSB + 0.43 KIN + 2 NAA	Evans, 1981
G. max (soybean)	Meristem 0.4–0.5 mm	MSB + 0.02 BA + 0.2 NAA 33% into whole plant	MSB + 2.3 BA		Kartha et al., 1981b
Grevillea rosmarinfolia	1-yr-old seedling nodal segment 1 cm	1/2 MS + 0.5 BA	Same 5–6fold/ mo	1/2 MS liq + 0.1 NAA paperbridge 85%	Ben-Jaacov & Dax, 1981
Gypsophila paniculata (baby's breath)	Shoot tip 1.5 cm	MS + 1 BA	Same 5.9fold	WH after auxin treatment	Kusey et al., 1980

Table 1. Cont.

SPECIES	EXPLANT	STAGE I (ESTABLISHMENT)	STAGE II (AXILLARY BUD PROLIFERATION)	STAGE III (ROOTING)	REFERENCE
Hosta decorata	Shoot tip 5 mm	MS + 5 BA + 0.01 or 0.1 NAA + 40 ADE sulfate	Same	MS or MS + 0.01 NAA	Papachatzi et al., 1981
Kalmia latifolia (mountain laurel)	3-yr-old seedling shoot tip	WPM liq + 2.8 2iP	Same + agar 8-10fold	Peat 73%	Lloyd & McCown, 1980
Lactuca sativa (head lettuce)	Shoot tip and nodal segment	M + 0.5 KIN + 5 IAA		M + 1 IBA	Koevary et al., 1978
Lolium multiflorum (Italian rye grass)	Meristem 0.3-0.9 mm	MS + 0.2 KIN	Same	Same	Dale, 1980
L. multiflorum, L. perenne, L. hybrid	Meristem 0.3 mm	MS + 0.2 KIN + 0.01 2,4-D 30-60% into whole plant			Dale, 1977
Lycopersicon esculentum (tomato)	Meristem 0.4-0.5 mm	MSB + 0.02 BA + 0.2 NAA into whole plant			Kartha et al., 1977
Malus domestica (Golden Delicious apple)	Shoot tip 1 cm	MS + 5 BA	MS + 3-5 BA 4.1-5.4fold/mo		Lundergan & Janick, 1980
M. domestica (4 cvs.) *M. pumila*	Shoot tip 0.1-0.5 mm	W + 5 BA	D + 0.2 BA + 40 ADE	QL + 5 CDPPN	Nemeth, 1981
M. sylvestris (M. 7)	Shoot tip 1 cm	1/2 MS + 0.5 BA	Same 13fold/mo	1/3 MS + 2 IBA + 0.27% agar 88%	Werner & Boe, 1980

SPECIES	EXPLANT	STAGE I (ESTABLISHMENT)	STAGE II (AXILLARY BUD PROLIFERATION)	STAGE III (ROOTING)	REFERENCE
M. sp. (M. 9)	Shoot tip 1-2 cm	LS + 1-2 BA + 0.1-0.5 IBA	Same 4.5fold/mo	LS + 3 IBA + 1620 PG[3]	James & Thurbon, 1981
M. sp. (M. 26)	Shoot tip 1-2 cm	MS + 1 BA + 1 IBA + 0.1 GA	Same 6 x 10 fold/8 mo	Same -BA 97%	Jones et al., 1977
M. sp. (MM 104, 106, 109)	Shoot tip 2-3 cm	MS + 1 BA + 1 IBA + 0.1 GA	Same 5fold/mo	1/2 MS + 1 IBA + AC[3] same -IBA 100%	Snir & Erez, 1980
Manihot utilissima (cassava)	Meristem dome 0.2-0.5 mm	MSB + 0.1 BA + 0.2 NAA + 0.035 GA >50% into whole plant			Kartha et al., 1974b
M. utilissima (cassava)	Meristem dome 0.2-0.5 mm	MSB + 0.1 BA + 0.2 ZEA + 2 NAA + + 0.035 GA 100% into whole plant			Nair et al., 1979
Phaseolus vulgaris (bean)	Meristem 0.4-0.5 mm	MSB + 0.2 NAA 73% into whole plant	MSB + 2.3 BA 15-30fold/2 wk	MSB + 0.23 BA + 0.2 NAA + 0.035 GA into MS + 1 IAA	Kartha et al., 1981
Phlox subulata (Ground phlox)	Shoot tip 0.5-1 mm	MS + 5 BA + .01 µM GA + 40 ADE sulfate	Same	MS + 0.01 or 0.5 NAA	Schnabelrauch & Sink, 1979
Pisum sativum (pea)	Seedling meristem dome 0.2-0.3 mm	B5 + 0.02 BA		1/2 B5 + 0.2 NAA	Kartha et al., 1974a

Table 1. Cont.

SPECIES	EXPLANT	STAGE I (ESTABLISHMENT)	STAGE II (AXILLARY BUD PROLIFERATION)	STAGE III (ROOTING)	REFERENCE
Populus nigra, *P. yunnanensis*, *P.* hybrid	Axillary bud	MS + 0.2 BA	MS + 0.1 BA + 0.02 NAA 10 fold/yr	MS + 0.01 BA + 0.01 NAA or Pumice + peat	Whitehead & Giles, 1977
Prunus amygdalus (almond) *P.* hybrid	Buds from clonal cv.	Knop's macro + MS org. + 1 BA	Same	Same limited	Tabachnick & Kester, 1977
P. armenaca (apricot)	Shoot tip from clonal cv. 0.5-1.5 cm	MS + 20 BA + 0.1 NAA	Same	Same limited	Skirvin et al., 1979
P. avium, *P. cerasus* (sour and sweet cherry)	Shoot tip from clonal cv. 0.5-1.5 cm	MS + 20 BA + 0.1 NAA	Same	Same limited	Skirvin et al., 1979
P. cistena	Meristem from clonal cv. 1-5 mm	MS + 5 BA	Same 35fold/ 3-4 wk	1/2 MS liq + 0.2 NAA	Lane, 1979
P. mariana, GF8-1, GF 1869 (plum rootstocks)	Nodal segment of mature plant	P7 + 0.1 BA + 0.2 IBA	Same	Same - BA	Seirlis et al., 1979
P. insititia (plum dwarf rootstocks)	Buds from clonal cv.	MS + 1 BA + 0.1 IBA + 0.1 GA + 162 PG^3	Same	MS + 3 IBA + 0.1 GA + 162 PG^3	Jones & Hopgood, 1979
P. persica (peach)	Shoot tip from clonal cv. 0.5-1.5 cm	MS + 20 BA + 0.1 NAA	Same 5-6fold/ 3 wk	Same limited	Skirvin et al., 1979
Pyrus communis (Bartlett pear)	Meristem 0.5-1.5 mm	MS + 1 BA + CH	MS + 1 BA 30fold/2-3 wk	1/2 MS + 2 NAA	Lane, 1979c

SPECIES	EXPLANT	STAGE I (ESTABLISHMENT)	STAGE II (AXILLARY BUD PROLIFERATION)	STAGE III (ROOTING)	REFERENCE
Rheum rhaponticum (rhubarb)	Meristem 0.2-0.5 mm	MS + 1 BA + 1 IBA	Same 5fold/mo	MS + 0.1 IBA 96%	Roggemans & Clae, 1979
Rhododendron sp.	Shoot tip	A + 2 2iP + 1 IAA	Same	1/4 A + AC[3]	Anderson, 1975, 1978
Ribes inebrians (wax currant)	Shoot tip 7-9 mm	Presoak in 150 GA to MS + 1 g/l NB[3]	MS + 0.5 BA + 2 IBA + 0.2 GA	Rootone F dip peat:perlite: sand	Wochok & Sluis, 1980b
Rosa hybrida (rose)	Shoot tip from clonal cv. 3-7 mm	MS + 3 BA + 0.3 IAA	Same 6fold/4 wk	1/2 or 1/4 MS + 0.1 NAA or IAA 55%	Hasegawa, 1979, 1980
R. sp. (rose)	Shoot tip 2 cm	MS + 2 BA + 0.1 NAA	Same 3fold/5 wk	1/4 MS 68%	Skirvin & Chu, 1979
Rubus idaeus (red raspberry)	Meristem of clonal plant 0.2-0.8 mm	A + 2 BA + 0.05 IAA + 0.1 GA	A + 1-3.5 BA + 0.05 IBA 3-5fold/2 mo	Pasteurized sand root powder & mist	Pyott & Converse, 1981
R. sp. (bramble raspberry)	Shoot tip of clonal plant 5-10 mm	NN + 1 IBA		Same	Kiss & Zatyko, 1978
R. rusinus x *R. idaeus* (raspberry)	Shoot tip 1 cm	MS + 1 BA + 0.1 GA	Same	MS + 1 IBA + 1000 PG[3] 100%	James, 1979
R. sp. (thornless blackberry)	Shoot tip from clonal plant 1-2 cm	MS + 2 BA + 0.1 NAA + 50 ascorbic acid	Same very rapid rate	WH or MS (1-1/16X) +/- NAA >80%	Skirvin et al., 1981
Salix matsudana x *S. alba* (willow)	Shoot nodal segment 2-3 cm	MS + 0.1 BA + 0.2 NAA	Same 4fold/mo	MS + 0.2 NAA	Bhojwani, 1980a

Table 1. Cont.

SPECIES	EXPLANT	STAGE I (ESTABLISHMENT)	STAGE II (AXILLARY BUD PROLIFERATION)	STAGE III (ROOTING)	REFERENCE
Santalum acuminatum (quandong)	Seedling shoot tip and nodal segment 5-10 cm	MS + 2 BA + 1 NAA	MS + 2 BA	MS + 1 IBA	Barlass et al., 1980
S. lanceolatum	Seedling shoot tip and nodal segment 5-10 cm	MS + 2 BA + 1 NAA	MS + 2 BA	MS + 1 IBA	Barlass et al., 1980
Saxifraga sp.	Shoot tip				see Murashige, 1974
Solanum etuberosum	Seedling shoot tip 0.5-1 mm	MS + 1 BA + 0.01 IAA	Same many	Same	Towill, 1981
S. tuberosum (potato)	Tuber sprout tip meristem 1 mm	1/3 MS	Same + .005 NAA; tuberization:MS liq + 10 BA + 8% sucrose		Wang, 1977; Wang & Hu, 1982
S. tuberosum (potato)	Meristem 0.3 mm	MS + 0.07 NAA + 0.04 GA 74% into whole plant			Pennazio & Vecchiat, 1976
Spinacia oleracea (spinach)	Seedling shoot tip 5 mm	Hoagland whole plants			Culafic, 1973
Spirea bumalda	Shoot tip 1-5 mm	MS + 1 BA	Same 300-400 fold/3-4 wk	1/2 MS liq + 0.1 NAA	Lane, 1979b
Stellaria media	Meristem 0.3-0.8 mm wide	LS liq + 8 IAA + 2.6 KIN stationary into plant	Same, on shaker	Same, stationary or shaker	Walkey & Cooper, 1976

SPECIES	EXPLANT	STAGE I (ESTABLISHMENT)	STAGE II (AXILLARY BUD PROLIFERATION)	STAGE III (ROOTING)	REFERENCE
Stevia rebaudiana	Seedling nodal segment	MS + 2 BA	Same, profuse	MS + 10 NAA 100% 8 root/plant	Yang, 1981
Sassafras randaiense	5-yr-old tree new bud 1 cm	1/2 LS + 60 KIN + 0.05 NAA	Same; shoot growth reduce to 5 BA	Same -KIN + 5 IBA sporadically	Wang & Hu, unpublished
Tectona grandis (teak tree)	100-yr-old tree bud 1 cm	MS liq + 0.1 BA + 0.1 KIN on shaker	MS + 1 BA + 0.5 KIN	WH + 2.5 IAA + 2.5 IBA + 2 IPA to WH liq 60%	Gupta et al., 1980
Vigna unguiculata (cowpea)	Seedling meristem 0.4-0.5 mm	MSB 100% into whole plant			Kartha et al., 1981b
Zea mays (corn)	Shoot tip 1-2 cm	MSW + 120 ADE sulfate	MSW + 3 BA + 1 IAA + 1 IBA 4.7-5.4fold/10 d	MSW + 5 NAA 51-94%	Raman, 1980

[1] A = Anderson, 1975; B5 = Gamborg et al., 1968; BS = Broad Spectrum Culture medium, deFossard, 1974a; D = Dudits et al., 1975; GD = Gresshof & Doy, 1972; GEW = Gamborg and Eveleigh, 1978; HB = Hartney and Barker, 1980; LS = Linsmaier and Skoog, 1965; M = Miller, 1963; MS = Murashige and Skoog, 1962; MSB = MS salt + B5 vit.; MSN = MS salt + NN vit.; MSW = MS salt + WH vit.; MT = Murashige and Tucker, 1969; NN = Nitsch and Nitsch, 1969; QL = see Nemeth, 1981 citation; W = Walkey, 1972; WH = White, 1963; WPM = Woody Plant Medium (Lloyd and McCown, 1980).

[2] The unit for ingredient concentration is mg/l unless indicated otherwise.

[3] AC = Activated charcoal; NB = Nutrient broth; PG = Phloroglucinol.

cinol, a phenolic compound (Jones and Hatfield, 1976). This discovery stimulated the current wave of micropropagation of woody fruit species. Successful examples of micropropagation in woody and semi-woody fruits through in vitro axillary bud proliferation include species in the genera of *Malus, Prunus, Pyrus, Ribes, Rubus,* and *Santalum;* and woody ornamentals and special crops including *Atriplex, Betula, Coffea, Grevillea, Kalmia, Rosa, Salix,* and *Spirea* (Table 1).

Most forest species do not readily root from cuttings. The standard method for their propagation is seeding. However, there are numerous problems associated with seed propagation: (a) It is extremely difficult to obtain large quantities of seeds from many economically important species for reforestation (Hu, 1979). (b) While heterosis is one of the hybrid qualities that should be taken advantage of during material selection for reforestation, the difficulty in mass cross-pollination of forest species makes the cost of hybrid seed production too high to be practical. (c) Tree species have long generation times which prohibit carrying out the type of breeding programs used in annual crop plants. Many selection programs are, therefore, carried out to search for genotypes with desirable traits among natural stands or first generation hybrids. Once final selections are made, they would have to be maintained as clones, but there is no effective conventional cloning method available for propagating forest species.

In vitro micropropagation through axillary bud proliferation may provide a practical solution to cloning of forest trees. In solving problems (a) and (b), a handful of seed will produce enough seedlings to serve as the explant sources. Whitehead and Giles (1977) estimated that more than 10^6 rooted plantlets can be produced per year from the culture of a single bud of *Populus nigra, P. yannanensis* and a *Populus* hybrid. Multiplication at such a rate is certainly feasible for plantlet production for reforestation. The solution of problem (c) is somewhat difficult since mature mother plants are involved. It is rather difficult to accomplish Stage III of micropropagation, namely rooting, when explants are not obtained at the juvenile stage. However, successful micropropagation with shoot tip or bud cultures have been accomplished with explants from mature plants of several *Eucalyptus* species (Cresswell and Nitsch, 1975; Gupta et al., 1981) and a 100-year-old teak tree (Gupta et al., 1980).

The potential usefulness of micropropagation methods for forest trees has long been recognized and discussed (Thorpe and Bianti, in press). However, it is only during the past decade that rigorous attempts have been made to commercially propagate important timber species. Very limited progress has been made in this area. The successful examples of axillary bud proliferation include species in the following genera: *Araucaria, Cryptomeria, Eucalyptus, Populus, Tectona, Thuja,* and *Sassafras* (Table 1).

Disease Elimination

When vegetatively propagated plants are systematically infected with a viral disease, the pathogen passes from one vegetative generation to the next. The entire population of a given clonal variety may, over years, be infected with the same pathogen. Especially with latent viruses, the symptoms are hardly detectable but the yield and/or quality of the crop may gradually decrease over generations. It is likely that all clonal crops cultivated today are harboring one or more viral diseases. In order to ensure the highest possible yield and quality, virus-free stock plants should be provided to growers. The use of certified seed potatoes is an impressive example.

It has been demonstrated that the shoot and root apices of virus-infected plants are frequently devoid of viral particles or contain very low viral concentrations (Kassanis, 1957; White, 1934). Morel was the first to demonstrate that virus-free plants can be recovered from infected plants through shoot meristem cultures (Morel and Martin, 1952). To date, chemotherapeutic agents or physical treatments have met with limited success in the eradication of viruses from infected plants. In vitro culture has become the only effective technique to obtain virus-free plants from infected stock. In vitro methods can also be used to produce pathogen-free plants from stocks systematically infected with pathogens such as mycoplasma (Fedotina and Krylova, 1976; Jacoli, 1978; Ulrychova and Petru, 1975), fungi (Bader and Phillips, 1962; Csinos, and Hendrix, 1977), and bacteria (Knauss, 1976; Theiler, 1977). The value of this method to the horticultural industry, therefore, is immeasurable.

A list of 49 species with virus-free plants regenerated in vitro was compiled by Wang and Hu (1980). The following additional species should be added to the list: The African cassava mosaic and cassava brown streak diseases were eliminated from the East African cassava (*Manihot utilissima*) cultivars (Kaiser and Teemba, 1979); rygrass mosaic from *Lolium multiflorum* (Dale, 1977a); cocksfoot streak virus, mild mosaic virus, and mottle virus from *Dactylis glomerata* (Dale, 1979); and various viral diseases from *Citrus* spp. (Navarro et al., 1975), *Pisum sativum* (Kartha and Gamborg, 1978), and *Trifolium repens* (Barnett et al., 1975). The elimination of viral diseases from *Citrus* spp. by Navarro presents an interesting variation from standard meristem culture method called in vitro micrografting. In this technique a meristem tip from viral infected scion cultivar is excised and placed on a decapitated, in vitro produced shoot of a rootstock. The entire process—grafting, healing of the union, growth of the scion, and rooting of the rootstock—is carried out in aseptic cultures. In vitro micrografting has also been applied successfully in regenerating virus-free apple plants (Huang and Millikan, 1980).

Although viral diseases are usually transmitted from generation to generation through asexual propagated organs, about 10% of the known plant viruses are also transmitted through seeds of infected host plants. In some cases, viruses are confined to the seed coat, such as TMV on tomato seeds (Broadbent, 1965). Others, such as seeds of leguminous crops, carry viruses internally. Meristem culture has further application in elimination of seedborne viruses. Pea seedborne mosaic, for example, was successfully eliminated in this way from over 100 breeding lines (Kartha and Gamborg, 1978; Kartha et al., 1979). Successful elimination of seedborne viruses would assist in the international exchange of genetic material.

Germplasm Preservation

Preservation of germplasm is a means to assure the availability of genetic materials as the needs arise. Since most seeds and vegetative organs have a limited storage life, research in germplasm preservation has concentrated on the development of procedures to extend usable life spans. For many crops, seed is the most appropriate storage form due to its low moisture content which enables the cells to survive low temperatures. However, shoot tips are preferred in low temperature storage of clonal lines. Since meristem cells are highly cytoplasmic and non-vacuolated, a high percentage of cells would be expected to survive cryopreservation. In addition, the meristems are genetically stable, readily regenerated into complete plants which bulk up after storage, and are able to yield pathogen-free plants. Meristems have also been identified as excellent material for germplasm preservation of crop species with seedborne viruses (Kartha et al., 1979).

COLD STORAGE. One strategy used for germplasm preservation is the maintenance of cultures under minimal growth conditions. Minimal medium and low temperature are used to slow down the metabolic rate. To save storage space and reduce the frequency of subculturing, Morel (1975) has been able to keep the meristem tip regenerated plantlets of grape (*Vitis vinifera*) alive at 9 C with only one transfer a year. Lundergan and Janick (1979) found that in vitro proliferated shoot tips of "Golden Delicious" apple could be stored at 1 or 4 C for at least one year with no loss of growth potential. They estimated that about 2000 culture tubes could be stored in an ordinary 0.28 m^3 refrigerator. It would require 5.7 ha to accommodate the same number of trees. Cold storage has been successfully employed to maintain in vitro virus-free strawberry plantlets derived from meristems (Mullin and Schlegel, 1976). More than 50 different strawberry cultivars have been maintained by them for up to 6 years at 4 C in darkness. Low temperature was also used to store *Lolium* (Dale, 1980).

Maintenance of cultures under minimal growth conditions does not stop the cellular processes. It merely reduces the frequency of culture transfer. Furthermore, to prevent desiccation, the addition of liquid medium, once every few months, is suggested. This not only is laborious and time consuming, but also subjects the culture to the possibility of microorganismal contamination.

CRYOPRESERVATION. During the past five years, considerable interest has been generated in exploring the possibilities of storing plant meristems or vegetative buds at cryogenic temperatures, i.e., in liquid nitrogen at -196 C. Since all the metabolic activities of meristem cells are likely to be totally arrested at this temperature, one could postulate that these meristems could be stored in liquid nitrogen indefinitely.

Seibert (1976) was the first to report shoot initiation of the carnation from cryopreserved shoot apices. Cryopreservation of strawberry meristems was reported by Sakai et al. (1978) and Kartha et al. (1980) with the demonstration of plantlet regeneration after thawing. Potato shoot tip cultures and the meristems of a nontuber-bearing potato, *Solanum etuberosum*, were frozen to -196 C and thawed by Grout and Henshaw (1978) and Towill (1981), respectively. The latter observed regeneration of a shoot mass as the predominant morphogenetic response. The cryopreservation of pea meristems (Kartha et al., 1979) and buds of hardy fruit trees (Sakai and Nishiyama, 1978) was also reported.

It has become apparent that the condition of plant tissue prior to freezing greatly influences survival. Seibert and Wetherbee (1977) found that cold treatment of donor carnation plants at 4 C for 3 days or more before meristem excision resulted in a doubling of the survival rate after freezing and up to seven fold increase in shoot regeneration. It was suggested that the reason for this increase was that the low temperature hardened the plant tissue, which was then capable of surviving at lower temperatures than unhardened plants. Freshly dissected shoot tips do not usually survive extensive cooling unless a period of in vitro conditioning is given before the freezing treatment. Grout and Henshaw (1978) first reported this requirement for the survival of rapid freezing in some potato species. In cryopreservation of strawberry meristems, Kartha et al. (1980) found that maximum viability and plant regeneration (95%) was obtained when the meristems were precultured on a culture medium, while the survival rate of the meristems frozen without preculturing was only about 5%. Similar observations were reported in pea meristems (Kartha et al., 1979) and shoot tips of *Solanum etuberosum* (Towill, 1981). The beneficial events that occur during this preculturing procedure, are obscure.

Explants, with an apical dome and two to three pairs of subjacent leaf primordia, are usually used in the cryopreservation of shoot meristems. Reports on the conditions of these explants after the freeze-and-thaw cycle were conflicting. Seibert and Wetherbee (1977) reported that, in general, the cells in the meristem dome remained viable after freezing while those in the leaf primordia did not. Similar observations were reported by Grout and Henshaw (1978). EM studies performed by Haskins and Kartha (1980), on the contrary, indicated that most of the actively growing cells in the pea meristems, which remained viable after cryotreatment, were located on primordial leaf tissues and in the axillary bud and stipule meristematic areas. These conflicting results might be due to species or procedural differences. Haskins and Kartha's work, however, indicated that the original meristematic dome does not have to be alive for the regeneration of whole plants after freezing and thawing.

In the majority of experiments performed, the testing tissues had been exposed to liquid nitrogen for only a few minutes. The assumption is that the critical periods for tissue survival are during freezing and thawing. Since cellular metabolism presumably is completely suspended, the length of time under liquid nitrogen should not affect the survival rate. Kartha et al. (1980), nevertheless, observed a reduction in regeneration rate of strawberry meristems from 95 to less than 65% after extending the period of cryostorage from 1 week to 2 or more (up to 8) weeks. With pea meristems, Kartha et al. (1979) reported that 73, 68, 62.5, and 61% of explants regenerated into whole plants after 1 hr, 1 wk, 7 wk, and 26 wk storage in liquid nitrogen, respectively. These reports suggest that long-range storage studies are needed to reveal whether cryostorage is indeed capable of preserving germplasms indefinitely.

Physiological Studies

The study of buds on an intact plant does not allow isolation of the specific physiological factors which affect bud development. Excised buds cultured in vitro, on the other hand, provide a simplified system, unaffected by correlative relationships with other parts of the plant, and by unwanted environmental factors. This technique offers an approach to study the control of growth, dormancy, reproductive cycles, and other physiological processes of buds. The following paragraphs include some examples of such in vitro studies.

In studying the time requirement for bud sprouting in vitro, Altman and Goren (1974) observed an innate characteristic of *Citrus* bud dormancy. They found that both the dormancy and the sprouting periods of buds in vitro corresponded to the natural periods occurring under field conditions. They postulated that culture experiments might help to elucidate more precisely the role of growth regulators in bud development. In *Citrus* summer bud culture they found the following: IAA delayed sprouting, while GA enhanced shoot elongation; cytokinins specifically induced the formation of numerous adventitious buds whereas ABA completely inhibited sprouting; this inhibition, however, was reversible. Using in vitro bud culture and chilling experiments, Borkowska and Powell (1979) were able to determine the dormancy status of apple buds.

Coleman and Thorpe (1978) were able to induce male strobili formation in western red cedar, *Thuja plicata*, from in vitro cultured vegetative shoot tips under high concentrations of various gibberellins and continuous illumination. No strobili were induced by GA under short-day conditions or continuous darkness. IAA (100 µM) completely inhibited the induction of the male strobili that had been induced by GA.

After the role of root tip in cytokinin biosynthesis was established, some workers suspected the possibility that the shoot apex was another site of cytokinin production. Koda and Okazawa (1980) demonstrated such production through *Asparagus* shoot apex culture. They found that the cultured shoot apices continued to diffuse a small but constant amount of cytokinin into the medium throughout five subcultures. The cytokinin content in the apices at the end of the subculture was not different from that at the start of the subculture. However, the root tip of *Asparagus* produced more cytokinin than the shoot apex.

THE PROTOCOLS

To illustrate the complete procedure of meristem, shoot tip, and bud culture, a protocol of meristem culture of a herbaceous species (potato)

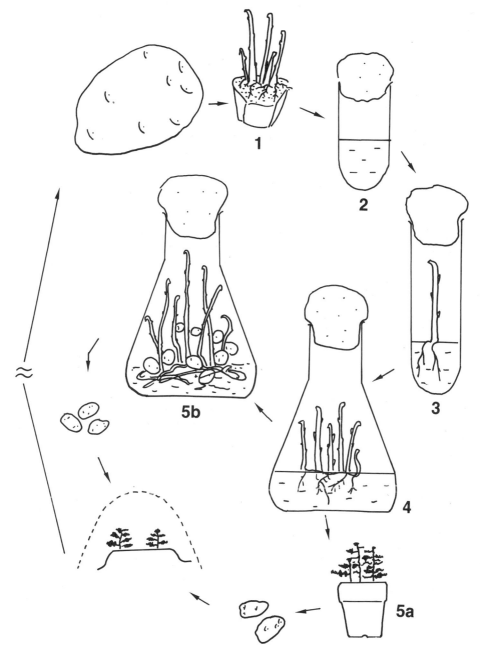

Figure 1. Schematic representation of the virus-free seed-potato production via meristem culture: (1) source of explants, (2) meristem explant in the primary culture, (3) growth of the explant, (4) in vitro layering—for axillary shoot proliferation, (5a) planting-out, (5b) in vitro mass tuberization.

and a protocol of shoot tip culture of a woody species (*Sassafras*) are provided. Since different species, different ages of the same species, as well as different parts of a plant, are likely to have different culture requirements, the readers are referred to the next section on the factors affecting success, for tips on modifying the procedure for other experimental plant materials.

Meristem Culture of Herbaceous Plant—*Solanum tuberosum* (see Fig. 1 for diagramatic representation).

1. Obtaining plant materials:
    a. Cut tubers into 20 g sections.
    b. Soak tuber sections in .03 mM GA for 1 hr to break bud dormancy.
    c. Sprout tuber sections on sterile moist vermiculite in growth chamber.
    d. Harvest shoot tips as they reach 3 to 5 cm length (Fig. 2a).
2. Surface disinfection is unnecessary.
3. Dissection:
    a. Place under 15–20X stereoscope in a Laminar flow hood.
    b. Dissect meristem domes with one subjacent primordium (approx. 75 μm, Fig. 2b). See "Dissection" subsection for details.
4. Stage I—The primary culture.
    a. Transfer each meristem explant onto solid MS medium plus 1 g/l bacto-tryptone in a 15 x 19 mm culture tube.
    b. Incubate cultures at 25 C, 12-hr photoperiod with an intensity of 150 lux during the first month and 500 lux during the second month. Lights are provided by a mixture of cool white fluorescent tubes and incandescent bulbs.
    c. Rooted plantlets of 3 cm long regenerate in two months (Fig. 2c).
5. Stage II—In vitro layering.
    a. Transfer regenerated plantlets horizontally on the surface of MS solid medium + 0.005 μM NAA in 250 ml flasks.
    b. Two-three axillary shoots, with adventitious roots, develop in 20 days (Fig. 2d).
    c. Press the newly developed axillary shoots to horizontal position on agar surface.
    d. Repeat step (c) at a 20–day interval until a mass of axillary shoots results (Fig. 2e).
    e. Harvest axillary shoots with a pair of sterile surgical scissors and subculture them onto fresh medium.
    f. Repeat step (e) at a 20–day interval with a two-three fold proliferation per subculture.

Figure 2. Virus-free seed-potato production via meristem culture. (a) Explant sources, (b) explant for meristem culture, including the apical dome and a subjacent leaf primordium, (c) 3 cm long shoot developed from explant after 40 d incubation, (d-e) in vitro layering, the shoots have been layered twice (d), and three times (e), (f) in vitro mass tuberization, minute dormant tubers form the shoots. They are ready to be harvested and stored outside the aseptic culture container.

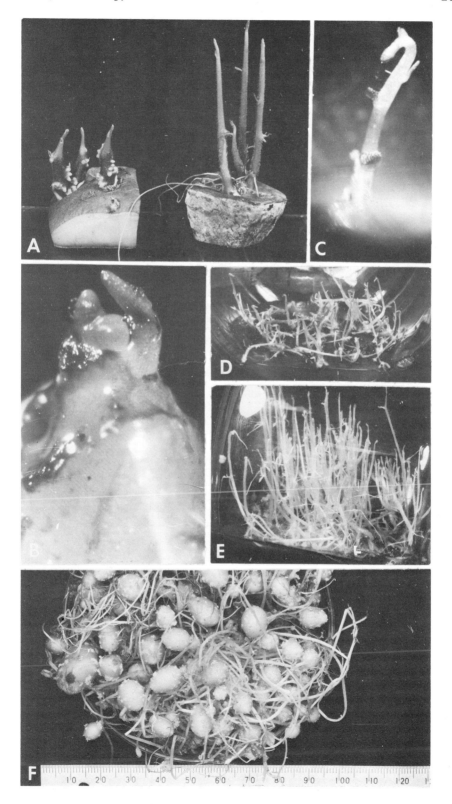

6. In vitro tuberization.
    a. Transfer harvested axillary shoots into MS medium with 22.0–
        44.0 μM BA and 0.23 M sucrose in 300 or 500 ml flasks.
    b. Incubate at 18–20 C, 8-hr photoperiod of 100–500 lux.
    c. Thirty to 50 dormant miniature tubers can be harvested from
        each 500 ml flask in a 4 month period.

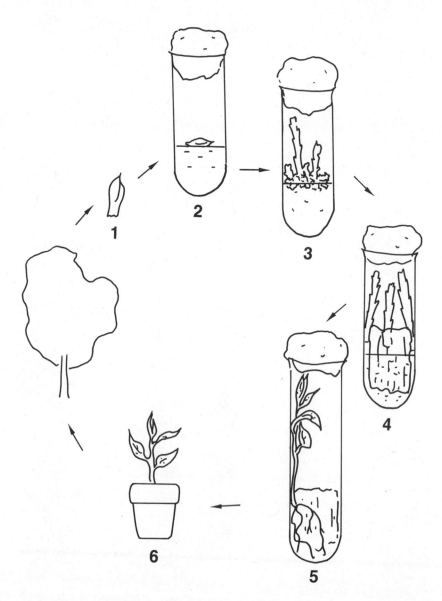

Figure 3. Schematic representation of the *Sassafras randaiense* shoot
tip culture. (1) Bud explant from 5-yr-old field grown tree, (2) primary
culture, (3) axillary shoot proliferation, (4) rooting of shoots in a liquid
medium, filter-paper-bridge system, (5) rooted plantlet, (6) planting-out.

Shoot Tip Culture of Woody Plant—*Sassafras randaiense* (Hay.) Rehd.
(see Fig. 3 for diagramatic representation)

1. Obtaining explants
   a. Spray the buds of a 5-year-old tree with a mixture of antibiotics and a systemic fungicide (Streptomycin 0.1% + Benlate 0.1%) at weekly intervals in early spring.
   b. Harvest tips of newly sprouted shoots (approx. 5 cm long) one week after the fourth spray.
   c. Dissect 8-10 mm long shoot tips with apical dome and seven-ten tightly packed leaf primordia.
2. Surface disinfection
   a. Dip explants in 75% ethanol for 2-3 sec.
   b. Rinse once with sterile distilled water.
   c. Soak explants in 0.5% NaClO with 0.01% Tween 20 under ultrasonic vibration for 5 min.
   d. Transfer into Laminar flow hood and rinse once with sterile distilled water.
3. Stage I—the primary culture
   a. Transfer shoot tip explants onto a modified LS (Linsmaier and Skoog, 1965) solid medium with 2x FeEDTA and the following additives: 0.15 M sucrose, 30% CW, 100 mg/l malt extract, 0.22 mM ADE sulfate, 0.34 mM glutamine, 0.3 mM arginine, 0.27 µM NAA, and 0.28 mM KIN.
4. Stage II—axillary shoot proliferation
   a. Subculture multiple bud mass onto fresh medium of the same composition in 250 ml flasks.
   b. Repeat step (a) at a monthly interval.
   c. Stimulate shoot growth by transferring multiple bud masses to the same medium with KIN reduced to 23.0 µM (Fig. 4a).
5. Stage III—rooting
   a. Transfer 1-cm-long shoots into culture tubes with filter-paper-bridge and LS liquid medium supplemented with 25.0 µM IBA.
   b. Sporadical rooting (up to 20%) results in 15-45 days (Fig.4b).
6. Planting-out
   a. Transfer rooted plantlets into vermiculite containing flats covered with transparent plastic sheets.
   b. Incubate flats under a 12-hr, 25 C day of 4 klx cool while fluorescent light and and 12-hr, 15 C night regime.
   c. Remove plastic sheet 1 month after transplanting. A 95% survival rate should result (Fig. 4c).

FACTORS AFFECTING SUCCESS

The ability of explants to survive, multiply, and regenerate is a consequence of a wide variety of factors such as the origin of cultures, history of explants, physiological state of explants, endogenous hormone concentrations, and general culture conditions, i.e., mineral salts, carbohydrate, light, and temperature. Only certain guidelines can be offered here. Since each species presents a somewhat unique problem

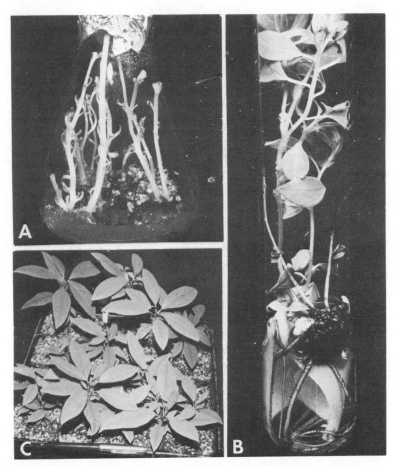

Figure 4. Micropropagation of *Sassafras randaiense*. (a) Axillary shoot
proliferation, (b) rooted plantlet, (c) planting-out.

for determining needs for in vitro culturing, trial and error is the best
available route to success.

Plant Materials

SIZE. When very small explants are used, the presence of leaf primordia appears to determine the capability of an explant to develop.
Using 0.1-mm-long potato meristems, Kassanis and Varma (1967) found
that the explants with one leaf primordium resulted in a more advanced
development than those lacking leaf primordia. For rhubarb it was
essential to dissect tips with two to three primordial leaves. Smaller
tips did not grow (Walkey, 1968). Working with in vitro micrografting
of apple, Huang and Millikan (1980) reported 15, 65, 75, and 90% success rates when apical domes with 0, 2, 4, and 6 subjacent leaf primordia were used, respectively. The lack of developmental capacity of

the apical domes is likely due to its dependence on the subjacent leaf and stem tissues for hormonal resources. A more sophisticated culture medium with balanced hormonal combination is required for growth (Shabde and Murashige, 1977; Kartha et al., 1974b).

The size of the explant determines the survival of the culture. In general, the larger the explant, the better the chance for survival. For example, Dale (1977b) reported that, in the meristem culture of eight grass species, variations in the survival of tips between species was highly correlated with the size of tips cultured; the longer the tip, the higher the survival rate. In cassava, Kartha and Gamborg (1975) demonstrated that only explants exceeding 0.2 mm in length formed complete plants. Those less than 0.2 mm produced either callus or roots.

Based on the foregoing discussions, it is evident that large explants, such as shoot tips and buds instead of the minute meristems should be selected for in vitro micropropagation. However, when eradication of viral infection is one of the objectives, meristems of the smallest size, while still within the regenerable range, should be used. Stone (1963) found that when excising carnation shoot meristems, explants smaller than 0.2 mm were unlikely to root while those larger than 0.75 mm produced plants that still contained mottle viruses. Tips between 0.2 and 0.5 mm most frequently produced virus-free plants. Mellor and Stace-Smith (1977) usually rejected meristems of potato less than 0.3 mm long because they were unlikely to root, and rarely used buds more than 0.7 mm long because they were prone to infection. Within the narrow size range, there is little variability in rooting ability. Although rare, there are examples for the use of large shoot tips and their success in viral elimination. For example, Vine and Jones (1969) obtained virus-free plantlets through cultivation of shoot tips of hops that were 5 mm in length.

BUD LOCATION. Explants taken from the tip of a shoot are in a younger stage of development than explants taken from the base. A young developmental stage has often been found to be optimum for shoot regeneration. Roest and Bokelmann (1981) reported that in carnation nodal segment culture, the percentage of shoot development between explants taken from the top and the base of the shoot was 88.6 and 69.8, respectively. In Chrysanthemum meristem culture, Hollings and Stone (1968) ascertained that the success rate of explants obtained from terminal buds was 32%, whereas from lateral buds it was 18%. The terminal buds presumably have a stronger growth potential than the lateral ones. Hasegawa (1979) observed that a higher percentage of shoot tip explants of rose developed multiple shoots than from lateral buds. Therefore, it is more desirable to use terminal explants in most cultures. Since there is only one terminal bud per shoot, many workers also use axillary buds. Satisfactory results have been reported with axillary buds (Ancora et al., 1981; Gupta et al., 1981; McComb and Newton, 1981).

SEASON. Like all asexual propagation methods, the success of meristem, shoot tip and bud cultures is affected by the season during which the explants are obtained. Altman and Goren (1974) studied the sprouting time required for excised *Citrus* buds in vitro and found that both the dormancy and the sprouting periods of buds in vitro corresponded to the natural periods occurring under field conditions. For plant species with a definite dormant period, the best results may be expected when the explants are dissected at the end of their dormancy period. Mellor and Stace-Smith (1969) found that, for most potato varieties, meristems excised right after dormancy, i.e., in spring and early summer, root more readily than those taken later in the year. When Tabachnik and Kester (1977) cultured dormant buds of almond in late October, some callus developed and the buds became chlorotic and eventually died. On the other hand, when the buds were excised in late December and January, after bud dormancy had already been broken by low temperature, cultures grew readily. The rate of shoot development appeared to be greater in material collected in January. In the in vitro micrografting studies, Huang and Millikan (1980) found there was a 60% success rate when apple meristems were obtained from the field between March and June. The rate gradually decreased to 10% during June to November and stayed at 10% throughout the winter months (November to February).

Actively growing shoot tips are recommended for meristem, shoot tip and bud cultures because of their strong growth potential, and their low virus concentration. Stone (1963) has found a better survival of carnation meristems excised and cultured during the active growing season of early spring and early autumn, compared to those taken in summer and winter. Poessel et al. (1980) reported that the best time for obtaining buds for in vitro micrografting was during the period of rapid shoot growth. Otherwise, a low success rate was observed. Actively growing shoots can be obtained by growing sterilized stem sections, bulbs, corms, or tubers in autoclaved vermiculite in greenhouse or laboratory (Kartha et al., 1974b; Wang, 1977) or from spring growth of the field plants.

Surface Disinfection

Shoot tips, buds, or nodal sections are usually cut to a size somewhat larger than the final explant. They are surface disinfected, then cut or dissected under an aseptic hood to the final size before being transferred into the culture vessel.

The most commonly used surface disinfectant is sodium hypochlorite (NaClO) which is often used as a 5-6% commercial bleach, e.g., Clorox or Javex. Plant segments are usually soaked in a 10-15% solution of such bleach (0.5-0.75% NaClO) for 5-15 min. The concentration as well as the duration of soaking may be reduced or increased according to need. As long as 45 min soaking in 10% Clorox was used by Kunisaki (1980). The use of concentrations of NaClO higher than 1% (Minocha, 1980; Tabachnik and Kester, 1977) and even up to 10% (Kuo and Tsay,

1977; Walkey and Cooper, 1976) have been reported. Tissue damage or cell death may result from such high concentrations. It is of interest to note that apple shoots, which are readily damaged by common surface disinfectants, become more resistant after a short period of incubation on a culture medium. Surface disinfection following such a procedure resulted in undamaged sterile apple shoot tips for in vitro micropropagation (Jones et al., 1977). It is not yet known whether such brief preculturing is also effective in reducing the sensitivity of other plant species to disinfectant damage.

To increase wettability, a small amount (0.01-0.1%) of surfactant such as Tween 20, Tween 80, Teepol, or Mannoxol should be added to the disinfecting solution. Magnetic stirring, ultrasonic vibration, or vacuum may be applied during soaking in order to reduce the possibility of trapping air bubbles on the explant surface. A quick dipping (5-30 sec) of the plant segments in 70-75% ethanol is frequently performed prior to bleach soaking. This extra step not only kills surface microorganisms but also acts as an effective surfactant. For explants with well protected outer structures, e.g., clove of garlic or bud of *Populus,* the ethanol dipping may be followed by a brief flaming (Bhojwani, 1980b; Whitehead and Giles, 1977).

Other commonly used surface disinfectants include calcium hypochlorite and mercury chloride. The concentration used for these disinfectants is similar to that of NaClO. One to three rinses with sterile distilled water are necessary to remove the disinfectant before the final excision of the explant.

Complete disinfection of many woody species can be extremely difficult, especially when the explants are taken from the field. To reduce the contamination rate in shoot tip culture of *Sassafras,* Wang sprayed the buds on a 5-year-old tree with a mixture of an antibiotic and a systemic fungicide (0.1% streptomycin + 0.1% Benlate) in weekly intervals early in the spring. Tips of the newly sprouted shoots were harvested one week after the fourth spraying, surface disinfected, then grown in vitro. Despite this effort, about 95% contamination rate was experienced in this species (Wang and Hu, unpublished). Similar rates were experienced when field grown papaya (Litz and Conover, 1978) and *Castanea* (Hu, unpublished) were used. Some recommendations were made by B.H. McCown (personal communication) for reducing the contamination rate in woody species:

1. Use plants produced in the greenhouse or phytotron if at all possible.
2. Use actively growing new shoots. The longer the material stays in the field, the more problems one will encounter with contamination (McGrew, 1980; Skirvin et al., 1979).
3. Some material works better if 2.5-5 cm nodal pieces are sterilized and placed in test tubes with 1 cm of liquid medium in the bottom. The new growth from the bud is used for further culturing on agar medium. The liquid medium in the test tube should be without sugar, thus considerably reducing the growth of contaminants.

Despite all disinfecting efforts, contamination may still persist in some species. This is possibly due to some endogenous microorganisms that are harbored within the explant tissue. Some of these contaminants are slow growing or latent and will not be visually apparent for several subcultures. To detect and weed out such contaminated cultures, some laboratories incorporate casein hydrolysate, bacto nutrient broth or yeast-peptone in the culture medium during certain stages of culturing to speed up the growth of the contaminants, if any, to a visible level (Dale, 1979; Lane, 1979b,c; Anderson, 1980; James and Thurbon, 1981).

Dissection

The dissecting operation is performed in a laminar flow hood or a UV-equipped aseptic transfer hood. Before dissection, the surfaces of the hood as well as the stereomicroscope and illuminator should be disinfected with 70 or 75% ethanol. Surface-sterilized culture containers and tools should be arranged in a handy fashion on the bench surface in the hood.

A single edged razor blade is used to excise the shoot tip, bud, or nodal explant. A pair of forceps are used to transfer the explant into a culture container. The blade and forceps should be dipped in ethanol and dried on a sterile paper towel or by flame after each operation.

A stereomicroscope, illuminator, and microscalpels are needed for the excision of meristems. Most spotlight illuminators on dissecting microscopes release intense heat energy. In order to prevent heat damage, the meristems should be exposed to light as briefly as possible during dissection. Illuminators with cool rays (fluorescent lamps) or glass fiber illumination are more desirable. Hypodermic needles (21 gauge) are used as microscalpels in our laboratory. They are cut into different lengths (2 and 3 cm) and fixed onto handles. To give each needle sufficient time to dry or cool down after ethanol-flame sterilization, several needles are used alternately. The short needles are employed to remove leaves and leaf primordia. It is not too difficult to excise the apical meristems of plants with loosely packed apices. Special care must be taken to avoid damaging the meristem in species whose primordia are tightly packed against the apical dome. As the dissecting operation approaches the apex, a different sterile needle should be used to cut each primordium—a precaution to avoid carrying microorganisms from the primordia to the sterile apex. This precaution becomes especially important when dissecting species with abundant hairs on the primordia, such as strawberries. The long needles are used for meristem excision. A smooth, clean cut is desired. The same needle is used to transfer the explant onto the surface of the culture medium.

Orientation of the explant on the medium surface does not appear to be critical as long as the explant is not buried underneath the medium surface. To create an ultimate contact between the nutrient and the frequently rough explant surface, we usually add a thin layer of sterile, distilled water on the surface of the agar medium. This extra layer of fluid may also facilitate an effective dispersion of toxic metabolic wastes released by the cultured tissue.

Culture Media

White's medium (1943) was the most widely used medium during the early days of meristem cultures. Many improvements have been made since then, the most noticeable of which are the enhancement of the N, P, and K levels, the reduction of the Ca level and the prevention of iron precipitation at high pH. There is no general purpose medium yet available for meristem, shoot tip or bud culture. Most of the commonly used media were originally developed for root cultures and other purposes. The optimum concentration of the ingredients, especially of micronutrients, have not been critically evaluated for meristem, shoot tip or bud culture. Among the media listed in Table 1, Murashige and Skoog (1962) medium (MS), with some modifications, is the one used most frequently and with greatest success. The growth additive and salt concentration requirement of the medium varies from species to species and, especially, from one stage of culture development to another. Some general tips are given in the appropriate sections below.

Stage I—Culture Establishment

In Stage I, the explants may develop either into single shoots, or into multiple shoot masses, or even into rooted plantlets. The first two cases are used in micropropagation, while the third case, the one-step plant regeneration, is usually preferred for virus-free plant regeneration work through meristem culture.

GROWTH REGULATORS. Although a small quantity of cytokinin may be synthesized by shoots grown in vitro (Koda and Okazawa, 1980), roots are the principal site of cytokinin biosynthesis. It is unlikely that the meristem, shoot tip, and bud explants have sufficient endogenous cytokinin to support growth and development. Thus, 85% of the Stage I culture media listed in Table 1 were supplemented with a cytokinin. There are three cytokinins frequently used: kinetin (KIN), N^6-benzyladenine (BA) and N^6-(2-isopentenyl)-adenine (2iP). BA is most effective for meristem, shoot tip, and bud cultures, followed by KIN. 2iP has been used less frequently (Nair et al., 1979). The range of effectiveness is reflected by the frequency of their usage. In Table 1, among the cytokinin containing Stage I media, 68% used BA, 23% used KIN and 9% used 2iP, zeatin (ZEA), and SD8339. The low percentage usage of the last three cytokinins is partially attributed to their high cost and limited availability. One should be aware of the fact that, although a given cytokinin may not work well in certain species, it may be quite effective in others. For instance, 2iP was found to be the cytokinin of choice for plants in *Ericaceae* (McCown, personal communication). Of the Stage I media listed in Table 1, 15% can be established without cytokinin. It is likely that a sufficient quantity of endogenous hormone is already present in the explants. Furthermore, some of the species in these cases, e.g., *Fragaria*, *Phaseolus*, *Rubus*, and *Solanum*, are known to regenerate adventitious roots readily. The regenerated roots may act as a new source of cytokinins before the residual hormone is exhausted.   None of these 15% Stage I media, however, supported axillary shoot proliferation.

Auxin is another hormone required for shoot growth.   Since the young shoot apex is an active site for auxin biosynthesis, exogenous auxin is not always needed in Stage I medium, especially when relatively large shoot tip explants from actively growing plants are used. Thus 40% of the Stage I media listed in Table 1 are auxin free. There are cases that exogenous auxin is not essential but is beneficial for the growth of Stage I cultures (Dale, 1975).   Resting buds and meristems 0.4 mm or less may not produce (or retain) enough endogenous auxins for shoot growth. In these cases, exogenous auxin needs to be added. IAA, IBA, NAA, and 2,4-D are the most frequently used auxins in plant tissue cultures.   IAA is considered as the weakest of the four and is readily inactivated by light and by tissues with high IAA–oxidase activity. Nevertheless, when effective, IAA shows minimum adversity on organ formation. IBA is slightly more potent than IAA and is not easily broken down.   In contrast, 2,4-D is the most potent auxin of the four.   2,4-D stimulates callus formation and strongly antagonizes organized development. Consequently, NAA is the auxin routinely used by most laboratories for meristem shoot tip, and bud cultures. A survey of auxin containing Stage I media in Table 1 indicated that NAA was used in 51%, IBA in 27%, IAA in 22%, and 2,4-D in 6% of the cases. The usual concentration range used in Stage I is between .045 to 10.0 μM.

Gibberellin (GA) has been included in only 17% of the Stage I media listed in Table 1.   Evidently, sufficient quantities of this hormone are synthesized by most explants. When GA is supplemented, its function is primarily for bud elongation (Schnabdrauch and Sink, 1979). The concentration used in Stage I media is 0.29 μM or less, which is exceedingly low compared to the levels used in other types of tissue culture experiments.

INCUBATION CONDITIONS.   Few studies have been carried out to reveal the optimum physical incubation conditions, such as temperature, light, etc.   The optimum incubation temperature varies with species. The climatic zone where the plant originated offers some clues as to what temperature should be selected. Most workers in meristem, shoot tip, and bud cultures select a constant incubation temperature ranging between 20-28 C, with the great majority in the middle of this range, i.e., 24-26 C. When the incubation temperature is raised above 28 C, water tends to condense on the plants and container walls, which may restrict growth (Lane, 1979b). A day:night temperature fluctuation may be desirable for some plants, especially for those adapted to temperate and desert climates.   Minocha (1980) and Roggemans and Claes (1979) used 25:20 C day:night temperature for *Betula papyrifera* nodal and *Rheum rhaponticum* meristem cultures, respectively; Ziv (1979) used 24:20 C day:night temperature for *Gladiolus* bud culture.   However, most workers who used a fluctuating day:night incubation temperature did so through arbitrary choice, without experimental support.   There are reports questioning the value of lowered night temperature (Hasegawa et al., 1973).

As in temperature selection, incubation light—duration, intensity, and quality—are also generally selected in an arbitrary way. In order to maximize the in vitro growth and prevent induction of dormancy, with few exceptions, long photoperiods (12-24 hr per day) are used. The most commonly used photoperiod regime is 16 hr day vs. 8 hr night. The light intensity is usually 1-10 klx. Since light stimulates tissue browning for explants with high polyphenol content, it is advisable to reduce light intensity below 1 klx or even incubate in darkness (see the next section). In most cases, light is provided by fluorescent tubes of the cool white or Gro-lux type. Small wattage incandescent bulbs are sometimes added to supplement the red and far-red region of the light spectrum.

Air humidity is infrequently controlled during incubation. Those who do control it usually set the relative humidity between 60-80%, with 70% being the most frequent setting.

POLYPHENOL OXIDATION. Many plants are rich in polyphenolic compounds. After tissue injury during dissection, such compounds will be oxidized by polyphenoloxidases and the tissue will turn brown or black. The oxidation products are known to inhibit enzyme activity, kill the explants, and darken the tissues and culture media. Such phenomena impose a serious block on the establishment of primary cultures, especially in woody plants. Some of the procedures used by various workers to combat this problem are (1) adding antioxidants to culture medium; (2) presoaking explants in antioxidant before inoculating into culture medium; (3) incubating the initial period of primary cultures in reduced light or darkness; and (4) frequently transferring explants into fresh medium whenever browning of the medium is observed.

The inclusion of antioxidants, such as ascorbic acid, polyvinylpyrrolidone (PVP), dithiothreitol, or bovine serum albumen into LS agar medium has been reported in Anigozanthos bud cultures (McComb and Newton, 1981) and cultures of immature Sassafras embryos (Hu and Wang, unpublished). The reason for the ineffectiveness might partially be attributed to the utilization of solid media. According to Ichihashi and Kako (1977), the browning of Cattleya shoot tips was most effectively inhibited by adding an antioxidant into a liquid medium in stationary condition. The inhibitor for polyphenoloxidase activity has little or no effect on the browning process of a solid medium. The inhibitors they used were 5 mM of potassium cyanide, ascorbic acid, cysteine, and thiourea. Stevenson and Harris (1980), however, reported reduction of agar medium discoloration with PVP-10 (0.01%) in Fuchsia shoot tip cultures.

In the preliminary experiments of Tactona granis bud cultures performed by Gupta et al. (1981), the medium turned black and all the explants died. In order to reduce blackening, the explants were suspended in a solution of different antibrowning agents in .058 M sucrose and agitated on a rotary shaker for 45 min before inoculation onto MS solid medium. The agents tested were H_2O_2 (5%), ascorbic acid (.28

mM), soluble PVP (0.7%) and polyclar AT (0.7%). Blackening was reduced by all the treatments. Multiple shoots, nevertheless, were formed only from explants treated with polyclar AT (an insoluble PVP) which is known to combine with phenolics by H-bonding. More recently, Polyclar AT has been replaced by insoluble polyvinylpyrrolidone (Sigma Chem. Co, USA) which gives similar results. In the culture of *Anigozanthos* buds, McComb and Newton (1981) soaked the explants in 0.55 M ascorbic acid while the material was being excised as well as after surface sterilization and before inoculation. Whether this pretreatment reduced tissue browning and explant loss was not reported.

Since the harmful phenolic oxidation products are formed under illumination, reduction of light intensity at the initial period of Stage I culture incubation should be beneficial. A less than 4% browning of the garlic meristem dome explants was obtained by Wang and Huang (1974) when the incubation light intensity during the first month was 150 lux, whereas 18% browning resulted under 500 lux. In order to reduce the accumulation of phenolic oxidates several workers carried out the initial incubation period (1-6 weeks) in darkness (Adams et al., 1979; McComb and Newton, 1981; Monaco et al., 1977). However, the degrees of effectiveness of those dark incubations were not reported.

The tissue and medium discoloration in thornless blackberry culture were effectively controlled by Broome and Zimmerman (1978) when the shoot tip explants were transferred to fresh medium 1-2 days after initial culturing. In shoot tip culture of mountain laurel, Lloyd and McCown (1980) transferred the explants into fresh liquid media 12 and 24 hr after the initial culturing, then continued the transfer process on a daily basis for one week before medium discoloration was completely under control. Axillary shoots were successfully produced from these explants 1-2 months later.

Stage II—Multiplication of the Propagule

The main objective of this stage is to produce the maximum number of useful propagule units. Among the three methods for achieving propagule multiplication, only the "axillary shoot proliferation" is covered in this chapter. Although the proliferation rate of this method may be slower than the other two in vitro methods, it provides genetic stability and is easily achievable by most plant species. Thus this method has rapidly gained popularity in recent years. For instance, Murashige (1974) cited only four species that were being multiplied using this method. By the end of 1981, there were at least 90 more species that could be added to the list (Table 1). With many genera, a million-fold enhancement of clonal plant increase in a year's time has been estimated using this method (Earle and Langhans, 1974a,b; Murashige, 1974; Wang, 1977; Whitehead and Giles, 1977).

GROWTH REGULATORS. In "axillary shoot proliferation," cytokinin is utilized to overcome the apical dominance of shoots and to enhance the branching of lateral buds from leaf axils. All, except one, of the Stage II media listed in Table 1 have cytokinin added. The effective concentration of exogenous cytokinin required to reverse apical dominance varies with the culture systems. Cytokinin concentration can be rather high compared to other types of in vitro cultures. Approximately three-quarters of the Stage II media listed in Table 1 were supplemented with 4.5 µM cytokinin or higher; one fourth with 25.0 µM or higher. Systems requiring as high as 90-270 µM cytokinin are occasionally encountered (Schnabelrauch and Sink, 1979; Skirvin et al., 1979; Wang and Hu, unpublished). In general, it appears that BA is the most effective cytokinin for stimulating axillary shoot proliferation, followed by, in decreasing order KIN and 2iP (Bhojwani, 1980a; Hasegawa, 1980; Kitto and Young, 1981; Lundergan and Janick, 1980; Papachatzi et al., 1981; Yang et al., 1981). A quite different order of cytokinin effectiveness may exist in certain species, such as Rhododendron (Anderson, 1975) and mountain laurel (Lloyd and McCown, 1980) in Ericaceae. The percentages of Stage II media in Table 1 using BA, KIN, 2iP, and ZEA are 7.6, 19.4, 3, and 3%, respectively.

Exogenous auxins do not promote axillary shoot proliferation; however, culture growth may improve by their presence. It was found that the multiplication potential of potato shoots during in vitro layering declined after 7-10 subcultures. This degeneration phenomenon could be completely reversed and eliminated by enriching the medium with 0.54 µM NAA (Wang and Hu, 1980). Of the Stage II media in Table 1, 48% were supplemented with this hormone. One of the possible roles of auxin in Stage II medium is to nullify the suppressive effect of high cytokinin concentrations on axillary shoot elongation and restore normal shoot growth (Lundergan and Janick, 1980). Too high a concentration of auxin may not only inhibit axillary bud branching (Hasegawa, 1980), but also induce callus formation, especially when 2,4-D is used. The percentages of auxin containing Stage II media in Table 1 using NAA, IBA, IAA, and 2,4-D are 58, 26, 16, and 0, respectively. With some exceptions (Hasegawa, 1979), this percentage data should reflect the rank of effectiveness of these auxins in Stage II medium.

Twelve percent of the Stage II media listed in Table 1 were supplemented with GA. The concentration range used was exceedingly low compared with other in vitro culture systems. Its role is essentially for axillary bud elongation (Schnabdrauch and Sink, 1979). Wochok and Sluis (1980a) observed that a topical treatment of Atriplex shoot explants with GA was effective not only in stimulating shoot elongation, but also in enhancing shoot multiplication beyond that of the most effective auxin-cytokinin combination. Lundergan and Janick (1980), however, detected no influence on either proliferation or growth of apple axillary buds when 0.29 µM GA was added. Ancora et al. (1981), working with globe artichoke, suggested that GA was not necessary to maintain shoot proliferation in vitro.

PROLIFERATION RATE. Since multiplication is the major economic criterion for successful commercial tissue culture propagation, the proliferation rate of Stage II determines the feasibility of in vitro propagation of a given species. This rate is affected by numerous factors. The chemical composition of culture medium and the physiological state of the plant material is of major importance. The chemical factors are largely determined by the concentration and types of exogenous growth regulators which have already been discussed in the previous section. Very little work has been done in exploring the effects of the physiological state of the plant material on culture response. The importance of this aspect on shoot proliferation, nevertheless, cannot be overlooked. For example, in carrizo citrange culture, the difference in the physiological states of the shoot tip vs. nodal explants resulted in a significantly different proliferation rate—3.0 vs. 1.7 shoots per explant, respectively (Kitto and Young, 1981).

The physiological state of the plant material can be modified through culture practices. In micropropagating a 20-year-old *Eucalyptus citriodora* tree, Gupta et al. (1981) found that bud explants gradually died when cultures were incubated under 23-25 C. However, when cultures were preincubated in 15 C for 3 days prior to transferring to 25 C, the explants grew and eventually led to a successful in vitro propagation.

Numerous examples indicated that the recalcitrant state of many species can be gradually modified through serial subculturing. In bud culture of *Betula platyphylla*, McCown and Amos (1979) observed that as the new growth was subcultured on a monthly basis, growth became more rapid and shoot multiplication was progressively easier to stimulate. Litz and Conover (1978) also experienced an increase in shoot multiplication rate with the increased number of subcultures. A proliferation rate of seven-fold was reached after nine subcultures.

Incubating the plant material in a liquid-shaking system for a brief period also appears to result in a modification of the physiological state of certain plant species and speeds up shoot proliferation. In the bud culture of Malling Merton apple rootstocks, a faster growth rate was observed when the in vitro produced shoots were cut and placed in liquid medium on an orbital shaker for four days. Intense proliferation became evident as these shoots were transferred back to a solid medium (Snir and Erez, 1980). Shaking the liquid culture seems advantageous. Walkey and Cooper (1976) reported that no multiple shoot proliferation was induced when the meristem tips of *Stellaria media* were cultured in a static liquid culture. As the meristem tips were placed in moving liquid culture, growth rate increased and the axillary buds divided to form a mass of proliferating shoots and roots. To induce bud proliferation from explants of mature *Eucalyptus* trees, Gupta et al. (1981) found that in addition to the 15 C preincubation, it was necessary to incubate the elongated explants in a shaking liquid medium for two weeks. In subsequent subcultures neither 15 C incubation nor liquid medium was necessary. They estimated that over 100,000 plants can be obtained by this method in a year from a single bud. It is, therefore, evident that the liquid-shaking system somehow reconditioned the plant and the newly acquired physiological state can be passed on for many subcultures on agar medium. However, the liquid-shaking system is not necessarily beneficial to all plant species (Kusey et al., 1980).

Few experiments have been designed to reveal the effects of incubation environments on shoot proliferation. Usually, the conditions used for Stage I are retained in Stage II cultures. Kitto and Young (1981) correlated the incubating light intensity and the number of shoots produced by each carrizo citrange shoot tip. They reported that under 0.0, 2.2, and 5.7 klx light intensity, 1.3, 3.0, and 2.1 shoots per explant resulted, respectively. This report indicates the need for knowledge in this neglected area.

Stage III—Root Regeneration

The purpose of this stage is de novo regeneration of adventitious roots from shoots obtained in Stage II or, in some cases, in Stage I and forming complete plants. Usually in vitro produced shoots of 10 mm or longer are cut and used in Stage III culture. Sometimes the elongation of shoots in Stage II is inhibited by high cytokinin level, thus, an intermediate shoot elongation stage becomes necessary. Wang (unpublished) reduced BA concentration from 264 to 22.0 µM to stimulate the elongation of *Sassafras* shoots from Stage II cultures before being transferred into rooting medium. Adventitious root formation can be induced quite readily in many herbaceous species, but it can be very recalcitrant in most woody species, especially from the mature trees. For these species Stage III probably is the most difficult of the three stages to accomplish. In this section, some helpful tips and techniques that have been reported will be examined.

Stage III does not always have to be carried out in vitro. A 100% rooting was reported by McCown and Amos (1979) when the shoots of *Betula platyphylla* were placed in 1:1 peat/perlite in a warm 30-35 C, high humidity (>80%) chamber. Kusey et al. (1980) obtained 60% rooting of *Gypsophila paniculata* by planting shoots in Jiffy 7 peat moss cylinders in greenhouse under intermittent mist. Instead of preparing auxin solution, commercial rooting powder, e.g., Rootone F, may be used to predip shoot bases before planting in rooting medium (Pyott and Converse, 1981; Wochok and Sluis, 1980b). Pyott and Converse (1981) found that rooting in vitro for red raspberry clones was unpredictable. Good rooting was obtained in greenhouse by placing shoots into pasteurized sand under intermittent mist.

CULTURE MEDIA. Growth Regulators. There are three phases involved in rhizogenesis: (a) induction, (b) initiation, and (c) elongation. Since it is rather difficult to isolate the induction phase in most experiments, this phase has usually been combined into the phase of initiation. It is well known, from the classic work of Skoog and Miller, that de novo root initiation depends on a low cytokinin to a high auxin ratio. Since the hormonal requirement for Stage II medium is opposite to this particular balance, Stage II medium is rarely used in Stage III.

In bud culture of Chinese cabbage, Kuo and Tsay (1977) demonstrated that an exogenous cytokinin/auxin ratio >1 was necessary for good shoot growth and <1 for root differentiation. Usually there is sufficient residual cytokinin in shoots from Stage II cultures; thus, little or no cytokinin is needed in Stage III medium. A survey of Table 1 reveals that there are only 27% Stage III media supplemented with cytokinin. The majority of the cytokinin containing Stage III media have reduced the cytokinin level to 4.5 μM or less. Too high an in vitro cytokinin content was shown to be deleterious to the initiation and the elongation of roots of both monocotyledonous and dicotyledonous plants (Henny, 1978; Lo et al., 1980; Nemeth, 1979; Pennazio, 1975). Sometimes, the residual cytokinin from Stage II cultures is high enough to suppress root formation. Transferring these unrooted shoots from the old rooting medium to fresh medium of the same composition might result in root formation (Ancora et al., 1981).

Since auxin is essential for root initiation, 86% of the Stage III media listed in Table 1 were supplemented with this hormone. Again, numerous studies have indicated that, among the common auxins, NAA is the most effective auxin for induction of root regeneration (Ancora et al., 1981; Kitto and Young, 1981; Johnson, 1978). The percentages of NAA, IBA, IAA, 2,4-D and other auxins used in Stage III media of Table 1 are 53, 29, 11, 3.6, and 3.6, respectively. Nemeth (1981) tested the capacity of several rare synthetic auxins on in vitro root induction in apple rootstocks, and found that 2-chloro-3-(2,3-dichloro-phenyl)propionitrile (CDPPN) applied at 5 μM was the most effective and produced up to 90% more roots than IBA which was the most effective common auxin in root induction of apple shoots.

Since the developing young shoots are a rich source of auxin production, the addition of exogenous auxin to Stage III media becomes unnecessary in many species (Hasegawa, 1980; Lee et al., 1977; Meredith, 1979; Papachatzi et al., 1981). When the auxin concentration is too high, callus will form at the shoot base which inhibits normal root development (Lane, 1979a). Another reason that too high an auxin level in Stage III media is undesirable is that the phase after root initiation, the "root elongation" phase, is very sensitive to auxin concentration, and will be inhibited by it high concentrations (Thimann, 1977). In order to provide the hard-to-root apple rootstocks a strong root induction stimulus, as well as, to avoid callusing and root growth inhibition, a two phase procedure was adopted for Stage III culture (James and Thurbon, 1979, 1981; Snir and Erez, 1980). The shoots were first cultured in an auxin containing "root initiation medium" for 4-8 days, then transferred to an auxin-free "root developing medium." This procedure effectively prevented callus formation, resulted in a 95% rooting, and led to a three-fold increase in root number, per rooted culture, compared to those in continuous contact with auxin. To stimulate root induction, Kusey et al. (1980) predipped the base of in vitro produced *Gypsophila* shoots in a 0.13 mM NAA or 0.12 mM IBA solution for 5 sec. and then transferred them into hormone-free White's medium for root initiation and development. A 48 hr soaking in an 11.0 μM NAA solution before being transferred to White's medium, successfully induced root formation in *Eucalyptus citriodora* (Gupta et al., 1981).

Five percent of the Stage III media listed in Table 1 were supplemented with GA. According to Pennazio, GA significantly improved the rooting percentage of potato meristem cultures (Pennazio and Redolfi, 1973; Pennazio and Vecchiati, 1976). Mosella Chancel et al. (1981), on the other hand, observed a suppressive effect by GA on the induction phase of in vitro peach rooting.

Auxin Synergists. The capacity of phenolic compounds to act as auxin synergists in the rooting process is well known, although the mode of action remains obscure. The difficulty in inducing in vitro rooting in woody fruit species led Jones to explore the phenolics and he obtained satisfactory results with phloroglucinol (PG) (Jones and Hatfield, 1976). Subsequently, the in vitro root inducing capacity of phenolics was demonstrated in apple rootstocks (James and Thurbon, 1979), *Fragaria* (James, 1979), *Prunus insititia* (Jones and Hopgood, 1979), *P. persica* (Mosella Chancel et al., 1980), and *Rubus* (James, 1979). With the exception of *Prunus persica*, all of these species responded to PG. However, the in vitro rooting percentage of *P. persica* (peach) improved greatly by the phenolics rutin and quercetin (Mosella Chancel et al., 1980).

According to Mosella Chancel et al. (1980) phenolics specifically act on the middle phase, the "initiation phase" of rhizogenesis. James and Thurbon (1981) reported that the presence of PG during the shoot proliferation stage (Stage II) significantly promoted root formation when measured as rooting percentage or number of roots per shoot in Stage III.

It appears that a given phenolic compound will show synergistic effect only with certain auxins. No synergistic effect between PG and non-indole auxin NAA was detected by James (1979). Mosella Chancel et al. (1980) found that rutin and quercetin combine best with IAA and NAA, respectively.

Salt Concentration. B5, LS, MS, and NN (see Table 1) media are all high N, P, and K salt media. Sometimes roots are unable to initiate in such high salt concentration regardless of the types of hormone present. When the salt concentration in the medium is lowered to one-half, one-third, or one-fourth of the standard strength, rooting becomes abundant (Kartha et al., 1974a, 1981b; Lane, 1979b; Skirvin and Chu, 1979). Successful rooting of *Eucalyptus citriodora* (Gupta et al., 1981) and *Gypsophila paniculata* (Kusey et al., 1980) was accomplished by transferring the in vitro produced shoots from Stage II MS high salt medium to the low salt WH medium.

Although lower salt concentration in a medium may be beneficial to root induction, it sometimes results in poor top growth. Wang (1978) observed that a one-ninth strength of MS medium stimulated 100% rooting in *Cryptomeria japonica*, but resulted in poor shoot growth. A one-third strength of MS medium, on the other hand, resulted in 87% rooting with good top growth. Although Gupta et al. (1981) successfully rooted *Eucalyptus citriodora* after transferring shoots from MS to WH medium, the rooted shoots failed to survive on this low salt medium and the leaves gradually turned yellow and dropped off. If the plants were transferred within two weeks, just after the emergence of the first roots, back into MS liquid medium, the yellowish leaves turned green and a well developed root system was formed.

Readers are referred to Table 1 for the list of plant species rooted successfully when salt concentrations were lowered in Stage III media.

Carrier Material. About nine-tenths of the Stage III media use agar as the carrier material to solidify the medium for supporting root growth. Physiologically, agar is not a completely inert material, and is a source of various types of substances which may affect growth. Agar, thus, may result in poor root growth in certain sensitive species. Lane (1979b) reported rooting inhibition in *Spirea* and *Prunus* when 0.6% agar was used. Kitto and Young (1981) observed an inverse relationship between the rooting ability and agar concentration in carrize citrange cultures.   In addition to the presence of possible growth inhibiting substances, the reason for poor rooting in agar medium may also be due to poor aeration and a slow rate of diffusion of the toxic metabolic wastes released by growing tissue.   Two methods are commonly used to circumvent the inhibitory effect of agar medium: (a) supplementing fine powder of activated charcoal (AC) to the agar medium; and (b) using a liquid and filter-paper-bridge system in place of agar.

AC may absorb toxic substances in the medium, thereby improving root regeneration and development (Ziv, 1979; Takayama and Misawa, 1980).   AC may also absorb residual cytokinin from Stage II medium. Takayama and Misawa (1980) reported an inhibition of root formation of *Lilium* by BA. Such inhibition was completely reversed by addition of AC.   Root formation and growth were even better in BA-free medium containing AC than in medium without AC.   AC is also capable of shading in vitro roots from light which, in high intensity, may inhibit root growth.

Liquid media are used in approximately 12% of the Stage III cultures listed in Table 1. It facilitates the free diffusion of toxic plant wastes. In combination with filter-paper-bridge system, it also provides excellent aeration for root development.   Its drawback is the amount of labor involved in preparing the bridge. It is not recommended unless agar or agar+AC media have proved to be unsatisfactory.

Polyurethane foam and vermiculite were used with satisfactory results as substitutes for high cost agar in Stage III media (McComb and Newton, 1981; Barnes, 1979). The use of liquid media with a vermiculite substrate for support and aeration in watermelon culture resulted in a significantly superior root system, with better branches and extensive root-hairs, than when grown on 0.4-1.2% agar (Barnes, 1979). Additionally, there was less damage to the roots when removed from the vermiculite during transplanting than from agar and better explant survival after transplanting.

JUVENILITY. For the hard-to-root species, especially the forest trees, the age of the plant plays a significant role in root regenerating capacity. deFossard et al. (1974b) observed that root development of *Eucalyptus grandis* nodal cultures was most frequent when explants were obtained from the basal end of the seedlings. The rooting data from the base cotyledon node to the apex were 21, 21, 11, 6, and 0 per 36 buds, respectively. Wang and Hu (unpublished) experienced a sporadic rooting in shoots resulting from bud cultures of a 5-year-old *Sassafras randaiense* tree, whereas the rooting of shoots resulting from embryo cultures of the same species was essentially 100%. In micropropagation of *Eucalyptus citriodora*, Gupta et al. (1981) reported a higher auxin (10.0 μM) requirement for rooting bud explants of a 20-year-old tree compared to 5.0 μM for seedling bud explants.

It is known that the subculturing process may change the physiological state and gradually rejuvenate the explants. Some of the juvenile characteristics were induced from the ex-buds of an adult clone of *Vitis vinifera* after two to three subcultures (Mullins et al., 1979). Rooting percentage increased from 10% in the primary cultures to 60% in the second and subsequent subcultures of the bud cultures of a 100-year-old *Tectora grandis* tree (Gupta et al., 1981), whereas the percentage of rooting from seedling explants was consistantly over 80%. In bud cultures of 20-year-old *Eucalyptus citriodora* trees, Gupta et al. (1981) reported that none of the Stage III treatments resulted in root formation in either the initial explants or in the first three subcultures. However, 35–40% of the shoots rooted at the fourth subculture. In the fifth and subsequent passages the percentage of rooting was about 45–50%. These reports clearly indicate that subculturing may modify the physiological state of the mature plant tissues and result in a return of some characteristics associated with juvenility. Subculturing thus may make root induction progressively easier.

INCUBATING CONDITIONS. As with the previous stages, few experiments have been designed to reveal the effects of incubating environments on root regeneration.

In general, high light intensity supports better plant formation (Bhojwani, 1980a; Ziv, 1979). Increased light intensity results in considerable culture growth, thus producing plantlets which could be successfully transferred to pots. Care should be taken, though, that the high intensity light does not directly impinge on the roots due to its inhibitory effect on root growth. Improved root growth was reported when the culture containers were wrapped with aluminum foil (Hennen and Sheehan, 1978) or 0.3% AC was added to the medium (Ziv, 1979). Light intensity used in this stage usually is within a 0.54 to 10 klx range.

A majority of investigators adopt the same incubating temperature as used in the previous stages. Schnabdrauch and Sink (1979) reported that rooting of shoot tip cultures of *Phlox* spp. could be hastened if the primordia development was promoted at 30 C for one week before being transferred to 22 C for root elongation.

Planting-Out

After rooting, the in vitro regenerated plantlets are ready to be transferred from the aseptic containers into pots. Factors that should be considered in transplantation are infections and desiccation. Steril-

izing the soil mixture eliminates serious infection problems. Desiccation usually is the last major block to be conquered in order to reach
the goal of micropropagation. Excessively high water loss was
recorded from the leaves of plants immediately after transplanting
(Brainerd and Fuchigami, 1981). Such a high rate of water loss is
related to: (a) the reduced quantities of epicuticular wax (Grout and
Aston, 1977; Sutter and Langhans, 1979), (b) the high volume of
mesophyll intercellular spaces (Brainerd et al., 1981), and (c) the
slowness of stomatal response to water stress (Brainerd and Fuchigami,
1981). To compound the problem, the xylem tissue in the regenerated
plants formed a closed system across the base of the shoot prior to
root formation. The de novo formed roots, arising from callus, have
poor connections to the main vascular system of the shoot (Grout and
Aston, 1977). Such a structuring is of no consequence in culture, when
plantlets are surrounded by high humidity, but it severely restricts
acropetal water transport after transplantation. "Transplant shock,"
thus results, which leads to tip dieback or death of the plantlets.

A period of humidity acclimatization is, therefore, required for the
newly transferred plantlets to adapt to the outside environment. During acclimatization, humidity is gradually reduced over a period of 2-3
weeks. In the meantime, the plantlets undergo morphological and
physiological adaptations enabling them to develop typical terrestrial
plant water control (Grout and Aston, 1977; Brainerd and Fuchigami,
1981).

Lane (1979c) maintained the newly transferred Bartlett pear plantlets
under a mist bed in the greenhouse for 2 weeks for hardening. The
subsequent growth in the greenhouse proceeded at the same rate as
seed derived plants. He established 500 plants in soil in this fashion.
Barnes (1979) attempted to maintain high humidity for the newly transferred watermelon plantlets under intermittent mist. The resulting
explant survival rate was poor. He subsequently found that plantlets
could be successfully established in the greenhouse by covering with a
clear plastic cup to maintain high humidity. The cups were partially
lifted one week later and tilted to allow air circulation, and later
removed for 5-6 hr daily. The cups were then completely removed
from those plantlets which exhibited vigorous growth and a lack of
wilting. Broome and Zimmerman (1978) obtained 60% survival rate with
blackberry by growing plantlets under inverted glass jars for 1-3 weeks.
It is interesting to note that acclimatization can be achieved with
direct low humidity exposure. Brainerd and Fuchigami (1981) left apple
plantlets in the Stage III culture jars with lids removed in a room with
a 30-40% relative humidity. To prevent drying of the medium, 10 ml of
distilled water was added daily. After the lids were opened for 5-6
days, 80% of the stomata of leaves closed within 15 min.—the same
rate as the greenhouse grown plants.

FUTURE PROSPECTS

Major breakthroughs have been accomplished in micropropagation of
woody fruits and, to a lesser extent, other woody species. The picture

of in vitro propagation in tree species is starting to emerge, but the details of the technique are still far from clear. Because of the long breeding cycle and few feasible conventional cloning methods available, without doubt, the in vitro propagation technique will become the method of the future for cloning desirable genotypes of forest, ornamental, and fruit trees. Although most woody species are recalcitrant in nature and working with them frequently results in frustration, we believe that the potential future rewards in this area will be too great to shy away from. We, therefore, strongly urge plant cell culture workers to devote efforts into this neglected area.

In vitro micropropagation is an extremely space saving propagation method (Boxus et al., 1977; Wang and Hu, 1982). The equipment and supplies for setting up a culture laboratory are not expensive. Despite these advantages, commercial adoption of this technique in urban areas is still limited to a few types of ornamental plants, even though culture requirements for many additional economically valuable plant groups are already available. This is mainly due to the intensive skilled labor required for subculturing the Stage II propagules and in transferring individual shoots or plantlets into and out of Stage III culture containers. The production costs, therefore, can not compete with the conventional propagation methods in the majority of ornamental and vegetable crops. In order to cut down the production costs, the development of a certain degree of automation in Stage II and III is essential.

It is evident that little effort has been directed to reveal the effects of incubation conditions on various phases of in vitro plant growth. The limited available data, nevertheless, suggests that sizeable potential gains can be made when optimal temperature, photoperiod, light intensity, and light quality levels have been identified. Experiments of these types are prohibitive mainly by the requirement for large numbers of expensive environmental control units. With the potential applications of plant tissue culture to genetic engineering and crop improvement, numerous industrial firms are investing in this field. Without doubt, the identification of the optimal incubation conditions is of crucial importance in culturing plant materials on an industrial scale. Hopefully, with the financial support from these industries, advances can be made in this neglected area of plant tissue culture.

It is important to be aware that in vitro cultures do not always produce pathogen-free plants. Large tips are frequently used by the commercial growers to propagate their plants. In such cases, if the source plants happened to be infected, the disease would spread in the clonal progenies. This probably is the case in the orchid industry. Before in vitro "mericloning" was developed, orchid viruses were a minor problem. However, orchid viruses are now generally widespread and becoming a costly problem. Obviously, "mericloning" of orchids without adequately carrying out virus indexing has resulted in efficient virus transmission. Since in no case is the meristem culture 100% effective in producing virus-free plants from infected plants, a rigid virus indexing must be used to test in vitro regenerated plants.

This work was supported in part by grants from National Science Council, Commission on Rural Reconstruction (ROC), and funds from the ART program of William Paterson College of New Jersey.

KEY REFERENCES

Anderson, W.C. 1980. Mass propagation by tissue culture: Principles and practices. In: Proceedings Conference on Nursery Production of Fruit Plants Through Tissue Culture—Applications and Feasibility. SEA, Agri. Res. Results, N.E. Series No. 11, pp. 1-10. USDA, Beltsville.

Kartha, K.K. 1981. Meristem culture and cryopreservation—Methods and application. In: Plant Tissue Culture (T.A. Thorpe, ed.) pp. 181-211. Academic Press, New York.

Murashige, T. 1974. Plant propagation through tissue cultures. Annu. Rev. Plant Physiol. 25:135-166.

Wang, P.J. and Hu, C.Y. 1980. Regeneration of virus-free plants through in vitro culture. In: Advances in Biochemical Engineering, Vol. 18, Plant Cell Cultures II (A. Fiechter, ed.) pp. 61-99. Springer-Verlag, Berlin, Heidelberg, New York.

REFERENCES

Adams, R.M., II, Koenigsberg, S.S., and Langhans, R.W. 1979. In vitro propagation of Cephalotus follicularis (Australian pitche plant). HortScience 14:512-513.

Altman, A. and Goren, R. 1974. Growth and dormancy cycles in Citrus bud cultures and their hormonal control. Physiol. Plant. 30:240-245.

Ancora, G., Belli-Donini, M.L., and Cuozzo, L. 1981. Globe artichoke plants obtained from shoot apices through rapid in vitro micropropagation. Sci. Hort. 14:207-213.

Anderson, W.C. 1978. Rooting of tissue cultured rhododendrons. Proc. Int. Plant Prop. Soc. 28:135-139.

_____ 1975. Propagation of rhododendrons by tissue culture: Part I. Development of culture medium for multiplication of shoots. Proc. Int. Plant Prop. Soc. 25:129-135.

Atanassov, A.I. 1980. Method for continuous bud formation in tissue cultures of sugar beet (Beta vulgaris L.). Z. Pflanzenzuchtg. 84:23-29.

Baker, P.K. and Phillips, D.J. 1962. Obtaining pathogen-free stock by shoot tip culture. Phytopathol. 52:1242-1244.

Barlass, M., Grant, W.J.R., and Skene, K.G.M. 1980. Shoot regeneration in vitro from native Australian fruit-bearing trees—Quandong and plum bush. Aust. J. Bot. 28:405-409.

Barnes, L.R. 1979. In vitro propagation of watermelon. Sci. Hort. 11:223-227.

Barnett, O.W., Gibson, P.B., and Sepo, A. 1975. A comparison of heat treatment and meristem-tip culture for obtaining virus-free plants of Trifolium repens. Plant Dis. Rep. 59:834-837.

Bayliss, M.W. 1973. Origin of chromosome number variation in cultured plant cells. Nature 246:529-530.

Ben-Jaacov, J. and Dax, E. 1981. In vitro propagation of Grevillea rosmarinifolia. HortScience 16:309-310.

Bhojwani, S.S. 1980a.   Micropropagation method for a hybrid willow (*Salix matsudana* x *alba* NZ-1002). New Zealand J. Bot. 18:209–214.

————— 1980b.   In vitro propagation of garlic by shoot proliferation. Sci. Hort. 13:47–52.

Borkowska, B. and Powell, L.E. 1979.   The dormancy status of apple buds as determined by an in vitro culture system. J. Am. Soc. Hort. Sci. 104:796–799.

Boxus, Ph., Quoirin, M., and Laine, J.M. 1977.   Large scale propagation of strawberry plants from tissue culture.   In: Applied and Fundamental Aspects of Plant Cell, Tissue, and Organ Culture (J. Reinert and Y.P.S. Bajaj, eds.) pp. 130–143.   Springer-Verlag, Berlin, Heidelberg, New York.

Brainerd, K.E. and Fuchigami, L.J. 1981.   Acclimatization of aseptically cultured apple plants to low relative humidity.   J. Am. Soc. Hort. Sci. 106:515–518.

Brainerd, K.E., Fuchigami, L.H., Kwiatkowshi, S., and Clark, C.S. 1981. Leaf anatomy and water stress of aseptically cultured "Pixy" plum grown under different environments. HortScience 16:173–175.

Broadbent, L.H. 1965.   The epidemiology of tomato mosaic.   XI. Seed transmission of TMV. Ann. Appl. Biol. 56:177–205.

Broome, O.C. and Zimmerman, R.H. 1978.   In vitro propagation of *Blackberry*. HortScience 13:151–153.

Chaturvedi, H.C., Sharma, A.K., and Prasad, R.N. 1978.   Shoot apex culture of *Bougainvillea glabra* "Magnifica". HortScience 13:36.

Coleman, W.K. and Thorpe, T.A. 1978.   In vitro culture of western red cedar (*Thuja plicata*). II. Induction of male strobili from vegetative shoot tips. Can. J. Bot. 56:557–564.

Cresswell, R. and Nitsch, C. 1975.   Organ culture of *Eucalyptus grandis*. Planta 125:87–90.

Csinos, A. and Hendrix, W. 1977.   Toxin produced by *Phytophythora cryptogea* active on excised tobacco leaves. Can. J. Bot. 55:1156–1162.

Culafic, L. 1973.   Induction of flowering of isolated *Spinacia oleracea* L. buds in sterile culture. Bull. DeL' Institut et du Jardin Botaniques de L'Universite de Beograd 53–56.

Dale, P.J. 1975.   Meristem tip culture in *Lolium multiflorum*. J. Exp. Bot. 26:731–736.

————— 1977a.   The elimination of ryegrass mosaic virus from *Lolium multiflorum* by meristem tip culture. Ann. Appl. Biol. 85:93–96.

————— 1977b.   Meristem tip culture in *Lolium*, *Festuca*, *Phleum* and *Dactylis*. Plant Sci. Lett. 9:333–338.

————— 1979. The elimination of cocksfoot streak virus, cocksfoot mild mosaic virus and cocksfoot mottle virus from *Dactylis glomerata* by shoot tip and tiller bud culture. Ann. Appl. Biol. 93:285–288.

————— 1980.   A method for in vitro storage of *Lolium multiflorum*. Ann. Bot. 45:497–502.

deFossard, R.A., Myint, A., and Lee, E.C.M. 1974a.   A broad spectrum tissue culture experiment with tobacco (*Nicotiana tabacum*) callus. Physiol. Plant. 30:125–130.

deFossard, R.A., Nitsch, C., Cresswell, R.J., and Lee, E.C.M. 1974b. Tissue and organ culture of *Eucalyptus*. New Zealand J. For. Sci. 4:267–278.

Dudits, D., Nemeth, G., and Haydu, Z. 1975. Study of callus growth and organ formation in wheat *Triticum aestivum* tissue culture. Can. J. Bot. 53:957–963.

Earle, E.D. and Langhans, R.W. 1974a. Propagation of *Chrysanthemum* in vitro. I. Multiple plantlets from shoot tips and the establishment of tissue culture. J. Am. Soc. Hort. Sci. 99:128–132.

_____ 1974b. Propagation of *Chrysanthemum* in vitro. II. Production, growth and flowering of plantlets from tissue cultures. J. Am. Soc. Hort. Sci. 99:352–358.

Edallo, S., Zucchinali, C., Petenzin, M., and Salamini, F. 1981. Chromosomal variation and frequency of spontaneous mutation associated with in vitro culture and plant regeneration in maize. Maydica 26:39–56.

Evans, D.A. 1981. Soybean tissue culture. Soybean Genetics Newsletter 8:27–29.

Fedotina, V.L. and Krylova, N.V. 1976. Ridding tobaccos of the mycoplasmic infection big bud by the method of tissue culturing. Dokl. Bot. Sci. 228:49–51.

Gamborg, O.L. and Eveleigh, D.E. 1968. Culture methods and detection of gluconates in suspension cultures of wheat and barley. Can. J. Biochem. 46:417–421.

Gamborg, O.L., Miller, R.A., and Ojima, K. 1968. Nutrient requirements of suspension cultures of soybean root cells. Exp. Cell Res. 50:151–158.

Gresshoff, P.M. and Doy, C.H. 1972. Development and differentiation of haploid *Lycopersicon esculentum* (tomato). Planta 107:161–170.

Grout, B.W.W. and Aston, M.J. 1977. Transplanting of cauliflower plants regenerated from meristem culture. I. Water loss and water transfer related to changes in leaf wax and to xylem regeneration. Hort. Res. 17:1–7.

Grout, B.W.W. and Henshaw, G.G. 1978. Freeze preservation of potato shoot-top cultures. Ann. Bot. 42:1227–1229.

Gupta, P.K., Mascarenhas, A.F., and Jagannathan, V. 1981. Tissue culture of forest trees—Clonal propagation of mature trees of *Eucalyptus citriodora* Hook, by tissue culture. Plant Sci. Lett. 20:195–201.

Gupta, P.K., Nadgir, A.L., Mascarenhas, A.F., and Jagannathan, V. 1980. Tissue culture of forest trees: Clonal multiplication of *Tectona grandis* L. (teak) by tissue culture. Plant Sci. Lett. 17:259–268.

Haines, R.J. and deFossard, R.A. 1977. Propagation of hoop pine (*Araucaria cunninghamii* Ait.) by organ culture. Acta Hort. 78:297–302.

Harper, P.C. 1978. Tissue culture propagation of blackberry and tayberry. Hort. Res. 18:141–143.

Hartney, V.J. and Barker, P.K. 1980. The vegetative propagation of *Eucalyptus* by tissue culture. Paper presented at IVFRO Symposium and Workshop on Genetic Improvement and Productivity of Fast-Growing Tree Species. August, 1980, Sau Paulo, Brazil.

Hasegawa, P.M. 1979. In vitro propagation of rose. HortScience 14:610–612.

Hasegawa, P.M. 1980. Factors affecting shoot and root initiation from cultured rose shoot tips. J. Am. Soc. Hort. Sci. 105:216–220.

Hasegawa, P.M., Murashige, T., and Takatori, F.H. 1973. Propagation of *Asparagus* through shoot apex culture. II. Light and temperature requirements, transplantability of plants, and cyto-histological characteristics. J. Am. Soc. Hort. Sci. 98:143-148.

Haskins, R.H. and Kartha, K.K. 1980. Freeze preservation of pea meristems: Cell survival. Can. J. Bot. 58:833-840.

Hennen, G.R. and Sheehan, T.J. 1978. In vitro propagation of *Platycerium stemaria* (Beauvois) Desv. HortScience 13:245.

Henny, R.J. 1978. In vitro propagation of *Peperomia* "Red Ripple" from leaf discs. HortScience 13:150-151.

Hollings, M. and Stone, O.M. 1968. Techniques and problems in the production of virus-tested plant material. Sci. Hort. 20:57-72.

Hu, C.Y. 1979. Propagation of *Sassafras randaiense* (Hay.) Rehd. Taiwan For. J. 5:30-31 (in Chinese).

Huang, S.-C. and Millikan, D.F. 1977. Production of apple plantlets by tip meristem culture. Trans. Mo. Acad. Sci. 10, 11:279 (abstract).

_____ 1980. In vitro micrografting of apple shoot tips. HortScience 15:741-743.

Ichihashi, S. and Kako, S. 1977. Studies on clonal propagation of *Cattleya* through tissue method. II. Browning of *Cattleya*. J. Jap. Soc. Hort. Sci. 46:325-330.

Jacoli, G.G. 1978. Sequential degeneration of mycoplasma-like bodies in plant tissue culture infected with aster yellows. Can. J. Bot. 56:133-140.

James, D.J. 1979. The role of auxins and phloroglucinol in adventitious root formation in *Rubus* and *Fragaria* grown in vitro. J. Hort. Sci. 54:273-277.

_____ and Thurbon, I.J. 1979. Rapid in vitro rooting of the apple rootstock M.9. J. Hort. Sci. 54:309-311.

_____ and Thurbon, I.J. 1981. Shoot and root initiation in vitro in apple rootstock M.9 and the promotive effects of phloroglucinol. J. Hort. Sci. 56:15-20.

Johnson, B.B. 1978. In vitro propagation of *Episcia cupreata*. HortScience 13:596.

Jones, O.P. and Hopgood, M.E. 1979. The successful propagation in vitro of two rootstocks of *Prunus*: The plum rootstock Pixy (*P. insititia*) and the cherry rootstock F12/1 (*P. avium*). J. Hort. Sci. 54:63-66.

_____, and O'Ferrell, D. 1977. Propagation in vitro of M.26 apple rootstocks. J. Hort. Sci. 52:235-238.

Kaiser, W.J. and Teemba, L.R. 1979. Use of tissue culture and thermotherapy to free East African cassava cultivars of African cassava mosaic and cassaava brown streak disease. Plant Dis. Rep. 63:780-784.

Kartha, K.K. and Gamborg, O.L. 1975. Elimination of cassava mosaic disease by meristem culture. Phytopathol. 65:826-828.

_____ 1978. Meristem culture techniques in the production of disease-free plants and freeze-preservation of germplasm of tropical tuber crops and grain legumes. In: Diseases of Tropical Food Crops (H. Maraite and J.A. Meyer, eds.) pp. 267-283. Universite Catholique de Louvain, Belgium.

_____, and Constabel, F. 1974a. Regeneration of pea (Pisum sativum L.) plants from shoot apical meristems. Z. Pflanzenphysiol. 72:172–176.

_____, Constabel, F., and Shyluk, J.P. 1974b. Regeneration of Cassava plants from apical meristems. Plant Sci. Lett. 2:107–113.

Kartha, K.K., Champoux, S., Gamborg, O.L., and Pahl, K. 1977. In vitro propagation of tomato by shoot apical meristem culture. J. Am. Soc. Hort. Sci. 102:346–349.

Kartha, K.K., Gamborg, O.L., and Leung, N.L. 1979. Freeze-preservation of pea meristems in liquid nitrogen and subsequent plant regeneration. Plant Sci. Lett. 15:7–15.

Kartha, K.K., Leung, N.L., and Pahl, K. 1980. Cryopreservation of strawberry meristems and mass propagation of plantlets. J. Am. Soc. Hort. Sci. 105:481–484.

Kartha, K.K., Mroginski, L.A., Pahl, K., and Leung, N.L. 1981a. Germplasm preservation of coffee (Coffea arabica L.) by in vitro culture of shoot apical meristems. Plant Sci. Lett. 22:301–307.

Kartha, K.K., Pahl, K., Leung, N.L., and Mroginski, L.A. 1981b. Plant regeneration from meristems of grain legumes: Soybean, cowpea, peanut, chickpea, and bean. Can. J. Bot. 59:1671–1679.

Kassanis, B. 1957. The use of tissue cultures to produce virus-free clones from infected potato varieties. Ann. Appl. Biol. 45:422–427.

_____ and Varma, A. 1967. The production of virus-free clones of some British potato varieties. Ann. Appl. Biol. 59:447–450.

Kehr, A.E. and Schaeffer, G.W. 1976. Tissue culture and differentiation of garlic. Hort. Sci. 11:422–423.

Kiss, F. and Zatyko, J. 1978. Vegetative propagation of Rubus species in vitro. Bot. Kozlem. 65:65–69.

Kitto, S.L. and Young, M.J. 1981. In vitro propagation of Carrizo citrange. HortScience 16:305–306.

Knauss, J.F. 1976. A tissue culture method for producing Dieffenbachia picta cv. Perfection free of fungi and bacteria. Proc. Fla. State Hort. Soc. 89:293–296.

Koda, Y. and Okazawa, Y. 1980. Cytokinin production by Asparagus shoot apex cultured in vitro. Physiol. Plant. 49:193–197.

Koevary, K., Rappaport, L., and Morris, L.L. 1978. Tissue culture propagation of head lettuce. HortScience 13:39–41.

Kunisaki, J.T. 1980. In vitro propagation of Anthurium andreanum Lind. HortScience 15:508–509.

Kuo, C.G. and Tsay, J.S. 1977. Propagation of Chinese cabbage by axillary bud culture. HortScience 12:459–460.

Kusey, W.E., Jr., Jammer, P.A., and Weiler, T.C. 1980. In vitro propagation of Gypsophila paniculata L. "Bristol Fairy." HortScience 15:600–601.

Lane, W.D. 1979a. The influence of growth regulators on root and shoot initiation from flax meristem tips and hypocotyls in vitro. Physiol. Plant. 45:260–264.

_____ 1979b. In vitro propagation of Spirea bumalda and Prunus cistena from shoot apices. Can. J. Plant Sci. 59:1025–1029.

_____ 1979c. Regeneration of pear plants from shoot meristem-tips. Plant Sci. Lett. 16:337–342.

Langhans, R.W., Horst, R.K., and Earle, E.D. 1977. Disease-free plants via tissue culture propagation. HortScience 12:149-150.

Lee, C.W., Skirvin, R.M., Soltero, A.J., and Janick, J. 1977. Tissue culture of *Salpiglossis sinuata* L. from leaf discs. HortScience 12:547-549.

Lester, D.T. and Berbee, J.G. 1977. Within-clone variation among black poplar trees derived from callus culture. For. Sci. 23:122-131.

Linsmaier, E.M. and Skoog, F. 1965. Organic growth factor requirement of tobacco tissue culture. Physiol. Plant. 18:100-127.

Litz, R.E. and Conover, R.A. 1978. In vitro propagation of papaya. HortScience 13:241-242.

Lloyd, G. and McCown, B. 1980. Commercially feasible micropropagation of mountain laurel, *Kalmia latifolia*, by use of shoot-tip culture. Proc. Int. Plant Prop. Soc. 30:421-427.

Lo, O.F., Chen, C.J., and Ross, J.G. 1980. Vegetative propagation of temperate foliage grasses through callus culture. Crop Sci. 20:363-367.

Lundergan, C. and Janick, J. 1979. Low temperature storage of in vitro apple shoots. HortScience 14:514.

_____ 1980. Regulation of apple shoot proliferation and growth in vitro. Hort. Res. 20:19-24.

Mahlderg, P.G., Turner, F.R., Walkinshaw, C., Venketeswaran, S., and Mehrotra, B. 1975. Observations on incomplete cytokinesis in callus cells. Bot. Gaz. 136:189-193.

McComb, J.A. and Newton, S. 1981. Propagation of Kangaroo paws using tissue culture. J. Hort. Sci. 56:181-183.

McCown, B. and Amos, R. 1979. Initial trials with commercial micropropagation of birch selections. Proc. Int. Plant Prop. Soc. 29:387-393.

McGrew, J.R. 1980. Meristem culture for reproduction of virus-free strawberries. In: Proceedings Conference on Nursery Production of Fruit Plants Through Tissue Culture—Application and Feasibility. SEA, Agri. Res. Results, N.E. Series No. 11, pp. 80-85. USDA, Beltsville.

Mellor, F.C. and Stace-Smith, R. 1969. Development of excised potato buds in nutrient medium. Can. J. Bot. 47:1617-1621.

_____ 1977. Virus-free potatoes by tissue culture. In: Applied and Fundamental Aspects of Plant Cell, Tissue and Organ Culture (J. Reinert and Y.P.S. Bajaj, eds.) pp. 616-637. Springer-Verlag, Berlin.

Meredith, C.P. 1979. Shoot development in established calli cultures of cultivated tomato (*Lycopersicon esculentum* Mill.). Z. Pflanzenphysiol. 95:405-411.

Miller, C.O. 1963. Kinetin and kinetin-like compounds. In: Modern Methods of Plant Analysis, Vol. 6 (H.F. Linskens and M.V. Tracey, eds.) pp. 194-202. Springer-Verlag, Berlin.

Minocha, S.C. 1980. Cell and tissue culture in the propagation of forest trees. In: Plant Cell Cultures: Results and Perspectives (F. Sala, B. Parisi, R. Cella, and O. Ciferri, eds.) pp. 295-300. Elsevier/North-Holland Biomedical Press, Amsterdam, New York, Oxford.

Monaco, L.C., Sondahl, M.R., Carvalho, A., Crocomo, O.J., and Sharp, W.R. 1977. Applications of tissue culture in the improvement of coffee. In: Applied and Fundamental Aspects of Plant Cell, Tissue and Organ Culture (J. Reinert and Y.P.S. Bajaj, eds.) pp. 109–129. Springer-Verlag, Berlin.

Morel, G.M. 1960. Producing virus-free *Cymbidiums*. Am. Orchid Soc. Bull. 29:495–497.

_____ 1965. Clonal propagation of orchids by meristem culture. *Cymbidium* Soc. News 20:3–11.

_____ 1975. Meristem culture techniques for the long-term storage of cultivated plants. In: Crop Genetics Resources for Today and Tomorrow (O.H. Frankel and J.G. Hawkes, eds.) pp. 327–332. Cambridge Univ. Press, Cambridge.

_____ and Martin, C. 1952. Guerison de dahlias atteints d'une maladie a virus. C. R. Acad. Sci. (Paris) 235:1324–1325.

Mosella Chancel, L., MacHeix, J.J., and Jonard, R. 1980. Les conditions du microbouturage in vitro du pecher (*Prunus persica* Batsch): Influences combinees des substances de croissance et de divers composes phenoliques. Physiol. Veg. 18:597–608.

Mullin, R.H. and Schlegel, D.E. 1976. Cold storage maintenance of strawberry meristem plantlets. Hort. Sci. 11:100–101.

Mullins, M.G., Nair, Y., and Sampet, P. 1979. Rejuvenation in vitro: Induction of juvenile characters in an adult clone of *Vitis vinifera* L. Ann. Bot. 44:623–627.

Murashige, T. and Skoog, F. 1962. A revised medium for rapid growth and bioassays with tobacco tissue cultures. Physiol. Plant. 15:473–497.

Murashige, T. and Tucker, D.P.H. 1969. Growth factor requirements of *Citrus* tissue culture. Proc. First Int. *Citrus* Symp. 3:1155–1161.

Nair, N.G., Kartha, K.K., and Gamborg, O.L. 1979. Effect of growth regulators on plant regeneration from shoot apical meristems of Cassava (*Manihot esculenta* Crantz) and on the culture of internodes in vitro. Z. Pflanzenphysiol. 95:51–56.

Navarro, L., Poistacher, C.N., and Murashige, T. 1975. Improvement of shoot tip grafting in vitro for virus-free *Citrus*. J. Am. Soc. Hort. Sci. 100:471–479.

Nemeth, G. 1979. Benzyladeninie–stimulated rooting in fruit-tree rootstocks cultured in vitro. Z. Pflanzenphysiol. 95:389–396.

_____ 1981. Adventitious root induction by substituted 2-chloro-3phenyl-propionitriles in apple rootstocks cultured in vitro. Sci. Hort. 14:253–259.

Nitsch, J.P. and Nitsch, C. 1969. Haploid plants from pollen grains. Science 163:85–87.

Novak, F.J. 1981. Chromosomal characteristics of long-term callus cultures of *Allium sativum* L. Cytologia 46:371–379.

Papachatzi, M., Hammer, P.A., and Hasegawa, P.M. 1981. In vitro propagation of *Hosta decorata* "Thomas Hogg" using cultured shoot tips. J. Am. Soc. Hort. Sci. 106:232–236.

Pennazio, S. 1975. Effects of adenine and kinetin on development of carnation meristem tips cultured in vitro. J. Hort. Sci. 50:161–164.

_____ and Redolfi, P. 1973. Factors affecting the culture in vitro of potato meristem tips. Pot. Res. 16:20–29.

_____ and Vecchiati, M. 1976. Effects of naphtaleneacetic acid on potato meristem tip development. Pot. Res. 19:257-261.

_____, Appiano, A., Vecchiati, M., and D'Agostino, G. 1976. Thiamine requirement of potato meristem tips cultured in vitro. Physiol. Veg. 14:121-131.

Poessel, J.-L., Martinez, J., Macheix, J.-J., and Jonard, R. 1980. Variations saisonnieres de l'aptitude au greffage in vitro d'apex de Pecher (*Prunus persica*, Batsch), Relations avec les teneurs en composes phenoliques endogenes et les activities peroxydasique et polyphenol-oxydasique. Physiol. Veg. 18:665-675.

Pyott, J.L. and Converse, R.H. 1981. In vitro propagation of heat-treated red raspberry clones. HortScience 16:308-309.

Raman, K., Walden, D.B., and Greyson, R.I. 1980. Propagation of *Zea mays* by shoot tip culture: A feasibility study. Ann. Bot. 45:183-190.

Roest, S., and Bokelmann, G.S. 1981. Vegetative propagation of carnation in vitro through multiple shoot development. Sci. Hort. 14:357-366.

Roggemans, J. and Claes, M.-C. 1979. Rapid clonal propagation of rhubarb by in vitro culture of shoot-tips. Sci. Hort. 11:241-246.

Sakai, A. and Nishiyama, Y. 1978. Cryopreservation of winter-vegetative buds of hardy fruit trees in liquid nitrogen. HortScience 13:225-227.

Sakai, A., Yamakawa, M., Sakata, D., Harada, T., and Yakuwa, T. 1978. Development of a whole plant from an excised strawberry runner apex frozen to -196 C. Low Temp. Sci. Ser. Bull. 36:31-38.

Schnabdrauch, L.S. and Sink, K.C. 1979. In vitro propagation of *Phlox subulata* and *Phlox paniculata*. HortScience 14:607-608.

Seibert, M. 1976. Shoot initiation from carnation shoot apices frozen to -196 C. Science 191:1178-1179.

_____ and Wetherbee, P.J. 1977. Increased survival and differentiation of frozen herbaceous plant organ cultures through cold treatment. Plant Physiol. 59:1043-1046.

Shabde, M. and Murashige, T. 1977. Hormonal requirements of excised *Dianthus caryophyllus* L. shoot apical. Am. J. Bot. 64:443-448.

Skirvin, R.M. and Chu, M.C. 1979. In vitro propagation of "Forever Yours" rose. HortScience 14:608-610.

_____, and Rukan, H. 1979. The culture of peach, sweet and sour cherry, and apricot shoot tips. Ill. State Hort. Soc. 113:30-38.

_____, and Gomez, E. 1981. In vitro propagation of thornless trailing Blackberries. HortScience 16:310-312.

Smith, R.H. and Murashige, T. 1970. In vitro development of isolated shoot apical meristems of angiosperms. Am. J. Bot. 57:562-568.

Snir, I. and Erez, A. 1980. In vitro propagation of Malling Merton apple rootstocks. HortScience 15:597-598.

Stevenson, J.H. and Harris, R.E. 1980. In vitro plantlet formation from shoot-tip explants of *Fuchsia hybrida* cv. Swingtime. Can. J. Bot. 58: 2190-2199.

Stone, O.M. 1963. Factors affecting the growth of carnation plants from shoot apices. Ann. Appl. Biol. 52:199-209.

Sutter, E. and Langhans, R.W. 1979. Epicuticular wax formation on carnation plantlets regenerated from shoot tip culture. J. Am. Soc. Hort. Sci. 104:493-496.

Tabachnik, L. and Kester, D.E. 1977.   Shoot culture for almond-peach hybrid clones in vitro. HortScience 12:545-547.

Takayama, S. and Misawa, M. 1980.   Differentiation in *Lilium* bulbscales grown in vitro.   Effects of activated charcoal, physiological age of bulbs and sucrose concentration on differentiation and scale leaf formation in vitro. Physiol. Plant. 48:121-125.

Theiler, R. 1977. In vitro culture of shoot tips of *Pelargonium* species. Acta Hort. 78:403-414.

Thimann, K.V. 1977.   Hormone Action in the Whole Life of Plant, Univ. of Massachusetts Press, Amherst.

Thorpe, T.A. and Biondi, S. 1983.   Conifers.   In: Vol. 2 of this series. Macmillan, New York.

Towill, L.E. 1981. *Solanum etuberosum*: A model for studying the cryobiology of shoot-tips in the tuber-bearing *Solanum* species.   Plant Sci. Lett. 20:315-324.

Ulrychova, M. and Petru, E. 1975.   Elimination of mycoplasma in tobacco callus tissues (*Nicotiana glauca* Grah.) cultured in vitro in the presence of 2,4-D in nutrient medium. Biol. Plant. 17:103-108.

Vine, S.J. and Jones, O.P. 1969.   The culture of shoot tips of hops (*Humulus lupulus* L.) to eliminate viruses. J. Hort. Sci. 44:281-284.

Walkey, D.G.A. 1968.   The production of virus free rhubarb by apical tip-culture. J. Hort. Sci. 43:283-287.

_____ 1972.   Production of apple plantlets from axillary-bud meristems. Can. J. Plant Sci. 52:1085-1087.

_____ and Cooper, J. 1976.   Growth of *Stellaria media, Capsella bursa-pastoris* and *Senecio vulgaris* plantlets from cultured meristem-tips. Plant Sci. Lett. 7:179-186.

_____, Neely, H.A., and Crisp, P. 1980.   Rapid propagation of white cabbage by tissue culture. Sci. Hort. 12:99-107.

Wang, P.J. 1977.   Regeneration of virus-free potato from tissue culture. In: Plant Tissue Culture and its Bio-technological Application (W. Barz, E. Reinhard, and M.H. Zenk, eds.) pp. 386-391.   Springer-Verlag, Berlin and Heidelberg.

_____ 1978.   Clonal multiplication of *Cryptomeria japonica* D. Don in vitro.   In: Studies and Essays in Commemoration of the Golden Jubilee of Academia Sinica, Taipei, Taiwan pp. 559-566.

_____ and Hu, C.Y. 1982.   In vitro mass tuberization and virus-free seed potato production in Taiwan. Am. Potato J. 59:33-39.

_____ and Huang, L.C. 1974.   Studies on the shoot meristem culture of *Allium sativum* L. Chinese Hort. 20:79-87.

Waterworth, P. and Kahn, R.P. 1978.   Thermotherapy and aseptic bud culture of sugar cane to facilitate the exchange of germplasm and passage through quarantine. Plant. Dis. Rep. 62:772-776.

Werner, E.M. and Boe, A.A. 1980.   In vitro propagation of Milling 7 apple rootstock. HortScience 15:509-510.

White, P.R. 1934.   Multiplication of the viruses of tobacco ancuba mosaic in growing excised tomato roots. Phytopathol. 24:1003-1011.

_____ 1963.   The Cultivation of Animal and Plant Cells, p. 59.   Ronald Press, New York.

Whitehead, H.C.M. and Giles, K.L. 1977.   Rapid propagation of poplars by tissue culture methods. New Zealand J. For. Sci. 7:40-43.

Wochok, Z.S. and Sluis, C.J. 1980a. Gibberellic acid promotes *Atriplex* shoot multiplication and elongation. Plant Sci. Lett. 17:363-369.

_____ 1980b. In vitro propagation and establishment of wax currant (*Ribes inebrians*). J. Hort. Sci. 55:355-357.

Yang, Y.-W., Hsing, Y.-I., and Chang, W.-C. 1981. Clonal propagation of *Stevia rebaudiana* Bertoni through axillary shoot proliferation in vitro. Bot. Bull. Academia Sinica 22:57-62.

Yie, S.T. and Liaw, S.I. 1977. Plant regeneration from shoot tips and callus of papaya. In Vitro 13:564-568.

Ziv, M. 1979. Transplanting *Gladiolus* plants propagated in vitro. Sci. Hort. 11:257-260.

CHAPTER 6
In vitro Production of Haploids

Y.P.S. Bajaj

As early as 1922 Blakeslee et al. (1922), while working with *Datura stramonium*, reported the natural occurrence of haploids. The term "haploid" refers to those plants which possess the gametophytic number of chromosomes in their sporophytes. Since this first report, the number of plant species in which haploids have been reported to occur spontaneously by parthenogenesis or have been induced experimentally by various techniques has increased substantially. Following the publication of extensive reviews on haploids by Kimber and Riley (1963) and Magoon and Khanna (1963), a tremendous amount of interest and progress in the field of experimental induction of haploids in higher plants developed, especially in the area of in vitro induction of androgenesis by anther culture (Guha and Maheshwari, 1964).

This interest in haploids stems largely from their considerable potential for plant breeding (Melchers and Labib, 1970; Kasha, 1974; Reinert and Bajaj, 1977; Hlasnikova, 1977). Haploids may be utilized to facilitate the detection of mutations and the recovery of unique recombinants. Since most mutations are recessive, they are difficult to detect in the presence of an unmutated dominant gene. Haploids possess only one set of alleles at each locus, so it is possible for recessive mutants to be detected. Furthermore, doubling of the chromosome number of haploids offers a method for the rapid production of homozygous plants, which in turn may be used for producing inbred lines for hybrid production.

Haploids may be grouped into two broad categories: (a) monoploids (monohaploids), which possess half the number of chromosomes from a diploid species, and (b) polyhaploids, which possess half the number of chromosomes (gametophytic set) from a polyploid species. To go into the details of classification of haploids is beyond the scope of this article (see Kimber and Riley, 1963). However, the general term "haploid" is applied to any plant originating from a sporophyte and containing half the number of chromosomes.

Haploids may occur spontaneously in nature, or they may be induced experimentally. Spontaneous haploids occur as a result of apomixis or parthenogenesis. In such cases the unfertilized egg, the sperm, or the synergids start to grow to form a haploid plant independent of any experimentally applied stimulus. Induced haploids can generally be obtained by the stimulation of the egg, synergids, or sperm by a number of methods, including ionizing irradiation and radioisotopes, thermal shocks, distant hybridization, delayed pollination, application of abortive pollen, spraying with various chemicals (see Kimber and Riley, 1963; Magoon and Khanna, 1963), the in vitro culture of the excised anthers, culture of isolated pollen and protoplasts, chromosome elimination by culture of young embryos, and in vitro parthenogenesis. The in vitro methods are discussed below.

IN VITRO METHODS FOR THE INDUCTION OF HAPLOIDS

Anther Culture

The technique of in vitro culture of anthers is relatively simple, quick, and efficient. By culturing excised anthers, (Fig. 1) haploid tissue or plants have been obtained in a number of species (Table 1).

In view of the critical importance of the stage of development of the anther for the induction of androgenesis, it is worthwhile to mention the structure of the anther and the development of the pollen. An Angiosperm stamen (Fig. 2a) comprises a filament, connective tissue, and anther. A typical anther (Maheshwari, 1950) in cross section (Fig. 2b) shows two anther lobes, and each lobe possesses two microsporangia or pollen sacs. At maturity the two pollen sacs of each lobe become confluent. During microsporogenesis in a young anther, there are four patches of primary sporogenous tissue which either function directly as pollen mother cells (PMCs) or undergo several divisions. The PMCs form pollen tetrads by meiosis. These are arranged in various patterns; tetrahedral and isobilateral are the most common. On secretion of a callose from the somatic tissue, the callose wall of the pollen tetrads dissolves and the four microspores are liberated. The newly released microspore is densely cytoplasmic, with a centrally located nucleus. As vacuolation occurs, the volume of the microspore increases rapidly. The nucleus is then displaced and pushed toward the periphery. At the first division (first mitosis) the microspore nucleus produces a large vegetative and a small generative nucleus. The second mitosis, which is restricted to the generative nucleus, forms two sperms and takes place in either the pollen or the pollen tube. The mode of division of the pollen nuclei is diagrammatically represented in Fig. 3a. The microspores in the microsporangia are surrounded (from inside to outside) by tapetum, middle layers, endothecium, and epidermis.

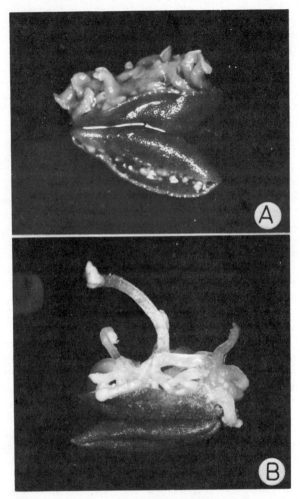

Figure 1.   In vitro culture of excised anthers of tobacco, undergoing androgenesis after 4 and 6 weeks, respectively.

PROTOCOL FOR ANTHER CULTURE.   Tobacco is the ideal material for the induction of haploids.   The immature anthers containing uni-nucleate pollen at the time of first mitosis are the most suitable material for the induction of androgenesis.   The flower buds are sterilised with 1% (w/v) solution of sodium hypochlorite, or 5% (v/v) solution of a commercially available disinfectant such as Clorox.   They are washed a couple of times with sterile distilled water, and the anthers are dissected out and cultured on agar-solidified medium.   The flower buds obtained from plants grown in greenhouses, if dissected carefully, need no sterilization.   An excision is made on one side of the flower bud, and the stamens are gently squeezed out and collected in a sterile petri dish.   Normally 5-10 anthers are inoculated in a culture vessel.   During excision of anthers special care should be taken to ensure that they are not injured in any way.   Damaged anthers

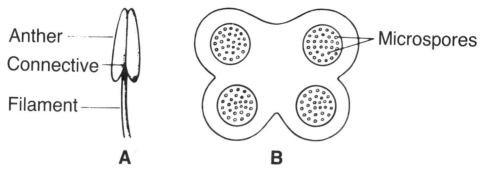

Figure 2.   (a) A stamen comprising filament, connective tissue and anther.   (b) An outline of a cross section of an anther showing four pollen sacs containing uninucleate microspores (haploid cells).

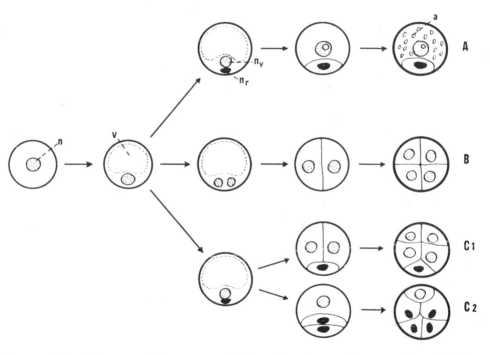

Figure 3.   Diagrammatic representation showing various modes of division of the microspores under in vivo (a) and in vitro (b, c_1, c_2) conditions (a = starch, n = nucleus, nr = generative nucleus, nv = vegetative nucleus, v = vacuole) (after Nitsch, 1972; courtesy of Professor E.C. Cocking).

should be discarded, as they often tend to produce callus from parts other than the pollen.

Basal medium of MS (Murashige and Skoog, 1962), White (1963), and Nitsch and Nitsch (1969), or their slight modification with various additives have been employed, but iron is indispensible (Nitsch, 1972).   The

Table 1. Plant Species in which Haploid Tissue, Embryos, or Plants Have Been Obtained by the Culture of Excised Anthers/Pollen

PLANT SPECIES	FAMILY	MODE OF DEVELOP-MENT[1]	REFERENCE
Aegilops caudata x A. umbellulata	Gramineae	C,P	Kimata & Sakamoto, 1972
Aesculus hippocastanum	Hippocastanaceae	E	Radojevic, 1978
Agropyron repens	Poaceae	C,E	Zenkteler et al., 1975
Anemone spp.	Ranunculaceae	C,E	Sunderland & Dunwell, 1977; Johansson & Eriksson, 1977; Georgiev & Chavdarov, 1974
Arabidopsis thaliana	Crucifereae	C,P	Gresshoff & Doy, 1972b; Amos & Scholl, 1978; Scholl & Amos, 1980
Arachis glabrata; A. hypogaea, A. villosa	Leguminoceae	C,E,P	Bajaj et al., 1980a, 1981; Mroginsky & Fernandez, 1980
Asparagus officinalis	Liliaceae	C,E,P	Pelletier et al., 1972; Hondelmann & Wilberg, 1973
Atropa belladonna	Solanaceae	C,E,P	Zenkteler, 1971; Narayanaswamy & George, 1972; Rashid & Street, 1973, 1974a; Bajaj et al., 1978
Beta vulgaris	Chenopodiaceae	C	Atanasov, 1973
Brassica campestris	Cruciferaceae	C,P	Keller et al., 1975; Keller & Armstrong, 1979
B. napus	Cruciferaceae	C,E,P	Thomas & Wenzel, 1975a; Keller & Armstrong, 1977
B. oleracea	Cruciferaceae	C,P	Kameya & Hinata, 1970; Quazi, 1978
B. oleracea x B. alboglabra	Cruciferaceae	C,P	Kameya & Hinata, 1970
Bromis inermis	Gramineae	E	Zenkteler et al., 1975
Cajanus cajan	Leguminoceae	C,E	Bajaj et al., 1980b

PLANT SPECIES	FAMILY	MODE OF DEVELOP-MENT[1]	REFERENCE
Capsicum annuum	Solanaceae	C	Kuo et al., 1973; Wang et al., 1973; George & Narayanaswamy, 1973; Novak, 1974
Capsicum frutescens	Solanaceae	C	Novak, 1974
Cassia fistula	Leguminoceae	C,E	Bajaj et al. (unpublished)
Cicer arietinum	Leguminoceae	C,E	Gosal & Bajaj (unpublished)
Citrus limon	Rutaceae	C,E	Drira & Bendalis, 1975
C. medica	Rutaceae	C,E	Drira & Bendalis, 1975
Coffea arabica	Rubiaceae	C,E	Sharp et al., 1973
Corchorus	Tiliaceae	C	Iyer & Raina, 1972
Chrysanthemum	Compositeae	C	Watanabe et al., 1972
Datura innoxia	Solanaceae	E,C,P	Guha & Maheshwari, 1964; Norreel, 1970; Engvild et al., 1972; Tyagi et al., 1979; Forche et al., 1981
D. metel	Solanaceae	E	Narayanaswamy & Chandy, 1971; Iyer & Raina, 1972; Gupta & Babbar, 1980
D. meteloides	Solanaceae	E,C,P	Kohlenbach & Geier, 1972; Nitsch, 1972; Geier & Kohlenbach, 1973
D. muricata	Solanaceae	E	Nitsch, 1972
D. stramonium	Solanaceae	E	Guha & Maheshwari, 1967
D. wrightii	Solanaceae	E	Kohlenbach & Geier, 1972
Digitalis purpurea	Scrophulariaceae	C,P	Corduan & Spix, 1975
Festuca arundinacea	Gramineae	C,P	Niizeki & Kita, 1974; Kasperbauer et al., 1980
F. pratensis	Gramineae	E	Nitzsche, 1970; Zenkteler & Misiura, 1974

Table 1. Cont.

234

PLANT SPECIES	FAMILY	MODE OF DEVELOP-MENT[1]	REFERENCE
Fragaria virginiana	Rosaceae	C	Fowler et al., 1971; Rosati et al., 1975
Freesia spp.	Iridaceae	C,P	Bajaj & Pierik, 1974
Gladiolus	Iridaceae	C	Bajaj et al., 1982
Glycine max	Leguminoceae	C	Ivers et al., 1974
Gossypium spp.	Malvaceae	C,E	Bajaj, 1982
Helleborus foetidus	Ranunculaceae	E	Zenkteler et al., 1975
Hevea brasiliensis	Euphorbiaceae	E,C	Satchuthananthavale & Irugalband-ara, 1972; Chen et al., 1979
Hordeum vulgare	Gramineae	E,C,P	Clapham, 1973, 1977; Zenkteler & Misiura, 1974; Kao, 1981; Xu et al., 1981
Hyoscyamus albus	Solanaceae	E	Raghavan, 1975; Dodds & Reynolds, 1980; Sunderland & Wildon, 1979
H. muticus	Solanaceae	E,C,P	Wernicke et al., 1979
H. niger	Solanaceae	E,C,P	Raghavan, 1975, 1978; Corduan, 1975
H. pusillus	Solanaceae	E	Raghavan, 1975
Iberis amara	Crucifereae	C,E,P	Babbar et al., 1980
Lilium longiflorum	Liliaceae	C,P	Sharp et al., 1971a
Lolium multiflorum	Gramineae	C,P	Clapham, 1971
L. multiflorum x *Festuca arundinacea*	Gramineae	C,P	Nitzsche, 1970
L. perenne	Gramineae	C,P	Clapham, 1971
Lotus corniculatus	Leguminoceae	C	Niizeki & Grant, 1971
Luffa cylindrica	Cucurbitaceae	C	Sinha et al., 1978a
L. echinata	Cucurbitaceae	F	Sinha et al., 1978b

PLANT SPECIES	FAMILY	MODE OF DEVELOPMENT[1]	REFERENCE
Lycium halimifolium	Solanaceae	E	Zenkteler, 1972
Lycopersicon esculentum	Solanaceae	C,P	Sharp et al., 1971b; Gresshoff & Doy, 1972a; Levenko et al., 1977; Zamir et al., 1980, 1981
L. peruvianum	Solanaceae	C	Gresshoff & Doy, 1972a
L. pimpinellifolium	Solanaceae	C,E	Nitsch & Nitsch, 1969; Debergh & Nitsch, 1973
L. esculentum x *L. peruvianum*	Solanaceae	C,P	Cappadocia & Ramulu, 1980
Malus	Rosaceae	C	Kubicki et al., 1975
Nicotiana affinis	Solanaceae	E,P	Nitsch & Nitsch, 1969
N. alata	Solanaceae	E	Nitsch, 1969, 1972
N. attenuata	Solanaceae	E	Collins & Sunderland, 1974
N. clevelandii	Solanaceae	E	Vyskot & Novak, 1974
N. glutinosa	Solanaceae	E	Nitsch, 1969
N. knightiana	Solanaceae	E	Collins & Sunderland, 1974
N. langsdorffii	Solanaceae	E	Durr & Fleck, 1980
N. otophora	Solanaceae	E	Collins et al., 1972
N. plumbaginifolia	Solanaceae	E,P	Tran Than Van & Trinh, 1980
N. raimondii	Solanaceae	E	Collins & Sunderland, 1974
N. rustica	Solanaceae	E	Nitsch & Nitsch, 1969
N. sanderae	Solanaceae	E	Vyskot & Novak, 1974
N. suaveolens x *N. langsdorfii*	Solanaceae	C	Guo, 1972
N. sylvestris	Solanaceae	E,C,P	Bourgin & Nitsch, 1967; Rashid & Street, 1974b; Butterfass & Kohlenbach, 1979; De Paepe et al., 1981
N. tabacum	Solanaceae	E,C,P	Nitsch & Nitsch, 1969, 1970; Bajaj, 1972, 1978b; and numerous others

235

Table 1. Cont.

PLANT SPECIES	FAMILY	MODE OF DEVELOP-MENT[1]	REFERENCE
Oryza sativa	Gramineae	C,E,P	Niizeki, 1968; Nishi & Mitsuoka, 1969; Wang et al., 1974; Chen & Chen, 1980; Chaleff & Stolarz, 1981; Chang & Hong-Yuan, 1981
Paeonia hybrida	Ranunculaceae	C,E	Sunderland, 1974
P. lactifolia	Ranunculaceae	E	Ono & Tsukide, 1978
P. lutea	Ranunculaceae	E	Zenkteler et al., 1975
P. suffruticosa	Ranunculaceae	E	Zenkteler et al., 1975
Pelargonium hortorum	Geraniaceae	C,P	Abo El-Nil & Hildebrandt, 1971, 1973; Abo El-Nil et al., 1976
Petunia axillaris	Solanaceae	C,P	Engvild, 1973; Doreswamy & Chacko, 1973
P. hybrida	Solanaceae	C,P	Iyer & Raina, 1972; Binding, 1972; Bajaj, 1978b; Mitchell et al., 1980
P. hybrida x *P. axillaris*	Solanaceae	C,P	Raquin & Pilet, 1972
Pharbitis nil	Convolvulaceae	E	Sangwan & Norreel, 1975
Phaseolus aureus	Leguminoceae	E,C	Bajaj & Singh, 1980
P. vulgaris	Leguminoceae	C	Peters et al., 1977
Phleum pratense	Gramineae	C	Niizeki & Kita, 1973
Pisum sativum	Leguminoceae	E,C	Bajaj & Gosal (unpublished)
Poinciana regia	Leguminoceae	E,C	Bajaj et al. (unpublished)
Populus spp.	Salicaceae	C,P	Sato, 1974
Primula obconica	Primulaceae	C,P	Bajaj, 1981b
Prunus amygdalus	Rosaceae	C	Michellon et al., 1974
P. avnals	Rosaceae	C	Seirlis et al., 1979
P. armenica	Rosaceae	C	Harn & Kim, 1972

PLANT SPECIES	FAMILY	MODE OF DEVELOPMENT[1]	REFERENCE
P. avium	Rosaceae	E,C	Jordan, 1974; Zenkteler et al., 1975
P. persica	Rosaceae	C	Michellon et al., 1974
Saccharum spontaneum	Gramineae	C,P	Fitch & Moore, 1981
Saintpaulia ionantha	Gesneriaceae	C,E,P	Hughes et al., 1975; Weatherhead et al., 1982
Scopolia carnicolica	Solanaceae	E	Wernicke & Kohlenbach, 1975
S. lurida		E	Wernicke & Kohlenbach, 1975
S. physaloides		E	Wernicke & Kohlenbach, 1975
Secale cereale	Gramineae	C,E,P	Wenzel & Thomas, 1974; Thomas & Wenzel, 1975b; Wenzel et al., 1975, 1977
S. montanum	Gramineae	E	Zenkteler & Misiura, 1974
Setaria italica	Gramineae	C	Ban et al., 1971
Solanum dulcamara	Solanaceae	E	Zenkteler, 1973
S. mamosum		C	Anand & Govindappa, 1979
S. melongena		C,P	Raina & Iyer, 1973; Guy et al., 1979
S. nigrum		C,P	Harn, 1971, 1972
S. tuberosum		C,E,P	Irikura & Sakaguchi, 1972, 1975; Dunwell & Sunderland, 1973; Sopory et al., 1978; Sopory & Tan, 1979
S. surattense		E,C	Sinha et al., 1978b, 1979
S. verrucosum		C,P	Weatherhead & Henshaw, 1979
Tradescantia reflexa	Commelinaceae	C	Yamada et al., 1963
Trifolium alexandrinum	Leguminoceae	C,P	Mokhtarzadeh & Constantin, 1978
T. pratense		C,P	Mokhtarzadeh & Constantin, 1978

Table 1. Cont.

PLANT SPECES	FAMILY	MODE OF DEVELOP-MENT[1]	REFERENCE
Triticale	Gramineae	C,P	Wang et al., 1973; Ono & Larter, 1976; Bernard, 1980
Triticum aegilopoides	Gramineae	C	Fujii, 1970
T. aestivum		C,P	Ouyang et al., 1973; Picard & de Buyser, 1973; Wang et al., 1973; Schaeffer et al., 1979; de Buyser & Henry, 1980; Shimada, 1981
T. dicocoides		C	Fujii, 1970
Ulmus americana	Ulmaceae	C	Redenbaugh et al., 1981
Vicia faba	Leguminaceae	C	Hesemann, 1980
Vigna unguiculata	Leguminaceae	C,S	Ladeinde & Bliss, 1977
Vitis vinifera	Vitaceae	C,P	Gresshoff & Doy, 1974; Hirabayashi et al., 1976; Rajasekaran & Mullins, 1979
Withania somnifera	Solanaceae	C,P	Vishnoi et al., 1979
Zea mays	Gramineae	C,P	Murakami et al., 1972; Opatrny et al., 1977; Nitsch, 1977

[1]C = Callus, E = Embryos, P = Plants.

usual level of sucrose is 2-4%; however, higher concentrations (8-12%) favor androgenesis in cereals (Clapham, 1971; Chen, 1978). Media rich in growth regulators encourage the growth of diploid tissue such as anther wall, connective tissue, and filament and should be avoided. The cultures are incubated at 24-28 C in a 14-hr daylight regime at about 2000 lux.

The anthers normally start to undergo pollen embryogenesis within 2 weeks, and either directly develop into haploid plants in about 6 weeks (in tobacco) and are capable of transferring to the soil, or undergo proliferation to form callus which can be induced to differentiate plants.

To obtain homozygous lines, the young plants that are still enclosed in or attached to the cultured anthers are treated with a 0.5% solution of colchicine for 24-28 hr, washed thoroughly, and separately planted.

INDUCTION OF ANDROGENESIS. The knowledge gained so far from anther and pollen culture has established that pollen at the uninucleate stage (Fig. 4a) just before the first mitosis or during mitosis are most susceptible to external stimuli for the induction of androgenesis. The formation of plantlets varies greatly between plant species. In tobacco, for instance, it takes about 3-5 weeks before the embryos are visible bursting out of the anthers, while in *Atropa* and rice it may take up to 8 weeks. Haploid plantlets are formed in two distinct ways, by direct androgenesis (i.e., embryos originating directly from the microspores without callusing) or by organogenesis from haploid callus tissue.

Androgenesis in tobacco is described here in some detail. The anthers (containing uninucleate pollen) cultured on basal medium (Nitsch and Nitsch, 1969) turn brownish within 2 weeks without any visible signs of growth. After 3-4 weeks in culture, small white protuberances, the embryos, appear which eventually develop into plantlets (Fig. 1). Frequently, the anthers may also show proliferation at their basal end which continues to increase and cover the whole of the anther. Occasionally this callus differentiates to form plantlets.

Microscopic observations reveal that within the excised anthers the microspores exhibit various modes of development (Figs. 3,4). The microspore nucleus either undergoes a normal mitosis and forms a vegetative and a generative nucleus or divides to form two "similar looking" nuclei (Fig. 3b). In some pollen the vegetative nucleus, while in others the two "similar looking" nuclei, undergo further repeated divisions to form multinucleate pollen. The generative nucleus remains quiescent or divides a couple of times and aborts; however, in *Hyoscyamus niger* (Raghavan, 1975, 1978) the generative nucleus is actively involved in androgenesis. As a result of repeated division (Fig. 4g,k,l), pollen with up to 30 nuclei may be formed with no walls separating the nuclei. Such pollen do not take part in androgenesis and generally abort.

In contrast to the formation of multinucleate pollen, the microspore may undergo direct segmentation (Fig. 4d), which continues until eventually a 40-50 celled proembryo is formed. The embryos, mostly at the globular stage, burst out of the confines of the exine and are released. The embryos undergo various stages of development (Fig. 4h-j) simulating those of the normal zygotic embryo. Finally, the cotyledons unfold, and the plantlets emerge out of the anthers in 4-5 weeks. The percentage of androgenic anthers and the number of plantlets per anther vary greatly, depending on the cultivar, medium, and various other endogenous and exogenous factors (discussed later). The haploid plantlets develop normally when transferred to pots, but are smaller than the diploids.

At the early stages of development it is not possible to differentiate between the multinucleate type of development and that which leads to the formation of callus. However, after about 2 weeks, pollen are observed to be larger than the multinucleate and the embryonal type and contain only a few cells. These cells increase in size, exerting pressure on the exine which bursts open and the contents are released in the form of callus (Fig. 4g). At the same time, in some cases, shoots differentiate from the callus. The plantlets originating from the callus are generally undesirable, as they exhibit various levels of ploidy.

The ultrastructural studies (Vazart, 1972; Dunwell and Sunderland, 1974a,b, 1975) on tobacco anthers have yielded some interesting insight into the sequence of early androgenesis. The embryogenesis begins after a short period of normal gametophytic differentiation, which proceeds at a similar rate to that in vivo during the first days of culture but thereafter the gametophytic processes gradually terminate. After 7-8 days in culture the vegetative cell of the embryonal pollen shows multivesiculate bodies resembling lysosomes. These zones increase with increasing age of the culture, and after about 12 days they are almost free of any contents. The ribosomes and many other organelles are virtually eliminated from the cells. The remaining organelles, mainly plastids, cluster around the vegetative nucleus. There is little change in the mitochondria, and the gametophytic cytoplasm seems to be destroyed before the first sporophytic division of the vegetative cell. The ultrastructural changes during androgenesis seem to correspond well with the observations on the pattern of nucleic acid changes (Bhojwani et al., 1973). It was observed that in embryonic pollen treated with specific stains for RNA and proteins, most of the stain taken up at the stages of the first sporophytic mitosis was located in the two nuclei and some cytoplasm surrounding them, while the rest of the pollen

Figure 4. (a) A uninucleate tobacco microspore (stage at inoculation). (b) Microspore showing first mitosis with two similar-looking nuclei, obtained from 1-week-old anthers. (c) A multinucleate pollen from 2-week-old culture; such pollen generally abort. (d) A bicelled embryonal microspore formed as a result of segmentation. (e) A multicellular pollen which appears to be of the callusing type. (f) A multicellular pollen (embryoid) obtained from 6-week-old anthers of *Triticum aestivum* cv. Kolibri. (g) A small mass of haploid callus obtained as a result of bursting of the pollen; note the small compact cells. (h-j) Various stages of embryogenesis in tobacco anthers. (k,l) Four and eight nucleate pollen obtained from 2-3-week-old cultures of anthers of *Paeonia suffruticosus* and *P. lutea* cv. Superba respectively (courtesy of Dr. M. Zenkteler).

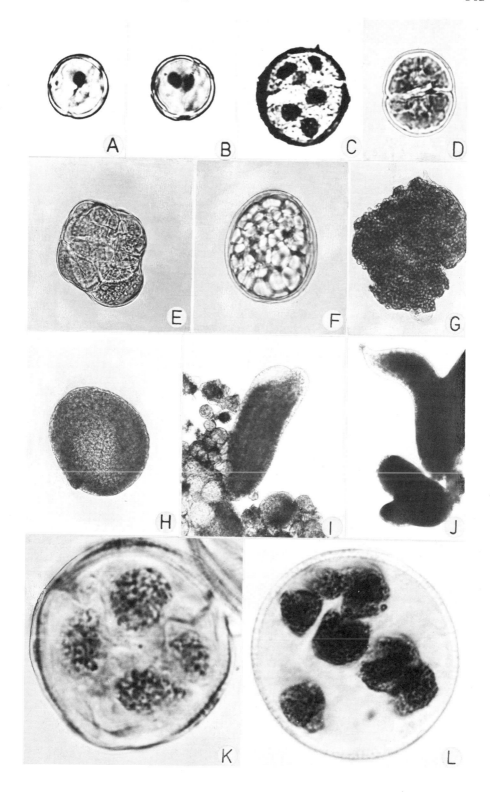

lacked stainable contents. Furthermore, there was less RNA per grain at the second mitosis than in a microspore at the first mitosis. The electron microscopic studies also confirmed that the generative cell is not involved in the process of androgenesis in tobacco, as it remains small, inert, and shows little change.

Culture of Isolated Pollen

Although anther culture has proved to be quite efficient for the induction of haploids, it has one main disadvantage: the plants not only originate from pollen but also from various other parts of the anther, with the result that a population of plants with various ploidy levels is obtained (Nishi and Mitsuoka, 1969; Devreux et al., 1971; Narayanaswamy and Chandy, 1971; Zenkteler, 1971; Engvild et al., 1972; Kohlenbach and Geier, 1972; Wagner and Hess, 1974; Thomas and Wenzel, 1975b; Bajaj et al., 1978; Bajaj, 1981a). This difficulty can be circumvented by the culture of isolated pollen. The culture of isolated pollen offers the following additional advantages:

1. Uncontrolled effects of the anther wall and other associated tissue are eliminated and various factors governing androgenesis can be better regulated.
2. The sequence of androgenesis can be observed starting from a single cell.
3. Pollen is ideal for uptake, transformation, and mutagenic studies, as pollen may be evenly exposed to chemicals or physical mutagens.
4. Unlike single callus cells, pollen is transformed directly into an embryo, and thus would be most suitable for understanding the physiology and biochemistry of androgenesis.
5. Higher yields of plants per anther could be expected.

Earlier work on the culture of pollen of Gymnosperms demonstrated their ability to divide. With the use of pollen culture haploid tissues have been obtained in *Ginkgo biloba* (Tulecke, 1953), *Ephedra foliata* (Konar, 1963), *Torreya nucifera* (Tulecke and Sehgal, 1963), *Cupressus* (Razmologov, 1973), and *Pinus* (Bonga, 1974; Bonga and McInnis, 1975). These experiments demonstrated that isolated pollen can be induced to change its primary role of germination to give rise to pollen tube and sperm. Instead, its mode of development can be diverted to produce haploid tissue. This work was extended to Angiosperms to yield some interesting results.

Kameya and Hinata (1970) used a hanging drop method to culture isolated pollen of *Brassica oleracea* and *B. oleracea* x *B. alboglabra*. Their method involved placing a drop of medium containing 50–80 grains on the cover glass, which is then inverted over a cavity slide and sealed with paraffin. Before inverting the cover glass, a column of paraffin is raised in the center, so that when inverted, it touches the bottom of the cavity slide. This facilitates the aeration as well as the movement of the pollen when the slide is rotated. Cell clusters were formed from isolated pollen after 4 weeks in a medium containing coco-

nut water. Later, Sharp et al. (1972) using a "nurse culture technique"
induced isolated pollen of *Lycopersicon esculentum* to form haploid
callus (Fig. 5).

Making use of the observations (Stow, 1930; Sax, 1935) that thermal
shocks given to plants can change the mode of division of the pollen
nuclei, Nitsch and Norreel (1973a) reported that trauma given to pollen
at the time of the first mitosis considerably enhanced the number of
pollen cells undergoing androgenesis. They succeeded in inducing embryo formation by subjecting *Datura innoxia* pollen to low temperatures
(4 C for 48 hr), and then growing them in a liquid medium. They also
claimed (Nitsch and Norreel, 1973b) to have improved the efficiency of
androgenesis by growing isolated pollen in a medium supplemented with
water extract from anthers undergoing androgenesis. Partial success
along the same lines was also reported in tomato (Debergh and Nitsch,
1973). Later Nitsch (1974a) substituted a fully synthetic medium for
the anther extract and obtained plantlets from *Nicotiana tabacum* cv.
Red Flowered and Coulo. This work was then extended to another
commercially important tobacco cultivar, Badischer Burley (Bajaj and
Reinert, 1975; Reinert et al., 1975; Bajaj et al., 1977; Bajaj, 1978a),
potato (Sopory, 1977; Weatherhead and Henshaw, 1979), *Hyoscyamus
niger* (Wernicke and Kohlenbach, 1977), *Datura innoxia* (Sangwan-
Norreel, 1977), *Cajanus cajan* (Bajaj et al., 1980b), and rye (Wenzel et
al., 1975).

The isolated pollen can be cultured by the methods which follow.

NURSE CULTURE TECHNIQUE. Sharp et al. (1972) induced the isolated pollen of *Lycopersicon esculentum* to form haploid callus. A
schematic protocol for the establishment of cultures is presented in
Fig. 5. In this method the anthers are placed horizontally on top of
the basal medium within a French square container. A filter paper
disk is placed over the intact anther, and about 10 pollen grains (in
suspension) are then placed on the filter paper disk. The controls are

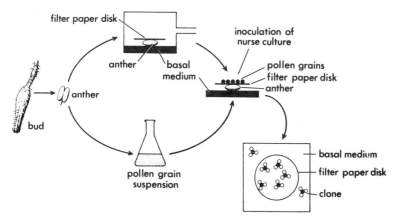

Figure 5. Schematic representation for the culture of isolated pollen
of tomato by nurse culture method (from Sharp et al., 1972; courtesy of
Dr. W.R. Sharp).

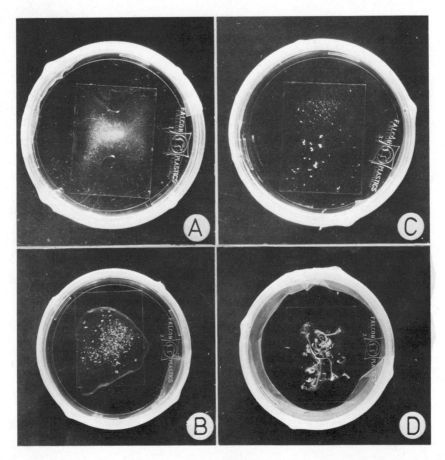

Figure 6. Various stages in the development of haploid tobacco plants from isolated pollen culture. (a) A drop of pollen suspension at the time of culture in a glutamine (800 mg/l) and serine (100 mg/l) rich medium. (b,c,d) Same, after 2-5 weeks showing pollen embryos and pollen plants (from Bajaj, 1978a).

prepared in exactly the same way, except that the pollen are placed on a filter paper disk kept directly on the agar. With this method the control did not grow at all, while the pollen kept on the nurse cultures had a plating efficiency of up to 60%, and clusters of green parenchymatous cells were formed on the filter paper disk in 2 weeks. These clones were observed to be haploid. Later, Pelletier (1973) successfully induced androgenesis in *Nicotiana tabacum* microspores using *Petunia* callus as a nurse tissue.

PRECULTURE METHOD. The technique for the culture of isolated pollen (Nitsch, 1974a,b, 1977) involves the following steps. The anthers are aseptically removed and cultured (Nitsch and Nitsch, 1969). After 4 days of incubation at 27 C, the pollen are squeezed out of the anther into the liquid medium. One liter of the medium contains KNO_3 (8.9 mM), $NH_4 NO_3$ (8.9 mM), $MgSO_4 \cdot 7H_2O$ (0.75 mM), $CaCl_2$ (1.42 mM), KH_2PO_4 (0.50 mM), FeEDTA (100 mM), glutamine (5.5 mM), serine (0.95 mM), Myoinositol (0.03 M), Zeatin (0.046 µM), indoleacetic acid (0.57 µM), and sucrose (0.06-0.23 M). After filtration the pollen are centrifuged to form a pellet. The pollen are washed twice, centrifuged again, and suspended in fresh medium with a density of about 5×10^4 pollen per ml. Aliquots of 2 ml are then dispensed in thin layers in small petri dishes or 25 ml Erlenmeyer flasks. However, if the material is available in microquantities, it can be grown in drop cultures (Bajaj 1978a; Fig. 6). To prepare such cultures, a drop of silicon is placed in the center of a small (5 cm) sterile plastic petri dish, and a cover slip (22 x 22 mm) is gently lowered onto the drop. Then a drop of 250-500 µl of the pollen suspension is pipetted onto the cover slip. To prevent dessication, the petri dishes are sealed with "Parafilm M" and the cultures incubated for the first 4-6 days under low light (500 lux), and then maintained in 14-hr daylight regime of 2000 lux at 28 C. The induced pollen undergo normal androgenesis and eventually produce haploid plants. The culture of excised anthers and pollen, and the formation of haploid and homozygous plants are schematically represented in Fig. 7.

Recently these techniques have been considerably modified and refined to obtain high percentages of pollen embryogenesis. For instance, rye pollen suspension may be purified to separate the viable microspores (Wenzel et al., 1975). Likewise, androgenesis can be induced in ab initio pollen cultures of tobacco (Rashid, 1981). The serial culture of anthers in the liquid medium enables the embryogenic pollen to discharge in the medium in good quantities. Such pollen can then be filtered and cultured (Sunderland and Roberts, 1977).

Although considerable success has been achieved with the culture of pollen, the induction of repeated nuclear and cellular divisions in isolated pollen mother cells (Bajaj, 1974a) and pollen tetrads (Bajaj, 1975a) of Atropa belladonna offers an alternative approach. The PMCs enclosed by the thick callose walls exhibit two types of behavior in culture. They either formed 6-8 additional microspores or they underwent repeated divisions to form a multicellular globular embryo-like structure (Fig. 8a). The frequency of these structures was increased if young anthers were subjected to low temperatures (4 C for 48 hr) before culture. It is presumed that in the pollen mother cells in which meiosis had already initiated, cytokinesis continued and gave rise to tetrads, and, by further mitosis, additional microspores were formed. However, if PMCs were cultured at an earlier stage (premeiotic) then perhaps mitosis was induced. This assumption is subject to cytological confirmation. If such behavior does occur in anther culture then diploid plants would be produced by these PMCs, and this would partially explain the origin of diploid plants obtained by anther culture.

Like the PMCs, pollen tetrads of Atropa in culture have also been observed to undergo divisions to form 5-8 additional microspores (Fig. 8b) with some microspores containing 2-3 nuclei. Even after 8 months in culture, the extra microspores were observed to be highly cytoplasmic and healthy. The culture of meiocytes and tetrads also offers

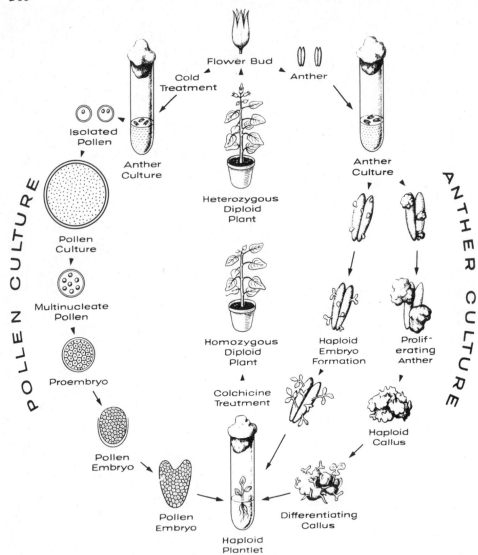

Figure 7.   Diagrammatic illustration showing various modes of andro-genesis and haploid plant formation by anther and isolated pollen culture. The homozygous plants are obtained by treating haploids with colchicine.

a unique tool for the study of the ontogeny of pollen.   The in vitro induction of repeated divisions also raises questions about the essential role of the tapetum during early microsporogenesis.

Following are some of the salient points which have emerged from the work on the culture of isolated pollen of *Nicotiana* and *Datura*.

1.   Growth and morphogenesis are adversely affected if the isolated pollen are not washed before culturing.   Unwashed pollen either fail to grow or have a tendency toward callus formation.

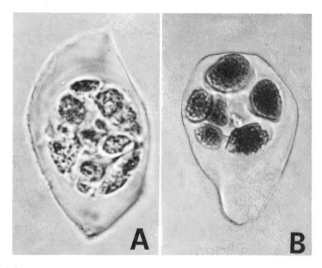

Figure 8.   (a) A 20-week-old culture of pollen mother cell of *Atropa belladonna* showing a multicellular embryo-like structure; note the intact callose (from Bajaj, 1974a).   (b) 10-week-old culture of a pollen tetrad of *Atropa belladonna* showing the formation of additional microspores (from Bajaj, 1975a)

2. Cold treatment has been used to increase the frequency of androgenesis (Fig. 9).   The cold treatment appears to inhibit spindle formation and in some cases may cause an abnormal first mitosis by triggering the pollen nucleus to form two similar-looking nuclei. However, close examination reveals that even in cold-treated pollen, normal mitosis usually takes place to form a vegetative and a generative nucleus.   The cold treatment does not induce androgenesis, but it enhances the viability of cultured pollen, and causes repression of the gametophytic differentiation which results in higher frequency of androgenesis.   The effect of cold treatment is indirect, and hence is not a prerequisite for the successful culture of isolated pollen.
3. Androgenesis in isolated tobacco pollen can be induced at various stages of development (ranging from uninucleate microspore, during and after first mitosis to late binucleate), but frequency of success is most frequently achieved using anthers taken at late uninucleate to binucleate pollen stage (Bajaj, 1978a).
4. The frequency of androgenesis can be enhanced considerably by raising the sucrose and myo-inositol concentrations in the medium.

Although the above techniques of pollen culture are an improvement over anther culture, the ideal situation would be to culture pollen, like single cells, directly in dishes or in suspension cultures without the intermediate step of anther culture.   Only limited success has so far been achieved using this approach.

HAPLOID PROTOPLASTS.   The removal of the cell wall and the release of the isolated plant protoplasts has become a routine procedure,

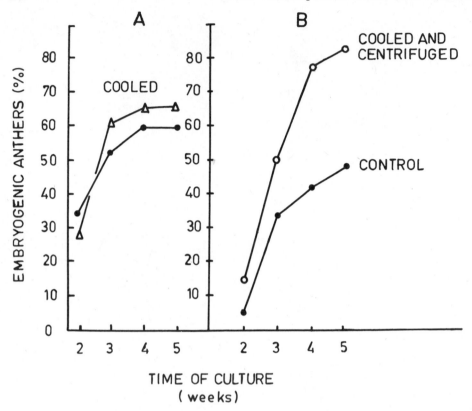

Figure 9.   Stimulatory effect of cooling and centrifugation on pollen embryogenesis in *Datura*.   Percentage of embryonic anthers from flower buds cultured (a) before and (b) after first mitosis.   Untreated, cooled at 3 C for 48 hr, zero-cooled and centrifuged (from Sangwan-Norreel, 1977).

and it is now possible to isolate protoplasts from almost every plant part (Bajaj, 1977b).   Mesophyll protoplasts isolated from haploid plants like those isolated from diploid plants readily divide in culture and give rise to regenerated plants.   In spite of the tendency for cells in culture to increase in ploidy, regenerated plants of tobacco, petunia, potato, atropa, and *Datura* have all been recovered that remain haploid (Ohyama and Nitsch, 1972; Bajaj, 1972; Binding, 1974a,b, 1978; Schieder, 1975; Furner et al., 1978; Bourgin and Missonier, 1978; Bajaj et al., 1978; Fig. 10).   If techniques can be extended to many plant species, the culture of mesophyll protoplasts could be a method for the large-scale multiplication of haploids.

The isolation of protoplasts from pollen and pollen tetrads would prove interesting from a number of aspects.   The behavior of pollen and tetrad protoplasts in culture might be useful for the study of pollen ontogeny (Bajaj et al., 1975).   In particular, the nature of cell wall regeneration by the protoplast when isolated from the surrounding somatic tissue would be of interest.   Apparently, pollen and tetrad

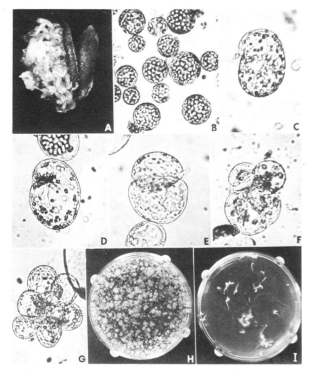

Figure 10.   Isolation and culture of haploid mesophyll protoplasts of *Atropa belladonna*.   (a) 5-week-old culture of an excised anther of *A. belladonna* undergoing androgenesis (the leaves from the haploid plants thus raised were employed for the isolation of protoplasts).   (b) Mesophyll protoplasts obtained from the peeled leaves incubated in an enzyme mixture containing cellulase (1.5%) and macerozyme (0.3%) in mannitol.   (c–e) Various stages in the elongation and division of an isolated protoplast cultured in a liquid nutrient medium 5 days after culture.   (f,g) One- and two-week-old cluster and a colony of callus cells obtained from protoplasts.   (h) Masses of calli obtained from cell colony (f,g) 3 weeks after plating on agar-solidified medium containing NAA.   (i) Differentiation of embryoids and plantlets from callus after transferring to an auxin-free medium (from Bajaj et al., 1978).

protoplasts isolated from the confines of the cell wall can respond to culture (Bajaj, 1974b,c,d).

Protoplasts from Maturing Pollen.   Pollen grains of Angiosperms are coated with sporopollenin (Zetsche, 1932) which is remarkably durable, chemically inert (Gherardini and Healey, 1969), and one of the most resistant materials in the organic world (Faegri and Iversen, 1964).   It is a polymer of carotenoids and carotenoid esters (Shaw, 1971) and can be dissolved only by treatment with KOH, strong oxidizing solutions and certain organic bases (Southworth, 1974), and also by certain microorganisms (Elsik, 1971).   As yet no enzyme preparation is known which can digest the exine; however, by combining both enzymatic and mechanical methods limited quantities of protoplasts have been obtained (Bajaj and Davey, 1974).

Protoplasts may be released in a variety of ways: by weakening of the germpore, by partial dissolution and sloughing off the exine, and by formation of subprotoplasts from the pollen tube. The freshly isolated protoplasts are spherical and vacuolated. When centrifuged, they tend to spontaneously fuse to form multinucleate giant protoplasts. In culture, the protoplasts occasionally showed elongation, division, and budding (Bajaj, 1974c, 1975b).

Protoplasts from Pollen Mother Cells and Pollen Tetrads. Somewhat better preparations are obtained from pollen taken at earlier stages of development, i.e., pollen mother cells and pollen tetrads. These are enclosed in a callose wall which is simple when compared to the complex nature of the exine. The callose is composed of unbranched β-1,3 glucan (Clowes and Juniper, 1968). The procedure for the isolation and culture of protoplasts from pollen mother cells and tetrads (Bhojwani and Cocking, 1972; Bajaj and Cocking, 1972; Bajaj, 1974c) involves a number of steps. One anther is aseptically removed from a young flower bud and stained with acetocarmine to ascertain the stage of development. The basal end of the anther is cut with a fine scalpel and the contents squeezed out with a glass spatula. This step is very important as squeezing with excess force will press out the diploid tapetal cells. The pollen mother cells and tetrads emerge as a milky fluid which is treated with an enzyme mixture derived from snail intestines (0.75-1% Helicase in 8-10% sucrose) for about 30-45 min; 1 ml of the enzyme mixture is used per anther. After incubation, the enzyme mixture is carefully replaced by a 10% sucrose solution and the protoplasts are allowed to settle. They are rinsed a couple of times with fresh medium and then resuspended in nutrient medium and cultured. Centrifugation is not advised, as the protoplast yields that are already low would be further reduced.

In culture, some pollen-derived protoplasts regenerate a wall and show budding and occasional division (Bajaj, 1974c). At places where they adhere, the tetrad protoplasts and developing microspores regenerate common walls giving rise to a clump of cells or a so-called microspore tissue. It is unfortunate that at present the culture of pollen protoplasts has yielded only limited success so that their potential for the induction of haploid plants has yet to be realized.

Elimination of Chromosomes by Bulbosum Technique

The induction of haploid barley by the novel method of Kasha and Kao (1970) of chromosome elimination has proved to be quite efficient for the large-scale production of these plants (Jensen, 1974, 1977). The method entails crossing *Hordeum vulgare* (2n = 14) with *Hordeum bulbosum* (2n = 14). In nature, the seed produced from such a cross develops for about 10 days, and then begins to abort. However, if the immature embryos are dissected out 2 weeks after pollination and cultured on B5 medium (Gamborg et al., 1968) without 2,4-D, they continue to grow. Almost all the plants originating from such embryos are monoploid (n = 7), and true hybrids are rare. This method for the formation of monoploid barley by crossing and embryo culture is diagrammatically represented in Fig. 11.

Kasha and Kao (1970) presented evidence to show that these monoploids are not caused by parthenogenesis, but by the elimination of *H. bulbosum* chromosomes. This elimination is under genetic control (Ho and Kasha, 1975). Their evidence comes from the following four sources: (1) The occurrence of cells in the embryo with more than seven chromosomes. (2) The percentage of seed set is higher than that observed by induced parthenogenesis. (3) When diploid *H. vulgare* is pollinated with tetraploid *H. bulbosum* the percentage of seed set is similar to that of the diploid cross. (4) Haploids of *H. vulgare* type are also obtained when *H. vulgare* is used as the male plant and *H. bulbosum* as the female.

Monoploid wheat has also been obtained by this technique (Barclay, 1975; Zenkteler and Straub, 1979) by crossing *Triticum aestivum* (2n = 6x = 42) with *H. bulbosum* (2n = 2x = 14) whereby the chromosomes of bulbosum are eliminated in the process and monoploid *Triticum aestivum* (n = 3x = 21) is recovered.

This technique represents a considerable advance in the production of barley haploids and it has a number of advantages over anther culture. In particular, haploids can be obtained from any cultivar of barley whereas with anther culture success is dependent on the genotype. Using this method, the frequency of haploid formation is very high. In addition, no haploid plants are obtained. In contrast, barley plants obtained from anthers via callus cultures include aneuploids (Clapham, 1971, 1977).

The mechanism for complete chromosome elimination in barley is gradual and perhaps these methods could be extended (Kasha and Kao, 1970) to plants like potato where dihaploids are obtained from tetraploid *Solalanum tuberosum* by crossing with pollen from diploid *S. phureja.*

INDUCTION OF PARTHENOGENESIS. The spontaneous occurrence of haploids in nature usually takes place through parthenogenesis (Chase, 1969), but in some cases it can also be induced by experimental treatments, especially by the application of inactivated pollen or distantly related pollen (see Maheshwari and Rangaswamy, 1965; Wardlaw, 1965). The in vitro induction of parthenogenesis is another approach that has not been exploited. The growth of isolated eggs or even unfertilized ovules have not been tested. However, Murgai (1959) attempted to culture flower buds and spikes of *Aerva tomentosa.* This is an apomict, the plant is dioecious and male plants are extremely rare in nature. The female flowers possess 5-8 perianth lobes enclosing a monocarpellary ovary and 8-10 staminoides. The unpollinated flowers and spikes cultured at the mature embryo sac stage yielded 40% seeds, and occasionally these seeds germinated in situ.

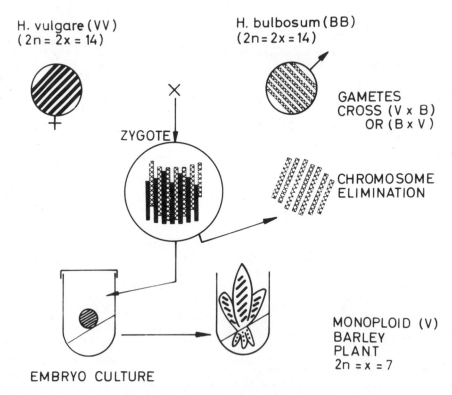

Figure 11. Diagrammatic representation showing the formation of monoploid barley by crossing diploid *Hordeum vulgare* x *Hordeum bulbosum* followed by culture of excised embryo; one set of chromosomes is eliminated (courtesy of Dr. C.J. Jensen).

Yet another approach worth considering is in vitro pollination by genetically unrelated pollen (Kanta, 1960; Kanta and Maheshwari, 1963; Rangaswamy, 1977). Haploid parthenogenesis can also be induced by the culture of unpollinated ovaries. Using this method, Hess and Wagner (1974) obtained haploids in *Mimulus*. The wall from the unpollinated pistil (ovary wall) of *Mimulus luteus* is removed and the pollen of

Torenia fournieri are directly placed on the ovules. With this method, haploid parthenogenesis was recorded in 1% of the cultures.

FACTORS INFLUENCING ANDROGENESIS

It has been suggested that in vitro induction of haploids by anther culture might be restricted to species within only a few families such as the Solanaceae and the Gramineae. However, it is now clear that although species belonging to these two families may be more prone to the induction of androgenesis, androgenesis can be achieved in many other species as well (Table 1), though with a lower frequency. Indeed, one of the main problems that a worker in the field of anther culture confronts is the low frequency of androgenesis. Thus attention is now being focused upon the factors which influence androgenesis with the goal of increasing its frequency. Some of the factors which have been implicated in the induction of androgenesis are discussed below. Manipulation of these factors has made it possible to increase the efficiency of the production of haploids (see Bajaj et al., 1977; Sangwan-Norreel, 1977; Chen, 1978; Maheshwari et al., 1980).

Genotype

One of the most important factors governing the success of in vitro induction of haploids is plant genotype. It has been repeatedly observed that various species and cultivars exhibit different growth responses in culture. Gresshoff and Doy (1972a,b) working with 43 cultivars of *Lycopersicon esculentum* and 18 lines of *Arabidopsis thaliana* could induce haploid tissues in only 3 cases each. The same trend was observed in rice (Guha-Mukherjee, 1973). The percentage of haploid induction in various species of *Nicotiana* (Nitsch, 1972; Tomes and Collins, 1976), *Solanum* (Irikura and Sakaguchi, 1975), and rye (Wenzel et al., 1977) differed remarkably. Likewise, out of 21 cultivars of *Triticum aestivum* (Table 2), haploid tissue could be obtained from anthers of only 10 cultivars (Bajaj, 1977a).

These studies make it clear that one of the reasons for failure in anther culture is that most workers restrict themselves to one cultivar and abandon the work if unsuccessful. It is highly desirable that a general survey of various cultivars be undertaken. Furthermore, for such a general survey, comparatively simple media should be used, as complex media rich in growth regulators tend to favor proliferation of somatic anther tissue giving rise to callus of various ploidy levels.

Physiological Condition and Age of the Plant

It has been a general experience that the age and physiological state of the plant considerably influences the efficiency of androgenesis. Flowers from relatively young plants at the beginning of the flowering season (Narayanaswamy and Chandy, 1971; Anagnostakis, 1974) are

Table 2.   Influence of Genotype on Growth Response in Excised
           Anthers of Various Cultivars of *Triticum aestivum*, Cultured
           at a Uninucleate Pollen Stage (from Bajaj, 1977a)

CULTIVAR	GROWTH RESPONSE	NUMBER ANTHERS CULTURED	NUMBER ANTHERS CALLUSING	PERCENTAGE ANTHERS CALLUSING
Maris Ensign	−	180		
Rothwell Sprite	+	120	2	1.6
Janus	+	310	5	1.6
Chinese Spring	+	250	4	1.6
Kolibri	+	800	14	1.7
Tilli	+	421	7	1.6
Cardinal	+	329	5	1.5
Maris Freeman	+	180	2	1.1
Maris Widgeon	−	75		
Maris Huntsman	+	129	1	0.8
Maris Templar	−	111		
Maris Dove	−	85		
Maris Nimrod	−	167		
Luna	−	129		
Jubilar 3	−	151		
Golden Valley	−	86		
Champlein	+	141	2	1.7
Nersee	+	175	1	0.6
Rampton Rivett	−	172		
1877	−	209		
Benno 780	−	114		

more suitable than the flower buds taken from plants at the end of their growing period. It has also been observed that flower vigor in tobacco is retained longer if the plants are irrigated with Hoagland's solution every 2 weeks. Moreover, the temperature at which the plants have been grown is important since *Datura* plants grown at 24 C gave a higher frequency of androgenesis (45%) than those grown at 17 C (8%) (Nitsch and Norreel, 1973a). However, *Brassica napus* yielded better results with anthers excised from the plants grown at lower temperatures (Keller and Stringam, 1978). The removal of the old flowers in *Datura* (Nitsch, 1975), and the apical portion of wheat inflorescence (Picard, 1973) caused an increase in the frequency of androgenesis. In *Triticum aestivum*, it was observed that anthers taken from field-grown plants were much healthier, more robust, and gave better results than those grown in poorly illuminated greenhouses (Bajaj, 1977a). Since the change in light and temperature are the result of variation in season, it is not suprising that the results vary between labs. It is, therefore, of prime importance that plants be grown under optimal growth conditions, watered with mineral salt solutions periodically, and that relatively young plants should be used.

Stage of Pollen

It has been established that not only the age and the environmental conditions under which the donor plants are grown but also the stage at which pollen are taken are most critical. Examples are known in tobacco (Nakata and Tanaka, 1968) and tomato (Gresshoff and Doy, 1972a) in which androgenesis has been induced at the pollen tetrad stage, and also in some gymnosperms where the grains were 4-5-celled. However, the data accumulated (see Table 1) has established the fact that uninucleate microspores (Table 3) are most prone to experimental treatment for anther culture just before or during first mitosis, while the culture of isolated pollen has yielded better results at the binucleate stage (Bajaj, 1978a). Moreover, the stage of pollen is also implicated in the ploidy level of the plants. The plants obtained from anthers of tobacco (Engvild, 1974), *Datura* (Engvild et al., 1972) and *Hyoscyamus* (Corduan, 1975) cultured at the uninucleate pollen were haploids, whereas those obtained from later stages showed various levels of ploidy.

Thermal Trauma

Haploids have been observed to occur in nature and to be induced experimentally following some trauma. By temperature manipulations, the egg could be induced to develop parthenogenetically. As early as 1922 Blakeslee et al. obtained *Datura stramonium* haploids by subjecting the plants to low temperatures at the time of fertilization. Povolochko (1937) obtained haploids by exposing tobacco plants to high as well as low temperatures. Similarly, Muentzing (1937) and later Noerdenskiold (1939) obtained rye haploids by low temperature (3 C) treatment. In all these cases the haploids originated from the egg. Both the egg and the pollen are prone to such trauma. Stow (1930) and Sax (1935) observed that thermal shocks given to plants can alter the mode of division of the microspore nucleus. It was further observed (Stow, 1930) that these cold-treated pollen of *Hyacinthus orientalis* lose male potency and resemble the 8 nucleate embryo sac. Floral buds or entire plants subjected to hot or cold treatment or perhaps alternate combinations of both might prove to be useful in enhancing androgenesis. Nuclear abnormalities such as budding and fusion have also been reported (Izhar, 1973) in *Petunia* plants kept at low temperatures for 24-48 hr. The stimulatory effect of thermal shocks for anther culture has been successfully extended to *Datura* (Nitsch and Norreel, 1973a), tomato (Debergh and Nitsch, 1973), *Atropa belladonna* (Bajaj, 1974a), and tobacco (Debergh and Nitsch, 1973; Bajaj and Reinert, 1975; Reinert et al., 1975; Bajaj, 1978a; Bajaj et al., 1977) to enhance androgenesis in anthers and isolated pollen. The combined effect of cold treatment plus centrifugation (Sangwan-Norreel, 1977), and cold treatment plus charcoal (Bajaj et al., 1977) are depicted in Fig. 9 and Table 4, respectively.

Table 3. Plant Regeneration from Pollen Callus Produced on Medium with Various Concentrations of Sucrose (from Chen, 1978)

ANTHER STAGE	SUCROSE CONCEN-TRATION (%)	NO. OF CALLUS CULTURED	NO. OF CALLUS PRODUCING PLANTS					GREEN/ALBINO PLANTS
			Green	Albino	Green + Albino	Total	(%)	
Early-uninucleate	3	16	6	2	2	10	62.5	2.0
	6	86	42	14	13	69	80.2	2.0
	9	94	40	17	17	74	78.7	1.7
Mid-uninucleate	3	67	29	8	3	40	59.7	2.9
	6	107	51	12	9	72	67.3	2.9
	9	65	20	14	8	42	64.6	1.3
Late-uninucleate	3	8	1	1	1	3	37.5	1.0
	6	30	2	9	0	11	36.7	0.2
	9	28	6	14	1	21	75.0	0.5

Table 4.   Effect of Various Exogenous Factors on Percentage of
            Anthers Undergoing Androgenesis and the Number of Plants
            per Anther (from Bajaj et al., 1977)

FACTOR	TREATMENT	EMBRYO-GENIC ANTHERS (%)	PLANTS PER ANTHER (MEAN)	SIGNIFI-CANTLY GREATER THAN CONTROL
Charcoal in the medium	Control	41.2	5.3	
	1% charcoal	83	14.6	**
	2% charcoal	91	17.5	**
Thermal shock (4 C) + charcoal medium	Control	78.6	12.9	
	24 hr	86	15.4	n.s.
	48 hr	91.6	23.2	**

** = statistically significant at $p < 0.01$; n.s. = nonsignificant.

Both cold treatment and brief exposure to higher temperature have
been reported to stimulate repeated division of pollen.   For example, in
Brassica pollen, embryogenesis is stimulated in anthers (Keller et al.,
1981) subjected to 30 C for 24 hr or 40 C for just 1 hr.   The temperature shock appears to cause dissolution of microtubules (Hepler and
Palevitz, 1974) and dislodging of the spindle which causes abnormal
division of the microspore nucleus.
    The effect of cold treatment is indirect.   The increase in androgene-
sis is mainly attributed to the fact that low temperature (3-5 C)
retains the pollen viability longer, delays senescence, and prevents the
abortion of pollen (Bajaj, 1978a) and thereby increases the number of
available viable pollen which are destined to form embryos.   This can
be verified by a simple experiment by studying the pollen viability in
cultured anthers incubated at various temperatures.
    Nemec (1898) observed that pollen grains in the petaloid anthers of
Hyacinthus orientalis sometimes assumed the 8 nucleate condition, resembling embryo sacs.   This observation, which later came to be known
as the Nemec Phenomenon, was reported in other cultivars of *Hyacinthus* by various workers (De Mol, 1923; Stow, 1930; Naithani, 1937),
Ornithogalum nutans (Geitler, 1961) and *Leptomeria billardierii* (Ram,
1959).   The culture of isolated pollen and excised anthers of plant
species showing the Nemec Phenomenon is worth consideration.

Chemical Treatment

    Various chemicals are known to induce parthenogenesis.   Yasuda
(1940) observed that by injecting the ovaries of *Petunia* with the chemical Belvitan, the egg was stimulated to divide repeatedly.   Likewise
by colchicine treatment, Smith (1943) and Levan (1945) obtained haploid

plants of *Nicotiana langsdorfii* and *Beta vulgaris*, respectively. The
efficiency of haploid induction in sweet corn was increased from 2.7/
1000 to 7/1000 by the application of 50 mg/l maleic hydrazide (Deanon,
1957). Spraying the plants with various growth regulators and hormones could also be taken into consideration. In this connection, Ethrel has yielded some interesting results. Following the report (Lower
and Miller, 1969) that 2-chloroethylphosphonic acid (commercially known
as Ethrel or Ethephon) acts as a gametocide and causes male sterility
(Rowell and Miller, 1971), Bennett and Hughes (1972) demonstrated that
plants of *Triticum aestivum* cv. Chinese Spring when sprayed with Ethrel just before meiosis in the pollen mother cells, undergo additional
mitosis to give rise to multinucleate pollen. Induction of additional
nuclear divisions and formation of nuclear bodies has also been
observed in *Medicago* and *Tradescantia* (MacDonald and Grant, 1974),
Petunia (Bajaj, 1975b), *Nicotiana* (Bajaj et al., 1977), and some
cultivars of wheat.

Flowers of tobacco plants sprayed at 1000-2000 ppm of Ethrel ab-
scissed within 2 days, while application of 250-500 ppm proved damaging as the flowers turned yellow and fell within a week. Flower buds
sprayed with 100 ppm occasionally underwent division in the vegetative
nucleus. Anthers excised from such buds (4 days after spray) showed
25% increase of androgenesis (Bajaj et al., 1977); however, anthers
tended to proliferate to form brownish callus, and the plantlets had a
tendency to turn yellow.

As compared to tobacco, wheat plants (Bajaj, 1975b) were relatively
resistant to Ethrel sprays. At high concentrations (8000 ppm) wheat
plants were badly damaged, and most of the leaves died, but later new
shoots appeared. The anthers were shriveled and contained a few
degenerating pollen. At low concentrations (500 ppm) no significant
effect was observed. In the plants treated with 4000 ppm Ethrel, the
anthers contained 4-6 nucleate pollen with fewer starch grains as compared with the controls. When cultured, the Ethrel treated anthers
occasionally showed proliferation. It was further observed that not
only before mitosis, but also plants sprayed immediately after the first
mitosis (binucleate stage) of pollen showed the 4-8 nucleate condition.
In some instances both vegetative and generative nuclei divided repeatedly. In one case four sperms were observed in a pollen (Bajaj,
1975b).

It is possible that multinucleate pollen isolated from Ethrel-treated
anthers might be induced to form embryos when cultured, and in this
way the yield of haploids could be enhanced.

Composition of the Medium

The composition of the medium is one of the most important factors
determining not only the success of anther culture but also the mode
of development. Pollen embryogenesis can be induced on a simple
mineral-sucrose medium in plants like tobacco (Nitsch, 1969) and *Hyoscyamus* (Raghaven, 1975), yet for androgenesis to be completed, addition of certain growth regulators is required. For instance, cereal

anthers require both auxins and cytokinins (Clapham, 1977), and optimal growth response depends on the endogenous level of these growth regulators. However, to promote direct embryogenesis, simple media with low levels of auxins are advisable. Complex media enriched with auxins such as 2,4-D encourage the formation of callus, which causes genetic instability and thus should be avoided.

Sugars are indispensable in the basal medium, as they are not only the source of carbon but are also involved in osmoregulation. Normally, 0.058-0.12 M sucrose is routinely used; however, higher concentrations (0.17-0.29 M) have yielded better results in wheat (Quyang et al., 1973), barley (Clapham, 1973), potato (Sopory et al., 1978), and rice (Chen 1978; Table 3).

In addition, the extract of androgenic anthers and potato tubers have been observed to stimulate androgenesis in tobacco (Nitsch and Norreel, 1973a) and wheat (Chinese Workers, 1976) respectively. Supplementing the media with serine and glutamine has also improved the culture of isolated pollen (Nitsch, 1974a).

Effect of Charcoal

It has been reported that activated charcoal stimulates the growth of various fungi (Parmentier, 1970; Butler and Bolkan, 1973), mosses (Proskauer and Berman, 1970; Klein and Bopp, 1971), fern (Kato, 1973), and orchids (Ernst, 1974). This work was extended to tobacco cv. Havana by Anagnostakis (1974), who observed an increase in the yield of androgenic anthers from 15% to a maximum of 45%. Further improvement has been achieved with cv. Badischer Burley by Bajaj et al. (1977). In this cultivar, the percentage of androgenic anthers was increased from 41 to 91% by supplementing the basal medium with 2% charcoal (see Fig. 12 and Table 4). In addition, the number of plants per anther was increased and the regeneration of plantlets from anthers was speeded up. Similar stimulatory effects of charcoal on androgenesis have been observed in *Anemone* (Johansson and Eriksson, 1977), rye (Wenzel et al., 1977), and potato (Sopory et al., 1978).

This method of charcoal stimulation to enhance the frequency of androgenesis is not understood; however, various conjectures have been proposed. Fridborg and Eriksson (1975) observed that charcoal enhances the embryogenesis in carrot, whereas normally it can not be induced without omission of the auxin from the medium. Similarly they obtained root formation in *Allium cepa* in a charcoal medium, although such cultures do not normally produce roots. This suggests that auxin is implicated. Weatherhead et al. (1978), on the other hand, believes that androgenic enhancement in tobacco is because of absorption of 5-hydroxymethylfurfural by the charcoal. This compound is produced during autoclaving of the sucrose and is deleterious.

It seems likely that in plants like tobacco, charcoal absorbs inhibitory substances and thereby reduces the number of potential pollen embryos that would normally have aborted. However, it is more probable that the level of growth substances (both endogenous and exogenous) is regulated by absorption into charcoal. Although no proof for this is

Figure 12. Effect of 1% activated charcoal on androgenesis in anthers grown for 5 weeks. (a) control, (b) charcoal medium.

yet available, the use of charcoal is very promising for the enhancement of the induction of haploid plants, at least in some species, and should be extended to others.

HOMOZYGOUS PLANTS

By conventional inbreeding and backcrossing it is possible to obtain pure lines, but it is a long and cumbersome process. However, homozygous plants can be obtained in a relatively short time by the production of haploids and by doubling their chromosomes. The duplication of the chromosomes can be achieved by a number of methods.

Endomitosis

Haploid cells are, in general unstable in culture and have a tendency to undergo endomitosis (chromosome duplication without nuclear division) to form diploid cells. This property of cell cultures has been exploited for obtaining homozygous tobacco plants (Nitsch, 1969; Kochhar et al., 1971). A small segment of stem from a haploid plant is grown on an auxin-cytokinin medium to induce callus formation. During callus growth and differentiation there is a doubling of the chromosomes by endomitosis to form diploid homozygous cells and ultimately plants.

Colchicine Treatment

Colchicine has been used extensively as a spindle inhibitor to induce chromosome duplication and to produce polyploid plants (Eigsti and Dustin, 1955). This has been employed for obtaining homozygous diploid plants from haploid cultures (Burk et al., 1972; Tanaka and Nakata, 1969; Kasperbauer and Collins, 1972). While still enclosed by the anther, the young plantlets are treated with 0.5% colchicine solution for 24-48 hr, washed thoroughly, and replanted. Mature haploid plants can also be used, in which case the colchicine (0.4% in lanolin paste) is applied to the axils of the leaves. The fertile homozygous plants thus obtained can be used for producing inbred lines used to produce hybrids.

Fusion of Pollen Nuclei

As mentioned earlier, a feature of anther culture is that the plants which are obtained exhibit various levels of ploidy. The diploid plants in general appear to arise from the somatic anther tissue, although some seem to arise from fusion of pollen nuclei. The embryos and plants obtained from such pollen would be completely homozygous as well as being diploid. The advantage of such spontaneously obtained homozygous plants is that the chances of nuclear aberrations are considerably reduced in 2n vs. n cells.

CYTOLOGICAL INSTABILITY OF HAPLOID CELL CULTURES

It has been repeatedly observed that cells in culture exhibit cytological instability. Moreover, it has been established that haploid cell lines have a greater tendency to increase in ploidy level to diploid than diploid cell lines have to increase to tetraploid (Sacristan, 1971). The use of this property for the production of homozygous diploid plants from haploid tissue has already been suggested. This tendency for cells in culture to undergo endomitosis is a major obstacle to the maintenance of haploid cells in culture. Indeed finding ways to maintain haploid cells in culture is as important in a number of areas of investigation as the induction of haploidy itself.

Parafluorophenylalanine (PFP) has been reported to induce haploidization in *Aspergillus* and *Ustilago* (Day and Jones, 1971). Gupta and Carlson (1972) extended this work to *Nicotiana tabacum* cv. Havana Wisconsin callus and observed that in mixoploid cultures, growth of diploid cells was progressively inhibited with increasing concentrations of PFP (1-9 mg/l) while the growth of haploid cells was unaffected. They further claimed that "our work demonstrates that it is possible to maintain stable cultures of haploid cells, and to select preferentially haploid cells from mixed populations of cells of varying ploidy." In contrast, Dix and Street (1974), although observing growth inhibition of diploid cells of *Nicotiana sylvestris* by PFP (37.5 mg/l), did not observe any preferential growth of haploid cells in mixed cell cultures. They concluded that it was genotype and not ploidy level that determines sensitivity to PFP. Similar negative results were reported for *Datura* by Evans and Gamborg (1979). Matthews and Vasil (1975) observed growth inhibition of both haploid and diploid tobacco pith–explants.

In experiments with *Nicotiana tabacum* and *Atropa belladonna* (Reinert and Bajaj, 1977; Bajaj and Grobler, unpublished) PFP (15–20 mg/l) caused growth inhibition of diploid callus while the haploid cultures were either stimulated or unaffected; however, a slight increase in the percentage of haploid cells was observed (Fig. 13). These observations support the conjecture that PFP may aid in stabilizing the haploid level. It was concluded that, although PFP may not routinely prevent the reversion of haploids to diploids, it may help to prolong the haploid phase in culture in some species.

CRYOPRESERVATION OF HAPLOID CULTURES

Freeze preservation of haploid cells and tissues (Bajaj, 1976a) at super–low temperatures (–196 C) in liquid nitrogen offers a novel approach to the preservation of genetic stability, and this method might prove especially useful for the establishment of haploid germplasm banks (Bajaj, 1976a, 1978b,d, 1979a,b, 1981a; Bajaj and Reinert, 1977). In view of the importance of haploid cells in plant breeding, mutation, and biochemical genetics research, it is highly desirable to develop methods

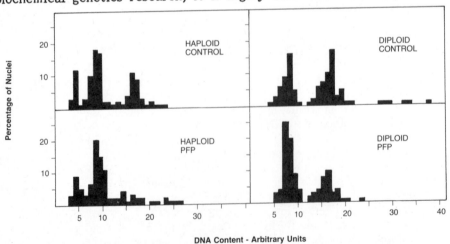

DNA Content - Arbitrary Units

Figure 13. Effect of parafluorophenylalanine (109 µM) on frequency distribution of values for the nuclear DNA contents of fresh stem callus of haploid and diploid *Atropa belladonna* grown for 4 weeks. DNA content was estimated by microspectrphotometry of 100 Feulgen stained nuclei (Bajaj and Grobler, unpublished).

by which genetic stocks can be conserved on a long-term basis. This can be brought about if cells are maintained in a metabolic inactive and nondividing state (i.e., at -196 C).

The limited work done on the regeneration of plants from haploid cell suspensions (Bajaj, 1976b), pollen embryos (Fig. 14; Bajaj, 1976a, 1977c, 1978c), and excised anthers of *Atropa, Petunia*, tobacco (Tables 5,6; Bajaj, 1978b), rice (Bajaj, 1980), and *Primula* (Bajaj, 1981b) stored in liquid nitrogen has greatly helped to enhance the optimism that cryogenic methods would be a promising approach for the conservation of haploids.

Pollen embryos and haploid meristem tips, which are generally known to be genetically stable, should be preferred over cell suspensions for freeze storage. The cell suspensions are usually heterogeneous, and in some cases it is difficult to induce the differentiation of an entire plant from callus, whereas pollen embryos and meristem tips develop into plants rather easily, and therefore, may prove to be better material for cryogenic studies.

INDUCTION OF MUTATIONS

One unique value of haploid cell cultures lies in the study of somatic cell genetics. In such studies mutant cell lines are particularly important. But the majority of mutations are recessive and, therefore, are not expressed in diploid cells in the presence of an unmutated dominant gene.

Haploid callus cells have been employed by various workers to study the effect of various mutagens, both irradiations as well as chemicals.

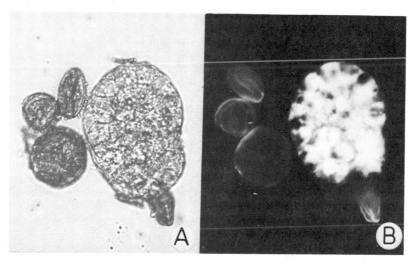

Figure 14. Survival of *Atropa belladonna* pollen–embryos subjected to -196 C in the presence of 5% DMSO. (a) Photographed in tungsten light, (b) in ultraviolet light; note the fluorescence of pollen–embryo in (b) after staining with fluorescein diacetate (from Bajaj, 1976a).

Table 5. Response of *Nicotiana tabacum* Anthers (4 weeks after culture) Subjected to Various Treatments and Recultured on Agar Solidified Medium (from Bajaj, 1978b)

EXPERIMENT	NO. ANTHERS CULTURED	NO. GROWING ANTHERS	SURVIVAL (%)	TOTAL NUMBER PLANTLETS	NO. PLANTS PER ANTHER
Untreated (control)	85	69	81.1	890	12.8
Treated with 7% DMSO	70	54	77.1	530	9.8
Treated with 7% DMSO and warmed at 37 C for 10 min	66	45	68.1	390	8.6
Treated with DMSO, cooled at the rate of 2 C/min, subjected to –196 C and thawed at 37 C	130	2	1.5	7	3.5

Table 6. Growth Response of Transversely Cut Halves of 4-week-old Cultured Anthers of *Nicotiana tabacum* Subjected to Cold Treatment and Recultured in Agitated Liquid Medium (from Bajaj, 1978b)

TREATMENT	NO. ANTHERS CULTURED	NO. GROWING ANTHERS	SURVIVAL (%)	TOTAL NUMBER PLANTLETS	NO. PLANTS PER ANTHER
Control (untreated)	45	30	66.6	623	20.7
Frozen at the rate 2 C/min to –196 C	60	4	6.6	10	2.5

Single cells and isolated pollen have the advantage over the entire plant in that they can be plated and screened in large numbers, in a similar way to the techniques used for microorganisms. Indeed, with the use of cell cultures a number of mutant cell lines have now been successfully isolated, and extensive work is being done to obtain lines that are resistant to various drugs (Widholm, 1977), pathotoxins (Bajaj, 1981c), salts (Nabors et al., 1980), chilling (Dix and Street, 1976), herbicides (Gressel et al., 1978), viruses (Heinz et al., 1977), and nematodes (Wenzel and Uhrig, 1981).

Tulecke (1960) isolated arginine-requiring strains from pollen of *Ginkgo* by substituting hormone in the medium with L-arginine (5.7 mM). With the availability of the technique of isolated pollen culture, it should be relatively easy to subject large numbers of pollen, like single cells, to various mutagens and to isolate mutants. This line of approach could be relatively efficient for the recovery of mutants.

Using haploid cells and protoplasts, mutations have been isolated in a number of plant species.   Cell lines, tissues, and complete plants resistant to streptomycin (Binding et al., 1970; Maliga et al., 1973a), 5-bromodeoxyuridine (Carlson, 1970; Maliga et al., 1973b), methionine sulfoximine (Carlson, 1973), potato cyst nematode (Wenzel and Uhrig, 1981), and various temperatures (Gebhardt et al., 1981) have been obtained.

Employing anther culture, Nitsch et al. (1969) claimed to have obtained mutants by subjecting young plantlets (while emerging from the anthers) to various doses (1500-3000 rads) of gamma irradiation. These plants showed a high proportion of abnormalities in shape, size, and color of the flowers, and some were definitely chimeras. In a further study, Nitsch (1972) observed a white-flowered mutant with marmoured leaves after growing tobacco anthers on a medium supplemented with 10^{-6} M N-3-nitrophenyl-N-phenylurea.   Devreux and Saccardo (1971) subjected flower buds of *Nicotiana tabacum* cv. Virginia Bright to X-rays (1000 R), and the excised anthers were subsequently cultured. About 50% of the haploid plants thus obtained were aberrant phenotypes, and nearly 6% showed chromosomal aberrations. Haploid tobacco protoplasts (Galun and Raveh, 1975) and cells (Eapen, 1976) showed differential radiosensitivity to ultraviolet and gamma radiations, and valine-resistant mutant plants were regenerated from such cultures (Bourgin, 1978).

INDUCTION OF GENETIC VARIABILITY

The success of any crop improvement program depends on the extent of genetic variability in the base population. However, because of depletion of the germplasm pools, there is a shrinkage of the genetic resources.   In this regard callus cultures are a rich source of genetic variability (D'Amato, 1977; Skirvin, 1978). By anther culture not only haploids but plants of various ploidy level and mutants can be regenerated.   Table 7 summarizes the results on the genetic variability for chromosome number obtained through the culture of excised anthers.

Table 7.   Variation in Chromosome Number of Anther-Derived Callus,
            Protoplasts, and Plants

PLANT SPECIES	RANGE IN CHROMO-SOME NUMBER	REFERENCE
Arachis hypogaea (2n = 40)	20, 21-24, 30, 40, 60-80	Bajaj et al., 1981
Atropa belladonna (2n = 72)	36, 72, 108	Zenkteler, 1971
A. belladonna (2n = 72)	36, 72, 132, 200	Bajaj et al., 1978
Cajanus cajan (2n = 22)	8, 10, 11-22, 23-28	Bajaj et al., 1980b
Cicer arietinum (2n = 16)	8, 9, 15, 16, 24, 25-64	Gosal & Bajaj (unpublished)
Hordeum vulgare (2n = 14)	7, 14, 28	Clapham, 1973
Nicotiana suaveolens x N. langsdorffi (2n = 50)	10-53	Guo, 1972
N. tabacum (2n = 48)	23, 46	Mattingly & Collins, 1974
N. tabacum (2n = 48)	24, 48, 96	Novak & Vyskot, 1975
N. otophora (2n = 24)	10, 11, 12, 13, 24	Collins et al., 1972
Oryza sativa (2n = 24)	12, 24, 36, 60	Nishi & Mitsuoka, 1969
O. sativa (2n = 24)	12, 24, 26, 36, 48, 72, 96	Chen & Chen, 1980
Petunia axillaris (2n = 14)	21, 28	Engvild, 1973
P. hybrida (2n = 14)	7, 14, 21	Wagner & Hess, 1974
Solanum tuberosum (4x = 48)	12, 14, 24, 48	Sopory & Tan, 1979

The anther-derived callus of *Arachis hypogaea* and *A. villosa* (Fig. 15; Bajaj et al., 1981), *Cajanus cajan* (Bajaj et al., 1980b), and *Cicer arietinum* (Gosal and Bajaj, unpublished) showed a wide range of genetic variability (Fig. 16), as the chromosome number varied from haploid to octoploid with abundant aneuploids. Thus, through the culture of anthers, desirable plants can be obtained and incorporated into the breeding programs.

SIGNIFICANCE OF HAPLOIDS IN THE EARLY RELEASE OF VARIETIES

The significance of the in vitro production of haploids in cell genetics and plant breeding and thus in agricultural crop and tree improvement programs has already been emphasized (Melchers, 1972; Winton and Stettler, 1974; Kasha, 1974; Collins, 1977; Reinert and Bajaj, 1977; Nitzsche and Wenzel, 1977; Hu et al., 1978; Vasil, 1980; Hermsen and

Figure 15. Regeneration of plants from anther-derived callus of *Arachis villosa* (2n = 20). (a) Excised anthers 3 weeks after culture on MS + IAA (22.0 μM) + KIN (9.3 μM) showing proliferation. (b,c) Differentiation of shoots from callus subcultured on MS+NAA (5.4 μM) + BAP (8.8 μM). (d) A complete plant obtained from anther callus and transferred to a pot (from Bajaj et al., 1981).

Figure 16 (see page 268). Induction of genetic variability in anther-derived callus of *Cajanus cajan* and *Cicer arietinum*; note the wide variation in chromosome number ranging from haploid to polyploid with abundant aneuploids (from Bajaj et al., 1980b).

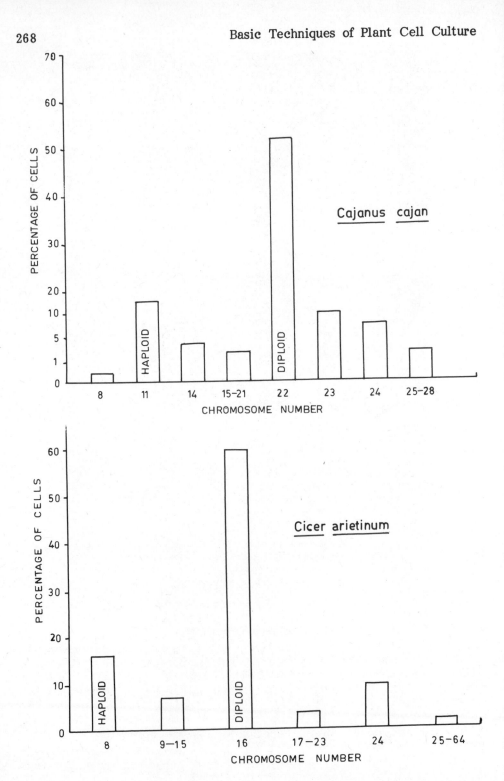

EXTENT OF GENETIC VARIABILITY IN ANTHER-DERIVED CALLUS

Ramanna, 1981), and to go into the details is beyond the scope of this article. Some aspects have, however, been discussed under the titles of homozygous plants, cryopreservation of haploid cultures, and the induction of mutation and genetic variability. Therefore, a summary of the most important feature, i.e., the early release of new varieties of crops follows.

The main and foremost advantage of the in vitro production of haploids over the conventional method is the saving of time. By the culture of anthers and pollen, homozygous plants and thus isogenic lines can be produced within a year as compared to the long inbreeding method which might take 4-6 years. Based on the anther culture method new varieties of rice namely *Huayu 1*, *Huayu 2*, and *Tanfong 1* (Yin et al., 1976), wheat *Lunghua 1* and *Huapei 1* (Hu et al., 1978), and tobacco *Tanyu 1*, *Tanyu 2*, and *Tanyu 3* have been released in China. In Japan, a superior tobacco variety *F 211*, resistant to bacterial wilt, has been obtained through anther culture. These reports have encouraged many a plant breeder to incorporate anther culture in breeding methods. It is envisaged that during the next decade the impact of anther culture work will be greatly felt in crop improvement programs.

CONCLUSIONS AND PROSPECTS

During the last decade, progress made in the vitro induction of haploids and their utilization in cell genetics and crop improvement programs have brought awareness among plant breeders of the use of these methods as a powerful tool for the induction of mutations and for the quick production of homozygous plants and isogenic lines that would result in the early release of crop varieties. Although considerable progress has been made, the main difficulty is still the low frequency of haploids produced by the anther culture of various crop plants. A logical step, therefore, is to study various factors, both endogenous and exogenous, that influence androgenesis. An ideal system would, however, be the direct culture and plating of isolated pollen and their protoplasts. This is an area that has potential, and more emphasis should be placed on it. This would not only yield large scale production of haploids and homozygous plants, but would also facilitate work in inducing desirable mutation, transformations, and biochemical genetics. It also has biotechnological potential. The method of monoploid production by chromosome elimination shows particular promise, especially if the technique can be extended to crops other than barley and wheat. The early release of superior cultivars of rice, wheat, barley, and tobacco through anther-derived plants has demonstrated the efficacy of the in vitro methods for these crops. This work should be seriously extended to other crops for their improvement. Moreover, mutations in haploid cultures could produce high yielding plants that are resistant to such elements as disease, salt, drought, and insects.

KEY REFERENCES

Bajaj, Y.P.S., Reinert, J., and Heberle, E. 1977. Factors enhancing in vitro production of haploid plants in anthers and isolated microspores. In: La Culture des Tissus et des Cellules des Vegetaux (R.J. Gautheret, ed.) pp. 47-58. Mason Press, Paris, New York.

Jensen, C.J. 1977. Monoploid production by chromosome elimination. In: Applied and Fundamental Aspects of Plant Cell, Tissue, and Organ Culture (J. Reinert and Y.P.S. Bajaj, eds.) pp. 299-330. Springer-Verlag, Berlin, Heidelberg, New York.

Nitsch, C. 1974b. Pollen culture—A new technique for mass production of haploid and homozygous plants. In: Haploids in Higher Plants—Advances and Potential (K.J. Kasha, ed.) pp. 123-135. Univ. Guelph Press, Guelph.

Reinert, J. and Bajaj, Y.P.S. 1977. Anther culture: Haploid production and its significance. In: Applied and Fundamental Aspects of Plant Cell, Tissue, and Organ Culture (J. Reinert and Y.P.S. Bajaj, eds.) pp. 251-267. Springer-Verlag, Berlin, Heidelberg, New York.

Sunderland, N. and Dunwell, J.M. 1977. Anther and pollen culture. In: Plant Tissue and Cell Culture (H.E. Street, ed.) pp. 223-265. Blackwell, Oxford.

REFERENCES

Abo El-Nil, M.M. and Hildebrandt, A.C. 1971. Differentiation of virus-symptomless geranium plants from anther callus. Plant Dis. Rep. 55: 1017-1020.
_____ 1973. Origin of androgenetic callus and haploid geranium plants. Can. J. Bot. 51:2107-2109.
_____, and Evert, R.F. 1976. Effect of auxin-cytokinin interaction on organogenesis in haploid callus of Pelargonium hortorum. In Vitro 12: 602-604.
Amos, J.A. and Scholl, R.L. 1978. Induction of haploid callus from anther of four species of Arabidopsis. Z. Pflanzenphysiol. 90:33-43.
Anagnostakis, S.L. 1974. Haploid plants from anthers of tobacco-enhancement with charcoal. Planta 115:281-283.
Anand, V.V. and Govindappa, A.K. 1979. In vitro culture of excised anthers of Solanum mammosum L. Indian J. Exp. Biol. 17:444-445.
Atanasov, A. 1973. Anther culture of tobacco and sugarbeet as a tool for producing haploid plants and callus of different ploidy levels. Genet. Sel. 6:501-507.
Babbar, S.B., Mittal, A., and Gupta, S.C. 1980. In vitro induction of androgenesis callus formation and organogenesis in Iberis amara Linn. anthers. Z. Pflanzenphysiol. 100:409-414.
Bajaj, Y.P.S. 1972. Protoplast culture and regeneration of haploid tobacco plants. Am. J. Bot. 59:647.
_____ 1974a. Induction of repeated cell division in isolated pollen mother cells of Atropa belladonna. Plant Sci. Lett. 3:309-312.

_____ 1974b.   The isolation, culture and ultrastructure of pollen pro-toplasts.   In: Haploids in Higher Plants: Advances and Potential (K.J. Kasha, ed.) pp. 139-140.   Univ. Guelph Pres, Guelph.

_____ 1974c.   Isolation and culture studies on pollen tetrad and pollen mother cell protoplasts.   Plant Sci. Lett. 3:93-99.

_____ 1974d.   Possibilities of haploid production from pollen proto-plasts.   Proceedings XIX International Horticultural Congress, p. 54. Warsaw.

_____ 1975a.   Formation of additional microspores in isolated tetrads of *Atropa belladonna* grown in microcultures.   Z. Pflanzenphysiol. 75: 464-466.

_____ 1975b.   Protoplast culture and production of haploids.   In: Form, Structure and Function in Plants, pp. 107-113.   Sarita Praka-shan Press, Meerut, India.

_____ 1976a.   Gene preservation through freeze-storage of plant cell, tissue and organ culture.   Acta Hortic. 63:75-84.

_____ 1976b.   Regeneration of plants from cell suspensions frozen at -20, -70 and -196 C.   Physiol. Plant. 37:263-268.

_____ 1977a.   In vitro induction of haploids in wheat (*Triticum aesti-vum* L.).   Crop Improv. 4:54-64.

_____ 1977b.   Protoplast isolation, culture and somatic hybridization. In: Applied and Fundamental Aspects of Plant Cell, Tissue, and Organ Culture (J. Reinert and Y.P.S. Bajaj, eds.) pp. 467-496. Springer-Verlag, Berlin, Heidelberg, New York.

_____ 1977c.   Survival of *Atropa* and *Nicotiana* pollen-embryos frozen at -196 C.   Curr. Sci. 46:305-307.

_____ 1978a.   Regeneration of haploid tobacco plants from isolated pollen grown in drop culture.   Indian J. Exp. Biol. 16:407-409.

_____ 1978b.   Effect of super-low temperature on excised anthers and pollen-embryos of *Atropa*, *Nicotiana*, and *Petunia*.   Phytomorphology 28:171-176.

_____ 1978c.   Regeneration of plants from pollen-embryos frozen at ultra-low temperature—A method for the preservation of haploids. In: Proceedings IV International Palynology Conference, Lucknow (1976-1977), Vol. 1, pp. 343-346.

_____ 1978d.   Mass production of haploids from isolated pollen and protoplasts, and their preservation by freezing.   In: Proceedings International Conference Cytogenetics and Crop Improvement, BHU, Varanasi.

_____ 1979a.   Establishment of germplasm banks through freeze-stor-age of plant tissue culture and their implications in agriculture.   In: Plant Cell and Tissue Culture—Principles and Applications (W.R. Sharp, P.O. Larsen, E.F. Paddock, and V. Raghavan, eds.) pp. 745-747.   Ohio State Univ. Press, Columbus.

_____ 1979b.   Technology and prospects of cryopreservation of germ-plasm.   Euphytica 28:267-285.

_____ 1980.   Induction of androgenesis in the rice anthers frozen at -196 C.   Cereal Res. Commun. 8:365-369.

_____ 1981a.   Plant genetic conservation through tissue culture.   Pro-ceedings International Workshop Improvement of Tropical Crops through Tissue Culture, Dacca Univ., Dacca, Bengladesh, pp. 41-46.

_____ 1981b. Regeneration of plants from ultra-low frozen anthers of *Primula obconica*. Sci. Hortic. 14:93–95.

_____ 1981c. Production of disease-resistant plants through cell culture—A novel approach. J. Nucl. Agric. Biol. 10:1–5.

_____ 1982. Survival of anther- and ovule-derived cotton callus frozen in liquid nitrogen. Curr. Sci. 51:139–140.

_____ and Cocking, E.C. 1972. The isolation and culture of microspore tetrads and microspore mother protoplasts of some angiosperms. Proceedings 3rd International Symposium Yeast Protoplast, (Salamanca) Spain, p. 79.

_____ and Davey, M.R. 1974. The isolation and ultrastructure of pollen protoplasts. In: Fertilization in Higher Plants (H.F. Linskens, ed.) pp. 73–80. North-Holland, Amsterdam.

_____ and Pierik, R.L.M. 1974. Vegetative propagation of *Freesia* through callus cultures. Neth. J. Agric. Sci. 22:153–159.

_____ and Reinert, J. 1975. Growth and morphogenesis in isolated pollen and their protoplasts. In: XII Botanical Congress, Leningrad, p. 278.

_____ and Reinert, J. 1977. Cryobiology of plant cell cultures and establishment of gene banks. In: Applied and Fundamental Aspects of Plant Cell, Tissue and Organ Culture (J. Reinert and Y.P.S. Bajaj, eds.) pp. 757–777. Springer-Verlag, Berlin, Heidelberg, New York.

_____ and Singh, H. 1980. In vitro induction of androgenesis in mung bean *Phaseolus aureus* L. Indian J. Exp. Biol. 18:1316–1318.

_____, Davey, M.R., and Grout, B.W. 1975. Pollen tetrad protoplasts: A model system for the study of ontogeny of pollen. In: Gamete Competition in Plants and Animals (D.L. Mulcahy, ed.) pp. 7–18. North-Holland, Amsterdam.

_____, Gosch, G., Ottma, M., Weber, A., and Groebler, A. 1978. Production of polyploid and aneuploid plants from anthers and mesophyll protoplasts of *Atropa belladonna* and *Nicotiana tabacum*. Indian J. Exp. Biol. 16:947–953.

_____, Labana, K.S., and Dhanju, M.S. 1980a. Induction of pollen-embryos and pollen-callus in anther cultures of *Arachis hypogaea* and *A. glabrata*. Protoplasma 103:397–399.

_____, Singh, H., and Gosal, S.S. 1980b. Haploid embryogenesis in anther cultures of pigeon pea (*Cajanus cajan*). Theor. Appl. Genet. 58: 157–159.

_____, Ram, A.K., Labana, K.S., and Singh, H. 1981. Regeneration of genetically variable plants from the anther-derived callus of *Arachis hypogaea* and *Arachis villosa*. Plant Sci. Lett. 23:35–39.

_____, Sidhu, M.M.S., and Gill, A.P.S. 1982. Some factors affecting the in vitro propagation of *Gladiolus*. Sci. Hortic. 17 (in press).

Ban, Y., Kokubu, T., and Miyaji, Y. 1971. Production of haploid plants by anther culture of *Setaria italica*. Bull. Fac. Agric. Kagoshima Univ. 21:77–81.

Barclay, I.R. 1975. High frequencies of haploid production in wheat (*Triticum aestivum*) by chromosome elimination. Nature 256:410–411.

Bennett, M.D. and Hughes, W.G. 1972. Additional mitosis in wheat pollen induced by Ethrel. Nature 240:566–568.

Bernard, S. 1980. In vitro androgenesis in hexaploid triticale—Determination of physical conditions increasing embryoid and green plant production. Z. Pflanzenzuchtung 85:308-321.

Bhojwani, S.S. and Cocking, E.C. 1972. Isolation of protoplasts from pollen tetrads. Nature New Biol. 239:29-30.

Bhojwani, S.S., Dunwell, J.M., and Sunderland, N. 1973. Nucleic acid and protein contents of embryogenic tobacco pollen. J. Exp. Bot. 24:863-871.

Binding, H. 1972. Nuclear and cell divisions in isolated pollen of Petunia hybrida in agar suspension cultures. Nature New Biol. 237:283-285.

_____ 1974a. Cell cluster formation by leaf protoplasts from axenic cultures of haploid Petunia hybrida L. Plant Sci. Lett. 2:185-188.

_____ 1974b. Mutation in haploid cell cultures. In: Haploids in Higher Plants—Advances and Potential (K.J. Kasha, ed.) pp. 323-337. Univ. Guelph Press, Guelph.

_____ 1978. Regeneration of mesophyll protoplasts isolated from dihaploid clones of Solanum tuberosum. Physiol. Plant. 43:52-54.

_____, Binding, K., and Straub, J. 1970. Selektion in Gewebekulturen mit haploiden Zellen. Naturwissenschaften 57:138-139.

Blakeslee, A., Belling, J., Farnham, M.E., and Bergner, A.D. 1922. A haploid mutant in Datura stramonium. Science 55:646-647.

Bonga, J.M. 1974. In vitro culture of microsporophylls and megagametophyte tissue of Pinus. In Vitro 9:270-277.

_____ and McInnis, A.H. 1975. Stimulation of callus development from immature pollen of Pinus resinosa by centrifugation. Plant Sci. Lett. 4:199-203.

Bourgin, J.P. 1978. Valine-induced inhibition of growth of haploid tobacco protoplasts and its reversal by isoleucine. Z. Naturforsch. 31:337-338.

_____ and Missonier, C. 1978. Culture of mesophyll protoplasts of haploid Nicotiana alata Link and Otto. Incompat. Newsl. 9:25-30.

_____ and Nitsch, J.P. 1967. Obtention de Nicotiana haploides a partir d'etamines cultivees in vitro. Ann. Physiol. Veg. (Paris) 9:377-382.

Burk, L.G., Gwynn, G.R., and Chaplin, J.F. 1972. Diploidized haploids from aseptically cultured anthers of Nicotiana tabacum. A colchicine method applicable to plant breeding. J. Hered. 63:355-360.

Butler, E.E. and Bolkan, H. 1973. A medium for heterokaryon formation in Rhizotonia solani. Phytopathology 63:542-543.

Butterfass, Th. and Kohlenbach, H.W. 1979. Monosomics of diploid Nicotiana sylvestris produced at will by androgenesis. Naturwissenschaften 66:162.

Cappadocia, M. and Ramulu, K.S. 1980. Plant regeneration from in vitro cultures of anthers and stem internodes in an interspecific hybrid, Lycopersicon esculentum L. x L. peruvianum Mill. and cytogenetic analysis of the regenerated plants. Plant Sci. Lett. 20:157-166.

Carlson, P. 1970. Induction and isolation of auxotrophic mutants in somatic cultures of Nicotiana tabacum. Science 168:487-489.

_____ 1973.   Methionine sulfoximine-resistant mutants of tobacco. Science 180:1366.

Chaleff, R.S. and Stolarz, A. 1981.   Factors influencing the frequency of callus formation among cultured rice (Oryza sativa) anthers. Physiol. Plant. 51:201-206.

Chang, Z. and Hong-Yuan, Y. 1981.   Induction of haploid rice plantlets by ovary culture. Plant Sci. Lett. 20:231-237.

Chase, S.S. 1969.   Monoploids and monoploid-derivatives of maize (Zea mays L.). Bot. Rev. 35:117-167.

Chen, C.-C. 1978.   Effects of sucrose concentration on plant production in anther culture of rice. Crop Sci. 19:905-906.

_____ and Chen, C.-M. 1980.   Changes in chromosome number in microspore callus of rice during successive subcultures. Can. J. Genet. Cytol. 22:607-614.

Chen, C.H.,   Chen, F.T., Chien, C.F., Wang, C.H., Chang, S.J., Hsu, H.E., Ou, H.H., Ho, Y.T., and Lu, T.M. 1979.   A process for obtaining pollen plants of Hevea brasilliensis Muell-Arg. Sci. Sinica 22:81-90.

Chinese Workers, 1976.   A sharp increase of the frequency of pollen plant induction in wheat with potato medium. Acta. Genet. Sinica 3:25-31.

Clapham, D. 1971.   In vitro development of callus from the pollen of Lolium and Hordeum. Z. Pflanzenzuchtg. 65:285-292.

_____ 1973.   Haploid Hordeum plants from anthers in vitro. Z. Pflanzenzuchtg. 62:305-310.

_____ 1977.   Haploid induction in cereals. In: Applied and Fundamental Aspects of Plant Cell, Tissue, and Organ Culture (J. Reinert and Y.P.S. Bajaj, eds.) pp. 279-298. Springer-Verlag, Berlin, Heidelberg, New York.

Clowes, F.L. and Juniper, B.E. 1968.   Plant Cell Botanical Monographs 8, Blackwell, Oxford.

Collins, G.B. 1977.   Production and utilization of anther-derived haploids of two Nicotiana species. J. Hered. 63:113-118.

_____ and Sunderland, N. 1974.   Pollen-derived haploids of Nicotiana knightiana, N. raimondii and N. attenuata. J. Exp. Bot. 25:1030-1039.

Corduan, G. 1975.   Regeneration of anther-derived plants from Hyoscyamus niger L. Planta 127:27-36.

_____ and Spix, C. 1975.   Haploid callus and regeneration of plants from anthers of Digitalis purpurea L. Planta 124:1-11.

D'Amato, F. 1977.   Cytogenetics of differentiation in tissue and cell cultures. In: Applied and Fundamental Aspects of Plant Cell, Tissue, and Organ Culture (J. Reinert and Y.P.S. Bajaj, eds.) pp. 343-357. Springer-Verlag, Berlin, Heidelberg, New York.

Day, A.W. and Jones, J.K. 1971.   p-Fluorophenylalanine-induced mitotic haploidization in Ustilago violacea. Genet. Res. 18:299-309.

Deanon, J.R. 1957.   Treatment of sweet corn silks with maleic hydrazide and colchicine as a means of increasing the frequency of monoploids. Phillipp. Agric. 41:364-377.

Debergh, P. and Nitsch, C. 1973.   Premiers resultans sur la culture in vitro de grains de pollen isoles chez la tomate. C.R. Acad. Sci. Ser. D. 276:1281-1284.

De Buyser, J. and Henry, Y. 1980.   Induction of haploid and diploid plants through in vitro anther culture of haploid wheat (n = 3x = 21).   Theor. Appl. Genet. 57:57-58.

De Mol, W.E. 1923.   Duplication of generative nuclei by means of physiological stimuli and its significance.   Genetica 5:225-227.

De Paepe, R., Bleton, D., and Gnangbe, F. 1981.   Basis and extent of genetic variability among doubled haploid plants obtained by pollen culture in Nicotiana sylvestris.   Theor. Appl. Genet. 59:177-184.

Devreux, M. and Saccardo, F. 1971.   Mutazioni spermimentali observate su piante aploidi di tabacco ottenute per colture in vitro di antere irradiate.   Atti. Assoc. Genet. Ital. 16:69-71.

Devreux, M., Laneri, U., and Brunori, A. 1971.   Plantes haploids et lignes isogeniques de Nicotiana tabacum obtenues par cultures d'anthers et tiges in vitro.   Caryologia 24:141-148.

Dix, P.J. and Street, H.E. 1974.   Effects of p-fluorophenylalanine on the growth of cell lines differing in ploidy and derived from Nicotiana sylvestris.   Plant Sci. Lett. 3:283-288.

_____ 1976.   Selection of plant cell lines with enhanced chilling resistance.   Ann. Bot. 40:903-910.

Dodds, J.H. and Reynolds, T.L. 1980.   A scanning electron microscope study of pollen embryogenesis in Hyoscyamus niger Z. Pflanzenphysiol. 97:271-276.

Dore Swamy, R. and Chacko, E.K. 1973.   Induction of plantlets and callus from anthers of Petunia axillaris (Lam.) B.S.P. cultured in vitro.   Hortic. Res. 13:41-44.

Drira, N. and Bendadis, A. 1975.   Analysis of the androgenetic potentialities of two Citrus species (C. medica L. and C. Limon (L) Burm.) by anther culture.   C.R. Acad. Sci. Ser. B 281:1321-1324.

Dunwell, J.M. and Sunderland, N. 1973.   Anther culture of Solanum tuberosum L.   Euphytica 22:317-323.

_____ 1974a.   Pollen ultrastructure in anther culture of Nicotiana tabacum.   I. Early stages of culture.   J. Exp. Bot. 25:352-361.

_____ 1974b.   Pollen ultrastructure in anther culture of Nicotiana tabacum.   II. Changes associated with embryogenesis.   J. Exp. Bot. 25:363-373.

_____ 1975.   Pollen ultrastructure in anther culture of Nicotiana tabacum.   III. The first sporophytic division.   J. Exp. Bot. 26:240-252.

Durr, A. and Fleck, J. 1980.   Production of haploid plants of Nicotiana langsdorffii.   Plant Sci. Lett. 18:75-79.

Eapen, S. 1976.   Effect of gamma- and ultraviolet-irradiation on survival and totipotency of haploid tobacco cells in culture.   Protoplasma 89:149-155.

Eigsti, O.J. and Dustin, P. 1955. Colchicine in Agriculture, Medicine, Biology and Chemistry. Iowa State College Press. Ames.

Elsik, W.C. 1971.   Microbiological degradation of sporopollenin.   In: Sporopollenin (J. Brocks, P.R. Grant, M. Muir, P. van Gijzel, and G. Shaw, eds.) pp. 480-511.   Academic Press, New York.

Engvild, K.C. 1973.   Triploid petunias from anther cultures.   Hereditas 74:144-147.

_____ 1974.   Plantlet ploidy and flower-bud size in tobacco anther cultures. Hereditas 76:320–322.

_____, Linde-Laursen, I., and Lundquist, A. 1972.   Anther cultures of *Datura innoxia*: Flower bud stage and embryoid level of ploidy. Hereditas 72:331–332.

Ernst, R. 1974.   The use of activated charcoal in asymbiotic seedling culture of *Paphiopedilum*. Am. Orchid Soc. Bull. 43:35–38.

Faegri, K. and Iversen, J. 1964.   Textbook of Pollen Analysis.   Munksgaard, Copenhagen.

Fitch, M.M. and Moore, P.M. 1981.   Anther culture production of haploids from *Saccharum spontaneum*, a tropical grass related to sugarcane. Plant Physiol. (Suppl.) 4:137 (abstract).

Forche, E., Kibler, R., and Neumann, K.-H. 1981.   The influence of developmental stages of haploid and diploid callus cultures of *Datura innoxia* on shoot initiation. Z. Pflanzenphysiol. 101:257–262.

Fowler, C.W., Hughes, H.G., and Janick, J. 1971.   Callus formation from strawberry anthers. Hortic. Res. 11:116–117.

Fridborg, G. and Eriksson, T. 1975.   Effects of activated charcoal on growth and morphogenesis in cell cultures. Physiol. Plant. 34:306–308.

Fujii, T. 1970.   Callus formation in wheat anthers.   Wheat Inf. Serv. 31:1–4.

Furner, I.J., King, J., and Gamborg, O.L. 1978.   Plant regeneration from protoplasts isolated from a predominantly haploid suspension culture of *Datura innoxia* (Mill). Plant Sci. Lett. 11:169–176.

Galun, E. and Raveh, D. 1975.   In vitro culture of tobacco protoplasts: Survival of haploid and diploid protoplasts exposed to X-ray radiation at different times after isolation. Radiat. Bot. 15:79–82.

Gamborg, O.L., Miller, R.A., and Ojima, O. 1968.   Nutrient requirements of suspension cultures of soybean root cell. Exp. Cell Res. 50:151–158.

Gebhardt, C., Schnebli, V., and King, P.J. 1981.   Isolation of biochemical mutants using haploid mesophyll protoplasts of *Hyoscyamus muticus* II.   Auxotrophic and temperature-sensitive clones.   Planta 153:81–89.

Geier, T. and Kohlenbach, H.W. 1973.   Development of embryos and embryonic callus from pollen grains of *Datura metaloides* and *D. innoxia*. Protoplasma 78:381–396.

Geitler, L. 1961.   Embryosacke aus Pollenkornern bei *Ornithocalum*. Ber. Dtsch. Bot. Ges. 59:419–423.

George, L. and Narayanaswamy, S. 1973.   Haploid *Capsicum* through experimental androgenesis. Protoplasma 78:467–470.

Georgiev, G. and Chavdarov, L. 1974.   Genet. Sel. 7:404–415.

Gherardini, G.L. and Healey, P.L. 1969.   Dissolution of outer wall of pollen grain during pollination. Nature 224:218–219.

Gressel, J., Zilkah, S., and Ezra, G. 1978.   Herbicide action, resistance and screening in cultures vs. plants. In: Frontiers of Plant Tissue Culture 1978 (T.A. Thorpe, ed.) pp. 427–436.   Univ. Calgary Press, Calgary.

Gresshoff, P.M. and Doy, C.H. 1972a.   Development and differentiation of haploid *Lycopersicon esculentum* (tomato). Planta 107:161–170.

_____ 1972b. Haploid *Arabidopsis thaliana* callus and plants from anther culture. Aust. J. Biol. Sci. 25:259-264.

_____ 1974. Derivation of a haploid cell line from *Vitus vinifera* and the importance of the stage of meiotic development of anthers for haploid culture of this and other genera. Z. Pflanzenphysiol. 73:132-141.

Guha, S. and Maheshwari, S.C. 1964. In vitro production of embryos from anthers of *Datura*. Nature 204:497.

_____ 1967. Development of embryoids from pollen grains of *Datura* in vitro. Phytomorphology 17:454-461.

Guha-Mukherjee, S. 1973. Genotypic differences in the in vitro formation of embryoids from rice pollen. J. Exp. Bot. 24:139-144.

Guo, C. 1972. Effects of chemical and physical factors on the chromosome number in *Nicotiana* anther callus cultures. In Vitro 7:381-386.

Gupta, N. and Carlson, P.S. 1972. Preferential growth of haploid plant cells in vitro. Nature New Biol. 239:86.

Gupta, S.C. and Babbar, S.B. 1980. Enhancement of plantlet formation in anther cultures of *Datura metel* L. by pre-chilling of buds. Z. Pflanzenphysiol. 96:465-470.

Guy, I., Raquin, C., and de Marley, Y. 1979. Haploid and diploid plants obtained in vitro from egg plant anthers (*Solanum melongena*). C.R. Acad. Sci. 288:987-989.

Harn, C. 1971. Studies on the anther culture in *Solanum nigrum*. SABRAO Newsl. 3:39-42.

_____ 1972. Production of plants from anthers of *Solanum nigrum* cultivated in vitro. Caryologia 25:429-437.

_____ and Kim, M.Z. 1972. Induction of callus from anthers of *Prunus armenica*. Korean J. Breed. 4:49-53.

Heinz, D.J., Krishnamurthi, M., Nickell, L.G., and Maretzki, A. 1977. Cell, tissue and organ culture in sugarcane improvement. In: Applied and Fundamental Aspects of Plant Cell, Tissue and Organ Culture (J. Reinert and Y.P.S. Bajaj, eds.) pp. 3-17. Springer-Verlag, Berlin, Heidelberg, New York.

Hepler, P.K. and Palevitz, B.A. 1974. Microtubules and microfilaments. Annu. Rev. Plant Physiol. 25:309-362.

Hermsen, J.G.T. and Ramanna, M.S. 1981. Haploidy in plant breeding. Philos. Trans. R. Soc. London. Ser. B 292:499-507.

Hesemann, C.U. 1980. Haploid cells in calli from anther culture of *Vicia faba*. Z. Pflanzenzuech. 84:18.

Hess, D. and Wagner, G. 1974. Induction of haploid parthenogenesis in *Mimulus luteus* by in vitro pollination with foreign pollen. Z. Pflanzenphysiol. 72:466-468.

Hirabayashi, T., Kozaki, I., and Akihama, T. 1976. In vitro differentiation of shoots from anther callus in *Vitis*. Hortic. Sci. 11:511-512.

Hlasnikova, A. 1977. Androgenesis in vitro evaluated from the aspects of genetics. Z. Pflanzenzuech. 78:44-56.

Ho, K.M. and Kasha, K.J. 1975. Genetic control of chromosome elimination during haploid formation in barley. Genetics 8:263-275.

Hondelmann, W. and Wilberg, B. 1973. Breeding all male varieties of asparagus by utilization of anther culture and tissue culture. Z. Pflanzenzuech. 69:19-24.

Hu, H., Hsi, T.Y., Tseng, C.C., Quyang, T.W., and Ching, C.K. 1978. Application of anther culture to crop plants. In: Frontiers of Plant Tissue Culture (T.A. Thorpe, ed.) pp. 123–130. Univ. Calgary Press, Calgary.

Hughes, K.W., Bell, S.L., and Caponetti, J.D. 1975. Anther–derived haploids of the African violet. Can. J. Bot. 53:1442–1444.

Irikura, Y. and Sakaguchi, S. 1972. Induction of 12 chromosome plants from anther culture in a tuberous *Solanum verrucosum*. Potato Res. 15:170–173.

_____ 1975. Induction of haploid plants by anther culture in tuber-bearing species and interspecific hybrids of *Solanum*. Potato Res. 18: 133–140.

Ivers, D.R., Palmer, R.G., and Fehr, W.R. 1974. Anther culture in soybeans. Crop. Sci. 14:891–893.

Iyer, R.D. and Raina, S.K. 1972. The early ontogeny of embryoids and callus from pollen and subsequent organogenesis in anther cultures of *Datura metel* and rice. Planta 104:146–156.

Izhar, S. 1973. Cell budding and fission in microspores of *Petunia*. Nature 244:35–37.

Jensen, C.J. 1974. Production of monoploids in barley: A progress report. In: Polyploidy and Induced Mutations in Plant Breeding, pp. 167–169. Int. Atomic Energy Agency, PL–503/24.

Johansson, L. and Eriksson, T. 1977. Induced embryo formation in anther cultures of several *Anemone* species. Physiol. Plant. 40:172–174.

Jordan, M. 1974. Multizellulare pollen bei *Prunus avium* nach in vitro. Z. Pflanzenzuech. 71:358–363.

Kameya, T. and Hinata, K. 1970. Induction of haploid plants from pollen grains of *Brassica*. Jpn. J. Breed. 20:82–87.

Kanta, K. 1960. Intra–ovarian pollination in *Papaver rhoeas*. Nature 188:683–684.

_____ and Maheshwari, P. 1963. Intraovarian pollination in some Papaveraceae. Phytomorphology 13:215–229.

Kao, K.N. 1981. Plant formation from barley anther cultures with Ficoll media. Z. Pflanzenphysiol. 103:437–443.

Kasha, K.J. (ed.) 1974. Haploids in Higher Plants: Advances and Potential. Univ. Guelph Press, Guelph.

_____ and Kao, K.N. 1970. High frequency haploid production in barley (*Hordeum vulgare* L.). Nature 225:874–876.

Kasperbauer, M.J. and Collins, G.B. 1972. Reconstitution of diploids from leaf tissue of anther–derived haploids in tobacco. Crop Sci. 12: 98–101.

Kasperbauer, M.J., Buckner, R.C., and Springer, W.D. 1980. Haploid plants by anther–panicle culture of tall Fescue. Crop Sci. 20:103.

Kato, Y. 1973. Active charcoal and vermiculite. Effective agents on growth and morphogenesis of fern gametophytes. Phytomorphology 23: 260–263.

Keller, W.A. and Armstrong, K.C. 1977. Embryogenesis and plant regeneration in *Brassica napus* anther cultures. Can. J. Bot. 55:1383–1388.

_____ 1979. Stimulation of embryogenesis and haploid production in *Brassica campestris* anther cultures by elevated temperature treatments. Theor. Appl. Genet. 55:65–67.

Keller, W.A. and Stringam, G.R. 1978. Production and utilization of microspore–derived haploid plants. In: Frontiers of Plant Tissue Culture (T.A. Thorpe, ed.) pp. 113–122. Calgary Univ. Press, Calgary.

Keller, W.A., Rajpathy, T., and Lacapre, J. 1975. In vitro production of plants from pollen in *Brassica campestris*. Can. J. Genet. Cytol. 17:655–666.

Keller, W.A., Armstrong, K.C., and de la Roche, I.A. 1981. The production and utilization of microspore–derived haploids in cruciferous crop species. Proceedings International Symposium Plant Cell Culture in Crop Improvement, Calcutta.

Kimata, M. and Sakamoto, S. 1972. Production of haploid albino plants of *Aegilops* by anther culture. Jpn. J. Genet. 47:61–63.

Kimber, G. and Riley, R. 1963. Haploid angiosperms. Bot. Rev. 29: 480–631.

Klein, B. and Bopp, M. 1971. Effect of activated charcoal in agar on the culture of lower plants. Nature 230:474.

Kochhar, T., Sabharwal, P., and Engelberg, J. 1971. Production of homozygous diploid plants by tissue culture technique. J. Hered. 62: 59–61.

Kohlenbach, H.W. and Geier, T. 1972. Embryonen aus in vitro kultivierten Antheren von *Datura meteloides* Dun, *Datura wrightii* Regel and *Solanum tuberosum*. L. Z. Pflanzenphysiol. 67:161–165.

Konar, R.N. 1963. A haploid tissue from the pollen of *Ephedra foliata* Boiss. Phytomorphology 13:170–174.

Kubicki, B., Telezynska, J., and Milewska-Pawliczuk, E. 1975. Induction of embryoid development from apple pollen grains. Acta Soc. Bot. Pol. 44:631–635.

Kuo, J.S., Wang, Y.Y., Chien, N.F., Ku, S.J., Kung, M.L., and Hsu, H.C. 1973. Investigations on the anther culture in vitro of *Nicotiana tabacum* L. and *Capsicum annuum* L. Actoa Bot. Sininca 15:36–52.

Ladeinde, T.A. and Bliss, F.A. 1977. A preliminary study on the production of plantlets from anthers of cowpea. Trop. Grain Legume Bull. 8:13.

Levan, A. 1945. A haploid sugar beet after colchicine treatment. Hereditas 31:399–410.

Levenko, B.A., Kunakh, V.A., and Yurkova, G.N. 1977. Studies on callus tissue from anthers. I. Tomato. Phytomorphology 27:377–383.

Lower, R.L. and Miller, C.H. 1969. Ethrel (2-chloroethanephosphonic acid) a tool for plant hybridizers. Nature 22:1072–1073.

MacDonald, I.M. and Grant, W.F. 1974. Anther culture of pollen containing ethrel induced micronuclei. Z. Pflanzenzuech. 73:292–297.

Magoon, M.L. and Khanna, K.R. 1963. Haploids. Caryologia 16:191–235.

Maheshwari, P. 1950. An Introduction to the Embryology of Angiosperms. McGraw-Hill, New York.

_____ and Rangaswamy, N.S. 1965. Embryology in relation to physiology and genetics. In: Advances in Botanical Research (R.D. Preston, ed.) p. 219. Academic Press, New York.

Maheshwari, S.G., Tyagi, A.K., and Malhotra, K. 1980. Induction of haploidy from pollen grains in angiosperms—The current status. Theor. Appl. Genet. 58:193-206.

Maliga, P., Breznovits, A.S., and Marton, L. 1973a. Streptomycin-resistant plants from callus culture of haploid tobacco. Nature New Biol. 244:29.

Maliga, P., Marton, L., and Breznovits, A.S. 1973b. 5-bromodeoxy-uridine-resistant cell lines from haploid tobacco. Plant Sci. Lett. 1:119-121.

Matthews, P.S. and Vasil, I.K. 1975. The dynamics of cell proliferation in haploid and diploid tissues of *Nicotiana tabacum*. Z. Pflanzenphysiol. 77:222-236.

Mattingly, C.F. and Collins, G.B. 1974. Use of anther-derived haploids in *Nicotiana* III. Isolation of nullisomics from monosomic lines. Chromosoma 46:29-36.

Melchers, G. 1972. Haploid higher plants for plant breeding. Z. Pflanzenzuech. 67:19-32.

_____ and Labib, G. 1970. Die Bedeutung haploider hoehrer Pflanzen fuer Pflanzenphysiologie und Pflanzenzuechtung. Durch Antherenkultur erzeugte. Haploide, ein neuer Durchbruch fuer die Pflanzenzuechtung. Ber. Dtsch. Bot. Ges. 83:129-150.

Michellon, R., Hugard, J., and Jonard, R. 1974. Sur l'isolement de colonies tissulaires de pecher (*Prunus persica* Batsch, cultivars Dixired et Nectared iv) et Amandier *Prunus amygdalus* Stokes cultivar 'A 1), a partir d' antheres cultivees in vitro. C.R. Acad. Sci. Ser. D 278:1719-1722.

Mitchell, A.Z., Hanson, M.R., Skvirsky, R.C., and Ausubel, F.M. 1980. Anther culture of *Petunia*: Genotypes with high frequency of callus, root or plantlet formation. Z. Pflanzenphysiol. 100:131-146.

Mokhtarzadeh, A. and Constantin, M.J. 1978. Plant regeneration from hypocotyl and anther-derived callus of Barseem clover. Crop Sci. 18:567-572.

Mroginski, L.A. and Fernandez, A. 1980. Obtainment of plantlets by in vitro culture of anthers of wild species of *Arachis* (Leguminosae). Oleagineux 35:89-92.

Muentzing, A. 1937. Note on a haploid rye plant. Hereditas 23:401.

Murakami, M., Takahashi, N., and Harada, K. 1972. Induction of haploid plant by anther culture in maize. I. On the callus formation and root differentiation. Kyoto Prefect Univ. Fac. Agric. Sci. Rep. 24:1-8.

Murashige, T. and Skoog, F. 1962. A revised medium for rapid growth and bioassays with tobacco tissue cultures. Physiol. Plant. 15:473-497.

Murgai, P. 1959. In vitro culture of the inflorescences, flowers and ovaries of an apomict, *Aerva tomentosa* Forsk. Nature 184:72-73.

Nabors, M.W., Gibbs, S.E., Bernstein, C.S., and Meis, M.E. 1980. NaCl-tolerant tobacco plants from cultured cells. Z. Pflanzenphysiol. 97:13-17.

Naithani, S.P. 1937. Chromosome studies in *Hyacinthus orientalis* L. III. Reversal of sexual state in the anthers of *H. orientalis* L. var. *Yellow Hammer*. Ann. Bot. 1:369-377.

Nakata, K. and Tanaka, M. 1968. Differentiation of embryoids from developing germ cells in anther culture of tobacco. Jpn. J. Genet. 43:65-71.

Narayanaswamy, S. and Chandy, L.P. 1971. In vitro induction of haploid, diploid and triploid androgenic embryoids and plantlets in *Datura metel* L. Ann. Bot. 35:535-542.

Narayanaswamy, S. and George (nee Chandy) L. 1972. Morphogenesis of belladonna (*Atropa belladonna* L.) plantlets from pollen in culture. Indian J. Exp. Biol. 10:382-384.

Nemec, B. 1898. Ueber den Pollen der petaloiden Antheren von *Hyacinthus orientalis* L. Rozp. Cesk. Akad. Prag. II 7:17.

Niizeki, H. 1968. Induction of haploid rice plant from anther culture. Jpn. Agric. Res. Q. 3:41-45.

Niizeki, M. and Grant, W.F. 1971. Callus, plantlet formation and polyploidy from cultured anthers of *Lotus* and *Nicotiana*. Can. J. Bot. 49:2041-2051.

Niizeki, M. and Kita, F. 1974. Studies on plant cell and tissue culture III. In vitro induction of callus from anther culture of forage crops. J. Fac. Agric. Hokkaido Univ. 57:293-300.

Nishi, T. and Mitsuoka, S. 1969. Plants from anther and ovary culture of rice plant. Jpn. J. Genet. 44:341-346.

Nitsch, C. 1974. La culture de pollen isole sur milieu synthetique. C.R. Acad. Sci. Ser. D 278:1031-1034.

_____ 1975. Single cell culture of a haploid cell: The microspore. In: Genetic Manipulations with Plant Material (L. Ledoux, ed.) pp. 297-310. Plenum Press, London.

_____ 1977. Culture of isolated microspores. In: Applied and Fundamental Aspects of Plant Cell, Tissue, and Organ Culture (J. Reinert and Y.P.S. Bajaj, eds.) pp. 268-278. Springer-Verlag, Berlin, Heidelberg, New York.

_____ and Norreel, B. 1973a. Effect d'un choc thermique sur le pouvoir embryogene du pollen de *Datura innoxia* cultive dans l'anthere ou isole de l'anthere. C.R. Acad. Sci. Ser. D 276:303-306.

_____ and Norreel, B. 1973b. Factors favouring the formation of androgenetic embryos in anther culture. In: Genes, Enzymes and Populations (Adrian M. Srb, ed.) pp. 129-144. Plenum, New York.

Nitsch, J.P. 1951. Growth and development in vitro of excised ovaries. Am. J. Bot. 38:566-577.

_____ 1969. Experimental androgenesis in *Nicotiana*. Phytomorphology 19:389-404.

_____ 1972. Haploid plants from pollen. Z. Pflanzenzuechtung. 67:3-18.

_____ and Nitsch, C. 1969. Haploid plants from pollen grains. Science 163:85-87.

_____, Nitsch, C., and Hamon, S. 1969. Production de *Nicotiana diploides* a partir de cals haploides cultives in vitro. C.R. Acad. Sci. Ser. D 269:1275-1278.

Nitzsche, W. 1970. Herstellung haploider Pflanzen aus *Festuca lolium* Bastarden. Naturwissenschaften 57:199-200.

_____ and Wenzel, G. 1977. Haploids in plant breeding. Verlag Paul Pary, Berlin.

Noerdenskiold, H. 1939.   Studies of a haploid rye plant.   Hereditas 28: 204-210.
Norreel, B. 1970.   Etude cytologique de l'androgenese experimentale chez *Nicotiana tabacum* et *Datura innoxia*. Bull. Soc. Bot. Fr. 117: 461-478.
Novak, F.J. 1974.   Induction of a haploid callus in anther cultures of *Capsicum* sp. Z. Pflanzenzuech. 72:46-54.
_____ and Vyskot, B. 1975.   Karyology of callus cultures derived from *Nicotiana tabacum* L. haploids and ploidy of regenerants.   Z. Pflanzenzuech. 75:62-70.
Ohyama, K. and Nitsch, J.P. 1972.   Flowering haploid plants obtained from protoplasts of tobacco leaves. Plant Cell Physiol. 13:423-428.
Ono, H. and Larter, E.N. 1976.   Anther culture of *Triticale*.   Crop Sci. 16:120-121.
Ono, K. and Tsukida, T. 1978.   Haploid callus formation from anther cultures in a cultivar of *Paeonia*. Jpn. J. Genet. 53:51-54.
Opatrny, Z., Dostal, J., and Martineck, V. 1977.   Anther culture of maize. Biol. Plant. 19:477.
Ouyang, T.-W., Hu, H., Chuang, C.-C., and Tseng, C.-C. 1973.   Induction of pollen plants from anthers of *Triticum aestivum* L. cultured in vitro. Sci. Sinica 16:79-95.
Parmentier, G. 1970.   L'incorporation de charbon actif aux milieux de culture de *Phytophthora infestans*. Parasitica 26:31-40.
Pelletier, G. 1973.   Les conditions et les premiers standes de l'androgenese in vitro chez *Nicotiana tabacum*. Mem. Soc. Bot. Fr. 1973: 261-268.
_____, Raquin, C., and Simon, G. 1972.   La culture in vitro d'anthere d'Asperge (*Asparagus officinalis*). C.R. Acad. Sci. Ser. D 274:848-851.
Peters, J.E., Crocomo, O.J., Sharp, W.R., Paddock, E.F., Tegenkamp, I., and Tegenkamp, T. 1977.   Haploid callus cells from anthers of *Phaseolus vulgaris*. Phytomorphology 27:79-85.
Picard, E. 1973.   Influence de modifications dans les correlations internes sur le devenir du gametophyte male de *Triticum aestivum* L. in situ et en culture in vitro. C.R. Acad. Sci. Ser. D 277:777-780.
_____ and de Buyser, J. 1973.   Obtension de plantules haploides de *Triticum aestivum* L. a partir de culture d'antheres in vitro. C.R. Acad. Sci. Ser. D 277:1463-1466.
Povolochko, P.A. 1937.   Experimental production of haploid plants in the genus *Nicotiana*. Bull. Appl. Bot. Genet. Plant Breed. Ser. II 7:175-190.
Proskauer, J. and Berman, R. 1970.   Agar culture medium modified to approximate soil conditions. Nature 227:1161.
Quazi, H.M. 1978.   Regeneration of plants from anthers of *Broccoli* (*Brassica oleracea* L.). Ann. Bot. 42:473-475.
Radojevic, L. 1978.   In vitro induction of androgenic plantlets in *Aesculus hippocastanum*. Protoplasma 96:369-374.
Raghavan, V. 1975.   Induction of haploid plants from anther culture of henbane. Z. Pflanzenphysiol. 76:89-92.
_____ 1978.   Origin and development of pollen embryoids and pollen calluses in cultured anther segments of *Hyoscyamus niger* (Henbane). Am. J. Bot. 65:982-1002.

Raina, S.K. and Iyer, R.D. 1973. Differentiation of diploid plants from pollen callus in anther cultures of *Solanum melongena* L. Z. Pflanzen-zuech. 70:275–280.

Rajasekaran, K. and Mullins, M.G. 1979. Embryos and plantlets from cultured anthers of hybrid grapevines. J. Exp. Bot. 30:399–407.

Ram, M. 1959. Occurrence of embryo sac-like structures in the micro-sporangia of *Leptomeria billardierri* R. Br. Nature 184:914.

Rangaswamy, N.S. 1977. Applications of in vitro pollination and in vit-ro fertilization. In: Applied and Fundamental Aspects of Plant Cell, Tissue, and Organ Culture (J. Reinert and Y.P.S. Bajaj, eds.) pp. 412–415. Springer–Verlag, Berlin, Heidelberg, New York.

Raquin, C. and Pilet, V. 1972. Production de plantules a partir d'an-theres de Petunias cultivees in vitro. C.R. Acad. Sci. Ser. D 274: 1019–1022.

Rashid, A. 1981. Induction of embryos in ab initio pollen cultures of *Nicotiana*. In: Proceedings International Symposium Plant Cell Cul-ture in Crop Improvement, Bose Institute, Calcutta.

_____ and Street, H.E. 1973. The development of haploid embryoids from anther cultures of *Atropa belladonna* L. Planta 113:263–270.

_____ and Street, H.E. 1974a. Growth embryogenic potential and sta-bility of a haploid cell culture of *Atropa belladonna* L. Plant Sci. Lett. 2:89–90.

_____ and Street, H.E. 1974b. Segmentations in microspores of *Nico-tiana sylvestris* and *Nicotiana tabacum* which lead to embryoid forma-tion in anther cultures. Protoplasma 80:323–334.

Razmologov, V.P. 1973. Tissue culture from the generative cell of the pollen grain of *Cupressus* spp. Bull. Torrey Bot. Club 100:18–22.

Redenbaugh, M.K., Klestfall, R.D., and Karnosky, D.F. 1981. Dihaploid callus production from *Ulmus americana* anthers. Bot. Gaz. 142:19–26.

Reinert, J., Bajaj, Y.P.S., and Heberle, E. 1975. Induction of haploid tobacco plants from isolated pollen. Protoplasma 84:191–196.

Rosati, P., Devreux, M., and Laneri, U. 1975. Anther culture of straw-berry. Hortic. Sci. 10:119–120.

Rowell, P.L. and Miller, D.G. 1971. Induction of male sterility in wheat with 2-chloroethylphosphonic acid (Ethrel). Crop Sci. 11:629–631.

Sacristan, M.D. 1971. Karyotypic changes in callus cultures from hap-loid plants of *Crepis capillaris* (L.) Wallr. Chromosoma 33:273–283.

Sangwan, R.S. and Norreel, B. 1975. Pollen embryogenesis in *Parbitis nil*. Naturwissenschaften 62:440.

Sangwan-Norreel, B.S. 1977. Androgenic stimulating factors in the an-thers and isolated pollen grain culture of *Datura innoxia* Mill. J. Exp. Bot. 28:843–852.

Satchuthananthavale, R. and Irugalbundara, Z.E. 1972. Propagation of callus from *Hevea* anthers. Q. J. Rubber Res. Inst. Ceylon 49:65–68.

Sato, T. 1974. Callus induction and organ differentiation in anther cul-ture of poplars. J. Jpn. For. Soc. 56:55–62.

Sax, K. 1935. The effect of temperature on nuclear differentiation in microspore development. J. Arnold Arbor. 19:301–310.

284   Basic Techniques of Plant Cell Culture

Schaeffer, G.W., Baenziger, P.S., and Worley, J. 1979.   Haploid plant development from anthers and in vitro embryo culture of wheat. Crop Sci. 19:696-702.

Schieder, O. 1975.   Regeneration von haploiden und diploiden *Datura innoxia* Mill. Mesophyll-Protoplasten zu Pflanzen.   Z. Pflanzenphysiol. 76:462-466.

Scholl, R.L. and Amos, J.A. 1980.   Isolation of doubled-haploid plants through anther culture in *Arabidopsis thaliana*.   Z. Pflanzenphysiol. 96:407-414.

Seirlis, G., Mouras, A., and Salesses, G. 1979.   In vitro culture of anthers and organ fragments.   *Prunus annals* Del L. Amelior. Plant. 29: 145-162.

Sharp, W.R., Raskin, R.S., and Sommer, H.E. 1971a.   Haploidy in *Lilium*. Phytomorphology 21:334-336.

Sharp, W.R., Dougall, D.K., and Paddock, E.F. 1971b.   Haploid plantlets and callus from immature pollen grains of *Nicotiana* and *Lycopersicon*.   Bull. Torrey Bot. Club 98:219-222.

_____ 1972.   The use of nurse culture in the development of haploid clones in tomato.   Planta 104:357-361.

Sharp, W.R., Caldas, L.S., and Crocomo, O.J. 1973.   Studies on the induction of *Coffea arabica* callus from both somatic and microsporogenous tissues; and subsequent embryoid and plantlet formation.   Am. J. Bot. 60:13.

Shaw, G. 1971.   The chemistry of sporopollenin.   In: Sporopollenin (J. Brocks, P.R. Grant, M. Muir, P. van Gijzel, and G. Shaw, eds.) pp. 305-350.   Academic Press, New York.

Shimada, T. 1981.   Haploid plants regenerated from the pollen callus of wheat (*Triticum aestivum* L.).   Jpn. J. Genet. 56:581-588.

Sinha, S., Jha, K.K., and Roy, R.P. 1978a.   In vitro development of callus from anthers in *Luffa cylindrica*.   Curr. Sci. 48:120-121.

_____ 1978b.   Segmentation pattern of pollen in anther cultures of *Solanum surattense, Luffa cylindrica* and *Luffa echinata*.   Phytomorphology 28:43-49.

_____ 1979.   Callus formation and shoot bud differentiation in anther culture of *Solanum surattense*.   Can. J. Bot. 57:2524-2527.

Skirvin, R.M. 1978.   Natural and induced variation in tissue culture. Euphytica 27:241-266.

Smith, H.H. 1943.   Studies on induced heteroploids of *Nicotiana*.   Am. J. Bot. 30:121-130.

Sopory, S.K. 1977.   Development of embryoids in isolated pollen culture of dihaploid *Solanum tuberosum*.   Z. Pflanzenphysiol. 84:453-457.

_____ 1979.   Effect of sucrose, hormones and metabolic inhibitors on the development of pollen embryoids in anther culture of dihaploid *Solanum tuberosum*.   Can. J. Bot. 57:2691-2694.

_____ and Tan, B.H. 1979.   Regeneration and cytological studies of anther and pollen calli of dihaploid *Solanum tuberosum*.   Z. Pflanzenzuecht.82:31-35.

_____, Jacobsen, E., and Wenzel, G. 1978.   Production of monoploid embryoids and plantlets in cultured anthers of *Solanum tuberosum*. Plant Sci. Lett. 12:47-54.

Southworth, D. 1974. Solubility of pollen exines. Am. J. Bot. 61:36-44.

Stow, I. 1930. Experimental studies on the formation of embryo sac-like giant pollen grain in the anther of *Hyacinthus orientalis*. Cytyologia 1:417-439.

Sunderland, N. 1974. Anther culture as a means of haploid induction. In: Haploids in Higher Plants: Advances and Potential (K.J. Kasha, ed.) pp. 91-122. Guelph Univ. Press, Guelph.

_____ and Roberts, M. 1977. New approach to pollen culture. Nature 270:236-238.

_____ and Wildon, D.C. 1979. A note on the pretreatment of excised flower buds in float culture of *Hyoscyamus* anthers. Plant Sci. Lett. 15:169-175.

Tanaka, M. and Nakata, K. 1969. Tobacco plants obtained by anther culture and the experiment to get diploid seeds from haploids. Jpn. J. Genet. 44:47-54.

Thomas, E. and Wenzel, G. 1975a. Embryogenesis from microspores of *Brassica napus*. Z. Pflanzenzuecht. 74:77-81.

_____ 1975b. Embryogenesis from microspores of rye. Naturwissenschaften 62:40-41.

Tomes, D.T. and Collins, G.B. 1976. Factors affecting haploid plant production from in vitro anther culture of *Nicotiana* species. Crop Sci. 51:139-140.

Tran Than Van, K. and Trinh, T.H. 1980. Embryogenetic capacity of anthers from flowers formed in vitro on thin cell layers and of anthers excised from the mother plant of *Nicotiana tabacum* L. and *N. plumbaginifolia*. Z. Pflanzenphysiol. 100:379-388.

Tulecke, W. 1953. A tissue derived from the pollen of *Ginkgo biloba*. Science 117:599-600.

_____ and Sehgal, N. 1963. Cell proliferation from pollen of *Torreya nucifera*. Contrib. Boyce Thompson Inst. 22:153-163.

Tyagi, A.K., Rashid, A., and Maheshwari, S.C. 1979. High frequency production of embryos in *Datura innoxia* from isolated pollen grains by combined cold treatment and serial culture of anthers in liquid medium. Protoplasma 99:11-17.

_____ 1980. Enhancement of pollen-embryo formation in *Datura innoxia* by charcoal. Physiol. Plant. 49:296-298.

Vasil, I.K. 1980. Androgenetic haploids. In: Perspectives in Plant Cell and Tissue Culture (I.K. Vasil, ed.) pp. 195-223. Academic Press, New York.

Vazart, B. 1972. The ultrastructure of the tobacco pollen cells destined to develop into embryoids. J. Microsc. (Paris) 14:98A.

Vishnoi, A., Babbar, S.B., and Gupta, S.C. 1979. Induction of androgenesis in anther cultures of *Withania somnifera*. Z. Pflanzenphysiol. 94:169-171.

Vyskot, B. and Novak, F.J. 1974. Experimental androgenesis in vitro in *Nicotiana clevelandii* Gray. and *N. sanderae* hort. Theor. Appl. Genet. 44:138-140.

Wagner, G. and Hess, D. 1974. Haploide, diploide und triploide Pflanzen von *Petunia hybrida* aus Pollenkorner. Z. Pflanzenphysiol. 73:273-276.

Wang, C.C., Chu, C.C., Sun, C.S., Wu, S.H., Yin, K.C., and Hsu, C. 1973. The androgenesis in wheat (*Triticum aestivum*) anthers cultured in vitro. Sci. Sinica 16:218-222.

Wang, C.C., Sun, C.S., and Chu, C.C. 1974. On the condition for the induction of rice pollen plantlets and certain factors affecting the frequency of induction. Acta Bot. Sinica 16:43-53.

Wang, Y.Y., Sun, C.S., Wang, C.C., and Chien, N. 1973. The induction of the pollen plantlets of *Triticale* and *Capsicum annum* from anther culture. Sci. Sinica 16:147-151.

Wardlaw, C.W. 1965. Physiology of embryonic development in Cormophytes. In: Encyclopedia of Plant Physiology (W. Ruhland, ed.) p. 932. Springer-Verlag, Berlin.

Watanabe, K., Nishi, Y., and Tanaka, R. 1972. Anatomical observations on the high frequency callus formation from anther culture of *Chrysanthemum*. Jpn. J. Genet. 47:249-255.

Weatherhead, M.A. and Henshaw, G.G. 1979. The production of homozygous diploid plants of *Solanum verrucosum* by tissue culture techniques. Euphytica 29:765-768.

Weatherhead, M.A., Burdon, J., and Genshaw, G.G. 1978. Some effects of activated charcoal as an additive to plant tissue culture media. Z. Pflanzenphysiol. 89:141-147.

Weatherhead, M.A., Grout, B.W.W., and Short, K.C. 1982. Increased haploid production in *Saintpaulia ionantha* by anther culture. Sci. Hortic. 17:137-144.

Wenzel, G. and Thomas, E. 1974. Observations on the growth in culture of anthers of *Secale cereale*. Z. Pflanzenzuecht. 72:89-94.

Wenzel, G. and Uhrig, H. 1981. Breeding methods and virus resistance in potato via anther culture. Theor. Appl Genet. 59:333-340.

Wenzel, G., Hoffmann, F., Potrykus, I., and Thomas, E. 1975. The separation of viable rye microspores from mixed populations and their development in culture. Mol. Gen. Genet. 138:293-297.

Wenzel, G., Hoffmann, F., and Thomas, E. 1977. Increased induction and chromosome doubling of androgenetic haploid rye. Theor. Appl. Genet. 51:81-86.

Wernicke. W. and Kohlenbach, H.W. 1975. Antherenkulturen bei *Scopolia*. Z. Pflanzenphysiol. 77:89-93.

_____ 1977. Experiment on the culture of isolated microspores in *Nicotiana* and *Hyoscyamus* Z. Pflanzenphysiol. 81:330-340.

Wernicke, W., Lorz, H., and Thomas, E. 1979. Plant regeneration from leaf protoplasts of haploid *Hyoscyamus muticus* L. produced via anther culture. Plant Sci. Lett. 15:239-250.

White, P.R. 1963. The Cultivation of Animal and Plant Cells. Ronald Press, New York.

Widholm, J.M. 1977. Selection and characterization of biochemical mutants. In: Plant Tissue Culture and its Biotechnological Applications (W. Barz, E. Reinhard, and M.H. Zenk, eds.) pp. 112-122. Springer-Verlag, Berlin.

Winton, L.L. and Stettler, R.F. 1974. Utilization of haploidy in tree breeding. In: Haploids in Higher Plants: Advances and Potential. pp. 259-273. Univ. Guelph Press, Guelph.

Xu, Z.H., Huang, B., and Sunderland, N. 1981.   Culture of anthers in conditioned media.   J. Exp. Bot. 32:767-778.

Yamada, T., Shoji, T., and Sinoto, Y. 1963.   Formation of calli and free cells in tissue culture of *Tradescantia reflexa*.   Bot. Mag. 76:332-339.

Yasuda, S. 1940.   A preliminary note on the artificial parthenogenesis induced by application of growth promoting substances.   Bot. Mag. 54:506-510.

Yin, K.-C., Hsu, C., Chu, C.-Y., Pi, F.-Y., Wang, S.-T., Liu, T.-Y., Chu, C.-C., Wang, C.-C., and Sun, C. 1976.   A study of the new cultivar of rice raised by haploid breeding method.   Sci. Sinica 19:227-242.

Zamir, D., Jones, R.A., and Kedar, N. 1980.   Anther culture of male sterile tomato (*Lycopersicon esculentum* Mill.) mutants.   Plant Sci. Lett. 17:353-361.

Zamir, D., Tanksley, S.D., and Jones, R.A. 1981.   Genetic analysis of the origin of plants regenerated from anther tissues of *Lycopersicon esculentum* Mill.   Plant Sci. Lett. 21:223-227.

Zenkteler, M. 1971.   In vitro production of haploid plants from pollen grains of *Atropa belladonna* L.   Experientia 27:1087.

_____ 1972.   Development of embryos and seedlings from pollen grains in *Lycium halimifolium* Mill. in the in vitro culture.   Biol. Plant. 14: 420-422.

_____ 1973.   In vitro development of embryos and seedlings from pollen grains of *Solanum dulcamara*.   Z. Pflanzenphysiol.69:189-192.

_____ and Misiura, E. 1974.   Induction of androgenic embryos and cultured anthers of *Hordeum*, *Secale* and *Festuca*.   Biochem. Physiol. Pflanzen 165:337-340.

_____ and Straub, J. 1979.   Cytoembryological studies on the progress of fertilization and the development of haploid embryos of *Triticum aestivum* L. (2n = 42) after crossing with *Hordeum bulbosum* (2n = 14).   Z. Pflanzenzuecht. 82:36-44.

_____, Misiura, E., and Ponitka, A. 1975.   Induction of androgenetic embryoids in the in vitro cultured anthers of several species.   Experientia 31:289-291.

Zetzsche, F. 1932.   Kork und Cuticularsubstanzen.   In: Handbuch der Pflanzenanalyse (G. Klein, ed.) pp. 205-215.   Springer-Verlag, Berlin.

PART B
SPECIALIZED CELL CULTURE TECHNIQUES

SECTION I
Genetic Techniques

CHAPTER 7
Protoplast Fusion

D. A. Evans

Sexual hybridization between closely related species has been used for years to improve cultivated crops (Bates and Deyoe, 1973). Unfortunately, sexual hybridization is limited, in most cases, to cultivars within a species or at best to a few wild species closely related to a cultivated crop (Smith, 1968). Species barriers thereby limit the usefulness of sexual hybridization for crop improvement. Sexual crossing can be limited by any of a number of factors usually related to the highly differentiated state of the specialized pollen and egg cells. In some cases these barriers may even be regulated by specific genes (Pandey, 1969). Despite these barriers, sexual hybridization has been used to incorporate a number of agriculturally important traits into numerous cultivated crops. Especially valuable has been the incorporation of disease resistance (Day, 1974) and pigment production (Rick, 1974). Sexual crosses between more distantly related plant species are desired, but are currently limited.

For a number of years the technique of protoplast fusion has been suggested as a method to overcome the species barriers to sexual hybridization (e.g., Gamborg et al., 1974). This technique, somatic cell hybridization, offers great promise for achieving wide crosses between species, with the hope to develop new crop varieties. Somatic hybridization, though, requires that after the fusion of two somatic plant cells that a complete plant is regenerated from the fused cell. While offering great promise, this technique has to date been successfully applied to only a small number of plant species.

CRITICAL REVIEW OF THE LITERATURE

Protoplast Fusion

When the plant cell wall is enzymatically removed, the resulting protoplasts often fuse spontaneously, thereby forming multinucleate fusion bodies. This occurrence is not surprising, as plasmodesmata that connect plant cells have been observed to expand in some cases rather than break (Withers and Cocking, 1972). It has been suggested that such spontaneous fusion products may divide and form plants. In some cases the occurrence of multinucleate protoplasts range from 8% for *D. innoxia* (Schieder, 1977) to 30% for *N. tabacum* (Power and Frearson, 1973). These intraspecific multinucleate protoplasts may divide synchronously (Fowke et al., 1974). Spontaneous fusion is particularly prevalent in protoplasts isolated from meiocytes (Ito, 1973). The spontaneous fusion of meiocyte-derived protoplasts has been used to produce intergeneric fusion products between *Lilium longiflorum* and *Trillium kamtschaticum* (Ito and Maeda, 1973).

While several methods were proposed to induce protoplast fusion, earliest success was achieved using sodium nitrate (e.g., Carlson et al., 1972). However, sodium nitrate is toxic to cells at fusogenic concentrations and only results in a small increase over spontaneous fusion frequencies (Keller and Melchers, 1973). Other methods have been proposed, including use of salt mixtures (Binding, 1974) or potassium dextran sulfate (Kameya, 1975). Keller and Melchers (1973) combined high pH (10.5) in medium containing Ca^{2+} ions to effectively induce cell fusion. Fusion frequencies of greater than 25% were reported after low-speed centrifugation of mixed protoplasts.

The method most frequently used today to achieve fusion is addition of polyethylene glycol (PEG) to a protoplast mixture. This method was first reported by Kao and Michayluk (1974). PEG agglutinates the plant protoplasts. High molecular weight PEG, 1540–6000, is added at 25–30% to cause agglutination. When eluted in the presence of high pH and Ca^{2+} concentration, a high frequency of protoplast fusion is obtained. In some cases up to 100% fusion has been reported using PEG to induce fusion (Vasil et al., 1975). The mechanism of action of PEG has not been reported. It has been suggested that PEG acts as a molecular bridge, thereby dissociating the plasmalemma. Other chemicals have been used to induce protoplast fusion, including polyvinyl alcohol (Nagata, 1978).

Identification and Selection of Hybrid Cells

Following fusion treatment, the protoplasts in liquid culture medium regenerate cell walls and undergo mitosis, resulting in a mixed population of parental cells, homokaryotic fusion products, and heterokaryotic fusion products or hybrids. Hybrid cells must be distinguished from the other cells present. Identification and recovery of protoplast fusion products have been based on the general observation that hybrid cells display genetic complementation for recessive mutations and physiological complementation for in vitro growth requirements (Table 1). Most somatic hybrid plants recovered to date have been identified by selection based on complementation. Carlson et al. (1972) first successfully used complementation to isolate auxin autotrophic somatic hybrids. Following fusion of two *Nicotiana* species, each with an auxin requirement for cell growth, somatic hybrids were isolated by growth on auxin-free culture medium. Auxin autotrophy was expressed as a result of the genetic combination of the two parental species used.

Melchers and Labib (1974), on the other hand, first used genetic complementation to isolate green somatic hybrids following fusion of two distinct homozygous recessive albino mutants of *Nicotiana tabacum*. This is the most frequently used method to isolate somatic hybrids. A population of protoplasts isolated from a genetically recessive albino is fused with (a) a population of protoplasts isolated from a second non-allelic albino mutation, or (b) with a population of normal green mesophyll protoplasts. For example, Schieder (1977) fused protoplasts of two diploid homozygous albino mutations of *Datura innoxia*, A1/5a and A7/1s, that had been induced by X-ray treatment. Intraspecific somatic hybrids were selected by isolating green regenerating shoots. Similarly, Douglas et al. (1981) isolated green interspecific hybrid shoots following fusion of chlorotic *N. rustica* protoplasts with albino *N. tabacum* protoplasts.

In most cases, though, it has not been necessary to use two albino mutants to recover somatic hybrid shoots. When used in combination with a morphological trait or a growth response, a single recessive albino mutation could be very useful in the isolation of somatic hybrid plants. For example, a culture medium can be selected that both favors regeneration of the albino species and prohibits regeneration of the green species. As the green species cannot regenerate, all green calli or shoots that are recovered represent putative somatic hybrids. This combination of genetic and physiological complementation using albino mutants has been used to recover interspecific somatic hybrids of *Datura, Daucus, Nicotiana,* and *Petunia,* as well as several intergeneric somatic hybrids (see Chapter 8). We have used a slight modification of this selection procedure to recover several interspecific somatic hybrids in the genus *Nicotiana*. By utilizing a semidominant albino mutation (Su/Su, *N. tabacum*) as one parent, each population of protoplasts can be uniquely identified when shoots are regenerated. The albino protoplasts produce only albino shoots when regenerated (Gamborg et al., 1979), while the mesophyll protoplasts of wild *Nicotiana* species produce only dark green shoots following protoplast regeneration (Evans, 1979). The protoplast fusion products that contain a mixture of green and albino genetic information have been visually distinguished from the parental regenerates as light green shoots. Such light green shoots have been verified as somatic hybrids (Evans et al., 1980).

In most cases use of a single recessive albino mutation as one parental line is insufficient to distinguish hybrid protoplasts from a second parent. Consequently, morphological markers have been used in combination with an albino mutation to distinguish putative somatic hybrid plants from plants derived from wild-type parental protoplasts. Dudits

Table 1. Phenotypes of Cells Used for Complementation or Partial Complementation To Recover Somatic Hybrid Cells or Plants[a]

PARENT 1	PARENT 2	HYBRID	GENUS	REFERENCE
Chlorophyll defect (s)	Chlorophyll defect (v)	Green	*Nicotiana*	Melchers & Labib, 1974
Chlorophyll defect	Chlorophyll defect	Green	*Datura*	Schieder, 1977
Chlorophyll defect (Su/Su)	Normal	Light green	*Nicotiana*	Evans et al., 1980
5MT resistant	AEC resistant	5MT resistant, AEC resistant	*Nicotiana*	White & Vasil, 1979
NR defect (nia)	NR defect (cnx)	Normal	*Nicotiana*	Glimelius et al., 1978
NR defect (nia)	Chlorophyll defect	Normal	*Nicotiana*	Glimelius et al., 1981a
Kanamycin	Unable to regenerate	Kanamycin	*Nicotiana*	Maliga et al., 1978
Actinomycin	Actinomycin	Actinomycin	*Petunia*	Power et al., 1976
Streptomycin	Normal	Streptomycin	*Nicotiana*	Medgyesy et al., 1980
Streptomycin	Auxin autotrophic	Streptomycin, auxin auto-trophic	*Nicotiana*	Wullems et al., 1980

[a]NR = Nitrate Reductase; 5MT = 5 Methyl tryptophan; AEC = Aminoethyl cysteine.

et al. (1977) fused albino *D. carota* protoplasts with wild-type *D. capillifolius*. As both *D. capillifolius* and the hybrid protoplasts were capable of regeneration, isolation of green shoots was insufficient to distinguish these two lines. However, origin of shoots could be detected as the morphology of the leaves in the hybrid plants more closely resembled *D. carota* leaves.

The development of more powerful selection methods, perhaps utilizing mutants induced in vitro, may be necessary to isolate more distant interspecies and intergeneric hybrids between more distantly related species. Utilization of two amino acid analogue resistance mutants (White and Vasil, 1979) and two nitrate reductase deficient mutants (Glimelius et al., 1978a) has been proposed, but has not yet resulted in the recovery of mature hybrid plants. On the other hand, some variants have been successfully used to recover somatic hybrid plants. Maliga et al. (1977) used a kanamycin resistant variant of *N. sylvestris*, KR103, isolated from cultured cells, as a genetic marker to recover fusion products between *N. sylvestris* and *N. knightiana*. Similarly, the SR1, streptomycin resistant mutant of *N. tabacum*, also isolated from cultured cells, was used to recover (1) intraspecific hybrids with *N. tabacum* (Wullems et al., 1980), (2) interspecific hybrids with *N. sylvestris* (Medgyesy et al., 1980), and (3) interspecific hybrids with *N. knightiana* (Menczel et al., 1981). The SR1 mutation is encoded in cytoplasmic DNA and those somatic hybrids that contained *N. tabacum* chloroplast DNA expressed streptomycin resistance. Variants isolated in vitro from carrot, *D. carota*, have also been used to identify somatic hybrid plants. Cycloheximide resistant plants of carrot were isolated using cultured cells. When these resistant lines were fused with albino lines of *D. carota*, somatic hybrids could be identified as being both cycloheximide resistant and green. Similarly Kameya et al. (1981) used cell line C123 of *D. carota* that simultaneously expressed 5-methyltryptophan (5MT) and azetidine-2-carboxylate (A2C) resistance to identify interspecific hybrids between *D. carota* and *D. capillifolius*. Selection for hybrid cells was based on resistance to 5MT. However, callus reinitiated from somatic hybrid plants expressed only intermediate resistance to 5MT and complete resistance to A2C.

While complementation of auxotrophic mutants has been successfully used to isolate somatic hybrids in *Sphaerocarpus* (Schieder, 1975) and *Physcomitrella* (Ashton and Cove, 1977), this method of selection has only been applied to higher plants on a limited scale. This limitation is due to the paucity of higher plant auxotrophs. There is only one report in which higher plant somatic hybrids have been selected using auxotrophic mutants. Glimelius et al. (1978a) fused the two different types of nitrate reductase deficient mutants of *N. tabacum* isolated by Muller and Grafe (1978). Neither mutant line could be grown with nitrate as sole nitrogen source, while hybrids could regenerate shoots in the nitrate media.

Plant regeneration from protoplasts has never been ascribed to a particular gene or group of genes. Evidence, though, from alfalfa suggests that regeneration from leaf explants is a selectable trait that can be stably transmitted (Bingham et al., 1975). In some cases, estimates have been made that 3-4 genes may control regeneration in cultured leaf explants. While evidence from fusion experiments has not been analyzed genetically, it is fortunate that the ability to regenerate from protoplasts behaves as a dominant trait in nearly all cell hybrids that have been examined. In most hybridization experiments, the hybrid line is capable of regeneration even if only one of the two parents is capable of regeneration. This observation has been extended to permit some researchers to develop culture media that permit growth of hybrid cells while prohibiting growth of at least one parent cell line. In some cases, hybrid cells have been produced that are capable of regeneration while neither parent line can be regenerated. For example, Maliga et al. (1977) fused KR103, the kanamycin resistant line of N. sylvestris that is incapable of plant regeneration, with N. knightiana protoplasts, also incapable of regeneration, and was able to recover interspecific somatic hybrid plants.

Metabolic complementation has also been proposed as a method to recover somatic hybrids. This method, first proposed by Wright (1978), has been applied to recover a number of animal cell hybrids. Parental cells are treated with an irreversible biochemical inhibitor, such as iodoacetate or diethylpyrocarbonate, and following treatment only hybrid cells are capable of cell division. Iodoacetate pretreatment has been used to aid recovery of somatic hybrids between N. sylvestris and N. tabacum (Medgyesy et al., 1980) and N. plumbaginifolia and N. tabacum (Sidorov et al., 1981). In each case the parent protoplasts treated with iodoacetate were unable to reproduce, while the newly formed hybrid protoplasts continued to develop and yield hybrid plants.

Numerous other methods have been proposed for selection of somatic hybrids. In some cases the growth pattern of hybrid callus is different from either parental line. In particular, a number of authors report that hybrid callus is often more vigorous than parental callus. Schieder (1978, 1980) has suggested that all interspecific Datura hybrids have much better callus growth than either parental line. Similarly, N. glauca + N. langsdorfii somatic hybrids could be preselected based on superior growth.

Perhaps the most efficient, yet most tedious method to select products of protoplast fusion is to visually identify hybrid cells and mechanically isolate individual cells. When morphologically distinct cells are used for protoplast fusion, microscopic observation can be used to visually distinguish fusion products from parental protoplasts. For example, following fusion of green chloroplast containing mesophyll protoplasts with colorless cell culture protoplasts that contain distinct starch granules due to growth on sucrose supplemented medium, fusion products can be distinguished shortly after fusion. Immediately after PEG treatment, the fusion products contain chloroplasts in one half of the cell and starch granules in the other half. Diffusion of chloroplasts throughout the cell occurs shortly after fusion. During first cell division in many hybrids the chloroplasts are clumped around the nuclear material. After 7-10 days of culture in protoplast medium, the chloroplasts appear as colorless proplastids and hence, usually leaf-cell culture hybrid cells can ony be distinguished for a short time after fusion. Similarly, Potrykus (1972) suggested using petal protoplasts to

visually identify hybrids. Petal + leaf fusion products and petal + cell
culture fusion products can be readily distinguished. The petal
pigment, usually vacuolar, is originally separated within the fused cell
but eventually becomes evenly distributed throughout the fused cell. In
some cases, this new mixture of protoplast contents produces cells with
unique coloration (Flick and Evans, 1983). Shortly after transfer to
protoplast culture medium, the color of the flower petals (usually
vacuolar) diffuses and as with leaf + cell culture fusion products can
only be used as a cell marker for a few days after fusion.

Visual markers can then be used as a basis to physically separate the
fusion products from parental protoplasts. Kao (1977) described a
method (see below) to mechanically isolate individual heterokaryocytes.
Using this method, he was able to isolate and then monitor the growth
of intergeneric and interfamilial cell hybrids. Gleba (1979) modified
this method to recover complete plants from microisolated plant protoplasts. Menczel et al. (1978) combined microisolation with nurse
culture techniques to develop a procedure to recover somatic hybrid
plants. In this method a single microisolated heterokaryocyte was
placed in a droplet containing albino protoplasts capable of rapid
growth. The albino protoplasts supported the growth of the single isolated heterokaryocyte which was subsequently identified as a regenerated green shoot among the albino shoots. While microisolation
methods are very tedious, these methods offer a unique opportunity to
monitor the cell and plant development of individual clonal lines
produced by protoplast fusion. Microisolation will no doubt be widely
applied in the future.

In most cases hybrid cells between distantly related plant species are
incapable of plant regeneration. However, stable intergeneric cell lines
have been frequently described (Table 2). Following protoplast fusion a
wide range of genetic products may be produced. Fusion products
could be either heterokaryotic or homokaryotic. The later would be
reflected in recovery of polyploid parental plants. Heterokaryocytes
can exist for a number of days before nuclei fuse. Up to 30-45% of the
heterokaryocytes contain only one nucleus from each parent (Constabel
et al., 1975). However, the remainder may be multinucleated. In most
cases multiple fusions will produce "giant" cells incapable of mitosis
and subsequent development. However, multiple fusion has been blamed
as the method of origin of a number of somatic hybrid plants with more
than the expected number of chromosomes, e.g., N. glauca + N. langsdorfii hybrids (Smith et al., 1976). More recently, some somatic hybrids
were reported between N. tabacum and N. sylvestris that contained 96
chromosomes (Medgyesy et al., 1980). It was suggested that these
plants arose by fusion of one N. tabacum nucleus (48 chromosomes) with
two N. sylvestris nuclei (2 x 24 chromosomes).

Nuclei of the heterokaryocytes can fuse to produce hybrids, segregate
producing cybrids, or undergo asynchronous mitosis. Szabados and
Dudits (1980) fused interphase protoplasts of one species with partially
synchronized metaphase protoplasts of a second species. In the
resulting dinucleate, the interphase nucleus was prematurely condensed,
resulting in some abnormal chromosomes. If such a phenomenon occurs
routinely, fusion of cells in different parts of the cell cycle could

Table 2. Viable Intergeneric Cell Fusion Products

FUSION PRODUCTS	REFERENCE
Glycine max (soybean) + *Nicotiana glauca* (wild tobacco)	Kao, 1977
G. max (soybean) + *Zea mays* (corn)	Kao et al., 1974
G. max (soybean) + *Hordeum vulgare* (barley)	Kao et al., 1974
Vicia faba (broad bean) + *Petunia hybrida* (petunia)	Binding & Nehls, 1978
Daucus carota (carrot) + *Hordeum vulgare* (barley)	Dudits et al., 1976
D. carota (carrot) + *Petunia hybrida* (petunia)	Reinert & Gosch, 1976
Nicotiana tabacum (tobacco) + *Lycopersicon esculentum* (tomato)	Evans et al., 1978
N. tabacum (tobacco) + *Solanum chacoense* (potato)	Gamborg et al., 1978
Sorghum bicolor (sorghum) + *Zea mays* (corn)	Brar et al., 1980

result in aneuploidy. Aneuploidy and specific chromosome elimination has been frequently reported among plant cell hybrids. Kao (1977) reported preferential loss of *N. glauca* chromosomes in hybrids between *N. glauca* and soybean. *Parthenocissus* + *Petunia* cell hybrids lost all *Petunia* chromosomes (Power et al., 1975), and *Daucus* + *Aegopodium* hybrids lost all *Aegopodium* chromosomes (Dudits et al., 1979). On the other hand, some stable hybrid cell lines have been produced that retain chromosomes from both species (e.g., *Vicia* + *Petunia*; Binding and Nehls, 1978).

Methods of Recovery and Characterization of Hybrid Plants

Protoplasts from any two species can be fused together. However, there are a number of limitations to widespread utilization of somatic hybridization in higher plants including aneuploidy, species barriers to hybridization, and the inability to regenerate plants from protoplasts. These limitations emphasize the need for more widespread research on plant protoplast fusion. (1) The primary limitation is certainly the restricted capability to regenerate plants from protoplasts. Plant regeneration from protoplasts has already been reported in a dozen plant genera, but at least seven of these genera are among the Solanaceae. This limitation temporarily precludes the use of somatic hybridization techniques for improvement of most of the important crop species, including legumes and cereals. Nonetheless, recent progress in protoplast regeneration in legumes and cereals is encouraging (Kao and Michayluk, 1980; Vasil and Vasil, 1980) and suggests that this limitation is not insurmountable. (2) Chromosome instability has been observed in most somatic hybrid plants recovered including intraspecific somatic hybrids (Melchers and Sacristan, 1977). Aneuploid somatic hybrid plants may be sexually sterile or even incapable of flower production (Gleba and Hoffmann, 1980) and, therefore, incapable of subsequent sexual propagation.   Aneuploidy has not been assessed for effect on the ability to utilize somatic hybrids to incorporate useful genetic information into a cultivated crop by repeated backcrossing. (3) Species limitations for successful production of somatic hybrids have not been adequately explored.   Interspecific sexual incompatibility should be carefully correlated to protoplast fusion to ascertain the limits to somatic compatibility (Zenkteler and Melchers, 1978), i.e., sexual hybrids should be carefully compared to somatic hybrids for ability to incorporate and express useful genetic traits.   Also, as our understanding of cytoplasmic genetics and our accumulation of useful cytoplasmic genetic markers increases, somatic hybridization could be used to produce unique nuclear-cytoplasmic combinations.   Protoplast fusion with PEG could then be used to transfer cytoplasmically controlled male sterility (e.g., Izhar and Tabib, 1980) or other useful cytoplasmic characters between breeding lines.

We have fused mesophyll protoplasts of Nicotiana glauca with suspension cultured protoplasts of N. tabacum (Evans et al., 1980).   Using the method of Kao (1976) a fusion frequency of 1-10% was obtained in three   separate   experiments.   Cell   growth   was   observed   when protoplasts were cultured in medium 8p of Kao and Michayluk (1975). After 3 weeks, unselected cells were plated onto solid agar containing MS medium (1962) with 5 µM 6BA.   Light green shoots representing putative hybrids were recovered in each experiment.   In 2 weeks shoots were transferred to rooting medium: one-half strength MS medium with 25 µM 3-aminopyridine.   Acclimated plantlets were subsequently transplanted to the greenhouse.   While selection of light green plants was insufficient to conclude that plants were somatic hybrids, hybridization was verified based on numerous criteria (Table 3), including morphological, biochemical, and cellular distinctions between N. glauca and N. tabacum.

Table 3.   Characteristics of Somatic and Sexual Hybrids of Nicotiana tabacum and Nicotiana glauca

PLANT	PETIOLE LENGTH[a]	FLOWER LENGTH[a]	COROLLA DIAMETER[a]	CHROMOSOME NUMBER
Nicotiana tabacum	0.0 mm	59 mm	25 mm	48
N. glauca	55.9 mm	38 mm	10 mm	24
N. tabacum x N. glauca	17.1 mm	43 mm	18 mm	36
N. tabacum + N. glauca somatic hybrid	12.9 mm	42 mm	18 mm	72

[a]Somatic hybrid is significantly different (p < 0.05) than either parent.

In   most   cases   reported   to   date,   morphological   characteristics   of either somatic or sexual hybrids were intermediate between the two parents (e.g., Carlson et al., 1972).   This is particularly true for both vegetative and floral morphology of N. tabacum + N. glauca somatic hybrids. Leaf shape and size, petiole size, and trichome (hairs on the leaf surface) density were all intermediate in the somatic hybrid plants (Table 3). In addition, flower shape, color, size, and structure were also all intermediate for these somatic hybrids. Using isoenzyme electrophoresis analysis, sexual or somatic hybrids usually contain the sum of isoenzyme bands found in the parents (Wetter, 1977). The sum of parental bands was observed for the N. tabacum + N. glauca somatic hybrids for alanylaminopeptidase using gel electrophoresis and for the small subunit of fraction-1 protein using isoelectric focusing. For both enzymes, different bands were observed in N. glauca and N. tabacum while the somatic hybrid contained the sum total of parental bands. For aspartate aminotransferase, which had been shown to be a dimeric enzyme, in addition to the sum of the parental bands, the somatic hybrid plants contained a unique hybrid band intermediate in mobility between a band of N. glauca and a band of N. tabacum. This probably represents formation of a hybrid dimeric enzyme unique to the somatic hybrids. Cytologically, the chromosome number of the somatic hybrids should be the sum of N. glauca (2n = 24) and N. tabacum (2n = 48). Of 25 separate somatic hybrid clones counted, all have 2n = 72 (Table 3). In addition, as some individual chromosomes can be distinguished, cells of both N. glauca, with large chromosomes and 1 pair of metacentric chromosomes, and N. tabacum, with small chromosomes and 9 pairs of metacentric chromosomes, were distinguishable from somatic hybrid plants that contained both large and small chromosomes and 10 pairs of metacentric chromosomes. Hence cytogenetic analysis coupled with enzyme analysis and morphological data clearly verify that the protoplast derived light green plants are indeed somatic hybrids.

Using this albino selection method, we have also produced somatic hybrid plants of N. tabacum with N. sylvestris and N. otophora, two species closely related to tobacco (Evans et al., 1983), and with N. nesophila and N. stocktonii, two species that cannot be crossed with tobacco using conventional breeding methods (Evans et al., 1981). Disease resistance has been incorporated into the N. tabacum + N. nesophila somatic hybrid plants, demonstrating the usefulness of these plants in crop improvement.

Most characteristics of other somatic hybrids that have been reported to date are intermediate between the two parents, except pollen viability and chromosome number, which are decreased and increased, respectively. Pollen viability is usually dependent on the taxonomic closeness of the two parental species used for hybridization; consequently, more distant somatic hybrids have lower pollen viability. Chromosome number, on the other hand, should equal the sum of somatic chromosome numbers of the two parents, but has been quite variable in most somatic hybrids reported to date. Most other morphological traits that can be quantified including both vegetative and floral characters are intermediate between the two parents. These include such vegetative characters as: leaf shape (Nagao, 1978), leaf area (Dudits et al., 1977), root morphology (Dudits et al., 1977), trichome length (Power et al., 1980), and trichome density (Carlson et al., 1972), and floral characters such as: floral length (Smith et al., 1976), corolla morphology (Schieder, 1978), intensity of floral pigment

(Power et al., 1980), and seed capsule morphology (Schieder, 1978). In many instances somatic hybrids with intermediate characters have been favorably compared to sexual hybrids already available. The genetic basis for most of these morphological traits has not been elucidated, but the intermediate behaviour in hybrids suggests the traits are controlled by multiple genes. Based on the sexual and somatic hybrids produced, intermediate morphology is the most frequently cited criterion to verify hybridity. Some traits, though, behave as dominant single gene traits as they are present in only one parent, but are also expressed in the somatic hybrids. Such traits include stem anthocyanin pigment (Evans et al., 1980), flower pigment (Schieder, 1977), heterochromatic knobs in interphase cells (Maliga et al., 1978), and leaf size (Schieder, 1978). Consequently, intermediate morphology is not observed for all characters in somatic hybrids. When possible, additional genetic data should be presented to support hybridity (cf., Chapter 8).

Isoenzyme analysis has been used extensively to verify hybridity (Chapter 18). Enzymes that have unique banding patterns for somatic hybrids versus either parental species include esterase (Wetter and Kao, 1976), aspartate aminotransferase (Evans et al., 1980), amylase (Lonnendonker and Schieder, 1980), and isoperoxidase (Carlson et al., 1972). Isoenzymes, though, are extremely variable within plant tissues (Bassiri and Carlson, 1979) and zymograms should therefore be prepared and interpreted cautiously. It is important to use the same tissue from each plant and to use plants at identical developmental ages when comparing parental plants with somatic hybrids.

A large number of intraspecific and interspecific somatic hybrid plants have been recovered to date. In most cases, intraspecific hybrids were produced as a method for verifying success of a predicted selection system. For example, 13 successful experiments have been reported (Table 4). Of these experiments, 9 were completed with *Nicotiana tabacum*. In addition to these 9 experiments, there are 3 other intraspecific somatic hybrids that have been produced with Solanaceous species. Hence only one species outside the Solanaceae, *D. carota*, has been used to produce an intraspecific somatic hybrid. These reports include the first successful fusion of two haploid protoplasts to recover hybrid plants (Melchers and Labib, 1974), first use of albino complementation (Melchers and Labib, 1974), and the first use of induced auxotrophic mutations to recover somatic hybrids (Glimelius et al., 1978a). In most cases at least some of the hybrid plants were recovered from each experiment that contained the summation chromosome number. Some hybrid combinations, though, are marked by a great deal of chromosome variability (Table 4). In some cases none or very few amphiploid plants were recovered (Bergounioux-Bunisset and Perennes, 1980; Gleba, 1979). Emphasis has not been placed on use of these hybrids for breeding programs. It would be expected that these aneuploid plants would have distorted segregation ratios. However, only two authors completed a detailed analysis of progeny. The authors each examined progeny following fusion of two haploid tobacco lines and reported that segregation ratios for the progeny were equal to or close to expected ratios (Kameya, 1975; Melchers, 1977).

Table 4. Intraspecific Somatic Hybrid Plants

PARENT 1 (pps source)	PARENT 2 (pps source)	SELECTION METHOD	CHROMOSOME NUMBER[a]	FERTILE PLANTS[a]	REFERENCE
Datura innoxia A1/5a	D. innoxia A7/1s	Albino complementation	34–108	NR	Schieder, 1977
Daucus carota CH	D. carota CH	Mutant and albino complementation	36	No	Lazar et al., 1981
Nicotiana debneyi	N. debneyi	Pgm isozymes	78–101	Yes	Scowcroft & Larkin, 1981
N. tabacum	N. tabacum	Albino complementation	48–96	Yes	Melchers & Labib, 1974
N. tabacum WS1,WS2	N. tabacum ws1,ws2	Progeny analysis	48	Yes	Kameya, 1975
N. tabacum nia	N. tabacum cnx	Mutant complementation	NR	NR	Glimelius et al., 1978a
N. tabacum nia	N. tabacum alb	Albino and mutant complementation	40–87	Fertile and sterile	Glimelius & Bonnett, 1981
N. tabacum nia	N. tabacum cms	Mutant complementation	37–83	No	Glimelius et al., 1981
N. tabacum p⁻	N. tabacum Su/Su	Albino complementation	60–154	NR	Gleba et al., 1975
N. tabacum p⁻	N. tabacum cms	Albino complementation	48, 96	Yes	Gleba, 1979
N. tabacum cms	N. tabacum	Morphology	48, 96	Fertile and sterile	Belliard et al., 1979
N. tabacum SR1	N. tabacum tumor	Media	NR	NR	Wullems et al., 1980

PARENT 1 (pps source)	PARENT 2 (pps source)	SELECTION METHOD	CHROMO- SOME NUMBER[a]	FERTILE PLANTS[a]	REFERENCE
Petunia hybrida	*P. hybrida* cms	Fertility and morphology	14->56	Yes	Bergounioux- Bunisset & Perennes, 1980

[a]NR = Not reported.

Many novel selection methods have been tested on intraspecific species combinations for the ability to identify hybrid cells. Plastome (p-), cytoplasmic male sterility (cms), and nuclear albino genes (e.g., ws1, ws2, Su, A1/5a, A7/1s, s, and v), have all been used as selectable markers. As discussed above, many mutations induced in cell culture have also been tested as selectable markers including the cnx and nia nitrate reductase deficient, cycloheximide (CH) resistant, and streptomycin resistant (SR1) mutants.

There have been 28 successful interspecific somatic hybrid experiments reported to date (Table 5). As with the intraspecific somatic hybrids, the majority of these hybrids were produced using solanaceous plant species. Of these 28 reports, 18 were completed using a total of 11 Nicotiana species. Of these 18 Nicotiana somatic hybrids, 14 were produced using N. tabacum as one parent line. As of now, 10 different Nicotiana species have been combined with N. tabacum via protoplast fusion. In addition to Nicotiana, 4 somatic hybrids have been achieved with Datura species, 3 with Petunia species, and 1 with Solanum species, all of which are members of the Solanaceae. Only 2 of the 28 successful interspecific hybridization experiments are outside the Solanaceae, and both of these publications report the recovery of somatic hybrids between Daucus carota + D. capillifolius. Hence while a large number of somatic hybrids have been reported, with one exception, this phenomenon has been restricted to the Solanaceae.

A wide range of chromosome numbers have been reported among somatic hybrid plants. Most of the hybrids that have been recovered are aneuploid. As aneuploidy may interfere with fertility, it may be very important to recover amphiploid plants in order to maximize usefulness of the hybrid plants. In some cases, no amphiploid plants were recovered, while in other cases only amphiploid plants were recovered. For example, Carlson et al. (1972) reported successful hybridization of N. glauca + N. langsdorfii. This hybrid is probably unstable as sexual hybrids between these two species produce genetic (Kostoff) tumors. In fact, each group that has selected hybrids between these two species has relied on a tumor-dependent characteristic, auxin autotrophy, to identify somatic hybrid lines. The small population of plants first reported contained only amphiploids (2n = 36) (Carlson et al., 1972). A second population of plants, later tested from an independent series of experiments, contained only aneuploid plants (none with 2n = 36) (Smith et al., 1976). Finally, Chupeau et al. (1978) examined a larger population of hybrid plants and found both amphiploid and aneuploid plants. The variation in chromosome number may reflect (a) the length of culture necessary to recover plants from fused protoplasts and (b) incompatibility between the nuclear and cytoplasmic genes that have been combined. Unfortunately, the critical experiments necessary to distinguish these two forces have not been performed.

Many of the recovered somatic hybrids are infertile. This is not surprising, as mostly aneuploid plants have been recovered and aneuploidy may result in sterility. In two cases infertile plants were recovered because a cms line was used as one source of protoplasts (Zelcer et al., 1978; Uchimiya, 1982). It appears that infertility is higher in more distantly related interspecies hybrids than in hybrids

Table 5. Interspecific Somatic Hybrid Plants

SPECIES PARENT 1 + PARENT 2	SELECTION METHOD	CHROMOSOME NUMBER[a]	FERTILE PLANTS[a]	REFERENCE
Datura innoxia + *D. discolor*	Albino complementation and media	48,46	Yes	Schieder, 1978
D. innoxia + *D. stramonium*	Albino complementation and media	46,48,72	Yes	Schieder, 1978
D. innoxia + *D. sanguinea*	Albino complementation and media	48,72,96	No	Schieder, 1980
D. innoxia + *D. candida*	Albino complementation and media	72	NR	Schieder, 1980
Daucus carota + *D. capillifolius*	Albino complementation and media	34–54	NR	Dudits et al., 1977
D. carota + *D. capillifolius*	Biochemical mutation and albino complementation	36–38	NR	Kameya et al., 1981
Nicotiana glauca + *N. langsdorffii*	Media	32	NR	Carlson et al., 1972
N. glauca + *N. langsdorffii*	Media	56–64	Yes	Smith et al., 1976
N. glauca + *N. langsdorffii*	Media	28–143	Yes	Chupeau et al., 1978
N. sylvestris + *N. knightiana*	Biochemical mutant complementation and media	NR	No	Maliga et al., 1977
N. tabacum + *N. alata*	Albino complementation	66,71	Yes	Nagao, 1979
N. tabacum + *N. glauca*	Partial albino complementation and media	72	Yes	Evans et al., 1980
N. tabacum + *N. glutinosa*	Albino complementation	50–88	Yes	Nagao, 1979
N. tabacum (cms) + *N. glutinosa*	Fraction–1–protein	NR	No	Uchimiya, 1982

Table 5. Cont.

SPECIES PARENT 1 + PARENT 2	SELECTION METHOD	CHROMOSOME NUMBER[a]	FERTILE PLANTS[a]	REFERENCE
N. tabacum + *N. knightiana*	Albino complementation and media	44–137	No	Maliga et al., 1978
N. tabacum + *N. nesophila*	Partial albino complementation	96	Yes	Evans et al., 1981
N. tabacum + *N. otophora*	Partial albino complementation	72	Yes	Evans et al., 1983
N. tabacum + *N. plumbaginifolia*	Metabolic complementation	NR	NR	Sidorov et al., 1981
N. tabacum + *N. rustica*	Albino complementation	60–91	Yes	Nagao, 1978
N. tabacum + *N. stocktonii*	Partial albino complementation	96	Yes	Evans et al., 1981
N. tabacum (cms) + *N. sylvestris*	Media + X-irradiation	24,38,76–80	No	Zelcer et al., 1978
N. tabacum (SR1) + *N. sylvestris*	Media	72–96	Yes	Medgyesy et al., 1980
N. tabacum + *N. sylvestris*	Partial albino complementation	72	Yes	Evans et al., 1983
Petunia parodii + *P. hybrida*	Actinomycin resistance	24–28	Yes	Power et al., 1976
P. parodii + *P. inflata*	Albino complementation and media	>48	Yes	Power et al., 1979
P. parodii + *P. parviflora*	Albino complementation	31–40	No	Power et al., 1980
Solanum tuberosum + *S. chacoense*	Morphology, chromosome number	60	NR	Butenko & Kuchko, 1980

[a]NR = not reported.

between closely related species.   In *Petunia* the two interspecific
hybrids between the most closely related species are fertile, while the
third interspecific hybrid is sterile (Power et al., 1980).   Similarly,
Schieder (1980) reported that his hybrids between distantly   related
species were infertile, while hybrids between closely related *Datura*
species were fertile. Comparisons on fertility are more difficult within
the genus *Nicotiana*, as most experiments were completed in different
laboratories.   However, it should be noted that in one laboratory
hybrids between the closely related species, *N. tabacum* + *N. sylvestris*, were fertile (Medgyesy et al., 1980) while hybrids between more
distantly related *N. tabacum* + *N. knightiana* were sterile (Maliga et
al., 1978).

Variability Among Hybrid Plants

   Populations of regenerated plants following protoplast fusion contain
more variability than comparable populations of plants produced by
sexual hybridization.   Variability has been observed between different
plants for phenotypic traits such as plant height (21-113 cm; Nagao,
1979), leaf shape (Smith et al., 1976), leaf size (Zelcer et al., 1978),
leaf petiole length (0-55.9 mm; Evans et al., 1980), flower length (Smith
et al., 1976), flower color (Maliga et al., 1978), pollen viability (6.0-
74%; Nagao, 1978), crossability (Smith et al., 1976), and isoenzyme
banding pattern (Maliga et al., 1977).
   Variability in a trait such as pollen viability could be very important
for use of somatic hybrid plants in breeding programs.   When 5 clones
of *N. tabacum* + *N. nesophila* somatic hybrid plants were compared,
differences were detected in the ability to collect seed from the hybrid
plants.   In vitro pollen germination varied from 12.2-46.7% for the 5
clonal lines (Evans et al., 1982).   The one line with the highest frequency of pollen germination had the greatest flexibility as it could be
used in backcrosses to either *N. tabacum* or *N. nesophila*. One clone,
on the other hand, has not been successfully backcrossed to either
parent.   A third clone was identified as the only clone that would self-
fertilize.   The variability for pollen viability regulates the usefulness of
each of these different clones.   Variability of hybrids may result from
any or all of three mechanisms.   (1) Genetic variability has been observed among plants regenerated from long-term cell cultures.   Populations of plants from protoplasts could be quite variable particularly in
light of the long period of in vitro growth required to obtain plant
regeneration from fused protoplasts.   This variation could be reflected
in recovery of aneuploid plants.   (2) The instability of certain nuclear
combinations may lead to loss of gene expression or physical loss of
part of the genetic information.   If the loss were sufficient, some of
these changes could be reflected in recovery of aneuploid plants.   (3)
Cytoplasmic or nuclear segregation following fusion results in unique
combinations of nuclear and cytoplasmic genetic information.   Some of
these unique combinations can be detected phenotypically.   Examination
of the published reports leads to the conclusion that each of these
three mechanisms may be responsible for the variability observed among

regenerated hybrids.   A more complete discussion of variability in
hybrid plants can be found in Chapter 8.

PROTOCOLS OF PROTOPLAST FUSION

A number of methods have been proposed to fuse protoplasts of two
different species.   The first successful fusion that produced a somatic
hybrid plant was accomplished using sodium nitrate, a treatment that is
not as effective as many recently developed methods.   Keller and
Melchers (1973) suggested treating isolated protoplasts with Ca^{2+} ions
and high pH (10.5).   The protoplasts are mixed and centrifuged at low
speeds (50xg) in the presence of fusion solution and then incubated for
30 min at 37 C.   The calcium ions and high pH method has been successfully applied with other species (Power et al., 1980).   The most
popular method now in use was developed by Kao and Michayluk (1974).
Rather than low-speed centrifugation, the protoplasts are brought in
physical contact by agglutination induced by polyethyleneglycol (PEG).
Protoplasts treated with PEG fuse during elution in the presence of
calcium and high pH.   The calcium ion concentration of the PEG eluting solution is less in the PEG method than in earlier reported methods
(Keller and Melchers, 1973).   The PEG method has achieved wide application and has been used for animal cell hybridization, thereby
replacing the famous, but less reliable Sendai virus-induced fusion
(Davidson et al., 1976).   The PEG method has also been successfully
applied to fungal (Das, 1980) and yeast protoplast fusion (Spencer et
al., 1980).   The PEG method of fusion is nonspecific and has been used
to produce intra- and interspecific somatic hybrids.   By manipulation of
steps of the procedure outlined below (Table 7), very high fusion frequencies have been demonstrated (Kao et al., 1974).   Solutions necessary to achieve protoplast fusion using the PEG method are listed in
Table 6.   Following protoplast release (Chapter 4), enzymes are removed by washing with enzyme wash solution via centrifugation.   The
PEG fusing solution includes PEG with molecular weight = 1540.   Concentrations and molecular weight of PEG has been varied considerably
in published reports.   After exposure of protoplasts to PEG, a toxic
chemical, the PEG must be diluted using the PEG eluting solution
which contains calcium and has pH = 10.5.   The basic protocol for
PEG-mediated fusion, based on Kao and Michayluk (1974), is summarized
in Table 7.
The fused protoplasts adhere tightly to the glass surface of the
coverslip and therefore can be monitored under the microscope.   If
protoplasts of two different sources are fused, such as leaf and cell
culture, fusion products can be visually distinguished from parental
protoplasts.   Heterokaryocytes can also be distinguished using differential staining (Keller et al., 1973); however, the process of staining kills
the protoplasts.   Attachment of protoplasts to a coverslip (Table 7) is
useful as the coverslip can be removed from the dish and the protoplasts fixed and stained while still attached to the coverslip.   The
coverslip is then inverted onto a microscope slide for examination of
fused protoplasts using a compound microscope.   It has been suggested

Table 6. Solutions Necessary for Protoplast Fusion

Enzyme Wash Solution
  Dissolve: 0.5 M Sorbitol (9.1 g)
               5.0 mM $CaCl_2 \cdot 2H_2O$ (75 mg)
  in 100 ml final volume.   pH = 5.8

PEG Fusing Solution
  Dissolve: 0.2 M Glucose (1.8 g)
               10 mM $CaCl_2 \cdot 2H_2O$ (73.5 mg)
               0.7 mM KH_2PO_4 (4.76 mg)
  in 50 ml final volume.
  To this 50 ml, add 25 g of PEG
  and dissolve.   pH = 5.8

PEG Eluting Solution
  Dissolve: 50 mM Glycine (375 mg)
               0.3 M Glucose (5.4 g)
               50 mM $CaCl_2 \cdot 2H_2O$ (735 mg)
  in 100 ml final volume.   pH = 10.5 using NaOH pellets.

Alternately, this solution can be prepared as two solutions that are
mixed at the time of fusion.

(A)   100 mM Glycine (750 mg)
      0.3 M Glucose (5.4 g)
in 100 ml final volume.   pH = 10.5 using NaOH pellets.

(B)   100 mM $CaCl_2 \cdot 2H_2O$ (1470 mg)
      0.3 M Glucose (5.4 g)
in 100 ml.

Using these two solutions permits storage without visible precipitation.
Mix A with B in 1:1 ratio at time of fusion

---

that protoplasts could be pretreated with fluorescent probes to identify
fusion products (Galbraith and Galbraith, 1979).   Using fluorescein and
rhodamine B conjugates, these authors were able to unambiguously
identify fusion products.   The fluorescence persists for at least 48 hr,
so that this method could be used to mechanically isolate fused protoplasts (see below). Galbraith and Mauch (1980) have also demonstrated
that fluorescent labeled protoplasts are still capable of plant regeneration.
    The PEG method of fusion has been used for a wide variety of plant
species with only slight modifications. (1) Many authors use PEG with
molecular weight 6000 rather than 1540.   Both of these appear to be
effective in agglutination plant protoplasts.   For fusion of animal cells
Klebe and Mancuso (1981) tested 7 different preparations of PEG with
molecular weight 200, 400, 600, 1000, 3000, 6000, and 20,000.   The
optimum molecular weight was 600 followed by 1000.   Both 200 and
20,000 MW were ineffective in induction of cell fusion. (2) The concen-

Table 7. Polyethylene Glycol Method of Protoplast Fusion

1. 0.5 ml of protoplasts from two sources are mixed and diluted to 8
    ml with enzyme wash solution (Table 6). This mixture is centri-
    fuged at 100 g for 4 min.
2. Excess enzyme solution is decanted and the mixed protoplasts are
    resuspended in enzyme wash solution. Step 1 and 2 are repeated
    once. After the second decantation, the precipitated protoplasts
    are resuspended in 1.0 ml of enzyme wash solution.
3. Put one drop of silicone fluid (Sigma Chemical Company) in a
    Falcon (1007) petri dish (60 x 15). Place a 22 x 22 mm coverslip
    on top of the silicone drop.
4. Pipette 0.15 ml of mixed protoplasts onto the glass coverslip.
    Allow the protoplasts to settle on the coverslip for ca. 5 min to
    form a thin layer of protoplasts.
5. Carefully add 0.45 ml of PEG solution (Table 6) to the protoplast
    mixture. To create a uniform single layer of protoplasts attached
    to the coverslip, the PEG should be added to one side of the
    protoplast droplet.
6. Incubate the protoplasts in the PEG solution at room temperature
    for 15-20 min.
7. Add 0.9 ml of PEG eluting solution (Table 6) to the mixture.
    After 10 min, add an additional 0.9 ml to the mixture.
8. After the second elution treatment, a protoplast culture medium
    (such as in Table 9) is added to the protoplasts to aid in removal
    of the PEG and the eluting solution.
9. Add 0.5 ml of culture medium to the mixture. An additional 0.5
    ml is added after 10 min.
10. After the final wash, the protoplasts should be in 1-2 ml of cul-
    ture medium. The petri dish can be sealed with a double layer
    of parafilm and examined in inverted microscope to ascertain the
    frequency of fusion.

tration of PEG has also been varied. Success has been reported with
concentrations of PEG between 10 and 50%. (3) Step 4 of the procedure appears to be very important and this step has been investigated
in detail (Weber et al., 1976). These authors varied the period of time
between the removal of the digestive enzymes from the isolated protoplasts and agglutination with PEG. Optimum preincubation was 5 min,
resulting in fusion frequency of 9%, with a decrease to 1% fusion
frequency after 2 hr of preincubation. (4) Careful preparation and
storage of the PEG eluting solution is also important (Table 6).
Difficulty is often encountered in precipitation of calcium hydroxide
from this solution.
    The PEG-method can reportedly be improved by adding concanavalin
A (con A) to the PEG solution (Glimelius et al., 1978b). The con A
strengthens the attachment of protoplasts induced by PEG; hence more
fusions remain after washing the fusion mixture. This treatment results
in a small but significant increase in heterokaryocytes. Addition of
dimethylsulfoxide (DMSO) also increases the fusion frequency (Haydu et
al., 1977). Addition of 15% DMSO to the PEG solution increases the

frequency of fusion from 3 to 13%. Presumably, DMSO makes the cells more susceptible to PEG. This treatment is also effective in animal cell fusion. The use of up to 20% DMSO in the fusion solution has no inhibitory effects on cell growth.

It has been stated that phospholipids in the protoplast membrane are important in regulating the fusion frequency. Hence it has been observed that variation in temperature, which is known to effect the concentration of phospholipids, alters the frequency of protoplast fusion (Yamada et al., 1980). Nagata et al. (1979) have identified synthetic phospholipids that are capable of inducing protoplast fusion. These have been proposed as being useful for studying the mechanism of protoplast fusion.

Other chemicals have also been observed to promote cell fusion. Nagata (1978) used a 15% solution of polyvinyl alcohol in combination with 0.05 M $CaCl_2$ and 0.3 M mannitol to fuse plant cells. Polyvinyl alcohol (PVA) is a nonionic surfactant and is apparently not harmful as cells can grow in up to 2% PVA. PVA is also available with different molecular weights, and Nagata (1978) has shown that for fusion PVA 500 is preferable to either 1400 or 2000. Dextran sulfate has also been used to induce protoplast fusion (Kameya, 1979). This chemical has had limited use because of its toxicity to plant cells. Klebe and Mancuso (1981) tested ability of 118 compounds, all purportedly membrane active, and found greater than 20 compounds that promoted animal cell fusion with nearly the same efficiency as PEG. Particularly active were a number of PEG derivatives, some commercially available in the pharmaceutical industry. Lectins are also known to agglutinate protoplasts (Larkin, 1978), but most of these have not been tested as fusogenic agents. A number of other treatments have been suggested for protoplast fusion. Of these treatments, most attention has been addressed to electrical stimulation. Senda et al. (1979) found that 2 glass capillary microelectrodes could be used to fuse adhering protoplasts. The microelectrodes are placed in contact with the protoplasts and a charge of 5-12 microamps for a few milliseconds is sufficient to induce fusion. Similarly, high intensity electrical fields have been used to fuse plant protoplasts (Vienken et al., 1981). It has even been suggested that plant viruses, e.g., potato yellow dwarf virus, may have cell fusion capability (Hsu, 1978).

A number of methods have been used to separate somatic hybrid protoplasts from mixtures of hybrid and parental protoplasts. Cell sorting methods vary in both complexity and efficiency. Galbraith and Mauch (1980) have suggested use of a cell sorter to separate differentially labeled protoplasts, but the high cost of such an apparatus precludes its widespread application. Harms and Potrykus (1978) used iso-osmotic density gradients to separate fusion products. By centrifuging a protoplast mixture for 2-4 min at 50-100 xg in KMC (potassium–magnesium–calcium) sucrose density gradients they were able to enrich for heterokaryocytes. The most reliable method of cell separation thus far applied to protoplast fusion is microisolation. A sample protocol is listed in Table 8, based on research of Kao (1977) on heterokaryocytes between *Nicotiana glauca* leaf protoplasts and soybean (SB-1) cultured cell protoplasts. Heterokaryocytes were easily distin-

Table 8.   Microisolation of Heterokaryocytes Following Protoplast
           Fusion

---

1. Fresh medium with lower osmolality is added to the protoplast mix-
    ture 24-48 hr after protoplast fusion. An appropriate solution
    would contain one part cell culture to two parts protoplast culture
    medium.
2. 24-48 hr later the protoplast mixture is diluted with the same cell
    culture:protoplast culture medium to 200-300 protoplast per ml.
3. The outer chamber of the Cuprak dish should be filled with 3 ml of
    protoplast medium. The protoplasts are transferred to the inner
    wells of a Cuprak dish (Costar 3268) with a disposable micropip-
    ette. The dishes are sealed with parafilm.
4. The inner wells of the Cuprak dish are examined and those wells
    containing a single heterokaryocyte are marked. Those wells that
    contain a mixture of protoplasts are reseparated.
5. When heterokaryocytes develop into colonies of 100-200 cells, they
    can be transferred to a 100-200 µl drop of culture medium (one
    part cell culture to two parts protoplast culture medium) in a
    petri dish.
6. Fresh cell culture medium is then added every 7-8 days. When
    culture has developed, it can be used to initiate a cell suspension
    culture or alternatively be transferred to plant regeneration
    medium.

---

guished from the parental protoplasts, as they contained both green
plastids and cytoplasmic strands. The microisolation procedure was
aided by a high frequency, 39%, of heterokaryocytes. It is important
to use a protoplast culture medium capable of supporting growth of a
single diluted heterokaryocyte. Kao and Michayluk (1975) developed
the protoplast medium listed in Table 9. While this medium is probably
unnecessarily complex, it was specifically developed for culture of cells
and protoplasts in very low densities. Gleba (1978) has suggested a
modification of the microisolation method in which heterokaryocytes
were cultured in very small volumes thereby resulting in an effective
density of 2-4 x 10^3 cells/ml. It was suggested that this method
resulted in very high plating efficiency.

FUTURE PROSPECTS

Somatic hybrids have been produced between a large number of plant
species, but most of these species combinations are restricted to the
Solanaceae. It is hoped that as efficient protoplast regeneration procedures become available for other economically important crop plants
that this technique can be extended to a wide range of plant families.
Agricultural application can also be expected by use of very wide
hybrids for intergeneric gene transfer (see Chapter 8), and by wider
use of cytoplasmic hybrids (see Chapter 9) to transfer useful traits. It
is necessary that these methods be more closely integrated with conventional breeding before it can be expected that new varieties may be
produced via protoplast fusion.

Table 9. Protoplast Culture Medium 8p of Kao and Michayluk (1975)[a]

MACRONUTRIENTS	PER LITER	
KNO_3	2500	mg
$CaCl_2 \cdot 2H_2O$	150	mg
$MgSO_4 \cdot 7H_2O$	250	mg
$(NH_4)_2SO_4$	134	mg
$NaH_2PO_4 \cdot H_2O$	150	mg
$FeSO_4 \cdot H_2O$	27.8	mg
$Na_2 \cdot EDTA$	37.3	mg

MICRONUTRIENTS	PER LITER	
KI	0.75	mg
H_3BO_3	3.0	mg
$MnSO_4 \cdot H_2O$	10.0	mg
$ZnSO_4 \cdot H_2O$	2.0	mg
$Na_2MoO_4 \cdot 2H_2O$	0.25	mg
$CuSO_4 \cdot 5H_2O$	0.025	mg
$CoCl_2 \cdot 6H_2O$	0.025	mg

VITAMINS	PER LITER	
Inositol	100	mg
Thiamine HCl	10	mg
Nicotinic acid	1	mg
Pyridoxine HCl	1	mg

	ADDITIVES	PER LITER	
SUGARS:	Sucrose	250	mg
	Glucose	68.4	g
	Fructose	250	mg
	Ribose	250	mg
	Xylose	250	mg
	Mannose	250	mg
	Rhamnose	250	mg
	Cellobiose	250	mg
	Sorbitol	250	mg
	Mannitol	250	mg
MINERALS:	$CaCl_2 \cdot 2H_2O$	450	mg
	$MgSO_4 \cdot 7H_2O$	50	mg
VITAMINS:	Ascorbic acid	2	mg
	Choline chloride	1	mg
	Calcium pantho- thenate	1	mg
	Folic acid	0.4	mg
	Riboflavin	0.2	mg
	Para-aminobenzoic acid	0.02	mg

Table 9.   Cont.

	ADDITIVES	PER LITER	
	Biotin	0.01	mg
	Vitamin A	0.01	mg
	Vitamin D_3	0.01	mg
	Vitamin B_{12}	0.02	mg
ORGANIC ACIDS:	Citric acid	40	mg
	Fumaric acid	40	mg
	Malic acid	40	mg
	Sodium pyruvate	20	mg
GROWTH REGULATORS:	Casamino acids	250	mg
	CW	20	ml
	2,4-D	0.2	mg
	6BA	0.5	mg
	NAA	1.0	mg

[a]pH = 5.6.   Medium must be filter sterilized.

Accumulating evidence also suggests that populations of somatic hybrids are more variable than conventional sexual hybrids.   It has already been established that within populations of somatic hybrid plants a wide range of chromosome numbers can be observed (Melchers and Sacristan, 1977), resulting in unique mixtures of interspecific nuclear DNA.   In addition, a range of unique nuclear-cytoplasmic combinations may be recovered following protoplast fusion if nuclear or cytoplasmic segregation occurs (Chapter 8).   As our knowledge of cytoplasmic genetics is increased, it may be possible to produce certain desirable nuclear-cytoplasmic combinations.   The production of unique mixtures of genetic material between species may aid in the selection and recovery of agriculturally useful plants following protoplast fusion.

KEY REFERENCES

Carlson, P.S., Smith, H.H., and Dearing, R.D. 1972.   Parasexual inter-specific plant hybridization.   Proc. Nat. Acad. Sci. 69:2292-2294.

Evans, D.A., Wetter, L.R., and Gamborg, O.L. 1980.   Somatic hybrid plants of *Nicotiana glauca* and *Nicotiana tabacum* obtained by proto-plast fusion.   Physiol. Plant. 48:225-230.

Kao, K.N. and Michayluk, M.R. 1974.   A method for high-frequency intergeneric fusion of plant protoplasts.   Planta 115:355-367.

_____ 1975.   Nutritional requirements for growth of *Vicia hajastana* cells and protoplasts at a very low population density in liquid media.   Planta 126:105-110.

REFERENCES

Ashton, N.W. and Cove, D.J. 1977. The isolation and preliminary characterisation of auxotrophic and analogue resistant mutants of the moss, *Physcomitrella patens*. Mol. Gen. Genet. 154:87-95.

Bassiri, A. and Carlson, P.S. 1979. Isozyme patterns in tobacco plant parts and their derived calli. Crop Sci. 19:909-914.

Bates, L.S. and Deyoe, C.W. 1973. Wide hybridization and cereal improvement. Econ. Bot. 27:401-412.

Belliard, G., Vedel, F., and Pelletier, G. 1979. Mitochondrial recombination in cytoplasmic hybrids of *Nicotiana tabacum* by protoplast fusion. Nature 281:401-403.

Bergounioux-Bunisset, C. and Perennes, C. 1980. Transfert de facteurs cytoplasmiques de la Fertilite male entre 2 lignees de *Petunia hybrida* par fusion de protoplasts. Plant Sci. Lett. 19:143-149.

Binding, H. 1974. Fusionsversuche mit isolierten Protoplasten von *Petunia hybrida* L. Z. Pflanzenphysiol. 72:421-426.

_____ and Nehls, R. 1978. Somatic cell hybridization of *Vicia faba* + *Petunia hybrida*. Mol. Gen. Genet. 164:137-143.

Bingham, E.T., Hurley, L.V., Kaatz, D.M., and Saunders, J.W. 1975. Breeding alfalfa which regenerates from callus tissue in culture. Crop Sci. 15:719-721.

Brar, D.S., Rambold, S., Constabel, F., and Gamborg, O.L. 1980. Isolation, fusion and culture of *Sorghum* and corn protoplasts. Z. Pflanzenphysiol. 96:269-275.

Butenko, R.G. and Kuchko, A.A. 1980. Production of interspecific somatic hybrids of potato by merging isolated protoplasts. Dok. Akad. Nauk. 247:491-495.

Chupeau, Y., Missonier, C., Hommel, M.-C., and Goujaud, J. 1978. Somatic hybrids of plants by fusion of protoplasts. Observations on the model system "*Nicotiana glauca-Nicotiana langsdorffii*". Mol. Gen. Genet. 165:239-245.

Constabel, F., Dudits, D., Gamborg, O.L., and Kao, K.N. 1975. Nuclear fusion in intergeneric heterokaryons. Can. J. Bot. 53:2091-2095.

Das, A. 1980. Parasexual hybridization and citric acid production by *Aspergillus niger*. Eur. J. Appl. Microbiol. 9:117-119.

Davidson, R.L., O'Malley, K.A., and Wheeler, T.B. 1976. Polyethylene glycol-induced mammalian cell hybridization: Effect of polyethylene glycol molecular weight and concentration. Som. Cell Genet. 2:271-280.

Day, P.R. 1974. The Genetics of Host-parasite Interaction. W.H. Freeman, San Francisco.

Douglas, G.C., Wetter, L.R., Nakamura, C., Keller, W.A., and Setterfield, G. 1981. Somatic hybridization between *Nicotiana rustica* and *N. tabacum*. Can. J. Bot. 59:220-227.

Dudits, D., Kao, K.N., Constabel, F., and Gamborg, O.L. 1976. Fusion of carrot and barley protoplasts and division of heterokaryocytes. Can. J. Genet. Cytol 18:263-269.

Dudits, D., Hadlaczky, Gy., Levi, E., Fejer, O., Haydu, Z, and Lazar, G. 1977. Somatic hybridization of *Daucus carota* and *D. capillifolius* by protoplast fusion. Theoret. Appl. Genet. 51:127-132.

Dudits, D., Hadlaczky, Gy., Bajszar, Gy., Koncz, Cs., Lazar, G., and Horvath, G. 1979. Plant regeneration from intergeneric cell hybrids. Plant Sci. Lett. 15:101–112.

Evans, D.A. 1979. Chromosome stability of plants regenerated from mesophyll protoplasts of *Nicotiana* species. Z. Pflanzenphysiol. 95:459–463.

_____, Gamborg, O.L., Shyluk, J.P., and Wetter, L.R. 1978. Somatic hybrid plants from protoplasts of *Nicotiana tabacum* and *Nicotiana glauca* and with *Lycopersicon.* International Association Plant Tissue Culture (Abstracts) p. 70, Univ. of Calgary, Calgary.

_____, Flick, C.E., and Jensen, R.A. 1981. Incorporation of disease resistance into sexually incompatible somatic hybrids of the genus *Nicotiana.* Science 213:907–909.

_____, Flick, C.E., Kut, S.A., and Reed, S.M. 1982. Comparison of *Nicotiana tabacum* and *Nicotiana nesophila* hybrids produced by ovule culture and protoplast fusion. Theor. Appl. Genet. 62:193–198.

_____, Bravo, J.E., Kut, S.A., and Flick, C.E. 1983. Genetic behavior of somatic hybrids in the genus *Nicotiana: N. otophora + N. tabacum* and *N. sylvestris + N. tabacum.* Theor. Appl. Genet. (in press).

Flick, C.E. and Evans, D.A. 1983. Isolation, culture, and plant regeneration of protoplasts isolated from flower petals of ornamental *Nicotianas.* Z. Pflanzenphysiol. 109:379–383.

Fowke, L.C., Bech-Hansen, C.W., Constabel, F., and Gamborg, O.L. 1974. A comparative study on the ultrastructure of cultured cells and protoplasts of soybean during cell division. Protoplasma 81:189–203.

Galbraith, D.W. and Galbraith, J.E.C. 1979. A method for identification of fusion of plant protoplasts derived from tissue cultures. Z. Pflanzenphysiol. 93:149–158.

Galbraith, D.W., and Mauch, T.J. 1980. Identification of fusion of plant protoplasts. II. Conditions for the reproducible fluorescence labeling of protoplasts derived from mesophyll tissue. Z. Pflanzenphysiol. 98:129–140.

Gamborg, O.L., Constabel, F., Fowke, L.C., Kao, K.N., Ohyama, K., Kartha, K.K., and Pelcher, L. 1974. Protoplast and cell culture methods in somatic hybridization in higher plants. Can. J. Genet. Cytol. 16:737–750.

Gamborg, O.L., Shyluk, J.P., Evans, D.A., and Wetter, L.R. 1978. Plant regeneration and hybridization with protoplasts from suspension cultures of the sulfur albino (Su/Su) mutant of *N. tabacum.* International Association Plant Tissue Culture (Abstracts) p.70. Univ. of Calgary, Calgary.

Gamborg, O.L., Shyluk, J.P., Fowke, L.C., Wetter, L.R., and Evans, D.A. 1979. Plant regeneration from protoplasts and cell cultures of *Nicotiana tabacum* sulfur mutant (Su/Su). Z. Pflanzenphysiol. 95:255–264.

Gleba, Yu.Yu. 1978. Microdroplet culture: Tobacco plants from single mesophyll protoplasts. Naturwissenschaften 65:158.

_____ 1979. Nonchromosomal inheritance in higher plants as studied by somatic cell hybridization. In: Plant Cell and Tissue Culture: Principles and Applications (W.R. Sharp et al. eds.) pp. 775–788. Ohio State Univ. Press, Columbus.

_____ and Hoffmann, F. 1980. "Arabidobrassica": A novel plant obtained by protoplast fusion. Planta 149:112-117.

_____, Butenko, R.G., and Sytnik, K.M. 1975. Fusion of protoplasts and parasexual hybridization in Nicotiana tabacum L. Dokl. Akad. Nauk. 221:1196-1198.

Glimelius, K. and Bonnett, H.T. 1981. Somatic hybridization in Nicotiana: Restoration of photoautotrophy to an albino mutant with defective plastids. Planta 153:497-503.

Glimelius, K., Eriksson, T., Grafe, R., and Muller, A.J. 1978a. Somatic hybridization of nitrate-deficient mutants of Nicotiana tabacum by protoplast fusion. Physiol. Plant. 44:273-277.

Glimelius K., Wallin, A., and Eriksson, T. 1978b. Concanavalin A improves the polyethylene glycol method for fusing plant protoplasts. Physiol. Plant. 44:92-96.

Glimelius, K., Chen, K., and Bonnett, H.T. 1981. Somatic hybridization in Nicotiana: Segregation of organellar traits among hybrid and cybrid plants. Planta 153:504-510.

Harms, C.T. and Potrykus, I. 1978. Enrichment for heterokaryocytes by the use of iso-osmotic density gradients after plant protoplast fusion. Theor. Appl. Genet. 53:49-55.

Haydu, Z., Lazar, G., and Dudits, D. 1977. Increased frequency of polyethylene glycol induced protoplast fusion by dimethylsulfoxide. Plant Sci. Lett. 10:357-360.

Hsu, H.T. 1978. Cell fusion induced by a plant virus. Virology 84:9-18.

Ito, M. 1973. Studies on the behavior of meiotic protoplasts. II. Induction of a high fusion frequency in protoplasts from Liliaceous plants. Plant Cell Physiol. 14:865-872.

_____ and Maeda, M. 1973. Fusion of meiotic protoplasts in Liliaceous plants. Exp. Cell. Res. 80:453-456.

Izhar, S. and Tabib, Y. 1980. Somatic hybridization in Petunia. Part II: Heteroplasmic state in somatic hybrids followed by cytoplasmic segregation into male sterile and male fertile lines. Theor. Appl. Genet. 57:241-245.

Kameya, T. 1975. Culture of protoplasts from chimeral plant tissue of nature. Jpn. J. Gen. 50:417-420.

_____ 1979. Studies on plant cell fusion: Effects of Dextran and Pronase E on fusion. Cytologia 44:449-456.

_____, Horn, M.E., and Widholm, J.M. 1981. Hybrid shoot formation from fused Daucus carota and D. capillifolius protoplasts. Z. Pflanzenphysiol. 104:459-466.

Kao, K.N. 1976. A method for fusion of plant protoplasts with polyethylene glycol. In: Cell Genetics in Higher Plants (D. Dudits et al. eds.) pp. 233-238. Akademiai Kiado, Budapest.

_____ 1977. Chromosomal behavior in somatic hybrids of soybean—Nicotiana glauca. Molec. Gen. Genet. 150:225-230.

_____ and Michayluk, M.R. 1980. Plant regeneration from mesophyll protoplasts of alfalfa. Z. Pflanzenphysiol. 96:135-141.

_____, Constabel, F., Michayluk, M.R., and Gamborg, O.L. 1974. Plant protoplast fusion and growth of intergeneric hybrid cells. Planta 120:215-227.

Keller, W.A. and Melchers, G. 1973. The effect of high pH and calcium on tobacco leaf protoplast fusion. Z. Naturforsch. 28B:737-741.

Keller, W.A., Harvey, B.L., Kao, K.N., Miller, R.A., and Gamborg, O.L. 1973. Determination of the frequency of interspecific protoplast fusion by differential staining. Protoplastes et fusion de Cellules Somatique Vegetales, Proc. C.N.R.S. 212:455-463.

Klebe, R.J. and Mancuso, M.G. 1981. Chemicals which promote cell hybridization. Som. Cell Genet. 7:473-488.

Larkin, P.J. 1978. Plant protoplast agglutination by lectins. Plant Physiol. 61:626-629.

Lazar, G.B., Dudits, D., and Sung, Z.R. 1981. Expression of cycloheximide resistance in carrot somatic hybrids and their segregants. Genetics 98:347-356.

Lonnendonker, N. and Schieder, O. 1980. Amylase isoenzymes of the genus *Datura* as a simple method for an early identification of somatic hybrids. Plant Sci. Lett. 17:135-139.

Maliga, P., Lazar, G., Joo, F., Nagy, A.H., and Menczel, L. 1977. Restoration of morphogenetic potential in *Nicotiana* by somatic hybridization. Mol. Gen. Genet. 157:291-296.

Maliga, P., Kiss, Z.R., Nagy, A.H., and Lazar, G. 1978. Genetic instability in somatic hybrids of *Nicotiana tabacum* and *Nicotiana knightiana*. Mol. Gen. Genet. 163:145-151.

Medgyesy, P., Menczel, L., and Maliga, P. 1980. The use of cytoplasmic streptomycin resistance: chloroplast transfer from *Nicotiana tabacum* into *Nicotiana sylvestris*, and isolation of their somatic hybrids. Mol. Gen. Genet. 179:693-698.

Melchers, G. 1977. Kombination somatischen und konventioneller Genetik fir die Pflanzenzuchtung. Naturwissenschaften 64:184-194.

_____ and Labib, G. 1974. Somatic hybridization of plants by fusion of protoplasts. I. Selection of light resistant hybrids of "haploid" light sensitive varieties of tobacco. Mol. Gen. Genet. 135:277-294.

_____ and Sacristan, M.D. 1977. Somatic hybridization of plants by fusion of protoplasts. II. The chromosome numbers of somatic hybrid plants of four different fusion experiments. In: La culture des tissus et des cellules des vegetaux (R.J. Gautheret ed.) pp. 169-177. Masson, Paris.

Menczel, L., Lazar, G., and Maliga, P. 1978. Isolation of somatic hybrids by cloning *Nicotiana* heterokaryons in nurse culture. Planta 143:29-32.

Menczel, L., Nagy, F., Kiss, Zs., and Maliga, P. 1981. Streptomycin resistant and sensitive hybrids of *Nicotiana tabacum* + *Nicotiana knightiana*: Correlation of resistance with N. tabacum plastids. Theor. Appl. Genet. 59:191-195.

Muller, A. and Grafe, R. 1978. Isolation and characterization of cell lines of *Nicotiana tabacum* lacking nitrate reductase. Mol. Gen. Genet. 161:67-76.

Murashige, T. and Skoog, F. 1962. A revised medium for rapid growth and bioassays with tobacco tissue cultures. Physiol. Plant. 15:473-497.

Nagao, T. 1978. Somatic hybridization by fusion of protoplasts. I. The combination of *Nicotiana tabacum* and *Nicotiana rustica*. Jpn. J. Crop Sci. 47:491-498.

_____ 1979. Somatic hybridization by fusion of protoplasts. II. The combinations of *Nicotiana tabacum* and *N. glutinosa* and of *N. tabacum* and *N. alata*. Jpn. J. Crop Sci. 48:385-392.

Nagata, T. 1978. A novel cell-fusion method of protoplasts by polyvinyl alcohol. Naturwissenschaften 65:263-264.

_____, Eibb, H., and Melchers, G. 1979. Fusion of plant protoplasts induced by a positively charged synthetic phospholipid. Z. Naturforsch. 34B:460-462.

Pandey, K.K. 1969. Elements of the S-gene complex. IV. S-allele polymorphism in *Nicotiana* species. Heredity 24:601-619

Potrykus, I. 1972. Fusion of differentiated protoplasts. Phytomorphology 22:91-96.

Power, J.B. and Frearson, E.M. 1973. The inter- and intraspecific fusion of plant protoplasts; subsequent development in culture with reference to crown gall callus and tobacco and petunia leaf systems. Protoplastes et fusion de Cellules Somatique Vegetales, Proc. C.N.R.S. 212: 409-415.

_____, Hayward, C., and Cocking, E.C. 1975. Some consequences of the fusion and selective culture of *Petunia* and *Parthenocissus* protoplasts. Plant Sci. Lett. 5:197-207.

_____, Hayward, C., George, D., Evans, P.K., Berry, S.F., and Cocking, E.C. 1976. Somatic hybridization of *Petunia hybrida* and *P. parodii*. Nature 263:500-502.

Power, J.B., Berry, S.F., Chapman, J.V., Cocking, E.C., and Sink, K.C. 1979. Somatic hybrids between unilateral cross-incompatible *Petunia* species. Theor. Appl. Genet. 55:97-99.

Power, J.B., Berry, S.F., Chapman, J.V., and Cocking, E.C. 1980. Somatic hybridization of sexually incompatible *Petunias* : *Petunia parodii*, *Petunia parviflora*. Theor. Appl. Genet. 57:1-4.

Reinert J., and Gosch, G. 1976. Continuous division of heterokaryons from *Daucus carota* and *Petunia hybrida* protoplasts. Naturwissenschaften 11:534.

Rick, C.M. 1974. High soluble-solids content in large-fruited tomato lines derived from wild green-fruited species. Hilgardia 42:493-510.

Schieder, O. 1975. Selection of a somatic hybrid between auxotrophic mutants of *Sphaerocarpos donnellii* using the method of protoplast fusion. Z. Pflanzenphysiol. 74:357-365.

_____ 1977. Hybridization experiments with protoplasts from chlorophyll-deficient mutants of some Solanaceous species. Planta 137:253-257.

_____ 1978. Somatic hybrids of *Datura innoxia* Mill. + *Datura discolor* Bernh. and of *Datura innoxia* Mill. + *Datura stramonium* L. var. tatula L. I. Selection and characterisation. Mol. Gen. Genet. 162:113-119.

_____ 1980. Somatic hybrids between a herbaceous and two tree *Datura* species. Z. Pflanzenphysiol. 98:119-127.

Scowcroft, W.R. and Larkin, P.J. 1981. Chloroplast DNA assorts randomly in intraspecific somatic hybrids of *Nicotiana debneyi*. Theor. Appl. Genet. 60:179-184.

Senda, M., Takada, J., Abe, S., and Nakamura, T. 1979. Induction of cell fusion of plant protoplasts by electrical stimulation. Plant Cell Physiol. 20:1141-1143.

Sidorov, V.A., Menczel, L., Nagy, F., and Maliga, P. 1981. Chloroplast transfer in *Nicotiana* based on metabolic complementation between irradiated and iodoacetate treated protoplasts. Planta 152:341–345.

Smith, H.H. 1968. Recent cytogenetic studies in the genus *Nicotiana*. Adv. Genet. 14:1–54.

———, Kao, K.N., and Combatti, N.C. 1976. Interspecific hybridization by protoplast fusion in *Nicotiana*. J. Hered. 67:123–128.

Spencer, J.F.T., Laud, P., and Spencer, D.M. 1980. The use of mitochondrial mutants in the isolation of hybrids involving industrial yeast strains. II. Use in isolation of hybrids obtained by protoplast fusion. Mol. Gen. Genet. 178:651–654.

Szabados, L. and Dudits, D. 1980. Fusion between interphase and mitotic plant protoplasts: Induction of premature chromosome condensation. Exp. Cell Res. 127:442–446.

Uchimiya, H. 1982. Somatic hybridization between male sterile *Nicotiana tabacum* and *N. glutinosa* through protoplast fusion. Theor. Appl. Genet. 61:69–72.

Vasil, I.K., Vasil, V., Sutton, W.D., and Giles, K.L. 1975. Protoplasts as tools for the genetic modification of plants. p. 82. In: Proceedings IV International Symposium on Yeast and Other Protoplasts. Univ. Nottingham, England.

Vasil, V. and Vasil, I.K. 1980. Isolation and culture of cereal protoplasts. Part 2: Embryogenesis and plantlet formation from protoplasts of *Pennisetum americanum*. Theor. Appl. Genet. 56:97–99.

Vienken, J., Ganser, R., Hampp, R., and Zimmerman, U. 1981. Electric field-induced fusion of isolated vacuoles and protoplasts of different developmental and metabolic provenience. Physiol. Plant. 53:64–70.

Weber, G., Constabel, F., Williamson, F., Fowke, L.C., and Gamborg, O.L. 1976. Effect of preincubation of protoplasts on PEG-induced fusion of plant cells. Z. Pflanzenphysiol. 79:459–464.

Wetter, L.R. 1977. Isoenzyme patterns in soybean-*Nicotiana* somatic hybrid cell lines. Mol. Gen. Genet. 150:231–235.

——— and Kao, K.N. 1976. The use of isozymes in distinguishing the sexual and somatic hybrids in callus cultures derived from *Nicotiana*. Z. Pflanzenphysiol. 80:455–462.

White, D.W.R. and Vasil, I.K. 1979. Use of amino acid analogue-resistant cell lines for selection of *Nicotiana sylvestris* somatic cell hybrids. Theor. Appl. Genet. 55:107–112.

Withers, L. and Cocking, E.C. 1972. Fine structural studies on spontaneous and induced fusion of higher plant protoplasts. J. Cell. Sci. 11:59–75.

Wright, W.R. 1978. The isolation of heterokaryons and hybrids by a selective system using irreversible biochemical inhibitors. Exp. Cell Res. 112:397–407.

Wullems, G.J., Molendijk, L., and Schilperoort, R.A. 1980. The expression of tumor markers in intraspecific somatic hybrids of normal and crown gall cells from *Nicotiana tabacum*. Theor. Appl. Genet. 56:203–208.

Yamada, Y., Hara, Y., Katagi, H., and Senda, M. 1980. Protoplast fusion: Effect of low temperature on the membrane fluidity of cultured cells. Plant Physiol. 65:1099–1102.

Zelcer, A., Aviv, D., and Galun, E. 1978.   Interspecific transfer of cytoplasmic male sterility by fusion between protoplasts of normal *Nicotiana sylvestris* and X-ray irradiated protoplasts of male-sterile *N. tabacum*. Z. Pflanzenphysiol. 90:397-407.

Zenkteler, M. and Melchers, G. 1978.   In vitro hybridization by sexual methods and by fusion of somatic protoplasts.   Theor. Appl. Genet. 52:81-90.

CHAPTER 8
Genetic Analysis of Somatic Hybrid Plants

Yu. Yu. Gleba and *D.A. Evans*

Distant interspecific sexual hybrids have been used for development of many new crop varieties. The breeding methods used to transfer useful genes from a wild species into a cultivated crop are time consuming and therefore limit the release of new varieties. Interspecific hybrids must be continually backcrossed to the cultivated parent to eliminate most of the wild species genome while retaining the useful gene(s) before a new variety can be released. It has been suggested in numerous review articles that protoplast fusion offers a unique method to transfer useful genes into cultivated crops (e.g., Melchers, 1977).

It is known that sexual crossing is a highly regulated system of hybridization, in which only certain organisms in restricted combinations can be used as parents. Moreover, the result of sexual hybridization is a single type of progeny possessing a restricted genetic constitution. Recognition of the limits of the sexual process will permit us to assess how protoplast fusion can be uniquely integrated with conventional breeding methods to produce novel plant varieties. These generalized limits to sexual hybridization include the following:

(1) The gametes involved in sexual crossing are highly specialized cells. Gametogenesis includes meiotic reduction and segregation of nuclear genetic material. Only plants that are capable of normal morphogenesis and gametogenesis can be sexually crossed. For example, in many cases prezygotic incompatibility has been based on limitations of the morphology of the specialized cells involved in sexual reproduction. In some cases protoplast fusion has already been shown to overcome prezygotic incompatibility (e.g., Power et al., 1977).

(2) Sexual crossing is symmetric in the sense that gametes of both parents equally contribute a gametophytic set of nuclear genetic material. The process of zygote formation includes the fusion of gametic nuclei and the restoration of the sporophytic set of nuclear genetic material. Each particular descendant is itself the genetic product of regular chromosome segregation of the parental chromosomes. The properties of nuclear segregation are strictly governed by Mendel's laws. These restrictive characteristics of sexual hybridization limit the mixture of genetic information that is recovered in hybrids. Products of protoplast fusion may result in many asymmetric hybrids. This could result in the most useful types of combinations that would have all chromosomal material from a cultivated crop, but only a few chromosomes or genes from the wild species parent.

(3) Extranuclear genetic determinants in most crop plants are inherited uniparentally, with maternal transmission expressed in progeny. In these species sexual crosses are restricted to only some nuclear-extranuclear genetic combinations out of many possible combinations (Evans, 1982). Fusion would permit researchers to produce unique mixtures of mitochondrial and chloroplast genetic information.

(4) Sexual crosses are limited to phylogenetically related plant species. Much research has been directed to use of protoplast fusion to produce hybrids between species that are too distant to hybridize sexually. These intergeneric somatic hybrids are most exciting as they permit new gene combinations that have previously been unexplored.

The most exciting distinction of parasexual hybridization in comparison with conventional sexual crossing is that this new method can be manipulated by experimentation to produce unique genetic mixtures. Cell hybridization with subsequent plant regeneration in vitro is a complicated process consisting of a number of variables that may all introduce genetic changes prior to plant regeneration. These changes in vitro can affect cell fusion, fusion/nonfusion of cell nuclei, or fusion/nonfusion of organelles. As the genotype of regenerated parasexual hybrids is affected by all of these phenomena, it is quite evident that by controlling some of the parameters of protoplast fusion that one can predetermine the genetic constitution of regenerated hybrid plants.

Although protoplast fusion can result in unique mixtures of plant cells, a number of requirements must be fulfilled before these plants can be utilized for crop improvement. First, hybrid cells must be regenerated into complete plants. Any two plant cells can be fused; however, regeneration of hybrid plants has been limited to only a few plant species. The techniques and successes to date in protoplast fusion and regeneration are discussed elsewhere in this volume (Chapter 7). If unique somatic hybrids are to be effectively used for crop improvement, the most important consideration is that the transferred genes must be stably inherited in somatic hybrid plants. Unfortunately, little attention has been directed to the fate of genetic information in somatic hybrids. The genetic behavior of intergeneric hybrids is particularly exciting, as all reported experiments have resulted in some asymmetric hybrids. By recovering controlled asymmetric hybrids, a great deal of time can be saved by circumventing otherwise necessary backcrosses to the cultivated crop. Many events may be extremely important in determining the fate of genetic material in somatic hybrid cells and regenerated plants. The frequency of cell fusion, nuclear fusion, nuclear segregation and recombination, organelle segregation, and recombination all impact on the recovery of novel somatic hybrid plants. In addition, the tissue culture conditions used to regenerate plants from fused protoplasts may introduce chromosome instability, gene mutations, or epigenetic changes that may further modify the appearance of somatic hybrids. While the potential variability of somatic hybrid plants is astonishing, it is important to control and regulate these phenomena. In this chapter we will attempt to summarize available information on the fate of genetic information in somatic hybrids and to speculate on the usefulness of protoplast fusion as a method of directed gene transfer.

REVIEW OF LITERATURE

Nuclear Genetics

Most somatic hybridization experiments have been designed to fuse two diploid protoplasts to produce a plant containing the summation chromosome number. Thus somatic hybrid plants would have a chromosome number equal to twice the chromosome number of comparable sexual hybrids. Hybrids that are directly comparable to sexual hybrids can be produced by fusing haploid cells of each species (Melchers and Labib, 1974); however, most experiments to date have used protoplasts isolated either from callus or from diploid somatic cells. In practice, though, most somatic hybrids that have been examined do not contain the amphiploid chromosome number.

Selection of hybrids has been based on complementation of mutations (Chapter 7). As these markers usually have a nuclear genetic basis, complementation represents evidence that the two parental nuclei have fused. Complementation has been obtained following fusion of protoplasts from two recessive chlorophyll deficiency mutants (Melchers and Labib, 1974; Schieder, 1977). If complementation is observed, it is expected that at least a part of the chromosomal material of each parent is present in regenerated somatic hybrid plants. While many aneuploid plants have been recovered following fusion, it is presumed that aneuploidy is the result of chromosome elimination following protoplast fusion. This elimination may reflect loss of chromosomes due to cellular incompatibility or due to chromosome variation induced by culture conditions prior to regeneration. In fact, the amphiploid chromosome number has been reported in at least some regenerated plants of a number of interspecific somatic hybrids (Chapter 7), representing further evidence that nuclear fusion has occurred.

MULTIPLE NUCLEAR FUSION. Microscopic examination of protoplast cultures shortly after fusion has shown that multiple fusions can result following polyethylene glycol treatment. It has been suggested that such multinuclear products are viable and may be able to form

whole hybrid plants. Among the intraspecific *Nicotiana tabacum* somatic hybrids produced following fusion of two chlorophyll defective, light
sensitive varieties (each with 2n = 24), a number of plants were recovered with 2n = 72 chromosomes (Melchers and Sacristan 1977). As
most hybrid plants contained the summation chromosome number of the
two haploid tobacco, 2n = 48, it was suggested that the triploid plants,
which in some cases were regenerated after only a short culture period, were the result of fusion of three nuclei. Similarly Smith et al.
(1976) counted chromosomes of 23 mature hybrid plants following fusion
of *N. langsdorfii* (2n = 18) with *N. glauca* (2n = 24). None of the
hybrid plants contained 42 chromosomes. The 2n number of regenerated plants was restricted to 56-63 chromosomes with a mean of 59.3.
As these plants were regenerated following extended culture periods, it
was proposed that the hybrid plants were derived from triple fusions
containing two glauca nuclei (GG + GG + LL = 66 chromosomes) or
two langsdorffii nuclei (LL + LL + GG = 60 chromosomes). Chupeau et
al. (1978) found a wider range of chromosome number when they fused
glauca with langsdorffii. These authors recovered some hybrids between these two species that had 42 chromosomes. As a few hybrids
that had 55-64 chromosomes were also isolated by these workers, it
has been suggested that in some species combinations, such as glauca-
langsdorffii, the growth conditions used to regenerate plants may favor
cells derived from multiple nuclear fusion.

Multiple cell fusion may be an important phenomenon during protoplast fusion. Evidence that plants are derived from multiple fusion
events also comes from experiments where intraspecific heteroplasmic
tobacco fusion products were isolated mechanically and cloned (Gleba
and Berlin 1979; Gleba, 1983). Among the cells derived from single
heterokaryons both hybrid and pure parental type cells have been
found. Such mosaic calli could only have originated from multiple
fusion events, in which two nuclei of different parents fused and a
third nucleus did not fuse. This third nucleus then segregated during
subsequent mitotic divisions. Similar results have been reported
following cloning of *N. tabacum* + *N. plumbaginifolia* interspecific fusion
products (Sidorov et al., 1981).

NUCLEAR SEGREGATION. In some cases nuclear fusion may not
occur following protoplast fusion. In protoplasts where nuclear fusion
does occur, the fusion of nuclei has been observed in interphase cells
or alternatively may not occur until fused protoplasts undergo the first
of later mitotic division (Kao, 1976). If nuclei do not fuse, it is
possible to establish "fused/nonfused" mosaic cell lines that may result
in regeneration of nonfused shoots. Plants resulting from nuclear segregation following protoplast fusion with subsequent cytoplasmic segregation (see below) could contain the nucleus from one parent and the
cytoplasm of a second parent.

In the early experiments of Gleba et al. (1975). the semidominant
genome mutation Su and plastome mutation both causing chlorophyll
deficiency were fused to follow the fate of nuclear and plastome genes.
It was demonstrated that some of the hybrids recovered on the basis of

genetic complementation and restoration of chlorophyll synthesis had wild nucleus type. The absence of Su gene in these products was due to nonfusion and segregation of nuclei in some heterokaryocytes.

While nuclear segregation may be dependent on the taxonomic relatedness of the two species being fused, it is evident from the literature that certain techniques can be modified to regulate the frequency of fused versus nonfused regenerated plants. Recently, Gleba (1983) used mechanical isolation and individual culturing (cloning) of heteroplasmic protoplast fusion products of tobacco to demonstrate that in the majority of the clones, nuclear fusion did not follow cell fusion. Hence nuclear segregants were obtained in high frequency. In other cases the nuclear fusion rate is high and sometimes is close to 100% (Sidorov et al., 1981; Menczel et al., 1981). When Sidorov et al. (1981) used mechanical isolation to separate putative N. tabacum + N. plumbaginifolia somatic hybrids approximately 100% of the regenerates had fused nuclei. However, when N. tabacum mesophyll cells were preirradiated with Co 60-rays (60 J/kg), the yield of hybrids with fused nuclei was reduced to 60%. Menczel et al. (1982) has demonstrated that the frequency of fused nuclei is correlated with the dose of preirradiation of tobacco cells. These authors observed a consistent reduction of hybrid nuclei from 100 to 22% as irradiation dose increased from 0 to 210 J/kg. Zelcer et al. (1978) first proposed use of preirradiation treatment to produce cybrids to permit transfer of cytoplasmic male sterility from N. tabacum to N. sylvestris. Zelcer et al. (1978) used 42 J/kg irradiation prior to fusing N. sylvestris + N. tabacum and only 73% of recovered male sterile plants had hybrid nuclei.

SEXUAL TRANSMISSION OF NUCLEAR GENETIC MARKERS. With the exception of the specific markers used to select hybrids, single nuclear genetic markers have not been followed in the sexual progeny of somatic hybrids. Certain morphological traits, presumed to be under nuclear genetic control, have been monitored in selfed and backcrossed progeny of somatic hybrids. Power et al. (1978) followed segregation of flower color in R_1 and BC_1 progeny of P. parodii + P. hybrida somatic hybrids and found that most R_1 plants were the same phenotype as the parent somatic hybrids. These authors observed that other vegetative and floral morphological characters segregated independently of flower color.

Chlorophyll deficient mutants have been used to visually select somatic hybrids. In most cases single gene recessive mutants were used to produce green somatic hybrids due to complementation. Intraspecific somatic hybrids are relatively easy to genetically analyze. Melchers and Labib (1974) fused protoplasts of two haploid light sensitive lines (s and v) of N. tabacum. These mutations were complemented following fusion and resulted in green somatic hybrid plants with 2n = 48. Similarly, sexual hybrids between these two lines could be produced. The s and v mutants segregated in the F_2 and R_1 generations of the sexual and somatic hybrids, respectively. The segregation ratio of the somatic hybrid was identical to the comparable sexual hybrid (Table 1). It is expected, and has been observed, that altered

segregation ratios occur in progeny of intraspecific tetraploids (Grant, 1975). Insufficient data has been published to permit complete analysis of genetic ratios in intraspecific hybrids although Melchers (1977) data suggest that nuclear segregation in somatic hybrids is not different from comparable sexual hybrids.

Table 1. Progeny Analysis of Intraspecific *Nicotiana tabacum* Somatic Hybrids (2n = 48) Derived by Fusing Two Haploid Nonallelic Chlorophyll-Deficient Mutants (s + v) (after Melchers, 1977)

F_2	GREEN	YELLOW-GREEN	LIKE-V	LIKE-S	LETHAL
Somatic hybrid	209	174	101	32	4
Sexual hybrid	108	77	64	20	5

Allotetraploids derived by protoplast fusion have the summation of chromosome complements from two diploid species. The mode of segregation of genes introduced into these hybrids depends on the presence or absence of active homologous genes in the different genomes. The degree of homology is reflected in chromosome pairing. It is presumed that most interspecies hybrids would be segmental allotetraploids, so that it is difficult to predict segregation ratios for specific genes in advance. It is likely that different segregation ratios would be observed for genes located in different regions of the genomes of a single amphiploid (Grant, 1975). Such ratios have been observed for progeny of interspecific amphiploid sexual hybrids in *Gossypium* (Gerstel and Phillips, 1958) and *Nicotiana* (Gerstel, 1963). In some *Gossypium* crosses it was demonstrated that the segregation ratios of a single gene varied dependent on the interspecies combination. Segregation ratios are correlated with the average degree of chromosome pairing between genomes carrying the genes in different hybrid combinations. Schieder (1980b) found that R_1, R_2, and R_3 progeny of *D. innoxia* + *D. discolor* and *D. innoxia* + *D. stramonium* somatic hybrid plants segregated for the albino Al/5a mutant derived from *D. innoxia*. Among seed of various generations, the number of chlorophyll-deficient seedlings varied from 0 to 10%. This percentage of chlorophyll-deficient seedlings is consistent with the limited chromosome pairing observed during meiotic analysis of the interspecific *Datura* somatic hybrids. Schieder also found one abnormal floral segregant among R_3 progeny that was later identified as an aneuploid variant plant (2n = 44). Similarly, Evans et al. (1981) observed albino seedlings among the progeny of *N. nesophila* + *N. tabacum* somatic hybrids by using the Su/Su albino genotype of *N. tabacum* (Table 2). The segregation ratio, about 2% albinos, is consistent with the limited chromosome pairing observed in embryo-culture produced hybrids of these two species (Reed and Collins, 1980). Backcrosses to either parent resulted in no albino progeny (Table 2). The light green trait (from Su/Su) could also be followed in these particular somatic hybrids (Evans et al., 1981). To this date genetic analysis has only been completed for these *Datura* and *Nicotiana* interspecific somatic hybrids. In each case the parent

species are distantly related and have very limited mitotic pairing,
thereby resulting in aberrant segregation ratios. As somatic hybrids
have twice the chromosome number of comparable sexual hybrids, the
genetic analysis of these plants is complicated.

Table 2. Progeny Analysis of Interspecific Somatic Hybrids (2n = 96)
Derived by Fusion Albino (Su/Su) N. tabacum with N.
nesophila (from Evans et al., 1981)

CROSS	DARK GREEN	LIGHT GREEN	ALBINO
(N. nesophila + N. tabacum) x N. tabacum	58	36	0
(N. nesophila + N. tabacum) x self	11	35	1

ANDROGENESIS.   Fortunately, anthers of somatic hybrids can be
cultured to recover plants that have a chromosome number that is
comparable to interspecific sexual hybrids.   Also, as androgenesis
produces plants from microspores, the anther–derived plants represent a
method to dissect the genome of the two species that were combined
by protoplast fusion.   Schieder (1978b) cultured anthers of intraspecific
D. innoxia somatic hybrids produced following fusion of two mutant
lines, A1/5a and A7/1s.   Among the 2n plants recovered following
anther culture, both albino mutant phenotypes could be distinguished.
In this autotetraploid a 25:10 ratio of green to albino is expected
when it is assumed that the double mutant is inviable.   Schieder
(1978b) recovered 45 wild type and 17 albino plants from anther
cultures (χ^2 = 0.040, 1 d.f., p > 0.95).

In interspecific hybrids, chromosome pairing and recombination
observed in somatic hybrids is reduced or eliminated, and it may therefore not be possible to recover comparable albinos via anther culture.
In the case of N. otophora + N. tabacum somatic hybrids, all plants
with 2n = 72 chromosomes are light green Su/Su/+/+.   As the tabacum
genome is Su/Su and the homologous allelic pair in the otophora
genome is presumed to be +/+, it follows that only one type of gamete
(Su/+) containing one homologous chromosome from each genome is
produced during microsporogenesis.   Hence, all N. otophora + N.
tabacum anther culture–derived plants examined to date contain 36
chromosomes and are light green.   Consequently, in this interspecies
somatic hybrid androgenic plants do not segregate for the Su trait.

MITOTIC RECOMBINATION AND GENE TRANSFER.   Leaf spot formation is associated with the Su locus of N. tabacum.   The frequently
observed double spots have been shown to be the result of mitotic
recombination in N. tabacum (Evans and Paddock, 1976).   The presence
of double spots implies an exchange between the two genomes.   In
these hybrids, there is no evidence that the exchange is due to reciprocal recombination; however, it is likely that spot formation reflects

exchange between the glauca and tabacum genomes. The frequency of spot formation can be increased in the somatic hybrids as in parental *N. tabacum*, by treating the shoot apex with X-irradiation (Evans et al., 1980). Similar spots have been observed when other *Nicotiana* species were fused to Su/Su *N. tabacum* (Evans et al., 1981). The Su/Su locus is a marker that permits us to monitor the frequency of intergenomic recombination. As the frequency of recombination is high and can be further increased with chemical or physical treatment, it is likely that gene exchange can be successfully manipulated in somatic hybrids.

There is already evidence that recombination has occurred in some very distant putative somatic hybrids to permit transfer of a small amount of genetic information from *Aegopodium podagraria* into carrot (Dudits et al., 1979). Similarly, it has been suggested that fusion could be used to facilitate gene transfer between widely divergent plant species by irradiating one nucleus prior to protoplast fusion (Dudits et al., 1980a). Incorporation of these methods would permit rapid introduction of useful, selectable genes from wild species into cultivated crops.

Cytoplasmic Genetics

BIPARENTAL TRANSMISSION AND MITOTIC SEGREGATION OF PLASMAGENES. Genes in the chloroplast and mitochondria have been reported to be inherited either biparentally or uniparentally. Sears (1980) lists those higher plant species with maternal or biparental organelle inheritance. In most crop species, strict maternal inheritance has been observed in sexual hybrids in which chloroplasts and mitochondria are excluded from sexual progeny. However, during protoplast fusion the contents of the cytoplasm of both cells are mixed together. The fate of mixed cytoplasms has been monitored in many somatic hybrids.

Gleba et al. (1975) reported presence of mixed cytoplasms in variegated plants obtained following fusion of two *N. tabacum* lines (see above), plastome mutant and Su nuclear mutant. Cytoplasmic heterozygosity rather than chimericity was confirmed by transmission of the variegated trait to sexual progeny. Protoplasts of *N. tabacum* plastome mutant were also fused with (a) cms tobacco analog that contained *N. tabacum* nuclear genome and *N. debneyi* cytoplasm and (b) protoplasts of *N. debneyi* (Gleba et al., 1978; Gleba 1978b; Gleba and Sytnik 1983). In both cases variegated plants have been recovered among progeny. Plants were at least cytoplasmic hybrids (heterozygotes), since the ribulose bisphosphate carboxylase large subunits were composed of both *N. tabacum* and *N. debneyi* polypeptides, and since variegation was transmitted maternally to sexual progeny. Also, "mixed" cells simultaneously containing both types of plastids were present in variegated plants. Heterozygosity for genes causing plastome chlorophyll deficiency has also been observed among fusion-derived hybrids in experiments of Sidorov et al. (1981), Glimelius and Bonnett (1981), and Hakata and Oshima (1982) i.e., in all experiments where plastome chlorophyll-less mutants have been used as one of the parents. Belliard et al. (1977) and Gleba and Piven (c.f. in Gleba and Sytnik 1983) observed segregation of flower malformation and male sterility/fertility characters in vegetatively propagated plants obtained after fusion between wild-type tobacco and either cms tobacco analog possessing *N. debneyi* cytoplasm or *N. debneyi*, respectively.

Because of mitotic segregation of plasmagenes, one or the other parental type of chloroplast is often observed after fusion. This segregation has been observed by monitoring the fate of various cytoplasmic markers in regenerated hybrid plants and in the progeny of somatic hybrids. Chen et al. (1977) followed segregation of the large subunit of Fraction-1 protein in *N. glauca* + *N. langsdorfii* somatic hybrids and observed that segregation to the langsdorfii or glauca-type chloroplasts occurred prior to meiosis in most of the hybrids containing mixed cytoplasms. In these hybrids and in most other cases segregation was random to one or the other type cytoplasm. In other reports nonrandom segregation to one parental type has been observed (Flick and Evans, 1982; Maliga et al., 1980). Nonrandom segregation may reflect nucleo-cytoplasmic incompatibility or may reflect the choice of cell types (e.g., cell suspension culture versus leaf-derived protoplasts) used for protoplast fusion. Detailed examination of *N. debneyi* chloroplast DNA in intraspecific somatic hybrids suggests that chloroplast segregation occurs shortly after protoplast fusion (Scowcroft and Larkin, 1981). In these hybrids where effort was made to insure that each parental protoplast had an equal chance of survival following protoplast fusion, segregation was random. Most likely segregation occurs during shoot regeneration, when the number of chloroplasts or proplastids is reduced to about 10-20 per cell.

CYTOPLASMIC MARKERS. There are a limited number of markers available to monitor the cytoplasmic inheritance in somatic hybrids. However, the number of cytoplasmic traits has been increasing rapidly within the past few years. Chloroplast DNA (cpDNA) is the site of coding for the large subunit of Fraction-1 protein that has been monitored in many hybrids (see Chapter 9). Insensitivity to tentoxin is also encoded in the cytoplasm as tentoxin binds to chloroplast coupling-factor-1 in sensitive species, a product of cpDNA. Maliga et al. (1980) has used streptomycin resistance as a chloroplast marker in several fusion experiments. The SR-1 trait was isolated from cultured tobacco cells and the streptomycin resistance of SR-1 is associated with an altered chloroplast ribosomal protein (Yurina et al., 1978). Plastome chlorophyll deficiency mutants encoded in cpDNA have also been used to follow the fate of chloroplasts following fusion. More recently restriction enzyme analysis of isolated cpDNA has been used to follow the fate of different parental chloroplasts (cf., Scowcroft and Larkin, 1981).

There is a paucity of mitochondrial genetic markers. In some cases it has been demonstrated that cytoplasmic male sterility (cms) is encoded in the mitochondria; however, no other useful phenotypic markers have been localized in the mitochondria. It is possible to follow the fate of mitochondrial DNA (mtDNA) by restriction enzyme maps of isolated mtDNA. However, in at least some species mitochondria contain a mixed population of many sizes of DNA that results in extensive rearrangement and recombination in hybrid cells (Quetier and Vedel, 1977). In most cases it has not been determined if cms is encoded in mitochondria or chloroplasts. Similarly, a number of potentially useful traits have been identified as being maternally inherited and eventually may add to the list of available cytoplasmic markers.

RECOMBINATION OF CYTOPLASMIC DNA. Recombination of mitochondrial DNA has been reported in several cytoplasmic hybrids. Recombination is monitored by examining restriction digest patterns of mtDNA. All reports of mtDNA recombination in somatic hybrids have been with species in the genus *Nicotiana* and include *N. tabacum* + *N. debneyi* (Belliard et al., 1979); *N. tabacum* + *N. sylvestris* (Galun et al., 1982); and *N. tabacum* + *N. knightiana* (Nagy et al., 1981). In some of these and other experiments, no recombination of cpDNA was detected using restriction digest analysis. No recombination of cpDNA was observed in the following cybrids: *N. tabacum* + *N. debneyi* (Belliard and Pelletier, 1978); *N. tabacum* + *N. knightiana* (Menczel et al., 1981); *N. tabacum* + *N. suaveolens* (Glimelius et al., 1981); or *N. debneyi* + *N. debneyi* (Scowcroft and Larkin, 1981). However, one preliminary report has suggested recombination of cpDNA in *N. glauca* + *N. langsdorffii* somatic hybrids (Conde, 1981).

Production and Gene Transfer of Intergeneric Hybrids

INTERFAMILIAL CELL HYBRIDS OF HIGHER PLANTS. There have been several reports on attempts to obtain cell hybrids between species that belong to different plant families. In most cases, however, observations have been limited to the first few cell divisions in fusion products. Visual observations were used when possible, particularly when morphologically different parental cell types were used, e.g., when callus protoplasts were fused with mesophyll protoplasts. The use of such visual observations is limited. However, the observations are of some importance as hybrid cells in interfamilial combinations have been able to complete several cell divisions following fusion.

One early successful experiment was aimed at isolation and study of clones of interfamiliar cell hybrids (Kao, 1977; Wetter, 1977; Kao and Wetter, 1976). These authors isolated cell hybrids of *Glycine max* + *Nicotiana glauca*. Individual dividing fusion products were mechanically isolated and then cultured in droplets of conditioned medium. Twenty cell lines were isolated, with cell division maintained for more than 8 months (about 100 cell generations). Chromosomes were analyzed and isozymes (aspartate aminotransferase and alcohol dehydrogenase) monitored for 2-8 months in culture. Chromosome analysis of the first cell division after fusion was also performed. In 48 hr mixed cultures, premitotic fusion of nuclei was observed. Karyokinesis in hybrid cells was nonsynchronous. The *N. glauca* chromosomes had a tendency to condense and fuse together.   Chromosome bridges were observed in anaphase of fusion products.

During the first division, superlong chromosomes (megachromosomes), ring chromosomes, fragmented chromosomes, and multiconstrictional chromosomes were observed.   After 1-2 months of isolated culture, clones from hybrid cells still revealed megachromosomes, chromosome fragments, and anaphase bridges.   During subsequent months the number of tobacco-type chromosomes was gradually reduced.   In addition, standard tobacco chromosomes were rarely observed, whereas chromosomes with two and more constrictions were frequently observed. Short chromosomes that were probably reconstituted tobacco chromosomes were also observed.   After 6 months of culture, some tobacco chromosomes were still preserved in 5 of the 20 hybrid lines.   During the same culture period, parental callus cultures of *N. glauca* and soybean had invariable chromosomal morphology.   The isozyme study confirmed the presence of isozyme forms of both tobacco and soybean in all the cell lines during the early stages (2-4 months) of culture. However, by the eighth month the specific isozyme forms of tobacco ceased to appear in most cultures.   The electrophoretic data were in good agreement with the data of chromosome analysis.   The authors did not report any morphogenesis, which is not surprising, as the soybean cultured cells are incapable of morphogenesis.   Hence the absence of morphogenesis in hybrid lines can be connected with the predominance of soybean chromosomes and loss of *N. glauca* chromosomes in cell hybrids.

These authors have used these unique hybrid cell lines in subsequent fusion experiments (Wetter and Kao, 1980). The hybrid cells of *Glycine max* + *Nicotiana glauca* that lost chromosomes of *N. glauca* were "backfused" twice by somatic hybridization with mesophyll protoplasts of *N. glauca.*   The first recurrent protoplast fusion was carried out 27 months after the initial fusion.   The second backfusion was performed 7 months after the first backfusion.   As a result, lines which had previously lost all evidence of tobacco chromosomes, as well as all isozyme forms of aspartate aminotransferase characteristic of tobacco, retained tobacco information.   The backfusion cell lines had a considerable increase of tobacco chromosome material, as was evidenced by normal and reconstructed chromosomes of *N. glauca.*   In addition, the isozyme bands of *N. glauca* were evident for at least 2 years after the first backfusion and at least 6 months after the second backfusion.   As numerous chromosomal rearrangements were visible in recurrent fusions, it is likely that increasing quantities of *N. glauca* genetic information were recombined with soybean chromosomes resulting in stable incorporations of *N. glauca* information into the new cell hybrids.   In order to alter the cell cycle stage of the *N. glauca* protoplasts during the second backcross fusion, the *N. glauca* leaves were preincubated in culture medium for 24 hr prior to protoplast isolation.   As fusion of different homophasic animal cells increases chromosome stability in resulting hybrid lines (Rao et al., 1975), it was suggested that these plant cells fused while at the same stage in the cell cycle would also have increased stability.   Additional studies with this well-characterized plant system could support this hypothesis and aid in understanding a number of basic questions regarding cell fusion. Chromosome elimination has been reported in cell hybrids of other distantly related plant species including *Vicia faba* + *Petunia hybrida* (Binding and Nehls, 1978), which represents another fusion of a legume and a solanaceous species. In this fusion some hybrid cell lines lost *Vicia* chromosomes, while other lines lost *Petunia* chromosomes.

Chien et al. (1982) have recently produced somatic cell hybrids of *Nicotiana tabacum* + *Glycine max* by fusion of tobacco mesophyll and soybean callus protoplasts. A total of 21 hybrid cell lines were obtained by mechanical isolation and culturing of individual fusion products. After 3 months of culturing, approximately two-thirds of the tobacco chromosomes were retained in some hybrid cells. After 6 to 7 months of culturing, 5 cell lines had retained more than half the tobacco chromosomes in the majority of cells, whereas another 6 cell clones studied had lost nearly all the *N. tabacum* chromosomes except for a very low number of cells which still retained a number of tobacco chromosomes. In many cells megachromosomes and chromosomal bridges have been observed. No morphogenetic activity has been reported for callus cells of these hybrids. These results provide another example of successful hybridization of species belonging to different families and indicate that greater chromosomal stability could be partially maintained in interfamiliar hybrids.

Using individual culturing of heteroplasmic fusion products, Gleba et al. (1983b) have been able to isolate somatic cell hybrids after fusion of mesophyll protoplasts of pea (*Pisum sativum*) and callus protoplasts of Chinese tobacco (*Nicotiana chinensis*). Three clones were obtained but only one fast growing clone was amenable to detailed analysis 6 months after hybridization. Electrophoretic separation and specific staining for esterase and amylase activities have demonstrated the presence of species specific multiple molecular forms of enzymes of both parents to be present in somatic cell hybrids. Karyological analysis has revealed the presence of both tobacco and pea chromosomal types in at least some of the dividing cells. Cells of the hybrid clone lose *Pisum* chromosomes. Nuclear material undergoes drastic rearrangements as indicated by fragmentation of nuclei, presence of multinuclear cells, occurrence of megachromosomes, and anaphase bridges, but these changes apparently do not affect morphology and number of Chinese tobacco specific chromosomes. Chromosomes of the two parents tend to group separately in a common metaphase indicating spatial separation of parental genomes in the hybrid nuclei. Until now, we were unable to induce any sort of morphogenetic activity in these cell clones.

The analysis of data obtained so far permits us to state that by using somatic hybridization, it is possible to obtain viable lines of cell hybrids between species belonging to different families and even orders. Interfamiliar plant cell hybrids are similar to animal somatic cell hybrids (Ringertz and Savage, 1976), as they typically display chromosome reconstruction and chromosome elimination of one of the parent species. It is still uncertain if morphogenesis can be induced in cell cultures of interfamiliar hybrids.

INTERGENERIC (INTERTRIBAL) HYBRID PLANTS.   Several works have dealt with attempts to produce somatic hybrids between distantly related plant species belonging to the same family.   Such intergeneric or intertribal hybrids have been reported in the Cruciferae, Solanaceae, and Umbelliferae families.   It should be mentioned that, as far as we know, all attempts to obtain hybrids of species belonging to different taxonomic tribes by sexual crossing have been unsuccessful.

In our work (Gleba and Hoffmann, 1978, 1979; Gleba et al., 1978; Komarnitsky and Gleba, 1981) the hybridization of callus protoplasts of *Arabidopsis thaliana* with mesophyll protoplasts of turnip, *Brassica compestris*, was reported.   The selection of hybrids was performed using a modified method of Kao (1977), that is, mechanical isolation of dividing heterokaryocytes using a micropipette followed by individual cultivation (cloning) in microdroplets of conditioned medium (Gleba, 1978).   With this method, more than 30 hybrid clones were obtained. Six *Arabidopsis-Brassica* clones were analyzed 4 months after hybridization.   Chromosome analysis was possible, as the parental chromosome types differed in size and morphology.   Detailed study of 6 cell hybrid clones confirmed in all the cases the presence of chromosomes of both parental types in the same metaphase cells.

Cytological study also revealed reconstructed chromosomes (chromosomes with two constrictions, ring chromosomes) that were absent in parental cells.   Mitotic abnormalities (anaphase bridges, multichromosome chains) were also observed.

Biochemical basis of the clones was evaluated using electrophoresis. Examination of multiple molecular forms of esterase, lactate and alcohol dehydrogenase, and peroxidase all revealed characteristic isozyme forms for both parents.   The analysis of *Arabidopsis* + *Brassica* cell hybrids has shown that chromosomes of both species were preserved in all lines for at least 7 months after fusion.   These data were in contrast to the *Glycine* + *Nicotiana* cell hybrids in which a rapid (during first months of culturing) reconstruction and elimination of tobacco chromosomes was reported (Kao, 1977; Wetter, 1977).   From our observation it was suggested that within the same plant family the genetic material of two distantly related species can be integrated and organized for subsequent cell reproduction without extensive elimination of genetic material of one of the parents.

Esterase isozymes were monitored in six cell lines for 15 months of culture, during which the cells had undergone at least 80-90 subsequent cell divisions.   It has been demonstrated that in four of the six cell lines, no changes in isozyme spectra occur.   This represents additional evidence in support of genetic stability of these intertribal cell hybrids. In one line, all molecular forms of turnip esterase were eventually lost.

The attempts to induce morphogenesis in cell cultures of *Arabidopsis* + *Brassica* intertribal cell hybrids were also successful.   Besides root organogenesis, in three clones shoots and abnormal plants were regenerated.   The regenerates from the stable cell lines were verified as hybrids based on analysis of chromosomes, esterase isozymes, and Fraction-1 protein.   All regenerated plants contained only *Brassica* plastids. The regenerated hybrid plants derived from chromosomally stable cell lines were highly homogenous and morphologically intermediate between

Arabidopsis and turnip. However, these regenerates would not flower and formed abnormal roots. On the other hand, regenerates from the segregating cell line that lost most of the turnip chromosomes were morphologically diverse and included both anomalous shoot-like teratomas as well as almost perfect flowering plants similar to *Arabidopsis* (Hoffmann and Adachi, 1981). The study of the most regular *Arabidopsis*-like plants has revealed that they contain only small amounts of turnip genetic material. In some regenerates, single large chromosomes characteristic of turnip are sometimes observed, as well as some morphological characters typical of turnip such as presence of chlorophyll, yellow petals, and simple nonbranching forms of trichomes. These morphological characters are absent in the parental strain of *Arabidopsis* that was used for fusion. All plastids of these "asymmetric" hybrids were derived from turnip. All the flowering hybrid plants were sterile. The above experiments demonstrate that at least in some hybrid cells morphogenesis of whole plants may be induced. Some morphogenesis is possible both for relatively symmetric hybrids and for asymmetric hybrid cells that contain only a portion of the genome of one of the parents. While these intergeneric cell hybrids were regenerated to hybrid plants, it should be noted that all the plants obtained in these experiments were morphologically abnormal and in all cases incapable of sexual reproduction.

Krumbiegel and Schieder (1979) reported the regeneration of hybrid shoots and teratomas from intertribal protoplast fusions of the Solanaceae species *Datura innoxia* and *Atropa belladonna*. These two species possess strikingly different chromosomes as *Datura* chromosomes are about twice as long as those from *Atropa*, thereby facilitating chromosome analysis. Protoplasts were isolated from the leaves of chlorophyll-defective *Datura* mutants and from the wild type leaves of *Atropa*. Hybrid calli were selected during shoot regeneration by isolating green (*Atropa* character) shoots that were covered with thick trichomes (*Datura* character). Eight independent fusion experiments were performed, resulting in isolation of 13 putative hybrids. The preliminary chromosome analysis has shown the presence of chromosomes of both parental species in the same metaphase cells.

One amphidiploid line has been isolated, but is incapable of shoot morphogenesis. Nevertheless, shoots were obtained in other lines carrying partial chromosome sets. Further analyses demonstrated that morphologically normal shoots could be obtained, but only after ca. 18 months of unorganized growth of the hybrid cells. Cytological studies revealed that in regenerated shoots, only ca. 40 of the original 72 *Atropa* chromosomes were retained (Schieder and Krumbiegel, 1980). In other instances, albino shoots reflecting *Datura* chlorophyll deficiency were regenerated and contained as many as 36 *Atropa* chromosomes, indicative that in this case, *Atropa* chromosomes controlling chlorophyll synthesis were eliminated (Krumbiegel and Schieder, 1981). In this experiment diploid *Atropa* (2n = 72) was fused with diploid *Datura* (2n = 24) cells. While the amphidiploid number of 96 chromosomes was expected, only one of the 13 hybrids had approximately 96 chromosomes. All other putative hybrid shoots exhibited lower or higher chromosome numbers.

Recently we were successful in obtaining (Skarzhynskaya et al., 1982) intertribal cell hybrids with regeneration of anomalous plants following protoplast fusion of *Nicotiana tabacum* with *Solanum tuberosum* and *Nicotiana tabacum* with *Solanum sucrense*. Mesophyll protoplasts of *Solanum* were fused with tobacco callus protoplasts. Hybrid cells were mechanically isolated and then cloned. As a result, 30 clones of potato + tobacco and 4 clones of tobacco + *Solanum sucrense* were maintained. The hybrid nature of several clones was verified by bio-chemical (study of multiple molecular forms of amylase and esterase) and cytochemical (cytophotometric determination of DNA content in nuclei) analysis of the initial cell lines, as well as the cell lines that were repeatedly cloned. Plant regeneration was induced in 9 cell lines of potato + tobacco 3 months after hybridization. Analysis of regen-erates has confirmed the hybrid origin of plants obtained from three of these clones. Plant regeneration is accompanied by a reduction of chromosome number. In these hybrids, tobacco chromosome material is preferentially eliminated.

In other experiments (Gleba et al., 1982, 1983) intertribal hybrids were obtained following fusion of mesophyll protoplasts of *Atropa bella-donna* with callus protoplasts of *Nicotiana chinensis*. Hybrid products were selected using mechanical cloning. To date, 12 cell clones were isolated, cloned, and then analyzed 4–6 and 11–13 months after fusion. Chromosome analysis was facilitated by the striking differences in chromosome size between *Nicotiana* (large) and *Atropa* (small, thin). Chromosome examination of cells of clones revealed both parental types in hybrid metaphases in most cases. Reconstructed (chain, ring) chromosomes were also observed. Species specific elimination of bella-donna chromosomes was observed in three lines. Biochemical analysis of multiple molecular forms of amylase and esterase revealed isozyme forms from both parents in a number of cell lines. The transfer of hybrid cells to regeneration medium has resulted in regeneration of anomalous shoots.

Dudits et al. (1979) have reported their attempts to obtain a para-sexual hybrid between albino carrot, *Daucus carota* (2n = 18) and gout-weed, *Aegopodium podagraria* (2n = 42). Protoplasts from suspension cultures of chlorophyll–deficient mutants of carrot were fused with mesophyll protoplasts of goutweed. Hybrid cells were produced when mixed protoplasts were treated with PEG. Unfortunately, putative green hybrid plants contained only 18 chromosomes and were morpho-logically similar to carrot, also with 2n = 18. Morphological markers for root development and root carotenoids derived from *A. podagraria* were also expressed in some regenerated plants. More recently, Dudits et al. (1980b) have extended these experiments to recover plants following fusion of protoplasts of their albino carrot mutant with leaf protoplasts of parsley, *Petroselinum hortense*, 2n = 22 (Dudits et al., 1980b). The highest frequency of regenerated plants contained 2n = 19 chromosomes representing incomplete elimination of the parsley chromo-somes. In addition, presence of parsley-derived isozyme bands and correction of the albino character indicate successful transfer of some parsley genetic information into carrot. Since neither group of plants were derived from cell lines that were isolated by microculture and as chromosome behavior shortly after fusion was not monitored, it may be too soon to state that these plants represent true hybrids. However, these experiments offer exciting possibilities for effective gene transfer between distantly related species by using protoplast fusion.

Based on these reports, a number of general characteristics of inter-tribal hybrids are evident. (1) The hybrid state is relatively stable. In some cases during months or even years of culturing hybrid cell lines, no chromosome elimination was reported. (2) In some cases intense genetic rearrangements consisting of reconstruction and partial elimination of parental chromosomes have been observed. Changes of DNA profiles or multiple molecular forms of enzymes have been re-ported. (3) Diverse genetic lines have been recovered from fusion experiments. These diverse lines may represent different fusion products that reflect preexisting diversity of parental protoplasts or may reflect changes in vitro that occur following fusion. (4) Intertribal fusion products are capable of morphogenesis but in all cases produce abnormal plants that are sterile. Functional sterility precludes genetic analysis.

Despite the increasing number of results obtained following cell hybridization of distantly related plant species, there are insufficient data to ascertain the cell mechanisms and processes that affect the fate of parental genetic material in fused cells. The nature of these fusion experiments has inadvertently introduced a number of variables that may change the fate of cell hybrids following protoplast fusion. (1) In most cases mitotically dormant mesophyll cells were used as one of the parents, whereas rapidly growing callus was used as the second parent. It is natural to expect that the fate of genetic material in heterokaryons depends on the state of nuclei at the moment of hybridi-zation. This has already been demonstrated for animal cell hybrids (Ringertz and Savage, 1976) and suggested in plant cells (Wetter and Kao, 1980). Detailed study of this cell cycle variable is, however, extremely difficult as cultured plant cells are seldom synchronized. (2) As a rule, cultured plant cells have a wide range of aneuploid chromo-some numbers. Consequently, the chromosome number of fusion pro-ducts is dependent on the degree of chromosome stability in the donor callus cultures. In some cases callus cultures that have been main-tained for many years were used to isolate protoplasts. (3) Finally, it should be mentioned that in most distantly related hybrids, the species-specific elimination of chromosomes is usually restricted to the species from which mesophyll protoplasts were isolated.

GENETIC ANALYSIS USING PROTOPLAST FUSION

Protoplast fusion can be used for genetic analysis in the same way that crossing has been used for years. Of special interest are those cases in which normal sexual processes cannot be performed. This would be important for analysis of nuclear genes in plants or plant cells defective in morphogenesis, the analysis of extranuclear genes of plant cells, the study of the fate of epigenetic traits in parental cells in the hybridization process, and analysis of the developmental genetics of cell hybrids.

Analysis of the Nature of Inherited Traits

Variability in plant cell cultures can be either genetic or epigenetic. Epigenetic changes are not stably transmitted through meiosis. Analysis of the nature (genetic versus epigenetic) of variability in cultured plant cells represents a special problem which in most cases cannot be discerned using available methods. The most obvious way to elucidate the nature of variability is to verify sexual transmission of induced characters using regenerated plants (Chaleff, 1981). Unfortunately, not all cultured cells are capable of morphogenesis. So, as an alternative method to assess the stability of variant cell lines, one can use protoplast fusion. Fusion may alter the epigenetic phenotype or may restore morphogenesis (e.g., Maliga et al., 1977). Restoration of fertility has been used to analyze stability of variants isolated in vitro. Gleba and Berlin (1979) fused 5-methyl-tryptophan (5MT) resistant tobacco protoplasts with normal tobacco protoplasts and regenerated hybrid plants, while the 5MT-resistant lines were incapable of regeneration (for details, see Gleba, 1983). Tissues of hybrid plants were used to induce callus that was partially resistant to 5MT. In this experiment the 5MT resistance was preserved despite fusion, regeneration, and callus reinitiation. So, while genetic analysis has not been completed, one can conclude that the 5MT resistance is most likely controlled by genetic factors. Similarly, we have fused albino, para-fluoro phenylalanine resistant protoplasts with normal tobacco protoplasts to demonstrate stability of the isolated variant (Flick et al., 1981; Flick and Evans, unpublished).

Plant cells obtained following mutagenesis are often defective for morphogenesis or gametogenesis. In these cases protoplast fusion and subsequent regeneration of intact hybrid plants can be used as a method to: (a) rescue and stably maintain mutations; (b) place recessive or codominant mutations into heterozygous state; (c) permit genetic analysis. Two works should be mentioned here as examples of the application of fusion for genetic analysis. First, protoplast fusion has been utilized for complementation analysis of auxotrophic mutants of the moss *Physcomitrella patens* (Grimsley et al., 1977a, b). The mutants obtained by mutagenic treatment in most cases were associated with developmental abnormalities and, as a result, were sterile. These included mutants that were auxotrophic for thiamine, para-amino benzoic acid, and nicotinic acid. By using protoplast fusion it was demonstrated that most auxotrophic isolates obtained by mutagenesis were nuclear recessive mutations. Complementation analysis using fusion of 6 nicotinic acid auxotrophic strains of independent origin revealed three nonoverlapping genes. The results obtained for complementation groups using fusion agreed with the limited data from sexual hybridization that was possible for some of these same mutant strains (Ashton and Cove, 1977). In these experiments heterozygous diploids were obtained by protoplast fusion. Production of heterozygotes permitted a more detailed genetic analysis of the mutational changes. Later, these authors performed complementation analysis and dominance tests for additional mutations, i.e., p-aminobenzoic acid auxotrophs and mutants causing abnormal development (Grimsley et al., 1980).

As a second example, the studies of tobacco cell lines defective for nitrate reductase (Glimelius et al., 1978) should be considered. A great number of cell lines defective for nitrate reductase enzyme activity were obtained using selection for resistance to chlorate (Muller and Grafe, 1978). The lines were resistant in the absence of the selective agent, but most mutants had lost the capacity for morphogenesis. The study of enzyme activity, selectivity, and reassociation of active enzymes in vitro (Mendel and Muller, 1976, 1978) of each of the variants suggested the existence of mutants with different blocks in activity of the same enzyme. Mutants of nia type were defective for the synthesis of nitrate reductase apoprotein whereas cnx mutants were defective in synthesis of the molybdenum-containing enzyme cofactor. Complementation analysis using protoplast fusion confirmed the biochemical data that already suggested that the nia and cnx mutants are controlled by recessive nonallelic genes.

More detailed study of the cnx mutants using somatic cell hybridization has been performed in the laboratory of P. Maliga (Sidorov and Maliga 1982) and three groups of cnx mutations have been identified that complement each other. This observation is not unexpected, because in other organisms, several complementation groups of cnx mutants have been found (see Cove 1979; Marzluf 1981).

Similar observations on the expression (recessiveness or codominance) of genes and their ability to complement in cell hybrids have been collected by other authors using protoplast mediated hybridization. In some cases these reports represent uncharacterized variants, including chlorophyll-deficient mutants (Schieder 1978, 1980a), resistance to amino acid analogues (Gleba and Berlin, 1979; White and Vasil, 1979; Kameya et al., 1981; Gleba 1983), and resistance to cycloheximide (Lazar et al., 1981).

A number of factors complicate the routine application of protoplast fusion for complementation analysis. (1) Some limits exist on the ability to isolate protoplasts and to microculture fused cells. (2) Most plants recovered by protoplast fusion are aneuploid. Aneuploidy complicates any genetic analysis. (3) In wide interspecific or intergeneric hybrids the plants recovered are often sterile. Sterility precludes sexual analysis.

Analysis of Mitotic Cycle Mechanisms

Studies with animal hybrid cells have permitted a detailed analysis of heterophasic fusion products. When G1 cells are fused with cells in S phase, the G1 nucleus enters S earlier than expected. Similarly, if two cells with different duration of G1 phase are fused, each nucleus enters the S phase following completion of the shortest G1 phase (Ringertz and Savage, 1976). Mitosis is equally well synchronized in G1/S or G1/G2 heterophasic fusions, suggesting that both the S and M stages of the cycle are triggered by cytoplasmic factors that even function in heterokaryocytes.

Studies with plant cells are more complicated, as synchronization of plant cells has not been reported to date. However, it is presumed that, as most experiments with plants represent fusion of mitotically active callus or suspension cells with mitotically dormant leaf mesophyll protoplasts, examination of plant experiments would permit an opportunity to observe striking mitotic cycle effects.

The results of a number of works suggest that upon hybridization of mesophyll cells (cells of G0 phase) with the protoplasts isolated from actively growing callus tissues containing a heterogenous mixture of cell cycles, mitoses in heterokaryons may be synchronous (Kao et al., 1974; Constabel et al., 1975; Gosch and Reinert, 1976; Constabel et al., 1977). These observations favor arguments for the theory of cytoplasmic control of initiation of mitosis. The cytoplasmic factors responsible for the mitotic process must not be species-specific, as the synchronization was also observed in heterokaryons obtained by cell fusion of distantly related plant species (Kao et al., 1974; Gosch and Reinert, 1976). In the distantly related hybrids of *Glycine* + *Nicotiana* (Kao, 1977) and *Arabidopsis* + *Brassica* (Gleba and Hoffmann, 1978) obtained by fusion of mesophyll and callus cells, specific elimination of chromosomes from the mesophyll and parental cells was observed. This observation may be connected with events similar to premature chromatin condensation observed in animal hybrid cells that results in preferential elimination of chromosomes of interphase parental cells (Ringertz and Savage, 1976).

Spatial Separation of Parental Genome in Hybrid Nuclei

Sexual hybridization results in formation of a common ("hybrid") nucleus in a zygote, but chromosomes of two parents are not necessarily randomly "mixed" within a common nucleus. On the contrary, there are indications that parental genomes remain spatially separated even after many successful mitotic divisions (cf. Finch et al., 1981). Possible mechanisms of this phenomonon are discussed in Bostock and Sumner (1978). Studies of primary cell divisions in somatic hybrid cells of different species (Kao, 1977; Constabel et al., 1977) show that chromosomes of the two parents are spatially arranged as separate groups during the first four cell divisions, forming a two-segment structure in metaphase and anaphase. This indicates preservation of strict spatial arrangement in the nucleus of hybrid cells might be a factor affecting the process of interspecific chromosomal rearrangements and subsequent chromosomal recombination. Therefore, Gleba et al. (1983c) have made an attempt to elucidate to what degree the arrangement of chromosomes from the two parental species is inherited in the hybrid nucleus and preserved in long-term cultured lines of *Atropa belladonna* + *Nicotiana chinensis*, after 12 months of unorganized growth. The information obtained in these experiments suggests that in cells of long-term cultures, chromosomes are to a large extent "mixed", reflecting an apparent random distribution in mitotic metaphase plates. At the same time statistical analysis demonstrates that in a definite number of metaphases rather large regions can be identified that contain chromosomes of only one species. It is difficult to explain this phenomenon by occasional fluctuations in the mutual distribution of chromosomes.

Moreover, statistical analysis of a number of preparations by calculation of Mahalonobis' generalized distance between two chromosome types in metaphase plates has demonstrated that in all metaphases studied, the distribution of the two parental chromosomes are statistically different (p = 0.95). This leads to the conclusion that nonrandom chromosomal arrangement is not an artifact due to squashing. Thus the spatial separation of the two genomes in cell cultures studied has been preserved for up to 12 months of culture. It is important to analyze the inheritance of the spatial arrangement of discrete chromosomes from the two parents in hybrid cells; for example, nucleolus organizer chromosomes could be monitored in progeny of somatic hybrids.

Spatial separation of chromosomes of the two parents has also been observed in metaphases of 6 month old cultures of interfamilial cell hybrids of *Pisum sativum* + *Nicotiana chinensis* (Gleba et al., 1983b), as well as in metaphases of somatic hybrid plants of *Nicotiana tabacum* + *N. glauca* (Evans, unpublished).

Inheritance of Chromosome Instability

After fusion of normal euploid somatic plant cells (mesophyll tobacco protoplasts) with the cells of the same species isolated from long-term chromosomally unstable callus cultures, hybrids exhibiting chromosome instability were usually obtained. In this sort of experiment the phenotype of chromosome instability behaves as a dominant trait (Gleba 1983). In this report intraspecific hybrids were recovered; hence the chromosome instability is most likely not a consequence of cell incompatibility. It is possible that chromosome instability could be the result of inheritance of parental chromosome instability.

Further study of a dominant or codominant character of chromosome instability in hybrid cells as well as studies of heterophasic fusions will elucidate which of the alternatives control chromosome instability: (a) the chromosome instability in hybrid cells may be due to genetic alterations or incompatibilities; (b) the instability may be determined by defects of the spindle apparatus. In the first case the instability should be restricted to chromosomes of one of the parents in vitro, i.e., the chromosomes of the parent with the instability character. In the second case all the chromosomes are equally involved in irregular distribution in mitosis. This basic genetic problem would best be examined by using a sufficient number of chromosome markers for each parent. Similar studies completed for animal hybrid cells (Bengtsson et al., 1975) demonstrated that chromosome instability of hybrid cells affects chromosomes of both parents suggesting a dominant trait involved in spindle formation or function.

Analysis of the Mechanisms of Differentiation and Morphogenesis

ALTERATION OF THE DIFFERENTIATED STATE. When leaf mesophyll protoplasts are fused with mitotically active cell culture protoplasts, the leaf cells are induced to divide. While in most cases leaf mesophyll protoplasts divide when cultured even if not fused, it is evident from a number of reports that hybrid cells undergo mitosis more rapidly than unfused leaf mesophyll cells following protoplast fusion. For example, Evans et al. (1981) found that somatic hybrid protoplasts divided more rapidly than unfused leaf mesophyll protoplasts following fusion of *N. tabacum* with *N. glauca*. While in most cases this mitotic activity of hybrid cells does not represent a reprogramming of the developmental fate of the cells, there is evidence in some species combinations that a reprogramming has occurred. Fusion of mesophyll protoplasts of the Gramineae that are incapable of mitosis with callus protoplasts has produced hybrid cells capable of cell division (Kao et al., 1974). Hence the developmental fate of the cereal leaf cells was altered by protoplast fusion. Similarly, Flick and Evans (1983) fused flower petal protoplasts incapable of cell division with suspension cultured protoplasts, and recovered mitotically active heterokaryocytes. It has been proposed that in both animals and plants many epigenetic characters of cells are altered following cell or protoplast fusion. This variation of epigenetic characters by fusion also has value in the analysis of mutants by protoplast fusion. Phenotypic variants that are equally expressed in cell lines before and after inter-specific protoplast fusion are less likely to be under epigenetic control and are more likely to be true mutants. However, this analysis alone is insufficient to ascertain that a trait is genetically controlled.

INHERITANCE OF MORPHOGENETIC POTENTIAL OF CELLS.   It had been demonstrated in our studies (Gleba and Sytnik, 1982; Gleba, 1983) that a hybrid cell obtained by fusion of a callus tobacco cell defective for morphogenesis with a morphologically potent mesophyll cell of the same species is capable of morphogenesis. In this case morphogenetic capacity behaves as a dominant trait following somatic hybridization.   The conditions of long-term culture in vitro are probably responsible for loss of morphogenetic ability in the callus parental line. Similar conclusions were made from the study of hybridization of mutant tobacco cells which were defective for nitrate reductase and for morphogenesis (Glimelius and Bonnett, 1981). Unlike either parent the hybrid cells were capable of morphogenesis. This work also demonstrates the complexity of the mechanisms examined. The nitrate reductase mutants of tobacco deficient for morphogenesis are defective in one enzyme that apparently has nothing to do with morphogenesis per se (Muller and Grafe, 1978).

Other workers have described the restoration of morphogenetic potential following fusion (Maliga et al., 1977). *Nicotiana knightiana*, used as one parent, was incapable of morphogenesis in the culture conditions utilized.   Kanamycin-resistant cells of *Nicotiana sylvestris* that were also incapable of morphogenesis, presumably due to the kanamycin mutation (Dix et al., 1977), were also used for protoplast isolation. Restoration of morphogenesis was found in these hybrids. In this particular work, it is difficult to ascribe regeneration to gene complementation as it is evident that the morphogenetic block of *N. sylvestris* is due to a mutation that is at best only indirectly connected with the mechanisms controlling morphogenesis. In addition, as a limited variation of culture media were tested, it is likely that some culture regime could have been identified that permitted regeneration of one of the parent protoplasts.

Restoration of morphogenetic potential has also been observed in distantly related intertribal combinations. In these reports the probability that this restoration is the result of gene complementation of a single pair of genes is even less probable. Restoration of morphogenesis was demonstrated to obtain *Arabidobrassica* hybrids (Gleba and Hoffmann, 1980), as well as in the tomato-potato somatic hybrids (Melchers et al., 1978).

As whole plant vigor is often reflected as vigor in vitro, it is not surprising that hybrids have been observed to grow more rapidly than parental cells, and in some cases to undergo morphogenesis when parental cells are incapable of regeneration. However, in interspecific or intergeneric hybrids it is unlikely that these phenomena are under the control of a single or small number of complementing genes.

INHERITANCE OF THE T-REGION OF AGROBACTERIUM Ti-PLASMID. Wullems, et al. (1980) studied cell hybridization of a hormone-independent tobacco strain derived from tumorous cells following infection with *Agrobacterium tumefaciens* with mesophyll protoplasts of a streptomycin-resistant tobacco mutant. The selection of hybrids was based on simultaneous hormone-independence and streptomycin resistance. Four cell colonies capable of shoot regeneration were recovered. In all cases lysopine dehydrogenase activity was identified in putative hybrids indicating the presence of *Agrobacterium* Ti-plasmid. Although the hybrids were capable of shoot regeneration, the initial hormone-independent strain was defective for shoot regeneration. Unfortunately, the regenerated shoots were unable to form roots. Similar inability for rhizogenesis in vitro was observed following transformation of tobacco protoplasts by *Agrobacterium* (Marton et al., 1979). Also in this experiment, hybrid cells capable of regeneration were obtained following fusion of morphogenetic defective hormone-independent callus cells that contained the Ti-plasmid of *Agrobacterium* with normal mesophyll protoplasts.

From the above-mentioned experiment, it follows that T-DNA is transferred to parasexual progeny, and the suppression of rhizogenesis in transformed cells seems to be a dominant character. Similar observations have recently been made by Gleba et al. (1983a) following fusion of hormone-independent Ti-transformed tobacco cells and mesophyll protoplasts of *Nicotiana plumbaginifolia*.

Analysis of Plasmagenes Using Cosegregation

If the production of heterozygotes for extranuclear genes is possible, these can be used to study mitotic segregation of extranuclear genes in double heterozygotes. If the characters coded for by the extranuclear genes segregate independently during mitosis, they must be controlled by different genophores. On the other hand, the cosegregation (joint segregation) of the characters coded for by extranuclear genes is evidence for the linkage of these genes. Cosegregation analysis is a way to ascertain linkage groups (groups of cosegregation) for genes outside the nucleus. These investigations seem to be especially valuable in the case of higher plants, whose cells contain at least two extranuclear genophores.

In the experiments of Gleba et al. (1978) and Gleba (1978b), cytoplasmic hybrids of *Nicotiana tabacum* + *N. debneyi* were used. These cybrids were heterozygous for genes coding for (a) plastome chlorophyll-deficiency, (b) polypeptide composition of large subunit of Fraction-1 Protein, and (c) cytoplasmic male sterility (cms), and therefore suitable for studies of the patterns of extranuclear gene segregation. Biochemical analysis of Fraction-1 Protein (F-1-P) has been performed for both green and white plants (segregants for chlorophyll deficiency). In all cases, the white plants possessed only *N. tabacum* large subunit, whereas in all green plants, only *N. debneyi* large subunit was observed. Thus, in all cases, the species specific plastome and large subunit F-1-P segregated together. The same phenomenon was observed in the analysis of green and white sexual progeny of variegated hybrids. The complete cosegregation of these two characters in both hybrids and their progeny confirms the assumption that the polypeptides of large subunit of F-1-P are coded for by the plastome.

The second important observation made in these experiments was that the heterozygosity and segregation of these two characters is independent of the genes coding for cytoplasmic male sterility (cms). On the contrary, plants identified as heterozygotes for sterility/fertility are frequently homoplastidic, i.e., they are complete segregants for plastome-encoded characters. These results have practically excluded the assumption that the genes coding for cms might be localized in the plastome and have indicated the presence of at least two independently segregating genophores outside the nucleus.

Extensive investigations of plasmagene cosegregation were subsequently performed by Belliard et al. (1978, 1979). They analyzed chloroplast (cp) and mitochondrial (mt) DNA using restriction endonucleases in parasexual plant hybrids that had novel morphological types of cms. Either of the tobacco varieties Samsun and Xanthi, and tobacco cms-analog, Techne variety, with cytoplasm of *N. debneyi*, were used as parental lines (Belliard et al., 1977a). The parental lines were morphologically distinct, as Samsun and Xanthi varieties had petiolate leaves and the Techne variety had sessile leaves (a polygenic Mendelian character). Biochemical analysis was completed, with regenerated plants possessing leaf morphology of one of the parents and morphologically intermediate flowers for both sterile and fertile (cybrid) types. Analysis of cp-DNA using Eco RI restriction enzyme patterns demonstrated the absence of cosegregation between the two groups of characters: sterility/fertility and modified flower shape, on one hand, and type of cp-DNA, on the other. In all cases cp-DNA of the hybrids was identical to one of the parental forms, and the type of parental cp-DNA was independent of the male sterility/fertility character.

Similarly, mt-DNA of the cybrids has been analyzed. The interpretation of these results has been complicated, because mt-DNA is usually split into a greater number of fragments than cp-DNA even using selective restriction enzymes. This phenomenon has been explained as a result of heterogeneity of mt-DNA molecular forms in higher plant cells (c.f. Quetier and Vedel, 1977). Each of the cybrids analyzed possessed mt-DNA that differed from either parental form as well as from the sum of parental forms. Moreover, mt-DNA of all but one cybrid revealed new bands. At the same time a correlation has been observed between the degree of male sterility and abnormal flower morphology and the number of *N. debneyi* mt-DNA bands. These results have demonstrated that in this system cms is most probably connected with mt-DNA.

Other investigators (Zelcer et al., 1978; Aviv et al., 1980; Galun and Aviv, 1979; Aviv and Galun, 1980; Galun et al., 1982) have studied segregation of plasmagenes in a more complex system. *Nicotiana sylvestris* was hybridized with a cms-analog of *N. tabacum* possessing cytoplasm of *N. suaveolens*. To select cybrids, cells of donor cytoplasm (tobacco with cms) were irradiated by X-rays and fusion products were cultured in medium containing mannitol in which protoplasts of *Nicotiana sylvestris* would not divide. Among the regenerates of seven callus colonies, three types of plants were identified: (a) *N. sylvestris* plants bearing the plastids of *N. suaveolens*; (b) nuclear hybrids between *N. sylvestris* + *N. tabacum* containing the plastids of *N. suaveolens*; and (c) parental plants of *N. sylvestris* bearing the plastome of *N. sylvestris*. The cosegregation of cp-DNA type and two other characters, (1) resistance to tentoxin and (2) isoelectric focusing of polypeptides of large subunit F-1-P, was verified for at least six different plants. In the same experiments the cms character segregated idependently of either nuclear or plastome markers. Independent segregation of cms and response to tentoxin were demonstrated for plants obtained in subsequent fusion (Aviv and Galun, 1980). Cybrids were obtained by fusing *N. sylvestris* with cms from the earlier fusion (type A plants) and *Nicotiana tabacum*. The authors also analyzed mt-DNA in segregants obtained in this second experiment (Galun et al., 1982). Complete correlation of the fertility/sterility character with presence of 4 (Sal I) and 8 (Xho I) fragments of parental mt-DNA was reported, as well as the absence of correlation between these characters, on one hand, and the type of cp-DNA and/or resistance to tentoxin, on the other.

Maliga and coworkers in Hungary have also studied segregation of extranuclear genes in interspecific hybrids of *Nicotiana*. Callus protoplasts of a streptomycin-resistant mutant of *Nicotiana tabacum* and mesophyll protoplasts of *N. knightiana* were hybridized. Selection of hybrids was performed using mechanical isolation and coculturing of fusion products. The Eco RI analysis of cp-DNA of plants regenerated from different subclones has demonstrated the complete correlation between resistance to streptomycin and *N. tabacun* cp-DNA type (Menczel et al., 1981a).

In other experiments different techniques for transfer of plastids from cells of plastome (Sidorov et al., 1981) or streptomycin resistant

(Menczel et al., 1982) tobacco mutants into the cells of *Nicotiana plumbaginifolia* were studied, using either irradiation of donor cells or the fusion of recipient protoplasts with donor cytoplasts. The transfer of plastids was confirmed by transfer of streptomycin resistance or plastome deficiency by cells of *N. plumbaginifolia* and the presence of tobacco specific EcoR-1 digests of cp-DNA. The analysis of cp-DNA in a total of 20 plants has confirmed cosegregation of the streptomycin-resistance character with cp-DNA, and the analysis of two plants has shown cosegregation of plastome chlorophyll–deficiency with cp-DNA type. Flick and Evans (1982) also reported cosegregation of two chloroplast encoded traits, large subunit F-1-P and tentoxin response, in *N. glauca* + *N. tabacum* somatic hybrids.

The fate of parental mt-DNA following cell hybridization has also been investigated (Nagy et al., 1981). The mt-DNA of *Nicotiana tabacum* + *N. knightiana* hybrids, obtained previously by joint culturing of individual heteroplasmic protoplast fusion products (Menczel et al., 1981) were studied. The restriction enzyme analysis of mt-DNA of eight nuclear hybrids (including five streptomycin–resistant and three streptomycin–sensitive types) has revealed unique types of mt-DNA in each hybrid.

Summarizing the data of published reports, one can conclude that: (1) Most characters encoded in plasmagenes sort to one or the other parental type. The exception is genes encoded in mt-DNA where recombination occurs. (2) In the process of segregation, characters coded for by plasmagenes segregate en bloc, i.e., cosegregate. (3) The higher plant cells comprise at least two cosegregation (linkage) groups of plasmagenes, the chloroplasts and mitochondria, and their segregation is independent. As can be seen, independent experiments in different laboratories have given noncontradictory results. That is, most characters studied, namely resistance to tentoxin and streptomycin, plastome chlorophyll deficiency and large subunit F-1-P cosegregate with the corresponding chloroplast DNA and thus are coded for by this genophore. This result is also consistent with earlier genetic and biochemical analyses. Mutation of plastome chlorophyll deficiency has been correlated with an alteration of cp-DNA (Wong-Staal and Wildman, 1973), large subunit F-1-P has been mapped in cp-DNA (Chan and Wildman, 1972), and resistance to tentoxin is connected with the changes of coupling factor I localized in chloroplasts and (partially) coded for by plasmagenes (Steele et al., 1976; Durbin and Uchytil, 1977). However, the studies using hybridization of somatic cells have for the first time permitted a confirmation of these reports by strict formal genetic analysis.

Two characters (cms and abnormal flower morphology) segregate independently of the cp-DNA and at the same time cosegregate with the type of mt-DNA. In this case, however, cosegregation must be discussed cautiously, as mt-DNA appears to undergo extensive recombination in cybrids. However, in cybrids a clear quantitative correlation between the relative content of fragments specific for each parental form and a prevalence of characters of this parent is observed. Subsequent cosegregation studies must consist of constructing multiple heterozygotes using all the available extranuclearly encoded characters (resistance to streptomycin, kanamycin, lincomycin, tentoxin, large subunit F-1-P, cms, 70 S ribosomes, and restriction spectra of mt and cp-DNA) and, based on the pattern of segregation in the mitotic process, to attribute them to definite linkage groups.

FUTURE PROSPECTS

Uniqueness of Somatic Hybrids

Parasexual hybridization by protoplast fusion permits the recovery of new combinations of parental genes to produce unique hybrid plants. Transmission genetics of parasexual hybridization by protoplast fusion is distinct from sexual hybridization (Fig. 1). Nuclear genes (designated as large circles) in nonsexual hybridization are inherited bi- and uniparentally, whereas in sexual hybridization only biparental inheritance is observed. Extranuclear genes (designated as small circles) in hybridization by protoplast fusion are inherited biparentally, whereas in the sexual process there is usually strict maternal uniparental inheritance. In plant cells there are at least two cytoplasmic genophores (chloroplasts and mitochondria), hence the scheme of transmission genetics of nonsexual hybridization becomes more complicated than in Fig. 1. In addition to two cytoplasmic genophores, protoplast fusion is further complicated, as mitotic segregation of extranuclear genes (genophores) has been reported (Fig. 2). Considering all these possibilities, it is easy to ascertain that by using protoplast fusion of two parental types it is possible to obtain 27 different combinations of one nuclear and two extranuclear gene systems, while sexual hybridization is limited to only two combinations, assuming that the reciprocal hybrids can be obtained. However, as segregation of cytoplasmic factors is usually complete by meiosis, it is unreasonable to assume that mixed chloroplast and mitochondrial genophores can be maintained. Hence we are left with 12 viable products of protoplast fusion, which is still significant when compared to sexual hybridization. By using parasexual hy-

SOMATIC HYBRIDIZATION WITH
SUBSEQUENT CYTOPLASMIC SEGREGATION

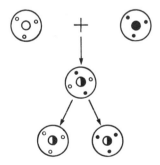

Figure 1. Comparison of nuclear genetic combinations after reciprocal sexual hybridization and following nuclear segregation during and after protoplast fusion.

RECIPROCAL SEXUAL HYBRIDIZATION

SOMATIC HYBRIDIZATION BY PROTOPLAST FUSION

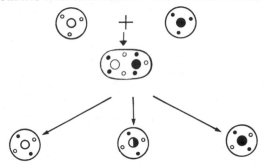

Figure 2.   Fate of cytoplasmic DNA following protoplast fusion.

bridization it is therefore possible to obtain: (a) plants containing the nucleus from one parent and the cytoplasm from the other parent and (b) plants containing chloroplast genes from one parent and mitochondria from the second parent.   Certain of these nuclear-cytoplasmic combinations are difficult or impossible to obtain using sexual hybridization.

The nuclear-cytoplasmic combinations diagrammed in Fig. 3 only depict the fate of complete nuclear or cytoplasmic genomes.   However, chromosome elimination has frequently been reported following protoplast fusion, suggesting that portions of nuclear genomes may be combined.   In addition, nuclear mitotic recombination has been reported following protoplast fusion.   While this phenomenon will most often result in sectors on the leaves of somatic hybrids, if recombination occurs prior to regeneration unique arrangements of nuclear genetic information could be recovered in regenerated plants.   Interspecific mitotic recombination may be important to permit gene transfer between wild species and cultivated crops.   Finally, mitotic recombination has also been reported in organelle DNA.   Extensive rearrangements of mt-DNA in somatic hybrids have been frequently reported (Chapter 9).   Less frequently, rearrangements have been reported in chloroplast DNA (Conde, 1981).

Study of Extranuclear Genetics

At present genetic localization and determination of linkage groups of extranuclear genes is limited by the number of extranuclear markers.

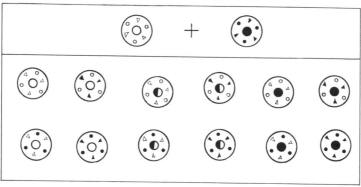

CHLOROPLAST AND MITOCHONDRIAL SEGREGATION
FOLLOWING PROTOPLAST FUSION

Figure 3. Possible mixtures of nuclear, chloroplast (o and ●), and mito-chondrial (Δ and ▲) DNA derived by protoplast fusion with subsequent segregation.

Cosegregation of extranuclear genic determinants in double cytoplasmic heterozygotes can be used to establish extranuclear linkage groups. Induction of new cytoplasmic mutations will permit a study of recombination of extranuclear genes.

Introduction of new cytoplasmic markers will aid in the use of proto-plast fusion for transfer of cytoplasm between wild species and culti-vated crops. For example, genes controlling cytoplasmic male sterility (cms) can be transferred from one species or variety to a second vari-ety within one cycle of hybridization, in contrast to the lengthy recur-rent backcrosses necessary to transfer cms using conventional methods. Transfer of cms into new varieties permits industrial production of hy-brid seeds.

As both chloroplasts and mitochondria are organelles responsible for energetic processes in the plant cell, it is likely that additional agriculturally useful traits encoded in these organelles will be identi-fied. Already traits controlling herbicide resistance, cms, disease resistance, response to toxin, and antibiotic resistance have been shown to be encoded in cytoplasmic organelles. Research programs concen-trating on crop improvement will most likely direct more attention to extranuclear inheritance. It is noteworthy that mitochondria of higher plants contain naturally occurring plasmids (Levings et al., 1980). As our knowledge of organelle genetics increases, it is possible that organelles may be used as vehicles for gene transfer (Chapter 14).

Somatic Hybrids Between Distantly Related Species

Parasexual hybridization permits crossing of distantly related plant species that cannot be crossed sexually. While symmetric hybridization resulting in plants with the amphiploid chromosome number have been produced between species incapable of conventional sexual hybridization (Evans et al., 1981), more distant intergeneric hybrids have produced anomalous hybrid plants that are sterile or incapable of subsequent sexual reproduction. Hence available intergeneric hybrid plants are useless for practical breeding. It is important to note that in some of these intergeneric hybrids stable chromosome number and morphology have been maintained in cell hybrids and is not lost until plants are regenerated (Gleba and Hoffmann, 1980). If procedures of regeneration are appropriately manipulated, it may be possible to retain chromosome stability in regenerated intergeneric somatic hybrids. Available intergeneric hybrid plants offer a unique opportunity to study the fate of genetic information when distantly related species are combined and suggest that even with extensive chromosome elimination that some genetic information can be stably transferred from one species to another. It is not unexpected that these intergeneric hybrids are morphologically abnormal and contain extensive chromosome elimination and rearrangements.

The limited coordination of the two genomes within distantly related plant cells may offer a new direction to achieve crop improvement using intergeneric hybrids. That is, hybridization of distantly related species is of greatest interest for practical breeding as a method to transfer small amounts of gene material between species (e.g., Dudits et al., 1980a). As was mentioned above, all distantly related parasexual hybrids obtained to date are morphologically abnormal, and it is unlikely that fertile amphiploid intergeneric hybrids will be produced in the future. However, attention can be directed toward production of asymmetric hybrids, in which the genome of one of the parents is intact and the second genome is represented by only a few chromosomes or genes. Already such methods have been proposed in the production of cybrids (Chapter 9). As has already been shown by a number of experiments, some of these asymmetric hybrids are morphologically more normal that symmetric hybrids. One can assume that further reduction of genetic material of the donor parent will permit production of plants which are fertile and at the same time retain certain genes from the second or donor species. This elimination of genetic material of the second species that occurs rapidly following wide somatic hybridization mimics conventional sexual hybridization. In conventional breeding desired traits are selected through a recurrent backcross procedure that may require 6-7 sexual generations. However, in wide somatic hybridization it may be possible to rapidly introduce a single chromosome or single gene by regulating somatic growth conditions. This somatic technique will become more useful as powerful cell selection methods are developed that permit identification of agriculturally useful traits.

The elaboration of the methods of induction of asymmetry in distantly related hybrids, as well as of the methods of selecting desired genes should permit intergeneric somatic hybridization to develop as a useful method for rapid transfer of genetic material between distantly related plant species.

KEY REFERENCES

Chaleff, R. 1981. Genetics of Higher Plants: Application of Cell Culture. Cambridge Univ. Press, Cambridge.

Evans, D.A. 1982. Plant regeneration and genetic analysis of somatic hybrid plants. In: Plant Regeneration and Genetic Variability (E.D. Earle, ed.) pp. 303-323. Praeger Press, New York.

Gleba, Yu.Yu. 1979. Nonchromosomal inheritance in higher plants as studied by somatic cell hybridization. In: Plant Cell and Tissue Culture (W.R. Sharp, P.O. Larsen, E.F. Paddock, and V. Raghavan, eds.) pp. 775-788. Ohio State Univ. Press, Columbus.

_____ and Sytnik, K.M. 1983. Protoplast Fusion and Genetic Engineering of Higher Plants. Springer, Berlin (in press).

REFERENCES

Ashton, N.W. and Cove, D.J. 1977. The isolation and preliminary characterization of auxotrophic and analogue resistant mutants of the moss, *Physcomitrella patens*. Mol. Gen. Genet. 154:87-95.

Aviv, D. and Galun, E. 1980. Restoration of fertility in cytoplasmic male sterile (CMS) *Nicotiana sylvestris* by fusion with X-irradiated *N. tabacum* protoplasts. Theor. Appl. Genet. 58:121-127.

Aviv, D., Fluhr, R., Edelman, M., and Galun, E. 1980. Progeny analysis of the interspecific somatic hybrids: *Nicotiana tabacum* (CMS) + *Nicotiana sylvestris* with respect to nuclear and chloroplast markers. Theor. Appl. Genet. 56:145-150.

Belliard, G. and Pelletier, G. 1978. Morphological characteristics and chloroplast DNA distribution in different cytoplasmic parasexual hybrids of *Nicotiana tabacum*. Mol. Gen. Genet. 165:231-237.

_____, and Ferault, M. 1977. Fusion de protoplastes de *Nicotiana tabacum* a cytoplasmes differents: Etude des hybrides cytoplasmiques neo-formes. C. R. Acad. Sci. Paris 284D:749-752.

Belliard, G., Vedel, F., and Pelletier, G. 1979. Mitochondrial recombination in cytoplasmic hybrids of *Nicotiana tabacum* by protoplast fusion. Nature 281:401-403.

Bengtsson, B.O., Nabholz, M., Kennett, R., Bodmer, W.R., Povery, S. and Swallow, D. 1975. Human intraspecific somatic cell hybrids: A genetic and karyotypic analysis of crosses between lymphocytes and D98/ AH-2. Somatic Cell Genet. 1:41-64.

Binding, H. and Nehls, R. 1978. Somatic cell hybridization of *Vicia faba* + *Petunia hybrida*. Mol. Gen. Genet. 164:137-143.

Blair, G.E. and Ellis, R.J. 1973. Protein synthesis in chloroplasts. I. Light-driven synthesis of the large subunit of Fraction-1 protein by isolated pea chloroplasts. Biochim. Biophys. Acta 319:223-237.

Chan, H. and Wildman, S.G. 1972. Chloroplast DNA codes for the primary structure of the large subunit of Fraction-1 protein. Biochim. Biophys. Acta 277:677-680.

Chen, K., Wildman, S.G., and Smith, H.H. 1977. Chloroplast DNA distribution in parasexual hybrids as shown by polypeptide composition of Fraction-1 protein. Proc. Nat. Acad. Sci. 74:5109-5112.

Chupeau, Y., Missonier, C., Hommel, M.-C., and Coujaud, J. 1978. Somatic hybrids of plants by fusion of protoplasts. Mol. Gen. Genet. 165:239-245.

Conde, M.R. 1981. Chloroplast DNA recombination in *Nicotiana* somatic parasexual hybrids. Genetics 97:s26.

Constabel, F., Dudits, D., Gamborg, O.L., and Kao, K.N. 1975. Nuclear fusion in intergeneric heterokaryons. Can. J. Bot. 53:2092-2095.

Constabel, F., Weber, G., and Kirkpatrick, J.W. 1977. Sur la chromosomes dans les hybrides intergeneriques de cellules de *Glycine max* + *Vicia hajastana*. C. R. Acad. Sci. 285:319-322.

Cove, D.J. 1979. Genetic studies of nitrate assimilation in *Aspergillus nidulans*. Biol. Rev. 54:291-337.

Dix, P.J., Joo, F., Maliga, P. 1977. A cell line of *Nicotiana sylvestris* with resistance to kanamycin and streptomycin. Mol. Gen. Genet. 157:285-290.

Dudits, D., Hadlaczky, Gy., Bajszar, Gy., Koncz, D.S., and Lazar, G. 1979. Plant regeneration from intergeneric cell hybrids. Plant Sci. Lett. 15:101-112.

Dudits, D., Fejer, O., Hadlaczky, Gy., Koncz, C., Lazar, G.B., and Horvath, G. 1980a. Intergeneric gene transfer mediated by protoplast fusion. Mol. Gen. Genet. 179:283-288.

Dudits, D., Hadlaczky, Gy., Lazar, G., and Haydu, Z. 1980b. Increase in genetic variability through somatic cell hybridization of distantly related plant species. In: Plant Cell Cultures: Results and Perspectives (F. Sala, B. Rarisi, R. cella, and O. Ciferri, eds.) pp.207-214. Elsevier/North-Holland Biomedical Press, Amsterdam.

Durbin, R.D. and Uchytil, T.F. 1977. Cytoplasmic inheritance of chloroplast coupling factor 1 subunits. Biochem. Gen. 15:1143-1146.

Evans, D.A. and Paddock, E.F. 1976. Comparisons of somatic crossing over frequency in *Nicotiana tabacum* and three other crop species. Can. J. Genet. Cytol. 18:57-65.

Evans, D.A., Wetter, L.R., and Gamborg, O.L. 1980. Somatic hybrid plants of *Nicotiana glauca* and *Nicotiana tabacum* obtained by protoplast fusion. Physiol. Plant. 48:225-230.

Evans, D.A., Flick, C.E., and Jensen, R.A. 1981. Disease resistance: Incorporation into sexually incompatible somatic hybrids of the genus *Nicotiana*. Science 213:907-909.

Flick, C.E. and Evans, D.A. 1982. Evaluation of cytoplasmic segregation in somatic hybrids in the genus *Nicotiana*: Tentoxin sensitivity. J. Hered. 73:264-266.

_____ 1983. Isolation, culture, and plant regeneration of protoplasts isolated from flower petals of ornamental *Nicotianas*. Z. Pflanzenphysiol. 109:379-383.

Flick, C.E., Jensen, R.A., and Evans, D.A. 1981. Isolation, protoplast culture, and plant regeneration of PFP-resistant variants of *Nicotiana tabacum* Su/Su. Z. Pflanzenphysiol. 103:239-245.

Galun, E., Arree-Gouen, P., Fluhr, R., Edelman, M., and Aviv, D. 1982. Cytoplasmic hybridization in *Nicotiana*: Mitochondrial DNA analysis in progenies resulting from fusion between protoplasts having different organelle constitutions. Mol. Gen. Genet. 186:50-56.

Gerstel, D.U. 1963. Segregation in new allopolyploids of *Nicotiana*. II. Discordant ratios from individual loci in 6x (*N. tabacum* x *N. sylvestris*). Genetics 48:677-689.

_____ and Phillips, L.L. 1958. Segregation of synthetic amphiploids in *Gossypium* and *Nicotiana*. Cold Spring Harbor Symp. Quant. Biol. 23: 225-237.

Gleba, Yu.Yu. 1978a. Microdroplet culture: Tobacco plants from single mesophyll protoplasts. Naturwissenschaften. 65:158.

_____ 1978b. Extranuclear inheritance investigated by somatic hybridization. In: Frontiers of Plant Tissue Culture 1978 (T.A. Thorpe, ed.) pp.95-102. IAPTC, Calgary.

_____ 1983. Transmission genetics of *Nicotiana* hybrids produced by protoplast fusion. I. Genetic constitution of the intraspecific *N. tabacum* progeny obtained by cloning individual fusion products. J. Hered. (submitted).

_____ and Berlin, J. 1979. Somatic hybridization by protoplast fusion in *Nicotiana*: Fate of nuclear genetic determinants. In: Abstracts V International Protoplast Symposium p. 73, Szeged, Hungary.

_____ and Hoffmann, F. 1978. Hybrid cell lines *Arabidopsis thaliana* + *Brassica campestris*: No evidence for specific chromosome elimination. Mol. Gen. Genet. 165:257-264.

_____ and Hoffmann, F. 1979. "Arabidobrassica": Plant-genome engineering by protoplast fusion. Naturwissenschaften 66:547-554.

_____ and Hoffmann, F. 1980. "Arabidobrassica": A novel plant obtained by protoplast fusion. Planta 149:112-117.

_____ and Sytnik, K.M. 1982. Protoplast Fusion and Genetic Engineering of Higher Plants. Naukova Dumka Pub., Kiev.

_____, Kohlenbach, H.W. and Hoffmann, F. 1978a. Root morphogenesis in somatic hybrid cell lines *Arbidopsis thaliana* + *Brassica campestris*. Naturwissenschaften 65:655-656.

_____, Piven, N.M., Komarnitsky, I.K., and Sytnik, K.M. 1978b. Cytoplasmic hybrids (cybrids) *Nicotiana tabacum* + *N. debneyi* obtained by protoplast fusion. Dokl. Akad. Nauk USSR 240:225-227.

_____, Momot, V.P., Cherep, N.N., Sharzhynskaya, M.V. 1982. Intertribal hybrid cell lines of *Atropa belladonna* + *Nicotiana chinensis* obtained by cloning individual protoplast fusion products. Theor. Appl. Genet. 62:75-79.

_____, Kanevsky, J.F., and Cherep, N.N. 1983a. Transmission genetics of *Nicotiana* hybrids produced by protoplast fusion. II. Genetic constitution of *N. tabacum* (x) *N. plumbaginiflolia* hybrids obtained using selective and non-selective screening methods. Theor. Appl. Genet. (in press).

_____, Momot, V.P., Okolot, A.N., and Cherep, N.N. 1983b. Isolation of somatic cell hybrid *Pisium sativum* (x) *Nicotiana chinensis*. Z. Pflanzenphysiol. (submitted).

_____, Momot, V.P., Okolot, A.N., Cherep, N.N., Skarrhynskagy, M.V., and Kotov, V. 1983c. Genetic processes in intergeneric cell hybrids *Atropa* x *Nicotiana*. I. Genetic constitution of cells of different clonal origin grown in vitro. Plant, Cell, Tissue and Organ Culture (submitted).

Glimelius, K. and Bonnett, H.T. 1981. Somatic hybridization in *Nicotiana*: Restoration of photoautotrophy to an albino mutant with defective plastids. Planta 153:497-503.

Glimelius, K., Eriksson, T., Grafe, R., and Muller, A. 1978. Somatic hybridization of nitrate reductase deficient mutants of *Nicotiana tabacum* by protoplast fusion. Physiol. Plant. 44:273-277.

Glimelius, K., Chen, K., and Bonnett, H.T. 1981. Somatic hybridization in *Nicotiana*: Segregation of organellar traits among hybrid and cybrid plants. Planta 153:504-510.

Gosch, G. and Reinert, J. 1976. Nuclear fusion in intergeneric heterokaryocytes and subsequent mitosis of hybrid nuclei. Naturwissenschaften 11:534.

Grant, W.F. 1975. Genetics of Flowering Plants. Columbia Univ. Press, New York.

Grimsley, N.H., Ashton, N.W., and Cove, D.J. 1977a. Complementation analysis of auxotrophic mutants of the moss, *Physcomitrella patens*, using protoplast fusion. Molec. Gen. Genet. 155:103-107.

_____ 1977b. The production of somatic hybrids by protoplast fusion in the moss, *Physcomitrella patens*. Molec. Gen. Genet. 154:97-100.

Grimsley, N.H., Featherstone, D.R., Courtice, G.R.M., Ashton, N.W., and Cove, D.J. 1980. Somatic hybridization following protoplast fusion as a tool for the genetic analysis of development in the moss *Physcomitrella patens*. In: Advances in Protoplast Research (L. Ferenczy, G.L. Farkar, and G. Lazar, eds.) pp. 363-376. Pergamon Press, Oxford.

Hoffmann, F. and Adachi, T. 1981. "Arabidobrassica": Chromosomal recombination and morphogenesis in asymmetric intergeneric hybrid cells. Planta 153:586-593.

Kameya, T., Horn, M.E., and Widholm, J.M. 1981. Hybrid shoot formation from fused *Daucus carota* and *D. capillifolius* protoplasts. Z. Pflanzenphysiol. 104:459-466.

Kao, K.N. 1976. Cytological studies on plant heterokaryocytes-nuclear behavior. In: Cell Genetics in Higher Plants (D. Dudits, G.L. Farkas, and P. Maliga, eds.) pp. 149-152. Akademiai Kiado, Budapest.

_____ 1977. Chromosomal behavior in somatic hybrids of soybean + *Nicotiana glauca*. Molec. Gen. Genet. 150:225-230.

_____ and Wetter, L.R. 1976. Advances in techniques of plant protoplast fusion and culture of heterokaryocytes. In: International Cell Biology (B.R. Brinkley and K.R. Porter, eds.) pp. 216-224. Academic Press, New York.

_____, Constabel, F., Michayluk, M.R., and Gamborg, O.L. 1974. Plant protoplast fusion and growth of intergeneric hybrid cells. Planta 120: 215-227.

Komarnitsky, I.K. and Gleba, Yu.Yu. 1981. Fraction I protein analysis of parasexual hybrids, *Arabidopsis thaliana* + *Brassica campestris*. Plant Cell Rep. 1:67-68.

Krumbiegel, G. and Schieder, O. 1979. Selection of somatic hybrids after fusion of protoplasts from *Datura innoxia* Mill. and *Atropa belladonna* L. Planta 145:371-375.

_____ 1981. Comparison of somatic and sexual incompatibility between *Datura innoxia* and *Atropa belladonna*. Planta 153:466-470.

Kung, S.D., Zhu, Y.D., and Shen, G.F. 1982. *Nicotiana* chloroplast genome. III. Chloroplast DNA evolution. Theor. Appl. Genet. 61:73-79.

Lazar, G.B., Dudits, D., and Sung, Z.R. 1981. Expression of cycloheximide resistance in carrot somatic hybrids and their segregants. Genetics 98:347-356.

Levings, C.S., Kim, B.D., Pring, D.R., Conde, M.R., Mans, R.J., Laughnan, J.R., and Gabay-Laughnan, S.J. 1980. Cytoplasmic reversion of CmS-S in maize: Associations with a transpositional event. Science 209: 1021-1023.

Maliga, P., Lazar, G., Joo, F., Nagy, A.H., and Menczel, L. 1977. Restoration of morphogenic potential in *Nicotiana* by somatic hybridization. Molec. Gen. Genet. 157:291-296.

Maliga, P., Nagy, F., Le Thi Xuan, Kiss, Z.R., Menczel, L., and Lazar, G. 1980. Protoplast fusion to study cytoplasmic traits in *Nicotiana*. In: Advances in Protoplast Research (L. Ferenczy and G.L. Farcas, eds.) pp. 341-348. Pergamon Press/Akademiai Kiado, Budapest.

Marton, L., Wullems, G.J., Molendijk, L., and Schilperoort, R.A. 1979. In vitro transformation of cultured cells from *Nicotiana tabacum* by *Agrobacterium tumefaciens*. Nature 277:129-131.

Marton, L., Sidorov, V., Biasini, G., and Maliga, P. 1982. Complementation in somatic hybrids indicates four types of nitrate reductase deficient lines in *Nicotiana plumbaginifolia*. Molec. Gen. Genet. 187:1-3.

Marzluf, G.A. 1981. Regulation of nitrogen metabolism and gene expression in fungi. Microbiol. Rev. 45:437-461.

Melchers, G. 1977. Kombination somatischer und konventioneller Genetik fur die Pflanzenzuchtung. Naturwissenschaften 64:184-194.

_____ and Labib, G. 1974. Somatic hybridization of plants by fusion of protoplasts. I. Selection of light resistant hybrids of "haploid" light sensitive varieties of tobacco. Molec. Gen. Genet. 135:277-294.

_____ and Sacristan, M.D. 1977. Somatic hybridization of plants by fusion of protoplasts. II. The chromosome numbers of somatic hybrid plants of four different fusion experiments. In: La Culture des Tissus et des Cellules des Vegetaus (G. Gautheret, ed.) pp. 169-177. Masson, Paris.

_____, Sacristan, M.D., and Holder, S.A. 1978. Somatic hybrid plants of potato and tomato regenerated from fused protoplasts. Carlsberg Res. Comm. 43:203-218.

Menczel, L., Nagy, F., Kiss, Z.R., and Maliga, P. 1981. Streptomycin resistant and sensitive somatic hybrids of *Nicotiana tabacum* + *Nicotiana knightiana*: Correlation of resistance of *N. tabacum* plastids. Theor. Appl. Genet. 59:191-195.

Menczel, L., Galiba, G., Nagy, F., and Maliga, P. 1982. Effect of radiation dosage on efficiency of chloroplast transfer by protoplast fusion in *Nicotiana*. Genetics 100:487-495.

Mendel, R.R. and Muller, A.J. 1976. Common genetic determinant of xanthine dehydrogenase and nitrate reductase in *Nicotiana tabacum*. Biochem. Physiol. Pflanzen. 170:538-541.

_____ 1978. Reconstitution of NADH-nitrate reductase in vitro from nitrate reductase-deficient *Nicotiana tabacum* mutants. Molec. Gen. Genet. 161:77-80.

Muller, A.J. and Grafe, R. 1978.   Isolation and characterization of cell lines of *Nicotiana tabacum* lacking nitrate reductase.   Molec. Gen. Genet. 161:67-76.

Nagy, F., Torok, I., and Maliga, P. 1981.   Extensive rearrangements in the mitochondrial DNA in somatic hybrids of *Nicotiana tabacum* and *Nicotiana knightiana*.   Molec. Gen. Genet. 183:437-439.

Nakata, K. and Oshima, H. 1982.   Cytoplasmic chimericity in the somatic hybrids of tobacco.   Abstr. V International Congress Plant Tissue Cell Culture.   p. 110.   July, 1982, Tokoyo.

Power, J.B., Berry, S.F., and Frearson, E.M. 1977.   Selection procedures for the production of interspecies somatic hybrids of *Petunia hybrida* and *Petunia parodii*.   I. Nutrient media and drug sensitivity complementation selection.   Plant Sci. Lett. 10:1-6.

Power, J.B., Sink, K.C., Berry, S.F., Burns, S.F., and Cocking, E.C. 1978.   Somatic and sexual hybrids of *Petunia hybrida* and *Petunia parodii*.   J. Hered. 69:373-376.

Quetier, F. and Vedel, F. 1977.   Heterogenous population of mitochondrial DNA molecules in higher plants.   Nature 268:365-368.

Rao, P.N., Hittleman, W.N., and Wilson, B.A. 1975.   Mammalian cell fusion.   VI. Regulation of mitosis in binucleate HeLa cells.   Exp. Cell Res. 90:40-46.

Reed, S.M. and Collins, G.B. 1980.   Chromosome pairing relationships and black shank resistance in three *Nicotiana* interspecific hybrids.   J. Hered. 71:423-426.

Ringertz, N.R. and Savage, R.E. 1976.   Cell Hybrids.   Academic Press, New York.

Schieder, O. 1977.   Hybridization experiments with protoplasts from chlorophyll-deficient mutants of some *Solanaceous* species.   Planta 137:253-257.

_____ 1978a.   Somatic hybrids of *Datura innoxia* Mill. + *Datura discolor* Bernh. and of *Datura innoxia* Mill. + *Datura stramonium* L. var. tatula L.   I. Selection and characterization.   Molec. Gen. Genet. 162: 113-119.

_____ 1978b.   Genetic evidence for the hybrid nature of somatic hybrids from *Datura innoxia* Mill.   Planta 141:333-334.

_____ 1980a.   Somatic hybrids between a herbaceous and two tree *Datura* species.   Z. Pflanzenphysiol. 98:119-127.

_____ 1980b.   Somatic hybrids of *Datura innoxia* Mill. + *Datura discolor* Bernh. and *Datura innoxia* Mill + *Datura stramonium* L. var. tatula L.   II. Analysis of progenies of three sexual generations. Molec. Gen. Genet. 139:1-4.

_____ and Krumbiegel, G. 1980.   Selection of somatic hybrids in plants.   In: International Cell Biology 1980-1981 (H.G. Schweiger, ed.) pp. 872-878.   Springer-Verlag, Berlin.

Schiller, B., Herrmann, R.G., and Melchers, G. 1982.   Restriction endonuclease analysis of plastid DNA from tomato, potato and some of their somatic hybrids.   Molec. Gen. Genet. 186:453-459.

Scowcroft, W.R. and Larkin, P.J. 1981.   Chloroplast DNA assorts randomly in intraspecific somatic hybrids of *Nicotiana debneyi*.   Theor. Appl. Genet. 60:179-184.

Sears, B.B. 1980.   Elimination of plastids during spermatogenesis and fertilization in the plant kingdom.   Plasmid 4:233-255.

Sidorov, V.A. and Maliga, P. 1982.   Fusion-complementation analysis of auxotrophic and chlorophyll-deficient lines isolated in haploid *Nicotiana plumbaginifolia* protoplast cultures.   Molec. Gen. Genet. 186:328-332.

Sidorov, V.A., Menczel, L., Nagy, F., and Maliga, P. 1981.   Chloroplast transfer in *Nicotiana* based on metabolic complementation between irradiated and iodoacetate treated protoplasts.   Planta 152:341-345.

Skarzhynskaya, M.V., Cherep, N.N., and Gleba, Yu.Yu. 1982.   Potato + tobacco hybrid cell lines and plants obtained by cloning individual protoplast fusion products.   Sov. Cytol. Genet. (in press).

Smith, H.H., Kao, K.N., and Combatti, N.C. 1976.   Interspecific hybridization by protoplast fusion in *Nicotiana*.   J. Hered. 67:123-128.

Steele, J.A., Mehytil, T.F., Durbin, R.D., Bhutuagar, P., and Rich, D.E. 1976.   Chloroplast coupling factor 1: A species specific receptor for tentoxin.   Proc. Natl. Acad. Sci. 73:2245-2248.

Wetter, L.R. 1977.   Isoenzyme patterns in soybean + *Nicotiana* somatic hybrid cell lines.   Molec. Gen. Genet. 150:231-235.

_____ and Kao, K.N. 1980.   Chromosome and isoenzyme studies on cells derived from protoplast fusion of *Nicotiana glauca* with *Glycine max-Nicotiana glauca* cell hybrids.   Theor. Appl. Genet. 57:273-276.

White, D.W.R. and Vasil, I.K. 1979.   Use of amino acid analogue-resistant cell lines for selection of *Nicotiana sylvestris* somatic cell hybrids.   Theor. Appl. Genet. 55:107-112.

Wong-Staal, F. and Wildman, S.G. 1973.   Identification of a mutation in chloroplast DNA correlated with formation of defective chloroplasts in a variegated mutant of *Nicotiana tabacum*.   Planta 113:313-326.

Wullems, G.J., Molendijk, L., and Schilperoort, R.A. 1980.   The expression of tumor markers in intraspecific somatic hybrids of normal and crown gall cells from *Nicotiana tabacum*.   Theor. Appl. Genet. 56:203-208.

Yurina, N.P., Odintsova, M.S., and Maliga, P. 1978.   An altered chloroplast ribosomal protein in a streptomycin resistant tobacco mutant.   Theor. Appl. Genet. 52:125-128.

Zelcer, A., Aviv, D., and Galun, E. 1978.   Interspecific transfer of cytoplasmic male sterility by fusion between protoplasts of normal *Nicotiana sylvestris* and X-ray irradiated protoplasts of male-sterile *N. tabacum*.   Z. Pflanzenphysiol. 90:397-407.

CHAPTER 9
Cytoplasmic Hybridization: Genetic and Breeding Applications

E. Galun and *D. Aviv*

This chapter will focus on a novel approach in plant cell genetics. In this approach, which may become of great relevance to crop improvement, the cytoplasms of two cells differing in either or both plastids and mitochondria are brought within the plasmalemma of one cell, thus producing a heteroplasmodic cell with one or two functional nuclei (a cybrid or hybrid cell). Such a cell can result from fusion between protoplasts that differ in their cytoplasmic organelles, or from the introduction of heterologous organelles in a plant protoplast. We shall be interested primarily in systems that enable the division of the cybrid cell and in the fate of the organelles during cell division and during plant regeneration. Our discussion will, with a few exceptions, be confined to angiosperms, and we shall bear in mind that crop improvement is the main objective of this treatise.

To comprehend the present and prospective use of cybrids as breeding tools, we shall briefly review current knowledge on basic chloroplast and mitochondrion features. We shall give short descriptions of the genomes of these organelles, the transcription of organelle DNA, the translation of organelle-coded proteins, and the sexual transmission of organelles to progenies. For details on these subjects the reader will be referred to specific reviews and recent relevant articles.

Most of this chapter will review the actual studies in which cybrid plants were produced. We shall critically look at experiments in which the fate and expression of organelles were followed during subsequent cell divisions and plant regeneration and in which sexual progenies were obtained from cybrid plants.

358

We shall finally evaluate the expected impact of cytoplasmic hybridization on future studies of plant organelle genetics and point out how cytoplasmic hybridization may be integrated with other breeding procedures to achieve crop improvement.

The reader is referred to other chapters of this book that deal with isolation, culture, and fusion of protoplasts; culture of calluses and cell suspension; and utilization of metabolic mutants in plant cell genetics and crop improvement.

EXTRACHROMOSOMAL INFORMATION IN ANGIOSPERMS, CHLOROPLASTS AND MITOCHONDRIA

The Chloroplast

Comprehensive reviews on chloroplast features are presented in several texts and reviews (e.g., Akoyunoglou and Argyroudi-Akoyunoglou, 1978; Gillham, 1978; Kirk and Tilney-Bassett, 1978; Tewari, 1979). The appreciation of the role of DNA in chloroplast genetics followed the detection of DNA in algal (Ris and Plaut, 1962) and higher plant (Kislev et al., 1965) chloroplasts. The recent availability of powerful biochemical methods vastly increased our familiarity with the chloroplast genome and caused a better understanding of its general organization, replication, coding capacity, and interaction with the nuclear genome.

To avoid ambiguity we would like to note that we use the term "plastome" as an equivalent to the chloroplast genome, while "plasmone" is used in a more general sense to include all extrachromosomal genomes. Furthermore, the term "chloroplast" includes both chloroplasts and plastids, but not vice versa; when "plastids" is mentioned, it should not include chloroplasts.

THE PLASTOME. The overall organization of the chloroplast genome appears to be rather uniform among many of the angiosperms (Herrmann and Possingham, 1980).   In all the plants tested, chloroplast DNA, which may be regarded as the chloroplast's genophore (analogous to the nuclear chromosome), was found to be a circular double-helix molecule with a contour length of approximately 35-60 μm, or 120-200kbp (e.g., Bedbrook and Bogorad, 1976; Kolodner and Tewari, 1979; Seyer et al., 1981; Van Ee et al., 1980).   The chloroplast DNA (cpDNA) of most presently analysed angiosperms can be viewed as being composed of four segments: two similar regions ("inverted repeats") that are separated by two single-copy regions of differing lengths. The inverted repeats include the ribosomal operon with the direction of transcription of the 16S, 23S, 4.5S, and 5S rRNAs pointing towards the short single-copy segment; Fig. 1 demonstrates a typical physical map of the chloroplast genophore. Exceptions to this general structure, in respect to the inverted repeat segment, were recently reported in legumes. Only one such segment was detected in the cpDNA of broad beans (Koller and Delius, 1980) and pea (Palmer and Thompson, 1981) while

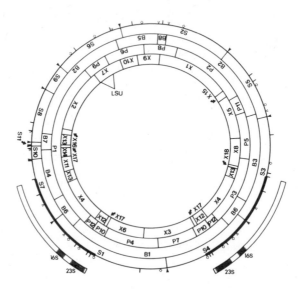

Figure 1.   The physical map of the *Nicotiana tabacum* chloroplast genophore.   The map is based on fragmentation with four restriction endonucleases: Sal I, Bgl I, Xho I, and Pvu II.   Fragments are designated by letter (marking the endonuclease) and number.   The bold regions and the expanded parts show the inverted duplications.   The positions of the genes for 16S and 23S rRNA (shaded areas) and the large subunit (LSU) of RUBPCase are indicated (from Seyer et al., 1981).

cpDNA of some other legume species seems to conform with the common inverted repeat structure.

Although the genophores of species from such widely separated families as mustard (Link et al., 1981) and wheat (Bowman et al., 1981) have many common structural features, notable intrageneric variations were detected in some genera such as *Oenothera* (Gordon et al., 1981) and *Nicotiana* (Fluhr and Edelman, 1981).   Figure 2 demonstrates the similarity as well as the variation in the physical maps of some Solanaceae genophores.   Such comparisons are obviously of interest for chloroplast evolution, but we should also bear in mind that they are very instructive in respect to our main thesis.   Knowledge of the molecular differences between specific genophores is essential for predicting the chances for the establishment of a viable plastome/nuclear genome relationship in cases where the endogenous plastome of a given species will be exchanged, via cytoplasmic hybridization, with an alien plastome.

RNAs AND PROTEINS CODED BY THE PLASTOME.   As mentioned above, chloroplast rRNAs were found to be transcribed from the plastome and were the first to be located on the physical map of this

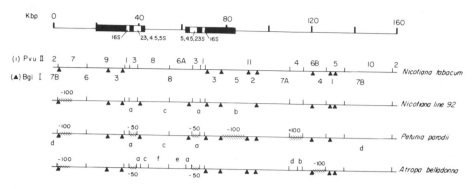

Figure 2. Differences in restriction endonuclease maps of representatives of the Solanaceae. The maps are based on fragmentation with Pvu II and Bgl I. Deletions and additions are shaded. All restriction fragments of *Nicotiana tabacum* are designated, while only those differing from the above are designated for the other species. The positions of the inverted repeats and their accompanying rRNA genes are indicated on the kbp scale (from Fluhr and Edelman, 1981).

genophore (Tewari, 1979). The plastome probably contains a full complement of tRNA species needed for protein synthesis in the chloroplast. Even though genes for some tRNA species (e.g., tRNA^{Asp}, tRNA^{Cys}, tRNA^{Glu}) were not yet identified in bean, spinach, and maize (Mubumbila et al., 1980), these authors have unpublished indications (personal communication) that these genes do exist on the plastome. As expected, the base sequence similarities between the respective spinach and bean chloroplast tRNAs seem to be greater than between either of these and chloroplast tRNAs in maize (Mubumbila et al., 1980). The locations of most chloroplast tRNA genes on the physical map of the spinach genophore have been determined (Steinmetz et al., 1980). The majority of these genes were located on the large single-copy segment, while others (e.g., genes for tRNA^{Ile1}, tRNA^{Ile2}, tRNA^{Leu2}, and tRNA^{Leu3}) were located on the inverted-repeat segment (Burkard et al., 1980; Steinmetz et al., 1980). A generally similar location of chloroplast tRNA genes emerges from studies with maize, broad beans, and other species (personal communication from J.H. Weil).

Contrary to chloroplast tRNAs, there is no indication that the chloroplast aminoacyl-tRNA synthetases are coded by the plastome. Evidence for the contrary comes from *Euglena* (Hecker et al., 1974), i.e., that chloroplast synthetases are coded by nuclear genes translated in the cytoplasm, and then imported into the chloroplast. However, direct evidence for such a situation in angiosperms is still not conclusive. If the plastome codes for tRNAs and the nuclear genome codes for their aminoacyl-tRNA synthetases a high degree of coordination would be required in respect to the temporal regulation of tRNAs and synthetases, and in respect to conformational aspects. This consideration brings us back to our main subject. Inserting an alien plastome into the domain of a given nuclear genome may cause a mismatch between plastome tRNA and its nuclear coded synthetase and thus prevent normal protein synthesis in the chloroplast.

The chloroplast proteins are either coded by the nuclear genome, translated on cytoplasmic ribosomes, and imported into the chloroplast; or they are coded by plastome genes and then transcribed and translated inside the chloroplast (Ellis, 1981). There is no unequivocal evidence for import of mRNA coded by the nuclear genome and then its translation inside the chloroplast. Although a large number of polypeptides are apparently synthesized by isolated chloroplasts, only a handful of polypeptide genes have as yet been identified and located on the physical map of the chloroplast genophore. One of these, the gene for the large subunit of ribulose 1,5-bisphosphate carboxylase, was located on the large single-copy segment of the maize chloroplast genophore (Bedbrook et al., 1979). Studies with other species have verified this finding (see Bottomley and Bohnert, 1981 for review and Westhoff et al., 1981 for a brief survey of recent studies). Another chloroplast polypeptide, the 32,000 MW membrane protein, which probably plays a role in regulating photosystem II-electron flow (Mattoo et al., 1981), is also coded by the plastome. This was verified by the isolation of its mRNA (Rosner et al., 1977; Reisfeld et al., 1978) from the chloroplasts as well as by location of its gene on the chloroplast genophore (Driesel et al., 1980).

A chloroplast specific elongation factor that is required for the addition of amino acids to the translated polypeptide chain was detected in spinach chloroplasts (Ciferri, 1975). Recent studies (O. Tiboni, L. Panzeri, G. di Pasquale, S.Sora, and O. Ciferri, communicated at the EMBO Workshop on Chloroplast DNA, Arolla, Switzerland, May 1981) mapped the gene for this factor on the spinach chloroplast DNA map.

Three polypeptides comprising subunits of the coupling factor (CF_1) in the chloroplast ATP synthetase complex were recently located on the spinach genophore physical map: the alpha, beta, and epsilon subunits (Westhoff et al., 1981). Other subunits (i.e., gamma and delta) of CF_1 are coded by the nuclear genome. In dealing with this protein we are faced again with a requirement for coordinated syntheses from plastome and nuclear genomes, and we should note this requirement in studies with cytoplasmic hybrids, resulting from fusion of cells from different species or genera.

LOCATION OF CHLOROPLAST DNA AND CHLOROPLAST REPRODUCTION. The number of genophores per chloroplast and the number of chloroplasts per cell should be of great interest in connection with somatic hybridization in plants, especially when cytoplasmic hybrids are considered. In the first phase of these hybridizations the chloroplasts of the fusion partners are included within one plasmalemma envelope and are expected to sort our during subsequent cell division. Whether or not this sorting out is directed or random, the velocity of sorting out should be coordinated with the number of chloroplasts and genophores contributed by each fusion partner to the fusion product. We should therefore like to know how genophores are "packed" in chloroplasts, i.e., does each chloroplast contain only one, a few, or many genophores? In addition, we should determine how many chloroplasts are in each cell. Kowallik and Herrmann (1972) and Herrmann and Kowallik (1970) furnished a good approach to the first question. By appropriate electron microscope techniques they detected several DNA regions in each chloroplast, indicating a multitude of genophores per chloroplast in green angiosperm leaves. Such DNA regions can now be revealed by relatively simple optical microscope techniques, with the aid of appropriate fluorescent dyes such as DAPI (4'6-diamidino-2-phenylindole). This technique was recently applied to several plant species (Coleman, 1979; Kuroiwa et al., 1981; Sellden and Leech, 1981). Although the results of these observations vary among species and types of cells, we at least now have a general idea on the number of DNA regions per chloroplast: etiolated leaf chloroplasts have only one or very few such regions, while mature chloroplasts (3 or more days after exposure of etiolated leaves to light) may have between 12 and 40 per chloroplast. Moreover, the increase of cpDNA is correlated with both light exposure and increase of chloroplast size (Gibbs et al., 1974). Since light also increases the number of plastids per cell (see below), the total number of genophores per cell is probably extremely variable. A model for the multiplication of the chloroplast genome is presented in Fig. 3.

The number of chloroplasts (or plastids) per cell and probable modes of chloroplast multiplication were discussed thoroughly by Butterfass (1979). The greatest numbers of chloroplasts per cell were commonly detected in mesophyll cells. Thus in *Raphanus sativus*—which belongs to a group of plants having a relatively high number of chloroplasts per cell—there are 200-350 chloroplasts in each palisade cell, while only 4 chloroplasts are usually found in each palisade cell of *Peperomia metalica* (Piperaceae). The common range of chloroplasts per mesophyll cell is 20-40. Lower numbers of chloroplasts per cell were recorded in epidermal and meristem cells while the lowest number of chloroplasts per cell and the least variation was found in guard cells: 2-15 with a mode of 7-8. In meristems there seems to be a positive correlation between nuclear division and plastome replication, while such a correlation is lacking in differentiated tissue. The lack of simple nuclear control over plastid development has already been demonstrated by Gibor and Granick (1962) in *Euglena*. Chloroplast differentiation in this alga was not impaired following UV irradiation which damaged the nucleus and its capability to divide. On the other hand it is noteworthy that leaf cells of polyploid plants contain much higher numbers of chloroplasts than comparable diploid cells. The control mechanisms which regulate the number of chloroplasts per cell are still unknown. We also lack knowledge on the regulation of genophore numbers per chloroplast. Moreover, except for evidence that chloroplast DNA is replicated by a nuclear-coded polymerase, we are ignorant of the molecular events of chloroplast DNA replication. Because of the requirement for close coordination between nuclei and chloroplasts, sustained culture of isolated chloroplasts, which could facilitate the molecular approach to cpDNA replication, is not expected to be achieved in the foreseeable future.

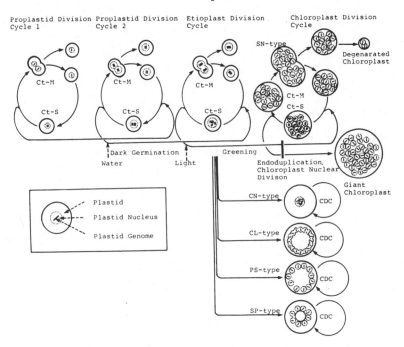

Figure 3.   A diagram of chloroplast nuclear events during chloroplast development and division.   A proplastid in land plants develops into a chloroplast with multiplication in the chloroplast genome (ct-genome) through at least four different division cycles: proplastid division cycle 1, the proplastid division cycle 2, the etioplast division cycle, and the chloroplast division cycle.   Nuclear division in the plastid takes place prior to plastidkinesis.   Most plants can be classified into one of five groups: SN-type, CN-type, CL-type, PS-type, or SP-type, based on differences in shape, size, and the distribution of the nucleoid region in their mature chloroplasts.   The differences in chloroplast nucleoid region patterns are explained by the distribution pattern of the replicated ct-genomes in mature chloroplasts which are distributed randomly (SN-type); gathered around the pyrenoid (SP-type); distributed along the periphery (PS-type); fused to form spherules in the central area (CN-type); or fused to form a circle along the periphery (CL-type) (from Kuroiwa et al., 1981).

Chloroplast Genetics.   Notable exceptions to Mendelian laws of inheritance were revealed soon after the "rediscovery" of these laws. One of the investigators who pioneered the detection of these exceptions was Carl Correns, who was also one of the three "rediscoverers" of Mendel's experiments.   Correns observed that reciprocal crosses between certain variegated and green plants resulted in different progenies.   His own studies, as well as studies of other investigators, on non-Mendelian inheritance were thoroughly reviewed (Correns, 1937). Lacking knowledge on genetic information in chloroplasts and mitochondria, Correns could not pinpoint cases of extranuclear inheritance to either of these organelles but rather attributed the genetic control to "cytoidioplasma" or "plasmone."   However, he did consider chloroplast transmission as being correlated with non-Mendelian inheritance.   For more recent reviews on chloroplast heredity, the reader is referred to reviews by Beale and Knowles (1978) and Gillham (1978).   Almost all the studies were based on morphological features (e.g., chlorophyll pigmentation) of whole plants, and genera expressing biparental transmission of chloroplasts (see below) as *Oenothera* and *Pelargonium* were the choice plant material.

A notable exception are the findings of Durbin and Uchytil (1977), who studied tentoxin sensitivity.   This fungal toxin binds to one of the subunits of the thylakoid CF_1 of some species but not to CF_1 subunits of other species (Steele et al., 1976).   The toxin sensitivity is maternally inherited in plants having uniparental chloroplast transmission. Tentoxin response can thus serve as a chloroplast marker (see Galun, 1982); since the genes of the three chloroplast-coded CF_1 subunits were recently located on the genophore's physical map, we have at hand a rare case (in angiosperms) in which we should be able to trace a change in expression in the mature leaf down to a specific site on the physical map of the chloroplast DNA.

Another chloroplast-controlled trait, streptomycin resistance, was revealed through cell culture.   Maliga et al. (1973) isolated a streptomycin tobacco cell line that could be regenerated into plants and found by reciprocal hybridization that this streptomycin resistance is maternally transmitted to the sexual progeny (Maliga et al., 1975).   Further studies by Maliga and his collaborators (Yurina et al., 1978) indicated that the resistant plants have an altered chloroplast ribosomal protein. This trait was rather useful in somatic hybridization both for the selection of fusion products and as a chloroplast marker—as shall be detailed below.

A conditioned lethal (thermal) tobacco mutant was detected by Matzinger and Wernsman (1973).   This mutant fails to become autotrophic at low temperatures but is normal in permissibe (e.g., 26/20—day/night) temperatures.   Further studies indicated that this trait is maternally inherited and the mutant has defective chloroplast (but normal mitochondria) when grown in nonpermissible conditions (Nessler et al., 1980; Nessler and Wernsman, 1980).

Resistance to triazine can probably also be traced to the 32,000 MW membrane protein (Pfister et al., 1981) and thus to a specific physical map location.   However, the intricate molecular mechanism causing resistance or sensitivity is still open to further clarification (Gressel, 1981).

It should be noted that while chloroplast genetics in algae (e.g., *Chlamydomonas*) is well advanced (Gillham, 1978), the molecular basis of many chloroplast genes has not yet been established.

CHLOROPLAST TRANSMISSION IN SEXUAL PROPAGATION. Genetic studies and the utilization of genetics in plant breeding are based on the mixing of parental genes in the zygote and their subsequent random segregation in the progeny. In considering chloroplast heredity and the employment of chloroplast traits in plant breeding, we face a rather complicated requirement. First, there should be a zygotic "mixing" of nuclear, chloroplast, and mitochondrial genomes of the two parents. Subsequently, these genomes should segregate in the progeny. Have such processes actually been identified in angiosperms? To answer this question we shall briefly survey chloroplast transmission in sexual propagation of angiosperms. This survey is based primarily on a recent detailed review (Sears, 1980) and on our own previous discussion (Galun, 1982). The reader is referred to these papers for specific literature.

There are two approaches to answer the above question. In one approach the fate of the plastids in the developing male and female gametophytes are followed by structural studies up to the stage of fertilization of the egg cell by the sperm cell. The transmission of plastids from both the maternal and paternal parents up to the fertilized egg cell will indicate (but not assure) biparental transmission of the plastome to the sexual progeny. The other approach is based on reciprocal crosses between plants having different recognizable plastome traits. Strict maternal inheritance indicates maternal transmission, and the rate of maternal to paternal transmission of a plastome trait serves to estimate the degree of maternal and paternal transmission of chloroplasts. However, the investigator should be careful: reciprocal differences in the phenotypes of F_1 plants are only an indication but not proof for extranuclear inheritance of a given trait. The striped leaf trait, *iojap*, in maize is one of several examples that the first generation resulting from reciprocal crosses may be misleading (Rhoades, 1955). The *iojap* trait is caused by a recessive gene (*ij*); plants having the *ij/ij* constitution produce defective chloroplasts which are transmitted maternally in maize, but *ij* is a nuclear gene which causes a programmed change to ribosome-less plastids (Walbot and Coe, 1979). Thus assigning a gene to the organelle genophore should be based on a proper genetic study and not merely on reciprocal crosses.

Because we do not have evidence for the elimination of plastids during female gametophyte development in angiosperms (such cases were recorded in gymnosperms), we shall consider only plastid transmission by the male gametophyte. The latter gametophyte passes through several decisive developmental stages (Frankel and Galun, 1977) which may constitute barriers for chloroplast transmission. The first is during the development of the diploid pollen mother cell into microspores; ultrastructural observation indicated that during this phase-transition there is a vast clearing of the cytoplasm. The second barrier is at the first asymmetrical mitosis leading into the larger vegetative and smaller generative cell. In some species the generative cell contains no, or only very few, plastids, while in other species the generative cell does contain an appreciable number of plastids (and mitochondria). A further barrier occurs during the discharge of the pollen tube into one of the embryo sac's synergid cells and the further penetration of the tube's cytoplasm into the egg cell. In some cases only the sperm nucleus enters the egg cell, while in other cases (e.g., in *Oenothera* species) both the nucleus and male gametophyte organelles are delivered into the egg cell. Ultrastructural observations are not always conclusive, as indicated by Sears (1980), because proplastids cannot unequivocally be differentiated from reduced mitochondria. Moreover, a failure to detect plastids is no conclusive evidence for their absence.

The other approach, namely deducing chloroplast transmission from genetic studies was followed in an appreciable number of species. Sears (1980, Table 1) should be consulted for details.   In summary, information is available from 22 monocot and 107 dicot species.   In most species the information is based on either genetic studies or structural investigation.   Only in a few species were both kinds of studies performed (e.g., in *Pelargonium* species—indicating biparental plastome transmission—and in tomato—indicating strict maternal plastome transmission).   There seems to be no clear correlation between phylogenetic evolution and the type of plastome transmission.   For example, biparental plastome transmission was detected in rye but not in wheat or barley.   On the other hand, in certain genera (e.g., *Pelargonium* and *Oenothera*) all the species studied had biparental plastome transmission, and in several of these paternal plastids were reported to enter the egg cell.   The list of examined species is thus still small and requires verification, but it indicates that in most of the economically important crops (e.g., wheat, rice, cotton, and tomato) the plastome is transmitted uniparentally (maternally).   This situation should be considered in plant breeeding; chloroplast characters in these crops can be transferred only in one direction and in order to establish a cultivar with a given nuclear genome but an alien plastome, the plastome donor should serve as female parent in the initial cross while the nuclear-genome donor should be the recurrent back-cross male parent.   We shall see below that this laborious and time consuming breeding procedure can now be speeded up by cytoplasmic hybridization.

The Mitochondrion

Mitochondria are consistent constituents of eukaryotic cells and have a central role in energy conversion and provision of ATP. Our discussion will focus on genetic properties of angiosperm mitochondria and on molecular characteristics of mitochondria related to these properties. We shall thus point out, right at the outset, that present information on these aspects in angiosperm mitochondria is rather poor as compared to that of mammalian and fungal mitochondria.   Nevertheless as we shall see below, we know already that the mitochondrial genome is involved in a number of traits (e.g., male sterility, phytotoxin sensitivity) that are of major importance to plant breeding. We shall therefore mention the major relevant mitochondrial characteristics of mammalian and fungal organisms, referring the reader to the appropriate literature for details on these characteristics.

The basic role of mitochondria is probably common in all eukaryotes (see Gillham, 1978 for review). Likewise, there are similar basic structural features in mitochondria of protozoa, metazoa, fungal organisms, and green plants. Although there are considerable size and shape variations even among cells of the same organism, the general structure of the mitochondrion is universal. It has a smooth and continuous outer membrane and an inner membrane which is also continuous but is highly

Table 1. Analysis of Chloroplasts, Mitochondria, and Male Sterility in Plants Resulting from Somatic Hybridization

FUSION PARTNERS	MEANS OF ANALYSIS	REFERENCES
N. glauca + N. langsdorffii	IEF of RUBPCase	Kung et al. 1975; Chen et al., 1977
L. esculentum + S. tuberosum	IEF of RUBPCase	Melchers et al., 1978; Poulsen et al., 1980
N. sylvestris + N. tabacum CMS (L-92)[a]	IEF of RUBPCase, tentoxin, cpDNA, mtDNA, male sterility	Zelcer et al., 1978; Galun & Aviv, 1979; Aviv et al., 1980
N. tabacum + N. tabacum CMS (debneyi)	cpDNA, mtDNA, male sterility	Belliard et al., 1978; Belliard et al., 1979
N. tabacum (P-) + N. tabacum CMS (debneyi, P+)	Variegation, IEF of RUBPCase, male sterility	Gleba, 1979
N. tabacum (P-) + N. sylvestris (P+)	Variegation, IEF of RUBPCase	Gleba, 1979
N. knightiana (P+) + N. sylvestris (P-, Kanam.-R)	Kanamycin, greening	Menczel et al., 1978
P. axillaris + P. hybrida CMS	Male sterility	Izhar & Power, 1979; Izhar & Tabib, 1980
P. hybrida + P. hybrida CMS	Male sterility	Bergounioux-Bunisset & Perennes, 1980
N. sylvestris + N. tabacum (Strept.-R)	Streptomycin	Medgyesy et al., 1980
N. sylvestris CMS (L-92) + N. tabacum[a]	Tentoxin, cpDNA, mtDNA, male sterility	Aviv & Galun, 1980a; Galun et al., 1981
N. tabacum + N. rustica	IEF of RUBPCase	Iwai et al., 1980; Iwai et al., 1981; Douglas et al., 1981
N. debneyi (type a) + N. debneyi (type b)	cpDNA	Scowcroft & Larkin, 1981
N. tabacum (Strept.-R) + N. knightiana	Streptomycin, cpDNA, mtDNA	Menczel et al., 1981a; Nagy et al., 1981
N. tabacum (Strept.-R) + N. plumbaginifolia[a]	Streptomycin, cpDNA	Sidorov et al., 1981; Menczel et al., 1981b

[a] X-irradiated protoplasts.

Abbreviations: CMS = cytoplasmic male sterility; IEF = isoelectric focussing; Kanam.-R = Kanamycin resistance; L-92 = Australian Nicotiana, cytoplasm of the Suaveolentes section; P- and P+ = deficient and normal chloroplasts; Strept. = streptomycin resistance. Parentheses following species names describe cytoplasmic characters used in the analyses of somatic hybrid plants (e.g., cytoplasm of alien species, deficient chloroplasts, antibiotic resistance).

convoluted into folds called cristae.   The lumen between the outer and the inner membranes is called the intermembrane space, while the space within the inner membrane is called the matrix.   It should be noted that the two membranes differ in structure, properties, and enzyme constituency.   In addition to those in or on the membranes, some enzymes are typically located in either the intermembrane space or the matrix.   Details on enzyme location and membrane permeabilities are outside the scope of our discussion, but we shall bear in mind that the outer membrane is relatively permeable (to molecules of up to 10,000 MW), the inner membrane is impermeable except to small uncharged molecules and the transfer across the inner membrane is regulated by specific transport system.   The inner membrane also contains the main respiratory chain.   We do, however, have a "methodological" interest in mitochondrial enzymes and their location.   Some of these enzymes are specific to mitochondria and can be located histochemically.   They can thus serve as mitochondrial markers (see Galun, 1981) in ultrastructural studies.   In physiological and biochemical respects the mitochondrion thus emerges as a highly organized, and rather complex, "power plant."

The mitochondrion has its own protein synthesis system based on endogeneous ribosomes and tRNAs (see Buetow and Wood, 1978, for review).   This system, as well as the mitochondrial DNA, will be discussed below in some detail because they have direct relevance to cytoplasmic hybridization.

THE MITOCHONDRIAL GENOME: PHYSICAL AND CHEMICAL CHARACTERISTICS.   The general properties of the mitochondrial genome were discussed in several texts and reviews (e.g., Gillham, 1978 for general review; Edelman, 1981 for plant mitochondria).   More often than not, the buoyant density (and the G + C percent) of the mitochondrial DNA (mtDNA) in a given organism differs considerably from that of its nuclear DNA.   This difference was pointed out by Luck and Reich (1964) for *Neurospora* and was subsequently verified in other eukaryotes, although there are exceptions (e.g., certain mammals and amphibia) to this generalization.   Angiosperms comprise an amazingly uniform group in respect to mtDNA buoyant density: their mtDNA bands at 1.706 ± 0.001 $g \cdot cm^{-3}$ (in neutral cesium chloride), while the buoyant density of their nuclear DNA varies considerably (e.g., 1.702 and 1.696 $g \cdot cm^{-3}$ for maize and cucumber, respectively).   Three main forms of mtDNA may be found in isolated preparation: linear, open circle, and duplex circular molecules.   It is not always clear whether the linear and open circle molecules represent the natural endogenous forms or result from breaks and nicks, respectively in duplex circular DNA during isolation.   Generally speaking, mammalian mtDNAs are circular molecules of 5-6 μm contour length (15-17 kbp; 9-12 x 10^6 apparent MW) while more complex and variable mtDNAs were found in invertebrates.   Among the latter some protozoa are outstanding in their apparent lack of circular mtDNA.   Considerable variations in length were reported for fungal organisms (e.g., ~ 6-25 μm in various yeasts, ~ 10 μm in *Aspergillus*, and 20-25 μm in *Neurospora*).   Among photo synthetic organisms, *Euglena* is outstanding, having circular mtDNA with only 1 μm contour length (although linear molecules up to 19 μm have also been found). The contour length of mtDNA from the green alga *Chlamydomonas* is about 5 μm.

Kolodner and Tewari (1972a, 1972b) described mtDNA of peas, lettuce, and spinach as 30 μm circles, having an approximate molecular weight of 70×10^6. Further studies on mtDNA of angiosperms reviewed in part by Levings and Pring (1978) indicated that the situation is probably more complex and variable. Thus in maize several discrete classes of mtDNA were detected by electron microscopy (16, 22, and 30 μm in length). More recently, additional small–size mtDNA populations were detected in certain fertile and male–sterile lines (Kemble and Bedbrook, 1980; Kemble et al., 1980) by both electron microscopy and electrophoretic analysis. Likewise, male–sterile and fertile sugarbeet lines were found to contain different mtDNA size classes (Powling, 1981). Different size–classes of mtDNA were also reported in tobacco (Sparks and Dale, 1980), cucumber, potato, Virginia creeper (Quetier and Vedel, 1977), and soybean (Syrenki et al., 1978). A detailed study of tobacco mtDNA classes indicated that the base–pair sequences of some classes are at least partially homologous (Dale, 1981). A yet unsolved paradox emerges from comparison of contour lengths of angiosperm mtDNA as determined by electron microscope techniques to length estimates based on the summing up of the apparent molecular weights of fragments obtained by electrophoretic separation of restriction endonuclease digested mtDNA; the latter estimates are constantly larger (e.g., Vedel et al., 1980). Moreover, completely different estimates for mtDNA size based on kinetic complexity were reported by Ward et al. (1981). These authors utilized reassociation kinetics and restriction analysis and suggested genome sizes of between 22 x 10 d and 1600 x 10 d (i.e., equivalent to about 60 and 300 μm contour length), respectively, for watermelon and muskmelon.

Information on the coding capacity of angiosperm mtDNA is only beginning to accumulate. Thus obvious questions as the relation between genome size and coding capacity and the existence of differences in coding capacities among angiosperms as well as the latter and other eukaryotes cannot be settled presently. Nevertheless, we should keep these considerations in mind, because they are of prime relevance in cytoplasmic hybridization. This relevance will become clearer during our discussion of these hybridizations. At this stage we like to note that restriction endonuclease analysis of mtDNA is presently the only direct method employed to characterize the mitochondrial population of hybrid progenies. Commonly the mtDNA restriction patterns obtained in cybrid plants are novel, being found in neither of the somatic hybrid parental plants.

THE MITOCHONDRIAL GENOME: CODING AND GENETICS. Research on the mitochondrial genome of angiosperms is still mostly at the level of DNA restriction patterns, physical properties, and fragmentary information on rRNA genes. On the other hand, our knowledge of the yeast (e.g., Linnane and Nagley, 1978; Borst and Grivell,

1978) and mammalian (e.g. Bibb et al., 1981) mitochondrial genomes is well advanced. Several researchers utilized genetic techniques (based mostly on the use of "petite" and antibiotic resistant mutants) to achieve a detailed understanding of yeast mitochondrial genome (cf. Borst and Grivell, 1978). Moreover, molecular approaches were added, resulting in an overlapping of the genetic and physical maps of the yeast mitochondrial genome. In brief yeast mtDNA codes for mitochondrial rRNA and most probably for all the tRNA required for the organelle translation system. The gene sequences for these RNAs as well as for a number of mitochondrial proteins were located on the genetic and physical maps of the yeast mitochondrial genome. About 5% of the proteins in the mitochondrion are coded by mitochondrial genes. Among these are subunits of cytochrome c oxidase, cytochrome bc_1 complex, ATPase complex, and ribosome associated proteins. As noted for chloroplasts, complete mitochondrial enzyme complexes (e.g., ATPase) are composed of some subunits coded by the nuclear genome and other subunits coded by the mitochondrial genome (e.g., *oli* genes). Interestingly, only two subunits of the mitochondrial ATPase are probably coded by mtDNA of *Neurospora*, while in yeast the mtDNA codes for three ATPase subunits. In this and probably other organelle membrane enzymes we face the possibility of a particular subunit gene being coded by different cellular genomes in different organisms (Borst and Grivell, 1978). Although comparable information on angiosperm mitochondrial enzymes does not yet exist, we can anticipate the subunit rule to hold true here as well. As a consequence, in cytoplasmic hybrids having mitochondria and nuclei from different plant species or genera we could face inadequate subunit coordination or lack of coordination between subunits and various cellular components such as the membrane. Alloplasmic male-sterility (see Gerstel, 1980) is a candidate for such considerations, which indicate the link between molecular aspects of mitochondrial biology and plant breeding.

Human (Anderson et al., 1981) and mouse (Bibb et al., 1981) mitochondrial genomes are almost identical in several respects. Both are short, being 16,569 and 16,295 base pairs in human and mouse, respectively (should we say that in respect to mtDNA, humans have a mere 274 base pair advantage over mice?) In spite of the short mammalian mitochondrial genome, the latter probably codes for the same major mitochondrial components as the yeast (and possibly the angiosperm) mitochondrial genome. The mammalian mitochondrial genome codes for the 12S and 16S rRNAs, 22 tRNAs, 3 subunits of the cytochrome c oxidase, 1 subunit of ATPase, one subunit of cytochrome b, and probably for an additional 8 polypeptides. The latter were predicted from "reading frames" of the mtDNA sequence. The genetic code of mammalian mitochondria differs from the universal code in several important respects. For example, mitochondrial UGA codes for tryptophan rather than for termination; AUA codes for methionine rather than for isoleucine, and AGA and AGG are termination rather than arginine codons.

Angiosperm mitochondria contain unique rRNA and tRNA species (see Edelman, 1981); and by analogy to other eukaryotes these are probably coded by the mitochondrial genome. Plant cells in culture are sensitive to most, if not all, the antibiotic drugs toward which mitochondrial resistant mutants were isolated in fungi and mammalian cells (e.g., chloramphenicol, erythromycin, oligomycin).   Therefore one possible approach to study mitochondrial genetics in angiosperms is to isolate antibiotic resistant mutants in these plants.   We have recently isolated an erythromycin resistant *Nicotiana* cell line (unpublished data); whether or not this resistance resulted from a mitochondrial mutation has not yet been determined.   Correlative evidence indicates that mitochondria are involved in cytoplasmic male sterility in certain angiosperms (e.g., maize, tobacco, wheat, and sugarbeet), such sterility resulted from an alloplasmic condition (Gerstel, 1980), i.e., plants contain nuclei of one species and cytoplasmic organelles of another species.   Various studies (e.g., Levings and Pring, 1976, 1978; Pring and Levings, 1978; Kemble and Bedbrook, 1980; Kemble et al., 1980) that characterized DNA mainly by restriction patterns indicated that cytoplasmic male sterile (CMS) lines of maize have unique mtDNA restriction patterns.   Differences in mtDNA restriction patterns were also reported between fertile and CMS sugarbeet cultivars (Powling, 1981) as well as a CMS tobacco line based on backcrossing *Nicotiana debneyi* with *N. tabacum* (as recurrent paternal parent; Belliard et al., 1979). We should note that these findings do not prove the causality of the correlation between unique mtDNA and male sterility.   Recent cytoplasmic hybridization, which will be detailed below, strengthen this correlation and furnish additional evidence for the involvement of the mitochondrial genome in CMS of *Nicotiana*.

Susceptibility to the fungus *Helminthosporium maydis* was found in maize CMS lines containing the Texas male sterile cytoplasm (T). Several studies indicated that mitochondria of T lines are specifically affected by the respective fungal toxin (see York et al., 1980 for details and literature).

Forde and Leaver (1980) reported that a 21,000 MW polypeptide is synthesized in mitochondria of normal fertile maize lines, while in mitochondria of CMS lines it appears to be replaced by a 13,000 MW polypeptide.   Moreover, nuclear genes that restore fertility to CMS lines also suppress the synthesis of the 13,000 MW polypeptide.   These findings may lead to an understanding of the nuclear and mitochondrial control of CMS and possibly to *H. maydis* sensitivity, especially if a firm correlation between synthesis of the 13,000 MW polypeptide and toxin sensitivity can be established.

SOMATIC AND CYTOPLASMIC HYBRIDS

Improved cultivars can be obtained by various breeding approaches. Frequently, especially when qualitative characters are at stake, the plant breeder knows which traits he intends to improve and has at hand one or several cultivars that are satisfactory but for the aforementioned traits.   The breeder is thus faced with the problem of "inserting" the required traits into existing cultivars.   If the traits are controlled by nuclear genes then cross breeding, followed by proper selection—in the selfed or backcrossed progenies—will commonly lead, through appropriate gene segregation, to the required improved cultivars. The selection phase, which usually requires several generations, may be shortened by androgenesis or other means of haploidization (see Chapter 6). If the required traits are not available in the cultivated species, interspecific crosses could be used, but these may encounter incongruity barriers of various kinds. In some cases the zygote is produced and the embryo starts to develop, but further development is stopped. In the latter cases in vitro embryo culture can be employed. However, when the traits required for cultivar improvement are organelle-controlled, this breeding scheme is not readily applicable. As outlined above, the hybrid will commonly contain only maternal organelles; thus no further segregation of organelle-controlled traits is expected. However, there are notable exceptions (e.g., rye; Melchers, 1980) which should be considered for specific crops. On the other hand, in somatic hybridization (see Schieder and Vasil, 1980; Chapter 5) we expect the first fusion product to contain organelles from both fusion partners. Such cells are termed heteroplasmonic. Evidence that this is actually the case was furnished by Gleba (1979) who fused tobacco protoplasts from plants with a chloroplast controlled albinism with tobacco protoplasts of a nucleus-controlled chlorophyll deficiency. Progeny plants with mixed chloroplast populations were obtained, and the detection of variegated plants in the sexual progeny of somatic hybrids indicated the heteroplastomic character of the fusion product. Other examples that indicate a mixture of organelles in the fusion product will be mentioned further on. Thus it is by now well established that somatic hybridization is a possible means to achieve new nuclear-gene/plasmone combinations. On the other hand, somatic hybridization also creates complications. Frequently we are not interested in changing the nuclear genome but merely in exchanging chloroplasts and/or mitochondria. In such cases our aim should be cytoplasmic hybrids (cybrids), i.e., to obtain fusion products, that will contain a specific nonhybrid nucleus but a heteroplasmonic cytoplasm. Such heteroplasmonic and homonuclear fusion products should yield the expected plants through organelle sorting-out and redifferentiation (Galun and Aviv, 1979). We shall deal with this approach later on in this chapter.

Organelles and Plasmone Traits in Somatic Hybridization

What is the relevance of organelles and plasmone traits in somatic hybridization of angiosperms? There are several answers to this question. First, morphological traits and biochemical characteristics, which are controlled by the plasmone, may serve to identify somatic hybrids. Second, certain traits such as streptomycin resistance and chloroplast deficiency are instrumental in the selection of heterofusion products among colonies or plants resulting from nonfused and autofused protoplasts. Third, for breeding purposes or for cell genetics studies we may be interested in exchanging the mitochondria and/or the chloroplasts of a given cultivar or genotype with organelles from another specific plant. The main available information concerning chloroplasts, mitochondria, and male sterility in plants resulting from somatic hybridization has been summarized in Table 1. In this section we shall discuss a few specific examples. The aspect of intentional transfer of organelles by cytoplasmic hybridization will be detailed in a separate section below.

IDENTIFICATION OF SOMATIC HYBRIDS BY PLASMONE TRAITS. As can be seen in Table 1, only a limited number of chloroplast traits were used for the identification of the plastome in plants resulting from somatic hybridization. Most of these, i.e., reaction to tentoxin, iso-electric focusing of the large subunit of RUBPCase, the restriction pattern of cpDNA, and streptomycin resistance, were detailed by Galun (1982). Fusion between a protoplast having normal chloroplasts and a protoplast having (plastome controlled) defective chloroplasts, may result in a green or variegated somatic hybrid having the nuclear geno-type of the second protoplast. Thus greening and variegation were used in some studies as chloroplast markers for the heteroplastomic state (Gleba, 1979; Menczel et al., 1978). The only mitochondrial marker presently available is the restriction pattern of mtDNA. We shall discuss the use of this marker as well as cytoplasmic male steril-ity when dealing with the appropriate examples.

Somatic hybridization of *Nicotiana langsdorffii* and *N. glauca* was the first reported somatic hybridization in angiosperms (Carlson et al., 1972). Chen et al. (1977) analysed the RUBPCase of 16 of these somatic hybrid plants. All these plants had the combined RUBPCase small subunits of both parental types, indicating nuclear hybridization. Of the 16 plants, 9 had only the large subunit of *N. langsdorffii* chloro-plasts while 6 plants had the large subunit of *N. glauca* chloroplasts. One plant had both kinds of large subunits, but the latter segregated in the sexual progeny. These results clearly indicated that the fusion caused a heteroplastomic state followed by sorting out of chloroplasts. Sorting out was rather abrupt in most cases. Rapid sorting out is common to most reported cytoplasmic hybridizations.

Isoelectric focusing was frequently used as chloroplast marker (Table 1), because both the plastome and the nuclear genome can be identified in one single test. On the other hand it lacks sensitivity. Thus using this method Iwai et al. (1980) identified only the large subunit RUBPC-ase of *N. tabacum* in a somatic hybrid resulting from *N. tabacum* + *N. rustica*, but some of the plants obtained from anther culture of this hybrid had *N. rustica* large subunit RUBPCase (Iwai et al., 1981). This somatic hybrid probably contained a mixture of *N. tabacum* and *N. rustica* chloroplasts, but the fraction of the latter was below detection by isoelectric focusing.

Another commonly used chloroplast marker is the cpDNA restriction pattern. It was first employed by Belliard et al. (1978) to character-ize *N. tabacum* and *N. debneyi* chloroplasts. Since the cpDNA of dif-ferent species is often fragmented differentially with at least some restriction endonucleases, this method is rather useful as a chloroplast marker in somatic hybridization. Furthermore even intraspecies differ-ences in cpDNA can be detected by the respective restriction pattern analysis.   Scowcroft and Larkin (1981) used this method to study chloroplast sorting out after fusion of protoplasts from two variants of N. debneyi.

Cytoplasmic male sterility (CMS) is strictly maternally inherited in several plants such as tobacco, maize, wheat, and Petunia (Frankel and Galun, 1977).   There is good circumstantial evidence that CMS is controlled in Nicotiana by a nuclear/mitochondrial interaction (Belliard et al., 1979; Galun et al., 1981).   CMS can therefore serve as an organelle (tentatively mitochondrial) marker in Nicotiana.   Belliard et al. (1978) utilized CMS to detect somatic hybrids between two tobacco lines.   One of these lines was CMS (with N. debneyi cytoplasm) and had sessile leaves (nuclear marker) while the other was fertile (with N. tabacum cytoplasm) and had petiolated leaves.   The leaf shape in the sexual hybrids between these lines was intermediate.   Consequently, these authors could distinguish somatic hybrids from cybrids.   Plants in which either leaf shape or male sterility/fertility differed from those of either parental type were presumptive somatic hybrids or cybrids. Chloroplast DNA restriction analysis was utilized to identify the chloroplasts of the tested plants, and no correlation was found between chloroplast type and CMS.   However, results of mtDNA restriction analysis were correlated with CMS.   Finally, male sterility was used by Izhar and coworkers to identify somatic hybrids among plants obtained after fusion of two Petunia species (Izhar and Power, 1979; Izhar and Tabib, 1980).

Restriction patterns of mtDNA were used in only two studies to help in the identification of somatic hybrids (or cybrids).   One of these (Galun et al., 1981) will be discussed in detail in the following section. The other study was performed by Nagy et al. (1981) on somatic hybrids between N. tabacum and N. knightiana.   In this study the mtDNA restriction patterns were analysed in 6 hybrid clones.   In addition to showing mostly either of the parental fragments, these hybrids also had unique fragments.   Based on probing the respective restriction patterns with E. coli DNA the authors suggested that a sequence rearrangement occurred in the mtDNA of the hybrids.   As mtDNA restriction patterns of Nicotiana species are remarkably stable even during in vitro cell culture (Sparks and Dale, 1980; Galun et al., 1981) the rearrangement should be attributed to the heteroplasmonic state following protoplast fusion.

Utilization of Plasmone Traits to Select Somatic Hybrids.   Because of the lack of appropriate mitochondrial mutants in angiosperms, our examples will be confined to plastome traits.   In practice two types of chloroplast mutants were used.   The first type have a malfunction in photosynthesis, e.g., lack of chlorophyll.   The second type is resistant to antibiotics (e.g., streptomycin, kanamycin), which affect translation inside the chloroplast but not in the cytosol.   Streptomycin resistance was used in several somatic hybridization studies by Maliga and coworkers, and in our laboratory to select hybrid plants (see Table 1). For example, Medgyesy et al. (1980) fused protoplasts from a streptomycin-resistant N. tabacum line with N. sylvestris protoplasts and transferred the resulting calli to a medium containing $1 \text{ mg} \cdot \text{ml}^{-1}$ streptomycin.   The *N. tabacum* protoplasts were pretreated by iodoacetate and therefore did not survive unless fused with *N. sylvestris* protoplasts.   Nonfused and autofused *N. sylvestris* protoplasts were expected to produce streptomycin–sensitive calluses.   The plants regenerated in the presence of streptomycin were, as expected, either somatic hybrids or cybrids.   The results confirmed this expectation.

Albinism and other chloroplast deficiencies may be caused by either the plastome or nuclear genes.   Gleba and coworkers (Gleba, 1979) used such traits to select somatic hybrids and cybrids.   Both fusion partners were protoplasts from *N. tabacum*, but one had a plastome mutation (P⁻) while the other was the "sulfur" mutant (Su/Su) which results in yellow and lethal plants (unless cultured in a sugar-containing medium).   Green or green/yellow variegated plants could thus be selected as fusion products.

Cytoplasmic Hybridization: A Tool for Organelle Exchange

We already mentioned above that for breeding and cell genetics purposes we may be interested in producing a plant of a certain nuclear genotype but with alien mitochondria and/or chloroplasts.   To achieve this purpose we should avoid the complication resulting from regular somatic hybridization, i.e., a heterozygous nuclear genome.   Fusion between protoplasts containing the full complement of nucleus, mitochondria, and chloroplasts with protoplasts that lack nuclei (e.g., Wallin et al., 1978; Hoffmann, 1981) or in which nuclear division is suppressed, can be helpful in this regard.   Another way could be the introduction of chloroplasts, mitochondria, or their respective DNAs into normal protoplasts followed by culture and plant regeneration.   The latter way was tried by several authors, but as yet did not yield the expected plants (see Binding, 1979).   Since our own study, using the "Donor-Recipient" technique, resulted in plants with the expected exchange of organelles, we shall detail our protocol and describe such experiments performed in our laboratory.

PROTOCOL

Only fusion and plant regeneration procedures by the "donor–recipient" method will be described.   This protocol will not detail procedures handled elsewhere in this book.   The reader should therefore consult other chapters for protoplast isolation, protoplast culture, protoplast fusion, and regeneration of plants from calli.   Figure 4 schematically illustrates the method.

Protoplasts

1.   Choose "donor" and "recipient" partners.   These should preferably differ in respect to both nuclear and cytoplasmic characters.   The required organelle characters should be in the "donor."

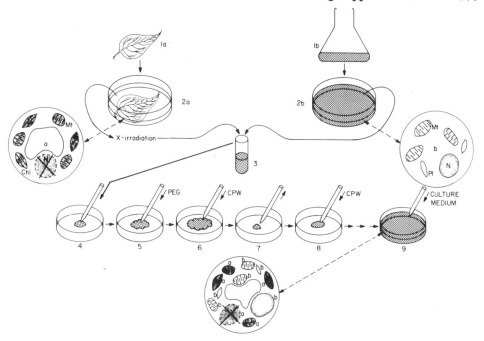

Figure 4.   The "donor-recipient" method of cytoplasmic hybridization: Scheme of procedure.   Legend for numbers: (1a) "donor" leaf as source of protoplasts; (1b) cell suspension culture as source of "recipient" protoplasts; (2a) maceration of "donor" leaf in enzyme solution; (2b) maceration of recipient cell suspension in enzyme solution; (3) mixing of X-irradiated "donor" and nonirradiated "recipient" protoplasts; (4) insertion of 0.25 ml mixed protoplast suspension to the center of a petri dish; (5) addition of 0.35 ml PEG solution; (6) addition of 0.5 ml CPW; (7) removal of solution, leaving agglutinated protoplasts on the dish; (8) addition of 0.5 ml CPW; (9) final dilution of fused protoplasts with liquid medium.   Protoplasts of "donor" (a) and "recipient" (b), containing nuclei (N), mitochondria and chloroplasts or plastids, as well as fusion product protoplast with only one functional nucleus are presented in a highly schematized manner.

2.   Prepare protoplasts from cell suspensions or leaf-mesophyll or from both.   There is an advantage of having "donor" and "recipient" protoplasts from different tissue sources, because then heterofusion can be observed and its percent can be evaluated.   Avoid strong light during isolation and fusion of protoplasts.

3.   X-irradiate "donor" protoplasts.   *Nicotiana* mesophyll protoplasts should be irradiated with an approximate dose of 5kr; other protoplasts may require different doses in order to arrest nuclear division.   Gamma, rather than X-irradiation may be applied.

Fusion

The fusion protocol is a modification of the method of Kao and Michay-luk (1974).

1. Mix "donor" and "recipient" protoplasts in a 1:1 proportion at a final density of about 5×10^5 protoplasts per ml.
2. Use a 6 cm diameter plastic petri dishes (tissue culture grade, e.g., Falcon 3002) and add 0.25 ml of the mixed protoplasts suspension in the center of each dish. Use several such dishes for each fusion experiment.
3. Add 0.35 ml of a PEG solution in the form of small drop at the periphery of each of the protoplast suspension "drops." Wait 15 min.
4. Add 0.5 ml CPW, in the form of small drops, to the periphery of each protoplast suspension "drop." Wait 10 min (you may observe that the aggregated protoplasts stay at the center of the "drop").
5. Use a Pasteur pipette to carefully remove the solution (but not the protoplasts) from the "drop" and add another portion of 0.5 ml CPW as above. Wait 10 min and repeat this step twice.
6. Add 1-3 ml liquid culture medium to each petri dish. Observe microscopically (use inverted microscope) to determine the proto-plast density for your specific protoplast system (for tobacco: optimal density is about 2×10^4 protoplasts per ml). Use culture medium that is appropriate to your specific protoplast system.
7. Close dishes with Parafilm and incubate (usually at 25-28 C). One or two days of dark or dim light incubation is probably of advant-age to most protoplasts. Thereafter shift to light of about 50-200 fc.

Growth and Selection

1. After 2-3 days add agar medium to reach final concentration of 0.8% agar (add a 42 C agar solution dropwise to avoid sudden heating of the protoplasts). If before adding agar medium you observe that the density of viable protoplasts is much below optimal, transfer the fused protoplast over a feeder layer (see Raveh et al., 1973; Cella and Galun, 1980, for principle; and Zelcer et al., 1978, for application).
2. Selection against unfused and autofused "recipient" protoplasts may start with the first incubation in liquid culture medium (e.g., man-nitol as osmoticum to select against N. sylvestris protoplasts) or it may be delayed to a later culture phase (e.g., transfer of young calli to a streptomycin containing medium to select against strep-tomycin sensitive cells). The selection procedure should thus be adopted to fit the specific system used.

Plant Regeneration

This stage will not be detailed, because the procedures to induce regeneration are specific for each system. Likewise, other means of identification of the cybrids and the determination of the plasmones will not be detailed here, because these subjects were detailed in previous sections of this chapter.

Specific Solutions

1. PEG solution: prepare a 50% polyethylene glycol 1500 (e.g., BDH Chemicals Ltd. Poole, England) solution in 10 mM $CaCl_2$, 0.1 M glucose.
2. CPW: 0.55 M mannitol (preferably from BDH), 0.19 mM KH_2PO_4, 0.01 M $CaCl_2$ 0.98 mM Mg SO_4. $7H_2O$, 0.98 mM KNO_3, 0.99 mM-KI, 0.16 mM $CuSO_4$ in 1000 ml H_2O.

The "Donor–Recipient" Method—Transfer of Male Sterility and Organelle Exchange

Because the "donor–recipient" method is potentially applicable to several systems, we shall describe it in some detail, using our own experiments as representative examples. These experiments were partially published (Zelcer et al., 1973; Galun and Aviv, 1979; Aviv and Galun, 1980a, 1980b; Aviv et al., 1980; Galun et al., 1981).

Our rationale for the first "donor–recipient" experiment was the following: Fusion of untreated "recipient" protoplasts having a given plasmone, with X-irradiated "donor" protoplasts having a different plasmone, and culturing under conditions that will prevent division of "recipient" protoplasts should result in cybrid plants having the "recipient" nuclei and plasmone traits of the "donor." We chose *N. sylvestris* as "recipient" because *N. sylvestris* protoplasts do not divide in a mannitol containing medium. The "donor" protoplasts were obtained from a CMS tobacco line (L-92) which was reported to have originated from a cross between *N. suaveolens* (female) and *N. tabacum* (male) followed by repeated back crossings to *N. tabacum* as pollinator. The morphology of L-92 was identical to *N. tabacum* cv. Xanthi but it is CMS, having stigmoid/petaloid anthers and no viable pollen. Because of uniparental transmission we assumed that the chloroplasts and the mitochondria of L-92 are identical or very similar to *N. suaveolens* (since we could not determine whether *N. suaveolens* or a related Australian *Nicotiana* species was in fact the original maternal progenitor of L-92, we shall term the organelles of L-92 as "L-92" rather than "sua"). Using abbreviations for nuclear genome, plastome, and male sterility (or fertility) respectively, the first fusion experiment can be written as:

$$syl/tbc/fert + tbc/L\text{-}92/ster \text{ (irr.).}$$

This means: *N. sylvestris* nucleus/*N. tabacum* chloroplasts/male fertile, plus *N. tabacum* nucleus/L-92 chloroplasts/male sterile; (irr.) denotes X-irradiation. The chloroplasts of *N. tabacum* and *N. sylvestris* are identical in respect to all tested characters (e.g., tentoxin resistance, cpDNA patterns after fragmentation with several restriction endonuclei). We therefore use "*tbc*" to denote the chloroplasts of both these species.

It was anticipated that the heterofused protoplasts have a selective advantage because the X-irradiated "donor" protoplasts should not produce colonies and the "recipient" protoplasts should not divide in our standard (mannitol) medium unless fused with the "mannitol resistant" *N. tabacum* protoplasts. The results can be summarized as follows. Most of the regenerated plants (20 plants from 5 calli), had *N. sylvestris* morphology but were male sterile like the "donor" plants. These 20 plants were termed Type A. Using three chloroplast markers (tentoxin sensitivity, isoelectric focusing of RUBPCase, and cpDNA restriction patterns) we found that Type A plants had L-92 chloroplasts. Thus chloroplasts and CMS were transferred from the CMS tobacco line (L-92) to *N. sylvestris*. The uniformity of Type A plants in respect to chloroplast composition and CMS was confirmed by out-crossings to *N. sylvestris* and *N. tabacum* as well as by androgenesis. Thus Type A plants can be denoted as: *syl*/L-92/ster.

These results urged us to ask the following questions: (a) Can the fertility of Type A plants be restored by fusion with a fertile donor?; (b) Is the restoration of fertility correlated with chloroplast transfer?

An affirmative answer to the first question would indicate the general applicability of the "donor-recipient" method to transfer male sterility and male fertility unidirectionally. Furthermore, if there is no correlation between the transfer of chloroplasts and male sterility, the latter trait might be mitochondrially controlled and consequently the "donor-recipient" method would be applicable in a more general manner: plastome traits could be transferred independently from mitochondrial traits.

To answer these questions we performed the following fusion:

syl/L-92/ster + *tbc*/*tbc*/fert (irr.).

Namely, we used Type A as "recipient" and normal *N. tabacum* cv. Xanthi as "donor." In one such experiment we obtained 63 calluses of which 13 were ascertained to be fusion products. These calli resulted in the following cybrid plants:

7 plants from 3 calli: *syl*/*tbc*/ster
19 plants from 8 calli: *syl*/*tbc*/fert
8 plants from 4 calli: *syl*/L-92/fert

In several cases one callus resulted in two kinds of plants (e.g., *syl*/*tbc*/fert and *syl*/L-92/fert). Such calli are listed twice in the above-mentioned data. In addition one plant, *syl*/L-92/ster, that had the same nuclear/chloroplast/sterility composition as Type A was obviously a fusion product, because it resulted from a callus that also produced *syl/tbc*/ster plants.   It is thus obvious that sorting out of plastids was still in progress at the callus level.   We rarely observed that both male sterile and male fertile plants differentiated from one fusion-product callus.   Could this mean that sorting out of mitochondria is completed much earlier than sorting out of chloroplasts?   We do not have an answer to this question but should note that the cytoskeleton may provide interconnections between the mitochondria in the cytosol. If this is the actual situation in angiosperms, the mitochondrial complement of a cell may not be regarded as composed of physically independent particles within a given cell.

Our experiments led us to ask the following further questions: (a) Is there a sorting out of mitochondria during cell division of a fusion product?   (b) Is there a correlation between the mitochondrial genome and male sterility/fertility?   (c)   Is there an independent assortment of chloroplasts and mitochondria in cybrids resulting from heteroplasmonic fusion?   We could answer these questions by employing mtDNA restriction patterns of the two fusion partners (e.g., Type A and *N. tabacum* cv. Xanthi) differ substantially.   Two additional observations are noteworthy.   There was no difference between the restriction patterns of *N. tabacum* and *N. sylvestris* mtDNA (after fragmentation with either Xho I or Sal I).   Furthermore, the mtDNA restriction patterns of cell suspensions were identical to those of intact plants.   Within each cybrid plant the mtDNA restriction pattern was stable: even the sexual progeny of a given cybrid retained the original cybrid pattern. The main two results of the analysis of the cybrids mtDNA were (1) most cybrids had some fragments found in neither of the parents and (2) male sterile cybrids always had mtDNA restriction patterns which were similar or identical to the sterile parent while male-fertile cybrids had restriction patterns which were similar or almost identical to the restriction pattern of the fertile parent.

The cybrids could therefore be grouped as follows:

$$
\begin{array}{ll}
syl/L\text{-}92/L\text{-}92* & \text{(ster)} \\
syl/L\text{-}92/tbc* & \text{(fert)} \\
syl/tbc/L\text{-}92* & \text{(ster)} \\
syl/tbc/tbc* & \text{(fert)}
\end{array}
$$

(The asterisks denote similarity or identity with the respective mtDNA).

The latter finding indicated a correlation between mitochondrial composition and male sterility/fertility.   These findings are in accordance with those of Belliard et al. (1979) and Nagy et al. (1981) working with different systems of *Nicotiana*.   The novel mtDNA restriction patterns obviously indicate changes in the base sequence of mtDNA following protoplast fusion which is in contradiction to the stability of mtDNA during sexual propagation and cell division.   Whether this change in sequence resulted from rearrangement or recombination should be determined in future studies.   Here we shall only note that if indeed there are interconnections between all or part of the mitochondria of an angiosperm cell there is a good chance for different mitochondrial genomes to meet in a cybrid cell—a chance which is probably rare or nonexisting in respect to plastomes, since we have no indications for interconnections between chloroplasts in angiosperm cells.

While we found a correlation between male sterility/fertility and mitochondrial type, no such correlation was found for chloroplasts: frequently two cybrids having identical chloroplast composition differed in male fertility and, vice versa, cybrids which differed in chloroplast composition were identical in respect to male fertility.

Our three questions were thus answered: there was a sorting out of mitochondria which was independent of chloroplast assortment and there was a correlation between mitochondrial assortment and segregation of male sterility/fertility.

The above described experiments were augmented in our laboratory with other similar "donor-recipient" studies. Table 2 indicates that although the number of calli obtained in six such experiments was variable these experiments consistently yielded a high proportion of cybrid calli. Furthermore, numerous cybrid plants were obtained from each experiment. Streptomycin resistance seems to be an efficient selection method resulting in a low number of "escapee" calli. Figure 5 demonstrates the transfer of the streptomycin-resistance character from the *N. tabacum* SR-1 "donor" to *N. sylvestris*. It should be noted that in this specific experiment most cybrids produced only streptomycin-resistant progenies while some cybrids produced both sensitive and resistant seedlings (details will be published elsewhere). Experiments 5 and 6 of Table 2 were designed to result in variegated plants. Such plants

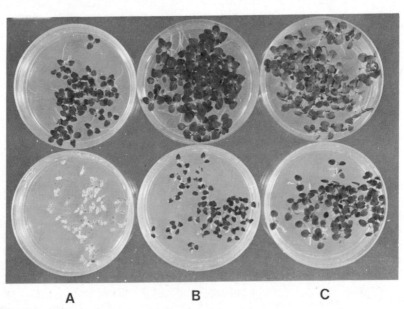

A   B   C

Figure 5. Verification of chloroplast transfer in a "donor-recipient" experiment by the streptomycin test. Surface-sterilized seeds were placed on nutrient agar without (upper row) or with 1 mg·ml^{-1} streptomycin (lower row); A-sensitive "recipient" parent: *N. sylvestris*, B-resistant cybrid (G-d-33), C-resistant "donor" parent: *N. tabacum* SR-1.

Table 2. Cybrid Plants Obtained in Six Different Fusion Experiments Employing Different Sources for "Donor–Recipient" Pairs

EXPERIMENT NUMBER	PARENTAL TYPES X-ray irrad. "Donor"	PARENTAL TYPES "Recipient"	MEANS OF SELECTION[a]	NUMBER OF CALLI OBTAINED	NUMBER OF CALLI PRODUCING CYBRID PLANTS	NUMBER OF CYBRID PLANTS[b]	REFERENCES
1	N. tabacum (CMS)	N. sylvestris	Mannitol	7	5	20	Zelcer et al., 1978
2	N. tabacum	N. sylvestris (CMS)	Mannitol	63	13	35	Aviv & Galun, 1980
3	N. tabacum (SR-1)[c]	N. sylvestris	Mannitol	21	13	22	Aviv, unpublished
4	N. tabacum (SR-1)	N. tabacum	Streptomycin	4	3	11	Fluhr & Aviv, unpublished
5	N. tabacum (CMS)	N. tabacum (albino)[d]	Green calli	3	2	27	Fluhr & Aviv, unpublished
6	N. tabacum	N. sylvestris (CMS)	Mannitol	8	3	20	Fluhr & Aviv, unpublished

[a]Selection against "recipient" ("donor" protoplasts were X-irradiated and will cease to divide).

[b]Identification of cybrids was based on morphological (nuclear) characters in addition to at least one cytoplasmic marker.

[c]Streptomycin-resistant mutant obtained from Dr. P. Maliga, Szeged, Hungary.

[d]Albino plantlets were derived by R. Fluhr, from a variegated N. tabacum line with defective plastids, originally obtained from Dr. L.G. Burk, Oxford, N.C.

were actually found among the respective cybrids, indicating that complete sorting out of chloroplasts is not an obligatory result of cytoplasmic hybridization.

MODIFICATION OF THE "DONOR-RECIPIENT" METHOD: USE OF IODOACETATE.   A notable modification to the "donor-recipient" method was reported by Maliga and coworkers (e.g., Sidorov et al., 1981).   These authors applied the iodoacetate inactivation technique which was suggested by Wright (1978) for the selection of fused mammalian cells.   This technique is based on metabolic inhibition of protoplasts which were pretreated with iodoacetate.   The pretreatment will cause the degeneration of nonfused and autofused protoplasts while fusion of pretreated protoplasts with nontreated protoplasts will cause metabolic complementation and result in viable hybrids.   In one experiment iodoacetate-treated *Nicotiana plumbaginifolia* cell-suspension protoplasts were fused with X-irradiated *N. tabacum* mesophyll protoplasts and 47 calli were obtained.   All regenerated (cybrid) plants from these calli had *N. plumbaginifolia* morphology, but most of them contained *N. tabacum* chloroplasts.   The iodoacetate treatment does not impair the nucleus of the treated protoplast.   Thus the latter can complement an X-irradiated protoplast.   Cybrid formation is therefore the expected result.   It is feasible that other metabolic inhibitors may be applicable in a similar manner.

CONCLUSIONS AND FUTURE PROSPECTS

Recent developments in molecular and somatic genetics can provide excellent tools for novel approaches in plant breeding.   A recent discussion of this subject by Cocking et al. (1981) provides numerous examples for such approaches.   In addition, these developments are useful for future studies on plant cell genetics.   Achievement in this latter area will obviously have an impact on crop improvement.

In this chapter we stressed the role of organelles in somatic hybridization and noted that protoplast fusion can provide new nuclear/plasmone combinations which cannot be obtained by sexual hybridization. We dealt in detail with the transfer of organelles from one plant species (or cultivar) to another by the "donor-recipient" method because this method was already shown to be applicable in practice and useful for the transfer of important plastome and mitochondrial traits.   Obviously there are other potential possibilities.   Thus "cytoplasts" (e.g., Hoffmann, 1981) or enucleated protoplasts (Bracha and Sher, 1981), rather than X-irradiated protoplasts may be applicable as organelle "donors" in specific cases.   Furthermore, it is conceivable that in the future efficient means will be available to introduce alien plastomes and mitochondrial genomes (or fractions of these genomes) into plant cells in a way which will assure the replication and expression of the introduced genomes.   Microinjection or encapsulation in liposomes could provide the appropriate "Trojan horse" in such manipulations.   The notable advantage of certain angiosperms over mammals is

the totipotency of higher plant cells from the former.   A single-injec-
ted (or otherwise manipulated) plant cell can regenerate a functional
plant, which can be further haploidized, subsequently diploidized, and
then self-pollinated to yield pure lines.

While it is expected that future achievements in molecular genetics
of microorganisms and animal cells could be adopted at least in part to
angiosperms, there is still one major limitation: only in a rather few
angiosperm species is a workable "protoplast system" available.   Therefore to fully utilize novel methodologies for crop improvement efficient
techniques should be established to regenerate functional plants from
isolated protoplasts.   Unless such techniques will be made applicable to
our major crops, the balance between "prospects" and "achievements"
will lean strongly to the former.

ACKNOWLEDGMENTS

We are grateful to Prof. Marvin Edelman and Robert Fluhr for help in
the preparation of this manuscript and for permission to use unpublished
experimental results.

KEY REFERENCES

Aviv, D. and Galun, E. 1980a.   Restoration of fertility in cytoplasmic
  male sterile (CMS) *Nicotiana sylvestris* by fusion with X-irradiated *N.
  tabacum* protoplasts.   Theor. Appl. Genet. 58:121-127.

Edelman, M. 1981.   Nucleic acids of chloroplasts and mitochondria.   In:
  The Biochemistry of Plants, Vol. 6 (A. Marcus, ed.) pp.249-301.   Academic Press, New York.

Frankel, R. and Galun, E. 1977.   Pollination mechanisms, reproduction
  and plant breeding.   Springer-Verlag, Berlin, Heidelberg, and New
  York.

Galun, E. 1982.   Screening for chloroplast composition in hybrids and
  cybrids resulting from protoplast fusion in angiosperms.   In: Methods
  in Chloroplast Molecular Biology (M. Edelman, R.G. Hallick, and
  N.-H. Chua, eds.) pp. 139-148.   Elsevier/North Holland Biochemical
  Press, Amsterdam.

Gillham, N.W. 1978.   Organelle Heredity.   Raven Press, New York.

Schieder, O. and Vasil, I.K. 1980.   Protoplast fusion and somatic
  hybridization.   In: Perspectives in Plant Cell and Tissue Culture (I.K.
  Vasil, ed.) International Reviews Cytology Suppl. 11B, pp. 21-46.
  Academic Press, New York.

REFERENCES

Akoyunoglou, G. and Argyroudi-Akoyunoglou, J.H. 1978.   Chloroplast
  Development.   Elsevier/North Holland, Amsterdam.

Anderson, S., Bankier, A.T., Barrell, B.G., de Bruijn, M.H.L., Coulson, A.R., Drouin, J., Eperon, I.C., Nierlich, D.P., Roe, B.A., Sanger, F., Schreier, P.H., Smith, A.J.H., Staden, R., and Young, I.G. 1981. Sequence and organization of the human mitochondrial genome. Nature 290:457-465.

Aviv, D. and Galun, E. 1980b. Biochemical and genetic analysis of plants derived from the fusion of X-irradiated male sterile *Nicotiana tabacum* protoplasts and *N. sylvestris* protoplasts. In: Advances in Protoplast Research (L. Ferenczy and G.L. Farkas, eds.) pp. 357-362. Pergamon Press, Oxford.

Aviv, D., Fluhr, R., Edelman, M., and Galun, E. 1980. Progeny analysis of the interspecific somatic hybrids: *Nicotiana tabacum* (CMS) + *Nicotiana sylvestris* with respect to nuclear and chloroplast markers. Theor. Appl. Genet. 56:145-152.

Arzee-Gonen, P. 1980. Isolation and characterization of mitochondrial DNA in *Nicotiana* and the utilization of mtDNA restriction patterns as cytoplasmic markers in *Nicotiana* somatic hybrids. M.Sc. Thesis, Submitted to the Feinberg Graduate School of the Weizmann Institute of Science, Rehovot.

Beale, G. and Knowles, J. 1978. Extranuclear Inheritance. Arnold, London.

Bedbrook, J.R. and Bogorad, L. 1976. Endonuclease recognition sites mapped on *Zea mays* chloroplast DNA. Proc. Nat. Acad. Sci. USA 73: 4309-4313.

Bedbrook, J.R., Coen, D.M., Beaton, A.R., Bogorad, L., and Rich, A. 1979. Location of the single gene for the large subunit of ribulose bisphosphate carboxylase on the maize chloroplast chromosome. J. Biol. Chem. 254:905-910.

Belliard, G., Pelletier, G., Vedel, F., and Quetier, R. 1978. Morphological characteristics and chloroplast DNA distribution in different cytoplasmic parasexual hybrids of *Nicotiana tabacum*. Molec. Gen. Genet. 165: 231-237.

Belliard, G., Vedel, F., and Pelletier, G. 1979. Mitochondrial recombination in cytoplasmic hybrids of *Nicotiana tabacum* by protoplast fusion. Nature 281:401-403.

Bergouniox-Bunisset, C. and Perennes, C. 1980. Transfer de facteurs cytoplasmiques de la fertilite male entre 2 lignees de *Petunia hybrida* par fusion de protoplastes. Plant Sci. Lett. 19:143-149.

Bibb, M.Y., Van Etten, R.A., Wright, C.T., Walberg, M.W., and Clayton, D.A. 1981. Sequence and gene organization of mouse mitochondrial DNA. Cell 26:167-180.

Binding, H. 1979. Subprotoplasts and organelle transplantation. In: Plant Cell and Tissue Culture—Principles and Application (W.R. Sharp, P.O. Larson, E.F. Paddock, and V. Raghavan, eds.) pp. 789-805. Ohio State Univ. Press, Columbus.

Borst, P. and Grivell, L.A. 1978. The mitochondrial genome of yeast. Cell 15:705-723.

Bottomley, W. and Bohnert, H.J. 1981. The biosynthesis of chloroplast proteins. In: Encyclopedia of Plant Physiology New Series (A. Pirson and M.H. Zimmerman, eds.) Springer-Verlag, Berlin, Heidelberg, New York (in press).

Bowman, C.M., Koller, B., Delius, H., and Dyer, T.A. 1981. A physical map of wheat chloroplast DNA showing the structural genes for ribosomal RNA and the large subunit of ribulose 1,5-bisphosphate carboxylase. Molec. Gen. Genet. 183:93-101.

Bracha, M. and Sher, N. 1981. Fusion of enucleated protoplasts with nucleated miniprotoplasts in onion (*Allium cepa* L.). Plant Sci. Lett. 23:95-101.

Buetow, D.E. and Wood, W.M. 1978. In: Subcellular Biochemistry (D.B. Roodyn, ed.) Vol. 5, pp. 1-85. Plenum, New York.

Burkard, G., Canaday, J., Crouse, E., Guillemaut, P., Imbault, P., Keith, G., Keller, M., Mubumbila, M., Osorio, L., Sarantoglou, V., Steinmetz, A., and Weil, J.H. 1980. Transfer RNAs and aminoacyl-tRNA synthetases in plant organelles. In: Genome Organization and Expression in Plants (C.J. Leaver, ed.) pp. 313-320. Plenum, New York.

Butterfass, T. 1979. Patterns of Chloroplast Reproduction—A Developmental Approach to Protoplasmic Plant Anatomy. Springer-Verlag, Wien, New York.

Carlson, P.S., Smith, H.H., and Dearing, R.D. 1972. Parasexual interspecific plant hybridization. Proc. Nat. Acad. Sci. 69:2292-2294.

Cella, R. and Galun, E. 1980. Utilization of irradiated carrot cell suspensions as feeder layer for cultured *Nicotiana* cells and protoplasts. Plant Sci. Lett. 19:243-252.

Chen, K., Wildman, S.G., and Smith, H.H. 1977. Chloroplast DNA distribution in parasexual hybrids as shown by polypeptide composition of fraction 1 protein. Proc. Natl. Acad. Sci. 74:5109-5112.

Ciferri, O. 1975. Mechanism of protein synthesis in higher plants. In: The Chemistry and Biochemistry of Plant Proteins (J.B. Harborne and C.F. Von Sumere, eds.) pp. 113-135. Academic Press, London.

Cocking, E.C., Davey, M.R., Pental, D., and Power, J.B. 1981. Aspects of plant genetics manipulation. Nature 293:265-270.

Coleman, A.W. 1979. Use of fluorochrome 4'6-diamidino-2-phenylindole in genetic and developmental studies of chloroplast DNA. J. Cell Biol. 82:299-305.

Correns, C. 1937. Nicht mendelnde Vererbung. In: Handbuch der Vererbungwissenschaft (E. Baur and M. Hartmann, eds.) Vol. IIH, pp. 1-59. Gebrueder Borntraeger, Berlin.

Dale, R.M.K. 1981. Sequence homology among different size classes of plant mtDNAs. Proc. Nat. Acad. Sci. 78:4453-4457.

Douglas, G.C., Wetter, L.R., Keller, W.A., and Setterfield, G. 1981. Somatic hybridization between *Nicotiana rustica* and *N. tabacum* IV. Analysis of nuclear and chloroplast genome expression in somatic hybrids. Can. J. Bot. 59:1509-1513.

Driesel, A.J., Speirs, Y., and Bohnert, H.J. 1980. Spinach chloroplast mRNA for a 32000 dalton polypeptide: Size and localization on the physical map of the chloroplast DNA. Biochim. Biophys. Acta 610:297-310.

Durbin, R.D. and Uchytil, T.F. 1977. Cytoplasmic inheritance of chloroplast coupling factor 1 subunits. Biochem. Gen. 15:1143-1146.

Ellis, R.J. 1981. Chloroplast proteins: Synthesis, transport and assembly. Annu. Rev. Plant Physiol. 32:111-137.

Fluhr, R. and Edelman, M. 1981. Conservation of sequence arrangement among higher plant chloroplast DNAs: Molecular cross hybridization among the Solanaceae and between *Nicotiana* and *Spinacia*. Nucl. Acid Res. 9:6841–6853.

Forde, B.G. and Leaver, C.J. 1980. Nuclear and cytoplasmic genes controlling synthesis of variant mitochondrial polypeptides in male-sterile maize. Proc. Natl. Acad. Sci. USA. 77:418–422.

Galun, E. 1981. Plant protoplasts as physiological tools. Annu. Rev. Plant Physiol. 32:237–266.

_____ and Aviv, D. 1979. Plant cell genetics in *Nicotiana* and its implications to crop plants. Monografia Genetica Agraria 4:153–175.

_____, Arzee-Gonen, P., Fluhr, R., Edelman, M., and Aviv, D. 1981. Cytoplasmic hybridization in *Nicotiana*: Mitochondrial DNA analysis in progenies resulting from fusion between protoplasts having different organelle constitutions. Molec. Gen. Genet. 186:50–56.

Gerstel, D.U. 1980. Cytoplasmic male sterility in *Nicotiana*. Tech. Bull. No. 263, North Carolina Agricultural Research Series, Raleigh.

Gibbs, S.P., Mak, R.N., and Slankis, T. 1974. The chloroplast nucleoid in *Ochromonas danica* II. Evidence for an increase in plastid DNA during greening. J. Cell Sci. 16:579–591.

Gibor, A. and Granick, S. 1962. Ultraviolet sensitive factors in the cytoplasm that affect the differentiation of *Euglena* plastids. J. Cell Biol. 15:599–603.

Gleba, Y.Y. 1979. Nonchromosomal inheritance in higher plants as studied by somatic cell hybridization. In: Plant and Tissue Culture—Principles and Applications (W.R. Sharp, P.O. Larsen, E.F. Paddock, and V. Raghavan, eds.) pp. 775–788. Ohio State Univ. Press, Columbus.

Gordon, K.H.J., Crouse, E.J., Bohnert, H.J., and Herrmann, R.G. 1981. Restriction endonuclease cleavage site map of chloroplast DNA from *Oenothera parviflora* (*Euoenothera* plastome IV). Theor. Appl. Genet. 59:281–296.

Gressel, J. 1981. Triazine herbicide interaction with a 32,000 M_r thylakoid protein—alternative possibilities. Plant Sci. Lett. 25:99–106.

Hecker, L., Egan, J., Reynolds, R.J., Nix, C.E., Schiff, J.A., and Barnett, W.E. 1974. The sites of transcription and translation for *Euglena* chloroplastic aminoacyl-tRNA synthetases. Proc. Natl. Acad. Sci. 71:1910–1914.

Herrmann, R.G. and Kowallik, K.V. 1970. Selective presentation of DNA-regions and membranes in chloroplasts and mitochondria. J. Cell Biol. 45:198–201.

Herrmann, R.G. and Possingham, J.V. 1980. Plastid DNA—The plastome. In: Results and Problems in Cell Differentiation, Vol. 10 (J. Reinert, ed.) pp. 45–96. Springer-Verlag, Berlin, Heidelberg, New York.

Hoffmann, F. 1981. Formation of cytoplasts from giant protoplasts in culture. Protoplasma 107:387–391.

Iwai, S., Nagao, T., Nakata, K., Kawashima, N., and Matsuyama, S. 1980. Expression of nuclear and chloroplastic genes coding for fraction 1 protein in somatic hybrids of *Nicotiana tabacum* + *rustica*. Planta 147:414–417.

Iwai, S., Nakata, K., Nagao, T., Kawashima, N., and Matsuyama, S. 1981. Detection of the *Nicotiana rustica* chloroplast genome coding for the large subunit of Fraction 1 protein in a somatic hybrid in which only the *N. tabacum* chloroplast genome appeared to have been expressed. Planta 152:478-480.

Izhar, S. and Power, J.B. 1979. Somatic hybridization in *Petunia*: A male sterile cytoplasmic hybrid. Plant Sci. Lett. 14:49-55.

Izhar, S. and Tabib, Y. 1980. Somatic hybridization in *Petunia*. Part 2: Heteroplasmic state in somatic hybrids followed by cytoplasmic segregation into male sterile and male fertile lines. Theor. Appl. Genet. 57:241-245.

Kao, K.N. and Michayluk, M.R. 1974. A method for high frequency intergeneric fusion of plant protoplasts. Planta 115:355-367.

Kemble, R.J. and Bedbrook, J.R. 1980. Low molecular weight circular and linear DNA molecules in mitochondria from normal and male-sterile cytoplasms of *Zea mays*. Nature 284:265-266.

Kemble, R.J., Gunn, R.E., and Flavell, R.B. 1980. Classification of normal and male-sterile cytoplasms in maize II. Electrophoretic analysis of DNA in mitochondria. Genetics 95:451-458.

Kirk, J.T. and Tilney-Bassett, R.A.E. 1978. The Plastid. Elsevier/North-Holland, Amsterdam.

Kislev, N., Swift, H., and Bogorad, L. 1965. Nucleic acids of chloroplasts and mitochondria in swiss chard. J. Cell Biol. 25:327-344.

Koller, B. and Delius, H. 1980. *Vicia faba* chloroplast DNA has only one set of RNA genes as shown by partial denaturation mapping and R-loop analysis. Molec. Gen. Genet. 178:261-269.

Kolodner, R. and Tewari, K.K. 1972a. Physicochemical characteristics of mitochondrial DNA from pea leaves. Proc. Natl. Acad. Sci. 69:1830-1834.

_____ 1972b. Genome sizes of chloroplast and mitochondrial DNAs in higher plants. In: Proceedings 30th Annual Meeting Electron Microscopy Society of America (C.J. Arceneaux, ed.) pp. 190-191. Claitors, Baton Rouge, Louisiana.

_____ 1979. Inverted repeats in chloroplast DNA from higher plants. Proc. Natl. Acad. Sci. 76:41-45.

Kowallik, K.V. and Herrmann, R.G. 1972. Variable amounts of DNA related to the size of chloroplasts IV. Three dimensional arrangement of DNA in fully differentiated chloroplasts of *Beta vulgaris*. J. Cell Sci. 11:357-377.

Kung, S.D., Gray, J.C., Wildman, S.G., and Carlson, P.S. 1975. Polypeptide composition of fraction 1 protein from parasexual hybrid plants in the genus *Nicotiana*. Science 187:353-355.

Kuroiwa, T., Suzuki, T., Ogawa, K., and Kawano, S. 1981. The chloroplast nucleus: Distribution, number, size and shape and a model for the multiplication of the chloroplast genome during chloroplast development. Plant Cell Physiol. 22:381-396.

Levings, C.S., III, and Pring, D.R. 1976. Restriction endonuclease analysis of mitochondrial DNA from normal and Texas cytoplasmic male sterile maize. Science 193:158-160.

_____ 1978. The mitochondrial genome of higher plants. In: Stadler Symposium Vol. V, pp. 77-93. University of Missouri, Columbia.

Link, G., Chambers, S.E., Thompson, J.A., and Falk, H. 1981. Size and physical organization of chloroplast DNA from mustard (*Sinapis alba* L.). Molec. Gen. Genet. 183:454–457.

Linnane, A.W. and Nagley, P. 1978. Mitochondrial genetics in perspective: The derivation of a genetic and physical map of the yeast mitochondrial genome. Plasmid 1:324–345.

Luck, D.J.L. and Reich, E. 1964. DNA in mitochondria of *Neurospora crassa*. Proc. Nat. Acad. Sci. 52:931–938.

Maliga, P., Breznovits, Sz., and Marton, L. 1973. Streptomycin resistant plants from callus of haploid tobacco. Nature New Biol. 224:29–30.

_____ 1975. Non–Mendelian streptomycin resistant tobacco mutant with altered chloroplasts and mitochondria. Nature 255:401–402.

Mattoo, A.K., Pick, U., Hoffman–Falk, H., and Edelman, M. 1981. The rapidly metabolized 32000–dalton polypeptide of the "proteinaceous shield" regulating photosystem II electron transport and mediating diuron herbicide resistance. Proc. Natl. Acad. Sci. 78:1572–1576.

Matzinger, D.F. and Wernsman, E.A. 1973. Extranuclear temperature–sensitive lethality in *Nicotiana tabacum* L. Proc. Nat. Acad. Sci. 70: 108–110.

Medgyesy, P., Menczel, L., and Maliga, P. 1980. The use of cytoplasmic streptomycin resistance: Chloroplast transfer from *Nicotiana tabacum* into *Nicotiana sylvestris* and isolation of their somatic hybrids. Molec. Gen. Genet. 179:693–698.

Melchers, G. 1980. The future. In: Perspectives in Plant Cell and Tissue Culture (I.K. Vasil, ed.) International Review Cytology Supplement 11B, pp. 241–253. Academic Press, New York.

_____, Sacristan, M.D., and Holder, A.A. 1978. Somatic hybrid plants of potato and tomato regenerated from fused protoplasts. Carlsberg Res. Commun. 43:203–218.

Menczel, L., Lazar, G., and Maliga, P. 1978. Isolation of somatic hybrids by cloning *Nicotiana* heterokaryons in nurse cultures. Planta 143:29–32.

Menczel, L., Galiba, G., Nagy, F., and Maliga, P. 1981a. Effect of radiation dosage on efficiency of chloroplast transfer by protoplast fusion in *Nicotiana*. Genetics 100:487–495.

Menczel, L., Nagy, G., Kiss, Zs.R., and Maliga, P. 1981b. Streptomycin resistant and sensitive somatic hybrids of *Nicotiana tabacum* + *Nicotiana knightiana*: Correlation of resistance to *N. tabacum* plastids. Theor. Appl. Genet. 59:191–195.

Mubumbila, M., Burkard, G., Keller, M., Steinmetz, A., Crouse, E., and Weil, J.H. 1980. Hybridization of bean, spinach, maize and *Euglena* chloroplast transfer RNAs with homologous and heterologous chloroplast DNAs. An approach to the study of homology between chloroplast tRNAs from various species. Biochem. Biophys. Acta 609:31–39.

Nagy, F., Torok, I., and Maliga, P. 1981. Extensive rearrangements in the mitochondrial DNA in somatic hybrids of *Nicotiana tabacum* and *Nicotiana knightiana*. Molec. Gen. Genet. 183:437–439.

Nessler, C.L. and Wernsman, E.A. 1980. Ultrastructural observations of extranuclear temperature–sensitive lethality in *Nicotiana tabacum* L. Bot. Gaz. 14:9–14.

Nessler, C.L., Long, R.C., and Wernsman, E.A. 1980. Physiological observations of extranuclear temperature-sensitive lethality in *Nicotiana tabacum* L. Z. Pflanzenphysiol. 99:27-35.

Palmer, J.D. and Thompson, W.F. 1981. Rearrangements in the chloroplast genomes of mung bean and Pea. Proc. Natl. Acad. Sci. 78: 5533-5537.

Pfister, K., Steinback, K.E., Gardner, G., and Arntzen, C.J. 1981. Photoaffinity labelling of an herbicide receptor protein in chloroplast membranes. Proc. Natl. Acad. Sci. 78:981-985.

Poulson, C., Porat, D., Sacristan, M.D., and Melchers, G. 1980. Peptide mapping of the ribulose bisphosphate carboxylase small subunit from the somatic hybrid of tomato and potato. Carlsberg Res. Commun. 45: 249-267.

Powling, A. 1981. Species of small DNA molecules found in mitochondria from sugar beet with normal and male-sterile cytoplasms. Molec. Gen. Genet. 183:82-84.

Pring, D.R. and Levings, C.S. III. 1978. Heterogeneity of maize cytoplasmic genomes among male sterile cytoplasms. Genetics 89:121-136.

Quetier, F. and Vedel, F. 1977. Heterogeneous population of mitochondrial DNA molecules in higher plants. Nature 268:365-368.

Raveh, D., Huberman, E., and Galun, E. 1973. In vitro culture of tobacco protoplasts: Use of feeder techniques to support division of cells plated at low densities. In Vitro 9:216-222.

Reisfeld, A., Gressel, J., Jakob, K.M., and Edelman, M. 1978. Characterization of the 32,000 dalton membrane protein I. Early synthesis during photoinduced plastid development of *Spirodela*. Photochem. Photobiol. 27:161-165.

Rhoades, M.M. 1955. Interaction of genetic and non-genic heredity units and the physiology of non-genic inheritance. In: Encyclopedia of Plant Physiology (W. Ruhland, ed.) Vol. 1, pp.19-57. Springer-Verlag, New York.

Ris, H. and Plaut, W. 1962. Ultrastructure of DNA-containing areas in the chloroplast of *Chlamydomonas*. J. Cell Biol. 13:383-391.

Rosner, A., Reisfeld, A., Jakob, K.M., Gressel, J., and Edelman, M. 1977. Shifts in the RNA and protein metabolism of *Spirodela* (Duckweed). In: Acids Nucleiques et Synthese de Proteines chez des Vegetaux (J. Weil and L. Bogorad, eds.) pp. 561-568. C.N.R.S., Paris.

Scowcroft, W.R. and Larkin, P.G. 1981. Chloroplast DNA assorts randomly in intraspecific somatic hybrids of *Nicotiana debneyi*. Theor. Appl. Genet. 60:179-184.

Sears, B.B. 1980. Elimination of plastids during spermatogenesis and fertilization in the plant kingdom. Plasmid 4:233-255.

Sellden, G. and Leech, R.M. 1981. Localization of DNA in mature and young wheat chloroplasts using the fluorescent probe 4'-6-diamidino-2-phenylindole. Plant Physiol. 68:731-734.

Seyer, P., Kowallik, K.V., and Herrmann, R.G. 1981. A physical map of *Nicotiana tabacum* plastid DNA including the location of structural genes for ribosomal RNAs and the large subunit of ribulose bisphosphate carboxylase/oxygenase. Curr. Gen. 3:189-204.

Sidorov, V.A., Menczel, L., Nagy, F., and Maliga, P. 1981. Chloroplast transfer in *Nicotiana* based on metabolic complementation between irradiated and iodoacetate treated protoplasts. Planta 152:341-345.

Sparks, R.B. and Dale, R.M.K. 1980. Characterization of ^{3}H labeled supercoiled mitochondrial DNA from tobacco suspension culture cells. Molec. Gen. Genet. 180:351-355.

Steele, J.A., Uchytil, T.F., Durbin, R.D., Bhutnagar, P., and Rich, D.E. 1976. Chloroplast coupling factor 1: A species specific receptor for tentoxin. Proc. Natl. Acad. Sci. 73:2245-2248.

Steinmetz, A., Mumbumbila, M., Keller, M., Burkard, G., Weil, J.-H., Driesel, A.J., Crouse, E.J., Gordon, K., Bohnert, H.-J., and Herrmann, R.G. 1980. Mapping of tRNA genes on the circular DNA molecule of *Spinacia oleracea* chloroplasts. In: Transfer RNA: Biological Aspects (D. Soll, J.N. Abelson, and P.R. Schimmel, eds.) pp. 281-286. Cold Spring Harbor, New York.

Syrenki, R.M., Levings, C.S. III, and Shah, D.M. 1978. Physicochemical characterization of mitochondrial DNA from soybean. Plant Physiol. 61: 460-464.

Tewari, K.K. 1979. Structure and replication of chloroplast DNA. In: Nucleic Acids in Plants I (T.C. Hall and J.W. Davis, eds.) pp. 41-108. CRC Press, Boca Raton, Florida.

Van Ee, J.H., Vos, Y.J., and Planta, R.J. 1980. Physical map of chloroplast DNA of *Spirodela oligorrhiza* analysis by the restriction endonucleases Pst I, Xho I and Sac I. Gene 12:191-200.

Vedel, F., Labacq, P., and Quetier, F. 1980. Cytoplasmic DNA variation and relationships in cereal genomes. Theor. Appl. Genet. 58: 219-224.

Walbot, V. and Coe, E.H. Jr. 1979. Nuclear gene *iojap* conditions a programmed change to ribosome-less plastids in *Zea mays*. Proc. Natl. Acad. Sci. 76:2760-2764.

Wallin, A., Glimelius, K., and Eriksson, T. 1978. Enucleation of plant protoplasts by cytochalasin B. Z. Pflanzenphysiol. 87:333-340.

Ward, B.L., Anderson, R.S., and Bendich, A.J. 1981. The mitochondrial genome is large and variable in a family of plants (Cucurbitaceae). Cell 25:793-803.

Westhoff, P., Nelson, N., Bunemann, H., and Herrmann, R.G. 1981. Location of genes for coupling factor subunits of the spinach plastid chloromosome. Curr. Gen. 4:109-120.

Wright, W.E. 1978. The isolation of heterokaryons and hybrids by a selective system using irreversible biochemical inhibitors. Exp. Cell. Res. 112:395-407.

York, D.W., Earle, E.D., and Gracen, V.E. 1980. Ultrastructural effects of *Helminthosporium maydis* race T toxin on isolated corn mitochondria and mitochondria within corn protoplasts. Can. J. Bot. 58: 1562-1570.

Yurina, N.P., Odintsova, M.S., and Maliga, P. 1978. An altered chloroplast ribosomal protein in a streptomycin resistant tobacco mutant. Theor. Gen. Genet. 52:125-128.

Zelcer, A., Aviv, D., and Galun, E. 1978. Interspecific transfer of cytoplasmic male sterility by fusion between protoplasts of normal *Nicotiana sylvestris* and X-ray irradiated protoplasts of male-sterile *N. tabacum*. Z. Pflanzenphysiol. 90:397-407.

CHAPTER 10
Isolation of Mutants from Cell Culture

C.E. Flick

Cell cultures offer many advantages for isolation of mutants in higher plants. Unlike the whole plant, a very large number of cells can be screened at one time for a desired trait. Because the cells are grown in a uniform cultural environment, reproducible selection schemes can be employed. The nature of the mutations can be more rigorously defined in a cell line for which stringent growth conditions can be imposed. Efficient mutagenesis of plant cell cultures is possible, as the cells can be uniformly treated with physical or chemical mutagens. The isolation of auxotrophic cell lines of higher plants has lagged behind the isolation of resistant mutants primarily because of the lack of stable haploid cell lines and the lack of selective systems. However, the isolation of auxotrophs by large scale screening of cell colonies descended from mutagenized protoplasts has been recently demonstrated (Sidorov et al., 1981).

Variant cell lines are usually isolated from protoplasts or more commonly from cell suspension cultures. Protoplasts are advantageous in that a population of single cells is available. Protoplasts, however, are more difficult to isolate and culture than cell suspension cultures. In some instances, e.g., *Zea mays*, plant regeneration is possible from suspension cultures but not from protoplasts. A cell suspension, however, may be filtered to remove large cell aggregates. Variant cell lines have been isolated by plating of cell suspension cultures on solid agar medium (e.g., Flick et al., 1981) and by direct selection in cell suspensions (e.g., Widholm, 1972a). Plating is advantageous in that cross-feeding of beneficial or toxic compounds is reduced. In addition, more than one variant colony can be retrieved from a single selective plate. Direct selection in cell suspension cultures can result in extensive cross feeding, as cells are bathed in liquid medium. Selection in liquid medium may produce a more heterogeneous cell population, because identification of individual colonies is not possible. Hence plating techniques have been used most frequently.

In almost all instances described mutants have been selected that are resistant to an antimetabolite. The most common selection has been for resistance to amino acid analogues and amino acids. Indeed, resistance to 13 different amino acid analogues and 4 amino acids has been reported. Other antimetabolites have included nucleic acid base analogues, antibiotics, herbicides, and phytotoxins. Most mechanisms of antimetabolite resistance arise through failure of the antimetabolite to interact with its target macromolecule. For example, when sensitivity to an herbicide or phytotoxin is due to the action of the compound at a single site, resistance can be selected using in vitro techniques. Simple positive selection has produced corn lines resistant to *Helminthosporium maydis* (Gengenbach and Green, 1975) and lines of *N. tabacum* resistant to the herbicide pichloram (Chaleff and Parsons, 1978a).

Mutant cell lines have been selected that have direct application in agriculture as well as those of more interest in basic research in plant genetics. The most common mutants isolated in vitro that are economically valuable are resistance to phytotoxins, (e.g., *Helminthosporium maydis*, Gengenbach and Green, 1975) herbicides (e.g., picloram resistance, Chaleff and Parsons, 1978a), NaCl tolerance, and chilling resistance. All but the last trait have been shown to be sexually transmitted. Many other mutants, while not of direct economic importance, may be useful as selective markers in somatic hybridization. The usefulness of amino acid analogue resistant cell lines for selection of somatic hybrids has been demonstrated in *N. sylvestris* (White and Vasil, 1979) and in *D. carota* (Harms et al., 1981).

In most instances positive selections will not lead to isolation of auxotrophic mutants, e.g., those mutants with a nutritional deficiency. Auxotrophic mutants are usually isolated by screening large numbers of cell lines regenerated from either protoplasts (e.g., isoleucine and uracil requiring lines, Sidorov et al., 1981) or single cells and small aggregates filtered out of suspension cultures (e.g., pantothenate and adenine requiring lines, Savage et al., 1979) for nutritional requirements. Carlson (1970) used bromodeoxyuridine (BudR) to enrich for nongrowing auxotrophic cells. Growing cells incorporate BudR into DNA. When transferred to light, photolysis of the BudR causes extensive DNA damage. Cells that are not growing do not incorporate BudR and are not light sensitive. Six cell lines were isolated with nutritional requirements. BudR was also used by Malmberg (1979) to enrich for temperature sensitive conditional lethal mutants of *N. tabacum*. Cells unable to metabolize nitrate (nitrate reductase deficient) have been isolated by selection for chlorate resistance (e.g., Muller and Grafe, 1978). Although these cell lines are technically auxotrophic, they are isolated by a straightforward positive selection for resistance to an antimetabolite, chlorate.

LITERATURE REVIEW

A great many variant cell lines (about 128) have been isolated through in vitro techniques. At present 51 phenotypes have been selected in 20 different species. Of these cell lines, 25 have been regenerated in 8 species. In only 13 instances has there been genetic analysis of mutant plants isolated from in vitro cultures. Of the genetic analyses, 11 involved *Nicotiana* species, most frequently *N. tabacum.* However, in more instances there has been some biochemical analysis of the variant cell lines. Many variant cell lines have not been analyzed at all. The current status of in vitro mutant isolation is reviewed here.

Amino Acid and Amino Acid Analogue Resistance

The majority of mutants isolated from in vitro cultures of higher plants have been selected for resistance to amino acid analogues or growth inhibitory naturally occurring amino acids. The reported amino acid analogue and amino acid resistant cell lines are summarized in Table 1. Most mutants have been isolated as resistant to either aromatic amino acid analogues, e.g., 5-methyltryptophan, p-fluorophenylalanine, and 6-fluorotryptophan or analogues of lysine and threonine, e.g., S-aminoethylcysteine, α-hydroxylysine, as well as lysine and threonine themselves.

Resistance to antimetabolites can arise through one or more of several mechanisms (see Fig. 1). First, overproduction of the naturally

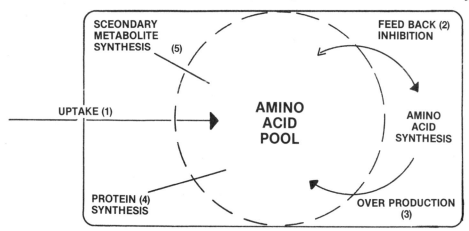

Figure 1. Mechanisms of resistance to amino acid analogs: (1) reduced uptake of analog, (2) decreased sensitivity of an amino acid biosynthetic enzyme to feedback inhibition leading to (3) overproduction of naturally occurring amino acid. (4) Excess of amino acid diluted effect of analog in protein synthesis or (5) led to increased synthesis of secondary metabolites may also be a method of degradation of amino acid analogs that are direct precursors of secondary metabolites.

Table 1. Amino Acid and Amino Acid Analog Resistant Cell Lines

SELECTIVE AGENT	SPECIES	MUTAGEN	PLANT REGEN.	RESISTANCE MECHANISM[a,b]	GENETICS[a]	REFERENCES
S-(aminoethyl)-L-cysteine	Arabidopsis thaliana	EMS	Yes	P,U	N.R.	Negrutiu et al., 1978
	D. carota	None	No	N.R.	N.R.	Widholm, 1978
	D. carota	None	No	A	N.R.	Matthews et al., 1980
	Hordeum vulgare	Azide	Yes	N.R.	Single recessive allele	Bright et al., 1980
	N. tabacum	EMS	No	C	N.R.	Widholm, 1976
	N. tabacum	UV or EMS	No	N.R.	N.R.	Widholm, 1977
	N. sylvestris	None	No	N.R.	N.R.	White & Vasil, 1979
	Oryza sativa	EMS	No	P	N.R.	Chaleff & Carlson, 1975
Azetidine-2-carboxylic acid	D. carota	None	No	P	N.R.	Nielsen et al., 1980
	D. carota	None	No	P,U	N.R.	Cella et al., 1982
Ethionine	D. carota	EMS	No	U	N.R.	Widholm, 1977
	D. carota	None	No	N.R.	N.R.	Widholm, 1978
	Medicago sativa	EMS	No	P	N.R.	Reish et al., 1981
p-Fluorophenyl-alanine	Acer pseudoplatanus	None	No	A,S	N.R.	Gathercole & Street, 1976

SELECTIVE AGENT	SPECIES	MUTAGEN	PLANT REGEN.	RESISTANCE MECHANISM[a,b]	GENETICS[a]	REFERENCES
	Acer pseudoplatanus	None	No	P,S	N.R.	Gathercole & Street, 1978
	D. carota	None	No	P	N.R.	Palmer & Widholm, 1975
	D. carota	None	No	N.R.	N.R.	Berlin & Widholm, 1977
	D. carota	None	No	N.R.	N.R.	Widholm, 1978
	D. innoxia	None	Yes	N.R.	N.R.	Evans & Gamborg, 1979
	N. tabacum	UV or EMS	No	N.R.	N.R.	Widholm, 1977
	N. tabacum	None	No	A	N.R.	Berlin & Vollmer, 1979
	N. tabacum	None	No	A	N.R.	Palmer & Widholm, 1975
	N. tabacum	None	No	S,U	N.R.	Berlin & Widholm, 1978b
	N. tabacum	None	No	A,S	N.R.	Berlin & Widholm, 1977
	N. tabacum	None	Yes	N.R.	N.R.	Flick et al., 1981
6-Fluorotryptophan	Petunia hybrida	NTG	No	N.R.	N.R.	Colijn et al., 1979
Glycine hydroxamate	N. tabacum	None	Yes	P	Single dominant allele	Lawyer et al., 1980
Δ-Hydroxylysine	N. tabacum	EMS	No	N.R.	N.R.	Widholm, 1977
	N. tabacum	UV or EMS	No	C	N.R.	Widholm, 1976

Table 1. Cont.

SELECTIVE AGENT	SPECIES	MUTAGEN	PLANT REGEN.	RESISTANCE MECHANISM[a,b]	GENETICS[a]	REFERENCES
Hydroxyproline	D. carota	EMS	No	U	N.R.	Widholm, 1976
	H. vulgare	Azide	Yes	P	Single semi-dominant allele	Kueh & Bright, 1981
Lysine plus threonine	Z. mays	Azide	Yes	P	Dominant nuclear allele	Hibberd & Green, 1982
	Z. mays	None	Yes	F,P	N.R.	Hibberd et al., 1980
Methione sulfoximine	N. tabacum	None	Yes	P	Single semi-dominant allele or two recessive alleles	Carlson, 1973
5-Methyl-tryptophan	D. carota	None	No	S	N.R.	Sung et al., 1979
	Catharanthus roseus	None	No	A	N.R.	Scott et al., 1979
	N. tabacum	UV or EMS	No	N.R.	N.R.	Widholm, 1977
	N. tabacum	None	No	F	N.R.	Widholm, 1972
	N. tabacum	None	Yes	N.R.	N.R.	Widholm, 1978

SELECTIVE AGENT	SPECIES	MUTAGEN	PLANT REGEN.	RESISTANCE MECHANISM[a,b]	GENETICS[a]	REFERENCES
5-Methyl-tryptophan	N. sylvestris	None	No	N.R.	N.R.	White & Vasil, 1979
	D. carota	None	No	F	N.R.	Widholm, 1974
	D. carota	None	No	F	N.R.	Widholm, 1978
	D. carota	None	No	F	N.R.	Widholm, 1972b
Seleno-amino acids	N. tabacum	None	No	P,U	N.R.	Flashman & Filner, 1978
Thienylalanine	N. sylvestris	None	No	N.R.	N.R.	Vunsh et al., 1980
Threonine	N. tabacum	None	Yes	U	N.R.	Heimer & Filner, 1970
Valine	N. tabacum (haploid)	UV	Yes	N.R.	Mendelian	Bourgin, 1978
	N. tabacum (haploid)	NTG or irradiation	No	N.R.	N.R.	Caboche & Muller, 1980

[a]N.R. = Not reported.

[b]Abbreviations for mechanisms: F = Enzyme resistant to feedback inhibition. S = Increase in synthesis of secondary metabolite. A = Increase in enzyme activity. P = Increased pools of natural compound. U = altered uptake. C = Cross resistant to other antimetabolites.

occuring compound may dilute the effect of the antimetabolite. Over-production of an amino acid, for example, is commonly a result of deregulation of enzymes normally subject to feedback inhibition. Wid-holm (1972a) has shown 5-methyltryptophan resistance in *N. tabacum* may arise through resistance of the first enzyme in the tryptophan pathway, anthranilate synthetase, to feedback inhibition by tryptophan or its analogue 5-methyltryptophan resulting in overproduction of tryp-tophan. Second, resistance may arise if the antimetabolite is effective-ly excluded from the cell, e.g., through a mutation effecting membrane permeability. Variant cell lines with altered transport have been iso-lated. Transport variants show resistance to more than one amino acid analogue (Berlin and Widholm, 1978a). Third, a compound may be detoxified through degradation. Berlin (1980) has shown that para-fluorophenylalanine (PFP) can be degraded by phenylalanine ammonia lyase. Such a mutation may lead to overproduction of phenolic com-pounds as was reported in some of Berlin's cell lines (Berlin, 1980). Fourth, increased synthesis of secondary metabolites can result in ana-logue resistant mutants. Increased activity of phenylalanine ammonia lyase (PAL, Berlin and Vollmer, 1979) in p-fluorophenylalanine resistant (PFP-res) cell lines can result in increased synthesis of phenolic com-pounds as in *N. tabacum* (Berlin and Widholm, 1977) and *Acer pseudo-platanus* (Gathercole and Street, 1976, 1978) or polyphenols as in *N. tabacum* (Berlin and Widholm, 1978b). Overproduction of a secondary metabolite can interfere with plant regeneration. A thirtyfold increase in IAA synthesis in 5-methyltryptophan resistant cell lines of *D. carota* was shown to inhibit embryogenesis (Sung, 1979).

In only a few instances has resistance to an amino acid analogue or amino acid been correlated with specific enzyme activities. In both *N. tabacum* (Widholm, 1972a) and *D. carota* (Widholm, 1974) resistance to 5-methyltryptophan has been correlated with a decrease in feedback inhibition of anthranilate synthetase by tryptophan. In *Catharanthus roseus* a 1.5-fold increase in anthranilate synthetase activity and a twofold increase in tryptophan synthetase activity were detected in lines with resistance to 5-methyltryptophan (Scott et al., 1979). In at least one variant cell line anthranilate synthetase was less sensitive than the wild type enzyme to feedback inhibition by tryptophan in cal-lus initiated from regenerated plants. However, anthranilate synthetase isolated from the whole plant was as sensitive to tryptophan as the wild type enzyme (Widholm, 1974). This emphasizes the developmental problem that cell cultures and the plants regenerated from cells may express different isoenzymes and hence respond differently to antimeta-bolites. Additional enzyme activities have been found to be altered in amino acid analogue resistant cell lines. For example, S-aminoethylcys-teine resistant cell lines of *D. carota* have altered aspartokinase and dihydropicolinic acid synthase activity (Matthews et al., 1980) and aspartokinase activity in lysine plus threonine resistant cell lines of *Zea mays* is resistant to feedback inhibition by lysine (Hibberd et al., 1980).

In many instances increases of amino acid pools have been reported in analogue or amino acid resistant cell lines (see Table 1). These increases may have agricultural value. Hibberd et al. (1980) used lysine plus threonine to select for high lysine lines of *Zea mays*, a crop usually deficient in lysine.   The methionine sulfoximine resistant mutants in *N. tabacum* isolated by Carlson (1973) overproduced methionine and had increased resistance to *Pseudomonas tabaci*, the causitive agent of wildfire disease.   Increased levels of proline in proline analogue resistant mutants (e.g., hydroxyproline resistant *Hordeum vulgare* plants isolated by Keuh and Bright, 1981) may lead to plants with increased resistance to water stress.   Although proline has been implicated in attracting locusts to drought–stressed, nitrogen–enriched plants, proline accumulating plants show no increased pest susceptibility (Bright et al., 1982).   Hence through use of amino acid analogues and amino acids normally toxic to plant cells it may be possible to select for economically important characteristics in higher plants which result from resistance to naturally occurring amino acids and their analogues.

Although more 5–methyltryptophan– and p–fluorophenylalanine–resistant cell lines have been isolated than lines resistant to any other analogue, in no instance has any genetic analysis of these variants been described.   Indeed, in only three instances has any of these types of variants been regenerated into plants.   While plants regenerated from a 5–methyltryptophan–resistant cell lines were sensitive to the analogue (Widholm, 1974), cell lines resistant to PFP in *D. innoxia* (Evans and Gamborg, 1979) and *N. tabacum* (Flick et al., 1981) regenerated plants in the presence of PFP.

A limited amount of genetic analysis of amino acid– or analogue–resistant mutants isolated in vitro has been reported.   Glycine hydroxamate resistance was shown to behave as a dominant Mendelian trait in *N. tabacum* (Lawyer et al., 1980).   Resistance to isonicotinic acid hydrazide (an inhibitor of the conversion of glycine to serine) in *N. tabacum* was also a dominant mutation (Berlyn, 1980).   Valine resistance in *N. tabacum* is inherited as either a single dominant or semidominant nuclear allele (Bourgin, 1978).   In *Zea mays* resistance to lysine plus threonine is attributed to a single dominant nuclear allele (Hibberd and Green, 1982).   Of the methionine sulfoximine–resistant cell lines isolated by Carlson, two are semidominant nuclear mutations and one is due to two recessive nuclear alleles (Carlson et al., 1973).   As one would expect for mutants isolated by direct positive selection, most of these resistance mutations are determined by dominant or semidominant alleles.

Whereas most amino acid– and analogue–resistant cell lines have been isolated from cell suspension cultures, callus, or protoplasts, a novel technique for mutant isolation has been devised in *Hordeum vulgare* (barley) by Bright et al. (1979).   Seeds were mutagenized with azide, planted, and grown to maturity.   Embryos were excised from harvested seeds and cultured on hormone free Murashige and Skoog medium (MS, Murashige and Skoog, 1962).   At this stage selection may be made for resistance to antimetabolites.   Plants resistant to S–aminoethylcysteine (Bright et al., 1979), lysine plus threonine (Bright et al., 1979), and hydroxyproline (Kueh and Bright, 1981) have been isolated.   This screening system for variants requires plant growth; hence genetic analysis of variants is possible.   Recessive (S–aminoethylcysteine resistant, Bright et al., 1979), semidominant (hydroxyproline resistant, Keuh and Bright, 1981), and dominant (lysine plus threonine resistant, Bright et al., 1979) mutations have all been identified. In addition, hydroxy-proline-resistant plants overproduce methionine, lysine, and threonine (Bright et al., 1979). Thus through a combination of conventional and in vitro culture techniques mutants with potential for crop improvement have been isolated.

Purine and Pyrimidine Analogues

Resistance to a number of purine and pyrimidine analogues has been reported in higher plants. Mutations of this nature have been extensively studied and used to identify somatic hybrids in animal cells, and thus their potential for use in higher plants has been explored. Biochemical or genetic analysis of purine- and pyrimidine analogue resistant mutants has been accomplished in a very limited number of cases.

Resistance to bromodeoxyuridine (BUdR, a thymidine analogue) has been most extensively studied. In *Glycine max* (soybean) BUdR resistance is a result of overproduction of deoxythymidilate monophosphate (Ohyama, 1974, 1976). This BUdR-resistant mutant is also resistant to fluorodeoxyuridine as well as aminopterin. As soybean does not regenerate from callus cultures, no genetic analysis was possible. However, in *N. tabacum* resistance to BUdR has been determined to be inherited as a dominant nuclear gene (Marton and Maliga, 1975), but no biochemical analysis has been reported. This cell line, however, was found to not be cytokinin habituated (Kandra and Maliga, 1977). BUdR-resistant cell lines have also been isolated in *Medicago sativa* (alfalfa) (Lo Schiavo et al., 1980). In some of these mutants thymidine, but not BUdR, could be incorporated into DNA. BUdR-resistant mutants were tested for growth in HAT (hypoxanthine, aminopterin, thymidine) medium, as described for animal cells (Lo Schiavo et al., 1980). Aminopterin inhibits folate reductase and hence inhibits purine and pyrimidine synthesis. Only cells that can incorporate exogenous hypoxanthine and thymidine into DNA (e.g., wild type cells) survive in HAT medium. Cells resistant to purine or pyrimidine analogues may not be able to incorporate exogenous hypoxanthine and thymidine and hence die in HAT medium. Five of seven BUdR-resistant cell lines and two of four azaguanine-resistant cell lines of *M. sativa* fail to grow in HAT medium. The HAT system is used extensively in animal cell culture to select for somatic hybrids based on complementation of analogue-resistant cell lines, i.e., only hybrids are able to incorporate both exogenous hypoxanthine and thymidine into DNA and survive. Based on results obtained in *M. sativa*, HAT selection may be useful in protoplast fusion experiments with higher plant cells.

Variants have been selected that are resistant to numerous other purine and pyrimidine analogues (Table 2). Cell lines of *Haplopappus gracilis* resistant to 8-azaguanine show a 30% reduction in hypoxanthine phosphoribosyltransferase (HPRTase) activity (Horsch and Jones, 1978), whereas 8-azaguanine-resistant mutants of *Acer pseudoplatanus* have a 50% reduction in HPRTase activity. Aminopterin resistance in two cell lines of *Zea mays* is due to increased dihydrofolate reductase activity,

Table 2. Purine and Pyrimidine Analog-Resistant Cell Lines

SELECTIVE AGENT	SPECIES	MUTAGEN	PLANT REGEN.	RESISTANCE MECHANISM[a,b]	GENETICS[a]	REFERENCES
Aminopterin	D. innoxia (haploid)	None	Yes	N.R.	N.R.	Mastrangolo & Smith, 1977
8-Azaguanine	A. pseudoplatanus	NTG	No	A	N.R.	Bright & Northcote, 1975
	Glycine max	EMS	No	N.R.	N.R.	Weber & Lark, 1979
	Haplopappus gracilis	EMS	No	A	N.R.	Horsch & Jones, 1978
	Medicago sativa	EMS	No	N.R.	N.R.	Loschiavo et al., 1980
6-Azauracil	N. tabacum	None	No	N.R.	N.R.	Lescure, 1973
	H. gracilis	None	No	C	N.R.	Jones & Hann, 1979
	Z. mays	EMS or NTG	No	F,A	N.R.	Shimamoto & Nelson, 1981
Bromodeoxy-uridine	Glycine max	NTG	No	P	N.R.	Ohyama, 1974, 1976
	M. sativa	EMS	No	N.R.	N.R.	Loschiavo et al., 1980
	N. tabacum (haploid)	None	Yes	N.R.	N.R.	Maisuryan et al., 1981
	N. tabacum	None	Yes	N.R.	Single dominant nuclear allele	Marton & Maliga, 1975

403

Table 2. Cont.

SELECTIVE AGENT	SPECIES	MUTAGEN	PLANT REGEN.	RESISTANCE MECHANISM[a,b]	GENETICS[a]	REFERENCES
5-Fluorouracil	*D. carota*	None	Yes	N.R.	N.R.	Sung & Jacques, 1980
Hydroxyurea	*N. tabacum*	None	Yes	N.R.	Single dominant nuclear allele	Chaleff & Keil, 1981; Chaleff, 1981
Thioguanine	*G. max*	EMS	No	N.R.	N.R.	Weber & Lark, 1979

[a]N.R. = Not reported.

[b]Abbreviations for mechanisms: F = Enzyme resistant to feedback inhibition. A = Increase in enzyme activity. P = Increased pools of natural compound. C = Cross resistant to other antimetabolites.

whereas in a third cell line increased resistance of the enzyme to aminopterin is observed (Shimamoto and Nelson, 1981). Only one mutant other than the BUdR-resistant line described above (Marton and Maliga, 1975) has been genetically analyzed. Chaleff (1981) selected cell lines of N. *tabacum* resistant to hydroxyurea which inhibits ribonucleotide reductase in animal cells and microorganisms. These mutants behaved as dominant nuclear alleles (Chaleff, 1981). In at least one instance an allele for hydroxyurea resistance was found to be tightly linked to a previously isolated allele for picloram resistance (Chaleff and Keil, 1981). Indeed, a large proportion of Chaleff's picloram-resistant mutants are also hydroxyurea resistant, indicating either a single gene involved in both resistances or enhancement of picloram resistance by hydroxyurea resistance (Chaleff and Keil, 1981). Picloram is probably not a mutagen, as it does not increase reversion of mutations in *Saccharomyces cerevisiae* (Chaleff and Keil, 1981). In addition, the recovery of hydroxyurea resistance in so many picloram-resistant mutants indicates that there must have been some selective pressure (Chaleff and Keil, 1981). This is the only reported instance of seemingly unrelated mutant genotypes arising simultaneously in a selected cell line of higher plants.

Antibiotic Resistance

Antibiotic-resistant cell lines of higher plants have been more exten-sively analyzed than any other mutants isolated in vitro in higher plants, especially those cell lines of N. *tabacum* isolated and character-ized by Maliga et al. (1973, 1977). Many antibiotics, e.g., streptomy-cin, kanamycin, and chloramphenicol, capitalize on differences in the physical structure of the translational machinery in prokaryotes and eukaryotes. Streptomycin and kanamycin, for example, are active on the prokaryotic 70S ribosome. The 70S ribosomes are only found in the mitochondria or chloroplasts in eukaryotes. Hence resistance to these compounds in eukaryotes may be inherited as a cytoplasmic trait. The isolation of streptomycin-resistant cell lines in N. *tabacum* was first reported by Maliga et al. (1973). As streptomycin causes bleaching of shoots regenerating from callus, resistant cell lines were isolated as green areas in regenerating callus. The streptomycin-resistant mutant SR1 has been analyzed genetically and biochemically. As expected, SR1 is inherited as a cytoplasmic gene, i.e., maternal inheritance (Maliga et al., 1975). Whereas streptomycin causes physical abnormali-ties in chloroplasts and mitochondria in sensitive cells, electron micro-graphs showed no effects of streptomycin on these organelles in SR1 (Maliga et al., 1975). Streptomycin is, however, taken up by these cells (Maliga et al., 1975). Two-dimensional electrophoresis of chloro-plast ribosomal proteins in SR1 and streptomycin-sensitive cells indi-cates an alteration in the electrophoretic mobility of one ribosomal protein in SR1 relative to the proteins from sensitive cells (Yurina et al., 1978). The cytoplasmic basis of this mutation was further demon-strated by the correlation of streptomycin resistance with chloroplast transfer from N. *tabacum* SR1 to N. *knightiana* via protoplast fusion

(Menczel et al., 1981).   In addition, the transfer of cytoplasmic streptomycin resistance from N. tabacum to N. sylvestris via protoplast fusion without creating a nuclear hybrid was demonstrated (Medgyesy et al., 1980).   Thus in the instance of SR1 a single gene mutation has been linked with changes in a single protein in the chloroplast.   A similar streptomycin-resistant mutant was isolated by Umiel et al. (1978).   This mutation is cytoplasmically inherited, and as SR1, plastids in the streptomycin-resistant cell line are normal in the presence of streptomycin, whereas plastids in sensitive cells are abnormal.

Resistance to streptomycin may also be inherited as a Mendelian trait.   Streptomycin resistance in a cell line isolated from haploid N. sylvestris, SR 180, is inherited as a recessive nuclear allele (Maliga, 1981).   A second streptomycin-resistant cell line of N. sylvestris (SR155) is sterile but not aneuploid.   Analysis of the karyotype of this cell line indicates a possible translocation (Maliga et al., 1979).

Cross resistance between antibiotics is also possible.   A kanamycin-resistant cell line of N. sylvestris (KR103) is resistant to streptomycin as well as neomycin (Dix, 1981b).   Maliga's SR1 cell line, however, is sensitive to kanamycin.   Two kanamycin-resistant cell lines of N. tabacum isolated by Owens (1981) are more resistant to streptomycin than to kanamycin.   These plants are partially sterile, and erratic segregation data makes genetic analysis difficult.

Several attempts have been made to isolate cycloheximide-resistant mutants in higher plants.   Resistance is lost in N. tabacum when callus is subcultured in the absence of cycloheximide (Maliga et al., 1976).   Gresshoff (1979) observed a similar instability of cycloheximide resistance in D. carota.   Sung et al. (1981) have shown that wild type embryos and plantlets of D. carota can detoxify cycloheximide.   However, a stable cycloheximide-resistant callus culture (WCH105) was isolated.   Plantlets regenerated from WCH105 are green, but they do not produce normal dissected leaves (Lazar et al., 1981).   This cycloheximide-resistant cell line was analyzed by somatic hybridization.   Although the chromosome number in somatic hybrids is as expected (2n = 36) callus initiated from somatic hybrids between cycloheximide sensitive and resistant cell lines is sensitive, indicating that cycloheximide resistance behaves as a recessive character (Lazar et al., 1981).   Reports of resistance to several other antibiotics, e.g., nystasin, amphotericin B, and chloramphenicol, are listed in Table 3.

Nitrogen Metabolism

Nitrogen metabolism involves several steps in higher plants.   (1) Nitrate must be taken into the cell, presumably mediated by a permease enzyme.   (2) Nitrate is reduced to nitrite via nitrate reductase.   (3) Nitrite reductase mediates the conversion of nitrite to ammonium.   (4) Finally, ammonium is combined with glutamate to yield glutamine, catalyzed by glutamine synthetase.   Mutants have been isolated in vitro that affect the first two steps.

When cell cultures of N. tabacum are grown on nitrate, certain amino acids or combinations of amino acids inhibit cell growth by

Table 3. Antibiotic Resistance

SELECTIVE AGENT	SPECIES	MUTAGEN	PLANT REGEN.	RESISTANCE MECHANISM[a,b]	GENETICS[a]	REFERENCES
Amphotericin B	N. tabacum	EMS	No	O	N.R.	Chiu et al., 1980
Chloramphenicol	N. sylvestris (haploid)	None	Yes[c]	N.R.	N.R.	Dix, 1980a
Colchicine	A. pseudoplatanus	None	No	C	N.R.[d]	Zryd, 1979
Cycloheximide	D. carota	None	Yes	D	N.R.[d]	Gresshoff, 1979
	D. carota	None	Yes	D	N.R.	Sung et al., 1981; Lazar et al., 1981
	N. tabacum	EMS	Yes	D	N.R.[d]	Maliga et al., 1976
Kanamycin	N. tabacum	None	Yes	C	Inconclusive[e]	Owens, 1981
	N. sylvestris (haploid)	None	Yes	R	N.R.	Dix et al., 1977
Nystasin	N. tabacum	EMS	No	O	N.R.	Chiu et al., 1980
Streptomycin	N. tabacum (haploid)	None	Yes	R	Single cytoplasmic allele	Maliga et al., 1973; Maliga et al., 1975; Yurina et al., 1978
	N. sylvestris (haploid)	None	Yes	N.R.	Single recessive nuclear allele	Maliga, 1981

Table 3. Cont.

SELECTIVE AGENT	SPECIES	MUTAGEN	PLANT REGEN.	RESISTANCE MECHANISM[a,b]	GENETICS[a]	REFERENCES
Streptomycin	N. sylvestris (haploid)	None	Yes	N.R.	Possible trans- location	Maliga et al., 1979
	N. tabacum	None	Yes	P	Single cyto- plasmic allele	Umiel & Goldner, 1976; Umiel, 1979

[a] Abbreviations for resistance mechanism: D = Detoxification. C = Cross resistant to streptomycin. O = Over-production natural compound. R = Alteration of chloroplast ribosomal protein. P = Normal plastid development.

[b] N.R. = Not reported.

[c] Possible chimera.

[d] Mutant not stable.

[e] Erratic segregation ratios implying undefined nuclear genetic control.

inhibiting the uptake of nitrate. Heimer and Filner (1970) isolated a cell line of *N. tabacum* (XDR-thr) that is resistant to threonine inhibition when grown on nitrate. The parent cell line is sensitive to threonine unless grown on nitrate-free culture medium, e.g., urea. The cell line is resistant to amino acids other than threonine that normally inhibit cell growth in nitrate medium, e.g., glycine, histidine, leucine, methionine, and valine (Heimer and Filner, 1970). Induction of nitrate reductase and nitrite reductase is inhibited by casein hydrolysate in XD or XDR-thr cells, indicating that the mutation does not affect either of these enzymes (Heimer and Filner, 1970). Since uptake of nitrate is not inhibited by threonine in the XDR-thr cell line, an alteration in the regulation of nitrate uptake must be the effect of the XDR-thr mutation. Unfortunately, no genetic analysis of the cell line has been reported.

Nitrate reductase has been extensively studied in *N. tabacum* through the use of mutants isolated in vitro (see Fig. 2). Nitrate reductase is a multisubunit enzyme that catalyzes the NADH–dependent reduction of nitrate to nitrite. In addition, $FADH_2$ or reduced benzyl viologen dye can be used as electron donors. These activities all depend on the presence of the molybdenum cofactor. In the absence of the cofactor the enzyme only has cytochrome c reductase activity.

In the absence of nitrate reductase activity chlorate is not converted to toxic chlorite; hence nitrate reductase deficient cell lines are chlorate resistant. Chlorate resistance has been used to select nitrate reductase deficient cell lines from haploid *N. tabacum* (Mendel and Muller, 1976; Muller and Grafe, 1978). Of nine original chlorate-resistant cell lines, one class, designated cnx, is also deficient in xanthine dehydrogenase. As nitrate reductase and xanthine dehydro-

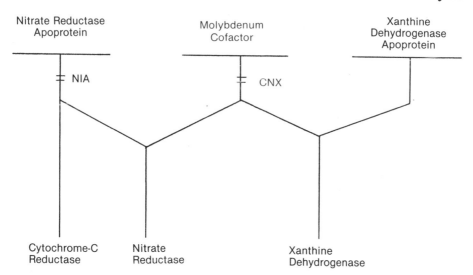

Figure 2. Genetic basis of nitrate reductase activity: (1) mutations in the nia allele inactivate the nitrate reductase apoprotein. (2) Mutations in the cnx allele inactivate the molybdenum cofactor necessary for both nitrate reductase and xanthine dehydrogenase activities.

genase use the same molybdenum cofactor, the cnx allele is believed to affect the molybdenum cofactor, but not the nitrate reductase apoprotein. Indeed, cnx lines of N. tabacum cannot complement a molybdenum cofactor deficient Neurospora crassa mutant (nit-1) in vitro to restore nitrate reductase activity, although wild type N. tabacum does complement the nit-1 mutant of Neurospora (Mendel et al., 1981). Partial nitrate reductase activity has been restored to cnx lines by addition of 1-10 mM molybdenum (Mendel et al., 1981). The cnx cell lines do have cytochrome c reductase activity, the molybdenum cofactor independent activity of nitrate reductase.

Other nitrate reductase mutants, designated nia, lack nitrate and cytochrome c reductase activities, but they do have xanthine dehydrogenase activity, indicating that the mutation affects the apoprotein (Mendel and Muller, 1979). In contrast to cnx cell lines of N. tabacum, crude extracts from nia cell lines can complement the N. crassa nit-1 mutant in vitro to restore nitrate reductase activity (Mendel et al., 1981). Further evidence has substantiated the premise that nia and cnx affect different subunits of nitrate reductase. For example, cnx and nia lines complement in vitro in crude extracts to restore nitrate reductase activity (Mendel and Muller, 1978), and hybrids produced by protoplast fusion between cnx and nia cell lines grow on nitrate. In vivo complementation implies that both cnx and nia mutations are recessive. Although most cnx and nia cell lines do not regenerate, shoots were produced from cnx + nia hybrids. Only leaky nia mutants (with reduced levels of $FADH_2$ and viologen dye dependent nitrate reductase activity but no NADH dependent activity) have regenerated shoots (Muller and Grafe, 1978).

Nitrate reductase deficient cell lines have been isolated in vitro in two additional species. King and Khanna (1980) selected nitrate reductase mutants as chlorate resistant in haploid Datura innoxia. The cell lines were shown to take up nitrate, but were deficient in nitrate reductase. Murphy and Imbrie (1981) selected chlorate-resistant strains of Rosa damascena. Of these cell lines, 15% fail to grow on nitrate, indicating a possible nitrate reductase deficiency. The remainder of these cell lines appear to have gained chlorate resistance through other means. Strauss et al. (1981) while screening for auxotrophs of haploid Hyoscyamus muticus discovered cell lines that depended on casein hydrolysate for growth. These mutants were subsequently discovered to be chlorate resistant and to lack in vivo nitrate reductase activity (Strauss et al., 1981).

Agriculturally Useful Phenotypes

Various phenotypes have been selected in cell cultures of higher plants that may have direct application in agriculture. Some of these phenotypes have been covered elsewhere in this chapter. For example, methionine sulfoximine-resistant cell lines of N. tabacum show increased resistance to wildfire disease (Pseudomonas tabaci), and lysine plus threonine resistant lines of Zea mays are potentially valuable in breeding high lysine lines of maize. Most agriculturally useful phenotypes have been selected for resistance to herbicides, pathotoxins and/or diseases, NaCl tolerance, or cold tolerance. Herbicide resistance is discussed elsewhere in this volume (see Chapter 12). Nabors (1976) reviewed the use of mutants to obtain agriculturally useful plants.

Selection for pathotoxin resistance in vitro promises immediate results for isolation of agriculturally valuable plants. Behnke (1979) isolated *Phytophthora infestans* toxin-resistant cell lines of *Solanum tuberosum* (potato). Regenerated plants as well as callus initiated from these plants were resistant to the toxin. Subsequently Behnke (1980a) showed that toxin-resistant plants are 25% more resistant to the late blight caused by *P. infestans* than parental lines. A genetic basis for this resistance has not been established, as *S. tuberosum* is commercially asexually propagated.

Plants have been regenerated from *H. maydis* race T toxin-resistant cell lines of *Zea mays*. Most of these plants (52/65) that retain resistance to the toxin have also reverted to male fertility. The remaining 13 plants are useful male steriles, but not of the cms-T type as the parent line. Thus resistance to the pathotoxin has only occurred through modification of the cms-T genes. Genetic analysis of these toxin-resistant plants indicates that pathotoxin resistance and male fertility are both cytoplasmically inherited.

Numerous instances of in vitro selection of NaCl-tolerant cell lines have been reported (see Table 4). In several of these reports plants have been regenerated. Sexual transmission of the NaCl-tolerance was reported in *N. tabacum* (Nabors et al., 1980), but segregation frequencies were not as expected. Regenerated plants and callus initiated from regenerated plants of NaCl-tolerant cell lines of *D. innoxia* were resistant to NaCl (Tyagi et al., 1981). In addition, plants of *Kickxia ramosissima* are resistant to NaCl up to 120mM (Mathur et al., 1980). As shown in Table 4, NaCl-tolerant cell lines have also been isolated in *N. sylvestris*, *Capsicum annuum* (Dix and Street, 1975; Dix and Pearce, 1981), *Medicago sativa* and *Oryza sativa* (Rains et al., 1979).

Cell lines with increased cold tolerance have been isolated in *N. sylvestris* (Dix and Street, 1975; Dix, 1977) and *D. carota* (Templeton-Somers et al., 1981). Although plants were regenerated from three lines of *N. sylvestris* with a low level of cold tolerance (Dix, 1977), the phenotype was not transmitted sexually. Templeton-Somers et al. (1981) reported that although cold-tolerant callus of *D. carota* was isolated, this trait was not expressed in embryos. This instance emphasizes the problem that not all traits selected in unorganized callus cultures will be expressed in the organized differentiated plant.

Auxotrophs

Auxotrophs have only rarely been isolated in vitro in higher plants. Polyploidy and aneuploidy in culture have made the isolation of recessive mutants almost impossible. In microorganisms and animal cells auxotrophs have helped elucidate biochemical pathways and are invaluable for selection of somatic hybrids by complementation as well as for transformation of cells with naked or liposome encapsulated DNA. Un-

Table 4. Agriculturally Useful Mutants

SELECTIVE AGENT	SPECIES	MUTAGEN	PLANT REGEN.	RESISTANCE MECHANISM[a,b]	GENETICS[a]	REFERENCES
Cold Tolerance						
	D. carota	EMS	Yes[e]	N.R.	N.R.	Templeton-Somers et al., 1981
	N. sylvestris	None	Yes	N.R.	E	Dix & Street, 1975; Dix, 1977
Herbicide Resistance						
Asulam	Apium graveolens	None	No	N.R.	N.R.	Merrick & Collin, 1981
Bentazone	N. tabacum (haploid)	γ-ray	Yes[c]	N.R.	Recessive nuclear allele	Radin & Carlson, 1978
2,4-D	Lotus corniculatus	None	Yes	A	N.R.	Swanson & Tomes, 1980
2,4-D; 2,4,5-T; 2,4-DB	Trifolium repens	None	No	N.R.	N.R.	Oswald et al., 1977
Isopropyl-N-phenylcarbamate	N. tabacum	EMS	Yes[c,d]	N.R.	N.R.	Aviv & Galun, 1977
Paraquat	N. tabacum	X-ray	Yes[c]	N.R.	N.R.	Miller & Hughes, 1980
Phenmedifarm	N. tabacum (haploid)	γ-ray	Yes[c]	N.R.	Recessive nuclear allele	Radin & Carlson, 1978

SELECTIVE AGENT	SPECIES	MUTAGEN	PLANT REGEN.	RESISTANCE MECHANISM[a,b]	GENETICS[a]	REFERENCES
Herbicide Resistance						
Picloram	*N. tabacum*	None	Yes[c]	N.R.	Dominant nuclear allele; 3 linkage groups	Chaleff & Parsons, 1978; Chaleff, 1980
Pathotoxin Resistance						
Fusarium oxysporum	*Solanum tuberosum*	None	Yes	N.R.	N.R.	Behnke, 1980b
Helminthosporium maydis	*Zea mays* T-cms	None[c]	Yes	C	Cytoplasmic allele	Gengenbach & Green, 1975; Gengenbach et al., 1977
Phytophthora infestans	*Solanum tuberosum*	None[c]	Yes	N.R.	N.R.	Behnke, 1979, 1980a,b
Salt Tolerance						
	Capsicum annum	None	No	N.R.	N.R.	Dix & Street, 1975
	D. innoxia	None	Yes	N.R.	N.R.	Tyagi et al., 1981
	Kickxia ramosissima	None	Yes	N.R.	N.R.	Mathur et al., 1980
	M. sativa	None	No	N.R.	N.R.	Croughan et al., 1978

Table 4. Cont.

SELECTIVE AGENT	SPECIES	MUTAGEN	PLANT REGEN.	RESISTANCE MECHANISM[a,b]	GENETICS[a]	REFERENCES
Salt Tolerance (Cont.)						
	O. sativa	None	No	N.R.	N.R.	Rains et al., 1979
	N. sylvestris	None	No	N.R.	N.R.	Dix & Pearce, 1981
	N. sylvestris	None	No	N.R.	N.R.	Nabors et al., 1980
	N. tabacum	EMS	Yes	N.R.	D	Templeton–Somers et al., 1981

[a] N.R. = Not reported.

[b] Abbreviations for resistance mechanisms: A = Alteration in 2,4-D conjugation. B = Regeneration in presence of selective agent. C = Reversion to male sterility. D = Transmitted sexually, but not at expected segregation ratios. E = Not transmitted sexually.

[c] Resistant plants.

[d] Sterile plants.

[e] Not expressed in plants.

til a collection of auxotrophic mutants is available in higher plants, these areas of research will lag far behind animal cell culture.

In contrast to other groups of mutants isolated in vitro, two features of the isolation procedures are always constant. First, haploids have been used in every instance. Most auxotrophic mutants are recessive and would not be detected in organisms of diploid or higher ploidy. As plant cell cultures are frequently unstable in chromosome number, the requirement for haploid cells is limiting, i.e., cell cultures initiated from haploid plant material may attain higher ploidies after brief periods of time in cell culture. Second, in all but one instance either a physical, e.g., gamma irradiation, or chemical, e.g., EMS or NG, mutagen has been used. In positive selection for resistance to antimetabolites the use of mutagens is unnecessary, as mutants can be isolated with or without mutagens. Only temperature–sensitive (ts) mutants isolated in *N. tabacum* were obtained without a mutagen (Malmberg, 1979). However, a BUdR enrichment was used, and BUdR itself may be mutagenic. Third, all mutants have been isolated from species of the Solanaceae. This is probably due to the amenability of solanaceous species to all aspects of tissue culture from anther to protoplast culture.

The first auxotrophs were reported in haploid *N. tabacum* by Carlson (1970). As *N. tabacum* is an amphidiploid, it is not surprising that these mutants are leaky. A positive selection scheme designed by Puck and Kao (1967) for mammalian cell cultures and previously used successfully with ferns by Carlson (1969) was used to select auxotrophs. After mutagenesis, cells were cultured in minimal medium so that cells with nutritional deficiencies would not grow. Bromodeoxyuridine, a thymidine analogue, was added, and the cultures were placed in the dark. BUdR is only incorporated into the DNA of actively growing cells. Because BUdR makes DNA light sensitive, when the culture is transferred to light photolysis of BUdR causes extensive damage to the DNA of cells that grew in minimal medium. Auxotrophic cells do not incorporate BUdR, and hence are not light sensitive. Surviving cells were then transferred to enriched culture medium and the mutant cells proliferated. With this enrichment scheme, Carlson isolated six auxotrophs that require biotin, paraaminobenzoic acid, arginine, hypoxanthine, lysine, and proline (Carlson, 1970). Plants were successfully regenerated from all but the proline auxotroph. Genetic analysis of these mutants showed that the biotin, paraaminobenzoic acid, and arginine auxotrophs resulted from changes in single recessive nuclear alleles, i.e., 3:1 segregation of dominants and recessives in the F_1 generation (Chaleff, 1981). The hypoxanthine auxotroph was more complex in that it resulted from mutations in two recessive nuclear alleles, i.e., a segregation ratio of 9:6:1 (dominant:heterozygous:recessive) (Chaleff, 1981). Although plants were regenerated from a lysine auxotroph and plant growth was shown to be improved by addition of lysine (Carlson et al., 1973), no genetic analysis has been reported. The sixth auxotroph, a proline auxotroph, was apparently never regenerated or analyzed.

An identical BUdR selection was used by Malmberg (1979, 1980) to isolate temperature–sensitive (ts) mutants of haploid *N. tabacum*. One such mutant (ts 4) has been biochemically analyzed.   The mutation is pleiotropic and decreases activities of ornithine decarboxylase, S-adenosylmethionine decarboxylase, and nitrate reductase, but it increases activity of phenylalanine ammonia lyase.   The ts 4 mutant also has a decreased level of chlorophyll.   A second site revertant of ts 4 (Rt 1) was isolated and has normal levels of ornithine decarboxylase and chlorophyll (Malmberg, 1980).

Two auxotrophs were isolated from a stable haploid cell suspension culture of *Datura innoxia*.   No selection scheme was used.   A large collection of cells were screened for nutritional requirements.   One mutant requires pantothenate (Savage et al., 1979), whereas the other requires adenine (King et al., 1980).   The pantothenate-requiring auxotroph failed to regenerate plants under conditions in which wild-type callus regenerates (Savage et al., 1979).   Neither auxotroph can be cross fed required nutrients by wild-type cells, an essential factor for designing new selection schemes.   No genetic or biochemical analysis of these cell lines has been reported.

Several auxotrophs have been isolated from haploid protoplasts of *Hyoscyamus muticus* (Gebhardt et al., 1981).   As with *Datura innoxia*, there was no enrichment method for auxotrophs, but large numbers of calli were analyzed for nutritional requirements.   Histidine, nicotinamide, and an undefined temperature-sensitive auxotroph have been regenerated into plants, but not yet analyzed.   A tryptophan auxotroph did not regenerate.   A nitrate reductase mutant, dependent on casein hydrolysate for growth, was isolated while screening for auxotrophs (see discussion above).

Two auxotrophs have been isolated from gamma-irradiated protoplasts isolated from leaf mesophyll of haploid *N. plumbaginifolia* (Sidorov et al., 1981).   As with *D. innoxia* and *H. muticus*, there was no method of enrichment.   One cell line is deficient in threonine deaminase, the first enzyme in the isoleucine pathway of biosynthesis.   Plants have been regenerated from the isoleucine auxotroph, but genetic analysis has not yet been reported.   A uracil auxotroph was also isolated, but it has neither been analyzed nor regenerated (Sidorov et al., 1981).

Only one report of failure to obtain auxotrophs has been published (Aviv and Galun, 1977b).   As is evident from Table 5, auxotrophs have only been isolated from haploids.   Although the FUdR enrichment scheme proposed by Aviv and Galun (1977b) is effective in animal cell cultures and should be equally effective as the BUdR enrichment used by Carlson (1969, 1970) and Malmberg (1979), Aviv and Galun used diploid tobacco.   As *N. tabacum* is an amphidiploid, the ploidy of somatic tissue may present insurmountable difficulties for auxotroph isolation. It is quite apparent from those published reports of auxotroph isolation that haploid cell lines and mutagenesis are necessary.   Present emphasis should be directed as did Aviv and Galun (1977b) to alternate methods for enrichment of auxotrophs.

Miscellaneous

Many other mutants have been isolated using in vitro cultures (see Table 6).   These phenotypes have involved carbohydrate metabolism,

Table 5. Auxotrophic Mutants

NUTRITIONAL REQUIREMENT	SPECIES	MUTAGEN	PLANT REGEN.	RESISTANCE MECHANISM[a,b]	GENETICS[a]	REFERENCES
Adenine	*D. innoxia* (haploid)	EMS	No	N.R.	N.R.	King et al., 1980
p-aminobenzoic acid	*N. tabacum* (haploid)	EMS	Yes	N.R.	Single recessive nuclear allele	Carlson, 1970; Chaleff, 1981
Arginine	*N. tabacum* (haploid)	EMS	Yes	N.R.	Single recessive nuclear allele	Carlson, 1970; Chaleff, 1981
Biotin	*N. tabacum* (haploid)	EMS	Yes	N.R.	Single recessive nuclear allele	Carlson, 1970; Chaleff, 1981
Histidine	*Hyoscamus muticus* (haploid)	NTG	Yes	N.R.	N.R.	Gebhardt et al., 1981
Hypoxanthine	*N. tabacum* (haploid)	EMS	Yes	N.R.	Two recessive nuclear alleles	Carlson, 1970; Chaleff, 1981
Isoleucine	*N. plumbaginifolia* (haploid)	γ-rays	Yes	A	N.R.	Sidorov et al., 1981

Table 5. Cont.

NUTRITIONAL REQUIREMENT	SPECIES	MUTAGEN	PLANT REGEN.	RESISTANCE MECHANISM[a,b]	GENETICS[a]	REFERENCES
Lysine	N. tabacum (haploid)	EMS	Yes	N.R.	N.R.	Carlson, 1970; Carlson et al., 1973
Nicotinamide	Hyoscamus muticus (haploid)	NTG	Yes	N.R.	N.R.	Gebhardt et al., 1981
Pantothenate	D. innoxia (haploid)	EMS	No	N.R.	N.R.	King et al., 1980
Proline	N. tabacum (haploid)	EMS	No	N.R.	N.R.	Carlson, 1970
Undefined temperature sensitive	N. tabacum (haploid)	None[c]	Yes	B	N.R.	Malmberg, 1980
Uracil	N. plumbaginifolia	γ-rays	No	N.R.'	N.R.	Sidorov et al., 1981

[a]N.R. = Not reported.

[b]Abbreviations for mechanisms: A = Threonine deaminase deficient. B = Decreased ornithine decarboxylase, S-adenosyl methione decarboxylase, and nitrate reductase activities, decreased chlorophyll content, increased PAL activity.

[c]BUdR selection.

Table 6. Miscellaneous Mutants

PHENOTYPE	SPECIES	MUTAGEN	PLANT REGEN.	RESISTANCE MECHANISM[a,b]	GENETICS[a]	REFERENCES
Carbohydrate Metabolism						
Glycerol utilization	N. tabacum	None	Yes	N.R.	Single[c] dominant nuclear allele	Chaleff & Parsons, 1978b
Maltose utilization	Glycine max	None	No	N.R.	N.R.[d]	Limberg et al., 1979
Plant Growth Regulator						
Abscisic acid resistant	N. tabacum N. sylvestris	EMS	No	N.R.	N.R.	Wong & Sussex, 1980
Auxin heterotrophic- and auxin-resistant	Sunflower Crown gall	γ-rays	No	A	N.R.	Atsumi, 1980
Miscellaneous						
Carboxin resistant	N. tabacum	None	Yes	N.R.	N.R.	Polacco & Polacco, 1977
Cell wall	D. innoxia	Azide	No	B	N.R.	Catt, 1981

419

Table 6. Cont.

PHENOTYPE	SPECIES	MUTAGEN	PLANT REGEN.	RESISTANCE MECHANISM[a,b]	GENETICS[a]	REFERENCES
Miscellaneous (Cont.)						
UV-resistant	*Rosa damascena*	None	No	C	N.R.	Murphy et al., 1979

[a] N.R. = Not reported.

[b] Abbreviations for mechanism: A = Regulation of auxin synthesis. B = Alkaline phosphotase secretor/β-glucosidase secretor. C = Increased polyphenolics.

[c] Homozygous dominant lethal.

[d] Stable in absence of selection.

growth regulators, and others. Only one of these mutants was genetically characterized. A line of *N. tabacum* able to utilize glycerol as a carbon source (Gut) was reported by Chaleff and Parsons (1978b). Preliminary genetic analysis indicated that the mutation was inherited as a dominant nuclear allele. As the Gut phenotype is not expressed in the regenerated plant, progeny analysis required initiation of callus from each seedling. The Gut mutation seemed to be inherited in a 2:1 ratio rather than the 3:1 ratio which would be expected for a dominant mutation. However, a 2:1 ratio is expected if Gut/Gut is lethal. Seed germination from self-fertilized Gut/+ is lower than from the Gut/ + x +/+ backcross, as would be expected if Gut/Gut is lethal. Further genetic analysis is necessary to establish the genetic base of this mutant.

Plants have been regenerated from various other cell lines. The 2,4-D-resistant *Lotus corniculatus* plants were more resistant to 2,4-D than the parent cell line, but not as resistant as some available varieties (Swanson and Tomes, 1980). Polacco and Polacco (1977) regenerated plants from a fungicide-resistant cell line (car-1) of *N. tabacum* in the presence of carboxin. Although carboxin inhibits mitochondrial succinic dehydrogenase in *N. tabacum*, the enzyme is equally sensitive to carboxin in car-1 (Polacco and Polacco, 1977). Other cell lines isolated in vitro and listed in Table 6 have not been well characterized.

PROTOCOLS

Selection of Starting Material

The selection of plant cell material from which to isolate mutants may determine the success or failure of in vitro mutant isolation. In some cases the initial cell culture is determined by the nature and aim of the selection, e.g., resistance to *Helminthosporium maydis* race T toxin is only practical in *Zea mays* T-cms cell lines. On the other hand, the choice of species is often a matter of convenience, i.e., a species is chosen because it is amenable to the desired selection, or because it is readily available in cell culture. In some cases the full potential of a selection scheme has not been realized because of inappropriate parental cell lines.

Several features of the initial plant cell material are crucial. If the parent cell line cannot regenerate plants, then the variant cell lines will likely not regenerate, and genetic analysis as well as biochemical analysis of plants are impossible. Hence in species such as *N. tabacum* for which cell lines with high regenerative capacity are available, cell lines that do not regenerate or regenerate poorly should be avoided. In addition, as cell lines with stable chromosome number are possible, aneuploid cell lines should be avoided, because aneuploidy can complicate or make impossible genetic and biochemical analysis. When the parent cell line is chosen only for convenience, a chromosomally stable regenerating cell population should be used. Such precautions may lead to regeneration of shoots from mutants that have not been previously regenerated, e.g., p-fluorophenylalanine-resistant cell lines of *N. taba-*

cum regenerated shoots (Flick et al., 1981). Chromosome stability and regenerative capacity are especially important if the cell line is to be used for selection of somatic hybrids. Deficiencies in either of these attributes can cause failure to produce hybrid plants or production of somatic hybrid plants that cannot be genetically analyzed or even maintained through sexual generations.

The use of haploid or diploid or higher ploidy cell lines must be considered when selecting a parent cell line. As indicated in the above discussion of auxotrophic cell lines, haploid cell lines are essential for identification of any recessive mutations. However, the chromosome stability of the haploid may affect the usefulness of a cell line. Haploids have been historically unstable in in vitro culture. For example, although *N. sylvestris* haploids are readily obtained and maintained in plants, polyploidy quickly results in cell cultures. In such an instance the use of a "haploid" cell line is really no better than a diploid cell line. A stable haploid cell line has only been reported in *Datura innoxia* (Evans and Gamborg, 1979), and this cell line has been used successfully to isolate auxotrophic mutants (e.g., Savage et al., 1979). Although haploid cell lines of *N. tabacum* have been used for isolation of auxotrophs (Carlson, 1970), it should be recalled that *N. tabacum* is an allotetraploid, and hence plants obtained through anther culture are not true haploids. As would be expected, this led to the isolation of leaky auxotrophs by Carlson (1970).

The use of haploid cell lines is probably unnecessary for the isolation of dominant mutations, e.g., stable mutations for resistance to most antimetabolites are possible in diploid cell lines, as has been documented repeatedly (see References). As cell cultures with ploidy level greater than haploid may be more stable in vitro, it is equally advantageous to use these cell lines whenever practical. Regeneration of diploid rather than haploid plants is desirable for genetic analysis. Regeneration of diploid plants from haploid cell cultures is not unusual, however. For genetic as well as biochemical analysis of mutants isolated from diploid or higher ploidy cell lines, it must be remembered that the variants isolated in vitro are heterozygous and will segregate in the R_1 generation.

The growth rate of the parental cell line can influence the success of in vitro mutant isolation. Some antimetabolites may have a short half-life in culture medium. If the cells involved also grow slowly, the antimetabolite may decay before sensitive cells die. Hence sensitive variants may survive under nonpermissive conditions. Cells growing slowly are also more prone to changes in chromosome number or structure. These changes may make plant regeneration more difficult or impossible, especially if aneuploids have accumulated in the culture. Even if plant regeneration is not desired, aneuploidy can complicate interpretation of biochemical analysis because of gene dosage effects. These problems are further complicated if selection is done directly in a suspension culture rather than by plating on solid culture medium (discussed below). Suspension cultures on 3-4 day subculture regimes and protoplasts isolated from such suspensions are ideal. These cultures, such as the Su/Su culture of *N. tabacum* (Gamborg et al., 1979), retain regenerative ability and chromosome stability over several years.

This cell culture, however, is limited in its usefulness, as it is an albino cell line and *N. tabacum* is allotetraploid.

Cell Source

A variety of cell sources may be used for in vitro isolation of mutants. Callus growing on solid culture medium will not be considered as a viable cell source for the following reasons: (1) The cells grow more slowly than suspension cultured cells. (2) Cells are not uniformly exposed to any selective agent included in the agar medium, i.e., cells growing closer to the agar surface will be exposed to higher concentrations of selective agent. (3) Cells are growing in large aggregates. Hence cells resistant to an antimetabolite may not be able to grow if surrounded by large numbers of sensitive cells that are not growing. (4) Finally, uniform application of either physical or chemical mutagens is not possible in callus cultures. Most of these limitations are due to cells growing in large aggregates.

The two most commonly used cell sources for mutant isolation are suspension cultures and protoplasts. Protoplasts are more versatile, in that cells can be directly isolated from plants, i.e., usually leaf mesophyll cells, or from suspension cultures. In either case a population of single cells is available. This results in uniform mutagenesis as well as application of selective agents. Cell colonies and regenerated plants result from single cells and hence should not be chimeric. Unfortunately, techniques of protoplast isolation and culture are not available in all species. In some species suspension cultures may be initiated and maintained, but protoplasts do not divide and proliferate or plating efficiencies of protoplasts are low. Isolation of very large quantities of protoplasts for mutant isolation may prove very costly. Protoplasts, therefore, are not always the ideal cell source.

Suspension cultured cells have been most commonly used for mutant isolation. Rapidly growing cell suspensions (at most 3-4 days between subcultures) can be manipulated similar to microorganisms. The most common problem is the size of cell aggregates in a suspension culture. Large cell clumps may present the same problems as are encountered with callus. However, large cell clumps may be removed from suspension cultures by filtration. (The Cellector tissue sieve available from Bellco Glass Co., Vineland, NJ, is convenient for filtration of plant cells.) Cell size will affect the filter size used; a 200-400 micron filter is useful for most plant species. Cell aggregates with up to 30 cells can be plated and will readily grow into large cell colonies. Under appropriate selective conditions it is possible to isolate single colonies with increased resistance to an antimetabolite. As each colony may have arisen from a group of cells rather than a single cell, a single cell isolation, i.e., protoplasts, may be advantageous to isolate a pure cell line.

Mutants have been isolated in vitro directly in suspension cultures (e.g., Widholm, 1972a; King and Khanna, 1980) as well as by plating cells on solid culture medium containing the selective agent (e.g., Flick et al., 1981). Numerous problems may be encountered when selecting directly in suspension cultures. (1) Because most cells will be sensitive to an antimetabolite, surviving cells will be cultured in a medium enriched by dead cells and any toxic products released by them. In solid agar medium diffusion of toxic products and effects of dead cells are reduced. (2) Individual colonies cannot be selected in a single cultured vessel, i.e., flask, in suspension cultures, since cell aggregates are constantly breaking apart. On solid medium more than one cell colony can be selected from a single culture vessel, i.e., petri plate, because growth on solid culture medium confines the dispersal of cells released from aggregates to the individual colony. (3) If cross feeding effects are important in survival of sensitive cells, then this effect will be greater in suspension cultures than on solid medium because of greater diffusion. Hence many sensitive cells may survive along with resistant cells in suspension cultures. (4) Slow growth in suspension cultures as when selecting for resistance to an antimetabolite may increase the probability of selecting aneuploid cells. As described above, aneuploidy may affect regenerative ability and complicate genetic analysis. (5) Removal of large cell clumps from suspensions is more difficult if the cells are to be returned to suspension cultures. Small cell aggregates seem to survive better when plated on solid medium than when transferred to liquid culture (Flick, unpublished). (6) Finally, the physical complications of handling large numbers of suspension cultures versus large numbers of petri plates must be considered. As the suspensions must be maintained shaking, a large investment in equipment is required. Overall, the advantages of plating on solid culture medium over direct selection in suspension culture seem obvious.

Mutations identified in callus, protoplasts, or suspension cultures may not be expressed in regenerated plants. For example, cell lines resistant to 5-methyltryptophan in suspension cultures regenerate plants sensitive to the analogue. Callus initiated from these plants is resistant to 5-methyltryptophan (Widholm, 1974). On the other hand, carrot embryos and plantlets are resistant to cycloheximide, but callus cultures are sensitive. Hence it is difficult to distinguish the regenerated mutant plant from wild type plants (Sung et al., 1981). The absence of the selected phenotype in regenerated plants may be unacceptable, especially for traits that were selected for agricultural usefulness.

Some phenotypes may not be apparent in callus culture, but only in plants. For example, as callus, suspension, and protoplast cultures are not active photosynthetically, it is not possible to select mutants involving photosynthesis. For example, although callus of D. carota is resistant to the herbicide metribuzin, embryos and plantlets are sensitive (Templeton-Somers et al., 1981). In this instance a novel selection scheme has been proposed. As photosynthesis-dependent embryo cultures are sensitive to metribuzin, resistance to the herbicide could be selected in embryo cultures (Templeton-Somers et al., 1981). Such a selection scheme should be viable with any inhibitor of photosynthetic activity. Indeed, Templeton-Somers (unpublished) has shown that nonlethal concentrations of streptomycin cause bleaching of photosynthetic embryos of D. carota with minimal growth inhibition. This could be used as the basis of a selection system for streptomycin-resistant green embryos. Maliga et al. (1973) used the bleaching effect of streptomycin on greening callus and regenerating shoots to select streptomycin-resistant cell lines.

Mutagenesis

The necessity of using mutagens when isolating mutants in vitro in higher plants has not been satisfactorily demonstrated. An almost equal number of mutants have been isolated in vitro with and without mutagens (see Table 7). However, in 14 of 15 instances reported, auxotrophic mutants have been isolated only after mutagenesis (Table 7). In the only selection from nonmutagenized cells a BUdR selection was used; BUdR itself, however, has mutagenic properties. Both chemical and physical mutagens appear effective in plant cell cultures for isolation of auxotrophs. Ethylmethane sulfonate, nitrosoguanidine, and gamma-irradiation have all been used with success. On the other hand, in 14 of 16 instances reported for in vitro isolation of antibiotic resistant mutants, mutagenesis has not been used. As many of the antibiotic resistance traits are cytoplasmically inherited, the conventional mutagens may not be as effective on organelle DNA as on nuclear DNA. Perhaps with the use of a mutagen that acts more effectively on organelle DNA than nuclear DNA, mutagenesis would enhance the success of isolation of mutants.

Table 7. Use of Mutagens for in vitro Mutant Isolation

CATEGORY	NUMBER OF MUTANT ISOLATED	
	With Mutagen	Without Mutagen
Amino acid analog resistant	19	28
Purine and pyrimidine analog resistant	11	6
Antibiotic resistant	2	14
Agriculturally useful	5	15
Auxotrophs	14	1[a]
Miscellaneous	3	6
Total	54	70

[a]Isolated with BUdR selection; BUdR itself is mutagenic.

If a mutagen is to be used, the choice between physical and chemical mutagens must be made. As neither one has been shown to be superior to the other in higher plants, convenience may be the overriding factor. Physical mutagens are advantageous in some respects, e.g., after mutagenesis the cells do not have to be manipulated to remove the mutagen. Unfortunately, physical mutagens require specialized equipment which is not always readily available, e.g., a gamma-ray source. Ultraviolet light (UV) is convenient for mutagenesis. Equipment is relatively inexpensive and mutagenesis can be carried out under sterile conditions, e.g., using a portable UV lamp in a laminar flow hood. However, after UV mutagenesis it is important to incubate cells in the dark so as to reduce photo-inducible DNA repair.

When using chemical mutagens it is necessary to wash the cells extensively following mutagenesis to remove all traces of the mutagen. This process may prove as deleterious to the cells as the mutagen, especially if the cells are fragile, e.g., protoplasts. In addition, as the medium used for washing the cells contains toxic compounds, waste disposal and decontamination of glassware may present problems.

In some selection procedures, e.g., selection of amino acid analogue-resistant cell lines, it is quite apparent that mutagenized as well as nonmutagenized cells can be used successfully. An effective mutagen must cause extensive DNA damage to cause a mutation in the gene of interest. Hence additional mutations may be present in the selected line. Secondary unselected mutations may cause effects such as reduced viability, interference with plant regeneration, pollen sterility, or innumerable other effects that could interfere with plant regeneration and/or genetic analysis of the mutant cell line. On the other hand, a spontaneous mutation may be expected to be the only mutation present in the cell line; unselected mutations should occur at an undetectable level in selected cell lines. Hence selection of mutants from unmutagenized cell lines is advantageous.

Before in vitro selection for the desired trait, cells should be grown for several generations after mutagenesis to allow segregation of the mutation and expression of the mutation in the cells. As a mutagen will only cause a modification in the DNA in one strand, it is essential that the DNA be replicated before selection for a phenotype so that at least both DNA strands will carry the mutation. If not present in both strands, a mutation may not be transcribed and expressed. In the absence of a growth period following mutagenesis, resistance to a selective agent may not be detected.

Methods of Selection

Ultimately the success of in vitro mutant isolation depends on the suitability of the selection scheme in use. The selection depends, of course, on the future use of the isolated mutants. Mutants with direct application are the most simple to isolate. For example, to study tryptophan biosynthesis Widholm (1972a,b) isolated 5-methyltryptophan-resistant cell lines. Herbicide, e.g., picloram (Chaleff and Parsons, 1978), resistance may be directly selected in cell culture. In these instances one need only identify the problem to be addressed and choose the appropriate selective agent. Most mutants that have been isolated in vitro in higher plants have been by direct selections (see Tables 1-6).

A large number of compounds have been shown to be toxic to higher plant cells in vitro, some of which are listed in Table 8. In several instances these are compounds that could produce mutants of agricultural interest. For example, resistance to substituted 2,6-dinitroaniline herbicides or metribuzin could be of immediate value. Resistance to heavy metals, e.g., zinc, copper, and aluminum, may increase the possibility of growing crops under conditions previously unsuitable for plant growth. Although attempts to isolate NaCl-tolerant cell lines have been made, Smith and McComb (1981) have examined the salt tolerance of salt-tolerant and sensitive species and shown that salt tolerance is expressed in callus cultures. Stress tolerance is discussed more thoroughly by Tal (Chapter 13). A large number of these compounds toxic to plant cells could be used as selective agents for in vitro mutant isolation.

Table 8. Potential Selective Agents for in vitro Isolation of Mutants

SELECTIVE AGENT	SPECIES	CELL SOURCE	REFERENCES
Aluminum	*Lycopersicon esculentum*	Callus	Meredith, 1978
Amino acid analogs (various)	*P. hybrida* *N. tabacum*	Protoplasts Protoplasts	Cocking et al., 1974
Drugs (various)	*P. hybrida* *N. tabacum*	Protoplasts Protoplasts	Cocking et al., 1974
G418 (antibiotic)	*N. tabacum*	Suspension	Ursic et al., 1981
Herbicides (substituted 2,6-dinitroanilines)	*N. tabacum*	Callus	Huffman & Camper, 1978
NaCl-tolerance	*Phaseolus vulgaris* *Beta vulgaris* *Atriplex vulgaris*	Callus Callus Callus	Smith & McComb, 1981
Metribuzin	*D. carota*	Embryos	Templeton-Somers et al., 1981
Phaseolotoxin	*D. carota*	Suspension	Jacques & Sung, 1981
Polyethylene glycol	*L. esculentum*	Suspension	Bressan et al., 1981
Zinc and copper	*Agrotis stolonifera*	Callus	Wu & Antonovics, 1978

Indirect methods may be of value for selection of certain phenotypes that are difficult or impossible to select directly in cell culture. For example, two methods have been used for in vitro isolation of disease-resistant cell lines. Carlson (1973) selected cell lines resistant to wildfire disease caused by *Pseudomonas tabaci*. Methionine sulfoximine is actually an analogue of the *P. tabaci* toxin. A more direct approach was taken by Gengenbach and Green (1975) for the isolation of southern corn blight-resistant cell lines of *Zea mays*. Cells resistant to

Helminthosporium maydis toxin were selected.   A similar approach was used by Behnke (1980a) for the isolation of potato cell lines resistant to *Phytophthora infestans*.   As direct in vitro selection for drought resistance is difficult, selection for overproduction of proline has been suggested (proline overproduction is observed as a response to stress). Mutants resistant to the analogue hydroxyproline may overproduce proline.   Indirect selections may in some instances produce the desired mutant phenotype.

Selection for resistance to antimetabolites is the most simple in vitro screening for mutants.   Several factors influence the success of a selection system.   (1) The action of the antimetabolite should be irreversible, i.e., the agent should be lethal, not just inhibitory to growth. (2) The selective agent should not be degraded in the culture medium either by factors in the medium itself or by wild-type cells.   If the compound is degraded, sensitive cells must be killed before its degradation.   In the case of a labile selective agent, cells may recover and begin to grow after the antimetabolite has been degraded.   Although some of the cell colonies isolated will be resistant variants, many will also be sensitive.

Selection for auxotrophic mutants is difficult.   Most schemes are based on the use of compounds that will kill growing cells, but not affect nongrowing cells.   BUdR selection, which has been successfully used with higher plants (Carlson, 1970), is based on this premise. Crossfeeding or leaky mutants that grow slowly should not be effectively recovered with this type of selection, because the toxic compound is incorporated into them as well as wild-type cells.   Without such a selection, however, auxotrophs can only be selected by massive screening of nonselected cells.   This has been the most common approach in selecting auxotrophs of higher plants.

GENERAL PROTOCOL FOR MUTANT ISOLATION

1. Subculture cells 4 days prior to plating or mutagenesis.   Cells should be grown to as high a density as possible while remaining in exponential growth.   As our tobacco cells are subcultured at least every 4 days, at this time they are still in exponential growth.
2. Cells should be grown for several generations following mutagenesis, whether it be by a chemical or physical agent.   For tobacco cells 4 days growth post-mutagenesis is adequate for expression of mutations.
3. Filter cells to remove large cell aggregates.   Clusters of up to 30 cells may remain, but smaller aggregates are better.   A 250–350 μm pore size filter may be used with tobacco suspensions.
4. Thoroughly mix filtered cells with an equal volume callus culture medium containing 1.2% agar held melted at 45 C (final agar concentration is 0.6%).
5. Evenly spread 2.5 ml cells in 0.6% agar on selective medium prepoured in 100 mm diameter petri plates.

6. After agar has solidified parafilm plates. Incubate in dark until individual cell colonies of 1-2 mm are visible. This will take 2-6 weeks for tobacco cells.
7. Transfer colonies to nonselective culture medium for rapid cell proliferation.

This protocol is designed for isolation of mutants from tobacco cell suspension cultures. Similar protocols can be efficiently used for other species, e.g., mutants of both *N. tabacum* and *D. carota* have been isolated in our lab using these procedures. The plating of suspension cultured cells on agar medium is an efficient method of screening cells for resistance to an antimetabolite. There are several ways to plate cells. (1) A volume of cells can be directly pipeted onto the prepoured solid culture medium in a petri plate. Under these conditions cell colonies may not remain attached to the agar surface or may break apart if the plate is moved. (2) Cells can be directly suspended in the solid culture medium and poured in a thin layer, e.g., 10 ml agar culture medium in a 100 mm petri plate. Unfortunately, thin layers will dry out quickly. We have observed different growth rates for cells submerged in agar medium versus those growing on the surface (Flick, unpublished). In addition, rescue of cell colonies laying beneath the agar is not always successful.

A method adapted from the microbiological technique of soft agar plating over prepoured agar plates is used in our lab. This method combines both direct spreading of suspension cultures on solid medium and plating in agar. As the cells are in a very thin layer of a low concentration agar, the cells are equally exposed to the atmosphere, firmly attached to the agar surface, and remain evenly dispersed.

The plating procedure is described below:

(1) Large cell aggregates must be removed to improve the efficiency of selection. Cells are filtered through a stainless steel tissue sieve (Bellco Glass Co. Cellector) that has been sterilized by flaming with ethanol. The mesh size depends on the plant species, but it should not permit passage of cell clumps containing more than 30 cells. For example, we have used 380 µm pore size for tobacco and 230 µm size for carrots. The filtered cells are collected in a sterile beaker.

(2) After filtration, tobacco cells must be concentrated before plating. This is achieved by centrifugation and resuspension of the cells in a minimal volume. Cells from more than one flask can be pooled and plated together. As the frequency of mutations is low, one should plate a continuous lawn of cells on selective medium. In contrast to tobacco cells, a single flask of carrot cells can provide sufficient cells.

(3) Filtered cells are diluted one-to-one with culture medium containing 1.2% melted agar held at 45-50 C (resulting in a final agar concentration of 0.6%). 2.5 ml cells in 0.6% agar is plated on prepoured agar plates (1.0% agar) containing 25 ml solid medium. After the soft agar solidifies, plates are parafilmed. As the plated cells constitute only 10% of the final volume in the plate, the selective agent can be included at 10% higher than desired concentrations so that following plating the desired concentration of antimetabolite will result.

The cultural conditions immediately following plating of cells on selective culture medium are important in the success of the screening. Senescent tobacco cells, for example, produce phenolic compounds when grown in the light but not in the dark. Hence it is advantageous to incubate selective plates in the dark. The nature of the selection may also necessitate culturing cells in the dark. p-Fluorophenylalanine resistant variants of *N. tabacum* Su/Su (Flick et al., 1981) produce large amounts of phenolic compounds and do not survive when cultured in light. Phenolic comounds are not a problem with some species, e.g., *D. carota*. Growth in the dark will also eliminate any possible photo degradation of the selective agent. All other cultural conditions, i.e., temperature, culture medium composition, etc., should be those that produce optimal cell growth. The only stress on the cells should be the selective agent.

Preliminary Analysis of Variant Cell Lines

A long period of time may result between plating of cells and appearance of cell colonies, e.g., up to several months. Dessication of the culture medium may cause growth inhibition. It is therefore important to remove cell colonies from the selective medium as soon as possible. Cells should be transferred to fresh culture medium without selective agent to allow rapid cell proliferation. When cell colonies reach a sufficient size to divide, a stage that will depend on the plant species, cells should be subcultured back onto selective medium and onto plant regeneration medium, as well as maintained on nonselective medium. As a loss of morphogenetic potential is possible after prolonged culture in vitro, especially under stressful conditions, it is essential that plant regeneration be attempted as soon as possible.

Only a portion of those variants isolated will be stable mutants. For example, of 250 p-fluorophenylalanine (PFP)-resistant cell lines isolated in *N. tabacum* Su/Su, only 21 were stably resistant to the phenylalanine analogue (Flick et al., 1981). The stability of variants can only be ascertained after prolonged subculturing in the absence of the selective agent. Variant cell lines isolated in vitro with any selection system should be verified as mutants. Several characteristics can be used to define a mutant, i.e., a cell bearing a stable change in its genetic makeup. (1) Mutants arise at low frequency, e.g., at a frequency of 10^{-5} -10^{-7} in the absence of mutagenesis. Up to a 100-fold higher mutation frequency may be possible following mutagenesis. In those instances where a selected phenotype arises at a very high frequency, e.g., 10^{-2} -10^{-3} a mutational basis for the variant is unlikely and physiological adaptation should be suspected. (2) A mutant must be stable in the absence of selection. Physiological adaptation may involve induction of certain biochemical functions in the presence of the antimetabolite. In the absence of selective pressure, variants may revert to sensitivity. Variant cell lines that are mutants will retain the selected phenotype after prolonged culture in the absence of selec-observed in cell cultures; the following criteria require plant regeneration. (3) A mutant cell line must remain stable after plant regeneration. Although plants may or may not show the selected phenotype, callus initiated from the regenerated plants must express the selected phenotype. (4) Sexual transmission of a selected phenotype assures its identity as a mutant. Chaleff and Parsons (1978) showed sexual transmission of picloram resistance and observed Mendelian segregation of sensitive and resistant phenotypes. Because in many instances plants regenerated from mutants are aneuploid and/or sterile, genetic analysis may be complicated.

Analogue sensitivity may be screened in either callus or cell suspension cultures. Testing for analogue sensitivity is more rapid in suspension culture, as cells are completely immersed in the antimetabolite. However, more than one subculture in the selective agent may be necessary before sensitive cells die. Diffusion of the selective agent into callus may prolong the time before all sensitive cells are killed. Cell lines that do not maintain resistance in the absence of selection should be discarded. Final analysis of the resistance of a cell line to an antimetabolite can be accomplished by either plating cell suspension cells on various concentrations of the antimetabolite or growth curves in the presence of the antimetabolite in suspension culture.

As a cell suspension culture composed of aggregates of cells has been used for plating, variant cell lines may be chimeric. Subsequent genetic and biochemical analysis of a regenerated plant could be complicated by the lack of homogeneity in the cell line. Protoplast isolation and culture from variant cell lines could be used to separate mosaics and produce homogeneous cell lines. In addition, the behavior of protoplasts isolated from a variant cell line in the presence of the selective agent is important if that variant is to be used as a parent in somatic hybridization.

Two approaches may be taken for the analysis of mutants isolated in vitro, i.e., the biochemical result of mutation and/or the genetic behavior of the mutation. Whether a biochemical and/or genetic analysis of a mutant is undertaken must depend on the aim of the mutant search. In some instances genetic analysis is not possible if no plants are regenerated. However, some genetic analysis is possible in cell cultures, e.g., Glimelius et al. (1978) showed by protoplast fusion that the nitrate reductase mutants of N. tabacum isolated by Muller and Grafe (1978) were recessive and defined two alleles, cnx and nia. In the case of plants that are conventionally asexually propagated, e.g., potato, genetic analysis, although interesting, may not be necessary for the usefulness of the mutant.

In almost all instances a biochemical analysis of a mutant is possible. Any mutant can be analyzed in cell culture or as a regenerated plant. Analysis of both cell culture and plant may contribute an understanding to the organization and/or development of the plant. However, biochemical analysis may not be necessary if the mutant is to be used only as a genetic marker.

FUTURE PROSPECTS

The success of in vitro mutant isolation has been limited by the status of tissue culture for species of interest. Only Solanaceous species, especially species of *Nicotiana*, have been fully exploited in tissue culture. In most species of agricultural interest at least one crucial aspect of tissue culture is not available, e.g., whereas plant regeneration from callus cultures of maize has been reported, plants have not been regenerated from protoplasts.

A species that would behave ideally as a model system has not yet been found. Several requirements for such a tissue culture system are: (1) availability of stable haploid plants or cell lines; (2) regeneration of plants from protoplasts, callus, and cell suspension cultures; (3) a reasonably short life cycle, e.g., flower first year from seed, so that genetic analysis is possible; (4) a background of genetic markers to use in genetic analysis; and (5) a selective system that will produce the desired mutants at a high frequency. No species has fulfilled all these criteria. Whereas auxotrophs were isolated from leaf mesophyll protoplasts of both haploid *N. plumbaginifolia* (Sidorov et al., 1981) and haploid *Hyoscyamus muticus* (Gebhardt et al., 1981; Strauss et al., 1981), in neither case do these species have a good genetic background for analysis of mutants, nor are they species of economic importance. Tissue culture systems need to be perfected in those species that have been extensively analyzed genetically, e.g., *Zea mays*, *Lycopersicon esculentum*, or other crop species.

The isolation of auxotrophs has been limited by the lack of a selection system. Selection for nongrowing cells, e.g., penicillin selection in bacteria or BUdR selection in eukaryotic cells, facilitates the isolation of auxotrophic cells by killing most of the wild-type cells. In the one instance when BUdR enrichment was used in higher plants (Carlson, 1970) a true haploid was not used, i.e., *N. tabacum*. Hence the success of the technique was limited. In the absence of a selection system, auxotrophs can only be identified by screening vast numbers of cell lines, a physically cumbersome project. The future of auxotroph isolation depends on development of appropriate selection systems.

The isolation of auxotrophs is of paramount importance to the full development of somatic cell genetics in higher plants. Much of the elucidation of biochemical pathways and their genetic basis in prokaryotic organisms was deduced using auxotrophic mutants. In higher plants a biochemical system based on information from auxotrophic cell lines has only been described for nitrate reductase, e.g., nitrate auxotrophs (Muller and Grafe, 1978). Metabolic complementation between auxotrophic cell lines can be used as a potent selection for identification of somatic hybrids constructed between two auxotrophic cell lines. Glimelius et al. (1978) demonstrated the potential application of such a selection with nitrate reductase mutants of *N. tabacum*. Both biochemical analysis of higher plants and somatic hybridization require a collection of auxotrophic cell lines.

The development of cytoplasmic genetics in higher plants may depend on the isolation of cytoplasmic mutants in vitro. As the number of organelle genetic markers is limited, each new cytoplasmic mutant expands the possibilities for analysis of the behavior of organelle genomes. As mutants resistant to a very limited number of antibiotics have been isolated up to now, new selective systems are necessary. In addition, efficient cytoplasmic mutagens may be necessary. Antibiotic resistance has already been shown to be valuable for selection of a particular cytoplasm in somatic hybridization (Medgyesy et al., 1980).

Agricultural research has barely realized the future applications of in vitro mutant isolation. Preliminary studies indicate that mutants resistant to diseases (e.g., *Helminthosporium maydis* race T in *Zea mays*, Gengenbach and Green, 1975) or herbicides (e.g., picloram resistance in *N. tabacum*, Chaleff and Parsons, 1978) can be selected directly in tissue cultures. Nutritional deficiencies in certain crops may be alleviated through the use of mutants isolated in vitro. For example, high lysine breeding lines of *Zea mays* with increased free lysine have been isolated by selection for lysine plus threonine resistance in cell cultures (Hibberd et al., 1980).

Perhaps the most promising applications of in vitro mutants in agriculture are those involving indirect selections for a desirable phenotype. For example, as proline accumulation is common in stress situations in higher plants, e.g., water stress, perhaps plants that overproduce proline would be more adaptable to stress. Proline overproduction has been demonstrated in mutants resistant to proline analogues (e.g., Keuh and Bright, 1981). As more selection systems are devised for indirect selection of agriculturally useful phenotypes, the potential applications of in vitro mutant isolation will be realized.

KEY REFERENCES

Chaleff, R.S. 1981. Genetics of higher plants. Applications of cell culture. Cambridge Univ. Press, Cambridge.

Handro, W. 1981. Mutagenesis and in vitro selection. In: Plant Tissue Culture. Methods and Applications in Agriculture (T.A. Thorpe, ed.) pp. 155–180. Academic Press, New York.

Maliga, P. 1980. Isolation, characterization and utilization of mutant cell lines in higher plants. In: International Review of Cytology, Suppl. 11A, Perspectives in Plant Cell and Tissue Culture (I.K. Vasil, ed.) pp. 225–251. Academic Press, New York.

Sala, F., Parisi, B., Cella, R., and Ciferri, O. (eds.) 1980. Plant Cell Cultures: Results and Perspectives, pp. 107–187. Elsevier/North-Holland Biomedical Press, Amsterdam.

REFERENCES

Atsumi, S. 1980. Induction, selection and isolation of auxin heterotrophic and auxin-resistant mutants from cultured crown gall cells irradiated with gamma rays. Plant Cell Physiol. 21:1041–1051.

Aviv, D. and Galun, E. 1977a. Isolation of tobacco protoplasts in the presence of isopropyl N-phenylcarbamate and their culture and regeneration into plants. Z. Pflanzenphysiol. 83:267–273.

_____ 1977b. An attempt at isolation of nutritional mutants from cultured tobacco protoplasts. Plant Sci. Lett. 8:299–304.

Behnke, M. 1979. Selection of potato callus for resistance to culture filtrates of *Phytophthora infestans* and regeneration of resistant plants. Theor. Appl. Genet. 55:69-71.

_____ 1980a. General resistance to late blight of *Solanum tuberosum* plants regenerated from callus resistant to culture filtrates of *Phytophthera infestans*. Theor. Appl. Genet. 56:151-152.

_____ 1980b. Selection of dihaploid potato callus for resistance to the culture filtrate of *Fusarium oxysporum*. Z. Pflanzenzuchtg. 85:254-258.

Berlin, J. 1980. Para-fluorophenylalanine resistant cell lines of tobacco. Z. Pflanzenphysiol. 97:317-324.

_____ 1981. Formation of putrescine and cinnamoylputrescines in tobacco. Z. Pflanzenphysiol. 20:53-55.

_____ and Mutert, U. 1978. Evidence for distinct amino acid transport systems in cultured tobacco cells. Z. Naturforsch. 33c:641-645.

_____ and Widholm, J.M. 1977. Correlation between phenylalanine ammonia lyase acitivty and phenolic biosynthesis in p-fluorophenylalanine-sensitive and -resistant tobacco and carrot tissue cultures. Plant Physiol. 59:550-553.

_____ and Widholm, J.M. 1978a. Amino acid uptake by amino acid analog resistant tobacco cell lines. Z. Naturforsch. 33c:634-640.

_____ and Widholm, J.M. 1978b. Metabolism of phenylalanine and tyrosine in tobacco cell lines resistant and sensitive to p-fluorophenylalanine. Phytochemistry 17:65-68.

_____ and Vollmer, B. 1979. Effects of α-aminooxy-β-phenylpropionic acid on phenylalanine metabolism in p-fluorophenylalanine sensitive and resistant tobacco cells. Z. Naturforsch. 34c:770-775.

Berlyn, M.B. 1980. Isolation and characterization of isonicotinic acid hydrazide-resistant mutants of *Nicotiana tabacum*. Theor. Appl. Genet. 58:19-26.

Bourgin, J.P. 1978. Valine-resistant plants from in vitro selected tobacco cells. Molec. Gen. Genet. 161:225-230.

Bressan, R.A., Hasegawa, P.M., and Handa, A.K. 1981. Resistance of cultured higher plant cells to polyethylene glycol-induced water stress. Plant Sci. Lett. 21:23-30.

Bright, S.W.J. and Northcote, D.H. 1975. A deficiency of hypoxanthine phosphoribosyltransferase in a sycamore callus resistant to azaguanine. Planta 123:79-89.

Bright, S.W.J., Norbury, P.B., and Miflin, B.J. 1979. Isolation of a recessive barley mutant resistant to S-(2-aminoethyl)L-cysteine. Theor. Appl. Genet. 55:1-4.

Bright, S.W.J., Lea, P.J., Kueh, J.S.H., Woodcock, C., Hollomon, D.W., and Scott, G.C. 1982. Proline content does not influence pest and disease susceptibility of barley. Nature 295:592-593.

Caboche, M. and Muller, J.F. 1980. Use of a medium allowing low cell density growth for in vitro selection experiments: Isolation of valine-resistant clones from nitrosoguanidine-mutagenized cells and gamma-irradiated tobacco plants. In: Plant Cell Cultures: Results and Perspectives (F. Sala, B. Parisi, R. Cella, and O. Ciferri, eds.) pp. 133-138. Elsevier/North-Holland Biomedical Press, Amsterdam.

Carlson, P.S. 1969. Production of auxotrophic mutants in ferns. Genet. Res. 14:337-339.

_____ 1970. Induction and isolation of auxotrophic mutants in somatic cell cultures of *Nicotiana tabacum*. Science 168:487-489.

_____ 1973. Methionine sulfoximine-resistant mutants of tobacco. Science 180:1366-1368.

_____, Dearing, R.D., and Floyd, B.M. 1973. Defined mutants in higher plants. In: Genes, Enzymes and Populations (A.M. Srb, ed) pp. 99-107. Plenum, New York.

Catt, J.W. 1981. Cell-wall mutants from higher plants: A new method using cell-wall enzymes. Phytochemistry 20:2487-2488.

Cella, R., Parisi, B., and Nielsen, E. 1982. Characterization of a carrot cell line resistant to azetidine-2-carboxylic acid. Plant Sci. Lett. 24:125-135.

Chaleff, R.S. 1980. Further characterization of picloram-tolerant mutants of *Nicotiana tabacum*. Theor. Appl. Genet. 58:91-95.

_____ and Carlson, P.S. 1975. In vitro selection for mutants of higher plants. In: Genetic Manipulations with Plant Material (L. Ledoux, ed.) pp. 351-363. Plenum, New York.

_____ and Keil, R.L. 1981. Genetic and physiological variability among cultured cells and regenerated plants of *Nicotiana tabacum*. Molec. Gen. Genet. 181:254-258.

_____ and Parsons, M.F. 1978a. Direct selection in vitro for herbicide-resistant mutants of *Nicotiana tabacum*. Proc. Natl. Acad. Sci. 75:5104-5107.

_____ and Parsons, M.F. 1978b. Isolation of a glycerol-utilizing mutant of *Nicotiana tabacum*. Genetics 89:723-728.

Chiu, P.L., Bottino, P.J., and Patterson, G.W. 1980. Sterol composition of nystatin and amphotericin B resistant tobacco calluses. Lipids 15: 50-54.

Cocking, E.C., Power, J.B., Evans, P.K., Safwat, F., Frearson, E.M., Hayward, C., Berry, S.F., and George, D. 1974. Naturally occurring differential drug sensitivities of culture plant protoplasts. Plant Sci. Lett. 3:341-350.

Colijn, C.M., Kool, A.J., and Nijkamp, H.J.J. 1979. An effective chemical mutagenesis procedure for *Petunia hybrida* cell suspension cultures. Theor. Appl. Genet. 55:101-106.

Croughan, T.P., Stavarek, S.J., and Rains, D.W. 1978. Selection of a NaCl-tolerant line of cultured alfalfa cells. Crop Sci. 18:959-963.

Dix, P.J. 1977. Chilling resistance is not transmitted sexually in plants regenerated from *Nicotiana sylvestris* cell lines. Z. Pflanzenphysiol. 84:223-226.

_____ 1981a. Inheritance of chloramphenicol resistance, a trait selected in cell cultures of *Nicotiana sylvestris*. Speg. and Comes. Ann. Bot. 48:315-319.

_____ 1981b. Cross-resistance in cell lines of *Nicotiana sylvestris* selected for resistance to individual antibiotics. Ann. Bot. 48:321-325.

_____ and Pearce, R.S. 1981. Proline accumulation in NaCl-resistant and sensitive cell lines of *Nicotiana sylvestris*. Z. Pflanzenphysiol. 102:243-248.

_____ and Street, H.E. 1975. Sodium chloride resistant cultured cell lines from *Nicotiana sylvestris* and *Capsicum annuum*. Plant Sci. Lett. 5:231-237.

_____, Joo, F., and Maliga, P. 1977. A cell line of *Nicotiana sylvestris* with resistance to kanamycin and streptomycin. Molec. Gen. Genet. 157:285-290.

Evans, D.A. and Gamborg, O.L. 1979. Effects of para-fluorophenylalanine on ploidy levels of cell suspension cultures of *Datura innoxia*. Environ. Exp. Bot. 19:269-275.

Flashman, S.M. and Filner, P. 1978. Selection of tobacco cell lines resistant to selenoamino acids. Plant Sci. Lett. 13:219-229.

Flick, C.E., Jensen, R.A., and Evans, D.A. 1981. Isolation, protoplast culture, and plant regeneration of PFP-resistant variants of *Nicotiana tabacum* Su/Su. Z. Pflanzenphysiol. 103:239-245.

Gamborg, O.L., Shyluk, J.P., Fowke, L.C., Wetter, L.R., and Evans, D.A. 1979. Plant regeneration from protoplasts and cell cultures of *Nicotiana tabacum* sulfur mutant (Su/Su). Z. Pflanzenphysiol. 95:255-264.

Gathercole, R.W.E. and Street, H.E. 1976. Isolation, stability and biochemistry of a p-fluorophenylalanine-resistant cell line of *Acer pseudoplatanus* L. New Phytol. 77:29-41.

_____ 1978. A p-fluorophenylalanine-resistant cell line of sycamore with increased contents of phenylalanine, tyrosine, and phenolics. Z. Pflanzenphysiol. 89:283-287.

Gebhardt, C., Schnebli, V., and King, P.J. 1981. Isolation of biochemical mutants using haploid mesophyll protoplasts of *Hyoscyamus muticus*. Planta 153:81-89.

Gengenbach, B.G. and Green, C.E. 1975. Selection of T-cytoplasm maize callus cultures resistant to *Helminthosporium maydis* race T pathotoxin. Crop Sci. 15:645-649.

_____, and Donovan, C.M. 1977. Inheritance of selected pathotoxin resistance in maize plants regenerated from cell cultures. Proc. Natl. Acad. Sci. 74:5113-5117.

Glimelius, K., Eriksson, T., Grafe, R., and Muller, A.J. 1978. Somatic hybridization of nitrate reductase-deficient mutants of *Nicotiana tabacum* by protoplast fusion. Physiol. Plant. 44:273-277.

Gresshoff, P.M. 1979. Cycloheximide resistance in *Daucus carota* cell cultures. Theor. Appl. Genet. 54:141-143.

Harms, C.T., Potrykus, I., and Widholm, J.M. 1981. Complementation and dominant expression of amino acid analogue resistance markers in somatic hybrid clones from *Daucus carota* after protoplast fusion. Z. Pflanzenphysiol. 101:377-390.

Heimer, Y.M. and Filner, P. 1970. Regulation of the nitrate assimilation pathway of cultured tobacco cells. II. Properties of a variant cell line. Biochim. Biophys. Acta 215:152-165.

Hibberd, K.A. and Green, C.E. 1982. Inheritance and expression of lysine plus threonine resistance selected in maize tissue culture. Proc. Natl. Acad. Sci. 79:559-563.

Hibberd, K.A., Walter, T., Green, C.E., and Gengenbach, B.G. 1980. Selection and characterization of a feedback-insensitive tissue culture of maize. Planta 148:183-187.

Horsch, R.B. and Jones, G.E. 1978. 8-Azaguanine-resistant variants of cultured cells of *Haplopappus gracilis*. Can. J. Bot. 56:2660-2665.

_____ 1980. The selection of resistance mutants from cultured plant cells. Mut. Res. 72:91-100.

Huffman, J.B. and Camper, N.D. 1978. Growth inhibition in tobacco (*Nicotiana tabacum*) callus by 2,6-dinitroaniline herbicides and protection by D-tocopherol acetate. Weed Sci. 26:527-530.

Jacques, S. and Sung, Z.R. 1981. Regulation of pyrimidine and arginine biosynthesis investigated by the use of phaseolotoxin and 5-fluorouracil. Plant Physiol. 67:287-291.

Jones, G.E. and Hann, J. 1979. *Haplopappus gracilis* cell strains resistant to pyrimidine analogues. Theor. Appl. Genet. 54:81-87.

Kandra, G. and Maliga, P. 1977. Is bromodeoxyuridine resistance a consequence of cytokinin habituation in *Nicotiana tabacum*? Planta 133:131-133.

King, J. and Khanna, V. 1980. A nitrate reductase-less variant isolated from suspension cultures of *Datura innoxia* (Mill.). Plant Physiol. 66: 632-636.

King, J., Horsch, R.B., and Savage, A.D. 1980. Partial characterization of two stable auxotrophic cell strains of *Datura innoxia* Mill. Planta 149:480-484.

Krumbiegel, G. 1979. Response of haploid and diploid protoplasts from *Datura innoxia* and *Petunia hybrida* L. to treatment with x-rays and a chemical mutagen. Environ. Exp. Bot. 19:99-103.

Kueh, J.S.H. and Bright, S.W.J. 1981. Proline accumulation in a barley mutant resistant to trans-4-hydroxy-L-proline. Planta 153:166-171.

Lawyer, A.L., Berlyn, M.B., and Zelitch. 1980. Isolation and characterization of glycine hydroxamate-resistant cell lines of *Nicotiana tabacum*. Plant Physiol. 66:334-341.

Lazar, G.B., Dudits, D., and Sung, Z.R. 1981. Expression of cycloheximide resistance in carrot somatic hybrids and their segregants. Genetics 98:347-356.

Lescure, A.M. 1973. Selection of markers of resistance to base-analogues in somatic cell cultures of *Nicotiana tabacum*. Plant Sci. Lett. 1:375-383.

Limberg, M., Cress, D., and Lark, K.G. 1979. Variants of soybean cells which can grow in suspension with maltose as a carbon-energy source. Plant Physiol. 63:718-721.

LoSchiavo, F., Mela, L., Nuti Ronchi, V. and Terzi, M. 1980. Use of HAT system on mutants isolated from cell cultures of *M. sativa*. In: Plant Cell Cultures: Results and Perspectives (F. Sada, B. Parisi, R. Cella, and O. Ciferri, eds.) pp 127-132. Elsevier/North-Holland, Amsterdam.

Maisuryan, A.N., Khadeeva, N.V., and Pogosov, V.Z. 1981. Isolation of 5-bromodeoxyuridine-resistant cell lines of haploid tobacco. Sov. Plant Physiol. 28:395-400.

Maliga, P. 1981. Streptomycin resistance is inherited as a recessive mendelian trait in a *Nicotiana sylvestris* line. Theor. Appl. Genet. 60:1-3.

_____, Breznovits, A.S., and Marton, L. 1973. Streptomycin-resistant plants from callus culture of haploid tobacco. Nature 244:29-30.

_____, Breznovits, A.S., Marton, L., and Joo, F. 1975. Non-mendelian streptomycin-resistant tobacco mutant with altered chloroplasts and mitochondria. Nature 255:401-402.

_____, Lazar, G., Svab, Z., and Nagy, F. 1976. Transient cycloheximide resistance in a tobacco cell line. Molec. Gen. Genet. 149:267-271.

_____, Kiss, Z.R., Dix, P.J., and Lazar, G. 1979. A streptomycin-resistant line of *Nicotiana sylvestris* unable to flower. Molec. Gen. Genet. 172:13-15.

Malmberg, R.L. 1979. Temperature sensitive variants of *Nicotiana tabacum* isolated from somatic cell culture. Genetics 92:215-221.

_____ 1980. Biochemical, cellular and developmental characterization of a temperature-sensitive mutant of *Nicotiana tabacum* and its second site revertant. Cell 22:603-609.

Martin, L. and Maliga, P. 1975. Control of resistance in tobacco cells to 5-bromodeoxyuridine by a simple mendelian factor. Plant Sci. Lett. 5:77-81.

Mastrangelo, I.A. and Smith, H.H. 1977. Selection and differentiation of aminopterin resistant cells of *Datura innoxia*. Plant Sci. Lett. 10:171-179.

Mathur, A.K., Ganapathy, P.S., and Johri, B.M. 1980. Isolation of sodium chloride-tolerant plantlets of *Kickxia ramosissima* under in vitro conditions. Z. Pflanzenphysiol. 99:287-294.

Matthews, B.F., Shye, S.C.H., and Widholm, J.M. 1980. Mechanism of resistance of a selected carrot cell suspension culture to S(2-amino-ethyl)-L-cysteine. Z. Pflanzenphysiol. 96:453-463.

Medgyesy, P., Menczel, L., and Maliga, P. 1980. The use of cytoplasmic streptomycin resistance: Chloroplast transfer from *Nicotiana tabacum* into *Nicotiana sylvestris*, and isolation of their somatic hybrids. Molec. Gen. Genet. 179:693-698.

Menczel, L., Nagy, F., Kiss, Zs.R., and Maliga, P. 1981. Streptomycin resistant and sensitive somatic hybrids of *Nicotiana tabacum* + *Nicotiana knightiana*: Correlation of resistance to *N. tabacum* plastids. Theor. Appl. Genet. 59:191-195.

Mendel, R.R. and Muller, A.J. 1976. A common genetic determinant of xanthine dehydrogenase and nitrate reductase in *Nicotiana tabacum*. Biochem. Physiol. Pflanzen 170:538-541.

_____ 1978. Reconstitution of NADH-nitrate reductase in vitro from nitrate reductase-deficient *Nicotiana tabacum* mutants. Molec. Gen. Genet. 161:77-80.

_____ 1979. Nitrate reductase-deficient mutant cell lines of *Nicotiana tabacum*. Molec. Gen. Genet. 177:145-153.

Mendel, R.R., Alikulov, Z.A., Lvov, N.P., and Muller, A.J. 1981. Presence of the molybdenum-cofactor in nitrate reductase-deficient mutant cell lines of *Nicotiana tabacum*. Molec. Gen. Genet. 181:395-399.

Meredith, C.P. 1978. Response of cultured tomato cells to aluminum. Plant Sci. Lett. 12:17-24.

Merrick, M.M.A. and Collin, H.A. 1981. Selection for asulam resistance in tissue cultures of celery. Plant Sci. Lett. 20:291-296.

Miller, O.K. and Hughes, K.W. 1980. Selection of paraquat-resistant variants of tobacco from cell cultures. In Vitro 16:1085–1091.

Muller, A.J. and Grafe, R. 1978. Isolation and characterization of cell lines of *Nicotiana tabacum* lacking nitrate reductase. Molec. Gen. Genet. 161:67–76.

Murashige, T. and Skoog, F. 1962. A revised medium for rapid growth and bioassays with tobacco tissue cultures. Physiol. Plant. 15:473–497.

Murphy, T.M. and Imbrie, C.W. 1981. Induction and characterization of chlorate-resistant strains of *Rosa damascena* cultured cells. Plant Physiol. 67:910–916.

Murphy, T.M., Hamilton, C.M., and Street, H.E. 1979. A strain of *Rosa damascena* cells resistant to ultraviolet light. Plant Physiol. 64:936–941.

Nabors, M. 1976. Using spontaneously occurring and induced mutations to obtain agriculturally useful plants. Bioscience 26:761–768.

_____, Gibbs, S.E., Bernstein, C.S., and Meis, M.E. 1980. NaCl-tolerant tobacco plants from cultured cells. Z. Pflanzenphysiol. 97:13–17.

Negrutiu, I., Cattoir-Reynaerts, A., and Jacobs, M. 1978. Selection and characterization of cell lines of *Arabidopsis thaliana* resistant to amino acid analogs. Arch. Soc. Belge Biochim. 422–443.

Nielsen, E., Rollo, F., Parisi, B., Cella, R., and Sala, F. 1979. Genetic markers in cultured plant cells: Differential sensitivities to amethopterin, azetidine-2-carboxylic acid and hydroxyurea. Plant Sci. Lett. 15:113–125.

Ohyama, K. 1974. Properties of 5-bromodeoxyuridine-resistant lines of higher plant cells in liquid culture. Exp. Cell Res. 89:31–38.

_____ 1976. A basis for bromodeoxyuridine resistance in plant cells. Environ. Exp. Bot. 16:209–216.

Oswald, T.H., Smith, A.E., and Phillips, D.V. 1977. Herbicide tolerance developed in cell suspension cultures of perennial white clover. Can. J. Bot. 55:1351–1358.

Owens, L.D. 1981. Characterization of kanamycin-resistant cell lines of *Nicotiana tabacum*. Plant Physiol. 67:1166–1168.

Palmer, J.E. and Widholm, J.M. 1975. Characterization of carrot and tobacco cell cultures resistant to p-fluorophenylalanine. Plant Physiol. 56:233–238.

Polacco, J.C. and Polacco, M.L. 1977. Inducing and selecting valuable mutations in plant cell culture: A tobacco mutant resistant to carboxin. Ann. N.Y. Acad. Sci. 287:385–400.

Puck, T.T. and Kao, F.-T. 1967. Genetics of somatic mammalian cells, V. Treatment with 5-bromodeoxyuridine and visible light for isolation of nutritionally deficient mutants. Proc. Natl. Acad. Sci. 58:1227–1234.

Radin, D.N. and Carlson, P.S. 1978. Herbicide-tolerant tobacco mutants selected in situ and recovered via regeneration from cell cultures. Genet. Res. 32:85–89.

Rains, D.W., Croughan, T.P., and Stavarek, S.J. 1979. Selection of salt-tolerant plants using tissue culture. In: Genetic Engineering of Osmoregulation (D.W. Rains, R.C. Valentine, and A. Hollaender, eds.) pp. 279–292. Plenum, New York.

Reish, B., Duke, S.H., and Bingham, E.T. 1981. Selection and characterization of ethionine-resistant alfalfa (*Medicago sativa* L.) cell lines. Theor. Appl. Genet. 59:89–94.

Savage, A.D., King, J., and Gamborg, O.L. 1979. Recovery of a pantothenate auxotroph from a cell suspension culture of *Datura innoxia* Mill. Plant Sci. Lett. 16:367–376.

Scott, A.I., Mizukami, H., and Lee, S.-L. 1979. Characterization of a 5-methyltryptophan resistant strain of *Catharanthus roseus* cultured cells. Phytochemistry 18:795–798.

Shimamoto, K. and Nelson, O.E. 1981. Isolation and characterization of aminopterin-resistant cell lines in maize. Planta 153:436–442.

Sidorov, V., Menczel, L., and Maliga, P. 1981. Isoleucine-requiring *Nicotiana* plant deficient in threonine deaminase. Nature 294:87–88.

Smith, M.K. and McComb, J.A. 1981. Effect of NaCl on the growth of whole plants and their corresponding callus cultures. Aust. J. Plant. Physiol. 8:267–275.

Strauss, A., Bucher, F., and King, P.J. 1981. Isolation of biochemical mutants using haploid mesophyll protoplasts of *Hyoscamus muticus*. Planta 153:75–80.

Sung, Z.R. 1979. Relationship of indole-3-acetic acid and tryptophan concentrations in normal and 5-methyltryptophan resistant cell lines of wild carrots. Planta 145:339–345.

_____ and Jacques, S. 1980. 5-Fluorouracil resistance in carrot cultures. Its use in studying the interaction of the pyrimidine and arginine pathways. Planta 148:389–396.

_____, Lazar, G.B., and Dudits, D. 1981. Cycloheximide resistance in carrot culture: A differentiated function. Plant Physiol. 68:261–264.

Swanson, E.B. and Tomes, D.T. 1980. In vitro responses of tolerant and susceptible lines of *Lotus corniculatus* L. to 2,4-D. Crop Sci. 20:792–795.

Templeton-Somers, K.M., Sharp, W.R., and Pfister, R.M. 1981. Selection of cold-resistant cell lines of carrot. Z. Pflanzenphysiol. 103:139–148.

Tyagi, A.K., Rashid, A., and Maheshwari, S.C. 1981. Sodium chloride resistant cell line from haploid *Datura innoxia* Mill. A resistance trait carried from cell to plantlet and vice versa in vitro. Protoplasma 105:327–332.

Umiel, N. 1979. Streptomycin resistance in tobacco. III. A test on germinating seedling indicates cytoplasmic inheritance in the St-R701 mutant. Z. Pflanzenphysiol. 92:295–301.

_____ and Goldner, R. 1976. Effects of streptomycin on diploid tobacco callus cultures and the isolation of resistant mutants. Protoplasma 89:83–89.

_____, Brand, E., and Goldner, R. 1978. Streptomycin resistance in tobacco: I. Variation among calliclones in the phenotypic expression of resistance. Z. Pflanzenphysiol. 88:311–315.

Ursic, D., Kemp, J.D., and Helgeson, J.P. 1981. A new antibiotic with known resistance factors, G418, inhibits plant cells. Bioch. Biophys. Res. Comm. 101:1031–1037.

Vagera, J. and Novak, F.J. 1979. Frequency of induced mutations at the haploid and diploid levels in *Nicotiana tabacum* L. Biol. Plant. 21:224–229.

Vunsh, R., Aviv, D., and Galun, E. 1980. An amino acid analogue resistant cell line derived from haploid *Nicotiana sylvestris.* In: Plant Cell Cultures: Results and Perspectives (F. Sala, B. Parisi, R. Cella, and O. Ciferri, eds.) pp. 145–150. Elsevier/North-Holland, Amsterdam.

Wang, T.L. 1979. The sensitivity of soybean tissue cultures to the thymidine analogue, 5-bromodeoxyuridine. Plant Sci. Lett. 17:123–128.

Weber, G. and Lark, K.G. 1979. An efficient plating system for rapid isolation of mutants from plant cell suspensions. Theor. Appl. Genet. 55:81–86.

White, D.W.R. and Vasil, I.K. 1979. Use of amino acid analogue-resistant cell lines for selection of *Nicotiana sylvestris* somatic cell hybrids. Theor. Appl. Genet. 55:107–112.

Widholm, J.M. 1972a. Cultured *Nicotiana tabacum* cells with an altered anthranilate synthetase which is less sensitive to feedback inhibition. Biochim. Biophys. Acta 261:52–58.

_____ 1972b. Anthranilate synthetase from 5-methyltryptophan-susceptible and -resistant cultured *Daucus carota* cells. Biochim. Biophys. Acta 279:48–57.

_____ 1974. Cultured carrot cell mutants: 5-methyltryptophan-resistant trait carried from cell to plant and back. Plant Sci. Lett. 3:323–330.

_____ 1976. Selection and characterization of cultured carrot and tobacco cells resistant to lysine, methionine, and proline analogs. Can. J. Bot. 54:1523–1529.

_____ 1977. Selection and characterization of amino acid analog resistant plant cell cultures. Crop Sci. 17:597–600.

_____ 1978. Selection and characterization of a *Daucus carota* L. cell line resistant to four amino acid analogues. J. Exp. Bot. 29:1111–1116.

Wong, J.R. and Sussex, I.M. 1980. Isolation of abscisic acid-resistant variants from tobacco cell cultures. I. Physiological bases for selection. Planta 148:97–102.

_____ 1980. Isolation of abscisic acid-resistant variants from tobacco cell cultures. II. Selection and characterization of variants. Planta 148:103–107.

Wu, L. and Antonovics, J. 1978. Zinc and copper tolerance of *Agrostis solonifera* L. in tissue culture. Am. J. Bot. 65:268–271.

Yurina, N.P., Odintsova, M.S., and Maliga, P. 1978. An altered chloroplast ribosomal protein in a streptomycin resistant tobacco mutant. Theor. Appl. Genet. 52:125–128.

Zyrd, J.P. 1979. Colchicine-induced resistance to antibiotic and amino-acid analogue in plant cell cultures. Experientia 35:1168–1169.

CHAPTER 11

Selection for Herbicide Resistance

K. Hughes

Pre- and post-emergent selective herbicides have been used to reduce weed populations that compete with agronomic crops for space and nutrients. The selective lethality of herbicides depends on differences between the crop and weed species with respect to one or more of the following: (1) uptake of the herbicide, (2) translocation of the herbicide, (3) inactivation of the herbicide, and (4) breakdown of the herbicide (Pinthus et al., 1972). Unfortunately, many of the herbicides in current use are not sufficiently selective, i.e., herbicide damage to the crop species occurs at herbicide concentrations necessary to kill weed species. Further, slight changes in dosages or herbicide-environment interaction may result in either crop damage or insufficient weed destruction (Pinthus et al., 1972). For many important crops there are no appropriate herbicides available. One way to solve this problem is to develop new and more selective herbicides, a lengthy and expensive process. Alternatively, crop plants can be selectively modified for resistance.

The use of cell culture systems in the isolation of potentially useful phenotypes in plants has been widely discussed in both the popular and scientific literature. Herbicide resistance has been mentioned as an area where this technology can be successfully used. The use of cell culture systems can offer several advantages in the isolation of herbicide resistant plants. (1) Many crop species in use today are highly inbred or derived from inbred lines. Genetic variability, as a result, may be reduced. Selection from culture systems provides new sources of genetic variability from which resistant phenotypes may be selected. Further, through a combination of conventional breeding techniques and in vitro mutant isolation, new herbicide resistant crops may be established.   (2) In vitro selection of herbicide resistance from cell culture systems may produce cell lines with differing forms of resistance. While the type of resistance obtained is dependent upon the specific mode of action of the herbicide, it is probable that for any given herbicide there are several ways of obtaining resistance (e.g., transport of the herbicide and detoxification of the herbicide).   Thus selection from culture can provide a convenient way to obtain a number of different types of herbicide resistant mutants.   If biochemical analysis indicates that the mutants are affected in different areas, then it may be possible to combine mutants using either conventional breeding techniques or somatic cell fusion to obtain plants that are more resistant than either mutant parent.   The development of highly resistant crop plants would increase the efficiency with which herbicides could be used in the elimination of weed species.   While biochemically different herbicide resistant variants have been isolated from culture systems (Hughes, unpublished data), no study to date has attempted to combine different phenotypes.

While the advantages are obvious, there are several constraints on the use of tissue culture technology that should be mentioned.   (1) Time required for development of resistance through tissue culture technology: several factors limit the functional life of a herbicide.   One such factor is the development of herbicide resistance in wild populations.   A paraquat-tolerant line of *Lolium perenne* was identified by Faulkner (1975).   Resistance to the S-triazine herbicides has appeared among wild species in several locations including: (a) Washington State, (b) Ontario, Canada, (c) England, and (d) Montpellier, France (Gressel, 1979).   Resistance of weed species to the S-triazine herbicides has appeared fairly rapidly, usually within a ten year period. This is considerably faster than the 20–30 year period expected for development of herbicide resistance in natural populations based on an annual spraying of the compound.   The rapid evolution of resistance in this instance may be due to the persistance of S-triazines in the soil and their widespread use (Gressel, 1979).   While the initial isolation of resistant phenotypes from cell culture systems is rapid, the regeneration of plants from culture, genetic analysis, and field trials will take several years to complete.   Thus herbicide-resistant crop species ideally should be developed and released concurrently with new herbicides for maximum use.   (2) Metabolic basis for herbicide action: a second constraint on the use of cell culture systems for the selection of, herbicide-resistant phenotypes lies in the physiological basis for herbicide action.   It is a prerequisite that the herbicidal action should be at the cell level.   Selection from cell cultures for resistance to several herbicides has been successful (Table 1).   In these cases the herbicidal action is apparent in both callus cultures and plants.

In some instances herbicidal action occurs only at the whole plant level and not in callus cultures.   Metribuzin, which acts on the photosynthetic electron transport system, is a highly effective herbicide on photosynthetic seedlings but has no effect on callus cultures (Ellis, 1978).   However, Templeton-Somers (1981) showed that metribuzin is toxic to asexual embryos of *D. carota*, and embryos could potentially

Table 1. Herbicide-Resistant Cell Lines and Plants Derived from Cell Cultures

HERBICIDE	SPECIES	REFERENCE	HERBICIDE RESISTANT PLANTS	R TRANSMISSION
Carbamates				
Asulam	Brassica campestris	Flack & Collins, 1978		
Isopropyl-N-carbamate	Nicotiana tabacum[b]	Aviv & Galun, 1977	+	+
Vernam	Glycine max	Howard & Meredith[3]		
Dipyridylium Compounds				
Diquat	Glycine max	Hughes, 1978b		
Paraquat	Glycine max	Hughes, 1978b		
	Lycopersicon peruvianum x L. esculentum[a]	Thomas & Pratt[4]	+	
	Nicotiana tabacum[b]	Miller & Hughes, 1980	+	+
Glycine Derivatives				
Glyphosate	Nicotiana tabacum[b]	McDaniel[1]	+	+
	Nicotiana tabacum	Bressan et al.[5]		
	Daucus carota	Bressan et al.[5]		
	Solanum tuberosum	Bressan et al.[5]		
Others				
Fluridon	Glycine max	Chu[6]		
Treflan	Nicotiana tabacum[b]	Chu[6]	+	
	Zea mays	Chu[6]		
Phenoxy Compounds				
2,4-D	Trifolium repens	Oswald et al., 1977		+
	Nicotiana sylvestris	Zenk, 1974		
	Nicotiana tabacum	Ono, 1979		
	Daucus carota	Gressel, 1974		

HERBICIDE	SPECIES	REFERENCE	HERBICIDE RESISTANT PLANTS	R TRANS-MISSION
Phenoxy Compounds				
2,4-DB	*Trifolium repens*	Oswald et al., 1977		
2,4,5-T	*Trifolium repens*	Oswald et al., 1977		
Pyridines				
Picloram	*Nicotiana tabacum*	Chaleff & Parsons, 1978	+	+
Triazines				
Atrizine	*Lycopersicon esculentum*	Locy[2]		
Triazoles				
Amitrol	*Nicotiana tabacum*[a]	Barg & Umiel, 1977	+	
	Nicotiana tabacum[b]	McDaniel[1]	+	+

[a] Plants were regenerated from herbicide-resistant cell cultures.
[b] Herbicide resistance was present in some R plants.

Personal communications are from:
[1] C.N. McDaniel, Dept. of Biol., Rensselaer Polytech, NY 12181.
[2] R. Locy, Dept. Horticulture, North Carolina State, Raleigh, NC 27650.
[3] J. Howard, Stauffer Chemical Co., 1200 S. 47th St., Richmond, CA 94804 and C. Meredith, Dept. of Viticulture and Enology, University of California, Davis, CA 95616. (Resistance in this line was not stable.)
[4] B. Thomas and D. Pratt, Dept. of Bacteriology, University of California, Davis, CA 95616.
[5] R. Bressan, M. Hasegawa, A. Hanada, and S. Weller, Dept. of Horticulture, Purdue University, West Lafayette, IN 47907.
[6] I. Chu, Lilly Research Labs., P.O. Box 708, Greenfield, IN 41640.

be used to select for metribuzin resistance in vitro. Radin and Carlson (1978) were unable to select for resistance to bentazone and phemedipherm in callus cultures; however, herbicide treatment of mutagenized whole plants revealed localized green areas of resistant cells. When these areas were isolated and plants regenerated via tissue culture, plants resistant to both herbicides were obtained.

CURRENT STATUS

Plant cell and tissue culture has been used with relatively few of the available herbicides to select for resistance. Resistant cell lines and/ or plants have been obtained for some of the phenoxy herbicides, the triazoles, the carbamates, the dipyridylium herbicides, and the herbicides picloram and glyphosate (Table 1). While it is relatively easy to obtain resistant cell lines, obtaining herbicide-resistant plants through cell culture procedures has been difficult. More mutants have been successfully isolated from N. tabacum than any other species.

There are many examples of herbicide resistance in cell cultures; however, in only a few cases have plants been regenerated from resistant cell lines. Where plants have been regenerated, they fall into three classes: (1) resistant to normally lethal levels of the herbicide, (2) tolerant to the herbicide (i.e., plants show enhanced survival at sublethal herbicide applications), and (3) not resistant. Herbicide-resistant plants have been regenerated from cell lines of N. tabacum resistant to amitrol and glyphosate (McDaniel, personal communication), isopropyl N-carbamate (Aviv and Galun, 1977), picloram (Chaleff and Parsons, 1978), and paraquat (Miller and Hughes, 1980).

In most studies, both resistant and nonresistant plants occurred in the regenerates (Aviv and Galun, 1977; Chaleff and Parsons, 1978; Miller and Hughes, unpublished). However, callus derived from both resistant and sensitive regenerated plants was herbicide resistant in most cases, indicating that the resistance trait was transmitted but not expressed in the whole plant. Callus initiated from regenerated tobacco plants was resistant to paraquat (Miller and Hughes, unpublished; Hughes, unpublished data). The only exception was callus initiated from a plant that was clearly chimeric for paraquat resistance. In this case the callus could have been derived from a sensitive sector of the plant.

Paraquat resistant cell lines have also been isolated from cell lines initiated from a Lycopersicon peruvianum x L. esculentum hybrid that has high regenerative ability (Thomas and Pratt, personal communication). Plants were regenerated from four of the resistant cell lines. Whereas one regenerate exhibited slight paraquat tolerance, all other regenerated plants were sensitive to paraquat. However, callus derived from the plants retained resistance (Thomas and Pratt, personal communication). Segregation for paraquat resistance in callus was observed in the R_1 generation.

In several instances plants and cell lines selected for resistance to one herbicide are cross resistant to another herbicide. Where the second herbicide is in a related category and has a similar metabolic

basis, this is to be expected. For example, a paraquat-resistant soybean cell line and paraquat-resistant tobacco cell lines are also resistant to the related dipyridylium herbicide diquat (Hughes, 1978b;
Hughes, unpublished data). Occasionally, cell lines that are resistant
to one herbicide are also found to be resistant to another apparently
unrelated herbicide. Two of five *N. tabacum* cell lines selected for
resistance to amitrol are also resistant to glyphosate. Conversely, two
of five lines selected for glyphosate resistance are also resistant to
amitrol. It is not yet known if the plants are cross resistant
(McDaniel, personal communication).

Chaleff and Keil (1981) found that more than half of their picloram-
resistant plants were also resistant to hydroxyurea. In two cases the
resistance trait segregated independently, indicating that at least two
separate and distinct mutations had occurred. As picloram is not
mutagenic in *Saccharomyces cerevisiae* and there is no apparent selection pressure for joint picloram-hydroxyurea resistance, the origin of
these double mutants is not understood (Chaleff and Keil, 1981). In a
third case no segregation of the traits occurred, indicating that either
the two genes are tightly linked or that they were identical.

Genetic analysis of herbicide-resistant mutants isolated in vitro has
only been reported by Chaleff (Chaleff and Parsons, 1978; Chaleff, 1980;
and Chaleff and Keil, 1981). Four picloram-resistant cell lines analyzed
defined three linkage groups. As the two mutants that were genetically linked arose in a single experiment, they may not be independent
mutations (Chaleff, 1980). Chaleff suggests that there must indeed be
a large number of possible alleles to produce a picloram resistant
phenotype. The subsequent discovery that more than half of his
picloram resistant cell lines are hydroxyurea resistant led Chaleff to
suggest that perhaps picloram resistance or hydroxyurea resistance
alleles are mutator alleles (Chaleff, 1980).

PROCEDURES

Mutagenesis

Selection procedures from cell culture do not require that the culture be mutagenized, and many investigators have successfully isolated
herbicide-resistant variants from cell cultures without mutagenesis;
however, we have observed that mutagenesis increases the recovery of
stable resistant cell lines. We have used a variety of mutagenic agents
with cell cultures of *N. tabacum*, including X-rays, short wavelength
UV irradiation (254 nm), and chemical mutagens, e.g., EMS. We now
use only UV irradiation for mutagenesis, because it is fast, convenient,
and easy to use. Most plastics and glass effectively block short wavelength UV irradiation; therefore, a large piece of glass or plastic may
be placed between the UV source and the operator to protect skin and
eyes. Safety glasses will provide additional protection for the eyes and
rubber gloves may be worn to prevent exposure of the hands.

All procedures should be performed under sterile conditions. All
manipulations following irradiation should be carried out in a darkened

room with as little light as possible to prevent light induced DNA repair.

EQUIPMENT NEEDED. The UV lamp and stir plate may be sterilized by wiping with 70% ethanol or by allowing the UV lamp to warm up in the work area, which will be sterilized by UV irradiation. Following UV irradiation, cultures should not be exposed to the light for several days to prevent photorepair. A fine suspension of cells is necessary. Cells in a \250 ml flask and 80 ml medium is sufficient for most studies. Developing a very fine suspension culture with primarily single cells and clusters of no more than three or four cells is technically difficult. It is often necessary to use 2,4-D, 2,4,5-T, or related compounds to increase the growth rate and produce fine suspensions. Either of these procedures may reduce the ability of cells to regenerate into plants at a later date. A technique that has been used successfully to prepare suspension cultures for direct plating is to force friable callus through a fine mesh screen (Pratt, personal communication; Bressan, personal communication). A new technique for producing primarily single cell suspension cultures using calcium alginate beads to immobilize cells in suspension cultures has recently been developed (Morris and Fowler, 1981). The suspensions may be started with either suspension cultures or sieved callus cultures. This procedure may eliminate the need for extended subculturing or the addition of 2,4-D to the medium.

Mutagenesis is carried out in a small, sterile beaker (50 ml). Using a sterile wide mouth pipette, 10 ml cells is transferred from a mid to late log phase culture to each beaker.

The ultraviolet light source is placed on a platform above the magnetic stirrer. In our setup, the distance between the bottom of the three lights and the stirring plate is 24 cm. At this distance, with the lower light on, the exposure is 140 ergs/mm^2/sec, and the loss due to absorbance of the medium is no more than 1.81%. With the lower and middle light on, the exposure rate is 155 ergs/mm^2/sec, and the UV absorbance by the medium is 6.45%. We use only the lower light for UV mutagenesis. If the equipment used is not the same as described above, it will be necessary to calibrate the UV fluence rate to obtain the same overall exposure. Using a radiometer specific for 254 nm light, place the sensor on the stirring plate and record the fluence.

Cells must be stirred during UV mutagenesis to expose the cells uniformly to UV light. Prior to mutagenesis the beaker is covered with a sterile glass plate. Mutagenesis must be carried out in the dark to prevent photorepair. The degree of mutagenesis is determined by the length of the exposure. Mutagenesis is initiated by removing the glass cover plate. The process is stopped by replacing the cover glass. We irradiated N. tabacum cultures for 2 min for a total exposure of 16,000 ergs/mm^2. This results in approximately a 30% reduction in viability as measured by exclusion of the dye bromphenol blue.

The irradiated cells are diluted up to 80 ml with fresh medium and cultured in the dark to prevent light-dependent DNA repair. Following mutagenesis, a period of growth is necessary to replicate the induced mutation as well as express the altered function. When the cells are growing well, they may be transferred to selective medium for isolation of resistant variants. An alternate procedure is to add 10 ml of medium containing twice the lethal concentration of irradiated cells and plate on selective medium. This procedure is described in more detail below. The advantage of direct plating is that the possibility of multiple isolates from a single mutation is reduced.

Selection Techniques

A major problem with selection for variants from cell cultures is the propensity for plant cells to grow in clumps. Except for protoplasts, single cell cultures are virtually nonexistent. When cells in clumps are cultured on selective medium (medium containing normally lethal concentrations of the selective agent), there is a possibility that wild-type sensitive cells can survive the selective pressure. This can occur if: (1) sensitive cells produce a detoxifying substance that can diffuse or be transported into adjacent sensitive cells, or (2) cells in the center of a clump of cells are protected from exposure to the selective agent (Chaleff, 1981).

The survival of sensitive cells in a clump of cells can lead to the formation of chimeric callus and plants. A stable chimeric callus line was identified among paraquat-resistant lines derived from callus of *Nicotiana tabacum*. Plants derived from this cell line were also chimeric and had leaf sectors that were clearly resistant or sensitive (Miller and Hughes, unpublished; Hughes, unpublished).

The survival of nonresistant cells in a clump of resistant cells can lead to the apparent loss of the resistance trait when the cells are transferred to nonselective medium. If the wild-type cells grow at a faster rate than the herbicide-resistant cells, wild-type cells can proliferate at the expense of mutants. After a few subcultures, only wild-type cells may remain in this type of culture. When retested, the resistance trait may be lost.

Determination of the herbicide level that causes death of cells in culture is an essential prerequisite to selection procedures. Several factors affect the toxicity of herbicides and other phytotoxic compounds on cells in culture including: (1) the size of the inoculum, (2) the growth phase of the culture (or plant) used as an explant source, and (3) hormonal levels in the medium (Zilkah and Gressel, 1977). We have observed that activities of enzymes which protect the cell against the toxic effects of paraquat vary sharply throughout the growth cycle, reaching a maximum level in the late log phase. Thus cells from certain stages of the growth cycle are more sensitive to paraquat than cells from other stages. A similar pattern of variability is seen in the whole plant and is associated with both age of the plant and position of the leaf on the plant. Here also, some leaves are more sensitive to the herbicide than others. Thus it is important to establish the conditions under which lethality occurs and to always carry out selection procedures within the bounds of these conditions.

A broad range of cell sources have been used in the isolation of cell lines that are resistant to herbicides. Selections of phenotypically stable cell lines have been successful from callus, suspension cultures, plated suspension cultures, and protoplasts. Both direct selection, i.e., selection for resistance to a lethal concentration of the herbicide, and stepwise selection procedures involving one or more initial selections at sublethal levels followed by selection at a lethal concentration of the herbicide have been used. However, in our experiments stepwise selection procedures did not increase the recovery of paraquat-resistant cell lines (Miller and Hughes, 1980). The major types of selective techniques are presented below.

SELECTION FROM CALLUS CULTURES. Selection of stable herbicide-resistant variants has been successful in a number of cases (Miller and Hughes, 1980; Hughes, 1978a). As noted above, there is a high probability of obtaining a callus that is a chimera of resistant and sensitive cells. Continued subculture on selective medium may or may not eliminate the sensitive cells. The low frequency of chimeric cell lines and plants that we recovered (1 in 24 isolates) using selection from callus cultures may be due to lack of cross feeding. None of our resistant cell lines so far tested has the ability to support growth of nonresistant cells (Hughes, unpublished).

Equipment Needed. All procedures should be performed in a sterile room or laminar flow hood. Callus is cultured in 125 ml Erlenmeyer flasks with 50 ml solidified medium. The flasks are plugged with cotton and the cotton covered with foil or a kaput (Bellco Glass). Cells are transferred with sterile stainless spoons.

Protocols. The concentration of herbicide that is lethal to cells in culture must be determined. Most herbicides are lethal to cultured cells at around 10^{-4} M. A range of herbicide concentrations between 10^{-5} M and 10^{-2} M should be tested. If the herbicide is heat labile, it will be necessary to add the herbicide sterilely to the medium after filter sterilization. Selection should be carried out at an herbicide concentration at which no callus growth occurs. Callus should be transferred from selective medium to nonselective medium to determine if the cells have been killed or simply arrested in growth. Lethality is essential. Cells should always be subcultured at the same stage in the growth cycle when determining lethality.
Callus is grown at the selective concentration of herbicide for 2-4 weeks or longer, depending on the growth rate of the callus. Transfer the callus from selective medium to nonselective medium to determine if cells have been killed or inhibited. A herbicide concentration and growth period that clearly kills wild-type cells should be chosen. This concentration will be used for future selections.
A small piece of callus (50-100 mg wet weight) is transferred onto selective medium. Cultures should be maintained in incubators under normal growing conditions. For N. tabacum, this is 25 C with 12-14 hr of light provided by incandescent and fluorescent lights. Cultures

should be observed at 2 week intervals. After 6-8 weeks, a few of the cultures (for paraquat resistance approximately 2% of the cultures) will develop small slow growing areas that can be seen as lighter colored tissue against a background of dead cells. The small growing areas are isolated and subcultured on medium containing the herbicide. Some of the calli will continue to grow and may be considered to be resistant. Failure to grow may result from a number of factors, including too small an explant and inviability of a given resistant variant. Continued subcultures in the presence of the selective agent may help to remove remaining sensitive cells. Alternately, callus may be transferred into suspension culture and plated on selective medium. If plants are to be regenerated, cells should be transferred as soon as possible to regeneration medium, before regenerative capacity of the cells is lost.

SELECTION FROM SUSPENSION CULTURES. Selection for resistant phenotypes may be carried out directly in suspension cultures. Widholm has successfully used this procedure to isolate a variety of analog-resistant variants (see Chapter 10). McDaniel has used this technique to isolate variants resistant to the herbicides amitrol and glyphosate (McDaniel, personal communication). The use of suspension cultures allows mutant cells in the culture to multiply under favorable growth conditions and increases the probability that a single mutant will be recovered. The possibility of recovering chimeras is also reduced. A disadvantage of this system is that one does not know if a single mutational event or several mutational events have occurred in the same flask. All colonies isolated from a single flask must be treated as a single mutation.

Equipment Needed. Cells are cultured in 250 ml Erlenmeyer or Bellco triple baffled shake flasks with 80 ml liquid medium on a rotary shaker. Sterile cotton plugged wide mouth 10 ml pipettes are used to transfer cells.

Protocol. The lethal concentration of herbicide must be determined as described above. For optimal results a fine suspension culture of cells should be established. Liquid selective medium is prepared and 80 ml of medium is dispensed into a 250 ml flask. Then 10 ml of cells is transferred from a suspension culture in exponential growth phase to the selective medium. Cells are placed on a rotary shaker at 100 rpm under standard lighting and temperature conditions. Cultures are examined every week. Cultures that appear to be growing should be reserved for further tests. Cultures may be plated out after 2-3 weeks of growth. The cells are plated onto selective medium as outlined below. Resistant cells should appear as small colonies which can be recovered for further analysis.

SELECTION FROM PLATED SUSPENSION CULTURES. Isolation of herbicide-resistant mutants from suspension cultures is summarized in Table 2. Cells are grown in liquid nonselective medium to the mid or late log phase of the growth curve. We maintain cultures in 250 ml flasks with 80 ml of medium. New flasks are inoculated with 10 ml of cells from 12-day-old cultures. Selective medium is prepared. Then 25 ml of selective medium is dispensed into petri dishes (20 mm x 155 mm). Medium and plates must be sterile. Selective medium in which to suspend cells for plating is prepared with two times the final concentration of both herbicide and agar and sterilized. The medium is reheated gently prior to use to melt agar and keep agar melted in an oven or waterbath at 40-45 C. A sample of the culture is examined. If the culture is a fine suspension, the culture may be used without filtering. If clumps of more than four to five cells are common, the culture may be filtered through nitex cloth to remove clumps. (Nitex cloth, 150 μm size from TECTO, Inc., 420 Sawmill River Rd., Elmsford, NY 10523. Cloth may be autoclaved.) We initially filter through cloth with a pore size of 500 μm, then through cloth with a pore size of 250 μm. Cells are collected on nitex with a pore size of 70 μm and resuspended in liquid medium. Cells may be concentrated if necessary by gently spinning them down in a clinical centrifuge and resuspending in a small volume of medium. To obtain fine suspension of cells from callus, cells are forced through a stainless steel screen (145 μm pore size). The mesh may be obtained from Fisher, alternately, a kitchen strainer with a fine mesh may be used. Cells are washed on a 70 μm mesh Nitex cloth to remove cell fragments and resuspended. Alternately, cells may be washed by spinning in a clinical centrifuge and resuspending (Bressan, personal communication).

Table 2. Isolation of Herbicide Resistant Mutants from Suspension
          Culture[a]

---

1. Establish a fine suspension of cells.
2. UV mutagenize with 16,000 ergs/mm
3. Grow cells for several generations in the dark to allow replication and expression of mutations.
4. Adjust cell concentration to 10^7–10^8 cells/ml.
5. Dilute cells with an equal volume of selective medium with double the final desired concentration of agar and selective agent.
6. Plate 10 ml with wide bore pipette on top of selective medium in petri dish. Seal dishes with parafilm and incubate until colonies of 2-3 mm diameter are visible.
7. Transfer resistant colonies to fresh culture medium with or without herbicide.
8. As soon as possible, transfer colonies to medium for shoot regeneration.

---

[a]Details given in text.

A small sample of the cell mixture is placed on a hemacytometer and the number of cells per ml estimated. The number of cells per ml is adjusted to 10^7–10^8. Alternately, 2.5 ml of packed cells may be resuspended in liquid medium (McDaniel, personal communication). The warm selective medium (2X) is combined with the cell suspension culture and quickly plated in 10 ml aliquots on top of the selective medium in a petri dish. Sterile wide-mouth pipettes can be useful here. The dishes are sealed with parafilm and placed in an incubator under standard conditions.

Plates should be examined at 2 week intervals for resistant colonies. Any colony larger than 2-3 mm can be subcultured on selective medium. Pratt has had better success with survival of his isolates when he subcultured initially on nonselective medium for a week to allow the culture to become established and then added the herbicide as a liquid which diffuses into the medium (Thomas and Pratt, unpublished).

SELECTION FROM PROTOPLASTS. In theory, selection from protoplast cultures for herbicide resistance should provide the best technique for avoiding chimeric callus and plants; however, protoplasts are fragile and require delicate and precise handling. No reports in the literature have used protoplasts for this purpose. If plating from protoplasts is desired, the investigator may use procedures outlined by D. Evans (see Chapter 7) to isolate and culture protoplasts.

REGENERATION OF PLANTS. Attempts to regenerate plants from herbicide-resistant cell lines should follow selection as rapidly as possible. The longer cells remain in a culture system, the more difficult they are to regenerate, and the more likely they are to accumulate chromosome aberrations (Smith and Street, 1974). Of course, care should be taken to initiate selection from cultures with morphogenetic potential.

Some investigators prefer to regenerate plants in the presence of the herbicide. If the callus is a chimera, i.e., composed of both resistant and sensitive cells, this procedure will reduce the chances of losing the trait if sensitive cells outgrow the resistant cells during plant development. On the other hand, this procedure may result in the failure of cell lines to regenerate in cases where the trait is expressed at the cell level only and not in the whole plant. We have shown that sensitive plants derived from callus produce resistant callus, indicating that the resistance trait is still present in the genome but is not expressed in the whole plant (Miller and Hughes, 1980). Similar observations were made by Thomas and Pratt in their studies with paraquat resistance in tomato hybrids (personal communication). On the other hand, picloram-resistant cell lines of *N. tabacum* regenerate picloram resistant plants (Chaleff and Parsons, 1978).

Regeneration procedures will vary with the species involved. The procedures given below are for *N. tabacum* var. *Wisconsin* 38.

Shoots are regenerated on MS medium with 4.6 µM KIN and 12.3 µM 2iP in 125 ml flasks or baby food jars. Some investigators regenerate in the presence of the herbicide. The necessary ratio of auxins and cytokinins in the medium to induce regeneration varies from species to species. A small piece of callus is transferred to the regeneration medium. The callus should form green areas which will develop into shoots. If sufficient shoot development has not occurred within a 6 week period, the green areas are transferred to fresh regeneration medium. This procedue is repeated until adequate shoot formation is obtained.

Small shoots (approximately 7.5-10 mm high) are transferred to rooting medium. This is usually a medium that lacks auxins and cytokinins. Rooting may be enhanced by dipping the end of the shoot in a sterile commercial rooting solution before transfer. When plants are rooted, the plant is carefully removed from the medium and transferred to damp potting soil. Care should be taken to not injure the root system. Plants from culture have very little cuticle and lack good vascular system and stomate development (Hughes, 1981). Thus it is important to prevent dehydration until the plant can "harden." Dehydration can be prevented by covering the pot with a plastic cone, glass, or plastic bag. Plants should be kept damp but not wet. When new leaves appear, the plant is gradually uncovered over a period of several days. After plants are sufficiently hardened, they can be transferred to a greenhouse for further studies.

TESTING PLANTS FOR HERBICIDE RESISTANCE. Techniques for assaying herbicide resistance are summarized in Table 3. Plants regenerated from herbicide-resistant cell cultures may or may not be resistant. In some cases the herbicide resistance is apparently lost during the developmental process. Unstable herbicide resistance may be due to either an epigenetic event, a temporary alteration of the genome, or a lack of expression of the phenotype in the whole plant. Epigenetic events may be defined functionally as "a stable cellular change of the type encountered in normal development" (Meins and Binns, 1977; Meins and Lutz, 1980). Epigenetic changes may possibly develop in response to selective pressure and may be reversed when pressure is removed.

Table 3. Analysis of Herbicide Resistance in Regenerated Plants[a]

A. Detached leaf assay (e.g., paraquat).
   1. Cut leaf disks from leaves of control plants and regenerates with cork borer.
   2. Float leaf disks on various levels of herbicide, e.g., both lethal and sublethal concentrations.
   3. Incubate in constant light for herbicides that are photosynthetic toxins.
   4. Observe disks at 24 and 48 hr. Resistant leaves will remain green whereas sensitive plants will show chlorosis or necrosis.
B. Whole plant assay (e.g., amitrol or glyphosate).
   1. Select your healthy control and regenerate plants about 30 cm tall.
   2. Apply solution of herbicide to leaf surface.
   3. Observe response at 1-2 weeks after inoculation. Chlorosis and/or wilting will occur in sensitive plants.

---

[a]Details given in text.

An alternate possibility is that the herbicide-resistant trait may be present in the genome but either not expressed or expressed at levels too low to protect the plant. Several studies have demonstrated that callus derived from sensitive regenerates can express the resistance trait (Miller and Hughes, 1980; D. Pratt, personal communication). Possibly, the genes for herbicide resistance are differentially activated, i.e., there may be genes that are expressed only in the callus state but not in differentiated plant tissues. It is also possible that the trait may be expressed in plant tissues but the metabolic change conferring herbicide resistance in vitro may not be sufficient to protect the plant. For example, paraquat toxicity is a function of free radical formation following reduction and reoxidation of paraquat. Reduction of paraquat occurs via electrons from photosynthesis. In cell cultures that photosynthesize at low levels, electron flow to paraquat is relatively low. Thus levels of detoxifying enzymes sufficient to protect the cell cultures may not protect a photosynthesizing plant where electron flow is much higher.

Resistance may be lost if plants are regenerated from a callus that is a chimera of resistant and sensitive cells. If sensitive cells have a selective growth advantage on nonselective medium, they may overgrow the resistant cell lines to produce a sensitive plant. Alternately, the mutant phenotype may not be morphogenic as a pleomorphic effect of the resistance trait. In this case only sensitive cells will form plant tissues.

Procedures for testing plants for resistance vary with the herbicide being tested. In some cases relatively simple assays for plant resistance have been developed that are nondestructive to the plant. In other cases whole plant assays must be used. Examples of assay systems for paraquat and for amitrol and glyphosate are given below; however, assay systems should be developed for each herbicide independently. It should be noted that plants and cell cultures may differ in their sensitivity to a herbicide (Watts and Collin, 1978) and that the selective level of the herbicide should be determined independently in each case.

Equipment Needed. The following items are necessary for determination of paraquat toxicity: (a) 50 cm glass petri dishes, (b) herbicide dissolved in water at lethal and sublethal levels, i.e., 10^{-6} M, 10^{-5} M, and 10^{-4} M, (c) leaves to be tested (leaves should always be taken from the same node of healthy plants as enzyme levels vary with leaf and plant age), and (d) 11 mm cork borer. In addition, for determination of amitrol and glyphosate resistance the following are required: (a) syringes, (b) amitrol at 2 g/l in water with 3% Tween 20, and (c) glyphosate provided as Roundup from Monsanto (McDaniel, personal communication).

Protocols for Determination of Paraquat Resistance. With a cork borer, leaf disks are cut from leaves of control plants and from putative paraquat-resistant plants. The leaf disks are floated on medium containing various levels of paraquat. The disks must be in contact with the solution. Covers are placed on the petri dishes, and the dishes are put in an incubator with 24 hr of fluorescent light. It is important that all dishes be exposed to the same light intensity.

The leaf disks are observed at 24 and 48 hr. Response will vary with the intensity of the light. Control leaves will bleach to white or brown (depending on the presence of nonchlorophyll pigments in the leaf), usually with 24 hr. Resistant leaves will remain green. Partially resistant (tolerant) plants will show some bleaching but will require a longer period of time to turn white. If desired, chlorophyll loss may be monitored by extracting chlorophyll with hot 95% ethanol.

Protocol for Determination of Amitrol Resistance (S.R. Singler and C.N. McDaniel, personal communication). Plants are selected approximately 30 cm high. A leaf near the top of the plant that is approximately 5 cm long is chosen. 0.5 mg of amitrol (0.25 ml) is applied to the leaf. Sensitive plants will exhibit chlorosis in the meristem area within 2 weeks. Eventually, the area will die. Resistant plants will not exhibit chlorosis.

Protocol for Determination of Glyphosate Resistance (S.R. Singer and C.N. McDaniel, personal communication). Plants and leaves are selected as outlined above. 10^{-6} moles of glyphosate (0.1 ml of Roundup) is applied to the leaf. Glyphosate kills the plant roots. Within 1-2 weeks, wilting will occur. Within a month, the sensitive plants are usually dead. Resistant plants do not wilt. The meristem area may become chlorotic, but the resistant plants will outgrow the chlorosis.

GENETIC ANALYSIS. To demonstrate conclusively that herbicide resistance from cell cultures has a genetic basis, it is necessary to show that the trait can be sexually transmitted to succeeding generations. Sexual transmission of herbicide resistance has been demonstrated for paraquat (Hughes, unpublished; Thomas and Pratt, personal communication), for amitrol (Singer and McDaniel, personal communication), and for picloram (Chaleff and Parsons, 1978; Chaleff, 1980); however, clear Mendelian segregation patterns have been demonstrated only for picloram. Picloram resistance was shown to be due to a single dominant gene in three mutants and to a semidominant gene in the fourth mutant (Chaleff, 1980). Thomas and Pratt (unpublished) have observed segregation of paraquat resistance and sensitivity in progeny from backcrosses to sensitive plants as well as in the progeny of self-fertilized plants of tomato. Normal Mendelian segregation was not observed. Dominant or semidominant nuclear alleles are believed to be involved in paraquat resistance in tomato (Thomas and Pratt, unpublished). Paraquat resistance has been transmitted sexually by four plants that exhibited paraquat resistance at both the callus and the whole plant level (Hughes, unpublished). However, the trait is transmitted to only about 2% of the progeny (Hughes, unpublished). In addition, more than 50% of the seeds from self-pollinations and cross pollinations were inviable. Both the high levels of inviable seed and the low segregation frequencies may be due to accumulation of additional mutations and chromosome aberrations as a result of extended periods in culture prior to regeneration.   Alternatively, plants may be aneuploid.   Thus plants should be regenerated from resistant cell lines as quickly as possible.   We are currently in the process of growing R_1 plants that are resistant to paraquat for further segregation test. We hope that the next sexual generation of plants will have lost some of the sterility barriers while retaining the trait.

The procedure we use for determining segregation of herbicide resistance from plants that express the trait is given below.   It should be noted when the trait is expressed in callus but not at the whole plant level, all progeny from a cross must be grown, and callus derived from each progeny plant must be tested for resistance.

Equipment Needed. The following items are recommended: (a) selective medium, 25 ml in 150 x 20 mm petri dishes (seed germination agar is half strength MS medium without growth regulators); (b) small funnel, 125 ml flask, and wash water; (c) sterile water; and (d) spatula.

Protocol.   The level of herbicide is determined that is lethal to seeds from control plants.   This may be different from the levels used to select for herbicide resistance in cell cultures.   For paraquat-resistance, the selective level is 10^{-3} M.

The cultures are observed every 2-3 days to determine the herbicide level in the medium that blocks germination of 100% of the control seeds.   For *N. tabacum*, a 2 week period on selective medium is sufficient to determine the selective level.

Selective medium is prepared containing the restrictive level of the herbicide.   Then 25 ml is dispensed to 150 x 20 mm petri dishes and sterilized.   Seed from a resistant plant (self-pollinated or cross pollinated) is washed in a 10% solution of Clorox for 10 min.   The seeds are poured through the nitex cloth suspended in a funnel.   The seeds are washed with sterile water.   With a spatula, the seeds are scraped off the nitex cloth and resuspended in a small volume of sterile water. The volume will depend on the number of seed, but no more than 1 ml of water should be distributed to any petri plate.   The beaker is swirled to keep the seed suspended and a small amount poured on each plate.   Each plate is sealed with parafilm and examined at weekly intervals until germination begins.   The number of surviving seedlings is counted.   Once seedlings are determined to be resistant, they may be transferred to seed germination medium for further growth and finally to soil following techniques given above.

FUTURE CONSIDERATIONS

The constantly expanding world population will insure a continuing need for improving crop species and increasing yields.   It will be necessary to use all technologies at hand in this effort, and tissue culture techniques will certainly play a major role.   Herbicide resistance is one of several areas that have been identified as important to future crop productivity.

If tissue culture techniques are to be effective in the development of herbicide-resistant plants, several problems must be addressed. These are problems that are shared by the general area of mutant isolation from plant cell culture. Among the most important of these problems is the question of variability from culture systems, i.e., where it comes from, the genetic and metabolic basis for the variability, and how it can be controlled.

Another problem is that little is known about the metabolic basis for herbicide resistance. For any herbicide there are probably numerous mechanisms by which cells and plants can avoid toxicity, including uptake, metabolism of the herbicide, and detoxification of the herbicide. Very few studies have been devoted to determining mechanisms of resistance (Harvey et al., 1978; Harper and Harvey, 1978). If a specific physiological basis for herbicide resistance can be determined and correlated with specific genetic elements, it may be possible to isolate and transfer genes for herbicide resistance via recombinant DNA technology. The next few years should see marked progress in this area of research.

Tissue culture techniques offer several advantages in developing herbicide-resistant plants. (1) The techniques for selection are relatively rapid. (2) Small volumes of cells represent the genetic potential of hectares of fields. (3) Metabolic studies are facilitated by uniform and sterile condition (Gressel et al., 1978). (4) Possible new types of resistance not found in nature may be obtained from plant cell cultures. A great deal of research will be required to solve some of the basic questions addressed above; however, the effort should be well worth the cost in terms of the potential benefits to be derived from genetically engineered herbicide resistance in higher plants.

ACKNOWLEDGMENTS

This work was supported in part by the U.S. Department of Agriculture Agreement No. 5901-0410-9-0347-0.

KEY REFERENCES

Chu, I. 1982. The use of tissue culture for breeding of herbicide tolerant varieties. In: Potentials of Cell and Tissue Culture Techniques in the Improvement of Cereals. International Rice Research Institute, Philippines, and The Chinese Academy of Science. (In press).

Meredith, C.P. and Carlson, P.S. 1982. Herbicide resistance in plant cell culture. In: Herbicide Resistance (H. LeBaron and J. Gressel, eds.) pp. 275-291. John Wiley and Sons, New York.

REFERENCES

Aviv, D. and Galun, E. 1977. Isolation of tobacco protoplasts in the presence of isopropyl-N-phenylcarbamate and their culture and regeneration into plants. Z. Pflanzenphysiol. 83:267-273.

Barg, R. and Umiel, N. 1977. Development of tobacco seedlings and callus cultures in the presence of amitrol. Z. Pflanzenphysiol. 83: 437-447.

Chaleff, R.S. 1980. Further characterization of picloram-tolerant mutants of Nicotiana tabacum. Theor. Appl. Genet. 58:91-95.

_____ 1981. Genetics of higher plants: Applications of cell culture. Cambridge Univ. Press, New York.

_____ and Keil, R.L. 1981. Genetic and physiological variability among cultured cells and regenerated plants of Nicotiana tabacum. Molec. Gen. Genet. 131:254-258.

_____ and Parsons, M.F. 1978. Direct selection in vitro for herbicide resistant mutants of Nicotiana tabacum. Proc. Natl. Acad. Sci. 75: 5104-5107.

Ellis, B.E. 1978. Non-differential sensitivity to the herbicide metribuzin in tomato cell suspension cultures. Can. J. Plant Sci. 58:775-778.

Faulkner, J.S. 1975. Heritability of paraquat tolerance in Lolium perenne L. Euphytica 23:281-288.

Flack, J. and Collin, H.A. 1978. Selection for resistance to asulam in seed rape. In: Abstracts, Fourth International Congress of Plant Tissue and Cell Culture. p. 171. Calgary, Canada.

Gressel, J. 1979. Genetic herbicide resistance: Projections on appearance in weeds and breeding for it in crops. In: Plant Regulation and World Agriculture. NATO Advanced Study Institute, Series A, Vol. 22 (T.K. Scott, ed.) pp. 85-109. Plenum, New York.

_____, Zilkah, S., and Ezra, G. 1978. Herbicide action, resistance and screening in cultures vs. plants. In: Proceedings Fourth International Congress on Plant Tissue Culture (T.A. Thorpe, ed.) pp. 427-436. Univ. of Calgary, Calgary, Canada.

Harper, D.B. and Harvey, B.M.R. 1978. Mechanism of paraquat tolerance in perennial rygrass. II. Role of superoxide dismutase, catalase and peroxidase. Plant Cell Environ. 1:211-215.

Harvey, B.M.R., Muldoon, J., and Harper, D.B. 1978. Mechanism of paraquat tolerance in perennial rygrass. I. Uptake, metabolism and translocation of paraquat. Plant Cell Environ. 1:203-209.

Hughes, K.W. 1978a. Isolation of a herbicide-resistant line of soybean cells. In: Plant Cell and Tissue Culture: Principles and Application (W.R. Sharp, P.O. Larson, E.F. Paddock, and V. Raghavan, eds.) p. 874. Ohio State Univ. Press, Columbus.

_____ 1978b. Diquat resistance in a paraquat resistant soybean cell line. In: Abstracts, Fourth International Congress of Plant Tissue and Cell Culture, p. 170. Calgary, Canada.

_____ 1981. Ornamental species. In: Cloning Agricultural Plants via In Vitro Techniques (B.V. Conger, ed.) pp. 5-50. CRC Press, Boca Raton, Florida.

Meins, F. and Binns, A. 1977. Epigenetic variation of cultured somatic cells: Evidence for gradual changes in the requirement for factors prompting cell division. Proc. Natl. Acad. Sci. 74:2928-2932.

Meins, F. and Lutz, J. 1980. The induction of cytokinin habituation in primary pith explants of tobacco. Planta 149:402-407.

Miller, O.K. and Hughes, K.W. 1980. Selection of paraquat-resistant variants of tobacco from cell culture. In Vitro 16:1085-1091.

Morris, P. and Fowler, M.W. 1981. A new method for the production of fine plant cell suspension culture. Plant Cell, Tissue Organ Culture 1:15-24.

Ono, H. 1979. Genetical and physiological investigations of a 2,4-D resistant cell line isolated from tissue cultures in tobacco. I. Growth responses to 2,4-D and IAA. Sci. Report Fac. Agr. Kobe Univ. 13:273-277.

Pinthus, M.J., Eshel, Y., and Shchari, Y. 1972. Field and vegetable crop mutants with increased resistance to herbicides. Science 177:715-716.

Oswald, T.H., Smith, A.E., and Phillips, D.V. 1977. Herbicide tolerance developed in cell suspension cultures of perennial white clover. Can. J. Bot. 55:1351-1358.

Radin, D.N. and Carlson, P.S. 1978. Herbicide-tolerant tobacco mutants selected in situ recovered via regeneration from cell culture. Genet. Res. 32:85-89.

Smith, S.M. and Street, H.E. 1974. The decline of embryogenic potential as callus and suspension cultures of carrot are serially subcultured. Ann. Bot. 38:223-241.

Templeton-Sommers, K.M. 1981. In vitro screening for metribuzin tolerance using somatic embryos of carrot. HortScience 16:87.

Watts, M.J. and Collin, H.A. 1978. The effect of asulam on the growth of tissue cultures of celery. Weed. Res. 19:33-37.

Zenk, M.H. 1974. Haploids in physiological and biochemical research. In: Haploids in Higher Plants (K.J. Kasha, ed.) pp. 339-353. Univ. of Guelph Press, Guelph, Canada.

Zilkah, S. and Gressel, J. 1977. Cell cultures vs. whole plants for measuring phytotoxicity. I. The establishment and growth of callus and suspension cultures; definition of factors affecting toxicity on calli. Plant Cell Physiol. 18:641-655.

CHAPTER 12
Selection for Stress Tolerance

M. Tal

Adaptation of plants to extreme environments can best be defined by applying Levitt's (1980) terminology, which provides a coherent frame of reference for the analysis of the response of plants to environmental stress. Extreme environments include those environments that potentially can cause stress to the organism exposed to them. A stress can be any environmental factor capable of eliciting from the plant a harmful chemical or causing a physical strain (change), which may be either reversible (elastic) or permanent (plastic). The resistance of the plant to a stress can operate through avoidance and/or tolerance mechanisms. In general, different kinds of stress resistance have been recognized (Levitt, 1980): (1) stress avoidance, i.e., excluding the stress and thus avoiding its potential strain, (2) stress tolerance due to strain avoidance, and (3) stress tolerance due to strain tolerance, which may result from an increased ability to repair the injury or elevation of the threshold level of tolerance of the affected constituents by increasing the elasticity of their response. In strain avoiders the stress does not induce any strain, even though the stress penetrates the plant, and the plant can, therefore, complete its normal life cycle. In strain-tolerant plants the normal activities are impaired under stress, and normal growth and development are renewed only when the stress is removed.

Blum (1980) suggested, in addition to the physiological or ecological definition of stress and resistance to stress, an agronomic definition which includes the individual physiological processes and the integrated plant responses leading to economic yield. His specific agronomic definition of drought stress and drought resistance can be used to define any stress and resistance. Accordingly, a stress can be any environmental factor capable of reducing the yield below the potential level, i.e., the highest possible yield for a given set of conditions.

In my opinion adaptation represents a whole spectrum of hereditary capabilities, which are the product of natural selection or breeding by man. These capabilities either enable the organism to resist a stress only after a gradual change during development or, if developmentally more "canalized" (Waddington, 1962), enable the organism to resist a stress relatively independently of developmental stage and environmental factors, i.e., independently of the immediate existence or absence of the stress. The position held by resistant plants between these two extremes depends on their evolutionary history.

The study of stress phenomena and the related resistance mechanisms in plants is important because large areas worldwide cannot be utilized for food production because of the limitations imposed by natural or man-made environmental stresses. The increase of population pressure requires maximizing the efficiency of the use of cultivated land and increasing the utilization of marginal soils. Because of the economic impact of stresses and the large amount of energy required to alter the environment to suit the plant, it is becoming increasingly important to utilize the existing technologies and to develop new ones in order to develop plants that are better adapted to stress. Such development can be made mainly by exploiting natural or induced genetic variability, along with a physiological study of the underlying mechanisms. Since plant resistance can be the product of many complex interrelated mechanisms, physiological studies can help the breeder identify those specific characteristics to be modified while planning efficient selection (Nabors, 1976; Kramer, 1980).

In the last decade tissue culture techniques have been recognized as a powerful tool for breeding work (Binding, 1974; Chaleff and Carlson, 1974, 1975; Street, 1975). One approach to genetic modification by using tissue culture is the selection of mutants. The possible contribution to agriculture of spontaneous or induced mutations selected through tissue culture has been previously reviewed (Nabors, 1976; Widholm, 1976; Chapter 10). To isolate mutations expressed in tissue culture, the use of haploid tissues would probably be most desirable (Zenk, 1974). However, since stable haploid cultures are still not easy to obtain, diploid tissues can be used for selection of dominant or semidominant mutations (Chaleff and Parsons, 1978). A given variant selected in tissue culture is established unequivocally as arising by mutation only when the following criteria are met: (a) low frequency of appearance, (b) stability when not in contact with the selective agent, (c) stability during the regeneration of whole plants, and (d) inheritance through the sexual cycle (Flick, Chapter 10). Although inheritance provides the most convincing and unequivocal evidence, only a few cases of altered phenotypes obtained in tissue culture were characterized in this way because of difficulties in the regeneration of whole plants, self-fertilization, and progeny testing for a particular trait (Chapter 10).

Another possible approach to genetic modification by using tissue culture may be the production of somatic hybrids by the fusion of naked protoplasts derived from plant species that cannot hybridize sexually. Practically nothing has yet been accomplished by this technique in improving plant adaptation to extreme environments. The only preliminary accomplishment was reported by Smillie et al. (1979). They suggested that somatic hybridization may be used for transferring chilling resistance from potato (*Solanum tuberosum* L.) to tomato (*Lycopersicon esculentum* Mill.), a chilling-sensitive species.

Physiological studies of the mechanisms of stress resistance can also benefit by using tissue culture. It may serve as an excellent model system in the study of mechanisms operating on the whole plant. Some of the advantages in using tissue or cell culture for physiological studies include (Babaeva et al., 1968; Ogolevets, 1976; Croughan et al., 1978; Singh, 1979, 1981) (1) homogeneity of the cell population, thus avoiding the complications that may result from the morphological variability and a highly differentiated state characteristic of the various tissues of the whole plant, (2) growth of tissues or cells on defined media in which they can be treated uniformly and in a controlled way, (3) the ability to perform experiments throughout the year since growth is independent of seasonal fluctuations of the environment, (4) the ability to study the response of tissues or cells isolated from different parts of the plant and thereby reveal the relative contribution of these parts to the resistance of the whole plant, (5) its use to differentiate between mechanisms which operate on the cellular level only and those which depend on the organization of the cells in the whole plant, and (6) the suitability of naked protoplasts for studying aspects of mechanisms related to changes in membrane characteristics.

The unique limitations of the tissue culture technique should be remembered during its application: (1) A variant that appears in tissues or cells in culture may be due to an inherited modification, which is expressed only in tissue culture, or to an epigenetic change resulting from the effect of stress on a process specific to culture without parallel in the whole plant. Bassiri and Carlson (1978), for example, demonstrated that callus from common bean (*Phaseolus vulgaris* L.) had a distinct isozyme pattern, which was independent of its tissue of origin. (2) Selection through tissue culture techniques cannot be applied for mechanisms that are expressed only on the differentiated multicellular level. Such mechanisms can be selected only by screening on the level of the whole plant (Nabors, 1976). (3) Cytological and nuclear changes together with progressive loss of totipotency frequently occur in culture (Street, 1975; Nabors, 1976).

Summing up the above discussion, it can be stated that an efficient program for improving plants in general and for adapting them to extreme environments should include genetic and physiological studies to be conducted on the whole plant level as well as on the level of isolated tissues and cells. In the following section the application of tissue and cell culture techniques in the study of the genetics and physiology of plant adaptation to extreme environment will be discussed.

LITERATURE REVIEW

The application of tissue and cell culture techniques to the improvement of the adaptation of plants to stresses is relatively recent, and the published data on this aspect, especially on its genetic applications, are very limited. Progress in the selection of cell lines resistant to various environmental stresses has been recently described very briefly (Dix, 1980). The discussion here will deal mainly with stresses that produce the most profound effects on plants in nature and that receive most attention, i.e., temperature, water, salt (including ions), and radiation stresses. With regard to temperature, only low-temperature stresses will be dealt with here, since nothing could be found in the literature on the effect of heat stress on tissue culture. Each stress will be discussed separately, including the main injuries incurred and the resistances against them, as in Levitt (1980). The review includes accomplishments related to genetic and physiological aspects of stress phenomenon in tissue and cell culture.

Low-Temperature (Chilling and Freezing) Stress

The stress due to low temperature is difficult to define quantitatively, since the threshold temperature under which a strain is induced in sensitive plants depends on the tissue. However, for most plants a chilling stress can be imposed by any temperature between 10-15 C and 0 C (Levitt, 1980). Only plants from tropical or subtropical regions are sensitive to this stress. Primary direct effects of this stress appear within hours from the initiation of the stress. These effects include changes in membrane permeability and consequent leakage. Primary indirect effects, which appear within days or weeks, are characterized by metabolic disturbances. Secondary effects of chilling may include water stress due to decreased permeability of the root to water. Direct or indirect, the two major symptoms of cells injured during chilling are the abrupt loss of the semipermeability of membranes and the alteration of the respiratory activity (Yoshida and Niki, 1979).

Practically, only tolerance mechanisms enable the plant to survive the effects of chilling strains. No avoidance-related changes have evolved, since the temperature of the plant is almost equal to that of its environment. Modern theory explains all kinds of chilling injuries by a temperature-dependent transition of membrane lipids from liquid crystalline to solid state and possibly also a temperature-dependent alteration in the hydrophobic nature of membranal proteins (Levitt, 1980). All these changes may lead to a secondary injury resulting in metabolic imbalance. The ability of membrane lipids in resistant plants to remain in a liquid state during chilling was explained by the presence of a high proportion of unsaturated fatty acids (Lyons and Raison, 1970) and/or an increased content of sterols (McKersie and Thompson, 1979).

Freezing stress is a shortened way of saying "freezing low-temperature stress," because freezing is not a stress but a strain produced by low-temperature stress, i.e., the plant can be exposed to temperatures below 0 C and remain unfrozen.

Freezing injury, in contrast to chilling injury, can occur in all plants. Because of its prevalence, freezing injury has been studied more intensively. One of the practical applications adopted in recent years from such studies is the freeze-storage of tissues for establishing germplasm banks (Chapter 28).

During freezing, ice is usually formed extracellularly. Because of efflux of water from the cell to the intercellular space as a result of the ice formed, a secondary water stress, which causes dehydration strain, is formed. Increase of solute concentration and collapse of the cell are the consequences of this dehydration. Freezing tolerance is the main component of freezing resistance, since the plant cannot avoid the freezing stress. According to Levitt (1980), the major kind of tolerance to freezing stress developed by the plant is the tolerance to the secondary water stress. Adaptation to freezing stress in most freezing-tolerant plants is built by gradual changes during the autumn (Quamme, 1978). The strain induced by temperatures below 0 C can be avoided by (1) the accumulation of antifreeze substances, (2) minimizing the amount of freezable water by dehydration, and (3) increasing the ability of supercooling (undercooling). According to Quamme (1978), the ultimate level of deep supercooling appears to be the major limitation to the northern distribution of many fruit species and for genetic improvement of freezing tolerance.

GENETIC STUDIES. Early attempts to select low-temperature-resistant lines in culture were made by Steponkus (in Dix, 1980), using callus of *Hedera helix* L. None of the selected lines displayed stable resistance to freezing temperatures.

Suspension cell cultures of *Nicotiana sylvestris* Speg and Comes and *Capsicum annuum* L. both exposed or unexposed to the mutagen ethyl methane sulfonate (EMS), have been submitted for 21 days to -3 C and 5 C, respectively (Dix and Street, 1976). The cell lines derived from the surviving cells were tested for their resistance to a chilling treatment. Some of the lines showed no increased survival when subjected again to the stress, while others retained their resistance after a long period of growth at 24 C. The treatment with the mutagen promoted the isolation of such stable resistant cell lines.

In another experiment (Dix, 1977), callus and suspension cell cultures of *Nicotiana sylvestris* were exposed to a chilling treatment of 21 days at 0 C and -3 C, respectively. Cell lines with enhanced chilling resistance were selected: two lines with high resistance and three with a low level of resistance. Attempts made to regenerate plants from these lines succeeded with the low-resistance lines only. Tissue culture developed from the sexually propagated progeny of the regenerated plants did not show any improvement of the chilling resistance when compared with cultures derived from sensitive control plants.

Tumanov et al. (1977) reported on the selection of callus cells resistant to freezing temperatures. Callus derived from seedlings of *Picea exoelsia* (Lam.) Link was passed through two phases of hardening treatment and then subjected to freezing temperatures. Frost-resistant cells were found capable of growing under optimal conditions after this treatment. No plants were regenerated from this callus, and no genetic study was conducted to clarify whether this capability is inherited.

None of the few reported experiments aimed at the selection of chilling or freezing-resistant mutations includes an unequivocal demonstration that the selected variant is a true gene mutation, i.e., being expressed in regenerated plants and transferred through the sexual cycle. The retention of the resistance characteristics in tissue culture grown under optimal temperatures and exposed again to chilling stress is not unequivocal proof of its genetic nature, since epigenetic changes might be retained in tissue culture for several passages without being exposed to the selection pressure (Dix, 1977). None of the reported variants obtained through selection in tissue culture has been the subject of a physiological study to identify the underlying mechanism(s) responsible for the acquired resistance.

PHYSIOLOGICAL STUDIES. Chilling. Considerable information exists on the problem of chilling injury, especially from the biochemical point of view. Most of this information is based on research made on the level of the whole plant.

Representative studies on the level of tissue or cell culture will be discussed here. Yoshida and Niki (1979) and Yoshida and Tagawa (1979) studied the changes that occur during early stages of cell injury by chilling in callus tissue derived from *Cornus stolonifera* Michx., a chilling-sensitive species, and *Sambucus sieboldiana* (Miq.) Graebn., a chilling-resistant species. They concluded that in the chilling-sensitive callus the early cellular responses associated with chilling injury include a change in the regulation of electron apportionment between the ordinary pathway of respiratory electron transport and an alternative pathway of respiration that is characterized by its insensitivity to cyanide and its inhibition by calicylhydroxamic acid.

Breidenbach and Wareing (1977) compared cells in suspension culture and seedlings of tomato, a chilling-sensitive plant, with respect to their response to chilling temperature of 10 C or below and found that growth and cellular activities related to cell viability were sharply lowered in both. They suggested, therefore, that cells in culture can be used as a model system instead of the complex tissues of the whole plant in the study of membrane modifications through cooling.

The first research designed to elucidate the ultrastructural responses of chilling-sensitive cultured cells within a short period of chilling treatment was performed by Niki et al. (1978). They concluded that the crucial event leading to irreversible cell decay under chilling stress may be the rupture of the tonoplast, thus disturbing the compartmentation of lytic enzymes. Such compartmentation is essential for maintaining normal cell functions.

Naked protoplasts were also used in the study of the effects of a chilling stress on plant cells. It was found that a pretreatment at a low temperature improved the vitality of protoplasts isolated from tobacco and cereals (in Muehlbach and Thiele, 1981). Tal and Watts

(1979) obtained a larger yield of viable protoplasts, as expressed by their plating efficiency, from tomato plants grown under temperature of 15 C and high humidity than from those grown at 25 C and low humidity. The possibility that changes occurring in membrane characteristics due to low temperatures are related to protoplast yield and viability was discussed. Yamada et al. (1980) found that cells cultured at a low temperature (10 C) contained a larger proportion of phospholipids of low transition point than those cultured at a normal temperature (25 C). Whether such a change may explain the increase of yield and viability of protoplasts isolated from plants grown under low temperatures remains to be elucidated.

Muehlbach and Thiele (1981) reported that the plating efficiency of fresh protoplasts isolated from tomato leaves and incubated in a medium containing mannitol at 29 C was raised if pretreated at 7 C for 12 hr. This positive effect of mannitol disappeared if the medium used during chilling was replaced by a fresh one. They suggested, therefore, that in the cold the protoplasts produce and excrete a factor that improves the vitality of the cells and/or stimulates cell division. Chilling of the protoplasts in a medium containing glucose, which, unlike mannitol, enhances the rate of cell division at 29 C, resulted in serious injury. Muehlbach and Thiele (1981) suggested, therefore, that chilling sensitivity of the protoplasts is related to their ability to divide.

Freezing. Callus tissue was used by Ogolevets (1976) in a study of the effects of freezing stress on plant cells. It was shown that cultured tissue of sour cherry (*Prunus cerasus* L.), apple (*Malus* species), and lemon (*Citrus limon* (L.) Burm.) behaved similarly to the whole plant upon hardening; although cherry and apple calli became resistant to freezing after this treatment, the lemon callus remained unhardened. The authors concluded that these tissues in cultures can be used for clarifying the resistance properties of the whole plant.

Chen et al. (1979) compared the adaptability of a callus derived from the freezing–sensitive cultivated potato with that of a callus derived from a resistant wild potato species *Solanum acaule* and *S. commersonii* Dunal. ex Poir. and found that only the callus of the latter species was capable of hardening to the same level as the mature leaf. They suggested, therefore, that the callus tissue can be used in studies of the adaptation of potato to freezing.

Toivio–Kinnucan et al. (1981) followed the ultrastructural changes that occurred in calli originated from the frost–sensitive cultivated potato and from its frost–resistant wild relative during the first 15 days of frost hardening. Aggregation of membrane protein particles was monitored during the first 10 days in the callus of the resistant species only, the aggregates disappearing later. They explained this phenomenon as an adaptive temperature–dependent mechanism that allows the alteration of the transition–point temperature of the membrane lipids.

Naked protoplasts were suggested as a very suitable material for the study of the involvement of surface membranes in freezing injury, because of the absence of the structural characteristics of plant tissues as well as the absence of a cell wall. Naked protoplasts can, therefore, be compared with cells of microorganisms and animals with respect to their reaction to freezing (Wiest and Steponkus, 1978). Singh (1979) suggested that the involvement of surface membranes in freezing tolerance may best be studied in protoplasts isolated from hardened and nonhardened winter rye (*Secale cereale* L.). He elaborated a technique to isolate a large amount of viable protoplasts, which retained the difference in sensitivity characteristic of the whole plants. A positive correlation between the reaction of isolated protoplasts to freezing stress and that of intact tissues from which the protoplasts were isolated was also found in black locust (*Robinia pseudoacacia* L.) trees by Siminovitch (1979). He also demonstrated that the resistance to freezing injury correlated to the resistance to plasmolysis injury, a finding that is in agreement with the suggestion that freezing injury mainly results from its dehydrating effect (Levitt, 1980).

The contribution of the physiological studies on the level of tissue or cell culture to the understanding of the mechanisms responsible for low-temperature resistance is still very limited. It was demonstrated, however, that in many cases tissue culture can be used as a relatively simple model system, which may facilitate the study of processes and mechanisms related to low-temperature resistance operating on both whole-plant and single-cell levels. The positive correlation between the responses of the two levels is not surprising, since, as mentioned above, the low-temperature stress penetrates the cells; hence the resistance to this stress depends in most cases on tolerance mechanisms operating at the cellular level.

Salt Stress

Excess salt, usually NaCl, is the most widespread chemical condition inhibiting plant growth in nature (Casey, 1972; Epstein, 1976).

The major efforts to circumvent salinity in the past have been directed toward soil reclamation and water desalination—practices that are becoming increasingly expensive. These efforts must, therefore, coincide with measures to improve the salt resistance of crops through genetic modification (Casey, 1972). In some species the diversity of salt resistance among cultivars seems quite extensive, and conventional breeding techniques are being used to improve their salt resistance (Epstein, 1976). In many species with little diversity for salt resistance, promising approaches would be either to use variation existing in wild relatives or to use tissue culture techniques for selection of mutations for salt resistance.

According to Levitt (1980), "if the salt concentration is high enough to lower the water potential appreciably (0.5-1.0 bar), the stress will be called a salt stress." Salt stress may have primary and secondary effects. Primary salt injuries may include direct, specific toxic effects as well as indirect effects, such as metabolic disturbances and inhibition of growth and development. Secondary salt effects include nutrient deficiency and osmotic dehydration. The estimation of the contribution of the primary and secondary effects to salt injury is still an open question.

The relative contribution of the specific toxic effect and the osmotic effect of salt to its general negative influence on the plant has been disputed for a long time. While Strogonov (1973) favors the idea that the specific toxicities of the ions are the most crucial factor, Bernstein (1963) claims that the negative osmotic effect of salt is the primary cause of salt damage.

Salt resistance includes both avoidance and tolerance mechanisms. The former may operate through passive exclusion of ions because of membrane permeability, active extrusion by ion pumps, or dilution by a rapid growth accompanied by an increase of water content (succulence). Osmotic stress is an unavoidable consequence of growth in a solution of high salt concentration and is followed by a loss of turgor. Tolerance to such a stress may operate either through dehydration tolerance, which permits the cell to survive without growing when turgor decreases, or through avoidance of dehydration by increasing the content of solutes in the cell following its rehydration, a process called osmoregulation. The solutes may be salt ions, which can be tolerated in the cytoplasm or be excluded into the vacuole, while being osmotically balanced by organic solutes in the cytoplasm. The solutes may also be organic substances. The accumulation of the latter as the major osmotica is accompanied by the prevention of salt ions from entering the cell. Greenway and Munns (1980) recommended that when breeding for salt resistance, concentration be placed on the ability of the plant to synchronize between an effective compartmentation of the absorbed solutes by the leaf cell and a high rate of transport of these ions to the shoot.

Some central, still unsolved aspects related to the mechanism of salt resistance are (1) the properties of the plasmalemma that control the uptake of the salt ions, (2) the nature and location of ion pumps, (3) the control of the synthesis of specific organic solutes under salinity, (4) the maintenance of the semipermeability of the plasma membrane in salt solutions, and (5) the control of salt-induced accumulation of toxic substances (Levitt, 1980).

GENETIC STUDIES. The advantage of the tissue culture technique for selecting salt-resistant mutants was discussed by Melchers (1972). This approach was applied successfully by Zenk (1974), who selected a cell line that was able to grow on medium containing 0.17 M NaCl, on which control cells could not grow, from haploid culture of *Nicotiana sylvestris*. The resistance of this line was stable in culture for many passages of saline medium.

Nabors et al. (1975) selected NaCl-resistant cell lines from *Nicotiana tabacum* L. cell-suspension culture treated by the mutagen EMS and then grown in a medium containing 0.03 M NaCl. Cells derived from these lines even resisted concentrations as high as 0.09 M NaCl in the medium. Most of the control cultures did not grow on the saline medium. Few of them, however, remained alive, and after several transfers started to grow fast. This growth was explained by the appearance of spontaneous dominant mutations.

Another successful selection was reported by Dix and Street (1975), who selected a number of cell lines of *Nicotiana sylvestris* and *Capsicum annuum* capable of growing in liquid medium containing up to 0.34 M NaCl. Some of these lines retained the resistance to salt after several subcultures in media lacking NaCl. Plants were regenerated from resistant cell lines obtained after either two or six passages in media containing 0.34 M NaCl (Dix, 1980). Calli initiated from the leaves of these two groups of plants were tested for their ability to grow in NaCl medium. While callus of the former plants showed only limited sectorial growth, if any, that of the latter grew as the original resistant cell lines.

Croughan et al. (1978) isolated NaCl-resistant cells, which could grow on a medium containing 0.17 M NaCl, from a cell culture of alfalfa (*Medicago sativa* L.). The selected line behaved like a halophyte in some respects, including the need for salt for optimal growth, maintenance of high level of K^+ in the presence of high levels of Na^+ and increased level of NO_3^- at low salt concentration, NO_3^- probably replacing Cl^-.

Similar results were obtained by Rains et al. (1980) with rice (*Oryza sativa* L.) cells selected for salt tolerance. Cells selected in the presence of 0.26 M NaCl, a concentration lethal for the unselected cells, required the presence of 0.09 M NaCl for optimal growth. They suggested that the selected and unselected cells of alfalfa and rice differed in the ion transport systems.

Kochba et al. (1980) reported the isolation of "Shamouti" orange (*Citrus sinensis* (L.) Osb.) callus lines with increased resistance to NaCl. The resistance was maintained in embryos obtained from these lines.

Nabors et al. (1980) reported the inheritance of salt resistance in plants regenerated from selected NaCl-resistant cells. Cell-suspension culture of *Nicotiana tabacum* was exposed to increasing levels of NaCl, and cell lines resistant to NaCl concentrations up to 0.88% were selected. The resistance was maintained in the progeny of plants regenerated from culture for the two subsequent generations examined. The level of resistance in the progeny of the regenerated plants was higher than in the original resistant cells in culture.

Bressan et al. (1981) found that cultured cells selected for resistance to drought imposed by polyethylene glycol were also more resistant to NaCl than unselected cells. The possible use of nonpenetrable osmotica for selecting mutations resistant to the osmotic effect of salt stress should, therefore, be taken into consideration.

Goldner et al. (1977) classified the possible mutations selected for resistance to salt stress in tissue culture into three major groups: (1) mutations resistant to osmotic stress, (2) mutations resistant to stress caused by high concentration of total salts, and (3) mutations resistant to stress caused by specific ions. They recommended that specific procedures be designed for the effective selection of mutants of each of these groups. They also suggested that the selection technique used by Zenk (1974), Dix and Street (1975), and Nabors et al. (1975), who subjected suspension culture to NaCl, is suitable for the selection of mutants of the first group.

PHYSIOLOGICAL STUDIES. The advantages of using tissue and cell cultures as model systems to answer questions related to the mechanisms of salt resistance operating at the whole plant level have been demonstrated in several investigations.

One of the earliest investigations of the effect of salt on plant tissue culture was reported by Babaeva et al. (1968). Tissues isolated from different parts of carrot root and calli derived from these tissues differed in their salt resistance. An additional interesting observation was that NaCl at growth-inhibitory concentrations induced ploidy changes.

Kulieva et al. (1975) suggested that *Crepis capillaris* Wallr. cells growing in culture can be used in studies of the cellular effects of salt, since they are very sensitive to salinity and are also genetically stable.

Some attempts to estimate the relative contributions of primary and secondary effects of salt to injury were recently made using tissue culture (Goldner et al., 1977). The investigators compared the effects of seawater, solutions of different inorganic salts, and mannitol on the growth and coloration of diploid callus originating from carrot root (*Daucus carota* L.). From these experiments they concluded that growth inhibition resulted mainly from the increase of the osmotic pressure, while discoloration and necrosis resulted from the toxicity of salt.

Chen et al. (1980) grew callus of *Nicotiana tabacum* in media containing various concentrations of seawater or NaCl. They found that a solution of a single salt (NaCl) was more toxic to the tissue than seawater of the same overall salt concentration. They recommended, therefore, in agreement with Epstein and Norlyn (1977), that the selection of salt-resistant mutants should be made using a stress caused not by a single salt but by seawater, which better represents the salt solutions existing in nature.

An experiment aimed at increasing the understanding of the role of osmotic adjustment in the adaptation of plant cells to salt was reported by Heyser and Nabors (1981). According to them, such an understanding is helpful in isolating salt-tolerant mutants. Callus of *Nicotiana tabacum* was adapted to NaCl by exposing it to increasing salt concentrations. Adapted and nonadapted calli were transferred to suspension culture and tested for their reaction to salt. The addition of NaCl to the medium reduced the growth of both kinds of cells equally. A specific toxic effect was suggested as the cause of NaCl inhibition of growth. No obvious qualitative differences were found in the osmotic adjustment of adapted and nonadapted cells.

The contribution of the mechanisms of salt resistance operating on the cellular level to the resistance of the whole plant has been examined in a series of works.

The response to NaCl of calli originated from four glycophytic and one halophytic species differing in their resistance was compared (Strogonov, 1973). While the response of calli originating from the glycophytic species was correlated to the response of the whole plants to salinity, no such correlation existed in the halophyte, i.e., callus originating from the latter was much more sensitive than that of the most salt-sensitive glycophyte.

The effect of NaCl on calli of four species, two glycophytes and two halophytes differing in salt resistance, was studied by Smith and McComb (1981). A positive correlation was found between the response of the two glycophytes and the calli derived from them; *Phaseolus vulgaris* was sensitive, while *Beta vulgaris* L. was more resistant to salt on both the whole plant and tissue levels. On the other hand, no such correlation was found in the two halophytes *Atriplex undulata* D. Dietr. and *Suaeda australis* (R. Br.) Moq. The investigators concluded that, in contrast to the glycophytes, salt resistance in the halophytic species depends on the anatomical and physiological integrity of the whole plant and not on cellular properties. Hence cells of these plants are not able to resist salt when they dedifferentiate to form callus.

A positive correlation between the response of the whole plant and callus derived from it was found in other halophytic and glycophytic species. Von Hedenstrom and Breckle (1974) examined the growth of callus derived from the halophyte *Suaeda maritima* (L.) Dumort. in media containing various concentrations of NaCl. As with the whole plant, the callus was able to grow in medium devoid of salt, and its growth was improved by salt added to the medium. Tal et al. (1978) found that cells which originated from leaves, stems, and roots of the cultivated tomato and two of its salt-tolerant wild relatives, *Lycopersicon peruvianum* (L.) Mill. and *Solanum pennellii* Corr., behaved in a similar fashion to the whole plant under salinity. They concluded, therefore, that the better osmotic adjustment, which characterizes the wild species under high salinity, is operating at the cellular level and does not depend on the organization of these cells in the whole plant. Accordingly, protoplasts isolated from leaves of *L. peruvianum* divided and grew better on a saline medium than those of the cultivated species (Rosen and Tal, 1981). A similar finding for callus was reported by Orton (1980), who compared the response to salt of calli derived from salt-sensitive and salt-resistant species of barley (*Hordeum vulgare* L.).

An attempt to use NaCl-resistant cell lines of *N. sylvestris* for investigating the underlying basis of resistance to high salinity was reported by Dix (1980). On the basis of X-ray microanalysis it was suggested that the accumulation, or sequestering, of Na^+ and Cl^- ions in a particular subcellular compartment cannot be the mechanism responsible for salt resistance in *N. sylvestris* cell cultures.

Drought Stress

The interest in the effect of drought stress on plants results mainly from the need to better understand the problems to which economically important crop plants are exposed when water is a limiting factor. Kramer (1980) suggests that the worldwide losses in yield caused by water shortage are greater than those caused by all other causes together.

Drought stress, which is commonly used for describing an environmental stress of sufficient duration to produce water stress in the plant (Kramer, 1980), is the least measurable of all environmental stresses (Levitt, 1980).

Drought stress can be a primary stress, produced by water deficit in the environment surrounding the plant, and a secondary stress, induced by either chilling, freezing, heat, or salt stresses (Levitt, 1980). Drought stress can induce various kinds of primary, direct, or indirect injuries, including growth inhibition, starvation, accumulation of protein and toxins, biochemical lesions, enzyme inactivation, and ion leakage. It can also induce secondary stresses, such as heat and nutrient deficiency.

Drought resistance can operate through avoidance and tolerance mechanisms (Levitt, 1980). Drought can be avoided by conservation of water or a fast supply of water, enough to compensate for a rapid loss. Avoidance protects the plant against all kinds of drought injuries by maintaining a high internal water potential in spite of a decrease in the water potential externally. In contrast to drought avoidance, which infers a general protection against drought injury, drought tolerance is highly specific and includes at least six possible types of tolerance. Levitt (1980) classified it into two major components: dehydration avoidance and dehydration tolerance. Dehydration avoidance includes the ability to lower the osmotic potential by increasing the net accumulation of solutes, i.e., osmotic adjustment, together with increased elasticity of the cell wall, and decreased cell size. These abilities enable the cell to maintain its turgor, which is essential for preserving normal metabolism, growth, and survival during a change in the water balance of the plant (Turner and Jones, 1980). Kramer (1980) suggested that the basis for differences among species in dehydration tolerance must be sought at the molecular level, including mainly membrane structure and enzyme activity.

In contrast to its minor role as a mechanism of resistance to temperature and salt stresses, avoidance is of great value for the survival of the higher plant under drought stress (Levitt, 1980). Because of the many possible combinations between the different kinds of drought avoidance and tolerance mechanisms, many kinds of adaptation to drought exist.

Most of the research on drought stress up to now has been concentrated on the responses of whole plants and of particular enzyme systems (Ruesink, 1978). Hardly any attention has been directed toward the application of tissue culture technology in studies of the physiological mechanisms of drought resistance or for the selection of resistant mutants.

GENETIC STUDIES. Heyser and Nabors (1979) were the first to report the selection of cell lines resistant to stress induced by polyethylene glycol (PEG). Bressan et al. (1981) selected cells of tomato resistant to drought stress by exposing callus to a medium containing PEG. The selected cells grew much better than the control unselected cells on media containing the nonpenetrating osmoticum. The resistance was stable, however, only in cells that were in continuous contact with PEG, becoming lost quickly on medium lacking in the osmoticum.

PHYSIOLOGICAL STUDIES.   Wahlstrom and Eriksson (1976) studied the effect of water stress imposed by various salts and organic osmotica for 30 min on protein synthesis in carrot cell culture.   The uptake and incorporation of labeled glutamic acid were not affected by the stress.   In contrast, a significant reduction of uptake and incorporation of labeled leucine into suspension-culture cells of *Convolvulus arvensis* L. exposed to water stress imposed by salts and organic osmotica for 0.5-4.5 hr was found by Ruesink (1978).   The difference in the time duration of the experiments was suggested as the most likely factor responsible for the sharp contradiction between the results of the two experiments.

Naked protoplasts isolated from bark of winter and summer black locust trees were subjected to strong osmotic dehydration by exposing them to a series of balanced salt solutions of increasing molarity.   The protoplasts showed almost the same response, i.e., susceptibility or resistance to osmotic dehydration, as the tissues from which they were originated (Siminovitch, 1979).

A serious limitation in using naked protoplasts as a model system in such studies is the continuous presence of osmotic stabilizers, incorporated in the protoplast-isolation medium, which may induce an osmotic stress in protoplasts in solution.   Various biochemical changes, which probably result from water stress induced by the osmotic stabilizers, were reported in isolated protoplasts, including a decrease of CO_2 fixation (Wehmann and Muehlbach, 1973), decrease of synthesis of protein and nucleic acids (Premecz et al., 1978), and an increase of RNase activity and proline content (Premecz et al., 1977).

Ion Stress

The determination of ion toxicity and ion resistance is very difficult because of the many complications resulting from the interactions either among ions or between ions and various other factors in the organism (Levitt, 1980).   In spite of these difficulties, there are many facts which show that plants differ in their resistance to ions.   The existence of genetic variability for resistance to ion toxicity in higher plants has been discussed by Epstein (1972).   The stress induced by heavy metals has received most attention because of its high occurrence in nature (Levitt, 1980).

Higher plants can resist ion stress by avoiding it, i.e., by means of decreased uptake or precipitation on the cell wall, or by tolerating it. Tolerance can operate through strain avoidance mechanisms, i.e., secretion into the vacuole, precipitation in the cytoplasm, or through strain tolerance mechanisms, for example, changes in enzyme structure that stabilize it against the effect of the toxic ion.

The use of tissue or cell culture techniques in physiological or genetic studies of ion toxicity is very limited.

GENETIC STUDIES.   Aluminum-resistant cell lines were selected from callus of tomato exposed to 200 μM Al^{3+} for several months

(Meredith, 1978a). Their frequency of occurrence was comparable to that reported for spontaneous mutations in eukaryotes (in Meredith, 1978a). The resistance was stable after growth in media lacking Al^{3+}. Although some facts suggested that the resistant variants were the result of dominant mutations, an unequivocal proof for that is still lacking, since inheritance and phenotypic expression were not demonstrated in regenerated whole plants. Such variants can, however, be used to facilitate the elucidation of the mechanism(s) of Al^{3+} toxicity and the resistance on the cellular level.

Colijn et al. (1979) selected cell lines resistant to mercury in culture of *Petunia hybrida* Vilm. The greatly enhanced frequency of resistant lines isolated after a treatment with a mutagen suggests the possibility that the resistance to mercury is genetically determined (Dix, 1980).

PHYSIOLOGICAL STUDIES. Meredith (1978b) demonstrated the effect of excessive Al^{3+} on callus originating from resistant and sensitive varieties of the cultivated tomato. Since the reaction of the calli to the ion excess was comparable to that of the whole plant, he concluded that the difference between the two varieties does not result from the differentiated state of the plant but depends on cellular characteristics. Meredith (1978b, in tomato) and Colijn et al. (1979, in *Petunia*) found that the resistant lines took up the toxic ions to the same extent as the sensitive lines.

Radiation Stress

Radiation, UV or ionizing, is not a natural stress in most habitats, since their intensities on earth are usually too low to injure plants. However, because of the increasing uses of such radiations, especially ionizing radiation, by man, injuries caused by this radiation might become a serious problem in the near future (Levitt, 1980).

Ultraviolet Radiation. When applied at high intensities, UV radiation may damage membranes, nucleic acids, and proteins, and it may also affect hormone balance and enzyme activity. Resistance to UV can result mainly from avoidance due to reflectance or absorbance by specific pigments at the leaf surface. The main advantage in using tissue culture in studies of UV effect on plant tissue is due to the very low penetration of UV into cells organized in the whole plant in multidimensional layers. Therefore, naked protoplasts and, to a lesser degree, cell suspensions are recommended.

The use of tissue culture at present is limited only to physiological and biochemical aspects of radiation. Wright and Murphy (1978) found that short UV radiation induced immediate specific leakage of Rb^+ from tobacco cells in suspension culture. Ohyama et al. (1974) found that protoplasts exposed to UV radiation of more than 192 ergs/mm^2 ceased to divide but remained viable, as shown by vital staining. UV intensity up to 1920 ergs/mm^2 inhibited nucleic acid synthesis to a greater degree than that of protein.

Ionizing Radiation.   The ionizing radiations most commonly investigated as environmental stresses are X-, γ-, and α-rays (Levitt, 1980). Differences in the response to radiation were found among different species and varieties, although plants are not generally subjected to selective force of radiation in nature.   The effects of high doses of ionizing radiation include many kinds of injuries and damage, which affect growth, metabolism, and differentiation.   Since this radiation is highly penetrating, the resistance against it is mainly based on tolerance mechanisms.

The effect of γ-rays on whole plants and isolated tissues was compared in a series of works.   Venketeswaren and Partanen (1966) were the first to compare the effect of γ radiation in the whole plant and in cell culture.   Similar comparisons were made among seeds, seedlings, and callus tissue in Pharbitis nil (L.) Choisy (Rao et al., 1976), Phaseolus vulgaris (Bajaj et al., 1970), and Petunia inflata R.E. Fries (Bapat and Rao, 1976).   The general conclusion drawn from these investigations is that in vitro culture is much more resistant to ionizing radiation than the organized plant.   They suggested that the higher resistance of cells in culture results from their structural simplicity and smaller cellular interdependence as compared with that of the shoot meristem, which is totally dependent upon the plant for its functioning.

PROTOCOLS FOR SELECTING STRESS-RESISTANT MUTANTS

All procedures designed to develop stress-resistant mutant plants from tissue or cell culture include five major steps: (1) preparation of callus and cell suspension culture from sensitive plants, (2) mutagen treatment, (3) selection, (4) test for retention of resistance in culture, (5) regeneration of resistant whole plants, and (6) genetic analysis. With the exception of selection (see Chapter 10) all the steps are identical for all procedures.

Preparation of Callus and Suspension Cell Culture

Callus can be derived from almost any organ of the plant.   The organ most frequently used for this purpose is the leaf.   The exact details of the optimal medium composition and incubation conditions for producing and growing the callus and the suspension culture are frequently specific to the species and sometimes even to the variety. General directions for preparing callus culture are given by Yeoman and MaCleod (1977) and for suspension culture by Street (1977).   The basal medium usually used and which is recommended as a starting point when no specific methods have been worked out for the respective species is either Murashige and Skoog's (1962) or Linsmaier and Skoog's (1965) medium.   The components, which usually vary among media for different plant types, are the quality and quantity of hormones and vitamins.

PROTOCOL FOR CULTURE FROM LEAVES

Callus Culture

1. Detach fully expanded young leaves from young plants.
2. Immerse the leaves in sodium hypochlorite (about 1% active chloride) for 5 min with frequent swirling. When leaves are taken from plants grown in the field instead of a greenhouse, more extreme measures of sterilization may be required, including higher concentration of and longer time in hypochlorite containing a few drops of detergent and additional treatment in 70% ethanol.
3. Rinse the leaves (3x) with sterile deionized water.
4. Cut small disks (1-1.5 cm in diameter) from the leaf and place them on 10 ml of agar-solidified (0.8-1%) medium in 100 x 15 mm plastic petri dishes.
5. Seal the dish with Parafilm and incubate. The incubation conditions, including light quality and intensity, photoperiod, and temperature, are usually specific to each species.
6. Excise and subculture callus, which usually develops at the circumference of the disk, and place it on a fresh medium of the same composition used for the incubation of the disks.
7. Every 2-3 weeks subculture the healthiest and fastest growing calli in a fresh medium. Repeat this subculturing three or four times.

Suspension Culture

1. Place 0.3 g or more (up to 2 g) callus into 50 ml of liquid medium in 250-ml Erlenmeyer flask rotating on a gyrotary shaker. The composition of the medium may be the same as that used for callus growth, but sometimes it may be varied slightly in hormones. The frequency of revolutions and the incubation conditions are usually specific to each plant type.
2. Maintain stock cultures by routine transfer of stationary-phase cells into fresh medium at the appropriate inoculum density.

PROTOCOL FOR CULTURE FROM SEEDS

1. Sterilize seeds for 20 min in sodium hypochlorite (about 1% active chloride containing a few drops of detergent).
2. Rinse the seeds (3x) in sterile deionized water.
3. Place the seeds on a wet sterilized sheet of Whatman No. 1 filter paper in a petri dish and incubate for germination. Light and temperature conditions may be specific to each species.
4. Excise the developing cotyledons, cut small disks or rectangles from them, place them on a solid medium, and incubate as described above.
5. Follow the same procedure (steps 5-7 and 1-2) as for callus and suspension cultures derived from the leaf.

Mutagen Treatment

1. Expose the suspension culture to a treatment with a mutagen, which is added aseptically to the suspension. The mutagen usually used is ethyl methane sulfonate at a concentration range of 0.075–1.5% (v/v) for 60 min.
2. Collect the cells on a fine nylon bolting cloth and wash twice with fresh culture medium.
3. Incubate for at least 3 days in fresh medium before submitting to selection.

Spontaneous mutations for stress resistance can also be obtained without mutagen treatment (see References).

Selection

Since the number of successful experiments for selecting stress-resistant mutants is still very small, no generalization can be made at this stage. The selection procedures described here can be used only as general guidelines. This will include, therefore, details which might be specific to the plant species used in each case. These methods of selection are based on principles used for general isolation of mutants in vitro (see Chapter 10).

Chilling Resistance (*Nicotiana sylvestris* and *Capsicum annuum*, Dix and Street, 1976)

1. Incorporate the fine suspension into melted agar medium (0.9% agar) held at 38 C to give cell densities in the range of $10 \cdot 10^3$ to $50 \cdot 10^3$ cells/ml.
2. Pour the medium immediately, 10 ml per 9 cm plastic petri dish.
3. Seal the dishes and incubate for 5 days at 25 C in the dark.
4. Expose the plates to chilling temperature (5 C for *C. annuum* and 0 C or less for *N. sylvestris*) for 21 days in darkness.
5. Return the plates to 25 C and maintain for 6 weeks (*C. annuum*) or 8–10 weeks (*N. sylvestris*).
6. Establish resistant cell lines by transferring individual growing colonies from the plates to an agar medium in bottles and incubate at 25 C.

Salt Resistance (*Medicago sativa*, Croughan et al., 1978 and *Nicotiana sylvestris* (N) and *Capsicum annuum* (C), Dix and Street, 1975)

Selection on Solid Medium

1. Allow the cells growing in suspension culture to settle for 5 min and decant most of the supernatant, which contains single cells

and small cell aggregates. An alternative (N and C) might be to filter cells of suspension culture (7-10 days after the final subculturing) through 0.6 mm mesh nylon bolting cloth and wash the cells retained with fresh medium.
2. Inoculate petri dishes containing agar-solidified medium supplemented by 0.17 M NaCl (w/v) by spreading aliquots of single cells and small cell aggregates. An alternative (N and C) might be to incorporate the fine suspension into melted agar medium (held liquid at 38 C) containing 0.17 or 0.34 M NaCl to give cell density in the range of $25 \cdot 10^3$ to $35 \cdot 10^3$ cells/ml, and to pour it into 9 cm petri dishes, 10 ml in each.
3. Seal the dishes with Parafilm and incubate under 60 μE/m^2.sec of continuous fluorescent light at 27 ± 1 C.
4. When most of the cells exhibit browning and arrested growth, characteristic of salt toxicity, subculture those pieces that look healthy onto fresh medium containing 0.17 M NaCl (about every 5-6 weeks).
5. Select salt-resistant lines that grow well and exhibit no discoloration during 6-7 such passages in the same saline medium.

Selection in Liquid Medium (*Nicotiana sylvestris* and *Capsicum annuum*, Dix and Street, 1975)

1. Transfer a fine suspension of single cells and small cell aggregates at a density of about $1 \cdot 10^5$ cells/ml to a fresh medium containing 0.17 or 0.34 M NaCl (w/v).
2. After 21 days of incubation, subculture in liquid medium containing the same level of NaCl as in the first passage and continue such subculturing every 21 days as long as the growth continues to improve in the presence of NaCl.
3. Establish salt-resistant cell lines by subculturing fragments of colonies developed from salt-resistant cells in bottles containing agar medium.

Sodium chloride-resistant cells may also be obtained from a cell culture treated with polyethylene glycol. For this alternative approach follow the procedure described for the selection of drought-resistant cells.

Drought Resistance (*Lycopersicon esculentum*, Bressan et al., 1981)

1. Dissolve polyethylene glycol (PEG) (mol. wt. 6000-7600) in double-strength medium prior to the adjustment with distilled water to the final volume.
2. Measure the initial water potential of media containing various amounts of PEG by the appropriate instrumentation. This step is required because the relationship between PEG content and water potential is not directly proportional and because other constituents of the medium, i.e., salts and sugars, influence the water potential.

3. Collect cells from a stationary-phase stock of suspension culture on a fritted glass funnel and resuspend them in a fresh medium containing 15% PEG at a density of approximately 0.2 g/ml in 125-ml Erlenmeyer flasks.
4. Incubate on a gyratory shaker (80–100 rpm) at constant temperature (26 C) under a 16 hr light (1500 lux Cool White fluorescent lamps)/8 hr dark regime.
5. Harvest the cultures at predetermined times by collecting the cells on a Whatman No. 4 filter paper in a Buchner funnel by aspiration.
6. Determine the amount of growth of cell samples by measuring their fresh weight and determining dry weight after drying in an oven at 80 C overnight.

Ion (Aluminum) Resistance (*Lycopersicon esculentum*, Meredith, 1978 ,b)

Callus Culture

1. Place callus on appropriate medium containing 200 μM aluminum as Al-EDTA, five callus pieces (approx. 300 mg each) in each petri dish.
2. Seal the dishes with Parafilm and incubate under the appropriate conditions of light and temperatures.
3. Transfer the callus pieces to fresh medium which contains Al^{3+} every 28 days and regularly examine for areas of growth.

Cell Suspension Culture

1. Spread 1-ml aliquots of cell suspension on agar-solidified medium containing 200 μM aluminum in petri dishes.
2. Incubate the plates at 30 C in the dark and regularly examine for growing colonies.

Test for Retention of Resistance in Culture

1. Grow calli of resistant lines without the selection agent.
2. After 2–4 passages expose the cells again to the selection agent and measure growth when maximal.

Regeneration of Plants

1. Induce shoot formation by transferring callus pieces to a shoot-inducing medium, usually the same basal medium used for culturing. The appropriate hormonal composition may be specific to each plant type.
2. Induce root formation by transferring the shoots to agar medium having hormonal composition suited for root induction specific to each plant type.

3.   Transfer the rooted plants to pots and after hardening transfer them to the greenhouse.

Genetic Analysis

1.   Grow the plants which were regenerated from resistant and sensitive cultures to the flowering stage and self them. Cross between the resistant and sensitive plants to obtain plants of F_1 generation, and self F_1 plants to obtain F_2 generation.
2.   Expose the plants of parental and hybrid generations to the respective stress and follow their response.

FUTURE PROSPECTS

The selection of mutant plants in tissue or cell culture and the production of somatic hybrids by protoplast fusion hold considerable potential for increasing the genetic diversity available to the breeder (Rains et al., 1980).

Inherited stress resistance selected in tissue culture results from rare mutations in few genes with major effects. Whether such monogenic mutations in genes with major effect can have significant potential contribution in breeding programs for stress resistance is a question of central importance. According to Johnson (1980), an advance in genetic improvement of valuable characteristics depends, necessarily, on a broad genetic basis provided by a diverse collection of germplasm. This view, which results from the generally held idea that inheritance of continuous characteristics, including those important for survival, depends on many genes with relatively small effects. However, this idea appears to be open to debate. It seems that the inheritance of many continuous characteristics is determined mainly by a few genes with major effects (Mayr, 1963; Wehrhahn and Allard, 1965; Tal, 1967). The phenotypic expression of continuous characteristics, as contrasted with the discrete expression of Mendelian characteristics, can be explained by a greater relative contribution of interactions among genes and between genes and environmental factors to the phenotype (Falconer, 1960). In my opinion the belief that a wide genetic basis is a prerequisite for genetic progress in improving valuable continuous characteristics results mainly from our ignorance of the complexity of the genetic and epigenetic levels of the multicellular organisms. An interdisciplinary study, including genetics, biochemistry, and physiology of mutations obtained in tissue culture, will help demonstrate the significance of genes with major effects in the control of stress-resistance mechanisms.

As discussed in the Introduction, resistance to stress usually includes tolerance and avoidance mechanisms, the former being probably more primitive adaptations depending mainly on cellular mechanisms, with the latter more advanced and depending more on the organization of cells in tissues and organs. The proportional contribution of avoidance or tolerance mechanisms to the resistance of the plant differs for the

482

various stresses; avoidance, for example, is the least important in low-temperature resistance and more important in salt resistance, while being a major component of drought resistance (Levitt, 1980). On the level of tissue or cell culture, which, to some extent, represents a reversal to a primitive stage, the mutations most expected are, therefore, those involved in mechanisms of tolerance. The contributions of tissue culture techniques to the increase of genetic diversity, through selection of stress-resistant mutations or by somatic hybridization, is expected, therefore, to be greater for resistance against low-temperature stress than for drought stress.

The present biochemical and physiological knowledge of the mechanisms controlling stress resistance is mainly based on studies on the level of whole plants and only very little on tissues or cell culture. These studies are varied and include different cellular aspects, i.e., general metabolic processes such as photosynthesis and respiration, level of various metabolites, activity of enzymes, and membrane characteristics. Based on such studies, which suggest the existence of common cellular targets for different stresses, and on the observation that hardening of the plant against a particular stress might also result in increased resistance to other stresses, Levitt (1980) has hypothesized the concept of a general stress tolerance. As the possible common targets Levitt (1980) suggested the membranes—in which the relevant changes are protein aggregation and lipid peroxidation—and/or the hormonal balance—in which the relevant changes may involve the activity of ethylene and ABA. The unified concept of stress tolerance has, however, been challenged by Steponkus (1980). Whether this hypothesis or its variation is correct will be verified by additional biochemical and physiological studies that will use the most efficient methodology and concentrate on the central questions. The use of stress-resistant cell lines originating from mutations in major genes is highly recommended for such a study. Central targets to be studied may include membrane characteristics, kinetics of enzymes that play a central role in the control of intermediary metabolism such as RuBP-carboxylase in the Calvin cycle and phosphofructokinase in the glycolysis pathway (Hochachka and Somero, 1973; Teeri, 1980), and hormonal regulation. There is an increasing evidence that ABA has a central role in the integration of plant responses to stress (Jones, 1980). According to Boyer (1980), the identification and understanding of the regulating mechanisms that control adaptation is one of the most important subjects to be studied in the area of stress adaptation.

Another question of central importance is whether identification and understanding of the mechanisms of stress resistance and the interrelationships among them will help in planning more efficient procedures designed to select stress-resistant mutations in tissue culture or to select stress-resistant hybrids produced by fusion of protoplasts of resistant and sensitive species. The general technique used at present for the selection of stress-resistant mutant cells is based on the application of behavioral criteria, i.e., survival or growth under the influence of the selection agent, irrespective of the stress-resistance mechanism involved. The understanding of the resistance mechanisms underlying a specific stress may offer an alternative approach to selection, which will be directed more specifically to the mode of action of the resistance. Such an approach was first recommended by Goldner et al. (1977) for the selection of salt-resistant mutations. It seems that the present knowledge of the mechanisms of stress resistance and the number of successful selections of resistant mutant cells in tissue culture are still too limited to be used as guidelines for detailed design of specific selection procedures.

The operation of mechanisms of stress resistance requires energy. The advantage of tissue culture for estimating the energy cost of such mechanisms was recommended by Rains et al. (1980). Nevertheless, the most meaningful evaluation of the price paid by the plant in terms of yield for being more resistant to a stress can only be made by incorporating the resistance mechanisms into commercial varieties. Here, their influence on the yield under the prevailing environmental conditions can be estimated. However, in most cases such proofs are lacking (Bidinger, 1980), since plant breeders are reluctant to breed and select for physiological mechanisms without this type of evidence, which depends, in its turn on breeding. Breaking this vicious circle is, thus, essential.

ACKNOWLEDGMENT

I thank Ms. Dorot Imber and Dr. Alan Witztum for their valuable comments, and Ms. Ruth Massil for her skillful typing.

KEY REFERENCES

Casey, H.E. 1972. Salinity problems in arid lands irrigation: A literature review and selected bibliography. Arid Lands Resource Information Paper No. 1, Univ. Arizona Office of Arid Lands Studies, Tucson, Arizona.

Dix, P.J. 1980. Environmental stress resistance, selection and plant cell cultures. In: Plant Cell Cultures: Results and Perspectives (F. Sala, B. Parisi, R. Cella, and O. Ciferri, eds.) pp. 183-186. Elsevier/ North-Holland Biomedical Press, Amsterdam, New York, and Oxford.

Levitt, J. 1980. Responses of Plants to Environmental Stresses, 2nd ed. Academic Press, New York.

Tal, M., Heiken, H., and Dehan, K. 1978. Salt tolerance in the wild relatives of the cultivated tomato: Responses of callus tissues of *Lycopersicon esculentum*, *L. peruvianum* and *Solanum pennellii* to high salinity. Z. Pflanzenphysiol. 86:231-240.

REFERENCES

Babaeva, Zh.A., Butenko, R.G., and Strogonov, B.P. 1968. Influence on salinization of the nutrient medium on the growth of isolated carrot tissue. Soviet Plant Physiol. 15:75-82.

Bajaj, Y.P.S., Saettler, A.W., and Adams, M.W. 1970. Gamma irradiation studies on seeds, seedlings and callus tissue cultures of *Phaseolus vulgaris* L. Rad. Bot. 10:119-124.

Bapat, V.A. and Rao, P.S. 1976. Differential radiosensitivity of seeds, seedlings and callus cultures of *Petunia inflata*. Plant Sci. Lett. 6: 291-298.

Bassiri, A. and Carlson, P.S. 1978. Isozyme patterns and differences in plant parts and their callus cultures in common bean. Crop Sci. 18: 955-958.

Bernstein, L. 1963. Osmotic adjustment of plants to saline media. II. Dynamic phase. Am. J. Bot. 50:360-370.

Bidinger, F. 1980. Breeding for drought resistance. In: Adaption of Plants to Water and High Temperature Stress (N.C. Turner and P.J. Kramer, eds.) pp. 452-454. John Wiley and Sons, New York.

Binding, H. 1974. Mutation in haploid cell cultures. In: Haploids in Higher Plants: Advances and Potentials (K.J. Kasha, ed.) pp. 323-327. Univ. Guelph Press, Guelph.

Blum, A. 1980. Genetic improvement of drought adaptation. In: Adaption of Plants to Water and High Temperature Stress (N.C. Turner and P.J. Kramer, eds.) pp. 450-452. John Wiley and Sons, New York.

Boyer, J.S. 1980. Physiological adaptations to water stress. In: Adaption of Plants to Water and High Temperature Stress (N.C. Turner and P.J. Kramer, eds.) pp. 443-444. John Wiley and Sons, New York.

Breidenbach, R.W. and Wareing, A.J. 1977. Response to chilling of tomato seedlings and cells in suspension cultures. Plant Physiol. 60: 190-192.

Bressan, R.A., Hasegawa, P.M., and Handa, A.K. 1981. Resistance of cultured higher plant cells to polyethylene glycol-induced water stress. Plant Sci. Lett. 21:23-30.

Chaleff, R.S. and Carlson, P.S. 1974. Somatic cell genetics of higher plants. Annu. Rev. Genet. 8:267-277.

_____ 1975. In vitro selection for mutants of higher plants. In: Genetic Manipulation with Plant Material (L. Ledoux, ed.) pp. 351-364. Plenum, New York and London.

Chaleff, R.S. and Parsons, M.F. 1978. Direct selection in vitro for herbicide-resistant mutants of *Nicotiana tabacum*. Proc. Natl. Acad. Sci. 75:5104-5107.

Chen, H.H., Gavinlertvatana, P., and Li, P.H. 1979 Cold acclimation of stem-cultured plants and leaf callus of *Solanum* species. Bot. Gaz. 140:142-147.

Chen, Y., Zahavi, E., Barak, P., and Umiel, N. 1980. Effects of salinity stresses on tobacco. I. The growth of *Nicotiana tabacum* callus cultures under seawater, NaCl and mannitol stresses. Z. Pflanzenphysiol. 98:141-153.

Colijn, C.M., Kool, A.J., and Nijkamp, H.J.J. 1979. An effective chemical mutagenesis procedure for *Petunia hybrida* cell suspension cultures. Theor. Appl. Genet. 55:101-106.

Croughan, T.P., Stavarek, S.J., and Rains, D.W. 1978. Selection of NaCl tolerant line of cultured alfalfa cells. Crop Sci. 18:959-963.

Dix, P.J. 1977. Chilling resistance is not transmitted sexually in plants regenerated in *Nicotiana sylvestris* cell lines. Z. Pflanzenphysiol. 84: 223-226.

_____ and Street, H.E. 1975. Sodium chloride-resistant cultured cell lines from *Nicotiana sylvestris* and *Capsicum annuum*. Plant Sci. Lett. 5:231-237.

_____ and Street, H.E. 1976. Selection of plant cell lines with enhanced chilling resistance. Ann. Bot. 40:903-910.

Epstein, E. 1972. Mineral Nutrition in Plants: Principles and Perspectives. John Wiley and Sons, New York.

_____ 1976. Adaptation of crops to salinity. In: Proceedings Workshop Plant Adaptaton to Mineral Stress in Problem Soils (M.J. Wright, ed.) pp. 73-82, Cornell Univ., Ithaca, New York.

_____ and Norlyn, J.D. 1977. Seawater based crop production: A feasibility study. Science 197:249-251.

Falconer, D.S. 1960. Quantitative Genetics. Ronald Press, New York.

Goldner, R., Umiel, N., and Chen, Y. 1977. The growth of carrot callus cultures at various concentrations and compositions of saline water. Z. Pflanzenphysiol. 85:307-318.

Greenway, H. and Munns, R. 1980. Mechanisms of salt tolerance in non-halophytes. Annu. Rev. Plant Physiol. 31:149-190.

Heyser, J.W. and Nabors, M.W. 1979. Osmotic adjustment of tobacco cells and plants to penetrating and non-penetrating solutes. Plant Physiol. Suppl. 63:77.

_____ 1981. Osmotic adjustment of cultured tobacco cells (*Nicotiana tabacum*. var. samsum) grown on sodium chloride. Plant Physiol. 67:720-727.

Hochachka, P.W. and Somero, G.N. 1973. Strategies of biochemical adaptations. Saunders, Philadelphia.

Johnson, D.A. 1980. Improvement of perennial herbaceous plants for drought-stressed western range lands. In: Adaptation of Plants to Water and High Temperature Stress (N.C. Turner and P.J. Kramer, eds.) pp. 419-434. John Wiley and Sons, New York.

Jones, H.G. 1980. Interaction and integration of adaptive responses to water stress. The implications on an unpredictable environment. In: Adaption of Plants to Water and High Temperature Stress (N.C. Turner and P.J. Kramer, eds.) pp. 353-365. John Wiley and Sons, New York.

Kochba, J., Spiegel-Roy, P., and Saad, S. 1980. Selection for tolerance to sodium chloride (NaCl) and 2,4-dichlorophenoxyacetic acid (2,4-D) in ovolar callus lines of *Citrus sinensis*. In: Plant Cell Cultures: Results and Perspectives (F. Sala, B. Parisi, R. Cella, and O. Ciferri, eds.) pp. 187-192. Elsevier/North-Holland Biomedical Press, Amsterdam, New York, and Oxford.

Kramer, P.J. 1980. Drought, stress and the origin of adaptations. In: Adaptation of Plants to Water and High Temperature Stress (N.C. Turner and P.J. Kramer, eds.) pp. 7-20. John Wiley and Sons, New York.

Kulieva, F.B., Shamina, Z.B., and Strogonov, B.P. 1975. Effect of high concentrations of sodium chloride on multiplication of cells of *Crepis capillaris* in vitro. Soviet Plant Physiol. 22:107-110.

Linsmaier, E.M. and Skoog, F. 1965. Organic growth factor requirements of tobacco tissue culture. Physiol. Plant 18:100-127.

Lyons, J.M. and Raison, J.K. 1970.   Oxidative activity of mitochondria isolated from plant tissue sensitive and resistant to chilling injury. Plant Physiol. 45:386-389.

Mayr, E. 1963.   Animal Species and Evolution.   Harvard Univ. Press, Cambridge.

McKersie, B.D. and Thompson, J.E. 1979.   Influence of plant sterols on the phase properties of phospholipid bilayers.   Plant Physiol. 63:802-805.

Melchers, G. 1972.   Haploid higher plants for plant breeding.   Z. Pflanzensuchtg. 67:19-32.

Meredith, C.P. 1978a.   Selection and characterization of aluminum-resistant variants from tomato cell cultures.   Plant Sci. Lett. 12:25-34.

_____ 1978b.   Response of cultured tomato cells to aluminum.   Plant Sci. Lett. 12:17-24.

Muehlbach, H.P. and Thiele, H. 1981.   Response to chilling of tomato mesophyll protoplasts.   Planta 151:399-401.

Murashige, T. and Skoog, F. 1962.   A revised medium for rapid growth and bioassays with tobacco tissue culture.   Physiol. Plant 15:473-497.

Nabors, M.W. 1976.   Using spontaneously occurring and induced mutations to obtain agriculturally useful plants.   BioScience 26:761-768.

_____, Daniels, A., Nadolny, L., and Brown, C. 1975.   Sodium chloride tolerant lines of tobacco cells.   Plant Sci. Lett. 4:155-159.

_____, Gibbs, S.E., Bernstein, C.S., and Meis, M.E. 1980.   NaCl-tolerant tobacco plants from cultured cells.   Z. Pflanzenphysiol. 97:13-17.

Niki, T., Yoshida, S., and Sakai, S. 1978.   Studies on chilling injury in plant cells.   I. Ultrastructural changes associated with chilling injury in callus tissues of *Cornus stolonifera*.   Plant Cell Physiol. 19:139-148.

Ogolevets, I.V. 1976.   Hardening of isolated callus tissue of woody plants with different frost resistances.   Soviet Plant Physiol. 23:115-119.

Ohyama, K., Pelcher, L.E., and Gamborg, O.L. 1974.   The effect of ultra-violet irradiation on survival and on nucleic acids and protein synthesis in plant protoplasts.   Rad. Bot. 14:343-346.

Orton, T.J. 1980.   Comparison of salt tolerance between *Hordeum vulgare* and *H. jubatum* in whole plants and callus cultures.   Z. Pflanzenphysiol. 98:105-118.

Premecz, G., Olah, T., Gulyas, A., Nyitrai, A., Palfi, G., and Farkas, G.L. 1977.   Is the increase in ribonnuclease level in isolated tobacco protoplasts due to osmotic stress?   Plant Sci. Lett. 9:195-200.

Premecz, G., Ruzicska, P., Olah, T., and Farkas, G.L. 1978.   Effect of "osmotic stress" on a protein and nucleic acid synthesis in isolated tobacco protoplasts.   Planta 141:33-36.

Quamme, H.A. 1978.   Breeding and selecting temperate fruit crop for cold hardiness.   In: Plant Cold Hardiness and Freezing Stress.   Mechanisms and Crop Implications (P.H. Li and A. Sakai, eds.) pp. 313-332.   Academic Press, New York.

Rains, D.W., Croughan, T.P., and Stavarek, S.J. 1980.   Selection of salt-tolerant plants using tissue culture.   In: Genetic Engineering of Osmoregulation.   Impact on Plant Productivity for Food, Chemicals, and Energy (D.W. Rains, R.C. Valentine, and A. Hollaender, eds.) pp. 279-292.   Plenum, New York and London.

Rao, P.S., Harada, H., and Bapat, V.A. 1976. A comparative study of the differential radiosensitivity of seeds, seedlings and tissue cultures of the Japanese morning glory (*Pharbitis nil*). Plant Cell Physiol. 17: 119–125.

Rosen, A. and Tal, M. 1981. Salt tolerance in the wild relatives of the cultivated tomato: Response of protoplasts isolated from leaves of *Lycopersicon esculentum* and *L. peruvianum* plants to NaCl and proline. Z. Pflanzenphysiol. 102:91–94.

Ruesink, A.W. 1978. Leucine uptake and incorporation by *Convolvulus* tissue culture cells and protoplasts under severe osmotic stress. Physiol. Plant. 44:48–56.

Siminovitch, D. 1979. Phrotoplastic surviving freezing to –196 C and osmotic dehydration in 5 molar salt solutions prepared from the bark of winter black locust trees. Plant Physiol. 63:722–725.

Singh, J. 1979. Freezing of protoplasts isolated from cold-hardened and non-hardened winter rye. Plant Sci. Lett. 16:195–201.

_____ 1981. Isolation and freezing tolerance of mesophyll cells from cold-hardened and non-hardened winter rye. Plant Physiol. 67:906–909.

Smillie, R.M., Melchers, G., and von Wettstein, D. 1979. Chilling resistance of somatic hybrids of tomato and potato. Carlsberg Res. Commun. 44:127–132.

Smith, M.K. and McComb, J.A. 1981. Effect of NaCl on the growth of whole plants and their corresponding callus cultures. Aust. J. Plant Physiol. 8:267–275.

Steponkus, P.L. 1980. A unified concept of stress in plants? In: Genetic Engineering of Osmoregulation. Impact on Plant Productivity for Food, Chemistry, and Energy (D.W. Rains, R.C. Valentine, and A. Hollaender, eds.) pp. 235–255. Plenum, New York and London.

Street, H.E. 1975. Plant cell cultures: Present and projected applications for studies in genetics. In: Genetic Manipulations in Plant Material (L. Ledoux, ed.) pp. 351–364. Plenum, New York and London.

_____ 1977. Cell (suspension) culture-techniques. In: Plant Tissue and Cell Culture (H.E. Street, ed.) pp. 61–102. Blackwell, Oxford.

Strogonov, B.P. 1973. Salt tolerance in isolated tissues and cells. In: Structure and Function of Plant Cells in Saline Habitats: New Trends in the Study of Salt Tolerance (B. Gollek, ed.) pp. 1–33. Israeli Program for Scientific Translations, Jerusalem.

Tal, M. 1967. Genetic differentiation and stability of some characters that distinguish *Lycopersicon esculentum* Mill. from *Solanum pennellii* Cor. Evolution 21:316–333.

_____ and Watts, J.W. 1979. Plant growth conditions and yield of viable protoplasts isolated from leaves of *Lycopersicon esculentum* and *L. peruvianum*. Z. Pflanzenphysiol. 92:207–214.

Teeri, J.A. 1980. Adaptation of kinetic properties of enzymes to temperature variability. In: Adaptation of Plants to Water and High Temperature Stress (N.C. Turner and P.J. Kramer, eds.) pp. 251–260. John Wiley and Sons, New York.

Toivio-Kinnucan, M.A., Chen, H.H., Li, P.H., and Stushnoff, C. 1981. Plasma membrane alterations in callus tissues of tuber-bearing *Solanum* species during cold acclimation. Plant Physiol. 67:478–483.

488   Specialized Cell Culture Techniques

Tumanov, I.I., Butenko, R.G., Ogolevets, I.V., and Smetyuk, V.V. 1977. Increasing the frost resistance of a spruce callus tissue culture by freezing out the less resistant cells. Soviet Plant Physiol. 24:728-732.

Turner, N.C. and Jones, M.M. 1980. Turgor maintainance by osmotic adjustment: A review and evaluation. In: Adaptation of Plants to Water and High Temperature Stress (N.C. Turner and P.J. Kramer, eds.) pp. 87-104. John Wiley and Sons, New York.

Venketeswaren, S. and Partanen, C.R. 1966. A comparative study of the effects of gamma radiation on organized and disorganized growth of tobacco. Rad. Bot. 6:13-20.

Von Hedenstrom, H. and Breckle, S.W. 1974. Obligate halophytes? A test with tissue culture methods. Z. Pflanzenphysiol. 74:183-185.

Waddington, C.H. 1962. New patterns in genetics and development. Columbia Univ. Press, New York.

Wahlstrom, D. and Eriksson, T. 1976. Uptake of ^{14}C-L-glutamic acid by *Daucus carota* cell suspension in different shock situations. Physiol. Plant. 38:138-140.

Wegmann, K. and Muehlbach, H.P. 1973. Photosynthetic CO_2 incorporation by isolated leaf cell protoplasts. Biochem. Biophys. Acta 314:79-82.

Wehrhahn, C.F. and Allard, R.W. 1965. The detection and measurement of the effects of individual genes involved in the inheritance of a quantitative character in wheat. Genetics 51:109-119.

Wiest, S.C. and Steponkus, P.L. 1978. Freeze-thaw injury to isolated spinach protoplasts and its simulation above freezing temperatures. Plant Physiol. 62:699-705.

Widholm, J.M. 1976. Selection and characterization of biochemical mutants. In: Plant Tissue Culture and its Bio-technological Application (W. Barz, E. Reinhard, and M.H. Zenk, eds.) pp. 112-122. Springer-Verlag, Berlin and New York.

Wright, L.A. and Murphy, T.M. 1978. Ultraviolet radiation-stimulated efflux of 86-rubidium from cultured tobacco cells. Plant Physiol. 61:434-436.

Yamada, Y., Hara, Y., Katagi, H., and Senda, M. 1980. Protoplast fusion. Effect of low temperature on the membrane fluidity of cultured cells. Plant Physiol. 65:1099-1102.

Yeoman, M.M. and MaCleod, A.J. 1977. Tissue (callus) cultures-techniques. In: Plant Tissue and Cell Culture (H.E. Street, ed.) pp. 31-60. Blackwell, Oxford.

Yoshida, S. and Niki, T. 1979. Cell membrane permeability and respiratory activity in chilling stressed callus. Plant Cell Physiol. 20:1237-1242.

Yoshida, S. and Tagawa, F. 1979. Alteration of respiratory function in chill-sensitive callus due to low temperature stress. I. Involvement of the alternate pathway. Plant Cell Physiol. 20:1243-1250.

Zenk, M.H. 1974. Haploids in physiological and biochemical research. In: Haploids in Higher Plants: Advances and Potential (K.J. Kasha, ed.) pp. 339-354. Univ. Guelph Press, Guelph.

CHAPTER 13
Selection for Photoautotrophic Cells

Y. Yamada and *F. Sato*

Cultured photoautotrophic cells for higher plants provide new materials for research on photosynthesis and for increasing the productivity of plant cells. They also have the economic advantage of using solar energy directly. Some cultured cells have chloroplasts with developed grana (Laetsch and Stetler, 1965; Seyer et al., 1975) and maintain photosynthetic activity (Bergmann, 1967; Hanson and Edelman, 1972). Light promotes the growth of green cells (Neumann and Raafat, 1973), but most green cells cannot grow without sugar. Many attempts to culture green tissues photoautotrophically have been made, with some success for short periods or with a low growth rate. Lack of vigorous photoautotrophic growth has been attributed to the low chlorophyll content and low photosynthetic activity of these green cells.

There are many studies on the physicochemical conditions and nutritional elements required for the growth and greening of green plant cells. These include research on auxins (Bergmann, 1967; Sunderland, 1966), cytokinins (Kaul and Sabharwal, 1971), sugars (Edelman and Hanson, 1971; Neumann and Raafat, 1973), inorganic nutrients (Vasil and Hildebrandt, 1966), light intensity (Bergmann and Balz, 1966), and the gas phase (Corduan, 1970; Dalton and Street, 1976). Although the number of studies has increased, only a few reports of successful photoautotrophic cultures have been made (Berlyn and Zelitch, 1975; Dalton, 1980; Husemann and Barz, 1977; Yamada and Sato, 1978; Yasuda et al., 1980).

Cultured cells consist of cells that are heterogeneous in their specific characters. Therefore, cells that show the particular character desired must be selected in addition to regulating the culture conditions.

489

490 Specialized Cell Culture Techniques

Yamada and Sato (1978) selected highly chlorophyllous cells of tobacco (*Nicotiana tabacum* var Samsun) and scotch broom (*Cytisus scoparius* L.) from photomixotrophic cultures and succeeded in culturing them photo-auxotrophically. Berlyn and Zelitch (1975) derived white and green haploid calli from an anther culture of yellow mutant tobacco, and their green calli had rapid net photosynthesis. Husemann and Barz (1977) made continuous selective subcultures of *Chenopodium rubrum* cells that had survived under photoautotrophic conditions until they obtained cell suspensions with high growth rates. These studies show that different types of cultured cells differ in their photosynthetic capacities.

The chlorophyll content of cultured cells usually is used as the criterion for the selection of photoautotrophic cells, but it is not the best index for photoautotrophy, as seen from the fact that we could not predict the low photoautotrophic growth of green amur cork tree (*Phellodendron amurense* Rupr.) cells. Measurement of photosynthetic O_2 evolution showed that there was very low photosynthetic activity in the amur cork tree cells, whereas the photosynthetic potential of tobacco and scotch broom cells in photomixotrophic culture was closely correlated with their photoautotrophic growth (Sato et al., 1979; Fig. 1). We confirmed that it is essential to select cells that have high photosynthetic potential if successful photoautotrophic cultures are to be established.

We here describe an efficient method for selecting cells that are capable of photoautotrophic growth. Photosynthetic potential is the criterion used to select these photoautotrophic cells.

MAJOR FACTORS FOR SUCCESSFUL PHOTOAUTOTROPHIC CULTURE

Plant Materials

Almost every part of the plant has been used for callus induction, but seeds and seedlings have been particularly common and easy materials. Whatever starting material is used, it must be sterilized in the usual way then inoculated on a culture medium.

Callus can be induced from many species of plants under light. The calli induced usually consist of cells that show different degrees of greening, even though they are derived from the same segments. Different plant species also have different potentials for the development of chloroplasts in callus. So far, only six varieties of plant cells have been reported to grow well in photoautotrophic culture: *Chenopodium rubrum*, *Cytisus scoparius* L., *Hyoscyamus niger* L., *Nicotiana tabacum* var JWB su/su, *Nicotiana tabacum* var Samsun, and *Spinacea oleracea*.

Culture Medium

The composition of the culture medium is important for the selection of green cells. 2,4-D is a common auxin used for callus induction and propagation, but 2,4-D inhibits chlorophyll synthesis in intact leaves

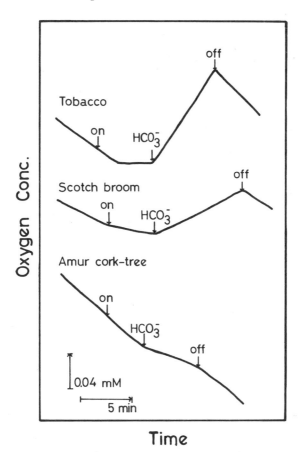

Figure 1. Oxygen exchange by cultured chlorophyllous cells. Photo-synthetic O_2 evolution and respiratory O_2 uptake were traced with an oxygen electrode. CO_2 was added as $NaHCO_3$ at a concentration of 1 mM. Cultured cells (0.1 g fresh weight) were suspended in 20 mM phosphate buffer, pH 7.8. On: light on; Off: light off.

and calli (Shewry et al., 1971; Sunderland, 1966). IAA, a natural aux-in, is favorable, but it decomposes easily under illumination. NAA is more effective than either IBA or 2,4-D for promoting active chloro-phyll synthesis (Yamada et al., 1979). Cytokinins, which regulate cytodifferentiation with auxins, are known to promote greening in non-green calli (Kaul and Sabharwal, 1971); hence we usually add BA or KIN to the culture medium. From our experience, a hormonal combina-tion of about 10 µM NAA and about 1 µM BA is best for use with the Linsmaier–Skoog basal medium (1965). This combination of basal medium and hormone may not necessarily be best for induction and propagation of all types of green callus, but so far it has been the most useful.

The addition of organic carbon sources to the medium at the time callus is induced is needed for the culture of green cells. But the addition of sugars, especially sucrose which is a product of photosynthesis, inhibits photosynthetic activity and the greening of cultured cells. After callus has been induced, selected green cells must be quickly transferred to conditions for photoautotrophic culture or the desired character will be lost.

Culture Conditions

The factors essential for establishment of successful photoautotrophic cultures are the selection of cell lines with high photosynthetic potential and the use of culture conditions favorable for photosynthesis; i.e., an adequate intensity of light and an enriched CO_2 concentration. A high CO_2 concentration (about 1-2% v/v) is maintained in cultures by bubbling CO_2-enriched air through distilled water then into culture flasks whose air inlets and outlets are plugged with cotton. Alternatively, CO_2 enriched air is led into transparent cabinets that contain petri dishes, or culture flasks plugged with cotton or silicone sponge (Fig. 2). Another way to maintain high levels of CO_2 in cultures is to use a 2 M $KHCO_3/K_2CO_3$ buffer solution as the CO_2 reservoir (Husemann and Barz, 1977). All cells then are cultured at a suitable temperature (about 26 C) and under an adequate intensity of light (6000 to 10,000 lux).

Large-Scale Culture

The culture conditions used for a jar fermenter differ from those used for flasks. Cultured cells have a high apparent viscosity and a sensi-

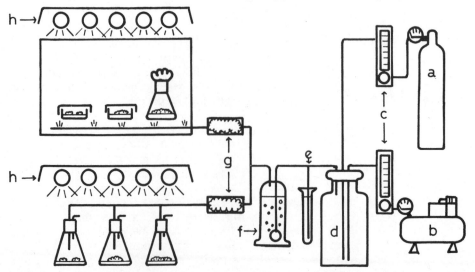

Figure 2. The system for the selection of cultured cells with high photosynthetic potential. (a) CO_2 gas, (b) air compressor, (c) flow control, (d) reservoir for mixed gas, (e) a safety valve, (f) distilled water (washing for gas), (g) air line filter (cotton filter, etc.), (h) illumination.

tivity to shear stress because of the relatively large volume of rigid cells (Wagner and Vogelmann, 1977). Mixing by slow agitation with a marine-type impeller and aeration has produced successful cultures of plant cells in a jar fermenter (Yamada et al., 1981). The culture conditions used for photoautotrophic culture in a jar fermenter also differ from those used for photomixotrophic culture. Because stimulated cell respiration compensates for the net photosynthesis, green cells cultured photoautotrophically in a jar fermenter in air enriched with 1% CO_2 have no apparent photosynthetic activity.

A low oxygen supply also is essential for photoautotrophic culture in a jar fermenter. A lowered oxygen supply has been produced by reducing the rate of aeration and by using air for which the partial pressure of oxygen has been lowered by the addition of N_2 gas; this enhances photoautotrophic growth. Similar results have been obtained by Dalton (1980), who cultivated *Spinacea* cells photoautotrophically in a continuous culture by controlling the dissolved oxygen concentration.

Measurements of Growth and Chlorophyll Content

Fresh weights should be determined only after excess water has been removed by blotting the cells with paper. After 2 days in an oven at 60 C the dry weights of the samples can be taken. The chlorophyll contents are determined spectrophotometrically in an 80% v/v acetone extract. Chlorophyll is extracted by the method of Sunderland (1966), and its concentration is calculated from the equations derived by Arnon (1949).

Measurements of Oxygen Exchange by Cultured Cells

Oxygen exchange is measured at 25 C with an oxygen electrode after 0.1 g cells (fresh weight) have been suspended in 1 ml of 50 mM phosphate buffer (pH 7.8). To measure the photosynthetic oxygen evolution, cell suspensions that contain 5 mM bicarbonate are illuminated with light from a projector that has been filtered through a 10 cm water layer at an intensity of about 100,000 lux.

ACTUAL EXAMPLE (YASUDA ET AL., 1980)

Three species from the family Solanaceae, *Atropa belladonna*, *Datura stramonium*, and *Hyoscyamus niger*, were used. Segments of excised leaves from redifferentiated and aseptically grown seedlings were inoculated on sugar-free, Linsmaier-Skoog agar media in petri dishes. The hormones used were combinations of 5 μM or 10 μM NAA and 0.05 μM, 0.5 μM or 5 μM BA. The petri dishes were placed in transparent 20 liter glass cabinets and, from the beginning of callus induction, were aerated with a mixture of 1% CO_2 in air at a flow rate of 1 liter per minute under continuous illumination with fluorescent lamps at an intensity of 3000-5000 lux. The temperature inside the cabinets ranged from 27 to 29 C.   After callus induction, subsequent transfers were maintained under these same photoautotrophic conditions.

After being cultured under the above photoautotrophic conditions for about 2 weeks, the inoculated leaf segments began to swell.   Green calli were induced in small areas along the edges of each segment in all the *Hyoscyamus* and *Datura* cultures, but the other parts of the tissues turned white and died (Fig. 3a).   The greenest cells from these calli were subcultured under photoautotrophic conditions (Fig. 3b, Fig. 4).   In the *Atropa* cultures, no callus induction was observed but the segments remained swollen.   Some segments of the *Atropa* leaves developed roots.

Data for photoautotrophic growth, chlorophyll content, and photosynthetic O_2 evolution in *Hyoscyamus*, and *Datura* cultures are summarized in Table 1.   The chlorophyll content of the *Datura* cells was much higher than that of the *Hyoscyamus* cells, but the increase in fresh weight for *Hyoscyamus* cells cultured photoautotrophically was higher than for *Datura* cells.   The photosynthetic activity of *Hyoscyamus* cells was also higher than for *Datura*, and the photosynthetic activity of *Hyoscyamus* cells (based on chlorophyll) was higher than that of regenerated *Hyoscyamus* seedlings.

The cultured *Hyoscyamus* cells were divided into small pieces, each of which was cultured photoautotrophically.   The relationship between the photosynthetic activity and photoautotrophic growth of each *Hyoscyamus* sample is shown in Fig. 5.   There was a clear correlation between photosynthetic activity and photoautotrophic growth, but no relationship between photoautotrophic growth and the chlorophyll content of *Hyoscyamus* cells.

DISCUSSION

Cells with high photosynthetic potential first must be obtained to establish photoautotrophic cultures.   As stated in the introduction, there

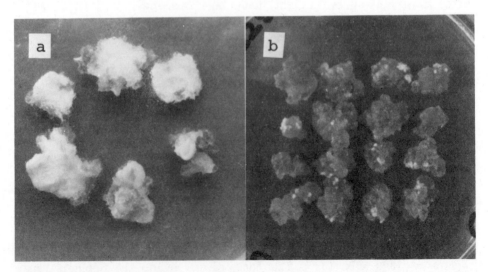

Figure 3.   Photoautotrophic culture of *Hyoscyamus* cells (a) at callus induction, (b) after selecting-subculture.

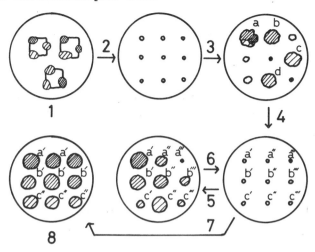

Figure 4.   Protocol for the selection of photoautotrophic cells.   (1)
Calli are induced under photoautotrophic or photomixotrophic conditions.   (2) Calli are broken into small pieces.   (3,5,7) Small pieces of
calli are cultured photoautotrophically.   (4,6) Green cells that have
grown well are selected and then broken into small pieces and recultured.   (8) Establishment of photoautotrophic cells which grow well
without sugars.   Steps (5) and (6) are repeated until homogeneous
photoautotrophic cells that grow well are produced and proliferate.
The number of oblique lines indicate the degree of greening.

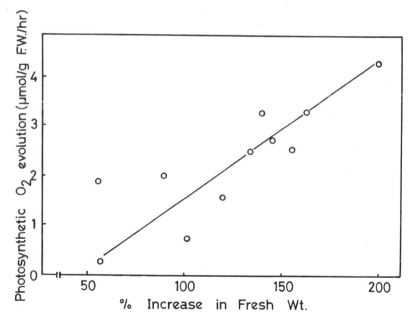

Figure 5.   Relationship between photosynthetic activity and the photoautotrophic growth of *Hyoscyamus* cells.   Each culture was plated with
1-2 g of cells, then harvested after 3 weeks of incubation.   Each value
is the mean of one to six replications.

Table 1. Photoautotrophic Properties in *Hyoscyamus niger* and *Datura stramonium* Cultures

MATERIALS	INCREASE IN F.W.	CHLOROPHYLL CONTENT	PHOTOSYNTHETIC O_2 EVOLUTION	
	(%)	(μg/g F.W.)	(μmol/g F.W./hr)	(μmol/mg Chl/hr)
Cultured *Datura* cells	55	129.6	3.4	26
Cultured *Hyoscyamus* cells	198	31.1	4.3	138
Regenerated *Hyoscyamus* seedlings	—	510.2	25.4	50

Hyoscyamus cultures were subcultured for 5 passages of 3 weeks, and *Datura* cultured for 7 passages of 3 weeks. Hormone concentrations were 10 μM NAA and 5 μM BA for *Hyoscyamus* and 5 μM NAA and 0.05 μM BA for *Datura* cells. For culture conditions see the text. Seedlings were grown with 2% sucrose in hormone-free medium under illumination.

have been many attempts to enhance photosynthetic activity by metabolic regulation. These attempts have been based on the belief that
all cells are homogeneous and that all have totipotency. Our experiment, however, started with the assumption that all cultured cells are
heterogeneous; thus they may function and respond differently to environmental selection. Therefore, cells must be selected that have high
photosynthetic potential if photoautotrophic culture is to be successful.

It was difficult to obtain highly chlorophyllous cells that had high
photosynthetic potential; however, newly derived calli illuminated from
the beginning of induction on a medium of simple composition, produced
green calli. For cultures maintained in media with sucrose, under conditions that are unsuitable for photosynthesis, cells which have obtained
most of their energy from sucrose are allowed to proliferate and to
predominate. After several years of subcultures under these conditions,
we have found that in some cases there has been a significant loss of
cells that have high photosynthetic potential. This is a problem often
encountered when plant cells are cultured, but natural selection most
likely occurs in any cell culture. One way to solve this problem is to
select only cells that have high photosynthetic potential from the first
culture and during each successive subculture. The effectiveness of
this method is, we believe, dependent on both the time and conditions
of selection.

Selected photoautotrophic cells have been cultured under conditions
favorable for photosynthesis under an adequate intensity of illumination
and aeration with CO_2-enriched air. However, it is difficult to culture
any type of green cell photoautotrophically under aeration with ordinary air (about 0.03% CO_2). In green tobacco cells the CO_2 compensa-
tion point for photosynthesis is high (Tsuzuki et al., 1981). This may
reflect the low concentration of CO_2 in cultured green cells, which is
due to low carbonic anhydrase activity and to the high diffusion resistance of cultured cells with large cell volumes, and/or high activity of
dark respiration in cultured green cells. Instead of aeration with CO_2-
enriched air, we may need to stimulate carbonic anhydrase activity or
to suppress respiration in cultured green cells.

To date no successful cultures of the green cells of cereals have
been reported. Callus induction from cereals requires a high concentration of auxin in the medium, but this inhibits chloroplast differentiation. Before this problem can be addressed, we need to know the detailed mechanism for the propagation and differentiation of chloroplasts
within cultured cells.

FUTURE PROSPECTS

The culture of photoautotrophic cells is a new method for potentially
increasing the productivity of plant cells. Some alkaloids (Hartmann et
al., 1980), a vitamin (Watanabe et al., 1982), and a volatile oil (Corduan and Rehard, 1972), all of which are produced in the green parts
of intact plants also have been produced by cultured green cells. If
the production of these metabolites is regulated by cellular differentiation, then photoautotrophic culture will increase the productivity of

cultured cells. The economic advantage of using solar energy directly and the decreased possibility of contamination by microorganisms make photoautotrophic cell culture a very useful technique for industrial purposes.

Intact plants probably consist of cells that are heterogeneous in their specific characters, and the culture of these cells enlarges the variation in specific characters. The selection of photoautotrophic cells is an effective new method that allows us to obtain cell lines that have high photosynthetic potential and that will, in turn, regenerate plants with enhanced photosynthetic potential. Photoautotrophically cultured green cells of C_3 plants incorporated large amounts of $^{14}CO_2$ into C_4 compounds (mainly as malate) in light, and they showed phosphoenol pyruvate (PEP) carboxylase activity (Sato et al., 1980). PEP carboxylase concentrates CO_2 in the bundle sheath cells of C_4 plants. Cultured green cells may have a new function as the concentrator of CO_2 for PEP carboxylase in C_3 plants.

KEY REFERENCES

Husemann, W. and Barz, W. 1977. Photoautotrophic growth and photosynthesis in cell suspension cultures of Chenopodium rubrum. Physiol. Plant 40:77–81.

Yamada, Y., Sato, F., and Hagimori, M. 1979. Photoautotrophism in green cultured cells. In: Frontiers of Plant Tissue Culture (T.A. Thorpe, ed.) pp. 453–462. IAPTC, Calgary.

Yasuda, T., Hashimoto, T., Sato, F., and Yamada, Y. 1980. An efficient method of selecting photoautotrophic cells from cultured heterogeneous cells. Plant Cell Physiol. 21:929–932.

REFERENCES

Arnon, D.I. 1949. Copper enzymes in isolated chloroplasts. Polyphenoloxidase in Beta vulgaris. Plant Physiol. 24:1–15.

Bergmann, L. 1967. Wachstum gruner Suspensionkulturen von Nicotiana tabacum var. "Samsun" mit CO_2 als Kohlenstoffquelle. Planta 74:243–249.

_____ and Balz, A. 1966. Der Einfluss von Farblicht auf Wachstum und Zusammensetzung pflanzlicher Gewebekulturen I. Nicotiana tabacum var. "Samsun". Planta 70:285–303.

Berlyn, M.B. and Zelitch, I. 1975. Photoautotrophic growth and photosynthesis in tobacco callus cells. Plant Physiol. 56:752–756.

Corduan, G. 1970. Autotrophe Gewebekulturen von Ruta graveolens und deren $^{14}CO_2$-Markierungs-Produkte. Planta 91:291–301.

_____ and Reinhard, E. 1972. Synthesis of volatile oil in tissue culture of Ruta graveolens. Phytochemistry 11:917–922.

Dalton, C.C. 1980. Photoautotrophy of spinach cells in continuous culture: Photosynthetic development and sustained photoautotrophic growth. J. Exp. Bot. 31:791–804.

_____ and Street, H.E. 1976. The role of the gas phase in the green-ing and growth of illuminated cell suspension culture of spinach (*Spinacea oleracea* L.). In Vitro 12:485-494.

Edelman, J. and Hanson, A.D. 1971. Sucrose suppression of chlorophyll synthesis in carrot tissue cultures. Planta 98:150-156.

Hanson, A.D. and Edelman, J. 1972. Photosynthesis by carrot tissue cultures. Planta 102:11-25.

Hartmann, T., Wink, M., Schoofs, G., and Teichmann, S. 1980. Bio-chemistry of lupin-alkaloid biosynthesis in leaf chloroplasts of *Lupinus polyphyllus* and photomixotrophic cell suspension culture. Planta Medica 39:282.

Kaul, K. and Sabharwal, P.S. 1971. Effects of sucrose and kinetin on growth and chlorophyll synthesis in tobacco tissue cultures. Plant Physiol. 47:691-695.

Laetsch, W.M. and Stetler, D.A. 1965. Chloroplast structure and func-tion in cultured tobacco tissue. Am. J. Bot. 52:798-804.

Linsmaier, E.M. and Skoog, F. 1965. Organic growth factor require-ments of tobacco tissue cultures. Physiol. Plant. 18:100-127.

Neumann, K.-H. and Raafat, A. 1973. Further studies on the photosyn-thesis of carrot tissue cultures. Plant Physiol. 51:685-690.

Sato, F., Asada, K., and Yamada, Y. 1979. Photoautotrophy and the photosynthetic potential of chlorophyllous cells in mixotrophic cul-tures. Plant Cell Physiol. 20:193-200.

Sato, F., Nishida, K., and Yamada, Y. 1980. Activities of carboxyla-tion enzymes and products of $^{14}CO_2$ fixation in photoautotrophically cultured cells. Plant Sci. Lett. 20:91-97.

Seyer, P., Marty, D., Lescure, A.M., and Peaud-Lenoel, C. 1975. Effect of cytokinin on chloroplast cyclic differentiation in cultured tobacco cells. Cell Differen. 4:187-197.

Shewry, P.R., Pinfield, N.J., and Stobart, A.K. 1971. The effect of 2,4-dichlorophenoxyacetic acid and (2-chloroethyl)-trimethylammonium chloride on chlorophyll synthesis in barley leaves. Planta 101:352-359.

Sunderland, N. 1966. Pigmented plant tissues in culture I. Auxins and pigmentation in chlorophyllous tissues. Ann. Bot. 30:253-268.

Tsuzuki, M., Miyachi, S., Sato, F., and Yamada, Y. 1981. Photosynthe-tic characteristics and carbonic anhydrase activity in cells cultured photoautotrophically and mixotrophically and cells isolated from leaves. Plant Cell Physiol. 22:51-57.

Vasil, I.K. and Hildebrandt, A.C. 1966. Growth and chlorophyll produc-tion in plant callus tissues grown in vitro. Planta 68:69-82.

Wagner, F. and Vogelmann, H. 1977. Cultivation of plant tissue cul-tures in bioreactors and formation of secondary metabolites. In: Plant Tissue Culture and Its Bio-technological Application (W. Barz, E. Reinhard, and M.H. Zenk, eds.) pp. 245-252. Springer-Verlag, Ber-lin, Heidelberg, and New York.

Watanabe, K., Yano, S., and Yamada, Y. 1982. The selection of cul-tured plant cell lines producing high levels of biotin. Phytochemistry 21:513-516.

Yamada, Y. and Sato, F. 1978. Photoautotrophic culture of chlorophyl-lous cultured cells. Plant Cell Physiol. 19:691-699.

Yamada, Y., Imaizumi, K., Sato, F., and Yasuda, T. 1981. Photoautotrophic and photomixotrophic culture of green tobacco cells in a jar-
fermenter. Plant Cell Physiol. 22:917–922.

CHAPTER 14
Genetic Transformation in Plants

K. Ohyama

Conventional plant breeding programs have introduced numerous improvements in crop yield during the past centuries. However, plant breeders may have reached a limit in the ability to introduce new genetic information into plants and to create new plant varieties through conventional plant breeding techniques. On the other hand, in the last 10 years two major achievements have been accomplished toward introducing foreign genetic information into plants: one was the isolation and culture of plant protoplasts; the other was the development of genetic engineering techniques for transferring and cloning genes. The introduction of foreign genetic material into recipient cells requires the following steps: (1) binding and uptake of genetic material, (2) stabilization and/or replication of genetic material, (3) expression of genetic material in the recipient cells, and (4) inheritance of genetic material (Fig. 1). The first two steps have been extensively investigated by feeding radioactive DNA to a variety of plant materials including seedlings (Ledoux et al., 1971), pollen (Hess, 1977), cultured cells (Lurquin and Hotta, 1975), protoplasts (Ohyama et al., 1972b), and isolated nuclei (Ohyama et al., 1977a; Ohyama, 1978). On the other hand, gene expression and its inheritance can be determined by the detection of phenotypic changes in the recipient cells. While a great deal of work on gene expression has been done (Lurquin, 1977), these results remain controversial because of lack of a suitable genetic marker for the selection of transformed cells. These problems necessitate the development of a plant host-vector system. Three types of vectors are being considered, including the bacterial plasmid (Ti plasmid) of *Agrobacterium tumefaciens*, plant virus DNA and plant organelle DNA.

EVENTS IN DNA UPTAKE BY PROTOPLASTS

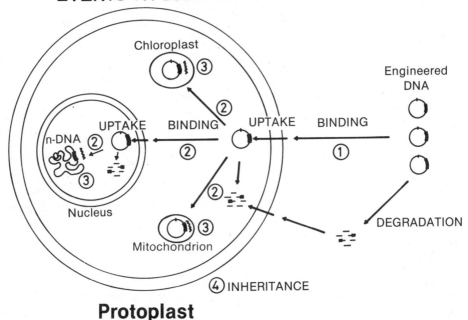

Figure 1.   Diagram of events in DNA uptake by plant protoplasts and nuclei.   Circled numbers 1, 2, 3, and 4 indicate each step of DNA uptake, its stabilization, gene expression, and gene inheritance, respectively.

The Ti plasmid of *A. tumefaciens*, an agent of crown gall induction in dicotyledonous plants, contains genes that are expressed in transformed plants.   These genes determine oncogenicity and opine synthesis (Schell and van Montagu, 1979).   Recently it was confirmed that a part of Ti plasmid (T region) is covalently linked to nuclear DNA in transformed cells (Willmitzer et al., 1980).   This system is now being considered as a model system for genetic transformation in plants. Alternative vectors include plant virus DNA, such as cauliflower mosaic virus (CaMV) DNA and organelle DNA from plant cells.   Information on genetic mapping of CaMV is being accumulated, and the transfection of CaMV to plant protoplasts is now under investigation.   Organelle DNA, on the other hand, such as chloroplast DNA and mitochondrial DNA, is extrachromosomal and can already replicate autonomously in plant cells. The first two vectors, Ti plasmid and CaMV DNA, represent genetic material that is exogenous to plant cells, but organelle DNA itself is native genetic material which is stably inherited and expressed in plants.   Therefore, organelle DNA can be a great candidate as a vector for plant genetic engineering studies (Ohyama et al., 1982).   In this chapter recent developments of DNA uptake by plant protoplasts and isolated nuclei will be described as well as the fundamental techniques of DNA uptake experiments.

LITERATURE REVIEW

Plant protoplasts appear to be very suitable recipient cells for DNA uptake studies, as protoplasts lack a thick cell wall and can be prepared in sufficient amounts by simple enzyme digestion.   Furthermore, protoplasts have several advantages over other plant materials; for example, (1) protoplasts provide a mild isolation procedure of cell organelles and extraction of DNA, (2) protoplasts can be cloned on agar plates as a single cell, and in some cases (3) protoplasts can be regenerated to whole plants.   Moreover, protoplasts can take up viral RNA such as tobacco mosaic virus (TMV) RNA and cowpea chlorotic mottle virus RNA; also, virus multiplication in the protoplasts has been detected (Aoki and Takebe, 1969; Motoyoshi et al., 1973).   These observations imply that protoplasts can be a potential host in plant genetic transformation experiments.

DNA Uptake by Plant Protoplasts

The published reports of DNA uptake using plant protoplasts are summarized in Table 1.   Ohyama et al. first reported on the utilization of protoplasts isolated from various suspension cultures in DNA uptake investigations (Ohyama et al., 1972a; Ohyama et al., 1973; Ohyama, 1975; Ohyama et al., 1978).   Protoplasts isolated from *Ammi visnaga, Glycine max* (soybean), and *Daucus carota* (carrot) cell suspension cultures were fed radioactive *Escherichia coli* DNA.   After DNase treatment to remove unbound DNA, about 0.6-2.8% of the exogenous *E. coli* DNA was taken up into the protoplasts, of which approximately 20% appeared to be acid-precipitable.   The time of incubation (0-4 hr), effects of incubation temperature (0-37 C), DNA concentration (10-30 μg/ml), and protoplast concentration (1.0-7.5 x 10^6 protoplasts/ml) were studied. The addition of polycations such as DEAE-dextran, poly-L-lysine, and poly-L-ornithine at a concentration of 5 μg/ml was found to markedly enhance DNA uptake.   The donor *E. coli* DNA was reextracted from incubated protoplasts and analysed by CsCl buoyant density gradient centrifugation.   After prolonged incubation with protoplasts, donor DNA appeared as a very broad band, indicating considerable degradation of donor DNA.   However, CsCl centrifugation analysis did not show the reutilization of degraded products of donor DNA, although active de novo DNA synthesis was observed in the protoplasts after a long period of incubation (24-96 hr).   Schaefer et al. extended their experiments using Ti plasmid DNA isolated from *Agrobacterium tumefaciens* (Ohyama et al., 1978; Schaefer et al., 1981).   They compared the results of plasmid DNA uptake with that of chromosomal DNA isolated from *A. tumefaciens*, and showed that plasmid DNA was degraded at slightly lower rates than chromosomal DNA.   They also subjected plasmid DNA that had been incubated with protoplasts, to agarose gel electrophoresis and detected exonuclease activity associated with protoplasts. They observed a variation of exonuclease activity in the protoplasts from 1-day-old and 4-day-old cell suspension cultures.   Protoplasts from 1-day-old and 4-day-old cells exhibited much less nuclease activity

Table 1. DNA Uptake by Plant Protoplasts

PROTOPLASTS	DONOR DNA	REMARKS	REFERENCE
Ammi visnaga, Daucus carota, Glycine max	E. coli	Acid-precipitation, effects of polycations	Ohyama et al., 1972, 1973
Ammi visnaga	E. coli	CsCl-centrifugation, DNA synthesis	Ohyama, 1975; Ohyama et al., 1978
Datura innoxia	Agrobacterium tumefaciens	Acid-precipitation, gel electrophoresis, comparison of plasmid with chromosomal DNA, nucleases associated with protoplasts	Ohyama et al., 1978; Schaefer et al., 1981
Petunia hybrida	Petunia hybrida	Autoradiography, nuclear fraction	Hoffmann & Hess, 1973
Petunia hybrida	Petunia hybrida (double-labeled)	CsCl-centrifugation, effects of metal ions	Hoffman, 1973
Petunia hybrida	E. coli	CsCl-centrifugation, nuclear fraction, sterile conditions	Liebke & Hess, 1977
Nicotiana tabacum	Bacteriophage fd	Acid-precipitation, sucrose density gradient centrifugation, inhibitors of DNA synthesis, effect of Zn^{+2} TMV-RNA transfection	Suzuki & Takebe, 1976
Nicotiana tabacum	Bacteriophage λ	DNase treatment, sucrose density gradient centrifugation, preincubation with polycations, distribution into subcellular fractions	Suzuki & Takebe, 1978

PROTOPLASTS	DONOR DNA	REMARKS	REFERENCE
Nicotiana tabacum Nicotiana glutinosa	N. tabacum E. coli	Acid-precipitation, sucrose density gradient centrifugation, DNA/DNA hybridization, distribution into intracellular componenets and subcellular fraction	Uchimiya & Murashige, 1977
Vigna sinensis, Daucus carota, Vinca rosea, Brassica campestris	Plasmid pBR313, plasmid pCR1	Acid-precipitation, sucrose density gradient centrifugation, polycations	Fernandez et al., 1978
Vigna sinensis	Plasmid pCK135, plasmid pBR313	Nuclear DNA extraction, sucrose density gradient-, CsCl-centrifugation, reconstitution of plasmid DNA with TMV coat protein	Kado & Lurquin, 1978
Hordeum vulgare, Nicotiana tabacum	Bacillus subtilis, Micrococcus luteus	Nuclei isolation, Sepharose 4B column chromatography, damaged and intact protoplasts, B. subtilis transformation activity	Hughes et al., 1978
Hordeum vulgare		Mannitol discontinuous gradient centrifugation, Sepharose 4B column chromatography, nuclear and nonnuclear fractions	Hughes et al., 1977
Glycine max	Glycine max	Autoradiography, cell wall contamination, addition of DEAE-dextran	Kool, 1977; Kool & Pelcher, 1978

indicating that the age of plant cells from which protoplasts are prepared may be an important factor in DNA uptake experiments.

Hoffmann and Hess (1973) reported on the uptake of radioactive homologous DNA by isolated protoplasts of *Petunia hybrida* using auto-radiography. Hoffmann (1973) also investigated the fate of double-labeled (^{3}H/^{14}C) homologous DNA in the protoplasts. In their experiments the majority of the radioactivity was found in the nuclear DNA fraction. Liebke and Hess (1977) extended earlier work to study the association of exogenous DNA with nuclei from protoplasts using CsCl density gradient centrifugation and demonstrated that the DNA from the nuclear fraction was recovered in a high molecular form. They also emphasized the importance of aseptic conditions in experiments on DNA uptake by plant protoplasts.

Suzuki and Takebe (1976) reported on DNA uptake by isolated tobacco mesophyll protoplasts using radioactive single-stranded DNA of the bacteriophage fd. They demonstrated in experiments with inhibitors of DNA synthesis that the radioactivity taken up by the protoplasts was not due to de novo DNA synthesis utilizing degradation products of added DNA. They investigated the effects of various divalent cations, polycations, such as poly-L-ornithine, poly-L-lysine, DEAE-dextran, and protamin sulfate, and other substances, such as cytochalasin B, amphtericin B, concanavalin A, and spermine, on DNA uptake by the protoplasts. They first reported that uptake of fd DNA was greatly enhanced by the presence of zinc ions in the incubation medium. They also attempted a study of DNA uptake by tobacco mesophyll protoplasts using bacteriophage λ double-stranded DNA and made a clear comparison with their previous work (Suzuki and Takebe, 1978). They reported that a major difference was found to be a much higher enhancement of double-stranded DNA uptake by preincubation of the DNA with poly-L-ornithine. They also found that the nuclear fraction (150 g pellet) consisted of 56% λ DNA taken up by the protoplasts, 14% by the chloroplast fraction (1000 g pellet). Little radioactivity was associated with the 12,000 g pellet (mitochondria, other smaller organelles, and fragments of membrane systems) and the 105,000 g pellet (ribosomes). The rest of the radioactivity was found in the 105,000 g supernatent fraction (ground cytoplasmic fraction).

Uchimiya and Murashige (1977) performed similar experiments using tobacco protoplasts and radioactive homologous DNA, and they reported that 3.5% of the input radioactivity was in acid-precipitable substances in protoplasts. They also reported that tobacco plants derived from protoplasts of a TMV-susceptible cultivar to which DNA from a resistant cultivar was fed did not show transfer of the virus resistant gene. They further studied the distribution of exogenous DNA using radioactive *E. coli* DNA and tobacco protoplasts. Distribution of the radioactivity was 60.9% in the soluble fraction and 28.4% in the nuclear fraction. They were the first to apply DNA/DNA hybridization techniques to protoplast-DNA uptake systems. The experiments revealed that 17.6% homology existed between reextracted radioactive DNA from tobacco protoplasts and *E. coli* DNA, 11.6% homology between the reextracted DNA and tobacco DNA. They suggested that the bulk radioactivity associated with protoplasts could be accounted for by reutilization of degraded *E. coli* DNA.

Lurquin and Kado (1977) used *E. coli* plasmid DNA to investigate the DNA uptake by *Vigna sinensis* (cowpea) protoplasts. They found that after a 15 min incubation period up to 3% of added plasmid DNA was incorporated into the protoplasts and that DNA incorporated into the protoplasts was distributed evenly to the cell fractions. They also reported that plasmid DNA incorporated into the protoplasts was not in the supercoiled form, indicating nuclease degradation of plasmid DNA during the incubation. Fernandez et al. (1978) extended the previous DNA uptake experiments using three kinds of bacterial plasmid DNA (pBR313, Col El from *E. coli*, and pCR1 plasmid from *A. tumefaciens*) and a variety of protoplast preparations such as *Daucus carota* (carrot), *Vinca rosea* (periwinkle), *Vigna sinensis* (cowpea), and *Brassica campestris* (turnip). They found that pBR313 fed to turnip protoplasts retained its original molecular size in the nuclear fraction for up to 4.5 hr incubation. Therefore, they concluded that this system could be a feasible vector for genetic engineering in plant cells. Kado and Lurquin (1978) reported on the successful uptake of *A. tumefaciens* plasmid pCK135 DNA and *E. coli* plasmid pBR313 DNA reconstituted with TMV protein by cowpea protoplasts. They demonstrated in sucrose density gradient and CsCl density gradient centrifugations that plasmid DNA coated with TMV protein was sufficiently protected from nucleases in plant cells and incorporated into protoplast nuclei in a short time.

Hughes et al. (1978) reported on DNA uptake by *Hordeum vulgare* (barley) and tobacco protoplasts. They compared uptake in intact and damaged protoplasts that were separated using discontinuous gradient centrifugation. Using this protoplast preparation, they studied various effects such as incubation time, temperature, DNA concentration, metabolic inhibitors, and polycations. They demonstrated that damaged protoplasts were rapidly saturated with relatively large amounts of added DNA, while intact protoplasts took up DNA very slowly. They also investigated depolymerization of DNA during uptake by using Sepharose 4B column chromatography. They first performed analysis of remaining transforming activity of recovered exogenous DNA by using a *Bacillus subtilis* transformation system. Transformation analysis demonstrated that DNA incubated with protoplasts for more than 3 hr completely lost transforming activity in a *B. subtilis* transformation assay. They also performed DNA uptake experiments using *E. coli* plasmids as donor DNA (Hughes et al., 1977). They compared degree of degradation of DNA associated with the nuclear fraction with that in the nonnuclear fraction, and demonstrated that plasmid DNA associated with the nuclear fraction was better protected than that of the nonnuclear fraction.

Kool and Pelcher reported on DNA uptake by protoplasts and cultured cells (Kool, 1977; Kool and Pelcher, 1978). By using autoradiography to examine protoplasts incubated with radioactive DNA, they concluded that the DNA was not associated with protoplasts, but only with aggregates of cell wall material contaminating the protoplast preparation. They also showed that most if not all DNase-resistant radioactive material was associated with cell wall materials. Therefore, they cautioned in interpreting experiments on the binding and uptake of DNA by plant protoplasts. However, they did not exclude the possibility that plant protoplasts can take up some exogenously supplied DNA.

DNA Uptake by Isolated Nuclei

As described above, most DNA uptake experiments indicated that DNA taken up by plant protoplasts had reached nuclei and remained in high molecular weight form, which implies large enough size to include a unit of genes. These promising results stimulated DNA uptake experiments using isolated nuclei to investigate the fate of DNA associated with the nucleus (summarized in Table 2).

Hotta and Stern (1971) briefly reported on DNA uptake by isolated nuclei from *Vicia faba* embryos. They demonstrated that once heterogenous DNA had been incorporated into a cell, it could reach the nuclear membrane and be associated with nuclei without extensive degradation.

Ohyama et al. extensively investigated DNA binding and uptake by isolated nuclei from soybean cell suspension cultures (Ohyama et al., 1977b; Ohyama, 1978). Using radioactive double-stranded soybean DNA (homologous), double-stranded *Salmonella typhimurium* DNA, and single-stranded bacteriophage fd DNA, they examined the competition for binding between double-stranded DNA and single-stranded DNA and demonstrated that soybean nuclei had a stronger affinity for single-stranded DNA than double-stranded DNA. Ohyama also studied the fate of single-stranded bacteriophage fd DNA bound to nuclei and associated with nuclei. Sucrose density gradient and CsCl density gradient centrifugation analyses revealed extensive degradation of DNA bound to nuclei in the absence of poly-L-lysine, and further revealed rapid cleavage of DNA associated with the nuclei into smaller fragments, even in the presence of poly-L-lysine.

Liebke et al. (1977) studied the uptake of *E. coli* DNA by isolated nuclei from *Petunia hybrida*. They demonstrated a stable association of the reextracted DNA with the nuclei by analyzing the reextracted DNA using CsCl density gradient centrifugation. They also observed transcription of exogenously supplied *E. coli* DNA by isolated nuclei from *Petunia hybrida* protoplasts (Blaschek and Hess, 1977). However, they did not investigate the translation activity of this exogenously supplied *E. coli* DNA in isolated nuclei, which would have been direct evidence for gene expression of foreign DNA.

Liposome-Mediated DNA Uptake by Protoplasts

Recent developments of liposome-mediated DNA uptake by plant protoplasts may make it possible to protect exogenously supplied DNA against intracellular nuclease activity and could therefore contribute to development of a genetic transformation system in plant cells (Table 3.)

Lurquin (1979) first reported on enhancement of plasmid DNA uptake into plant protoplasts by liposome encapsulation. It was demonstrated

Table 2. DNA Uptake by Isolated Nuclei

SOURCE OF NUCLEI	DONOR DNA	REMARKS	REFERENCE
Vicia faba	*Microccocus lysodeikticus*	Acid-precipitation	Hotta & Stern, 1971
Glycine max	*Glycine max*	Acid precipitation, pronase, DNA competition, ATP generating system	Ohyama et al., 1977
Glycine max	Bacteriophage fd, *Salmonella typhimurium*	Acid-precipitation, SDG-, CsCl-centrifugation, comparison single-stranded DNA with double-stranded DNA, preincubation with nonradioactive DNA	Ohyama, 1978
Petunia hybrida	*E. coli*	DNase-treatments, SDG-, CsCl-centrifugation, competition between *E. coli* and calf thymus DNA	Liebke et al., 1977
Petunia hybrida	*E. coli*	RNA transcription, DNA/DNA hybridization	Blaschek & Hess, 1977

Table 3. Liposome–Mediated DNA Uptake by Plant Protoplasts

PROTOPLAST SOURCES	DONOR DNA	REMARKS	REFERENCE
Vigna sinensis	Plasmid pBR322, plasmid pCR1	Acid–precipitation, gel electrophoresis, Sepharose 4B column chromatography, PEG–mediated fusion	Lurquin, 1979
Nicotiana tabacum	Plasmid pBR322	Affinity and molecular sieving column chromatography, SDG–centrifugation, association with nuclei and chromatin	Lurquin, 1981
Daucus carota	*Bacillus subtilis*	Molecular sieving column chromatography, autoradiography, PEG–mediated fusion	Rollo et al., 1981

Table 4. Gene Expression of Foreign DNA in Plant Protoplasts

PROTOPLASTS	DONOR DNA	REMARKS	REFERENCE
Glycine max	*Azotobacter vinelandii*	Mannitol utilization, callus formation on mannitol medium	Holl et al., 1974
Nicotiana tabacum (virus susceptible)	*Nicotiana tabacum* (virus resistant)	Virus resistance, no evidence of virus resistance in plants	Uchimiya & Murashige, 1977
Nicotiana tabacum	Plasmid Col El (kanamycin)	Kanamycin resistance, no expression of kanamycin resistance in tissues	Owens, 1979
Petunia hybrida x *Petunia parodii*	Plasmid pTi-Ach5	Cell growth on hormone free medium	Davey et al., 1980

that DNA-filled liposomes strongly interact with plant protoplasts under conditions that induce protoplast fusion, and that plasmid DNA in liposomes is subsequently transferred to protoplast nuclei. Lurquin (1981) extended studies on the binding of liposomes carrying pBR322 DNA to tobacco protoplasts. The results indicated that a substantial proportion of the DNA was associated with the nucleus and with chromatin. A minimal amount of degradation was observed.

Rollo et al. (1981) studied the interaction of liposomes loaded with DNA from *B. subtilis* with carrot protoplasts. Using molecular sieving and autoradiographic analyses, they demonstrated that liposomes can efficiently protect DNA from nucleases present in the protoplast suspension and can facilitate transfer of DNA into the protoplasts.

Gene Expression of Foreign DNA in Plant Protoplasts

There have been few reports on expression of foreign genes in protoplasts treated with exogenous DNA (Table 4). Holl et al. (1974) reported that regenerated callus from soybean protoplasts grew slowly on medium containing mannitol as a carbon source following treatment with DNA extracted from *Azotobacter vinelandii*. The bacterium is able to metabolize mannitol as carbon source, while soybean cells do not grow on mannitol medium. However, they demonstrated, with radioactive mannitol and glucose, that the slow growth of the cells on mannitol medium was due to limited uptake of mannitol by the cells, and suggested that callus growth on mannitol medium may not be controlled by expression of the exogenously supplied DNA.

Owens (1979) reported on binding of *E. coli* Col El-kan plasmid DNA by tobacco protoplasts. He demonstrated that 2.9% of the bound or irreversibly bound DNA was in an acid-insoluble form. He also tested kanamycin-resistance of plantlets or shoots obtained from DNA-treated protoplasts. However, there was no evidence for expression of the kanamycin-resistance gene derived from the Col El-kan plasmid in tobacco tissues.

Davey et al. (1980) carried out transformation experiments using *Petunia hybrida* protoplasts and Ti plasmid DNA from *A. tumefaciens*. They stated that isolated Ti plasmid DNA can transform *Petunia hybrida* protoplasts. Evidence for transformation included continued proliferation of *Petunia* protoplasts on hormone-free medium, overgrowth formation, octopine synthesis, and lysopine dehydrogenase activity in the transformed cells. While mentioning that preliminary DNA-hybridization tests indicate the presence of T-DNA in callus that is still synthesizing octopine, these tests have not been reproduced.

PROTOCOLS FOR DNA UPTAKE BY PLANT PROTOPLASTS

DNA Uptake

1. Transfer 0.1 ml of a protoplast suspension (approximately 10^5 protoplasts in culture medium containing 0.275 M sorbitol) into a 1.5 ml

Eppendorf centrifuge tube. (See Gamborg and Wetter, 1975, for protoplast preparation and medium. Protoplast concentration can be determined by a Coulter electronic counter or a haemocytometer.)

2. Add 50 µl of a protoplast culture medium containing 0.55 M sorbitol.

3. Add 50 µl of radioactive DNA solution (approximately 15,000 cpm for ^{3}H- or ^{14}C-labeled DNA) in 0.1 X SSC (15 mM NaCl, 1.5 mM trisodium citrate, pH 7.0) buffer. (See Marmur, 1961, for bacterial radioactive DNA preparation, and Ohyama et al., 1972a,b, for radioactive plant DNA preparation.)

4. Incubate at 28 C for 4 hr.

DNase-Treatment, Washing, and Lysis of Protoplasts

1. After incubation add 1.0 ml of protoplast culture medium containing 0.275 M sorbitol. Centrifuge at 300 g for 3 min in International model HN centrifuge. (Place the Eppendorf centrifuge tube into a 15 ml Corex centrifuge tube with a cushion of cotton.)

2. Suspend protoplasts in 0.9 ml of the protoplast culture medium containing 0.275 M sorbitol, and incubate the protoplast suspension with 0.1 ml of DNase solution (1 mg/ml, Worthington) containing 0.3 M $MgCl_2$ at 37 C for 5–10 min.

3. Wash protoplasts three times with 1.0 ml of the culture medium containing 0.275 M sorbitol by centrifugation at 300 g for 3 min.

4. Lyse protoplasts with 1.0 ml x SSC (0.15 M NaCl, 15 mM trisodium citrate, pH 8.0) buffer containing 2% Sarkosyl NL 30 (Ciba-Geigy).

Determination of Total Uptake

1. Transfer 0.5 ml of the lysate to a scintillation vial.

2. Add 10 ml of toluene-Triton X-100 (2:1) containing 10 g per liter of butyl-PBD (2-(4-tert-butylphenyl)-5-(4-biphenylyl)-1,3,4-oxadiazole).

3. Measure radioactivity in a scintillation counter for 10 min.

Determination of Acid-Precipitable Uptake

1. Add 0.5 ml of cold 10% trichloroacetic acid (TCA) to the rest of the lysate (0.5 ml from section A, 2-4) in an Eppendorf centrifuge tube.

2. After 30 min in ice-bath, collect precipitate on a glass fiber filter (Whatman GF/C, 2.1 cm diameter) attached to a Millipore filter apparatus.

3. Wash precipitate three times with 5 ml of cold 5% TCA and once with cold 95% ethanol, and dry filter under an infrared lamp.

4. Place filter in a scintillation vial, and add 5 ml of toluene containing butyl-PBD (10 g/l). Count radioactivity for 10 min.

PROTOCOLS FOR DNA BINDING AND UPTAKE BY ISOLATED NUCLEI

Preparation of Nuclei from Soybean Protoplasts

1. Suspend soybean protoplasts in 20 ml of B5 medium containing 0.275 M sorbitol, 10 mM $MgCl_2$, 1 mM 2-mercaptoethanol, and 0.5% Triton X-100 (medium A). (See Gamborg and Wetter, 1975, for B5 medium).
2. Disrupt the protoplasts using a Dounce homogenizer with 10 gentle strokes.
3. Pass the homogenate once through a layer of Miracloth and twice through triple layers of Miracloth.
4. Layer the filtrate on 4 ml of medium A in a 15 ml Corex centrifuge tube, and centrifuge at 400 g for 5 min.
5. Suspend nuclei pellets in 2 ml of medium A, layer the suspension on 4 ml of medium A, and centrifuge at 400 g for 5 min. Repeat this step twice to remove debris, cytoplasmic organelles, and starch granules.
6. Suspend nuclei in 2 ml of medium A, layer the suspension on gradients of 5 ml of 0.5 M sorbitol in medium A and 2 ml of 1 M sorbitol in medium A, and centrifuge at 100 g for 3 min.
7. Take out the top layer containing the nuclei fraction and spin down the nuclei at 400 g for 5 min. Repeat steps 6 and 7 at least twice to ensure complete removal of larger particles such as unbroken protoplasts. (Check under a light microscope.)
8. Suspend the purified nuclei in 1-2 ml of 0.01 M Tris-HCl buffer (pH 7.5) containing 0.275 M sorbitol, 10 mM $MgCl_2$, 1 mM 2-mercaptoethanol, and 0.1% Triton X-100. (The nuclei concentration can be determined by using a Coulter electronic counter or a haemocytometer.)

DNA Binding and Uptake

1. Transfer 0.1 ml of soybean nuclei suspension (approximately 10^6 nuclei) into a 1.5 ml Eppendorf centrifuge tube.
2. Add 0.1 ml of 1 M sorbitol, 0.1 ml of 0.05 M Tris-HCl (ph 7) containing 10 mM $CaCl_2$, and 0.1 ml of distilled water.
3. Add 0.1 ml of radioactive DNA (approximately 20,000 cpm for ^{3}H- or ^{14}C-labeled DNA). Incubate at 30 C for 20 min.

Determination of DNA Binding

1. After incubation, centrifuge at 400 g for 5 min in an International model HN centrifuge. (Place the Eppendorf centrifuge tube onto a 15 ml Corex centrifuge tube with a cushion of cotton.)
2. Wash nuclei three times with 1.0 ml of B5 medium containing 0.275 M sorbitol, 10 mM $MgCl_2$, 1 mM 2-mercaptoethanol, and 0.5% Triton X-100. (See Gamborg and Wetter, 1975, for B5 medium).

3. Suspend nuclei in 0.8 ml of the same medium and add 0.2 ml of 10% Sarkosyl NL 30 (Ciba–Geigy).
4. Transfer the whole lysate to a scintillation vial.
5. Add 10 ml of toluene–Triton X-100 (2:1) containing butyl-PBD (10 g/l). Count radioactivity in a scintillation counter for 10 min.

Determination of DNA Uptake

1. Suspend the washed nuclei (from above) in 0.8 ml of 0.01 M Tris-HCl (pH 7.0) containing 0.275 M sorbitol, 10 mM $MgCl_2$, 1 mM 2-mercaptoethanol, and 0.1% Triton X-100.
2. Add 0.2 ml of DNase solution (1 mg/ml, Worthington) and incubate at 30 C for 5-10 min.
3. Wash nuclei three times with 1.0 ml of B5 medium containing 0.275 M sorbitol.
4. Lyse nuclei by addition of 0.5 ml of 2% Sarkosyl NL 30.
5. Transfer 0.1 ml of the lysate to a scintillation vial and count radioactivity with 10 ml of toluene–Triton X-100 (2:1) containing butyl-PBD (10 g/l).

Determination of Acid–Precipitable Uptake

1. Add 0.4 ml of cold 10% TCA to the rest of the lysate (from previous section, step (4)).
2. Collect precipitate on glass fiber (Whatman GF/C, 2.1 cm diameter) attached to a Millipore filter aparatus.
3. Wash the filter three times with cold 5% TCA and once with cold 95% ethanol, and dry filter under infrared lamp.
4. Place filter in a scintillation vial, and add 5 ml of toluene containing butyl-PBD (10 g/l). Count radioactivity for 10 min.

FATE OF EXOGENOUS DNA TAKEN UP BY PROTOPLASTS AND ISOLATED NUCLEI

DNA Analysis by Sucrose Density Gradient Centrifugation

(Important: use ^{3}H-labeled DNA for DNA analysis because of higher specific radioactivity.)

1. Transfer the lysate (250 μl from previous protocols approximately 10,000 cpm) into a 1.5 ml Eppendorf centrifuge tube.
2. Immediately heat at 80 C for 5 min to destroy nuclease activity.
3. Apply the lysate with a size marker ([32]P-, or ^{14}C-labeled DNA of a known size) onto 8-20% linear sucrose density gradient formed in a Beckman SW 40 rotor centrifuge tube.
4. Centrifuge at 23,000 rpm at 20 C for 17 hr in a Beckman model L5-65B ultracentrifuge.
5. Attach the tube to a ISCO density fractionator model 640, and puncture the tube bottom.

6. Collect each 12-drop fraction into a scintillation vial. (Drop size may change during fractionation because of detergents in the lysate. A 12-drop fractionation gives approximately 60 fractions.)
7. Count radioactivity for 10 min with 10 ml of toluene–Triton X-100 (2:1) containing butyl-PBD (10 g/l).

DNA Analysis by CsCl Buoyant Density Gradient Centrifugation

1. Transfer the lysate (250 µl from previous protocols, approximately 5,000 cpm) into a 15 ml Corex centrifuge tube. (Important: Lyse nuclei with neutral detergents like Sarkosyl.)
2. Make total volume up to 3.0 ml 1 x SSC buffer, and add a ^{32}P-, ^{14}C-labeled DNA (1,000 cpm), or nonradioactive *Microccocus lysodeikticus* DNA (approximately $OD_{260 nm}$ = 2) as a density marker.
3. Add 3.75 g of solid CsCl. (If any insoluble material is formed, remove by centrifugation at 20,000 rpm for 20 min.)
4. Adjust refractive index to 1.400 at 25 C.
5. Transfer the solution into a Beckman SW 56 rotor centrifuge tube, and fill up with a good-quality mineral oil.
6. Centrifuge at 23,000 rpm for 68 hr in a Beckman model L5-65B ultracentrifuge.
7. Attach the tube to a ISCO density fractionator model 640. Puncture the tube bottom. (When nonradioactive density marker DNA used, monitor absorbance at 254 nm with ISCO model UA-5 monitor.)
8. Collect each 6-drop fraction into a scintillation vial. (A 6-drop fractionation gives approximately 50 fractions.)
9. Count radioactivity for 10 min with 10 ml of toluene–Triton X-100 (2:1) containing butyl-PBD (10 g/l).

FUTURE PROSPECTS

The methods described here provide information on DNA binding and uptake by plant protoplasts as well as isolated nuclei. The DNA uptake was enhanced by the addition of polycations such as poly-L-lysine, poly-L-ornithine, and some factors such as zinc ions. In most cases there is extensive DNA degradation by nucleases when protoplasts and nuclei are incubated for a longer period of time. This indicates that donor DNA must be protected against nuclease degradation until reaching the host genome. Recent developments of liposome-mediated DNA uptake by plant protoplasts offer an unique method to solve this problem (Lurquin, 1981).

In DNA uptake experiments caution must be paid to bacterial contamination that gives false information of DNA uptake and the fate of DNA in recipient cells (Kleinhofs and Behki, 1977). Furthermore, it is suggested that several methods such as acid-precipitable uptake determination, autoradiography, chromatography, ultracentrifugation analysis, DNA-DNA hybridization, and agarose gel electrophoresis all be used to investigate the uptake and fate of donor DNA. Intact and clean protoplast preparations must be used, as dead protoplasts or membrane debris preferentially absorbed exogenous DNA molecules (Hughes et al., 1977; Kool, 1977; Kool and Pelcher, 1978).

We are facing three major barriers to accomplish routine genetic transformation in plants:

(1) Appropriate host cells must be identified. Plant protoplasts are believed to be the most suitable recipient cells, as protoplasts were found to take up exogenously supplied DNA. In some cases protoplasts can be regenerated into whole plants. It would be appropriate to use cell lines with genetic markers.

(2) It is necessary to develop appropriate vectors. Donor DNA from various sources has been used in DNA uptake studies. For donor DNA to replicate in recipient cells, it must be incorporated into the replication system of the recipient cells or it must have its own DNA replication system that can function in the recipient cells. Vector systems capable of autonomous replication in plant cells are being developed to deliver donor DNA into recipient cells. Presently, several potential vectors are being studied: (a) the Ti plasmid of *Agrobacterium tumefaciens.* The T region of the Ti plasmid was found to integrate into nuclei of transformed cells. As genes adjacent to the T region are also integrated, this system could be used to introduce genes into recipient cells. (b) DNA from plant DNA viruses such as cauliflower mosaic virus also replicates in plant cells. This DNA can replicate in the cytoplasms of susceptible cells. Protoplast infection with CaMV DNA is being investigated. (c) Plant organelle DNA such as chloroplasts and mitochondria are found naturally in plant cells. The DNA from organelles is autonomously replicating and has several functional genes, including the large subunits of Fraction 1 protein, 32,000 dalton membrane protein, ribosomal RNA genes, and transfer RNA genes. Yet the development of this DNA as a vector has not been explored (Ohyama et al., 1982).

(3) It is necessary to use donor DNA with a marker that can be selected and expressed in recipient cells. To be useful for introduction of new traits into plant cells, genes of exogenously supplied DNA must be phenotypically expressed in recipient plant cells. To select transformed cells, normal recipient cells must be recessive to the gene which is used for the selection of transformants. That is, recipient cells must be recessive mutants and the donor gene must be dominantly expressed in the recipient cells. From this point of view, availability of recessive auxotrophic mutants in recipient plant cells, such as amino acid-, carbohydrate-, base-, and nucleoside-requiring cells, would be excellent markers for DNA transformation in plants. Unfortunately, there are few auxotrophic or metabolic deficient mutants in plant cells. Alternatively, donor DNA with dominant resistance mutants could be used. The need for plant genetic markers suggests that efforts must be directed to get a variety of metabolic mutants to facilitate DNA tranformation experiments in plants.

Development of efficient transformation systems in higher plants would permit introduction of genetically engineered DNA into plants. This engineered DNA could be cloned in plant cells and would contribute to crop improvements in the near future.

ACKNOWLEDGMENTS

I am grateful to Dr. Y. Yamada for his encouragements in my contribution to this chapter. I also thank Mrs. K. Morita for her competent typing of this manuscript.

KEY REFERENCES

Gamborg, O.L. and Wetter, L.R. 1975. Plant Tissue Culture Methods. National Research Council of Canada, Saskatoon.

Ohyama, K. 1978. DNA binding and uptake by nuclei isolated from plant protoplasts: Fate of single-stranded bacteriophage fd DNA. Plant Physiol. 61:515-520.

_____, Gamborg, O.L., and Miller, R.A. 1972b. Uptake of exogenous DNA by plant protoplasts. Can. J. Bot. 50:2077-2080.

_____, Pelcher, L.E., and Horn, D. 1977a. A rapid, simple method for nuclei isolation from plant protoplasts. Plant Physiol. 60:179-181.

_____, Pelcher, L.E., and Horn, D. 1977b. DNA binding and uptake by nuclei isolated from plant protoplasts: Factors affecting DNA binding and uptake. Plant Physiol. 60:98-101.

_____, Pelcher, L.E., and Schaefer, A. 1978. DNA uptake by plant protoplasts and isolated nuclei: Biochemical aspects. In: Frontiers of Plant Tissue Culture (T.A. Thorpe, ed.) pp. 75-84. University of Calgary, Calgary.

REFERENCES

Aoki, S. and Takebe, I. 1969. Infection of tobacco mesophyll protoplasts by tobacco mosaic virus ribonucleic acid. Virology 39:439-448.
Blaschek, W. and Hess, D. 1977. Transcription of bacterial DNA by isolated plant nuclei. Experientia 33:1594-1595.
Davey, M.R., Cocking, E.C., Freeman, J., Pearce, N., and Tudor, I. 1980. Transformation of Petunia protoplasts by isolated Agrobacterium plasmid. Plant Sci. Lett. 18:307-313.
Fernandez, S.M., Lurquin, P.F., and Kado, C.I. 1978. Incorporation and maintenance of recombinant-DNA plasmid vehicles pBR313 and pCR1 in plant protoplasts. FEBS Lett. 87:277-282.
Hess, D. 1977. Cell modification by DNA uptake. In: Applied and Fundamental Aspects of Plant Cell, Tissue Culture, and Organ Culture (J. Reinert and Y.P.S. Bajaj, eds.) pp. 506-535. Springer-Verlag, Berlin.
Hoffmann, F. 1973. Uptake of double labelled DNA into isolated protoplasts of Petunia hybrida. Z. Pflanzenphysiol. 69:249-261.
_____ and Hess, D. 1973. Uptake of radioactively labelled DNA into isolated protoplasts of Petunia hybrida. Z. Pflanzenphysiol. 69:81-83.

Holl, F.B., Gamborg, O.L., Ohyama, K., and Pelcher, L.E. 1974. Genetic transformation in plants. In: Tissue Culture and Plant Science (H.E. Street, ed.) pp. 301-327. Academic Press, New York.

Hotta, Y. and Stern, H. 1971. Uptake and distribution of heterologous DNA in living cells. In: Informative Molecules in Biological Systems (L. Ledoux, ed.) pp. 176-186. North Holland, Amsterdam.

Hughes, B.G., White, F.G., and Smith, M.A. 1977. Fate of bacterial plasmid DNA during uptake by barley protoplasts. FEBS Lett. 79:80-84.

_____ 1978. Fate of bacterial DNA during uptake by barley and tobacco protoplasts. Z. Pflanzenphysiol. 87:1-23.

Kado, C.I. and Lurquin, P.F. 1978. Reconstitution of plasmid DNA with tobacco mosaic virus protein for introducing plasmids into higher plant cells. In: Microbiology—1978 (D. Schlessinger, ed.) pp. 231-234. American Society Microbiology, Washington, D.C.

Kleinhofs, A. and Behki, R. 1977. Prospects for plant genome modification by nonconventional methods. Annu. Rev. Plant Physiol. 11:79-101.

Kool, A.J. 1977. An autoradiographic study of binding and uptake of DNA by cultured soybean SB1 cells and protoplasts. Genen. Phaenen. 19:61-63.

_____ and Pelcher, L.E. 1978. Conditions that affect binding of DNA by cultured plant cells and protoplasts: An autoradiographic analysis. Protoplasma 97:71-84.

Ledoux, L., Huart, R., and Jacobs, M. 1971. Fate of exogenous DNA in *Arabidopsis thaliana*: Translocation and integration. Eur. J. Biochem. 23:96-108.

Liebke, B. and Hess, D. 1977. Uptake of bacterial DNA into isolated mesophyll protoplasts of *Petunia hybrida*. Biochem. Physiol. Pflanzen. 171:493-501.

Liebke, B., Blaschek, W., and Hess, D. 1977. Studies on the uptake of exogenous DNA into isolated nuclei of *Petunia hybrida*. Z. Pflanzenphysiol. 84:265-274.

Lurquin, P.F. 1977. Integration versus degradation of exogenous DNA in plants: An open question. In: Progress in Nucleic Acid Research and Molecular Biology (W.E. Cohn, ed.) Vol. 20, pp. 161-207. Academic Press, New York.

_____ 1979. Entrapment of plasmid DNA by liposomes and their interactions with plant protoplasts. Nucl. Acids Res. 6:3773-3784.

_____ 1981. Binding of plasmid loaded liposomes to plant protoplasts: Validity of biochemical methods to evaluate the transfer of exogenous DNA. Plant Sci. Lett. 21:31-40.

_____ and Hotta, Y. 1975. Reutilization of bacterial DNA by *Arabidopsis thaliana* cell in tissue culture. Plant Sci. Lett. 5:103-112.

_____ and Kado, C.I. 1977. *Escherichia coli* plasmid pBR313 insertion into plant protoplasts and into their nuclei. Molec. Gen. Genet. 154:113-121.

Marmur, J. 1961. A procedure for the isolation of DNA from microorganisms. J. Molec. Biol. 3:208-218.

Motoyoshi, F., Bancroft, J.B., Watts, J.W., and Burgess, J. 1973. The infection of tobacco protoplasts with cowpea chlorotic mottle virus and its RNA. J. Gen. Virol. 20:177-193.

Ohyama, K. 1975. Introduction of genetic information into plant proto-plasts. In: Plant Tissue Culture Methods (O.L. Gamborg and L.R. Wetter, eds.) pp. 28–35. National Research Council of Canada, Saskatoon.

_____, Gamborg, O.L., and Miller, R.A. 1972a. Isolation and proper-ties of DNA from protoplasts of cell suspension cultures of *Ammi visnaga* and carrot (*Daucus carota*). Plant Physiol. 50:319–321.

_____, Gamborg, O.L., Shyluk, J.P., and Miller, R.A. 1973. Studies on transformation: Uptake of exogenous DNA by plant protoplasts. In: Protoplastes et Fusion de Cellules Somatiques Vegetales (J. Tempe, ed.) Vol. 212, pp. 423–428. C.N.R.S., Paris.

_____, Wetter, L.R., Yamano, Y., Fukuzawa, H., and Komano, T. 1982. A simple method for the isolation of chloroplast DNA from *Marchantia polymorpha* L. cell suspension cultures. Agric. Biol. Chem. (in press).

Owens, L.D. 1979. Binding of Col El-kan plasmid DNA by tobacco protoplasts: Nonexpression of plasmid gene. Plant Physiol. 63:683–686.

Rollo, F., Galli, M.G., and Parisi, B. 1981. Liposome-mediated transfer of DNA to carrot protoplasts: A biochemical and autoradiographic analysis. Plant Sci. Lett. 20:347–354.

Schaefer, A., Ohyama, K., and Gamborg, O.L. 1981. Detection by agar-ose gel electrophoresis of nucleases associated with cells and proto-plasts from suspension cultures using *Agrobacterium tumefaciens* Ti plasmid. Agric. Biol. Chem. 45:1441–1445.

Schell, J. and van Montagu, M. 1979. The Ti plasmids of *Agrobac-terium tumefaciens* and their role in crown gall formation. In: Genome Organization and Expression in Plants (C. Leaver, ed.) pp. 453–470. Plenum, New York.

Suzuki, M. and Takebe, I. 1976. Uptake of single-stranded bacterio-phage DNA by isolated tobacco protoplasts. Z. Pflanzenphysiol. 78:421–433.

_____ 1978. Uptake of double-stranded bacteriophage DNA by iso-lated tobacco leaf protoplasts. Z. Pflanzenphysiol. 89:297–311.

Uchimiya, H. and Murashige, T. 1977. Quantitative analysis of the fate of exogenous DNA in *Nicotiana* protoplasts. Plant Physiol. 59:301–308.

Wilmitzer, L., De Bueckeleer, M., Lemmers, M., van Montagu, M., and Schell, J. 1980. DNA from Ti plasmid present in nucleus and absent from plastids of crown gall plant cells. Nature 287:359–361.

CHAPTER 15
Liposome-Mediated Delivery of DNA to Plant Protoplasts

B. Matthews

INTRODUCTION

Early Uses of Liposomes

Liposomes are artificial lipid vesicles used to study membrane-membrane interactions and to deliver macromolecules including DNA and RNA, proteins, and therapeutic drugs to plant and animal cells. The highly ordered structure of liposomes consists of alternating lipid sheets and parallel aqueous compartments readily visible under the light microscope. Each bimolecular lipid sheet forms a complete, discrete compartment, sequestering its aqueous components such that they can escape only through diffusion across the bimolecular membrane.

Liposome technology rapidly developed when preparative techniques became available that furnished phospholipids in high quantity for the first time. Early pioneers examined dried phospholipids on glass slides using the light microscope and watched while membranous mazes formed when a drop of water was added. Lipids such as phosphatidylcholine, phosphatidylserine, and cholesterol were observed to form closed membrane systems in the presence of water. These membranes simulated cell membranes, thus becoming interesting model systems to examine membrane physiology and transport.

One early interest was the study of the effects of ions, such as sodium and potassium, on the permeability of these artificial membranes (Bangham et al., 1965). The effects of charge and ionic forces upon the spacing of the concentric lamella of the lipid vesicles were also investigated. The spacing of the concentric lamella could be altered by changing the composition of the aqueous medium and the phospholipid composition being examined (Papahadjopoulos and Miller, 1967).

As liposomes were used increasingly as artificial membrane models, new lipid compositions were employed, and the interaction and effects of ions on these liposomes were investigated. The transition temperatures of several different lipids were determined, i.e., the temperature at which the lipid was no longer in a gel state but became fluid. Cholesterol was observed to act as a membrane stabilizing agent decreasing permeability and fluidity of liposomes. These early experiments led to a better understanding of cell-cell fusion and membrane transport.

More recently liposomes have been used to sequester a wide variety of molecules for delivery into cells. Papahadjopoulos et al. (1974) encapsulated cyclic AMP within unilamellar vesicles and examined the effects of the delivered cyclic AMP upon cultured mouse cells. The delivery of cyclic AMP by unilamellar liposomes into these cells was demonstrated by inhibition of cell proliferation. A variety of enzymes have also been introduced into cells via liposomes (Weissmann et al., 1975; Cohen et al., 1976; Finkelstein and Weissmann, 1978). Many human diseases, including Tay-Sachs, Hunter, and Hurler disease, are caused by genetic deficiencies resulting in the aberrance or absence of specific cytoplasmic enzymes. Several investigators are examining the effects of enzyme replacement by delivering the appropriate enzyme to enzyme-deficient cells. Injection of unencapsulated enzyme directly into the bloodstream may induce an immune response. Also, the enzyme disappears rapidly from the bloodstream and accumulates in the liver rather than reaching specific cells. The injected enzyme may interact undesirably with molecules in the circulatory system or liver to form toxic or potentially harmful compounds. Liposome-mediated delivery allows the enzyme to be sequestered and protected from components in the bloodstream to minimize unwanted interactions. The enzyme can be directed or "targeted" by liposomes to specific sites in the body. Site-specific targeting of liposomes is accomplished by attaching antibodies, specific for antigenic sites of the target cell, to the outside of the liposome (Weissmann et al., 1975; Heath et al., 1980; Van Houte et al., 1979; Jansons and Mallett, 1981). Although much more basic research is needed before liposome technology is widely utilized for delivering enzymes, antineoplastic agents, antibiotics, and other chemotherapeutic and pharmaceutical compounds to target cells, this area of research promises to be exciting and rewarding for the treatment of several types of diseases.

Entrapment and Delivery of Nucleic Acids by Liposomes to Mammalian Cells

Nucleic acids encapsulated within liposomes are protected from degradation by extraliposomal nuclease activity. Ostro et al. (1977) encapsulated ^{3}H-RNA from *E. coli* within unilamellar vesicles formed by the ether infusion method of Deamer and Bangham (1976). The liposome mixture was treated with RNase which degraded the unencapsulated 4S, 16S, and 23S bacterial RNA, and the liposomes were separated from the unencapsulated fragments by column chromatography. The liposome–encapsulated RNA was preserved intact, as indicated by gel electrophoresis. Similar experiments have shown that DNA is protected from external DNase degradation by liposome encapsulation (Hoffman et al., 1978; Dimitriadis, 1978, 1979; Mannino et al., 1979). Because liposomes can encapsulate and protect nucleic acids from nuclease attack, liposomes have been used to insert mRNA, rRNA, DNA, and chromosomes into a variety of cells. Ostro et al. (1978) encapsulated mRNA coding for rabbit globin within liposomes and incubated the liposomes with nonglobin producing HEp-2 cells. ^{3}H-amino acids were added to label newly synthesized proteins. Analysis of the extracted proteins revealed that the HEp-2 cells produced a globin–like protein as characterized by SDS polyacrylamide gel electrophoresis and sucrose density gradient centrifugation. A similar experiment was described by Dimitriadis (1978), whereby liposome–encapsulated rabbit globin mRNA was delivered to mouse lymphocytes; once again the mRNA was translated.

Liposomes have been used to deliver the E. coli plasmid, pBR322, to competent E. coli cells in the presence of high amounts of DNase (Fraley et al., 1979). Transformation occurred as indicated by the formation of tetracycline–resistant colonies when liposome–encapsulated plasmid was used. No colonies were formed when E. coli were treated with unencapsulated plasmid. Similar experiments using Simian virus 40 (SV40) DNA were conducted in which encapsulated viral DNA enhanced the infection of permissive monkey cells 100–fold over controls using unencapsulated SV40 DNA (Fraley et al., 1980). Thus liposome encapsulation protects DNA from extraliposomal DNase degradation, preserving the biological activity of the DNA until the contents of the liposomes are inserted into cells.

Evidence indicating the transfer of a restriction fragment from the E. coli plasmid pBR322 to HeLa cells and chick embryo cells using liposome techniques has been reported by Wong et al. (1980). An 875 bp restriction fragment coding for β–lactamase activity was encapsulated within phosphatidylcholine:phosphatidylserine vesicles (molar ratio = 9:1). Unencapsulated DNA was removed by chromatographing the liposome mixture using a Sepharose 4B column. The purified vesicles were incubated with murine LM fibroblasts, chick embryo cells, and HeLa cells. Assays of extracts from each of these cells indicated the presence of β–lactamase activity, suggesting that this prokaryotic gene could be expressed in eukaryotic cells. However, hybridization with probes to determine where the fragment was located, how much was present, and its stability and integration were not reported.

Liposomes have been used successfully to coat human X chromosomes to form "lipochromosomes" and to insert closely linked genes on these chromosomes into A9 mouse cells (Mukherjee et al., 1978). Metaphase chromosomes from hypoxanthine guanine phosphoribosyltransferase (HGPRTase) positive cells were encapsulated, forming lipochromosomes, and were fused with HGPRTase-negative cells. Transformed cells were selected for survival in hypoxanthine-aminopterinthymidine (HAT) medium. A tenfold increase in frequency of HGPRT transfer occurred using lipochromosomes as compared to uncoated chromosomes.   Two other enzyme markers, human glucose-6-phosphate dehydrogenase and phospho-glycerate kinase, were also transferred, as indicated by electrophoretic analysis of the transformants.   The frequency of transfer of the HGPRT-ase gene to the A9 mouse cells was approximately 10^{-5}.   In each of four clones examined, a fragment of presumptive human chromatin was present that was either free within the nucleus or was associated with the centromere of a mouse chromosome (Hoffman et al., 1981).   These data indicate that intact chromosomes can be transferred using lipo-chromosomes and that chromosome fragments can survive either free or attached to a chromosome in the nucleus of the recipient cells after delivery.   When the fragment is integrated into another chromosome it may do so at a preferential site.   However, the stability of the trans-ferred trait is generally poor in the absence of selection pressure.

Liposome-Mediated Delivery of DNA to Plant Protoplasts

Conventional plant breeding has relied upon the sexual breeding pro-cess to produce new gene combinations.   Combinations that cannot be achieved through this process may be derived by plant breeders in the future through the evolving field of plant genetic engineering.   There still is a great gap in knowledge of gene structure, regulation, func-tion, and expression as well as a lack of well-defined and understood gene transfer vehicles, selectable and characterized markers, and DNA uptake and integration processes.

DNA uptake by plant protoplasts has been examined by numerous laboratories and has been the topic of several extensive reviews (Klein-hofs and Behki, 1977; Lurquin and Kado, 1979).   When radioactively labeled DNA is added to protoplasts, the DNA is rapidly degraded by DNase in the medium.   Thus transformation of plant protoplasts by naked DNA, such as occurs in bacterial transformation experiments, does not appear to be feasible.   Precautions must be taken to preserve the integrity of the DNA so that it can be taken up by the proto-plasts.   Varying degrees of protection are afforded by manipulating the protoplast medium.   Low temperature and sodium citrate have been shown to retard degradation of the DNA in the protoplast medium, but no experiments have been conducted to demonstrate transfer of intact DNA to protoplasts (Slavik and Widholm, 1978).   High pH, $ZnSO_4$, and kinetin can protect exogenously added DNA, but treatment with these agents decrease protoplast wall formation (Fernandez et al., 1978; Hughes et al., 1977).   Currently, only poly-L-ornithine appears to have significant value in protecting DNA from exogenous DNase activity.   It does not appear to damage protoplasts; however, much of the exogen-ously added DNA is still degraded in the protoplast incubation medium (Hughes et al., 1977).

An alternative method for delivering intact DNA to protoplasts is through the sequestration of the DNA within liposomes and subsequent fusion with protoplasts.   In 1979, liposomes were shown to encapsulate and protect RNA from nuclease degradation and to deliver substantial amounts of the RNA to carrot protoplasts (Matthews et al., 1979).   E.

coli ^{3}H-RNA was encapsulated within liposomes of various lipid compositions and fused with carrot protoplasts. The RNA was then extracted from the protoplasts and examined on formamide-polyacrylamide gels and on sucrose gradients. These analyses indicated that the 23S RNA was degraded after delivery to the protoplasts, but some 16S and 4S prokaryotic RNA was still intact. No degradation of the *E. coli* RNA sequestered within liposomes was apparent even after treatment of the liposomes with RNase. Comparison of liposome-inserted RNA with naked RNA delivered to protoplasts indicated that liposome-mediated insertion provided much better protection of the RNA and delivered more intact RNA than did delivery attempts using unencapsulated RNA.

Several investigators have reported the liposome-mediated delivery of DNA to protoplasts. Matthews and Cress (1981) successfully used liposomes to deliver pBR322 plasmid DNA to carrot protoplasts. After incubation of the protoplasts with liposomes containing pBR322, the protoplasts were thoroughly washed to remove unincorporated liposomes, ruptured in the presence of an excess amount of rat DNA to prevent binding of undelivered pBR322, and the protoplast nuclei were isolated. The nuclei were treated briefly with DNase to remove any pBR322 adhering to the outside of the nuclei, lysed, and subjected to electrophoresis followed by hybridization with ^{32}P-probe for pBR322. This analysis indicated that between 200 and 1000 intact copies of pBR322 were present in each nucleus after liposome-mediated delivery.

Uchimiya and Harada (1981), utilizing a similar approach, examined the fate of pBR322 DNA 5 and 20 hr after insertion by liposomes into *D. carota* protoplasts. After 5 hr both open and closed circular pBR-322 molecules were present within the protoplasts; after 20 hr many of the open circular molecules present after 5 hr were degraded. Covalently closed circular pBR322 molecules appeared to be more stable and after 20 hr were still distinct on electrophoretic gels, indicating a preferential degradation of the open circular molecules.

Independent experiments by Lurquin (1979, 1981) demonstrated encapsulation of DNA within liposomes and its subsequent delivery to cowpea and tobacco protoplasts. ^{3}H-pBR322 plasmid was encapsulated within multilamellar vesicles (MLV) and incubated with protoplasts. Analysis of the plasmid DNA after delivery to the protoplasts indicated that most of the pBR322 remained highly polymerized, suggesting delivery of the plasmid to the protoplasts with little degradation.

Delivery of liposome contents to protoplasts has been monitored using fluorescence techniques. The use of fluorescein diacetate (Uchimiya, 1981; Cassells, 1978) and acridine orange-DNA complex (Lurquin and Sheehy, 1982) have been reported. However, fluorescein diacetate must be used with care when monitoring delivery of liposome contents to protoplasts, because low molecular weight compounds readily leak out of many types of liposomes. Rigorous controls must be performed to validate actual delivery. Liposomes encapsulating *E. coli* chromosomal DNA treated with the fluorescent dyes ethidium bromide or 4'-6-diamidino-2-phylindole (DAPI) have been used to monitor the interaction of liposomes with cowpea protoplasts (Lurquin, 1979). Individual and clusters of liposomes were observed in association with the protoplasts.

With the use of autoradiographic techniques, Rollo et al. (1980, 1981) also demonstrated the insertion of liposome-encapsulated DNA into protoplasts. When ^{3}H-DNA from *Bacillus subtilis* was entrapped within liposomes and incubated with *D. carota* protoplasts, approximately 4% of the protoplasts were labeled, indicating the presence of ^{3}H-DNA within the cytoplasm or nucleus. Control experiments in which unencapsulated ^{3}H-DNA was incubated with protoplasts indicated no association of the DNA within the protoplasts.

The effects of sequestration and insertion of nucleic acids into protoplasts utilizing liposomes prepared by several techniques and using several different lipid compositions have been examined by Lurquin et al. (1981). Positively charged vesicles sequestered more DNA than did neutral and negatively charged vesicles, but it remains to be established which lipid composition is optimal for insertion of liposome contents. Autoradiographic studies indicate that DNA delivery occurs when either positively or negatively charged vesicles are used (Lurquin and Sheehy, 1982).

The contents of liposomes are probably delivered to the protoplast by fusion and endocytosis. Scanning electron micrographs taken by Lurquin and Sheehy (1982) suggest that liposomes fuse with the plasma membrane of protoplasts, delivering their contents to the cytoplasm and nucleus. In addition to fusion, endocytosis may also play a role. When cytochalasin B, an inhibitor of endocytosis in mammalian cells, was added to carrot protoplast-liposome incubation mixtures, the delivery of the liposome contents to the protoplasts was reduced by 35-60% (Matthews et al., 1979). The amount of inhibition was partially dependent upon the lipid composition of the liposome. The amount of delivery occurring in the presence of cytochalasin B probably represents the amount delivered by fusion.

Biological Activity of Liposome-Delivered DNA and RNA

Several investigators are attempting to document the delivery of biologically active RNA and DNA molecules to plant protoplasts. The investigations discussed in this section are very recent and in some cases only preliminary results are available, but they serve as a guide to future applications of liposome technology.

Infection of plant protoplasts has been achieved through liposome-mediated delivery of biologically active viral RNA molecules. Tobacco mosaic virus (TMV) RNA was encapsulated within large, phosphatidylserine, unilamellar vesicles (LUV) and fused with protoplasts from *Vinca rosea* suspension cultures (Fukunaga et al., 1981). TMV production was monitored using fluorescent antibody against TMV. The viral RNA was protected by the liposomes from nuclease attack and delivered to the protoplasts. Over 50% of the protoplasts were infected by this method. TMV RNA has also been encapsulated within rotary evaporated vesicles (REV) by the method of Szoka and Papahadjopoulos (1978) and delivered to tobacco protoplasts (Fraley et al., 1982). Over 400 μg of TMV RNA per 10^5 protoplasts was detected 48 hr after delivery. Rollo and Hull (personal communication) have encapsulated turnip rosette virus (TRosV) and its RNA within multilamellar (MLV) and unilamellar (REV) vesicles. The TRosV encapsulated within MLVs infected turnip leaves but not protoplasts. The TRosV RNA entrapped within REVs infected turnip protoplasts, while unencapsulated TRosV RNA did not. Cauliflower mosaic virus (CaMV) and its DNA have also been entrapped within liposomes where it is protected from DNase degradation (Ulrich, Goodman, Widholm, and Matthews, unpublished; Rollo and Hall, personal communication). Local lesions can be formed on turnip leaves through the infection of the leaves by liposome-encapsulated CaMV DNA; however, neither group has yet reported the successful infection of turnip protoplasts by liposome-sequestered CaMV DNA.

Liposomes have also been used to deliver the crown gall tumor inducing (Ti) plasmid from *Agrobacterium tumefaciens* to protoplasts. Evidence for liposome-mediated delivery of the Ti plasmid to tobacco cells has been obtained by two independent groups (Dellaporta and Fraley, 1981; Owens, Matthews, and Cress, unpublished). In the latter studies, octopine-producing clones which displayed hormone-independent growth were isolated. These putative transformants have proved to be unstable upon successive subculturing and both traits were eventually lost. Although many of these reports are preliminary, it is clear that liposomes can deliver biologically active molecules to plant protoplasts under appropriate conditions.

METHODS OF LIPOSOME FORMATION

Parameters Influencing Encapsulation

Liposomes can be composed of an infinite number of varied proportions of phospholipids. Lipids frequently used for liposome studies involving plant protoplasts include phosphatidylcholine, phosphatidylserine, stearylamine, β-sitosterol, dicetyl phosphate, and cholesterol. Depending upon solubility, the phospholipids can be stored as stock solutions either in chloroform or methanol under nitrogen at -20 C for several weeks.

The phospholipid composition of liposomes is a major factor influencing encapsulation, permeability, fluidity, and surface charge. Ostro et al. (1980) demonstrated that within one hour 50% or more of ^{3}H-uridine was lost from liposomes, depending upon the lipid composition. These compositions included various proportions of phosphatidylcholine, dicetyl phosphate, lysophosphatidylcholine, and cholesterol. Cholesterol lended rigidity to the liposomes, making them less fluid and less permeable, thus helping to retain the encapsulated material. When ^{3}H-RNA was encapsulated within liposomes of identical compositions, only 3-13% of the radioactivity was lost after 24 hr, indicating that although low molecular weight molecules readily permeate through the liposome membrane, high molecular weight molecules remain entrapped for a substantial period of time.

Entrapment of DNA by liposomes increases when larger amounts of phospholipid are used (Hoffman et al., 1978; Szoka and Paphadjopoulos, 1978; Matthews and Cress, 1981) and is dependent upon the encapsulation technique.   The size of the DNA does not appear to influence encapsulation; even large structures such as chromosomes (Mukherjee et al., 1978; Matthews, unpublished) and chloroplasts (Giles et al., 1980) have been encapsulated or at least lipid–coated under the appropriate conditions.

Liposome Encapsulation Techniques

Several methods of liposome formation are available, and the choice depends upon the use of the liposomes and the material to be encapsulated.   There are at least a dozen general methods for making liposomes which differ in ease of formation, efficiency of encapsulation, vesicle size, trapped volume, homogeneity, and desired phospholipid composition.   Several of these methods are compared in Table 1.

Table 1.   Techniques for Liposome Formation

METHOD[a]	TYPE VESICLE	SIZE (Å)	ENCAPSU-LATION (%)
Hand–shaken	MLV	Vary	15
Ether infusion	LUV	500–2500	10
Sonication	SUV	200–500	1
Calcium–induced fusion	LUV	2,000–10,000	10
Reverse–phase evaporation	REV	2,000–10,000	40

[a] Data were compiled from Szoka and Papahadjopoulos (1978), Szoka (1980), Ostro et al. (1980), and Papahadjopoulos et al. (1975).

1.   HAND–SHAKEN MULTILAMELLAR VESICLES (MLV).   This method entails the rotary evaporation of lipids in an organic phase until the round bottom flask is evenly coated with a thin film of lipid and no trace of organic solvent remains.   The material to be encapsulated is then added in an aqueous phase solution, and the flask is shaken by hand or sonicated.   The lipid material forms multilamellar vesicles entrapping portions of the aqueous phase.   The suspension is allowed to incubate at room temperature for 2 hr.   These multilamellar vesicles generally trap less volume than large unilamellar liposomes, but have a much lower passive diffusion rate (Scieren et al., 1978).   This methodology may find applications in the encapsulation or coating of large particulate matter such as mitochondria, chloroplasts, or chromosomes for transfer to protoplasts, because it avoids contact of the material to be encapsulated with an organic phase, such as when the reverse phase evaporation technique is used (Method 5).   Also, it avoids high temperatures that are necessary with the ether infusion technique (Method 2). However, the heterogeneity in liposome size, the multilamellar structure, and low encapsulation efficiency of the liposomes limit the uses of liposomes made by this method.

2. ETHER INFUSION TECHNIQUE FOR UNILAMELLAR VESICLES (LUV). This technique, first described by Deamer and Bangham (1976), can be used to efficiently entrap high molecular weight molecules by injecting an organic phase, usually petroleum ether, containing phospholipids into a heated aqueous phase containing the material to be encapsulated. The temperature of the aqueous phase, the rate of injection of the organic phase, and the concentration of the phospholipid are important in obtaining good vesicle formation. Unilamellar liposomes of fairly uniform size are produced with a high trapped volume. The ether infusion technique was the first available method that could be used to routinely encapsulate high molecular weight compounds such as RNA and DNA within unilamellar vesicles. It is more difficult to produce sterile liposomes by this method than by the reverse phase evaporation technique, because more manipulations are involved. Furthermore, heat-labile proteins are denatured by this method. Although the either infusion technique forms liposomes which entrap 10% of the available aqueous volume, the amount of material which can be entrapped using the reverse phase evaporation technique is higher. The former method provides liposomes of a more uniform size suitable for studies involving membrane model systems.

3. SONICATION TO FORM SMALL UNILAMELLAR VESICLES (SUV). Large multilamellar vesicles (MLV) encapsulating desired macromolecules can be sonicated to form small unilamellar vesicles (SUV). Sonication can be achieved using either a probe sonicator or a bath sonicator. The probe sonicator introduces contamination and shears DNA more readily, so the bath sonicator is usually preferred. The small size of these vesicles limits the size of the molecules which can be encapsulated. However, they are useful when a homogeneous population of vesicles is required.

4. CALCIUM-INDUCED FUSION (LUV). Small unilamellar vesicles made from acidic phospholipids can be induced to fuse to produce large cochleate cylinders by adding calcium (Papahadjopoulos et al., 1975, 1976). Upon the addition of EDTA, these multilamellar structures form large unilamellar vesicles. This method enhances the range of size of molecules that can be encapsulated within unilamellar vesicles and encapsulates macromolecules under mild conditions, thus reducing the risk of mechanical damage.

5. REVERSE-PHASE EVAPORATION TECHNIQUE (REV). This procedure produces large liposomes which encapsulate a high percentage of the available aqueous material. In 1978 Szoka and Papahadjopoulos described their newly devised system using reverse-phase evaporation which routinely produces liposomes encapsulating approximately 40–50% of the aqueous material presented. These liposomes possess a high aqueous space to lipid ratio and are unilamellar and oligolamellar. This method is useful for encapsulating high molecular weight molecules

such as DNA and intact mitotic plant chromosomes, but it destroys chloroplasts and mitochondria (Matthews and Cress, 1981; Matthews, unpublished). The amount of encapsulation of aqueous material is strongly dependent upon the amount of lipid material present in the organic phase. With this method, the aqueous material to be encapsulated is added to the organic phase containing the phospholipids. The mixture is sonicated (or hand shaken if the material of be encapsulated is sensitive to sonication) to form an emulsion which is then rotary evaporated under nitrogen until the liposomes are formed and no trace of organic solvent is present. The protocol for making liposomes by this method is presented below. Because this method of producing liposomes is very simple, it lends itself to maintaining liposome sterility for use in DNA and RNA transfer studies. Within 3 years it has become the method of choice for many investigators conducting genetic engineering studies.

PROTOCOL FOR REVERSE-PHASE EVAPORATION VESICLES (REV) MADE OF PHOSPHATIDYLCHOLINE (PC), DICETYL PHOSPHATE (D), AND LYSOPHOSPHATIDYLCHOLINE (LPC) (8PC:2D:0.4LPC).

1. Lipids, stored at -20 C under nitrogen in chloroform, are allowed to warm up to room temperature.
2. Pipet the following lipids into a 50 ml round bottom flask: 12.5 ml D (100 mg/100 ml), 0.5 ml PC (100 mg/ml), 5.0 LPC (50 mg/100 ml).
3. Rotary evaporate under vacuum until all chloroform is removed.
4. Redissolve lipid in 5 ml anhydrous ether.
5. Add 1.5 ml HEPES buffer containing ^{3}H-DNA in HEPES buffer: 2 mM HEPES, 2 mM NaCl, 2 mM KCl, pH 7.4.
6. Layer nitrogen over surface.
7. Hand shake 5 min or until a good dispersion is formed.
8. Rotary evaporate 15-20 min at 200 rpm to remove ether.
9. Add 2.0 ml HEPES, shake to disperse lipids and rotary evaporate 5-10 min.
10. Remove liposomes and rinse flask with 1.0 ml HEPES to obtain remaining liposomes.

Purification of Liposomes from Unencapsulated Materials

Several methods are available for removing unencapsulated DNA or RNA from the liposome preparation. The removal of unencapsulated materials is needed to calculate the amount of a compound encapsulated within the liposomes. Unencapsulated DNA and RNA are degraded by nuclease added to the liposome preparation; the resultant nucleic acid fragments are separated from the liposomes by column chromatography (Ostro et al., 1977; Matthews et al., 1979). Elution of the liposomes is monitored by change in absorbance at 650 nm and is observed as a milky white suspension eluting with the void volume. Autoclavable column matrices, such as Sepharose CL-2B, are available

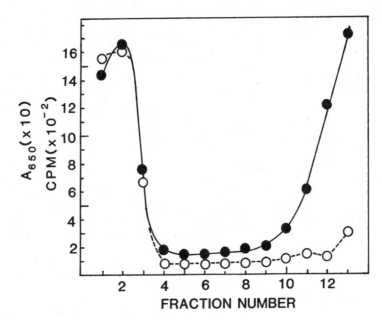

Figure 1. Profile of Ficoll gradient used to separate liposomes from unencapsulated materials. Radioactivity (—) and absorbance at 650 nm (---) are depicted. Gradient was fractionated from top to bottom. Procedure is as described in the Protocol.

for sterile column chromatography. Liposomes can also be separated from unencapsulated fragments by high speed centrifugation which pellets the liposomes while the fragments remain in the supernatant (Fraley et al., 1980). An efficient method used routinely in several laboratories utilizes Ficoll, as described in detail in the following protocol. This method is simple, rapid, provides good separation of the liposomes from unencapsulated material (Fig. 1), and allows sterility to be easily maintained. If a liposome preparation is suspected of being contaminated, the preparation can be sterilized by passing it through a 0.2 micron filter (Szoka et al., 1980). However, 30–40% of the entrapped molecules are released by this procedure.

PROTOCOL FOR FICOLL GRADIENT SEPARATION OF LIPOSOMES FROM UNENCAPSULATED DNA (PROCEDURE DERIVED FROM FRALEY ET AL., 1980)

1. Mix 1.5 ml liposomes with 3 ml of 30% Ficoll in a test tube.
2. Layer 9 ml of 10% Ficoll on top of mixture.
3. Layer 3 ml of HEPES on top of 10% Ficoll layer.
4. Centrifuge gradient for 30 min at 16,300 x g.

5. Liposomes are at the top of the gradient, while unencapsulated material is in the lower portion of the gradient (see Fig. 1).

LIPOSOME-PROTOPLAST INTERACTION

Liposome Membrane Composition

Conditions optimizing encapsulation of material by liposomes, insertion of liposome contents into protoplasts, and protoplast viability after treatment are necessary to maximize the probability of success of gene transfer experiments. By altering the phospholipid composition of the liposome membrane, the overall charge of the liposome surface can be made positive, neutral, or negative. Stearylamine is commonly used with neutral lipids such as cholesterol and phosphatidylcholine to form positively charged vesicles, while dicetyl phosphate or phosphatidylserine are used in combination with neutral lipids to form negatively charged vesicles. Lurquin et al. (1981) reported that positively charged multilamellar vesicles (MLV) and unilamellar vesicles (LUV) possess a higher affinity for nucleic acids, although negatively charged vesicles formed by the reverse phase evaporation technique (REV) also successfully encapsulated DNA and RNA (Lurquin and Sheehy, 1982). Experiments by several investigators (Ostro et al., 1980; Rollo et al., 1981; Fraley et al., 1982; Lurquin and Sheehy, 1982) indicate that the charge of the liposome also affects the interaction of the liposome with the protoplast. However, the conditions for encapsulation and incubation of liposomes with protoplasts appear to play a large role in determining which charge is most effective in promoting liposome-protoplast interaction. Most investigators agree that liposomes containing cholesterol are more stable and tend to retain the encapsulated molecules longer without leakage, thus providing more protection for encapsulated DNA and RNA. Several investigators working with mammalian systems have modified the external liposome membrane by the addition of proteins and immunoglobins in order to "target" liposomes for interaction with specific cells. Similar methods may increase liposome-protoplast interaction.

Liposome-Protoplast Incubation Conditions

By manipulating liposome-protoplast incubation conditions, delivery of liposome contents to protoplasts can be varied from less than 1% to over 50%. Many parameters, including the chemical composition and pH of the incubation medium, proportion of liposomes to protoplasts, temperature, and the use of membrane fusing agents, augment liposome-mediated deliver. For example, incorporation of liposome contents into protoplasts is greatly dependent upon the temperature of the incubation medium (Fig. 2). As the temperature of the fusion medium is increased the fluidity of the liposome membrane and the protoplast membrane increases, allowing the membranes to fuse more easily. *Daucus carota* protoplasts can be incubated with liposomes at 37 C for more than 2

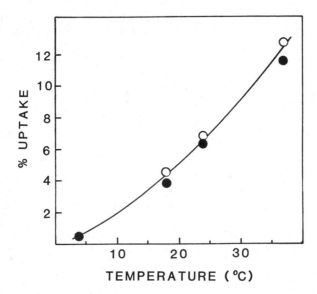

Figure 2. Temperature effect on uptake of liposomes by carrot proto-plasts. Liposomes (A_{650} = 0.5) were added to 2 x 10^7 protoplasts and incubated 1 hr.

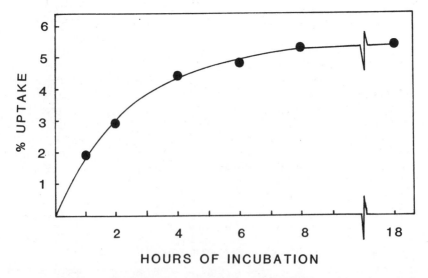

Figure 3. Uptake of lipsomes by carrot protoplasts with time. Proto-plasts (6 x 10^5) and liposomes (A_{650} = 0.761) were incubated for 2 hr at 37 C in an incubation volume of 6.0 ml. Each point represents the average of three experiments.

hr to increase delivery while retaining high protoplast viability; other protoplasts may be more sensitive to the effects of temperature. The pH of the fusion medium also influences liposome-protoplast interaction (Matthews and Cress, 1981). When the pH of the medium is decreased to 5 there is an increase in fusion of protoplasts with liposomes composed of phosphatidylcholine, dicetyl phosphate, and lysolecithin (8PC: 2D:0.4 Ly).

Liposome uptake by *D. carota* protoplasts occurs in a nearly linear fashion during the first two hours of incubation (Fig. 3) and then decreases. Insertion of liposome contents into the protoplasts is also proportional to the number of protoplasts available for interaction. Although liposome incorporation can be increased using a high protoplast to liposome ratio, the number of molecules inserted per protoplast is higher when a high liposome to protoplast ratio is used.

An increase in the insertion of liposome contents into protoplasts can be achieved by using various chemical fusing agents (Matthews and Cress, 1981; Lurquin, 1979; Fukunaga et al., 1981; Fraley et al., 1982). Polyethylene glycol (PEG) 4000, 6000, and 8000 have been very effective as fusing agents as has polyvinyl alcohol (PVA). Insertion of the contents of over 40% of the available liposomes has been achieved using PEG; however, protoplast viability 24 hr after treatment was less than 10% (Matthews, unpublished). Although high amounts of fusion can be achieved, moderate delivery conditions tend to maintain higher protoplast viability (Table 2). The protocol achieving these results is given in Fig. 4 and below. Maximum percentage incorporation of liposome contents by protoplasts must be balanced with protoplast viability. Under current conditions such a balance occurs within the range of 20-30% incorporation. Cells with a higher percentage incorporation do not form cell walls as quickly, and many were still protoplasts or were inviable after 7 days.

Table 2. Effect of Polyethylene Glycol (PEG 8000) on Liposome-Mediated Delivery of ^{3}H-*E. coli* DNA to *D. carota* Protoplasts and on Protoplast Viability

PEG (%)	CPM	INCORPORATION (%)	VIABILITY (%)[a]
0	100	1.1	90.4
6	393	4.6	45.2
11	3061	35.6	9.4
15	2193	25.5	14.4

[a]Viability was determined by staining cells with phenosafranin or Evans Blue 1 week after plating on agar medium.

LIPOSOME-PROTOPLAST INCUBATION PROTOCOL

Carrot Protoplast Formation

1. Carrot protoplasts are prepared by incubating 5-day-old *D. carota* cells in 4% cellulysin and 1% macerase in 0.3 M sorbitol, 0.3 M mannitol, 3 mM 2[N-morpholino]ethanesulfonic acid (MES), 6 mM CaCl$_2$·2H$_2$O, 0.7 mM NaH$_2$PO$_4$·H$_2$O, pH 5.7.

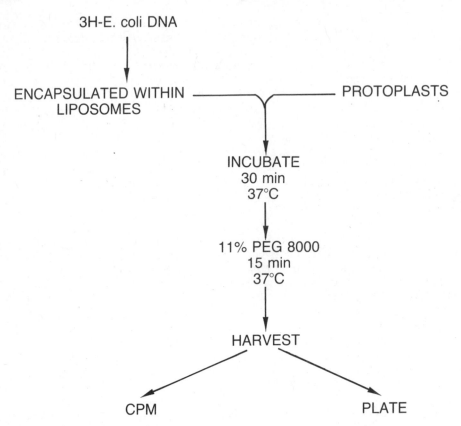

Figure 4. Protocol for determining uptake of liposomes by protoplasts and protoplast viability.

2. After 4 hr wash the protoplasts using the same incubation medium, but lacking enzymes.
3. Dilute the protoplasts to a concentration of 10^6/ml.

Liposome Preparation

1. Liposomes are prepared as described in the Protocol for REV using desired lipid composition.
2. Dilute the liposomes with HEPES buffer to $A_{650} = 0.6$.

Liposome–Protoplast Incubation

1. Mix in a capped, sterile test tube: 0.5 ml liposomes; 0.5 ml fusion medium consisting of 0.6 M Mannitol, 0.6 M Sorbitol, 6 mM MES, 12 mM $CaCl_2 \cdot 2H_2O$, 1.4 mM NaH_2PO_4, pH 4.7; 1 ml carrot protoplasts (10 /ml).
2. Incubate for 30 min at 37 C with test tube inverting at 2 rpm.

3.   Slowly add 1.1 ml of PEG Solution (37 C) containing: 30% PEG 8000, 5 mM $CaCl_2$, 0.1 M Glucose, pH 4.7.   This yields a final PEG concentration of 11%.
4.   Gently mix and incubate for 15 min at 37 C.
5.   Dilute slowly with protoplast incubation solution (see text).
6.   Centrifuge at 200 x g for 5 min and resuspend for plating or wash twice more and determine amount of uptake.

FUTURE PROSPECTS FOR LIPOSOME-MEDIATED DELIVERY

Several new frontiers in molecular biology and genetics will develop in conjunction with liposome encapsulation and delivery techniques. The use of liposomes for delivering DNA and mRNA to plant protoplasts is particularly exciting and virtually unexplored. A range of experiments examining in vivo regulation of mRNA synthesis and translation can be envisioned through the delivery of different mRNAs to plant protoplasts.   It may be possible to encapsulate and insert different mRNAs into protoplasts to study in vivo translation under different cellular environments.   Furthermore, the processing of the resultant translated protein product may be studied as well as its transport to specific cellular locations.   Protein assembly could be examined from a variety of perspectives.   Also, in vivo processing of liposome-delivered mRNA may be investigated, whereby factors involved in mRNA recognition by ribosomes and regulatory enzymes could be identified.   This could involve the analysis of mRNA half-lives and the possible recognition of mRNA as self or foreign.   How long and how effectively would mRNA from one species be translated in closely related and distantly related species?   For instance, if biologically active mRNA coding for the small subunit of RuBPCase was isolated from two related species and these mRNAs were delivered individually and together to a mesophyll protoplast, which one would be preferentially synthesized, how would this affect the synthesis of the large subunit, would there be a difference in the half-life of the two mRNAs, would the protein product be transferred into the chloroplast and be assembled with the large subunit, and what effect would this have on carbon fixation?   Many other questions could be asked and perhaps answered using similar systems to study basic molecular mechanisms.

Liposome encapsulation is currently a promising method for delivering genetically engineered DNA to plant protoplasts. Transformation experiments utilizing concentrated DNA with high gene copy number have high probability for success using liposome-mediated transfer.   Resistance to herbicides, pathogens, environmental stresses, and other important traits, especially those coded for by single genes, may be inserted into plants via liposomes in the future.   Matthews and Cress (1981) have reported liposome-mediated delivery of between 200 and 1000 intact copies of pBR322 per carrot nucleus.   Significant gains in optimizing liposome delivery have been obtained since then, improving the chances for successful delivery and insertion of biologically active DNA into protoplast nuclei.   Within the next few years it is likely that several laboratories will be able to deliver genes to plant protoplasts using liposomes. Continued use of liposomes in this capacity is dependent upon the success and limitations of other techniques which show potential. One exciting alternative method is that of direct fusion of bacterial spheroplasts with protoplasts (Hasezawa et al., 1981). This method would allow genetically engineered plasmid DNA to be replicated within the bacteria, rather than extracting and purifying the plasmid DNA for liposome-mediated insertion. The bacterial wall would then be enzymatically removed and the resultant bacterial spheroplast fused with the plant protoplast. If this method can be developed into an efficient mode of DNA insertion and protoplast transformation, liposomes may be less useful for certain genetic engineering applications.

It is also likely that liposome-mediated chromosome transfer will be successful as it has been in mammalian systems. Although plant cell cultures are less easy to synchronize than mammalian cell cultures, intact chromosomes can be isolated in high enough numbers to initiate chromosome transfer experiments using plant materials (Malmberg and Griesbach, 1980; Griesbach et al., 1981; Matthews, 1981). Liposomes can protect the isolated chromosomes from extracellular degradation. Although much of the chromosome may be degraded once it is inside the cell, fragments that contain gene clusters are likely to remain intact, thus conferring new traits. The success of such insertions may be enhanced by determining optimum conditions for liposome encapsulation, liposome-mediated delivery, and insertion. Such factors as the influence of cell cycle stage on uptake and insertion, the use of liposomes in combination with other DNA-protecting agents, and the form, complexity, and base sequence of the DNA to be inserted may all contribute to the success or failure of the insertion and expression of biologically active DNA. When these parameters are optimized for insertion while maintaining protoplast viability, liposomes promise to be a useful tool to aid in the modification of plant genomes for basic and applied research.

ACKNOWLEDGEMENT

The author would like to thank Drs. R.T. Fraley, P.F. Lurquin, L.D. Owens, F.U. Rollo, and H. Uchimiya, for providing communications and papers in press.

Mention of trademark, proprietary product, or vendor does not constitute a guarantee or warranty of the product by the U.S. Department of Agriculture and does not imply its approval to the exclusion of other products or vendors that may also be suitable.

KEY REFERENCES

Fraley, R., Subramani, S., Berg, P., and Papahadjopoulos, D. 1980. Introduction of liposome-encapsulated SV40 DNA into cells. J. Biol. Chem. 255:10431-10435.

Matthews, B.F. and Cress, D.E. 1981. Liposome-mediated delivery of DNA to carrot protoplasts. Planta 153:90-94.

Szoka, F., Jr. 1980. Comparative properties and methods of preparation of lipid vesicles (liposomes). Annu. Rev. Biophys. Bioeng. 9:467-508.

Szoka, F. and Papahadjopoulos, D. 1978. Procedure for preparation of liposomes with large internal aqueous space and high capture by reverse-phase evaporation. Proc. Natl. Acad. Sci. 75:4194-4198.

REFERENCES

Bangham, A.D., Standish, M.M., and Watkins, J.C. 1965. Diffusion of univalent ions across the lamellae of swollen phospholipids. J. Molec. Biol. 13:238-252.
Cassels, A.C. 1978. Uptake of charged lipid vesicles by isolated tomato protoplasts. Nature 275:760.
Cohen, C.M., Weissmann, G., Hoffstein, S., Awasthi, Y.C., and Srivastava, S.K. 1976. Introduction of purified hexosaminidase A into Tay-Sachs leukocytes by means of immunoglobin-coated liposomes. Biochemistry 15:452-460.
Deamer, D. and Bangham, A.D. 1976. Large volume liposomes by an ether vaporization method. Biochim. Biophys. Acta 443:629-634.
Dellaporta, S.L. and Fraley, R.J. 1981. Delivery of liposome-encapsulated nucleic acids into plant protoplasts. Plant Molec. Biol. Newsletter 2:59-66.
Dimitriadis, G.J. 1978. Entrapment of ribonucleic acids in liposomes. FEBS Lett. 86:289-293.
_____ 1979. Entrapment of plasmid DNA in liposomes. Nucl. Acids Res. 6:2697-2705.
Fernandez, S.M., Lurquin, P.F., and Kado, C.I. 1978. Incorporation and maintenance of recombinant-DNA plasmid vehicles pBR313 and pCR1 in plant protoplasts. FEBS Lett. 87:277-282.
Finkelstein, M. and Weissmann, G. 1978. The introduction of enzymes into cells by means of liposomes. J. Lipid Res. 19:289-303.
Fraley, R.T., Fornari, C.S., and Kaplan, S. 1979. Entrapment of a bacterial plasmid in phospholipid vesicles: Potential for gene transfer. Proc. Natl. Acad. Sci. 76:3348-3352.
Fraley, R.T., Dellaporta, S.L., and Papahadjopoulos, D. 1982. Liposome-mediated delivery of tobacco mosaic virus RNA into tobacco protoplasts: A sensitive assay for monitoring liposome-protoplast interactions. Proc. Natl. Acad. Sci. 79:1859-1863.
Fukunaga, Y., Nagata, T., and Takebe, I. 1981. Liposome-mediated infection of plant protoplasts with tobacco mosaic virus RNA. Virology 113:752-760.
Giles, K.L., Vaughan, V., Ranch, J.P., and Emery, J. 1980. Liposome-mediated uptake of chloroplasts by plant protoplasts. In Vitro 16:581-584.
Griesbach, R.J., Koivuniemi, P.J., and Carlson, P.S. 1981. Extending the range of plant genetic manipulation. BioScience 31:754-756.
Hasezawa, S., Nagata, T., and Syono, K. 1981. Transformation of Vinca protoplasts mediated by Agrobacterium spheroplasts. Molec. Gen. Genet. 182:206-210.

Heath, T.D., Fraley, R.T., and Papahadjopoulos, D. 1980. Antibody targeting of liposomes: Cell specificity obtained by conjugation of F(ab')$_2$ to vesicle surface. Science 210:539–541.

Hoffman, R.M., Margolis, L.B., and Bergelson, L.D. 1978. Binding and entrapment of high molecular weight DNA by lecithin liposomes. FEBS Lett. 93:365–368.

Hoffman, W.K., Lalley, P., Butler, J.DeB., Orloff, S., Schulman, J.D., and Mukherjee, A.B. 1981. Lipochromosome mediated gene transfer: Identification and probable specificity of localization of human chromosomal material and stability of the transferents. In Vitro 17:735–740.

Hughes, B.G., White, F.G., and Smith, M.A. 1977. Fate of bacterial plasmid DNA uptake by barley protoplasts. FEBS Lett. 79:80–84.

Jansons, V.K. and Mallett, P.L. 1981. Targeted liposomes: A method for preparation and analysis. Anal. Biochem. 111:54–59.

Kleinhofs, A. and Behki, R. 1977. Prospects for plant genome modification by nonconventional methods. Annu. Rev. Genet. 11:79–101.

Lurquin, P.F. 1979. Entrapment of plasmid DNA by liposomes and their interactions with plant protoplasts. Nucl. Acids Res. 6:3773–3784.

_____ 1981. Binding of plasmid loaded liposomes to plant protoplasts: Validity of biochemical methods to evaluate the transfer of exogenous DNA. Plant Sci. Lett. 21:31–40.

_____ and Kado, C.I. 1979. Recent advances in the insertion of DNA into higher plant cells. Plant Cell Environ. 2:199–203.

_____ and Sheehy, R.E. 1982. Binding of large liposomes to plant protoplasts and delivery of encapsulated DNA. Plant Sci. Lett. 25:133–146.

_____, Sheehy, R.E., and Rao, N.A. 1981. Quantitative aspects of nucleic acids sequestration in large liposomes and their effects on plant protoplasts. FEBS Lett. 125:183–187.

Malmberg, R.L. and Griesbach, R.J. 1980. The isolation of mitotic and meiotic chromosomes from plant protoplasts. Plant Sci. Lett. 17:141–147.

Mannino, R.J., Allebach, E.S., and Strohl, W.A. 1979. Encapsulation of high molecular weight DNA in large unilamellar phospholipid vesicles. FEBS Lett. 101:229–232.

Matthews, B.F. 1981. Isolation of chromosomes from carrot cell cultures. Plant Physiol. 67(suppl.):656.

_____, Dray, S., Widholm, J., and Ostro, M. 1979. Liposome-mediated transfer of bacterial RNA into carrot protoplasts. Planta 145:37–44.

Mukherjee, A.B., Orloff, S., Butler, J.DeB., Triche, T., Lalley, P., and Schulman, J.D. 1978. Entrapment of metaphase chromosomes into phospholipid vesicles (lipochromosomes): Carrier potential in gene transfer. Proc. Natl. Acad. Sci. 75:1361–1365.

Ostro, M.J., Giacomoni, D., and Dray, S. 1977. Incorporation of high molecular weight RNA into large artificial lipid vesicles. Biochem. Biophys. Res. Comm. 76:837–842.

Ostro, M.J., Giacomoni, D., Lavelle, D., Paxton, W., and Dray, S. 1978. Evidence for translation of rabbit globin mRNA after liposome-mediated insertion into a human cell line. Nature 274:921–923.

Ostro, M.J., Lavelle, D., Paxton, W., Matthews, B., and Giacomoni, D. 1980. Parameters affecting the liposome-mediated insertion of RNA into eukaryotic cells in vitro. Arch. Biochim. Biophys. 201:392-402.

Papahadjopoulos, D. and Miller, N. 1967. Phospholipid model membranes. I. Structural characteristics of hydrated liquid crystals. Biochim. Biophys. Acta 135:624-638.

Papahadjopoulos, D., Poste, G., and Mayhew, E. 1974. Cellular uptake of cyclic AMP captured within phospholipid vesicles and effect on cell-growth behavior. Biochim. Biophys. Acta 363:404-418.

Papahadjopoulos, D., Vail, W.J., Jacobson, K., and Poste, G. 1975. Cochleate lipid cylinders: Formation by fusion of unilamellar lipid vesicles. Biochim. Biophys. Acta 394:483-491.

Papahadjopoulos, D., Hui, S., Vail, W.J., and Poste, G. 1976. Studies on membrane fusion. I. Interactions of pure phospholipid membranes and the effect of myristic acid, lysolecithin, proteins and dimethylsulfoxide. Biochim. Biophys. Acta 448:245-264.

Rollo, F., Sala, F., Cella, R., and Parisi, B. 1980. Liposome-mediated association of DNA with plant protoplasts: Influence of vesicle lipid composition. In: Plant Cell Cultures: Results and Pespectives (F. Sala, B. Parisi, R. Cella, and O. Ciferri, eds.) pp. 237-246. Elsevier/North Holland Biomedical Press, Amsterdam.

Rollo, F., Grazia Galli, M., and Paresi, B. 1981. Liposome-mediated transfer of DNA to carrot protoplasts: A biochemical and autoradiographic analysis. Plant Sci. Lett. 20:347-354.

Schieren, H., Rudolph, S., Finkelstein, M., Coleman, P., and Weissmann, G. 1978. Comparison of large unilamellar vesicles prepared by a petroleum ether vaporization method with multilamellar vesicles. Biochim. Biophys. Acta 542:137-153.

Slavik, N.S. and Widholm, J.M. 1978. Inhibition of deoxyribonuclease activity in the medium surrounding plant protoplasts. Plant Physiol. 62:272-275.

Szoka, F., Olson, F., Heath, T., Vail, W., Mayhew, E., and Papahadjopoulos, D. 1980. Preparation of unilamellar liposomes of intermediate size (0.1-0.2 µm) by a combination of reverse phase evaporation and extrusion through polycarbonate membranes. Biochim. Biophys. Acta 601:559-571.

Uchimiya, H. 1981. Parameters influencing the liposome-mediated insertion of fluorescein diacetate into plant protoplasts. Plant Physiol. 67:629-632.

_____ and Harada, H. 1981. Transfer of liposome-sequestering plasmid DNA into Daucus carota protoplasts. Plant Physiol. 68:1027-1030.

Van Houte, A.J., Snippe, H., and Willers, J.M.N. 1979. Characterization of immunogenic properties of haptenated liposomal model membranes in mice. I. Thymus independence of the antigen. Immunology 37:505-514.

Weissmann, G., Bloomgarden, D., Kaplan, R., Cohen, C., Hoffstein, S., Collins, T., Gotlieb, A., and Nagle, D. 1975. A general method for the introduction of enzymes, by means of immunoglobulin-coated liposomes, into lysosomes of deficient cells. Proc. Natl. Acad. Sci. 72:88-92.

Wong, T.K., Nicolau, C., and Hofschneider, P.H. 1980.   Appearance of
   β-lactamase activity in animal cells upon liposome-mediated gene
   transfer.   Gene 10:87-94.

CHAPTER 16

Chromosome Number Variation and Karyotype Stability in Cultures and Culture-Derived Plants

A.D.Krikorian, S.A. O'Connor, and *M.S. Fitter*

INTRODUCTION

Some Generalities

In this chapter emphasis will be placed on the fact that it will *not* be possible to provide details of rigorous experimental techniques and protocols that can be applied without modification to a broad spectrum of material. One of the central problems is that it is often very difficult to extend techniques used on intact plants or plant organs such as root tips to cultured tissues and cells. Even if the methods have been developed for cultures of a given species and have been published, extrapolation is frequently not easy, because of variation among species, cultivars, and even genotypes within a species. Therefore, variation in responsiveness to a given method is to be expected. In outbred populations it is sure to be especially wide. Our approach, therefore, will necessarily be a generalized one, and by far the greater emphasis will be placed on certain principles. The most one can expect from such a summary are guidelines!

This chapter tacitly assumes that the study of karyology or chromosomes of cultured plant cells and tissues can provide a realistic bridge between the need for a better understanding of the nutrition of cultured plant tissues and cells, the factors which attend the release and expression of their morphological competence, and crop plant improvement. At the same time, the study of chromosomes in cultures may be an end in itself against the day when sophisticated cytogenetics can routinely be carried out using cell or protoplast cultures. Because definitive studies on chromosomal stability or lack thereof in cultured systems are, in our view, still at a preliminary stage of design and development, it is fair to say that before the use of cultures for crop improvement are fully realized, it will be critical to control and understand better the culture conditions in the broadest sense possible. The karyology and cytogenetics of cultured plant cells have not been nearly as extensively developed as one would have expected, and plant workers lag far behind those who work with animal cells (Yunis, 1974). This state is due in part to the relatively few investigators that have concerned themselves seriously with the problems, but it is also due to the very real technical difficulties that exist and reflect the state of the art.

Establishment of Cultures and Procurement of Cell Lines

Unlike some mammalian cell lines that are certified and may be purchased from places like the American Type Culture Collection, "standard" plant tissue or cell lines are not available through a centralized facility. Except through the courtesy of individual investigators who may be willing to provide others with already established materials (Withers, 1981), those working with cultured plant systems are forced to initiate and maintain cultures of their own. Whatever the reason, this is highly desirable, since it is the only way to begin to know and fully understand what might be occurring in a culture under a given set of conditions. There are sure to be plenty of unforeseen difficulties confronting a worker in this field without adding to the problems of the additional unknowns entailing subtle (and some not so subtle) intricacies of culture methodology, etc. Throughout these volumes one will repeatedly appreciate various peculiarities of methodology and seeming or real idiosyncracies of a given laboratory. In some cases these may affect or seriously alter the outcome of a procedure. Occasionally, they may be the sole difference between success and failure! One cannot exaggerate the overriding need for establishing cultures under as carefully defined conditions as possible. We even insist on processing our own supplies of so poorly defined a supplement as coconut water. References may also be made to publications which show that the composition of even a nominally well-established culture medium such as that of White may not be universally agreed upon (Singh and Krikorian, 1981). The same is true for the amount of chelate in the widely used iron EDTA complex in the medium of Murashige and Skoog (Singh and Krikorian, 1980).

We hope that all will agree that there is little to be gained by working with cultured cells and tissues that are inadequately defined or described as to their origin. There is a critical need to know as much biology (including the genetics) as possible of the plant with which one is working. Equally important is the need to know the anatomical origin or histological site from which the primary culture was originally derived. Attention must also be given to the medium on which it was initiated and subsequently maintained and the number of passages that it underwent and the environmental parameters to which the culture was exposed.   In short, as obvious and elementary as this may all seem, only occasionally does one know the precise "cultural history" of the material which is being examined.

Baseline Information

A PLEA FOR WORK USING IDENTIFIED GERMPLASM.   Investigators at various research institutes that specialize in a particular crop plant will perhaps not need to be concerned with the first criterion as much as others, because substantial amounts of information are sure to be available to them.   They will be familiar with the botany or biology of the plant and its agronomic needs.   The specific research goals will be carefully delineated, and a rather large body of data will often be available to them on the cytology and cytogenetics of the plant(s) in question.   If the organization is fortunate enough to have individuals who are skilled in culture techniques, then a collaborative effort between "tissue culturist" and breeder is sure to be satisfying.   But it is equally certain that there will be many individuals who are skilled to varying degrees in one or another of the disciplines.   Such individuals can play a major role in the development and expansion of the knowledge and expertise that ultimately will permit the anticipated potential of crop improvement via tissue culture to be realized if they adhere strictly to a few guidelines.

A listing of the national and international centers responsible for germplasm collection of specific crop plants may be found in Thurston (1977, p. 225) and Sprague (1980, p. 154).

LITERATURE SOURCES FOR PRELIMINARY BASELINE DATA.   Although it would be very valuable, but because it could never be even marginally comprehensive and would take too much space, primary literature citations will not be given here to the many sources on the cytology and cytogenetics of crop plants available to workers.   It would be unpardonable, however, to omit a few secondary sources. Simmonds (1976) is an excellent place to start for anyone interested in tracking down the karyotype or chromosomal characteristics of a specific crop plant.   The contributors to that volume, all experts on their respective plants, have included valuable information and key references on the chromosomal relationships of virtually all the major and many of the minor crops.   Ferwerda and Wit (1969) cover a wide range of tropical plants in substantial detail.   Simmonds (1979) should also be consulted.   In addition to the many journals in which cytogenetic information is available, mention should be made of *Field Crop Abstracts* and the now-standard books in the Leonard Hill *World Crops* series and the Longman *Tropical Agriculture* series.   The American Society of Agronomy and the Crop Science Society of America occasionally publish monographs on selected plants.   Invariably these include comprehensive cytogenetic treatments, see *Helianthus*, the sunflower, for instance in Whelan (1978).   If a crop is so "minor" or so "exotic" that it has not yet been included in these conventional sources, we recommend for a start that one consult Fedorov (1969).

GENERATION AND PRESENTATION OF BASELINE DATA. The study of chromosome morphology in some crop plants has advanced to the point where oftentimes it is possible to establish a detailed profile of the nucleus or karyotype. A karyogram may be defined as a systemized or orderly arrangement of the chromosomes of a single cell prepared with the aid of photography or by drawing. The implication is that the number and morphology of the chromosomes shown is not only representative of the cell, but typifies or karyotypes a given individual, cultivar, species, etc. The term idiogram is ordinarily reserved for the diagrammatic representation of a karyotype based on measurements taken from several or many cells.

Unfortunately, the chromosomes of crop plants may not always necessarily be the most favorable for detailed study. In some cases the chromosome number is relatively high and the size very small. In others the chromosomes may have few, if any, clear-cut characteristics or distinguishing features. As a result, published karyograms or karyotypes of certain crop plants may differ in a number of their details. In many crop plants the karyotypes are mutually compatible, but sometimes they can cause real confusion. Such difficulties could be minimized if a standard system of nomenclature were agreed upon for each of the many crops grown, but specialists rarely convene to discuss and ultimately agree upon terminology. Guidelines and systems for uniform terminology, and even linkage groups, have been set forward for investigators working with some of the major crops like maize (Neuffer et al., 1968; Neuffer and Coe, 1974; Sprague, 1977), rice (Khush, 1974; Chang and Li, 1980), wheat (Sears, 1972, 1974), and barley (Nilan, 1964, 1971, 1974). Most crops, however, have no such accepted conventions.

Counting of chromosomes in metaphase in squash preparations is the most reliable method of determining the chromosome number and degree of ploidy of a plant. In preparations in which the spread is good and the chromosomes are favorably condensed, it is merely a matter of counting. It is important, however, to make sure that counts are only taken from cells that have entire or intact membranes. Counting is subject to great error if membranes are destroyed or broken, and spreads are likely to intermix. Photographs should be taken at appropriate magnification using a fine-grain film. (We use Kodak Technical Pan Film TP 135.) After prints are made, the chromosomes from one print may be cut out using fine, curved, cuticle scissors. (In the event it is not possible to get a uniform spread, it may be necessary as a last resort to photograph chromosomes at different levels of focus. In such cases prints must be patched or pieced together. This is by no means a recommendation, however, for it indicates the methodology needs to be improved.) Care should be taken in cutting to follow the outline of the chromosome shape. A few millimeters of margin should be retained. Thus, using the index finger or a pair of tweezers, grouping of the chromosomes should be attempted on a clean, flat, smooth surface.

The diagnostic features of chromosomes include form, size, points of spindle fiber attachment (centromere location or primary constriction), secondary constrictions (nucleolar oranizer regions, called NORs), and presence or absence of satellites. Aside from size, the single most obvious feature characterizing an individual chromosome is the location of the centromere. Nomenclatural relationships and the means of determining centromere position have been discussed in detail by Levan et al. (1965). While in some cases one can readily associate homologous pairs of chromosomes, measurements are often needed to achieve accurate pairing. Quantifying specific chromosome characteristics is, of course, highly desirable when it can be achieved. Investigators are urged to try to quantify wherever possible.

Measurements are, perhaps, most easily made with a compass on photographic enlargements. Compass spans may then be translated to millimeters against a good-quality ruler. (A metal ruler with the finest gradations or subdivisions available seems to be best.) Needless to say, all enlargments must be fastidiously made and precautions taken to ensure uniform magnification and measurement. In view of the apparent morphological variation of satellites, they and their connecting strands are ordinarily not included in computing arm ratios or centrometric indices.

While ratios are often the endpoint in reporting certain results, and hence the actual numbers used for computation should not affect the final values, measurements are best made on preparations that have been as uniformly treated as possible. Theoretically, the degree of contraction should, for instance, not affect the ratio of the long arm to the short arm, but it will be appreciated that the potential for error in measuring increases if the chromosomes are overly or underly contracted. Two very important diagnostic features of each chromosome are the length of each chromosome relative to the total haploid set and the position of the centromere. Centromere position may be conveniently expressed in either the arm ratio of the chromosome, expressed as the length of the longer arm relative to the shorter one, or the centromere index, expressed as a percentage of the length of the shorter arm of the whole length of the chromosome. Relative chromosomal length is very useful in the preparation of the idiogram, for it permits a chromosome to be related in terms of its size to the overall complement. More precisely, relative length is the length of each chromosome relative to the total length of a normal haploid set. This is calculated by the formula:

(chromosome length/total of the chromosome lengths) x 100
= relative chromosome length

The arm ratio (r) is expressed as the length of the longer arm (l) relative to the shorter one (s) or r = l/s. (Arm index is another parameter and is the ratio of the short arm to the long arm, i.e., short arm/long arm = arm index.) Sometimes centromere position is calculated as a difference (d) between the length of long and short arms (l and s) respectively, i.e., d = l - s. With the use of the terminology recommended by Levan et al. (1965), an arm ratio of 1.00 indicates a precise median centromere position (M) in the strict sense; an arm ratio ranging from 1.00 to 1.70 delimits a metacentric chromosome (m); 1.7 to 3.0 delimits a chromosome with a submedian centromere (sm);

3.0 to 7.0 delimits a chromosome with a subterminal centromere (st) or an acrocentric chromosome.

Centromere index (i) is the percentage of the shorter arm of the whole length of the chromosome. This is calculated by the formula:

(length of the short arm/total length of the chromosome) x 100 = centromeric index or i = 100 s/c.

Often by eye alone or certainly by use of the measurements and computed values, it should be possible to arrange the chromosomes into coherent groups. All those chromosomes which fall into a given category as to centromere position should be grouped together. These in turn should be paired in pairs or homologues according to their length, i.e., long homologue, short homologue. The paired homologues should then be arrayed in descending order of length. While there is no convention that dictates it, it is infinitely more pleasing to the eye, in our view, if the shorter arms, secondary constrictions, or satellites where present, point upward on a page in those cases where centromeres are not median. If rules exist for chromosome groupings in a given plant, then they should, of course, be followed. If they do not exist, the preferred order when viewed from top to bottom is median, submedian, subterminal, and terminal. Allosomes or sex chromosomes are not very common in plants (Allen, 1940; Westergaard, 1958; Dronamraju, 1965), but where they do exist they should be retained for the last in any karyogram. After the groupings have been firmly established, numbers may be assigned to the chromosome pairs. This can be done in descending order of length. Chromosomes can be referred to by group, e.g., those with median centromeres, rather than assigned to a given pair; this permits flexibility until the identities of individual chromosomes can be further resolved.

If chromosomes are lined up by centromere position, identification becomes much easier for the observer. Final illustrations can be prepared by permanently affixing chromosomes to a mounting board. Although rubber cement can be conveniently used, affixing dry mounting tissue with a tacking iron to the back of a photographic print prior to cutting out the chromosomes makes the job considerably easier. After the prints are arranged in appropriate order, they can be permanently attached with heat in a dry mount press. Figures 1 and 2 provide a karyogram and an idiogram respectively of a diploid cultivar of sunflower (Helianthus annuus) (n = 17) with which we have had considerable experience. Both were prepared as outlined above by taking into consideration the length relations, arm ratios, etc. from 10 metaphases from different preparations. This example is useful not only because it relates to an economically important plant but also because is emphasizes several points we have mentioned.

The metaphase chromosomes of the dwarf cultivar in question are separable into three distinct groups (see Table 1; Figs. 1 and 2). The centromeres of chromosome pairs 1 through 7, (group 1) are median (m). The arm ratios of each of these chromosomes varies between 1.15 and 1.39. The longest chromosome is 3.47 μm and the shortest one is 2.58 μm. The centromeres of chromosome pairs 8 through 12 (group 2) are

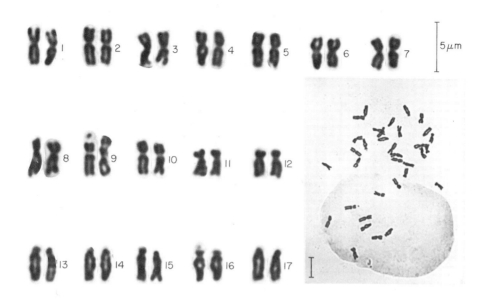

Figure 1. Karyogram of *Helianthus annuus* cv. Teddy Bear. The inset shows a representative cell stopped in metaphase. The rest of the figure shows the chromosomes from that cell cut out and rearranged as an associated diploid karyogram. Arrangement of homologous pairs is according to their centromere position and according to decreasing size. Scale bar of both, 5 μm.

submedian (sm). The arm ratio of each of these chromosomes varies between 1.87 and 2.31. The longest chromosome is 3.28 μm, and the shortest is 2.35 μm. Although pairs 11 and 12 have the same arm ratio (2.31), the difference between their lengths (relative length of 11 is 5.08 and that of 12 is 4.68) helps one easily identify each pair. Also pair 9 has microsatellites. Pairs 13, 14, 15, 16, 17 (group 3) have subterminal centromeres (st). The arm ratio of each of these chromosomes varies between 3.77 and 8.09. The longest chromosome is 3.25 μm and the shortest is 2.69 μm. Pair 16 also has microsatellites.

Despite the relative uniformity of overall length within group 3 and its great similarity to group 2, the overriding characteristic that permits distinct grouping is the very different arm ratios. But here a decision must be made. Chromosome pair 16, in particular, has an arm ratio of 3.77 but 13, 14, 15 and 17 have arm ratios ranging from 6.47 to 8.09. Following Levan et al. (1965), an arm ratio of 3.0 is the generally accepted "cut-off" for a submedian centromeric position, and hence the pair belongs in group 3. This naturally emphasizes the need to have good measurements but also shows that a group does not necessarily have a narrow range of homogeneity in terms of centromere position. Groups 1 and 2 do have a rather restricted range in terms of arm ratio, but this is not so in group 3. Figure 3 provides an example of yet another means of helping an investigator to match and pair homologous chromosomes. In that preparation of sunflower, the heterochromatic regions are particularly evident.

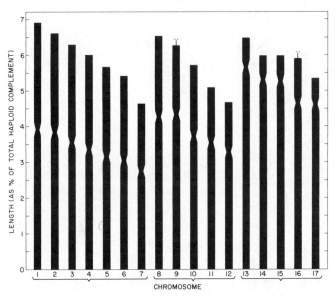

Figure 2.   Idiogram showing length, position of centromere, and satellites of chromosomes of diploid cells in metaphase from *Helianthus
annuus* cv. Teddy Bear.   The ordinate represents the length of the
chromosomes as a percentage of the haploid genome length (50.19 μm).
The diagrammatic representation of the genome is arranged according,
first to centromere position, and second, to length.   Chromosome pairs
1-7 are metacentric; 8-12 are submetacentric; and 13-17 subterminal.
Microsatellites on pairs 9 and 16 are shown as dots to emphasize their
very small size (cf., Fig. 1).

A BIT ABOUT TERMINOLOGY.   Frequently, deviation from the diploid number is encountered in plants.   The most commonly observed
change involves the number of chromosomes.   A simple glossary of
terms adapted from Schulz-Schaeffer (1980) is provided.   Attention is
drawn to two broad categories of change.   The first involves change in
chromosome number in only one pair of the diploid; the second involves
change in chromosome number in all sets or pairs.

1. Change in chromosome number in only one pair of the diploid.
    Monosomic = 2n - 1 (also called haplo forms).
    Trisomic = 2n + 1 extra chromosome:
     Primary trisomic = 2n + 1 (extra is complete homologue of one
     pair).
     Secondary trisomic = 2n + 1 (extra has two similar arms).
    Tetrasomic = 2n + 2 (both extras are homologous with one origin-
    al pair).
2. Change in chromosome number in all sets or pairs.
    Polyploid (heteroploid)—a form with chromosome number other than
    the true haploid (monoploid) or diploid number.
    Euploid—an exact multiple of the haploid (triploid, tetraploid,
    penta-, hexa-, hepta-, octoploid, etc.).

Autopolyploid (autoheteroploid)—a multiple chromosome complement of a single kind of the haploid set.   An autotetraploid, for example, has four similar chromosomes in each set.

Allopolyploid (alloheteroploid)—a multiple chromosome complement of dissimilar sets of chromosomes.

Aneuploid—a chromosome number other than exact multiple of the haploid.

Hypoploid—a little lower than some multiple.

Hyperploid—a little higher than some multiple.

Table 1.   Mean Relative Length and Arm Ratios of Chromosomes from Ten Root Tip Metaphases of *Helianthus annuus* cv. Teddy Bear.

CHROMOSOME PAIR NO.	RELATIVE LENGTH[1]	ARM RATIO[2]	TYPE[3]
1	6.9	1.31	m
2	6.6	1.39	m
3	6.3	1.29	m
4	6.0	1.26	m
5	5.7	1.27	m
6	5.4	1.29	m
7	5.1	1.15	m
8	6.5	1.90	sm
9	6.3	2.27	sm
10	5.7	1.87	sm
11	5.1	2.31	sm
12	4.7	2.31	sm
13	6.5	6.93	st
14	6.0	8.09	st
15	6.0	7.57	st
16	5.9	3.77	st
17	5.4	6.47	st

[1]Percentage of the length of the total haploid genome.
[2]Ratio of the long arm to the short arm.
[3]m = median
 sm = submedian
 st = subterminal centromeres

In the context of the times when they were invented and first adopted, the terms given above referred only to individuals or populations.   With the advent of cell culture technology and the discovery that cultured animal cells (especially human and other mammalian lines) displayed irregularities and frequently underwent changes that were almost unknown in classical animal genetics, it became clear that pains should be taken to adopt a standardized terminology so as to avoid confusion.   The initiative was taken by Levan and Muentzing (1963), who recommended uniform adoption of the following designations especially in terms of chromosome number:

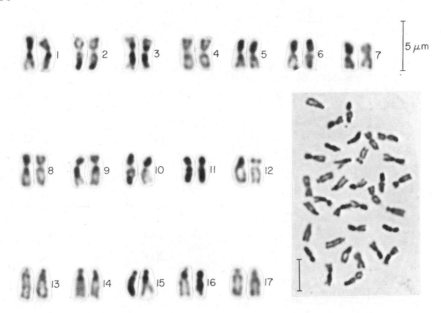

Figure 3. Karyogram of *Helianthus annuus* cv. Teddy Bear. The inset shows a representative cell stopped in metaphase. The rest of the figure shows the chromosomes from that cell cut out and rearranged as an associated diploid karyogram. Arrangement of homologous pairs is according to their centromere position and according to decreasing size. When compared with the chromosomes of Fig. 1, these show discrete distribution of heterochromatin and confirms the pairing as made and proves the reliability of the homology. Scale bar of both, 5 μm.

Haploid
1. The basic number of a polyploid series (symbol: x). In this use of the word, haploid means monoploid.
2. The chromosome number of the so-called haplophases (symbol: n). In this use of the word, haploid means the reduced, gametic chromosome number.
Diploid, triploid, tetraploid, etc.
The double, triple, quadruple, etc. basic number (symbols: 2x, 3x, 4x, etc.)
Polyploid
A general designation for multiples of the basic number, higher than diploid.
Heteroploid
1. In organisms with the predominating diplophase: all chromosome numbers deviating from the normal chromosome number of the diplophase (i.e., the unreduced zygotic number).
2. In organisms with predominantly haplophase: all chromosome numbers deviating from the normal chromosome number of the haplophase (i.e., the reduced, gametic number).
Euploid
All exact multiples of x.

Aneuploid

All numbers deviating from x and exact multiples of x.

Mixoploidy

The presence of more than one chromosome number in a tissue or cell population. This term covers all cases of mosaicism and chimeric constitution.

In all the above the use of x is potentially the most confusing. Usually, 2n is identical with the somatic chromosome number, except in instances of somatic polyploidy, somatic elimination of chromosomes, chromosome number variation due to abnormal mitoses, etc. A couple of examples, taken from Levan and Muentzing (1963) will help illustrate the point that n designates chromosome numbers in specific types of cells and that x means an exact numerical entity. In the case of a particular triploid species of *Allium* with the basic number $x = 8$, the somatic cells are designated $2n = 3x = 24$. This plant forms gametes with chromosome numbers ranging from $n = x = 8$ to $n = 2x = 16$. The most frequent number is $n = x + 4 = 12$. In the case of rye, very rarely a diploid plant, with $2n = 2x = 14$ gives rise to a haploid plant with $2n = x = 7$. Hexaploid timothy with $2n = 6x = 42$ forms polyploids with $n = 3x = 21$.

Attention will be directed in the next section and elsewhere in this chapter to a slightly extended discussion of terminology using specific examples, but for the time being the above will orient the reader.

LITERATURE REVIEW

Many years ago it was thought that mitotic irregularities were frequent in cells of malignant plant tissues and that tumor cells often contained abnormal numbers of chromosomes. At one time there was even a school of thought that attempted to correlate chromosomal abnormalities with the tumor state in a causal way (Levine, 1931). As work on the culture of aseptically cultured tissues expanded, it became clear that abnormal behavior of the nucleus, chromosome inconstancy, and heteroploidy characterized many normal cell lines as well. This feature has become so well publicized that it has virtually become a rule that cultured plant tissues and cells that have gone through the callus state are very likely to deviate in some way or other from the norm in their chromosomal complement. In primary cultures the deviation seems to be minimal, but as continuous cultivation through subculture is achieved, substantial differences from the normal somatic condition may be observable.

One of the earliest attempts to monitor changes in karyology and chromosome behavior in cultured tissues was made by Straus (1954). Using 12 days post-pollination maize endosperm, he accumulated evidence for many kinds of irregular chromosomal duplication and distribution, e.g., polyploidy, aneuploidy, hypoploidy, polyteny, chromosomal bridges, and lagging chromosomes. Straus (1954) counted as many as 105 chromosomes in cultures of ordinarily triploid endosperm ($3n = 30$). He fully appreciated the potential affect of the medium on his cultures and suggested that nucleic acid-rich yeast extract (0.5% w/v filter sterilized) could have been responsible, since nucleic acid salts or "nucleates" were known to cause similar problems in onion root tip "assays." Straus questioned whether the yeast extract could be fully responsible, however, since he knew an original abnormal condition of the tissue (mosaic or variegated endosperms do occur in maize) could have accounted for the seemingly atypical behavior. Straus also wondered if the failure to regenerate complete plantlets from maize endosperm could be due to chromosomal inconstancy and erratic chromosome behavior in the cultures. In the course of preparing and growing a large number of cultures, LaRue (1947) encountered fewer than 1 in a 1000 cultures that formed roots and only one that "formed a shoot root axis and miniature leaves." Straus (1954, p. 837) speculated, "If it were possible to get endosperm to grow in culture some time before 12 days post-pollination, it might be possible to escape the effects of the postulated influences and then the tissue might be able to demonstrate its inherent totipotency." Straus also urged that "similar studies of other plant tissue cultures should be undertaken in an effort to determine whether this property [chromosomal irregularity] is generally acquired by all plant tissue cultures" (Straus, 1954, p. 838).

A few years later, Torrey (1959) reported that although predominantly composed of diploid cells during the first few days after culture initiation, pea root callus cultures gradually changed so that tetraploid (or near tetraploid) cells came to predominate in the population. Two very early workers in this area, Mitra and Steward (1961), appreciated the need for a system that could be readily cultured, had low chromosome number, or had chromosomes with distinctive morphology. *Haplopappus gracilis* (Compositae) fit the needs admirably (Jackson, 1957). The normal diploid complement of this weed consists of two chromosomes with a submedian centromere and two shorter chromosomes with subterminal centromeres, bearing satellites on the short arm. Mitra and Steward (1961) noted the presence of polyploidy, aneuploidy, and anaphase abnormalities as well as morphologically altered chromosomes in cells of predominantly normal diploid *Haplopappus* cultures. A number of distinctive clones with various chromosome numbers, even up to 64, were isolated and maintained serially. A bit later and in the same laboratory, Blakely (1963) described a strain of *Haplopappus* that remained predominantly diploid for approximately 3 years. Two other strains, composed primarily of diploid cells at the outset, underwent an alteration in karyotype, leading to chromosome complements moderately to drastically different from the normal. Cell lines or strains with obvious differences in phenotype (friability, coloration, etc.) had abberant karyotypes from the outset. Cells of the strains with more than the normal chromosome number generally did not contain euploid cells. In those strains with variant karyotype, chromosomes with normal morphology were identifiable in most cells, but obviously abnormal chromosomes in terms of number and morphology were present as well. Strains with aberrant numbers also showed a high level of anaphase disturbances. Blakely (1963) found that some *Haplopappus* strains could remain predominantly diploid without diminution in vigor. Thus although he encountered many chromosomal aberrations in his cultures, there seemed no a priori reason why cells with normal chromosomal features could not be maintained in long-term cell cultures. Unfortunately, the cultured *Haplopappus* cells did not regenerate, and hence no correlations could be made between chromosomal constitution and regenerative capacity. Even so, it is clear that *Haplopappus* represents an extreme example of the sort of culture that may tolerate cytological aberrations. Also, since haploidy was seen to occur in free cell cultures of *Haplopappus*, the question has been raised whether an opportunity may have been missed to study meiosis in this system (Steward and Krikorian, 1979). In our view, the most significant single observation made on *Haplopappus* cultures during that early period was that chromosomal arrangements strongly resembling chiasmata could be seen in somatic cells in mitosis (Mitra and Steward, 1961). If this represented mitotic crossing over or segregation, and had it been predictable or controllable, it could have opened the way for cytogenetic analysis in cultured plant cells long ago.

Over 20 years ago there was evidence that cells of carrot could be maintained in suspension in such a way that they would preferentially give rise to normal plantlets. Based on the fidelity of phenotype expressed by the regenerated carrots they produced, Steward and his coworkers suspected very early that clones in the strict sense of the term, were being regenerated. Although there was some diversity in the nuclear constitution of some of the population of carrot cells multiplied in liquid, no cytological aberrations could be detected in the suspension-derived plantlets. In their hands, apparently only those cells with normal genotype led to somatic embryos. After an initial study of root tips from 200 plants showed that the regenerated plants seemed to be karyologically normal diploids, the original plan to examine a total of 1000 tips was abandoned (cf., also Mitra et al., 1960). Although the question of slightly altered genotype without detectable consequences in phenotype remained unanswered in the case of the "Cornell carrots," it is unlikely that there were serious problems, since all phenotypic details were recapitulated accurately—even to the presence of anthocyanin-containing florets in an otherwise white inflorescence (cf. Steward et al., 1964). Some years later Mok et al. (1976) reported recovering, albeit on a much smaller scale, normal diploid plants from carrot root callus, although cells with varying chromosome numbers also existed in their cultures.

Even on the strength of the early work alone, therefore, some interesting questions can be raised. These include: What is the basis of different ploidy levels in cell cultures? How do chromosomal aberrations arise in cultured tissues and cells? Are variant karyotypes stable upon selection and subculture? Do cells with identifiable altered chromosomes persist "indefinitely" in those strains? Are structurally altered chromosomes more common in polyploid cells than in diploid cells? Can particular phenotypic traits of tissue and cell cultures be associated with particular features of karyotype? Can plantlets be reared from cells that are not predominantly euploid or that have altered karyotypes? Is there any evidence that chromosomal rearrangements have serious effects on phenotype of cultured cells or plants to which they may give rise? Are certain genotypes more amenable to culture, and/or manipulations and do they express morphogenetic competence more readily than others? We can examine these questions only in a superficial way, because in many cases the answers are only now just beginning to emerge.

There have been a number of reviews summarizing the work of a given period. There is little point in recapitulating these, since they are all readily available and all serious investigators will wish to consult them in detail (Partanen, 1957, 1963a, 1965, 1973; Muir, 1965; Torrey, 1959, 1965; Sheridan, 1975; Torrey and Landgren, 1977; Sunderland, 1973, 1977; D'Amato, 1975, 1977a,b, 1978; D'Amato et al., 1980; Meredith and Carlson, 1979; Bayliss, 1980; Evans and Reed, 1981; Constantin, 1981).

Problems of Sampling

Before addressing these questions, a few words are in order about sampling.

CALLUS CULTURES VERSUS LIQUID SUSPENSION CULTURES. In cases where cultures form coherent masses but are friable and easily fragmented, sampling is relatively easy. But one must constantly ask whether the means of cell selection for counting or examination introduces a bias. Are the cells in the sample actually observable in mitosis representative of the entire culture? Even in cases where sampling is carefully carried out and multiple replicates are taken, there is the very real problem of variation in results. These variations may reflect so-called multimodal development of different lines or even clones of cells in different parts of the culture. Irregular polyploidy and aneuploidy may occur in localized areas, and this, of course, reflects that replication or multiplication of chromosomes may have proceeded in different cells at different rates. There seems little doubt that mixoploidy can and does occur in callus masses. Since there is little evidence of what the selective forces are that operate and control differential proliferation of distinct cell clones, it is imperative to accept that certain cell types may be selectively favored. Thus it is necessary to sample different portions of callus cultures that have been established under as carefully controlled conditions as possible. Only in this way can one establish whether cultures have undergone changes during maintenance through subculture. A few chromosome counts randomly taken are insufficient to establish the status of a given culture. It is essential to screen for possible mosaics. If a strain or line of callus has a firm, compact growth habit, meristematic activity is frequently confined to the sheath of cells which surrounds or envelopes the culture mass. Peripheral growth in this sort of callus was described many years ago by Caplin (1947). In lines that are more friable, meristematic activity may be confined to more localized nodules scattered throughout the mass.

When differences in karyotype exist between replicate cultures grown on agar or between different portions of the same culture, especially if they can be localized to specific sectors, there is little doubt that the cause of diversity must be examined in greater depth. When a culture containing portions of "normal" and "abnormal" cells is used as the source of the inoculum for making subcultures, the replicate subcultures derived from it (if they grow) are sure to reflect its diversity in karyotype. At the time of the next subculturing, a culture containing entirely abnormal cells might be selected by chance as the source of subculture inocula. In this way the normal or characteristic karyotype might be entirely lost, although normal selection pressure might ordinarily not favor, or could even discriminate against, the abnormal karyotype. The problem is perhaps greatest in cultures maintained on agar media and was, in fact, one of the reasons why Steward and his early co-workers decided that they would attempt to work almost exclusively in liquid, although the tide of opinion at the time was against their approach (White, 1953).

The problem of sampling is of lesser importance and, in general, not as critical with cell suspension cultures. If random samples of cells and cell clusters are taken up in a pipette during subculture, as long as the inoculum is representative of the culture as a whole, shifts in the characteristics of a culture should not be favored by the subculturing process per se. Even so, this does not preclude evolution by mutation and natural cell selection. In this laboratory, routine use has been made since 1966 of filtration techniques through screens of known dimension. Passage of suspensions through these screens permit very uniform inocula to be utilized for subculture or further manipulation (see Fig. 4). While personal prejudices and other considerations have caused us to emphasize liquid suspension methods (Krikorian, 1982), we recognize that in some cases it is not feasible or is impractical to do so. Workers using callus, therefore, are urged to exercise particular caution in reporting their results (see Demoise and Partanen, 1969). Against the above comments, attention may now be directed to some of the problems posed.

Polysomaty and Polyploidy

The condition of having cell populations whose component cells differ in their chromosome numbers in the same individual is not as unusual as one might suppose. In addition to diploid cells with the 2n number of chromosomes, some plants contain many polyploid cells with 4n, 8n, and occasionally even 16n chromosomes. This phenomenon is called polysomaty. One process leading to this condition is called endopolyploidization and is due to successive chromosome multiplication within the nucleus, with one mitosis intervening at each new level of polyploidy. Unfortunately, there is not nearly enough information as to the full range of tissues in which polysomaty occurs. One fact is clear, however; it is fairly common and it is particularly common in those plants that are propagated mainly through vegetative means. The role, if any, of euploid, aneuploid, and polyploid cells within an otherwise diploid individual from a developmental viewpoint has yet to be rigorously determined, but it is clear that somatic alterations result in

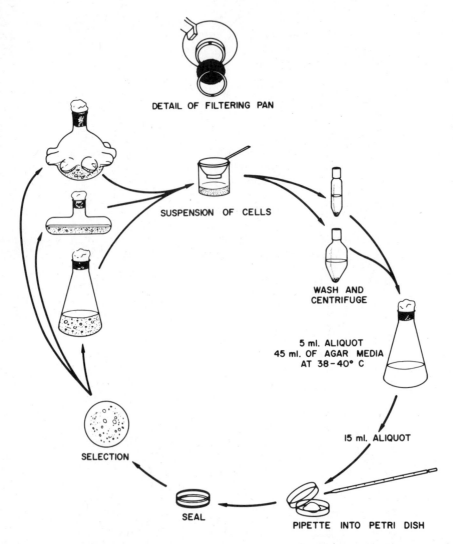

DETAIL OF FILTERING PAN

SUSPENSION OF CELLS

WASH AND
CENTRIFUGE

5 ml. ALIQUOT
45 ml. OF AGAR MEDIA
AT 38-40° C

15 ml. ALIQUOT

SELECTION

SEAL

PIPETTE INTO PETRI DISH

Figure 4.   Scheme of procedures that can be followed in preparing cells and cell clusters of known dimension from suspensions for distribution in agar.   Their subsequent growth and selection for subculture or analysis is also shown.   Erlenmeyer flasks, so-called nipple culture flasks (which can accommodate 225 ml of medium) or T-tubes (which can hold 10 ml of medium) can be conveniently used to grow cells. The point of "selection" after growth of distributed cells is, understandably, the most arbitrary and demanding step.

genotypically different individuals. This phenomenon is particularly well documented in certain vegetatively propagated edible aroids like *Colocasia* (Sharma and Das, 1954; Sharma and Sarkar, 1963).   Long ago, D'Amato (1952), Geitler (1953), and Chiarugi (1954) reviewed the occurrence and modes of origin of polysomaty.   D'Amato in particular concentrated on the relationship of endopolyploidy to differentiation in otherwise 2n plants (D'Amato, 1964, 1977b).

D'Amato wrote as far back as 1964 that "It is hoped that the successful development of techniques for isolation and culture of single plant cells (Muir, Hildebrandt and Riker, 1954) will represent a new stage of research in the cytology and genetics of the plant cell." Unfortunately the cloning of higher plant cells has not been easily extended from its early successful use with tobacco to other plants. In those cases where it has been possible to use tissue cloned from single cells, higher ploidy levels and aneuploids have been encountered (Cooper et al., 1964). This indicates that preexisting polyploid nuclei in the initial explant are not the only cause of chromosomal variability (Chouinard, 1955; Partanen, 1963b). Endomitosis may occur and persistent chromosome bridges may also result in polyploid cells (Cf. Nagl, 1978, for a detailed treatment of endopolyploidy, etc.).

Understandably, most investigators are distressed by the production of polyploids when they are unwanted (cf., the production of asparagus tetraploids in Malnassy and Ellison, 1970, and the production of carrot tetraploids in Bayliss, 1973, 1975). On the other hand, it has occasionally provided a substantial opportunity for potential improvement of a given crop. Potato tissue culture may be used to produce tetraploids (Hermsen et al., 1981). Tissue culture has made a real contribution to research in the sugarcane industry. Doubling of chromosome number was not possible through conventional methods but callus techniques have permitted doubling. This has also opened the way to the development of an aneuploid series to check on yield potential at chromosome numbers not previously available (cf. Nickell, 1977, and references there cited).

Aneuploidy

In some cell lines the majority of cells in each population is represented by a zone of connecting chromosome numbers or gradations in numbers. If we invoke the term used in reference to animal tumor cell populations, cells with chromosome numbers above or below the stemline may be common in cultures, even though clonal cultures are derived from single cells. Although some investigators initially assumed that cloning is the most effective means of ensuring cell populations with a relatively homogeneous makeup, this amounts to near fantasy and at best is a gross oversimplification. Methods for establishing single cell colonies are, as stated above, still problematical, and the culture of single cells of higher plants is still one of the greatest challenges facing investigators.

Deviation from the euploid or diploid chromosome number might arise in a number of ways. Perhaps the simplest is by elimination of chromosomes through failure of chromosomes to assemble at metaphase and disjoin properly at anaphase. Ordinarily when properly aligned, the chromosomes are pushed into a single plane. The centromere of each chromosome is attached to each of the opposite poles of the cells. Also, lagging can occur at anaphase. Nondisjunction is another potential cause of difficulty.   If this is sufficiently frequent, elimination alone could give origin to the weird numbers often encountered, but micronuclei would more than likely form.   Loss of an entire chromosome in plants is frequently sufficient to prevent a cell reappearing in mitosis; it is very deleterious (Khush, 1973).

There is substantial literature on the occurrence of aneuploid cells in plant cell and callus cultures (cf. Sunderland, 1973; D'Amato, 1977a; Skirvin, 1978; Novak, 1981, and references there cited).   There are fewer papers that reported aneuploid plants that have been regenerated by means of culture techniques.   Aneuploidy in tobacco has been particularly well documented in plants derived from both normal and tumorous tissue (Murashige and Nakano, 1967; Sacristan and Melchers, 1969). Other cases of aneuploidy in culture-derived plants have been described as well (cf. Hiraoka and Tabata, 1974, for *Datura innoxia*).   Mosaicism is encountered more often (e.g., Pelletier and Pelletier, 1971; Bennici, 1974; Lester and Berbee, 1977; Bennici and D'Amato, 1978).

In this laboratory, several primary trisomic daylily plants have been isolated from suspension cultures.   Primary trisomics provide an excellent cytogenetic tool for testing the independence of linkage groups and for assigning linkage groups to particular chromosomes.   Since there are three homologous chromosomes in a primary trisomic instead of two for a particular member of the complement, the genetic ratios in segregating progenies for genes that are located in this chromosome are very different from the 3:1 or 1:1 ratios obtained in the F_2 and backcross generations of a normal disomic heterozygous for a recessive gene.   Primary trisomics can also be of use for assigning unlocated genes to linkage groups.   In addition to their usefulness in a genetic sense, the various primary, secondary, and tertiary trisomics that we have obtained can be employed in studying the effect of duplication of a whole chromosome, a chromosome arm or parts of chromosome arms on the morphology, anatomy, and physiology of the organism.   Access to trisomics can throw light on the basic nature of the genome of a species.   There is every reason to believe that the study of the development of chromosomally abnormal plants from cell clones would be productive.

Translocations

Clones of cells with rearranged chromosomes arise relatively frequently in cell cultures, but they are usually reciprocal, since in the cases most studied, there is no evidence of change in chromosome number.   In cultures where there are extreme mitotic irregularities during growth and cell multiplication, it appears that most of these disturbances render the cell less able to compete with the other cells in the population. They are not maintained during subculture.

An extreme example of culture stability following chromosome rearrangement involves the maintenance by Sheridan (1974, 1975) of morphogenetically competent cell lines of *Lilium* in liquid.   These cultures retained their normal ploidy and apparently formed normal plants longer than any continuous culture.   Sheridan (1975) did not detect unexpected phenotypic differences in either his cultures or the regenerated plants. In this laboratory translocations have been detected in suspension cultures of daylily. While we are certain of the constancy of this condition in some lines, we are not yet in a position to state categorically that there is no phenotypic consequence in regenerated plants. There is clearly no deleterious effect on the morphogenetic competence.

It was recognized long ago that translocations could occur in cultured cell populations (Cf. Sacristan, 1971, on *Crepis capillaris*), but those cultures most amenable to cytogenetic analysis were not then amenable to regeneration. Bayliss (1975) encountered frequent chromosomal rearrangements in carrot cells, and Novak (1974, 1981) has described them in long-term callus cultures of *Allium sativum* L. There is no doubt that many translocations occur but have gone undetected, and culture conditions can play a substantive role in selection (Kao et al., 1970; Singh et al., 1975).

When one considers the various points in the mitotic cycle at which aberrations might arise, it is all the more remarkable that normalcy is so often maintained. It is a moot point whether the aberrations encountered in cells in the cultured state reflect adequately and accurately the variation that occurs in nature in situ. The data of Verma and Mittal (1978) on garlic suggests that more karyotype changes may be going on than has been hitherto appreciated. In spite of this, however, one needs to reconcile the observation that in many cases chromosome stability is maintained throughout culture and regeneration (Evans, 1979; Krikorian and Kann, 1980; Krikorian et al., 1981), whereas in many others karyotypic heterogeneity is frequently reported.

A modal chromosome number can be identified in many cultures, and the implications of the presence of absence of numerical variability about that mode is something that cannot be discussed with precision. If marker chromosomes were present, it would make the task easier in those cases where the chromosomes are small and numerous. Much of the published literature suggests, however, that organization occurs preferentially from cells that do not show severe cytologic abnormalities. There is some evidence that points to the site of morphogenetically competent cells from which organogenesis ultimately derives as controlling the ploidy of the offspring (cf. Murray et al., 1977; Tran Thanh Van and Trinh, 1979, and references there cited). In short, the problems seem to become broader as more information is gathered.

Advances in Methodology

The fact is that the classical methods to be covered in this chapter will more than likely be adequate for the needs of most investigators at this time. Microspectrophotometric measurements of DNA per nucleus in dividing meristems can, on occasion, be helpful in establishing the proportions of diploid and polyploid cells (Conger and Carabia, 1978, and references there cited) but it must be recognized that small alterations in chromosomal number and structure are not detectable by these means. It is at best a crude measure.

Specific attention will be given to cytostatics in the next section, but one should be constantly on the lookout for new pretreatment chemicals, no matter how unlikely they might seem at first glance (cf. Lavania, 1978b, for chlorflurenol) and other means for help in establishing the cytotype of proverbially difficult plant tissues (cf., Lin, 1977 for maize endosperm).   Fluorescent stains are becoming increasingly useful in the study of small chromosomes or under conditions where conventional technique is fraught with many problems (Galbraith et al., 1981).

Nucleolar organizer regions (NORs) can be visualized quite well in many cases using silver staining techniques (Schubert et al., 1979; Hizume et al., 1980), and in some cases one can simultaneously stain the NORs and the centromeres (Brat et al., 1979).   These are all rather specialized techniques, however, and are only mentioned here.

SOME SELECTED PROTOCOLS

Before presenting some sample protocols, it is essential to reemphasize that different investigators will invariably have their preferred methods to achieve the same end.   We also restate that different materials, by virture of their nature or their particular stage or degree of development in culture, often require quite different procedures.   There is an enormous literature to draw upon as far as methodology applicable to intact plants or plant parts (see, e.g., the journal Stain Technology and the key references in this chapter).   There is much less to provide detailed guidelines geared to unorganized or callus cultures or cells in suspension (Partanen, 1973; Kao, 1975; Evans and Reed, 1981).

Selection of Material to be Sampled

When plants have been regenerated in vitro via any of the several culture methods, the task is generally straightforward.   The cells that are arrested in metaphase in root tip preparations are ideal.   Even so, readers may be surprised to learn that it is usually quite difficult to use agar-grown or even liquid-grown material directly for root tip squashes.   It is generally far more productive to remove rooted plantlets from aseptic culture and to transfer them to a vermiculite-perlite or vermiculite (i.e., nonaseptic) environment.   The fresh roots formed under these conditions are, in our experience, more amenable to study, since they will have many cells in division.   Because one is often concerned with determining whether a mosaicism exists in a plant produced via a given culture technique, it is also often useful to examine parts other than roots.   Young leaves are frequently good sources of mitotic figures (Baldwin, 1939; Meyer, 1943; Sharma and Sharma, 1957; Bezbaruah, 1968).   In plants like Nicotiana, young corolla tissue may be used (Burns, 1964; Collins, 1979).   In plants like palms where microspore mother cells or roots may be hard to obtain or to process, pollen culture may provide an alternative avenue for chromosome study (Read, 1964, 1965).   One should always be aware that root tip analysis alone does not necessarily constitute reliable evidence for the karyological constitution. Sectorial and periclinal chimeras are not uncommon (Dermen, 1960; Neilson-Jones, 1969; Arisumi, 1972), and meiotic analysis is particularly important in these cases.

Selection of the Appropriate Protocol

Investigators have available to them a wide range of chemicals and protocols. The task is to ascertain which is the best one to use and which regimen to follow. One has to ascertain, study, and optimize the following variable conditions: (1) the "best" cytostatic or prefixation chemical; (2) the best concentration; (3) the length of exposure; (4) the volume of cytostatic to which the test materials are exposed; and (5) the best environment in which to use the cytostatic, e.g., light vs. darkness; high, medium, or low temperature; the osmotic environment; etc.

One can list in skeleton form the steps important in working with metaphase chromosomes suitable for karyotype analysis.

1. Prefixation
2. Fixation
3. Hydrolysis and maceration
4. Staining
5. Squashing

Several cytostatics have been used in this laboratory with success, but there is no way of knowing whether one is better than another before actually testing it. Sharma and Bhattacharya (1960) carried out a comparative study using a number of pretreatment chemicals on several groups of plants and investigators are referred to that paper. Colchicine, 1-bromonaphthalene, vinblastine, 8-hydroxyquinoline, p-dichlorobenzene, coumarin, aesculin, etc. are only a few of the many available. In our work on intact roots we have found that 1-bromonaphthalene to be quite satisfactory, although preparation of the prefixative can be a bit cumbersome (see Krikorian et al., 1981, and protocols). We provide a sample protocol.

We have also used 1-bromonaphthalene for investigating callus cultures, but a longer period of exposure to the prefixative is generally necessary—usually 5.5-8 hr. Treatment in a cold (ca., 4 C) environment can be helpful as well. Hence caution must be exercised in collecting and interpreting data because there is a danger of polyploidy being induced by prolonged treatment. Another general feature of cells in callus culture is that they are, in general, much more weakly staining than cells from organized meristems. Also, because some callus cultures can be rather hard and nonfriable, it may be helpful to squash the cells a bit before hydrolysis. We have found it useful to "map" a callus by carefully cutting it into component parts, distributing them among different vials, labelling them, and processing them. In general, the outermost parts of the callus have the greatest number of division figures. In older subcultures, the innermost cells may even become necrotic and devoid of mitotic figures. In general, good mitotic figures are rather hard to find in callus cultures.

Cell suspension cultures provide better opportunities for chromosome study. At the same time one must expend a substantial amount of effort in establishing the most appropriate time of collection for processing. It is by no means a casual affair to determine that period of time when cells are undergoing maximal division.

We have used 1-bromonaphthalene, colchicine, vinblastine, and CIPC as cytostatics in suspension cultures. At the time of this writing, we have been most pleased with vinblastine. Even so, we suggest that interested investigators test several cytostatics before reaching any conclusions. For instance, a major drawback to the use of 1-bromonaphthalene in liquid suspension culture derives from the difficulty and inconvenience encountered in washing out this "oily" prefixation chemical. Otherwise it can be used with good success. We again stress that differences in sensitivity are invariably encountered depending on the culture system employed, the taxonomic group, age of material samples, etc.

Colchicine has been used for many years as a cytostatic agent. Dermen (1940), Eigsti and Dustin (1955), and Mehta and Swaminathan (1957) all contain valuable information on its use. As a cytostatic it is generally used at 0.01%-0.5% w/v in aqueous solution. While some batches seem to be quite pure, others may need to be partially purified (see Davidson, 1968, p. 190). It is always a good idea to prepare solutions of colchicine just prior to use. However, some investigators maintain that it can be stored in a dark container in a refrigerator for up to a year or so without losing activity. In suspension cultured cells of daylily, 0.5% w/v colchicine has produced reasonably good results (see Fig. 5b and e). Colcemid, a trade name for demecolcine or N-methylcolchicine, is usually more active than colchicine and a rule of thumb seems to be to use about one-third the concentration of colchicine. It is very expensive, however, and one cannot recommend its routine use. Both colchicine and democolcine are highly toxic chemicals and should, of course, be used with appropriate precautions.

Vinblastine sulfate has been widely used by investigators working with animal cells to arrest cells in metaphase. It is readily available, albeit with shipping restrictions since it is a highly hazardous chemical. Like democolcine, it is quite expensive. Levels that may be tested with plant cells are on the order of 10-40 µg/ml. Higher concentrations usually produce overcontracted chromosomes. In daylily suspension, 20 µg/ml has been optimal in our hands (see Fig. 5c, d and f). We recommend that an aqueous solution be made up fresh just before use since this is consistent with storage instructions (0-5 C). Some say, however, that it is stable at room temperature for a few days (cf. also Segawa and Kondo, 1980).

The herbicide O-isopropyl-N-phenylcarbamate (IPC) and its m-chlorophenyl derivative, N-3-chlorophenylcarbamate (CIPC) are mitotic inhibitors (Hepler and Jackson, 1969) and have been used to contract chromosomes. Storey and Mann (1967) first drew attention to the use of this compound to cytologists.

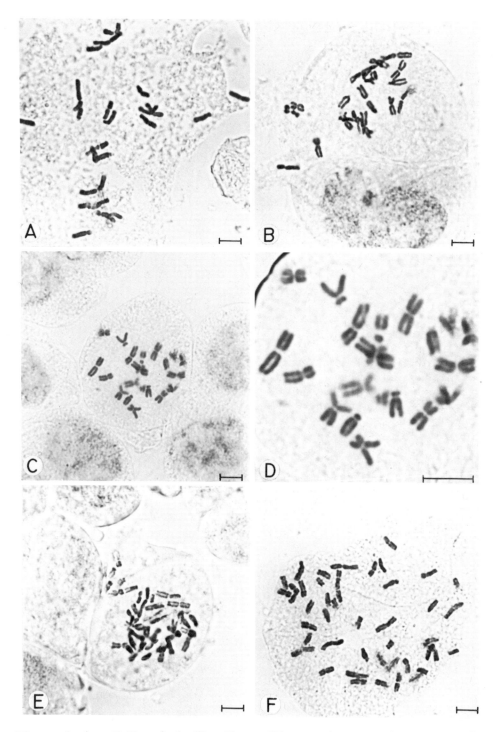

Figure 5a-f. Cells of daylily *Herocallis* cv. Autumn Blaze grown in suspension culture arrested in metaphase by different cytostatic agents.

Figure 5a-f (Cont.).   The length of time in a given cytostatic is given in brackets.   The number of easily counted chromosomes is given in parentheses.   (a) [3.5 hr in CIPC, N-3-chlorophenylcarbamate] (23 chromosomes) Description of chromosome shape, morphology, contraction, and spread: The morphology of the chromosomes is poor.   Little useful detail.   There is too much spread which could lead to confusion when counting the chromosome number of a cell.   (Note: This is a characteristic problem encountered when working with suspension culture cells—the cells rupture very easily!   The chromosome shapes are very poor making it extremely difficult to distinguish the chromosome arms, the centromere, and other distinquishing characteristics.   This picture could not be successfully used in making a karyotype.)   (b) [3.5 hr in colchicine] (22 chromosomes) Description: The chromosome morphology is better than in (a).   The spread is fair but not good.   The chromosome shapes are good.   It would be difficult, unless one was already very familiar with this genotype, to make a karyotype from this picture.   Some of the chromosomes have broken out and away from the cell.   This could lead to confusion or error.   (c) [3.5 hr in vinblastine] (21 chromosomes) Description: Notice that in contrast to the previous pictures the cell is intact.   Therefore, there is no doubt that all of this cell's chromosomes are present.   The chromosomes are well contracted in this preparation.   The morphology, chromosome shapes, and spread are all excellent.   This preparation is by far better than that at (a) and (b).   A precise karyotype can be made from an enlargement of this picture (cf., d).   (d) An enlargement of a preparation such as that at (c).   This enlargement could easily be used to prepare a karyotype. (Note: Some of these chromosomes lie out of the plane of focus due to the three dimensionality of the cell itself.   However, this problem can easily be alleviated by taking photographs in different planes.)   With an increased number of chromosomes, a good spread becomes even more important as well as more difficult to obtain.   Fig. 5e-f illustrates this point.   (e) [3.5 hr in colchicine] (44 chromosomes?) Description: These chromosomes show a great deal of overlap.   There is a poor spread which makes the individual chromosomes indistinguishable from each other.   Even those which are not overlapping have only a fair morphology and shape.   The poor spread makes one unable to make an accurate count.   This is especially true of cells cultured in suspension since they are often very heterogenous.   In this case one can only venture a guess that there are 44 chromosomes in the cell. (f) [3.5 hr in vinblastine] (44 chromosomes) Description: The chromosomes are reasonably contracted in this preparation and this helps to make for a better spread.   Scale bars on each 5 μm.

Cycloheximide has been known for some time to affect chromosome contraction but its use in combination with pretreatment chemicals is relatively recent.   Tlaskal (1980) found that prefixation with 8-hydroxy-quinoline and cycloheximide (250 mg/l and 70 mg/l respectively) was useful in sugar cane (Sisodia, 1968).   It seems that the method is especially useful in those plants that have a large number of small chromosomes since it causes substantial contraction.   To that extent, reliable counts can be readily made.   In our hands, cycloheximide has been most useful when used alone (see Fig. 6a), but it tends to overcontract

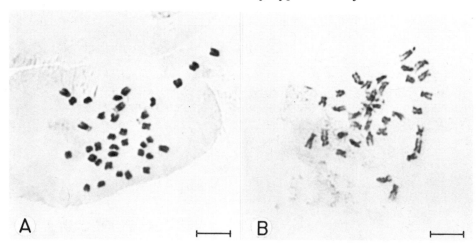

Figure 6. Cells of carrot root arrested in metaphase by 1-bromonaph-thalene or cycloheximide. These photographs derive from a study of carrot plants regenerated from protoplasts prepared from cells grown in suspension. The length of time in the cytostatic is given in brackets. The number of chromosomes is given in parentheses. (a) [3 hr in cycloheximide, 70 µg/ml in distilled water at room temperature] (4n = 36) This metaphase has an excellent spread. There is substantial detail in the chromosome morphology. Note the satellites. Although the larger chromosomes are well contracted, some of the smaller ones look like little more than large dots. (b) [4 hr in 1-bromonaphthalene at 4-5 C] (4n = 36) The spread is good and the contraction is such that the morphology can be described in considerable detail. Although the specimen was exposed for a longer time to this cytostatic than to cycloheximide, the chromosomes are not as contracted as those in (a). It would be possible from an enlargement of this photograph to estab-lish a karyotype. (Scale bars, 5 µm).

small chromosomes making it virtually useless for making karyotypes. Lowering the levels of cycloheximide does not, moreover, seem to result in a diminution of contraction.

Sharma (1956) reviewed fixation in great depth and Sharma and Sharma (1980) provide 40 pages on fixation and many recipes have been compiled. Farmer's fixative, however, has been used very successfully for a long time by many investigators. It is prepared by mixing abso-lute ethyl alcohol and glacial acetic acid in a 3:1 ratio (v/v). Some find that using a 3:1 mixture of ethanol (95%) and propionic acid helps in the staining of small chromosomes (Hyde and Gardella, 1953). Fixa-tives should be freshly prepared. If samples are to be kept for any length of time prior to hydrolysis and staining, it is essential to re-place the fixative with a couple of changes of 70% ethanol. Samples should then be stored in the cold at 4 C. In most cases it is best, however, if the specimens can be processed promptly.

Subsequent processing is more or less routine and need not be dealt with here. Suffice to say that progress is being made all the time and plants with rather small chromosomes are beginning to be effectively karyotyped (e.g., Kurata and Omura, 1978; Tsuchiya and Nakamura, 1979; Pal and Sharma, 1980).

The final protocol is a sample protocol for use in suspension cultured cells.

PROTOCOL FOR 1-BROMONAPHTHALENE PREFIXATIVE

1. Measure out 100 ml of glass distilled water into a 250 ml Erlenmeyer flask with a ground glass stopper.
2. Carefully add 50 mg of saponin, making sure the saponin does not stick to the sides of the vessel being used. Allow the saponin to dissolve more or less by itself. This will take about 5 min. Do not shake. When it has dissolved, gently swirl the vessel so that a few suds appear.
3. Add 64 mg of MES [2-(N-morpholino)ethanesulfonic acid].
4. Add 30 ml of 1-bromonaphthalene.
5. Shake the vessel well to saturate the aqueous solution.
6. If the above procedure is carried out properly, the dispersion will separate out only a few mm of 1-bromonaphthalene at the bottom of the container.
7. Always make sure that the mixture is well dispersed before using and that it is made up freshly for use.

PROTOCOL FOR METAPHASE ANALYSIS OF ROOT TIP CELLS

Roots may be induced directly in aseptic culture or produced in greenhouse or growth chamber from propagules placed in vermiculite moistened periodically with diluted fertilizer (e.g., the mineral salt medium of MS). Even deionized water seems to be adequate for short periods. It usually takes anywhere from a few days to about 2 weeks to produce a sufficient number of roots of appropriate size for examination. Unfortunately the designation appropriate is not very helpful and needs to be determined empirically (see Davidson, 1968 for a discussion of various parameters important to root physiology).

The stepwise procedure of specimen preparation for squashes of chromosomes arrested in metaphase is:

1. Collect roots washed free of vermiculite by cutting tip portions with scissors or, preferably, use whole plants or plantlets. In the latter case a much greater number of division figures are generally encountered, particularly if they are kept in bright light and at a temperature at which the plant in question flourishes. Presumably this is because cell divisions progress more readily in intact plants than in root systems severed from their shoots.
2. Prefix in 1-bromonaphthalene solution for about 4.5–5.5 hr at 10 C in the light. The mixture is comprised of 30 ml of 1-bromonaphthalene, 50 mg of saponin, and 64 mg of MES[2-(N-morpholino)ethanesulfonic acid] in 100 ml of water (see preceding protocol). The timing is important and will vary with the plant. In general,

it should not be allowed to remain in the prefixative for longer than 6 hr, because polyploidization may occur.

3. Remove the roots from the above and rinse 3 or 4 times in deionized water at room temperature. Shake off and blot the excess water promptly.

4. Rinse for a couple of minutes in absolute alcohol—glacial acetic acid (2:1 or 100 and 50 ml, respectively, or 3:1) that has been prechilled to 4 C. Caution: Glacial acetic acid is corrosive and must be handled with extreme care! Note: Prechill the absolute alcohol-acetic acid fixative to 4 C before washing.

5. Sever the roots transversely with razor blades or scalpels (root tips 3-4 cm or so long are adequate) from the plant and plunge firmly but gently into fixative at 4 C. Material may be safely kept in a fixative for a couple of weeks, but deterioration in terms of morphology usually occurs within 96 hours. If it is certain that prolonged storage is unavoidable, it is important to replace the fixative with 70% ethyl alcohol (a couple of changes is advisable) and store at 4 C.

6. Place root tips in 1N HCl (82.2 ml of concentrated HCl and 917.8 ml of water) for 5-20 min at 58-60 C for hydrolysis. Decant. The timing is important and will vary with the plant. In general, 8-10 min is sufficient for hydrolysis.

7. Stain roots in modified Feulgen for 5 min to 24 hr and then soak in tap water for 5 min or so at room temperature. Note: The length of exposure to the stain is important. An hour or so is the more usual case for staining (see the following protocol).

8. Squash in aceto-carmine by applying firm pressure with a suitable object (eraser end of pencil, blunt end of dissecting needle) through several thicknesses of filter paper or paper toweling. Note: Properly prepared roots will fall apart with only gentle pressure, and the cells will spread over the slide.

9. In this condition these preparations are temporary and will dry out in a few hours. The edges of the cover glass can be sealed with a melted mixture of equal parts of paraffin and gum mastic and the slides will keep in this condition for several days.

10. (Optional.) The slides may be made permanent by the following prodedure: (a) If necessary, carefully scrape the paraffin-gum mastic mixture from around the edge of cover glass, taking special pains not to move the cover glass. (b) When this has been done the solution remaining under the cover glass can be quickly frozen by discharging CO_2 under the slide. (c) When the slide is sufficiently frosty, carefully pry under a corner of the cover glass with a razor blade. The cover glass will usually pop right off. Cells ordinarily will adhere to both slide and cover glass. (Alternatively, Cryokwik, dichlorodifluoromethane (CCl_2F_2) packaged by Shield Chemical Company, Inc. for Damon/IEC Division 300 Second Ave., Needham Heights, MA 02194, is a convenient way to "pop off" the cover slip.) (d) Quickly but carefully put both slide and cover glass in a Coplin jar containing 95% ethyl alcohol for 5-10 min. (e) 100% (absolute) ethyl alcohol, 5-10 min. (f) Mount in a drop of Permount or other suitable mounting medium.

PROTOCOL FOR CHROMOSOME STAINS

A. Modified Feulgen (Unpublished)

1. Boil 800 ml distilled water, pour over 6 g basic fuchsin, stir with a glass rod, and leave it to cool to 50 C (check with a thermometer).
2. Filter through two layers of Whatman No. 1 filter paper premoistened with a few drops of water and then add 12 g potassium metabisulfite. Stir to dissolve the $K_2S_2O_5$.
3. Add 320 ml glacial acetic acid. Place in a tightly stoppered amber glass container and store in the refrigerator at least overnight before use. Note: The final color is a faint pink. The solution should be stored in the cold and away from light. Even so, it is good for about 2 months only.

B. Feulgen (cf. also DeTomasi, 1936 and Van't Hof, 1968)

1. Add 1 g basic fuchsin and 1.9 g potassium (or sodium) metabisulfite to 100 ml hydrochloric acid (0.15 N) solution and allow to stand for 24 hr.
2. Add a heaping teaspoonful (ca 500 mg) of activated charcoal for each 100 ml of mixture.
3. Filter through a Whatman No. 1 filter paper on a Buchner funnel premoistened with a few drops of 1 N HCl. The solution after filtration should be virtually colorless.
4. Store in refrigerator in an amber or otherwise darkened, tightly stoppered glass container. There are a number of variations in formulation of the Fuelgen stain. Variability in the basic fuchsin is the major problem. The solution is good for about 2-4 months and should be discarded if it turns pink.

Aceto-Carmine (cf. also Belling, 1921, 1926)

1. Combine 0.5 g carmine, 45 ml glacial acetic acid, and 55 ml distilled water, stir well with a glass rod, and heat the mixture on a water bath until it begins to boil. Continue to boil for a couple of minutes until the carmine dissolves (there is a sudden change in color).
2. Filter while hot through two layers of Whatman No. 1 paper.
3. Add a few crystals of ferric acetate or other ferric salt (to excess).
4. Filter again through double Whatman and refrigerate in an amber or darkened bottle. The solution is very stable and can be used for a long time. Some investigators prefer to use this full strength. In many cases, however, it is better to use a 1:1 dilution with 45% acetic acid, since it generally gives a virtually unstained cytoplasm.

SAMPLE PROTOCOL FOR MITOTIC ANALYSIS OF CELLS GROWN IN SUSPENSION (cf., Fig. 4 for a schematic outline of some of the procedures used in this laboratory for growing cells)

1. Prepare a fresh solution of cytostatic agent on the day of anticipated use (see text).
2. Aseptically remove samples of cell suspension (ca. 20 ml each) using a widemouthed pipet and place each in a 35 ml conical centrifuge tube. (The tubes have metal closures.)
3. Allow the tubes to remain in a holding rack without any disturbance for about 5 min to let the cells settle.
4. Decant the supernatants, allowing a tiny amount of liquid to remain so as not to expose the cells to the drying affects of air.
5. Carefully wash the cells by adding about 10 ml of "wash solution" (see text) to each tube.
6. Repeat steps 4 and 5 two to three times to effect a good wash.
7. Select a vessel, such as an Erlenmeyer flask, that will permit exposure of the cells to a broad surface area. Do not use a centrifuge or test tube.
8. Transfer the cells from the conical centrifuge tube to this container using a Pasteur pipet. Some of the solution of cytostatic agent will have to be used to effect the transfer. It is very useful to retain for study some cells that have not been exposed to cytostatic agent. These may be treated with a fixative at about the same time cytostatic is added to the "test" samples. Fixative, made fresh and prechilled, should be added and the cells retained in it for a minimum of 2 hr at 4 C. No interpretation of the experiment should be made without adequately studying this "control."
9. Add enough additional solution of cytostatic agent to cover the cells. (The amount of cytostatic solution should be approximately equal to the volume of the suspension initially removed for treatment, i.e., if 20 ml of medium containing cells in suspension is the starting point, 20 ml cytostatic will be enough to cover the washed cells.
10. Place the samples on a gently rotating shaker for the necessary period. The speed should be very slow (order of 25 rpm). This step is frequently best carried out in a cold room (5-10 C). The variable time period (see text) will usually be about 3.5 hr but may range from a half-hour to 4.5 hr.
11. Transfer the contents of the flasks containing samples into clean centrifuge tubes.
12. Allow the tubes to rest in a rack without any disturbance for 5 min or so to let the cells settle.
13. Decant the supernatants. Discard. Wash the cells several times. Decant the final wash.
14. Slowly drip with a Pasteur pipet prechilled (4 C) fixative down the wall of the centrifuge tubes, taking care not to harshly "churn up" the cells. Even so, they need to be released from the "pellet" to be uniformly fixed.

15. Refrigerate the tubes for 10 min.
16. Carefully remove the fixative with a pipette. (Use a bulb!) Repeat step 14 and place the tubes in the refrigerator for about 1.5 hr.
17. Carefully withdraw the fixative with a pipet.
18. In this, the final fixation, it is often convenient and instructive to subdivide individual samples among several (five or so) containers.
19. Keep the samples in the refrigerator for a minimum of 12 hr. (Quality of final preparations will deteriorate, however, if they are kept longer than 48 hr.)
20. To prepare for hydrolysis, decant the fixative from each sample. Remove excess fluid from each by inverting the vial over layers of filter paper or toweling to absorb the liquid.
21. Add a few ml of 1 N HCl to each of the samples and place in an oven at 58-60 C for 12-15 min or so. (The length of hydrolysis will vary and tends to be longer in cultured cells than in intact organs.)
22. Decant the hydrolyzing fluid and add a Feulgen stain (see previous Protocol).
23. After about an hour or so, the cells are ready for squashing in a drop of 45% acetic acid.

FUTURE PROSPECTS

Hesitancy to use free cell systems for cloning plants derives primarily from the lack of expertise on the part of some, but it also stems from the view held by many that chromosomal aberrations, especially aneuploidy and polyploidy, are very common in cultured plant cells. We have made an attempt to refute this much too broad generalization. Suffice to say that the incidence of these changes in morphologically competent cell systems has not been adequately studied. Few seem to appreciate that such changes are often induced, even fostered, by the culture procedures adopted. To subscribe to the view that too great a risk of chromosomal change occurs when callus is induced and that to contemplate induction of somatic embryos via cell suspension cultures for the purposes of clonal micropropagation is courting disaster is ill-advised for serious investigators. The first rebuttal should be that there is insufficient critical data on the genetic stability or lack thereof in totipotent cell suspensions capable of producing plantlets, and hence there is no adequate base for generalizations. Much of the literature deals with callus, cultured cells or callus-derived roots rather than the plantlets (or the roots thereof) to which they give rise.

One can find several tables in the literature summarizing the chromosomal instability encountered in tissue cultures. Readers are urged to consult Evans and Reed (1981, pp. 215-216) and D'Amato (1977a, pp. 350-351) for the tables and their accompanying citations.

A major obstacle to meaningful chromosome analysis is that in many cases it is not yet possible to initiate morphogenetically competent callus or suspension cultures within a short period of time. If too much time elapses between taking of the primary explant and the satisfactory establishment of the morphogenetically competent cell suspension state, so much selection and variation may have gone on that the culture might in all reality only slightly resemble the cells in the initial explant.   Efforts to disclose the requirements for inducing rapid cell divisions, especially in more recalcitrant cases, still is a laudable goal and investigators should not rest comfortably in a state of false security thinking that one knows everything there is to know about initiating and maintaining cultures.   We do not (cf. Krikorian and Kann, 1981; Krikorian, 1982, and refs. there cited).

There is a vast body of information that exists or is beginning to surface on the genetic toxicology of various substances used by tissue culture workers as growth regulators.   Whereas the greater amount of attention has been given to the phenoxy acids (Shoji et al., 1960; Seiler, 1978; Ehrenberg, 1978; Grant, 1979), there is some information on other growth regulators as well (e.g., Sawamura, 1964, 1965).   It will be a long time before we really can authoritatively advise investigators to refrain from the use of any given compound.

For one reason or other, chromosomal changes, including mutations that carry over into regenerated plantlets, might be utilized as potentially valuable sources of genetic variations (Larkin and Snowcroft, 1981).   Banding techniques would permit investigators to routinely screen morphogenetically competent cell populations prior to their use in specific procedures or their use in regenerating plantlets.   If the cells contained chromosome abnormalities, then the cultures would be discarded and fresh ones initiated.   In this laboratory, it is a routine procedure to establish fresh cultures on a continuing basis so as to minimize chromosomal abnormalities.   Unpublished studies carried out several years ago revealed that the phenotype of cultured carrot plants throughout their development—from young plants to mature ones complete with flowers and seeds—is normal provided the suspensions, from which the somatic embryo-initiating cells were derived, were less than one-year-old.   Plants from cells kept in suspension culture for longer periods showed a number of phenotypic aberrations that could be correlated with their chromosomal condition.

The high level of success enjoyed by animal cell workers in terms of Giemsa banding seems more remote for those working with plant cells. Even so, progress is being made (Vosa, 1975; Mok and Mok, 1976; Lavania, 1978a; Dumas de Vaulx, 1980).   Similarly, use of substances such as the fluorochrome 'Hoechst 33258' to disclose constitutive heterochromatin and/or repetitive DNA and even to detect sister chromatid exchange has also been used in plants with varying degrees of success (Perry and Wolff, 1974; Filion et al., 1976; Cesarone et al., 1979; Peacock et al., 1981).

Presumed epigenetically-determined responses in culture-derived plantlets are only one type of problem that cannot be addressed adequately by only counting chromosomes.   It is well known that epigenetic effects can be noted in products of cells that are karyologically normal (Sibi, 1976; Siminovitch, 1976; Larkin and Scowcroft, 1981).   One example of an epigenetic effect being expressed in culture involves production of juvenile or adult leaf forms according to the source of the primary explant (Stoutemeyer and Britt, 1965; Marcavillaca and Montaldi, 1967; Banks, 1979). Another involves shoot plagiotropism (i.e., having the longer axis inclined away from the vertical) in culture-derived plants. This is particularly true of conifers. Although plants with orthotropic shoots (i.e., those that elongate vertically) can be retrieved from culture-derived plants that show plagiotropism, it is often a time-consuming and labor-intensive procedure (e.g., Franclet, 1977; Franclet et al., 1980). In this laboratory, tiny daylily plantlets that have a phyllotactic arrangement normally encountered in, and characteristic of, the more mature forms (Fitter and Krikorian, unpublished) have been encountered.

As time goes on, plant cells are sure to be amenable to more and more sophisticated techniques such as satellite DNA localization (Peacock et al., 1981). Even so, there is no doubt that use of the existing, even classical, methodology covered in this chapter can go far to disclose many of the kinds of events occurring under a given set of culture conditions. With the intense interest in somatic cell hybridization via use of protoplasts for genetic engineering (Kleinhofs and Behki, 1977; Chaleff, 1981; Cocking et al., 1981), one is seeing increased interest in careful application of cytological techniques. One cannot fail to see the limitations of metaphase chromosome analysis, but there seems little point in attempting more advanced technology when even the available methods have not been fully exploited. Had adequate karyological and chromosomal analysis been carried out on all the materials that have thus far been cultured, it would have been possible to treat the subject matter of this chapter with a greater degree of certainty.

ACKNOWLEDGEMENTS

Investigations providing the background for this chapter were supported by various grants. Prominant among them has been support from the National Aeronautics and Space Administration (Grant NSG 7270) Plant Cells, Embryos, and Morphogenesis in Space. This support is gratefully acknowledged.

KEY REFERENCES

Darlington, C.D. and La Cour, L.F. 1975. The Handling of Chromosomes, 6th ed. John Wiley & Sons, New York.

Dyer, A.F. 1979. Investigating Chromosomes. Edward Arnold, London; John Wiley & Sons, New York.

Loeve, A. and Loeve, D. 1975. Plant Chromosomes. J. Cramer, Vaduz Lichtenstein. (Available in the U.S. from Lubrecht & Cramer, RFD 1, Box 227, Monticello, NY. 12701.)

Phillips, R.L. 1981. Plant cytogenetics. In: Staining Procedures, 4th ed. (G. Clark, ed.) pp. 341-359. Biological Stain Commission, Williams & Wilkins, Baltimore and London.

Sharma, A.K. and Sharma, A. 1980. Chromosome Techniques. Theory and Practice, 3rd ed. Butterworths, London, Boston, Sydney, Wellington, Durban, and Toronto.

REFERENCES

Allen, C.E. 1940. The genotypic basis of sex-expression in angiosperms. Bot. Rev. 6:227-300.

Anonymous. 1978. The National Plant Germplasm System. U.S. Government Printing Office, 1978 0-264-791.

Anonymous. 1978. National Seed Storage Laboratory, Fort Collins, Colorado. U.S. Government Printing Office, 1978 - 793-035.

Arisumi, T. 1972. Stabilities of colchicine-induced tetraploid and cytochimeral daylilies. J. Hered. 63:15-18.

Baldwin, J.T. 1939. Chromosomes from leaves. Science 90:240.

Banks, M.A. 1979. Plant regeneration from callus from two growth phases of English Ivy, *Hedera helix* L. Z. Pflanzenphysiol. 92:349-353.

Bayliss, M.W. 1973. Origin of chromosome number variation in cultured plant cells. Nature 246:529-530.

_____ 1975. The effects of growth in vitro on the chromosome complement of *Daucus carota* (L.) suspension cultures. Chromosoma (Berl.) 51:401-411.

_____ 1980. Chromosomal variations in plant tissues in culture. In: Perspectives in Plant Cell and Tissue Culture (I.K. Vasil, ed.) Int. Rev. Cytol. Suppl. 11A:113-144.

Belling, J. 1921. On counting chromosomes in pollen-mother cells. Am. Nat. 55:573-574.

_____ 1926. The iron-acetocarmine method of fixing and staining chromosomes. Biol. Bull. 50:160-162.

Bennici, A. 1974. Cytological analysis of roots, shoots and plants regenerated from suspension and solid in vitro cultures of haploid *Pelargonium*. Z. Pflanzenzuchtg. 72:199-205.

_____ and D'Amato, F. 1978. In vitro regeneration of *Durum* wheat plants. 1. Chromosome numbers of regenerated plantlets. Z. Pflanzenzuchtg. 81:305-311.

Bezbaruah, H.P. 1968. An evaluation of preparatory procedures for leaf-tip chromosome spreads of the tea plant (*Camellia sinensis*). Stain Technol. 43:279-282.

Blakely, L.M. 1963. Growth and Variation of Cultured Plant Cells. Ph.D. Thesis, Cornell University, Ithaca, New York.

Brat, S.V., Verma, R.S., and Dosik, H. 1979. A simplified technique for simultaneous staining of nucleolar organizer regions and kinetochores. Stain Technol. 54:107-108.

Burns, J.A. 1964. A technique for making preparations of mitotic chromosomes from *Nicotiana* flowers. Tobacco Sci. 8:1-2.

Caplin, S.M. 1947. Growth and morphology of tobacco tissue cultures in vitro. Bot. Gaz. 108:379-393.

Cesarone, C.F., Bolognesi, C., and Santi, L. 1979. Improved microfluorometric DNA determination in biological material using 33258 Hoechst. Analyt. Biochem. 100:188-197.

Chaleff, R.S. 1981. Genetics of Higher Plants. Applications of Cell Culture. Cambridge Univ. Press, Cambridge.

Chang, T.-T. and Li, C.-C. 1980. Genetics and breeding. In: Rice: Production and Utilization (B.S. Luh, ed.) pp. 87-146. AVI Pub., Westport, Connecticutt.

Chiarugi, A. 1954. La poliploidia somatica nelle piante. Caryologia 6 (suppl.):488-520.

Chouinard, A.L. 1955. Nuclear differences in *Allium cepa* root tissues as revealed through induction of mitosis with indoleacetic acid. Can. J. Bot. 33:628-646.

Cocking, E.C., Davey, M.R., Pental, D., and Power, J.B. 1981. Aspects of plant genetic manipulation. Nature 293:265-269.

Collins, G.B. 1979. Cytogenetic techniques. In: *Nicotiana*: Procedures for Experimental Use (R.D. Durbin, ed.) pp. 17-27. USDA Tech. Bull. 1586, Washington, D.C.

Conger, B.V. and Carabia, J.V. 1978. Proportions of 2C and 4C nuclei in the root and shoot of dormant and germinated embryos of *Festuca arundinacea* and *Dactylis glomerata.* Environ. Exp. Bot. 18:55-59.

Constantin, M.J. 1981. Chromosome instability in cell and tissue cultures and regenerated plants. Exp. Environ. Bot. 31:359-368.

Cooper, L.S., Cooper, D.C., Hildebrandt, A.C., and Riker, A.J. 1964. Chromosome numbers in single cell clones of tobacco tissue. Am. J. Bot. 51:284-290.

D'Amato, F. 1952. Polyploidy in the differentiation and function of tissues and cells in plants. Caryologia 4:311-358.

_____ 1964. Endopolyploidy as a factor in plant tissue development. Caryologia 17:41-52.

_____ 1975. The problem of genetic stability in plant cell and tissue cultures. In: Crop Genetic Resources for Today and Tomorrow (IBP 2) pp. 335-348. Cambridge Univ. Press, Cambridge.

_____ 1977a. Cytogenetics of differentiation in tissue and cell cultures. In: Applied and Fundamental Aspects of Plant Cell, Tissue and Organ Culture. (J. Reinert and Y.P.S. Bajaj, eds.) pp. 343-357. Springer-Verlag, Berlin, Heidelberg, and New York.

_____ 1977b. Nuclear Cytology in Relation to Development. Cambridge Univ. Press, Cambridge.

_____ 1978. Chromosome number variation in cultured cells and regenerated plants. In: Frontiers of Plant Tissue Culture 1978 (T.A. Thorpe, ed.) pp. 287-295. Univ. of Calgary Press, Calgary.

_____, Bennici, A., Cionini, P.G., Baroncelli, S., and Lupi, M.C. 1980. Nuclear fragmentation followed by mitosis as mechanism for wide chromosome number variation in tissue cultures: Its implications for plantlet regeneration. In: Plant Cell Cultures: Results and Perspectives (F. Sala, R. Parisi, R. Cella, and D. Ciferri, eds.) pp. 67-72. Elsevier/North-Holland Biomedical Press, Amsterdam, New York, and Oxford.

Davidson, D. 1968. Physiological studies of cells in root meristems. In: Methods in Cell Physiology, Vol. III, pp. 171-211. Academic Press, New York.

Demoise, C.F. and Partanen, C.R. 1969. Effects of subculturing and physical condition of medium on the nuclear behavior of a plant tissue culture. Am. J. Bot. 56:147-152.

Dermen, H. 1940. Colchicine polyploidy and technique. Bot. Rev. 6: 599-635.

Dermen, G.M. 1960. Nature of plant sports. Am. Hort. Mag. 39:123-173.

DeTomasi, J.A. 1936. Improving the technic of the Feulgen Stain. Stain Technol. 11:137-144.

Dronamraju, K.R. 1965. The function of the Y chromosome in man, animals and plants. Adv. Genet. 13:227-310.

Dumas de Vaulx, R. 1980. Technique de coloration ou de surrcoloration des chromosomes par le "Giemsa R." Ann. L'Amelior. Plantes 30:218-219.

Ehrenberg, L. 1978. Genetic effects of chlorophenoxyalkanoic acids on plants. In: Chlorinated Phenoxy Acids and then Dioxins (C. Ramel, ed.) pp. 186-189. Ecol. Bull. No. 27, Stockholm.

Eigsti, O.J. and Dustin, P. 1955. Colchicine in Agriculture, Medicine, Biology and Chemistry. Iowa State College Press, Ames.

Evans, D.A. 1979. Chromosome stability of plants regenerated from mesophyll protoplasts of Nicotiana species. Z. Pflanzenphysiol. 95:459-463.

_____ and Reed, S.M. 1981. Cytogenetic techniques. In: Plant Tissue Culture: Methods and Applications in Agriculture (T.A. Thorpe, ed.) pp. 213-240. Academic Press, New York.

Fedorov, A. (ed.) 1969. Chromosome Numbers of Flowering Plants by Z. Bolkhovskikh, V. Grif, T. Matvejeva, O. Zaharyeva. Nauka Pub., Leningrad.

Ferwerda, F.P. and Wit, F. (eds.) 1969. Outlines of Perennial Crop Breeding in the Tropics. H. Veenman & Zonen, NV., Wageningen.

Filion, W.G., MacPherson, P., Blakey, D., Yen, S., and Culpeper, A. 1976. Enhanced Hoechst 33258 fluorescence in plants. Exp. Cell Res. 99:204-206.

Franclet, A. 1977. Manipulation des pieds-mères et amélioration de la qualité des boutures. Études et Recherches, Association Forêt-Cellulose No. 8-12/77, Nangis, France.

_____, David, A., David, H., and Boulay, M. 1980. Premiere mise en evidence morphologique d'un rajeunissement de méristemes primaires caulinaires de pin maritime âgé (Pinus pinaster Sol.). Comptes Rendus a L' Academie des Sciences, Paris 290 D:927-930.

Galbraith, D.W., Mauch, T.J., and Shields, B.A. 1981. Analysis of the initial stages of plant protoplast development using 33258 Hoechst: Reactivation of the cell cycle. Physiol. Plant. 51:380-386.

Geitler, L. 1953. Endomitose und endomitotische Polyploidisierung. Protoplasmatologia VIC, Springer-Verlag, Vienna.

Grant, W.F. 1979. The genotoxic effects of 2,4,5-T. Mut. Res. 65:83-119.

Hepler, P.K. and Jackson, W.T. 1969. Isopropyl-N-phenylcarbamate affects spindle microtubule orientation in dividing endosperm cells of Haemanthus katherinae Baker. J. Cell Sci. 5:727-743.

Hermsen, J.G.Th., Ramanna, M.S., Roest, S., and Bokelmann, G.S. 1981. Chromosome doubling through adventitious shoot formation on in vitro cultivated leaf explants from diploid interspecific potato hybrids. Euphytica 30:239-246.

Hiraoka, N. and Tabata, M. 1974. Alkaloid production by plants regenerated from cultured cells of *Datura innoxia*. Phytochemistry 13: 1671-1675.

Hizume, M., Sato, S., and Tanaka, H. 1980. A highly reproducible method of nucleolus organizing regions staining in plants. Stain Technol. 55:87-90.

Hyde, B.B. and Gardella, C.A. 1953. A mordanting fixation for intense staining of small chromosomes. Stain Technol. 28:305-308.

Jackson, R.C. 1957. New low chromosome number plants. Science 126:1115-1116.

Kao, K.N. 1975. A chromosomal staining method for cultured cells. In: Plant Tissue Culture Methods (O.L. Gamborg and L.R. Wetter, eds.) pp. 63-65. National Research Council of Canada, Prairie Regional Laboratory, Saskatoon.

_____, Miller, R.A., Gamborg, O.L., and Harvey, B.L. 1970. Variations in chromosome number and structure in plant cells grown in suspension cultures. Can. J. Genet. Cytol. 12:297-301.

Khush, G.S. 1973. Cytogenetics of Aneuploids. Academic Press, New York.

_____ 1974. Rice. In: Handbook of Genetics (R.C. King, ed.) pp. 31-58. Plenum Press, London and New York.

Kleinhofs, A. and Behki, R. 1977. Prospects for plant genome modification by non-conventional methods. Annu. Rev. Genet. 11:79-101.

Krikorian, A.D. 1982. Cloning higher plants from aseptically cultured tissues and cells. Biol. Rev. 57:151-218.

_____ and Kann, R.P. 1980. Mass blooming of a daylily clone reared from culture tissues. Hemerocallis J. 34:35-38.

_____ and Kann, R.P. 1981. Plantlet production from morphogenetically competent cell suspensions of daylily. Ann. Bot. 47:679-686.

_____, Staicu, S.A., and Kann, R.P. 1981. Karyotype analysis of a daylily clone reared from aseptically cultured tissues. Ann. Bot. 47: 121-131.

Kurata, N. and Omura, T. 1978. Karyotype analysis in rice 1. A new method for identifying all chromosome pairs. Jap. J. Genet. 53:251-255.

LaRue, C.D. 1947. Growth and regeneration of the endosperm of maize in culture. Am. J. Bot. 34:585-586.

Larkin, P.J. and Snowcroft, W.R. 1981. Somaclonal variation. A novel source of variability from cell cultures for plant improvement. Theor. Appl. Genet. 60:197-214.

Lavania, U.C. 1978a. Differential staining and plant chromosomes—A progress in cytogenetics. Curr. Sci. (Bangalore) 47:255-260.

_____ 1978b. Chlorflurenol—A new pretreating agent for chromosome work. Curr. Sci. (Bangalore) 47:632-633.

Lester, D.T. and Berbee, J.G. 1977. Within-clone variation among black poplar trees derived from callus culture. For. Sci. 23:122-131.

Levan, A. and Muentzing, A. 1963. Terminology of chromosome numbers. Portugaliae Acta Biologica, Series A. Morfologia, Fisiologia, Genetica Biologica 7:1-16.

Levan, A., Fredga, K., and Sandberg, A.A. 1965. Nomenclature for centromeric position in chromosomes. Hereditas 52:201-220.

Levine, M. 1931. Studies in the cytology of cancer. Crown gall disease. Am. J. Cancer 15:1410-1494.

Lin, B.-Y. 1977. A squash technique for studying the cytology of maize endosperm and other tissues. Stain Technol. 52:197-201.

Malnassy, P. and Ellison, J.H. 1970. Asparagus tetraploids from callus tissue. HortScience 5:444-445.

Marcavillaca, M.C. and Montalde, E.R. 1967. Diferentes formas de lojas producidas por yemas adventicias inducidas experimentalmente en lojas aisladas de Nicotiana tabacum L. y Passiflora coerulea L. Rev. Investigaciones Agropecuarias INTA (Buenos Aires) Ser. 2, 4:1-7.

Mehta, R.K. and Swaminathan, M.S. 1957. Studies on induced polyploids in forage crops. Ind. J. Genet. Plant Breed. 17:27-57.

Meredith, C. and Carlson, P.S. 1979. Genetic variation in cultured plant cells. In: Propagation of Higher Plants through Tissue Culture (K.W. Hughes, R. Henke, and M. Constantin, eds.) pp. 166-176. USDOE, Technical Information Center, Springfield, Virginia.

Meyer, J.R. 1943. Colchicine-Feulgen leaf smears. Stain Technol. 18: 53-56.

Mitra, J. and Steward, F.C. 1961. Growth induction in cultures of Haplopappus gracilis. II. The behavior of the nucleus. Am. J. Bot. 48:358-368.

Mitra, J., Mapes, M., and Steward, F.C. 1960. Growth and organized development of cultured cells. IV. The behavior of the nucleus. Am. J. Bot. 47:357-368.

Mok, D.W.S. and Mok, M.C. 1976. A modified Giemsa technique for identifying bean chromosomes. J. Hered. 67:187-188.

Mok, M., Gabelman, W.H., and Skoog, F. 1976. Carotenoid synthesis in tissue cultures of Daucus carota L. J. Am. Soc. Hort. Sci. 101:442-449.

Muir, W.H. 1965. Influence of variation in chromosome number on differentiation in plant tissue cultures. In: Proceedings International Conference Plant Tissue Culture (P.R. White and A.R. Grove, eds.) pp. 485-492. McCutchan Pub., Berkeley.

———, Hildebrandt, A.C., and Riker, A.J. 1954. Plant tissue cultures produced from single isolated cells. Science 119:877-878.

Murashige, T. and Nakano, R. 1967. Chromosome complement as a determinanat of the morphogenetic potential of tobacco cells. Am. J. Bot. 54:963-970.

Murray, B.E., Handyside, R.J., and Keller, W.A. 1977. In vitro regeneration of shoots on stem explants of haploid and diploid flax (Linum usitatissimum). Can. J. Genet. Cytol. 19:177-186.

Nagl, W. 1978. Endopolyploidy and Polyteny in Differentiation and Evolution. Towards an Understanding of Quantitative and Qualitative Variation of Nuclear DNA in Ontogeny and Phylogeny. North-Holland, Amsterdam and New York.

Neilson-Jones, W. 1969. Plant Chimeras. Methuen and Co., London and New York.

Neuffer, M. and Coe, E.H. 1974. Corn (maize). In: Handbook of Genetics (R.C. King, ed.) pp. 3-30. Plenum Press, London and New York.

Neuffer, M., Jones, L., and Zuber, M. (eds.) 1968. The Mutants of Maize. Crop Science Society of America, Madison.

Nickell, L.G. 1977. Crop improvement in sugarcane: Studies using in vitro methods. Crop Sci. 17:717-719.

Nilan, R.A. 1964. The cytology and genetics of barley, 1957-1962. Research Studies, Vol. 32, No. 1. Washington State Univ. Press, Pullman.

_____ (ed.) 1971. Barley Genetics II. Proceedings 2nd International Barley Genetics Symposium, Washington State Univ. Press, Pullman.

_____ 1974. Barley (Hordeum vulgare). In: Handbook of Genetics (R.C. King, ed.) pp. 93-110. Plenum Press, London and New York.

Novak, F.J. 1974. The changes of karyotype in callus cultures of Allium sativum L. Caryologia 27:44-54

_____ 1981. Chromosomal characteristics of long term callus cultures of Allium sativum L. Cytologia 46:371-379.

Pal, A. and Sharma, A.K. 1980. Analysis of cytotypes of Discorea and scope of increasing the diosgenin content. La Cellule 73:116-134.

Partanen, C.R. 1957. Quantitative chromosomal changes and differentiation in plants. In: Developmental Cytology (D. Rudnick, ed.) pp. 21-45. 16th Symposium of the Society for the Study of Development and Growth. Ronald Press, New York.

_____ 1963a. Plant tissue culture in relation to developmental cytology. Int. Rev. Cytol. 15:215-243.

_____ 1963b. The validity of auxin-induced divisions in plants as evidence of polyploidy. Exp. Cell. Res. 31:597-599.

_____ 1965. Cytological behavior of plant tissues in vitro as a reflection of potentialities. In: Proceedings International Conference Plant Tissue Culture (P.R. White and A.R. Grove, eds.) pp. 463-471. McCutchan Pub., Berkeley.

_____ 1973. Karyology of cells in culture. In: Tissue Culture Methods and Applications (P.F. Kruse, Jr., and M.K. Patterson, Jr., eds.) pp. 791-797. Academic Press, New York.

Peacock, W.J., Gerlach, W.L., and Dennis, E.S. 1981. Molecular aspects of wheat evolution: Repeated DNA sequences. In: Wheat Science— Today and Tomorrow (L.T. Evans and W.J. Peacock, eds.) pp. 41-59. Cambridge Univ. Press, Cambridge.

Pelletier, G. and Pelletier, A. 1971. Culture in vitro de tissus de trefle blanc (Trifolium repens): Variabilite des plantes regenerees. Ann. Amelior. Plantes 21:221-233.

Perry, P. and Wolff, S. 1974. New Giemsa method for differential staining of sister chromatids. Nature 251:156-158.

Read, R.W. 1964. Palm chromosome studies facilitated by pollen culture on a colchicine-lactose medium. Stain Technol. 39:99-105.

_____ 1965. Chromosome numbers in the Coryphoideae. Cytologia 30:385-391.

Sacristan, M.D. 1971. Karyotypic changes in callus cultures from haploid and diploid plants in Crepis capillaris (L.) Wallr. Chromosoma (Berl.) 33:273-283.

_____ and Melchers, G. 1969. The caryological analysis of plants regenerated from tumorous and other callus cultures of tobacco. Molec. Gen. Genet. 105:317-333.

Sawamura, S. 1964. Cytological studies on the effect of herbicides on plant cells in vivo. I. Hormonic herbicides. Cytologia 29:86-102.

_____ 1965. Cytological studies on the effect of herbicides on plant cells in vivo. II. Non-hormonic herbicides. Cytologia 30:325-348.

Schubert, I., Anastassova-Kristeva, M., and Rieger, R. 1979. Specificity of NOR staining in *Vicia faba*. Exp. Cell Res. 120:430-435.

Schulz-Schaeffer, J. 1980. Cytogenetics. Springer-Verlag, New York.

Sears, E.R. 1972. Chromosome engineering in wheat. Stadler Symp. 4: 23-38.

_____ 1974. The wheats and their relatives. In: Handbook of Genetics (R.C. King, ed.) pp. 59-91. Plenum Press, London and New York.

Segawa, M. and Kondo, K. 1980. Root fasciation in *Cucumis sativus* and *Vicia faba* treated with vinblastine. Environ. Exp. Bot. 20:201-206.

Seiler, J.P. 1978. The genetic toxicology of phenoxyacids other than 2,4,5-T. Mut. Res. 55:197-226.

Sharma, A.K. 1956. Fixation of plant chromosomes. Bot. Rev. 22:665-695.

_____ and Bhattacharya, N.K. 1960. An investigation on the scope of a number of pre-treatment chemicals for chromosome studies in different groups of plants. Jap. J. Bot. 17:152-162.

Sharma, A.K. and Das, N.K. 1954. Study of karyotypes and their alterations in aroids. Agron. Lusitana 16:23-48.

Sharma, A.K. and Sarkar 1963. Cytological analysis of different cytotypes of *Colocasia antiquorum*. Bot. Soc. Bengal. Bull. (Calcutta) 17: 16-22.

Sharma, A.K. and Sharma, A. 1957. Permanent smears of leaf tips for the study of chromosomes. Stain Technol. 32:167-169.

Sheridan, W.F. 1974. Long term callus cultures of *Lilium*: Relative stability of the karyotype. J. Cell. Biol. 63:313a.

_____ 1975. Plant regeneration and chromosome stability in tissue cultures. In: Genetic Manipulations with Plant Material. (L. Ledoux, ed.) pp. 263-293. Plenum, New York.

Shoji, T., Shoji, T., Oda, Y., and Matsuura, Y. 1960. Effects of agricultural chemicals on mitosis I. Herbicides: Phenoxyacetic acid derivatives. Kromosomo (Institute for Chromosome Research) Tokyo 46/47:1531-1540 (in Japanese with English summary).

Sibi, M. 1976. La notion de programme genetique chez les vegetaux superieurs. II. Aspect experimental. Obtention de variants par cultures de tissus in vitro sur *Lactuca sativa* L. Apparition de vigeur chez les croisements. Annales de l'Amelioration des Plantes (Paris) 26:523-547.

Siminovitch, L. 1976. On the nature of heritable variation in cultured somatic cells. Cell 7:1-11.

Simmonds, N.W. (ed.) 1976. Evolution of Crop Plants. Longman, London and New York.

_____ 1979. Principles of Crop Improvement. Longman, London and New York.

Singh, B.D., Harvey, B.L., Kao, K.N., and Miller, R.A. 1975. Karyotypic changes and selection pressure in *Haplopappus gracilis* suspension cultures. Can. J. Genet. Cytol. 17:109-116.

Singh, M. and Krikorian, A.D. 1980. Chelated iron in culture media. Ann. Bot. 46:807-809.

_____ 1981. White's standard nutrient solution. Ann. Bot. 47:133-139.

Sisodia, N.S. 1968. A root squash technique for counting somatic chromosomes in sugar cane. Stain Technol. 43:129-135.

Skirvin, R.M. 1978. Natural and induced variation in tissue culture. Euphytica 27:241-266.

Sprague, G.F. 1977. Corn and Corn Improvement. American Society of Agronomy, Madison.

_____ 1980. Germplasm resources of plants: Their preservation and use. Annu. Rev. Phytopathol. 18:147-165.

Steward, F.C. and Krikorian, A.D. 1979. Problems and potentialities of cultured plant cells in retrospect and prospect. In: Plant Cell and Tissue Culture: Principles and Applications (W.R. Sharp, R.O. Larsen, E.F. Paddock, and V. Raghavan, eds.) pp. 221-262. Ohio State Univ. Press, Columbus.

Steward, F.C., Mapes, M.O., Kent, A.E., and Holsten, R.D. 1964. Growth and development of cultured plant cells. Science 143:20-27.

Storey, W.B. and Mann, J.D. 1967. Chromosome contraction by o-isopropyl-N-phenylcarbamate (IPC). Stain Technol. 42:15-18.

Stoutemeyer, V.T. and Britt, O.K. 1965. The behavior of tissue cultures from English and Algerian ivy in different growth phases. Am. J. Bot. 52:805-810.

Straus, J. 1954. Maize endosperm tissue grown in vitro. II. Morphology and cytology. Am. J. Bot. 41:833-839.

Sunderland, N. 1973. Nuclear cytology. In: Plant Cell and Tissue Culture (H.E. Street, ed.) pp. 161-190. Univ. of California Press, Berkeley and Los Angeles.

_____ 1977. Nuclear cytology. In: Plant Tissue and Cell Culture, 2nd ed. (H.E. Street, ed.) pp. 177-205. Univ. California Press, Berkeley and Los Angeles.

Thurston, H.D. 1977. International crop development centers: A pathologists' perspective. Annu. Rev. Phytopathol. 15:223-247.

Tlaskal, J. 1980. Combined cycloheximide and 8-hydroxyquinoline pretreatment for study of plant chromosomes. Stain Technol. 54:313-319.

Tran Thanh Van, K. and Trinh, H. 1979. Plant propagation: Non-identical copies. In: Propagation of Higher plants through Tissue Culture (K.W. Hughes, R. Henke, and M. Constantin, eds.) pp. 134-153. USDOE, Technical Information Center, Springfield, Virginia.

Torrey, J. 1959. Experimental modification of development in the root. In: Cell, Organism and Milieu (D. Rudnick, ed.) pp. 189-222. Ronald Press, New York.

_____ 1965. Cytological evidence of cell selection by plant tissue culture media. In: Proceedings International Conference Plant Tissue Culture (P.R. White and A.R. Grove, eds.) pp. 473-484. McCutchan Pub., Berkeley.

_____ and Landgren, C.R. 1977. Mitosis and cell division in cultures of higher plant protoplasts and cells. In: La Culture des Tissus et des Cellules des Vegetaux. Travaux dedies a la memoire de Georges Morel (R.J. Gautheret, ed.) pp. 146-168. Masson, Paris.

Tsuchiya, T. and Nakamura, C. 1979. Acetocarmine squash method for observing sugar beet chromosomes. Euphytica 28:249-256.

Van't Hof, J. 1968. Experimental procedures for measuring cell population kinetic parameters in plant root meristems. In: Methods in Cell Physiology, Vol. III, pp. 95-117. Academic Press, New York.

Verma, S.C. and Mittal, R.K. 1978. Chromosome variation in the common garlic, *Allium sativum* L. Cytologia 43:383-396.

Vosa, C.G. 1975. The use of Giemsa and other staining techniques in karyotype analysis. Curr. Adv. Plant Sci. 6:495-510.

Westergaard, M. 1958. The mechanism of sex determination in dioecious flowering plants. Adv. Genet. 9:217-281.

Whelan, E.D.P. 1978. Cytology and interspecific hybridization. In: Sunflower Science and Technology (J.F. Carter, ed.) pp. 339-360. Agronomy Monographs 19, American Society of Agronomy, Madison.

White, P.R. 1953. A comparison of certain procedures for the maintenance of plant tissue cultures. Am. J. Bot. 40:517-524.

Withers, L.A. 1981. Institutes Working on Tissue Culture for Genetic Conservation. International Board for Plant Genetic Resources. Consultant Report, AGP:IBPGR/81/30. IBPGR Secretariat, Rome.

Yunis, J. (ed.) 1974. Human Chromosome Methodology. Academic Press, New York.

SECTION II
Molecular Techniques

CHAPTER 17

Fraction-1 Protein and Chloroplast DNA as Genetic Markers

S.D. Kung

Somatic cell fusion and recombinant DNA technology form the basis of the intense current interest in plant genetic manipulation. Somatic cell fusion overcomes certain restrictions on gene flow between organisms that are sexually incompatible. Recombinant DNA technology permits gene transfer even between prokaryotes and eukaryotes (Cocking et al., 1981). Together, they offer an opportunity not available before to improve plants genetically. The success of both approaches depends, to a large degree on the refinement and success of tissue culture techniques. One limitation confronting us today in using tissue culture, particularly the somatic cell fusion, as a way of genetic manipulation, is the lack of proper and effective markers for selection. Considerable effort should be directed toward development of efficient selection schemes.

One of the many nonselective genetic markers currently used in plant systems is the Fraction-1 (F1) protein (Kung, 1976b; von Wettstein et al., 1978). This protein has many unique properties. One of them is the variability of the isoelectric focusing patterns that have been widely used to identify the parentage of many progenies from a variety of plant species, especially in tobacco (Kung, 1976b). Tobacco, in turn, is the most commonly used material for tissue culture. Therefore, this chapter is confined to *Nicotiana* F1 protein and its application as a genetic marker for both chloroplast and nuclear genomes. Because recent studies of *Nicotiana* chloroplast DNA (cpDNA) have revealed a higher degree of variability in its restriction sites, it appears to be more useful than the isoelectric focusing patterns of F1 protein as a genetic marker for the chloroplast genome. Therefore, the potentials and limitations of using cpDNA as a genetic marker will also be discussed.

FRACTION-1 PROTEIN AS A UNIQUE GENETIC MARKER

F1 protein is a major chloroplast protein comprising more than 50% of the total soluble leaf protein (Fig. 1). Historically it was discovered from the study of plant hormones (Wildman and Bonner, 1947) and subsequently identified as the major photosynthetic enzyme, ribulose 1,5-bisphosphate carboxylase-oxygenase (RuBPCase) (Dorner et al., 1957). It is a unique enzyme in many aspects. For example, it catalyzes both reactions of photosynthesis and photorespiration (Andrews et al., 1973); it requires both chloroplast and nuclear genomes for its synthesis (Kung, 1977); and it consists of two types of subunits and has separate sites for activation and catalysis (Lorimer, 1981). Because of

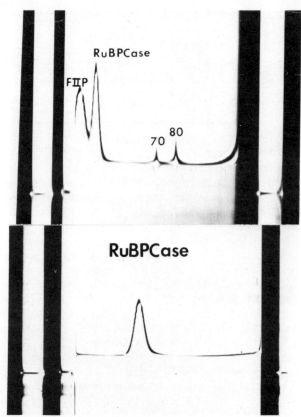

Figure 1.   Analytical ultracentrifuge schlieren pattern of total soluble protein and free ribosomes extracted from tobacco leaves (top). The four distinct components are Fraction II protein (FIIP), F1 protein (RuBPCase), 70S and 80S ribosomes. The crystalline F1 protein is homogeneous and has a sedimentation coefficient of 18S (below).

these unusual properties, F1 protein has been the subject of intensive study by many laboratories in recent years.   One of the practical applications of F1 protein is the wide adaptation of this protein as a genetic marker.   In this chapter, the molecular biology, the procedure for crystallization, and the application of F1 protein as a genetic marker of *Nicotiana* will be described.

Characterization of *Nicotiana* Fraction-1 Protein

STRUCTURE AND FUNCTION.   F1 protein from higher plants, including tobacco, is composed of eight large and eight small subunits, each of which has an estimated molecular weight of 5.5×10^4 and 1.4×10^4 d, respectively (Fig. 2).   The eight large subunits (LS) are situated at the four corners of a two-layered structure with the eight small subunits (SS) on the surface.   This proposed L_8L_8 structure for F1 protein fits well with all evidence obtained from X-ray diffraction (Baker et al., 1975), electron micrographic (Bovien and Mayer, 1978), and chemical (Rutner, 1970) investigations.   The subunits can be dissociated under a variety of denaturing conditions, such as urea, sodium dodecyl

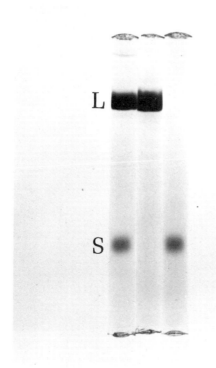

Figure 2.   SDS-gel electrophoretic separation of the large (L) and small (S) subunits of carboxymethylated F1 protein; left—F1 protein, center—large subunit, right—small subunit.

sulfate, and extreme pH, and then can be separated by gel filtration
(Rutner and Lane, 1967). The large and small subunits have different
amino acid compositions and tryptic maps. The LS is also much more
hydrophobic than the SS. Upon the digestion with trypsin about 55
and 28 peptides were released from the large and small subunits
respectively (Kung et al., 1974). Recent DNA sequence analysis of the
gene coding for the LS of F1 protein from maize revealed that there
are a total of 475 amino acid residues per LS with a molecular weight
of 52,682 (McIntosh et al., 1980). It contains 11 sulfhydryl groups, 26
lysine, and 29 arginine residues similar to that of tobacco (Kung et al.,
1974).

Isoelectric focusing of S-carboxymethylated F1 protein from *N. tabacum* in polyacrylamide gels has resolved the subunits into their component polypeptides (Kung et al., 1974). The 8 LS are resolved into three
polypeptides each having a molecular weight of 5.5×10^4 d (Fig. 3).

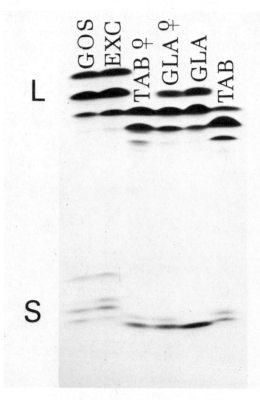

Figure 3. Genetic analysis of the carboxymethylated *Nicotiana* F1 protein by isoelectric focusing techniques. The isoelectric focusing pattern of large (L) and small (S) subunits of F1 protein are from (left to
right) *N. gossei*, *N. excelsior*, *N. tabacum* (female) x *N. glauca* (male),
N. glauca (female) x *N. tabacum* (male), *N. glauca*, and *N. tabacum*.
The large subunits consist of three polypeptides and the small subunits
vary from one to four polypeptides.

The 8 SS are resolved into two polypeptides each having an identical molecular weight of 1.3 x 10^4 and a different N-terminal amino acid sequence (Gibbon et al., 1975). In this case the two types of SS are randomly distributed in each molecule (Hirai, 1973).

F1 protein is the important photosynthetic enzyme RuBPCase, a unique enzyme having dual functions that either fix or lead to the evolution of CO_2. This enzyme catalyzes the addition of CO_2 and the cleavage of ribulose 1,5-bisphosphate (RuBP) to form two molecules of 3-phosphoglycerate. Since this enzyme also possesses oxygenase activity it produces one molecule each of phosphoglycolate and 3-phosphoglycerate (Jensen and Bahr, 1977; Lorimer, 1981). The phosphoglycolate serves as the substrate of photorespiration.

GENETICS AND SYNTHESIS. The reciprocal interspecific hybridization studies of *Nicotiana* spp. have been used to localize the genetic information of F1 protein (Fig. 3). The basis for this approach is that organelle DNA (chloroplasts and mitochondria) of many higher plants including tobacco is transmitted maternally. Therefore, any protein coded by organelle DNA is likely inherited uniparentally via the female parent. By using this approach it was demonstrated that the genes that code for the LS are inherited only from the maternal line and are therefore located in the chloroplast genome (Chan and Wildman, 1972). A similar genetic analysis showed that the coding information for SS of F1 protein is transmitted biparentally and is therefore contained in the nuclear DNA (Kawashima and Wildman, 1972). This has been confirmed by many studies using in vitro systems in which the LS of F1 protein was either synthesized by isolated chloroplasts (Blair and Ellis, 1973), purified cpDNA (Bottomley and Whitfeld, 1979), or a cloned chloroplast DNA fragment (Link et al., 1978; Gatenby et al., 1981). Furthermore, the chloroplast gene for the LS of F1 protein has been sequenced from maize (McIntosh et al., 1980), whereas the nuclear gene for the SS of F1 protein has been sequenced from pea (Bedbrook et al., 1980). Thus the separate location of the coding information for F1 protein has been firmly established.

The evidence for the site of synthesis of subunits of F1 protein was first provided by studies with inhibitors (Criddle et al., 1970). Chloramphenicol was reported to specifically inhibit the synthesis of the LS, whereas cycloheximide preferentially inhibits the synthesis of the SS. This suggested that the LS of F1 protein was synthesized within the chloroplast and the SS was synthsized at a separate site, probably in the cytoplasm. This suggestion was later confirmed by the evidence obtained from the cell-free system and immunological approaches (Gooding et al., 1973). It is now clearly established that the LS is coded for by a single chloroplast gene, its messenger RNA is a chloroplast RNA, relatively non-polyadenylated, and it is synthesized on chloroplast ribosomes (70S). The SS is coded by nuclear genes; its mRNA is a cytoplasmic RNA, and it is synthesized on cytoplasmic ribosomes (80S). The SS is synthesized in the form of a precursor polypeptide of about 20,000 d, and then transported into the chloroplast where it is processed and assembled into the holoenzyme. It has been postulated that a transit sequence on the N-terminal end of the SS directs transport and assembly with the LS (Schmidt et al., 1979). Many details of the sequence, uptake and processing of the precursor, have been elucidated (Chua et al., 1978). However, recent evidence also indicates that the mRNA for the LS of F1 protein can be translated in both eukaryotic and prokaryotic in vitro protein synthesizing systems (Bottomley and Whitfeld, 1979).

Since separate coding information as well as different sites of synthesis are required for F1 protein, proper control mechanisms must be exerted in such a fashion that the synthesis of large and small subunits are coordinated. Although there is no clear view of this controlling mechanism, the study of assembly of this protein inside the chloroplast and the free subunit pool may provide some insight. Currently, the suggestion of control mechanisms operating to coordinate the regulation of the synthesis of large and small subunits is primarily based on the results obtained from the studies using inhibitors. For example, the specific inhibition of cytoplasmic protein synthesis not only blocks synthesis of the SS but also synthesis of the LS of F1 proteir.. This is generally viewed as the evidence suggesting the existence of an in vivo control mechanism. When the pool of SS is depleted because of inhibition of synthesis, it in turn shuts down the synthesis of the LS. In this case it would imply that a nuclear gene remains in control of the overall rate of synthesis of F1 protein.

MOLECULAR EVOLUTION. The evolution of the subunit structure of F1 protein during speciation of *Nicotiana* has been systematically studied (Chen et al., 1976). The analysis of F1 protein from new species by isoelectric focusing shows that the LS is invariably identical to that of the female parent, as is expected because of the maternal inheritance of chloroplast genomes, and the SS is composed of polypeptides similar to both parents. Therefore, hybridization of two species each with a single but different SS polypeptide could give rise to a new F1 protein with two combined SS polypeptides, and a second round of interspecific hybridization could give rise to a F1 protein with four different SS. For example, *N. tabacum* F1 protein is known to have originated from a hybridization of *N. sylvestris* as the female parent, with *N. tomentosiformis* as the male parent (Gray et al., 1974); therefore, it is composed of the LS identical to *N. sylvestris* and two SS, one from each parent. Likewise, the F1 protein of amphidiploid *N. digluta*, the interspecific hybrid between *N. glutinosa* and *N. tabacum*, contains LS identical to *N. glutinosa* and four SS, two from each parent (Kung et al., 1976). This information permits inferences of the origin of existing F1 protein and the prediction of new F1 protein derived from breeder assisted interspecific hybridization.

Many cytoplasmic male sterile (CMS) lines of *N. tabacum* have F1 proteins consisting of different large but identical small subunits (Fig. 4). The LS are identical to the female parents from which these CMS lines were derived by interspecific hybridization, with *N. tabacum* as the male parent. The SS is identical to that of *N. tabacum* because of repeated backcrosses (Kung and Rhodes, 1978). Recently, a series of

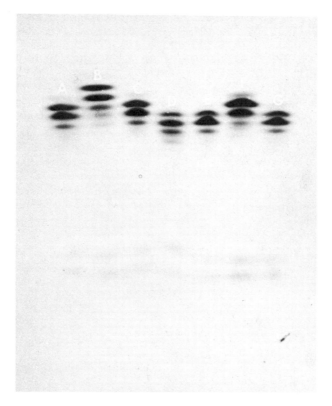

Figure 4.   Isoelectric focusing patterns of carboxymethylated F1 protein polypeptides.   Representing the variable large (L) and identical small (S) subunit combinations are: N. *tabacum* cv. Maryland 609 (A and C); cv. Burley 21 genome in N. *megalosiphon* cytoplasm (B); cv. Burley 21 (C); cv. Burley genome in N. *plumbaginifolia* cytoplasm (D and G); and cv. Hicks genome in N. *undulata* cytoplasm (F).

F1 protein has been genetically constructed in such a fashion that they all have identical LS by using a single species, N. *sylvestris*, as the female parent in all crosses.   Since the male parent in each cross has different SS, the resultant F1 protein is characterized by having identical large but different small subunits, the opposite situation as the CMS lines.

Physicochemical Properties and Procedure of Crystallization

PHYSICOCHEMICAL PROPERTIES.   Tobacco F1 protein possesses many unique physicochemical properties.   It is only soluble in the presence of its substrate, RuBP, or high salt (NaCl) and requires heat treatment to reach the maximum enzymatic activity.   These two properties provide the basis for the development of a simpler crystallization procedure (Lowe, 1977) than the one originally devised by Chan et al.

(1972). This procedure consists of three steps; namely, the breakage of chloroplasts in high salt concentration to release the protein, heating the crude filtrate to precipitate the undesirable material, and removal of the salt by gel filtration to yield crystalline protein. To date, tobacco F1 protein is the only known plant protein that can be prepared in crystalline form by such a simple procedure. The crystalline F1 protein is homogeneous as determined by sedimentation velocity (Fig. 1) and by polyacrylamide gel electrophoresis (Fig. 2). The crystalline protein contains no carbohydrate (Sakano et al., 1973), no tightly bound metals (Chollet et al., 1975), and no unusual amino acids. It has a sedimentation coefficient ($S_{20,w}^{o}$) of 18S, corresponding to the most commonly reported molecular weight of 5.6×10^5 d (Kung, 1976a). This protein assumes spherical shape in solution and has a density near 1.0 in a crystalline form. These and other physicochemical properties of the F1 protein are listed in Table 1 (Kung et al., 1980).

Table 1. Physicochemical Properties of F1 Protein

PHYSICOCHEMICAL PROPERTY		PLANT SOURCE
Sedimentation coefficient $S_{20,w}^{o}$	= 18.3 S	Tobacco
Molecular diameter (Å)	= 112	Tobacco
Molecular weight (daltons)	= 5.6×10^5	Tobacco
Partial specific volume (cm/g)	= 0.7	Tobacco
Wet crystal density (g/cm^3)	= 1.058-1.095	Tobacco
Water content of wet crystal (%)	= 80	Tobacco
$A_{280\,nm}/A_{260\,nm}$	= 1.92	Tobacco
Diffusion constant $d_{20,w}^{o}$(cm^2/sec)	= 2.93×10^{-7}	Spinach
Frictional coefficient (f/f^o)	= 1.11	Spinach
Extinction coefficient ($E_{1cm}^{1\%}$)	= 14.1	Spinach

PROCEDURE OF CRYSTALLIZATION. The crystallization procedure described below is for large-scale operation. This is based on the method developed by Lowe (1977) and requires 200-400 g of leaves.

Young tobacco leaves grown in a greenhouse (4-5 weeks after germination) are homogenized with 0.01 ml 2.0 M NaCl plus 2-mercaptoethanol per gram leaf in a Waring blender. With the blender at low speed, leaves are gradually added as pieces. After all leaves are added, the slurry is blended at high speed for 30-50 sec. The resulting slurry is carefully squeezed through four layers of cheesecloth and one layer of Miracloth. The green filtrate is immediately heated in a 50 C water bath with occasional stirring until the filtrate reaches 45-48 C. The suspension is rapidly cooled in an ice bath to 15-20 C. Then 2 ml of 10% Na$_2$EDTA is added for every 100 ml of filtrate and the pH adjusted to 7.5 with 1.0 M Tris (about 4-5 ml per 100 ml of filtrate). The green suspension is centrifuged for 20 min at 16,000 x g. The supernatant is carefully decanted and passed through a Sephadex G-50 column equilibrated with 25 mM Tris-HCl pH 7.9, 0.2 mM EDTA. Two fractions are collected when protein starts to appear (as detected by with 5% TCA) in the effluent. A large fraction of effluent is collected first (about 70-80% of the protein fraction). The remainder or the protein fraction of effluent is collected separately. The protein concentration in the small fraction is rather low, and sometimes this fraction fails to yield crystals. Usually, crystals are formed in the column or will be formed overnight at 4 C. The crystals appear having a hexagonal shape (Fig. 5). A yield of 2-3 mg of crystals per gram of fresh weight of leaves can be expected. Protein concentration, as mg per ml, is calculated by the factor $A_{280\,nm}$ x 0.7 (1 cm light path). The column can be cleaned by washing with distilled water to remove salt and pigments.

In some cases, only a small quantity of leaf material can be obtained. If the material is less than 50 g, the crystallization procedure originally developed by Chan et al. (1972) can be employed. In a small Waring blender, or by using a pestle and mortar, 20-30 g of leaves are homogenized in 1-2 volumes of 0.05 M Tris-HCL (pH 7.4) buffer containing 1.0 M NaCl. Filter the paste through cheesecloth and Miracloth and squeeze to obtain juice, which is centrifuged at 17,000 g for 5 min. Remove any floating material; decant the supernatant, and centrifuge it at 17,000 g for 30 min. Concentrate the supernatant to about 5 ml by ultrafiltration (Amicon). The resulting concentrate is placed in a collodion dialysis bag and subjected to 0.025 M Tris-HCl (pH 7.4) buffer containing no NaCl. Crystals usually appear overnight at 4 C.

A procedure for microscale preparation of *Nicotiana* F1 protein was developed by Uchimiya et al. (1979a). F1 protein contained in as little

Figure 5. The photomicrograph of crystals of tobacco F1 protein. The largest crystal is about 0.33 mm in diameter.

as 300-500 mg of leaves can be precipitated by antiserum. The anti-body-protein complex can then be carboxymethylated and subjected to isoelectric focusing. This method is described briefly as follows:

Leaf tissues weighing 1-2 g are homogenized manually with 1-2 ml of 0.05 M Tris-HCl (pH 7.4) buffer using a Teflon tissue grinder. The resulting slurry is centrifuged at 8,000 x g for 15 min to remove the debris. To every ml of buffer used 0.3 ml of the anti-F1 protein serum is added to the supernatant. After incubation for 1 hr at 37 C it is incubated at 4 C overnight. The antibody-protein complex is collected by centrifugation at 3,000 x g for 15 min. The precipitate is washed twice by resuspension in the same buffer and centrifugation at 3,000 x g for 15 min. The washed antibody-F1 protein pellet is solubilized in 0.1 ml of 0.5 M Tris-HCl buffer (pH 8.5) containing 8 M urea, 1 mM EDTA and 5 mg dithiothreitol (DTT). This solution or suspension is placed in a 1 cm x 7.5 cm test tube sealed with a rubber cap to maintain an N_2 atmosphere for carboxymethylation.

Analysis of F1 Protein by Isoelectric Focusing

PRINCIPLE. The principle of isoelectric focusing is straightforward. Proteins migrate through a stable pH gradient under the influence of an electric potential. Migration ceases when a protein reaches its isoelectric point in the gradient. Every protein has an isoelectric point, pl, which is the pH value at which its net charge is zero. If the protein is added to a solution with a pH higher than pl, it loses protons and becomes negatively charged. Conversely, if the protein is in an environment with a pH lower than pl, it will capture protons and become positively charged. In an electric field, positively charged proteins migrate toward the cathode and negatively charged proteins toward the anode.

When a sample of proteins with different isoelectric points within the pH range of the gradient is applied to the system, the protein molecules acquire different charges according to the pl of each protein. Applying a voltage across such a gradient results in each protein molecule migrating toward the pH value in the gradient where its net charge is zero. The proteins are thus exactly focused at the point where the pH is equal to the pl. Each protein zone remains sharpened in its position as long as the pH gradient is stable and the voltage is maintained. In gel isoelectric focusing, therefore, no matter how each protein is distributed at the beginning of a run, it always ends up at its isoelectric point. Thus the proteins can be applied either by mixing with the gel solution or by loading on to the top of the gel.

PROTOCOLS

1. S-Carboxymethylation of F1 Protein. Approximately 3-5 mg of protein is dissolved with 0.6 ml of 0.05 M Tris-HCl pH 8.5, 18.0 M urea, 0.001 M EDTA in a 1.0 x 7.5 cm tube. The tube is then

sealed, and flushed with N_2 for 5 min. Then 0.2 ml of Tris-urea medium containing 3-5 mg of DTT is carefully injected into the sealed tube and the solution is incubated for 2 hr at 25 C. The tube is then covered with aluminum foil to exclude light. Next, 0.2 ml of Tris-urea medium, containing 9-15 mg of iodoacetic acid or iodoacetamide, is injected. After 15 min further incubation at 25 C the reaction mixture is passed through a 1 x 15 cm Sephadex G-25 column previously equilibrated with Tris-urea medium. The protein fraction is eluted with Tris-urea medium and collected using 5% TCA as an indicator (Kung et al., 1974).

2. Isoelectric Focusing. After installing the gel in a vertical slab apparatus, 200 ml of 0.2% H_2SO_4 are placed in the bottom trough and 200 ml of 0.4% triethanolamine in the top. The pH gradient is established during a 2 hr prerun at 5 mA. Since electrical resistance rises during formation of the pH gradient, the voltage must be increased gradually to maintain a constant current until the gradient is established. This takes about 15 min. After 2 hr prerun samples can be applied to each slot of the gel. Isoelectric focusing may be carried out for 4 hr or overnight at 25 C, maintaining a maximum current of 5 mA and up to a maximum voltage of about 300 V (Kung et al., 1974).

3. Application of Sample. Top loading is preferred to mixing samples with the bulk acrylamide solution. The protein concentration of each sample after S-carboxymethylation is determined, and 15-20 µg are applied in 10-20 µl per slot. No sucrose solution is needed, because the proteins are in the Tris-urea medium. A Hamilton syringe is used for sample application.

4. Staining of Protein. The gel is removed at the conclusion of the run and stained with 0.2% bromphenol blue in ethanol-acetic acid-water (10:1:9, v/v/v) for 1 hr and destained in ethanol-acetic acid-water (90:15:195, v/v/v). Protein dyes such as amido black should not be applied directly to the gel, since the carrier ampholytes stain strongly. However, ampholytes can first be removed by repeated washing with 5% TCA.

Application of F1 Protein as a Unique Genetic Marker

PRACTICAL APPLICATIONS. The requirement of both nuclear and chloroplast genomes for the biosynthesis of F1 protein not only offers the opportunity to study the cooperative interaction between nuclear and cytoplasmic systems, but also provides a unique genetic marker for both genomes. This is particularly useful in identifying hybrid plants derived from somatic cell fusions as well as in studying the fate and distribution of organelles in the fused product where the coexistence of two cytoplasms is a distinct possibility (Izhar and Tabib, 1980). In the absence of an effective selective marker to identify somatic hybrids, the analysis of the isoelectric focusing pattern of F1 protein has become a standard procedure to verify the hybrid nature of the regenerated plants. This technique was first applied to identify the parasexual hybrid plants produced by the fusion of protoplasts from N. glauca and

N. langsdorffii (Chen et al., 1977), as well as the combination of *N. suaveolens* nuclei with chloroplasts of *N. tabacum*. The expression of both nuclear and chloroplast genomes in these parasexual hybrid plants was examined by an analysis of the isoelectric focusing pattern of F1 protein. Figure 6 shows the polypeptide composition of the large and small subunits of F1 protein prepared from the leaves of the parasexual hybrid of *N. glauca* and *N. langsdorffii*, produced by protoplast fusion. This is compared with the two parental species and an artificial mixture of the parental proteins. The results suggest that the nuclear genes for the small subunits of both species are equally expressed in this parasexual hybrid, whereas only the chloroplast genome for the large subunit of *N. glauca* is expressed. This confirms the hybrid nature of the plant produced by somatic cell fusion. It is also evident from the analysis of F1 protein that the nuclear and chloroplast genomes from both *N. suaveolens* and *N. tabacum* are present and equally expressed in the hybrid plant (Kung et al., 1975). Whether the two different chloroplast populations exist in the same cell or in different cells of chimeral tissue is not certain in this case. However, it has later been demonstrated, by using F1 protein, that different populations of chloroplast do not coexist within the same cell (Chen et al., 1977). The value of F1 protein as a unique genetic marker for both nuclear and chloroplast genomes has thus been clearly illustrated.

Figure 6. The isoelectric focusing pattern of F1 protein prepared from the parasexual hybrid of *N. glauca* and *N. langsdorffii*. (A) *N. langsdorffii*, (B) parasexual hybrid of *N. glauca* and *langsdorffii*, (C) *N. langsdorffii* F1 protein, (D) equal mixture of *N. glauca* and *N. langsdorffii* F1 protein.

The application of F1 protein as a genetic marker is not limited in identifying the sexual and somatic hybrid plants but also can be applied to determine the origin, evolution, and speciation of many plants. In the case of identifying hybrid plants, the application is widely adapted. For example, it has been successfully used in recent years to demonstrate the hybrid nature of plants regenerated from the interspecific and intergeneric somatic cell fusion of protoplasts. Somatic hybrid plants produced by fusion of protoplasts between different species of tobacco (Kung et al., 1975), different species of petunia (Kumar et al., 1980), as well as between different genera such as tomato and potato (Melchers et al., 1978) have been identified by using F1 proteins. It is very clear from analysis of the LS of F1 protein in the somatic hybrid that only chloroplasts from one species of tobacco (Chen et al., 1977) and petunia (Kumar et al., 1980) are present in the hybrid plants and are being expressed. However, both species have an equal chance for presence and expression (Chen et al., 1977). A similar observation was reported from the potato–tomato somatic hybrids (Melchers et al., 1978). It seems true that although it is possible to combine two distinct chloroplast populations in one cytoplasm, it is rather difficult to keep them functionally together.

There is a large body of evidence in using F1 protein to study the evolution of many plant species. The study of the origin of *N. tabacum* (Gray et al., 1974) and other new *Nicotiana* species (Kung et al., 1976) illustrated the first successful application. Subsequently, it has been employed to study the origin of polyploid wheats (Chen et al., 1975b), of European potato (Gatenby and Cocking, 1978a), of amphidiploid Brassica (Uchimiya and Wildman, 1978), the evolution of the genus Lycopersicon (Gatenby and Cocking, 1977; Kumar et al., 1980), and some other Angiosperm groups (Uchimiya et al., 1977; Chen and Wildman, 1981). Furthermore, F1 protein has also been employed as a powerful genetic marker for studying chloroplast uptake, inheritance, and distribution in the somatic hybrids as well as possible mechanisms of cytoplasmic male sterility (Chen et al., 1975a; Kung, 1976b; Chen and Meyer, 1979; Glimelius et al., 1981).

POTENTIAL AND LIMITATIONS. The success of using F1 proteins as genetic markers depends entirely on differences in the isoelectric focusing pattern of the large and small subunits. Without such distinct differences, the value of F1 protein as a genetic marker is diminished. In *Nicotiana* there are four types of large and thirteen types of small subunits among the 64 species that have been examined (Chen et al., 1976). The limited differences in isoelectric focusing pattern among the LS of F1 protein confined its usefulness as a genetic marker for chloroplast genomes to differences between four groups. However, the identical isoelectric points of polypeptides do not necessarily reflect the identity of their primary structure. This is because changes in amino acid composition or sequence without a change in overall net charge of the polypeptides would not alter the isoelectric point. This can be illustrated by demonstrating the differences in the amino acid composition and peptide maps of four F1 proteins exhibiting identical isoelectric focusing patterns (Kung et al., 1977).

The advantages of using F1 protein as a genetic marker in *Nicotiana* are (1) the simple crystallization procedure whereby a large quantity of pure crystalline F1 protein can be easily obtained and (2) a one-step operation for comparison of the isoelectric focusing patterns of the large and small subunits.   Such advantages are nullified when the difference between the large and small subunits of two species can be detected only by amino acid analysis or peptide mapping.   Furthermore, the instability and variation of the isoelectric focusing pattern of the LS of F1 protein can make the identification of two closely related LS difficult.   It is known that there is a single gene for the LS of F1 protein per chloroplast genome and that the multiple polypeptide bands resolved from each LS are the results of a combination of post-translational modifications and oxidation of the thiol groups (Kung, 1976a). Consequently, the number of polypeptides detected from the LS of different plant species or from different preparations of the same species can vary from two to as many as twenty depending on how the F1 protein was pretreated (Kung et al., 1974).   This phenomenon deserves serious consideration when the isoelectric focusing pattern of F1 protein is used as genetic marker for chloroplast genomes.   However, the isoelectric focusing pattern of the SS of F1 protein is reproducible. Each band represents the product of a single nuclear gene.   Since there are only a few thiol groups per small subunit, the randomized oxidation in this case does not constitute a serious problem.

CHLOROPLAST DNA AS GENETIC MARKER

Physicochemical Properties and Organization

Chloroplast DNA is a double-stranded structure having a high molecular weight of $95-100 \times 10^6$ d (Kung, 1977; Bedbrook and Kolodner, 1979).   It has an average buoyant density of $1.698-1.700$ g/cm^3 corresponding to a GC content of 38-41% (Kung, 1977).   In all cases the buoyant density of cpDNA is indistinguishable from that of the nuclear DNA.   However, cpDNA possesses some unique features.   To date, no detectable 5-methyl cytosine can be found in the cpDNA from higher plants, whereas the corresponding nuclear DNA invariably contains 4-5%.   Another unique property of cpDNA is the ease with which it can be renatured.   This is a reflection of the degree of simplicity and homogeneity of cpDNA which is a much less complicated molecule than nuclear DNA, both in structural complexity and in genetic content.

Chloroplast DNA is circular in form and has a contour length of 50 µm.   There are multiple copies per organelle, ranging from a few to over 100, existing in different forms; circular, supercoil circles and circular dimers (Bedbrook and Kolodner, 1979).   Each circle can be divided into four regions, the inverted repeat regions which are separated by a small and a large single-copy sequence with only a few exceptions (Koller and Delius, 1980).   The inverted repeat is usually 20-25 kilobases (Kb) in length and contains the genes for ribosomal and transfer RNAs (Bedbrook and Kolodner, 1979).   The single-copy regions contain the genes for many chloroplast proteins and transfer RNAs (Bedbrook and Kolodner, 1979). Its coding capacity may well exceed 100 poly-peptides of 4×10^4 d in size; over 90 of them have been recently demonstrated (Ellis, 1981). However, only a few polypeptides have been identified as the product of chloroplast genes, including the LS of RuBPCase (Kung, 1976b, 1977), the 32,000 d thylakoid membrane protein (Bedbrook et al., 1978), some subunits of coupling factor (Nelson et al., 1980), cytochrome f (Doherty and Gray, 1979), some ribosomal proteins (Mets and Bogorad, 1972), and a protein associated with function of the triazine herbicides (Steinbach et al., 1981).

Chloroplast DNA Diversity and Evolution

Based on a number of biochemical and functional considerations cpDNA of higher plants has long been thought to be highly conserved. This concept originated from the observations that cpDNA exists in multiple copies per chloroplast, contains similar coding information, and exhibits uniform physicochemical properties. This view should be reex-amined in the light of the current evidence. The results obtained from restriction enzyme analysis revealed a wide range of variability of cpDNA. Tobacco cpDNA exhibits a high degree of diversity while re-taining similarity for many bands. For instance, the individual restric-tion pattern is species specific for any given species, while the general configuration is characteristic of the genus *Nicotiana* (Kung et al., 1981).

Among the many restriction enzymes used, EcoR1 produces the larg-est number of fragments and therefore has the highest resolving power to uncover differences. Of the 40 fragments generated by EcoR1 enzyme, the first 10 are certainly worth noting (Kung et al., 1982). Fragments 1-3 are stable and present in all 40 *Nicotiana* species exam-ined so far. In contrast, fragments 4 and 5 are extremely variable. One or both can be altered; one or both may be absent. Thus, a com-bination of diversity and similarity in fragment pattern forms the basis of species specificity and overall identity of tobacco cpDNA (Fig. 7).

The fragment pattern generated by BamH1 revealed a similar degree of diversity of tobacco cpDNA. For example, *N. gossei* and *N. otopho-ra* cpDNAs differ in 13 out of the 27 BamH1 bands (Fig. 8). This matches the extent of variation in their EcoR1 fragments in which 18 of the 40 bands are different. Therefore, both EcoR1 and BamH1 can be used to accurately measure the degree of diversity in tobacco cpDNA.

Alteration of DNA is the main force of evolution. Judged from the wide diversity of tobacco cpDNA, it is evident that there have been considerable changes through the course of evolution. It has been demonstrated in tobacco that the mechanism of cpDNA alteration in-volves point mutation, inversion, deletion, duplication, and recombina-tion (Kung et al., 1981). Point mutations are primarily responsible for the observed gain and elimination of many restriction sites. They occur rather frequently in relation to other mechanisms and are clus-tered in one region which can be described as a hot spot (Kung et al., 1982). A large insertion involving a segment of about 7×10^6 d has

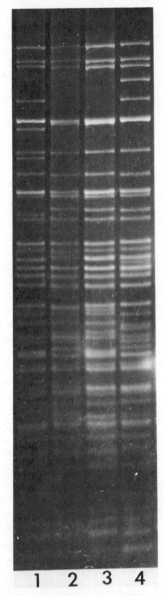

1 2 3 4

Figure 7.   EcoR1 restriction fragment pattern of cpDNA showing the
large number of fragments generated (Kung et al., 1981).   (1) *N. gossei*,
(2) *N. tabacum*, (3) *N. langsdorffii*, and (4) *N. otophora*.

also been discovered (Shen et al., 1982).   Whether this segment, which
accounts for 7% of the tobacco chloroplast genome, contains any structural genes is not known.   The existence of inverted and tandem repeats suggests that there is some cpDNA duplication (Bedbrook and
Kolodner, 1979).   Recombination has also been suggested but only as a
rare event (Kung et al., 1981).

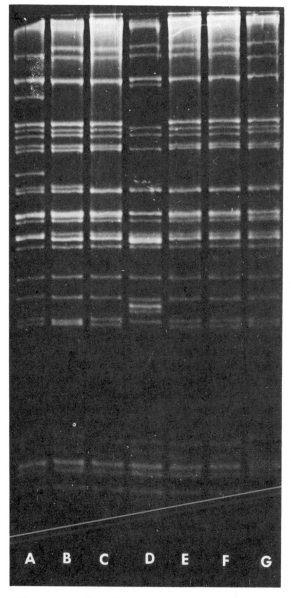

Figure 8.   BamH1 restriction fragment patterns of cpDNA illustrating the species specificity within a common realm of similarity (Kung et al., 1981).   (A) *N. gossei*, (B) *N. glauca*, (C) *N. tabacum*, (D) *N. otophora*, (E) *N. sylvestris*, (F) *N. langiflora*, and (G) *N. langsdorffii*.

The limited differences in the restriction patterns generated by Sma 1 provide excellent markers for phylogenetic studies.   It is a better system than the isoelectric focusing patterns of F1 protein for identification of evolutionary steps in cpDNA.   By using a single restriction enzyme, Sma 1, eliminations and sequential gains of recognition sites during the course of cpDNA evolution are clearly demonstrated (Fig. 9).

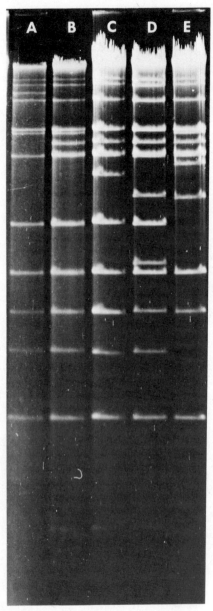

Figure 9.   SmaI restriction fragment patterns of cpDNA demonstrating the gains and eliminations of its restriction sites.   The limited differences of SmaI site are very valuable in the study of phylogeny (Kung et al., 1982).   (A) *N. plumbaginifolia,* (B) *N. langsdorffii,* (C) *N. tabacum,* (D) *N. otophora,* and (E) *N. tomentosa.*

Purification and Restriction Analysis of Chloroplast DNA

Following filtration through one layer of miracloth and centrifugation at 1500 x g for 15 min in a GSA rotor, the crude pellets were gently resuspended in isolation buffer and layered onto discontinuous silica sol gradients for isopycnic banding of intact chloroplasts. These gradients consisted of four layers of 0.25 M sucrose, 50 mM Tris-HCl (pH 8.0), 5 mM $MgCl_2$, 1% bovine serum albumin, and 10 mM mercaptoethanol containing Ludox AM at 52.6, 36.8, 18.4, and 9% (v/v) for densities of 1.20, 1.16, 1.11, and 1.08 g/cm^3 respectively. Prior purification of Ludox AM by passage over 14 mesh charcoal was not necessary. The silica sol gradients were centrifuged at 2,600 x g for 30 min in an HB-4 rotor. The resulting chloroplast bands were collected, diluted with isolation buffer, and centrifuged at 1,500 x g for 10 min to pellet the organelles. Chloroplasts recovered from these gradients were then lysed in the presence of 2% sarkosyl, and DNA was isolated as described by Kolodner and Tewari (1975).

The cpDNA can be digested with various restriction enzymes, as directed by the supplier. In order to achieve a complete digestion of cpDNA, it is recommended that 2-3 times more restriction enzymes than required be used to digest the same amount of λDNA. At a given concentration of cpDNA, each molecule will be cut by a particular restriction enzyme at precisely the same sites, yielding as many subsets of identical molecular fragments as there are cleavage sites on the original molecule. It is known that cpDNA contains a variable amount of restriction sites for different enzymes ranging from 10-40 (Kung et al., 1982). These subsets of fragments can be conveniently separated from one another according to size by gel electrophoresis. Generally, the cleaved DNA is placed in a rectangular slot near one end of a slab of agarose gel (0.8 or 1.5%). When an electric current is applied, the DNA fragments migrate through the gel toward the positive electrode at a rate inversely proportional to the log of their molecular length. Each subset of identical fragments forms a narrow band whose position can be visualized by staining the gel with a fluorescent dye such as ethidium bromide and photographing it under UV light.

Since the large fragment cannot penetrate the agarose gel or migrate very slowly, a low percentage gel should be used. It is a general guide that if there are fewer than 15 fragments generated by a given restriction enzyme from a higher plant chloroplast genome, then a 0.7-0.8% gel is adequate. If the number of fragments are over 25, then a 1.0-1.5% gel is suitable. The intensity of some bands are higher than others because of the existence of multiple copies having a similar or identical size. The exact number of copies in such bands can be estimated from the densitometer tracings.

The Potential and Limitation of Applying Chloroplast DNA as Genetic Marker

Chloroplast DNA is potentially a better genetic marker than F1 protein, at least in the genus of *Nicotiana*. There are only four groups of isoelectric focusing patterns of the LS of F1 protein. The restriction fragment pattern of *Nicotiana* cpDNA is almost species specific. Therefore, this species specificity can be used as genetic marker to identify each species. However, leaves contain only a very small amount of cpDNA, and a large quantity of material and tedious work are required to prepare sufficient amount of sample for any effective analysis. Therefore, in some cases, it is not practical to use it as a genetic marker.

CONCLUDING REMARKS

The existence of easily distinguishable isoelectric focusing patterns of the large and small subunits of F1 protein makes it a unique genetic marker for both chloroplast and nuclear genomes. It has been widely used to identify the hybrid nature of plants produced by somatic cell fusion as well as to trace the evolutionary origin of many plant species. However, the usefulness is not without limitations. It can be used only for leaf tissues that contain chloroplasts. This confines its application to the identification of plants derived from protoplast fusion and cannot be used as the type of selective marker that is urgently needed in somatic cell fusion studies. Moreover, it is a valuable genetic marker only in the cases where differences in the large and small subunits of F1 protein exist. Without such differences its usefulness is diminished, which is true in many instances. In the case of *Nicotiana* there are over 60 species, and only four groups of isoelectric focusing patterns of the LS of F1 protein have been identified. Likewise, the degree of variations of LS of F1 protein in other genera is even more limited. However, this deficiency may be compensated by using cpDNA as an additional genetic marker for the chloroplast genome in cases where no difference between two LS of F1 proteins can be detected. It has been demonstrated recently that *Nicotiana* cpDNAs exhibit a high degree of variability in their restriction fragment patterns. The limitation for this approach is restricted by the fact that a large quantity of leaf material is required for DNA isolation and purification, and the procedure is rather time consuming when compared with the analysis of F1 protein. In this chapter the potential and limitations of both F1 protein and cpDNA as genetic marker are discussed.

ACKNOWLEDGMENT

The work described here was supported by NIH grant GM 27746-01 and USDA cooperative agreement 58-3244-0-157 from the tobacco laboratory.

KEY REFERENCES

Chen, K., Wildman, S.G., and Smith, H.H. 1977. Chloroplast DNA distribution in parasexual hybrids as shown by polypeptide composition of fraction 1 protein. Proc. Natl. Acad. Sci. 74:5109-5112.

Kolodner, R. and Tewari, K.K. 1975. The molecular size and conformation of the chloroplast DNA from higher plants. Biochem. Biophys. Acta 402:372-390.

Kung, S.D. 1976b.   Tobacco fraction 1 protein: A unique genetic marker.   Science 191:429–434.

Rhodes, P.R. and Kung, S.D. 1981.   Chloroplast DNA isolation: Purity achieved without nuclease digestion.   Can. J. Biochem. 59:911–915.

REFERENCES

Andrews, T.J., Lorimer, G.H., and Tolbert, N.E. 1973.   RuBP oxygenase.   I. Synthesis of phosphoglycolate by fraction 1 protein of leaves. Biochemistry 12:11–18.

Baker, T.S., Eisenberg, D., Eiserling, F.A., and Weissman, L. 1975.   The structure of form I crystals of RuBPCase.   J. Molec. Biol. 91:391–399.

Bedbrook, J.R. and Kolodner, R. 1979.   The structure of chloroplast DNA.   Annu. Rev. Plant Physiol. 30:593–620.

Bedbrook, J.R., Link, G., Coen, D.M., Bogorad, L., and Rich, A. 1978. Maize plastid gene expressed during photoregulated development. Proc. Natl. Acad. Sci. 75:3060–3064.

Bedbrook, J.R., Smith, S.M., and Ellis, J.R. 1980.   Molecular cloning and sequence of cDNA encoding the precursor to the small subunit of chloroplast RuBPCase.   Nature 287:692–697.

Blair, G.E. and Ellis, R.J. 1973.   Protein synthesis in chloroplasts I. Light-driven synthesis of the large subunit of fraction 1 protein by isolated pea chloroplasts.   Biochem. Biophys. Acta 319:223–234.

Bottomley, W. and Whitfeld, P.R. 1979.   Cell-free transcription and translation of total spinach chloroplast DNA.   Eur. J. Biochem. 93: 31–39.

Bovien, B. and Mayer, F. 1978.   Further studies on the quartenary structure of ribulose 1,5-bisphosphate carboxylase from *Alcoligenes eutrophus*.   Eur. J. Biochem. 88:97–107.

Chan, P.H. and Wildman, S.G. 1972.   Chloroplast DNA codes for the primary structure of the large subunit of fraction 1 protein.   Biochem. Biophys. Acta 277:677–680.

Chan, P.H., Sakano, K., Sigh, S., and Wildman, S.G. 1972.   Crystalline fraction 1 protein: Preparation in large yield.   Science 176:1145–1146.

Chen, K. and Meyer, V.G. 1979.   Mutation in chloroplast DNA coding for the large subunit of fraction 1 protein correlated with male sterility in cotton.   J. Hered. 70:431–433.

Chen, K. and Wildman, S.G. 1981.   Differentiation of fraction 1 protein in relation to age and distribution of Angiosperm groups.   Plant Syst. Evol. 138:89–113.

Chen, K., Kung, S.D., Gray, J.C., and Wildman, S.G. 1975a.   Polypeptide composition of fraction 1 protein from *Nicotiana glauca* and from cultivars of *Nicotiana tabacum*, including a male sterile line.   Biochem. Genet. 13:771–778.

Chen, K., Gray, J.C., and Wildman, S.G. 1975b.   Fraction 1 protein and the origin of polyploid wheats.   Science 190:1304–1306.

Chen, K., Johal, S., and Wildman, S.G. 1976.   Role of chloroplast and nuclear genes during evolution of fraction 1 protein.   In: Genetics and Biogenesis of Chloroplasts and Mitochondria. (Th. Bucher, W. Neupert, W. Sebald, and S. Werner, eds.) pp. 3-11. Elsevier/North-Holland Biomedical Press, Amsterdam.

Chollet, R., Anderson, L.L., and Hovspian, L. 1975. The absence of tightly bound copper, iron, and flavin nucleotides in crystalline RuBPCase/oxygenase from tobacco. Biochem. Biophys. Res. Comm. 64:97-107.

Chua, N.H., Schmidt, G.W., and Maltin, K.S. 1978. In vitro synthesis transport and assembly of RuBPCase subunits. In: Photosynthetic Carbon Assimilation (H.W. Siegelman and G. Hind, eds.) pp. 325-346. Plenum, New York.

Cocking, E.C., Davey, M.R., Pental, D., and Power, J.B. 1981. Aspects of plant genetic manipulation. Nature 293:265-270.

Criddle, R.S., Dann, B., Kleinkopf, G.E., and Huffaker, R.C. 1970. Differential synthesis of RuBPCase subunits. Biochem. Biophys. Acta 41: 621-627.

Doherty, A. and Gray, J.C. 1979. Synthesis of cytochrome f by isolated pea chloroplasts. Eur. J. Biochem. 108:131-136.

Dorner, R.W., Kahn, A., and Wildman, S.G. 1957. The protein of green leaves VII. Synthesis and decay of the cytoplasmic protein during the life of tobacco leaf. J. Biol. Chem. 229:945-951.

Ellis, J. 1981. Chloroplast proteins: Synthesis, transport and assembly. Annu. Rev. Plant Physiol. 32:111-137.

Gatenby, A.A. and Cocking, E.C. 1977. Polypeptide composition of fraction 1 protein subunits in the genus *Petunia*. Plant Sci. Lett. 10: 97-103.

_____ 1978a. Fraction 1 protein and the origin of the European potato. Plant Sci. Lett. 12:177-181.

_____ 1978b. The polypeptide composition of the subunit of Fraction 1 proteins in the genus *Lycopersicon*. Plant Sci. Lett. 13:171-176.

Gatenby, A.A., Castleton, J.A., and Saul, M.W. 1981. Expression in *E. coli* of maize and wheat chloroplast genes for large subunits of RuBPCase. Nature 291:117-121.

Gibbon, G.C., Strokbaek, S., Haslet, B., and Boulter, D. 1975. The N-terminal amino acid sequence of the small subunit of RuBPCase from *Nicotiana tabacum*. Experientia 31:1040-1041.

Glimelius, K., Chen, K., and Bonnett, H.T. 1981. Somatic hybridization in *Nicotiana*: Segregation of organellar traits among hybrid and cybrid plants. Planta 153:504-510.

Gooding, L.R., Roy, H., and Jagendorf, A.T. 1973. Immunological identification of nascent subunits of wheat RuBPCase on ribosomes of both chloroplast and cytoplasmic origin. Arch. Biochem. Biophys. 159:324-355.

Gray, J.C., Kung, S.D., Wildman, S.G., and Sheen, S.J. 1974. Origin of *Nicotiana tabacum* L. detected by polypeptide composition of fraction 1 protein. Nature 252:226-227.

Hirai, A. 1977. Random assembly of different kinds of small subunit polypeptides during formation of fraction 1 protein macromolecules. Proc. Natl. Acad. Sci. 74:3443-3445.

Izhar, S. and Tabib, Y. 1980. Somatic hybridization in petunia. Theor. Appl. Genet. 57:241-245.

Jensen, R.G. and Bahr, J.T. 1977. Ribulose, 1,5-bisphosphate carboxyl-ase-oxygenase. Annu. Rev. Plant Physiol. 28:379-400.

Kawashima, N. and Wildman, S.G. 1972. Studies on fraction 1 protein VI. Mode of inheritance of primary structure in relation to whether chloroplast or nuclear DNA contains the code for chloroplast pro-teins. Biochem. Biophys. Acta 262:42-49.

Koller, B. and Delius, H. 1980. *Vicia faba* chloroplast DNA has only one set of ribosomal RNA genes as shown by partial denaturation mapping and R-Loop analysis. Molec. Gen. Genet. 178:261-269.

Kumar, A., Wilson, D., and Cocking, E.C. 1980. Polypeptide composi-tion of fraction 1 protein of the somatic hybrid between *Petunia parodii* and *Petunia parviflora*. Biochem. Genet. 19:255-261.

Kung, S.D. 1976a. Isoelectric points of the polypeptide components of tobacco fraction 1 protein. Bot. Bull. Acad. Sinica 17:185-191.

_____ 1977. The expression of chloroplast genomes in higher plants. Annu. Rev. Plant Physiol. 28:401-437.

Kung, S.D. and Rhodes, P.R. 1978. Interaction of chloroplast and nu-clear genomes in regulating RuBPCase activity. In: Photosynthetic Carbon Assimilation (H.W. Siegelman and G. Hind, eds.) pp. 307-324. Plenum, New York.

Kung, S.D., Sakano, K., and Wildman, S.G. 1974. Multiple peptide com-position of the large and small subunits of *N. tabacum* fraction 1 protein ascertained by finger printing and electrofocusing. Biochem. Biophys. Acta 365:138-147.

Kung, S.D., Gray, J.C., Wildman, S.G., and Carlson, P.S. 1975. Polypep-tide composition of Fraction 1 protein from parasexual hybrid plants in the genus *Nicotiana*. Science 187:353-355.

Kung, S.D., Sakano, K., Gray, J.C., and Wildman, S.G. 1976. The evo-lution of fraction 1 protein during the origin of a new species of *Nicotiana*. J. Molec. Evol. 7:59-64.

Kung, S.D., Lee, C., Wood, D.D., and Moscarello, M.A. 1977. Evolution-ary conservation of chloroplast genes coding for the large subunit of Fraction 1 protein. Plant Physiol. 60:89-94.

Kung, S.D., Chollet, R., and Marsho, T.V. 1980. Crystallization and assay procedures of tobacco RuBPCase/oxygenase. Meth. Enzym. 69: 326-336.

Kung, S.D., Zhu, Y.S., Chen, K., Shen, G.F., and Sisson, V. 1981. *Nic-otiana* chloroplast genome II. Chloroplast DNA alteration. Molec. Gen. Genet. 183:20-24.

Kung, S.D., Zhu, Y.S., and Shen, G.F. 1982. *Nicotiana* chloroplast gen-ome III. Chloroplast DNA evolution. Theor. Appl. Genet. 61:73-79.

Link, G.L., Bogorad, L., Bedbrook, J.R., Coen, D.M., and Rich, A. 1978. The expression of the gene for the large subunit of RuBPCase in maize. In: Photosynthetic Carbon Assimilation (H.W. Siegelman and G. Hind, eds.) pp. 349-362. Plenum, New York.

Lorimer, G.H. 1981. The carboxylation and oxygenation of RuBP: The primary events in photosynthesis and photorespiration. Annu. Rev. Plant Physiol. 32:349-383.

Lowe, R.H. 1977. Crystallization of fraction 1 protein from tobacco by a simplified procedure. FEBS Lett. 78:98-100.

McIntosh, L., Poulsen, C., and Bogorad, L. 1980. Chloroplast gene sequence for the large subunit of ribulose bisphosphate carboxylase of maize. Nature 288:556-560.

Melchers, G., Sacristan, M.D., and Holder, A.A. 1978. Somatic hybrid plants of potato and tomato generated from fused protoplasts. Carlsberg Res. Comm. 43:203-218.

Mets, L. and Bogorad, L. 1972. Altered chloroplast ribosomal proteins associated with erythiomycin-resistant mutants in two genetic systems of *Chlamydomonas reinhardtii*. Proc. Natl. Acad. Sci. 69:3779-3783.

Nelson, N., Nelson, H., and Schatz, G. 1980. Biosynthesis and assembly of the proton-translocating adenosine triphosphatase complex from chloroplasts. Proc. Natl. Acad. Sci. 77:1361-1364.

Rutner, A.C. 1970. Estimation of molecular weights of RuBPCase. Biochem. Biophys. Res. Comm. 39:923-929.

_____ and Lane, M.D. 1967. Non-identical subunits of RuBPCase. Biochem. Biophys. Res. Comm. 28:531-537.

Sakano, K., Partridge, J.E., and Shannon, L.M. 1973. Absence of carbohydrate in crystalline fraction 1 protein isolated from tobacco leaves. Biochem. Biophys. Acta 329:339-341.

Schmidt, G.W., devillers-Thiery, A., Desruisseaux, H., Blobe, G., and Chua, N.H. 1979. NH2-Terminal amino acid sequences of precursors and nature forms of the RuBPCase small subunit from *Chlamydomonas reinhardtii*. J. Cell Biol. 83:615.

Shen, G., Chen, K., Wu, M., and Kung, S.D. 1982. *Nicotiana* chloroplast genome IV. *N. accuminata* has larger inverted repeats and genome size. Molec. Gen. Genet. 187:12-18.

Steinbach, K., McIntosh, L., Bogorad, L., and Arntzen, C.J. 1981. Identification of the triazine receptor protein as a chloroplast gene product. Proc. Natl. Acad. Sci. 78:7463-7467.

Uchimiya, H. and Wildman, S.G. 1978. Evolution of Fraction 1 protein in relation to origin of amphidiploid *Brassica* species and other members of the Cruciferae. J. Hered. 69:299-303.

Uchimiya, H., Chen, K., and Wildman, S.G. 1977. Polypeptide composition of Fraction 1 protein as an aid in the study of plant evolution. Stadler Symp. 9:83-100.

Uchimiya, H., Chen, K., and Wildman, S.G. 1979a. A micro electrofocusing method for determining the large and small subunit polypeptide composition of fraction 1 protein. Plant Sci. Lett. 14:387-394.

Uchimiya, H., Chen, K., and Wildman, S.G. 1979b. Evolution of Fraction 1 protein in the genus *Lycopersicon*. Biochem. Genet. 17:333-338.

von Wettstein, D., Poulsen, C., and Holder, A.A. 1978. Ribulose 1,5-bisphosphate carboxylase as a nuclear and chloroplast marker. Theor. Appl. Genet. 53:193-197.

Walbot, V. 1977. Use of silica sol step gradients to prepare bundle sheath and mesophyll chloroplasts from panicum maximum. Plant Physiol. 60:102-108.

Wildman, S.G. and Bonner, J. 1947. The proteins of green leaves. I. Isolation, enzymatic properties and auxin content of spinach cytoplasmic proteins. Arch. Biochem. 14:381-413.

CHAPTER 18

Isoenzyme Analysis of Cultured Cells and Somatic Hybrids

L. Wetter and J. Dyck

Electrophoretic techniques are useful in many protein problems. Initially it was utilized to assess the electrical complexity of protein mixtures and to compare similar protein mixtures from various sources, e.g., blood serum from different species of animals, fish, etc. These early studies employed very sophisticated equipment and were carried out in liquid medium. The development of starch as a stable support for electrophoresis (Smithies, 1955) was the forerunner of what has turned out to be a very useful technique. This application now allows one to study protein mixtures and, by adapting histological methods, also to reveal specific proteins, i.e., enzymes or isozymes (Hunter and Markert, 1957). Other solid support systems, such as polyacrylamide gels and agarose, have been introduced. At the present time very detailed investigations can be carried out utilizing relatively inexpensive equipment and materials.

For isozyme studies one will find that both starch and polyacrylamide gels are employed. Polyacrylamide gel electrophoresis (PAGE) can be used for many kinds of studies. The electrical properties of proteins (enzymes) can be used to obtain mobility values for characterization information and the cross-linking of the gel can give information on the molecular weight. Gels can be prepared to establish a pH-gradient and thus yield information on the isoelectric point of proteins. Finally, specially treated gels can be used to determine the molecular weight of dissociated proteins (constitutive polypeptide chains), although this technique is limited in its application.

Polyacrylamide gel electrophoresis can be employed for a variety of applications and situations. In our hands it has proven to be an invaluable technique for establishing the success of somatic hybridization experiments at an early stage (Wetter, 1977). Very small samples are required for an assay which is particularly important where supplies are limited. A large number of samples can be analyzed on a routine basis. The results usually are easy to interpret and assess. The study of cell cultures, as well as callus and various tissues of the complete plant, lend themselves well to electrophoretic investigations.

REVIEW OF THE LITERATURE

Electrophoretic and isozyme techniques are powerful and useful tools in plant investigations. Numerous publications and reports can be found on its application; however, only a few will be cited in this short review. Several excellent reviews pertaining to higher plants are available for background information (Scandalios, 1977; Scandalios and Sorenson, 1977). One should be aware that much of the methodology and application of gel electrophoresis can be found in the fields of medicine and animal biology (Markert, 1975).

The technique can be employed to determine at a relatively early stage whether cell hybridization has been achieved. Wetter (1977) was able to show within 2 months after fusion that somatic hybridization had taken place between *Glycine max* (L.) Merr. and *Nicotiana glauca* Grah. (Kao, 1977). Callus tissue from a fusion of *Parthenocissus tricuspidata* and *Petunia hybrida* was assayed for peroxidase and the results established that hybridization had been realized (Power et al., 1975).

The technique has been employed more extensively on plant material derived from regenerated plants. The first report of parasexual interspecific hybridization of *Nicotiana glauca* Grah. and *Nicotiana langsdorffii* Weinm. employed peroxidase isozymes among other characters to verify a successful hybridization (Carlson et al., 1972). There are a number of other fusions that have been verified by the utilization of isozymes and electrophoresis, e.g., *Nicotiana rustica* L. + *Nicotiana tabacum* L. (Douglas et al., 1981b); *Petroselinum hortense* Hoffm. + *Daucus carota* L. (Dudits et al., 1980); *Arabidopsis thaliana* (L.) Heynh. + *Brassica campestris* L. (Gleba and Hoffman, 1978); interspecific somatic hybrids from 9 species of *Datura* (Lonnendonker and Schieder, 1980); *Nicotiana tabacum* L. + *Nicotiana knightiana* (Maliga et al., 1978); *Nicotiana sylvestris* + *Nicotiana knightiana* (Maliga et al., 1977); *Nicotiana tabacum* L. + *Nicotiana glauca* Grah. (Evans et al., 1980); *Petunia parodii* + *Petunia hybrida* (Power et al., 1976); 2 genotypes of *Nicotiana debneyi* (Scowcroft and Larkin, 1981). The isozyme systems employed to assess somatically derived hybrid material from plants are listed in Table 1.

The electrophoretic properties of ribulose-1,5-bisphosphate (RuBP) carboxylase, also referred to as Fraction-1 protein, can be used to establish whether hybridization has been achieved. Since RuBP carboxylase is only found in green chloroplasts, one must have green plant material available to make this study. The technique, which requires the isolation, chemical modification, and isoelectric focusing of RuBP carboxylase, was developed by Wildman and his coworkers for investiga-

Table 1.   Isozyme Systems Studied in Somatically-Derived Hybrid Plants

ENZYME	TISSUE	REFERENCE
Acid phosphatase	Cell suspension	Dudits et al., 1980
Alanyl aminopeptidase	Leaf	Evans et al., 1980
Alcohol dehydrogenase	Cell suspension	Dudits et al., 1980; Wetter, 1977
	Callus	Gleba et al., 1978; Maliga et al., 1977
	Leaf	Maliga et al., 1978
Amylase	Shoots	Lonnendonker & Schieder, 1980
Aspartate aminotransferase	Cell suspension	Dudits et al., 1980; Wetter, 1977
	Shoots	Douglas et al., 1981b
	Leaf	Evans et al., 1980
Esterase	Cell suspension	Dudits et al., 1980
	Callus	Gleba et al., 1978; Maliga et al., 1977
	Shoots	Douglas et al., 1981b
	Leaf	Maliga et al., 1978
Glucose-6-PO$_4$ dehydrogenase	Cell suspension	Dudits et al., 1980
	Callus	Maliga et al., 1977
Lactate dehydrogenase	Callus	Gleba & Hoffmann, 1978
Leucine aminopeptidase	Cell suspension	Dudits et al., 1980
Malate dehydrogenase	Cell suspension	Dudits et al., 1980
	Callus	Maliga et al., 1977
Peroxidase	Cell suspension	Dudits et al., 1980
	Callus	Power et al., 1975; Gleba et al., 1978; Maliga et al., 1977
	Leaf	Carlson et al., 1972; Power et al., 1976
Phosphodiesterase	Cell suspension	Dudits et al., 1980
	Leaf	Maliga et al., 1978
Phosphoglucomutase	Shoots	Scowcroft & Larkin, 1981
Superoxide dismutase	Shoots	Douglas et al., 1981b

tions on the genus *Nicotiana* (Wildman et al., 1974).   Electrophoretic studies of RuBP carboxylase isolated from somatic hybrids of *Nicotiana rustica* and *Nicotiana tabacum* verified that hybridization had been obtained (Douglas et al., 1981a).   Investigations are not limited to *Nicotiana*, as they can be utilized wherever RuBP carboxylase can be isolated.   The somatic hybridization products of *Solanum tuberosum* L. and *Lycopersicon esculentum* Mill. were identified by using the isoelectric focusing patterns of the RuBP carboxylase as phenotypic markers (Melchers et al., 1978; Poulsen et al., 1980).

A survey of the Solanaceae (Gatenby et al., 1980) shows that this methodology could be usefully employed for the identification of nuclear genomes in somatic hybrid plants. The isolation and study of RuBP carboxylase, isolated from polyploid wheats, has helped in establishing their origin (Chen et al., 1975).

Isozymes can be employed as effective markers particularly in studies on differentiation and genetics. There are several excellent reviews on these subjects, on differentiation (Scandalios, 1974, 1977), and on genetics (Jacobs, 1975a,b). A few reports describing the isozyme changes that occur in the initial stages of differentiation, i.e., from cells in a cell culture to callus and then to plant, can be cited. Thorpe and Gaspar (1978) report on the changes in peroxidase isozyme patterns that occur when tobacco callus is induced to form shoots.

Further, studies comparing the isozyme patterns or changes of callus and cell suspensions of plants with organized tissue are very limited. Such a comparison of peroxidase in *Solanum melongena* L. demonstrates that there are differences (del Grosso and Alicchio, 1981). Comparative studies using a number of isozyme systems on suspension cell cultures of *Phaseolus vulgaris* derived from different tissues of a single seedling also indicate some differences (Arnison and Boll, 1975).

Isoelectric focusing would appear to have additional value, particularly as an adjunct to conventional electrophoresis. For example, in cultured human cells the separation of glucose-6-phosphate dehydrogenase by conventional gel electrophoresis produced relatively simple banding patterns; however, when examined by isoelectric focusing it became obvious that there were several isozymes which were useful for further identification (Hunter, 1979). A review by Righetti and Glanazza (1980) serves as a good starting point. The technique has been widely utilized in animal cell culture studies but is only beginning to be applied to plant investigations. Isoelectric focusing patterns of esterase and peroxidase in *Brassicoraphanus* clearly showed that it was a hybrid (Kato and Tokumasu, 1979). Isoelectric focusing of esterases demonstrated that differences could be detected between haploid green plants and mutants of *Nicotiana sylvestris* Speg. et Comes (Szilagyi and Nagy, 1977). A detailed study using isoelectric focusing has been done on peroxidase and indole acetic acid oxidase (Hoyle, 1977).

PROTOCOLS

Preparation of Extract and Protein Assay

Many different extraction media are available for proteins. The investigator may want to experiment with several before settling on one; however, one should keep in mind that the recovery of undenatured enzymes is essential. For comparative studies, the use of the same extraction system throughout is important. We employ the one described here because it is simple and effective.

Extraction

1. Select appropriate plant material, e.g., cell suspension cultures, callus, and plant tissue.
2. The cells (from suspension culture) or callus are collected on Miracloth, cut to fit a Millipore filter holder (15 ml).
3. Wash the material exhaustively with mannitol solution (Table 2).
4. The cells or callus can be extracted in one of several ways. (i) In a cold miniature French pressure cell (3.7 ml). The ratio of extraction mixture (Table 2) to plant material is 1 ml to 1 g. The pressure is 140 MPa. (ii) In a Polytron homogenizer with a probe generator PT 10 ST. One ml is the smallest volume one can grind. One drop of n-octanol is employed to inhibit foaming. (iii) In glass tissue homogenizers of any convenient size.
5. The extraction is done at 5 C.
6. The homogenized tissue is centrifuged at 30,000 x g for 1 hr. The extract is stored under N_2 gas at 4 C.

Table 2. Reagents for Extraction and Protein Assay

EXTRACTION REAGENTS
1. Washing solution
   0.16 M mannitol (w/v)
2. Extraction solution
   1 M sucrose and 0.056 M 2-mercaptoethanol in 0.2 M Tris-HCl buffer, pH8.5

PROTEIN ASSAY REAGENTS
1. Precipitation solutions
   2% sodium deoxycholate (w/v)
   24% trichloroacetic acid (TCA) (w/v)
2. Color development solutions

Reagent A	Reagent B
10 parts 10% Na_2CO_3 (w/v) in 0.5 N NaOH	1 part phenol reagent (2 N)
	10 parts distilled water
0.5 parts 2% potassium tartrate (w/v)	
0.5 parts 1% $CuSO_4$ x $5H_2O$ (add 1 drop conc. H_2SO_4 for every 100 ml)	

Protein Assay

The protein assay is based on a precipitation step (Bensadoun and Weinstein, 1976) and a modification of the Lowry method (Miller, 1959). The precipitation step is very important, as it removes the protein from the materials that interfere with the subsequent color development step.

Protein Precipitation

1. Place sample (10–50 µl, depending on protein content) in 15 ml conical centrifuge tube.
2. Add in order 3 ml of distilled water and 25 µl deoxycholate solution (Table 2).
3. Shake well and let stand for 15 min.
4. Add 1 ml TCA solution (Table 2).
5. Stir well and centrifuge for 30 min at 1300 x g. A centrifuge with a horizontal head is recommended.
6. Pour off supernatant and let drain.

Color Development

1. Add 1 ml of Reagent A (Table 2) to tube containing precipitate. Mix well; vortex mixer is recommended.
2. After dissolution of precipitate add 3 ml of Reagent B (Table 2) and heat at 50 C for 15 min.
3. Cool, transfer to suitable cuvettes, and read at 625 nm in a spectrophotometer.
4. Compare readings to a calibration curve prepared from crystalline bovine serum albumin.

Conventional Polyacrylamide Gel Electrophoresis

One buffer system is described in detail in this chapter. A survey of the literature will reveal a number of variations; however, the most frequently used is the alkaline buffer system described by Davis (1964). If an acidic buffer system is desired, one can use the one described by Reisfeld et al. (1962). As for apparatus we employ cylindrical or, preferably, slab gel formers. We use a Bio-Rad Model 220 slab electrophoresis cell. The size and type of gel electrophoresis equipment depends on the researcher, as many different models are available commercially. For the power supply we recommend one that can be operated at one of three regulation modes: constant current, constant voltage, or constant power. Again, there are several different commercial suppliers.

Preparation of Gels

1. Select two clean glass plates, 140 mm x 120 mm x 1.5 mm. A slab gel will be cast between the plates. The quantities of reagent specified below are for this size slab. The assembly of plates for gel casting are usually with the electrophoresis unit.

2. Prepare a 5% separating gel by mixing together, 3.0 ml of Solution 1 (all solutions used in this section are described in Table 3), 4.0 ml of Solution 6, and 5.0 ml of distilled water.

3. Degas with water aspirator for 1 min and then add 12 ml of Solution 7.

4. Pour the slab immediately, filling to the desired height, approximately 100 mm. Quickly push the gel-forming plate into position above the gel so that no bubbles are trapped between it and the gel. Overlay with water by carefully pouring water into tray above the gel. Do not disturb the gel surface. Let stand for at least 1 hr before adding the stacking gel.

5. The stacking gel is prepared by mixing 0.75 ml of Solution 2, 1.5 ml of Solution 3, 3.0 ml of Solution 5, and 0.75 ml of Solution 4. Degas as above for 1 min.

6. Pour off the water above the separation gel and remove the last trace with filter paper strips. Add stacking gel solution to the top of the separation gel and insert the well-forming comb. Overlay immediately with water by gently pouring water into tray above gel.

7. Place gel slab before a fluorescent light source (within 8 cm) to polymerize the stacking gel and leave for 1 hr.

Table 3. Solutions Required for Gel Electrophoresis at pH 8.9

SOLUTION	INGREDIENTS[a]
1	36.6 g Tris, 0.23 ml TEMED, 48 ml of 1 N HCl made up to 100 ml of solution
2	5.7 g Tris, 0.08 ml of TEMED, add enough H_3PO_4 to bring pH to 6.9 in 100 ml of solution
3	10 g acrylamide, 2.5 g Bis in 100 ml of solution
4	4 mg riboflavin in 100 ml of solution (stable for 2 weeks)
5	40 g sucrose or glycerol in 100 ml of solution
6	30 g acrylamide, 0.74 g Bis in 100 ml of solution
7	140 mg ammonium persulfate in 100 ml of solution (stable for 1 week in a brown bottle)
8	6.0 Tris, 28.8 g glycine in 1 liter of solution

[a]All solutions except 8 are stored in the refrigerator. All solutions are made up to volume with distilled water. Abbreviations: Tris = tris (hydroxymethyl) aminomethane; TEMED = N,N,N',N'-tetramethylenediamine; Bis = N,N'-methylenebisacrylamide.

Electrophoresis of Samples

1. Just prior to electrophoresis of the samples, the well-forming comb is carefully removed so as not to disturb the sample wells. The water is then poured off. Wash the sample wells thoroughly with electrode buffer (see below), best done by employing a syringe.

2. Add electrode buffer, Solution 8 diluted by one-tenth, to correct levels in both the anode and cathode compartments of the electrophoresis cell.
3. Add approximately 1 μl of bromphenol blue (0.5% (w/v) in water) to 100 μl of protein extract.
4. Place the protein extract (usually 100 μl protein) at the bottom of the well under the upper tray buffer. The sample is best applied with a Hamilton syringe.
5. The electrophoresis unit is now attached to a circulating cooling bath (4 C) for the duration of the run.
6. The power supply is connected and turned on, 0.5 mA of current is passed through the slab gel for approximately 1 hr. The power is increased to 10 watts (constant power) and the electrophoresis run is continued until the bromphenol blue has moved to within 1 cm of the bottom of the slab (usually about 90 min).
7. Turn off power. Remove the gel from the glass plates. Remove the stacking gel from the separating gel with a spatula and mark the gel by notching upper right hand corner.
8. The position of the bromphenol blue often can be seen after the gel is dried. However, if the dye fades its position can be identified by carefully trimming the gel at that point.
9. The gel is then transferred to a tray containing the appropriate staining solution. Disposable vinyl examination gloves should be worn whenever the gel is touched with the hands.

Staining Gels

1. The isozyme stains that are routinely employed in our laboratory are listed along with their composition in Table 4.
2. The gels are placed in the staining solution, which consists of substrate, cofactors, and dye, until the bands appear. This can vary from 10 to 90 min or more. Aminopeptidase is an exception, as in this case the gel is incubated at 37 C for 30 min in substrate (Solution A), then stained in Solution B.
3. When the staining is judged to be complete, the solution is discarded and the gel is rinsed and stored in 7% acetic acid (v/v). The exceptions are aminopeptidase and aspartate aminotransferase which are stored in water.
4. Protein is detected by staining the gel for 90 min in a 3.5% (w/v) perchloric acid solution containing 0.04% (w/v) Coomassie Blue G-250. The gel is destained overnight in at least two changes of the following solution: 100 ml ethyl acetate, 70 ml ethyl alcohol, 50 ml acetic acid, and 780 ml of distilled water.

Gel Preservation and Records

1. After staining, the gels are photographed against a frosted glass light box. A permanent record is then kept on 35 mm black and white negatives, from which a photographic record can be obtained for publication.

Table 4. Some General Stains Employed for the Detection of Isozymes on Gels

ENZYME	STAIN INGREDIENTS	QUANTITY OF INGREDIENT
Dehydrogenases; Solution A	NAD or NADP	40 mg
	Phenazine methosulfate	1 mg
	Nitro blue tetrazolium	20 mg
	0.1 M Tris buffer, pH8.5	46 ml
Dehydrogenases; Solution B	One M substrate in 0.1 M Tris buffer, pH 8.5 (for the more expensive substrates 0.1 M substrate is adequate). Immediately before staining add 4 ml of solution B to 46 ml of solution A.	
Esterase	α-Naphthyl acetate	20 mg
	Acetone	0.5 ml
	Dissolve	
	Water	0.5 ml
	0.2 M phosphate buffer, pH 6.0	50 ml
	Filter	
	Fast Blue RR salt	38 mg
Aminopeptidase; Solution A	L-Alanyl-β-naphthylamide	25 mg
	N,N-dimethyl formamide	1 ml
	Dissolve	
	0.2 M acetate buffer, pH 4.8	50 ml
Aminopeptidase; Solution B	Fast Garnet GBC salt	100 mg
	0.2 acetate buffer, pH 4.8	50 ml
Aspartate amino-transferase[a]	Pyridoxal-5'-phosphate	0.5 mg
	Bovine serum albumin	50 mg
	L-Aspartic acid	200 mg
	α-Ketoglutarate	32 mg
	Polyvinylpyrrolidone (M.W. 40,000)	1.12 g
	0.2 M phosphate buffer, pH 7.5	41 ml
	Adjust pH to 7.4 if necessary. Prior to staining, add following to above	
	Fast Violet B salt	68 mg
	Water	9 ml
Peroxidase	o-Dianisidine	49 mg
	β-Naphthol	29 mg
	Acetone	20 ml
	0.1 M Tris-acetate buffer, pH 4.0	10 ml
	Water	50 ml
	3% Hydrogen peroxide	1 ml
	Water	to 100 ml
Acid phosphatase	α-Naphthyl acid phosphate sodium salt	50 mg

Table 4. Cont.

ENZYME	STAIN INGREDIENTS	QUANTITY OF INGREDIENT
Acid phosphatase	Fast Garnet GBC salt	50 mg
	1.0 M acetate buffer, pH 4.8	50 ml

[a]Recently a more sensitive method for detecting this enzyme has been proposed. In our hands it works very well; however, the substrate is L-cystine sulfinate and not L-aspartate (Yagi et al., 1981).

2. One can also calculate relative mobilities from the wet gel by calculating the ratio of the distance the band has moved to the distance the bromphenol blue has moved.
3. The gel can be preserved by drying it on a commercial gel slab drier. We use a Bio-Rad Model 224 dryer. The instructions on how to use it come with the dryer.

The procedure outlined here applies to our particular situation. We employ a vertical slab electrophoresis assembly; however, there is no reason why a horizontal slab assembly or a cylindrical unit cannot be employed. Many parameters can be varied and should always be considered in any study. For example, the size and thickness of the slab can be modified or the composition of the separating gel can be changed from 5 to 12%. The stacking gel is employed by us because it results in sharper bands and superior separation.

The protein samples are usually added as a solution, but samples can also be added to filter paper and then inserted into the well for electrophoresis. Recently we have demonstrated that 5-10 mg pieces of tissue, e.g., leaves, can be crushed between buffer-soaked filter paper and then employed for gel isozyme studies. We are now using this technique for the study of isozymes in *Hordeum vulgare* L.

The isozyme staining solutions are usually modeled on histological staining techniques. We have listed only a few enzymes in Table 4 but many more can be investigated. For a list, one should consult Scandalios (1974) and Scandalios and Sorenson (1977). Although the methods are well standardized, one should always be on the lookout for improvements to existing methods. A case in point is aspartate aminotransferase; recently a much improved method has been proposed which is more sensitive and permanent (Yagi et al., 1981).

Isoelectric Focusing of Ribulose-1,5-Bisphosphate Carboxylase

The utilization of this technique is dependent on the availability of green tissue from which the carboxylase can be isolated, subjected to isoelectric focusing, and then used to identify hybrids by studying the patterns of the large and small subunits. The methodology employed here is based on those developed by Wildman and his coworkers (Kung et al., 1974). The equipment is the same as that described for conventional electrophoresis.

Preparation of Leaf Extracts

1. Collect 0.5 g of healthy green leaves, rinse well in distilled water.
2. The leaves are cut into small pieces and placed in a glass conical-tipped tissue grinder.
3. To the grinder add 1.3 ml of extraction reagent (see Table 5). Grind for at least 3 min keeping the temperature at 4 C.
4. Centrifuge sample at 8000 x g for 15 min, then transfer supernatant to 15 ml heavy-walled conical tubes.

Precipitation of RuBP Carboxylase

1. To a 15 ml conical-tipped tube containing approximately 1 ml of leaf extract add 0.3 ml of anti-RuBP carboxylase serum. This ratio of extract to antiserum gives a satisfactory yield of precipitate for *Nicotiana tabacum*; however, a preliminary test should be done to ascertain the best ratio for precipitation in other systems.
2. Stir the mixture gently and then incubate for several hours at 37 C. It is during this time that the insoluble RuBP carboxylase-antibody complex usually forms. The incubation is continued at 4 C overnight.
3. Next morning the precipitate is collected as a pellet after centrifugation at 3000 x g for 15 min in a centrifuge with a horizontal head.
4. Pour off the supernatant and discard. Let the precipitate drain for several minutes. Suspend precipitate in 1 ml of cold borate-saline buffer (see Table 5). Centrifuge as described above.
5. Repeat step 4.
6. Let precipitate drain well. The tube containing the precipitate is immediately flushed with N_2 gas, after which it is sealed with a rubber serum stopper. Proceed onto the next step immediately.

Carboxymethylation of Protein Complex

1. The precipitate is dissolved in 0.12 ml of Tris-urea buffer (Table 5) containing 5 mg dithiothreitol (Cleland's reagent). It is important to add this reagent and all subsequent reagents with a hypodermic syringe through the serum stopper.
2. Let mixture stand at room temperature for 2 hr.
3. Wrap the tube in aluminum foil to keep out the light.
4. The carboxymethylated mixture is now passed through a small Sephadex G-25 (medium grade) column (for preparation see 6 below). The protein is eluted with Tris-EDTA buffer (Table 5).
5. The elute from the column is closely monitored for protein by testing with 24% trichloroacetic acid (w/v). When protein appears, a 0.7 ml sample is collected for isoelectric focusing.
6. Preparation of Sephadex G-25 is carried out as follows: 4 g are equilibrated in degassed distilled water in the refrigerator overnight or in a boiling water-bath for 1 hr. The column is poured and equilibrated with Tris-EDTA buffer. This amount will make three 5 ml columns. We employ Mohr 5 ml pipettes for columns.

Table 5.   Composition of Solutions Required for Preparation of RuBP Carboxylase for Isoelectric Focusing

---

Extraction Reagent (from Gray and Wildman, 1976)

0.05 M Tris	0.01 M sodium metabisulfite
0.10 M NaCl	0.1% bovine serum albumin
0.001 M KCN	Dowex-1, 200 mg per 0.5 g leaf
Adjust the pH to 7.8	

Borate-Saline Buffer
0.0025 M Borax (Na$_2$B$_4$O$_7$ x 10H$_2$O)
0.04 M Boric acid
0.12 M NaCl
  Adjust to pH 7.3

Tris-Urea Buffer
0.5 M Tris-HCl
8.0 M Urea
0.001 M EDTA
  Adjust pH to 8.5

Tris-EDTA Buffer
0.5 M Tris-HCl
0.001 M EDTA
  Adjust pH to 8.5

Isoelectric Focusing

1. Prepare the 5.25% polyacrylamide gel by dissolving 25 mg of ammonium persulfate in 25 ml of Solution 2 (Table 6), then add 25 μl of TEMED and 1.25 ml of mixed Ampholines (i.e., 0.25 ml of pH 3-10 and 1.00 ml of pH 4-6). Then 3.75 ml of Solution 1 (Table 6) is degassed and added to the urea solution. Mix and pour immediately into a slab. Place the well-forming comb in position and carefully cover the top of the gel with distilled water. Let stand for at least 1 hr to gel.
2. When gel is to be used, the distilled water is poured off and the comb is removed. The wells are washed with water and filled with sucrose-Ampholine solution (Table 6).
3. The electrode solutions are now added. In our apparatus care must be taken not to disturb the sucrose-Ampholine solution in the wells when adding the catholyte.
4. Add sufficient urea to the protein sample to make it 8 M, then transfer samples (approximately 20 μl) to the well, placing them below the sucrose-Ampholine solution.
5. The focusing run is performed at 4 C using the following electrical conditions: 1 hr at 1 watt, 2 hr at 2.5 watts, and finally 2 hr at 4.5 watts. Care should be taken that during the last 2 hr the voltage does not exceed 800 volts. If it does, set the power supply at a constant voltage of 800.

6. After 5 hr the power is switched off. The gel is marked by notch-
    ing the upper right hand corner. A pH profile can be obtained in
    one of two ways, either by using a specially designed micro-elec-
    trode or by removing a 1 cm wide strip (top to bottom) from gel
    and cutting into 0.5 cm segments. Each segment is equilibrated in
    1 ml of 0.1 M NaCl after which the pH is determined.
7. The gel is stained for protein in Coomassie Blue G-250 exactly as
    described earlier for conventional electrophoresis.
8. The recording of results is the same as that employed in the previ-
    ous protocol.

Table 6. Solutions for Isoelectric Focusing

SOLUTION	INGREDIENTS
1	42% acrylamide and 0.8% Bis (w/v) made up in distilled water
2	10 M urea (ultrapure) freshly prepared in degassed distilled water
Sucrose-Ampholine	50 mg sucrose and 0.05 ml pH 3-10 Ampholine per ml of solution
Electrode	The catholyte is 0.4% triethanolamine; the ano-lyte is 0.01 M glutamic acid

The initial extraction of the leaves could lead to some serious problems. The yield of RuBP carboxylase could be very low if the extraction mixture is inadequate. Some extraction reagents will not permit
the precipitation of the protein complex. Finally, the leaf extract
could become very dark in color, because of the action of polyphenol
oxidases, yielding a very dark precipitate that results in isoelectric
focusing patterns that are very difficult to interpret because of variant
bands. After trying a number of extraction mixtures we prefer the one
described here (Table 5).

The direct precipitation method described here is basically the one
developed by Uchimiya et al. (1979). The anti-RuBP carboxylase serum
is obtained by immunizing rabbits to crystalline RuBP carboxylase obtained from healthy (i.e., TMV-free) *Nicotiana tabacum*. We employed a
method described by Lowe (1977) for the crystallization of the enzyme.
The antiserum to the enzyme was prepared by the method described by
Gray and Wildman (1976). The specificity of the antibody-carboxylase
appears to be very broad and therefore can be employed to isolate the
carboxylase from a wide source of green plant material (Chen et al.,
1975, 1976). It is important to centrifuge in conical tubes and in a
horizontal head in order to obtain a well-packed pellet of protein in
the bottom of the tube.

The carboxymethylation step is straightforward, but the following
precautions should be observed. It is important to exclude O_2 during
the operation, and therefore, the N_2 gas flush is essential. The precipitate must be dissolved in the Tris-urea buffer immediately, as any
drying of the precipitate makes its solution extremely difficult. The
urea employed in this assay must be highly purified, as urea can often

be contaminated with appreciable concentrations of cyanate ions. This will lead to carbamyl derivatives of protein and thus produce extraneous bands. Because of this contamination with cyanate ion it is important to complete this assay as quickly as possible, i.e., within days.

Polyacrylamide Gel Isoelectric Focusing of Isozymes

The combination of isoelectric focusing on polyacrylamide gels and enzyme recognition by histological stains does not appear to be generally used, particularly in plant studies. In some investigations much sharper separation can be achieved by isoelectric focusing than by conventional electrophoresis. This is illustrated by a study conducted on glucose-6-phosphate dehydrogenase obtained from cultured human cells (Hunter, 1979).

The equipment employed is the same as for conventional polyacrylamide electrophoresis.

Preparation of Extracts and Protein Assay

These procedures are identical to those described for conventional gel electrophoresis.

Preparation of Gels

1. A 6% polyacrylamide gel is prepared by diluting 4.3 ml of Solution 1 (Table 6) with 20.7 ml of distilled water. This solution is degassed. To this mixture add the Ampholine which consists of 0.25 ml of wide range pH 3-10 and 1.0 ml of an applicable narrow pH range Ampholine and 3.75 ml of Solution 4 (Table 3). Finally add 25 µl of TEMED.
2. Pour the gel immediately; insert the well-forming comb and overlayer completely with distilled water to exclude air. The gel is then photopolymerized under a fluorescent lamp, usually overnight or until completely gelled.
3. In preparation for electrophoresis the water is removed and the wells are flushed before filling with sucrose-Ampholine solution (Table 6). The protein samples are then placed at the bottom of the well below the sucrose-Ampholine solution.
4. The electrode chambers are filled with the appropriate solution (Table 6), triethanolemine in the cathode chamber, and glutamic acid in the anode chamber. The electrical conditions were the same as those described for the isoelectric focusing of RuBP carboxylase.

Staining of Gels

1. After the run has been completed the gel is removed, rinsed in distilled water, and incubated in the appropriate enzyme buffer (there is no substrate in this solution) for 30 min with gentle shaking.

2. The gel is then placed in the staining solution and the patterns developed as previously described. The composition of the various staining solutions are given in Table 4.

3. The pH profile of the gel and its preservation have been described previously.

Isoelectric focused gels are usually developed with Coomassie Blue to visualize the proteins; however, the intent here is to identify the various enzyme systems. To do this it is important to remove the Amphol-ines, thereby changing the pH environment to that which will give an optimal enzyme reaction. This is done by equilibrating the gel with the buffer employed in the staining mixture; usually 30 min is suffici-ent.

SDS–Polyacrylamide Gel Electrophoresis

Sodium dodecyl sulfate (SDS) polyacrylamide gel electrophoresis is employed to separate and estimate the molecular weight of constituent polypeptide chains (subunits) of proteins. For basic information the re-searcher is referred to two excellent papers (Laemmli, 1970; Weber and Osborn, 1975). This technique has been employed to determine the molecular weight of the two subunits derived from the enzyme, ribu-lose-1,5-bisphosphate carboxylase in *Nicotiana* (Douglas et al., 1981a).

Once again the technique is performed on standard gel electrophore-sis equipment.

Preparation of Extracts

1. Protein samples are extracted and prepared in the usual manner, as has already been described.

2. Just prior to electrophoresis 100 µl of protein extract is mixed with an equal volume of dissociation medium (Solution 1, see Table 7) to which 7.7 mg of dithiothreitol is added. Heat in a boiling water bath for one and a half min. For an assay approximately 20–30 µg of protein is used.

Preparation of Gel

1. The running gel is prepared by dissolving 3.8 g of acrylamide and 76 mg of Bis in 21.4 ml of distilled water. Then add 9.7 ml of buffer (Solution 2, Table 7), 0.15 ml ammonium persulfate (Solution 5, Table 7), 0.39 ml SDS (Solution 6, Table 7), and finally 30 µl of TEMED.

2. The gel slab is poured immediately to the desired height (approxi-mately 3 cm from top), the gel-forming plate positioned on top and covered with overlay solution. This solution consists of 5 ml of Solution 2 (Table 7), 15 ml distilled water, and 0.2 ml Solution 6 (Table 7).

3. Allow at least 1 hr for gel to set.

4. The stacking gel is prepared and poured just prior to the experiment. It is prepared by dissolving 0.51 g acrylamide and 0.12 g Bis in 10.1 ml of distilled water. To this mixture add 3.75 ml of Solution 3 (Table 7), 0.18 ml of Solution 5 (Table 7), 0.20 ml of Solution 6 (Table 7), and 15 μl of TEMED.

5. Remove the overlay solution from top of running gel and wash thoroughly with distilled water, then dry gel surface with filter paper. The stacking gel is added to the appropriate level. Insert the well-forming comb into the stacking gel and cover with electrode buffer (Solution 7, Table 7). Allow 20–30 min to gel, slowly remove the comb, and fill wells with electrode buffer.

6. Add the specially prepared protein samples to the well. Every run must include a molecular weight standard (these can be purchased) for comparison purposes. The standard protein is prepared exactly as the unknown, i.e., in the dissociation medium (Solution 1, Table 7).

7. Run the gel at 50 ma constant current until the voltage reaches 200 V, then continue run at 200 V constant voltage until the dye front approaches the bottom of gel (about 5 hr). The run is performed at 15 C; otherwise the SDS precipitates.

8. When run is completed, shut off power and disassemble apparatus. Mark one corner of gel for orientation purposes.

9. Transfer gel to Solution 8 (Table 7) for at least 1 hr at 50 C. The gel can remain in this solution overnight without harmful effects. Remove the staining solution and rinse several times at room temperature with Solution 9 (Table 7). The gel is finally destained at 50 C against several changes of Solution 10 (Table 7).

10. The molecular weight of unknown protein bands is determined by comparing with the standards. A standard curve is obtained by plotting R_fs versus molecular weight of standard proteins on semilogarithmic paper, as described by Weber and Osborn (1975).

11. The gel can be preserved by drying, and a photographic record should be kept.

The SDS technique is straightforward and simple. Protein standards must be run in conjunction with the unknown samples in every experiment. It is the standards that serve as markers to which protein bands in the unknown sample can be compared. One must also keep in mind that the protein is a dissociation product and not one in its native state. This is one reason why this method cannot be employed to look at isozymes except on rare occasions when the dissociated molecule can be renatured and then assessed.

FUTURE PROSPECTS

The utilization of solid-support electrophoresis, whether it is conventional, isoelectric focusing, or SDS electrophoresis, has a great potential in plant cell and tissue culture investigations. The equipment is relatively inexpensive, and easy to operate and maintain. The tech-

Table 7. Solutions for SDS-Polyacrylamide Gel Electrophoresis

SOLUTION	INGREDIENTS
1	To 15.5 ml of distilled water add 1.0 ml of 1.5 Tris (Solution 2); then add 5.0 ml 10% SDS (Solution 6), 2.5 g Ficoll, 0.5 ml 0.1 M Na_2EDTA, and 100 µl 1.0% bromphenol blue
2	Dissolve 18.15 g Tris in distilled water and adjusted to pH 8.8 by adding 2.6 ml concentrated HCl; make up to 100 ml
3	Dissolve 60.6 g Tris in distilled water and adjusted to pH 6.8 with 1 M HCl; make up to 100 ml
4	Dissolve 60.6 g Tris and 285 g glycine in distilled water to 2 liter; filter
5	Ammonium persulfate, 10% (w/v) (make up fresh)
6	SDS (sodium dodecyl sulfate), 10% (w/v)
7	To 200 ml of Solution 4 add 10 ml of Solution 6 and 1.0 ml mercaptopropionic acid; dilute to 1 liter with distilled water
8	Dissolve 1 g of Coomassie Blue R 250 in 455 ml of methanol and 90 ml glacial acetic acid; make up to 1 liter with distilled water; filter
9	To 500 ml of methanol add 100 ml glacial acetic acid and 500 ml distilled water
10	Acetic acid, 7% (v/v)

niques can be learned quickly and are adaptable to the screening of large numbers of samples. What is perhaps more important to the plant cell culturist is that investigations can be carried out on small quantities of tissue. Sample sizes of a few milligrams can yield usable results.

Most of the studies to date have been carried out on only a few isozyme systems, most frequently the peroxidase system. It is questionable whether this is the most desirable system to employ because of its susceptibility to subtle environmental changes, which leads to a great deal of variation. Thus it would seem advisable to consider other enzyme systems. Some of these are aspartate aminotransferase, esterase, a few dehydrogenases, and acid phosphatase which give excellent results. Other enzyme systems should be surveyed with a view to identifying those that reflect more accurately the genetic make-up of the cell and not the environmental changes. It is imperative to remember that when investigating hybrids, enzyme systems that give distinctly different patterns for each parent must be used.

Since enzymes are concerned with fundamental life processes and are products of gene expression, techniques associated with enzymes should be a useful measure of what is happening in plant cells and the effect of their environment. Several areas come to mind in which electrophoretic techniques might be employed to advantage. For example, the changes that might occur in isozyme patterns or specific enzyme markers with the initiation of differentiation, i.e., from nondifferentiated cell to callus to shoot or root.   Might one employ this technique to predict the beginning of embryo formation?   What changes take place in a meristem tip as it begins its growth?   What happens in protoplasts as they begin to produce a cell wall and are cultured as a suspension? The synthesis of ribulose-1,5-bisphosphate carboxylase in tobacco protoplasts (Fleck et al., 1979) is an attempt at answering the last question.

More information is required as to how effectively electrophoretic results that are obtained in cell and callus cultures can be transferred to the whole plant situation and vice versa.   There is evidence that the basic isozyme pattern is retained (Khavkin and Sukhorzhevskaia, 1979), but there also is evidence that would suggest that this is not the case (Arnison and Boll, 1975).

Most of the enzyme systems employed in isozyme studies of plants are those found in all living cells.   A real need exists to search for and investigate enzyme systems that are unique to plants.   Some are already being investigated, e.g., ribulose-1,5-bisphosphate carboxylase (Chen et al., 1975).   The identification of unique plant enzymes is particularly applicable in the field of secondary product production in plant cell cultures.   At the present time the end (or desired) product is the only marker one can employ, and very frequently its identification requires very expensive and sophisticated equipment.   The identification of the enzyme and its adaptation to the isoenzyme electrophoretic technique could be an invaluable aid to these investigations.

Recent reports suggest that electrophoretic techniques could be usefully employed in the study of stress.   Protein synthesis patterns change dramatically when soybean seedlings are subjected to heat stress (Key et al., 1981).   The same observation was made when soybean and tobacco suspension cultures were heat stressed (Barnett et al., 1980).   These changes were followed by employing the SDS electrophoresis technique.   These results suggest that electrophoretic techniques could be used to study and assess a whole variety of stresses, for example, cold, salt, and chemical stress as well as stress initiated by plant diseases.   This would appear to be an exciting prospect.

Can existing methodology be modified or adapted in such a manner that more information can be obtained in the investigations of plant tissue cultures?   There would appear to be several.   One intriguing possibility arises out of the two-dimensional electrophoresis concept designed to fingerprint the proteins of an organism (O'Farrell, 1975).   In this technique, isoelectric focusing is performed in the first dimension and SDS electrophoresis in the second.   We propose the following modification: isoelectric focusing in the first dimension but conventional electrophoresis in the second.   One could now stain for specific enzymes and thus obtain a two-dimensional fingerprint of isozymes.   In some preliminary experiments we have found it works very well for esterases.   The second area of investigation, arising out of the one just discussed, relates to the utilization of isoelectric focusing for separation and the subsequent staining for specific enzymes.   Several areas require investigations.   How seriously do Ampholines interfere with enzyme reactions and their staining?   Do Ampholines complex with cofactors?   How can these compounds be removed from the gel?   Finally, what enzyme systems can be studied by this technique?

We certainly have by no means exhausted or even attempted to cover the areas where the electrophoretic technique could aid plant tissue culturists in their investigations. We hope we have pointed out the possibilities and that this chapter will encourage them to utilize this technique.

KEY REFERENCES

Allen, R. and Maurer, H.R. (eds.) 1974. Electrophoresis and Isoelectric Focusing in Polyacrylamide Gel: Advances of Methods and Theories, Biochemical and Clinical Applications. Walter de Gruyter, Berlin.

Brewer, G.J. and Sing, C.F. 1970. An Introduction to Isozyme Techniques. Academic Press, New York.

Markert, C.L. (ed.) 1975. Isozymes. Molecular Structure (Vol. I), Physiological Function (Vol. II), Developmental Biology (Vol. III), Genetics and Evolution (Vol. IV). Academic Press, New York.

Righetti, P.G. and Gianazza, E. 1980. New developments in isoelectric focusing. J. Chromatogr. 184:415-456.

Scandalios, J.G. and Sorenson, J.C. 1977. Isozymes in plant tissue culture. In: Plant Cell, Tissue, and Organ Culture (J. Reinert and Y.P.S. Bajaj, eds.) pp. 719-730. Springer-Verlag, Berlin, New York.

Vesterberg, O. 1973. Isoelectric focusing of proteins in thin layers of polyacrylamide gel. Sci. Tools 20:22-29.

Weber, K. and Osborn, M. 1975. Proteins and sodium dodecyl sulfate: Molecular weight determination on polyacrylamide gels and related procedures. In: The Proteins (H. Neurath and R.L. Hill, eds.) pp. 179-223. Academic Press, New York.

REFERENCES

Arnison, P.G. and Boll, W.G. 1975. Isoenzymes in cell cultures of bush bean (Phaseolus vulgaris cv. Contender): Isoenzymatic differences between stock suspension cultures derived from a single seedling. Can. J. Bot. 53:261-271.

Barnett, T., Altschuler, M., McDaniel, C.N., and Mascarenhas, J.P. 1980. Heat shock induced proteins in plant cells. Dev. Genet. 1:331-340.

Bensadoun, A. and Weinstein, D. 1976. Assay of proteins in the presence of interfering materials. Anal. Biochem. 70:241-250.

Carlson, P.S., Smith, H.H., and Dearing, R.D. 1972. Parasexual interspecific plant hybridization. Proc. Natl. Acad. Sci. 69:2292-2294.

Chen, K., Gray, J.C., and Wildman, S.G. 1975. Fraction 1 protein and the origin of polyploid wheats. Science 190:1304-1306.

Chen, K., Kung, S.D., Gray, J.C., and Wildman, S.G. 1976. Subunit polypeptide composition of fraction 1 protein from various plant species. Plant Sci. Lett. 7:429-434.

Davis, B.J. 1964. Disc electrophoresis. II. Method and application to human serum proteins. Ann. N.Y. Acad. Sci. 121:404–427.

del Grosso, E. and Alicchio, R. 1981. Analysis in isozymatic patterns of *Solanum melongena*: Differences between organized and unorganized tissues. Z. Pflanzenphysiol. 102:467–470.

Douglas, G.C., Wetter, L.R., Keller, W.A., and Setterfield, G. 1981a. Somatic hybridization between *Nicotiana rustica* and *N. tabacum*. IV. Analysis of nuclear and chloroplast genome expression in somatic hybrids. Can. J. Bot. 59:1509–1513.

Douglas, G.C., Wetter, L.R., Nakamura, C., Keller, W.A., and Setterfield, G. 1981b. Somatic hybridization between *Nicotiana rustica* and *N. tabacum*. III. Biochemical, morphological, and cytological analysis of somatic hybrids. Can. J. Bot. 59:228–237.

Dudits, D., Fejer, O., Hadlaczky, G., Koncz, C., Lazar, G.B., and Horvath, G. 1980. Intergeneric gene transfer mediated by plant protoplast fusion. Molec. Gen. Genet. 179:283–288.

Evans, D.A., Wetter, L.R., and Gamborg, O.L. 1980. Somatic hybrid plants of *Nicotiana glauca* and *Nicotiana tabacum* obtained by protoplast fusion. Physiol. Plant 48:225–230.

Fleck, J., Durr, A., Lett, M.C., and Hirth, L. 1979. Changes in protein synthesis during the initial stage of life of tobacco protoplasts. Planta 145:279–285.

Gatenby, A.A., Zapata, F.J., and Cocking, E.C. 1980. Molecular markers for the identification of nuclear and organelle genomes in somatic hybrid plants of the Solanaceae. Z. Pflanzenzuecht. 84:1–8.

Gleba, Yu.Yu. and Hoffmann, F. 1978. Hybrid cell lines *Arabidopsis thaliana* + *Brassica campestris*: No evidence for specific chromosome elimination. Molec. Gen. Genet. 165:257–264.

Gray, J.C. and Wildman, S.G. 1976. A specific immunoabsorbent for the isolation of fraction 1 protein. Plant Sci. Lett. 6:91–96.

Hoyle, M.C. 1977. High resolution of peroxidase–indole acetic acid oxidase isoenzymes from horseradish by isoelectric focusing. Plant Physiol. 60:787–793.

Hunter, L. 1979. Glucose-6-phosphate dehydrogenase isoenzymes in cultured human cell lines: Separation by isoelectric focusing. Anal. Biochem. 101:78–87.

Hunter, R.L. and Markert, C.L. 1957. Histochemical demonstration of enzyme separated by zone electrophoresis in starch gels. Science 125:1294–1295.

Jacobs, M. 1975a. Isozymes and a strategy for their utilisation in plant genetics. I. Isozymes: Genetic and epigenetic control. In: Genetic Manipulation in Higher Plants (L. Ledoux, ed.) pp. 365–378. Plenum Press, New York.

_____ 1975b. Isozymes and a strategy for their utilisation in plant genetics. II. Isozymes as a tool in plant genetics. In: Genetic Manipulation in Higher Plants (L. Ledoux, ed.) pp. 379–389. Plenum Press, New York.

Kao, K.N. 1977. Chromosomal behavior in somatic hybrids of soybean—*Nicotiana glauca*. Molec. Gen. Genet. 150:225–230.

Kato, M. and Tokumasu, S. 1979. An electrophoretic study of esterase and peroxidase isozymes in *Brassicoraphanus*. Euphytica 28:339–350.

Key, J.L., Lin, C.Y., and Chen, Y.M. 1981. Heat shock proteins of higher plants. Proc. Natl. Acad. Sci. 78:3526–3530.

Khavkin, E.E. and Sukhorzhevskaia, T.B. 1979. Maintenance of isoenzyme spectra in callus and suspension cultures derived from internodes of maize (Zea mays L.). Biochem. Physiol. Pflan. 174:431–437.

Kung, S.D., Sakano, K., and Wildman, S.G. 1974. Multiple peptide composition of the large and small subunits of Nicotiana tabacum fraction 1 protein ascertained by fingerprinting and electrofocusing. Biochim. Biophys. Acta 365:138–147.

Laemmli, U.K. 1970. Cleavage of structural proteins during the assembly of the head of bacteriophage T4. Nature (London) 227:680–685.

Lonnendonker, N. and Schieder, O. 1980. Amylase isoenzymes of the genus Datura as a simple method for an early identification of somatic hybrids. Plant Sci. Lett. 17:135–139.

Lowe, R.H. 1977. Crystallization of fraction 1 protein from tobacco by a simplified procedure. FEBS Lett. 78:98–100.

Maliga, P., Lazar, G., Joo, F., Nagy, A.H., and Menczel, L. 1977. Restoration of morphogenic potential in Nicotiana by somatic hybridization. Molec. Gen. Genet. 157:291–296.

Maliga, P., Kiss, Z.R., Nagy, A.H., and Lazar, G. 1978. Genetic instability in somatic hybrids of Nicotiana tabacum and Nicotiana knightiana. Molec. Gen. Genet. 163:145–151.

Melchers, G., Sacristan, M.D., and Holder, A.A. 1978. Somatic hybrid plants of potato and tomato regenerated from fused protoplasts. Carlsberg Res. Comm. 43:203–218.

Miller, G.L. 1959. Protein determination for large number of samples. Anal. Chem. 31:964.

O'Farrell, P.H. 1975. High resolution two-dimensional electrophoresis of proteins. J. Biol. Chem. 250:4007–4021.

Poulsen, C., Porath, D., Sacristan, M.D., and Melchers, G. 1980. Peptide mapping of the ribulose bisphosphate carboxylase small subunit from the somatic hybrid of tomato and potato. Carlsberg Res. Comm. 45:249–267.

Power, J.B., Frearson, E.M., Hayward, C., and Cocking, E.C. 1975. Some consequences of the fusion and selective culture of petunia and Parthenocissus protoplasts. Plant Sci. Lett. 5:197–207.

Power, J.B., Frearson, E.M., Hayward, C., George, D., Evans, P.K., Berry, S.F., and Cocking, E.C. 1976. Somatic hybridisation of Petunia hybrida and P. parodii. Nature (London) 263:500–502.

Reisfeld, R.A., Lewis, U.J., and Williams, D.E. 1962. Disk electrophoresis of basic proteins and peptides on polyacrylamide gels. Nature (London) 195:281–283.

Scandalios, J.G. 1974. Isozymes in development and differentiation. Annu. Rev. Plant Physiol. 25:255–258.

_____ 1977. Regulation of isozyme patterns in plant cell differentiation. In: Cell Differentiation in Microorganisms, Plants and Animals, International Symposium (L. Nover and K. Mothes, eds.) pp. 467–483. North-Holland, Amsterdam.

Scowcroft, W.R. and Larkin, P.J. 1981. Chloroplast DNA assorts randomly in intraspecific somatic hybrids of Nicotiana debneyi. Theor. Appl. Genet. 60:179–184.

Specialized Cell Culture Techniques

Shaw, C.R. and Prasad, R. 1970. Starch gel electrophoresis of enzymes—a compilation of recipes. Biochem. Genet. 4:297-320.

Smithies, O. 1955. Zone electrophoresis in starch gels: Group variations in the serum proteins of normal human adults. Biochem. J. 61: 629-641.

Szilagyi, L. and Nagy, A.H. 1977. Comparative analysis of green and spontaneous albino mutant lines of *Nicotiana sylvestris* Speg. et Comes. Acta Bot. Acad. Sci. Hung. 23:219-227.

Thorpe, T.A. and Gaspar, T. 1978. Changes in isoperoxidases during shoot formation in tobacco callus. In Vitro 14:522-526.

Uchimiya, H., Chen, K., and Wildman, S.G. 1979. A micro electrofocusing method for determining the large and small subunit polypeptide composition of fraction 1 protein. Plant Sci. Lett. 14:387-394.

Wetter, L.R. 1977. Isoenzyme patterns in soybean—*Nicotiana* somatic hybrid cell lines. Molec. Gen. Genet. 150:231-235.

Wildman, S.G., Chen, K., Gray, J.C., Kung, S.D., Kwanyuen, P., and Sakano, K. 1974. Evolution of ferredoxin and fraction 1 protein in the genus *Nicotiana*. In: Genetics and Biogenesis of Mitochondria and Chloroplasts (C.W. Birky, Jr., P.S. Perlman, and T.J. Byers, eds.) pp. 309-329. Ohio State Univ. Press, Columbus.

Yagi, T., Kagamiyama, H., and Nozaki, M. 1981. A sensitive method for the detection of aspartate: 2-oxoglutarate aminotransferase activity on polyacrylamide gels. Anal. Biochem. 110:146-149.

CHAPTER 19
DNA Replication in Cultured Plant Cells

D. Cress

The major experimental advances leading to our understanding of the basic mechanics, enzymology, and regulation of the replication of DNA have come primarily from investigations on the prokaryotic chromosome and bacterial viruses and plasmids. There are several reasons for the success of these systems.

First, bacterial plasmids, viruses, and even the bacterial chromosome all are units of structural simplicity. Initiation of replication occurs usually at one site on the circular molecule. Bidirectional propagation of the replication forks usually follows until the forks join at termination, and the semiconservatively replicated daughter DNA molecules separate after completion of replication. The replication unit and the segregation unit are identical.

Second, these genomes are also units of genetic simplicity. They are normally haploid, facilitating isolation of recessive mutants; are manipulable in large numbers; have extremely short generation times; and are subject to genetic analysis as a consequence of genetic exchange by transduction, conjugation, and transformation. These factors have led to the isolation of a large number of mutants altered in replication properties, those particularly useful being conditional lethal (temperature sensitive) for DNA synthesis. The power of the genetic approach

---

[1] Mention of trademark, proprietary product, or vendor does not constitute a guarantee or warranty of the product by the U.S. Department of Agriculture and does not imply its approval to the exclusion of other products or vendors that may also be suitable.

has been in the identification and analysis of enzymes involved in the replication process and in the reconstitution of the replication complex by complementation in in vitro DNA synthesis.

Third, prokaryotes have considerable physiological flexibility. They are capable of growth under a variety of experimental conditions and can be easily radioactively labeled using nucleic acid precursors, and DNA in various stages of replication can be isolated in native form for physical analysis.

The extension of experimental focus to DNA replication in eukaryotes came in the 1960s subsequent to the development of tissue culture methods for mammalian cells. Such methods permitted for the first time quantitative physiological studies on the replication of DNA of higher organisms. It was quickly apparent that DNA replication in eukaryotes was substantially more complex than in prokaryotes. Such a discovery should come as no surprise in view of the substantial differences in the organization of the genetic material and the multicellular environment of the eukaryotic cell (Hand, 1978).

First, the genome of eukaryotes is several orders of magnitude larger than in prokaryotes, typically around 10^{10} base pairs (bp) in total length. If such DNA from a single cell were organized as one continuous chromosome, it would extend to over 1 meter in length. If this DNA were contained within a single replication unit, or replicon as termed by Jacob et al. (1963), it would require weeks for complete replication of the molecule, even at the fastest known rates of replication fork movement. The discoveries that replication is initiated at different relative times on the eukaryotic chromosome (Taylor, 1960) and that there are multiple sites for initiation of replication (Cairns, 1966; Huberman and Riggs, 1968) explain how such a large genome could be replicated in sufficient time to permit observed rates of cell division and growth. The complexity of the process of initiation and replication at multiple sites on multiple chromosomes immediately suggests additional levels of control or regulation not apparent in prokaryotes.

Second, the chromosomes of eukaryotes are replicated during a defined period of the cell division cycle, the S phase of interphase (Howard and Pelc, 1953). The control of entry of a cell into this phase constitutes a major regulatory point for commitment to cellular division as the consequence of the duplication of the cell's genetic material.

Third, a substantial portion of the eukaryotic genome (30–70%) consists of DNA sequences that are repeated from several hundred to several hundred thousand times in the single cell (Britten and Kohne, 1968). The discovery that these sequences are sometimes replicated at defined times within the S phase suggests that the replication of DNA at multiple subchromosomal sites might be subject to regulation by sequence specificity as well as by temporal and spatial relationships.

The advantages of tissue cultured cells of mammals for experimental purposes lies primarily in technical parameters. The development of techniques for synchronization of cell populations resulted in opportunities to study the replication process at different time points within S phase, as well as by simply increasing the proportion of cells in the population actively engaged in replication at any given moment. Further, the synchrony of entry into S phase by cell populations blocked at the G1/S boundary also resulted in increased synchrony of initiation of individual replicons. Such synchrony is extremely important for the analysis of replication units whose physical parameters change as a consequence of the replication process. Tissue cultured cells also proved quite amenable to radioisotopic labeling with tritiated thymidine, permitting analysis of replicating molecules by autoradiographic and ultracentrifugation techniques. The isolation of DNA could be performed by gentle lysis of the cell and nuclear membranes resulting in large, relatively unsheared molecules containing many replication units. In spite of these advantages, mammalian tissue culture cells still do not exhibit the physiological flexibility of bacteria. They grow well only in a narrow temperature range and require nutritionally complex media. Such physiological limitations have contributed to limiting the success of attempts to isolate mutant cell lines defective in various components of DNA synthetic process.

Because of these limitations and the increasing potential for the applications of genetic engineering techniques to higher plants, tissue cultured plant cells are promising experimental material for the study of DNA replication. Higher plants in the differentiated state grow in environments that vary widely in temperature and nutritional quality, and they respond to a wide variety of physical and biological stresses. This physiological flexibility is also exhibited in some undifferentiated tissue cultures, permitting growth over a wide temperature range and in nutritionally defined media. Generation times are often comparable to some animal cell cultures (1-2 days). Cultures can grow as suspensions of cells or clumps of cells approaching a relatively homogeneous population with respect to physical proximity of the culture environment.

LITERATURE REVIEW

Experimental Approaches

The utilization of higher plants for studies of DNA replication has been rather limited, despite use in the early and classic experiments that demonstrated that DNA synthesis occurs during a particular period within the cell division cycle (Howard and Pelc, 1953) and that chromosomes (and hence DNA, since chromosomes are unineme) are replicated in a semiconservative fashion (Taylor et al., 1957).

The development of methodologies for the sterile in vitro culture of plant cells has led to a large number of applications for many plant species in areas of clonal propagation, morphogenesis, and genetic modification. The totipotency of higher plant cells provides remarkable opportunities for correlation between physiology and genetics in undifferentiated versus differentiated tissues. The ability to culture cell suspensions has also led to a number of desirable properties which make cell suspension cultures good experimental systems for studying DNA replication.

In designing a model system for the study of DNA replication, several characteristics are desirable: It should be possible to isolate large DNA molecules, undegraded by shear or by nucleases. This isolation should proceed under conditions in which synthesis is inhibited and replication intermediates preserved. Cells should be easily radioactively labeled with DNA precursors or analogues of precursors. To accomplish this, it should be possible to reduce the endogenous precursor supply in order to increase the efficiency with which exogenous label is incorporated. Moreover, pools of exogenous precursors should be rapidly exchanged or emptied to facilitate pulse-chase experiments. One would like to be able to synchronize the system and align cells at the G1/S boundary so that replication of specific sequences can be studied temporally within S phase and synchronous initiation of replication at numerous replication units at the beginning of S phase can be analyzed. Moreover, since such synchronization increases the number of cells in which replication is occurring, it increases the efficiency of the system for studying DNA replication. These characteristics are met to one extent or another by using either cultured cells, protoplasts, or nuclei derived from them.

Figure 1 illustrates the several alternate approaches that can be used and their interrelationship as far as the final DNA product to be analyzed. Depending on the method of analysis and the experimental regime, any one of the three might be the method of choice. The simplest experimental regime involves the radioactive labeling of cells directly in the cell suspension culture. This requires minimum perturbation of cells that are actively growing and hence replicating their DNA. One problem with this approach is that it is difficult to extract

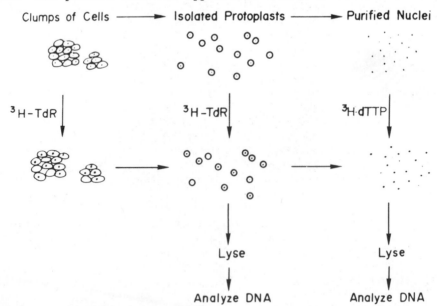

Figure 1.   Experimental approaches in studying DNA replication in plant tissue cultures.

DNA directly from cells without shear degradation. High molecular weight DNA is essential for the analysis of the size and spatial relationship of replication intermediates. One solution to this problem is the enzymatic conversion of cells to protoplasts after labeling and gentle lysis of protoplasts by osmotic shock or detergent to permit purification of nuclei. Nuclei can then be lysed with detergent and DNA extracted for purification or analysis. The major limitation to this approach is the time period required for conversion of cells to protoplasts and for isolation of nuclei. There is a real possibility that metabolic processing of replication intermediates might occur prior to extraction. Such processing could either be processed into more mature products, or be degraded as the consequence of plant nucleases active in vivo during the protoplasting and isolation.

A solution to this problem is to convert cells to protoplasts prior to labeling. This would permit quick termination of a radioactive pulse by lysis of protoplasts with alkali to extract DNA directly or with detergent for nuclei isolation. The processing factor would be minimized by this protocol. Difficulties with this approach lie in the effect of the protoplasting procedure on the metabolic state of the isolated protoplasts.

Macromolecular Synthesis in Protoplasts

The early observations (Nagata and Takebe, 1970) that isolated tobacco leaf mesophyll protoplasts had the capability after wall regeneration to undergo subsequent cell division led to a number of studies on macromolecular synthesis or uptake by the incorporation of radioactive precursors of protein, RNA, and DNA. Sakai and Takebe (1970) demonstrated that RNA and protein synthesis occur in isolated tobacco mesophyll protoplasts. Other studies with cucumber leaf mesophyll protoplasts (Robinson and Mayo, 1975) indicated a low rate of macromolecular synthesis immediately after protoplast isolation. This low rate increased gradually within the next several days. Watts and King (1973) failed to detect any significant incorporation of thymidine into tobacco mesophyll protoplasts. Ohyama et al. (1972) observed little or no thymidine incorporation into protoplasts freshly isolated from soybean suspension cells.

Zelcer and Galun (1976) later examined the kinetics of incorporation of radioactive thymidine into acid-insoluble material in tobacco mesophyll protoplasts isolated under conditions to assure at least 60% cell division within 3 days culture. They found that leucine and uridine incorporation began immediately after culture, with rates increasing after 12 hr. Thymidine incorporation was very low initially, began increasing after about 20 hr, and peaked 40 hr after culture. This peak preceded the increase in cell division by several hours. Rubin and Zaitlin (1976) further showed that high concentrations of protoplast suspensions during culture could decrease leucine and uridine incorporation as a result of dilution of labeled precursors in the incubation medium by nonlabeled precursors of cellular origin. Incorporation of radioactive leucine and uridine into tobacco protoplasts was reduced by the degree of osmotic stress experienced during isolation (Premecz et al., 1978).   Similar inhibition of leucine uptake and incorporation was reported for nondividing protoplasts isolated for *Convolvulus* suspension cultured cells (Ruesink, 1978).

While these studies were concerned primarily with protein and RNA synthesis, it is clear that rates of synthesis for all macromolecules in isolated protoplasts could be influenced by a number of factors, that rates could change during culture, and in particular that DNA synthesis appeared to be quiescent for quite some time after protoplast isolation. These facts all suggested that the isolated protoplasts in these systems might not reflect the same physiological condition as the cells from which they were isolated.   No attempt was made in these studies to compare rates of synthesis in protoplasts to rates in intact cells, a factor of critical importance in assessing the physiological significance of the incorporation that was occurring.   One study (Kulikowski and Mascarenhas, 1978) did show that protoplasts enzymatically derived from suspension cultures of *Centaurea cyanus* incorporated radioactive uridine into total RNA with kinetics similar to that of whole cells. However, during the first hours of culture, protoplasts showed an increased proportion of poly(A) containing RNA and faulty processing of ribosomal RNA precursor.   Within 24 hr, during which time cell wall regeneration was initiated, RNA profiles resembled those of whole cells.

There have been few studies designed primarily toward increasing the rates of macromolecular synthesis in freshly isolated protoplasts.   Numerous experiments with oat mesophyll protoplasts have been carried out, with the goal of improving protoplast viability, the hope being that increased viability might lead to sustained cell division of the cultured protoplasts, a serious roadblock to many tissue culture approaches with cereals.   Rates of macromolecular synthesis were used as assays to monitor protoplast viability (Fuchs and Galston, 1976).   They found thymidine incorporation into acid-insoluble material to proceed uninterrupted for at least 21 hr, while uridine and leucine incorporation stopped about 6 hr after isolation.   Efforts were made to lengthen the extent and increase the rate of precursor incorporation.   Leaf pretreatment with cycloheximide or kinetin was shown to increase protoplast yields as well as rates of leucine and uridine incorporation (Kaur-Sawhney et al., 1976).   Additions of polyamines were found to stabilize protoplasts against spontaneous lysis (Altman et al., 1977), decrease the postexcision rise in RNase and protease activity (Kaur-Sawhney et al., 1977), and increase the incorporation of amino acids and nucleosides into the acid-insoluble fraction of the protoplasts (Altman et al., 1977; Kaur-Sawhney et al., 1977).   Recent studies (Kaur-Sawhney et al., 1980) have indicated that most of the previously observed incorporation of thymidine into acid-insoluble material was not into a DNase sensitive component, indicating little or no net DNA synthesis in oat leaf protoplasts.   The same study, however, shows that the addition of 1 mM spermine or spermidine increased the rate of total thymidine incorporation, and more importantly, the amount converted into the DNase sensitive fraction.   Some increase in the frequency of binucleate protoplasts was observed in 72-168 hr cultures, perhaps the consequence of improved DNA synthetic capacity in these protoplasts.   No such stimulatory effect has been observed on fully viable protoplasts isolated from soybean or *Datura innoxia* suspension cultured cells (Cress, unpublished data).

Radojevic and Kavoor (1978) investigated the rate of DNA synthesis during the initial period of culture of protoplasts isolated from callus cultures of *Corylus avellana*. They normalized the rate of synthesis by prelabeling callus cells with long exposure to ^{3}H thymidine prior to protoplast isolation and labeling with ^{32}P orthophosphate. The rate of synthesis was estimated by the ^{32}P/^{3}H ratio of radioactivity of RNase-resistant material isolated by agarose gel chromatography instead of by acid-precipitation. Their results indicated a fairly linear rate of DNA synthesis from 16 hr after isolation up to over 100 hr; however, no comparison was made to rates of synthesis in the callus culture from which the protoplasts were isolated.

The most detailed study examining DNA synthesis in freshly isolated protoplasts was carried out by Cress et al. (1978). A short (1 hr) pro-toplasting procedure in sucrose-glucose medium resulted in preparations of protoplasts which were able to synthesize DNA without delay and at the same rate as the cell suspension from which they were prepared. Kinetics of incorporation into DNA were reproducibly linear for several hours and permitted the study of replication intermediates by immediate extraction in the cold with alkali. Some results of that study will be presented below. Subsequent studies (Roman et al., 1980; Cress, unpublished data) have demonstrated that use of low osmotic medium is the key factor in obtaining high initial rates of DNA synthesis in fresh-ly isolated soybean protoplasts. Optimal conditions for high initial rates of DNA synthesis involved 0.2 M mannitol or KCl as osmotic sta-bilizers. High osmotic strength medium inhibited incorporation of thy-midine into DNA with little effect on uptake or thymidine pool size.

The above studies on DNA synthesis in isolated protoplasts in one way or another measured net rates of synthesis. The net rate is pro-portional not only to the actual synthetic rate but could be influenced by degradation of either parental or nascent DNA. The observations of increased RNase levels during culture of oat leaf protoplasts (Altman et al., 1977; Kaur-Sawhney et al., 1977; Galston et al., 1978) and tobacco mesophyll protoplasts (Lazar et al., 1973; Premecz et al., 1977) sugges-ted a possible correlation between protoplast vitality and nuclease levels in vivo. The RNase levels in the oat protoplasts could be re-duced by antisenescence treatments such as addition of polyamines. The tobacco protoplast experiments suggested that osmotic stress in-duced the synthesis of new protein responsible for the increase in RNAse activity (Premecz et al., 1977, 1978).

Nuclease degradation of DNA was observed in cultured protoplasts of *Corylus avellana* (Radojevic and Kovoor, 1978). There was some corre-lation between the extent of degradation of preexistent or parental DNA, the integrity of the protoplast preparation, and the extent of synthesis of new DNA during culture. Similar observations have been made for protoplasts isolated from soybean suspension cultured cells (Cress, unpublished data).

An additional factor influencing the measurement of rate of synthesis is the capacity for uptake of the exogenous radioactive precursor.

Howland and Yette (1975) observed that the low level of incorporation of thymidine into wild carrot cells or protoplasts resulted from degradative catabolism of the exogenous thymidine rather than loss of DNA synthetic capacity. The addition of 10^{-5} M 5-fluorodeoxyuridine (FUdR) in conjunction with radioactive thymidine inhibited this degradation as well as enhancing thymidine incorporation into DNA. Subsequent studies using wild carrot suspension cultures (Slabas et al., 1980) and sugarcane suspension cultures (Lesley et al., 1980) confirmed catabolism of exogenously supplied thymidine to β-aminoisobutyric acid. The addition of FUdR or thymine was effectual in reducing thymidine degradation but without complete success. The authors concluded that degradation was most seriously a problem during long incubation times in radioactive thymidine. Proper selection of cell concentration, incubation time, and exogenous thymidine concentration were critical factors in establishing thymidine incorporation as a valid measure of the rate of DNA synthesis in cultured carrot cells (Slabas et al., 1980).

FUdR was utilized to enhance thymidine incorporation into suspension cultured cells of soybean (Cress et al., 1978). The kinetics of uptake as well as analysis of isolated labeled DNA showed that incorporation under these culture conditions was into DNA and that rates of incorporation were valid measures of DNA synthesis. The pool of endogenous thymidine within these cells was small, since uptake of radioactive thymidine proceeded within 15 min without an observable lag and was stopped immediately when radioactive thymidine was replaced by non-radioactive precursor. That permitted the authors to carry out pulse-chase experiments to analyze the precursor-product relationships of replication intermediates and replicating units.

Cell Synchronization

A large number of different experimental systems have been used to study the plant cell cycle and its regulation. The only detailed studies on DNA replication used suspension cultured cells partially synchronized by accumulation at the G1/S boundary (Chu and Lark, 1976; Cress et al., 1978; Lark and Cress, 1978). Treatment with FUdR for one generation results in alignment of about 70% of the cell population at the G1/S boundary, with the majority of the remainder arrested in the various portions of the S phase in which they were when inhibitor was added. Upon removal of FUdR and addition of thymidine, the synchronized cells are viable and undergo replication and a subsequent wave of mitoses at times expected from the established periods of their cell cycles. Protoplasts (Cress et al., 1978) or nuclei (Roman et al., 1980) populations derived from synchronized cell populations exhibited replication properties similar to those of the cell cultures.

DNA Polymerases

The capacity for controlling experimental conditions by using suspension cultured cells has been exploited to study the relationship between

DNA synthesis and DNA polymerases of plant cells. The major DNA polymerase (α-like) present in rice suspension cells was isolated and its activity correlated with cell proliferation (Amileni et al., 1979). Inhibition by aphidicolin of in vivo DNA synthesis and of in vitro α-like DNA polymerase activity suggests that this enzyme is the nuclear replicating enzyme of plant cells (Sala et al., 1980).

DNA Fiber Autoradiography

The characteristic feature of DNA replication in eukaryotes is the spatial organization of the chromosome into multiple discrete units of replication that are simultaneously active. This feature was first discovered by Cairns (1966) using the technique of DNA fiber autoradiography to study DNA replication in mammalian tissue culture cells. The technique consists of labeling DNA to high specific activity with ^{3}H thymidine, releasing DNA from the cell by very gentle lysis, and spreading out the fibers linearly on the surface of a glass slide. The preparations are fixed and coated with suitable photographic emulsion and exposed until tandem arrays of silver grains of sufficient density are produced. These arrays represent daughter DNA helices replicated during the period of labeling.

The pattern of tandem arrays reflects the activities of replication origins at and during the time of the radioactive pulse. If adjacent units initiate replication simultaneously, as in FUdR synchronized cell cultures (Fig. 2), then the distance between those origins would be equivalent to the measured distances between the centers of adjacent autoradiographic tracks. If, however, units were in various stages of bidirectional replication, the two adjacent tandems would reflect the two symmetrical replicating forks on either side of the origin (Fig. 2, exponential). In that case the distance between adjacent origins would not be equivalent to center-to-center distances, but instead would be measured by the distance between the midpoints of two pairs of tandem tracks of identical length. Both of these patterns require at least two adjacent replicating units to be active during the radioactive pulse. In practice, fibers with numerous tandemly active units are found, partly because of the subjectivity of selection by the investigator but in large part because active replicating units tend to be clustered, with almost simultaneous activation of neighboring origins.

More sophisticated analyses can be performed by altering the specific activity of the radioactive precursor during the pulse, producing changes in grain intensity within an autoradiographic track. Such protocols have clearly demonstrated that bidirectionality of replication from the origin of a replicating unit is the norm for eukaryotic cells. It should be noted that for purposes of measurements of interorigin distances the period of labeling should be short enough to prevent adjacent tandems from fusing.

With the use of autoradiographic techniques, the same general pattern of replicating units has been observed in most eukaryotes, including *Drosophila*, yeast, animal cells, and plants. The first and most extensive use of fiber autoradiography to study replication units in higher

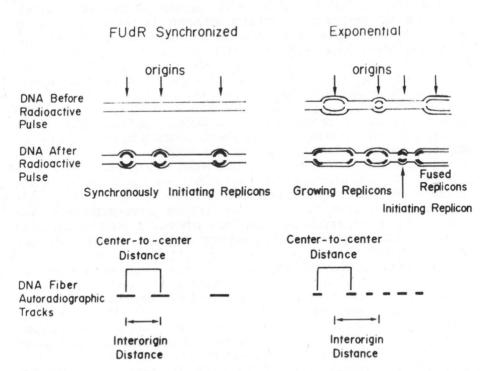

Figure 2.   Schematic representation of DNA replication analysis by
DNA fiber autoradiography.

plants was by Van't Hof and colleagues.   They showed that the interorigin distance between replicating units in pea root meristem ranged
from 20 to 140 μm, with an average of 55 μm.   Replication was bidirectional in most units, and the average rate of fork movement was
29 μm/h at 23 C (Van't Hof, 1975).   In nutritionally synchronized root
meristem cells, replicon size averaged 38 μm for cells terminating S
phase (Van't Hof, 1976a), while two size classes of 40 μm and 50 μm
were observed for cells in early S phase (Van't Hof, 1976b).   Cultured
explants of stelar and cortical mature pea root tissues were found to
have a replicon size of 18 μm, indicating that some origins not activated in embryonic cells are active in differentiated cells of pea (Van't
Hof and Bjerknes, 1977).   Analysis of seedlings of *Arabidopsis thaliana*,
a plant of much smaller genome size, revealed two replicon families
initiating replication at different times in S, with average replicon size
of 24 μm (Van't Hof et al., 1978a).   Root tip meristem cells of sunflower grown at different temperatures varied in the length of the S
phase of the cell cycle; however, the average replicon size of 22 μm
was invarient (Van't Hof et al., 1978b).   Rather, variations in rates of
fork movement were observed.
    In summary, the analysis of several plant species including undifferen-
tiated cultured cells suggests a basic replicon spacing of about 18 μm

(58 kb), with certain tissues displaying spacings of approximately two or three times that in length, indicating supression of activable replication origins under those conditions. Replication appears to proceed bidirectionally in the vast majority of replicons, and both replicon spacing and the rate of fork movement can vary throughout S. As yet there is little evidence for the existence of defined termini of replicating units, and the nature of the regulation of individual origin activation and regional cluster activation is not known. The power of DNA fiber autoradiography lies in the simplicity, directness, and precision of measurements. Three aspects of autoradiography prevent a detailed analysis of replication intermediates: By this method, the resolution of the technique does not permit a quantitative analysis of replicated or nonreplicated regions shorter than 3 μm; adjacent regions which are not covalently linked cannot be distinguished from those that are; the technique is, by its nature, subjective and greatly influenced by the microscopic fields available. Thus aggregates must be avoided, and long, labeled regions tend to confuse shorter, tandem regions. Long regions are often belittled in analysis, since they represent a small proportion of the DNA fibers, although they may represent large amounts of replicated material (i.e., one hundred 10 μm pieces are equal to 1 mm piece).

Sedimentation Velocity Analyses

The above drawbacks of fiber autoradiography can be avoided by complimentary analysis of replicated DNA using rate zonal sedimentation in neutral or alkaline sucrose density gradients. These methods involve the entire population of replicating or radioactively labeled molecules and can utilize pulse-chase techniques to determine the fate of a replicated piece of DNA as to its relationship with neighboring replicating units, i.e., ligation or fusion of adjacent replicons.

Figure 3 presents a schematic representation of actively replicating units in FUdR synchronous or asynchronous soybean suspension cells in which replication intermediates exist that are smaller than replicon-sized intermediates. If such DNA is sedimented in neutral sucrose gradients, the newly replicated (labeled) DNA will sediment with the velocity characteristics of the larger parental strands. That velocity is proportional to the size of the isolated fragments. Under gentle lysis conditions, parental DNA of length many times that of replicon-sized fragments can be obtained, permitting analysis of several replicating units in localized regions. If the DNA is denatured by alkali, the separated single strands assume the structure of a random coil, and replicating fragments will sediment in alkaline sucrose gradients independently of the parental strands. The quantitative relation of sedimentation rate to molecular weight (hence, length) has been empirically determined (Studier, 1965), and therefore lengths of labeled pieces can be calculated from sedimentation rates. If cells have been prelabeled with a different radioactive percursor (e.g., ^{14}C) for a generation prior to pulse labeling with ^{3}H thymidine, then parental DNA and replicating DNA can be distinguished in the same analysis.

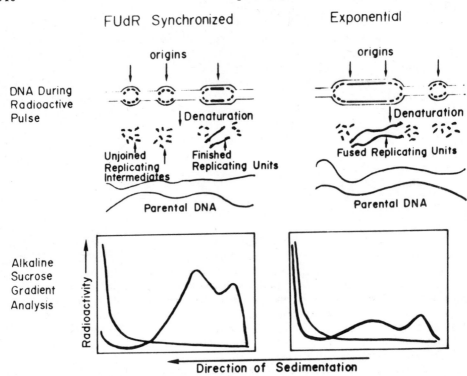

Figure 3. Schematic representation of DNA replication by analysis by alkaline sucrose density gradient centrifugation.

Few such studies have been carried out with plant tissue. Experiments with *Vicia* embryos (Sakamaki et al., 1975, 1976) and cotton radicles (Clay et al., 1976) indicate that a unit sedimenting at 25S (corresponding to about 50 kb in length) is synthesized during long pulses, but that with brief pulses a much smaller intermediate sedimenting at 4-6S (200-400 nucleotide pairs long) can be observed. The relationship of these intermediates to each other was not determined. Such small intermediates have been observed in other eukaryotic systems (Sheinen et al., 1978) and can be shown by pulse-chase experiments to be precursors of the larger units.

With the use of protoplasts from asynchronous or FUdR synchronized suspension cultures of soybean (Cress et al., 1978; Lark and Cress, 1978) it was shown that such small intermediates do occur and rapidly become associated with higher molecular weight DNA. In such quantitative experiments it is easily shown that during the chase period all of the radioactive label is conserved, but no residual incorporation occurs. These small "Okazaki" pieces occur in both exponential and synchronized cultures. However, the rate at which they are joined is much faster in exponential cultures.

Moreover, in synchronous cultures the product of joining corresponds to the tandem units seen in autoradiograms, whereas a much broader size distribution is observed in asynchronous exponential cultures. Why should the two be different?   Two explanations are likely: The different patterns are an effect of the synchronizing treatment, i.e., FUdR treatment somehow inhibits joining.   If this is the case, the inhibition is not immediate, since no such effect is observed after short treatments with this analogue.   The spacing of replicating units at the beginning of S is different from that at later times, regularly interspersing replicating and nonreplicating units at the onset of S.   Similarly, small "Okazaki" intermediates may be separated by unreplicated "Okazaki" regions.   Evidence exists consistent with both of these possibilities.   We have observed an inhibition of DNA replication by the continued presence of FUdR despite the presence of exogenous thymidine and uridine.   Thus it is possible that FUdR slows the synthesis of "Okazaki" piece intermediates which would prevent joining of both larger and smaller pieces.

A more interesting set of observations support the second alternative.   In both mammalian cells (Hori and Lark, 1974) and *Physarum* (Funderud et al., 1978) it has been observed that clustering of replicating units occurs such that units synthesized early in S are separated by units that will only be synthesized later.   Moreover, in *Physarum*, there is evidence that small "Okazaki" units join more slowly at the beginning of S than at the end (and are thus observed for longer times).   The *Physarum* data are particularly strong, since they take advantage of a naturally synchronous system.   There is some evidence for similar differences in clustering of replicons in higher plants, since Van't Hof (1976a,b) observed much longer basic replicating units late in S as compared to the onset of S.   If the larger units are aggregates of simultaneously replicating smaller units, the conclusion could be drawn that replicating units are clustered and therefore joining more rapidly in S.

Quantitative analysis of replicating DNA by sedimentation velocity was performed on the same experimental material as used for fiber autoradiography (Cress et al., 1978) and revealed that the maximum sizes of labeled fragments measured by autoradiography corresponded to those observed by alkaline sedimentation, indicating that the lengths measured in autoradiograms are representative of the entire population of labeled DNA molecules.

Buoyant Density Analyses

Suspension cell cultures also permit the use of density analogues of thymidine such as bromodeoxyuridine (BUdR).   Since this analogue can be substituted for thymidine in the medium and since FUdR can be used to inhibit the synthesis of endogenous thymidine, complete substitution of BUdR for thymidine in DNA can be obtained.   An immediate advantage of this is that DNA replicated in BUdR can be separated from the rest of the genome by isopycnic centrifugation in CsCl.   In this manner it becomes possible to use synchronized cultures to enrich for and isolate DNA synthesized at particular periods in S.   The density separation is particularly good with soybean, since the DNA is normally AT rich and therefore light, whereas substitution with BUdR results in DNA rich in BUdR or extra heavy (the density shift upon substitution goes from 1.694 to 1.747 gm/cm^3).

In addition to separating DNA that is replicated at a particular time, density labeling can be used to estimate the number of replicating units per nucleus. This is done by labeling cells for short periods with BUdR and measuring the amount of nonreplicated adjoining DNA banding in CsCl in a region of heavier density. If the size of the pieces approaches the size of replicating units and if the size of the replicated regions is small by comparison, then the amount of unreplicated DNA pulled to a region of heavier density represents the size of the replication units multiplied by the number of BUdR centers (i.e., the number of units in the process of replication). Knowing the genome size allows a calculation of the number of replication units. For soybean cells beginning S, this number is about 5000 (Cress et al., 1978).

Another aspect of replication which can be inferred from experiments with density labeling is the spacing or interspersion of small "Okazaki" intermediates. We have just discussed the existence of these very small 4-6S intermediates and suggested that they may not all be immediately contiguous, especially if synthesized after FUdR synchronization. If this is so, we expect a density label to be incorporated into small pieces separated by regions not yet replicated. If these other regions are subsequently replicated in BUdR, the whole region will become uniformly dense corresponding to the expected density shift. If, however, subsequent replication occurs in thymidine, we expect "finely divided" regions of heavy and light density and the fineness of the division should correspond to the 4-6S subdivision. This result has been observed, leading to the conclusion that the "Okazaki" piece intermediates are separated by unreplicated space of similar dimensions (Cress et al., 1978; Lark and Cress, 1978).

The ability to enrich for replicating DNA synthesized at different times in S by use of FUdR-synchronized cultures and BUdR labeling has been applied to the study of ribosomal (rDNA) cistrons in soybean (Jackson and Lark, 1982). DNA replicated at different intervals following release from inhibition was separated from nonreplicated DNA on the basis of buoyant density increase and hybridized to ribosomal RNA to measure replication of ribosomal cistrons. More than one-third of the cistrons replicated within the first 90 min of the 8 hr S period. These cistrons constituted the "major" type of rDNA cistron in soybean, while rDNA sequences of the "minor" type were replicated later in S.

DNA Stability

The long-standing question of the relation of changes of genomic DNA content and composition to development has been studied in carrot tissues along with DNA from whole carrot plants for comparison. A heavy satellite DNA band was observed in carrot explants that could not be detected in the DNA of the whole carrot plant, and pulse-chase experiments indicated the occurrence of a metabolic turnover of DNA (Schafer et al., 1978). Such changes in DNA composition, particularly in the intermediate repeated DNA fraction, could be induced by treatment of explants with gibberellic acid (Schafer and Neumann, 1978). Further studies showed that another satellite consisting of highly repeated DNA sequences was preferentially synthesized during a 24 hr labeling period in carrot tissue cultures growing with high cell division activity (Duhrssen and Neumann, 1980).

A more detailed study of the genomic DNA sequence alteration has been performed in the ribosomal DNA cistrons of soybeans (Jackson and Lark, 1982). Soybean SB-1 cells grow rapidly in sucrose with a 1 day doubling time. Upon transfer to medium containing maltose, the doubling time is reduced to 8 days (Limberg et al., 1979). After several generations of slow growth in maltose, one-third of the rDNA cistrons has been lost. These sequences represent the minor type of ribosomal cistron which is replicated later in S phase. The major type of cistron, replicated early in S, is maintained during growth in maltose. These sequences cannot be essential for cell growth in culture because subsequent, prolonged, rapid growth in sucrose medium fails to completely restore the lost sequences. It has not been determined whether this loss results from active removal of specific sequences or from the failure to replicate these sequences during the first generations growth in selective medium. Likewise, the significance of the correlation between sequence loss and timing of replication in S phase is unclear at present.

In vitro DNA Synthesis

The use of cells or protoplasts in studies of in vivo DNA synthesis has limitations in the detail with which replication process can be examined. For ultimate purposes of biochemical reconstitution of the replication complex and complementation using mutants defective in different aspects of replication, an in vitro system of DNA replication is required. A first step in that direction has been the in vitro replication of DNA by isolated soybean nuclei (Roman et al., 1980). The rate of DNA synthesis in those nuclei was as high as 30% of the in vivo rate, yet ceased after 10-15 min. Several lines of evidence supported the hypothesis that such synthesis was replication rather than repair synthesis. However, the DNA replicated in vitro was in large part comprised of specific repeated DNA sequences (Caboche and Lark, 1981), suggesting that different components of the replication machinery might be involved in replicating single copy and repeated DNA sequences. Quantitative changes in the pattern of replication of these repeated sequences in vitro were observed in nuclei isolated from cells cultured for 16 hr in the presence of cytokinins. However, the addition of cytokinins to the in vitro reaction mixture had no effect on the pattern of replication. While it has not yet been demonstrated that initiation of replication at replication origins occurs in this in vitro system, the system offers great potential for elucidating the biochemical machinery of replication and the process of regulation of that machinery.

PROTOCOLS

Cell Culture

The cell line SB-1 of *Glycine max* (L.) Merr. cv. Mandarin (Gamborg et al., 1968) is grown in suspension in liquid medium. Either B5 (Gamborg et al., 1968) or MS (Murashige and Skoog, 1962) medium containing 4.5 μM 2,4-D and 0.4% CH (vitamin and salt free, ICN Pharmaceutical Inc., Cleveland, Ohio) is suitable for rapid cell growth. Stock cultures can be maintained at temperatures from 20 C to 33 C. Cells are routinely cultured in the dark at 33 C as 50 ml batch suspension in 250 ml baffled Delong shake flasks on a gyrotary shaker (110-150 rpm). Cell growth is measured either by converting cells to protoplasts and counting these in a Neubauer hemacytometer or by measuring the packed volume of centrifuged cells. Cell density is maintained between 6×10^5/ml and 3×10^6/ml by directly diluting 10 ml cells into 40 ml fresh medium every 2 days. Under these conditions, exponential growth is maintained and cell number doubles every 20-24 hr. Cell viability is determined using the vital stain trypan blue (0.4% in normal saline, Grand Island Biological Company, Grand Island, N.Y.).

Synchronization

Populations of SB-1 cells can be partially synchronized by treatment with the DNA synthesis inhibitor fluorodeoxyuridine (FUdR) (Chu and Lark, 1976). FUdR (filter-sterilized in aqueous solution) is added directly to suspension cultures on the second day following subculture to a final concentration of 1 μg/ml, along with 0.5 μg/ml uridine. Cultures are incubated a further 24 hr, during which time cells accumulate at the G1/S boundary of the cell division cycle. Cells are released from inhibition by washing twice with fresh filter-sterilized medium containing 2 μg/ml thymidine.

Protoplast Isolation

For subsequent nuclei isolation:

1. Prepare enzyme solution containing desalted 2% cellulase Onozuka R-10 and 0.5% Macerozyme R-10 (both enzymes from Kinki Yakult, Nishinomiya, Japan) in digestion buffer containing 0.6 M mannitol, 5 mM $CaCl_2$, and 2 mM KH_2PO_4. Pass the solution through filter paper (Whatman No. 1), adjust to pH 5.4, filter sterilize, store at 4 C.
2. Mix equal volumes of cell suspension and enzyme solution in sterile petri dish or centrifuge tube. Incubate with gentle shaking (ca., 40 rpm on gyrotary shaker) at 33 C for 30 min to 2 hr.
3. Monitor protoplast production by examination on an inverted stage light microscope. Protoplast production should be 95% complete within 2 hr.

4. To remove incompletely digested clumps of cells, filter the proto-
    plast suspension twice by gravity through a stainless–steel or nylon
    screen (50 µm pore size) into sterile centrifuge tube (e.g., Falcon
    2037).
5. Centrifuge (IEC Clinical model) the protoplasts at 100 x g for 5
    min, pour off supernatant and gently resuspend protoplasts in 10 ml
    protoplast digestion buffer.
6. Repeat step 5 two times.
7. Protoplast concentration is determined by counting using a hema-
    cytometer.

For subsequent isolation of nuclei for in vitro DNA synthesis (Roman
et al., 1980):

1. Centrifuge exponentially growing cells at 100 x g for 5 min and
    resuspend at the. same concentration in digestion buffer (200 mM
    KCl, 50 mM sucrose, 3 mM $CaCl_2$, 2 mM NH_4NO_3, and 1 mM
    KH_2PO_4, pH 5.5) containing 1% cellulase Onozuka R10, 0.02%
    Rhozyme HP150 (Rohm and Haas, Philadelphia). Incubate at 33 C
    with gentle shaking for 3 hr.
2. Remove cell debris and cell clumps by filtration through stainless
    steel screen (50 µm pore size).
3. Wash once with digestion buffer. Resuspend protoplasts at a con-
    centration of 2–4 x 10^6/ml.
4. Layer 10 ml protoplast suspension onto 10 ml digestion buffer con-
    taining 20% (w/v) sucrose in a 30 ml centrifuge tube. Gently mix
    the interface of the two layers to form a localized gradient.
5. Centrifuge in a swinging bucket rotor at 800 x g for 5 min.
6. Collect the material at the top of the 20% sucrose layer, dilute
    with digestion buffer.
7. Centrifuge the protoplasts at 500 x g for 5 min. Resuspend in di-
    gestion buffer.
8. Repeat steps 4 through 7 if necessary to eliminate residual contam-
    ination with cells.

For DNA synthesis in isolated protoplasts:

1. Centrifuge cell suspension (1 to 2 x 10^6 cells/ml) at 100 x g for 3
    min. Wash once with filter–sterilized culture medium.
2. Resuspend cells in digestion buffer (0.2 M mannitol, 25 mM glucose,
    5 mM $CaCl_2$, 2 mM KH_2PO_4, pH 5.4) containing 2% cellulase Ono-
    zuka R–10 and 0.2% Macerozyme R–10.
3. Incubate at 33 C with gentle shaking (50 rpm) for 1 hr.
4. Filter four times through nylon screen (50 µm pore size).
5. Wash twice with digestion buffer.
6. Resuspend at a concentration of 10^6/ml in filter–sterilized cell cul-
    ture medium with the replacement of sucrose by 0.2 M mannitol.

Nuclei Isolation

For subsequent DNA extraction:

1. Pellet washed protoplasts and resuspend at a concentration of 1-2 x 10^7/ml on ice in lysis buffer containing 0.25 M sucrose, 25 mM KCl, 2 mM $CaCl_2$, 1 mM 2-mercaptoethanol, and 25 mM Tris-HCl (pH 7.8).
2. Add an equal volume of lysis buffer containing 0.4% (v/v) of the nonionic detergent Triton X-100.
3. Shake by inversion by hand for 1 min.
4. Dilute with equal volume of lysis buffer.
5. Filter by gravity through two layers of nylon screen (50 μm pore size), and collect filtrate.
6. Filter through one layer of nylon screen (10 μm pore size) three times in succession, each time collecting filtrate and agitating by inversion by hand.
7. Centrifuge the filtrate at 750 x g for 10 min, and resuspend the pellet in lysis buffer containing 0.4% Triton X-100.
8. Set on ice for 5 min.
9. Repeat step 7, washing the pellet in lysis buffer.
10. If further purification is required because of contaminating plastids, membranes, or fragments of nuclei, continue with steps 11 through 13.
11. Layer the nuclei suspension over an equal volume of lysis buffer containing 2.3 M sucrose and 0.2% Triton X-100. Mix the interface gently.
12. Centrifuge for 1 hr at 75,000 x g in a swinging bucket rotor at 4 C.
13. Resuspend the pellet in lysis buffer and wash twice by centrifugation at 800 x g for 10 min.

For subsequent in vitro DNA synthesis (Roman et al., 1980) in isolated nuclei:

1. All following steps should be carried out at 4 C.
2. Resuspend protoplasts (purified as before) on ice at a concentration of 2-4 x 10^7/ml in buffer containing 0.2 M sorbitol, 60 mM KCl, 3 mM $CaCl_2$, and 25 mM Tris-HCl (pH 8.0).
3. Add Triton X-100 to a final concentration of 0.1% (v/v) and mix.
4. After 1 min, 2-ml aliquots in 15 ml tubes are shaken by hand five times at 2 second intervals.
5. Add 5 ml lysis buffer.
6. Centrifuge at 800 x g for 10 min.
7. Wash pellets and resuspend nuclei at concentration of 5 x 10^7/ml.
8. Examine nuclei by phase-contrast microscopy. Contamination by cells or protoplasts should be less than 0.5%.

Autoradiography

Of labeled protoplasts:

1. Prepare labeled protoplasts as above and wash twice in a fixative solution composed of 1 part glacial acetic acid, 3 parts 95% ethanol, 4 parts of 0.56% KCl in distilled water.
2. Resuspend protoplasts to a concentration of 1 x 10^7/ml.
3. Place two drops of the protoplast-fixative mixture on a microscope slide and spread by tilting the slide. Air-dry.
4. Wash the slide in cold 5% (w/v) trichloroacetic acid (TCA) for 5 min and then in 95% ethanol for 2 min. Air-dry.
5. Dip the slide into Kodak NTB-3 liquid emulsion (in the dark, of course) and allow to expose in light-proof boxes for an appropriate period of time (1-10 days).
6. Develop at 25 C with Kodak D-19 developer for 5 min and fix in Kodak fixer for 7 min.
7. Wash with distilled water and air-dry.
8. For staining, dip the slide in Giemsa stain solution (Fisher) for 20 min at room temperature. Destain by dipping in acetone.

Of labeled nuclei:

1. Resuspend labeled nuclei in lysis buffer containing 20% sucrose.
2. Fix by gentle addition of fixative solution consisting of 1 part glacial acetic acid, 3 parts ethanol, 1 part sucrose-containing lysis buffer.
3. Prepare for autoradiography as above.

Of labeled mitoses:

1. Prepare modified carbol fuchsin stain (Kao, 1975) as follows: Dissolve 2 g basic fuchsin in 50 ml of 70% ethanol. Filter through Whatman No. 1 filter paper. Add 2 ml of filtrate to 18 ml of 5% aqueous phenol in distilled water. Mix and add 2.5 ml acetic acid plus 2.5 ml of 37% formaldehyde. Dilute 5 ml of this solution to 100 ml with 45% acetic acid and 2 g mannitol. Age at room temperature for several weeks before use. This final staining solution is stable at room temperature for at least several years.
2. Wash appropriate cells once in fixative solution of acetic acid-ethanol (1:3, v/v).
3. Resuspend in fixative and incubate overnight at 4 C.
4. Wash once with fixative and once with 45% acetic acid.
5. Place about 20 µl of the fixed cells suspension on a clean glass microscope slide.
6. Slowly add about 50 µl of 5% modified carbol fuchsin.
7. Expose to air for 2 min, then lower cover slip onto the drop.
8. After 10-30 min, place two layers of absorbant paper over the cover slip and press firmly and evenly with the thumb.
9. Wash with 5% TCA and 95% ethanol, and prepare for autoradiography by dipping in photographic emulsion as described above.

Of DNA fibers:

1. Purify nuclei from protoplasts as described above.
2. Wash twice with cold phosphate–buffered saline by centrifugation at 1000 x g for 5 min.
3. Place a drop containing 10^3–10^4 nuclei on a 1" x 3" glass microscope slide (Gold–Seal, precleaned) previously coated with either 0.5% serum albumin or a solution of 0.01% $CrK(SO_4)_2$ x 12 H_2O, 0.1% gelatin.
4. Place a drop of solution containing 2% sodium dodecyl sulfate (SDS), 0.05 M EDTA adjacent to the drop of nuclei.
5. Slightly tilt the microscope slide to permit the drops to run together.
6. Spread the gently mixing solutions down over the surface of the slide by pulling the drop with a clean glass stirring rod by surface tension.
7. Air–dry the slide at room temperature.
8. Fix the slide by dipping in cold 5% trichloroacetic acid for 5 min.
9. Wash by dipping in water once and 95% ethanol twice. Air–dry.
10. Coat with film by dipping into a 1/1.6 dilution of Ilford L4 liquid emulsion.
11. Place in light-proof boxes to expose at 4 C for the appropriate period of time (2-6 months).
12. Develop the exposed slides with Kodak D19 developer for 5 min at 25 C and fix with Kodak fixer for 7 min.
13. Observe tracks of silver grains under a microscope with brightfield optics. Typical magnification required for measurement of tandem arrays is 6000–12,000x.

Extraction and Analysis of DNA

Isopycnic centrifugation in CsCl:

1. Resuspend nuclei thoroughly in solution containing 20% (w/v) sucrose, 100 mM NaCl, 50 mM EDTA, and 100 mM Tris-HCl, pH 7.8.
2. Incubate at 65 C for 2 min.
3. Add an equal volume of the same lysis buffer containing freshly dissolved 2% (w/v) sodium dodecyl sulfate and 100 μg/ml proteinase K (EM Laboratories); mix gently.
4. Incubate at 65 C for 30-60 min.
5. Slowly dilute the lysate to 6.7 gm with distilled water.
6. Pour this solution onto 8.72 gm CsCl in a polyallomer centrifuge tube. Dissolve the CsCl slowly by gentle inversion.
7. Overlay the solution with clean mineral oil, seal the tube with centrifuge cap, and centrifuge in a Beckman Ti50 rotor at 40,000 rpm at 15 C for 40-60 hr.
8. Fractionate the gradient by piercing the bottom of the tube with a needle and collecting fixed volume aliquots onto Whatman 3MM filter (2.4 cm circles). Air–dry the filters.

9. Wash the filters once in cold 10% TCA, twice in 5% TCA, once in 70% ethanol, and twice in 95% ethanol (200 ml in each wash). Air-dry.
10. Immerse filters in scintillation vials containing a toluene based counting fluid and assay for radioactivity in a liquid scintillation spectrometer.

Velocity sedimentation through neutral sucrose gradients:

1. Prepare a neutral sucrose gradient as follows: Pipet a 1 ml cushion of sterile 70% (w/v) sucrose directly into the bottom of a polyallomer centrifuge tube. Filter-sterilize 5% (w/v) sucrose and 20% (w/v) sucrose solutions containing 20 mM Tris-HCl (pH 7.3), 1 M NaCl, and 4 mM EDTA. Form the linear 5-20% sucrose gradient of 10-11 ml over the 1 ml sucrose cushion using a mixing chamber and peristaltic pump (Buchler Instruments, Fort Lee, New Jersey).
2. Resuspend nuclei in 60 μl neutral lysis buffer as for isopycnic centrifugations. Incubate at 65 C for 2 min.
3. Lyse by the addition of 240 μl buffer containing 100 mM NaCl, 100 mM EDTA, 100 mM Tris-HCl (pH 7.8), 1% SDS, and 100 μg/ml proteinase K. A radioactive marker of ^{32}P bacteriophage DNA (λ, S = 32, or fd, S = 16.5) is mixed with the lysate prior to layering.
4. After incubation at 65 C for 15 min, gently pour the viscous lysate directly onto the top of the neutral sucrose gradient.
5. Centrifuge in a Beckman SW41 rotor at 38,000 rpm at 15 C for 3-6 hr.
6. Fractionate the gradient and assay for radioactivity as for CsCl gradients.
7. If protoplasts are to be lysed directly rather than nuclei, 1% (v/v) diethylpyrocarbonate (Sigma) is added to the lysing buffer.

Velocity sedimentation through alkaline sucrose gradients:

1. Linear 5-20% alkaline sucrose gradients are formed as above for neutral gradients except the sucrose solutions contain 0.2 N NaOH and 10 mM EDTA.
2. Resuspend nuclei in 60 μl of buffer containing 20% (w/v) sucrose, 100 mM NaCl, 50 mM EDTA, 10 mM Tris-HCl (pH 8.0), and proteinase K (100 μg/ml).
3. Incubate at 65 C for 2 min.
4. Lyse by the addition of 240 μl of solution containing 0.25 N NaOH, 20 mM EDTA. Mix gently.
5. Pour directly onto the top of the alkaline sucrose gradient.
6. Centrifuge and fractionate as above for neutral sucrose gradients.
7. For direct lysis, labeled protoplasts are resuspended on ice in 60 μl of buffer containing 20 mM Tris (pH 8.0), 50 mM EDTA, 0.2 M KCl, and 1% diethylpyrocarbonate, and lysed by the addition of 200 μl of solution containing 0.25 N NaOH, 50 mM EDTA, and 1% diethylpyrocarbonate. The lysate is mixed gently, poured directly onto the alkaline sucrose, and centrifuged as above.

FUTURE PROSPECTS

Basic research on DNA replication in higher plants leads to the conclusion that the primary features of organization of the genome for replication are quite similar in plants and animals. The size of the basic replication unit or replicon, the rate of replication fork movement, the nature of transient replication intermediates, the replication enzymes, and the temporal and spatial structure of replication within the S phase of the cell cycle, all appear to have been fundamentally conserved within higher eukaryotic organisms.

Differences, however, exist in cellular control mechanisms in plants and animals, because of differences in the nature of the organism's capacity to respond to the environment. It is reasonable to expect these differences to be also manifested in some aspects of the regulation of DNA synthesis, and data from some of the studies discussed previously do point in that direction. Future basic research in DNA replication in higher plants will likely focus in the following areas: (a) The isolation of conditional lethal (cold-sensitive or temperature-sensitive) mutant cell lines in tissue culture. This will permit the genetic dissection of the replication process through genetic complementation studies by somatic cell hybridization. Mutant cell lines will also be important in studying biochemical complementation in an in vitro DNA synthesis system. (b) The improvement of methods for in vitro DNA synthesis. DNA synthesis in isolated nuclei permeable to proteins or in concentrated nuclear lysates will be critical in identifying and characterizing the gene products essential for replication. Progress will be required to increase the extent (length) of total synthesis, the capacity for initiation at replication origins, and the full genomic sequence participation in such in vitro systems. (c) Sequence specificity initiation of replication. It is not at all clear whether the regulation of initiation of replication is exercised simply at the level of DNA sequence specificity or whether other levels such as higher order periodic features of chromatin structure are involved.

While progress in the above areas will greatly increase our knowledge of the basic biochemistry of DNA replication, the question of DNA stability is one of practical importance in crop improvement programs utilizing plant tissue culture methods. The increasing base of data on genomic variability in plants leads to concern over the effects of tissue culture regimes on genetic stability. It is clear that plants can respond to environmental stress by selective amplification or deamplification (loss) of certain DNA sequences. The mechanisms behind such a response are unknown, but must involve alteration in the regulation of the replication of the DNA sequences involved. The ability to control precisely the environment of cultured cells should facilitate the study of this important phenomenon in hopes that it can be exploited for selectively amplifying desired selectable genes (and hence, gene products) or for controlling (limiting) the loss of unselectable genes or sequences. In any event, further knowledge on the relationship between DNA replication and genetic stability will be most useful in understanding and exploiting tissue culture and genetic engineering methods for crop improvement.

KEY REFERENCES

Cress, D.E., Jackson, P.J., Kadouri, A., Chu, Y.E., and Lark, K.G. 1978. DNA replication in soybean protoplasts and suspension-cultured cells: Comparison of exponential and fluorodeoxyuridine synchronized cultures. Planta 143:241-253.

Hand, R. 1978. Eucaryotic DNA: Organization of the genome for replication. Cell 15:317-325.

Lark, K.G. and Cress, D.E. 1978. Cell division and DNA synthesis in plant cells. In: Frontiers of Plant Tissue Culture 1978 (T.A. Thorpe, ed.) pp. 179-189. International Association for Plant Tissue Culture, Calgary.

Roman, R., Caboche, M., and Lark, K.G. 1980. Replication of DNA by nuclei isolated from soybean suspension cultures. Plant Physiol. 66:726-730.

Van't Hof, J. and Bjerknes, C.A. 1979. Chromosomal DNA replication in higher plants. Bioscience 29:18-22.

REFERENCES

Altman, A., Kaur-Sawhney, R., and Galston, A.W. 1977. Stabilization of oat leaf protoplasts through polyamine-mediated inhibition of senescence. Plant Physiol. 60:570-574.

Amileni, A., Sala, F., Cella, R., and Spadari, S. 1979. The major DNA polymerase in cultured plant cells: Partial purification and correlation with cell multiplication. Planta 146:521-527.

Britten, R.J. and Kohne, D.E. 1968. Repeated sequences in DNA. Science 161:529-540.

Caboche, M. and Lark, K.G. 1981. Preferential replication of repeated DNA sequences in nuclei isolated from soybean cells grown in suspension culture. Proc. Natl. Acad. Sci. USA 78:1731-1735.

Cairns, J. 1966. Autoradiography of HeLa cell DNA. J. Molec. Biol. 15:372-373.

Chu, Y.E. and Lark, K.G. 1976. Cell-cycle parameters of soybean (Glycine max L.) cells growing in suspension culture: Suitability of the system for genetic studies. Planta 132:259-268.

Clay, W.F., Bartels, P.G., and Katterman, F.R.H. 1976. Mechanism of nuclear DNA replication in radicles of germinating cotton. Proc. Natl. Acad. Sci. USA 73:3220-3223.

Coutts, R.H.A., Barnett, A., and Wood, K. 1975. Ribosomal RNA metabolism in cucumber leaf mesophyll protoplasts. Nucleic Acid Res. 2:1111-1112.

Duhrssen, E. and Neumann, H.-K. 1980. Characterization of satellite-DNA of Daucus carota L. Z. Pflanzenphysiol. 100:447-454.

Fuchs, Y. and Galston, A.W. 1976. Macromolecular synthesis in oat leaf protoplasts. Plant Cell Physiol. 17:475-482.

Funderud, S., Andreassen, R., and Haugli, F. 1978. Size distribution and maturation of newly replicated DNA through the S and G_2 phases of Physarum polycephalum. Cell 15:1519-1526.

Galston, A.W., Altman, A., and Kaur-Sawhney, R. 1978. Polyamines, ribonuclease and the improvement of oat leaf protoplasts. Plant Sci. Lett. 11:69-79.

Gamborg, O.L., Miller, R.A., and Ojima, K. 1968. Nutrient requirements of suspension cultures of soybean root cells. Exp. Cell Res. 50:151-155.

Hori, T.-A. and Lark, K.G. 1974. Autoradiographic studies of the replication of satellite DNA in the kangaroo rat. Autoradiographs of satellite DNA. J. Molec. Biol. 88:221-232.

Howard, A. and Pelc, S. 1953. Synthesis of DNA in normal and irradiated cells and its relation to chromosome breakage. Heredity (Suppl.)6:261-273.

Howland, G.P. and Yette, M.L. 1975. Simultaneous inhibition of thymidine degradation and stimulation of incorporation into DNA by 5-fluorodeoxyuridine. Plant Sci. Lett. 5:157-162.

Huberman, J.A. and Riggs, A.D. 1968. On the mechanism of DNA replication in mammalian chromosomes. J. Molec. Biol. 32:327-341.

Jackson, P.J. and Lark, K.G. 1982. Inherited changes in frequencies of different ribosomal RNA cistrons which occur in soybean (*Glycine max*) suspension cultures as a result of altered growth conditions. Molec. Gen. Genet. (in press).

Jacob, F., Brenner, S., and Cuzin, F. 1963. On the regulation of DNA replication in bacteria. Cold Spring Harbor Symp. Quant. Biol. 28:329-348.

Kao, K.N. 1975. A chromosomal staining method for cultured cells. In: Plant Tissue Culture Methods (O.L. Gamborg and L.R. Wetter, eds.) pp. 63-64. National Research Council of Canada, Saskatoon.

Kaur-Sawhney, R., Rancillac, M., Staskawicz, B., Adams, W.R., and Galston, A.W. 1976. Effect of cycloheximide and kinetin on yield, integrity and metabolic activity of oat leaf protoplasts. Plant Sci. Lett. 7:57-67.

Kaur-Sawhney, R., Adams, W.R., Tsang, J., and Galston, A.W. 1977. Leaf pretreatment with senescence retardants as a basis for oat protoplast improvement. Plant Cell Physiol. 18:1309-1317.

Kaur-Sawhney, R., Flores, H.E., and Galston, A.W. 1980. Polyamine-induced DNA synthesis and mitosis in oat leaf protoplasts. Plant Physiol. 65:368-371.

Kulikowski, R.R. and Mascarenhas, J.P. 1978. RNA synthesis in whole cells and protoplasts of *Ceutaurea*—a comparison. Plant Physiol. 61:575-580.

Lazar, G., Borbely, G., Udvardy, J., Premecz, G., and Farkas, G.L. 1973. Osmotic shock triggers an increase in ribonuclease level in protoplasts isolated from tobacco leaves. Plant Sci. Lett. 1:53-57.

Lesley, S.M., Maretzki, A., and Nickell, L.G. 1980. Incorporation and degradation of ^{14}C and ^{3}H-labeled thymidine by sugarcane cells in suspension culture. Plant Physiol. 65:1224-1228.

Limberg, M., Cress, D., and Lark, K.G. 1979. Variants of soybean cells which can grow in suspension with maltose as a carbon-energy source. Plant Physiol. 63:718-721.

Murashige, T. and Skoog, F. 1962. A revised medium for rapid growth and bioassays with tobacco tissue cultures. Physiol. Plant 15:473-497.

Nagata, T. and Takebe, I. 1970. Cell wall regeneration and cell division in isolated tobacco mesophyll protoplasts. Planta 92:301-308.

Ohyama, K., Gamborg, O.L., and Miller, R.A. 1972. Uptake of exogenous DNA by plant protoplasts. Can. J. Bot. 50:2077-2080.

Premecz, G., Olah, T., Gulyas, A., Nyitrai, A., Palfi, G., and Farkas, G.L. 1977. Is the increase in ribonuclease level in isolated tobacco protoplasts due to osmotic stress? Plant Sci. Lett. 9:195-200.

Premecz, G., Ruzicska, P., Olah, T., and Farkas, G.L. 1978. Effect of "osmotic stress" on protein and nucleic acid synthesis in isolated tobacco protoplasts. Planta 141:33-36.

Radojevic, L. and Kovoor, A. 1978. Characterization and estimation of newly synthesized DNA in higher plant protoplasts during the initial period of culture. J. Exp. Bot. 29:963-968.

Robinson, D.J. and Mayo, M.A. 1975. Changing rates of uptake of [^{3}H] leucine and other compounds during culture of tobacco mesophyll protoplasts. Plant Sci. Lett.8:197-204.

Roman, R., Caboche, M., and Lark, K.G. 1980. Replication of DNA by nuclei isolated from soybean suspension cultures. Plant Physiol. 66: 726-730.

Rubin, G. and Zaitlin, M. 1976. Cell concentration as a factor in precursor incorporation by tobacco leaf protoplasts or separated cells. Planta 131:87-89.

Ruesink, A.W. 1978. Leucine uptake and incorporation by Convolvulus tissue culture cells and protoplasts under severe osmotic stress. Physiol. Plant 44:48-56.

Sakai, F. and Takebe, I. 1970. RNA and protein synthesis in protoplasts isolated from tobacco leaves. Biochim. Biophys. Acta 224:531-540.

Sakamaki, T., Fukuei, K., Takahashi, N., and Tanifuji, S. 1975. Rapidly labeled intermediates in DNA replication in higher plants. Biochim. Biophys. Acta 395:314-321.

Sakamaki, T., Takahashi, N., Takaiwa, F., and Tanifuji, S. 1976. Double strandedness of nascent DNA of a higher plant (Vicia faba). Biochim. Biophys. Acta 447:76-81.

Sala, F., Parisi, B., Burroni, D., Amileni, A.R., Pedrali-Noy, G., and Spadari, S. 1980. Specific and reversible inhibition by aphidicolin of the α-like DNA polymerase of plant cells. FEBS Lett. 117:93-98.

Schafer, A. and Neumann, K.-H. 1978. The influence of gibberellic acid on reassociation kinetics of DNA of Daucus carota L. Planta 143:1-4.

Schafer, A., Blaschke, J.R., and Neumann, K.-H. 1978. On DNA metabolism of carrot tissue cultures. Planta 139:97-101.

Sheinen, R., Humbert, J., and Pearlman, R.E. 1978. Some aspects of eukaryotic DNA replication. Annu. Rev. Biochem. 47:277-316.

Slabas, A.R., MacDonald, G., and Lloyd, C.W. 1980. Thymidine metabolism and the measurement of the rate of DNA synthesis in carrot suspension cultures. Plant Physiol. 65:1194-1198.

Studier, F.W. 1965. Sedimentation studies of the size and shape of DNA. J. Molec. Biol. 11:373-390.

Taylor, J.H. 1960. Asynchronous duplication of chromosomes in cultured cells of Chinese hamster. J. Biophys. Biochem. Cytol. 7:455-464.

_____, Woods, P., and Hughes, W. 1957.   The organization and duplica-
tion of chromosomes as revealed by autoradiographic studies using
³H-labeled thymidine.   Proc. Natl. Acad. Sci. USA 43:122-128.

Van't Hof, J. 1975.   DNA fiber replication in chromosomes of a higher
plant (*Pisum sativum*).   Exp. Cell Res. 93:95-104.

_____ 1976a.   DNA fiber replication of chromosomes of pea root cells
terminating S.   Exp. Cell Res. 99:47-56.

_____ 1976b.   Replicon size and rate of fork movement in early S of
higher plant cells (*Pisum sativum*).   Exp. Cell Res. 103:395-403.

_____ and Bjerknes, C.A. 1977.   18 μm replication units of chromosomal DNA fibers of differentiated cells of pea (*Pisum sativum*).
Chromosoma 64:287-294.

_____, Bjerknes, C.A., and Clinton, J.H. 1978a.   Replicon properties
of chromosomal DNA fibers and the duration of DNA synthesis of sunflower root-tip meristem cells at different temperatures.   Chromosoma
66:161-171.

_____, Kuniyuki, A., and Bjerknes, C.A. 1978b.   The size and number
of replicon families of chromosomal DNA of *Arabidopsis thaliana*.
Chromosoma 68:269-285.

Watts, J.W. and King, J.M. 1973.   The metabolism of proteins and nucleic acids in freshly isolated protoplasts.   In: Protoplasts et fusion
de cellules somatiques vegetales (J. Tempe, ed.) pp. 119-123.
Editions I.N.R.A., Paris.

Zelcer, A. and Galun, E. 1976.   Culture of newly isolated tobacco protoplasts: Precursor incorporation into protein, RNA and DNA.   Plant
Sci. Lett. 7:331-336.

CHAPTER 20
Biochemistry of Somatic Embryogenesis

V. Raghavan

One of the finest endeavors in the history of plant tissue culture has been the cultivation of free cells and cell groups derived from higher plants, especially the angiosperms, in a chemically defined liquid medium as a suspension (suspension culture) and manipulation of their regenerative potential by changes in the hormonal balance of the medium. One type of regeneration frequently observed in suspension cultures is the development, in large numbers, of embryo-like structures that recapitulate with a high degree of precision the typical stages in the embryogenesis of a fertilized egg cell. Since the embryos are formed from the sporophytic or somatic cells of the plant, as opposed to gametophytic or germ cells, the phenomenon is conveniently referred to as somatic embryogenesis. The cryptic potentiality of cells of angiosperms to grow as callus, the callus to dissociate into free cells, and free cells to regenerate whole plants by recapitulating stages of embryogenesis is so widespread that it must be considered as a general property of this group of plants. Where it has not been possible to demonstrate this capacity by existing techniques, it is probably due to special inhibiting conditions within the system, rather than to its unspecialized or primitive state.

Somatic embryogenesis was first clearly described in domestic carrot, *Daucus carota* L., and to date the carrot system is the most comprehensively studied with respect to culture conditions and developmental physiology and biochemistry of somatic embryogenesis. A callus that yields free cells is the starting point for the study of somatic embryogenesis in carrot, and one must be familiar with basic manipulative techniques and experimental conditions that allow embryogenic type of

growth in these cells in order to appreciate the questions asked in biochemical research. That is the goal of the paragraphs that immediately follow. Next, this chapter describes the directions in which biochemical investigations have been pressed to gain an insight into the
mechanism by which an undifferentiated parenchymatous cell is transformed into a fully differentiated bipolar embryo-like structure. Finally, a brief account of the biochemical aspects of differentiation of
pollen grains into embryos is also given. For a complete discussion of
somatic embryogenesis, see Chapter 3; for the generation of haploid
plants, refer to Chapter 5.

INDUCTION OF EMBRYOGENESIS IN SUSPENSION CULTURES

The procedures for developing and maintaining a suspension of carrot
cells of high embryogenic potential available continuously for use in
physiological and biochemical research may be described as follows. A
callus is initiated from seedling hypocotyl or root segments obtained
from carrot seeds germinated under aseptic conditions. The excised
segment is placed on the surface of a nutrient medium solidified with
agar and incubated in the dark to foster growth of a callus. The
medium used is a high-nitrogen-containing mineral salt formulation such
as that of Murashige and Skoog, supplemented with sucrose, an organic
addendum including myo-inositol, a source of iron like sodium ferric
ethylenediaminetetraacetate, a cytokinin (KIN or ZEA), and an auxin,
2,4-D. A suspension culture is initiated by transferring a piece of the
callus into a liquid medium of the same composition, but with a reduced amount of 2,4-D, and incubated on a horizontal shaker. Under
these conditions the callus proliferates, releasing the superficial cells
into the medium, where they multiply and cohere to form small cell
aggregates. The liquid in which the free cells and cell aggregates
grow is the suspension culture. The cellular units of an aliquot of this
suspension, serially transferred at appropriate intervals to fresh media
possess the capacity to differentiate into somatic embryos when challenged in a medium of a different composition.
Embryogenic type of growth is induced in the suspension culture
when it is transferred to a medium from which 2,4-D is omitted. Under these conditions, the first sign of embryogenesis are observed in
about 4-5 days when clumps of cell aggregates are transformed into
globular embryos. To obtain a homogenous inoculum, the suspension is
graded as to unit size by filtration on stainless steel or nylon screens
of different pore sizes and washed with the basal medium before transferring to fresh media with or without 2,4-D (Fujimura and Komamine,
1975; Sengupta and Raghavan, 1980a). This technique allows analysis
of cell aggregates of the same size and age during growth in the undifferentiated state or during embryogenic development. In another
method to obtain high yields of somatic embryos of particular stages of
development, Warren and Fowler (1977) separated globular, heart-
shaped, and torpedo-shaped embryos by filtering the suspension culture
successively through glass beads of different sizes. Any contaminating
cells and cell clumps were removed from the fractions by centrifugation

through Ficoll to yield nearly homogenous bulk samples of viable embryos. This method is, however, of limited value, since homogenous fractions of embryogenic cells prior to the globular stage cannot be obtained by this method. Recently, Fujimura and Komamine (1979) established conditions to obtain high yields of synchronously developing carrot somatic embryos of different stages of development. In this method cells and cell clusters from a stationary phase suspension culture retained on a 31-47 μm screen were initially subjected to density gradient centrifugation in Ficoll solution. Next, the heaviest fraction from the Ficoll gradient was repeatedly centrifuged at a low speed (50 g) for a short time (5 sec) until most of the contaminating vacuolate cells were removed. When the final fraction was transferred to an embryo-inducing medium, the cell clusters exhibited a high degree of synchrony in embryogenesis, with more than 90% frequency.

Although it would appear from the above description that a cell suspension of embryogenic cells is an excellent model of a differentiating system in which all biochemical steps of embryogenesis beginning with the single progenitor cell can be studied in a clonal population, this goal has hardly been achieved. Moreover, since most of the results have come from work done with carrot cell suspension, their validity to other systems cannot be taken for granted.

CHANGES IN MACROMOLECULE COMPOSITION

In determining the changes in the composition of macromolecules during growth cycle in a tissue, the usual aim is to obtain a quantitative measure of the rates of synthesis of the macromolecules and of the increase or decrease in their amounts due to synthesis minus degradation. The macromolecules on which most attention has been focused are nucleic acids and proteins. Although different groups of workers have followed the metabolism of DNA, RNA, and proteins during growth and embryogenic development in carrot cell suspension, as shown below, because of differences in the nature of the starting materials used, and because of the lack of synchrony in the cultures, some discrepancies have become apparent in the data obtained.

Verma and Dougall (1978) found that DNA content of carrot cell clumps in the range of 63-125 μm in diameter grown in a medium containing auxin, increased exponentially until the ninth day, after which it slowed down, apparently because of limiting nutrient conditions of the culture. A similar pattern of DNA accumulation was observed in the embryogenic cells grown in the absence of auxin, except that in these cells DNA continued to increase beyond 10 days. On the other hand, using a synchronized cell culture, Fujimura et al. (1980) have established that compared to nonembryogenic cells of carrot growing in a medium enriched with auxin, embryogenic cells growing in the absence of auxin exhibited a higher DNA content at 3-4 days after culture and coincident with the appearance of the first globular embryos. As shown by ^{3}H-thymidine incorporation (Fig. 1) a rapid replication of chromatin and nuclear materials in cells growing in a medium lacking 2,4-D after about 3 days is a biochemical transition point in anticipation of embryogenic induction.

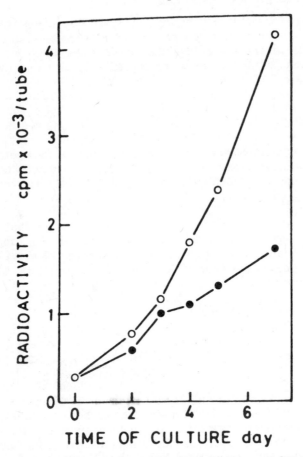

Figure 1.   Changes in the rate of incorporation of ³H-thymidine into perchloric acid-insoluble fraction of carrot cells subcultured in the presence (●) or absence (○) of 2,4-D in the medium.   Cells were pulsed with 10 μCi ³H-thymidine in 2 ml medium for 1 hr; radioactivity is expressed as cpm/tube (from Fujimara et al., 1980).

From the work of Verma and Dougall (1978) it appeared that RNA and protein contents of embryogenic and nonembryogenic cells of carrot increased at similar rates up to 6 days after transfer to new media; thereafter the nonembryogenic cells accumulated more RNA and proteins than the embryogenic cells.   In contrast, other workers have noted slight increases in RNA (Sengupta and Raghavan, 1980a) and protein (Fujimura et al., 1980; Sengupta and Raghavan, 1980a) contents of embryogenic cells even at earlier periods after transfer to new media. Although the reasons for this discrepancy remain unclear, the results have provoked further analysis of the role of RNA and protein metabolism during embryogenic induction in carrot cell cultures.

It seems clear that more RNA and protein are synthesized in the embryogenic cells than in the nonembryogenic cells during the period of their culture preparatory to the appearance of embryos.   The pulse-labeling experiments of Fujimura et al. (1980) using ^{3}H-uridine and ^{3}H-leucine as precursors of RNA and protein synthesis, respectively, have demonstrated that embryogenic cells of carrot incorporate the precursors into respective macromolecules at a higher rate than nonembryogenic cells beginning about 2 days after culture. Since the increased RNA and protein accumulation observed in the embryogenic cultures is too small to be accounted for by increased synthesis, it has been suggested that there is an active turnover of RNA and protein in the embryogenic cells. This is in agreement with the results of Sengupta and Raghavan (1980a), who found that in short-term labeling experiments the rate of RNA and protein synthesis in the embryogenic cells increased appreciably over that of nonembryogenic cells as early as 2 to 4 hr after transfer to new media (Fig. 2,3). The increased RNA synthetic activity of the embryogenic cells continued up to 12 hr after their transfer to a medium lacking auxin. Since no new cells are formed in the suspension culture during the first 12 hr after transfer, the augmentation of RNA synthesis is, in all likelihood, due to cellular RNA synthesis. Another small peak in RNA synthetic activity of the embryogenic cells was seen at 96 hr after transfer, coincident with the formation of the first embryos. Although overall these results indicate that an increased synthesis of RNA and protein is associated with induction of somatic embryos in carrot cell suspension, interpretation of the data is also complicated by the well-known effects of 2,4-D non-specifically stimulating RNA and protein synthesis in plants. Thus the

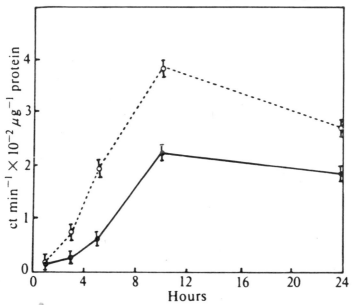

Figure 2.   Rates of protein synthesis in embryogenic (o) and non-embryogenic (•) carrot cell suspension during the first 24 hr of sub-culture. At specific times, samples were pulsed with ^{3}H-leucine (2.5 µCi/ml in 5 ml medium) for 1 hr. Vertical lines represent 2x standard error of 3 replicates (from Sengupta and Raghavan, 1980a).

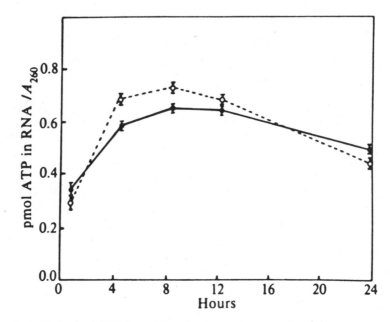

Figure 3. Rates of RNA synthesis in embryogenic (o) and nonembryogenic (•) carrot cell suspension during the first 24 hr of subculture. At specific times, samples were pulsed with 2.5 µCi/ml ^{3}H-adenosine for 1 hr and mol of ATP in RNA determined according to the method of Emerson and Humphreys (1971). Vertical bars represent 2x standard error of 3 replicates (from Sengupta and Raghavan, 1980a).

effect of an auxin-free medium on carrot cell suspension could be partly attributed to the promotion of synthesis of specific RNA and proteins involved in embryogenesis, and partly to the inhibition of the synthesis of certain RNA and proteins necessary for undifferentiated growth.

Sengupta and Raghavan (1980b) have attempted to characterize the nature of RNA synthesized during the early hours of transformation of somatic cells of carrot into embryogenic cells by acrylamide gel electrophoresis and affinity chromatography on an oligo (dT) cellulose column of single- and double-labeled RNA. In the double-labeling experiment, embryogenic carrot cells growing in a medium lacking 2,4-D were labeled with ^{3}H-adenosine, while nonembryogenic cells of the same age growing in a medium containing auxin were labeled with ^{14}C-adenosine. The control experiment consisted of labeling nonembryogenic cells in two separate sets with ^{3}H-adenosine and ^{14}C-adenosine. After incubation in the isotopes, the two sets of cells were mixed together and used for extraction of RNA. Electrophoretic separation of RNA was carried out on acrylamide gels containing 2.5% acrylamide, 0.125% bisacrylamide, and 0.5% agarose. Following electrophoresis, the gel was cut into 1 mm slices and radioactivity due to ^3H and ^{14}C in each slice determined by scintillation counting. The ratios of ^3H to ^{14}C calculated were plotted against gel slice number corresponding to UV absorption profile.

To isolate poly(A)-rich RNA, labeled total RNA was dissolved in a binding buffer (0.01 M Tris-HCl, pH 7.5, 0.4 M NaCl; 0.5% NaDodSO$_4$) along with yeast carrier RNA. It was then mixed with oligo(dT)-cellulose that was previously equilibrated with the binding buffer. After centrifugation, the supernatant was removed. The oligo(dT)-cellulose was washed with the binding buffer until no more radioactive RNA was eluted, combined with supernatants constituting poly(A)-lacking RNA. The poly(A)-rich RNA was eluted with 0.01 M Tris-HCl (pH 7.5) and 0.05% NaDodSO$_4$ and was precipitated in ethanol.

To isolate poly(A) segments, bound RNA eluted from oligo(dT)-cellulose column was initially incubated in pancreatic RNase. RNase T$_1$ was then added, and RNA resistant to the enzymes was precipitated with trichloroacetic acid. Radioactivity in the precipitate due to poly(A) segments was determined by scintillation counting.

Data derived from these experiments suggest the following effects of auxin omission from the medium on the type of RNA synthesized by cells. As early as 6 hr after transfer of cells to a medium lacking auxin, there was a decrease in the rate of synthesis of rRNA, concomitant with an increased synthesis of minor species of RNA in the 12S-18S region (Fig. 4). On the other hand, embryogenic cells synthesized poly(A) containing RNA (Poly A + RNA) at a higher rate than the non-embryogenic cells during an experimental period of 96 hr (Fig. 5). Since it is generally assumed that addition of poly(A) residues to mRNA is a prerequisite for its use as an active message, these results point to the hypothesis that removal of auxin from the medium controls the biochemical events of embryogenesis by eliciting the synthesis of poly(A) containing mRNA.

Yet the evidence does not ascribe a triggering role for newly synthesized mRNA in embryogenic induction. We have seen earlier that transfer of cells to a medium lacking auxin is followed by a stimulation of protein synthesis, and the question is whether these proteins are translated on newly synthesized mRNA templates. In other words, to what extent do mRNAs synthesized by cells while they are bathed in the auxin-enriched medium direct embryogenesis, as opposed to those that are made after cells are transferred to a medium lacking auxin? Results of studies using cordycepin, an inhibitor of polyadenylation, have pointed to the tentative conclusion that embryogenic induction in carrot cells growing in a medium devoid of auxin is controlled both at the transcriptional and translational levels (Sengupta, 1978). For example, it was noted that when cells were transferred to an auxin-free medium containing 10.0 mg/l cordycepin, which inhibited polyadenylation by 90%, somatic embryo formation proceeded to an arrested globular or heart-shaped stage. Since inhibition of polyadenylation does not inhibit protein synthesis in the embryogenic cells, it seems that proteins synthesized by the cells during the early hours of their growth in an auxin-free medium are not translated on newly transcribed poly(A) + RNA. Presumably, mRNAs made by the cells while they are still in

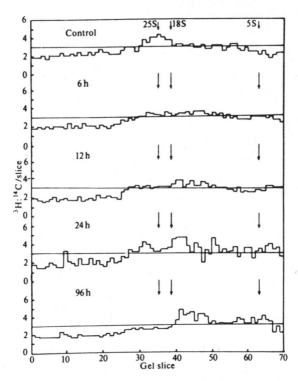

Figure 4. ^{3}H:^{14}C ratios of double-labeled RNA from acrylamide gel
slices.   Carrot cells growing in an auxin-free medium were labeled for
2 hr with ^{3}H-adenosine (3.0 μCi/ml) at 6, 12, 24, and 96 hr.   Cells
growing in an auxin-enriched medium were labeled for 2 hr with ^{14}C-
adenosine (1.0 μCi/ml) at the same time intervals.   RNA extracted from
the combined cell mixture was analyzed electrophoretically and radioactivity due to ^{3}H and ^{14}C in the gel slices determined.   The ratios
of ^{3}H:^{14}C calculated from the data are plotted against gel slice
number corresponding to UV absorption profile.   Controls consisted of
cells growing in the auxin-enriched medium labeled in 2 separate sets
with ^{3}H-adenosine and ^{14}C-adenosine (from Sengupta and Raghavan,
1980b).

contact with auxin are able to sustain protein synthesis during the
early period of their growth in the auxin-free medium.   The fact that
the cell clumps differentiating in the auxin-enriched medium do not become fully embryogenic suggests that the message component of RNA
synthesized is masked in some way until the cells are transferred to a
medium lacking auxin.   If this assumption is correct, and this is not
unreasonable, a major contribution to further our knowledge of the biochemistry of somatic embryogenesis would be an analysis of the mechanism by which mRNA is unmasked.
    The complexity of the biochemical events during somatic embryogene-
sis in carrot thus makes it seem probable that rather than being solely
involved in promoting undifferentiated growth, 2,4-D might be directing

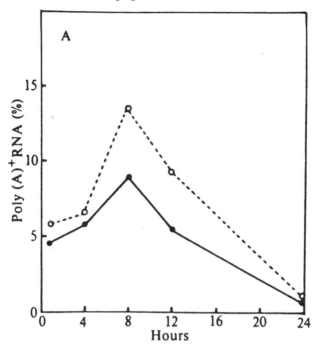

Figure 5.   Rates of poly(A) + RNA synthesis in embryogenic (o) and nonembryogenic (•) carrot cell suspension.   At specified times, cells were labeled with ^{3}H-adenosine (2.5 µCi/ml) for 2 hr.   RNA was extracted and poly(A) + RNA isolated.   Results are expressed as percentage of counts due to poly(A) + RNA in total RNA.   Points represent mean of 4 replicates (from Sengupta and Raghavan, 1981b).

part of a repertoire of changes leading to embryogenesis.   From an ultrastructural study of embryogenesis in carrot cell suspension, Halperin and Jensen (1967) have contended that embryogenic induction probably occurs during isolation and growth of tissues in the auxin-containing medium, although formation of more organized structures reminiscent of zygotic embryos is prevented as long as auxin is present in the medium.   Recently, Sung and Okimoto (1981) have studied the profile of newly synthesized proteins of nonembryogenic and embryogenic cells of carrot by two-dimensional gel electrophoresis.   In this work, using ^{3}H-methionine to label the proteins, the authors have identified two proteins, designated as embryonic proteins, in 12-day-old embryogenic cells (Fig. 6).   The surprising finding was that regardless of the presence or absence of 2,4-D in the medium, these proteins were synthesized by the cells during the early days of their growth in fresh media, but in the presence of 2,4-D the proteins gradually diminished and completely disappeared after 12 days.   Since the cell suspension grown in the presence of 2,4-D had globular embryogenic masses of cells, it seems that synthesis of embryonic proteins is an early event of embryogenesis triggered by auxin, but by its very presence in the medium auxin also prevents the continued synthesis of these proteins necessary for embryogenesis coming to fruition.   Thus the development-

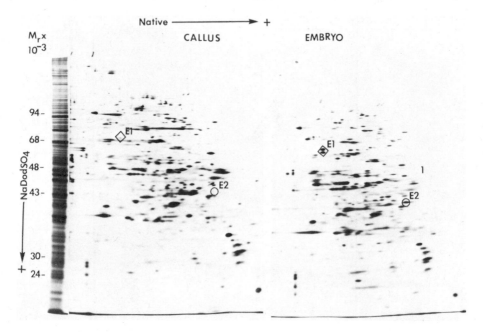

Figure 6.   Autoradiographs of soluble polypeptide profiles of carrot cells grown in the nonembryogenic (left) and embryogenic (right) media for 12 days.   Cells grown in media with or without 2,4-D were labeled with ^{35}S-methionine and extracts separated by native NaDodSO$_4$ poly-acrylamide gel electrophoresis.   The one dimensional NaDodSO$_4$ gel of the nonembryogenic cells is also shown on the left.   E1 and E2 are embryogenic proteins preferentially synthesized in the cells growing in an auxin-free medium (from Sung and Okimoto, 1981; print kindly sup-plied by Dr. Z. R. Sung).

al versatility of the carrot cell suspension may stem from the action of embryonic proteins that serve as early markers of differentiation.   Sig-nificant changes in protein composition as indicated by increasing or decreasing staining intensity of bands on acrylamide gels have also been noted during embryogenesis in petiole explants of Chinese celery, *Apium graveolens* L. (Zee et al., 1979).

ROLE OF THE GENOME

It appears from the preceeding that the genome exerts a major im-pact on the transition of somatic cells of carrot into embryo mother cells.   Since the genes for embryogenesis are apparently transcribed under special cultural conditions that inhibit undifferentiated growth of the cells, can we assume that embryogenesis is associated with changes in the quality or quantity of chromosomal proteins?   One way to an-swer this question is to isolate chromatin (a complex mixture of DNA, histones, acid proteins, and possibly some RNA) and study the proper-ties of histones and nonhistone nuclear proteins as a function of embryogenic development. Of the two classes of nuclear proteins, histones have long been favored as regulators of gene activity, particularly as a mechanism for silencing gene expression by inhibiting transcription (Huang and Bonner, 1962). However, other studies have tended to minimize the role of histones in gene expression because they seem to lack specificity. Consequently, considerable attention has been directed toward nonhistone chromosomal proteins as regulators of gene expression (Stellwagen and Cole, 1969). Although no qualitative differences were noted in the histone composition between embryogenic and nonembryogenic cells of carrot, the percentage of histone H_1 to the total histone was found to be lower in the embryogenic than in the nonembryogenic cells (Gregor et al., 1974; Fujimara et al., 1981). It has been suggested that this is related to structural changes in the chromatin of embryogenic cells to facilitate gene expression (Fujimara et al., 1981). On the other hand, changes in several minor bands were characteristically observed in the electrophoretic profile of nonhistone proteins of embryogenic cells. Moreover, preparations of nonhistone proteins from 14-day-old embryogenic cells were more effective in restoring histone-inhibited, DNA-directed, RNA synthesis than similar preparations from nonembryogenic cells (Matsumoto et al., 1975). It is interesting to note that the increased capacity of nonhistone proteins to make chromatin accessible to transcription occurs just at about the same time as two embryonic proteins are detected in the cells (Sung and Okimoto, 1981). From these experiments the extent to which changes in histones or nonhistone proteins poise or arrest embryogenesis in carrot cells by controlling the activity of DNA template remains uncertain. We still have to learn more about the interaction of DNA and chromosomal proteins of the intact nuclei of embryogenic and nonembryogenic cells before precise correlations of changes in the properties of histones and nonhistone proteins to embryogenic processes can be attempted.

OTHER BIOCHEMICAL CHANGES

Apart from nucleic acid and protein synthesis, many other areas of metabolism are also undoubtedly involved in embryogenic differentiation. This is particularly true of a suspension culture, since the somatic embryo mother cell, unlike a zygote, is a simple parenchymatous cell which is relatively undifferentiated for the anticipated developmental task. Unfortunately, very few studies have been carried out on other biochemical events involved in somatic embryogenesis, and where information is available the contribution of such events to the processes of determination and subsequent cellular differentiation is uncertain. Montague et al. (1978) have shown that there is a significant difference in polyamine metabolism between embryogenic and nonembryogenic cells of carrot. For example, putrescine levels in the embryogenic cells were enhanced nearly twofold over the control within 24 hr of their transfer to a medium lacking 2,4-D. In line with this observation the activity of arginine decarboxylase (arginine carboxylase, EC 4.1.1.19), an enzyme in the synthesis of putrescine from arginine, was found to be higher in the embryogenic cells than in their nonembryogenic counter-part; moreover, the difference in the enzyme activity was noticeable within 6 hr after transfer of cells to the embryogenic medium (Monta-gue et al., 1979). Although these results might indicate a role for arginine decarboxylase in somatic embryogenesis of carrot, specifically in the regulation of putrescine levels, the question of whether it repre-sents an effect on terminal gene expression requires further study.

Electrophoretic variations have been reported for several isozymes of carrot cells grown in embryogenic and nonembryogenic media (Lee and Dougall, 1973). Greatest differences were observed in the pattern of glutamate dehydrogenase (1-glutamate: NAD oxidoreductase, EC 1.4.1.2 and L-glutamate: NADP oxidoreductase, EC 1.4.1.4) which was repre-sented by only slowly migrating bands in the embryogenic cells while the nonembryogenic cells had in addition faster migrating bands. Ac-cording to Kochba et al. (1977) an embryogenic line of Shamouti orange callus showed an upsurge of peroxidase activity concomitant with the appearance of embryos. A particularly interesting finding was that at the time when the rise in activity was under way, a new band typical of the embryogenic cell line was also detected in the isozyme profile of the enzyme. From a causal point of view, the pattern of enzyme changes during embryogenesis remains obscure. The small amount of evidence available suggests that isozyme changes in glutamate dehydro-genase are connected with nitrogen metabolism of the embryogenic cells.

In another biochemical approach, lipid and fatty acid profiles of car-rot cells were found to exhibit quantitative rather than qualitative changes during embryogenesis (Warren and Fowler, 1979). Such changes were seen particularly during the transition of small meristematic groups of cells grown in a medium lacking 2,4-D into later stages of somatic embryos. In this system a decrease in the level of dissolved oxygen in the cell suspension which led to an increase in the cellular levels of ATP was found to induce embryogenesis (Kessell et al., 1977). A histochemical analysis of embryogenesis in callus cultures of *Corylus avellana* L. and *Paulownia tomentosa* Stued. has shown a significant in-crease in starch content of the cells from which somatic embryos dif-ferentiate; the starch content of cells also remained high during embryo development (Radojevic et al., 1979).

BIOCHEMICAL ASPECTS OF POLLEN EMBRYOGENESIS

In contrast to the somatic cells, eggs and sperm, and the associated cells of the gametophyte, are designated as sexual reproductive cells or germ cells. Germ cells have their origin in simple diploid parenchyma-tous cells that in other circumstances differentiate into specialized somatic cells. Since germ cells are born out of a reduction division of somatic cells, embryos formed from germ cells in the absence of fertili-zation are haploids at the cellular level.

The pollen grain, or microspore, is the first cell of the male gameto-phytic generation of angiosperms in which two distinct phases can be recognized. First, while still enclosed in the microsporangium, the pollen grain divides asymmetrically to yield a small generative cell and a large vegetative cell. Next, in the natural environment of the stigma, or under artificial culture conditions in the laboratory, the pollen grain enters the germination phase, characterized by the formation of a pollen tube. Upon germination of the pollen grain, the generative cell and the nucleus of the vegetative cell loose from its cytoplasm move into the emerging pollen tube. During subsequent growth of the pollen tube, the generative cell divides to form two sperms which are involved in the act of double fertilization in the embryo sac. On the other hand the nucleus of the vegetative cell disintegrates or remains as a vestigial organelle in the pollen tube. Although pollen grains are thus programmed for terminal differentiation, culture of anthers of certain plants at an appropriate stage of development in a simple mineral salt medium has been shown to evoke repeated divisions in a small proportion of the enclosed pollen grains. The multicellular pollen grains thus formed subsequently differentiate into haploid embryos and plantlets. This phenomenon is known as pollen embryogenesis or anther androgenesis. It is recognized that the transformation of pollen grains into embryos does not fall within the preview of somatic embryogenesis. However, in both somatic embryogenesis and pollen embryogenesis, cells that have not embarked upon a pathway of irreversible differentiation within an already committed plan exhibit their inherent genetic potential in the same way and to the same extent as a zygote; for this reason an account of our present status of knowledge of the biochemistry of pollen embryogenesis is included here to complement the information on the biochemistry of somatic embryogenesis presented in the preceding pages.

Biochemical changes associated with the transformation of pollen grains into embryos have been largely inferred from cytochemical and autoradiographic studies. In plants such as *Nicotiana tabacum* L. where the embryo is formed by the repeated division of the vegetative cell there is an apparent lack of RNA and protein accumulation in the embryogenic pollen grains, while pollen grains that mature and complete the gametophytic program in culture accumulate relatively large amounts of these macromolecules (Bhojwani et al., 1973). On the basis of these results it has been claimed that the initial event of embryogenic induction is a suppression of the gametophytic program. This will ensure that genes coding for proteins involved in embryogenic divisions can be expressed fully without being masked by the simultaneous expression of genes for pollen germination and pollen tube growth. Similarly, in embryogenic pollen grains of *Datura innoxia* Mill. there was a period of reduced RNA accumulation immediately after culture of anthers, although formation of vegetative and generative cells was followed by an increase in the pyroninophily of the cytoplasm. During the same time span the histone content of the pollen first increased and then decreased (Sangwan–Norreel, 1978). According to Sangwan (1978) there was a rapid turnover of amino acid metabolism during pollen embryogenesis in *D. metel* L.; however, since the analysis was carried with cultured anthers, it is not certain whether the changes in amino acid titre occurred in the somatic tissues of the anther or in the enclosed pollen grains. In *Hyoscyamus niger* L., where a proportion of embryos originate by the division of the generative cell of the bicellular pollen, Raghavan (1979a,b) has shown that the pollen grains become embryogenically determined as early as the first hour of culture of anthers and that this is accompanied by the synthesis of autoradio-graphically detectable RNA. After the first haploid mitosis, embryo-genic divisions are initiated in pollen grains in which the generative cell nucleus or both vegetative and generative cell nuclei synthesize RNA, whereas those in which RNA synthesis occurs exclusively in the nucleus of the vegetative cell become starch-filled and nonembryogenic. Thus pollen embryogenesis in *H. niger* appears to be correlated with an initial level of RNA synthesis in the uninucleate pollen grain and in the generative cell formed afterward.

Further work was directed to determine whether RNA synthesized by the embryogenic pollen grains contains any nonribosomal RNA that carries information for embryogenic divisions (Raghavan, 1981). In this work the distribution of poly(A) + RNA during pollen embryogenesis was followed by in situ hybridization with ^{3}H-polyuridylic acid ^{3}H-poly(U) in histological preparations of anthers at different times after culture. Results of these studies have shown that appreciable binding of the iso-tope occurred in a small number of uninucleate, embryogenically deter-mined pollen grains within a few hours after culture of anthers, while the majority of pollen grains, which are nonembryogenic, do not bind any ^{3}H-poly(U). Lack of ^{3}H-poly(U) binding in the uninucleate, embry-ogenic pollen grains of anthers cultured in a medium containing actino-mycin D, has led to the conclusion that a poly(A) containing mRNA is newly synthesized by pollen grains as they establish contact with the nutrient medium. This mRNA is probably concerned with embryogene-sis, since uninucleate pollen grains do not bind ^{3}H-poly(U) at the time of culture or at earlier stages in their ontogeny. Comparative analysis of ^{3}H-poly(U) binding activity in the nuclei of the generative and vege-tative cells during gametogenesis and during induced pollen embryo-genesis has provided some insight into how they respond to different developmental signals. These data have established that during the normal ontogeny of the pollen the nuclei of the generative and vegeta-tive cells are only transiently active in binding ^{3}H-poly(U). The strik-ing feature of the potentially embryogenic binucleate pollen grains is the continued transcriptional activity of the nucleus of the generative cell, as indicated by ^{3}H-poly(U) binding in the cytoplasm in the vicin-ity of this nucleus. Overall, these studies have led to the conclusion that as a result of the trauma of excision and culture of the anther, a certain proportion of the enclosed pollen grains change their develop-mental program by the synthesis of poly(A)-containing RNA which prob-ably codes for enzymes necessary to induce the first haploid mitosis. Subsequent embryogenic divisions of the pollen grains are mediated by the synthesis of additional and perhaps new mRNA by the nucleus of the generative cell. On the other hand, synthesis of mRNA by the nucleus of the vegetative cell perpetuates part of the gametophytic program that leads to starch accumulation.

From the above account, the parallels between the biochemical changes that result in the deflection of a somatic cell and a pollen grain in the embryogenic pathway are obvious. The precise course of biochemical changes may vary, depending upon the experimental system and the conditions under which embryogenic induction is achieved.

CONCLUDING COMMENTS

At the present time, biochemical studies on somatic embryogenesis are restricted to the so-called model system, carrot. Similar studies should now be extended to other crop plants in which somatic embryogenesis is accomplished by simple manipulative techniques. Even in plants where somatic embryos are produced with difficulty, knowledge of the molecular basis of differentiation can help to devise protocols to induce a high frequency of somatic embryogenesis.

A formidable obstacle to our understanding of the biochemistry of pollen embryogenesis is the very low percentage of pollen grains in cultured anthers that become embryogenic. Therefore, biochemical data on embryogenic differentiation of pollen grains based on analysis of whole anthers must be viewed against an overwhelming background of nonembryogenic pollen. In the author's opinion, the challenge of unraveling the biochemical and molecular changes during pollen embryogenesis can only be met by formulation of techniques to induce pollen grains, routinely and in large numbers, in the embryogenic pathway.

KEY REFERENCES

Fujimura, T., Komamine, A., and Matsumoto, H. 1980. Aspects of DNA, RNA and protein synthesis during somatic embryogenesis in a carrot cell suspension culture. Physiol. Plant. 49:255-260.

Halperin, W. and Jensen, W.A. 1967. Ultrastructural changes during growth and embryogenesis in carrot cell cultures. J. Ultrastruct. Res. 18:428-443.

Raghavan, V. 1979a. Embryogenic determination and synthesis of ribonucleic acid in pollen grains of *Hyoscyamus niger* (henbane). Am. J. Bot. 66:36-39.

Sengupta, C. and Raghavan, V. 1980a. Somatic embryogenesis in carrot cell suspension I. Pattern of protein and nucleic acid synthesis. J. Exp. Bot. 31:247-258.

REFERENCES

Bhojwani, S.S., Dunwell, J.M., and Sunderland, N. 1973. Nucleic acid and protein contents of embryogenic tobacco pollen. J. Exp. Bot. 24:863-871.

Fujimura, T. and Komamine, A. 1975. Effects of various growth regulators on the embryogenesis in a carrot cell suspension culture. Plant Sci. Lett. 5:359-364.

_____ 1979. Synchronization of somatic embryogenesis in a carrot cell suspension culture. Plant Physiol. 64:162-164.

_____, and Matsumoto, H. 1981.   Changes in chromosomal proteins during early stages of synchronized embryogenesis in a carrot cell suspension culture. Z. Pflanzenphysiol. 102:293-298.

Gregor, D., Reinert, J., and Matsumoto, H. 1974. Changes in chromosomal proteins from embryo induced carrot cells. Plant Cell Physiol. 15:875-881.

Huang, R.C. and Bonner, J. 1962. Histone, a suppressor of chromosomal ribonucleic acid synthesis. Proc. Natl. Acad. Sci. USA 48:1216-1222.

Kessell, R.H.J., Goodwin, C., Philip, J., and Fowler, M.W. 1977. The relationship between dissolved oxygen concentration, ATP and embryogenesis in carrot (*Daucus carota*) tissue cultures. Plant Sci. Lett. 10:265-274.

Kochba, J., Lavee, S., and Spiegel-Roy, P. 1977. Differences in peroxidase activity and isoenzymes in embryogenic and nonembryogenic 'Shamouti' orange ovular callus lines. Plant Cell Physiol. 18:463-467.

Lee, D.W. and Dougall, D.K. 1973. Electrophoretic variation in glutamate dehydrogenase and other isozymes in wild carrot cells cultured in the presence and absence of 2,4-dichlorophenoxyacetic acid. In Vitro 8:347-352.

Matsumoto, H., Gregor, D., and Reinert, J. 1975. Changes in chromatin of *Daucus carota* cells during embryogenesis. Phytochemistry 14:41-47.

Montague, M.J., Koppenbrink, J.W., and Jaworski, E.G. 1978. Polyamine metabolism in embryogenic cells of *Daucus carota*. I. Changes in intracellular content and rates of synthesis. Plant Physiol. 62:430-433.

Montague, M.J., Armstrong, T.A., and Jaworski, E.G. 1979. Polyamine metabolism in embryogenic cells of *Daucus carota*. II. Changes in arginine decarboxylase activity. Plant Physiol. 63:341-345.

Radojevic, L.J., Kovoor, J., and Zylberberg, L. 1979. Etude anatomique et histochimique des cals embryogenes du *Corylus avellana* L. et du *Paulownia tomentosa* Stued. Rev. Cytol. Biol. Veg. 2:155-167.

Raghavan, V. 1979b. An autoradiographic study of RNA synthesis during pollen embryogenesis in *Hyoscyamus niger* (henbane). Am. J. Bot. 66:784-795.

_____ 1981. Distribution of poly(A)-containing RNA during normal pollen development and during induced pollen embryogenesis in *Hyoscyamus niger*. J. Cell. Biol. 89:593-606.

Sangwan, R.S. 1978.   Amino acid metabolism in cultured anthers of *Datura metel*. Biochem. Physiol. Pflanz. 173:355-364.

Sangwan-Norreel, B.S. 1978.   Cytochemical and ultrastructural peculiarities of embryogenic pollen grains and of young androgenic embryos in *Datura innoxia*. Can. J. Bot. 56:805-817.

Sengupta, C. 1978. Protein and nucleic acid metabolism during somatic embryogenesis in carrot. Ph.D. Dissertation, Ohio State Univ., Columbus.

_____ and Raghavan, V. 1980b. Somatic embryogenesis in carrot cell suspension II. Synthesis of ribosomal RNA and poly(A) + RNA. J. Exp. Bot. 31:259-268.

Stellwagen, R.H. and Cole, R.D. 1969.   Chromosomal proteins.   Annu. Rev. Biochem. 38:951-990.

Sung, Z.R. and Okimoto, R. 1981.   Embryonic proteins in somatic embryos of carrot. Proc. Natl. Acad. Sci. USA 78:3683-3687.

Verma, D.C. and Dougall, D.K. 1978.   DNA, RNA and protein content of tissue during growth and embryogenesis in wild-carrot suspension cultures. In Vitro 14:183-191.

Warren, G.S. and Fowler, M.W. 1977.   A physical method for the separation of various stages in the embryogenesis of carrot cell cultures. Plant Sci. Lett. 9:71-76.

_____ 1979.   Changing fatty acid composition during somatic embryogenesis in cultures of *Daucus carota*. Planta 144:451-454.

Zee, S.-Y., Wu, S.C., and Yue, S.B. 1979. Morphological and SDS-polyacrylamide gel electrophoretic studies of pro-embryoid formation in the petiole explants of Chinese celery.   Z. Pflanzenphysiol. 95:397-403.

CHAPTER 21
Biochemical Mechanisms of Plant Hormone Activity

H.-J. Jacobsen

The most relevant breakthrough in the development of plant tissue culture techniques was the discovery of the morphogenetic activities of phytohormones, particularly those of the native and synthetic auxins and cytokinins. The identification of IAA as a native auxin, the investigations on its multiple effects on plants, and the discovery of the synthetic auxins NAA and the other compounds from the phenoxy group, stimulated tissue culture research. In addition, the observation that the cytokinins are the active growth-regulating substances present in coconut water, together with intensive studies on the nutritional requirements of in vitro cultivated plant cells, led to the formulation of a number of culture media and procedures. These media and procedures proved to be a successful basis for serious tissue culture studies. The use of undefined extracts in tissue culture medium, e.g., coconut water, yeast, or malt extracts ended a rather long lag period in the development of plant tissue culture toward an application for crop improvement. The use of defined concentrations of phytohormones permitted better reproducibility of experiments and the initiation of investigations leading to a more precise understanding of developmental processes in higher plants.

Throughout this chapter, the term phytohormone is used instead of the term growth regulator for compounds having auxin and/or cytokinin activities, irrespective of their origin as native or synthetic. It is insinuated by the author that auxins and cytokinins, as well as other plant hormone or growth regulator compounds, control the induction and maintenance of developmental processes and probably differ from

672

growth promoting compounds e.g., vitamins, certain sugars, and some amino acids as well as essential minerals and other compounds. In this sense, the term "phytohormone" is used to obtain more precision. Since the basic mode of phytohormone action is yet uncertain, use of the term does not anticipate a mode of action in a strict analogy to the animal hormone system, although the author hypothesizes such an analogy for the initial steps of hormone action.

PLANT TISSUE CULTURE: RELEVANT HORMONES

The aim of this chapter is to outline biochemical mechanisms of plant hormone action.   Thus it seems worthwhile to restrict the text to those hormones that are frequently used in in vitro studies of plant growth and development. The examples cited in this chapter are mainly based on work with auxins (IAA, NAA, and 2,4-D) and cytokinins (ZEA, BA, 2iP, and KIN). In some instances, other auxins or auxin-like compounds have been found to be advantageous. The number of reports on the application of hormones other than auxins and cytokinins is rather limited, however, and the relevance of these hormones will be outlined briefly.   There may be a wider application of these hormones in the future, as was indicated very recently by studies on the dynamics of endogenous hormones (Weiler, 1981).

Auxins

Analysis of the effects and mode of action of auxins is difficult, as conflicting results of auxin action have been reported for different tissues. One can generalize that some effect of auxins can be found in any living plant tissue, but it is very often impossible to prove whether this effect is a direct one or a secondary consequence of auxin application.

Auxins are thought to control the following procedures: (a) apical dominance, (b) cell elongation in roots and shoots (with much lower concentrations in roots than in shoots), (c) H^+-extrusion and permeability changes of the plasmalemma, (d) formation of ethylene, (e) induction of adventitious root formation, (f) enhancement of the respiration rate, (g) induction of disorganized growth at higher (herbicidal) concentrations, (h) inhibition of embryo formation in cell suspension cultures, (i) formation of parthenocarpic fruits in some species, and (j) mitotic irregularities in long term tissue cultures.

This still incomplete list of auxin effects may give an idea of the complexity required for the development of a model that would explain all these divergent observations in a comprehensive way. In fact, it appears to be unlikely that all auxin effects are mediated by a single mechanism.

The auxins most frequently used in tissue culture are listed in Fig. 1.   For a detailed history of auxin research, see Soeding (1961) and Thimann (1972).

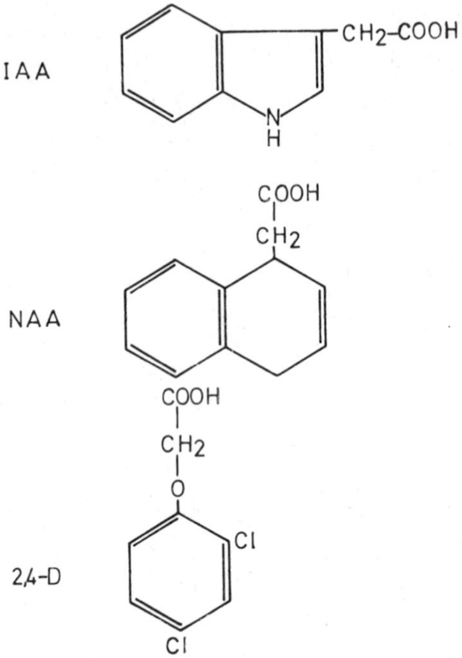

Figure 1.   The major auxins used in plant tissue culture: IAA (native auxin), NAA, and 2,4-D (synthetic auxins).

Cytokinins

The cytokinins have been used very successfully in plant tissue culture prior to the discovery of their being the essential compounds in coconut water (van Overbeek et al., 1941). Cytokinins have been characterized by Miller et al. (1955) in immature seeds of *Zea mays*, and subsequently, in other plant tissues.   The chemical nature of cytokinins indicates that they are derivatives of the purine base adenine.   Kinetin, the most widely used cytokinin, however, is not a native but a synthetic compound isolated by Miller et al. (1955) from yeast extract and old herring sperm after autoclaving.   Kinetin has not been found in higher plant extracts to date.   In contrast to the auxins and auxin-like substances, the cytokinins chemically represent a rather homogenous group (Fig. 2).

In plant tissues, the cytokinins act as: (a) stimulators of cell division, (b) retardants of senescence, and (c) stimulators of seed germination.

Moreover, the cytokinins counteract the role of auxins in the control of apical dominance, and, in certain combinations with auxins, are necessary for the commitment of cultured cells to organogenesis (Skoog, 1971).   From the large number of known cytokinin-active purine derivatives, only the four shown in Fig. 2 are frequently used in tissue culture media.

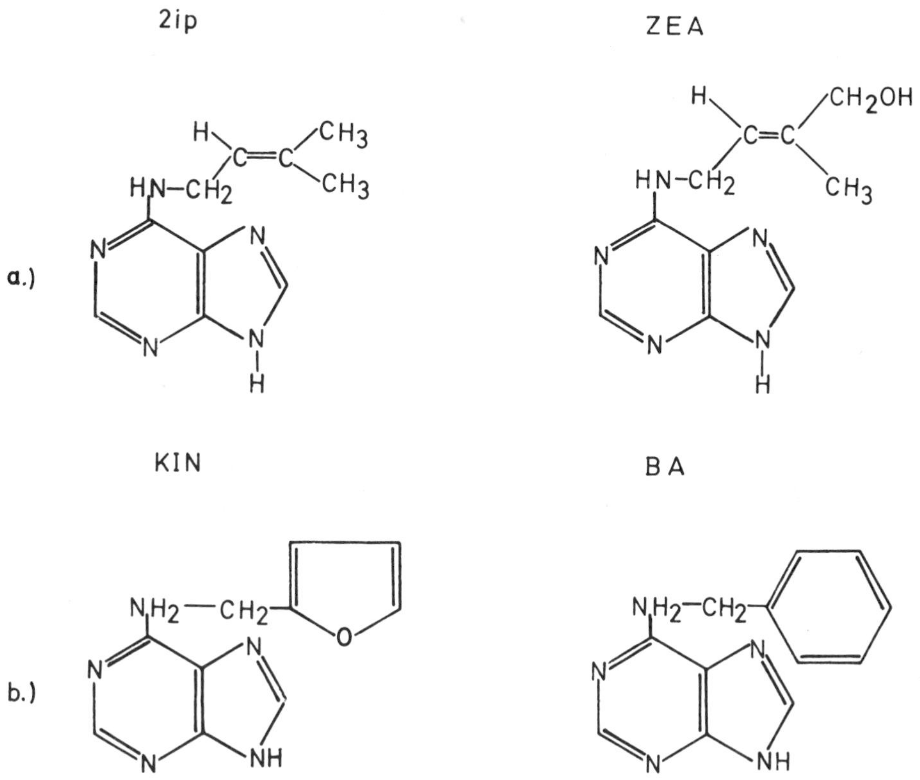

2ip   ZEA

a.)

KIN   BA

b.)

Figure 2. Cytokinins used in plant tissue culture: 2-isopentenyl adenine and zeatin (native cytokinins), kinetin and benzylamino purine (synthetic cytokinins).

Gibberellins

Gibberellins have no significant effects of practical value in tissue
culture experiments, although tobacco callus weight was reported to be
enhanced by this hormone (Murashige, 1965; Lance et al., 1976).   Most
probably, this positive influence was due to cell size enlargement, as
has been reported by Singh and co-workers (1974) for a number of
plant species.   Also from a report of Skoog (1971), it is known that
GA_3 exhibits some influence on tissue growth and organ formation in
combination with auxins and cytokinins in tobacco, but generally the
use of gibberellins in plant tissue culture was rather limited when compared to other phytohormones.   However, recent sensitive and selective
radioimmunoassays for the principle hormones in hormone-autotrophic
cultures of Beta vulgaris revealed that both abscisic acid and gibberellins may play a role in the tissues during growth (Weiler, 1981).

AVAILABILITY OF PHYTOHORMONES

Since the first and rather trivial prerequisite for phytohormone action
is the stable and defined availability of the hormones in the cells and

tissues, it seems worthwhile to pay some attention to mechanisms affecting availability.   Generally the availability of phytohormones depends on (a) concentration and stability in the medium during preparation, sterilization, and the culture period, and (b) uptake, translocation, and metabolism in the tissues during the culture period.

Physical and Chemical Factors Affecting Phytohormone Concentrations in the Media

In aqueous solutions, the native auxin IAA is degraded by acids, ionizing radiations, UV-light, visible light in the presence of sensitizing pigments, and oxygen or peroxides in the presence of suitable redox systems (Sembdner et al., 1980).   The most frequently observed destruction process for IAA is decarboxylation, as has been reported by a number of authors.

While the other widely used auxins like NAA and 2,4-D are rather stable against the above-mentioned factors, none of the auxins is affected by autoclaving, except IAA under certain conditions: Loewenberg (1965) found some destruction of IAA in the presence of Mn^{2+}, citric acid, and oxygen.

According to Dekhuijzen (1971), the cytokinins KIN, ZEA, and 2iP show no breakdown products using chromatography in aqueous solutions up to 1 hr after autoclaving.   On the other hand, when natural extracts with nonactive 1-, 3-, or 9-substituted purines are autoclaved, callus growth stimulating N^6-substituted compounds can be found.   For this reason, the use of undefined autoclaved natural extracts should be avoided, since nonreproducible conditions in respect to cytokinin concentrations may occur, thus influencing tissue growth in an unpredictable way.

In conclusion, it seems to be advisable, to prefer the use of NAA, 2,4-D or other more stable auxins whenever possible instead of IAA, or special control of the IAA concentration after media preparation should be used.

Biochemical Factors Affecting Phytohormone Concentration in the Media

Exogenously applied IAA in tissue culture is subject to enzymatic decarboxylation by excreted IAA-oxidases and peroxidases with rates and degrees that vary with tissue and physiological conditions (Davies, 1972; Thimann, 1972).   Generally, enzymatic decarboxylation of IAA is higher in older than in younger tissues (Epstein and Lavee, 1975, 1977).

It is not likely that biochemical modification of the other phytohormones, i.e., NAA, 2,4-D or cytokinins occurs as a "cell-free" process in the medium.   Nonetheless, the synthetic auxins and cytokinins are much less labile than IAA.

UPTAKE, TRANSLOCATION AND METABOLIZATION OF PHYTOHORMONES IN TISSUES

Auxins

An important factor controlling the morphogenetic responses of phytohormones is the availability and nonavailability of phytohormones in the cells and tissues. There have been a large number of reports on the uptake, translocation, and metabolism of phytohormones in higher plants. In this section the author will consider only those reports that have an impact on in vitro growth processes.

In reading papers or protocols about tissue culture experiments, one often is confronted with reports of different effects of a single tissue when exposed to the same hormones. Generally one would expect the hormones of the same type to act in the same way in identical tissues, but practical experience shows that this is not necessarily the case. In our laboratory we have conducted careful studies on the actions of the auxins IAA, NAA, and 2,4-D on intact pea seedlings (*Pisum sativum* cv. Dippes Gelbe Victoria) and dramatic differences have been observed with respect to uptake, mobility, and metabolism. These differences could be correlated to differences in the morphogenetic activities of the respective auxins (Ingensiep et al., 1981; Ingensiep, 1982). Figure 3 shows the different responses of pea seedlings to these auxins. The data on uptake, metabolism, and mobility are compiled in Table 1. Only the application of 2,4-D exhibited marked effects on the pea seedlings, i.e., short roots, formation of callus-like structures on the roots, and swollen tissues in the shoots. In contrast to the 2,4-D effects, following application of IAA and NAA only a slight stimulation of adventitious root formation could be observed. A comparison of these observations with the data from Table 1 shows that only 2,4-D is present in the tissues as free auxin at relatively high concentrations. This accumulation of 2,4-D occurs because 2,4-D has a high mobility and only a limited rate of oxidation and conjugation. IAA and NAA, on the other hand, have higher degradation rates and lower mobility, and consequently are present in the tissues as free auxins at much lower concentrations (Fig. 3).

Summarizing these results, it seems worthwhile to argue that the repeatedly higher efficiency of 2,4-D in the induction of callus formation apparently depends on its greater availability at least in the tissues of most dicotyledonous plants. In monocots and some dicots, however, resistance against the application of phenoxy-type auxins can be observed, probably because of an enhanced detoxification of the hormones by the enzymatic formation of less auxin-active compounds. Wilcox et al. (1963) found evidence for the formation of ring-labeled metabolites by excised roots of the resistant cereals oat (*Avena sativa*), barley (*Hordeum vulgare*), and corn (*Zea mays*), while the susceptible legumes peanut (*Arachis hypogea*), soybean (*Glycine max*), and alfalfa

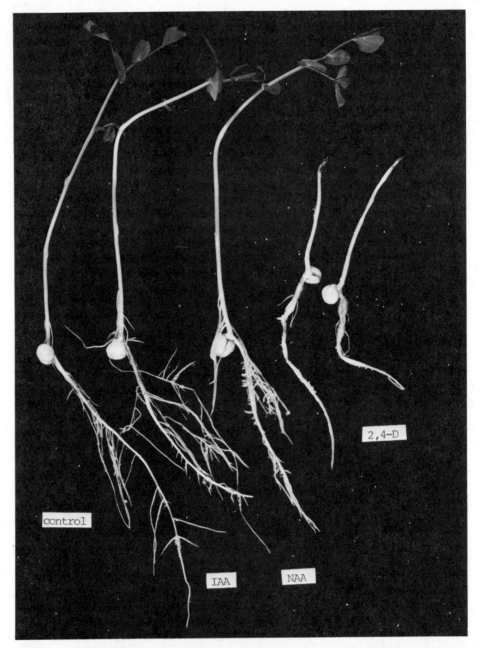

Figure 3. The effect of various auxins on intact pea seedlings (7 days old etiolated pea seedlings were treated with 10^{-4} M solutions of IAA, NAA, and 2,4-D for 24 hr, washed and grown for 7 days in a light/dark regime of 14/10 hr).

(*Medicago sativa*), did not show this type of metabolite. These results were confirmed by Feung et al. (1975) for callus tissues of five species,

Table 1. Comparison of the Fate of Root Applied 2,4-D, IAA, and NAA in Intact Pea Seedlings

AUXIN (root-applied, 10^{-4} M)	UPTAKE	METABOLISM Oxidation	METABOLISM Conjugation	METABOLISM Free Auxin	MOBILITY
2,4-D	Low	±	−	+++	High
IAA	Very high	+++	+	±	Low
NAA	Low	−	+	+	Low

where only the callus derived from maize possessed a high concentration of hydroxylated metabolites, whereas susceptible species showed mainly conjugate formation with amino acids. The phenomenon of resistance in dicots seems to have a genetic basis, since in different varieties of identical species different behavior against the application of 2,4-D was observed: Luckwill and Lloyd-Jones (1960) found decarboxylation of applied 2,4-D at a rate of 57% in 92 hours in the resistant variety Cox, while the susceptible variety Bramley's seedling only metabolized 2% during the same time period.

Another factor influencing the concentration of active auxin in the cells is the interaction between exogenously applied hormones with the pathways of the endogenous auxins. Venis (1972) reported that any active auxin applied to pea stem tissues induced the formation of an enzyme, which formed aspartate conjugates with the hormones. The auxin 2,4-D at higher concentrations is known to inhibit the transport of IAA, while at the lower growth-promoting concentrations, 2,4-D behaves like any other auxin in the promotion of IAA transport (Hay, 1956).

Besides the existence of pathways for metabolism of phytohormones, the mode of reaction to hormone application of a plant cell may also be determined by its ability to release hormones after changes of the hormonal equilibrium. Montague et al. (1981a) reported that carrot cells contain high amounts of primarily free 2,4-D when grown under maintenance conditions. When carrot cells are transferred to 2,4-D-free medium, 2,4-D is excreted, coinciding with the formation of somatic embryos. Carrot cells release much more 2,4-D than cultured soybean cells grown in the same manner. Soybean cells conjugate 2,4-D with amino acids and lack the ability to form embryos under conditions that are embryogenic for carrots. It was demonstrated by the authors that the higher retention of 2,4-D in soybean cells was due to the higher amino acid-conjugation rate in this species, since young cells of soybean are able to release unmetabolized 2,4-D in the same way as carrot cells do following a short exposure to 2,4-D. The authors suggest that the incapability of soybean cells to form embryos is at least partly related to the high endogenous level of 2,4-D-amino acid conjugates, which, according to Feung et al. (1974) possess weak auxin activity. Similar mechanisms were found by Spiegel-Roy and Kochba (1980) for *Citrus* callus, where the ability to form auxin-amino acid conjugates is correlated with positive embryogenic responses. Moreover, addition of inhibitors of endogenous IAA synthesis stimulated embryogenesis. Interestingly, Montague et al. (1981b) found an inhibitory influence of various cytokinins on 2,4-D conjugation in soybean, but since no new morphological structures could be observed in this system, more unknown factors must be related to the failure of soybean cells to regenerate embryos in the same way that carrot cells do.

Cytokinins

The metabolism of cytokinins is closely linked to the RNA-metabolism, when both biosynthesis and metabolic fate are examined (Letham,

1978; Sembdner et al., 1981). As was stated earlier, cytokinins represent chemically a rather homogenous group as compared to all the various compounds showing auxin activity. As a general feature, active cytokinins possess an intact purine ring and an N^6-substituent (Fig. 2), while the structure of auxin-active compounds exhibit much greater heterogeneity.

The availability of cytokinins in cells depends much more than that of the auxins, on the species- and cell-specific enzymatic interconversion rates, leading to new active or inactive derivatives. For example, leaves from *Populus alba* and root nodules from *Alnus glutinosa* were able to convert zeatin or zeatin conjugates to dihydro analogs (Letham et al., 1977; Henson and Wheeler, 1977). This conversion could not be observed in seedlings of radish, *Raphanus sativum* (Parker and Letham, 1973; Gordon et al., 1974).

Besides enzymatic interconversion processes, the pattern of cytokinins is changed in the direction of inactive derivatives by formation of nucleosides or nucleotides of bases with hormone activity as well as by glycosylation (Sembdner et al., 1981).

CONCLUSION

As a general conclusion it may be stated that the biological activity of phytohormones in cells is a function of its stable availability at a certain concentration in an active form. The concentration of hormones in the cells depends on the endogenous level, the exogenous supply, and both likewise are influenced by the uptake-, translocation-, and metabolization patterns. However, as was mentioned earlier, the tissue availability of hormones is only the first prerequisite for understanding these actions.

BIOCHEMICAL INTERACTIONS OF PHYTOHORMONES

The biochemically accessible effects of phytohormones are numerous and sometimes controversial. Only those that may contribute to an understanding of the mechanisms of morphogenetic phytohormone activity will be included in this section. Particularly important are long-term hormone effects.

In this respect research has been focused on phytohormone effects correlated with protein and/or nucleic acid syntheses. It has been demonstrated and reviewed by various authors that long-term treatment or higher (herbicidal) dosages of auxins strongly enhance both protein and nucleic acid synthesis (Key and Ingle, 1969). Most of the RNA produced after auxin treatment was shown to be rRNA (Melanson and Ingle, 1978) in artichoke tissue. Sen (1975) found an enhancement of DNA and histone precursor incorporation after 2,4-D-treatment in diploid and polytenic nuclei of *Allium cepa* roots by means of autoradiography. In Jerusalem Artichoke tissue, Yajima et al. (1980) found the induction of DNA synthesis by 2,4-D during callus induction. Moreover, these authors reported the incorporation of ^{14}C-2,4-D into chromatin, especially into the moderately lysine-rich fraction of the histones. They concluded that electrophoretically detectable changes in the histone contents depend on the binding of the acidic 2,4-D to the chromatin proteins, which are closely related to gene activation for RNA-synthesis (Yajima et al., 1980). Some reports are available showing the existence of proteins as intermediate factors in the interaction of hormones with the genome (Matthysse and Phillips, 1969), or as the hormone-dependent enhancement of RNA transcription or RNA-polymerase activity (Venis, 1971; Hardin et al., 1972; Teissere et al., 1975).

The unequivocal demonstration of newly and specifically synthesized proteins following hormone application, however, is still rare. Mozer (1980) reported the post-transciptional control function of gibberellic acid on the synthesis of α-amylase and other, GA-dependent polypeptides in the barley aleurone system. In addition, these syntheses were inhibited by the simultaneous addition of ABA to the tissues. Moreover, direct or indirect modification of proteins (methylation, phosphorylation) may occur as an expression of phytohormone activity. With the use of 2D-electrophoresis, Zurfluh and Guilfoyle (1980) found decreases and increases in the amount of certain polypeptides as well as changes in the electric charges of polypeptides in soybean hypocotyls following 2,4-D treatment. Following the application of herbicidal concentrations of 2,4-D (10^{-4} M) to intact pea seedlings, the pattern of the major acidic cytoplasmic proteins was found to be conserved, whereas de novo synthesis of some basic proteins could be detected (Fig. 4; Ingensiep, personal communication). Presently it is unclear whether these basic proteins are required for the formation of ribosomes or nucleosomes during the process of dedifferentiation.

In a recent paper Shaefer and Kahl (1981) reported an enhancement of DNA-dependent RNA polymerase I and II and a drastic stimulation of phosphorylation of both high- and low-molecular weight chromosomal proteins by 2,4-D in white potato tuber tissues after wounding. The effect on protein kinases occurred after a lag of 10 hr and was not observed after in vitro application of the hormone. This indicated the necessity for intermediate steps during the process of hormone action. In addition, these results indicate that phytohormone activity may depend not only on transcriptional control, but also on protein modification. Since tissues grown under in vitro conditions generally are wounded ones, these results are of great importance for an understanding of phytohormone effects in plant tissue culture.

Models for Phytohormone Action

The models required for an understanding of hormone actions in plants must be categorized in respect to the time course of hormone-dependent responses. Usually hormone effects are separated into (a) short-term effects and (b) long-term effects.

The short-term effects, which in the case of auxins are observed after a lag phase of 8-15 min (Stoddart and Venis, 1980), are suggested not to involve gene activations, since cycloheximide inhibition does not occur until 35-45 min after application. This second cycloheximide-

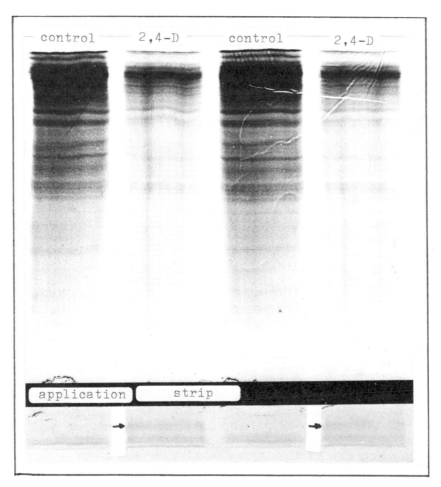

Figure 4. Autoradiograph of cytoplasmic proteins from pea roots treated with 2,4-D after isoelectric focusing and ^{14}C-amino acid labelling (autoradiograph with courtesy of H.W. Ingensiep).

inhibited response has a timing, which is much more consistent with the temporal requirements for macromolecular synthesis (Evans, 1974). Whether the rapid effects on fatty acid synthesis reported by Kull and Ultes (1980) caused by KIN after a time period of 45 min in isolated petunia protoplasts belong to the first or second group of rapid hormone effects cannot be answered at present. In the pattern of RNA and protein synthesis during the continuation of auxin-induced cell elongation, only minor qualitative changes can be found, so the discussion of whether gene activation is required for this process is controversial (Zurfluh and Guilfoyle, 1980; Bates and Cleland, 1980; Vanderhof, 1980; Jacobsen, 1977; Key, 1969). Moreover, there is no evidence that these short-term effects are functionally correlated with the morphogenetic long-term responses such as callus induction, shoot formation, rhizogenesis, or somatic embryogenesis, which occur after time lags of days or even weeks. These appear to fit long-term effect models, which have been developed in analogy to nonplant systems and which are thought to be independent from short-term responses (Libbenga, 1978).

From animal hormone research, two basic models are known and widely accepted for the action of the two main classes of animal hormones, reflecting the nature of the respective hormones.

First, peptide hormones, neurotransmitters and proteohormones bind to primary receptor sites, located at the outer cell membrane, thus identifying the cells as target cells. Binding activates an enzyme, adenylate cyclase, on the inner side of the membrane that forms cyclic AMP (cAMP) from AMP. cAMP acts as second messenger for the induction of a number of cellular processes. However, in plants neither the enzyme nor the cAMP has been found in noteworthy amounts (Amrhein, 1977), so this mode of action most likely is not realized in higher plants.

Second, steroid hormones enter their target cells and are recognized by soluble cytoplasmic receptor proteins. Newly formed dimeric hormone-receptor complexes bind to acceptor sites in the chromatin, where specific RNA syntheses are induced. Neither the receptor protein nor the hormone alone are able to bind to the acceptor sites in the nucleus. This clear causality between hormone application, hormone recognition, and hormone-induced gene activation has yet to be demonstrated in plants, although particular steps of this model have experimental documentation. This indicates some similarities between the action of plant hormones and steroids. The degree of similarity between plant and animal hormones is still uncertain.

A number of papers published in the field of phytohormone research in the last ten years have shown that hormone binding in plants has been found both at membrane-located sites in *Zea mays* coleoptiles, *Cucumis sativus* fruits, *Pisum sativum* epicotyls, and *Avena* roots (Hertel et al., 1972; Narayanan et al., 1981; Bhattacharyya and Biswas, 1978; Doellstaedt et al., 1976) and as soluble binding sites in the cytoplasm.

However, as was stated above, the membrane-bound sites have no cAMP-synthesizing activities, a result that once more confirms the nontransferability of the animal second messenger model to the existing conditions in higher plants. These binding sites on plant membranes may have some connection to the rapid auxin responses in the initial phase of cell elongation, as these responses apparently do not require protein synthesis.

An indication for a possible analogy with the steroid model is that cytoplasmic binding sites have been reported by various authors for auxins (Oostrom et al., 1975, 1980; Ihl, 1976; Wardrop and Polya, 1977; Jacobsen, 1981), cytokinins (Hecht, 1980), gibberellins (Konjevic et al., 1976; Stoddart et al., 1974), and ethylene (Sisler, 1979). Moreover, in the case of auxins macromolecular factors influencing RNA-synthesis in vitro has been reported (Matthysse and Phillips, 1969; Hardin et al., 1972; Venis, 1971; Teissere et al., 1975; Roy and Biswas, 1977). However, evidence is rare that these factors, which in most cases have been identified as proteins, have any relevant in vivo function.

Detection of Specific Soluble Auxin-Binding Proteins

Evidence for the existence of specific soluble phytohormone-binding proteins has been obtained by (a) equilibrium dialysis or (b) analysis of uptake of radiolabeled hormones. In the latter, radiolabeled hormone is incubated with increasing concentrations of cold hormone at adequate conditions for the proteins to be tested. After the incubation period the "free" hormone is separated from the "bound" hormone by dextran-coated charcoal (which binds the "free" hormone molecules) or by gently pelleting the proteins containing specific and unspecific hormone-protein complexes with a saturated solution of $(NH_4)_2SO_4$. Since equilibrium dialysis requires high specific activities of the labeled phytohormones or highly purified receptor proteins, most results on soluble hormone binding in plants have been obtained with the other method (Oostrom et al., 1975, 1980; Wardrop and Polya, 1977; Jacobsen, 1981). Since the author has had most experience with the pelleting assay, a protocol of this method is presented.

Criteria for Hormone Receptors in Plants

Although the biological functions of the hormone-binding proteins have not been elucidated unequivocally, some theoretical properties for proteins relating to their possible functions as receptors have been suggested (Kende and Gardner, 1976): (a) A receptor should be highly specific, and can be identified by a high affinity to all hormones of identical type, i.e., for all auxins or all active cytokinins. The K_a or its reciprocal (the K_d) should be the same order of magnitude as the concentration range of active hormones in the cells. (b) The capacity of a receptor should be limited and expressed by a rapid saturation with the hormone. The saturation rate should be parallel to the range of concentration of the biological response. Both parameters, K_a (or K_d) and the number of binding sites (R_t) can be derived by a kinetic analysis of experimental data according to Scatchard (1949). (c) The binding should be reversible and competitive. Nonactive or less active structural analogs of the hormone should bind in approximate accordance with their biological activity.

PROTOCOL

Preparation of Plant Material

Seeds of *Pisum sativum* were surface-sterilized and grown in moist vermiculite in the dark for 7 days at 22 C.

Preparation of Soluble Proteins

Roots or shoots of the seedlings were homogenized in a minimum of Tris-HCl (50 mM, pH 7.8) in a Waring blender (1 min, maximum speed), filtered through several layers of pure glass wool (Merck 4086), and centrifuged for 2 hr at 105,000 x g. The resulting supernatants were concentrated by ultrafiltration (Amicon UM-10 membrane) and separated from low-molecular weight compounds by Sephadex G-75 filtration. The protein fractions were collected, again concentrated, and directly used for the binding assays.

Binding Assays

The binding assays were performed according to Wardrop and Polya (1977). Tris-HCl (50 mM, pH 8.0) was adjusted with ^{14}C-labeled hormone to a hormone concentration of $5-6 \times 10^{-8}$ M. Using this labeled buffer, a stock solution of cold hormone is stepwise diluted, resulting in hormone concentrations between 5.005×10^{-5} M and the dilution buffer alone. The radioactivity of the subsequent fractions was checked for possible dilution errors by counting 10 μl samples in triplicate. With 2-3 replications/auxin concentration, aliquots of the proteins were mixed, and incubated in an icebath for 5 min. After this incubation, the proteins were precipitated with a saturated solution of $(NH_4)_2SO_4$, left some minutes in the ice and then centrifuged (15 min, 50,000 x g). The supernatant, containing "free" hormone, was carefully removed from the tubes, and the pellets resuspended in 1 ml 0.1% SDS and counted in Nuclear Chicago Liquid Scintillation Counter with Quickszint 212. From the cpm-readings, the dpm-rates were calculated by means of tabulated correction coefficients from a computerized quench correction curve based on the channel ratio method. From these data a displacement curve can be obtained, showing a decrease of radioactivity with increasing amounts of cold hormone (Fig. 5, upper graph). It was found that subtraction of the background observed in the presence of 5×10^{-5} M cold hormone, resulting Δ dpm-values (Fig. 5, mid graph) gave a much clearer picture of the displacement. It could be demonstrated by the kinetic analysis that this background was due to events other than specific or unspecific binding.

Calculation of Kinetic Data from Binding Experiments

The amount of total bound hormone was calculated by the known relation of cold and labeled hormone in every sample by the following formula:

$$B_{tot} = \Delta \text{ dpm} \times A \left[1 + (^{12}C\text{-hormone}/^{14}C\text{-hormone})\right]$$

Δ dpm = radioactivity corrected for the plateau value in the presence of 5×10^{-5} M cold hormone

A = factor depending on the specific activity of the labeled hormone

Figure 5.   Schematic representation of the different steps for the evaluation of kinetic data from phytohormone-binding studies (for explanation see text).

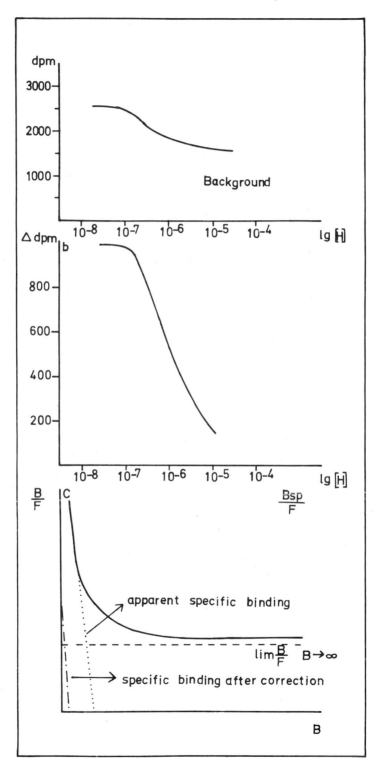

The Scatchard plots were constructed from these data (Scatchard, 1949). In this plot, specific binding is represented by a steeply descending line, while unspecific binding is shown by a line almost parallel to the abscissa, resulting in a hyperbolic curve (Fig. 5, graph at the bottom). According to Chamness and McGuire (1975), nonspecific binding present at all ligand concentrations has to be separated from specific binding by multiplying the limiting B/F-ratio (lim B/F, B→∞) with the free ligand concentration F at each point and subtracting this nonspecific binding from total binding:

$$B_{sp} = B_{tot} - F \ (lim \ B/F, \ B\to\infty).$$

The corrected binding line (–.–.–.) was calculated by plotting the values for B_{sp} versus B_{sp}/F. The kinetic parameter K_d and R_t can be calculated with relative accuracy by using the linear regression function, although mathematical laws are violated in this case (Quednau and Jacobsen, 1981). All calculations should be based on a protein content of 1 mg/ml of assay volume after addition of the pelleting agent. In this plot, the negative reciprocal of the slope equals K_d, and the intercept with the abscissa gives the number of binding sites, R_t.

The same test can be used to ascertain the specificity of the binding sites by performing competition experiments with active and nonactive analogs of the respective hormone tested. If the binding is specific, active analogs should displace the labeled hormone in the same way as the unlabeled one, or at least much better than inactive compounds. Figure 6 shows the displacement of ^{14}C-NAA by cold NAA and cold 2,4-D in the soluble cytoplasmic protein fraction of etiolated pea epicotyls.

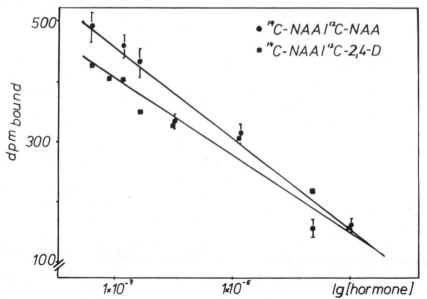

Figure 6: Displacement of ^{14}C-labelled NAA by cold NAA and cold 2,4-D.

CRITICAL REMARKS

The preceding section outlines general methodology for obtaining evidence on the suggested primary level of interaction of plant hormones with the cell. It appears very likely that plant cells possess proteins that are able to recognize hormone molecules in in vitro assays under artificial conditions. But, as was shown in the case of auxins, nonplant proteins like BSA are able to bind auxins under acidic conditions in a similar way as the tentative receptors (Murphy, 1979). Hence some doubts on the relevance of this binding has been suggested (Venis, 1980). A recent paper of Wardrop and Polya (1980a) shows auxin-binding by the soluble proteins of pea and bean leaves. In a variety of chromatographic procedures this auxin-binding protein co-purifies with RuBPCase, indicating that this enzyme is the auxin-binding protein. However, the authors could not find the same auxin-binding in Fraction-1 protein from spinach or sugar beet. As they point out, the observed binding by RuBPCase does not mean that this enzyme is an auxin-receptor, because the concentration of this enzyme in the cells is about 100 times higher than the K_d for the binding, and potent auxins like 2,4-D bind with a much lower K_d (Wardrop and Polya, 1980a,b). From this, the authors concluded that the function of this binding might be auxin sequestration and/or auxin translocation. Heilmann et al. (1981) found high accumulation of fed ^{14}C-IAA in mesophyll chloroplasts of *Spinacia oleracea* (47% of the endogenous IAA was found in the chloroplasts, while the chloroplasts represent only 7% of the tissue volume). It is not clear whether these results have any physiological relevance in respect to the morphogenetic activities of auxins.

In the case of soluble auxin-binding proteins from etiolated pea epicotyls (Jacobsen, 1981), since SDS-pherograms exhibited the absence of RuBPCase in the proteins investigated, this binding apparently is different from that reported from the green tissues (Wardrop and Polya, 1977, 1980a).

In all other cases where soluble cytoplasmic proteins have been reported to bind specifically phytohormones, the relevance of the binding must be questioned in respect to the relationship between the observed binding phenomena and the hormone effects in the systems in question. This relationship is only tentative, and there exists no clear evidence for the presence of steroid-like mechanisms in higher plants. Some comments may illustrate this hypothesis.

First, the results of Yajima et al. (1980) demonstrate that 2,4-D directly binds to chromatin, a clear contradiction to an analogy to the steroid paradigm. As a hypothesis, evolutionary differences between chromatin composition in plants and animals may be assumed. This may have also been indicated by Muller et al. (1980), who showed some atypical properties of barley chromatin as having a surprisingly low solubility at low ionic strength, which was attributed to the presence of some acidic nonhistone proteins. In addition, Greilhuber (1977) demonstrated that plant metaphase chromosomes have a much higher degree of condensation than vertebrate chromosomes, which during mitotic metaphase are only about 2.3 times shorter than in pachytene. The higher contraction of plant chromosomes, which do not show Giemsa-banding, may be an indication for a significantly altered chromatin composition during interphase.

Second, in a refreshingly provocative article, Trewavas (1981) introduced two interesting aspects of phytohormone research: (a) plants
possess developmental processes that are unique, and (b) the concept of
growth substances as *hormones* may be a misleading one.   Moreover,
Trewavas (1981) criticizes the fact that tissue sensitivity generally is
interpreted as a matter of change of growth substances instead of a
change of the number of receptor molecules.   In this respect two results on auxin-binding proteins may be of great importance for future
work: Oostrom et al. (1980) obtained evidence on a change of soluble
high-affinity binding of IAA, extractable from tobacco-pith callus after
different periods of subculture, with a peak at 10–12 days after transfer
to fresh medium.   Vreugdenhil et al. (1981) found modulation of the
number of membrane-located auxin-binding sites during the growth of
batch-cultured tobacco cells.

As a conclusion, it can be argued that morphogenetic activities of
phytohormones depend on the specific recognition of the morphogenetic
information carried by the plant hormone molecule.   The most relevant
phytohormones and proteins have been reported which may fulfill this
criterion in a satisfactory manner.   However, the next steps toward a
biochemical realization of the hormonal stimulus are yet unclear.   It is
not known whether transport of possible hormone/hormone-receptor
complexes to the nucleus, followed by the initiation of specific transcription, occurs or whether the hormones or hormone/receptor-complex-
es exhibit an influence on the processing of the primary gene transcript
or on the control of translation.   A direct modification of proteins can
not be excluded.   The present state of the art may be illustrated by
Fig. 7, where question marks indicate the lack of proven evidence.
Hopefully, the question marks will be reduced in the next few years as
a result of continued phytohormone research.

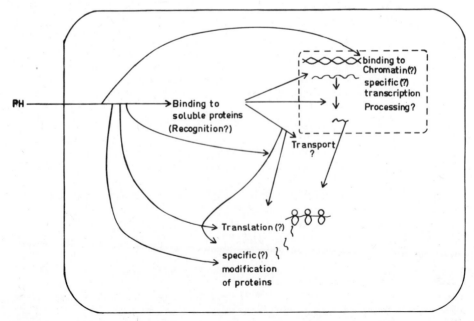

Figure 7.   The flow of the morphogenetic information of a phytohormone and probable control points (?) as indicated by available results.

ACKNOWLEDGMENT

The author gratefully acknowledges the helpful criticism of H.W. Ingensiep during the preparation of the manuscript as well as permission to use unpublished data (Fig. 4).

KEY REFERENCES

Evans, M.L. 1974. Rapid responses to plant hormones. Annu. Rev. Plant Physiol. 25:195-224.

Jacobsen, J.V. 1977. Regulation of ribonucleic acid metabolism by plant hormones. Annu. Rev. Plant Physiol. 28:537-564.

Kende, H. and Gardner, G. 1976. Hormone binding in plants. Annu. Rev. Plant Physiol. 27:267-290.

Letham, D.S. 1978. Cytokinins. In: Phytohormones and Related Compounds: A Comprehensive Treatise (D.S. Letham, J. Higgins, and P.Z. Goodwin, eds.) Vol. I, pp. 205-293. Elsevier, Amsterdam.

Libbenga, K.R. 1978. Hormone receptors in plants. In: Frontiers of Plant Tissue Culture 1978 (T.A. Thorpe, ed.) pp. 325-333. International Association of Plant Tissue Culture, Calgary.

Sembdner, G., Gross, D., Liebisch, H.-W., and Schneider, G. 1980. Biosynthesis and metabolism of plant hormones. In: Encyclopedia of Plant Physiology, New Series, Vol. 9: Hormonal Regulation of Development I (J. McMillan, ed.) pp. 336-444. Springer-Verlag, Berlin, Heidelberg, New York.

Soeding, H. 1961. Die Auxine-Historische Uebersicht. In: Encyclopedia of Plant Physiology (W. Ruhland, ed.) pp. 450-484. Springer-Verlag, Berlin, Goettingen, Heidelberg.

Stoddart, J.L. and Venis, M.A. 1980. Molecular and subcellular aspects of hormone action. In: Encyclopedia of Plant Physiology, New Series, Vol. 9: Hormonal Regulation of Development I (J. McMillan, ed.) pp. 445-510. Springer-Verlag, Berlin, Heidelberg, New York.

Thimann, K.V. 1972. The natural plant hormones. In: Plant Physiology: A Treatise (F.C. Steward, ed.) pp. 1-359. Academic Press, New York.

Trewavas, A. 1981. How do plant growth substances work? Plant Cell Environ. 4:203-228.

REFERENCES

Amrhein, N. 1977. The current status of cyclic AMP in higher plants. Annu. Rev. Plant Physiol. 28:123-132.
Bates, G.W. and Cleland, R.E. 1980. Protein patterns in the oat coleoptile as influenced by auxin and by protein turnover. Planta 148: 429-436.
Bhattacharyya, K. and Biswas, B.B. 1978. Membrane-bound auxin receptors from *Avena* roots. Indian J. Biochem. Biophys. 15:445-448.

Chamness, G.C. and McGuire, W.L. 1975. Scatchard plots: Common errors in correction and interpretation. Steroids 26:538-542.

Davies, P.J. 1972. The fate of exogenously applied IAA in light grown stems. Physiol. Plant. 27:262-270.

Dekhuijzen, H.M. 1971. Sterilization of cytokinins. In: Effects of Sterilization on Components in Nutrient Media. pp. 129-132, Wageningen.

Deverall, B.J. 1965. Apparently spontaneous decarboxylation of indole-3-acetic acid. Nature 207:828-829.

Doellstaedt, R., Hirschberg, K., Winkler, E., and Huebner, G. 1976. Bindung von Indolylessigsaeure an Fraktionen aus Epikotylen und Wurzeln von *Pisum sativum*. Planta 130:105-111.

Epstein, E. and Lavee, S. 1975. Uptake and fate of IAA in apple callus tissue using IAA-1-[14]C. Plant Cell Physiol. 16:553-561.

_____ 1977. Uptake, translocation and metabolism of IAA in the olive (*Olea europea*). I. Uptake and translocation of 1-^{14}C-IAA to detached "Manzanilla" olive trees. J. Exp. Bot. 28:619-628.

Feung, C., Mumma, R.O., and Hamilton, R.H. 1974. Metabolism of 2,4-dichlorophenoxyacetic acid. VI: Biological properties of amino acid conjugates. J. Agric. Food Chem. 22:307-309.

Feung, C., Hamilton, R.H., and Mumma, R.O. 1975. Metabolism of 2,4-dichlorophenoxyacetic acid. VII: Comparison of metabolites from five species of plant callus tissue cultures. J. Agric. Food Chem. 23:373-376.

Gordon, M.E., Letham, D.S., and Parker, C.W. 1974. The metabolism and translocation of zeatin in intact radish seedlings. Ann. Bot. 38:809-825.

Greilhuber, J. 1977. Why plant chromosomes do not show G-bands. Theor. Appl. Genet. 50:121-124.

Hardin, J.W., Cherry, J.H., Morre, D.J., and Lemby, C.A. 1972. Enhancement of RNA polymerase activity by a factor released by auxin from plasma membrane. Proc. Natl. Acad. Sci. USA 69:3146-3150.

Hay, J.R. 1956. The effect of 2,4-D and 2,3,5-tri-iodobenzoic acid on the transport of IAA. Plant Physiol. 31:118-120.

Hecht, S.M. 1980. Probing the cytokinin receptor site(s). In: Plant Growth Substances 1979 (F. Skoog, ed.) pp. 144-158. Springer-Verlag, Berlin.

Heilmann, B., Hartung, W., and Gimmler, H. 1981. Subcellular compartmentalization of indole-3-acetic acid in mesophyll cells of *Spinacia oleracea*. Z. Naturforschung. 36c:679-685.

Henson, I.E. and Wheeler, C.T. 1977. Hormones in plants bearing nitrogen-fixing root nodules: Metabolism of (8-^{14}C)-zeatin in root nodules of *Alnus glutinosa* (L.) Gaertn. J. Expt. Bot. 28:205-214.

Hertel, R., Thomson, K.-S., and Russo, V.E.A. 1972. In vitro auxin binding to particulate cell fractions from corn coleoptiles. Planta 107:325-340.

Ihl, M. 1976. Indole-acetic acid binding proteins in soybean cotyledon. Planta 131:223-228.

Ingensiep, H.W. 1982. The morphogenetic response of intact pea seedlings with respect to translocation and metabolism of root applied auxin. Z. Pflanzenphysiol. 105:149-164.

—————, Herlt, M., and Jacobsen, H.-J. 1981. Morphogenetic response, translocation and metabolism of root applied auxins in pea seedlings. Pisum Newsl. 13:21-23.

Jacobsen, H.-J. 1981. Soluble auxin-binding proteins in pea. Cell Biol. Int. Rep. 5:768.

Key, J.L. 1969. Hormones and nucleic acid metabolism. Annu. Rev. Plant Physiol. 20:449-474.

—————— and Ingle, J. 1969. Effects of auxin on RNA metabolism: RNA metabolism in response to auxin. In: Biochemistry and Physiology of Plant Growth Substances (F. Wightman and G. Setterfield, eds.) pp. 711-722. Runge Press, Ottawa.

Konjevic, R., Grubisic, D., Markovic, R., and Petrovic, J. 1976. Gibberellic acid-binding proteins from pea stems. Planta 131:125-128.

Kull, U. and Ultes, U. 1980. Rapid kinetin effects on lipid synthesis in isolated mesophyll protoplasts of *Petunia*. Naturwissenschaften 67:97.

Lance, B., Reid, M., and Thorpe, T.A. 1976. Endogenous gibberellins and growth of tobacco callus cultures. Physiol. Plant. 36:287-292.

Letham, D.S., Parker, C.W., Duke, C.C., Summons, R.E., and MacLeod, J.K. 1977. O-Glucosylzeatin and related compounds: A new group of cytokinin metabolites. Ann. Bot. 41:261-263.

Loewenberg, J.R. 1965. Promotion of indoleacetic acid destruction by citric acid and L-alanine. Physiol. Plant. 18:31-40.

Luckwill, L.C. and Lloyd-Jones, C.P. 1960. Metabolism of plant growth regulation. II. Decarboxylation of 2,4-D in leaves of apple and strawberry. Ann. Appl. Biol. 48:626-636.

Matthysse, A.G. and Phillips, C. 1969. A protein intermediary in the interaction of a hormone with the genome. Proc. Natl. Acad. Sci. USA 63:897-903.

Melanson, D.L. and Ingle, J. 1978. Regulation of ribosomal RNA-accumulation by auxin in artichoke tissue. Plant Physiol. 61:190-198.

Miller, C.O., Skoog, F.V., Saltza, M.H., and Strong, F.M. 1955. Kinetin, a cell division factor from deoxyribonucleic acid. J. Am. Chem. Soc. 77:1329.

Montague, M.J., Enns, R.K., Siegel, N.R., and Jaworski, E.G. 1981a. A comparison of 2,4-dichlorophenoxyacetic acid metabolism in cultured soybean cells and in embryogenic carrot cells. Plant Physiol. 67:603-607.

—————— 1981b. Inhibition of 2,4-dichlorophenoxyacetic acid conjugation to amino acids by treatment of cultured soybean cells with cytokinin. Plant Physiol. 67:701-704.

Mozer, T.J. 1980. Control of protein synthesis in barley aleurone layers by the plant hormones gibberellic acid and abscisic acid. Cell 20:479-485.

Muller, A., Philipps, G., and Gigot, C. 1980. Properties of condensed chromatin in barley nuclei. Planta 149:69-77.

Murashige, T. 1965. Effects of stem elongation retardants and gibberellin on callus growth and organ formation in tobacco tissue culture. Physiol. Plant 8:665-673.

Murphy, G.J. 1979. Plant hormone receptors: Comparison of naphtaleneacetic acid binding by maize extracts and by a non-plant protein. Plant Sci. Lett. 15:183-191.

Narayanan, K.R., Mudge, K.W., and Poobaiah, B.W. 1981. In vitro auxin binding to cellular membranes of cucumber fruits. Plant Physiol. 67: 836-840.

Oostrom, H. 1980. Investigations on a soluble auxin receptor. Ph.D. Thesis, Leiden, The Netherlands.

_____, van Loopik-Detme M.A., and Libbenga, K. 1975. A high affinity receptor for indoleacetic acid in cultured tobacco pith explants. FEBS Lett. 59:194-197.

_____, Kulescha, Z., van Vliet, T.B., and Libbenga, K. 1980. Characterization of a cytoplasmic auxin receptor from tobacco pith callus. Planta 149:44-47.

Parker, C.W. and Letham, D.S. 1973. Regulators of cell division in plant tissues. XVI. Metabolism of zeatin by radish cotyledons and hypocotyls. Planta 114:199-218.

Quednau, H.D. and Jacobsen, H.-J. 1981. Biometric evaluation of Scatchard plot data from phytohormone binding studies. EDV Biol. Med. 12:83-88.

Roy, P. and Biswas, B.B. 1977. A receptor protein for indolacetic acid from plant chromatin and its role in transcription. Biochem. Biophys. Res. Comm. 74:1597-1606.

Scatchard, G. 1949. The attractions of proteins for small molecules and ions. Ann. N. Y. Acad. Sci. 51:660-672.

Schaefer, W. and Kahl, G. 1981. Auxin-induced changes in chromosomal protein phosphorylation in wounded potato tuber parenchyma. Plant Mol. Biol. 1:5-17.

Sen, S. 1975. Effect of 2,4-D on nucleic acid and histone-precursor incorporation into nuclei. Naturwissenschaften 62:184.

Singh, B.D., Thomas, E.T., and Harvey, B.L. 1974. Effects of gibberellic acid on cell suspension culture of higher plants. Indian J. Exp. Biol. 12:213-215.

Sisler, E.C. 1979. Measurement of ethylene binding in plant tissue. Plant Physiol. 64:538-542.

Skoog, F. 1971. Aspects of growth factor interactions in morphogenesis of tobacco tissue culture. In: Les Cultures de Tissus de Plantes. Colloques Internationaux du C.N.R.S., No. 193. Paris

Spiegel-Roy, P. and Kochba, J. 1980. Embryogenesis in *Citrus* tissue cultures. In: Advances in Biochemical Engineering, Vol. 16: Plant Cell Cultures (A. Fletcher, ed.) pp. 27-48. Springer-Verlag, Berlin, Heidelberg, New York.

Stoddart, J., Breidenbach, W., Nadeau, R., and Rappaport, L. 1974. Selective binding of (^{3}H) gibberellin A^1 by protein fractions from dwarf pea epicotyls. Proc. Natl. Acad. Sci. USA 71:3255-3259.

Teissere, M., Penon, P.M., van Hustee, R.B., Azou, Y., and Ricard, J. 1975. Hormonal control of transcription in higher plants. Biochem. Biophys. Acta 402:391-402.

Vanderhoef, L.N. 1980. Auxin-regulated cell enlargement: Is their action at the level of gene expression? In: Genome Organization and Expression in Plants (C.J. Leaver, ed.) pp. 159-173. Plenum, New York.

van Overbeek, J., Conklin, M.E., and Blakeslee, A.F. 1941. Factors in coconut milk essential for growth and development of very young *Datura* embryos. Science 94:350-351.

Venis, M.A. 1971. Stimulation of RNA transcription from pea and corn DNA by protein retained on Sepharose coupled to 2,4-dichlorophenoxacetic acid. Proc. Natl. Acad. Sci. USA 68:1824-1827.

_____ 1972. Auxin-induced conjugation systems in peas. Plant Physiol. 49:24-27.

_____ 1980. Cellular recognition of plant growth regulators. In: Aspects and Prospects of Plant Growth Regulators (joint DPGRG and BPGRG symposium) Monograph 6.

Vreugdenhil, D., Burgers, A., Harkes, P.A.A., and Libbenga, K. 1981. Modulation of the number of membrane-bound auxin-binding sites during growth of batch cultured tobacco. Planta 152:415-419.

Wardrop, A.J. and Polya, G.M. 1977. Properties of a soluble auxin-binding protein from dwarf bean seedlings. Plant Sci. Lett. 8:155-163.

_____ 1980a. Co-purification of pea and bean leaf soluble auxin-binding proteins with ribulose-1,5-bisphosophate carboxylase. Plant Physiol. 66:105-111.

_____ 1980b. Ligand specificity of bean leaf soluble auxin-binding protein. Plant Physiol. 66:112-118.

Weiler, E.W. 1981. Dynamics of endogenous growth regulators during the growth cycle of a hormone-autotrophic plant cell culture. Naturwissenschaften 68:377-378.

Wilcox, M., Moreland, D.E., and Klingman, G.C. 1963. Aryl hydroxylation of phenoxy aliphatic acids by excised roots. Physiol. Plant. 16:565-571.

Yajima, Y., Yasuda, T., and Yamada, Y. 1980. Induction of DNA-synthesis by 2,4-dichlorophenoxyacetic acid during callus induction in Jerusalem artichoke tuber tissue. Physiol. Plant. 48:564-567.

Zurfluh, L.L. and Guilfoyle, T.J. 1980. Auxin-induced changes in the patterns of protein synthesis in soybean hypocotyl. Proc. Natl. Acad. Sci. USA 77:357-361.

CHAPTER 22
Transport of Ions and Organic Molecules

C. McDaniels

The literature on solute transport in intact plants is enormous (Zimmerman and Dainty, 1974; Luttge and Higinbotham, 1979; Pate, 1980; Spanswick et al., 1980; Spanswick, 1981) while the literature on transport in cultured plant cells is limited. Maretzki and Thom (1978) have reviewed the latter and indicated some of the reasons for using cultured cells in transport studies. Cultured cells are free of contaminating organisms, and the external environment can be easily controlled. Cells for transport studies can be grown in continuous culture systems to achieve relatively homogeneous phenotypic populations of cells. The possibility of isolating variants from cultured cells is an exciting prospect. For some species, e.g., *N. tabacum*, plants can be regenerated from single cells (Vasil and Hildebrandt, 1965), haploids can be produced by anther culture (Nitsch, 1969), and genetic variants produced during culture can be regenerated into whole plants (Carlson, 1973). Thus in many respects cultured plant cells can be treated like microorganisms, and the genetic variants selected in culture can be proven to be mutants via Mendelian analysis. An additional advantage of cell cultures, e.g., protoplasts, single cells, or small clumps of cells, is that essentially all cell surfaces are exposed to the incubation solution. Perhaps the major limitation to the use of cultured cells is the fact that it is impossible to gain a complete understanding of the transport phenomenon in plants at the tissue, organ, or plant levels of organization by studying transport in cultured cells. However, if one is in fact interested in characterizing transport at the cellular level, suspension cultured cells, including protoplasts, may offer an ideal system.

REVIEW OF THE LITERATURE

It has been just a little more than a decade since investigators began to employ cultured cells for transport studies (Hart and Filner, 1969; Maretzki and Thom, 1970; Heimer and Filner, 1971). In the early 1970s there were few studies, but within the past several years the use of cultured cells has increased significantly (e.g., Berlin and Widholm, 1978; Blackman and McDaniel, 1980; Guy et al., 1978, 1980, 1981; King and Khanna, 1978; Ruesink, 1978; Smith, 1978a; Cheruel et al., 1979; Mettler and Leonard, 1979a; Thoiron et al., 1979, 1980; Courtois et al., 1980; Morris and Thain, 1980; Rubinstein and Tattar, 1980; Harrington et al., 1981; Jones and Smith, 1981; Kaiser and Hartung, 1981; Thom et al., 1981). The primary aim of many of these as well as the earlier studies was to establish credibility and repeat what had already been done in other systems. I believe that credibility has been established and that, in most respects, the basic parameters of transport are similar in intact plant cells and in cultured cells. Uptake of amino acids, ions, and sugars at physiological concentrations is carrier mediated and active. Uptake kinetics are most often complex, but the underlying basis of this complexity is unclear (Nissen, 1974; Bange, 1979; Borstlap, 1981). The carriers exhibit specificity and can, for example, distinguish between Na^+ and K^+ (Mettler and Leonard, 1979b), the L and D isomers of sugars (Guy et al., 1978), and the D and L isomers of amino acids (McDaniel et al., 1983). Current evidence supports the hypothesis that amino acids and sugars are co-transported with H^+ and that the driving force for uptake is the membrane potential and the proton gradient across the cell membrane (Etherton and Rubinstein, 1978; see various papers in Spanswick et al., 1980). Some molecules appear to enter the cell by diffusional processes. Several recent papers have examined the uptake of organic compounds in cultured cells which appear to enter the cell via diffusional, noncarrier mediated processes, such as abscisic acid (Kaiser and Hartung, 1981), amitrole (Singer and McDaniel, 1982), and tryptamine (Courtois et al., 1980). Transport can occur across several membranes (e.g., plasma membrane, tonoplast, or other plastid membranes), and it has been difficult to establish which membrane is being studied. The solution to this problem may require the use of isolated organelles or artificial vesicles with reconstituted transport systems.

A variety of assays have been used to measure solute transport in suspension cultured cells and protoplasts. There are no standard methods, and each laboratory has developed its own procedure. In developing an assay to measure solute transport, many variables must be considered, for example, preparation of the cells, contents of the incubation solution, number of points used to estimate the uptake rate, standardization parameters (e.g., number of cells, fresh weight, dry weight, protein and homogenized cell volume), and length of transport period. In selecting or developing an assay, an investigator needs to take into account the assumptions being made and how the selection of conditions will influence the precision and possibly the accuracy of the measurement being made.

PROTOCOLS

There are no standard methods for analyzing the transport of solutes in suspension cultured cells or protoplasts. Our laboratory has developed assays for measuring the uptake of organic molecules in suspension cultured tobacco cells and tobacco leaf protoplasts. I will present in detail two protocols: our assay for measuring solute uptake in suspension cultured cells (Blackman and McDaniel, 1978, 1980) and with minor modifications, Rubinstein's assay for measuring solute uptake in oat leaf protoplasts (Rubinstein, 1978; Rubinstein and Tattar, 1980). However, before considering these protocols, I will discuss developing an assay in considerable detail. This discussion will be for an assay employing suspension cultured cells, but essentially all of the comments are pertinent to transport studies in general.

Developing an Assay

TRANSPORT MEASUREMENTS. Radioactive isotopes are most often used to follow the movement of a molecule, and all of the discussion will be related to the use of these tracers. In using a radioactive isotope one assumes that the membrane does not distinguish between the tracer (radioactive isotope) and the unlabeled molecule. Although this is a reasonable assumption, it may not be completely valid. In the case of ^3H_2O being used as a tracer for the movement of H_2O, the interpretation of some data suggests that the membrane may distinguish between ^3H_2O and H_2O (Shafer and Andreoli, 1977).

The objective of a transport assay is to measure, quantitatively, the movement of a solute into or out of a cell. To insure that this is what is being measured, one must consider binding, metabolism, and trapping in the extracellular free space. Binding to cell walls and cell membranes can be measured in several ways (Blackman and McDaniel, 1978). One approach is to measure the adsorption of solute to cell wall material by exposing homogenized cells to the labeled solute and measure the amount bound. A second approach is to expose the cells to the solute and then, after a period of time, homogenize the cells and fractionate the components through a series of centrifugations at increasing g's to estimate binding to various cellular components (e.g., cell walls, membranes, organelles). The first approach measures passive binding, while the latter measures binding dependent upon cell integrity. We have made these binding measurements for several amino acids and the herbicide, amitrole. The binding to homogenized cells was negligible for these solutes when compared to the amount transported, but the amount of amitrole bound to the walls of intact cells was significant. Depending upon the assay employed, this binding could influence the results. That is, if the whole cell is solubilized for scintillation counting, then the well-bound isotope would be counted, and higher rates would result.

Removing the isotope from the free space is generally accomplished by rinsing the cells with a volume of incubation solution that does not contain isotope. The presence or absence of the transported solute in the rinse solution may influence the retention of the isotope. This possibility can be checked by comparing the counts retained in cells washed with rinse solution containing or lacking the solute. The rinse procedure should be accomplished with a volume of solution and in a manner such that essentially no counts appear in the final rinse solution.

The purity of the isotope should be monitored. We have found L-leucine (^{3}H) to be more than 98% pure and to be stable over several months of storage. This has not been true for labeled L-aspartic acid, L-histidine, and L-phenylalanine. After several months of storage, 10-30% of the counts are no longer associated with the authentic amino acid. Metabolism or spontaneous breakdown of the solute before or after entering the cell may also influence results. Metabolism of organic isotopes can be estimated by standard chromatographic methods. If the solute can be incorporated into a large molecule (protein, nucleic acid, carbohydrate, lipid), one should check for this using standard isolation methods for the molecule in question. In general, chromatographic methods should reveal metabolism into macromolecules, since most large molecules will remain at the origin in solvent systems which separate into small molecules.

DURATION OF UPTAKE STUDIES. Some workers have used uptake time periods of an hour or more (Heimer and Filner, 1971; Harrington and Smith, 1977). These long-term assays measure net flux ($F_{net} = F_{in} - F_{out}$), and make it necessary to correct for metabolism. Additionally, inaccuracies are also possible, since the uptake rate may change over the uptake period. Some workers have used short time periods to reduce the influence of metabolism, efflux, and changes in uptake rate (Maretzki and Thom, 1970; Blackman and McDaniel, 1978). However, even with a short uptake period one must still measure metabolism, efflux, and changes in uptake rate over the uptake time period to ascertain the degree to which these processes influence the rate being calculated.

Efflux can be estimated by placing preloaded cells in unlabeled medium and following the loss of counts from the cells or the appearance of counts in the medium. Efflux may involve exchange diffusion (Shtarkshall et al., 1970). For this reason, efflux into bath media with and without the transported solute should be calculated. Influx and efflux can be compared by calculating the influx or efflux rate constant using the flux equation $\ln[A] = \ln[Ao] - kt$, where $[A]$ is the internal radioactivity in cpm, $[Ao]$ is the initial internal radioactivity in cpm, k is the flux constant, and t is the time. The ln of the cpm in the cells plotted against time gives a straight line when efflux is from a single compartment and the absolute value of the slope of this line equals the flux rate constant, k. The rate constant is concentration independent, and thus a direct comparison between influx and efflux can be made.

Uptake rate changes can be measured in several ways. In one approach a large population of cells can be sampled continuously over a several hour period. On a plot of uptake versus time, a straight line would indicate a constant net flux, provided that the external concentration remained constant.   This could result from a constant influx rate and virtually no efflux, or from increasing influx as well as increasing efflux.   Other shaped curves are possible and can be difficult to interpret.   For example, we have measured uptake over a 12 hr period by sampling the same cell population as well as the bath at hourly intervals (Blackman and McDaniel, 1980).   Three different initial bath concentrations were studied.   At an initial bath concentration of 20 μM L-leucine, net flux became zero after 6 hr indicating that influx equaled efflux after 6 hr.   At 100 μM and 1.0 mM uptake continued for 12 hr, but it was clearly not linear.   One can easily see the difficulty in estimating time-dependent rate changes using this method.   An alternative method is to measure uptake over a short time period in cells that have been exposed to unlabeled solute.   We have measured the initial rate before and 3 hr after cells were exposed to 1.0 mM L-leucine using our standard 9 min assay (Table 1).   The rate was depressed about 30% in cells that had accumulated L-leucine for 3 hr.

Table 1.   Changes in the L-leucine Uptake Rate Over a 3 hr Uptake
            Period[a]

CONDITIONS	PERCENTAGE OF INITIAL (0 HR) RATE ± SE
3-hr control	110 ± 3
3-hr uptake period	76 ± 2

[a] Two cell samples were taken from two separate flasks of cells and the uptake rate measured over a 9 min uptake period in each of these four samples.   Then 0.5 ml of sterile distilled water was added to one flask (control flask) and 0.5 ml of a sterile 100 ml L-leucine solution was added to the other flask, thereby making the extracellular concentration approximately 1.0 mM.   After 3 hr the L-leucine uptake rate was measured in two samples of cells from each flask.   Prior to measuring uptake, cells were washed to remove L-leucine and control cells were washed in a similar fashion.

INCUBATION AND BUFFER SOLUTIONS.   Some researchers use dilute incubation solutions (Harrington and Smith, 1977; Harrington and Henke, 1981), while others employ an incubation solution that is similar to the growth medium (Maretzki and Thom, 1970; Smith, 1975; Blackman and McDaniel, 1978; King and Khanna, 1978).   The advantage of a dilute incubation solution is that the influence of other molecules on transport can be studied.   The disadvantage is that the cells are in an abnormal environment with a greatly reduced osmolarity.   The question of buffering is a difficult one.   If a buffer is used, it insures that uptake is measured at a constant pH, and this may be important for some solutes such as sugars or amino acids where uptake is pH-dependent.   However, cells are not usually grown in strongly buffered medium, and the buffer used can influence the uptake rate.

Significantly different results can be obtained when different incubation solutions are employed.   Different buffers result in very different uptake rates.   In citrate buffer, cysteine uptake rates are 3 times higher at pH 5.0 than in MES buffer (Harrington and Smith, 1977). When a dilute, buffered incubation solution is used, there is a dramatic time, protein synthesis, and Ca^{2+} dependent stimulation of amino acid (Smith, 1978a; Lyons et al., 1980; Harrington et al., 1981) and sulfate (Jones and Smith, 1981) uptake rates.   The stimulation is greatly reduced or absent in an incubation solution which is the same as the culture medium.   We use unbuffered culture medium (Linsmaier and Skoog, 1965) as our incubation solution, believing that this enables us to measure uptake under more physiological conditions.

PREPARATION OF CELLS.   Manipulation of cultured cells has been reported to influence uptake (Thoiron et al., 1979, 1980).   A study by Thom et al. (1981) has shown that cells from different species respond differently to manipulation and that the depressing effects of some types of manipulation result from a compound released as a result of cell damage.   The use of vacuum filtration for collecting cells for an uptake assay appears to be very damaging for some cultured cells and probably should be avoided.   Collection via centrifugation is preferable (Thom et al., 1981).

We have observed that thoroughly washing cells prior to subculturing can influence the uptake rates measured during the growth cycle (Fig. 1).   Although washing has increased the lag phase about a day, the relationship between uptake rate and growth phase is different for washed and unwashed cells.   These examples dramatically illustrate the importance of evaluating the influence various types of manipulation have on solute uptake.

NUMBER OF SAMPLES.   Some assays have only one point from which a rate is calculated (Harrington and Smith, 1977; Harrington and Henke, 1981), while others have many (Maretzki and Thom, 1970; Blackman and McDaniel, 1978).   We originally used an eight point assay but switched to four points when we determined that four points gave rates as precise as eight (Blackman and McDaniel, 1980).   With the help of a computer we have taken data from 39 of the four point assays and calculated rates using a single sample point, pair of points, and groups of three points.   Analysis of these data indicated that more precise measurements are not obtained by using more than two points (Table 2).   In fact, one point plus the origin gives rates that are as precise as four points, provided the time of the single point is greater than several minutes.   However, using more than two points makes it possible to determine the fit of a line to the points.   In this way the linearity of uptake is checked in each assay, and bad sample points can be detected.   Before selecting the number of sample points to use, the linearity of uptake over at least the time period of the assay should be confirmed with numerous sample points.

STANDARDIZATION PARAMETERS.   In measuring uptake, the rate should be related to a unit area of membrane.   Unfortunately, this is

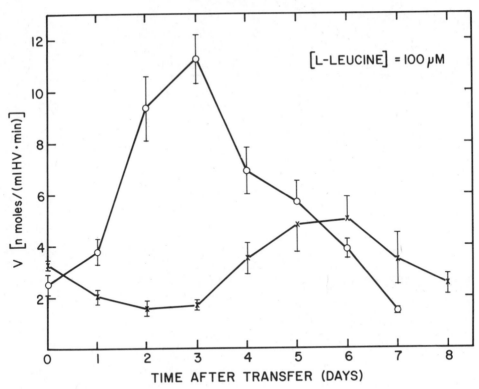

Figure 1. L-leucine uptake rate as a function of time after subculture
in cells which had been washed or not washed at the time of subculturing.   Cells from a flask of stationary-phase cells were divided in
half, washed twice with 100 ml of fresh Linsmaier and Skoog (1965)
medium supplemented with KIN (0.47 μM), NAA (11.0 μM), and sucrose
(40 g/l) (L and S medium), and subcultured in fresh L and S medium.
Unwashed cells represent our regular subculturing procedure.   In this
case 50 ml of fresh L and S medium were added to a flask of stationary cells and the resultant suspension divided between two flasks.
Final cell density was similar for washed and unwashed cells.   L-leucine uptake rate was measured at a L-leucine concentration of 100 μM.
Means ± SE are plotted.   Sample number = 4. x – x = washed.   o – o
= unwashed.

virtually impossible for cultured cells.   Thus researchers have standardized the uptake measurements to various parameters that relate to the
quantity of biological material in a sample (e.g., g fresh weight, mg dry
weight, number of cells, mg protein, or volume of homogenized cell
material).   None of the standardization parameters is a direct measurement of membrane area, and the variation in some is not related to
changes in membrane area.   Owens and Poole (1979) have shown that
several physical parameters change during the growth of suspension cultures of bean cells.   For example, protein content (mg/g fresh weight)
and average cell weight vary more than twofold.   Thus standardizing
by fresh weight or mg protein in these cultures would lead to inaccur-
ate rate measurements over the culture cycle.

Table 2. Number of Sample Points in an Assay and Precision of the Uptake Measurement[a]

	NUMBER OF SAMPLE POINTS								
	1	1	2	2	2[b]	2[b]	3	3	4
Time used for rate determination	0, 0.25	0, 9	6, 9	0.25, 9	0, 0.25, 9	0, 6, 9	3, 6, 9	0.25, 3, 6	0.25, 3, 6, 9
Average rate	19.86	12.76	12.43	12.57	14.02	12.80	12.07	12.66	12.49
SD	5.26	2.83	5.6	2.80	3.38	2.77	3.27	2.92	2.78
Maximum rate	32.97	19.49	26.85	19.39	22.19	19.07	19.21	18.16	18.80
Minimum rate	9.47	7.67	4.54	7.35	8.36	7.51	6.90	6.87	7.20

[a] Data from 39 L-leucine uptake assays was analyzed with the aid of a computer to determine if fewer than four sample points would result in data as precise as that given by four sample points. All of the four point assays had line correlations of 0.97 or better as determined by linear regression. The standard assay has sample points at 0.25, 3, 6, and 9 min. The computer program gave rates for one point assays and various combinations of 2 and 3 point assays. A "0" time indicates that the origin was used as a time point. For assays with the same number of sample points, adjacent columns indicate the least precise and the most precise uptake measurements.

[b] Two point assays were forced through the origin.

We have elected to relate solute taken up to the volume of homogenized cell material in a sample for two reasons: (1) the amount of cell wall should be directly related to the area of surface membrane, provided there are no wall ingrowths and that the thickness of the wall is constant, and (2) it is easy to make this measurement precisely. We simply homogenize a sample in a tissue grinder, centrifuge the homogenate in a table top centrifuge, and measure the pellet volume. Centrifuge tubes calibrated to 0.01 ml enable us to measure pellet volume to within several thousandths of a ml.

After selecting a standardization parameter, one should publish conversion factors so that rates can be expressed in terms of other parameters for comparison (Blackman and McDaniel, 1978). I do not know of a comparative analysis of standardization parameters. Thus the choice appears to be one of personal preference. However, one should determine that rate variation is not a reflection of parameter variation (McDaniel and Wozniak, 1982). This can be accomplished by standardizing rates with several different parameters under conditions where rate variation has been observed.

RATE DETERMINATION. The rate of uptake can be calculated by plotting the cpm's/standardization parameter versus time for each sample point. The slope of the line generated is the uptake rate in cpm/(std. parameter unit·time). If one knows the specific activity of the incubation solution, the rate can be converted to moles/(std. parameter·unit time). One can estimate the specific activity by calculating the number of curies that have been added to the known amount of unlabeled solute in the incubation solution. A more accurate method is to count a small sample of each incubation solution and thereby precisely know the specific activity in each assay. In plotting cpm versus time, only rarely does a perfect line result. Best fit lines and line correlations can be determined by linear regression. A poor line correlation will indicate a problem, and one might elect to use data only if the line correlation is better than some value. Different compounds can give, on average, very different line correlations using the same assay. In our laboratory, line correlations for L-leucine uptake are almost always better than 0.98, while L-aspartic acid line correlations average about 0.85. Thus the uptake of some compounds is more variable than that of others.

Growth Phase and Culture Viability

The uptake rate of various solutes changes as a function of the culture growth condition (King and Oleniuk, 1973; Harrington and Smith, 1977; Ruesink, 1978; Blackman and McDaniel, 1980; Jones and Smith, 1981). We have found the uptake of amino acids and the herbicide, amitrole, to be greatest in rapidly growing cells and lowest in stationary phase cells (Blackman and McDaniel, 1980; McDaniel et al., 1981, 1983; Singer and McDaniel, 1982). The almost continual rate changes occurring over the growth cycle of suspension cultured cells (see Fig.

1) are a constant source of variability. This variability is illustrated in two ways. First, we have collected a large number of uptake measurements and calculated the range, SD, and SE (Table 3). Second, we have made six L-leucine uptake measurements on cells from the same flask and six measurements on cells from six flasks that have been treated identically (Table 4). It is clear that we observe large variations in uptake rate. Others have also reported similar variations (King and Oleniuk, 1973; Ruesink, 1978). Thus interflask variability can make comparisons of uptake rates difficult to interpret and can lead to inaccuracies. This problem of variability can be dramatically reduced either by making measurements or by comparing rates for cells from the same flask. For example, if the influence of a metabolic inhibitor is being determined, assays with and without the inhibitor would be conducted for cells from the same flask at the same time. In this way the influence of the metabolic inhibitor can be estimated without interflask variability. However, even though several replicates may be made from a single flask, only one population of cells has been tested. To have more confidence in the values obtained, the same set of assays should be performed on cells from additional flasks of cells.

The general health of a culture can drastically influence the uptake rate. L-leucine uptake rates for a culture of tobacco cells that has turned just slightly brown can be 0.5-0.2 times the normal rate. As a general rule we only measure uptake in cells that appear to have the growth characteristics and color of the normal phenotype.

Table 3. Variability in Uptake Rates[a]

L-LEUCINE RATES	AMITROLE RATES
12.49 ± 2.78 ± 0.45 [7.20-18.80]	1.35 ± 0.24 ± 0.04 [0.97-1.79]

[a] L-leucine and amitrole uptake rates were measured in cells from 39 and 30 different flasks, respectively. All measurements were made on cells 3 days after subculturing at an L-leucine concentration of 0.1 mM and an amitrole concentration of 0.2 mM. One person ran all of the amitrole assays and another person ran the L-leucine assays. Rates in nmol/(ml HV·min) ± SD ± SE [range].

Table 4. Interflask and Intraflask Variability[a]

CELLS FROM SIX DIFFERENT FLASKS	CELLS FROM THE SAME FLASK
12.0 ± 2.4 [9.5-16.2]	14.9 ± 0.9 [14.2-15.9]

[a] Cells had their L-leucine uptake rate measured 3 days after subculturing. Six measurements were made at an L-leucine concentration of 0.1 mM. Rate in nmol/(ml HV·min) ± SD [range].

Transport in Protoplasts

Although the above is pertinent to transport studies in general, it has focused on transport measurements in suspension cultured cells. The reader wishing to use protoplasts may encounter some unique problems associated with isolating and measuring uptake in protoplasts. Protoplasts can be made from either cultured cells or leaves. The considerable literature on the production and characterization of protoplasts has recently been reviewed (Galun, 1981; Gamborg et al., 1981). Protoplasts from cultured cells (Briskin and Leonard, 1979; Mettler and Leonard, 1979a,b) and from leaves (Taylor and Hall, 1976; Robinson and Mayo, 1977; Guy et al., 1978, 1980; Rubinstein and Tattar, 1980; Guy et al., 1981; Volokita et al., 1981) have been employed in transport studies. The reader may wish to consult some of these sources when developing a protoplast transport assay.

PROTOCOL FOR MEASURING UPTAKE IN SUSPENSION CULTURED CELLS

Equipment

1. Tabletop clinical centrifuge.
2. Vortex mixer.
3. Wire or cloth mesh basket with a pore size of about 1.0 mm.
4. 15 ml graduated centrifuge tubes.
5. Millipore suction filter apparatus with 15 ml chimney. The filter should be a screen with a mesh size that traps the cells but does not clog when cells are trapped. We make the screens by gluing wire mesh cloth between two metal washers.
6. 40 ml heavy-duty plastic tubes that fit snugly in the top of the vortex mixer.
7. Special centrifuge tubes for albumin and total protein; 6.5 ml graduated in 0.01 ml from 0 to 0.4 ml; in 0.1 ml from 0.4 to 1.0 ml; in 0.5 ml from 1 to 2 ml; and at 3, 4 and 6.5 ml (Fisher Scientific catalog #05-663. Kimble catalog #46800, listed under tubes, sedimentation, McNaught and MacKay-Shevky-Stafford).
8. Tissue grinders [Bellco Glass, Inc. Vineland, N.J. 08360; catalog #1979-00005 (3 ml) or 1979-00003 (1 ml)]. Either size will work, but the 1 ml size breaks more easily.
9. Widemouth pipets for taking samples which can be made by cutting off the tip of a 10 ml pipet. Mechanical pipetters like Pipetmen (Rainin Inst. Co.) can be used by cutting off the end of a disposable plastic tip.
10. Liquid scintillation counter.
11. Pasteur pipets, flasks, beakers, spatula, protective gloves, and other general lab supplies.

Assay

1. Make 6 ml of incubation solution using complete culture medium (sterile), appropriate amount of unlabeled solute (a filter-steri-

lized, stock solution can usually be stored at 5-10 C for several weeks), and several μCi of isotope. The amount of isotope used depends upon the uptake rate. Each counted sample should have a minimum of 100 cpm. The pH should be checked prior to the addition of the isotope and adjusted if necessary. Pipet 0.05 ml of the incubation solution into a scintillation vial. This is the assay standard used for estimating the specific activity of the incubation solution.

2. Prepare four 100 ml wash solutions. We use Linsmaier and Skoog (1965) minus KH_2PO_4 and $FeSO_4$.

3. In a 250 ml beaker containing about 20 ml of complete culture medium place a wire mesh basket. Into the basket pour from a culture flask several ml of cell suspension. Gently swirl the basket and then pour the filtered cell suspension into a 15 ml graduated centrifuge tube.

4. Centrifuge the cell suspension for 1 min at maximum speed in a tabletop clinical centrifuge (about 1600 x g).

5. If fewer than 3 ml of packed cells are recovered, pour off supernatant, add more filtered cell suspension, and repeat. If more than 3 ml of cells have been pelleted, gently stir with a spatula just above the pelleted cells until 3 ml of pelleted cells remain.

6. Pour off supernatant and by gently squirting wash fluid into the test tube, transfer the cells to a 40 ml plastic incubation tube. Centrifuge the cell suspension for 1 min at maximum speed in a clinical centrifuge. Pour off the supernatant.

7. Place tube on Vortex and add 6 ml of incubation solution. Stir solution and start Vortex.

8. At 15 sec, 3, 6, and 9 min stop Vortex and take a 1.0-1.2 ml sample with a widemouth pipet.

9. Place sample in the chimney of a Millipore suction filtration apparatus. Suck 100 ml of wash fluid through the cell pellet and then suck the cell pellet dry.

10. Homogenize the pellet with a tissue grinder in about 1 ml of water. Transfer the homogenate to a special centrifuge tube and centrifuge for 1 min at maximum speed in a tabletop centrifuge.

11. Measure total volume and pellet volume (HV). Put 0.1 ml of the supernatant in a scintillation vial.

12. Count the standard and the sample points. Plot cpm/ml HV versus time. Calculate the slope of the resulting line and determine the line correlation. The slope in cpm/(ml HV·min) is the uptake rate. Using the specific activity of the standard, this rate can be converted to moles/(ml HV·min). The line correlation will indicate if there are "bad" points in an assay.

PROTOCOL FOR MEASURING UPTAKE IN PROTOPLASTS

Equipment

1. 30 C and 25 C incubators with fluorescent lamps.
2. 15 ml centrifuge tubes graduated in 0.1 ml units.

3. Variable, slow-speed centrifuge.
4. Hemocytometer.
5. Inverted microscope or other microscope for counting protoplasts.
6. Liquid scintillation counter.
7. Pasteur pipets, flasks, beakers, protective gloves, and other general lab supplies.

Protoplast Preparation

1. Select uniform healthy leaves and peal the lower epidermis. Cut the peeled leaf parts into strips several mm in width.
2. Plasmolyze leaf pieces in 600 mM sorbitol, 29 mM sucrose, 5 mM $MgCl_2$, 20 mM MES-KOH pH 5.5 at 25 C for 20 min. This removes the plasma membrane from the wall, so it is less likely to be attacked by enzymes in cellulysin.
3. Add cellulysin (Calbiochem) and dithiothreitol to a final concentration of 0.5% (w/v) and 2 mM, respectively. Incubate for 2-3 hr, in the light, at 30 C without agitation.
4. Filter the mixture through a 74 μm polypropylene screen and centrifuge at 400 x g for 2.5 min.
5. Resuspend pellet in 1.2 ml of 600 mM sorbitol, 29 mM sucrose, 5 mM $MgCl_2$, 50 mM Hepes-KOH pH 8.0 at 4 C.
6. Place the suspension on a chilled solution of 5.5% (w/v) PEG (6000), 2.5 mM dextran (T40), 400 mM Na-phosphate buffer (pH 7.5), and 0.6 M sorbitol, and centrifuge at 300 x g for 5 min. This procedure removes damaged protoplasts and other debris as well as contaminating microorganisms. The rest of the procedure should be done in aseptic solutions for short-term assays and aseptically for long-term assays (>1 hr). Remove the purified protoplasts banding at the interface and wash twice by centrifuging at 400 x g for 1 min in the solution to be used for uptake experiments [600 mM sorbitol, 29 mM sucrose, 1 mM glucose, 1 mM $Ca(NO_3)_2$, 0.25 mM $MgSO_4$, 5 mM each of Tris, HEPES and MES (pH 7.0)].
7. Purified protoplasts resuspended in the uptake solution (minus uptake solute) can be used immediately or stored for about 18 hr at 4 C.

Uptake Measurement

1. Determine the density of protoplasts in the protoplast suspension by counting a small volume using a hemocytometer.
2. Prepare 1.2 ml of incubation solution which is the solution above (step 6) plus labeled and unlabeled solute. Put 0.05 ml of the incubation solution and a volume of protoplast suspension equal to that to be used in each sample point into a scintillation vial. This will be used to determine the specific activity of the incubation solution.
3. Add ca. 3×10^5 protoplasts to each of four incubation solutions so that the total volume in each is 250 μl. Gently swirl solution and then incubate without swirling at 25 C in the light.

4. Take one sample point as quickly as possible by removing a 200 μl aliquot from one incubation tube and placing it in a calibrated centrifuge tube containing 10 ml of ice cold incubation solution minus isotope. Centrifuge at 45 x g for 7 min, aspirate to 0.2 ml above the pellet, and repeat. Add 0.2 ml of incubation solution (minus isotope) and vortex the tube. Transfer 0.3 ml to a scintillation vial for counting. The dpm/10^6 protoplasts in this sample are considered to represent background binding and are subtracted from uptake samples taken at later times.
5. Stop the uptake in the other three incubation tubes as above after 5, 10, and 15 min.
6. Convert cpm/10^6 protoplasts to dpm/10^6 protoplasts with a quench curve prepared using protoplasts.
7. Subtract zero time dpm/10^6 protoplasts from other sample dpm/10^6 protoplasts and calculate the rate from the three sample points.
8. Using the specific acitivity of the incubation solution, convert the rate to pmol/10^6 protoplast · min.

Other Considerations

1. Sorbitol and sucrose are not pure and may contain heavy metals and other compounds which reduce the effectiveness of cellulysin and damage the protoplasts. Sugars should be purified. Chelex-100 (Bio · Rad, Richmond, Ca.) can be used. Agitate a 500 ml of sugar solution with 5 g of Chelex-100 overnight at 4 C. The solution can be centrifuged to remove the Chelex-100.
2. Protoplasts are very delicate and should be handled very gently. At various times the condition of the protoplasts should be checked. Visual inspection and an exclusion dye like Evans blue can be used.

FUTURE PROSPECTS

There are a multitude of reasons why a biologist will want to characterize the transport of a solute in a cell or an organism. Cultured cells, including protoplasts, provide an excellent system for studying transport into or out of a cell. Advantages of the cultured cell system were mentioned in the introduction. Two of these are of particular importance to the field of transport. The first involves the exposure of all cells to the incubation solution. Diffusion of solutes into the cell free space may present a serious problem for the study of transport in tissues or organs. Ehwald et al. (1979) have shown that low substrate concentrations (below 0.1 M for potato slices 0.7-1.5 mm thick and below 0.001 M for slices of sugarbeet root 0.6-2.4 mm thick) and the rate of sugar uptake by plant tissue is limited by diffusion in the free space. If this is true for other tissue and organ systems, then much of the work reported for tissue slices and organs is imprecise at best. This problem of diffusion in the free space can be totally eliminated by using protoplasts or reduced markedly by using cell suspensions that have been filtered through a fine mesh filter. Along these same lines, the cell wall is a charged, unstirred layer where solute movement is diffusional. The movement of ions and charged molecules like amino acids may be influenced by the cell wall, and this influence may significantly alter transport. Again the wall-free protoplast may be the ideal solution to this problem. Various reports have shown that protoplasts are similar to tissue cells (Guy et al., 1978; Rubinstein, 1978; Briskin and Leonard, 1979; Mettler and Leonard, 1979a,b; Rubinstein and Tattar, 1980; Morris and Thain, 1980; Galun, 1981). However, too little work has been done with protoplasts to establish their real value. We cannot say for certain how the transport systems are effected by the enzymatic removal of the wall or if they function normally in the medium required to keep protoplasts healthy. Although the results have been encouraging, some questions have been raised. Transport rates appear to be lower in protoplasts than in the source tissue, one-sixth to one-third lower for α-aminoisobutyric acid in oat leaf protoplasts (Rubinstein and Tatter, 1980), and about 50% lower for 3-0-methyl glucose and α-aminoisobutyric acid in pea leaf protoplasts (Guy et al., 1978). Rubinstein (1978) reported that the membrane potential in oat leaf protoplasts was -62 mV, while that of oat leaf cells was -140 to -150 mV. Only additional study will indicate whether these and other reported differences will prove to be detriments to the use of protoplasts in transport studies.

The second advantage of major future importance is the ability to use cultured cells for the isolation of transport mutants. Microbial transport is understood far better than transport in any higher organism. This is primarily because transport mutants have been isolated and characterized. Mutants are powerful tools, as they provide us with a way of dissecting normal functions into component parts. Although several preliminary reports have suggested that amino acid transport mutants have been isolated (Widholm, 1974; Berlin and Widholm, 1978), transport mutants in plants and their use in dissecting transport phenomena are for the future. Only when considerable effort has been expended will we know if permeability mutants can be isolated and if they will be of significant value in elucidating transport phenomena.

It has only been within the last decade or two that the study of genetics became more than an academic pursuit. The importance of genes was never questioned, but the practical usefulness of genetics was lacking. I believe that the history of transport may turn out to be similar to that of genetics. No one questions that transport is a vital cellular and organismal function. For this reason we study it. But will transport knowledge enable us to improve crop yields? No one knows, but we can speculate. What is hybrid vigor? Could it be the uptake of critical nutrients like nitrate? Are the quantities of specific seed proteins produced, at least in part, a function of the amino acids taken up by the cotyledons or the endosperm? Transport studies can provide information on such questions and that information may enable us to design or select better crop plants. Cultured cells certainly have a place in the transport field and their contribution in the future will be significant.

ACKNOWLEDGMENT

I would like to thank Buddy Blackman, Robert Lyons, Susan Singer, and Paul Wozniak for their many contributions to our transport studies, Bernie Robinson for his protoplast protocol, and NIH (1R01 GM 25838-03), NSF (PCM78-11793), and USDA (CRGO #5901-0410-8-0075-0) for their financial support.

KEY REFERENCES

Luettge, U. and Higinbotham, N. 1979. Transport in Plants. Springer-Verlag, New York.

Maretzki, A. and Thom, M. 1978. Transport of organic and inorganic substances by plant cells in culture. In: Frontiers of Plant Tissue Culture 1978 (T.A. Thorpe, ed.) pp. 463-473. The International Association for Plant Tissue Culture, Calgary.

Spanswick, R.M., Lucas, W.J., and Dainty, J. (eds.). 1980. Plant Membrane Transport: Current Conceptual Issues. Elsevier, Amsterdam, New York, Oxford.

REFERENCES

Bange, G.G.J. 1979. Multiphasic kinetics in solute absorption—intrinsic property of the transport-system. Z. Pflanzenphysiol. 91:75-78.
Berlin, J. and Widholm, J.M. 1978. Amino acid uptake by amino acid resistant tobacco cell lines. Z. Naturforsch. 33c:634-640.
Blackman, M.S. and McDaniel, C.N. 1978. Amino acid transport in suspension cultured plant cells. I. Methods and kinetics of L-leucine uptake. Plant Sci. Lett. 13:27-34.
_____ 1980. Amino acid transport in suspension-cultured plant cells. II. Characterization of L-leucine uptake. Plant Physiol. 66:261-266.
Borstlap, A.C. 1981. Concept of multiphasic uptake in plants rejected. Naturwissenschaften 68:41-43.
Briskin, D.P. and Leonard, R.T. 1979. Ion transport in isolated protoplasts from tobacco suspension cells. III. Membrane potential. Plant Physiol. 64:959-962.
Carlson, P.S. 1973. Methionine sulfoximine-resistant mutants of tobacco. Science 180:1366-1368.
Cheruel, J., Jullien, M., and Surdin-Kerjan, Y. 1979. Amino acid uptake into cultivated mesophyll cells from Asparagus officinalis L. Plant Physiol. 63:621-626.
Courtois, D., Kurkdjian, A., and Guern, J. 1980. Tryptamine uptake and accumulation by Catharanthus roseus cells cultivated in liquid medium. Plant Sci. Lett. 18:85-96.
Ehwald, R., Meshcheryakov, A.B., and Kholodova, V.P. 1979. Hexose uptake by storage parenchyma of potato and sugar beet at different

concentrations and different thicknesses of tissue slices. Plant Sci. Lett. 16:181-188.

Etherton, B. and Rubinstein, B. 1978. Evidence for amino acid-H$^+$ cotransport in oat coleoptiles. Plant Physiol. 61:933-937.

Galun, E. 1981. Plant protoplasts as physiological tools. Annu. Rev. Plant Physiol. 32:237-266.

Gamborg, O.L., Shyluk, J.P., and Shahin, E.A. 1981. Isolation, fusion, and culture of plant protoplasts. In: Plant Tissue Culture: Methods and Applications in Agriculture (T.A. Thorpe, ed.) pp. 115-153. Academic Press, New York.

Guy, M., Reinhold, L., and Laties, G.G. 1978. Membrane transport of sugars and amino acids in isolated protoplasts. Plant Physiol. 61:593-596.

Guy, M., Reinhold, L., and Rahat, M. 1980. Energization of the sugar transport mechanism in the plasmalemma of isolated mesophyll protoplasts. Plant Physiol. 65:550-553.

Guy, M., Reinhold, L., Rahat, M., and Seiden, A. 1981. Protonation and light synergistically convert plasmalemma sugar carrier system in mesophyll protoplasts to its fully activated form. Plant Physiol. 67:1146-1150.

Harrington, H.M. and Henke, R.R. 1981. Amino acid transport into cultured tobacco cells. I. Lysine transport. Plant Physiol. 67:373-378.

Harrington, H.M. and Smith, I.K. 1977. Cysteine transport into cultured tobacco cells. Plant Physiol. 60:807-811.

Harrington, H.M., Berry, S.L., and Henke, R.R. 1981. Amino acid transport into cultured tobacco cells. II. Effect of calcium. Plant Physiol. 67:379-384.

Hart, J.W. and Filner, P. 1969. Regulation of sulfate uptake by amino acids in cultured tobacco cells. Plant Physiol. 44:1253-1259.

Heimer, Y.M. and Filner, P. 1971. Regulation of nitrate assimilation pathway in cultured tobacco cells. III. The nitrate uptake system. Biochim. Biophys. Acta 230:362-372.

Jones, S.L. and Smith, I.K. 1981. Sulfate transport in cultured tobacco cells. Effects of calcium and sulfate concentration. Plant Physiol. 67:445-448.

Kaiser, W.M. and Hartung, W. 1981. Uptake and release of abscisic acid by isolated photoautotrophic mesophyll cells, depending on pH gradients. Plant Physiol. 68:202-206.

King, J. and Khanna, V. 1978. The effect of ammonium ion on uptake of glutamine and other amino compounds by cultured cells of rapeseed. Planta 139:193-197.

King, J. and Oleniuk, F.H. 1973. The uptake of alanine-^{14}C by soybean root cells grown in sterile suspension culture. Can. J. Bot. 51:1109-1114.

Linsmaier, E.M. and Skoog, F. 1965. Organic growth factor requirements of tobacco tissue cultures. Physiol. Plant. 18:100-127.

Lyons, R.A., Blackman, M.S., and McDaniel, C.N. 1980. Incubation solution dependent stimulation of L-leucine uptake rate in suspension cultured Nicotiana tabacum cells. Plant Physiol. (Supp.)65:58.

Maretzki, A. and Thom, M. 1970. Arginine and lysine transport in sugarcane cell suspension cultures. Biochemistry 9:2731-2736.

McDaniel, C.N., Lyons, R.A., and Blackman, M.S. 1981. Amino acid transport in suspension-cultured plant cells. IV. Biphasic saturable uptake kinetics of L-leucine in isolates from six *Nicotiana tabacum* plants. Plant Sci. Lett. 23:17-23.

McDaniel, C.N., Holterman, R.K., Bone, R.F., and Wozniak, P.M. 1983. Amino acid transport in suspension cultured plant cells. III. Common carrier system for the uptake of L-arginine, L-aspartic acid, L-histidine, L-leucine, and L-phenylalanine. Plant Physiol. (in press).

Mettler, I.J. and Leonard, R.T. 1979a. Ion transport in isolated protoplasts from tobacco suspension cells. I. General characteristics. Plant Physiol. 63:183-190.

———— 1979b. Ion transport in isolated protoplasts from tobacco suspension cells. II. Selectivity and kinetics. Plant Physiol. 63:191-194.

Morris, P. and Thain, J.F. 1980. Comparative studies of leaf tissue and isolated mesophyll protoplasts. II. Ion relations. J. Exp. Bot. 31:97-104.

Nissen, P. 1974. Uptake mechanisms: Inorganic and organic. Annu. Rev. Plant Physiol. 25:53-79.

Nitsch, J.P. 1969. Experimental androgenesis in *Nicotiana*. Phytomorphology 19:389-404.

Owens, T. and Poole, R.J. 1979. Regulation of cytoplasmic and vacuolar volumes by plant cells in suspension culture. Plant Physiol. 64:900-904.

Pate, J.S. 1980. Transport and partitioning of nitrogenous solutes. Annu. Rev. Plant Physiol. 31:313-340.

Robinson, D.J. and Mayo, M.A. 1977. Changing rates of uptake of [³H] leucine and other compounds during culture of tobacco mesophyll protoplasts. Plant Sci. Lett. 8:197-204.

Rubinstein, B. 1978. Use of lipophilic cations to measure the membrane potential of oat leaf protoplasts. Plant Physiol. 62:927-929.

———— and Tattar, T.A. 1980. Regulation of amino acid uptake into oat mesophyll cells: A comparison between protoplasts and leaf segments. J. Exp. Bot. 31:269-279.

Ruesink, A.W. 1978. Leucine uptake and incorporation by *Convolvulus* tissue culture cells and protoplasts under severe osmotic stress. Physiol. Plant. 44:48-56.

Schafer, J.A. and Andreoli, T.E. 1977. Action of antidiuretic hormone on water and non-electrolyte transport processes in mammalian collecting tubules. In: Disturbances in Body Fluid Osmolality (T.E. Andreoli, J.J. Grantham, and F.C. Rector, Jr., eds.) pp. 57-83. American Physiological Society, Bethesda.

Shtarkshall, R.A., Reinhold, L., and Harel, H. 1970. Transport of amino acids in barley leaf tissue. I. Evidence for a specific uptake mechanism and the influence of "aging" on accumulating capacity. J. Exp. Bot. 21:915-925.

Singer, S.R. and McDaniel, C.N. 1982. Transport of the herbicide 3-amino-1,2,4-triazole by cultured tobacco cells and leaf protoplasts. Plant Physiol. 69:1382-1386.

Smith, I.K. 1975. Sulfate transport in cultured tobacco cells. Plant Physiol. 55:303-307.

_____ 1978a.   Role of calcium in serine transport into tobacco cells. Plant Physiol. 62:941-948.

_____ 1978b.   Effect of plant growth regulators on calcium-stimulated serine transport into tobacco cells. Plant Physiol. 62:949-953.

Spanswick, R.M. 1981.   Electrogenic ion pumps. Annu. Rev. Plant Physiol. 32:267-289.

Taylor, A.R.D. and Hall, J.L. 1976.   Some physiological properties of protoplasts isolated from maize and tobacco tissues. J. Exp. Bot. 27: 383-391.

Thoiron, B., Thoiron, A., Le Guiel, J., Luettge, U., and Thellier, M. 1979.   Solute uptake of *Acer pseudoplatanus* cell suspensions during recovery from gas shock. Physiol. Plant. 46:352-356.

Thoiron, B., Thoiron, A., Espego, J., Le Giuel, J., Luettge, U., and Thellier, M. 1980.   The effects of temperature and inhibitors of protein biosynthesis on the recovery from gas-shock of *Acer pseudoplatanus* cell cultures. Physiol. Plant. 48:161-167.

Thom, M., Komor, E., and Maretzki, A. 1981.   Transport studies with cell suspensions of higher plants: Effect of cell manipulation. Plant Sci. Lett. 20:203-212.

Vasil, V. and Hildebrandt, A. 1965.   Differentiation of tobacco plants from single, isolated cells in microcultures. Science 150:889-892.

Volokita, M., Kaplan, A., and Reinhold, L. 1981.   Evidence for mediated HCO_3^- transport in isolated pea mesophyll protoplasts. Plant Physiol. 67:1119-1123.

Widholm, J.M. 1974.   Cultured carrot cell mutants: 5-methyltryptophan-resistance trait carried from cell to plant and back. Plant Sci. Lett. 3:323-330.

Zimmermann, U. and Dainty, J. 1974.   Membrane Transport in Plants. Springer-Verlag, Berlin.

PART C
MODIFICATIONS AND APPLICATIONS

CHAPTER 23

Production of Useful Compounds in Culture

Y. Yamada and *Y. Fujita*

The production of useful compounds by plant cell cultures has become increasingly significant in the field of biotechnology. There are two important problems that have to be overcome for in vitro production of useful compounds. These are the selection of specific cells that produce high amounts of the desired compounds and the development of an adequate culture medium for the production of such useful compounds. In this chapter we discuss the selection method of cell aggregate cloning as used with *Euphorbia millii* and the establishment of the most suitable medium for production of shikonin derivatives by *Lithospermum erythrorhizon* cells in suspension culture.

SELECTION FOR THE CELL WITH A SPECIFIC CHARACTER

Usually plant cell cultures produce only small amounts of secondary metabolites when we induce callus and then subculture cells. Recently, however, cell strains containing amounts of secondary metabolites greater than those found in intact plants have been isolated by clonal selection. The successful selection of cells producing high amounts of secondary metabolites has been made possible because of the heterogeneity associated with cultured plant cells. It has therefore been possible to select cells and develop cell lines with desirable characteristics. For example, photoautotrophic cells (Huesemann and Barz, 1977; Yamada and Sato, 1978; Yasuda et al., 1980), resistant cells (Palmer and Widholm, 1975; Maliga et al., 1976; Nabors et al., 1980), high vitamin-producing cells (Yamada and Watanabe, 1980; Matsumoto et al.,

1980; Watanabe and Yamada, 1982), high pigment-producing cells (Kinnersley and Dougall, 1980; Yamamoto et al., 1982), and high alkaloid-producing cells (Zenk et al., 1977; Ogino et al., 1978; Yamada and Hashimoto, 1982) have each been obtained in various plant species.

The selection of cells producing high amounts of a useful metabolite is an important method for the production of useful compounds using cultured cells. In selecting specific cells, the two methods generally used are single-cell cloning and cell-aggregate cloning. The difficulties associated with the isolation and culture of single cells limit application of this method. The latter method of cell-aggregate cloning may appear to be more time consuming but so far it is easier and has been more successful.

The cell-aggregate cloning method has been successfully used to obtain high anthocyanin-producing cells from *Euphorbia millii* by Yamamoto et al. (1981, 1982).

Method of Selection

There are two questions with respect to clonal selection: (1) How can we determine whether the high productivity for secondary metabolites in selected cell strains is stable? (2) How long must we continue clonal selection in order to obtain stable cell strains that produce secondary metabolites?

The outline of the selection method, cell-aggregate cloning, is shown in Fig. 1. Original *Euphorbia millii* calli which produced a red pigment, cyanidin monoglucoside, were induced from leaves. The color of the induced calli was mottled red and white. These calli were cut into 128 segments (cell mass, ca. 3 mm) with a scalpel. Each segment was coded and placed on agar medium (25 ml) in a sectioned petri dish 9 cm in diameter. The agar medium consisted of Murashige and Skoog's basal solution (Murashige and Skoog, 1962) supplemented with 1 μM 2,4-D, 0.2% (w/v) malt extract, 2% (w/v) sucrose and 0.8% (w/v) agar. The segments were cultured at 28 C under fluorescent light of 6000 lux for 10 days. Each of the nine segments on a petri dish was cut into two cell aggregates; one (D_1) for subculture and the other (D_2) for quantitative analysis of the pigment. From the analysis of D_2 Yamamoto et al. selected the reddest D_1 cell aggregate from each petri dish. These selected cell aggregates were cut into several segments (cell mass, ca. 3 mm). All these segments were coded and transplanted onto fresh medium in a nine section petri dish. Segments selected at each transplantation were cultured on the same medium using the above conditions. This selection procedure was repeated 28 times.

The reddest cell-aggregates were selected continuously from each of the 28 subcultures of the original mottled-red *Euphorbia millii* callus. Two characteristic terms ($\bar{C}$ and C^{max}) in the frequency distribution of cell-aggregates from the 16th to 28th passages were plotted against the pigment content (Fig. 2).

The mean value ($\bar{C}$) for the pigment content increased nearly threefold from the 16th to 22nd passages. It came within the limit of 7.46 ± 0.56 after the 23rd passage which was seven times higher than the

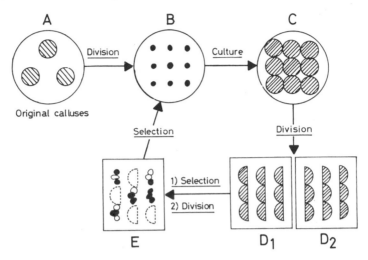

Figure 1. Outline of the selection method.   (A) The original *E. millii* calli were divided into 128 segments, and a segment was placed on agar-medium in one section of a nine-section petri dish and coded.   (B) Segments were cultured at 28 C under light (6000 lux) for 10 days. (C) Each grown segment was then divided into two cell aggregates; one (D_1) for subculture and the other (D_2) for quantitative analysis of the pigment.   (D_1) The reddest of the nine aggregates were removed and placed in an empty petri dish (E).   (E) Each of these red aggregates was divided into several segments, of which the reddest pieces were removed, then coded and placed on agar medium.

pigment content (1.05) of the original callus.   The maximum value (C^{max}) increased nearly threefold from the 16th to 19th passages, and was within the limit of 12.96 ± 2.36 after the 20th passage.

The 9A and 9F cell lines originated from the 9A and 9F cell aggregates of the 9th subculture.   In this subculture, the 9A and 9F lines produced pigment-rich descendants at high frequencies.   The distribution rates of these two cell lines in the population of cell aggregates from the 16th to 24th passages is shown in Fig. 3.

The distribution rate of the 9F cell line was lower than that of the 9A cell line at the 16th subculture, but equalled it at the 20th subculture.   This rate gradually increased and reached 100% at the 24th subculture.   Yamamoto et al. found that the mean value ($\bar{C}$) for the pigment content in cell aggregates of *Euphorbia millii* became stable after 24 clonal selections (Fig. 2).   In addition, the distribution rate for cell aggregates of the 9F cell line reached 100% in the population after the 24th passage (Fig. 3).   All the aggregates after the 24th passage were derived from the 9F cell aggregate.   If the mean value for the content of secondary metabolites is stable in a population of cell aggregates chosen by successive clonal selection, and if all the cell aggregates are derived from one cell aggregate, they should consist of cells with high and stable productivity of secondary metabolites.   These results show that it is possible to isolate and culture a cell strain containing a high

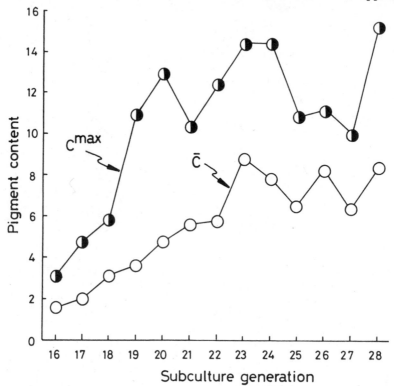

Figure 2.   Trends of the two characteristic terms; the mean value ($\bar{C}$) and maximum value (C^{max}) in the frequency distribution of cell aggregates with various pigment contents from the 16th to 28th passages. Cell aggregates were subcultured on Murashige and Skoog's agar medium with 2,4-D (1 μM), malt extract (0.2% w/v), and sucrose (2% w/v) at 28 C under light (6000 lux) every 10 days.

and stable pigment content from *Euphorbia millii* callus after 24 successive clonal selections.

Summary of Selection of Specific Cells

From the preceeding experiment it is our belief that cultured cells are heterogeneous in their production of pigments during early passages and that high pigment-producing variant cells also exist.   Results show that one could select more cell aggregates of the 9F cell line than of the 9A cell line at each subculture and that the 9F cell aggregate had higher pigment-producing variant cells (Yamamoto et al., 1982).

In preliminary experiments, Yamamoto et al. (unpublished) failed to obtain a high pigment-producing strain of cultured *E. millii* cells by single cell cloning because culture of a single cell was very difficult. The cell-aggregate cloning used in this study appears superior to single cell cloning.

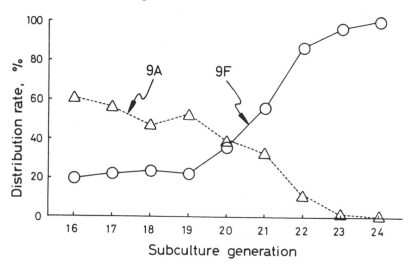

Figure 3. Distribution rates of the 9A and 9F cell lines in the population of cell aggregates from the 16th to 22nd passages. The 9A and
9F cell lines were derived from the 9A and 9F cell aggregates of the
ninth subculture. Distribution rate (the number of cell aggregates of a
specific cell line/the number of total cell aggregate) x 100.

CULTURE MEDIA FOR PIGMENT FORMATION BY *Lithospermum erythrorhizon*

Lithospermum erythrorhizon Sieb et Zucc. (Boraginaceae) is a perennial plant native to Japan, China, Korea, and Southeast Asia. Its root
is called shikon in Chinese medicine and contains red pigments. The
red pigments comprise shikonin possessing a structure of naphthazarin,
its lower aliphatic acid ester, and the deoxy derivative of shikonin
(Fig. 4).

The red pigments have long been used in Japan as a medicine for
wounds, burns, etc. and as dyestuffs. In Japan it is still being used as
a medicine for hemorrhoids.

Tabata et al. (1974) studied the production of shikonin derivatives by
the cell cultures of *L. erythrorhizon*. Their study has shown that callus derived from the seedling could produce shikonin derivatives on LS
agar medium supplemented with 1 μM IAA and 10 μM KIN. Mizukami
et al. (1977) have also reported the effects of nutritional factors such
as sucrose, nitrogen sources, etc., on shikonin production in *L. erythrorhizon*.

Based on the research of Tabata et al., Fujita et al. (1981a,b) took
up an investigation of the production of shikonin derivatives by cell
suspension cultures in large tank cultures. They aimed at improving
the production of shikonin derivatives by giving particular attention to
the medium composition. A cell line, M-18 (Mizukami et al., 1978) received from Tabata was cultured in known liquid medium, where cell
growth and amount of shikonin derivatives produced were studied by
Fujita et al. (1981a,b).

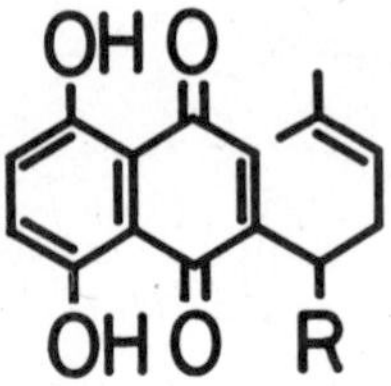

R=H   Deoxyshikonin
R=OH   Shikonin
R=OCOCH₃   Acetylshikonin
R=OCOCH(CH₃)₂   Isobutylshikonin
R=OCOCH=C(CH₃)₂   β,β-Dimethyacryl-
                          shikonin
R=OCOCH₂CH(CH₃)₂   Isovalerylshikonin
R=OCOCH(CH₃)CH₂CH₃   α-Methyl-n-butyl-
                          shikonin
R=OCOCH₂C(OH)(CH₃)₂   β-Hydroxyisovaleryl-
                          shikonin

Figure 4. Structures of shikonin derivatives.

No shikonin derivatives were produced in cell suspension cultures grown in Linsmaier-Skoog medium, a suitable medium for cell growth. Although White medium favored the production of shikonin derivatives, it could not support good cell growth. Fujita et al., however, considered that a culture combining these two media might effectively yield shikonin derivatives and support cell growth because of mutual supplements of each other's deficiencies. Fujita et al. evaluated a two-stage culture method consisting of proliferation of cells using LS medium, suitable for cell growth as the first stage culture, and obtained shikonin derivatives from the resulting cells using White medium suitable for the production as the second stage culture. The results obtained are given in Table 1. Although the culture duration was as long as 23 days because of the two-staged culture method, the growth rate attained a twentyfold increase. The production of shikonin derivatives in the second stage was not affected by the preculture with LS medium as the cells produced 130 mg/l shikonin derivatives. Consequently, Fujita et al. decided to adopt a two-staged culture method for the production of shikonin derivatives.

Table 1. Effects of Two-Staged Culture Method on Cell Growth and Production of Shikonin Derivatives

	CULTURE FOR CELL GROWTH	CULTURE FOR PRODUCTION	TOTAL
Medium	LS[a]	White	
Culture duration (days)	9	14	23
Growth rate (times)	6.7	3.0	20
Shikonin derivatives formed (mg/l)	0	130	130

[a]Linsmaier-Skoog.

Development of Medium for Production of Shikonin Derivatives

As shown in Table 2, cell growth and production of shikonin derivatives were greatly influenced by the medium used. The results listed in Table 1 were obtained by the combination of LS medium and White medium that were selected from the known media listed in Table 2. Therefore, the best media for cell growth and shikonin derivative production might be compositions other than the two media mentioned above. Fujita et al. (1981a,b) investigated the relationships between the different levels of all the components in LS medium and cell growth and between those of the White medium and the production of shikonin derivatives in order to identify the optimum medium for secondary product synthesis.

Table 2. Effects of Type of Media on Cell Growth and Production of Shikonin Derivatives[a]

MEDIUM	CONCENTRATIONS OF NUTRIENTS (mM)				CELL YIELDS (g DW/l)	SHIKONIN DERIVATIVES FORMED (mg/l)
	NH_4^+	NO_3^-	PO_4^{-3}	K^+		
LS	21	39	1.3	20	16.8	0
Gamborg et al. B5	2.0	25	1.1	25	12.2	0
White	0	2.5	0.12	1.7	6.8	130
Nitsch- Nitsch	9.0	18	0.5	9.9	4.6	0
Blaydes	13	25	2.2	13	4.0	0

[a]Cells of 0.5 g fresh weight were inoculated into 100 ml flasks containing 27 ml of medium, and cultured for 14 days in the dark at 25 C. All the media were supplemented with 1 µM IAA and 10 µM KIN.

The results indicated that most of the unique components in White medium had strong effects on the production of shikonin derivatives as well as on cell growth (Fujita et al., 1981a,b). The effects of inorganic and organic components on the production of shikonin derivatives were as follows:

INORGANIC COMPONENTS. White medium contains 15 kinds of inorganic components, of which NO_3^-, Cu^{+2}, and SO_4^{-2} demonstrated the strongest effects on the production of shikonin derivatives. Both the yield of shikonin derivatives and the cell growth increased as the NO_3^- concentration increased but decreased above 10 mM NO_3^- (Fig. 5). The optimum concentration for both growth and shikonin derivative production was 6.7 mM, which was different from the concentration in White

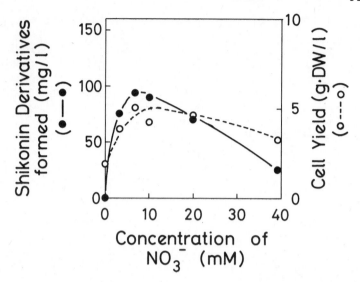

Figure 5. Effect of the concentration of NO₃⁻ on the production of shikonin derivatives and on cell growth. The concentration of NO₃⁻ in White medium was 3.3 mM.

medium (3.3 mM). The yield of shikonin derivatives increased with an increase in the concentration of Cu^{2+}, but became constant when the concentration exceeded 0.8 µM. Based on these results, Fujita et al. adopted the Cu^{2+} concentration of 1.2 µM, which corresponds to 30 times the concentration of Cu^{2+} in White medium. The concentration of Cu^{2+} had almost no effect at all on the cell growth. The yield of shikonin derivatives steeply rose with an increase in the concentration of SO_4^{-2}, and the optimum concentration was found to be 13.5 mM (4.5 mM in White medium). Cell growth was slightly inhibited by SO_4^{-2}.

ORGANIC COMPONENTS. The yield of shikonin derivatives increased with the concentration of sucrose up to 2% and was inhibited when concentration was over 5%. Cell growth increased as the concentration of sucrose rose to 4%, above which it became constant. From these results, Fujita et al. fixed the concentration at 3%.
   The optimum medium for the production of shikonin derivatives was obtained by replacing the concentration of each component in White medium by the corresponding optimum concentration of each medium component. It was named the M-9 medium by Fujita et al. Its composition is compared with that of White medium in Table 3. The significant characteristics of M-9 medium as compared with White Medium are that the concentration of Cu^{2+} is very high, the number of inorganic components are reduced, and no vitamins or amino acids are added to the culture medium.
   As a result of the studies, Fujita et al. developed the MG-5 medium for optimum cell growth (Table 4) and the M-9 medium for the optimum production of shikonin derivatives. The two-staged culture of the cells

Table 3. Compositions of White Medium and M-9 Medium

COMPONENT	WHITE MEDIUM (mg/l)	M-9 MEDIUM (mg/l)
$Ca(NO_3)_2$ x $4H_2O$	300	694
KNO_3	80	80
NaH_2PO_4 x $2H_2O$	21	19
KCl	65	65
$MgSO_4$ x $7H_2O$	750	750
Na_2SO^4	200	1,480
$MnSO_4$ x $4H_2O$	5	0
$ZnSO_4$ x $7H_2O$	3	3
$Fe(SO^4)$	2.5	0
NaFe-EDTA	0	1.8
H_3BO_3	1.5	4.5
KI	0.75	0
$CuSO_4$ x $5H_2O$	0.01	0.3
MoO_3	0.001	0
Sucrose	20,000	30,000
Glycine	3	0
Nicotinic acid	0.5	0
Thiamine x HCl	0.1	0
Pyridoxin x HCl	0.1	0
IAA	0.00018	0.0018
Kinetin	0.0022	0

Table 4. Compositions of LS Medium and MG-5 Medium

COMPONENT	LS (mg/l)[a]	MG-5 (mg/l)
NH_4NO_3	1650	500
KNO_3	1900	1900
$NaNO_3$	0	2480
$CaCl_2$ x $2H_2O$	440	150
$MgSO_4$ x $7H_2O$	370	120
$MgCl_2$ x $6H_2O$	0	203
KH_2PO_4	170	170
$FeSO_4$ x $7H_2O$	27.8	27.8
Na_3 x EDTA	37.3	37.3
$MnSO_4$ x $4H_2O$	22.3	22.3
$ZnSO_4$ x $7H_2O$	8.6	8.6
H_3BO_3	6.2	1.9
KI	0.83	0
Na_2MoO_4 x $2H_2O$	0.25	0.25
$CuSO$ x $5H O$	0.025	0.025
Sucrose	30,000	30,000
Inositol	100	100
Thiamine x HCl	0.4	0.4

[a] Linsmaier-Skoog

was performed by combining these new media, and the results are presented in Table 5.

Table 5.   Results of Two-Staged Culture by Combining MG-5 and M-9 Media

	CULTURE FOR CELL GROWTH	CULTURE FOR PRODUCTION OF SHIKONIN DERIVATIVES	TOTAL
Medium	MG-5	M-9	
Culture duration (days)	9	14	23
Growth rate (times)	7.5	3.6	27
Shikonin derivatives formed (mg/l)	0	1,500	1,500
Content of shikonin derivatives in the cells (%)	0	13.6	13.6

Thus, by using a two-staged culture, as much as 1500 mg/l of shikonin derivatives were produced (the content of these derivatives reached 13.6%), and the yield was about 11.5 times that obtained when LS and White media were combined. In addition, growth rate was improved 1.1 times. The productivity of shikonin derivatives per inoculum therefore increased about 13 times, certainly a significant improvement.

Summary of Culture Medium Manipulations

These studies were aimed at the production of shikonin derivatives with higher yield by the cell suspension cultures of *Lithospermum erythrorhizon*. It was observed that the respective media suitable for cell growth and for production were different. To get a higher production of shikonin derivatives, a two-staged culture method was adopted. In this procedure the first culture was designed to promote cell growth, while the second culture was designed for production of shikonin derivatives. By the consequent development of media most suitable for each culture, the two-staged culture method made it possible to produce 1500 mg/l shikonin derivatives, about 13 times higher than before. Compared with shikon requiring 2-3 years for harvest of the plant, cultured cells permit harvesting within about 3 weeks, thereby greatly shortening the production period. The content of shikonin derivatives in the cultured cells was about 14%, which was extremely high compared with 1-2% in the normal field grown shikon. The chemical composition of shikonin derivatives obtained by the cultured cells were similar to those of the shikon, and the composition in the cultured cells was found to have less fluctuation when compared with that of the shikon.

FUTURE PROSPECTS

The production of useful compounds by cultured plant cells is possible if cells with specific characteristics produce the useful compounds that can be successfully selected. It is also possible to increase the amount of the compounds by establishing the most suitable medium for production of those useful compounds. Recently, the number of scientific publications on in vitro production of high amounts of useful compounds has increased. It is our belief that with the successes obtained so far in experimental studies, industrial production of such compounds using these techniques is a distinct possibility in the near future.

KEY REFERENCES

Fujita, Y., Hara, Y., Ogino, T., and Suga, C. 1981a. Production of Shikonin derivatives by cell suspension cultures of *Lithospermum erythrorhizon*. I. Effects of nitrogen sources on the production of shikonin derivatives. Plant Cell Rep. 1:59-60.

Fujita, Y., Hara, Y., Suga, C., and Morimoto, M. 1981b. Production of Shikonin derivatives by cell suspension cultures of *Lithospermum erythrorhizon*. II. A new medium for the production of shikonin derivatives. Plant Cell Rep. 1:61-63.

Kinnersley, A.M. and Dougall, D.K. 1980. Increase in anthocyanin yield from wild-carrot cell cultures by a selection system based on cell-aggregate size. Planta 149:200-204.

Tabata, M., Mizukami, H., Hiraoka, N., and Konoshima, M. 1974. Pigment formation in callus cultures of *Lithospermum erythrorhizon*. Phytochemistry 13:927-932.

Watanabe, K. and Yamada, Y. 1982. The selection of cultured plant cell lines producing high levels of biotin. Phytochemistry 21:513-516.

Yamamoto, Y., Mizuguchi, R., and Yamada, Y. 1982. Selection of a high and stable pigment-producing strain in cultured *Euphorbia millii* cells. Theor. Appl. Genet. 61:113-116.

REFERENCES

Huesemann, W. and Barz, W. 1977. Photoautotrophic growth and photosynthesis in cell suspension cultures of *Chenopodium rubrum*. Physiol. Plant. 40:77-81.

Maliga, P., Lazat, G., Svab, Z., and Nagy, F. 1976. Transient cycloheximide resistance in a tobacco cell line. Molec. Gen. Genet. 149:267-271.

Matsumoto, T., Ikeda, T., Kanno, N., Kisaki, T., and Noguchi, M. 1980. Selection of high ubiquinone 10-producing strain of tobacco cultured cells by cell cloning technique. Agric. Biol. Chem. 44:967-969.

Mizukami, H., Konoshima, M., and Tabata, M. 1977. Effect of nutritional factors on shikonin derivative formation in *Lithospermum* callus cultures. Phytochemistry 16:1183-1186.

_____ 1978. Variation in pigment production in *Lithospermum erythrorhizon* callus cultures. Phytochemistry 17:95–97.

Murashige, T. and Skoog, F. 1962. A revised medium for rapid growth and bioassays with tobacco tissue cultures. Physiol. Plant. 15:473–497.

Nabors, M.W., Gibbs, S.E., Bernstein, C.S., and Meis, M.E. 1980. NaCl-tolerant tobacco plants from cultured cells. Z. Pflanzenphysiol. 97:13–17.

Ogino, T., Hiraoka, N., and Tabata, M. 1978. Selection of high nicotine-producing cell lines of tobacco callus by single-cell cloning. Phytochemistry 17:1907–1910.

Palmer, J.E. and Widholm, J. 1975. Characterization of carrot and tobacco cell cultures resistant to p-fluorophenylalanine. Plant Physiol. 56:233–238.

Yamada, Y. and Hashimoto, T. 1982. Production of tropane alkaloids in cultured cells of *Hyoscyamus niger*. Plant Cell Rep. 1:101–103.

Yamada, Y. and Sato, F. 1978. The photoautotrophic culture of chlorophyllous cells. Plant Cell Physiol. 19:691–699.

Yamada, Y. and Watanabe, K. 1980. Selection of high vitamin B_6 producing strains in cultured green cells. Agric. Biol. Chem. 44:2683–2687.

Yamamoto, Y., Mizuguchi, R., and Yamada, Y. 1981. Chemical constituents of cultured cells of *Euphorbia tirucalli* and *E. millii*. Plant Cell Rep. 1:29–30.

Yasuda, T., Hashimoto, T., Sato, F., and Yamada, Y. 1980. An efficient method of selecting photoautotrophic cells from cultured heterogeneous cells. Plant Cell Physiol. 21:929–932.

Zenk, M.H., El-Shagi, H., Arens, H., Stoeckgt, J., Weiler, E.W., and Dues, B. 1977. Formation of the indole alkaloids serpentine and ajmalicine in cell suspension cultures of *Catharanthus roseus*. In: Plant Tissue Culture and its Biotechnological Application (W. Barz, E. Reinhard, and M.H. Zenk, eds.) pp. 27–51. Springer-Verlag, Berlin, Heidelberg, New York.

CHAPTER 24
Flavor Production in Culture

H.A. Collin and *M. Watts*

Much of the research and development on secondary product formation in plant tissue cultures has concentrated on compounds such as the pharmaceuticals which can be classified as high-cost, low-volume, high-demand compounds. However, a number of other secondary compounds also come under this category. These are the essential oils, food flavorings, colorings, and gums which are all used by the convenience food, ice cream, and confectionary industries. Table 1 provides a number of examples of these compounds (including some fragrant oils), grouped in order of cost. Some of the most expensive compounds are the fragrant oils used by the cosmetic industry, but also included in Group I is quassin which is a bittering agent for drinks. Saffron, a coloring agent, is in Group II, as well as gum acacia and the flavoring angelica. In the less expensive Group III are the herbs and their essential oils, such as cardamom oil, chamomile, the fruit flavor of buchu oil, black pepper, the bittering flavor of hops, spices such as ginger, and the food flavors of onion and vanilla. With the less-expensive compounds, such as onion flake and powder, the category changes to one of lower cost and higher volume. This change is seen more clearly in the next group, which contains the beverages tea, coffee, and cocoa; the flavoring of peppermint and spearmint; and the coloring turmeric. Less-well-known compounds are the polysaccharide gums which have a very large market in all aspects of the food industry as stabilizers and thickeners. Examples are gum arabic, which is a tree exudate, guar derived from seeds, and carrageenan from marine algae.

One of the major problems with a plant source of flavors and colorings is ensuring that there is a constant supply. Many of the com-

Table 1. A Selection of Plants that Produce an Oil, Flavor, Coloring or Gum

GROUP £/kg	ENGLISH NAME	BOTANICAL NAME	PART USED	AVERAGE PRICE (£/kg)	WORLD DEMAND (£ million)
I 1000+	Quassin	Quassia amara	Wood	4500	Unknown
	Jasmin	Jasmin grandiflorum	Flowers	3300	0.19
II 500–1000	Rose otto	Rosa damascena	Flowers	2200	9.0
	Gum arabic	Acacia senegal	Gum	780	3.85
	Angelica	Angelica sylvestris	Root oil	670	11.5
III 100–500	Buchu oil	Barosina betulina	Leaves	170	117.9
	Cardamon oil	Elletoria Cardamomum	Herbs, leaves	330	19.5
	Cassia	Cinnamonium cassia	Herbs, leaves	260	8.4
	Chamomile	Anthemisia nobilis	Herbs/flowers	350	Very high
	Cinnamon	Connamomum	Bark (oil)	150	3.2
	Hops	Humulus lupulus	Flowers	200	Unknown
	Ginger	Zingber officinale	Rhizome	100	25.0
	Black pepper	Piper nigrum	Berries	150	12.0
	Onion	Allium sativum	Bulb	300	Very high
	Vanilla	Vanilla planifolia	Beans (oil)	350	10.0
IV Below 100	Celery	Apium graveolens	Seeds	90 (oil)	0.9
	Coffee	Coffea arabica	Beans	8.7 (instant)	1700
	Tea	Thea sinensis	Leaves	1.5	335 (UK)
	Turmeric	Cucurna longa	Rhizomes	0.7	51
	Mints (peppermint and spearmint)	Mentha peperita and M. viridis	Leaves, flowers	11.0 (oil)	40

GROUP	£/kg	ENGLISH NAME	BOTANICAL NAME	PART USED	AVERAGE PRICE (£/kg)	WORLD DEMAND (£ million)
IV	Below 100	Cocoa	*Theobroma cacao*	Seeds	3.3	40
		Capsicum	*Capsicum annuum*	Fruit	25	1.1 (UK)
		Coriander	*Coriandrum sativum*	Seeds	16 (oil)	3.4

pounds are obtained from plants that are not grown under large-scale or controlled cultivation. This instability, combined with climatic, harvesting, transport difficulties, and possible political problems in the country of origin, often leads to considerable fluctuations in the supply and therefore price of the compound. With each of these compounds there is interest in a more stable and easily controlled source. In addition to a stable source, new regulations governing the use of synthetic flavor and coloring additives to food have stimulated interest in replacing the synthetic sources with a plant or naturally derived source. Large-scale tissue culture has been seen as a way of providing a constant supply of compounds as well as going some way toward overcoming the toxicological objections to compounds derived from synthetic sources.

Although the potential for secondary product formation in large-scale tissue culture has been recognized for some time (Staba, 1963; Carew and Staba, 1965; Constabel et al., 1974; Zenk, 1978; Dougall, 1980; Yeoman et al., 1980), there has been only a limited investigation of the production of food flavors. One of the main reasons for this small amount of research effort has been the prolonged lack of success in stimulating flavor synthesis in undifferentiated tissue cultures. Most of the flavors are usually present in plants as a component of the essential oil, so investigators have looked for the presence of the essential oil in the culture and the effect of culture on the composition of the oil. One of the earliest investigators was Staba, who in a review (Carew and Staba, 1965) suggested that the essential oils were not produced from cultures of flavor sources such as lemon, peaches, avocado, and mint because the tissue lacked the oil glands that are necessary for their synthesis and accumulation. This concept that the undifferentiated cultures of oil-producing plants were unable to produce the oil was reinforced by Becker (1970), who analyzed the tissue cultures of a variety of herbs (anise, fennel, sage, and peppermint) and found none of the volatile oils. However, Becker found that following redifferentiation of his cultures into roots and shoots, the normal capacity for synthesis was regained. In both celery and onion cultures the normal pattern of flavor also reappeared in the differentiated tissue (Al-Abta et al., 1979; Selby et al., 1979; Turnbull et al., 1981). However, the production of essential oils and flavors does not always appear to require redifferentiated organs. Partially differentiated callus or suspension cultures of *Ruta graveolens* (Nagel and Reinhard, 1975), chamomile (Szoeke et al., 1978), perilla (Sugisawa and Ohnishi, 1976), coriandrum (Sardesai and Tipnis, 1969), peppermint (Kireeva et al., 1978), and cocoa (Townsley, 1974; Jalal and Collin, 1979) all show synthesis of some or all of the flavor compounds. The reformation of secondary products in redifferentiated or partially differentiated tissue cultures indicates that cells in undifferentiated cultures still retain the capacity to synthesize secondary products, but that the path of synthesis is blocked in some way. The following is a discussion of the attempts to resolve this problem.

BIOSYNTHESIS OF FLAVOR COMPOUNDS

Supply of Precursors

Although it is generally accepted that tissue culture cells are totipotent for their secondary pathways, there has been very little basic research on why the tissue culture cell shows a reduced activity of secondary pathways when compared to the normal plant. Overton and Picken (1977) appropriately state that too little is known about factors that inhibit secondary metabolite synthesis in tissue cultures or cause its reemrgence upon redifferentiation. Zenk (1978) has given many examples supporting this view. In one approach it has been assumed that the blocked secondary pathways are deficient in one or more intermediates. The intermediates are then supplied to the tissue culture in the nutrient medium. This approach has been largely unsuccessful unless the intermediate is at a position in the pathway which is close to the final product.

In *Capsicum annuum* the flavor component capsaicin is synthesized from valine and phenylalanine. The immediate precursors of capsaicin from these two pathways are 8-methylnonenoic acid, which is similar to isocapric acid, and vanillylamine. Thus supplying vanillylamine and isocapric acid raised the level of capsaicin considerably above that achieved with the primary amino acid precursors (Yeoman et al., 1980). With the use of combinations of precursors to establish whether the limitation on production lay between phenylalanine to vanillylamine, or valine to isocapric acid, it was shown that as long as isocapric acid was present, capsaicin was produced, but if only valine was supplied, the production of capsaicin was an order of magnitude less. This indicated that the rate-limiting step for the production of capsaicin lay between valine and isocapric acid.

In onion tissue cultures the rate-limiting step was established in a similar way. The flavor in onion, *Allium cepa* L., is produced from three flavor precursors, s-methyl, and s-propyl-l-cysteine sulphoxides, and trans-prop-l-enyl-l-cysteine sulphoxide which are hydrolyzed by an enzyme alliinase when the tissue is cut or crushed to give a variety of volatile flavor components (Schwimmer and Guadagni, 1962; Schwimmer and Mabelis, 1963). The s-methyl and s-propyl-l-cysteine sulphoxides are formed originally from serine, but the principal flavor precursor s-trans-prop-l-enyl-l-cysteine sulphoxide is synthesized from valine and cysteine (Granroth, 1970; Whitaker, 1976). The pathway of synthesis from valine to the intermediate, methacrylic acid, is part of a primary metabolic sequence. Methacrylic acid then combines with cysteine to form (s-2-carboxypropyl) cysteine, which is the start of a secondary pathway largely restricted to the *Allium* species, and which leads to trans-s-prop-l-enyl-l-cysteine and finally to the principal flavor precursor. When intermediates on the primary pathway such as sulphate, valine, methacrylic acid, and cysteine were fed to the onion callus, there was no production of the flavor precursor (Selby et al., 1980).

However, when intermediates in the secondary pathway were used, i.e., (s-2-carboxypropyl) cysteine and trans-s-prop-1-enyl-1-cysteine, there was active synthesis of the major precursor compound. It was suggested that the block to precursor synthesis in the callus was where methacrylic acid combined with cysteine at the beginning of the secondary pathway.

When radioactive intermediates were used it was found that, in both *Capsicum* (Yeoman et al., 1980) and onion (Turnbull et al., 1980), even with intermediates which were at an early stage of the primary pathways, e.g., C^{14} valine in *Capsicum* and C^{14} cysteine in onion tissue cultures, some radioactivity was detected in the capsaicin and in the main onion flavor precursor. The presence of radioactivity in the flavor compounds suggested that the secondary pathways were in fact operating in the tissue culture, but at an extremely low level. However, when the cultures were fed intermediates, either in the primary pathway or at an early stage in the secondary pathway, no significant increase in synthesis of the secondary products occurred. This suggested that the block in synthesis could be due to two possible factors. (1) The intermediates in the primary, or early stages of the secondary pathways, are more likely to be drawn off into other pathways than intermediates at a final stage in the secondary pathways. (2) A key enzyme determining the entry of intermediates into a secondary pathway may be inactive in the undifferentiated culture (Yeoman et al., 1980, 1981). Hence, intermediates supplied to the tissue cultures after this rate-limiting step stimulate synthesis of the final product, but not those intermediates before this stage.

Evidence for reduced activity of key enzymes in intact tissue culture cells was provided by Banthorpe et al. (1976), who compared the synthesis of monoterpenes from a ^{14}C labeled precursor, isopentyl pyrophosphate, in cell-free extracts from callus and intact plants of *Tanacetum vulgare* (tansy). The activities of the enzyme system in the callus extracts was higher than from the plants, yet none of the callus contained detectable levels of these monoterpenes. The enzymes required for the biosynthesis of the monoterpenes, though present in the callus, were inactive or inhibited. An enzyme that has a regulatory role on another component of flavor, the polyphenols, is phenylalanine ammonialyase (PAL). In cultures where polyphenol synthesis occurs at the end of the growth cycle, the appearance of the polyphenols is always preceded by an increase in activity of PAL (Davies, 1971, 1972). The factors responsible for determining the activity of the regulatory enzymes need to be analyzed in more detail.

Composition of the Medium

Another approach to overcoming the block in the secondary product synthesis in tissue cultures is to alter the composition of the nutrient medium. This approach is based on the assumption that the pathway can be activated either by (1) a specific effect of a growth regulator, resulting in release of the rate-limiting stage of the secondary pathway or, (2) an indirect effect that releases the limitation on the supply of an intermediate caused by the activity of the primary pathways. The supply of available intermediates can be increased either by increasing the supply of basic nutrients to the culture or restricting the demand of the primary pathways by reducing the growth rate of the culture.

In celery the essential oil contains the main flavor compounds the phthalides, of which 3-isobutylidene 3a,4-dihydrophthalide, 3-isovalidene 3a,4-dihydrophthalide, 3-n-butylphthalide, and sedanolide are the most important. In a study of the effect of different auxin and cytokinin treatments on flavor production in undifferentiated celery tissue cultures, it was found that replacing 2,4-D with IAA initiated a low level of flavor production. Replacing 2,4-D with 2,3- and 3,5-dichlorophenoxyacetic acids, 3,5-dimethylphenoxyacetic acid, and chloroisobutyric acid stimulated flavor production to a much greater extent, while chlorophenoxyacetic acid was ineffective. All these treatments except 2,4-D and chlorophenoxyacetic acid caused the cultures to produce green, round aggregates; 3,5-dichlorophenoxyacetic acid produced both the highest flavor and greening levels. Thus the effects of growth factor changes on flavor production are difficult to separate from the effects of differentiation in the form of greening.

In tissue cultures of *Capsicum* the supply of macronutrients in the medium was reduced to inhibit growth of the tissue cultures and stimulate the synthesis of capsaicin. The cultures were placed in a situation of nutrient stress in medium with 5% of the normal nitrogen supply and no sucrose. Detectable levels of capsaicin were found in the tissue and in the medium. When cultures under nutrient stress were incubated with the intermediates vanyllalamine and isocapric acid at the same time, the synthesis of the capsaicin was further increased to a level comparable with the concentration in the intact fruit (Yeoman et al., 1980). The same inverse correlation between growth and secondary product formation has been noted for alkaloids (Neumann and Mueller, 1974; Yeoman et al., 1981).

THE ROLE OF MORPHOLOGICAL AND CELLULAR DIFFERENTIATION

In an early review of secondary product formation in tissue cultures Constabel et al. (1974) stated that whenever the synthesis and accumulation of secondary metabolites is dependent upon special biochemical and structural modifications in the plant cells there is no chance to exploit cell cultures unless those modifications can be induced. It is known that the essential oils and flavor compounds in plants accumulate in specialized tissues, such as glands in mint, oil ducts in celery, and swollen leaf bases in onion. Most of the work on essential oil and flavor production in tissue cultures has shown that synthesis of the complete oil or flavor only occurs when the cultures have been morphologically differentiated to produce shoots and roots, or the culture contains specialized cells. Thus Becker (1970) found no essential oil synthesis in tissue cultures of a variety of plant herbs until redifferentiation. Fridborg (1971), Freeman et al. (1974), and Turnbull et al. (1981) demonstrated that redifferentiation of onion tissue cultures into roots or shoots was essential for the production of onion flavor. In

celery cultures embryogenesis was essential before the celery flavor
was formed (Al-Abta et al., 1979).

Some tissue cultures do produce an essential oil, but an examination
of the culture generally reveals the presence of specialized cells or
ducts or a slow growth rate of the plant cells. *Ruta graveolens* cultures for instance produce an essential oil that is similar in composition to the essential oil in the plant (Nagel and Reinhard, 1975), but
the callus also contains the specialized oil-producing schizogenous passages found in the differentiated plant (Reinhard et al., 1968). Tissue
cultures of perilla also synthesize an essential oil, but this may be
because of the aggregated state of the suspension. It is possible that
aggregates have internal cell differentiation, or contain cells with a
slow growth rate (Sugisawa and Ohnishi, 1976). Townsley (1974) found
that to produce a cocoa aroma it was necessary to allow the cocoa
suspension culture to become senescent. Oil-producing mint cultures
also contained partially differentiated cells (Bricout and Paupardin,
1975). In later studies on peppermint an essential oil was produced by
culture, but the oil did not have the same composition as the oil from
the plant. In these cultures oil production appeared to be derived
from specialized giant cells in the callus (Kireeva et al., 1978). Chamomile cultures only showed greening and morphological differentiation
in the light, and only these light-grown cultures had the same essential
oil as the plant (Szoeke et al., 1977). In all these cases of essential
oil production by tissue cultures, a level of cell or organ differentiation
was found to be necessary before synthesis and accumulation of the
secondary product occurred. It does suggest that cells need to attain
a specific stage of differentiation before the secondary pathways are
activated.

In an investigation of the effect of stage of differentiation on the
initiation of flavor production in celery, tissue cultures were analyzed
for flavor production at various stages of differentiation. These included undifferentiated cells; globular, heart, and torpedo embryoids; and
differentiated plants (Al-Abta et al., 1979). The early globular and
heart-shaped embryoids produced no flavor compounds, whereas the torpedo stage which was a more differentiated stage, possessing chloroplasts and containing an internal cell differentiation did contain the
phlthalide flavor compounds. There were no oil ducts in torpedo embryos (Al-Abta and Collin, 1978), so the synthesis of the flavor compounds in the essential oil was not dependent on the presence of highly
specialized cell types. However, the torpedo-stage embryos were green
which raised the possibility that flavor synthesis may be correlated to
the greening.

The effect of greening on flavor production in celery plant petioles
showed a positive correlation (Table 2). Subsequent greening in celery
tissue cultures was initiated by altering the composition of the medium
and also by selecting for green clones. By replacing 2,4-D in the
medium with a range of other phenoxyacetic acids of which the most
effective was 3,5-dichlorophenoxyacetic acid, greening could be induced.
All the green cultures had the odor of celery, and an analysis of the
media in which a selected green clone was being maintained showed
the presence of phlthalides (Fig. 1). Light microscope studies of the

Table 2. The Relationship between Chlorophyll and Phthalide (Sedanonic Anhydride Measured as µg Ligustilide Equivalents) in Intact Celery Tissue and Cell Suspension Medium

TISSUE SOURCE	CHLOROPHYLL (mg g fwt^{-1})	SEDANONIC ANHYDRIDE (µg ligustilide equiv. g fwt^{-1})
Blanched petioles	0.003 ± 0.001	0.085 ± 0.01
Green petioles	0.041 ± 0.001	0.091 ± 0.02
Green leaves	0.650 ± 0.100	0.920 ± 0.33
Green suspension	–	0.064 (medium)
Nongreen suspension	–	–

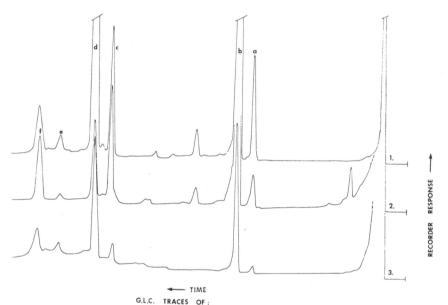

G.L.C. TRACES OF :
1 = SEED OIL ≡ 0.02 GRAM FRESH WEIGHT ; 2=HERB OIL ≡ 1 GRAM FRESH WEIGHT ; 3=GREEN CLONE MEDIUM EXTRACT ≡ 0.2 GRAM FRESH WEIGHT

Figure 1. GC trace of extract of seed oil, celery leaf tissue, and a nutrient medium that had contained a green celery tissue culture clone (a, pinene; b, limonene; c, caryophyllene; d, sesquiterpene; e, 3-butylphthalide; f, sedanonic anhydride).

green celery aggregates showed no differentiation into organized meristem or embryoid structures, and under the electron microscope there appeared to be no obvious difference in the starch containing plastids in cells from both green and nongreen cultures. However, the presence of chlorophyll in the green cells did imply a more advanced chloroplast development, although the role this might play in phlthalides synthesis is unknown. Greening has also been found to stimulate alkaloid production (Garve et al., 1980; Yeoman et al., 1981), so it is possible that the greening of plastids is merely an expression of an overall cell differentiation, and an increase in capacity for biosynthesis is associated with this differentiation as before.

ENHANCEMENT OF YIELD

Where cultures show a limited synthesis of the secondary product, it should be possible to select for a higher yield by making use of natural genetic variations that exist in tissue cultures. There are a number of approaches to select for high yields. Some of these methods have been successful, particularly with the pharmaceuticals.

(1) One of the most sensitive methods to detect presence of chemicals is radio immuno assay (RIA). This technique is described in detail by Weiler (1977). This method is ideal, since there is a great deal of literature on the use of the technique in medicine and pharmacology. Also, the technique is rapid, very sensitive, and can be used to test a very large number of small samples. It should be possible to develop an RIA system for detection of many interesting compounds.

(2) An alternative to the RIA method is to use more simple, direct, semi-quantitative methods of selection. Thus tissue culture clones can be assessed on the basis of smell, since the nose is very sensitive (Townsley, 1974). If yield is related to color, such as in the production of food colorings, then selection is straightforward. Similarly where synthesis is proportional to greening, intensity of greening can be used as a basis of selection. A broad general chromatographic location reagent has proved useful in selecting for alkaloids in crude cell extracts, and direct squashes on the chromatogram. It may be possible to develop a similar technique for detecting terpenes in a cell squash since the presence of these compounds are good indicators of the cells' capacity to synthesize related flavor compounds.

(3) The direct methods of selection are time consuming and laborious, involving the analysis of many samples. A possible alternative which has not yet been explored is to select for resistance of cells to analogs in the culture medium. Cell clones that are resistant to amino acid analogs in the medium may survive by overproduction of the specific amino acid (Widholm, 1976). The method would be to establish analogs for intermediates in the secondary pathway, then select for cells which are resistant to the presence of these analogs in the medium. In theory, resistance will be conferred in the cells by an increased activity of the secondary pathway and therefore increased levels of secondary products.

PROTOCOL FOR ASSESSMENT OF FLAVOR PRODUCTION IN ONION

Thin Layer Chromatography and Electrophoresis of Onion Flavor Precursor Compounds

Thin layer chromatography and electrophoresis provide a rapid method of separating the amino acids and flavor precursor compounds. The methods described are modifications used by Bieleski and Turner (1966), Stahl (1969), and Granroth (1970).

1. Place 100 mg fwt. of tissue in a prechilled homogenizer containing 2 cm^3 methanol/chloroform/water (12 + 5 + 3 v/v) (MCW), at -20 C for 30 min and then homogenize.
2. Filter the homogenate and then centrifuge at high speed in a small bench centrifuge.
3. Transfer supernatant to another centrifuge tube, homogenize the residue in a furthe 2 cm^3 of MCW, centrifuge, and add the supernatant to the first extract.
4. Add 1 cm^3 chloroform and 1.5 cm^3 water to the combined extracts and the extract mixed using a vortex mixer.
5. Separate the two phases by centrifuging for 15 min at high speed. Transfer the upper aqueous layer to a 50 cm^3 volume pear-shaped flask.
6. Dry in a vacuum oven at a temperature not exceeding 30 C and then store at -20 C.
7. Dissolve extract in 10% isopropanol to give a concentration of 400 mg fwt. per cm^3.
8. Homogenize 15 g cellulose powder and 2.5 g silica gel H (Merck) with 100 ml distilled water for 30 sec, then allow the mixture to stand for 30 sec, mix again for 30 sec, then allow to stand for 60 sec before spreading on grease-free glass plates (4 mm, 20 x 20 cm) at a thickness of 300 μm using a motorized TLC coater.
9. Load the equivalent of 10-20 mg fresh weight of tissue onto a plate in a thin band 2.5 cm long and 2.5 cm from the edges of the plate using a microcapillary pipette. Place a spot of marker amino acid, (2,4-dimetrophenyl)-l-lysine hydrochloride at the origin (Fig. 2).
10. Spray the plate with electrophoresis buffer pH 2 (15.3 ml of 98-100% formic acid, 57 cm^3 acetic acid in 1 liter distilled water). Remove excess buffer from plate by blotting plate with chromatography paper. Hold plates horizontally to prevent streaming from the origin.
11. Place plate on water-cooled base plate of Shandon electrophoresis chamber with the origin at the anode. Fold wicks of buffer-soaked Whatman No. 3 chromatography paper over the edges of the plate and secure with glass strips and a glass plate. This ensures good contact between the wicks and the thin layer (Fig. 3).
12. Carry out electrophoresis at 1000 V and 20-40 mA for approximately 25 min until the yellow marker spot reaches the 4 cm mark. This ensures optimum electrophoretic separation of the amino acids on a 20 x 20 cm plate.
13. Dry the plate in a stream of cool air, parallel with the amino acid bands. Reduce the amino acids to spots by a water run then dry the plates as before.
14. Develop the plates twice in the same direction, first with methyl ethyl ketone/pyridine/water/acetic acid, (70 + 15 + 15 + 2, v/v),

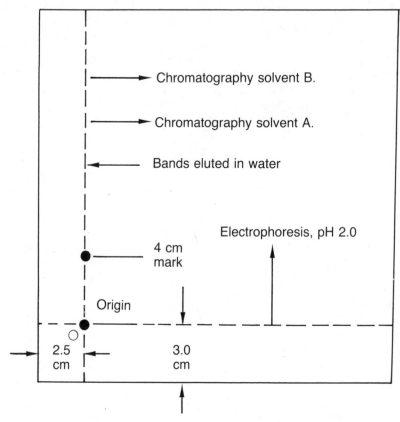

Figure 2.   Diagram of procedure used in two–dimensional separation of amino acids by high voltage electrophoresis and chromatography.

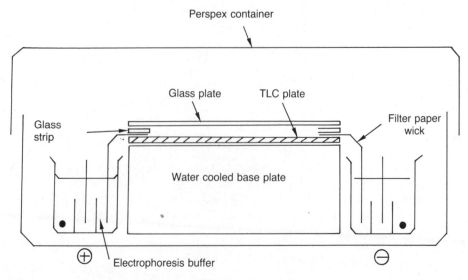

Figure 3.   Schematic diagram of electrophoresis apparatus.

then with n-propanol/water/n-propyl acetate/acetic acid/pyridine (120 + 60 + 20 + 4 + 1, v/v). All solvents to be of Analar and Aristar grade. Saturate chromatography tanks 1 hr before use.
15. Spray the dried plates with 0.2% ninhydrin in acetone and allow the color to develop overnight at room temperature.
16. Run all plates in duplicate and record the pattern by tracing onto paper and by photography.
17. Run standards of amino acids to produce a map of their distribution under these conditions.

Assay for Total Flavor Precursor

One of the first products of the reaction between the alkyl cysteine sulphoxides (the flavor precursors) and the enzyme alliinase is pyruvate. Hence a measure of pyruvate can be used to estimate the total flavor precursor level in the tissue. The pyruvate was measured according to the method of Schwimmer and Guadagni (1962) in which both the endogenous level of pyruvate (Pc) and the level produced by tissue disintegration (PT) are determined.

1. For a measurement of PT, grind 2.5 g tissue using a pestle and mortar with 2.5 ml 0.1 M sodium pyrophosphate buffer, pH 9.0. Stand at room temperature for 30 min.
2. Dilute the homogenate to 25 ml with distilled water and filter through Whatman No. 1 filter paper.
3. Repeat for the measurement of Pc, but prepare the initial homogenate in 2 M HCl to prevent enzyme activity.
4. Add 1 ml 0.0125% 2:4 dimitro-phyl-hydrozine in 2 M HCl to each 2 ml filtrate.
5. After incubating for 10 min at 27 C, add 5 ml 0.6 M NaOH, and measure the optical density at 420 nm.
6. Construct a calibration curve using standard solutions of sodium pyruvate up to a maximum of 2 μg per ml.
7. Express the results as μmol pyruvate per ml or estimate protein in filtrate and express results as a specific activity, $\mu mol\ mg^{-1}$ protein min^{-1}.
8. For the protein estimates, shake a small known volume of filtrate with 25 volumes of ice-cold acetone and centrifuge the mixture at high speed in a small bench centrifuge. This will precipitate the protein.
9. Discard the acetone and redissolve the protein in a known volume of 2% sodium carbonate in 0.1 M NaOH.
10. Estimate the protein content using the Folin Ciocalteau method.
11. Prepare a calibration curve using bovine serum albumin with a concentration range of 10–100 mg/l. Express the results as mg protein per ml.

Extraction and Assay of Alliinase Activity

1. Homogenize 10 g tissue using a pestle and mortar with 10 ml 0.2 M potassium phosphate buffer pH 6.8 in 0.3 M sucrose for 2 min, then filter through five layers of muslin.

2. Centrifuge the filtrate for 15 min 3,500 x g in an MSE superspeed 65 centrifuge and recentrifuge the supernatant for 20 min at 30,000 x g to produce a clear yellow supernatant.

3. Precipitate the soluble proteins from this supernatant by the addition of solid ammonium sulphate to 75% v/v saturation. After stirring for 4 hr at 4 C, centrifuge the mixture for 20 min at 30,000 x g. Store the resulting pellet at -20 C for assay.

Assay for Alliinase Activity

This is modification of the method of Schwimmer and Mabelis (1963) where synthetic s-propyl-l-cysteine sulphoxide was used as a substrate.

1. Assay alliinase activity at two protein concentrations of approximately 0.1 and 0.05 mg protein per ml reaction mixture. Determine the concentrations accurately by the Folin method.

2. Prepare reaction mixture containing 0.4 ml 50 mM s-propyl-l-cysteine sulphoxide, 0.1 ml 0.5 mM pyridoxal phosphate, 0.4 ml 0.1 M sodium pyrophosphate buffer, pH 9.0 and 0.1 ml sample.

3. Incubate the mixture in a water bath at 25 C. Remove 0.1 ml samples at various times and stop the reaction by the addition of 1 ml 10% TCA.

4. Make each sample up to a total volume of 2 ml and estimate the pyruvate as before.

5. Express the results as activity mg/protein.

PROTOCOL FOR SMALL-SCALE FLAVOR AND CHLOROPHYLL EXTRACTION OF CELERY

Flavor Extraction from Whole Celery or Filtered Cells from Suspension Culture

1. Homogenize 50 g celery (fresh weight) in 50 ml ethyl acetate in a rotary blade homogenizer.

2. Add 75 ml distilled water, mix thoroughly, and decant to a 500 ml spherical flask, adding a few antibumping granules.

3. Distill homogenate in a modified Stahl apparatus (1969) for 12 hr, running off the distillate every 2 hr into a foil capped flask. (Note: use of parafilm will result in contamination of the extract with phthalate plasticisers.) See Fig. 4.

4. Saturate the aqueous phase of the distillate with sodium chloride and partition, reserving the ethyl acetate fraction.

5. Partition the remaining aqueous phase an additional three times with its equal volume of ethyl acetate, shaking the aqueous and solvent thoroughly before each separation.

6. Pool the ethyl acetate fractions, and dry over anhydrous magnesium sulphate (0.5-1.0 g $MgSO_4$ per 100 ml ethyl acetate).

7. Filter, and evaporate under vacuum to 100-200 µl.

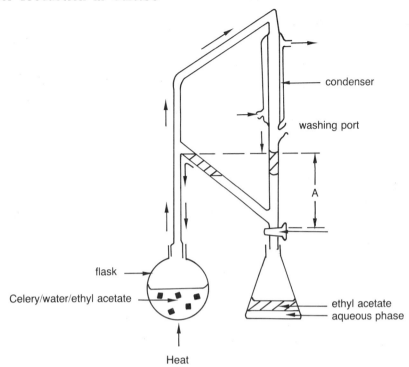

Figure 4. Modified Stahl's apparatus for the distillation of essential oils.

8. Analyze 0-10 μl of this extract by gas liquid chromatography (9 ft x 4 mm internal diameter glass column, fitted in a Pye 105 gas chromatograph; column packing: 10% SP2100 on a 60-100 mesh chromosorb; carrier gas: nitrogen, flow rate 60 ml/min; temperature program: 60-250 C at 6 C/min).

9. Phthalide levels are calculated from GLC peak areas and expressed as equivalents to a pure phthalide standard, ligustilide (3-butylid-ene-4,5-dihydrophthalide) kindly provided by Prof. H. Mitsuhashi, Faculty of Pharmaceutical Science, Hokkaido University, Sapporo, Japan.

From Suspension Culture Media

1. 120 ml of media are shaken thoroughly with 50 ml ethyl acetate and the procedure as for extraction of celery followed from steps 3-8.

Chlorophyll Extraction

1. Grind 1 g fwt. celery in a mortar and pestle with 3 ml 80% v/v acetone.

2. Pour homogenate into a 50 ml centrifuge tube and centrifuge at 1500 x g for 3-5 min.
3. Decant the supernatant and repeat steps 1-2 until no further green coloring is extracted with the acetone.
4. Pool the supernatants and note the volume.
5. Read the absorbance of a portion of the acetone extract at 663 nm and 645 nm in a spectrophotometer.
6. Calculate the chlorophyll content from Arnon's formula:

$$\text{mg cm}^{-3} \text{ chlorophyll a} = 0.0127D_{663} - 0.00269D_{645}$$

$$\text{mg cm}^{-3} \text{ chlorophyll b} = 0.0229D_{645} - 0.00468D_{663}$$

where D_{663} and D_{645} are the absorbances of the extract at 663 nm and 645 nm, respectively.

FUTURE PROSPECTS

The situation for large-scale tissue culture of plant cells is looking much more hopeful than it was even 8 years ago (Zenk, 1978; Dougall, 1980). Up till now there has been only limited attention given to the essential oils and flavors, although the cost of some of these compounds could justify further investigation. It is suggested the following areas are important:

(1) Where essential oil and flavor production is inhibited in tissue culture, the rate-limiting step should be established. At the moment it may not be possible to activate a specific regulatory stage in biosynthesis, but it is important that this basic research should proceed.

(2) The minimum cell or organ differentiation at which the secondary pathways are initiated must be identified. If maximum yield of oils and flavors are obtained from partially differentiated cultures such as slow-growing cells, then the large-scale culture system must accommodate this fact. The new developments in immobilized cells, either in liquid medium or attached to a column, have a contribution to make to the support and maintenance of slow-growing but biosynthetically active units.

(3) The capacity of undifferentiated cells to transform simple, cheap precursors to the final product has been largely unsuccessful. The capacity for biotransformation of the more biosynthetically active and partially differentiated cells must be investigated, as in *Mentha* cultures (Aviv et al., 1981).

(4) The development of rapid, sensitive, direct, and indirect selection techniques for isolating high-yielding clones must be expanded.

(5) The mechanism of control of differentiation of cultures in liquid suspension must be established. It may be possible to consider growing cultures on a large scale to accumulate biomass, then alter the medium or cultural conditions so as to switch the cells to a differentiation phase to initiate secondary product formation.

KEY REFERENCES

Carew, D.P. and Staba, E.J. 1965. Plant tissue culture: Its fundamentals, application and relationship to medicinal plant studies. Lloydia 28:1-26.

Constabel, F., Gamborg, O.L., Kurz, W.G.W., and Steck, W. 1974. Production of secondary metabolites in plant cell cultures. Planta Med. 25:158-165.

Dougall, D.K. 1980. Production of biologicals by plant cell cultures. In: Advances in Experimental Medicine and Biology (J.C. Petricciana, H.E. Hopps, and D.J. Chafpte, eds.) pp. 136-151. Plenum, New York and London.

Staba, E.J. 1963. The biosynthetic potential of plant tissue cultures. Dev. Microbiol. 4:193-198.

Yeoman, M.M., Miedzybrodyka, M.B., Lindsey, K., and McLauchlan, W.R. 1980. The synthetic potential of cultured plant cells. In: Plant Cell Cultures: Results and Perspectives (F. Sala, B. Parisi, R. Cella, and O. Ciferri, eds.) pp. 327-343. Elsevier/North-Holland Biomedical Press, Amsterdam.

Yeoman, M.M., Lindsey, K., Miedzybrodyka, M.B., and McLauchlan, W.R. 1981. Accumulation of secondary products as a facet of differentiation in plant cell and tissue culture. In: Differentiation In Vitro, 4th Symposium British Society Cell Biology, pp. 65-81. Cambridge Univ. Press, Cambridge.

Zenk, M.H. 1978. The impact of plant cell culture on industry and agriculture. In: Frontiers of Plant Tissue Culture (T.A. Thorpe, ed.) pp. 1-13. Calgary Univ. Press, Calgary.

REFERENCES

Al-Abta, S. and Collin, H.A. 1978. Cell differentiation in embryoids and plantlets of celery tissue cultures. New Phytol. 80:517-521.
Al-Abta, S., Galpin, I.J., and Collin, H.A. 1979. Flavour compounds in tissue cultures of celery. Plant Sci. Lett. 16:129-134.
Aviv, D., Krochmal, E., Dantes, A., and Galun, E. 1981. Biotransformations of monoterpenes by mentha cell lines: Conversion of menthone to neomenthol. Planta Med. 42:236-243.
Banthorpe, D.V., Bucknall, G.A., Doonan, H.J., Doonan, S., and Rowan, M.G. 1976. Biosynthesis of veraniol and nerol in cell-free extracts of Tanacetum vulgare. Phytochemistry 15:91-100.
Becker, H. 1970. Untersuchungen zur Frage der Bildung fluechtiger Stoffwechset produkte in Calluskulturen. Biochem. Physiol. Pflanz. 161:425-441.
Bieleski, R.L. and Turner, N.A. 1966. Separation and estimation of amino acids in crude plant extracts by thin layer chromatography and electrophoresis. Anal. Biochem. 17:278-293.

Bricout, J. and Paupardin, C. 1975.   Sur la composition de l'huile essentielle de *Mentha piperita* L. cultivee in vitro: Influence de quelque facteurs sur sa synthese.   C.R. Acad. Sci. Ser D 281:383-386.

Davies, M. 1971.   Multi-sample enzyme extraction from cultured plant cell suspensions.   Plant Physiol. 47:38-42.

_____ 1972.   Polyphenol synthesis in cell suspension cultures of Pauls Scarlet Rose.   Planta 104:50-65.

Freeman, G.G., Whenham, R.J., Mackenzie, I.A., and Davey, M.R. 1974. Flavour components in tissue cultures of onion (*Allium cepa* L.). Plant Sci. Lett. 3:121-125.

Fridborg, G. 1971.   Growth and organogenesis in tissue cultures of *Allium cepa* v. proliferum.   Physiol. Plant. 25:436-440.

Garve, R., Luckner, M., Vogel, F., Trewes, A., and Novev, L. 1980. Growth morphogenesis and cardenolide formation in long term cultures of *Digitalis lanata*.   Planta Med. 40:92-103.

Granroth, B. 1970.   Biosynthesis and decomposition of cysteine derivatives in onion and other *Allium* species.   Ann. Acad. Sci. Fenn. Ser. A2 152:1-71.

Jalal, M.A.F. and Collin, H.A. 1979.   Secondary metabolism in tissue cultures of *Theobroma cacao*.   New Phytol. 83:343-349.

Kireeva, S.A., Melnikov, U.N., Reznikov, S.A., and Meshcheryakova, N.I. 1978.   Essential oil accumulation in a peppermint callus culture. Soviet Plant Physiol. 25:438-443 (translation).

Nagel, M. and Reinhard, E. 1975.   Das Aetherische Oel der Callusculturen von *Ruta graveolens* L.   I. Die Zusammensetzung des Oeles. Planta Med. 27:151-158.

Neumann, D.V. and Muller, E. 1974.   Beitraege zur Physiologie der Alkaloide.   IV. Alkaloidebildung in Kalluskulturen von Macleaya. Biochem. Physiol. Pflan. 165:271-282.

Overton, K.H. and Picken, D.J. 1977.   Studies in secondary metabolism with plant tissue culture.   Prog. Chem. Org. Nat. Prod. 34:249-298.

Reinhard, E., Corduan, G., and Volk, O.H. 1968.   Uber Gewebekulturen von *Ruta graveolens* L.   Planta Med. 16:8-16.

Sardesai, D.L. and Tipnis, H.P. 1969.   Production of flavouring principles by tissue culture of *Coriandrum sativum*.   Curr. Sci. 38:345.

Schwimmer, S. and Guadagni, D.G. 1962.   Relation between refactory threshold concentration and pyruvic acid content of onion juice.   J. Food Sci. 27:94-97.

Schwimmer, S. and Mabelis, M. 1963.   Characterisation of alliinase of *Allium cepa* (onion).   Arch. Biochem. Biophys. 100:66-73.

Selby, C., Galpin, I.J., and Collin, H.A. 1979.   Comparison of the onion plant (*Allium cepa*) and onion tissue culture.   I. Alliinase activity and flavour precursor compounds.   New Phytol. 83:351-359.

Selby, C., Turnbull, A., and Collin, H.A. 1980.   Comparison of the onion plant (*Allium cepa*) and onion tissue culture.   II. Stimulation of flavour precursor synthesis in onion tissue cultures.   New Phytol. 84: 307-312.

Stahl, E. 1969.   Thin Layer Chromatography.   A Laboratory Handbook, 2nd ed.   George Allen and Unwin, London.

Sugisawa, H. and Ohnishi, Y. 1976.   Isolation and identification of monoterpenes from cultured cells of *Perilla* plants.   Agric. Biol. Chem. 40:231-232.

Szoeke, E., Kuzovkina, I.N., Verzar-Petri, G., and Smirnov, A.M. 1977. Cultivation of wild chamomile tissue. Fiziol. Rast. 24:832-840 (translation).

Szoeke, E., Kuzovkina, I.N., Verza-Petri, G., and Savarda, A.L. 1978. The influence of coconut milk on the synthesis of ethereal oils in plant tissue cultures. In: Proceedings 18th Hungary Annual Meeting Biochem. Salgotarjan. pp. 189-190.

Townsley, P.M. 1974. Chocolate aroma from plant cells. J. Inst. Can. Sci. Tech. Alim. 7:76-78.

Turnbull, A., Galpin, I.J., and Collin, H.A. 1980. Comparison of the onion plant (*Allium cepa*) and onion tissue culture. III. Feeding of ^{14}C labelled precursors of the flavour precursor compounds. New Phytol. 85:483-487.

Turnbull, A., Galpin, I.J., Smith, J.L., and Collin, H.A. 1981. Comparison of the onion plant (*Allium cepa*) and onion tissue culture. IV. Effect of shoot and root morphogenesis on flavour precursor synthesis in onion tissue. New Phytol. 87:257-268.

Weiler, E.W. 1977. Radio immuno assay for the determination of digitoxin and related compounds in *Digitalis lanata*. In: Plant Tissue and its Biotechnological Application (W. Barry, E. Reinhard, and M.H. Zenk, eds.) pp. 3-16. Springer-Verlag, Berlin, New York.

Widholm, J.M. 1976. Selection and characterisation of cultured carrot and tobacco cells resistant to lysine, methionine and proline analogs. Can. J. Bot. 54:1523-1529.

Whitaker, J.R. 1976. Development of flavour odour and pungency in onion and garlic. Adv. Food Res. 22:73-133.

CHAPTER 25
Genetic Variability in Regenerated Plants

B. Reisch

Plants regenerated from undifferentiated tissue cultures have become a new and useful source of genetic variation. Variation generated by the use of a tissue culture cycle has been termed somaclonal variation by Larkin and Scowcroft (1981). They defined a tissue culture cycle as a process that involves the establishment of a "dedifferentiated cell or tissue culture under defined conditions, proliferation for a number of cell generations, and the subsequent regeneration of plants." Even though this interest in somaclonal variability is rather recent, this phenomenon was reported more than a decade ago among callus culture regenerates (Morel, 1971; Sacristan and Melchers, 1969) and even among the first plants reportedly regenerated directly from isolated protoplasts.

Plants regenerated from shoot apex cultures may occassionally include variants (Bush et al., 1976; Denton et al., 1977); however, most reports describing plants from such cultures contend that genetic stability is preserved (D'Amato, 1975; Morel, 1975; Murashige, 1974). The shoot apex provides a structure in which (1) cell division and DNA replication are strictly controlled, and (2) those cells with impaired reproductive ability (i.e., those with genetic defects) are eliminated because of competition (D'Amato, 1977).

Cultured cells, on the other hand, are not usually genetically stable. Polyploidy, aneuploidy, and chromosome structure changes in cells in culture under various conditions have been extensively described (see reviews such as Bayliss, 1980; D'Amato, 1977; Larkin and Scowcroft, 1981; Skirvin, 1978; Sunderland, 1977). Variation in genetic composition is also known in whole plant tissues (Murashige and Nakano, 1966;

Neilson-Jones, 1969). Thus it is not surprising that many of the plants regenerated from undifferentiated cultures (callus, cell suspension, and protoplast cultures) are altered. For the purposes of germplasm preservation and plant micropropagation the inherent stability of shoot apex cultures is required. But the diverse variation characteristic of the plants obtained from undifferentiated cell cultures might be of great use to plant breeders, and it is this type of variation that will be discussed herein.

This chapter is not intended to be a review of every report of somaclonal variation. Rather, it will focus upon representative examples of variation among plants derived from tissue cultures originating from sporophytic tissue in which no effort was made to induce variation via mutagenesis. Since technical details vary considerably from case to case, no effort will be made to elaborate upon specific techniques. Instead, this chapter will be organized around a discussion of major techniques used to produce variation. Readers are referred to the original sources for technical details. The scope of this chapter will be limited to the variation obtained from cultures derived from sporophytic tissues. Variation obtained from gametophytic cultures will not be discussed, since culture-produced variation is probably confounded with residual heterozygosity in autogamous species (Collins and Legg, 1980). This topic is fully discussed elsewhere (Larkin and Scowcroft, 1981). A critical review of recent progress in this area and an analysis of major techniques used to enhance somaclonal variation will be followed by a discussion of the value of somaclonal variation in a plant breeding program.

CRITICAL REVIEW

Major Techniques

Plant variation generated by use of a tissue culture cycle may be encouraged by seven different techniques. The seven specific manipulations, employed to generate somaclonal variation are as follows: (a) a long-term culture cycle, (b) a protoplast culture cycle, (c) a callus culture cycle, (d) the use of explants from specified tissues, (e) the generation of random variation concomitant with the selection of a specific nutrient medium or hormone formulation, (g) the use of certain genotypes that tend to produce increased amounts of variation.

Explants used in this process may come from virtually any tissue, including leaves, internodes, ovaries, roots, and inflorescences. But because these classifications are broad in scope, they are not mutually exclusive. For example, the use of a long-term tissue culture or a specific explant source must be done in conjunction with either protoplast or callus culture.

A summary of variation obtained in 23 plant genera and the techniques used to obtain this variation are presented in Table 1. In most cases only those plants derived directly from tissue culture and their vegetative propagules have been examined, but in at least 10 reports the sexual transmittance of such traits has been verified (Table 1). In all of these cases the variation was stably transmitted to the progeny.

Table 1. Documented Variation among Plants Regenerated from Sporophytic Tissue Cultures

SPECIES	ALTERED CHARACTERISTICS	ASSOCIATED TECHNIQUES[1]	INHERITANCE EXAMINED[2]	REFERENCE
Allium sativum L.	Bulb size and shape, clove number, failure to flower, plant height, plant vigor, chromosome number	a,c,g	–	Novak, 1980
Ananas cosmosus (L.) Merr.	Spine type, leaf color, leaf waxiness, foliage density, albino stripes	d	–	Wakasa, 1979
Avena sativa L.	Heteromorphic pairs, trisomics, monosomics, interchanges	a,h	–	McCoy, 1980; McCoy & Phillips, 1982
Brassica L. spp.	Fasciation, variegation, chromosome number	c,g	–	Dunwell, 1981
Brassica napus L.	Chromosome number	a	–	Sacristan, 1981
Chrysanthemum x *morifolium* Ramat.	Flower color	c	–	Ben-Jaacov & Langhans, 1972
Chrysanthemum x *morifolium* Ramat.	Plant vigor, leaf morphology and hairiness, lateral bud growth, phyllotaxy	a	–	Sutter & Langhans, 1981
Daucus carota L.	Erect stems, leaf dissection, leaf thickness, leaf color	f	–	Ibrahim, 1969
Haworthia setata Haw.	Chromosome number, vigor, leaf shape, leaf color, esterase zymogram, chromosome associations, pollen fertility	c,f	–	Ogihara, 1981
Hordeum L. spp.	Isoenzyme intensity, growth habit, head morphology, auricle size, chromosome number	c	–	Orton, 1980

SPECIES	ALTERED CHARACTERISTICS	ASSOCIATED TECHNIQUES[1]	INHERITANCE EXAMINED[2]	REFERENCE
Hordeum vulgare L.	Albinism, leaf shape, fertile tillers	f	+	Deambrogio & Dale, 1980
Lactuca sativa L.	Leaf shape, leaf color, axillary bud, plant vigor	c	+	Sibi, 1976
Lilium L. spp.	Plant vigor, leaf variegation	a,c	–	Stimart et al., 1980
Lolium L. spp.	Chromosome number, leaf shape and size, floral development, growth vigor, survival, pereniality	c	–	Ahloowalia, 1978
Medicago sativa L.	Chromosome number, leaf shape, petiolule length, herbage yield, plant height, shoot length, rooting response	e	–	Reisch & Bingham, 1981
Medicago sativa L.	Cotyledon number and shape, leaf morphology, plant vigor	c	+	Johnson et al., 1980
Nicotiana suaveolens Lehm. x *N. glutinosa* L.	CO_2 absorption, chlorophyll content	c	+	Mousseau, 1970
Nicotiana sylvestris Speg. et Comes.	Leaf shape, photoperiod response, peroxidase isoenzyme	e	–	Maliga et al., 1979
Nicotiana tabacum L.	Chromosome number, male fertility, plant vigor, leaf shape	a	–	Butenko et al., 1967
Nicotiana tabacum L.	Chromosome number, plant morphology, fertility	b	–	Takebe et al., 1971
Nicotiana tabacum L.	Plant vigor, flower morphology, male and female fertility	a,c	–	Syono & Furuya, 1972
Nicotiana tabacum L.	Chromosome number, leaf shape, leaf color, pollen and seed fertility, plant vigor	c	+	Ogura, 1976
Nicotiana tabacum L.	Plant vigor, root development, leaf shape, growth habit	e	+	Berlyn, 1980

751

Table 1. Cont.

SPECIES	ALTERED CHARACTERISTICS	ASSOCIATED TECHNIQUES[1]	INHERITANCE EXAMINED[2]	REFERENCE
Nicotiana tabacum L.	Leaf color, leaf shape	a,b,c,d	+	Barbier & Dulieu, 1980
Nicotiana tabacum L.	Chromosome number, leaf morphology, sterility, leaf color	a	−	Sacristan & Melchers, 1969
Nicotiana tabacum L.	Tricotyly, hydroxyurea resistance	e	+	Chaleff & Keil, 1981
Oryza sativa L.	Seed fertility, plant height, heading date, morphology, chlorophyll deficiency	c	+	Oono, 1978
Panicum L. spp.	Plant size, leaf shape, tillering	c	−	Bajaj et al., 1981
Pelargonium L'Her ex Ait spp.	Plant and organ size, leaf and flower morphology, oil constituents, fasciation, pubescence, anthocyanin pigmentation	a,c,g	−	Skirvin & Janick, 1976a
Pelargonium graveolens L'Her ex Ait	Chromosome number, leaf shape and type	c	−	Janick et al., 1977
Saccharum L. spp.	Plant morphology, chromosome number, isozyme systems	c,g	−	Heinz & Mee, 1971
Saccharum L. spp.	Auricle length, leaf attitude, hairiness, sugar content, chromosome number	c,g	−	Liu & Chen, 1976
Saccharum officinarum L.	Resistance to *Drechslera sacchari* (But.) Subr. et Jain, *Solerospora sacchari* Miyake and Fiji virus; cane and sugar yield	c	−	Heinz et al., 1977

SPECIES	ALTERED CHARACTERISTICS	ASSOCIATED TECHNIQUES[1]	INHERITANCE EXAMINED[2]	REFERENCE
Saccharum officinarum L.	Cane and sugar yield; stalk number, length, diameter, volume and density; fiber percent	c	-	Liu & Chen, 1978
Saccharum officinarum L.	Resistance to *H. sacchari* toxin	c	-	Larkin & Scowcroft, 1981
Solanum tuberosum L.	Resistance to *Alternaria solani* (Ell. et Mart.) Sor.	b	-	Matern et al., 1978
Solanum tuberosum L.	Tuber shape	a,b	-	Shepard et al., 1980
Solanum tuberosum L.	Tuber shape, yield, maturity date, photoperiod requirement, plant morphology, resistance to *A. solani* and *Phytophthora infestans* (Mont.) de Bary	b	-	Secor & Shepard, 1981
Solanum tuberosum L.	Leaf morphology, leaf color, glossiness, hairiness, plant growth habit	b	-	Thomas, 1981
Solanum tuberosum L.	Tuber shape, tuber color, leaf shape	d	-	Van Harten et al., 1981
Sorghum bicolor (L.) Moench	Seed fertility, leaf morphology, plant growth habit	c	-	Gamborg et al., 1977
Trifolium incarnatum L.	Male and female fertility, leaf shape, flower head structure	c	-	Beach & Smith, 1979
Triticum turgidum L. Durum Group	Chromosome number	c	-	Bennici & D'Amato, 1978
Zea mays L.	Leaf arrangement, plant height, node number	c	-	Green, 1977

753

Table 1. Cont.

SPECIES	ALTERED CHARACTERISTICS	ASSOCIATED TECHNIQUES[1]	INHERITANCE EXAMINED[2]	REFERENCE
Zea mays L.	Chromosome number, pollen fertility, endosperm and seedling mutations e.g., opague, germless, etched, yellow, yellow green, pale green, viviparous, virescent	c,g	+	Edallo et al., 1981

[1] a = a long-term culture cycle; b = a protoplast culture cycle; c = a callus culture cycle; d = the use of explants from specified tissues; e = the generation of random variation concomitant with the selection for a resistance trait in tissue culture; f = the use of a specific nutrient medium or hormone formulation; g = the use of certain genotypes which tend to produce increased amounts of variation.

[2] + = sexual transmission of altered characteristics has been confirmed; - = sexual transmission of altered characteristics has not been examined.

LONG-TERM CULTURE CYCLE. The regeneration of plants from long-term cultures is becoming possible with an increasing number of species (e.g., *Medicago sativa* L., Stavarek et al., 1980) but has been generally difficult in the past. Relatively few reports have presented information concerning the morphology and genetic stability of plants regenerated from long-term cultures. Changes in karyotypic structure are known to occur with increasing time in culture (Bayliss, 1980); hence it follows that the frequency of plant abnormalities would likely increase with increasing time in culture. But regenerated plants do not usually reflect the full range of abnormalities reported in cultured cells, indicating that some selection prior to regeneration usually takes place (Edallo et al., 1981; Sacristan and Melchers, 1969).

Flower and leaf abnormalities among tobacco (*Nicotiana tabacum* L.) plants regenerated from callus subcultured for prolonged periods were common but were not found among plants regenerated from short-term cultures or grown from seed under the same cultural conditions (Syono and Furuya, 1972). However, it was noted that abnormal type callus may have been inadvertently and selectively maintained in the long-term culture.

In the chimeral plant *Chrysanthemum morifolium* Ramat. a few abnormalities were observed among plants regenerated from short-term shoot tip cultures (Bush et al., 1976). But plants regenerated from another cultivar maintained as leafy callus for 9 years were highly variable: plants were stunted, leaf shapes were altered, flowers were smaller and irregularly shaped, and lateral shoot growth was excessive as compared to plants regenerated from leafy callus maintained in culture for 1 month (Sutter and Langhans, 1981). Though chimeral rearrangements are a possibility, residual hormone effects and genetic instability cannot be excluded as explanations.

McCoy (1980) determined that the frequency of cytogenetically abnormal plants increased with culture age for the oat cultivars Lodi and Tippecanoe. Abnormal plants from young cultures usually possessed a single alteration; abnormal plants from cultures up to 20 months old frequently had more than one chromosome alteration. By contrast, long-term cultures of corn were stable, 94% of the plants analyzed after 8 months were normal.

PROTOPLAST CULTURE CYCLE. The first plants regenerated from protoplast cultures were altered in morphology, chromosome number, and fertility (Takebe et al., 1971). In other reports of regeneration from protoplasts, however, phenotypic stability is often observed. Plants obtained from protoplasts of four *Nicotiana* L. species were unaltered (Evans, 1979). Vegetative and floral morphology and chromosome number of all regenerated plants were normal. However, these observations were based on a total of only 20 plants.

Wenzel et al. (1979) also observed remarkable uniformity among potato plants derived from protoplast cultures. Some plants varied in tuber shape, but only after extended culture periods. The slight amounts of observed variation could be ascribed to culture-induced aneuploidy. Such results are difficult to reconcile with those of Shepard et al.

(1980) who, using the cultivar Russet Burbank, screened 10,000 clones
derived from protoplasts and found great amounts of horticulturally significant variation.   Variation in tuber shape, yield, plant morphology,
and resistance to early and late blight was identified in some of these
clones.   The genetic basis of variation was not determined.   Of these
clones, 65 were analyzed in detail for 35 characteristics, including leaf,
tuber, flower, and whole plant traits (Secor and Shepard, 1981).   The
65 clones in this study were selected from an original population of
1700.   Grossly aberrant types were eliminated; the selected clones
were relatively normal in appearance and possessed acceptable vigor,
vine type, and tuber conformation.   Each clone differed from the parent
Russet Burbank in at least one trait, with three clones varying in only
one trait and one clone in 17 traits.   Further testing of potato clones
derived from protoplasts would be necessary to determine variant trait
stability and the effect of the environment on such variability.   An
excellent, detailed account of the Russet Burbank protoplast culture
technique can be found in the article by Shepard (1980).
    Several possible explanations could be offered to reconcile the dis-
crepancies between the works of Shepard and Wenzel.   Wenzel et al.
(1979) worked with potato lines that were completely unrelated to Russet Burbank.   That different genotypes produce different amounts and
types of variation from culture has already been verified (see below).
In addition, Wenzel screened relatively few (211) clones, while Shepard
screened over 10,000 clones.   Other differences between their techniques include (1) source of protoplasts (Shepard—leaf mesophyll; Wenzel—stem tip culture) and (2) ploidy level (Shepard—4x; Wenzel—2x).
Interestingly, Thomas (1981) reported that among plants regenerated
from stem tip culture-derived protoplasts of the tetraploid British potato cultivar Maris Bard, large amounts of variation were observed.
Changes in growth habit, leaf morphology, color, degree of gloss, and
hairiness were noted.   Only 2 of 25 groups of plants derived from different protoplasts resembled the parental cultivar Maris Bard.   So it
appears that variation can be detected among plants from either shoot
tip culture protoplasts or leaf mesophyll protoplasts, but somaclonal
variation among dihaploid potato clones has not yet been observed.

CALLUS CULTURE CYCLE.   The most widely employed technique
for creating genetic variation via tissue culture has been the use of a
callus culture cycle.   That plants regenerated from callus cultures can
be highly variable has been well documented by Barbier and Dulieu
(1980), Sibi (1976), Edallo et al. (1981), Oono (1978), and others (Table
1).   Sibi (1976) found that lettuce plants regenerated from callus cultures displayed altered morphology, and their selfed progeny showed
altered leaf characteristics, axillary bud development, and leaf color.
Certain leaf traits were clearly the result of an altered cytoplasm,
since inheritance was maternal.   A diallel cross between three of the
variants and the control plant indicated that outcrosses showed more
vigor than selfed progeny, even though the starting material was homozygous.   Other changes that never segregated in three generations of
selfing were also assumed to be maternal.   But the possibility of chro-

mosome substitution, which would also cause such true breeding mutants, cannot be excluded (Peloquin, 1981).

By contrast, Edallo et al. (1981) regenerated plants from corn tissue cultures derived from immature embryos of two inbred lines and found changes only in nuclear DNA. Most variants were due to visible gene mutations, as demonstrated by inheritance studies. As earlier reported by McCoy (1980), corn appears to be cytogenetically stable; Edallo et al. (1981) found that only 2 of 110 regenerated plants differed from diploidy, but intrachromosomal changes were not examined. On the average, there appeared to be approximately one single gene mutation per plant.

Barbier and Dulieu (1980) used nuclear genes to trace the pattern of plant variation that arose from tobacco plants regenerated directly from (1) cotyledons, (2) cotyledon-derived protoplasts, and (3) callus cultures induced to form plants at varying time intervals. Deletions and reversions could account for the high frequencies of leaf color variants observed. The authors conclude, by comparison of the variation obtained from the three different tissue culture treatments that "regeneration allows expression of potential variability accumulated in cotyledon resting cells after the last cell cycle; variability also accumulates in cultured cells." (For a detailed review of this work, see Larkin and Scowcroft, 1981).

Among 800 somaclones derived from rice (*Oryza sativa* L.) callus only 28.1% were considered to have normal inheritance patterns for all characters observed (Oono, 1978, 1981). Altered traits were analyzed through two generations of selfing. Plants were altered in chlorophyll content, flowering date, plant height, fertility, and morphology. Variation was believed to be due to mutations occurring in cultured cells. Hence there is a belief that plants from callus can be variable because of either preexisting whole plant variation or variation that accumulates during the growth of undifferentiated cells.

CULTURE OF A SPECIFIC PLANT TISSUE. Under identical conditions the explant used can greatly affect the quantity and type of variation produced. Pineapple (*Ananas cosmosus* (L.) Merr.) plants which were regenerated from callus of four types of explants (syncarp, slip, crown, and axillary buds) were altered according to the explant source (Wakasa, 1979). Only spine characters were altered in plants from the slip, crown, and axillary buds, but variants from syncarp were altered in leaf color, spine, wax, and foliage density. By choice of the proper organs, it was concluded that tissue culture could be useful for rapid propagation of nonvariant plantlets or for the induction of useful variants. The work of Van Harten et al. (1981) showed that careful explant choices could be used to facilitate the increased production of somaclonal variants. Phenotype alterations were found in 12.3% of plants from leaflet discs, while 50.3% of plants from rachis- and petiole-derived callus were altered.

VARIATION CONCOMITANT WITH BIOCHEMICAL CELLULAR SELECTION. Somatic cell selection schemes have been discussed at length as a method of selecting variants of value at the cellular level (Maliga,

1978, 1980). Such a system has been proposed to take advantage of microbial-type cellular manipulations, since large numbers of plant cells can be grown quite easily in a defined medium. Initial reports were limited by the inability of selected cells to regenerate into plants. More recently, plants have been regenerated and examined, and their progeny have also been checked for the inheritance of the selected trait (e.g., Berlyn, 1980; Bourgin, 1978; Chaleff and Parsons, 1978; Chaleff, 1980). Several plants from variant cultures were reportedly altered for traits other than those being sought (e.g., Chaleff and Keil, 1981; Maliga et al., 1979), suggesting either pleiotropic effects from simple alterations or multiple alterations within the selected variant cells.

This type of somaclonal variation is perhaps different in that cultures are usually mutagenized prior to the application of cellular selection systems. It cannot usually be determined whether variation is induced by mutagenesis in tissue cultures or randomly via processes more in line with other reports on somaclonal variation. Berlyn (1980) selected variant lines resistant to isonicotinic acid hydrazide (INH) in irradiated cell cultures of haploid *Nicotiana tabacum*. Plants regenerated from these cultures were altered in vigor, root development, leaf shape, and growth habit. The slow growth trait associated with INH resistance could be separated by backcrossing hybrids to the parental stock, indicating that these traits were genetically independent.

Plants that were regenerated from diploid alfalfa (*Medicago sativa* L.) cultures selected for growth on ethionine-containing medium were altered when compared directly to the original plant (Reisch and Bingham, 1981). Plants from 23 of 91 cell lines were altered in chromosome number, leaf shape, petiolule length, herbage yield, plant height, shoot length, and ease of rooting. Alterations in some of these characteristics could be of importance in alfalfa breeding. Hence it was suggested that agronomically desirable clones might be improved via a tissue culture cycle.

Most of the variants examined in this study were not resistant to ethionine toxicity, despite selection for growth in its presence. This suggests that both resistant and nonresistant cells were carried in many cell lines. It also indicates that the wide variation observed is completely independent of ethionine resistance. Also, one tetraploid plant in this study, NS1, was regenerated from unmutagenized cells plated in noninhibitory medium. This plant yielded significantly more dry matter of herbage than the control tetraploid line and showed differences for other traits as well. A low frequency of variants is usually observed among plants regenerated from callus cultures of alfalfa (E.T. Bingham, personal communication). So it appears that distinct variants can arise from alfalfa tissue cultures, but at a much higher frequency when associated with selection for growth on ethionine.

The idea that plants regenerated in these experiments were altered because of exposure to ethionine or both EMS and ethionine was suggested by Reisch and Bingham (1981). Few plants (3 out of 61) regenerated from a control that was EMS-mutagenized but not plated in ethionine were altered. Ethionine may indeed be mutagenic to eukaryotic cells and enhance the mutation frequency when used in combination with known mutagens (Talmud and Lewis, 1974; Lewis and Tarrant, 1971; Davies and Parry, 1978). Ethionine treatments of soybean seed and corn caryopses indicated that, even alone, ethionine can be mutagenic (Fujii, 1981). Further tests are warranted to determine whether useful variation can be consistently induced via ethionine treatments of cultured cells with or without mutagenic treatments.

GROWTH MEDIUM AND HORMONE EFFECTS. The effects of media and hormones, particularly 2,4-D, on cell cultures are well described by Bayliss (1980). But few reports detailing the effects of culture medium components on regenerated plants can be found. Plants were regenerated from tissue cultures of barley exposed to 4.5-18.0 µM 2,4-D for 4-56 days (Deambrogio and Dale, 1980). Genetic variation was detected by examining the progeny of regenerated plants. Variation occurred only at 18.0 µM 2,4-D for such traits as albinism, leaf shape, and tiller fertility.

In an extensive characterization of plants from *Haworthia* callus cultures, cytogenetic changes were examined in detail (Ogihara, 1981). Two media that differed only in hormone content were used to regenerate plants: the NK medium (26.8 µM NAA and 0.46 µM KIN) and the I medium (0.57 µM IAA). The NK medium tended to regenerate more tetraploids and fewer plants carrying translocations than the I medium.

GENOTYPE RESPONSE. Under carefully controlled procedures for culture manipulation and regeneration, variability among regenerated plants can be genotype-dependent. Liu and Chen (1976) tested 8 sugarcane cultivars and found that among a total of 4600 plants examined, many of the plants differed from their donors; for clone F146 the frequency of morphological changes was 1.8%, while for F156 the frequency was 34.0%.

Skirvin and Janick (1976a) detailed the changes among plants regenerated from callus cultures of *Pelargonium* sp. cultivars. The type of variation observed was dependent upon the original cultivar used. They proposed that such variability might be useful for intraclonal plant improvement especially with polyploid, asexually propagated clones.

The Role of Somaclonal Manipulations in Plant Breeding

Plant breeders have long recognized the value of selectively improving a popular cultivar rather than creating a new one. Growers are more willing to plant older cultivars because of proven value and familiar growth characteristics under local conditions. New cultivars are unproven, and their consumer and marketing acceptance is unknown, though the possibility of improved performance is attractive. There are basically five techniques which can be used to enhance the quality of popular cultivars while maintaining their basic identity: (1) backcross breeding (Allard, 1960), (2) clonal selection or sport selection (Van

Oosten and Van der Borg, 1977), (3) mutation breeding (IAEA, 1977), (4) production of variants from tissue cultures (Larkin and Scowcroft, 1981; Maliga, 1978, 1980), and (5) single gene transformation (Kado and Kleinhofs, 1980).

Improvement of homozygous, autogamous crops might be accomplished by any of the above methods. However, where simply inherited traits are desired in an acceptable cultivar deficient in only one or two traits, backcross breeding is usally the method of choice. In heterozygous, vegetatively propagated crops, backcross breeding fails to recreate the elite combinations of genes and also fails to maintain the heterozygous nature of the recurrent parent if a single clone is to be improved. To selectively improve such crops, other methods must be used.

Such other methods are often limited in their application. Single gene transformation in higher plants has not yet been sufficiently developed to be a useful tool. The selection of occasional sports requires large-scale plantings, keen observation, and the chance occurrence of a rare event. Pest-resistant sports cannot be spotted in pesticide-sprayed plantings. Mutation breeding is an alternative in such cases, along with the newer techniques of somaclonal manipulations. Mutation breeding techniques are compared with somaclonal manipulation and backcross breeding methods in Table 2.

Somaclonal Variation as a Form of Mutation Breeding

Mutation breeding and plant improvement via somaclonal variation may actually be equivalent techniques. Somaclonal variation may be exploiting genetic changes that preexist in the whole plant or changes that occur in cultured cells (Barbier and Dulieu, 1980). Where changes occur during a tissue culture cycle, somaclonal variations can be considered a logical and useful extension of the mutation breeding process which broadly encompasses any process by which genetic variability is induced.

Research on mutation breeding in potato (*Solanum tuberosum* L.) led to the suggestion that in vitro techniques might replace mutagenic treatments as a method for inducing variation. An in vitro adventitious bud technique for mutation breeding of potato was examined for efficiency of mutation induction and chimerism (Van Harten et al., 1981). Chimerism can be a formidable problem when vegetative tissues are mutated. Plants derived from adventitious buds are not usually chimeric, but plants obtained in this study following X-irradiation from in vitro-produced adventitious buds exhibited a very high mutation frequency, a wide mutation spectrum, and a very low rate of chimerism. Unexpectedly, nonirradiated controls derived from rachis and petiole explants (but not from leaf explants) had a mutation frequency of 50.3% which was almost as high as the irradiated series. In previous experiments with the same cultivar, the induced mutation frequency was only as high as 24-38% (Van Harten and Bouter, 1973). Therefore, it was suggested that "one could even think of mutation breeding of potato without irradiation, thus avoiding the undesired side-effects of the

Table 2. Comparison among Three Methods Designed to Produce Single Changes in Popular Cultivars or Breeding Stock: Backcross Breeding, Mutation Breeding, and Somaclonal Manipulations

	BACKCROSS BREEDING	MUTATION BREEDING	SOMACLONAL MANIPULATIONS
Source of variation	Natural population	Induced	Spontaneous and induced
Safety hazard	None	Slight (due to use of mutagens)	None
Likelihood of success	Guaranteed except where undesirable linkages are not easily broken	None (variation cannot be directed)	None (variation cannot be directed)
Alteration of quantitative traits	Rarely successful	Possible	Possible
Rate of progress	One trait in 5-7 sexual generations	One to several traits improved in each cycle	One to several traits improved in each cycle
Chimerism	None	A major drawback	Low frequency
Diplontic selection	None	A major drawback	Frequency limited
Species limitations (1)	Useful only with sexually reproducing species, especially autogamous species	Useful in sexually- and asexually reproducing species	Limited to species and genotypes that can regenerate plants from tissue cultures
(2)	Not useful for the improvement of specific asexually propagated cultivars	Useful especially with asexually propagated cultivars	Useful especially with asexually propagated cultivars
Evaluation	Less extensive than for advanced lines with more completely altered genotypes	Same as backcross breeding	Same as backcross breeding

mutagenic treatment" (Van Harten et al., 1981). Apparently, these researchers were unaware of the literature describing wide variation
among plants regenerated from potato protoplasts and suspension cultures (Shepard et al., 1980; Thomas, 1981; Behnke, 1980) and arrived
independently at the idea of utilizing the variation occurring among
plants regenerated from tissue culture for the purpose of crop improvement.

Current Applications

As a result of research on somaclonal variants a number of valuable
breeding lines have been developed. High-yielding and smut-resistant
lines of sugarcane developed from callus cultures are currently under
test in Taiwan (Liu, 1981). Work on other somaclonal variants of
sugarcane has led to the development of a potentially important clone,
identical to its parent in most respects, but with the addition of being
resistant to downy mildew and Fiji disease, a virus of widespread importance (Heinz et al., 1977). While several Fiji virus-resistant clones
were developed and determined to be free from obvious deleterious
abnormalities, in replicated trials only one was as good as Pindar, the
original clone, in other respects such as cane yield and percentage of
sucrose. Replicated trials are a necessity to verify the qualities of
advanced selections. As with any other breeding technique where new
characteristics are incorporated into popular cultivars, extensive testing
is still necessary to determine a genotype's response over time and
space. Techniques for producing somaclonal variation may only be advantageous in producing novel variation within a specified genetic
background, but then the traditional breeder's approach to assessing the
relative performance of cultivars and selections must be used for a
final determination.

Skirvin and Janick (1976b), following extensive investigations of somaclonal variation among *Pelargonium* sp., developed an improved scented
geranium that they named Velvet Rose. This is believed to be the
first named cultivar developed by any type of tissue culture technique.

That somaclonal variation might be useful in enhancing the exchange
of genetic material required in wide sexual hybrids for the introgression
of desirable alien genes was discussed by Larkin and Scowcroft (1981).
Plants from hybrid *Lolium* and *Hordeum* embryo cultures were highly
variable and were agronomically more valuable than those sexual hybrids that had not been through a tissue culture cycle (Ahloowalia,
1978; Orton, 1980). Frequently, attempts to introgress alien genes into
crop plants are frustrated by a lack of exchange between crop and
alien genomes in the hybrids. Results with hybrid embryo cultures suggest that genetic factors that limit the exchange in germ line cells
appear to break down in cell culture; thus a callus phase of the hybrid
embryo may overcome barriers to genetic exchange.

POTENTIAL AND FUTURE PROSPECTS

Genetic variation via somaclonal techniques should become more
widely available for cultivar improvement in the near future. Already,

one cultivar has been developed (Skirvin and Janick, 1976b) and many species have been identified in which somaclonal variants have been observed. Until improvements in this technique are made, it is at best only creating random genetic variation that differs widely between genotypes and species. Directed selection systems (Maliga, 1978, 1980; Thomas et al., 1979) would seem preferable but are limited in application, since they depend on the correlation between a cellular selection system and the expression of a whole plant trait often within a specific tissue. Without a doubt, the development of novel selection techniques that would facilitate rapid cellular selection for a wide variety of traits would seem to be of paramount importance. The potential applications for both directed cellular selection systems and somaclonal manipulations can only be realized when regeneration from cells can be easily accomplished in a wide range of genotypes and species.

More knowledge will be required to be able to increase variation or direct variation toward a desired goal. Research in medium, hormone, and physical environment effects might prove to be useful in defining optimum conditions for enhancing somaclonal variation. Possibly, certain classes of genetic changes might be more frequent than others following subtle alterations in the culture medium (Ogihara, 1981) or other variables. Perhaps, as seen with work on protoplast culture (Shepard, 1980), the successful production of variation might be dependent upon the environment of the plant used to initiate tissue cultures. Further work in this area is sorely needed. Such information will be necessary if somaclonal manipulations are to make a lasting contribution to crop improvement.

It will be more difficult, but of great importance, to determine the genetic basis for variation among somaclones. Peloquin (1981) suggests that chromosome substitutions may be responsible for the changes observed in plants regenerated from tissue cultures. Cytogenetic changes have been observed in plants regenerated from tissue cultures (McCoy and Phillips, 1982; Ogihara, 1981) and are no doubt responsible for some of the observed changes. Other reports verify that single gene changes occur (Edallo et al., 1981; Barbier and Dulieu, 1980) while still others implicate cytoplasmic alterations (Sibi, 1976; Brettell et al., 1980). Virtually any type of genetic variation might contribute to the basis for somaclonal variation and should be considered in any research examining such variation. Larkin and Scowcroft (1981) divided the possible origins of variation into seven categories: (1) karyotypic changes, (2) cryptic chromosome rearrangements, (3) transposable elements, (4) somatic gene rearrangements, (5) gene amplification and depletion, (6) somatic crossing over and sister chromatid exchange, and (7) cryptic virus elimination. In addition to these possibilities, where sexual transmittance of somaclonal variation has not been identified, the possibility of epigenetic changes cannot be excluded (Meins and Binns, 1977). Reports concerning the basis for somaclonal variation are summarized in Table 3.

It will be important to determine the types of plants that will most readily produce variants following a tissue culture cycle. It appears that both diploids and polyploids can be altered easily (Table 1). Variation occurs at a higher frequency among somatically unstable geno-

Table 3.   Possible Sources of Variation among Plants Regenerated from
            Tissue Cultures

SOURCE OF VARIATION	REFERENECE
Genetic	
Point mutation	Edallo et al., 1981; Oono, 1978
Cytoplasmic	Brettell et al., 1980; Sibi, 1976
Transposable element[1]	
Gene amplification[1]	
Somatic crossing over[1]	
Cytogenetic	
Aneuploidy	McCoy, 1980; Sacristan & Melchers, 1969
Polyploidy	Murashige & Nakano, 1966; Janick et al., 1977
Translocation	McCoy, 1980; Orton, 1980
Deletion	Barbier & Dulieu, 1980
Inversion	McCoy, 1980
Duplication[1]	
Chromosome substitution[1]	
Nongenetic	
Virus elimination[1]	
Epigenetic change	Ibrahim, 1969

[1] No conclusive documentation among reports of somaclonal variants.

types than among stable genotypes (Heinz and Mee, 1971). It will also
be of considerable interest to determine whether old asexually propagated cultivars in which cellular variation presumably accumulates will
produce a greater frequency of variation than recently bred cultivars.

It should be noted that while enthusiasm is justified, the production
of somaclonal variants should be considered merely a useful adjunct to
a sexual breeding program. More immediate use will likely be found
with asexually propagated crops where the variation obtained can be
used directly. In programs where the desired variation does not exist
or exists only in relatively undomesticated germplasm, somaclonal variation should be of value. With the current state of the art, somaclonal
variants are still just a shot in the dark until general methods can be
developed to enhance selectively the frequency of a desired form of
plant variation. The genetic variation obtained will be of no use unless handled skillfully in accordance with standard plant breeding practices for either asexually propagated or seed-propagated crops.

KEY REFERENCES

Larkin, P.J. and Scowcroft, W.R. 1981. Somaclonal variation—A novel
    source of variability from cell cultures for plant improvement.
    Theor. Appl. Genet. 60:197-214.

Shepard, J.F. 1980. Mutant selection and plant regeneration from pota-
    to mesophyll protoplasts. In: Genetic Improvement of Crops: Emer-

gent Techniques (I. Rubenstein, B. Gengenbach, R.L. Phillips, and C.E. Green, eds.) pp. 185-215. Univ. Minn. Press, Minnesota.

_____, Bidney, D., and Shahin, E. 1980. Potato protoplasts in crop improvement. Science 28:17-24.

Skirvin, R.M. 1978. Natural and induced variations in tissue cultures. Euphytica 27:241-266.

REFERENCES

Ahloowalia, B.S. 1978. Novel ryegrass genotypes regenerated from embryo-callus culture. In: Frontiers of Plant Tissue Culture (T.A. Thorpe, ed.) Abstr. 1723, p. 162. Univ. Calgary Press, Calgary.

Allard, R.W. 1960. Principles of Plant Breeding. John Wiley and Sons, New York.

Bajaj, Y.P.S., Sidhu, B.S., and Dubey, V.R. 1981. Regeneration of genetically diverse plants from tissue cultures of forage grass—*Panicum* spp. Euphytica 30:135-140.

Barbier, M. and Dulieu, H.L. 1980. Genetic changes observed on tobacco (*Nicotiana tabacum*) plants regenerated from cotyledons by in vitro culture. Ann. Amelior. Plant. 30:321-344.

Bayliss, M.W. 1980. Chromosomal variation in plant tissue culture. In: International Review of Cytology, Suppl. 11A: Perspectives in Plant Cell and Tissue Culture (I.K. Vasil, ed.) pp. 113-144. Academic Press, New York.

Beach, K.H. and Smith, R.R. 1979. Plant regeneration from callus of red clover and crimson clover. Plant Sci. Lett. 16:231-238.

Behnke, M. 1980. General resistance to late blight of *Solanum tuberosum* plants regenerated from callus resistant to culture filtrates of *Phytophthora infestans*. Theor. Appl. Genet. 56:151-152.

Ben-Jaacov, J. and Langhans, R.W. 1972. Rapid multiplication of *Chrysanthemum* plants by stem tip proliferation. HortScience 7:289-290.

Bennici, A. and D'Amato, F. 1978. In vitro regeneration of durum wheat plants. I. Chromosome numbers of regenerated plants. Z. Pflanzenzuecht. 81:305-311.

Berlyn, M.B. 1980. Isolation and characterization of isonicotinic acid hydrazide resistant mutants of *Nicotiana tabacum*. Theor. Appl. Genet. 58:19-26.

Bourgin, J.P. 1978. Valine-resistant plants from in vitro selected tobacco cells. Molec. Gen. Genet. 161:225-230.

Brettell, R.I.S., Thomas, R., and Ingram, D.S. 1980. Reversion of Texas male-sterile cytoplasm maize in culture to give fertile T-toxin resistant plants. Theor. Appl. Genet. 58:55-58.

Bush, S., Earle, E., and Langhans, R.W. 1976. Plantlets from petal segments, petal epidermis and shoot tips of the periclinal chimera, *Chrysanthemum morifolium* Indianapolis. Am. J. Bot. 63:729-737.

Butenko, R.G., Shemina, Z.B., and Frolova, L.V. 1967. Induced organogenesis and characteristics of plants produced in tobacco tissue culture. Genetika 3:29-39.

Chaleff, R.S. 1980. Further characterization of picloram-tolerant mutants of *Nicotiana tabacum*. Theor. Appl. Genet. 58:91-95.

_____ and Keil, R.L. 1981. Genetic and physiological variability among cultured cells and regenerated plants of *Nicotiana tabacum*. Molec. Gen. Genet. 181:254-258.

_____ and Parsons, M.F. 1978. Direct selection in vitro for herbicide-resistant mutants of *Nicotiana tabacum*. Proc. Natl. Acad. Sci. USA 75:5104-5107.

Collins, G.B. and Legg, P.D. 1980. Recent advances in the genetic application of haploidy in *Nicotiana*. In: The Plant Genome (D.R. Davies and D.A. Hopwood, eds.) pp. 197-213. The John Innes Charity, Norwich.

D'Amato, F. 1975. The problem of genetic stability in plant tissue and cell cultures. In: Crop Genetic Resources for Today and Tomorrow (O.H. Frankel and J.G. Hawkes, eds.) pp. 349-358. Cambridge Univ. Press, Cambridge.

_____ 1977. Cytogenetics of differentiation in tissue and cell cultures. In: Applied and Fundamental Aspects of Plant Cell, Tissue and Organ Culture (J. Reinert and Y.P.S. Bajaj, eds.) pp. 343-357. Springer-Verlag, Berlin.

Davies, P.J. and Parry, J.M. 1978. The modification of induced genetic changes in yeast by an amino acid analogue. Molec. Gen. Genet. 162:183-190.

Deambrogio, E. and Dale, P.I. 1980. Effect of 2,4-D on the frequency of regenerated plants in barley (*Hordeum vulgare*) cultivar 'Akka' and on genetic variability between them. Cereal Res. Commun. 8:417-424.

Denton, I.R., Westcott, R.J., and Ford-Lloyd, B.V. 1977. Phenotypic variation of *Solanum tuberosum* L. cv. Dr McIntosh regenerated directly from shoot-tip culture. Potato Res. 20:131-136.

Dunwell, J.M. 1981. In vitro regeneration from excised leaf-disks of 3 *Brassica* species. J. Exp. Bot. 32:789-799.

Edallo, S., Zucchinali, C., Perenzin, M., and Salamini, F. 1981. Chromosomal variation and frequency of spontaneous mutation associated with in vitro culture and plant regeneration in maize. Maydica 26:39-56.

Evans, D.A. 1979. Chromosome stability of plants regenerated from mesophyll protoplasts of *Nicotiana* species. Z. Pflanzenphysiol. 95:459-463.

Fujii, T. 1981. Mutagenic effect of L-ethionine in soybean and maize. Environ. Exp. Bot. 21:127-131.

Gamborg, O.L., Shyluk, J.P., Brar, D.S., and Constabel, F. 1977. Morphogenesis and plant regeneration from callus of immature embryos of sorghum. Plant Sci. Lett. 10:67-74.

Green, C.E. 1977. Prospects for crop improvement in the field of cell culture. HortScience 12:131-134.

Heinz, D.J. and Mee, G.W.P. 1971. Morphologic, cytogenetic and enzymatic variation in *Saccharum* species hybrid clones derived from callus tissue. Am. J. Bot. 58:257-262.

Heinz, D.J., Krishnamurthi, M., Nickell, L.G., and Maretzki, A. 1977. Cell, tissue and organ culture in sugar cane improvement. In: Applied and Fundamental Aspects of Plant Cell, Tissue and Organ Culture (J. Reinert and Y.P.S. Bajaj, eds.) pp. 3-17.   Springer-Verlag, Berlin.

Ibrahim, R.K. 1969.   Normal and abnormal plants from carrot root tissue cultures.   Can. J. Bot. 47:825-826.

International Atomic Energy Agency. 1977.   Manual on Mutation Breeding, 2nd Ed.   Report Series No. 119.   IAEA, Vienna.

Janick, J., Skirvin, R.M., and Janders, R.B. 1977.   Comparison of in vitro and in vivo tissue culture systems in scented geranium.   J. Hered. 68:62-64.

Johnson, L.B., Stuteville, D.L., and Skinner, D.Z. 1980.   Phenotypic variants in progeny of selfed alfalfa regenerated from calli.   Plant Physiol. (Suppl.) 65:36.

Kado, C.I. and Kleinhofs, A. 1980.   Genetic modification of plant cells through uptake of foreign DNA.   In: International Review of Cytology, Suppl. 11B: Perspectives in Plant Cell and Tissue Culture (I.K. Vasil, ed.) pp. 47-80.   Academic Press, New York.

Lewis, C.M. and Tarrant, G.M. 1971.   Induction of mutation by 5-fluorouracil and amino acid analogues in Ustilago maydis.   Mutat. Res. 12:349-356.

Liu, M.C. 1981.   In vitro methods applied to sugarcane improvement. In: Plant Tissue Culture.   Methods and Applications in Agriculture (T.A. Thorpe, ed.) pp. 299-323.   Academic Press, New York.

_____ and Chen, W.H. 1976.   Tissue and cell culture as aids to sugar cane breeding.   I. Creation of genetic variation through callus culture.   Euphytica 25:393-403.

_____ and Chen, W.H. 1978.   Tissue and cell culture as aids to sugar cane breeding.   II. Performance and yield potential of callus-derived lines.   Euphytica 27:273-282.

Maliga, P. 1978.   Resistance mutants and their use in genetic manipulations.   In: Frontiers of Plant Tissue Culture 1978 (T.A. Thorpe, ed.) pp. 381-392.   Univ. Calgary Press, Calgary.

_____ 1980.   Isolation, characterization and utilization of mutant cell lines in higher plants.   In: International Review of Cytology, Suppl. 11a: Perspectives in Plant Cell and Tissue Culture (I.K. Vasil, ed.) pp. 225-250.   Academic Press, New York.

_____, Zsuzsa, R.-K., Dix, P.J., and Lazar, G. 1979.   A streptomycin resistant line of Nicotiana sylvestris unable to flower.   Molec. Gen. Genet. 172:13-15.

Matern, U., Strobel, G., and Shepard, J.F. 1978.   Reactions to phytotoxins in a potato population derived from mesophyll protoplasts. Proc. Natl. Acad. Sci. USA 75:4935-4939.

McCoy, T.J. 1980.   Cytogenetic stability of tissue cultures and regenerated plants of oats (Avena sativa L.) and corn (Zea mays L.). Ph.D. Thesis, Univ. Minnesota, St. Paul.

_____ and Phillips, R.L. 1982.   Cytogenetic variation in tissue culture regenerated plants of Avena sativa: High frequency of chromosome loss.   Can. J. Genet. Cytol. 24:37-50.

Meins, F. and Binns, A. 1977.   Epigenetic variation of cultured somatic cells: Evidence for gradual changes in the requirement for factors promoting cell division.   Proc. Nat. Acad. Sci. USA 74:2928-2932.

Morel, G. 1971. The impact of plant tissue culture in plant breeding.
    In: The Way Ahead in Plant Breeding (F.G.H. Lupton, G. Jenkins, and
    R. Johnson, eds.) pp. 185-194. Sixth Congress, Eucarpia, Cambridge.
_____ 1975. Meristem culture techniques for the long term storage of
    cultivated plants. In: Crop Genetic Resources for Today and Tomor-
    row (O.H. Frankel and J.G. Hawkes, eds.) pp. 327-332. Cambridge
    Univ. Press, Cambridge.
Mousseau, J. 1970. Fluctuation induite par la neoformation de bour-
    geons in vitro. Colloq. Int. CNRS 193:234-239.
Murashige, T. 1974. Plant propagation through tissue culture. Annu.
    Rev. Plant Physiol. 25:135-166.
_____ and Nakano, R. 1966. Tissue culture as a potential tool in ob-
    taining polyploid plants. J. Hered. 57:115-118.
Neilson-Jones, W. 1969. Plant Chimeras, 2nd Ed. Methuen, London.
Novak, F.J. 1980. Phenotype and cytological status of plants regener-
    ated from callus cultures of Allium sativum L. Z. Pflanzenzuecht.
    84:250-260.
Ogihara, Y. 1981. Tissue culture in Haworthia. IV. Genetic charac-
    terization of plants regenerated from callus. Theor. Appl. Genet. 60:
    353-363.
Ogura, H. 1976. The cytological chimeras in original regenerates from
    tobacco tissue cultures and in their offsprings. Jpn. J. Genet. 51:
    161-174.
Oono, K. 1978. Test tube breeding of rice by tissue culture. Proceed-
    ings Symposium Methods in Crop Breeding. Tropical Agriculture Re-
    search Series No. 11. Ministry of Agriculture and Forestry, Ibaraki,
    Japan.
_____ 1981. In vitro methods applied to rice. In: Plant Tissue Cul-
    ture. Methods and Applications in Agriculture (T.A. Thorpe, ed.) pp.
    273-298. Academic Press, New York.
Orton, T.J. 1980. Chromosomal variability in tissue cultures and re-
    generated plants of Hordeum. Theor. Appl. Genet. 56:101-112.
Peloquin, S.J. 1981. Chromosomal and cytoplasmic manipulations. In:
    Plant Breeding II (K.J. Frey, ed.) pp. 117-150. Iowa State Univ.
    Press, Ames.
Reisch, B. and Bingham, E.T. 1981. Plants from ethionine-resistant al-
    falfa tissue cultures: Variations in growth and morphological charac-
    teristics. Crop Sci. 21:783-788.
Sacristan, M.D. 1981. Regeneration of plants from long-term callus cul-
    tures of haploid Brassica napus. Z. Pflanzenzuecht. 86:248-253.
_____ and Melchers, G. 1969. The caryological analysis of plants re-
    generated from tumorous and other callus cultures of tobacco. Molec.
    Gen. Genet. 105:317-333.
Secor, G.A. and Shepard, J.F. 1981. Variability of protoplast-derived
    potato clones. Crop Sci. 21:102-105.
Sibi, M. 1976. Genetic program in higher plants. Part 2: Experimen-
    tal aspects, production of variants by in vitro tissue culture of Lac-
    tuca sativa. Increase in vigor in outcrosses. Ann. Amelior. Plant.
    26:523-547.
Skirvin, R.M. and Janick, J. 1976a. Tissue culture-induced variation in
    scented Pelargonium spp. J. Am. Soc. Hort. Sci. 101:281-290.

_____ 1976b. 'Velvet Rose' *Pelargonium*, a scented geranium. Hort-Science 11:61-62.

Stavarek, S.J., Croughnan, T.P., and Rains, D.W. 1980. Regeneration of alfalfa plants from long term cultures of alfalfa cells. Plant Sci. Lett. 19:253-261.

Stimart, D.P., Ascher, P.D., and Zagorski, J.S. 1980. Plants from callus of the interspecific hybrid *Lilium* 'Black Beauty.' HortScience 15: 313-315.

Sunderland, N. 1977. Nuclear cytology. In: Plant Tissue and Cell Culture (H.E. Street, ed.) pp. 177-205. Univ. California Press, Berkeley.

Sutter, E. and Langhans, R.W. 1981. Abnormalities in *Chrysanthemum* regenerated from long term cultures. Ann. Bot. 48:559-568.

Syono, K. and Furuya, T. 1972. Abnormal flower formation of tobacco plants regenerated from callus cultures. Bot. Mag. Tokyo 85:273-284.

Takebe, I., Labib, G., and Melchers, G. 1971. Regeneration of whole plants from isolated mesophyll protoplasts of tobacco. Naturwissenschaften 58:318-320.

Talmud, P.J. and Lewis, D. 1974. The mutagenicity of amino acid analogues in *Coprinus lagopus*. Genet. Res. 23:47-61.

Thomas, E. 1981. Plant regeneration from shoot culture-derived protoplasts of tetraploid potato (*Solanum tuberosum* cv. 'Maris Bard'). Plant Sci. Lett. 23:81-88.

_____, King, P.J., and Potrykus, I. 1979. Improvement of crop plants via single cells in vitro—An assessment. Z. Pflanzenzuecht. 82:1-30.

Van Harten, A.M. and Bouter, H. 1973. Dihaploid potatoes in mutation breeding: Some preliminary results. Euphytica 22:1-7.

_____, and Broertjes, C. 1981. In vitro adventitious bud techniques for vegetative propagation and mutation breeding of potato (*Solanum tuberosum* L.). II. Significance for mutation breeding. Euphytica 30: 1-8.

Van Oosten, H.J. and Van der Borg, J.H. (eds.). 1977. Symposium on Clonal Variation in Apple and Pear. Acta Hortic. 75.

Wakasa, K. 1979. Variation in the plants differentiated from tissue culture of pineapple. Jpn. J. Breed. 29:13-22.

Wenzel, G., Schieder, O., Przewozny, T., Sopory, S.K., and Melchers, G. 1979. A comparison of single cell derived *Solanum tuberosum* plants and a model for their application in breeding programs. Theor. Appl. Genet. 55:49-55.

CHAPTER 26

Herbicide Tolerance in Regenerated Plants

O.J. Crocomo and *N. Ochoa-Alejo*

Classical plant breeding programs require a large amount of space and are time consuming, because of the long biological cycles of most plant species (Devine et al., 1975). Moreover, the conditions necessary for selection pressure in the field are often not easy to establish and maintain for a long period of time (Gressel and Segel, 1978). The recent advances in the isolation and culture of cell, tissue, and protoplasts allow for rapid selection of natural mutants that exist in a certain cell population or alternatively to increase the frequency and select mutations through the use of mutagens. This new technology, indeed, permits the manipulation of very large cell populations in small spaces and in a shorter period of time.

The cells can be cultivated in chemically defined media and the mutants selected using physical or chemical factors. Toxic substances added to the medium have been used to select resistant cell lines (Widholm, 1974; Meredith, 1978; Brettell and Ingram, 1979; Behnke, 1979).

Some of the problems and advantages of the use of cell and tissue culture as systems to isolate tolerant lines to specific compounds have already been discussed in general by Maliga (1978), and in the case of herbicide tolerance by Gressel (1980). In general, a cell population with rapid growth and high plating efficiency at low cell densities is necessary to select cell lines showing specific characteristics. Ideally, cells under selection pressure should show similar responses as those of the intact plant in the presence of the toxic compounds and should be able to regenerate intact plants. Finally, the genetic information selected at the cellular level should be stable in the cell lines, in the regenerated plants, and be transmissible to the progeny.

Suspension culture is the most frequently used system for selection of resistant cell lines. The conditions for selection are established through tests with different concentrations of the compound, focusing the effects on growth and survival of the cells. Cells or cell colonies are selected that are capable of growing and dividing when exposed to toxic concentrations of the compound. Alternatively, the cells can be inoculated directly into a liquid or solid medium that completely inhibits the growth of the major part of the cell population. Selection of cell lines can also be made by imbedding cells or protoplasts in a semisolid medium containing the toxic substance (Maliga, 1978).

Callus cultures have been utilized to select cell lines resistant to several chemical compounds (Widholm, 1977). In this case the resistant cells can be located in some points or even regions of the callus maintained in the selective medium. The cells are then subcultured for subsequent characterization.

Frequently the terms resistant and tolerant are used interchangeably when describing the response of plants to a toxic compound. However, as was stressed by Gressel and Segel (1978), these terms can be differentiated by response to effective doses. Populations of resistant plants completely survive an effective dose while the susceptible type does not. Tolerant plants express variations between biotypes as the dose is varied.

LITERATURE REVIEW

In recent years cell and tissue cultures have been utilized as systems to test phytotoxicity and to study the metabolism of herbicides (Hardlie et al., 1977; Zilkah and Gressel, 1977; Oswald et al., 1978; Smith, 1979). Some progress has been made toward establishment of cell lines and cultivars that are tolerant to herbicides. The methodology used and the problems faced by the researchers during the procedures of selection have already been reported (Chapter 11). For instance, Zenk (1974) reported a cell suspension tolerant to 2,4-D when the cells were subcultured for a period of 6 months in liquid medium in the presence of increasing amounts of the herbicide. These cells were able to grow in 1 mM 2,4-D, while the nontreated suspension was completely inhibited at 0.3 mM 2,4-D. The proposed mechanism of tolerance involves an enhancement in the metabolism of 2,4-D. Under the same conditions as this previous work, Widholm (1977) observed that carrot cells lost the tolerance to the herbicide when transferred to a 2,4-D-free medium, suggesting that tolerance was the result of induction of the enzymatic systems responsible for the degradation of 2,4-D.

Oswald et al. (1976) reported that soybean cell suspension cultures derived from cultures that are susceptible or tolerant to metribuzin continued to show these characteristics when cultured in presence of the herbicide.

Gressel (1979) selected carrot cells resistant to 2,4-D by treating cell suspensions with 4.5 µM of the herbicide for 3 hr followed by inoculation in 2,4-D-free medium. The cell colonies were subcultured

in liquid medium in the presence of 2,4-D. The treated cells were
more viable in the presence of the herbicide than the wild type cells.
One year later, two cell lines isolated from 2,4-D medium still expressed resistance to the herbicide suggesting that genetic stability had
been obtained.

Oswald et al. (1977) observed the development of tolerance in cell
suspensions of *Trifolium repens* to 2,4-D, 2,4,5-T (trichlorophenoxyacetic
acid), and 2,4-DB (4-(2,4-dichlorophenoxy) butyric acid). The pretreatment of the cells with 2,4-D (18.0 µM), or 2,4,5-T (31.3 µM) or 2,4-DB
(8.0 µM) for 5 days increased the tolerance to 2,4-D 3.4 times, to
2,4,5-T 4.2 times, and to 2,4-DB 3 times. According to the authors,
the adaptability to toxic levels of each herbicide was followed by heritability of the tolerance.

Ellis (1978) established tomato cell suspension cultures from 4 cultivars showing different degrees of tolerance to metribuzin. Cell growth
was not affected by levels of the herbicides that were lethal to seedlings, but was inhibited at 150 ppm. The cells in suspension did not
show differences when derived from sensitive or tolerant cultivars. No
positive correlation was observed between this system and the response
of intact plants to the herbicide. This was probably because metribuzin inhibits photosynthesis, a process that was not active in the cell
suspensions. Results to the contrary were described previously by
Oswald et al. (1976), indicating variability dependent on the cell system utilized.

Chaleff and Parsons (1978) reported the most complete work on herbicide selection published to date. Seven cell lines of *Nicotiana tabacum* resistant to picloram were isolated from cell suspensions inoculated
in medium containing 500 µM of the herbicide. One or two months
later resistant callus was obtained, transferred to a selective medium,
and subsequently to plantlet induction medium. Out of the seven selected lines, six lines regenerated plants. Later explants from those
regenerated plants were inoculated to obtain callus that was grown in
the presence of 500 µM picloram. The derived callus of five lines still
showed resistance. Callus from one line exhibited sensitivity, although
it was obtained from plants derived from a cell line resistant to the
herbicide. This observation shows that stability of one phenotype
maintained in the absence of the selection factor is not always a good
criterion to define the genetic basis of an alteration. Moreover it
cannot be a definitive criterion to distinguish between genetic and
epigenetic events. An analysis of the segregation of the tolerance in
plants that produced resistant callus showed that plants derived from
one of the lines were sterile, while plants from another line were sensitive to the herbicide. So, the expression of an altered phenotype is
not general to all the tissues or organs of a plant. Only in four lines
was the resistance to picloram transferred sexually; three of the phenotypes behaved as dominant alleles and one as a semidominant allele of
single nuclear genes.

Swanson and Tomes (1980) isolated lines from callus of *Lotus corniculatus* cv. Leo resistant to 2,4-D. Two lines of callus were selected:
one showing rapid growth and one slow growth in the presence of 4.5
µM 2,4-D. On the other hand, cell lines resistant to 2,4-D were ob-

tained when callus from the wild type was inoculated into liquid medium with 60 μM of the herbicide. The cells were subcultured 14 days later in liquid medium containing the same concentration of the herbicide. The suspension was then plated in a semisolid medium with 4.5 μM 2,4-D. The green spots of the callus obtained 2 weeks later were excised and inoculated onto a plantlet induction medium. According to the authors, the genotypes that were obtained could be propagated through meristem cultures.

Selection for resistant cell lines of celery was reported by Merrick and Collin (1981). Cells in suspension were incubated for 1 week in medium containing 80 μM Asulam. Single cells from small clumps of cells were inoculated in a herbicide-free semisolid medium. The cell colonies were again selected in 80 μM Asulam. After 15 subcultures in a herbicide-free medium, the isolated cell lines still showed tolerance to Asulam. One cell line showed stable tolerance after 3 years of continuous subculture (Collin, 1981).

With the use of the plating technique described by Bergman (1960), a suspension of celery embryoids was cultured in Asulam (Collin, 1981). The surviving embryoids, however, regenerated plants whose seeds were not viable.

Seven variants of *Nicotiana tabacum* cells resistant to amitrol were obtained by McDaniel (1981). The regenerated plants and their progenies showed tolerance to the herbicide. Variants resistant to amitrol or to glyphosphate have been recently obtained and are being subjected to characterization.

PROTOCOLS

The effects of two herbicides and one ripener on the growth of callus derived from internodal leaf explants of three sugarcane (*Saccharum* spp.) varieties have been reported by Crocomo et al. (1981) using the protocol outlined first. Different varietal responses to two herbicides ametryn (a triazine) and dalapon (an aliphatic) and to the ripener mefluidide, which were used at the following concentrations: 0, 20, 40, 80, and 160 ppm, have been noted. The callus of var. CB41-76 showed greater sensitivity toward the three compounds than those from var. IAC48-65 and var. NA56-79, which exhibited a similar degree of tolerance (Fig. 1). Of the three compounds, ametryn was the one that showed greater deleterious effects on the three sugarcane varieties.

Sugarcane callus when cultivated in special media can differentiate and regenerate plants (Crocomo et al., 1979; third protocol). This system has been utilized in our laboratory to obtain sugarcane plants directly with different degrees of tolerance to ametryn and dalapon (Ochoa-Alejo et al., 1981). Callus from sugarcane (cv. NA56-79) was inoculated for plant regeneration in media containing 0, 20, 40, 80, and 160 ppm of either ametryn, dalapon, or mefluidide. Plants tolerant to 20 and 40 ppm ametryn were obtained; however, as the herbicide concentration increased the frequency of regeneration decreased (Fig. 2a). Callus inoculated in medium with dalapon showed very low frequency of regeneration at 20 ppm, as the plants displayed symptoms of toxicity (Fig. 2b). The callus inoculated with mefluidide did not show signifi-

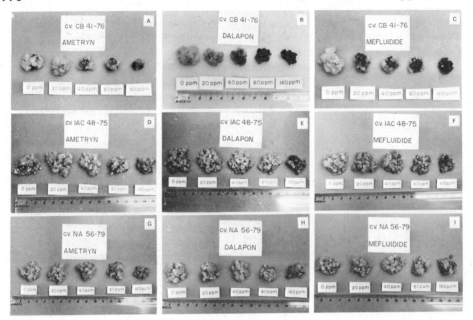

Figure 1.   Influence of ametryn, dalapon, and mefluidide on the growth of sugarcane callus derived from cv. CB41-76 (a,b,c), cv. IAC 48-65 (d,e,f), and cv. NA56-79 (g,h,i).

cant plant regeneration at the concentration employed.   The regenerated plants are being propagated in the presence of the herbicides.

Techniques for culturing cells, tissues, and protoplasts of sugarcane (*Saccharum* spp.) have been established in our laboratories (Crocomo et al., 1979; Evans et al., 1980).   Callus, cell suspensions, and/or plant regeneration are obtained with inoculation of the tissue in specific media (liquid or solid medium).   These techniques were adapted for phytotoxicity studies and for the selection of plants showing tolerance to herbicides.   The general protocol is illustrated in the diagram in Figure 3, and each step is described in the following Protocols.

Protocol for the Induction of Callus in Sugarcane Tissue and Treatment with Herbicides

1.   Buds from internodal cuttings of sugarcane stalks are germinated in a growth chamber, in vermiculite containing trays, at 28 C, 12 hr light.
2.   Leaves from the seedlings are removed 15–20 days later and segments ca. 5 cm long are excised from the shoot.
3.   The segments are washed 4–5 times (3 min each time) in a commercial sodium hypochlorite solution (20% v/v).
4.   In the laminar flow, the segments are transferred to 250 ml flasks containing 200 ml of commercial sodium hypochlorite solution (20% v/v) and kept in a shaker (180 rpm) for 20 min.

Figure 2.   Sugarcane plant regeneration from callus of cv. NA56-79 in IM medium containing ametryn (a), or dalapon (b).

(Note: All the subsequent steps are conducted in a laminar flow under aseptic conditions.)

5.   The segments are washed three times in sterile distilled water, in Petri dishes.

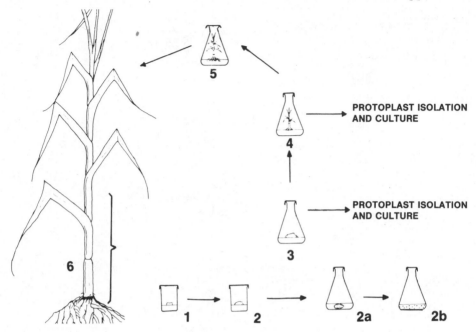

Figure 3.   General protocol to obtain callus (solid medium), cell
suspension, protoplasts, and plant regeneration of sugarcane (*Saccharum*
spp.). (1) internodal meristematic tissue innoculated in the conditioning
medium (CM); (2) cell proliferation and callus formation; (a) callus
transferred to CM liquid medium; (b) cells in suspension; (3) callus is
subcultured in the shoot induction medium (IM); (4) shoot regeneration;
(5) transferrence to root induction medium (RM); (6) entire plant ready
to be transferred to greenhouse after adaptation in a humid chamber.

6.   The segments are cut in 1-2 cm long pieces.
7.   The explants are inoculated in glass flasks (e.g., "French square"
      type) containing 10 ml of conditioning medium (CM), modified from
      Murashige and Skoog (1962), for callus formation. The basal medi-
      um has MS mineral salts and (per liter): sucrose 0.06 M, inositol
      0.55 mM, 2,4-D 13.5 µM, thiamine 2.9 µM, arginine 0.34 mM, 100
      ml CW, 8 g agar (Crocomo et al., 1979). The pH is adjusted to
      5.8.
8.   The flasks containing the explants are incubated in the dark for
      3-4 weeks at 24 C.
9.   The calli produced at both ends of the explants are excised and
      reinoculated in CM. The flasks are exposed to fluorescent (Gro-
      lux type lamps) and incandescent light for 4 weeks at 24 C. The
      callus can be transferred to CM medium at least one more time
      without apparent loss of the capacity to regenerate intact plants.
      Tissues derived from some sugarcane cultivars are more stable and
      can regenerate intact plants even when in CM medium for a long-
      er period (e.g., 1 year).
10.   The calli are divided into 1-2 cm diameter portions and trans-
      ferred to CM medium containing the desired concentration of the
      herbicide.

11.   The flasks are exposed to light at 24 C for 4 weeks.
12.   Surviving cells are removed for reselection or lyophilization (chemical analysis).

Protocol to Establish Sugarcane Cells in Suspension and Treatment with Herbicide

1.   Callus portions (1-2 g fwt.) derived from sugarcane internodal explants (preceding Protocol) are transferred to 125 ml flasks containing 50 ml CM liquid medium, in the laminar flow hood.
2.   The flasks are shaken on a rotary shaker (180 rpm) for 2 weeks at 24 C.
3.   The dispersed cells (25 ml) are transferred to 125 ml flasks containing 50 ml CM liquid medium.
4.   The cells are subcultured as in the previous step every 2 weeks.
5.   The cell mass is determined in a 1 or 2 ml suspension culture collected in Miracloth discs and dried in an oven at 85 C for 2 hr.
6.   An equal mass of cells is inoculated in the CM liquid medium containing the desired concentration of herbicide in order to study metabolic alterations. The CM liquid medium is not suitable for the selection of herbicide-tolerant cells, since many subculture steps can (and usually do) lead to loss of the capacity to regenerate intact plants. This problem could most probably be addressed by using shoot induction medium (IM) with different hormones.

Protocol for Regeneration of Sugarcane Plants Tolerant to Herbicide

1.   Callus pieces (1-2 g, see first Protocol) are transferred to 250 ml flasks, containing 50 ml of shoot induction medium (IM) and different concentrations of the herbicide. All the steps are made in the laminar flow. The IM medium has the MS mineral salts, plus (in 1 liter) sucrose 0.06 M, thiamine 2.9 μM, inositol 0.55 mM, KIN 4.6 μM, NAA 5.4 μM, 400 mg hydrolysed (enzymatic) casein, 100 ml CW, 8 g agar (Crocomo et al., 1979). The pH is adjusted to 5.8.
2.   The flasks are incubated for 4-6 weeks at 24 C under light (first Protocol).
3.   The regenerated shoots are transferred to 250 ml flasks containing 50 ml root induction medium (RM), which corresponds to the SH medium of Schenk and Hildebrandt (1972). The sucrose content can be increased up to 0.2 M (Larkin, personal communication; Maretzki and Hiraki, 1980).
4.   The regenerated plantlets are left under light as in step 2 above. The regenerated material can be reselected by repeating the whole process: callus induction from its explants and plant regeneration in the presence of herbicides.
5.   The regenerated plants are transferred to pots containing a mixture of vermiculite and sand (4:1), irrigated with nutrient solution (Hoagland and Arnon, 1950), and maintained in a humid chamber.

FUTURE PROSPECTS

Utilization of protoplasts as a system to select cell lines tolerant to herbicides has not been extensively explored. Using protoplasts, Aviv and Galun (1977) isolated lines of *Nicotiana tabacum* resistant to isopropyl N-phenylcarbamate (IPC). A high proportion of plants regenerated in the presence of the herbicide were sterile. Only one variant produced protoplasts more resistant to IPC than the control plants.

Protoplast technology can be used to increase the possibility to obtain monoclonal lines and offers the opportunity for interspecific transfer of cytoplasmic factors of resistance to some types of herbicides. Such factors were observed in *Brassica,* and were transferred using the classical methodology of crossing (Beversdorf et al., 1980). The technique of protoplast fusion has already been used with success to transfer cytoplasmic factors of male sterility between *Nicotiana sylvestris* and *N. tabacum* (Zelcer et al., 1978; Chapter 9).

Protoplasts from one species carrying the herbicide tolerance factor can be fractionated into enucleated microplasts or enucleated subprotoplasts using the methodology developed by Bilkey and Cocking (1980) providing enucleate units for fusion with protoplasts from another species. This technique will avoid irradiating protoplasts to inactivate nuclei before fusion which could cause side effects to the cytoplasm. The novel range of cybrids so formed would carry the tolerance factor (Cocking, 1981).

A potential application for the transfer of herbicide tolerance factor is the transplant of cytoplasmic organelles. Differential tolerance to atrazine in several *Chenopodium album* has been observed (Souza Machado et al., 1977, 1978). In this case, the mechanism of tolerance is determined by the differential tolerance of the chloroplasts. Therefore, as has been demonstrated in other systems (Bonnett and Eriksson, 1974; Davey et al., 1976), there is a possibility to transfer intact chloroplasts of resistant species to protoplasts of sensitive species.

In nature, microorganisms harboring enzymatic systems to degrade herbicides genetically coded by plasmids can be found (Fisher et al., 1978). Karns et al. (1981) were able to isolate a *Pseudomonas* strain that utilizes 2,4,5-T and other chlorophenols as a sole carbon source. This new degradative function of this bacteria could be evolved by recruitment of genes from various other plasmids. The incorporation of plasmids responsible for herbicide degradation can open new perspectives for the transference of resistance/tolerance in plant cells in the future.

The treatment of cells with mutagens to increase the frequency of phenotypes has not been used often to isolate cell lines resistant to herbicides. Radin and Carlson (1978) isolated mutants tolerant to phenmedipharm and to bentazone. Gamma-irradiated leaves of *N. tabacum* were treated and plants were regenerated resulting in 10 stable mutants.

Selection of variants at a frequency from 10^{-5} to 10^{-8} can be obtained in some cases without mutagen treatment (Maliga, 1978). However, the frequency of some phenotypes can be increased up to 10 times when cells are treated with mutagens. On the other hand, lesions can be introduced in other genes unrelated to the resistance by the mutagens leading to undesirable characteristics in the selected phenotype (Aviv and Galun, 1977; Gressel, 1980).

ACKNOWLEDGMENT

The authors wish to thank the Brazilian Nuclear Commission (CNEN, Rio) and PLANALSUCAR, Araras-SP, for financial support for the experiments discussed in this review.

KEY REFERENCES

Chaleff, R.S. and Parsons, M.F. 1978. Direct selection in vitro for herbicide resistant mutants of *Nicotiana tabacum*. Proc. Nat. Acad. Sci. USA 75:5104-5107.

Cocking, E.C. 1981. Opportunities from the use of protoplasts. Phil. Trans. R. Soc. London Ser. B. 292:557-568.

Gressel, J. 1980. Uses and drawbacks of cell cultures in pesticide research. In: Plant Cell Cultures: Results and Perspectives (F. Sala, B. Parisi, R. Cella, and O. Ciferri, eds.) pp. 379-388. Elsevier, Amsterdam.

REFERENCES

Aviv, D. and Galun, E. 1977. Isolation of tobacco protoplasts in the presence of isopropyl N-phenylcarbamate and their culture and regeneration into plants. Z. Pflanzenphysiol. 83:267-274.
Behnke, M. 1979. Selection of potato callus for resistance to culture filtrates of *Phytophtora infestans* and regeneration of resistant plants. Theor. Appl. Genet. 55:69-72.
Bergmann, L. 1960. Growth and division of single cells of higher plants in vitro. J. Gen. Physiol. 43:841-851.
Beversdorf, W.D., Weiss-Lerman, J., Erickson, L.R., and Souza Machado, V. 1980. Transfer of cytoplasmically inherited triazine resistance from bird's rape to cultivated oilseed rape (*Brassica campestris* and *B. napus*). Can. J. Genet. Cytol. 22:167-172.
Bilkey, P.C. and Cocking, E.C. 1980. Isolation and properties of plant microplasts: Newly identified subcellular units capable of wall synthesis and division into separate micro cells. Eur. J. Cell Biol. 22: 502.
Bonnett, H.T. and Eriksson, T. 1974. Transfer of algal chloroplasts into protoplasts of higher plants. Planta 120:71-79.
Brettell, R.I.S. and Ingram, D.S. 1979. Tissue culture in the production of novel disease-resistant crop plants. Biol. Rev. 54:329-345.
Collin, H.A. 1981. Selection for herbicide resistance. Newsletter International Association for Plant Tissue Culture, No. 35, pp. 13-14, IAPTC, Kyoto, Japan.

Crocomo, O.J., Sharp, W.R., and Carvalho, M.T.V. 1979. Controle da morfogenese e desenvolvimento de plantas em cultura de tecido de cana-de-acucar, resultados experimentais. An. 1 Congr. Nac. Soc. Tecn. Acuc. (STAB) Brasil pp. 241-243.

Crocomo, O.J., Ochoa-Alejo, N., Goncalves, C.H.R.P., and Bacchi, O.S. 1981. Tolerancia de cultivares de cana-de-acucar a herbicidas utilizando a tecnica de cultura de tecidos. An. 2 Congr. Nac. Soc. Tecn. Acuc. (STAB) pp. 21-40.

Davey, M.R., Frearson, E.M., and Power, J.B. 1976. Polyethylene glycol induced transplantation of chloroplast into protoplasts: An ultrastructural assessment. Plant Sci. Lett. 7:7-16.

Devine, T.E., Seaney, R.R., Linscott, D.L., Hagin, R.D., and Brace, N. 1975. Results of breeding for tolerance to 2,4-D in birdsfoot trefoil. Crop. Sci. 15:721-726.

Ellis, B.E. 1978. Non-differential sensitivity to the herbicide metribuzin in tomato cell suspension cultures. Can. J. Plant Sci. 58:775-778.

Evans, D.A., Crocomo, O.J., and Carvalho, M.T.V. 1980. Protoplast isolation and subsequent callus regeneration in sugarcane. Z. Pflanzenphysiol. 98:355-358.

Fisher, P.R., Appleton, J. and Pemberton, J.M. 1978. Isolation and characterization of the pesticide degrading plasmid pJP1 from *Alcaligenes paradoxus*. J. Bacteriol. 135:798-804.

Gressel, J. 1979. Genetic herbicide resistance: Projections on appearance in weeds and breeding for it in crops. In: Plant Regulation and World Agriculture (T.K. Scott, ed.) pp. 85-109. Plenum, New York.

_____ and Segel, L.A. 1978. The paucity of plants evolving genetic resistance to herbicides. Possible reasons and implications. J. Theor. Biol. 75:349-371.

Hardlie, L.C., Widholm, J.M., Slife, F.W. 1977. Effect of glyphosate on carrot and tobacco cells. Plant Physiol. 60:40-43.

Hoagland, D.R. and Arnon, D.I. 1950. The water-culture method for growing plants without soil. California Agricultural Experiment Station Circular 347.

Karns, J.S., Kilbane, J.J., Chaterjee, D.K., and Chakrabarty, A.M. 1981. Laboratory breeding of a bacterium for enhanced degradation of 2,4,5-T. In: Genetic Engineering for Biotechnology (O.J. Crocomo, F.C.A. Tavares, and D. Sodrzeieski, eds.) pp. 37-40. PROMOCET, SP, Brazil.

Maliga, P. 1978. Resistance mutants and their use in genetic manipulation. In: Frontiers of Plant Tissue Culture (T.A. Thorpe, ed.) pp. 381-392. International Association for Plant Tissue Culture, Calgary.

Maretzki, A. and Hiraki, P. 1980. Sucrose promotion of root formation in plantlets regenerated from callus of *Saccharum* spp. Phyton 38:85-88.

McDaniel, C.N. 1981. Herbicide resistant mutants. Newsletter International Association for Plant Tissue Culture, No. 35, p. 21, IAPTC, Kyoto, Japan.

Meredith, C.P. 1978. Selection and characterization of aluminum-resistant variants from tomato cell cultures. Plant Sci. Lett. 12:25-34.

Merrick, M.M.A. and Collin, H.A. 1981. Selection for asulam resistance in tissue cultures of celery. Plant Sci. Lett. 20:291-296.

Murashige, T. and Skoog, F. 1962. A revised medium for rapid growth and bioassays with tobacco tissue cultures. Physiol. Plant. 15:473-497.

Ochoa-Alejo, N., Machado, I.S., Oliveira, E.T., and Crocomo, O.J. 1981. Sugarcane cell and tissue culture systems as fundamental step in the study of the herbicide tolerance factor. In: Genetic Engineering for Biotechnology (O.J. Crocomo, F.C.A. Tavares, and D. Sodrzeieski, eds.) pp. 106-107. PROMOCET, SP, Brazil.

Oswald, T.H., Smith, A.E., and Phillips, D.V. 1976. Phytotoxicity and intercultivar detoxification of the herbicide metribuzin in soybean cell suspension cultures. Abstracts of Crop Science Society of America Annual Meeting, p. 59. American Society Agronomy, Madison.

_____ 1977. Herbicide tolerance developed in cell suspension cultures of perennial white clover. Can. J. Bot. 55:1351-1358.

_____ 1978. Phytotoxicity and detoxification of metribuzin in dark-grown suspension cultures of soybeans. Pestic. Biochem. Physiol. 8:73-83.

Radin, D.N. and Carlson, P.S. 1978. Herbicide-tolerant tobacco mutants selected in situ and recovered via regeneration from cell culture. Genet. Res. 32:85-90.

Schenk, R.U. and Hildebrandt, A.C. 1972. Medium and techniques for induction and growth of monocotyledonous and dicotyledonous plant cell cultures. Can. J. Bot. 50:199-204.

Smith, A.E. 1979. Metabolism of 2,4-DB by white clover (*Trifolium repens*) cell suspensions. Weed Sci. 37:392-396.

Souza Machado, V., Bandeen, J.D., Stephenson, G.R., and Jensen, K.I.N. 1977. Differential atrazine interference with the Hill reaction of isolated chloroplasts from *Chenopodium album* L. biotypes. Weed Res. 17:407-413.

Souza Machado, V., Arntzen, C.J., Bandeen, J.D., and Stephenson, G.R. 1978. Comparative triazine effects upon system II photochemistry in chloroplasts of two common lambsquarters (*Chenopodium album*) biotypes. Weed Res. 26:318-322.

Swanson, E.B. and Tomes, D.T. 1980. Plant regeneration from cell cultures of *Lotus corniculatus* and the selection and characterization of 2,4-D tolerant cell lines. Can. J. Bot. 58:1205-1209.

Widholm, J.M. 1974. Cultured carrot cell mutants: 5-methyl-tryptophan-resistance trait carried from cell to plant and back. Plant Sci. Lett. 3:323-330.

Zelcer, A., Aviv, D., and Galun, E. 1978. Interspecific transfer of cytoplasmic male sterility by fusion between protoplasts of normal *Nicotiana sylvestris* and X-ray irradiated protoplasts of male sterile *N. tabacum*. Z. Pflanzenphysiol. 90:397-407.

Zenk, M.H. 1974. Haploids in physiological and biochemical research. In: Haploids in Higher Plants: Advances and Potential (K.J. Kasha, ed.) pp. 339-354. Univ. Guelph Press, Guelph.

Zilkah, S. and Gressel, J. 1977. Cell cultures vs. whole plants for measuring phytotoxicity. III. Correlations between phytotoxicities in cell suspension cultures, calli and seedlings. Plant Cell Physiol. 18:815-820.

CHAPTER 27
Germplasm Preservation

W. Nitzsche

The conservation of living material is of interest to many scientific workers. Prokaryotes, plants, and animals have been included in such experiments. The aim of germplasm conservation is to insure the availability of useful germplasm at any time. Therefore the storage of callus tissue should address the problem of conservation of special phenotypes, and not of well-defined genotypes. This is of interest in the following fields: (1) The conservation of callus and cell lines for tissue culture purposes, which have lost the ability of morphogenesis or at least to undergo mitosis and to produce seeds. This may be the case in cell lines with or without a special enzyme activity (Schroeder et al., 1979), or in animal cells producing monoclonal antibodies (Koehler and Milstein, 1975). (2) In open-pollinated, self-incompatible species, special breeding methods must be used, such as polycross or topcross (Frandsen, 1951) to estimate the breeding value of single plants. After progeny testing, the best original plants are used for multiplication. This is only possible in perennial plants, as annuals do not survive their progeny. Tissue culture and storage can overcome this problem. (3) In vegetatively propagated species, maintenance breeding has to produce healthy, mainly virus-free plants. With the use of tissue culture, along with stored calli, there is a high protection against infection, and healthy clones can be maintained for years. A presupposition for all fields is the need for methods that enable safe conservation and stable regeneration of cells, calli, or meristems to plants after several years of storage.

LITERATURE REVIEW

Theoretical Basis of Conservation

For long-term storage, the living material must be fixed invariably. Changes of chemical composition or physical structure will lead to cell death or permit accumulation of mutations. Only reversible processes can be tolerated.

Biochemical reactions in the cells require liquid water as substrate, otherwise processes are stopped. Water can be eliminated by transferring the liquid water, including the "bulk water" and the "vicinal water" (Løvtrup and Hansson Mild, 1981) to the solid or gaseous state, or to substitute water with other substances. There are therefore different principles for preparing the callus for storage.

Transfer of Water to the Solid State

Freezing of water occurs at a wide range of temperatures. While pure water becomes ice at 0 C, the cell water needs much lower temperatures because of freezing point depression by salts or organic molecules. The lowest temperature, where liquid water has been demonstrated is -68 C (Scheurmann, 1967). This implies that storage temperatures of hydrated cells and tissues must be lower than -70 C. Type of crystal formation during freezing process and during storage controls survival of the tissue.

Transfer of Water to the Gaseous State

The transfer of water to the gaseous state results in dehydration. In most cases this leads to cell death. However, in some specially adapted organisms or tissues, dehydration occurs naturally. Dehydrated seeds of cereals can stay alive for more than 100 years (Aufhammer and Fischbeck, 1964). Tiller buds of *Vitis riparia* can dehydrate and thereby overcome freezing damage during midwinter (Pierquet and Stushnoff, 1980). Also, animals, e.g., nematodes, can naturally dehydrate without dying. Artificial drying of mammalian cells is done successfully with bull sperm down to 6% residual water. After storage for 2 days at temperatures above zero, the sperm survived (Jeyendran et al., 1981).

The Substitution of Water

Sometimes water can be substituted by a lot of organic solvents in cells. Little is known about the biochemical background of this conservation method. Iwanami (1972a) successfully used 19 different organic solvents for the storage of *Camellia japonica* pollen (Table 1). These three principles form the basis for methods of cryopreservation. Physical structures may be altered reversibly, but the biochemical

Table 1. Pollen Protectants Used Successfully by Iwanami (1972b)

Acetone (0.4% H_2O)	Carbon tetrachloride
n–Amyl acetate	1,1,1-Trichloroethane
Ethyl acetate	iso–Amyl ether
n–Amyl alcohol	Diethyl ether
n–Butyl alcohol	Petroleum ether
Ethyl alcohol	n–Heptane
iso–Butyl alcohol	n–Pentane
Benzene	iso–Pentane
Paraldehyde	Toluene
Petroleum benzene	Xylene

structures, bearing the genetic information, undergo changes that are stable.

In addition to these storage principles, the use of slow growth is also possible. By this method living material is maintained with as low a growth activity as possible while retaining some activity. Water is maintained in the liquid condition, but all biochemical processes are delayed. This phenomenon occurs in nature with the dormancy of bulbs and buds.

METHODS OF CRYOPRESERVATION

Freezing

The type of crystal water within stored cells is very important for survival of the tissue. For this reason three different types of freezing procedures have been developed: (1) slow freezing with a temperature decrease of 0.1-10 C/min, (2) the rapid freezing method, with a temperature decrease of 50-1200 C/min, and (3) the stepwise freezing method, combined from a slow freezing method to -20 to -40 C, then a stop for a period of time and an additional rapid freezing to -196 C (liquid nitrogen) (Withers, 1978a). A scheme of the temperatures used for each method is given in Fig. 1. Storage temperatures for all these methods are between -70 C and -196 C. Changes of temperatures may lead to an additional increase in crystal size.

PRETREATMENT. The effect of temperature on plants depends equally on genotype and environment, as well as on physiological conditions. While genotype for practical purposes is an unchangeable factor, the other two factors can be altered according to circumstances. Some plants are always killed at the same temperature, e.g., sunflowers at -2 C. Other plant species are killed according to the environmental temperature that was present before freezing. For instance, cabbage dies at -2 C when it is grown at 20-30 C; when grown at 5 C for 1 week it is killed at -7 C, and after a second week at 0 C it is killed at -12 C. This dependence of killing temperature on the culture temperature is known as hardening. Levitt and Dear (1970) call the first

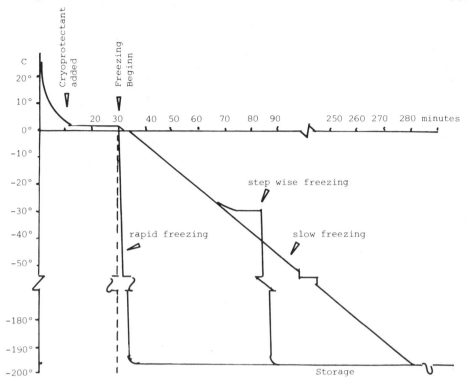

Figure 1.   Temperature flow at different freezing procedures.

group tender plants, and the second group, which can be hardened, hardy plants.   During the hardening process, structures in the plasma membrane may be altered, molecular linkages within proteins may be changed from SS (sulphur-sulphur) to SH (sulphur-hydrogen), and additionally, sugars and similar substances, which have the function of cryoprotectants, are accumulated.

The process of hardening is also important as a pretreatment for freezing in tissue culture.   As both genetic and environmental factors are influential, no general rules can be stated.   Each situation has to be investigated specially.   Photoperiodic processes may additionally complicate the process.

Appearance of ice crystals in cells is controlled by adding cryoprotectants, substances which in the solid state are amorphous instead of crystalline.   A number of different cryoprotectants are known, with different efficiencies (Table 2).   Most frequently used is DMSO (dimethylsulfoxide), which was discovered in 1959 by Lovelock and Bishop.

A diluted solution of the cryoprotectant (e.g., 5-10% DMSO) has to be added gradually to prevent plasmolysis in intervals of at least 5 min.   This procedure has to be done in an ice bath, as room temperature affects the viability of cells and tissues.   After the last addition of the cryoprotectant, but before cooling, there should be an interval of 20-30 min.

Mixtures of several cryoprotectants may also be used.   Ulrich et al. (1979) and Finkle et al. (1980) put pieces of embryonic palm callus into

Table 2.   Efficiency of Different Cryoprotectants (Cinatl and Tolar, 1971)

| | SURVIVAL | |
80%	60%	20%
Glycerol	Acetamide	Dimethylsulfone
Dimethylsulfoxide	Dimethylacetamide	Polyvinylpyrrolidone
Ethylene glycol	Formamide	
Diethylene glycol	Monoacetine	
Propylene glycol	Mannose	
Pyridoxine-n-oxide	Ribose	
Hexamethylentetra- mine	Glucose	

a mixture of 10% polyethylene glycol, 0.44 M glucose, and 10% DMSO, froze the material and subsequently grew it into plants after thawing.

THE SLOW-FREEZING METHOD.   This method is the most common one.   Survival of cells frozen at slow-freezing rates of -0.1 to -10 C/min may involve some beneficial effects of dehydration, which minimize the amount of intercellular water that freezes intercellularly.
A lot of temperature-controlling instruments, including computer programmed freezers, have been developed (Cinatl and Tolar, 1971; Bajaj and Reinert, 1977).   Technically these systems are expensive, but in many species success has been achieved.   In *Oryza* Sala et al. (1979) have used the method of slow freezing with an inexpensive variant utilizing Dewar bottles.

THE RAPID-FREEZING METHOD.   The quicker the freezing is done, the smaller the intracellular ice crystals are.   When plant material is placed in vials directly into liquid nitrogen, a decrease of -300 to -1000 C/min or more will occur.   This method is technically simple and easy to handle.   A somewhat slower temperature decrease is achieved when the vial containing plant material is put in the atmosphere over liquid nitrogen (-10 to -70 C/min).   Dry ice (CO_2) instead of nitrogen probably can be used similarly.
In the rapid-freezing method cryoprotectants are used in the same manner as in the slow-freezing method.   For deciding between both methods the water content of the samples is very important.   For the rapid freezing method, small specimens with low water content seem to be more appropriate.   None of the reports of preservation by the rapid freezing method describes a postthaw deplasmolysis injury effect in contrast to the more widely used slow freezing method (Withers, 1978b).

THE STEPWISE FREEZING METHOD.   This method combines the advantages of both the other methods.   A slow freezing procedure down to -20 to -40 C permits a protective dehydration of the cells.   An additional rapid freezing in liquid nitrogen prevents the growing of big ice crystals in the biochemically important structures. Cryoprotectants are needed as with the other techniques. The expense is similar to both above-mentioned methods.

The stepwise freezing method gives excellent results with suspension cultures, but is unsatisfactory for organized structures.

Storage of Dried Callus

The basic idea for the drying method originated from the totipotency of cells: every plant cell has the potential to develop into a complete plant. Consequently, it must be assumed that each callus cell can be transferred to conditions that are similar to those in stored seeds.

Cells of embryos are meristematic; however, they lose moisture during the ripening process. This process is controlled by hormones; in addition chemicals that facilitate storage substances are also often accumulated. This process may be simulated in vitro. The ripening process is not well understood, as a ripened cell has not been chemically defined. Hormones influencing ripening include abscissic acid (ABA) and ethylene. Other hormones are also important. Storage substances include carbohydrates, lipids, and proteins that may be influential as well.

PRETREATMENT. In several investigations the influence of ABA and sucrose concentration in the medium has been tested. ABA acts as an inhibitor of many processes in the cell, especially those biochemical events that are initiated by GA (Trewavas, 1976). One effect of both ABA and Rifampicin is the decrease of RNA-polymerase activity; however, in contrast to ABA, a pretreatment with Rifampicin does not influence the survival of cells. Hence the significance of the action of ABA during the pretreatment is not yet understood.

Increasing sucrose concentration up to 0.15 M or more has a positive effect on the survival of dried callus. Sucrose may act as a cryoprotectant in altering the structure of the cell water or alternately may increase dry matter content which in turn increases the survival rate.

From our investigations, optimum pretreatment is from 12 to 20 days. Apparently, a significant number of biochemical transformations are necessary for effective treatment. It is of both biological and biochemical interest that nucleic acids, proteins, etc. can be dehydrated without losing their information, and that the crystal water is not necessary for the preservation of their structure.

The pretreatment seems to be a necessary prerequisite in many dicotyledonous species, although calli of *Gramineae* survive without it.

DRYING. Contamination of the calli has to be prevented during the drying and storage processes. This has been achieved by using either of two different methods. In the first method, callus pieces were placed in sterile gelatin capsules (0.5 cm^3) and closed. The capsules may then be dried under unsterile conditions in the laboratory for several days. Survival of *Daucus carota* L. callus was not affected by 4-7 days drying (Nitzsche, 1978). In the second method, callus pieces were placed on sterile filter paper and dried in a laminar air flow chamber in the stream of sterile air. The dried callus pieces are then stored in sterile Petri dishes or similar vessels (Nitzsche, 1980).

STORAGE. In contrast to the freezing method, superlow temperatures are not necessary for the storage of dried calli. The storage conditions can be chosen according to those that are used for seeds. Low humidity is necessary while low temperatures are useful. Temperatures above 0 C should be used. Oxygen content of the surrounding atmosphere may also influence the survival, but this has not yet been investigated. As storage requirements are not as rigid for dried callus, this method does not require as much energy as the freezing method.

Substitution of Water

The substitution of water with other chemicals has not been tried in tissue culture. Chemical substitution has been investigated in the storage of pollen (Iwanami and Nakamura, 1972). They have been able to store pollen of *Chrysanthemum pacificum*, *Lilium longiflorum*, *L. speciosum*, *L. auratum*, *Camellia japonica*, *C. sasanqua*, *Impatiens balsamina*, and Roggen (1974) has stored *Brassica* spp. The substitution is completed by submerging the pollen in organic solvents without any pretreatment. Most environmental effects that have been investigated are unimportant. Successfully used organic solvents are listed in Table 1. It seems to be important that the solvents are free of water. For example, 1.5% water in ethanol prevents germination of pollen. Acetone may contain 0.4% H_2O which does not impede storage, but an increase to 0.9% leads to a decrease in subsequent pollen germination capacity (Iwanami, 1972a,b, 1975).

For storage in organic solvents, low temperatures (0 C) are sufficient; pollen of *Lilium* could be stored under these conditions up to 180 days.

It is unknown if the method of chemical substitution of water also can be used in the preservation of tissues or cells in culture.

Combination of the Methods

The three different methods of cryopreservation may also be used in combinations. The use of cryoprotectants can be considered as a combination of the freezing method and the water substitution method, but with an incomplete substitution of the water.

The combination of substitution and drying methods is also done in pollen. In this case the organic solvent is evaporated, and the pollen is used for fertilization of flowers (Iwanami, 1972a).

Drying and freezing have been combined in the cold storage of dried callus as a control in the experiments of Nitzsche (1980). The dry-freezing method of Withers (1978a, 1979) cannot be mentioned here, because the cells contained water and only the embryos were blotted dry of superficial moisture, prior to the freezing procedure.

Slow-Growing Method

In addition to those methods already discussed, in which the liquid water is replaced by other states or substances, cells may be stored in a slow-growing state. This technique is comparable to the traditional technique of Japanese gardeners known as Bonsai, in which trees are kept alive in small pots over hundreds of years. Growing processes are reduced to a minimum by limitation of a combination of growth factors. This technique can be applied to tissue cultures mainly by using temperatures from 1-4 C, but also up to 26 C, by using low oxygen pressure, or by reducing the growing capacity by hormones.

TEMPERATURE. The use of this technique has been demonstrated by Lundergan and Janick (1979). They cultured shoot tips of apples (*Malus domestica* Borkh. cv. Golden Delicious) on artificial medium and stored it at 1 and 4 C. The tissues survived the 12 month storage. After returning to 26 C the shoot tips proliferated new shoots.

LOW OXYGEN PRESSURE. The oxygen pressure can be regulated by atmospheric or partial pressure. The possibilities and applications of this method can be found in Chapter 29.

Alternatively, Caplin (1959) investigated the influence of a mineral oil overlay for tissue conservation. Callus of *Daucus carota* cv. Long Chantenay were preserved in test tubes overlayed with 0.5-4 cm mineral oil and the oxygen excluded. At 26 C the tissues could be maintained viable by subculturing at 5 month intervals for 3 years.

HORMONES. It is also possible to use plant growth regulators to delay callus growth. One of the substances that has been tested is succinic acid 2,2 dimethyl hydrazide (B-9). This chemical causes a reduced growth, mainly shorter internodes of the plants. It has been tested on potatoes (*Solanum tuberosum* L.) in addition to reducing temperature to 10 C and the light intensity to 4000-5000 lux. Using this method, the interval of transfer to new medium is prolonged up to 2 years (Mix, 1981). The genetic stability seems to be unaltered. Genotypic differences between cultivars have been observed.

DETERMINATION OF SURVIVAL

There is only one realistic test of survival of plant material: the regrowth of plants from stored tissues or cells. This test needs a lot of time and work, and therefore it is unsuitable in basic investigations.

For basic research purposes, the test is stopped at any stage during thawing and plant regrowth. The regrowth of callus after thawing also gives evidence of cell survival.

Staining methods are much more rapid than the more time-consuming methods mentioned above. Two different techniques are available. The first technique, fluorescein diacetate staining (Widholm, 1972) works only for living cells. One drop of fluorescein diacetate (0.1%) is mixed with one drop of the thawed cell suspension. The cells are counted microscopically in normal light and in UV light. The relation of both gives the percentage of survival.

In the second technique, the triphenyl tetrazolium chloride method (TTC), the cell survival is estimated by the amount of formazen produced as a result of reduction of TTC. This reaction results in a pink color. The procedure involves the following steps (Bajaj and Reinert, 1977):

1. Buffer solution: 78% Na_2HPO_4 x $2H_2O$ solution (0.05 M):22% KH_2PO_4 solution (0.05 M).
2. TTC-solution: 0.18 M TTC dissolved in buffer solution.
3. About 150 mg of cell sample is put into 3 ml of TTC solution and incubated for 15 hr at 30 C.
4. The TTC solution is drained off and the cells washed with distilled water.
5. Cells are centrifuged and extracted with 7 ml of ethanol (95%) in a water bath at 80 C for 5 min.
6. The extract is cooled and made to 10 ml volume with 95% ethanol.
7. The absorbance (pink color) is then recorded with a spectrophotometer at 530 μm.

It should be noted that neither of these staining methods can replace the plant regrowth experiments. Staining is useful only to give a preliminary indication of viability.

SPECIFIC SECTION

The methods used for different plant genera are listed in Table 3. This list implies that the slow or stepwise freezing method excludes the rapid freezing method and vice versa. The only exception is *Fragaria*, where both methods have been successfully applied. However, upon examination of different genera within a family, some generalizations can be made. In the Solanaceae both methods are successful, and there is no reasonable explanation that a method working in *Datura* or *Nicotiana* should not work in *Lycopersicum* or *Solanum*. Furthermore, from looking at the whole literature it can be concluded that the success of the methods depends more or less on where the investigations were conducted. Therefore the absence of the + sign does not necessarily mean that the method does not work. The better interpretation is that the special case has probably not yet been investigated.

Table 3. Genera and Methods Used for Tissue Conservation Studies

FAMILY	GENUS	SLOW AND STEP-WISE FREEZING	RAPID FREEZING	DRYING	SLOW GROWING
Araliales	Daucus	+			+
Caryophyllales	Dianthus		+	+	+
Fabales	Glycine	+			
	Lotus				+
	Medicago				+
	Trifolium				+
	Pisum	+			
Geraniales	Linum	+			
Primulales	Primula		+		
Rosales	Fragaria	+	+		+
	Malus		+		+
	Prunus	+			
Rhamnales	Vitis	+			
Sapindales	Acer	+			
Campanulales	Chrysanthemum	+			+
	Haplopappus	+			
Solanales	Atropa	+			
	Capsicum	+			
	Datura	+			
	Ipomoea	+			
	Lycopersicum		+		
	Nicotiana	+			+
	Solanum	+	+	?	+
Salicales	Populus	+			
Gentianales	Coffea	+			
Palmales	Phoenix	+			
Poales	Lolium			+	+
	Oryza	+			
	Zea	+			

Araliales

Daucus carota L.   This is one of the classical objects of tissue culture.   Therefore, it has also been used for basic investigations in germplasm conservation studies.   The most frequently used medium for carrot tissue culture is MS supplemented with 2,4-D (2.3 μM).

The basic investigations of the stepwise freezing method have been reported by Latta (1971), Nag and Street (1973), Dougall and Wetherell (1974), and Bajaj (1976).   In these publications the influence of genotype, cryoprotectant, cooling rate, storage temperature, and duration and thawing methods have been tested.

An influence of genotype is not mentioned in any of the publications, unless wild versus cultivated carrots have been used.   The cryoprotectants tested include DMSO, glycerol, and sucrose.   Best results were obtained with 5-7% (up to 10%) DMSO.   Optimal cooling rate is -1 to -2 C down to -40 C or more and subsequent storage in liquid nitrogen. Cells can be stored at -78 C, but there is a progressive decline in cell survival as storage is prolonged.   At -196 C no decline has been reported (Nag and Street, 1973).   Rapid thawing yielded the best results of all thawing methods tested.   All these results were confirmed by Popov et al. (1978) and Withers (1979).   Using synchronized cells, Withers and Street (1977) determined that best survival was obtained using cells from the lag or early exponential growth phase.

The dry-freezing method was introduced by Withers (1978a, 1979). After cryoprotectant treatment, plantlets (1-3 cm) derived from embryos were excised and blotted dry of superficial moisture.   Subsequently the plantlets were frozen using the slow-freezing method.   The dry frozen and thawed plantlets showed the induction of unorganized growth from shoot and root meristem regions.   Addition of activated charcoal to the medium promoted organized growth.

The callus-drying method was developed by Nitzsche (1978, 1980) for carrots.   While callus grown on the medium of Gamborg and Eveleigh (1968) did not survive drying, those grown for 16 days on the same medium with 0.15 M sucrose instead of 0.58 M and supplemented with 37.8 μM ABA can be dried without loss of viability.   The low residual water content of the dried callus allows freezing to -80 C without cryoprotectants as well as storage under low humidity conditions at room temperature.   At 25% relative humidity and 15 C callus can be stored more than 2 years without loss of viability (Nitzsche, unpublished).   On the original medium regrowth occurs after 4 weeks, and plants can be grown from them.

Caryophyllales

Dianthus caryophyllus L.   Shoot apices, including two leaf primordia of *Dianthus caryophyllus* can be grown on the inorganic parts of MS medium completed with 1.1 μM thiamine-HCl, 0.55 mM myo-inositol, 0.57 μM IAA, 2.3 μM KIN, and .087 M sucrose.   Rapid freezing has been investigated with this species (Seibert, 1976; Seibert and Wetherbee, 1977).   From the cultivars Scania and Ellen Marie, 33% and 15%, respectively, of cells survived freezing in liquid nitrogen and rapid thawing. This material was stored up to 2 months in liquid nitrogen. In further investigations with the cultivars Scania and Linda a cooling rate of -50 C/min was considered optimum. An acclimation of the shoot apices to 4 C for 4 days prior to freezing increased the survival rate to 60%. These results could be confirmed in experiments started at different seasons of the year. Only short day conditions during January and February gave viable shoot tips in cv. Pink Sim. Scania needed a 6-7 week short days treatment and Ellen Marie a 9 week treatment. It is uncertain if these differences reflect genotypic differences.

Fabales

Glycine max L. The conservation of *Glycine max* has been investigated in cell suspension cultures by Bajaj (1976). The cells were cultured in a White medium supplemented with 11.0 µM NAA and 0.47 µM KIN. Pretreatment was best with 5% DMSO using a freezing procedure with a rate of -2 C/min to -196 C. The procedure resulted in 20% survival determined by the TTC reduction method.

Lotus corniculatus L. The slow-growing method has been described in *Lotus corniculatus* by Tomes (1979). The cultivar Leo was used in node cultures on B5 medium containing 0.22 µM benzyladenine (BA-4) and B5-H, which contained neither cytokinin nor auxin. A total of 11 genotypes on BA-4 and 19 genotypes on B5-H survived at a temperature of 2-4 C for 4 weeks representative of a 91-100% survival rate.

Trifolium sp., *Medicago sativa* L. These forage legumes have been examined under slow-growing conditions by Cheyne and Dale (1980) and by Bhojwani (1981). The species and varieties used were: *Trifolium repens* L. var. S100, S184, and Huia, *Trifolium pratense* L. var. S123, Norseman (tetraploid), and *Medicago sativa* L. var. Sabilt.

Shoots, shoot tips, and meristems were cultured on B5, completed with 1.1 µM IAA and 0.98 µM 2iP (*Trifolium*) or 1.1 µM NAA (*Medicago*) or on a medium containing the inorganic salts of MS and 4.0 µM nicotinic acid, 0.4 µM pyridoxin-HCl, 0.3 µM thiamine-HCl, 0.55 mM inositol, .09 mM adenine sulfate, .06 M sucrose, and 8 g/l agar. This medium is called BM. In some cases the medium of Blaydes (1966), which is identical with the medium of Miller (1963), was used successfully.

The storage temperatures were 2-6 C at 300 lux/8 hr or in complete darkness. Survival was tested by using regrowth after 4-8 weeks. The survival rates after 15-18 months storage were: for *Trifolium repens* var. S100, 92%; for *T. repens* var. S184, 92%; for *T. repens* var. Huia, 100% (after 10 months); for *Trifolium pratense* var. S123, 83%; for *T. pratense* var. Norseman, 83%; for *Medicago sativa* var. Sabilt, 81%.

No real differences between dark and light treatment could be obtained. The number of shoots per culture decreased depending on storage time from 15 shoots after 1 month to 10 shoots after 10 months.

Pisum sativum L. Leaf promordia meristems were the starting material for freeze preservation experiments done by Kartha et al. (1979) with *Pisum sativum*. They were grown on B5 medium supplemented with 0.5 μM BA. DMSO, glycerol and ethylene glycol at concentrations from 0-20% each were tested as cryoprotectants. The general procedure requires a cryoprotectant treatment at 0 C for 30 min and a 10 min gap followed by a freezing rate from -0.5 to -1 C/min down to -40 C, followed by storing in liquid nitrogen for at least 1 hr to 26 weeks. After rapid thawing the meristems were washed four times and cultured for 3 weeks. Survival was based on callus formation and shoot production. A 100% shoot production was observed using 0-5% DMSO, 0% ethylene glycol, or 0-15% glycerol. A freezing rate of -0.6 C/min was optimum. There were no differences in survival at storage between 1 and 26 weeks.

Geraniales

Linum usitatissimum L. The only investigated species of the Geraniales is *Linum usitatissimum* (Quatrano, 1968). Cell cultures were established in MS medium supplemented with 0.12 M sucrose, 1.0 mM L-glutamine, 0.7 mM cysteine–HCl, 3.0 μM thiamine–HCl, 4.5 μM 2,4-D. Cryoprotection was achieved with 10% DMSO using a freezing rate of -5 to -10 C/min. After storage at -50 C for up to 1 month the cell suspension was thawed rapidly and washed three times. The survival rate was 14%, determined by the tetrazolium test.

Primulales

Primula obconica Hance. The MS medium completed with 2.9 μM IAA, 2.3 μM 2,4-D, and 9.1 μM ZEA has been used for anther culture in *Primula obconica* (Bajaj, 1981b). After 3 weeks cultivation at 26 C no growth of callus or embryos could be obtained. The anthers were treated with 7% DMSO and 0.2 M sucrose for 2 hours, then the cryoprotectant solution was removed, and the anthers were blotted on sterile filter paper and quickly frozen to -196 C. After rapid thawing and further cultivation, haploid plantlets were regenerated. The efficiency rate was not given.

Rosales

Fragaria annanasis Duch. Tissue culture is well established in the breeding and multiplication of *Fragaria annanasis*. The maintenance of tissue in vitro has become of interest for pratical purposes. The slow-

growing method was used by Mullin and Schlegel (1976). Sterile plantlets were grown in test tubes on a medium containing 1 l Knop solution (macroelements), 0.5 ml Berthelot solution (microelements), 3.0 μM
thiamine-HCl, 4.1 μM nicotinic acid, 2.4 μM pyridoxine-HCl, .03 mM
glycine, 0.55 mM myoinositol, 14.2 μM IAA, 0.47 μM KIN. The plantlets were grown on filter paper with 2.5 ml solution in a test tube and
stored in a refrigerator at 1 and 4 C. A total of 58 different genotypes were tested. The material was checked every 3 months and 1-2
drops of the solution were added to those cultures showing evidence of
desiccation. The genotypes could be stored up to 6 years. Storage at
1 C yielded better results than at 4 C. This storage method is very
inexpensive.

Freezing methods have been tested by Kartha et al. (1980). Strawberry meristems were isolated from plantlets propagated in vitro and
precultured on MS with 10 μM BA, supplemented with either 5% DMSO
for 2 days or different concentrations of glycerol for 1-3 days. Individual cryprotectants were added for 60-120 min at 0 C. Stepwise
freezing was done to -40 C with additional storage in liquid nitrogen,
or rapid freezing by plunging into liquid nitrogen.

Optimum cooling velocity was -0.84 C/min which resulted in a survival rate of 95%, the tested range was -0.56 to -0.95 C/min. Rapid
freezing and rapid dry freezing resulted in a survival of only 5-7%.
Prefreezing treatment with glycerol resulted in a lower survival rate
than cells with 5% DMSO as a cryoprotectant. After 1 week's storage
in liquid nitrogen, the plant survival and regeneration was 95%, but
after 2-8 weeks survival was reduced to 50-60%.

Malus domestica Borkh. The low-temperature storage of apple shoots
in vitro has been investigated by Lundergan and Janick (1979). *Malus
domestica* Borkh. cv. Golden Delicious shoot tips have been cultivated
on MS medium supplemented with 0.3 μM thiamine-HCl, 2.4 μM pyridoxin, 4.0 μM nicotinic acid, 0.3 mM glycine, 22 μM BA, .087 M sucrose,
and 10 g/l agar. After a 12 months' storage, there was a survival at
-17 C of 0%, at 1 C of 100%, at 4 C of 70%, and at 26 C of 3%. At
the low temperatures the average number of shoots per culture tube
increased with storage time.

Prunus cerasus L. Turmanov et al. (1968) investigated freezing survival of plum callus. On White or MS medium they cultivated callus
tissues at 26 C and varied the sucrose content for 10 days from 2% to
15%. Calluses were tested at -4, -7, -10, -13, -15, -20, -25, -30, -35,
-40, and -50 C. At low sugar concentrations callus survived at -25 C,
at the high concentrations callus survived at -35 to -40 C. Up to 50%
of the dry weight of the callus was sugar.

Rhamnales

Vitis rupestris Scheele. Slow-growing experiments have been done by
Glazy (1969) in *Vitis rupestris*. Sterile tillers were grown on a medium

composed of 0.5 liter Knop's solution, 0.5 ml Berthelot's micronutrients, thiamine, pyridoxin, nicotinic acid, ca-pantothenate, inositol, biotin, (all given without quantification), and 0.04 M sucrose. The survival after 24 days were at 2 C 13%, at 7 C 46%, at 9 C 100%, and after 42 days at 9 C 100%. At the highest temperature the tiller could be stored up to 300 days without loss of viability and without elongation of the tillers.

Sapindales

Acer pseudoplatanus L. One of the species more frequently investigated for tissue conservation is *Acer pseudoplatanus*. *Acer* can be cultured on WH medium with 27.0 µM 2,4-D, 10% CW, on LS or on a medium developed by Stuart and Street (1969). The optimum storage condition was determined by Sugawara and Sakai (1974) using 24% DMSO (v/v) and 0.55 M glucose (w/v) in distilled water, slow freezing to -40 to -50 C and storage in liquid nitrogen. After rapid thawing, survival in the TTC-test was 20-30%. It seems essential to use the cells in late lag phase or in the early cell division phase. Nag and Street (1975a) reported a survival of 2% after using 5% DMSO, a cooling rate of -2 C/min to -65 C and rapid freezing in liquid nitrogen with rapid thawing.

Slow freezing was shown to be optimum by Withers and Davey (1978) and Withers (1978b) for *Acer pseudoplatanus*. They used a cooling rate of -1 C/min and thawed it rapidly for electron microscopic purposes. Slow freezing in the presence of cryoprotectants was associated with a reduction in cell size by dehydration, reduced intracellular ice formation and good preservation of organelle integrity. Only cryoprotected cells survived, but even in these cells some ultrastructural modifications were evident.

Campanulales

Chrysanthemum morifolium Ramat. The only investigated species of the Campanulales is *Crysanthemum morifolium*. It has been tested in acclimation experiments by Banner and Steponkus (1976) and in slow-growing experiments (Chapter 29). The acclimation was investigated on MS medium with 4.6 µM KIN, 5.4 µM NAA, 0.55 mM inositol, 200 mg/l casein hydrolysate, and 100 ml/l CW. After subculturing at 27 C for 10, 17, 31, 45, and 59 days the callus was acclimated at 4.5 C for 0, 1, 2, 4, and 6 weeks. Freezing procedure was done stepwise to -4, -5, -6, -9, -12, and -15 C. The best combination occurs at 10 days preculture and 4-6 weeks acclimation. The cultivars Shining Light, Jessamine Williams, and John Milbrath survived at 4 weeks acclimation at a temperature of -15 C, the cultivars Chiquita and Larry at 6 weeks acclimation at a temperature of -16.1 C.

Haplopappus sp. Cass. *Haplopappus* is extremely sensitive to freezing in contrast to other Compositae. Without cryoprotectants and starting with cell suspension cultures of *H. gracilis*, Towill and Mazur (1976) observed a 10% survival of cells, frozen in Erikkson medium from -1 C with a cooling of -0.1 C/min down to -10 C. Cells frozen in distilled water had a lower survival rate. Hollen and Blakely (1975) used *H. ravenii* for their experiments. At a cooling rate of -1.3 to -3 C and using 10% DMSO they observed 3-5 times greater viability than the control that was transferred from -20 C by rapid thawing. However, a prolonged storage at -20 C was not possible. The best stage for starting such experiments seems to be the early log phase, because these rapidly growing cells are most resistant to freezing.

Solanales

Atropa belladonna L. Solanales are of interest in freezing experiments, because this order contains many economically important species, and most of these species are tropical or subtropical. Therefore acclimation by low temperatures to freezing conditions cannot be expected.

Experiments with *Atropa belladonna* had been done by Nag and Street (1975a,b). Using callus cultures of *Atropa* on the SSM medium of Thomas and Street (1972) with 11.0 µM NAA, different cryoprotectants (DMSO, glycerol) were tested; also the cooling rate was varied from 0.5-65 C/min. Best results were obtained with 5% DMSO and a cooling rate of 1-2 C/min. In the absence of cryoprotectants cell death occurred between -10 and -30 C, with DMSO cell survival decreased from 90% at 2 C to 28% at -60 C and was constant at lower temperatures. From the rapidly thawed cultures on the mentioned medium without NAA embryoids were formed.

Capsicum annuum L. One report is given on the cryopreservation of *Capsicum annuum* by Withers and Street (1977). The influence of molarity was investigated by increasing the concentration of mannitol. The cultures were grown on LS medium supplemented with 1.8 µM 2,4-D, and 0.14 µM KIN. Mannitol was added at concentrations of 0 M, .05 M, .09 M, and 0.29 M (w/v). After 7 days subculture cell diameter decreased from 112 to 85 then to 72 µm. Freezing procedure was done using glycerol (10%) and DMSO (5%) as cryoprotectants and a freezing rate of 1-2 C/min down to -100 C with subsequent storage at -196 C.

A 40 C water bath was used for thawing. The cell suspensions were then diluted 1:50 with new medium. *Capsicum* cells were shown to be highly sensitive to chilling injury by storage at 2 C, but 70% of cells grown in presence of 5% mannitol and treated directly with cryoprotectants survived, and the survival rate was decreased to 25-30% with additional freezing procedures.

Slow growth of *Capsicum annuum* was also examined by Withers (1978a). The callus cultures survived a temperature of 4 C for 25 days.

Datura stramonium L.   Like other Solanaceae, *Datura* sp. have excellent response to tissue culture procedures.   Tissue conservation of *Datura* has only been investigated by Bajaj (1976).   He cultured *Datura stramonium* on MS medium with 2.3 µM 2,4-D.   After a pretreatment with 7% DMSO, the cells were frozen at cooling rates from 1-3 C/min to -196 C.   Cell survival was determined by absorption at 546 nm caused by tetrazolium chloride reduction.   Highest survival rate was obtained at a cooling rate of 1 C/min.

Ipomoea sp. L.   From the Convolvulacea, Latta (1971) investigated the freezing possibilities of sweet potatoes.   He started with 3-year-old cell cultures on 67V medium and compared it to carrots.   Addition of cryoprotectants showed that *Ipomoea* cells are much less permeable than those of carrots, mainly because of the sucrose content.   So *Ipomoea* cells did not survive the technique successfully used in carrots.   However, preculture on a medium containing 65% sucrose and the addition of 2.5% glycerol and 2.5% DMSO resulted in good survival of the cells at temperatures of -40 C.

Lycopersicum esculentum L.   Another Solanaceous species that has been investigated is tomato (Grout et al., 1978).   Seedlings with radicle length of 18-20 mm were surface sterilized and put into MS medium with .08 M sucrose (w/v) at 0 C.   As a cryoprotectant DMSO was added stepwise at 20 min intervals for 2 hr.   Different DMSO concentrations and freezing methods were tested.   The best survival was obtained using 5-10% DMSO, using a rapid freezing method, and using vapor cooling in a Dewar bottle over liquid nitrogen with a temperature decrease of -20 to -55 C/min.   The best survival was obtained at 10-15% DMSO.   Under the slow-freezing condition all the seedlings died.   Of the cultures, 40-45% produced shoots immediately.   This is about half the frequency of the nonfrozen control.   By adding GA to the nonsurvivors, all material showed morphogenesis.   Based on this result it was concluded that the hormonal regulation of organized growth has been altered by the freezing and/or thawing process.

Nicotiana tabacum L.   Freezing experiments in *Nicotiana tabacum* have been reported (Chapter 28; Bajaj, 1976, 1978).   Culture medium is MS with 1.1-11.0 µM   IAA and .047-0.93 µM KIN.   Cell suspensions, callus or anthers could be frozen.   DMSO, ethylene glycol, glucose, and glycerol have been tested as cryoprotectants.   The best results obtained were with 10% glycerol or a mixture of 0.22 M glucose, 3% DMSO, and 2.5% ethylene glycol.   With glycerol, cells survived a 6 months storage at -196 C.   Using this mixture of 3 cryopreservation compounds only 20% of the cells recovered from a temperature of -23 C.   After thawing the cells remained in lag phase for up to 4 months.   The stepwise freezing method was used with a freezing rate of 2 C/min.   The slow-growing method has also been applied to tobacco and Chrysanthemum (Chapter 29).

Solanum sp. L.   The rapid–freezing method has been successful in *Solanum* sp., as reported by Grout and Henshaw (1978) for *S. goniocalyx* and Bajaj (1981a) for *S. tuberosum*.   The media used were modified MS with various concentrations and combinations of sucrose, IAA, 2,4-D, BA, GA, and KIN.   In addition, 10% DMSO or 0.15 M sucrose, 5% glycerol, and 5% DMSO were used as cryoprotectants.   Shoot tips, axillary buds, or tuber sprouts could all be preserved.

After rapid freezing, storage from 4 weeks to 24 months, and rapid thawing, the survival rate was determined.   Combinations of the three cryoprotectants gave better results than DMSO alone.   Tuber sprouts survived at a rate of 11-14%, while 19-27% of axillary buds and shoot tips survived.   Regeneration of plants occurred after a lag period.

The slow–growing method was used by Mix (1981).   Nodal segments were grown on liquid medium on filter paper.   The medium is given in Table 4.   At 10 C and 4000-5000 lux the tissues may survive for 2 years.   Some genotypic differences have been observed.

Storage of dried callus was successful in only one line of potato. Here the dihaploid clone HH 258 was grown on MS with 37.8 µM ABA and 0.15 M sucrose, dried in sterile air, and cooled to -80 C for 36 hr. Regrowth occurred with shoot formation after transfer to new medium (Nitzsche, unpublished).

Table 4.   Medium for Slow-Growing Culture of Potatoes (from Mix, 1981)

KNO_3	.019 M
KH_2PO_4	1.25 mM
NH_4NO_3	.02 M
$MgSO_4 \cdot 7H_2O$	1.79 mM
$CaCl_2 \cdot 2H_2O$	3.0 M
H_3BO_3	0.1 mM
$MnSO_4 \cdot 4H_2O$	0.1 mM
$FeSO_4 \cdot 7H_2O$	.05 mM
$Na_2EDTA \cdot 2H_2O$	0.1 mM
$CoCl_2 \cdot 6H_2O$	0.1 µM
$CuSO_4 \cdot 5H_2O$	0.1 µM
$ZnSO_4 \cdot 7H_2O$	.03 mM
$Na_2MoO_4 \cdot 2H_2O$	1.0 µM
KI	5.0 µM
Succinic acid 2.2 dimethyl hydrazide 5% 1 ml/l	
Sucrose	.06 M
pH5.8	

Salicales

Populus euamericanus Guinier.   As with all woody species, cold hardiness is important in *Populus*, and cryopreservation was therefore investigated by Sakai and Sugawara (1973).   They cultured popular callus of *Populus euamericanus* cv. Gelrica on MS medium with 16.1 µM NAA at temperature regimes of 23/19, 12/0, 8/5, 15, 12, and 0 C for

25 days. After this pretreatment the callus tissue was cooled to -5 C
and later on in daily intervals in -5 C steps to -30 C.   This temperature was maintained for 16 hr during the test.   A part of the
tissue was also frozen in liquid nitrogen for 2 hr.   Thawing procedure
was done slowly in air at 0 C.   Survival of material was determined by
regrowth on MS medium for 50 days.   All treatments except the 23/19
treatment survived -5 C.   The 12/0 and the 8/5 treatments were hardy
to -15 C.   When the callus was subsequently subjected to 0 C for 20
days to induce acclimation, callus of the 12/0 regime survived -70 C or
below.

Gentianales

Coffea arabica L.   Coffee seeds retain their viability at low-tempera-
ture/low moisture conditions for only 4 months to 2.5 years.   Therefore, tissue storage is very interesting for germplasm conservation.
Kartha et al. (1981) investigated this problem using the cultivars Cat-
urra Rojo and Catuai.   Seedling meristem tips were grown on MS medium supplemented with B5 vitamins.   The addition of 0.44-4.4 µM BA or
0.46-4.6 µM ZEA led to single shoot formation, while the addition of
22.0-44.0 µM BA or ZEA caused multiple shoot formation.   Root formation occurred at half concentration of MS without sugar after 6-8
weeks for 90-100% of the shoots.   The shoots regenerated from meristems have been maintained in vitro at 26 C, using a 16 hr photoperiod with 7500 lux intensity for over 2 years as plantlets.   During
this period the regenerated trees are only 3-4 cm high.

Palmales

Phoenix dactylifera L.   The possibility of freeze conservation of a
tropical palm tree, date palm, has been investigated by Finkle et al.
(1979) and by Ulrich et al. (1979) and is described in Chapter 28.

Poales

Lolium sp. L.   The callus-drying method has been tested in *Lolium
temulentum* L. by Nitzsche (unpublished).   Callus was grown on B5
medium with 11.3 µM 2,4-D.   A pretreatment for 16 days with the
same medium using 0.06, 0.15, and 0.35 M sucrose and 0, 0.37, 3.8 and
37.8 µM ABA was tested.   After drying for 7 days the callus was put
on the original medium.   Regrowth was obtained at a rate of 51-96%,
independent of the media variation, but no plants could be regrown
from the material.
   Dale (1980) used the slow-growing method for *L. multiflorum* Lam.,
starting with shoot tip culture.   On MS supplement and with 0.93 µM
KIN plantlets were recovered from shoots that were stored at 2-4 C in
an 8 hr day at 300 lux.   After one year the plantlets were transferred
to 25 C.   The regeneration rate obtained was 67%.

Oryza sativa L.   The tropical gramineae *Oryza sativa* L. has been examined in freezing experiments.   Sala et al. (1979) used the cultivar Ronsarolo for the initiation of suspension cultures on the R-2 medium of Ohira et al. (1973).   Cells treated with DMSO and glycerol were put in thermos flasks (Dewar bottles) and these were kept at -70 C for 18 hr. The lack of an apparatus for controlled freezing could be overcome by this simple method.   Thawing was done rapidly.   In the tetrazolium chloride test the survival was determined as 58% after the 5% DMSO pretreatment, and the dry weight increase was the same as of the un-frozen control.   Glycerol and other concentrations of DMSO failed in these experiments.

Zea mays L.   *Zea mays* has also been investigated (Withers, 1978a). The stepwise freezing method and dry freezing method have been used, the latter with mature and immature zygotic embryos.   The experiments were successful, but no survival rates are given.

CONCLUSIONS

Methods and techniques for cell and tissue conservation have in-creased in recent years, and more and more success has been gained. An examination of the literature reveals that there are a lot of gaps in the available information.   It is hoped that we will learn more about cryopreservation in the future.

The knowledge of cooling rate and cryoprotectants offers the possi-bility to start special systematic investigations with every cultured species.   It may be expected that these experiments will enjoy great success.   Our understanding of acclimation and action of hormones is even more fragmented.   Many biochemical investigations in basic re-search may be necessary for planning reasonable further investigations. Probably this field will offer a lot of useful new possibilities in understanding the living material in action and dormancy.   The survival rate is an important parameter, but it must not be overestimated in its value.   A 5-10% survival with an immediate start of growth after thaw-ing may offer better results than an 80% survival with a lag phase of several weeks after thawing.   The 5% survival of a 1000 cell charge offers enough growth potential for regaining the stored genotype.   For practical purposes the tissue conservation will compete with the stor-age of seeds.   These have the advantage of being naturally qualified for storage, while the tissues have to be altered in such conditions by artificial manipulations.   Therefore, the seeds will have in most cases an advantage and should be preferred whenever possible.   But in spe-cial cases, as mentioned in the introduction, tissues and cells can not be replaced by seeds, and new possibilities are offered by the methods of tissue conservation.

KEY REFERENCES

Bajaj, Y.P.S. and Reinert, J. 1977.   Cryobiology of plant cell cultures
and establishment of gene-banks.   In: Applied and Fundamental Aspects of Plant Cell, Tissue, and Organ Culture (J. Reinert and Y.P.S.
Bajaj, eds.) pp. 757-777.   Springer-Verlag, Berlin, Heidelberg, New
York.

Cinatl, J. and Tolar, M. 1971.   Technik der Zellkultivation.   In: Aktu-
elle Probleme der Zellzuechtung (B. Mauersberger, ed.) pp. 63-129.
Gustav Fischer Verlag, Stuttgart.

Withers, L.A. 1978a.   Freeze preservation of cultured cells and tissues.
In: Frontiers of Plant Tissue Culture 1978 (T.A. Thorpe, ed.) pp. 297-
306.   International Association for Plant Tissue Culture, Calgary.

REFERENCES

Aufhammer, G. and Fischbeck, G. 1964.   Ergebnisse von GefaeB- und
Feldversuchen mit dem Nachbau keimfaehiger Gersten- und Haferkorn-
er aus dem Grundstein des 1832 errichteten Nuernberger Stadt-
theaters.   Z. Pflanzenzuecht.   51:354-373.
Bajaj, Y.P.S.   1976.   Regeneration of plants from cell suspensions frozen at -20, -70, and -196 C.   Physiol. Plant. 37:263-268.
_____ 1978.   Effect of super low temperature on exised anthers and
pollen embryos of Atropa belladonna, Nicotiana tabacum and Petunia
hybrida.   Phytomorphology 28:171-176.
_____ 1981a.   Regeneration of plants from potato meristems, freeze-
preserved for 24 months.   Euphytica 30:141-145.
_____ 1981b.   Regeneration of plants from ultra-low frozen anthers of
Primula oboenica.   Sci. Hort. 14:93-95.
Bannier, L.J. and Steponkus, P.L. 1976.   Cold acclimation of Chrysanthemum callus cultures.   J. Am. Soc. Hort. Sci. 101:409-412.
Bhojwani, S.S. 1981.   A tissue culture method for propagation and low
temperature storage of Trifolium repens genotypes.   Physiol. Plant.
52:187-190.
Blaydes, D.F. 1966.   Interaction of kinetin and various inhibitors in the
growth of soybean tissue.   Physiol. Plant. 19:748-753.
Caplin, S.M. 1959.   Mineral oil overlay for conservation of plant tissue
cultures.   Am. J. Bot. 46:324-329.
Cheyne, V.A. and Dale, P.J. 1980.   Shoot tip culture in forage legumes.
Plant Sci. Lett. 19:303-309.
Dale, P.J. 1980.   A method of in vitro storage of Lolium multiflorum
Lam.   Ann. Bot. 45:497-502.
Dougall, D.K. and Wetherell, D.F. 1974.   Storage of wild carrot cul-
tures. Cryobiology 11:410-415.
Finkle, B.J., Ulrich, J.M., Rains, D.W., Tisserat, B.B., and Schaefer,
G.W. 1979.   Survival of alfalfa, Medicago sativa, rice, Oryza sativa,
and date palm, Phoenix dactylifera, callus after liquid nitrogen freez-
ing.   Cryobiology 16:583.

Finkle, B.J., Ulrich, J.M., and Tisserat, B.B. 1980. Regeneration of date palm trees from callus stored at -196 C. Plant Physiol. 65(Suppl.):36.

Frandsen, K.J. 1951. Methodische Fragen der daenischen Futterpflanzenzuechtung. Vortraege ueber Pflanzenzuchtung. Land- and Forstwirtschaftl. Forschungsrat e.V. Bonn.

Gamborg, O.L. and Eveleigh, D.E. 1968. Culture methods and detection of glucanases in suspension cultures of wheat and barley. Can. J. Biochem. 46:417.

Glazy, R. 1969. Recherches sur la croissance de *Vitis rupestris* Scheele sain et court noue cultive in vitro a differentes temperatures. Ann. Phytopathol. 1:149-166.

Grout, B.W.W. and Henshaw, G.G. 1978. Freeze preservation of potato shoot tip cultures. Ann. Bot. 42:1227-1229.

Grout, B.W.W., Westcott, R.J., and Henshaw, G.G. 1978. Survival of shoot meristems of tomato seedlings frozen in liquid nitrogen. Cryobiology 15:478-483.

Hollen, L.B. and Blakely, L.M. 1975. Effects of freezing on cell suspension cultures of *Haplopappus ravenii*. Plant Physiol. 56(Suppl.):39.

Iwanami, Y. 1972a. Viability of pollen grains in organic solvents. Botanique (Nagpur) 3:61-68.

_____ 1972b. Retaining the viability of *Camellia japonica* pollen in various organic solvents. Plant Cell Physiol. 13:1139-1141.

_____ 1975. Absolute dormancy of pollen induced by soaking in organic solvents. Protoplasma 84:181-184.

_____ and Nakamura, N. 1972. Storage in organic solvent as a means for preserving viability of pollen grains. Stain Technol. 47:137-139.

Jeyendran, R.S., Graham, E.F., and Schmehl, M.K.L. 1981. Fertility of dehydrated bull semen. Cryobiology 18:292-300.

Kartha, K.K., Leung, N.L., and Gamborg, O.L. 1979. Freeze-preservation of pea meristems in liquid nitrogen and subsequent plant regeneration. Plant Sci. Lett. 15:7-16.

Kartha, K.K., Leung, N.L., and Pahl, K. 1980. Cryopreservation of strawberry meristems and mass propagation of plantlets. J. Am. Soc. Hort. Sci. 105:481-484.

Kartha, K.K., Mroginski, L.A., Pahl, K., and Leung, N.L. 1981. Germplasm preservation of coffee (*Coffea arabica* L.) by in vitro culture of shoot apical meristems. Plant Sci. Lett. 22:301-303.

Koehler, G. and Milstein, C. 1975. Continuous cultures of fused cells secreting antibody of predefined specificity. Nature 256:495-497.

Latta, R. 1971. Preservation of suspension cultures of plant cells by freezing. Can. J. Bot. 49:1253-1254.

Levitt, J. and Dear, J. 1970. The role of membrane proteins in freezing injury and resistance. In: The Frozen Cell (G.E.W. Wolstenhome and M. O'Connor, eds.) J.A. Churchill, London.

Lovelock, J.E. and Bishop, M.W.H. 1959. Prevention of freezing damage to living cells by dimethyl sulfoxide. Nature (London) 183:1394-1395.

Løvtrup, S. and Hansson Mild, K. 1981. Permeation, diffusion, and structure of water in living cells. In: International Cell Biology 1980-1981 (H.G. Schweiger, ed.) pp. 889-903. Springer-Verlag, Berlin, Heidelberg, New York.

Lundergan, C. and Janick, J. 1979. Low temperature storage of in vitro apple shoots (*Malus domestica* cv. Golden Delicious). Hort-Science 14:514.

Miller, C.O. 1963. Kinetin and kinetin-like compounds. In: Molderne Methoden der Pflanzenanalyse (K. Paech and M.V. Tracy, eds.) Vol. VI, pp. 194–202. Springer-Verlag, Berlin, Heidelberg, New York.

Mix, G. 1981. Kartoffelsorten aus dem Reagenzglas—Bedingungen zur Langzeitlagerung. Der Kartoffelbau 32:198–199.

Mullin, R.H. and Schlegel, D.E. 1976. Cold storage maintenance of strawberry meristem plantlets. HortScience 11:100–101.

Nag, K.K. and Street, H.E. 1973. Carrot embryogenesis from frozen cultured cells. Nature 245:270–272.

_____ 1975a. Freeze preservation of cultured plant cells. I. The pretreatment phase. Physiol. Plant. 34:254–260.

_____ 1975b. Freeze preservation of cultured plant cells. II. The freezing and thawing phase. Physiol. Plant. 34:261–265.

Nitzsche, W. 1978. Erhaltung der Lebensfaehigkeit in getrocknetem Kallus. Z. Pflanzenphysiol. 87:469–472.

_____ 1980. One year storage of dried carrot callus. Z. Pflanzen-physiol. 100:269–271.

Ohira, K., Ojima, K., and Fujiwara, A. 1973. Studies on the nutrition of rice cell culture. I. A simple, defined medium for rapid growth in suspension culture. Plant Cell Physiol. 14:1113–1121.

Pierquet, P. and Stushnoff, C. 1980. Relationship of low temperature exotherms to cold injury in *Vitis riparia*. Am. J. Enol. Vitic. 31:1–6.

Popov, A.S., Butenko, R.G., and Glukhova, I.N. 1978. Effect of pretreatment and condition of deep freezing on renewal of suspension cultures of *Daucus carota* cells. Fiziol. Rast. 25:1227–1236.

Quatrano, R.S. 1968. Freeze-preservation of cultured flax cells utilizing dimethyl sulfoxide. Plant Physiol. 43:2057–2061.

Roggen, H.P.J.R. 1974. Fertilization in Higher Plants (H.F. Linsken, ed.) ASP, Amsterdam.

Sakai, A. and Sugawara, Y. 1973. Survival of poplar callus at superlow temperatures after cold acclimation. Plant Cell Physiol. 14:1201–1204.

Sala, F., Cella, R., and Rollo, F. 1979. Freeze-preservation of rice cells grown in suspension culture. Physiol. Plant. 45:170–176.

Scheuermann, E.A. 1967. Fluessiges Wasser bei Minusgraden. Kosmos (Stockholm) 63:331.

Schroeder, J., Heller, W., and Hahlbrook, K. 1979. Flavanone synthase: Simple and rapid assay for the key enzyme of flavonoid biosynthesis. Plant Sci. Lett. 14:281–286.

Seibert, M. 1976. Shoot initiation from carnation shoot apices frozen to -196 C. Science 191:1178–1179.

_____ and Wetherbee, P.J. 1977. Increased survival and differentiation of frozen herbaceous plant organ cultures through cold treatment. Plant Physiol. 59:1043–1046.

Stuart, R. and Street, H.E. 1969. Studies on the growth in culture of plant cells. IV. The initiation of division in suspension of stationary-phase cells of *Acer pseudoplatanus* L. J. Exp. Bot. 20:556–571.

Sugawara, Y. and Sakai, A. 1974. Survival of suspension-cultured syca-more cells cooled to the temperature of liquid nitrogen. Plant Physiol. 54:722-724.

Thomas, E. and Street, H.E. 1972. Factors influencing morphogenesis in excised roots and suspension cultures of *Atropa belladonna*. Ann. Bot. 36:239-247.

Tomes, D.T. 1979. A tissue culture procedure for propagation and maintenance of *Lotus corniculatus* genotypes. Can. J. Bot. 57:137-140.

Towill, L.E. and Mazur, P. 1976. Osmotic shrinkage as a factor in freezing injury in plant tissue cultures. Plant Physiol. 57:290-296.

Trewavas, A.J. 1976. Plant growth substances. In: Molecular Aspects of Gene Expression in Plants (J.A. Bryant, ed.) Academic Press, London, New York, San Francisco.

Tumanov, I.I., Butenko, R.G., and Ogolevets, I.V. 1968. Application of isolated tissue technique for studying hardening processes in plant cells. Fiziol. Rast. 15:749-756.

Ulrich, J.M., Finkle, B.J., Moore, P.H., and Ginoza, H. 1979. Effect of a mixture of cryoprotectants in attaining liquid nitrogen survival of callus cultures of a tropical plant *Saccharum* sp. cultivar H-50-7209. Cryobiology 16:550-556.

Widholm, J.M. 1972. The use of fluorescein diacetate and phenosafranin for determining viability of cultured plant cells. Stain Technol. 47:189-194.

Withers, L.A. 1978b. A fine structural study of the freeze-preservation of plant tissue cultures. II. The thawed state. Protoplasma 94:235-249.

_____ 1979. Freeze-preservation of somatic embryos and clonal plant-lets of carrot (*Daucus carota* L.). Plant Physiol. 63:460-467.

_____ and Davey, M.R. 1978. A fine structural study of the freeze-preservation of plant tissue cultures. I. The frozen state. Protoplasma 94:207-219.

_____ and Street, H. 1977. Freeze-preservation of cultured plant cells. III. The pregrowth phase. Physiol. Plant. 39:171-178.

CHAPTER 28
Protocols of Cryopreservation

B. Finkel and *J. Ulrich*

There is an urgent need to preserve crop plants through improved methods of vegetative propagation. This is especially so with tropical crops. Successful agricultural practice requires that many of the tropical species be reproduced vegetatively. In addition many lines that are, or may be, valuable as gene pools for breeding are lost each year by neglect or by mass eradication of their habitats for alternative uses. Crop cultivars, parental lines, and experimental lines (e.g., mutants) often require long-term preservation. Preservation is needed, also, for endangered wild species of all types. Improved methods of preservation might be extended to include any tropical plant, including those from tropical forests.

The practice of using tissue cultures for the propagation of plant lines has proven useful. Tissue culture has several advantages over field propagation in the study of plant growth and the preservation of genotypes, such as a greatly decreased demand for land and manpower and decreased exposure, in axenic culture, to crop-borne diseases and pests. But there are also problems with the maintenance of tissue cultures. These include the time commitment of skilled personnel for repeated transfers, the danger of contamination during each transfer operation which, unfortunately, can cause elimination of a line, and most serious of all, selective genetic changes in the culture that take place during repeated subculturing (Sunderland, 1977; Torrey, 1967). Loss of morphogenic potential, and the other problems associated with tissue culturing, would be minimized by our being able to freeze tissues of the desired lines in a living condition at very cold temperatures such as in liquid nitrogen (boiling point, -196 C). Successful cryogenic methods for doing this are now being developed.

Freezing followed by growth of the cells after thawing is now possible for a number of plant species, including several which have been regenerated into whole plants (Finkle et al., 1980; Kartha et al., 1979; Sakai et al., 1978; Withers and Street, 1977). So far, success has been attained in viably freezing a variety of tissue types, including callus cultures, suspension cultures, growing points, anthers, and pollen. Even protoplasts—single, membrane-bound cells whose cell walls have been enzymatically removed—can be treated to survive freezing (Siminovitch, 1979). A great amount of experimentation and several successful accomplishments have been reported on the freezing of callus cultures to very cold temperatures, e.g., -196 C. At this temperature they can be stored without further care, and with little or no change in their metabolic or genetic characteristics (Nag and Street, 1973; Ulrich et al., 1981) for many months or possibly for hundreds of years (Ashwood-Smith and Friedman, 1979). Thus the highly desirable idea of a cell repository or bank of many frozen lines of genetically identified plant cells is slowly becoming feasible.

In most cases the freezing of plant tissues has been performed in the presence of cryoprotective chemicals such as dimethylsulfoxide (DMSO), glycerol, sugars, large molecular weight polymers, or combinations of these (Finkle and Ulrich, 1979; Ulrich et al., 1979). These compounds protect plant cells from damage by freezing, even at very cold temperatures, so that growth can continue after thawing. We do not understand very well how the cryoprotective chemicals prevent freezing damage, but we are able to use their stabilizing effect, even if the compounds themselves sometimes cause a degree of irreversible damage to plant cells. The following protocol makes use of a mixture of compounds whose combination gives a number of species of plant cultures excellent cryoprotection at -196 C while minimizing adverse effects (Ulrich et al., 1979).

METHODS

Plant Material

Within a species, a number of types of tissue can be used for freezing. Callus tissue, particularly among tropical species, has often responded to our present method of freezing (Finkle et al., 1980; Tisserat et al., 1981; Ulrich et al., 1979). Callus tissue is most resistant to freezing damage when at a vigorous, rapidly growing stage shortly after transplanting (1 or more weeks, depending on the growth rate). When selecting experimental tissue, old cells at the top of the callus, blackened areas, etc., should be avoided. Callus cultures display different growth habits, so that various transfer techniques are useful for experimental manipulation. Calli grow as large, friable lumps that can be cut into small pieces, or they may proliferate as loosely associated lumps or grow as sheet-like films on the agar surface or as aggregates of a watery consistency. With highly hydrated lines (such as many rice lines) the callus can be scraped together and transferred with a spatula or spoon. Alternatively, for some cultures or situations the plant material may be handled better as suspension cultures in flasks or tubes and transferred with wide-tipped pipettes.

Medium and Wash Solutions

Murashige and Skoog (MS) medium, or whatever medium is suitable to the tissue of interest, is used. In our laboratory for culturing rice the standard MS components, namely, macronutrients, micronutrients, and organic compounds, are supplemented with 2,4-D (4.5 µM) and IAA (5.7 µM) as auxins, KIN (4.6 µM), L-glutamine (1 mM), and sucrose (0.087 M) (Schaeffer and Sharpe, 1981). To prepare solid medium, agar is dissolved at 1% concentration in a hot solution of medium before sterilization.

A simplified sterile wash solution is used on thawed samples to remove cryoprotective chemicals. It contains only the MS macronutrients plus 0.087 M sucrose.

Cryoprotective Solution

Many individual chemical substances or combinations of them have been used. A particularly successful mixture consists of PGD, a solution consisting of polyethylene glycol (Carbowax PEG 6000, Union Carbide Corp., 10% solution by weight), glucose (0.44 M), and dimethylsulfoxide (DMSO, 10%).

Discussion of Special Equipment

A sterile, or laminar flow, hood ("clean bench") is used for aseptic transfers.

CONTROLLED TEMPERATURE CULTURE CHAMBER. For suspension cultures, one needs shaking equipment to aerate and stir the solutions in flasks or tubes.

TUBES FOR FREEZING THE TISSUE. Sterile heavy-walled Pyrex borosilicate glass conical centrifuge tubes, 12 ml, graduated, with screw caps or capped with heavy aluminum foil (100 µm thick) or plastic screw-capped freezing vials obtainable in several sizes (A/S Nunc, Denmark) are required. The graduations of the glass tubes have the advantage that the approximate settled or packed volume of cells can be estimated for experimental purposes, while the plastic vials require less storage space.

SUCTION PIPETTES. Sterile drawn-tip (Pasteur) pipettes, about 20 cm long, with a rubber bulb at the top, are used for adding and withdrawing solutions while manipulating the callus samples.

DISPENSING BOTTLE.   This should be autoclavable, for adding small portions of sterile solution (e.g., wash solution) to samples.

FREEZING FACILITY.   Freezing may be performed either stepwise in baths of different temperatures, e.g., -10, -15, -23, and -30 C, 4 min at each temperature (Ulrich et al., 1979) or in a continuous-change freezing bath that may be assembled from an immersion cooler and other widely available, simple components (Withers and King, 1980), or it may be a more elaborate programmed unit (e.g., Cryo-Med, Mt. Clemens, Michigan).   A temperature curve that approximates a temperature drop of 1-3 C/min, along with a step to nucleate (initiate) ice formation in the temperature region -4 to -10 C, gives high recoveries.

A tank of liquid nitrogen, small accessory liquid nitrogen containers (double-walled Dewar beakers or bottles), and a liquid nitrogen sample-storage container are needed to facilitate the processing and storage of tissue at -196 C.

PRETREATMENT OF TISSUE

Rice callus tissue is used at 1-2 weeks after transfer.   In a sterile hood, scrape the loose tissue together and transfer an adequate amount by spatula to a small, cold petri dish standing on cracked ice.   Harvest 5 ml or more of cells, depending on the amount needed for the experiment or for propagation.   For different experimental treatments, a minimum convenient amount for handling is about 0.2-0.5 ml volume of settled cells per sample.   Much larger amounts may be difficult to handle or may not respond well to the freezing and thawing program. (The cracked ice need not be sterile if used carefully.)   Mix the cells in the dish gently, divide them into uniform piles, and remove some of them to separate dishes if they are to be used for different experimental pretreatments.

To each volume of tissue in a cold 6 cm petri dish, add cryoprotective solution as follows: At first add about 3 volumes (compared to settled cell volume) of 4x diluted PGD at ice temperature and stir gently.   After 10 min, remove the liquid by suction pipette. Subdivide the cells further into separated portions, one for each sample.   Transfer the portions into labeled freezing vials chilled in an ice bucket. Add 1 ml cold full-strength PGD to each vial.

The Freezing Operation

Starting with the samples at ice temperature, place them either at -10 C in the bath, or in the programmed freezer.   Initiate ice crystallization manually by pressing a wedge of dry ice (solid carbon dioxide) to the outside of the freezing tube just until crystals of ice begin to form inside, or, if using a programmed freezer, set the program to activate a short-duration drop in temperature (e.g., to -50 C).   The transient large flow of cold nitrogen vapor cools the samples for just enough time to initiate ice formation in the solution inside the sample tube

PROCEDURE

CRYOPROTECTIVE ADDITIONS:
PEG-GLUCOSE-DMSO (P-G-D, 10-8-10% W/V)
ADDED TO CELLS

THAWING:
SWIRL IN +40°C BATH
WASH WITH 3% SUCROSE MEDIUM AT R.T.

FREEZING:
3°/ MIN. TO −4°C
SEEDING AT −4°C
1°/MIN. TO −30°C
THEN INTO L.N. (−196°C)

DETERMINING SURVIVAL:
TRANSFER TO GROWTH MEDIUM.
OR, ADD TRIPHENYLTETRAZOLIUM CHLORIDE (TTC)
TO DETERMINE VIABILITY INDEX
(530nm ABSORBANCE)

Figure 1. Outline of steps for freezing, thawing, and evaluating plant cells. (PEG, polyethyleneglycol; DMSO, dimethylsulfoxide; L.N., liquid nitrogen; R.T., room temperature.)

without excessively cooling the callus cells. Following the initiation of freezing, continue to cool the tubes, according to the freezing program. When the samples reach -30 C, transfer to a Dewar beaker containing liquid nitrogen and hold at this temperature (-196 C) until the program calls for thawing, either soon or after a period of storage. During the course of experimental runs, the program may require the withdrawal and thawing of some samples at temperatures above -196 C, e.g., at -15 amd -30 C.

The Thawing Operation

Attention to several precautionary details is required for the thawing operation. Plunge the frozen tips of the sample tubes into warm water with a vigorous swirling wrist action just to the point of ice disappearance. Protect the tops to avoid contamination from splashing. When using glass tubes a bath at 40-45 C is best; for plastic, 60 C. It is important for survival of the tissue that the tubes not be left in the very warm bath after the ice melts. Just at the point of thawing, quickly transfer the tubes to a water bath at room temperature (R.T., 20-25 C) and continue the swirling action for 15 sec to cool the warm walls of the tubes. Then leave the tubes at R.T. for the short time until ready for washing and culture transfers.

Reculturing

With tubes containing tissue suspended in 1 ml PGD at R.T. (again in a sterile hood) flame-sterilize the tubes and, using a suction pipette or calibrated dispensing bottle, add 0.5 ml of sucrose wash solution at R.T. (Persidsky et al., 1980) to each tube, in turn, with gentle stirring by rotation (50% dilution). After 10-15 min, remove the liquid by suction. Again, add 1 ml wash and remove. Using the pipette tip, push cells up onto the side of the tube for drainage and later ease of removal.

Transfer the callus cells onto agar plates using two spatulas. One of the spatulas may be spoon shaped, beaten from stainless steel or aluminum sheet metal, to make tissue transfer simpler and so that a level measured volume of callus can be transferred with experimental accuracy. A deep petri dish (25 mm height) with 50 ml of nutrient agar for culturing often gives better growth of callus cultures than the ordinary shallow (15 mm) type, especially when several callus samples are transferred onto each plate. A double layer of stretched Parafilm sheet (American Can Co., Greenwich, Conn.) or equivalent gas-permeable film can be wound around the petri dish lip to hold the plates together, minimize contamination, and diminish the rate of water loss from the cultures. Alternatively, with suspension cultures, the washed cells are cultured in nutrient medium in tubes or flasks, on a shaker.

Evaluation of Survival and Recovery

Although there are colorimetric and other chemical methods by which the viability of treated cells can be evaluated (Finkle and Ulrich, 1979; Nag and Street, 1973), renewed growth is the ultimate measure of survival. With callus cultures this can be determined by size, by weight, or in some cases, by the number of growing cells or plantlets. With cell suspension cultures growth can be determined similarly, measuring the packed volume of cells after centrifugation in a volume-graduated tube.

When the experiment is designed to compare the sizes and weights of calli formed during a growth period, it is an advantage to have uniform-sized callus pieces for all treatments and then to subculture these in a uniform, favorable environment. In this way differences in growth capability can be easily recognized. For example, four treatments in duplicate can be equally spaced in a circle around the dish (eight total calli), with one or two untreated control pieces placed at the center (Figure 2). See note on the use of deep petri dishes in the previous section (above). Sufficient mathematic reliability for statistical analysis (e.g., of final harvest weights) can be obtained by inoculating at least three petri dishes in this manner.

Evaluate the changes in size after sufficient growing time has elapsed for comparison of the effects of the treatments. Size can be judged by eye or by more precise methods. The size attained by callus pieces may be outlined on the surface of the petri dish or traced on paper or photographed, or the callus shapes may be copied with a photocopying machine (e.g., Xerox). Determine the area of each callus.

Alternatively, the callus can be removed for weighing. Remove each callus quantitatively and aseptically if you wish to continue the experiment by reimplanting the callus on nutrient agar for further growth treatment. The callus can be replaced on the original dish for continued growth, or transferred to another dish for different treat-

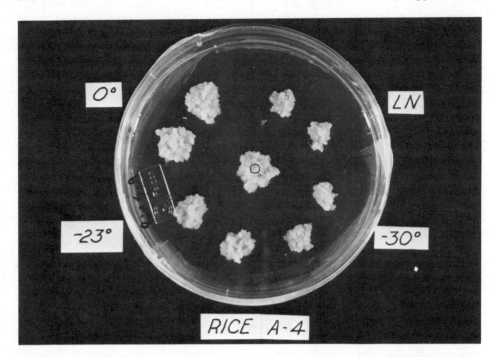

Figure 2. Rice culture dish showing the grouping of callus tissue after
treatments, and subsequent growth. Treatments: PGD added at 0 C,
then callus brought to indicated temperature. Tissue treatments are in
duplicate. Untreated control tissue at center; black circle indicates
original size of callus implant. Growth period, 3 weeks.

ment, e.g. to differentiation medium (Ulrich et al., 1979). Weigh the
calli directly (fresh weight, recoverable), or dry them in an oven before
weighing (dry weight, not recoverable).

The calli can also be observed under a microscope to evaluate their
cell types and condition. With some types of callus, e.g., with date
palm meristemoid callus (Tisserat et al., 1981; Ulrich et al., 1982), one
can count the number of individual plantlets that have survived, by
direct visual examination (Fig. 3) or after screening at an early growth
stage.

Any young plantlets obtained may be transplanted to porous solid
medium (e.g., vermiculite) and, after a time, to soil, thereby completing
the tissue culture cycle (Fig. 4). Plantlets obtained after freezing
treatments can be further examined for their form and growth habit
and constituents. Changes in genetic makeup can be determined from
chromosome squashes and isozyme patterns (Finkle et al., 1980; Ulrich
et al., 1982), and eventually, progeny characteristics can be evaluated.

ACKNOWLEDGMENTS

Acknowledgments are made to Dr. G. W. Schaeffer and Mr. F. T.
Sharpe for experimental rice tissue cultures, Dr. B. Tisserat for date
palm cultures, and Dr. P. H. Moore and the Hawaiian Sugar Planters'
Association for sugarcane cultures.

Figure 3. Development of date palm plantlets from callus tissue exposed to different temperatures. Growth period, 4 months.

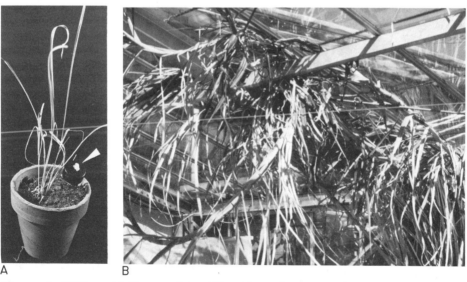

A   B

Figure 4. Differentiation and proliferation of sugarcane plant from callus frozen at -23 C. (a) Typical size of frozen callus piece (on black cap at right, arrow; cap diameter, 23 mm) and growth of callus into a soil-rooted plant are shown. Thawed callus was first implanted on nutrient agar and after 3 months transferred to agar medium lacking 2,4-D; transplanted to soil after 9 months. (b) Plant after 30 months growth.

KEY REFERENCES

Finkle, B.J. and Ulrich, J.M. 1979. Effects of cryprotectants in combination on the survival of frozen sugarcane cells. Plant Physiol. 63:
598-604.

Ulrich, J.M., Finkle, B.J., and Tisserat, B. 1982. Effects of cryogenic
treatment on plantlet production from frozen and unfrozen date palm
callus. Plant Physiol. 69:624-627.

REFERENCES

Ashwood-Smith, M.J. and Friedman, G.B. 1979. Lethal and chromosomal
effects of freezing, thawing, storage time, and X-irradiation on mammalian cells preserved at -196 C in dimethylsulfoxide. Cryobiology
16:132-140.
Finkle, B.J., Ulrich, J.M., Tisserat, B., and Rains, D.W. 1980. Regeneration of date palm and alfalfa plants after freezing callus tissues to
-196 C in a combination of cryoprotective agents. Cryobiology 17:
625-626.
Kartha, K.K., Leung, N.L., and Gamborg, O.L. 1979. Freeze-preservation of pea meristems in liquid nitrogen and subsequent plant regeneration. Plant Sci. Lett. 15:7-15.
Nag, K.K. and Street, H.E. 1973. Carrot embryogenesis from frozen
cultured cells. Nature 245:270-272.
Persidsky, M., Ulrich, J., and Finkle, B. 1980. Effect of washing temperature on the survival of animal and plant cells treated with cryoprotectants containing DMSO. Cryobiology 17:616.
Sakai, A., Yamakawa, M., Sakata, D., Harada, T., and Yakuwa, T. 1978.
Development of whole plant from an excised strawberry runner apex
frozen to -196 C. Low Temp. Sci. Ser. B 36:31-38.
Schaeffer, G.W. and Sharpe, F.T. 1981. Lysine in seed protein from S-
aminoethylcysteine resistant anther-derived tissues of rice. In Vitro
17:345-352.
Siminovitch, D. 1979. Protoplasts surviving freezing to -196 C and osmotic dehydration in 5 molar salt solutions prepared from the bark of
winter black locust trees. Plant Physiol. 63:722-725.
Sunderland, N. 1977. Nuclear cytology. In: Plant Tissue and Cell Culture, 2nd ed. (H.E. Street, ed.) pp. 177-205. Univ. California Press,
Berkeley.
Tisserat, B., Ulrich, J.M., and Finkle, B.J. 1981. Cryogenic preservation and regeneration of date palm tissue. Hort. Sci. 16:47-48.
Torrey, J.G. 1967. Morphogenesis in relation to chromosomal constitution in long-term plant tissue cultures. Physiol. Plant. 20:265-275.
Ulrich, J.M., Finkle, B.J., Moore, P.H., and Ginoza, H. 1979. Effect of
a mixture of cryoprotectants in attaining liquid nitrogen survival of
callus cultures of a tropical plant. Cryobiology 16:550-556.
Withers, L.A. and King, P.J. 1980. A simple freezing unit and routine
cryopreservation method for plant cell cultures. CryoLetters 1:213-
220.

Withers, L.A. and Street, H.E. 1977.   The freeze-preservation of plant cell cultures.   In: Plant Tissue Culture and Its Biotechnological Application (W. Barz, E. Reinhard, and M.H. Zenk, eds.) pp. 226-243. Springer-Verlag, New York.

CHAPTER 29
Protocols of Low-Pressure Storage

M.P. Bridgen and *G.L. Staby*

Two forms of tissue culture storage have recently been studied as alternatives to dry storage (Chapter 27) and cryopreservation (Chapter 28). These include low-pressure (hypobaric) and low-oxygen storage (Bridgen and Staby, 1981). The low-pressure system (LPS) functions by decreasing the atmospheric pressure surrounding the tissue cultures, as a result decreasing the partial pressure of all gases that are in contact with the plant material (Fig. 1). The low-oxygen system (LOS) functions at atmospheric pressure (760 mm Hg) by combining an inert gas, such as nitrogen, with oxygen to create the desired partial pressure of oxygen (Fig. 1). Experimental results obtained from plant tissue culture experiments with LPS and LOS are similar regardless of plant part or species used (Table 1).

Table 1. Conclusions Derived from Experiments with Plant Tissue
Cultures Subjected to Low-Pressure and Low-Oxygen
Treatments

1. Partial pressures of oxygen below 50 mm Hg reduce the rate and the amount of growth of plants stored in vitro.
2. Low partial pressures of oxygen cause similar effects in plant tissue cultures regardless if obtained by LPS or LOS.
3. Both organized and unorganized plant tissues are affected by low partial pressures of oxygen.
4. No phenotypic growth differences are observed after the tissue cultures are removed from storage and then grown to maturity in vivo.

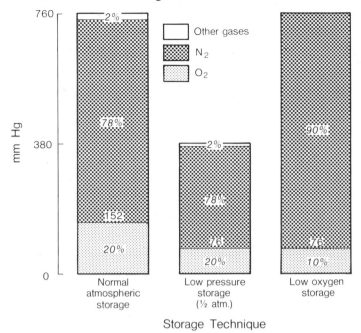

Figure 1.   Comparison between normal atmospheric storage, low-pressure storage, and low-oxygen storage.

LITERATURE REVIEW

No universal storage method has been found to be suitable for all plant material in the tissue-cultured state (Caplin, 1959; Withers, 1978). However, short-term storage by low temperatures (4 C) is successful with many plants (Mullin and Schlegel, 1976; Seibert and Wetherbee, 1977; Withers, 1978).   Refrigerated storage is already being used for the storage of *Fragaria* spp. meristem plantlets (Mullin and Schlegel, 1976), but this method has several disadvantages that limit its usefulness.   The medium still needs periodic replenishment (Mullin and Schlegel, 1976); there is possible cell deterioration through dehydration (Bajaj and Reinert, 1977); selection of plant material may occur (Withers, 1978); and cultures must be removed from storage before cellular damage occurs (Bannier and Steponkus, 1976; Withers, 1978).

Low-pressure storage is useful in extending the shelf life of meat, poultry, shrimp, fish, vegetables, fruits, cut flowers, potted plants, cuttings, and other metabolically active products (Anonymous, 1975).   It is also successful in controlling physiological and pathological disorders of many horticultural crops (Anonymous, 1975; Dilley et al., 1975).   Low-pressure or hypobaric systems are based on the following principles. First, the commodity must be placed in an atmosphere of controlled temperatures (Anonymous, 1975).   Second, the atmosphere must be at reduced pressures so that the partial pressures of each gas within the storage and commodity are reduced proportionately to the pressure causing an increase in the gas exchange in the commodity (Lougheed et al., 1976). Third, a continuous air exchange is used to flush away any toxic vapors released into the storage area and last, high humidity prevents shrinkage, weight loss, and desiccation of the commodity (Anonymous, 1975; Gaffney, 1978).

Previous work has indicated that low pressures may be a potentially useful tool in the long-term storage of plant tissue cultures. It was first shown that normally short-lived seeds such as onion, celery, and cabbage exhibit increased germination after low-pressure storage when compared to atmospheric storage (Lougheed et al., 1976). It was then demonstrated that tomato plant growth was inhibited at low pressures (Rule and Staby, 1981).

In addition to the potential for long-term storage, low pressures also have the added advantage of reducing the activity of culture medium pathogens in aseptic material (Covey and Wells, 1970). Spore germination, mycelial growth, and sporulation of *Penicillium digitatum*, *Alternaria alternata*, *Botrytis cinerea*, *Diplodia natalensis*, and *Sclerotinia sclerotiorum* are reduced under low pressures (Adair, 1971; Apelbaum and Barkai-Golan, 1977). Subatmospheric pressures also have a fungistatic effect on *Penicillium expansum*, *Rhizopus nigricans*, *Aspergillus niger*, *Botrytis alli*, and *Alternaria* sp. (Wu and Salunkhe, 1972).

Low-oxygen storage is the combination of different gases to create a desired atmosphere at atmospheric pressure. Originally reported for the storage of apples and pears under low oxygen and high carbon dioxide, it is still being used in commercial operations for the storage of various fruit crops (Dewey et al., 1969; Smock, 1979).

There are several theories as to why low-pressure storage and low-oxygen storage delay senescence of horticultural crops. One theory is that by decreasing the partial pressure of oxygen in the atmosphere, the amount of CO_2 evolved is also reduced (Kessel and Carr, 1972; Parkinson et al., 1974; Siegel, 1961), and with the low temperatures, respiration is decreased. In addition senescence may also be delayed in low-pressure storage because of the continuous flow of air which flushes away toxic gases such as ethylene that may accumulate (Gamborg and LaRue, 1971; LaRue and Gamborg, 1971).

Most growth studies of plants with oxygen have been related to the measurements of oxygen uptake of CO_2 release in the light. This has allowed the proposal of several theories explaining why low oxygen has an effect on plants growing in vivo. Researchers now know that photosynthesis is increased as O_2 in the atmosphere is reduced (Forrester et al., 1965; Hesketh, 1967; Ludwig and Canvin, 1971; Servaites and Ogren, 1978; Takabe and Akazawa, 1977) by a direct inhibitory effect of O_2 on the RuBP carboxylase of the photosynthetic carbon cycle (Challet and Ogren, 1975). It is also possible that photorespiration decreases as the partial pressure of oxygen is lowered (Ehleringer and Bjorkman, 1977; Forrester et al., 1965; Tregunva et al., 1964). This would inhibit CO_2 production and possibly stimulate CO_2 fixation (Tjepkema and Yocum, 1973; Yentur and Leopold, 1976). Hesketh (1967) found that both the increase in photosynthesis and the decrease in photorespiration are dependent upon species and temperature. Since O_2 is necessary for opening and closing of stomates, as was shown with wheat and barley (Akita and Moss, 1973), transpiration can also be affected by low partial pressures of oxygen (Regehr et al., 1975).

The idea that low partial pressures of oxygen may be advantageous for the storage of plant tissue cultures arose from Caplin (1959). He noticed that liquid petrolatum, commonly known as mineral oil, was widely used for the conservation of cultures of various microorganisms. The mineral oil was used to reduce the rate of growth and to decrease the amount of evaporation from the agar medium. Caplin's experiments with carrot tissue cultures demonstrated that the amount of growth under oil or nutrient solution is controlled by the supply of oxygen to the tissue.

The only reported experiments examining low partial pressures of oxygen with plant tissue cultures have been completed by Bridgen and Staby (1981), using low-pressure storage (LPS) and low-oxygen storage (LOS). Differentiated cultures of *Nicotiana tabacum* L. Wisconsin 38 and *Chrysanthemum* x *morifolium* Ramat. Nob Hill and undifferentiated cultures of *Nicotiana tabacum* L. Wisconsin 38 were studied. The exact LPS and LOS procedures are described in the next section; however, the results follow in this section.

Growth of chrysanthemum shoots was measured by fresh weight gain (Table 2), height increases (Fig. 2), and total number of leaves (Table 2). Plantlet growth was not totally inhibited by any of the treatments over the 6 week period; however, there was a difference in the amount of growth among treatments. Treatments having a partial pressure of oxygen (PO_2) of 50 mm Hg or higher were not different from the controls after 6 weeks in storage. Growth of treatments less than 50 mm Hg was less than the controls with plantlets grown at a PO_2 of approximately 8 mm Hg increasing the least over the 6 weeks. Plantlets grown under LOS and LPS at corresponding PO_2 had similar growth patterns.

Table 2.   Average Fresh Weight Gain and Number of Leaves of Chrysanthemums after 6 Weeks in Storage

TREATMENT		AVERAGE FRESH WT. GAIN (mg)[a]	AVERAGE NUMBER OF LEAVES[b]
Atmospheric Pressure (mm Hg)	PO_2 (mm Hg)		
760	152.0	383.3	7.58
760	54.0	316.6	6.56
300	60.8	325.0	7.50
760	28.1	151.6	5.78
150	30.4	233.3	6.64
760	8.4	55.0	1.73
70	8.0	50.0	2.29

[a]S.E. = 0.01.
[b]S.E. = 1.81.

Growth of tobacco shoot tips was measured by counting the number of leaves and roots and by measuring plant height. The visual rating

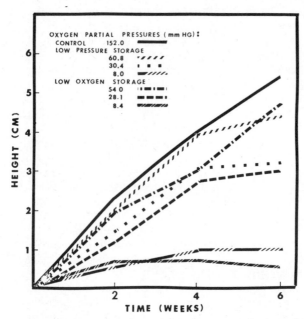

Figure 2.   Increase in height of chrysanthemum plants after 6 weeks in storage.   S.E. = 1.71.

Table 3.   Visual Rating System for Evaluating Tissue Responses

RATING	VISUAL DESCRIPTION
1	≤0.5 cm high, ≤6 leaves
2	0.5–1.0 cm high, 6–10 leaves
3	1.0–3.0 cm high, 10–15 leaves
4	≥3.0 cm high, ≥15 leaves
5	Culture bottle completely filled; could not calculate without opening the bottle

system described in Table 3 was used to express results.   Growth trends were similar to those observed for the chrysanthemums; as the PO_2 was reduced, the rate of growth decreased.   Similarly, the lower the PO_2, the greater the reduction in growth (Table 4).   Medium desiccation was observed in this experiment after 2 weeks only with the LPS treatment which was held at 40 mm Hg.   This caused the plantlets to dehydrate in 50% of the bottles and prevented growth measurements of one replication.

Tobacco callus growth was evaluated by measuring the increase in height, width, and length to estimate volume increase (cm^3) from the initial 125 mm^3 masses (Figs. 3 and 4).   The growth curves for the callus tissue were similar to differentiated chrysanthemum and tobacco tissue; however, differences among treatments were more evident. Growth decreased as the PO_2 was lowered, and there was no difference between LPS and LOS at similar PO_2.

Table 4. Visual Rating of Tobacco Shoots after 6 Weeks in Storage

	TREATMENT	
Atmospheric Pressure (mm Hg)	PO$_2$ (mm Hg)	Rating
760	152.0	3.46
760	48.6	2.97
300	60.8	2.93
760	26.6	2.34
150	30.4	1.92
760	9.1	1.74
40	8.0	1.29[1]
		S.E. = 0.83

[1]Value represents data from only 1 replication.

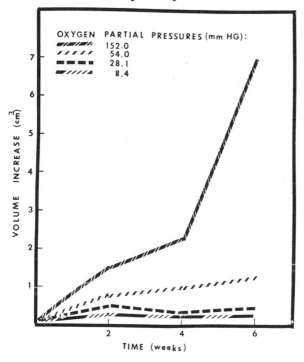

Figure 3. Tobacco callus volume increase after low-oxygen storage for 2, 4, and 6 weeks.

One-third of the chrysanthemum and tobacco plantlets were grown to flowering following each 6 week experiment. To do this plantlets were first transferred onto MS medium supplemented with 1.1 µM IAA and 0.93 µM KIN. Then after 4 weeks these plantlets were potted up in a Metro Mix 200 soil formula and placed in the greenhouse under misting and long days. After 2 weeks plants were grown under standard greenhouse conditions. Little difference was noticed between the treatments flowering, growth habits, and final heights (Tables 5 and 6).

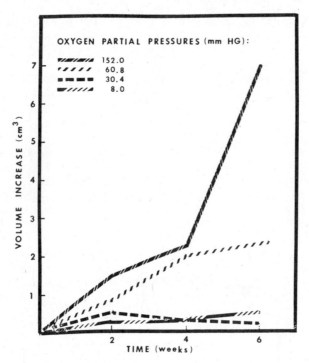

Figure 4. Tobacco callus volume increase after low-pressure storage for 2, 4, and 6 weeks. S.E. = 2.55.

Table 5.   Average Height, after 40 Days in the Greenhouse, of Chrysanthemum Plants Which Were Previously Stored for 6 Weeks
Under Low Oxygen and Low Pressure Conditions

TREATMENT		
Atmospheric Pressure (mm Hg)	PO$_2$ (mm Hg)	Height (cm)
760	152.0	32.7
760	54.0	27.0
300	60.8	35.2
760	28.1	32.4
150	30.4	29.7
760	8.4	32.3
70	8.0	27.7
		S.E. = 2.96[1]

[1]S.E. = Standard error.

PROTOCOLS

In Bridgen and Staby's experiments (1981), the plant material was prepared in the following manner. Chrysanthemum plants were grown

Table 6.   Average Height, after 150 Days in the Greenhouse, of
           Tobacco Plants Which Were Previously Stored for 6 Weeks
           Under Low Oxygen and Low Pressure Conditions

| TREATMENT | | |
Atmospheric Pressure (mm Hg)	PO$_2$ (mm Hg)	Height (cm)
760	152.0	49.25
760	48.6	47.75
300	60.8	50.00
760	26.6	44.50
150	30.4	40.30
760	9.1	52.75
40	8.0	48.00
		S.E. = 4.04[1]

[1]S.E. = Standard error.

under greenhouse conditions with 16 hr days, and pinched when they
reached a height of 15.0 cm.   Lateral bud breaks were then removed,
soaked in 1.05% sodium hypochlorite for 15 min, rinsed in 50% ethanol
for 1 min, and then allowed to remain in 0.26% sodium hypochlorite
until ready to culture.   These lateral, vegetative buds were placed in
vitro and used as stock plants.   Each experiment commenced with 5.0
mm shoot tips removed from the stock plants.   Tobacco stock cultures
were obtained from existing cultures.

All cultures were grown in 30 ml French square glass bottles on a
modified MS medium supplemented with 0.11 µM IAA and 0.93 µM KIN
for chrysanthemum sections, 1.1 µM IAA and 0.93 µM KIN for tobacco
shoots, and 4.5 µM 2,4-D and 100 ml/l CW for tobacco callus.   Culture
medium was sterilized using a steam pressure autoclave at 121 C for
12-15 min.   Immediately before each experiment, each bottle cap was
completely unscrewed and set on top of the bottles to allow adequate
moisture exchange.

All experiments were performed in a randomized block design, with
each treatment being replicated twice and with 16 bottles of plantlets
per treatment.   The plant material in each experiment was grown at
uniform conditions under 26-28 C with 16 hr daylengths and 2.0-2.2
kilolux light intensity supplied by cool-white fluorescent lights.   Each
treatment was maintained in a 10 liter desiccator which was placed inside a clear polyethylene bag.   The desiccators were scrubbed in 1.05%
sodium hypochlorite before the onset of each experiment.

All low-pressure systems were run from a Precision Scientific Model
75 vacuum pump which pulled the air through potassium permanganate
filters to remove various hydrocarbons including ethylene (Scott et al.,
1970), then through Matheson Model 49 pressure regulators to air flow
meters and a water bath, before reaching the desiccators (Fig. 5a).   A
relative humidity of 94-96% was maintained by passing the atmospheres
through the water bath which consisted of 1 liter side-arm Erlenmeyer
flasks filled with 850 ml of distilled water.   The temperature of this
water was raised 3 C above room temperature to allow maximum humi-

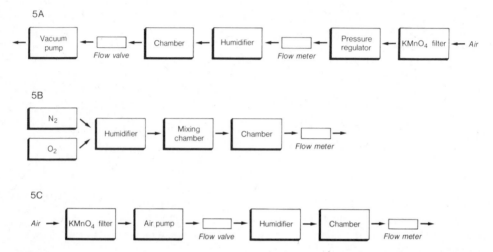

Figure 5.   Schematic of low pressure (a), controlled atmosphere (b), and atmospheric pressure systems (c).

dity in the chamber (Gaffney, 1978).   The gas flow rates of all treatments were monitored daily along with the pressures of each LPS treatment which were estimated with a mercury manometer.

Controlled atmosphere systems were comprised of various combinations of oxygen and nitrogen, which were humidified to 60-72% by bubbling the gases through a tank of water (Fig. 5b).   Gas concentrations were obtained by maintaining constant pressures of each gas and by using glass tube orifices of different sizes to control the percentage of each gas that was entering the system.   Gas composition of the atmosphere was measured on a Packer thermal conductivity gas chromatograph at an oven temperature of 100 C and an injector and detector temperature of 170 C.   A 3 mm x 90 cm stainless steel column was packed with a 5 Å 60/80 mesh molecular sieve for O_2 and N_2.   Initially, each experiment had 3 ml gas samples evacuated for daily analysis; however, after an atmosphere was established, samples were tested weekly.

Controls were set up at atmospheric pressures and atmospheric oxygen concentrations at a relative humidity of approximately 94-96% created by the same system as described for the low pressures.   All air was pulled through a potassium permanganate filter and air flows were maintained by a Universal 1.3 amp air pump (Fig. 5c).

Contamination in each experiment was never greater than 10% after the 6 week period.   This was due in part to the low-oxygen environments, but also due to the fairly aseptic conditions that were maintained.

FUTURE PROSPECTS

Although the tissue culture research with LPS and LOS has been very successful to date, several aspects of the procedures should be examined before commercial applications can be made. Of major concern, particularly if these techniques were to be used for plant germplasm preservation, is the examination of long-term effects of low partial pressures of oxygen on the plants. If these low PO_2 cause subtle genotypic variations in the cultures, these systems may not be feasible.

An aspect of the LPS system that should be examined is the medium desiccation at low pressures. This was exhibited in the tobacco shoot tip experiments and somewhat in experiments with tomato root tips grown in liquid medium (Bridgen, 1979). Medium desiccation for germplasm preservation would limit the storage time of cultures and could possibly be controlled by elevating the relative humidity within the growth chamber or by decreasing the number of air exchanges per hour.

Another aspect of LPS and LOS storage systems to be examined would be the effects of C_3 and C_4 plants under the various PO_2. There may be additional advantages to storing C_4 plants under the low-oxygen conditions over C_4 plants.

Comparisons should be made between the LPS and LOS systems to determine which one is the easiest to use and the most economical. The LOS system may be relatively costly on a large scale, whereas an efficient vacuum pump in the LPS system would be less expensive. Once set up, the LPS should be relatively easy to run and monitor.

These experiments and the theories backing them demonstrate that partial pressures of oxygen below 50 mm Hg reduce the amount of both organized and unorganized plant tissue growth. This can be accomplished by using either LPS or LOS and does not create phenotypic growth differences. With these facts in mind, it appears that these techniques may be feasible to use in the future for plant tissue culture germplasm banks.

KEY REFERENCES

Bridgen, M.P. 1979. Low pressure and controlled atmosphere storage of plant tissue cultures. M.S. Thesis, Ohio State Univ., Columbus.

_____ and Staby, G.L. 1981. Low pressure and low oxygen storage of *Nicotiana tabacum* and *Chrysanthemum* x *morifolium* tissue cultures. Plant Sci. Lett. 22:177-186.

Caplin, S.M. 1959. Mineral oil overlay for conservation of plant tissue cultures. Am. J. Bot. 46:324-329.

REFERENCES

Adair, C.N. 1971. Influence of controlled-atmosphere storage conditions on cabbage postharvest decay fungi. Plant Dis. Rep. 55:864-868.

Akita, S. and Moss, D.N. 1973. The effect of an oxygen-free atmosphere on net photosynthesis and transpiration of barley and wheat leaves. Plant Physiol. 52:601-603.

Anonymous. 1975. Hypobaric storage and transportation of perishable commodities. Gruman Allied Ind., Inc., Garden City, New York

Apelbaum, A. and Barkai-Golan, R. 1977. Spore germination and mycelial growth of postharvest pathogens under hypobaric pressure. Phytopathology 67:400-403.

Bajaj, Y.P.S. and Reinert, J. 1977. Cryobiology of plant cell cultures and establishment of gene-banks. In: Applied and Fundamental Aspects of Plant Cell, Tissue, and Organ Culture (J. Reinert and Y.P.S. Bajaj, eds.) pp. 757-777. Springer-Verlag, Berlin.

Bannier, L.J. and Steponkus, P.L. 1976. Cold acclimation of *Chrysanthemum* callus cultures. J. Am. Soc. Hort. Sci. 101:409-412.

Challet, R. and Ogren, W.L. 1975. Regulation of photorespiration in C_3 and C_4 species. Bot. Rev. 41:137-179.

Covey, H.M. and Wells, J.M. 1970. Low-oxygen or high-carbon dioxide atmospheres to control postharvest decay of strawberries. Phytopathology 60:47-49.

Dewey, D.H., Herner, R.C., and Dilley, D.R. 1969. Controlled Atmospheres for the Storage and Transport of Horticultural Crops. Michigan State Univ., East Lansing.

Dilley, D.R., Carpenter, W.J., and Burg, S.P. 1975. Principles and applications of hypobaric storage of cut flowers. Acta Hort. 41:249-262.

Ehleringer, J. and Bjorkman, O. 1977. Quantum yields for CO_2 uptake in C_3 and C_4 plants—Dependence on temperature, CO_2 and O_2 concentrations. Plant Physiol. 59:86-90.

Forrester, M.L., Krotkov, G., and Nelson, C.D. 1965. Effect of oxygen on photosynthesis, photorespiration and respiration on detached leaves. I. Soybean. Plant Physiol. 41:422-427.

Gaffney, J.J. 1978. Humidity: Basic principles and measurement techniques. HortScience 13:551-555.

Gamborg, O.L. and LaRue, T.A.G. 1971. Ethylene production by plant cell cultures. Plant Physiol. 48:399-401.

Henshaw, G.G., Westcott, R.J., and Roca, W.M. 1978. Tissue culture methods for the storage and utilization of potato germplasm. In: Frontiers of Plant Tissue Culture (T. Thorpe, ed.) p. 507. International Association for Plant Tissue Culture, Calgary, Canada.

Hesketh, J. 1967. Enhancement of photosynthetic CO_2 assimilation on the absence of oxygen, as dependent upon species and temperature. Planta 76:371-374.

Kessel, R.H.J. and Carr, A.H. 1972. The effect of dissolved oxygen concentration on growth and differentiation of carrot (*Daucus carota*) tissue. J. Exp. Bot. 28:996-1007.

LaRue, T.A.G. and Gamborg, O.L. 1971. Ethylene production by plant cell cultures. Plant Physiol. 48:394-398.

Lougheed, E.C., Murr, D.P., Harney, P.M., and Sykes, J.T. 1976. Low pressure storage of seeds. Experientia 32:1159-1161.

Ludwig, L.J. and Canvin, D.T. 1971. The rate of photorespiration during photosynthesis and the relationship of the substrate of light respiration to the products of photosynthesis in sunflower leaves. Plant Physiol. 48:712-719.

Mullin, R.H. and Schlegel, D.E. 1976. Cold storage of meristem plantlets of *Fragaria* spp. HortScience 11:100-101.

Parkinson, F.J., Penman, H.L., and Tregunna, E.B. 1974. Growth of plants in different oxygen concentrations. J. Exp. Bot. 25:132-145.

Regehr, D.L., Bazzaz, F.A., and Boggess, W.R. 1975. Photosynthesis, transpiration and leaf conductance of *Populus deltoides* in relation to flooding and drought. Photosynthetica 9:52-61.

Rule, D.E. and Staby, G.L. 1981. Growth of tomato seedlings at subatmospheric pressures. HortScience 16:331-332.

Scott, K.J., McGlasson, W.B., and Roberts, E.A. 1970. Potassium permanganate as an ethylene absorbent in polyethylene bags to delay ripening of bananas during storage. Aust. J. Exp. Agric. Anim. Husb. 10:237-240.

Seibert, M. and Wetherbee, P.J. 1977. Increased survival and differentiation of frozen herbaceous plant organ cultures through cold treatment. Plant Physiol. 59:1043-1046.

Servaites, J.C. and Ogren, W.L. 1978. Oxygen inhibition of photosynthesis and stimulation of photorespiration in soybean leaf cells. Plant Physiol. 61:62-67.

Siegel, S.M. 1961. Effects of reduced oxygen tension on vascular plants. Physiol. Plant. 14:554-557.

Smock, R.M. 1979. Controlled atmosphere storage of fruits. In: Horticultural Reviews, Volume 1 (J. Janick, ed.) pp. 301-336. AVI Pub., Westport, Connecticut.

Takabe, T. and Akazawa, T. 1977. A comparative study on the effect of O_2 on photosynthetic carbon metabolism by *Chlorobium thiosulfatophilum* and *Chromatium vinosum*. Plant Cell Physiol. 18:753-765.

Tjepkema, J.D. and Yocum, C.S. 1973. Respiration and oxygen transport in soybean nodules. Planta 115:59-72.

Tregunva, E.B., Krotkov, G., and Nelson, C.D. 1964. Effect of oxygen on the rate of photorespiration in detached tobacco leaves. Physiol. Plant. 19:723-733.

Withers, L.A. 1978. Freeze-preservation of cultured cells and tissues. In: Frontiers of Plant Tissue Culture (T. Thorpe, ed.) pp. 297-306. International Association for Plant Tissue Culture, Calgary, Canada.

Wu, M.T. and Salunkhe, D.K. 1972. Fungistatic effects of sub-atmospheric pressures. Experientia 28:866-867.

Yentur, S. and Leopold, A.C. 1976. Respiratory transition during seed germination. Plant Physiol. 57:274-276.

CHAPTER 30
Nitrogen Fixation

A.P. Ruschel and *P.B. Vose*

The genetic information for nitrogen (N_2) fixing ability is found in prokaryotes, and one of the most important tasks is to try to transfer this character to eukaryotes. However, this possibility is far from being achieved, and at present we have to use symbiotic and associative N_2-fixation systems (diazotrophic biocoenoses) already existing in nature. Cell culture can be used to support N_2-fixing microorganisms and aid in understanding the effect of the macrosymbiont on N_2-fixation. Conditions under which a microsymbiont can express N_2-fixing ability in vitro have been established.

Although Raggio et al. (1957, 1959) used cultures of excised pea roots for nodulation studies 25 years ago, there was no immediate follow-up or trend set by this work. The reason is obvious: the idea was ahead of both plant tissue culture methodology and our knowledge of nitrogen-fixing mechanisms. Moreover, development of plant tissue culture in relation to nitrogen fixation has been comparatively slow because, as Giles and Vasil (1980) have pointed out, workers have had rather diverse aims, and there has been little concentration on either material or objectives. Additionally, the fact that plant regeneration from legume callus has proven very troublesome has blocked various in vitro plant selection procedures that might otherwise be possible.

The demonstration of nitrogen fixation by free-living *Rhizobia* appeared after studies of nitrogen fixation in tissue culture. Historically, it was believed that *Rhizobia* have to interact with the legume to perform N_2-fixation through symbiosis. Active *Rhizobia* outside nodules were not known until Holsten et al. (1971) provided evidence for the establishment of *Rhizobium japonicum* on cell suspensions of soybean roots with structure similar to infection channels in 1-10% of cells with intracellular microorganisms.   The paper caused excitement in the scientific community, since active *Rhizobia* were obtained in absence of nodules, leghemoglobin, and bacteroids, which were apparently necessary in intact plants.   In other laboratories development of N_2ase (nitrogenase) in soybean callus-*Rhizobium* system growing in solid media (Phillips, 1974a; Child and LaRue, 1974) was observed concomitantly.

This chapter attempts to highlight the important developments and possibilities in relation to plant tissue culture.

BACKGROUND OF N_2-FIXING SYSTEMS

The N_2-fixing process in different systems is similar and is dependent on nitrogenase enzyme energy supply (ATP), anaerobic conditions, and strong reductant (Haaker et al., 1980).   Photoautotrophs use light energy and are able to fix CO_2 as well as N_2 (Gallon, 1980).   Heterotrophs use carbohydrates as a source of energy and as electron donor.

Nitrogenase is a key enzyme in the nitrogen cycle and catalyses the ATP-dependent 6 electron reduction of dinitrogen (N_2) into ammonia. Physicochemical properties of nitrogenase isolated from different microorganisms are very similar (Eady and Smith, 1979), with a catalyst dinitrogenase (formerly nitrogenase component I or MoFe protein), and another protein-dinitrogenase reductase (formerly N_2ase component II or Fe-protein) (Haaker et al., 1980).

Nitrogenase reduces N_2 and H^+ (Schrauser, 1977) at the same time using 75% of energy in N_2-reduction and 25% in H^+-reduction. N_2-fixation is a biological process with high energy requirement and therefore energy losses should be minimized to enhance its efficiency.   Results indicate that hydrogen evolution, via nitrogenase, is the major factor affecting efficiency of N_2-fixation.   However, there are microorganisms that through hydrogenase can recycle part of the hydrogen evolved through nitrogenase, providing an ATP generating H_2 uptake process (Hup+) (Dixon, 1972, 1978).   Another hydrogenase type, called reversible hydrogenase, was described in *Clostridium pasteurianum*.   This hydrogenase could recycle and dispose of excess reducing power in this fermentative bacteria (Gray and Gest, 1965).   Production of H_2 by nodules of soybean was observed by Hoch et al. (1957).

One of the most important points to be elucidated is the means of protecting nitrogenase against oxygen damage in vivo (Yates, 1977), since the growth and N_2ase activity of N_2-fixing obligate aerobes is inhibited by excess oxygen.   The mechanisms that microorganisms use to control access of oxygen are respiratory protection, morphological change, association with other biological macromolecules, and possession of superoxide dismutase, catalase, and peroxidase.

Specific points can be considered for the overall improvement of nitrogen fixation efficiency: (1) microorganisms with highly efficient N_2 utilization and low H_2 evolution; (2) microorganisms that can recycle the H_2 evolved; (3) plants with highly efficient nitrogen utilization of N from biological nitrogen fixation, especially in the production of seed protein; (4) N_2-fixing systems tolerant to mineral N, especially NH_4^+, as fertilizers; (5) plants with improved carbohydrate production.

N₂-FIXATION AND TISSUE CULTURE

The nature and factors that affect symbiotic associations, especially *Rhizobium-Leguminoseae* interactions, were better understood after studies of tissue culture inoculated with this bacteria were started by Veliky and LaRue (1967). In their experiments the stimulation of plant cell differentiation and lignification were apparent; however, there was no proof of intracellular bacterial nodule-like tissue.

An attempt to establish a symbiotic association with plant cell and *Rhizobium* in vitro showing active nitrogenase was made by Holsten et al. (1971), using root cells obtained from surface-sterilized seed grown in specific media, forming undifferentiated cells (callus) inoculated with two strains of *R. japonicum*. Examination by the light microscope showed structures resembling the infection threads, with *Rhizobia* as well as bacteria inside intracellular spaces within the cell mass and which multiplied inside cells. Active *Rhizobia* were assayed by acetylene reduction, in light or in the dark and was found to be more active in the dark.

Rhizobium is primarily known to produce effective nodules in *Leguminoseae*, but a nonlegume, *Trema aspera* fam. *Ulriaceae* has been reported to produce active nodules with that bacteria (Trinick, 1973). He concluded that genetic information as well as conditions for nodule formation or N₂-fixing ability are independent of the host plant.

Table 1 summarizes the evidence for and factors affecting nitrogen fixation in tissue culture. Based on research using soybean callus, it is clear that variety and strain affect N₂-fixation. Child and LaRue (1974) tested five different culture media in order to observe nitrogenase activity and obtained positive results using Gamborg's medium with lower inorganic nitrogen, indicating that low levels of mineral N might improve the symbiotic process. Testing the soybean varieties Acme, Mandarin, and Norman, they observed that Acme callus showed higher N₂ase activity than the others, with Mandarin showing the lowest activity. Differences were observed between strains, indicating a variety x bacteria relationship. The nitrate:ammonium ratio showed that less ammonium must be used to increase bacterial activity.

Morphological evidence of callus/*Rhizobium* symbiosis was found by Holsten et al. (1971) and Reporter et al. (1975). The Holsten group noted structures similar to infection threads in initial stages of infected roots. These pseudo-infection threads cross the intracellular spaces into cells where the *Rhizobium* multiplies. Reporter et al. (1976) noticed bacteria only in certain cells that did not differ morphologically from the others in tissue culture. However, entry of bacteria into cells was done at specific sites of initial bacterial mass formation. Bacteria were oriented normally to the curvature of the plant cell. The plant cell selected the bacteria to host. Electron microscope examination showed that bacteria are trapped by the extracellular filaments, and they clump the bacteria by invagination and differentiation for symbiosis.

Light has been found to be a factor affecting N₂-fixation by tissue culture, as reported by Holsten et al. (1971) who noticed that incubation in the dark increased bacterial activity when compared with incubation in the light.

Table 1. Factors Affecting Nitrogen Fixation in Soybean, *Glycine max*, Tissue Culture as Measured by Acetylene Reduction

		C_2H_2/g FRESH WEIGHT·24 HR (nmol)	REFERENCE
EFFECT OF LIGHT			
Soybean callus +	Light	72.8	Holsten et al.,
Rhizobium strain 61A76	Dark	318.6	1971
EFFECT OF PLANT VARIETY			
var. Acme	Exp. 1	274.6	Child & LaRue,
	Exp. 2	87.3	1974
var. Mandarin	Exp. 1	3.8	Child & LaRue,
	Exp. 2	2.5	1974
var. Norman	Exp. 1	5.0	Child & LaRue,
	Exp. 2	20.7	1974
EFFECT OF STRAIN			
var. Mandarin	83[1] USA	10.2	Child & LaRue,
	10324 (ATCP)	8.4	1974
	61A76 (Nitragin)	3.3	
COMPATIBILITY OF STRAIN $NO_3:NH_4$ (mg/l)			
Effect of NH_4 x	1000:150	3.9	Reporter et al.,
NO_3	1000:50	16.6	1975
MORPHOLOGICAL EVIDENCE			Holsten et al., 1971; Reporter et al., 1975

Induction of *Rhizobium* Nitrogenase in Tissue Culture

An in vitro experiment to study the effect of soybean meal and whether any effect observed could be due to host plant cells or *Rhizobia* was carried out by Anderson and Phillips (1976). They observed that the soybean extract had a direct effect on the bacteria by observing increased N_2ase activity in *Rhizobium*–soybean associations supplied with aqueous extracts of hexane obtained from soybean meal. They observed that it was not necessary for the extract to be a protein in structure and that soybean meal is the active agent or is transformed into a promoting compound.

Succinates promote N_2ase activity of the *Rhizobium*–soybean cell association (Phillips, 1974a,b; Anderson and Phillips, 1976). These authors observed that when succinate affected N_2ase activity, it was due to the effect on the plant cell, since succinate does not affect the activity of free–living *Rhizobia*.

As mentioned before, *Rhizobia* incubated together with plant cells showed activity even in the absence of any kind of structure; however, the techniques used did not make it possible to independently study the steps involved in N_2-fixation. Reporter and Hermina (1975) described bacterial activity in transfilter suspension cultures of *R. japonicum*.

Actively nodulating strains of *R. japonicum* were found (Reporter, 1976) to exhibit acetylene reduction in vitro under conditions termed synergistic, because the plant cell activated nitrogenase activity of *Rhizobium* that could not be activated when grown in agar. Using a transfilter apparatus and studying the effect of substrate concentration it was possible to observe that high O_2 level (22%), when produced in various substrates, i.e., pyruvate (py), glucose (gl), α-hetoglutarate (α-hg), and succinate (succ), increased acetylene reduction. Pyruvate results were higher than the others. However, α-hg and succ affected N_2ase activity only after 40 hr incubation. If carbon monoxide was used as an inhibitor, α-hg and succ showed high activity, and py and gl increased N_2ase activity after 70 hr incubation, indicating that oxygen inactivates unprotected nitrogenase and CO inhibits N_2ase activity. When *Rhizobium* was reisolated from the culture in the transfilter apparatus, it was surrounded by a pellicule presumably obtained from plant cells.

Reporter (1976) was able to demonstrate with the transfilter apparatus that activation of *Rhizobia* showing N_2ase activity was related to: (1) *Rhizobia* activity in nodulated plants; (2) plants that affect activation are those that nodulate; (3) invasion can happen directly; and (4) activity of activated bacteria is similar to that of bacteroids.

Viable cells from root hairs maintained in tissue culture (Hermina and Reporter, 1977) showed (1) C_2H_2 reduction needs more O_2 than cells grown in transfilter apparatus, or free-living *Rhizobia*; (2) washed preparations reduced acetylene; (3) tissue cultures of root hairs have lectin; and (4) there were differences between plant varieties in root hair cell differentiation (Acme greater than Harosoy).

The first paper showing factors affecting electron transfer from bacteria to nitrogenase of soybean nodules was that of Phillips et al. (1973). It is known that the leghemoglobin in nodules acts as an equilibrator of oxygen level which permits the process of anaerobic fixation of N_2. Therefore, the role of proteins is very important both in controlling biological nitrogen fixation and as electron carriers. Growing *R. japonicum* in culture media (KNO_3 under anaerobiosis), Phillips and co-workers observed proteins that could transfer electrons to nitrogenase, similar to those later found by Phillips (1974a). This indicated increased nitrogenase activity from adding the protein precursors succinic acid and glutamine.

Difficulties in observing nitrogenase activity of fast-growing *Rhizobium* were overcome by using rhizobial strains under free-living conditions with a plant cell conditioned medium, produced by coculture of legume cell cultures and *Rhizobia* (Mohapatra et al., 1980). They found a reproducible technique, using clover-cell conditioned media, for assessing the activity of slow- and fast-growing bacteria and mutants.

N$_2$-Fixation by Free-Living *Rhizobia*

The first report of N$_2$-fixation by *Rhizobium* without contact with plant cells was made by Reporter and Hermina (1975). N$_2$-fixation was obtained in transfilter suspension culture using a defined medium to observe the effect of strain and carbon source. The soybean variety Harosoy was the donor of a factor that could affect N$_2$-fixation (C$_2$H$_2$ reduction). When inoculated with the same strain and ammended with glucose, the culture showed different levels of N$_2$ase activity. With the var. Acme, the same strain and different substrates showed particularly high activity in the presence of beta-hydroxybutyrate. However, as *Klebsiella* was added to the media, there arose some doubt as to whether N$_2$-fixation was solely due to *Rhizobium*, since reduction of acetylene was noted only after the oxygen concentration was reduced below 1%.

The observation that diffusible factors from plants are responsible for expression of nitrogenase activity (Child, 1975; Scowcroft and Gibson, 1975) induced research toward finding active *Rhizobium* in culture media. The search for knowledge of which factors could affect N$_2$-fixation by free-living *Rhizobium* resulted in near simultaneous publication of three papers on the subject.

Pagan et al. (1975) found that only 15 of 27 strains (from *Rhizobium* sp.—cowpea group; *R. japonicum*; *R. lupin*; *R. meliloti*; *R. trifolii* and *R. leguminosarum*) were activated, and those were from just *Rhizobium* sp. and *R. japonicum*. They also observed that arabinose increased activity in media with normal salts plus glutamine, inositol, and succinate.

Kurz and LaRue (1975) observed that addition of extracts of carrots and rice permitted expression of N$_2$ase activity by *R. cowpea*. Studying carbohydrates, they observed that xylose, arabinose, and galactose also stimulated N$_2$ase activity when *Rhizobia* were grown with carrot cells. However, using each of these plus saccharose, activity was found in the absence of plant cells but only with strains that produced slime, indicating that protection from oxygen was required.

Working with one of four strains of *Rhizobium* that were able to show N$_2$ase activity in the presence of plant callus and growing it in specific media, McComb et al. (1975) observed that addition of (glutamic acid + succinic acid) and (asparagine + succinic acid) increased acetylene reduction. According to their conclusions, genes for nitrogenase are present in bacteria, although this invalidates the theory of Dilworth and Parker (1969) that plant DNA contributes to part of the N$_2$ase in legume nodules.

It appears that factors affecting free-living fixation could be carbohydrate (5-carbon enhancing activity), strain (not all strains expressed activity), fixed nitrogen (ammonium and nitrate), and organic-C (amine-C). Keister (1975) observed that free-living *Rhizobia* require low cell density to ensure activity, that removal of O$_2$ was important, and that NO$_3$ could not be used as the terminal electron acceptor. There was also an effect of strain and source of combined-N. However, with stationary cultures 20% O$_2$ was required for reduction of C$_2$H$_2$, while 10

M NH_4Cl inhibited activity. Other inhibitors were CO, nitriles, cyan-
ite, and nitrate. Shaken cultures that had O_2 concentration above 1%
showed an inhibitory effect (Evans and Keister, 1976).

Tjepkema and Evans (1975) observed fixation under microaerophilic
conditions with a rate of fixation similar to soybean nodules. However,
Criswell et al. (1976) observing the effect of altered pO_2 on the rhizosphere of intact soybean nodules, concluded that the system is able to
adapt to a wide range of external pO_2 through an undefined mechanism. Gibson et al. (1976a) observed that a pO_2 concentration of 0.20-
0.25 in the medium was necessary for good nitrogenase activity of Rhizobium sp., but as bacteria were grown in solid media, the possibility
of internal protection against oxygen is not excluded. Work on factors
affecting free-living N_2-fixation of Rhizobium is summarized in Table 2.

Table 2. Factors that Affect Free-Living N_2-Fixation by Rhizobium

FACTORS	BACTERIA	REFERENCE
Oxygen tension	Rhizobium sp.	Gibson et al., 1976
Incubation time	Rhizobium sp.	Gibson et al., 1976
Temperature incubation	Rhizobium sp.	Gibson et al., 1976
N concentration	Rhizobium sp.	Gibson et al., 1976
C source	Rhizobium sp.	Gibson et al., 1976
	Rhizobium sp.	Pagan et al., 1975
O_2 concentration	Rhizobium sp.	Tjepkema & Evans, 1975
	Rhizobium sp.	Evans & Keister, 1976
Rhizobium species	Several	Pagan et al., 1975
Rhizobia strain	Rhizobium sp.	McComb et al., 1975
Inhibitor of N_2ase	Rhizobium sp.	Keister, 1975

Regulation of N_2-Fixation in Rhizobium sp.

The ability to culture nitrogen-fixing bacteria without the presence
of plant cells has given us additional insight into the regulation of N_2-
fixation. A good early summary of regulation of N_2-fixation in bacteria
was made by Tubb (1976).

Studying the regulation of N_2ase activity in Klebsiella, Tubb and
Postgate (1973) observed that synthesis of N_2ase is repressed by excess
free ammonium, as was later found for other bacteria (Tubb, 1976).
Shanmugam and Valentine (1975a) obtained mutants of Klebsiella derepressed for nitrogenase synthesis growing in the presence of ammonium.
Such strains have derepressed levels of glutamine synthetase and lack
glutamate dehydrogenase activity. These mutants retain only 30% of
N_2ase activity, but if mutants are obtained without pathways of NH_4^+
assimilation, such as glutamate or glutamine-requiring and without glutamate synthetase activity, then they have 100% N_2ase activity in the
presence of NH_4^+, and NH_4^+ excretion is obtained in Azotobacter and
cyanobacteria when methionine sulphoxamine is used as an inhibitor of
glutamine synthetase.

There is evidence that bacteroids do not assimilate NH_4^+ via glutamine synthetase as *Rhizobia* cultures do.   Very low synthetase is found
in bacteroids during nodule development, but a high concentration is
found in plant nodule supernatant fractions (Tubb, 1976).   This observation suggests that the bacteroid is not the site of NH_4^+ assimilation.
Tubb et al. (1976) noticed that nitrogenase activity in cultures of *Rhizobium* grown in defined media with NH_4^+ was only 14-36% of glutamate
grown cultures.   The excreted NH_4^+ from bacteroids is subsequently
assimilated via plant enzymes, and glutamate may play a key role in
promoting NH_4^+ excretion (Tubb, 1976).   However, Scowcroft et al.
(1976) observed that although N_2ase activity was rapidly inhibited by
NH_4^+, it was reactivated with increasing O_2-tension.   Ammonium may
increase O_2 consumption and lower tension (and hence ATP supply) or
may stimulate N-assimilation and protein synthesis thereby decreasing
ATP.   Scowcroft et al. (1976) concluded that NH_4^+ inhibition of nitrogenase activity is not affected through glutamine synthetase regulation
of nitrogenase synthesis.   On the other hand, Ranga Rao (1977) observed that the activity of *Rhizobium* was dependent on glutamine in
the media, while glutamine plus asparagine increased activity.

Appropriate nutritional conditions are needed for N_2ase synthesis by
Rhizobium in culture media (Pankhurst, 1981).   Some strains can show
N_2ase activity, and others do not.   Pankhurst also noted that the slow-
growing *Rhizobium* strains 32HI and CB627 showed activity but not
CB744, with CB627 having more specific requirements than 32HI, mainly because of N source.   Otherwise, no differences were observed in
requirements for vitamins and trace elements.   Nucleotide AMP (adenosine 3'5'-cyclic monophosphate) accelerated derepression of N_2ase synthesis of both strains that showed activity.

Appleby et al. (1981) observed that continuous culture retains cytochrome o and aa_3 (which are terminal oxidases functioning at moderate
to high concentration of dissolved oxygen) in the absence of leghemoglobin, and consider these oxidases as a useful marker technique for
selection of strains tolerant to functioning in the absence of leghemoglobin.

By using an inhibitor of bacterial RNA polymerase and a DNA inter-
calating agent (Pankhurst et al., 1981), it was possible to verify that
those products repress N_2ase synthesis and a few other proteins, and
there are suggestions that to show N_2ase acitivity, *Rhizobium* may be
dependent on a structurally distinctive or modified RNA polymerase.

Tissue Culture of Legumes for N_2-Fixation

As already noted, although pea roots and soybean callus were cultured in vitro fairly easily, the problems involved in plant regeneration
from legume callus have held up many potential studies relating to
nitrogen fixation.

Some progress is now being made.   Bajaj and Dhanju (1979) have
been able to regenerate plants from apical meristems of varieties of
Cicer arietinum, Lens esculentum, Pisum sativum, Phaseolus aureus, and

Phaseolus mungo.   Martin et al. (1979) have rooted meristem cultures
and callus culture of field bean, *Vicia faba.*   Working with alfalfa, *Medicago sativa,* Walker et al. (1979) have been able to regenerate plants
from callus tissue.   It appears therefore that there is something of a
breakthrough in legume tissue culture.

As more legumes can be regenerated a number of plant culture possibilities exist; for example, the use of haploid (Sharp et al., 1983) and
protoplast (Cailloux, 1983) techniques in breeding and selection.   Plant
breeding is now clearly inseparable from the quest for improved N_2-fix-
ation, whether in legumes or nonlegumes, although as Rennie (1981) has
pointed out, most work aimed at improving N_2-fixation has concentrated on nitrogenase synthesis and expression (nif) controlled in the bacterium in response to the plant, and very little work, i.e., plant breeding, has been carried out for the N_2-fixation supportive traits (nis).

In many legumes it is now apparent that in some cases there is a
notable plant genotype x *Rhizobium* reaction, and consequently there is
no best universal strain but that optimum results are obtained when
legume variety and *Rhizobium* strain are uniquely matched, e.g., in field
bean, *Vicia faba* (Mytton et al., 1977), cowpea, *Vigna unguiculata* (Minchin et al., 1978), and white clover, *Trifolium repens* (Jones and Hardarson, 1979; Jones and Morley, 1981).

Coevolution of host and rhibosomal genotypes in soybean has been
suggested by Devine and Breithaupt (1980), based on a study of 851
genotypes.   They found that Rj_4 was common in 31% of the lines (those
originating from south-east Asia), while Rj_2 was noted in only 2% of
the lines (Those originating from China and Japan).   Presuming coevolution of host and rhizobial genotypes to be a general phenomenon, this
suggests that coculture of *Rhizobium* and legume tissue in vitro could
lead to the development, through mutation and selection, of mutually
compatible types.   This offers the possibility of achieving through relatively few subculture and selection routines in vitro what might otherwise take generations in the field.

Similarly, there are a number of practical requirements for improving
legume culture and N_2-fixation in modern agriculture that might also
be approached through coculture in vitro of *Rhizobium* and host plant.
These include: (1) tolerance to mineral nitrogen given as fertilizer
because mineral-N, especially NH_4, tends to inhibit N_2-fixation; (2) tolerance to high levels of available Al and Mn, such as are found in acid
soils; (3) tolerance to seed-applied pesticides, which inevitably come
into contact with seed-applied *Rhizobium* inoculum; (4) tolerance and
adaptation to more selective herbicides.

General experience, from whole plant studies, from selection of mutant plant cell lines in vitro, and from selection work on mutant strains
of *Rhizobia* tolerant to various factors, suggests that in principle such
coselection is possible.   Hitherto, such work has been carried out with
plant culture (i.e., whole-plant) and *Rhizobium* cultures independently,
but coculture of plant host cells and *Rhizobium* could ensure optimum
compatibility of bacteria and host genotypes.   The development of such
procedures clearly depends on the routine regeneration of whole plants
from cell cultures.

Tissue Culture of Nonlegumes for N_2-Fixation

In the last few years there has been growing recognition that N_2-fixation by nonlegumes is widespread, in addition to the long-recognized symbiotic systems of *Alnus* (*Frankia alni*), *Azolla* sp. (*Anabaena azollae*) and *Gunnera* (*Nostoc nuscorum*). Associative N_2-fixing systems (diazotrophic biocoenoses) involving a number of microorganisms have been demonstrated in very many tropical grasses, sugarcane, maize and sorghum, rice, wheat, *Spartina alterniflora* Loisel, etc. Vose and Ruschel (1981) should be consulted for a general view of this area.

A certain amount of work has been concerned with the interaction of *Rhizobium* on substrates of nonlegume cells, and this will be mentioned first. Although such work is at present academic, it does show that the *Rhizobium* system is not as inflexible as first thought, and it offers the hope that its capacity to fix atmospheric nitrogen might be transferable to species hitherto lacking this ability. Trinick (1973) observed that the nonlegume *Trema aspera* was able to produce effective nodules, when with *Rhizobium*-like bacteria, that infect and form nodules in four legume species. The endophytic nature of the bacteria is the necessary criterion for classifying them as *Rhizobium*. Attempts to obtain infection and activity of *Rhizobium* were made by Child (1975) with great success, with association of *Rhizobia* x callus of nonlegumes (brome grass, rapeseed, wheat) and nonnodule forms of legumes (sweet clover and *Vicia hajastana*), as well as with five varieties of pea. Nitrogenase activity was observed with all the species tested, and the symbiotic N_2-fixing association was shown to be aerobic. Bacteria populated the outside of cells and in cracks and intracellular spaces. Nodule–like structure was not observed. Bacteria were round in shape, without any membrane surrounding them.

The *Rhizobium*–plant callus association differs from the nodule system morphologically and in its reaction to plant regulators. The first subculture of a *Rhizobium* culture from callus showed N_2ase activity; however, the second subculture did not show N_2ase activity and hence the effect of plant tissue on fixation. Activity with plants other than legumes suggests that the same common plant product may be involved. It was concluded that in symbiotic N_2-fixation the species barriers are at the stages of infection and nodule formation, rather than the expression of nitrogenase by microorganisms that populate these plants. Using $^{15}N_2$ enriched atmosphere and the association of *Rhizobium*–tobacco cells, isotope enrichment was observed after 28 hr and 42 hr incubation, confirming that acetylene reduction is a real indication of N_2ase activity under these conditions (Scowcroft and Gibson, 1975).

Nitrogen fixation of a cowpea strain of *Rhizobium* associated with a tobacco cell culture was noted (Gibson et al., 1976a). They observed maximization of nitrogenase activity at 20% O_2-tension and 30 C. Glutamine as a nitrogen source could be replaced by other sources, such as glutamic acid, aspartic acid, and asparagine. Nitrate and ammonium suppressed nitrogenase.

The effect of *Rhizobium* strain on nitrogenase activity in association with tobacco cells was observed. Gibson et al. (1976a) found activity with a few strains of *R. cowpea*, *R. japonicum*, and *R. lupini*, but no activity with strains of *R. trifolii*, *R. meliloti*, and *Cicer* strains. Using the same type of substrate of tobacco cells, Gibson et al. (1975) noticed that some strains tested (*R. trifolii*, *R. meliloti*, and *Cicer* strains) did not show N_2ase activity, while almost 100% of *R. japonicum*, *R. lupini*, and *R. cowpea* showed activity, indicating that some affinity is needed. Child and LaRue (1974) and Scowcroft and Gibson (1975) found activity with *R. cowpea*.

Schetter and Hess (1977) studied association of *Rhizobium* with callus tissue of horticultural plants (*Portulaca grandiflora*, *Petunia hybrida*, *Ipomoea bicolor*, and *Nemesia strumosa*). Except for *Ipomoea* all the callus associations showed N_2ase activity.

Compared with plant tissue culture of legumes x *Rhizobium* the basic difficulty of studying diazotrophic biocoenoses in nonlegumes is somewhat different. In the former we now have good knowledge of *Rhizobium* but have problems in regenerating whole plants from legume callus, whereas with the latter the main problems lie with the bacteria, as plant regeneration methods are available for most of the associated species involved. A summary of the main associative N_2-fixation systems is given in Table 3, together with major references.

A note of caution should be sounded here: the study of associative N_2-fixing systems is very recent, and in some cases there is still uncertainty as to the identification of the associated bacterial species. Also, we do not yet have knowledge as to whether species- or genotype-specific strains of the bacteria are involved. For example, in sugarcane it is still not entirely clear as to whether there is a different dominant bacterial population in the stalk and roots as compared with the rhizosphere soil; although it appears that this is so (Ruschel, 1981a,b; Rennie et al., 1982).

We do not have information as to whether some associative bacterial populations are adventitious, depending on location and soil. In some cases where more than one bacteria is involved, it is not clear as to which is the major N_2-fixer. There is some evidence in the sugarcane system (Ruschel and Vose, 1977) and in sorghum (Dart and Subba Rao, 1981) for commensalism, i.e., bacteria dependent on each other in some way, possibly for growth factors.

Almost certainly these associative systems are not optimal at the present time, and there is the possibility of improving the nis factors by plant breeding, either by conventional means or through plant tissue culture. This is shown by the work with sugarcane that has indicated that some varieties support much greater nitrogenase activity than others, and that this is heritable (Ruschel and Ruschel, 1981). Dart and Subba Rao (1981) found wide variation in nitrogenase activity of 334 field-grown sorghum lines, and 15 of them showed consistently high nitrogenase levels over three seasons. Brill and his group (personal communication), working with maize, have been able to identify, cross, and select progeny lines with improved capacity to support an association with *Azotobacter vinelandii*.

In these associative N_2-fixing systems the same limitations on nitrogen fixation are present as in the *Rhizobium* nodule systems, i.e., the bacteria must receive a source of carbohydrates, and the anaerobic

Table 3. The Major Associative N_2-Fixing Systems (diazotrophic biocoenoses)

PLANT SPECIES	PRINCIPLE MICROORGANISM	NOTES	REFERENCES
GRASSES			
Paspalum notatum var. batatais	*Azotobacter paspali*	Specific	Dobereiner, 1966
Panicum maximum	*Azospirillum lipoferum*[1]	Specific	Dobereiner & Day, 1974
Cynodon dactylon	*Azospirillum lipoferum*	Specific	Dobereiner & Day, 1974
Digitaria decumbens	*Azospirillum lipoferum*	Specific	Dobereiner & Day, 1974
Pennisetum purpureum	*Azospirillum lipoferum*	Specific	Dobereiner & Day, 1974
Spartina alterniflora Loisel	*Campylobacter*	In roots	Patriquin et al., 1981
WHEAT			
Triticum spp.	*Bacillus polymixa*	Genotype specific with certain spring wheats	Nelson et al., 1976; Larson & Neal, 1978; Rennie, 1981
RICE			
Oryza sativa	*Achromobacter*	Wetland rice	Watanabe & Barraquio, 1979
	Enterobacteriaceae *Azospirillum brasilense*[1]	Wetland rice	Balandreau et al., 1975
SUGARCANE			
Saccharum spp.	*Azotobacter, Beijerinckia, Bacillus, Klebsiella, Derxia, Vibrio, Azospirillum*	Rhizosphere	Ruschel, 1981b; Singh et al., 1981
	Enterobacteriaceae Bacillaceae	In roots	Rennie et al., 1982
PEARL MILLET AND SORGHUM			
Pennisetum purpureum	*Azospirillum, Bacillus polymyxa, Klebsiella, Azotobacter, Derxia, Enterobacter*	Rhizosphere and roots	Dart and Subba Rao, 1981

Table 3. Cont.

PLANT SPECIES	PRINCIPLE MICROORGANISM	NOTES	REFERENCES
MAIZE *Zea mays*	*Azospirillum lipoferum*[1], *Azotobacter vinelandii*	Inoculation experiments with selected genotypes have shown promise	Brill, personal communication

[1] *Azospirillum* spp. seem ubiquitous in many associative N_2-fixing systems, but *A. brasilense* seems to have a primary affinity for C-3 plants, i.e., barley, oat, rice, rye, and wheat, while *A. lipoferum* has affinity for C-4 plants, i.e., maize, sorghum, and most tropical grasses (Boddey and Doebereiner, 1982).

840

nitrogen-fixing mechanism must be protected from oxygen. In *Azotobacter* it is recognized that this is achieved by its intense respiratory activity and protectory conformational change of the enzyme, while in the case of *Azospirillum* microaerophyllism is necessary in culture medium, but sometimes these points are not properly established. It should be noted that in some cases N_2-fixation may not be the only or even the prime effect of the bacteria, which may also produce hormonal effects.

It seems likely, even from the few references given above, that in most cases the maximum possible level of associative N_2-fixation has not yet been achieved, and to do this we need to have better knowledge of these systems. For example, future investigations of N_2-fixing systems will want to establish: (1) positive identification of the bacteria, (2) the specificity of bacterial strains vis a vis associated plant species and variety, (3) the nature of the carbohydrate source, (4) the degree of bacterial commensalism and synergy, (5) whether the effect of the bacteria is solely through N_2-fixation or if there is an effect of growth factors, (6) the infection mechanism and the nature of the affinity, and (7) obtain evidence that microorganisms in the plant are really active in N_2-fixation.

We need to remain open-minded as to what these associations represent, i.e., they show us the possibility of useful biological N_2-fixing systems not dependent on *Rhizobium*. Some discussions on the possibility of transference of N_2-fixing ability from legumes + *Rhizobium* to nonlegumes have centered on the possibility of adaptation of *Rhizobia* or the transfer of its nif genes. It could be that this is not the only approach; indeed it may be a less favorable approach than either improving existing associative N_2-fixing systems or trying to transfer nif genes from bacteria already closely adapted to nonlegumes.

Plant tissue culture could have a useful role in a number of areas. These might include investigations on the affinity of specific bacteria to the host plant and attempts to obtain, by mutation and selection, optimal compatibility of plant genotype and bacterial strain. This could be especially important in sugarcane, where conventional breeding is troublesome and slow.

Limited ad hoc inoculated tissue culture work has already been attempted. Vasil et al. (1979) and Berg et al. (1979) studied combined cultures of sugarcane callus tissue and *Azospirillum brasilense*, and found appreciable rates of acetylene reduction activity. They were able to maintain the cultures for 18 months, even though A. *brasilense* is not a colonizer of sugarcane roots and stem. Carlson and Chaleff (1974) reported an association between carrot tissue and *Azotobacter vinelandii*, and Vasil et al. (1979) reported tobacco tissue cultures inoculated with A. *vinelandii* and *Azospirillum brasilense*, but the cultures did not develop an intracellular association and eventually died.

Plant tissue culture and plant regeneration should not now be a limiting factor in studying associative N_2-fixing systems in vitro. Many procedures now exist, and Green (1978) reviewed in vitro plant regeneration in grasses and cereals, while Brar et al. (1979) reviewed regeneration in maize and sorghum. Sugarcane has long (Heinz, 1973) been very sucessfully cultured in vitro.

METHODOLOGY

The techniques used to study biological nitrogen fixation are quite routine. Burris (1975) has described standard methods for nitrogen fixation studies. Vincent (1970) has described practical methods for root nodule bacteria, and Bergersen and Gibson (1978) have described the techniques of nitrogen fixation by *Rhizobium* spp. in laboratory culture media.

General methods for evaluating nitrogen fixation have been described extensively (Bergersen, 1980). Nitrogen fixation is usually determined routinely via nitrogenase activity as evaluated by acetylene reduction (Hardy et al., 1973) samples being incubated with 10% acetylene and ethylene subsequently analyzed by gas chromatography. The stable isotope ^{15}N offers an exact method of determining nitrogen fixation, and methodology and problems have been discussed by Vose et al. (1981). Knowles (1981) has discussed the problems of measuring N_2-fixation.

LaRue et al. (1975) described a continuous and nondestructive method to determine N_2ase activity in microbial cultures, using low and noninhibitory concentrations of acetylene (10^{-7} mol/liter), mixed with the gas flow aerating the microbial cultures, which can be carried out in situ and automated. Direct evidence can be achieved by exposing the system being studied to $^{15}N_2$-dinitrogen (Scowcroft and Gibson, 1975).

Callus Culture

Plant cell culture can be obtained from different plant parts, generally using Gamborg's method (1968) as used by many authors (Child and LaRue, 1974; Holsten et al., 1971). However, great care is always necessary, since contamination with N_2-fixing free-living microorganisms is very common. Aseptic tissue culture of sugarcane has been obtained (Chapter 26) from very newly expanding leaves from surface sterilized buds, since after 3-4 days bacteria situated inside the stalk quickly move to every plant part and, without being harmful have, of course, active nitrogenase.

The effect of diffusible factors has been studied by growing bacteria and plant cells close to but not in direct contact, using special containers with membranes that facilitate separation of plant cells from bacteria in the same media (LaRue et al., 1975; Reporter and Hermina, 1975; Anderson and Phillips, 1976).

It is difficult to study free-living N_2-fixers in association with tissue culture. Bearing in mind that bacteria can show activity in culture media and that callus has to develop on N-enriched media, the following points should be stressed: (1) N_2ase activity can be repressed by an excess of N in the system. (2) N_2ase activity can be due to the development of bacteria in media depleted by previous growth of callus or tissue culture and not by the association tissue culture x bacteria. (3) When plants are obtained from tissue culture differentiation, it is almost impossible to maintain them without some contamination by other microorganisms.

As an example, we can cite an experiment carried out with sugar-cane plants obtained from tissue culture inoculated with different N_2-fixing microorganisms (*Azotobacter*, *Bacillus*, *Azospirillum*, *Erwinia*) and maintained in closed vessels for 1 month with all precautions that nevertheless showed contamination by other bacteria. Even so, treatments with *Azospirillum* and *Bacillus* showed N_2ase activity when compared with the others (Graciolli, unpublished).

The above difficulties might be overcome through: (1) separation of the bacteria in the callus from the callus medium, (2) the use of $^{15}N_2$ to provide more reliable evidence of N_2-fixation, and (3) the use of a known enriched culture of plant tissue, so that subsequent isotope dilution might show incorporation of new nitrogen following inoculation.

FUTURE PROSPECTS

Future prospects must be related to the need to improve biological nitrogen fixation, either to reduce the need for nitrogen fertilizers in developed agriculture or to increase nitrogen inputs to crops in developing countries where little or no nitrogen fertilizer is used. Many points have been emphasized in the text and will merely be summarized here.

Work with legumes will concentrate on improving the existing *Rhizobium* symbiosis. This will require improved compatibility of *Rhizobium* and legume, improved competitiveness of the *Rhizobium* inoculum with wild type soil *Rhizobia* and other bacteria, increased tolerance to factors such as soil acidity and seed-applied pesticides, and an ability to function in the presence of mineral-N. As regards the legume plant, it is necessary to seek those genotypes with efficient photosynthesis and improved partitioning of carbohydrate to the nodules, the latter being necessary even if this results in slightly lower yields. Achieving maximum possible yields may not always be a desirable objective, especially in countries where fertilizer costs are high and require foreign currency (Ruschel and Vose, 1980; Vose, 1981). In some cases legumes are needed with better adaptation to Al and Mn toxicities of acid soils, and for all legumes, varieties with tolerance to selective herbicides that still retain the capacity for efficient symbiosis.

The real challenge lies in achieving greater inputs of biologically fixed nitrogen into nonlegume crops. Conceivably this might be done by developing much improved versions of the existing associative N_2-fixing systems; by conferring *Rhizobium*-fixing capacity (nodules) on non-fixing crops by intergeneric hybridization with legumes; or by transfer of nif genes from bacteria. Improvement of associative N_2-fixation by sugarcane, wheat, and the crops associated with *Azospirillum* spp. will have essentially the same objective as *Rhizobium* work: improved compatibility of plant and bacteria, improvement of the association and the carbohydrate supply, and making the systems tolerant to mineral fertilizer. Additionally, in sugarcane it would be useful to eliminate the repressor system (inhibitors?) that prevent N_2-fixing bacteria present in the stems from actively fixing nitrogen in situ. This is because, as far as is known, bacteria do not fix nitrogen in the stems, only in the cut stem pieces (germinating sets used for planting), in the roots, and in the rhizosphere soil.

Protoplast techniques, in particular the intergeneric fusion of protoplasts offer, at least theoretically, the possibility of crossing legume with nonlegume crops. It has to be admitted that there are still substantial difficulties, even though intergeneric protoplast fusion products are known. Constabel (1978) noted 24 cases of viable interspecific somatic hybrids, but in virtually all these cases the species could also be crossed by sexual means.

In the case of legumes, Kao et al. (1974) achieved fusion and cell division with barley and soybean, and maize and soybean. Vasil et al. (1977) approached the problem via nodule protoplasts, with the argument that this should both overcome the problem of the infection barrier and also permit protection of the bacteroids and nitrogenase system within the original membranes. Protoplasts from *Lupinus augustifolius* were fused with mesophyll protoplasts of *Nicotiana tabacum*. The fusion products were viable for a few days, and it was not possible to demonstrate acetylene reduction activity, although such activity had been present in the isolated nodule protoplasts. Binding and Nehls (1978a) obtained fusion products of *Vicia faba* and *Petunia hybrida*, but hybrid cells tended to eliminate *Vicia* chromosomes. The same authors (1978b) achieved regeneration of isolated *Vicia faba* protoplasts.

In general, plant regeneration from protoplast fusion of plants that normally cannot be crossed sexually has not been very successful. The plants have tended to grow slowly, show malformation, and be sterile. However, this relative lack of success at this early stage of research should not rule out the possibility of ultimate success of such methods.

There is considerable interest at the present time in the possibility of transferring the nif genes for nitrogen fixation from bacteria to higher plants. It has been known for a long time that genes for nitrogen fixation can be transferred between bacteria (Dunican and Tierney, 1974; Cannon et al., 1974), the latter workers successfully transferring the nitrogen fixing genes of *Klebsiella pneumoniae* into *Escherichia coli* and conferring on it the capacity to fix atmospheric nitrogen. The method adopted was to incorporate the nif genes in a plasmid, i.e., extrachromosomal DNA, followed by introduction of the plasmid into the *E. coli* cells.

Although the transfer of bacterial nif genes to plants is theoretically possible, it presents many difficulties, because of the presence of an apparent genetic barrier between prokaryotes and eukaryotes. Mere simple transfer of nif genes to nonlegume plants seems unlikely to give them the immediate capacity to fix nitrogen. A consideration of plant evolution indicates that nitrogen fixation capacity has not yet jumped the gap, say, between bluegreen and green algae, bacteria and yeasts, nor is it present in the simplest higher plant. The problems to be solved have been formulated by many workers (e.g., Shanmugan and Valentine, 1975a; Postgate, 1977).

It must be noted that it is not only the nif genes that must be transferred, but probably also those for leghaemoglobin synthesis (an oxygen carrier in the case of legume-type nodules) and for oxygen protection for nitrogenase. This requires the introduction of foreign DNA into plant chromosomes, or at least to ensure its maintenance as stable plasmids. The basic problem is that genetic information of higher plants and bacteria is expressed in different ways. Thus bacteria possess the language for regulating, transcribing, and translating information on nif and the other genes that must support it. As higher plants cannot read this language, an intermediary must be sought. Two intermediaries have been suggested as vectors: some viruses and the bacteria *Agrobacterium tumefaciens.*

Viruses have the advantage that they carry plasmid sites that can be read by the plant. Thus a DNA virus might prove to be a good intermediary. However, it seems not to be properly known whether bacterial RNA sequences can replicate by means of plant DNA polymerase or whether they can use plant RNA polymerase for transcription. One possibility is to construct hybrid molecules between a virus such as cauliflower mosaic virus, CMV (Langridge, 1976) and bacterial DNA sequences, so that bacterial genes for nif might be given a capacity for replication that can be expressed within plant cells. Of course, it is not sufficient to insert genetic information into a cell; it must thereafter show functional expression.

Agrobacterium belongs with *Rhizobium* in the two-genera family of Rhizobiaceae. Both genes produce cell proliferation, which in the case of *Rhizobium* finds expression in the nodules of legumes and in the case of *Agrobacterium* produces crown galls or growth on plants which are nonlegumes. Theoretically then, *Agrobacterium* could be used to transfer the nitrogen fixing ability of *Rhizobium* to nodule-like structures on nonlegumes. Davey et al. (1980) have shown the possibility of transforming isolated plant protoplasts with *Agrobacterium* plasmids. Plasmids of *Agrobacterium* and *Rhizobium* apparently share a high degree of homology (Drummond and Chilton, 1978; Jouanin et al., 1981; Hooykaas et al., 1981). Genetic studies in *Rhizobium* have indicated that at least some of the genes for symbiosis are plasmid carried (Johnson et al., 1978). Nuti et al. (1979) have put forward evidence for nif genes on *Rhizobium* plasmids, while Hooykaas et al. (1981) have reported a plasmid, Sym (biosis) in *R. trifolii* which determines host specificity, and most steps to root nodule formation and nitrogen fixation.

Although taken together these factors may be positive for the use of *Agrobacterium tumefaciens* as a nif vector, the fact remains that we have no means of controlling the gall production and, as Postgate (1977) noted, although *A. tumefaciens* accepts nif genes on a plasmid, it does not fix nitrogen; but if the plasmid is moved into *E. coli*, then derivatives will fix nitrogen. There are obvious problems. Although there are exciting possibilities, clearly we cannot guess how nif will ultimately be transferred to nonlegume plants.

In the meantime, more efficient rhizosphere bacteria may be obtained. Gordon and Brill (1972) found mutant strains of *Azotobacter* that will fix nitrogen even in the presence of NH_4^+. Kleeberger and Klingmueller (1980) have suggested transferring nif genes to bacteria common in the rhizosphere that hitherto do not have the capacity to fix nitrogen. They were successful in achieving acetylene reduction activity in a normally nonnitrogen-fixing strain of *Enterobacter cloacae*, after transferring plasmid pRD1 carrying nif genes from *Escherichia coli*.

Azotobacter paspali is specific for the batatais variety of *Paspalum notatum* (Dobereiner, 1976). There seems to be no theoretical reason why different strains of this bacteria should not be made specific for various Gramineae, with corresponding increased efficiency.

KEY REFERENCES

Eady, R.R. and Smith, B.E. 1979. In: Dinitrogen Fixation (R.W.F. Hardy, F. Bottomley, and R.C. Burns, eds.) pp. 399–490. John Wiley and Sons, New York.

Giles, K.L. and Vasil, I.K. 1980. Nitrogen fixation and plant tissue culture. Int. Rev. Cytol. (Suppl.)11B:81–99.

Vose, P.B. and Ruschel, A.P. (eds.) 1981. Associative N_2-Fixation, Vols. I and II. CRC Press, Boca Raton, Florida.

REFERENCES

Anderson, S.J. and Phillips, D.A. 1976. Effect of protein additives on acetylene reduction (nitrogen fixation) by *Rhizobium* in the presence and absence of soybean cells. Plant Physiol. 57:890–893.

Appleby, C.A., Bergersen, F.J., Ching, T.M., Gibson, A.H., Gresshoff, P.M., and Trinick, M.J. 1981. In: Current Perspectives in Nitrogen Fixation (A.H. Gibson and W.E. Newton, eds.) p. 369. Proceedings 4th International Symposium on Nitrogen Fixation, Canberra. Australian Academy of Science, Canberra.

Bajaj, Y.P.S. and Dhanju, M.S. 1979. Regeneration of plants from apical meristem tips of some legumes. Curr. Sci. 48:906–907.

Balandreau, J., Rinaudo, G., Fares-Hamad, I., and Dommergues, Y. 1975. N_2 fixation in paddy soils. In: Nitrogen Fixation and the Biosphere (W.D.P. Stewart, ed.) pp. 57–70. Cambridge Univ. Press, Cambridge and London.

Berg, R.H., Vasil, V., and Vasil, I.K. 1979. The biology of *Azospirillum*-sugarcane association. II. Ultrastructure. Protoplasma 101:143–163.

Bergersen, F.J. and Gibson, A.H. 1978. Nitrogen fixation by *Rhizobium* spp. in laboratory culture media. In: Limitations and Potentials for Biological Nitrogen Fixation in the Tropics (J. Dobereiner, R.H. Burris, and A. Hollaender, eds.) pp. 263–274. Plenum Press, New York.

Bergersen, F.J. (ed.) 1980. Methods for Evaluating Biological Nitrogen Fixation. John Wiley and Sons, New York.

Binding, H. and Nehls, R. 1978a. Somatic cell hybridization of *Vicia faba* and *Petunia hybrida*. Molec. Gen. Genet. 164:137–143.

_____ 1978b. Regeneration of isolated protoplasts of *Vicia faba* L. Z. Pflanzenphysiol. 88:327–332.

Boddey, R.M. and Dobereiner, J. 1982. Symposium Papers, pp. 28–47, Proceedings 12th International Congress of Soil Science, New Delhi.

Brar, D.S., Rambold, S., Gamborg, O., and Constabel, F. 1979. Tissue culture of corn and *Sorghum*. Z. Pflanzenphysiol. 95:377-389.

Burris, R.H. 1975. Methodology. In: The Biology of Nitrogen Fixation (A. Quispel, ed.) pp. 9-36. Elsevier, New York.

Cailloux, M. 1983. Protoplast methods. In: Contemporary Bases for Crop Breeding (P.B. Vose and S. Blixt, eds.). Pergamon Press, Oxford and New York. In press.

Cannon, F.C., Dixon, R.A., Postgate, J.R., and Primrose, S.B. 1974. Chromosomal intergration of *Klebsiella* nitrogen fixation genes in *Escherichia coli*. J. Gen. Microbiol. 80:227-239.

Carlson, P.S. and Chaleff, R.S. 1974. Forced association between higher plant and bacterial cells. Nature 252:393-394.

Child, J.J. 1975. Nitrogen fixation by *Rhizobium* sp. in association with non-leguminous plant cell cultures. Nature 253:350-351.

_____ and LaRue, T.A. 1974. A simple technique for the establishment of nitrogenase in soybean callus culture. Plant Physiol. 53:88-90.

Constabel, F. 1978. Development of protoplast fusion products, heterokaryocytes and hybrid cells. In: Frontiers of Plant Tissue Culture (T.A. Thorpe, ed.) pp. 141-150. International Association Plant Tissue Culture, Univ. Calgary Press, Alberta.

Criswell, J.G., Havelka, U.D., Quebedeaux, B., and Hardy, R.F. 1976. Adaptation of nitrogen fixation by intact soybean nodules to altered rhizosphere pO_2. Plant Physiol. 58:622-625.

Dart, P.J. and Subba Rao, R.V. 1981. Nitrogen fixation associated with sorghum and millet. In: Associative N_2-Fixation, Vol. I (P.B. Vose and A.P. Ruschel, eds.) pp. 169-177. CRC Press, Boca Raton, Florida.

Davey, M.R., Cocking, E.C., Freeman, J., Pearce, N., and Tudor, I. 1980. Transformation of *Petunia* protoplasts by isolated *Agrobacterium* plasmids. Plant Sci. Lett. 18:307-313.

Devine, T.E. and Breithaupt, B.H. 1980. Significance of incompatibility reactions of *Rhizobium japonicum* strains with soybean host genotypes. Crop Sci. 20:269-271.

Dilworth, M.J. and Parker, C.A. 1969. Development of the nitrogen-fixing system in legumes. J. Theor. Biol. 25:208-218.

Dixon, R.O.D. 1972. Hydrogenase in legume root nodule bacteroids: Occurrence and properties. Arch. Microbiol. 85:193-201.

_____ 1978. Nitrogenase-hydrogenase interrelationships in *Rhizobia*. Biochemie 60:233-236.

Dobereiner, J. 1966. *Azotobacter paspali* n. sp. uma bacteria fixadora de nitrogenio na rhizosphera de *Paspalum*. Pesq. Agropec. Bras. 1:357-365.

_____ and Day, J. 1974. Associative symbioses in tropical grasses: Characterization of microorganisms and dinitrogen-fixing sites. In: Proceedings International Symposium Nitrogen Fixation, Vol. 2 (W.E. Newton and C.J. Nyman, eds.) pp. 518-538. Washington State Univ. Press, Pullman.

Drummond, M.H. and Chilton, M.D. 1978. Tumor-inducing (Ti) plasmids of *Agrobacterium* share extensive regions of DNA homology. J. Bacteriol. 136:1178-1183.

Dunican, L.K. and Tierney, A.B. 1974. Genetic transfer of nitrogen fixation from *Rhizobium trifolii* to *Klebsiella aerogenes*. Biochem. Biophys. Res. Comm. 57:62-72.

Eady, R. 1981. In: Current Perspectives in Nitrogen Fixation (A.H.
Gibson and W.E. Newton, eds.). Proceedings 4th International Symposium on Nitrogen Fixation, pp. 172-181. Australian Academy Science, Canberra.

Evans, W.R. and Keister, D.L. 1976. Reduction of acetylene by stationary culture of free-living *Rhizobium* sp. under atmospheric oxygen
levels. Can. J. Microbiol. 22:949-952.

Gallon, J.R. 1980. Nitrogen fixation of photoautotrophs. In: Nitrogen
Fixation (W.D.P. Stewart and J.R. Gallon, eds.) pp. 197-238. Academic Press, London.

Gibson, A.H., Child, J.J., Pagan, J.D., and Scowcroft, W.R. 1976a. The
induction of nitrogenase activity in *Rhizobium* by non-legume plant
cells. Planta 128:233-239.

Gibson, A.H., Scowcroft, W.R., Child, J.J., and Pagan, J.D. 1976b. Ni-
trogenese activity in cultured *Rhizobium* sp. Strain 32H1. Arch.
Microbiol. 108:45-54.

Gordon, J.K. and Brill, W.J. 1972. Mutants that produce nitrogenase in
the presence of ammonia. Proc. Nat. Acad. Sci. USA 69:3501.

Green, C.E. 1978. In vitro plant regeneration in cereals and grasses.
In: Frontiers of Plant Tissue Culture (T.A. Thorpe, ed.) pp. 414-419.
International Association of Plant Tissue Culture, Univ. Calgary
Press, Alberta.

Gray, C.T. and Gest, H. 1965. Biological formation of molecular hydrogen. Science 148:186-192.

Haaker, H., Laane, C., and Veeger, C. 1980. Dinitrogen fixation and
the proton-motive force. In: Nitrogen Fixation (W.D.P. Stewart and
J.R. Gallon, eds.) pp. 113-183. Academic Press, London and New
York.

Hardy, R.W.F., Burns, R.C., and Holsten, R.D. 1973. Application of the
acetylene reduction assays for nitrogenase. Soil Biol. Biochem. 5:47-
81.

Heinz, D.J. 1973. Sugarcane improvement through induced mutations
using vegetatively propagated plants. In: Induced Mutations in Vegetatively Propagated Plants, pp. 53-59. FAO/IAEA Panel Proc. IAEA,
Vienna.

Hoch, G.E., Little, H.N., and Burris, R.H. 1957. Hydrogen evolution
from soybean root nodules. Nature 179:430-431.

Holsten, R.D., Burns, R.C., Hardy, R.W.F., and Herbert, R.R. 1971. Establishment of symbiosis between *Rhizobium* and plant cells in vitro.
Nature 232:173-176.

Hooykaas, P.J.J., van Brussel, A.A.N., den Dulk-Ras, H., van Slogteren,
G.M.S., and Schilperoort, R.A. 1981. Sym plasmid of *Rhizobium trifolii* expressed in different rhizobial species and *Agrobacterium tumefaciens*. Nature 291:351-353.

Johnston, A., Berynon, J., Buchanan-Wollaston, A., Setchell, S., Hirsch,
P., and Beringer, J. 1978. High frequency transfer of nodulating
ability between strains and species of *Rhizobium*. Nature 276:634-
636.

Jones, D.G. and Hardarson, G. 1979. Variation within and between white clover varieties in their preference for strains of *Rhizobium trifolii*. Ann. Appl. Biol. 92:221-228.

Jones, D.G. and Morley, S.J. 1981. In: Current Perspectives in Nitrogen Fixation (A.H. Gibson and W.E. Newton, eds.) p. 429. Proceedings 4th International Symposium on Nitrogen Fixation. Australian Academy of Science, Canberra.

Jouanin, L., DeLajudie, P., Bazetoux, S., and Huguet, T. 1981. Molec. Gen. Genet. 183:189-196.

Kao, K.N., Constabel, F., Michayluk, M.R., and Gamborg, O.L. 1974. Plant protoplast fusion and growth of intergeneric hybrid cells. Planta 120:245-227.

Keister, D.L. 1975. Acetylene reduction by pure cultures of *Rhizobia*. J. Bacteriol. 123:1265-1268.

Kleeberger, A. and Klingmueller, W. 1980. Plasmid-mediated transfer of nitrogen-fixing capability of bacteria from the Rhizosphere of grasses. Molec. Gen. Genet. 180:621-627.

Knowles, R. 1981. In: Current Perspectives in Nitrogen Fixation (A.H. Gibson and W.E. Newton, eds.) pp. 327-333. Proceedings 4th International Symposium on Nitrogen Fixation, Australian Academy of Science, Canberra.

Kurz, W.G.W. and LaRue, T.A. 1975. Nitrogenase activity in absence of plant host. Nature 256:407-409.

Langridge, J. 1977. Genetic engineering in plants. Search 8:13-15.

Larson, R.I. and Neal, J.L. 1978. Selective colonization of the rhizosphere of wheat by nitrogen-fixing bacteria. Ecol. Bull. 26:331-342.

Martin, C., Carre, M., and Duc, G. 1979. Note sur les cultures de tissues de feverole (*Vicia faba* L.) bouturage, culture de cals, culture de meristems. Ann. Amelior. Plantes 29:227-289.

McComb, J.A., Elliot, J., and Dilworth, M.J. 1975. Acetylene reduction by pure *Rhizobium* in culture. Nature 256:409-410.

Minchin, F.R., Summerfield, R.J., and Eaglesham, A.R.J. 1978. Plant genotype x Rhizobium strain interactions in cowpea (*Vigna unguiculata* (L.) Walp.). Trop. Agric. (Trinidad) 55:107-115.

Mohapatra, S.S., Reporter, M., Rolfe, B.G., and Gresshoff, P.M. 1980. Abstract 238, Programme and Abstracts, 4th International Symposium on Nitrogen Fixation. Australian Academy of Science, Canberra.

Mytton, L.R., El Sherbeeny, M.H., and Lawes, D.A. 1977. Symbiotic variability in *Vicia faba*. 3. Genetic effects of host plant, *Rhizobium* strain and of host x strain interaction. Euphytica 26:785-791.

Nelson, A.D., Barker, L.E., Tjepkema, J., Russell, S.A., Powelson, R., Evans, H.J., and Seidler, R.J. 1976. Nitrogen fixation associated with grasses in Oregon. Can. J. Microbiol. 22:523-530.

Nuti, M.P., Lepidi, A.A., Prakash, R.K., Schilperoot, R.A., and Cannon, F.C. 1979. Evidence for nitrogen fixation (nif) genes on indigenous *Rhizobium* plasmids. Nature 282:533-535.

Pagan, J.D., Child, J.J., Scowcroft, W.R., and Gibson, A.H. 1975. Nitrogen fixation by *Rhizobium* cultured on a defined medium. Nature 256:406-407.

Pankhurst, C.E. 1981. Nutritional requirement for the expression of nitrogenase activity by *Rhizobium* sp. in agar culture. J. Appl. Bacteriol. 50:45-54.

Patriquin, D.G., Boyle, C.D., Livingstone, D.C., and McLung, C.R. 1981. In: Associative N₂-Fixation, Vol. II (P.B. Vose and A.P. Ruschel, eds.) pp. 11-26. CRC Press, Boca Raton, Florida.

Phillips, D.A. 1974a. Factors affecting the reduction of acetylene by Rhizobium-soybean cell association in vitro. Plant Physiol. 53:67-72.

_____ 1974b. Promotion of acetylene reduction by Rhizobium-soybean cell association in vitro. Plant Physiol. 54:654-655.

_____ 1976. The effect of combined nitrogen on Rhizobium-soybean cell associations in vitro. In: Proceedings 1st International Symposium on Nitrogen Fixation (W.E. Newton and C.J. Wyman, eds.) Vol. 2, pp. 367-373. Washington State Univ. Press, Pullman.

_____, Daniel, R.M., Appleby, C.A., and Evans, H.J. 1973. Isolation from Rhizobium of factors which transfer electrons to soybean nitrogenase. Plant Physiol. 51:136-138.

Postgate, J.R. 1977. Possibilities for the enhancement of biological nitrogen fixation. Phil. Trans. R. Soc. London 281:249-260.

Raggio, N., Raggio, M., and Torrey, J.G. 1957. The nodulations of isolated leguminous roots. Am. J. Bot. 44:325-334.

Raggio, N., Raggio, M., and Burris, R.H. 1959. Nitrogen fixation by nodules formed on isolated bean roots. Biochem. Biophys. Acta 32: 274.

Ranga Rao, V. 1977. Nitrogenase activity of Rhizobium sp. Strain 1552 on defined medium (Stylosanthes gracilib nodule bacteria). Plant Sci. Lett. 8:363-366.

Rennie, R.J. 1981. In: Induced Mutations—A Tool in Plant Research, pp. 293-321. IAEA, Vienna.

_____ and Larson, R.I. 1981. Dinitrogen fixation associated with disomic chromosome substitution lines of spring wheat in the phytotron and in the field. In: Associative N₂-Fixation, Vol. I (P.B. Vose and A.P. Ruschel, eds.) pp. 145-154. CRC Press, Boca Raton, Florida.

Reporter, M. 1976. Synergetic culture of Glycine max root cell and Rhizobia separated by membrane filters. Plant Physiol. 57:651-655.

_____ and Hermina, N. 1975. Acetylene reduction by transfilter suspension cultures of Rhizobium japonicum. Biochem. Biophys. Res. Comm. 64:1126-1133.

_____, Raveed, D., and Norris, G. 1975. Binding of Rhizobium japonicum to cultured soya root cells: Morphological evidence. Plant Sci. Lett. 5:73-76.

Ruschel, A.P. 1981a. Associative N₂-fixation by sugarcane. In: Associative N₂-Fixation, Vol. II (P.E. Vose and A.P. Ruschel, eds.) pp. 81-90. CRC Press, Boca Raton, Florida.

_____ 1981b. Sugarcane—Position paper. In: Associative N₂-Fixation, Vol. II (P.B. Vose and A.P. Ruschel, eds.) pp. 249-251. CRC Press, Boca Raton, Florida.

_____ and Vose, P.B. 1977. Present situation concerning studies on associative N₂-fixation in sugarcane. Boletim Cientifico No. 45, CENA (CP 96, 13400 Piracicaba, SP, Brasil).

_____ and Vose, P.B. 1980. Nitrogen fixation as a source of energy in tropical agriculture. In: FACO/SIDA Workshop on Organic Recycling in Agriculture, Costa Rica. FAO, Rome.

Ruschel, R. and Ruschel, A.P. 1981. Inheritance of N_2-fixing ability in sugarcane. In: Associative N_2-Fixation, Vol. II. (P.B. Vose and A.P. Ruschel, eds.) pp. 133-140. CRC Press, Boca Raton, Florida.

Schetter, C. and Hess, D. 1977. Nitrogenase activity in in vitro associations between callus tissues of non-leguminous horticultural plants and *Rhizobium*. Plant Sci. Lett. 9:1-5.

Schrauser, G.N. 1977. In: Recent Developments in Nitrogen Fixation (W. Newton, J.R. Postgate, and C. Rodriguez-Barrueco, ed.). Academic Press, London and New York.

Scowcroft, W.R. and Gibson, A.H. 1975. Nitrogen fixation by *Rhizobium* association with tobacco and cowpea cell cultures. Nature 253:351-352.

_____, and Pagan, J.D. 1976. Nitrogen fixation in cultured cowpea rhizobia. Inhibition and regulation of nitrogenase activity. Biochem. Biophys. Res. Commun. 73:516-523.

Shanmugam, K.T. and Valentine, R.C. 1975a. Microbial production of ammonium ion from nitrogen. Proc. Nat. Acad. Sci. USA 72:136-139.

_____ 1975b. Molecular biology of nitrogen fixation. Science 137:919-924.

Sharp, W.R., Reed, S.M., and Evans, D.A. 1983. Production and application of haploid plants. In: Contemporary Bases for Crop Breeding (P.B. Vose and S. Blixt, eds.). Pergamon Press, Oxford and New York. In press.

Singh, K., Sinha, A.P., and Agnihotri, V.P. 1981. Role of Azotobacter in sugarcane culture and the effect of pesticides on its population in soil. In: Associative N_2-Fixation, Vol. II (P.B. Vose and A.P. Ruschel, eds.) pp. 103-108. CRC Press, Boca Raton, Florida.

Tjempkema, J. and Evans, H.J. 1975. Nitrogen fixation by free living *Rhizobium* in a defined liquid medium. Biochem. Biophys. Res. Commun. 65:625-628.

Trinick, M.J. 1973. Symbiosis between *Rhizobium* and the non-legume, *Trema aspera*. Nature 24:459-460.

Tubb, R.S. 1976. Regulation of nitrogen fixation in *Rhizobium* sp. Appl. Environ. Microbiol. 32:483-488.

_____ and Postgate, J.R. 1973. Control of nitrogenase synthesis in *Klebsiella pneumoniae*. J. Gen. Microbiol. 79:103-117.

Vasil, I.K., Vasil, V., and Hubbell, D.H. 1977. Engineered plant cell or fungal association with bacteria that fix nitrogen. In: Genetic Engineering for Nitrogen Fixation (A. Hollaender et al., eds.) pp. 197-211. Plenum Press, New York.

Vasil, V., Vasil, I.K., Zuberer, D.A., and Hubbell, D.H. 1979. The biology of *Azospirillum*-sugarcane association. I. Establishment of the association. Z. Pflanzenphysiol. 95:141-147.

Veliky, I. and LaRue, T.A. 1967. Changes in soybean root culture induced by *Rhizobium japonicum*. Naturwissenschaften 54:96-97.

Vincent, J.M. 1970. A manual for the practical study of root nodule bacteria. IBP Handbook, No. 15. Blackwell Sci. Pub., Oxford.

Vose, P.B. 1981. Crops for all conditions. New Sci. 89:688-690.

_____, Ruschel, A.P., Victoria, R.L., Tsai Saito, S.M., and Matsui, E. 1981. 15-Nitrogen as a tool in biological nitrogen fixation research. In: Proceedings International Workshop Biological Nitrogen Fixation Technology for Tropical Agriculture, CIAT, Colombia.

Walker, K.A., Wendeln, M.L., and Jaworski, E.G. 1979. Organogenesis in callus tissue of *Medicago sativa*. The temporal separation of induction processes from differentiation processes. Plant Sci. Lett. 16: 23-30.

Watanabe, I. and Barraquio, W.L. 1979. Low levels of fixed nitrogen required for isolation of free-living N_2-fixing organisms from rice roots. Nature 277:565.

Yates, M.G. 1977. Physiological aspects of nitrogen fixation. In: Recent Developments in Nitrogen Fixation (W. Newton, J.R. Postgate, and C. Rodriguez-Barrueco, eds.) pp. 219-270. Academic Press, London and New York.

CHAPTER 31
Evaluation of Disease Resistance

S.A. Miller and *D.P. Maxwell*

Tissue culture systems have been utilized in the past 15 years to investigate numerous aspects of host–pathogen interactions (Ingram, 1977; Ingram and Helgeson, 1980). Emphasis in this chapter is placed on the application of tissue culture methods to studies of plant disease resistance, particularly the role of phytoalexins in resistance. Tissue cultures (i.e., callus tissues, suspension cultures, isolated cells, and protoplasts) offer many advantages over intact plants for these studies, including (a) exclusion of contaminating microorganisms, (b) control of environmental parameters, e.g., temperature, light, and nutrients, (c) ability to inoculate host cells without wounding, (d) control of pathogen inoculum levels and number of host cells, (e) ability to alter the nature of the host–pathogen interaction by altering componenets of the growth medium, (f) the presence of only one or a few major host cell types, and (g) the ease of application or removal of materials, e.g., labeled precursors, from cultured cells. The major disadvantage of tissue culture systems is that plant cells grown in culture may be, in some instances, genetically and physiologically different from cells in intact plants (Ingram, 1980). Nonetheless, valuable information has been acquired using tissue cultures, and they should continue to be useful as model systems in studies of plant disease resistance.

REVIEW OF THE LITERATURE

Expression of Disease Resistance in Tissue Cultures

The tissue culture system that has been most carefully defined and presents the best evidence for the applicability of in vitro systems to the study of host-parasite interactions is the tobacco (*Nicotiana tabacum* L.)/*Phytophthora parasitica* Dast. var. *nicotianae* (Breda de Haan) (Ppn) system (Helgeson et al., 1972; Helgeson and Haberlach, 1980). Monogenic resistance in tobacco to race O of Ppn is determined by a dominant genetic factor (Goins and Apple, 1970) that is expressed in tissue culture (Helgeson et al., 1976; Maronek and Hendrix, 1978; Deaton et al., 1982).   Helgeson et al. (1976) compared rooted cuttings from 185 tobacco plants with callus tissues derived from these plants for their reaction to race O of Ppn.   The plants tested were cuttings from a homozygous resistant parent, a homozygous susceptible parent, the F_1 plants from a cross of these parents, the S_1 plants of selfed resistant and susceptible parents, the F_2 plants from selfed F_1 plants, the F_3 plants from crosses of homozygous susceptible with heterozygous resistant F_2 plants and progeny of an outcross between an F_1 plant and a susceptible plant from the cultivar Wisconsin 38.   In each case, the segregation pattern reflected the presence of a single, dominant genetic factor for resistance to race O of Ppn.   In addition, callus tissues derived from all of the resistant plants were resistant, and those derived from the susceptible plants were susceptible.   Since there were no deviations from the correlation of callus tissue reactions to Ppn with the reactions of the plants from which they were derived, these results indicate that the resistance gene that is expressed in intact tobacco plants is also expressed in tissue culture.   Quantitatively inherited resistance to Ppn is also expressed in tissue culture.   Deaton et al. (1982) found that the response of callus tissue from field resistant burley cultivars to Ppn was highly correlated with the whole plant response to the pathogen.   The degree of resistance of these calluses was intermediate between monogenic resistance and susceptibility.

The expression of resistance in tobacco callus cultures is affected by several factors, including temperature, inoculum concentration, callus morphology, and particularly, the balance of phytohormones in the tissue culture medium (Helgeson et al., 1972).   Resistant-responding callus tissues are characterized by a hypersensitive reaction in the tissue beneath the point of inoculation, accompanied by reduced colonization by the pathogen.   In susceptible tissues no hypersensitive reaction is observed, and colonization is extensive within 7 days after inoculation (Haberlach et al., 1978).   Cytological studies demonstrate that within 48 hr after inoculation, only 5-8 cell layers are penetrated in resistant tissues, while in susceptible tissues more than 50 cell layers are colonized (de Zoeten et al., 1982).   Ultrastructural changes in compatible and incompatible interactions in callus tissue are strikingly similar to those occurring in intact roots infected with Ppn (Hanchey and Wheeler, 1971).

The expression of race-specific resistance in tissue culture has also been demonstrated for potato (*Solanum tuberosum* L.) in response to *Phytophthora infestans* (Mont.) de By.   Tissue culture aggregates from the variety Majestic, with no known R genes for resistance to *P. infestans*, stimulate growth of race 4 of the pathogen, whereas aggregates from the variety Orion, which contains the R_1 gene, do not (Ingram and Robertson, 1965).   Tissue culture aggregates derived from other *Solanum* lines resistant to race 4 are also resistant, while those derived from susceptible lines are susceptible (Ingram, 1967). Thus Ingram (1967) concluded that R genes for resistance to *P. infestans* are expressed in *Solanum* tissue cultures. As in the tobacco/Ppn system, the expression of resistance to *P. infestans* in potato callus tissue is influenced by the balance of plant growth regulators in the growth medium. By manipulating the auxin:cytokinin concentration ratios in chemically defined growth media, Campbell (1979) was able to demonstrate the expression of resistance in callus tissue to *P. infestans*. The resistant response is not accompanied by tissue browning, although *P. infestans* does cause browning in tobacco callus tissue.

Other systems in which the expression of disease resistance in tissue cultures has been investigated include (a) tomato (*Lycopersicon esculentum* Mill.)/*P. infestans* (Warren and Routley, 1970), (b) soybean (*Glycine max* (L.) Merr.)/*Phytophthora megasperma* Drechs. var. *sojae* (Holliday and Klarman, 1979), (c) tobacco/*Pseudomonas* spp. (Huang and Van Dyke, 1978), and (d) tobacco/tobacco mosaic virus (Beachy and Murakishi, 1971).

In our laboratory we have utilized an alfalfa (*Medicago sativa* L.) callus tissue culture system to study host and nonhost resistance in alfalfa to *Phytophthora megasperma* f. sp. *medicaginis* (Pmm), an alfalfa root pathogen, and *P. megasperma* f. sp. *glycinea* (Pmg, syn. *P. megasperma* var. *sojae*), a pathogen of soybean but not of alfalfa, respectively. Cultivated alfalfa is an autotetraploid with considerable genetic diversity, and inheritance is complex (Busbice et al., 1972). There appears to be more than one genetic system governing the resistance of alfalfa to Pmm (Lu et al., 1973; Irwin et al., 1981a); in the system we have studied resistance is determined by two incompletely dominant complementary genes, requiring at least a duplex genotype at one locus and a simplex genotype at the other locus for the expression of resistance (Irwin et al., 1981b). Resistance to Pmm and to the nonpathogen, Pmg, is expressed in callus tissue culture (Miller et al., 1981; Miller, 1982). The resistant interactions are characterized by a reduction in tissue colonization and an increase in tissue browning, relative to the susceptible interaction (Table 1, Figs. 1 and 2). The difference between resistant and susceptible interactions, based on the presence of aerial hyphae on inoculated callus tissue pieces, is apparent within 2 days after inoculation (Table 1).

Studies of Phytoalexin Induction and Biosynthesis in Tissue Culture

Phytoalexins are low molecular weight, antimicrobial compounds formed by plants in response to infection and exposure to various toxic and nontoxic compounds and are thought to play a role in disease resistance (Kuc, 1976; Stoessel, 1980; Keen, 1981; Heath, 1981a). Tissue culture systems, especially those utilizing suspension cultures, have been used recently to investigate various aspects of the induction and biosynthesis of phytoalexins. Plant cells grown in culture retain the capacity to produce phytoalexins in response to biotic or abiotic phytoalexin-inducing agents (elicitors) (Table 2). In some cases phytoalexins are produced in the absence of added elicitors, but in general, the

Table 1. Aerial Hyphal Development and Callus Discoloration in Two Alfalfa Callus Tissue Lines after Inoculation with *Phytophthora megasperma* f. sp. *medicaginis* (Pmm) or *P. megasperma* f. sp. *glycinea* (Pmg).[1]

TREATMENT	AERIAL HYPHAL DEVELOPMENT[2,3]			TISSUE DISCOLORATION[2,3]		
	Day 1	Day 2	Day 3	Day 1	Day 2	Day 3
M269/Pmm	0.0 a	1.5 b	2.7 c	2.0 bc	3.3 def	3.0 de
M194/Pmm	0.0 a	0.3 a	0.3 a	1.3 b	2.7 cd	4.0 f
M269/Pmg	0.0 a	0.0 a	0.3 a	1.3 b	1.7 b	4.0 f
M194/Pmg	0.0 a	0.0 a	0.0 a	1.3 b	1.3 b	3.7 ef
M269/Control	0.0 a	0.0 a	0.0 a	0.0 a	0.0 a	0.0 a
M194/Control	0.0 a	0.0 a	0.0 a	0.0 a	0.0 a	0.0 a

[1] Line M269 is susceptible to Pmm; line M194 is resistant. Both are nonhosts of Pmg.

[2] Mean of three replications. For each parameter (aerial hyphal development or tissue discoloration), means followed by the same letter are not significantly different at P = 0.01, according to Duncan's multiple range test.

[3] Rating system for aerial hyphal development and callus tissue discoloration:

0 = no aerial hyphae or tissue discoloration
1 = aerial hyphae or tissue discoloration on top 25% of callus piece
2 = aerial hyphae or tissue discoloration on top 50% of callus piece
3 = aerial hyphae or tissue discoloration on top 75% of callus piece
4 = aerial hyphae or tissue discoloration on entire callus piece

Table 2. Tissue Culture Systems Used to Study Disease Resistance Expression and Phytoalexin Accumulation for Various Host-Pathogen Interactions (from Dixon, 1980)

HOST	PATHOGEN/ELICITOR	PHYTOALEXIN	MEDIUM	REFERENCE
Nicotiana tabacum	Phytophthora parasitica var. nicotianae (Ppn)	1	Linsmaier & Skoog	Helgeson et al., 1972; Haberlach et al., 1978
N. tabacum	Ppn	Rishitin; Capsidiol	Linsmaier & Skoog	Budde, 1981
N. tabacum	Ppn	1	Modified MS	Maronek & Hendrix, 1978
N. tabacum	Pseudomonas tabaci P. pisi P. fluorescens	1	SH	Huang & Van Dyke, 1978
N. tabacum	Tobacco mosaic virus (TMV)	1	Modified MS	Beachy & Murakishi, 1971
N. glutinosa N. tabacum	TMV TMV	1	Modified MS	Russell & Halliwell, 1974
Solanum tuberosum	Phytophthora infestans (Pi)	1	see reference	Ingram & Robertson, 1965
S. tuberosum	Cell-free hyphal homogenates of Pi	Rishitin; Phytuberin	Modified White's	Ersek & Sziraki, 1980
S. tuberosum	Pi	1	Modified MS; Lam's medium	Campbell, 1979
Lycopersicon esculentum	Pi	1	Modified MS	Warren & Routley, 1970
Ipomoea batatas	2	Furano-terpenes	see reference	Oba & Uritani, 1979
Pisum sativum	2	Pisatin	see reference	Bailey, 1970

Table 2. Cont.

HOST	PATHOGEN/ELICITOR	PHYTOALEXIN	MEDIUM	REFERENCE
Canavalia ensiformis	Pithomyces chartarum	Medicarpin	Modified Miller's medium	Gustine et al., 1978
Glycine max	Phytophthora megasperma var. sojae (Pms)[3]	Glyceollin	Linsmaier & Skoog	Keen & Horsch, 1972
G. max	Pms[3] glucan elicitor, Nigeran	Glyceollin	B-5	Ebel et al., 1976
G. max	Pms[3]	1	B-5	Holliday & Klarman, 1979
Phaseolus vulgaris	Botrytis cinerea culture filtrate	Phaseollin	Modified SH	Dixon & Fuller, 1976, 1978
P. vulgaris	Bean hypocotyl extract	Phaseollin; Phaseollidin; Kievitone; Phaseollin-isoflavan	Modified SH	Hargreaves & Selby, 1978
Trifolium repens	Sulfhydryl reagents	Medicarpin	B-5	Gustine, 1981
Medicago sativa	Phytophthora megasperma f. sp. medicaginis, Pmg	Medicarpin; Sativan	Modified SH	Miller & Maxwell (see Protocol)

[1] Phytoalexin levels were not determined.
[2] Phytoalexins produced in culture in absence of added elicitors.
[3] Phytophthora megasperma var. sojae (Pms) = P. megasperma f. sp. glycinea (Pmg) (Kuan and Erwin, 1980).

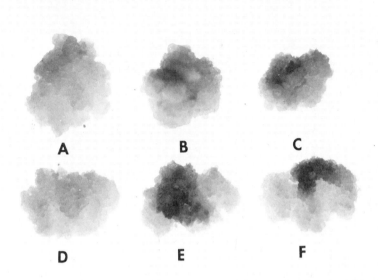

Figure 1. Response of two alfalfa callus tissue lines to *Phytophthora megasperma* f. sp. *medicaginis* (Pmm) and *P. megasperma* f. sp. *glycinea* (Pmg) 3 days after inoculation. Line M269 is susceptible to Pmm, line M194 is resistant; both are nonhosts of Pmg. (a) M269, noninoculated control, (b) M269/Pmm, (c) M269/Pmg, (d) M194, noninoculated control, (e) M194/Pmm, (f) M194/Pmg. Note aerial hyphal development on M269/ Pmm (b) and browning reaction on M269/Pmg (c), M194/Pmm (e) and M194/Pmg (f).

level of constitutive, or noninduced, phytoalexin production in culture is low (Dixon, 1980) and decreases with continual subculture (Bailey, 1970; Hargreaves and Selby, 1978; Dixon, 1980).

Most of the studies of phytoalexin induction in tissue culture have involved the production of isoflavonoid-derived phytoalexins by members of the Leguminosae. Changes in levels of various enzymes of phenylpropanoid metabolism in plant cultures, after treatment with spore suspensions or other elicitors, have been examined in *Phaseolus vulgaris* L. (Dixon and Fuller, 1976; Lamb and Dixon, 1978; Dixon and Bendall, 1978; Dixon and Lamb, 1979; Lawton et al., 1980; Dixon et al., 1981), *Glycine max* (Ebel et al., 1976; Zaehringer et al., 1981), and *Canavalia ensiformis* (L.) DC. (Gustine et al., 1978). The activity of L-phenylalanine ammonia-lyase (PAL), which catalyzes the first step in the biosynthesis of phenylpropanoid phytoalexins from phenylalanine, increases in elicitor-treated callus or suspension cultures, and precedes increases in phytoalexin concentration in these systems. In *P. vulgaris* suspension cultures treated with an elicitor preparation from the bean pathogen, *Colletotrichum lindemuthianum* (Sacc. & Magn.) Br. & Cav., the marked but transient increase in PAL activity observed is the result of increased de novo synthesis of the enzyme (Dixon and Lamb, 1979).

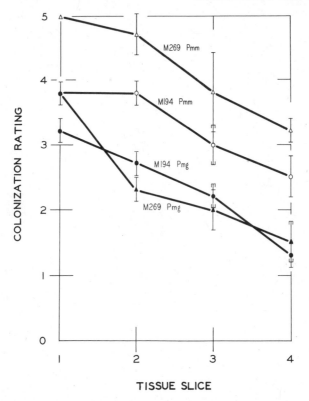

Figure 2.   Colonization of alfalfa callus tissue 72 hr after inoculation
with *Phytophthera megasperma* f. sp. *medicaginis* (Pmm) or *P. megasperma* f. sp. *glycinea* (Pmg).   Callus pieces are sliced horizontally into
four equal slices (numbered one to four from top to bottom) that are
rated for colonization on a zero to five scale.   A value of zero indicates that no hyphae are visible in the slice, while a value of five is
assigned to slices in which the hyphae form a thick mat.   Alfalfa line
M269 is susceptible to Pmm; line M194 is resistant.   Both are nonhosts
of Pmg.   No hyphae are observed in noninoculated controls.

However, it is still not clear whether PAL plays a regulatory role in
phytoalexin biosynthesis (Partridge and Keen, 1977; Zaehringer et al.,
1978).   Other enzymes of phenylpropanoid metabolism which have been
shown to increase in induced callus or suspension cultures include (a)
o-methyl transferase in *C. ensiformis* (Gustine et al., 1978), (b) dimethylallylpyrophosphate:3,6a,9-trihydroxy-pterocarpan dimethylallyl transferase in *G. max* (Zaehringer et al., 1981), and (c) cinnamic acid 4-hydroxylase, p-coumaric acid-coenzyme A ligase, flavanone synthase, and
chalcone-flavanone isomerase in *P. vulgaris* (Dixon and Bendall, 1978).

The Role of Phytoalexins in Resistance Expression in Tissue Cultures

Several careful studies with intact plants have indicated that phytoalexins accumulate at the proper time, place, and concentration to re-

strict growth of incompatible races of a pathogen, and thus are presumed to be responsible for the expression of race–specific resistance
(see review by Keen and Bruegger, 1977; Yoshikawa et al., 1978).
However, studies of resistance in tobacco callus tissue to incompatible
races of Ppn have failed to provide evidence which supports this concept (Helgeson et al., 1978; Budde, 1981; Budde and Helgeson, 1981).
Rishitin, capsidiol, and two related sesquiterpenoid compounds with
antifungal activity accumulate in tobacco callus tissues challenged with
compatible or incompatible races of Ppn, but the levels are higher and
detected earlier in the compatible interactions than in the incompatible
ones.   Phytoalexins are not detected in the incompatible interactions
until 60–70 hr after inoculation, even though the restriction of hyphal
growth occurs within 24 hr after inoculation (de Zoeten et al., 1982).
Thus it is not obvious how phytoalexins could be responsible for the
expression of race–specific resistance in this system.

We have observed similar results in our studies of the role of phytoalexins in the resistance of alfalfa callus tissue to Pmm.   Our efforts
have focused on the pterocarpanoid phytoalexin, medicarpin.   It is the
predominant phytoalexin that accumulates in alfalfa roots (L.B. Graves,
Jr. and D.P. Maxwell, unpublished data) and seedlings (Vaziri et al.,
1981) challenged with Pmm, and in leaves inoculated with *Helminthosporium carbonum* Ullstrup, a nonpathogen of alfalfa (Higgins, 1972).
Although colonization of resistant callus tissue (M194) by Pmm is significantly lower than colonization of susceptible tissue (M269) (Fig. 2),
levels of medicarpin in the resistant interaction are lower than in the
susceptible interaction (Fig. 3).   In addition, the highest medicarpin
concentration reached in the resistant interaction (approx. 20 µg/g
fresh weight callus tissue) is much lower than the ED_{50} value for Pmm
(>90 µg/ml; Miller, 1982) determined in in vitro bioassays of hyphal
growth inhibition.

Two other isoflavonoid phytoalexins, sativan and vestitol, are reported to accumulate in alfalfa leaves following inoculation with *H. carbonum* (Ingham and Millar, 1973; Ingham, 1979).   Vestitol has not been
detected in alfalfa callus tissue challenged with Pmm or Pmg or in noninoculated controls.   Low levels of sativan (2.5 µg/g fresh weight callus
tissue) are detected in Pmm- and Pmg-challenged callus tissue within 48
hr after inoculation and increase slightly by 72 hr after inoculation.
Accumulation of sativan, like medicarpin accumulation, is highest in the
susceptible interaction.   Thus phytoalexin accumulation does not appear
to be responsible for the expression of resistance to Pmm in alfalfa callus tissues.   However, phytoalexins may play a role in nonhost resistance in alfalfa callus tissues to Pmg.   In vitro tests of hyphal growth
inhibition by medicarpin indicate that Pmg is highly sensitive to this
phytoalexin (ED_{50} <20 µg/ml).   By 72 hr after inoculation with Pmg,
medicarpin is present in both M269 and M194 callus tissues in concentrations that surpass the ED_{50} value of medicarpin for Pmg (Fig. 3).

The inability to demonstrate that phytoalexins are primarily responsible for the expression of host resistance in the tobacco/Ppn and
alfalfa/Pmm tissue culture systems should not be interpreted as conclusive evidence against the role of phytoalexins in the expression of resistance in intact plants.   Parallel studies of disease resistance in

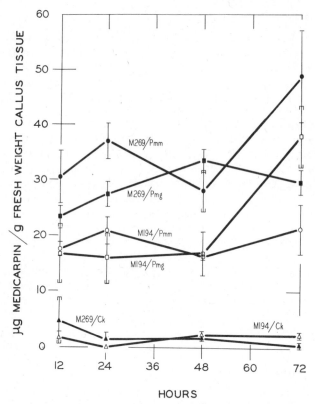

Figure 3.   Level of the phytoalexin, medicarpin, in alfalfa callus tissue after inoculation with *Phytophthora megasperma* f. sp. *medicaginis* (Pmm) or *P. megasperma* f. sp. *glycinea* (Pmg).   Line 269 is susceptible to Pmm; line M194 is resistant.   Both are nonhosts of Pmg.

plants and tissue cultures derived from them are needed to determine whether the same mechanisms of resistance operate in tissue culture and intact plants.   However, the failure of studies utilizing tissue culture systems to support the phytoalexin concept indicates that the concept should be examined more carefully.   Tissue culture systems may provide a means whereby other possible mechanisms of disease resistance might be investigated.

PROTOCOLS

There are many factors to consider in establishing a tissue culture system to study plant disease resistance (see Helgeson and Haberlach, 1980).   Ultimately, the choice of a system will depend upon the aspects of resistance that are being investigated.   For example, studies of the biochemical basis of race-specific resistance require a system in which the genetics of host and pathogen are well defined and near-isogenic lines of the host and races of the pathogen are available.   Alternatively, the biochemical basis of nonhost resistance might be investi-

gated with a less-well-defined system.   In both of these cases it is
critical that conditions (e.g., media composition, temperature, light, and
inoculum concentration) are defined that allow the expression of disease
resistance in culture.   In other instances, e.g., studies of the induction
and biosynthesis of phytoalexins, expression of disease resistance is not
a major factor as long as phytoalexins are produced.   In fact, many of
these investigations are carried out using nonspecific elicitors of
phytoalexin biosynthesis.   In these instances factors that influence
phytoalexin accumulation (e.g., media composition, age of cultured cells,
types of elicitor and concentration of elicitor) must be considered.
These have been recently reviewed (Dixon, 1980) with an emphasis on
isoflavonoid phytoalexins that are characteristic of members of the
Leguminosae.

Type of Tissue or Cell Culture

One of the first considerations in developing a system to study disease resistance is the type of tissue or cell culture to be utilized, i.e.,
callus tissue, suspension cultures, isolated cells, or protoplasts.   Each
has advantages for particular areas of study, and detailed methods have
been published for the initiation and maintenance of callus cultures
(Yeoman and Macleod, 1977) and suspension cultures (Street, 1977; Helgeson, 1980), the isolation of cells (Takebe et al., 1968; Servaites and
Ogren, 1977; Callow and Dow, 1980), and the isolation and culture of
protoplasts (Bajaj, 1977; Evans and Cocking, 1977; Eriksson et al.,
1978; Hanke, 1980; Shepard, 1980).   Callus tissue culture systems have
been used widely for studies of the expression of race-specific and nonhost resistance (Ingram and Robertson, 1965; Warren and Routley, 1970;
Helgeson et al., 1972; Keen and Horsch, 1972; Maronek and Hendrix,
1978; Huang and Van Dyke, 1978; Holliday and Klarman, 1979; Campbell, 1979; Miller, 1982) and offer several advantages over suspension
cultures, isolated cells, or protoplasts.   These include (a) the ease of
initiation and maintenance of cultured tissues, (b) the ability to add
inoculum (spores, zoospores, etc.) directly to callus tissue, so that the
tissue culture medium is not a direct source of nutrients for the pathogen, and (c) the ability to follow, cytologically, the progress of infection and colonization of callus tissue by the pathogen and the host's
response.   Phytoalexin accumulation can be determined in pathogen-
challenged callus tissue and related to the extent of colonization
(Budde and Helgeson, 1981; Miller, 1982).   In addition, resistance that
is expressed by callus tissue may not be expressed in suspension culture (Ingram and Robertson, 1965).
Suspension cultures commonly are used in studies of phytoalexin induction and biosynthesis (Bailey, 1970; Ebel et al., 1976; Dixon and
Fuller, 1977, 1978; Hargreaves and Selby, 1978; Oba and Uritani, 1979).
Phytoalexin-inducing agents (elicitors) and labeled precursors can be
applied easily and uniformly to suspension cultures, and sampling is
facilitated.   In addition, since a small number of cell types are present
in suspension cultures, interpretation of results (e.g., changes in enzyme
activity after addition of elicitor) should be simplified (Dixon, 1980).

However, the possibility that changes may occur in cultured cells that alter their reaction to exogenously applied compounds must be considered (Larken and Skowcroft, 1981). Isolated plant cells have the advantage of being more representative of cells in an intact plant (Callow and Dow, 1980). Suspension cultures, isolated cells, and protoplasts can provide excellent model systems for the study of various aspects of host-pathogen specificity, particularly the interaction of elicitors (Callow and Dow, 1980; Doke and Tomiyama, 1980a) and suppressors (Doke and Tomiyama, 1980b) of the hypersensitive response and/or phytoalexin accumulation, with plant cells. These systems have also been useful in studying the toxic effects of phytoalexins on plant cells (Shiraishi et al., 1975; Skipp et al., 1977; Lyon and Mayo, 1978).

Media Composition

A variety of culture media have been used in studies of the expression of disease resistance and phytoalexin accumulation in culture (Table 2). The best approach in choosing a culture medium is to select one that allows good growth of callus tissue (or suspension cells), then to modify the medium as necessary. For example, one may wish to modify the levels of plant growth regulators in the growth medium, because both the expression of disease resistance (Helgeson et al., 1972; Haberlach et al., 1978; Holliday and Klarman, 1979; Campbell, 1979) and production of phytoalexins (Dixon and Fuller, 1976, 1978) are influenced by the concentrations of auxins and cytokinins in the medium. In tobacco callus tissue a particular balance is required between the cytokinin and auxin concentrations in the culture medium for resistance to Ppn to be expressed. Resistance to Ppn can be eliminated by increasing the concentrations of the cytokinins kinetin (KIN) or benzyladenine (BA), but not Δ^2-isopentenyl adenine (2-ip), relative to the concentration of the auxin indoleacetic acid (IAA) (Haberlach et al., 1978). Thus it may be necessary to test several plant growth regulators at several concentrations to develop a suitable culture medium that will allow good growth of the host tissue and expression of disease resistance.

Several investigators have found that expression of disease resistance is affected by callus morphology, which is also influenced by the concentrations of plant growth regulators in the growth medium (Helgeson et al., 1972; Haberlach et al., 1978; Holliday and Klarman, 1979; Campbell, 1979). In tobacco and potato, resistance to an incompatible race of Ppn and P. infestans, respectively, is expressed in compact callus tissue, but not in callus tissue more friable in appearance. Resistance in alfalfa to Pmm and Pmg is expressed in rather friable callus tissue; however, callus morphology is also important in this system. Callus tissues in which the surface appears wet or glossy are colonized more rapidly and extensively than callus tissues that are dry in appearance. The dryness of alfalfa callus tissue appears to increase with increasing concentrations of kinetin in the growth medium (Miller, 1982).

The Pathogen

It is important to work with genetically uniform isolates of the pathogen, especially in studies of race-specific resistance. This can be accomplished by using isolates that have been grown from single, selected spores or isolated bacterial cells. Methods for isolating individual spores have been published elsewhere (Tuite, 1969).

Most facultative plant pathogens can be grown on undefined media such as V-8 medium (Tuite, 1969) and maintained by frequent transfers to fresh media. However, prolonged maintenance on artificial media may result in a reduction of aggressiveness or loss of pathogenicity in the pathogen. Isolates should be checked periodically for aggressiveness and pathogenicity by inoculating host plants and observing symptom development. If aggressiveness has been reduced in culture, several cycles of inoculation and reisolation from diseased plants may restore it to an acceptable level.

Inoculum

Bacterial cells, purified virus preparations, fungal spores, zoospores, sporangia, and hyphae have been used successfully to infect callus tissues. Where quantification of inoculum is important, fungal spores, zoospores, or sporangia are preferable to hyphae, although hyphal inoculum is generally easier. The results of several studies indicate that the expression of resistance in callus tissue is relatively insensitive to changes in inoculum levels (Ingram and Robertson, 1965; Holliday and Klarman, 1979; Campbell, 1979; Budde, 1981). For example, resistance to *P. infestans* is expressed in potato tissue culture aggregates at inoculum levels of 100 and 1000 sporangia/aggregate, but begins to break down at 5000 sporangia/aggregate. Phytoalexin accumulation in callus tissue is affected by large changes in inoculum levels (Gustine et al., 1978; Budde, 1981); medicarpin accumulation in jack bean (*Canavalia ensiformis*) callus tissue increases with increasing concentrations of spores (10^4-10^7 spores/ml) of *Pithomyces chartarum* (Berk. & Curt.) M. B. Ellis, a nonpathogen of jack bean.

Temperature

An easily discernable difference between the resistant and susceptible reactions in callus tissue may be evident over a somewhat limited range of temperatures. In tobacco callus tissue, resistance to Ppn is expressed at 20, 22, and 24 C, but not at 28 C, and the difference between resistant and susceptible interactions is optimal at 20 C (Helgeson et al., 1972). Soybean callus tissues express resistance to an incompatible race of Pmg at 16 and 20 C, but not at 24 or 28 C (Holliday and Klarman, 1979). In the potato/*P. infestans* system the greatest difference in tissue reactions occurs at 20 C, while at 12 and 16 C, susceptible callus tissues appear resistant, and at 24 C, resistant callus tissues are colonized (Campbell, 1979). Therefore, the host-pathogen interaction should be tested at a range of temperatures to determine the temperature at which the best differential between resistant and susceptible host tissue is expressed.

Little information is available concerning the effect of temperature on phytoalexin accumulation, although a recent report (Ersek and Sziraki, 1980) indicates that phytoalexin accumulation in callus tissue is influenced by temperature. In that study production of rishitin and phytuberin by potato callus tissue in response to cell-free hyphal homogenates of *P. infestans* was higher at 20 C than at 28 C, a phenomenon also observed in potato tuber slices (Currier and Kuc, 1975).

Light

The effects of light regime on the expression of disease resistance in tissue cultures has not been fully documented. In the tobacco/Ppn system, several light regimes were tested and found not to be an important factor (Helgeson et al., 1972). For studies on the induction of isoflavonoid phytoalexins, cultures should be grown in the dark, since many of enzymes of flavonoid and isoflavonoid biosynthesis are induced by light (Grisebach and Hahlbrock, 1974; Heller et al., 1979).

Protocol for Evaluation of Disease Resistance and Phytoalexin Accumulation in Alfalfa Callus Tissue Culture

TISSUE CULTURE GROWTH MEDIUM. Tissue cultures are grown on modified SH medium (Schenk and Hildebrandt, 1972; Table 3). Only freshly prepared growth regulator solutions are used in each batch of medium. After appropriate amounts of concentrated solutions of inorganic elements, vitamins, and growth regulators are combined, inositol and sucrose are weighed and dissolved in the medium. The final volume is adjusted with distilled water, and the pH corrected to 5.9 with 1 N NaOH. The agar is added and dissolved by steaming at 100 C for a length of time determined by the volume of the medium. Then 40 ml aliquots of molten medium are dispensed into 150 ml medicine bottles that are capped, then sterilized at 121 C, 20 psi for 20 min. The bottles are placed horizontally so that the medium is approximately 1.5 cm thick and allowed to cool in a laminar flow hood to prevent contamination. Alternatively, the medium (40 ml) may be dispensed into 100 x 20 mm plastic petri dishes. After the medium has cooled and tissue cultures initiated, the dishes should be wrapped in parafilm to prevent desiccation and contamination.

INITIATION AND MAINTENANCE OF TISSUE CULTURES. Alfalfa tissue cultures are initiated from immature ovaries (Saunders and Bingham, 1972) from two alfalfa lines, one resistant (M194), the other susceptible (M269) to Pmm. Flower buds 2-4 mm in length are surface

Table 3.   Modified Schenk-Hildebrandt Medium for Growth of Alfalfa
            Callus Tissue Cultures

	CONCENTRATION		CONCENTRATION
Major Elements		Iron–EDTA	
KNO_3	0.025 M	$FeSO_4 \cdot 7H_2O$	0.05 mM
$MgSO_4 \cdot 7H_2O$	1.6 mM	Na_2EDTA	0.05 mM
$NH_4H_2PO_4$	2.6 mM	Vitamins	
$CaCl_2 \cdot 2H_2O$	1.8 mM	Thiamine-HCl	0.015 mM
Minor Elements		Nicotinic Acid	0.04 mM
$MnSO_4 \cdot H_2O$	0.06 mM	Pyridoxine-HCl	2.4 µM
H_3BO_3	0.08 mM	Growth Regulators	
$ZnSO_4 \cdot 7H_2O$	3.5 µM	2,4-D	9.0 µM
KI	0.012 mM	KIN	9.3 µM
$CuSO_4 \cdot 5H_2O$	0.8 µM	NAA	11.0 µM
$NaMoO_4 \cdot 2H_2O$	0.4 µM	Inositol	5.5 mM
$CoCl \cdot 6H_2O$	0.4 µM	Agar	10 g/l
Sucrose	0.088 M		

sterilized by soaking 3 min in 95% ethanol, then 3 min in 1.1% sodium
hypochlorite (a 1:5 dilution of commercial bleach), followed by two
rinses in sterilized distilled water.   Ovaries 1-2 mm in length are removed aseptically from the buds and transferred to SH medium.   Cultures are grown in the dark at 21 ± 1 C for 4-6 weeks, then subdivided into callus pieces 3-5 mm in diameter and transferred to fresh
media.   Cultures are grown under similar conditions for 4-6 weeks,
until callus pieces are approximately 2 cm in diameter.   These callus
pieces are used in studies of fungal colonization and phytoalexin accumulation.

ISOLATION AND MAINTENANCE OF FUNGAL CULTURES.   The
Pmg isolate (race 1, isolate 16) used in our system was obtained from
C. R. Grau, University of Wisconsin, Madison.   A pure culture of the
alfalfa pathogen, Pmm, was isolated from Pmm-infested soil by baiting
(Marks and Mitchell, 1970).   This culture was induced to form zoospores
(Irwin et al., 1979) which were plated on 2% water agar medium.   Single, germinated cysts were selected and transferred to V-8 agar medium (Tuite, 1969); the isolate used in these studies (isolate 5b4) was
selected from that population.   Both Pmm and Pmg are maintained on
V-8 agar at 22 ± 2 C in ambient light.   Periodically, Pmm and Pmg
are tested for pathogenicity on alfalfa and soybean, respectively (Klarman and Gerdemann, 1963; Irwin et al., 1979).

INOCULATION PROCEDURES.   Callus pieces of M269 or M194, 4-6
week old, are inoculated with hyphae of Pmm or Pmg.   Blocks of inoculum 1 mm^2 are taken from the edge of 4-6-day-old colonies growing on
V-8 medium.   The excess agar is removed, and the inoculum blocks are
placed on the top of callus pieces using a long-handled transfer needle.
Inoculated callus tissues are maintained at 21 ± 1 C in the dark for

12, 24, 48, or 72 hr, when the callus pieces are removed from the bottles and hyphal development and phytoalexin accumulation are determined. Noninoculated callus tissue pieces serve as controls.

RATING WHOLE CALLUS PIECES.   Callus pieces are evaluated daily for hyphal colonization and tissue discoloration.   For hyphal colonization, the following rating system is used:

0 = no aerial hyphae visible
1 = aerial hyphae present on top 25% of callus piece
2 = aerial hyphae present on top 50% of callus piece
3 = aerial hyphae present on top 75% of callus piece
4 = aerial hyphae present on entire callus piece

A similar rating system is used to evaluate tissue discoloration. Ratings for colonization and discoloration in a typical experiment are presented in Table 1.

DETERMINATION OF HYPHAL DEVELOPMENT IN CALLUS TISSUE. Callus pieces (three replicates/treatment) are trimmed to 1 cm^2, then sliced in half.   One half is used to determine the extent of colonization, while colonized tissue from the other half is used in medicarpin quantification.   To determine the level of colonization, the halved callus pieces are divided horizontally into four equal slices; each slice is mounted on a glass slide and stained with lactophenol cotton blue (10 g phenol, 10 ml glycerine, 10 ml lactic acid, 0.02 g aniline blue, and 10 ml distilled water).   Slides are examined using a dissecting microscope (20x magnification), and colonization is rated subjectively on a zero to five scale.   A value of zero is assigned to tissue slices where no hyphae are present, and a value of five to slices in which the hyphae form a thick mat.   Figure 2 depicts the level of colonization in slices of M269 and M194 callus tissue inoculated with Pmm or Pmg and in noninoculated controls, 72 hr after inoculation.

Phytoalexin Quantification

EXTRACTION.   All organic solvents used in phytoalexin extraction and analysis (ethanol, chloroform, and methanol) are purified by filtration through aluminum oxide, followed by redistillation.   Water is purified by reverse osmosis and subsequent filtration in a Milli-Q Water Purification System (Millipore Corp., Bedford, MA, 10730).
After the extent of colonization has been determined in half of each callus piece, the colonized portion of the remaining half (100–500 mg) is placed in a preweighed test tube, and the fresh weight is determined.   Then 5 ml 95% ethanol are added to the tissue, and the samples are allowed to stand at 22 ± 2 C for 48 hr.   All samples are then stored in the freezer until the extraction procedure is continued.

Samples are allowed to warm to room temperature and the tissue and ethanol are transferred to a 90 ml cup of a Sorvall Omni Mixer (DuPont Instruments, Newtown, CT, 06470). An additional 10 ml of ethanol are used to rinse the sample tube (five 2 ml rinses) and are added to the cup. The tissue is ground at high speed for 15 sec, in three, 5 sec intervals. Samples are filtered by vacuum through a 5.5 cm Hirsch funnel containing Whatman #50 filter paper directly into a 100 ml evaporating flask; the cup is rinsed with five 2 ml volumes of ethanol, which are added to the filtrate. Five ml of water are added to the filtrate, which then is reduced to the aqueous phase under vacuum at 40 C. The aqueous phase is partitioned three times with three volumes of chloroform (total of nine volumes) in a 250 ml separatory funnel equipped with a Teflon stopcock. The chloroform phase is reduced to dryness under vacuum at 40 C; the residue is redissolved in five 0.2 ml aliquots of chloroform, which are transferred to a 1 ml glass syringe attached to a silica Sep-pak cartridge (Waters Associates, Inc. Milford, MA 01757) that has been rinsed previously with 10 ml chloroform. The Sep-pak cartridge is loaded with the sample, then eluted slowly (3 ml/min) with 10 ml chloroform. The first 2 ml are discarded, and the remaining 8 ml collected and reduced to dryness under vacuum at 40 C. The residue is dissolved in 1 ml methanol for analysis by high performance liquid chromatography (HPLC). The elution profiles from a silica Sep-pak cartridge for medicarpin, sativan, and vestitol are presented in Fig. 4. Medicarpin and sativan elute from the Sep-pak in the first 10 ml, while vestitol remains on the cartridge. By collecting fractions 3 through 10, medicarpin and sativan can be separated from numerous interfering compounds (Fig. 5). The percentage recovery of

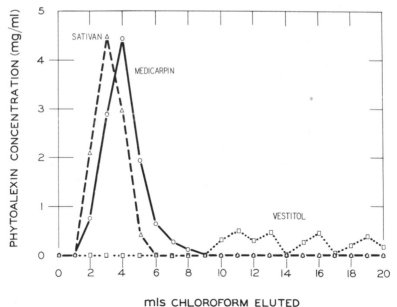

Figure 4. Elution of medicarpin, sativan, and vestitol from a silica Sep-pak cartridge using chloroform as solvent.

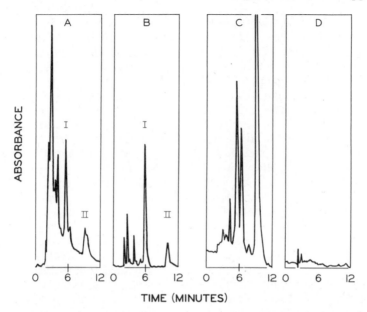

Figure 5. HPLC chromatograms of an extract from alfalfa callus tissue (line M269) 72 hr after inoculation with *Phytophthora megasperma* f. sp. *medicaginis*. (a,c) Extract prior to purification by a silica Sep-pak cartridge; (a) absorbance at 280 nm; (c) absorbance at 365 nm; (b,d) extract after purification by silica Sep-pak cartridge; (b) absorbance at 280 nm; (d) absorbance at 365 nm. I = medicarpin, II = sativan. Solvent system: methanol/dilute (1%) acetic acid (65/35, v/v); solvent flow rate: 1.5 ml/min; column: μBondapak C18 (Waters).

medicarpin, determined by adding a known quantity of the phytoalexin to noninoculated callus tissue and proceeding with the procedure as described above, is approximately 85%. A higher percentage recovery of both medicarpin and sativan can be obtained by collecting fractions 2 through 10 from the Sep-pak cartridge.

Since any vestitol present in callus tissue will not be detected with this procedure, crude chloroform extracts are reduced to dryness, redissolved in methanol, and analyzed by HPLC. No vestitol (retention volume = 8.0 ml) is detected in any of the callus tissues 72 hr after inoculation with Pmm or Pmg, or in noninoculated controls.

HPLC ANALYSIS. Medicarpin and sativan are quantified from callus tissue extracts by HPLC using Waters Associates, Inc. (Milford, MA 01757) instruments (Model M45 and 6000A pumps, Model 660 solvent programmer, Model U6K universal injector) equipped with a 30-cm μBondapak-C18 analytical column. Absorbance is measured simultaneously at 280 and 365 nm with a Model 440 UV detector. The system is run isocratically, with methanol/dilute (1%) acetic acid (65/35;v/v) as the solvent, at a flow rate of 1.5 ml/min. Medicarpin and sativan are resolved successfully using this system; the medicarpin peak appears 5.8

min after injection (retention volume = 14.7 ml). Retention volumes may vary somewhat, depending on the condition of the column and/or the length of connective tubing in the system. Medicarpin and sativan concentrations are calculated from the area under their respective peaks using standard curves. Three injections are made per replicate for each treatment.

Using these procedures, we have been able to detect medicarpin and sativan in callus tissues in concentrations as low as 1 µg/g fresh weight tissue and 0.4 µg/g fresh weight tissue, respectively. At the highest sensitivity, nanogram amounts (10-20 ng/injected sample) of medicarpin and sativan can be detected.

Bioassay for Antifungal Compounds

Compounds in addition to medicarpin and sativan with antifungal activity can be detected in colonized tissue using a thin layer chromatography (TLC) - Cladosporium bioassay (Homans and Fuchs, 1970; Bailey and Burden, 1973). Colonized callus tissue is extracted as described above for HPLC analysis, except that after partitioning with chloroform, the chloroform is reduced to dryness and the residue is dissolved in 30-50 µl ethanol. Alternatively, ethanol extracts of the tissue may be reduced to a small volume (30-50 µl) and assayed directly (Helgeson and Haberlach, 1980). Ethanol extracts are spotted on TLC plates (Silica gel 60, E. Merck, Darmstadt, Germany) and developed in hexane/ethyl acetate/methanol (60/40/1, v/v/v). Plates are allowed to dry overnight, then are sprayed with a thin layer of molten potato dextrose agar (PDA). Care should be taken to keep the PDA layer relatively thin; the plates should be sprayed slowly and evenly, until the surface glistens and appears somewhat grainy. The plates then are sprayed with a thin film of spores of Cladosporium cucumerinum Ellis and Arth. The spore suspension is prepared using 4-6-day-old cultures of the fungus grown on PDA slants in the dark at 20 C. Approximately 10 ml of a sucrose/salts medium (Table 4) is poured over each of five cultures, and the surface is rubbed with a glass rod or wire loop to dislodge the spores. The resulting suspension is filtered through three layers of cheesecloth and sprayed on agar-coated TLC plates. The plates are incubated horizontally for 3 days in covered plastic

Table 4. Medium Used to Suspend Spores of Cladosporium cucumerinum for TLC-Cladosporium Bioassay (Homans and Fuchs, 1970)

	CONCENTRATION
KH_2PO_4	0.05 M
Na_2HPO_4	0.017 M
KNO_3	0.04 M
$MgSO_4 \cdot 7H_2O$	4.0 mM
NaCl	0.017 M
Sucrose	0.15 M

boxes lined with moist paper towels.   Zones of inhibition appear as
white areas in an olive green background.

FUTURE PROSPECTS

Each type of tissue or cell culture has provided excellent model systems for the study of specific aspects of plant disease resistance.   Suspension cultures have been particularly advantageous in studies on the
induction and regulation of phytoalexin biosynthesis; further studies
should result in a clearer understanding of these processes and help de-
lineate the enzymes and intermediates involved in phytoalexin biosynthetic pathways.   These systems should also be useful in the isolation
and characterization of pathogen-produced compounds that induce
accumulation of phytoalexins.   However, as previously noted, care must
be taken to guard against possible changes in cultured cells that may
alter their responses to these compounds.
Studies with callus tissue culture systems have demonstrated that
resistance to plant pathogens is expressed in vitro, and that, in at
least one system (tobacco/Ppn), the same resistance gene that is expressed in the intact plant is expressed in callus tissue.   These findings
indicate that callus tissue culture systems are quite suitable for studies
on the factors critically important in the expression of disease resistance.   One such factor is thought to be the accumulation of inhibitory
levels of phytoalexins in infected tissue; but in the tissue culture systems that have been examined, phytoalexins do not appear to be primarily responsible for the expression of host resistance.   Therefore,
these systems should be utilized to investigate other, perhaps as yet
unknown, mechanisms of disease resistance.
Protoplasts have been utilized recently to investigate two areas of
host-pathogen specificity: the interaction of race-specific toxins with
plant cells and the effects of various pathogen-produced compounds on
the induction of the hypersensitive response.   Ultrastructural and biochemical studies of corn (*Zea mays* L.) protoplasts treated with HmT
toxin, a host-specific toxin produced by *Helminthosporium maydis* Nisik.
and Miyake race T, have contributed strong evidence that mitochondria
are the primary site of action of HmT toxin (Earle and Gracen, 1981).
Additional studies with this system and others that utilize host-specific
toxins should provide valuable insights into the mode of action of these
toxins and mechanisms of resistance to them in plant cells.   In the
latter case Doke and Tomiyama (1980a,b) have utilized potato protoplasts to investigate the mechanism(s) of race specificity in potato/*P.
infestans* interactions.   Hyphal wall components nonspecifically elicit a
hypersensitive reaction in potato protoplasts similar to the response observed in tuber cells infected with an incompatible race of *P. infestans*
(Doke and Tomiyama, 1980a).   Water-soluble glucans isolated from hyphae of compatible, but not incompatible races of *P. infestans* suppress
the hypersensitive reaction (Doke and Tomiyama, 1980b).   The authors
suggest that the suppressive effect of the glucans may be due to their
occupation of elicitor binding sites in the protoplasmic membrane or to
changes in the conformation of the receptor sites caused by the reac-

tion of the glucans with the membranes. The use of protoplasts should facilitate further studies of host-pathogen specificity in this system, e.g., the demonstration of membrane binding by elicitors and the characterization of putative receptors in the plasma membrane.

The expression of race–specific resistance in callus tissue cultures provides the opportunity to utilize simplified, controlled experimental systems to investigate factors critically important in the interaction of plants and plant pathogens. These interactions are likely, in many instances, to be highly complex (Heath, 1981b; Bushnell and Rowell, 1981), and in vitro systems may provide a means to separate and identify the components of host-pathogen interactions that ultimately lead to compatibility or incompatibility.

ACKNOWLEDGMENTS

We thank Dr. Lynn B. Graves, Jr. and Ms. Candice K. Elliot for their part in developing and carrying out these procedures, Dr. Graves and Dr. J. P. Helgeson for helpful discussions and critically reviewing the manuscript, and Ms. Elliot and Mr. Steve Vicen for preparing the figures.

KEY REFERENCES

Grisebach, H. and Ebel, J. 1978. Phytoalexins, chemical defense substances of higher plants. Angew. Chem. Int. Ed. Eng. 17:635-647.

Ingram, D.S. 1977. Applications in plant pathology. In: Plant Tissue and Cell Culture (H.E. Street, ed.) pp. 463-500. Univ. California Press, Berkeley.

_____ and Helgeson, J.P. (eds.) 1980. Tissue Culture Methods for Plant Pathologists. Blackwell Scientific, Oxford.

REFERENCES

Bailey, J.A. 1970. Pisatin production by tissue cultures of Pisum sativum L. J. Gen. Microbiol. 61:409-415.
_____ and Burden, R.S. 1973. Biochemical changes and phytoalexin accumulation in Phaseolus vulgaris following cellular browning caused by tobacco necrosis virus. Physiol. Plant Pathol. 3:171-177.
Bajaj, Y.P.S. 1977. Protoplast isolation, culture and somatic hybridization. In: Applied and Fundamental Aspects of Plant Cell, Tissue, and Organ Culture (J. Reinert and Y.P.S. Bajaj, eds.) pp. 467-496. Springer-Verlag, Berlin.
Beachy, R.N. and Murakishi, H.H. 1971. Local lesion formation in tobacco tissue culture. Phytopathology 61:877-878.
Budde, A.D. 1981. Determination of the localization of phytoalexins in tobacco callus after infection by Black Shank of tobacco. M.S. Thesis, University of Wisconsin, Madison.

_____ and Helgeson, J.P. 1981. Chronology of phytoalexin production and histological changes in tobacco callus infected with *Phytophthora parasitica* var. nicotianae. (abstr.). Phytopathology 71:864.

Busbice, T.H., Hill, R.R., Jr., and Carnahan, H.L. 1972. Genetics and breeding procedures. In: Alfalfa Science and Technology (C.H. Hanson, ed.) pp. 283-318. American Society of Agronomy, Madison.

Bushnell, W.R. and Rowell, J.B. 1981. Supressors of defense reactions: A model for roles in specificity. Phytopathology 71:1012-1014.

Callow, J.A. and Dow, J.M. 1980. The isolation and properties of tomato mesophyll cells and their use in elicitor studies. In: Tissue Culture Methods for Plant Pathologists (D.S. Ingram and J.P. Helgeson, eds.) pp. 197-202. Blackwell Scientific, Oxford.

Campbell, J.S. 1979. Potato tissue culture and the expression of resistance to *Phytophthora infestans*. M.S. Thesis, University of Wisconsin, Madison.

Currier, W.W. and Kuc, J. 1975. Effect of temperature on rishitin and steroid glycoalkaloid accumulation in potato tuber. Phytopathology 65:1194-1197.

Deaton, W.R., Keyes, G.J., and Collins, G.B. 1982. Expressed resistance to black shank among tobacco cell cultures. Theor. Appl. Genet. 63:65-70.

de Zoeten, G.A., Gaard, G.R., Haberlach, G.T., and Helgeson, J.P. 1982. Infection of tobacco callus by *Phytophthora parasitica* var. nicotianae. Phytopathology 72:743-746.

Dixon, R.A. 1980. Plant tissue culture methods in the study of phytoalexin induction. In: Tissue Culture Methods for Plant Pathologists (D.S. Ingram and J.P. Helgeson, eds.) pp. 185-196. Blackwell Scientific, Oxford.

_____ and Bendall, D.S. 1978. Changes in the levels of enzymes of phenylpropanoid and flavonoid synthesis during phaseollin production in cell suspension cultures of *Phaseolus vulgaris*. Physiol. Plant Pathol. 13:295-306.

_____ and Fuller, K.W. 1976. Effects of synthetic auxin levels on phaseollin production and phenylalanine ammonia-lyase (PAL) activity in tissue cultures of *Phaseolus vulgaris* L. Physiol. Plant Pathol. 9: 299-312.

_____ and Fuller, K.W. 1977. Characterization of components from culture filtrates of *Botrytis cinerea* which stimulate phaseollin biosynthesis in *Phaseolus vulgaris* cell suspension cultures. Physiol. Plant Pathol. 11:287-296.

_____ and Fuller, K.W. 1978. Effects of growth substances on non-induced and *Botrytis cinerea* culture filtrate-induced phaseollin production in *Phaseolus vulgaris* cell suspension cultures. Physiol. Plant Pathol. 12:279-288.

_____ and Lamb, C.J. 1979. Stimulation of de novo synthesis of L-phenylalanine ammonia lyase in relation to phytoalexin accumulation in *Colletotrichum lindemuthianum* elicitor-treated cell suspension cultures of French bean (*Phaseolus vulgaris*). Biochim. Biophys. Acta 586:453-463.

_____, Dey, P.M., Murphy, D.L., and Whitehead, I.M. 1981. Dose responses for *Colletotrichum lindemuthianum* elicitor-mediated enzyme induction in French bean cell suspension cultures. Planta 151:272-280.

Doke, N. and Tomiyama, K. 1980a. Effect of hyphal wall components from *Phytophthora infestans* on protoplasts of potato tuber tissues. Physiol. Plant Pathol. 16:169-176.

_____ 1980b. Suppression of the hypersensitive response of potato tuber protoplasts to hyphal wall components by water soluble glucans isolated from *Phytophthora infestans*. Physiol. Plant Pathol. 16:177-186.

Earle, E.D. and Gracen, V.E. 1981. The role of protoplasts and cell cultures in plant disease research. In: Plant Disease Control: Resistance and Susceptibility (R.C. Staples and G.H. Toenniessen, eds.) pp. 285-297. John Wiley and Sons, New York.

Ebel, J., Ayers, A.R., and Albersheim, P. 1976. Host pathogen interactions XII. Response of suspension cultured soybean cells to the elicitor isolated from *Phytophthora megasperma* var. sojae, a fungal pathogen of soybeans. Plant Physiol. 57:775-779.

Eriksson, T., Glimelius, K., and Wallin, A. 1978. Protoplast isolation, cultivation and development. In: Frontiers of Plant Tissue Culture 1978, Proceedings 4th International Congress of Plant Tissue and Cell Culture (T.A. Thorpe, ed.) pp. 131-139. International Association for Plant Tissue Culture, Calgary.

Ersek, T. and Sziraki, I. 1980. Production of sesquiterpene phytoalexins in tissue culture callus of potato tubers. Phytopathol. Z. 97:364-368.

Evans, P.K. and Cocking, E.C. 1977. Isolated plant protoplasts. In: Plant Tissue and Cell Culture, 2nd ed. (H.E. Street, ed.) pp. 103-135. Univ. California Press, Berkeley.

Goins, R.B. and Apple, J.L. 1970. Inheritance and phenotypic expression of a dominant factor for black shank resistance from *Nicotiana plumbaginifolia* in a *Nicotiana tabacum* milieu. Tob. Sci. 14:7-11.

Grisebach, H. and Hahlbrock, K. 1974. Enzymology and regulation of flavonoid and lignin biosynthesis in plants and plant cell suspension cultures. In: Recent Advances in Phytochemistry, Vol. 8 (V.C. Runeckles and E.E. Conn, eds.) pp. 21-52. Academic Press, New York.

Gustine, D.L. 1981. Evidence for sulfhydryl involvement in regulation of phytoalexin accumulation in *Trifolium repens* callus tissue cultures. Plant Physiol. 68:1323-1326.

_____, Sherwood, R.T., and Vance, C.P. 1978. Regulation of phytoalexin synthesis in jackbean callus cultures. Stimulation of phenylalanine ammonia-lyase and O-methyltransferase. Plant Physiol. 61:226-230.

Haberlach, G.T., Budde, A.D., Sequeira, L., and Helgeson, J.P. 1978. Modification of disease resistance of tobacco callus tissues by cytokinins. Plant Physiol. 62:522-525.

Hanchey, P. and Wheeler, H. 1971. Pathological changes in ultrastructure: Tobacco roots infected with *Phytophthora parasitica* var. nicotianae. Phytopathology 61:33-39.

Hanke, D.E. 1980. The preparation, manipulation and culture of plant protoplasts. In: Tissue Culture Methods for Plant Pathologists (D.S. Ingram and J.P. Helgeson, eds.) pp. 27-31. Blackwell Scientific, Oxford.

Hargreaves, J.A. and Selby, C. 1978.   Phytoalexin formation in cell suspensions of *Phaseolus vulgaris* in response to an extract of bean hypocotyls. Phytochemistry 17:1099-1102.

Heath, M.C. 1981a.   Nonhost resistance.   In: Plant Disease Control: Resistance and Susceptibility (R.C. Staples and G.H. Toenniessen, eds.) pp. 201-217.   John Wiley and Sons, New York.

_____   1981b.   A generalized concept of host-parasite specificity. Phytopathology 71:1121-1123.

Helgeson, J.P. 1980.   Plant tissue and cell suspension culture.   In: Tissue Culture Methods for Plant Pathologists (D.S. Ingram and J.P. Helgeson, eds.) pp. 19-25.   Blackwell Scientific, Oxford.

_____   and Haberlach, G.T. 1980.   Disease resistance studies with tissue cultures.   In: Tissue Culture Methods for Plant Pathologists (D.S. Ingram and J.P. Helgeson, eds.) pp. 179-184.   Blackwell Scientific, Oxford.

_____, Kemp, J.D., Haberlach, G.T., and Maxwell, D.P. 1972.   A tissue culture system for studying disease resistance: The black shank disease in tobacco callus cultures. Phytopathology 62:1439-1443.

_____, Haberlach, G.T., and Upper, C.D. 1976.   A dominant gene conferring disease resistance to tobacco plants is expressed in tissue cultures. Phytopathology 66:91-96.

_____, Budde, A.D., and Haberlach, G.T. 1978.   Capsidiol: A phytoalexin produced by tobacco callus tissues.   Plant Physiol. 61(Suppl.): 58.

Heller, W., Egin-Buehler, B., Gardiner, S.E., Knobloch, K.-H., Matern, U., Ebel, J., and Hahlbrock, K. 1979.   Enzymes of general phenylpropanoid metabolism and of flavonoid glycoside biosynthesis in parsley. Plant Physiol. 64:371-373.

Higgins, V.J. 1972.   Role of the phytoalexin medicarpin in three leaf spot diseases of alfalfa.   Physiol. Plant Pathol. 2:289-300.

Holliday, M.J. and Klarman, W.L. 1979.   Expression of disease reaction types in soybean callus from resistant and susceptible plants.   Phytopathology 69:576-578.

Homans, A.L. and Fuchs, A. 1970.   Direct bioautography on thin-layer chromatograms as a method for detecting fungitoxic substances.   J. Chromatogr. 51:327-329.

Huang, J.-S. and Van Dyke, C.G. 1978.   Interaction of tobacco callus tissue with *Pseudomonas tabaci*, *P. pisi*, and *P. fluorescens*.   Physiol. Plant Pathol. 13:65-72.

Ingham, J.L. 1979.   Isoflavonoid phytoalexins of the genus *Medicago*. Biochem. Syst. Ecol. 7:29-34.

_____   and Millar, R.L. 1973.   Sativin: An induced isoflavan from the leaves of *Medicago sativa* L.   Nature 242:125-126.

Ingram, D.S. 1967.   The expression of R-gene resistance to *Phytophthora infestans* in tissue cultures of *Solanum tuberosum*.   J. Gen. Microbiol. 49:99-108.

_____   1980.   Tissue culture methods in plant pathology.   In: Tissue Culture Methods for Plant Pathologists (D.S. Ingram and J.P. Helgeson, eds.) pp. 3-9.   Blackwell Scientific, Oxford.

_____   and Robertson, N.F. 1965.   Interaction between *Phytophthora infestans* and tissue cultures of *Solanum tuberosum*.   J. Gen. Microbiol. 40:431-437.

Irwin, J.A.G., Miller, S.A., and Maxwell, D.P. 1979. Alfalfa seedling resistance to *Phytophthora megasperma*. Phytopathology 69:1051-1055.

Irwin, J.A.G., Maxwell, D.P., and Bingham, E.T. 1981a. Inheritance of resistance to *Phytophthora megasperma* in diploid alfalfa. Crop Sci. 21:271-276.

_____ 1981b. Inheritance of resistance to *Phytophthora megasperma* in tetraploid alfalfa. Crop Sci. 21:277-283.

Keen, N.T. 1981. Evaluation of the role of phytoalexins. In: Plant Disease Control: Resistance and Susceptibility (R.C. Staples and G.H. Toenniessen, eds.) pp. 155-177. John Wiley and Sons, New York.

_____ and Bruegger, B. 1977. Phytoalexins and chemicals that elicit their production in plants. In: Host Plant Resistance to Pests, ACS Symposium Series, No. 62 (P.A. Hedin, ed.) pp. 1-26. American Chemical Society, Washington, D.C.

_____ and Horsch, R. 1972. Hydroxyphaseollin production by various soybean tissues: A warning against use of "unnatural" host-parasite systems. Phytopathology 62:439-442.

Klarman, W.L. and Gerdemann, J.W. 1963. Resistance of soybeans to three *Phytophthora* species due to the production of a phytoalexin. Phytopathology 53:1317-1320.

Kuan, T.-L. and Erwin, D.C. 1980. Formae speciales differentiation of *Phytophthora megasperma* isolates from soybean and alfalfa. Phytopathology 70:333-338.

Kuc, J. 1976. Phytoalexins and the specificity of plant-parasite interaction. In: Specificity in Plant Diseases (R.K.S. Wood and A. Graniti, eds.) pp. 253-268. Plenum, New York.

Lamb, C.J. and Dixon, R.A. 1978. Stimulation of de novo synthesis of L-phenylalanine ammonia-lyse during induction of phytoalexin biosynthesis in cell suspension cultures of *Phaseolus vulgaris*. Fed. Eur. Biochem. Soc. Lett. 94:277-280.

Larkin, P.J. and Scowcroft, W.R. 1981. Somaclonal variation—A novel source of variability from cell cultures for plant improvement. Theor. Appl. Genet. 60:197-214.

Lawton, M.A., Dixon, R.A., and Lamb, C.J. 1980. Elicitor modulation for the turnover of L-phenylalanine ammonia-lyase in french bean cell suspension cultures. Biochem. Biophys. Acta 633:162-175.

Lu, N.S.-J., Barnes, D.K., and Frosheiser, F.I. 1973. Inheritance of *Phytophthora* root rot resistance in alfalfa. Crop Sci. 13:714-717.

Lyon, G.D. and Mayo, M.A. 1978. The phytoalexin rishitin affects the viability of isolated plant protoplasts. Phytopathol. Z. 92:298-304.

Marks, G.C. and Mitchell, J.E. 1970. Detection, isolation, and pathogenicity of *Phytophthora megasperma* from soils and estimation of inoculum levels. Phytopathology 60:1687-1690.

Maronek, D.M. and Hendrix, J.W. 1978. Resistance to race O of *Phytophthora parasitica* var. nicotianae in tissue cultures of a tobacco breeding line with black shank resistance derived from *Nicotiana longiflora*. Phytopathology 68:233-234.

Miller, S.A. 1982. Cytological and biochemical factors involved in the susceptible, host resistant, and nonhost resistant interactions of alfal-

fa with *Phytophthora megasperma*.   Ph.D. Thesis, University of Wisconsin, Madison.

_____, Graves, L.B., Jr., and Maxwell, D.P. 1981.   Hyphal development
and phytoalexin accumulation in an alfalfa tissue culture/*Phytophthora
megasperma* system (abstr.).   Phytopathology 72:895.

Oba, K. and Uritani, I. 1979.   Biosynthesis of furano-terpenes by sweet
potato cell culture.   Plant Cell Physiol. 20:819-826.

Partridge, J.E. and Keen, N.T. 1977.   Soybean phytoalexins: Rates of
synthesis are not regulated by activation of initial enzymes in flavonoid biosynthesis.   Phytopathology 67:50-55.

Russell, T.E. and Halliwell, R.S. 1974.   Response of cultured cells of
systemic and local lesion tobacco hosts to microinjection with TMV.
Phytopathology 64:1520-1526.

Saunders, J.W. and Bingham, E.T. 1972.   Production of alfalfa plants
from callus tissue.   Crop Sci. 12:804-808.

Schenk, R.U. and Hildebrandt, A.C. 1972.   Medium and techniques for
induction and growth of monocotyledonous and dicotyledonous plant
cell cultures.   Can. J. Bot. 50:199-204.

Servaites, J.C. and Ogren, W.L. 1977.   Rapid isolation of mesophyll
cells from leaves of soybean for photosynthetic studies.   Plant
Physiol. 59:587-590.

Shepard, J.F. 1980.   Abscisic acid-enhanced shoot initiation in protoplast-derived calli of potato.   Plant Sci. Lett. 18:327-333.

Shiraishi, T., Oku, H., Isono, M, and Ouchi, S. 1975.   The injurious effect of pisatin on the plasma membrane of pea.   Plant Cell Physiol.
16:939-942.

Skipp, R.A., Selby, C., and Bailey, J.A. 1977.   Toxic effects of phaseollin on plant cells.   Physiol. Plant Pathol. 10:221-227.

Stoessel, A. 1980.   Phytoalexins—a biogenetic perspective.   Phytopathol.
Z. 99:251-272.

Street, H.E. 1977.   Cell (suspension) cultures—techniques.   In: Plant
Tissues and Cell Culture, 2nd ed. (H.E. Street, ed.) pp. 61-102.
Univ. California Press, Berkeley.

Takebe, I., Otsuki, Y., and Aoki, S. 1968.   Isolation of tobacco mesophyll cells in intact and active state.   Plant Cell Physiol. 9:115-124.

Tuite, J.T. 1969.   Plant Pathological Methods: Fungi and Bacteria.
Burgess Publ., Minneapolis.

Vaziri, A., Keen, N.T., and Erwin, D.C. 1981.   Correlation of medicarpin production with resistance to *Phytophthora megasperma* f. sp.
medicaginis in alfalfa seedlings.   Phytopathology 71:1235-1238.

Warren, R.S. and Routley, D.G. 1970.   The use of tissue culture in the
study of single gene resistance of tomato to *Phytophthora infestans*.
J. Am. Soc. Hort. Sci. 95:266-269.

Yeoman, M.M. and Macleod, A.J. 1977.   Tissue (callus) cultures—techniques.   In: Plant Tissue and Cell Culture, 2nd ed. (H.E. Street, ed.)
pp. 31-59.   Univ. California Press, Berkeley.

Yoshikawa, M., Yamauchi, K., and Masago, H. 1978.   Glyceollin: Its
role in restricting fungal growth in resistant soybean hypocotyls infected with *Phytophthora megasperma* var. sojae.   Physiol. Plant Pathol. 12:73-82.

Zaehringer, U., Ebel, J., and Grisebach, H. 1978. Induction of phyto-
    alexin synthesis in soybean. Elicitor-induced increase in enzyme ac-
    tivities of flavonoid biosynthesis and incorporation of mevalonate into
    glyceollin. Arch. Biochem. Biophys. 188:450-455.
Zaehringer, U., Schaller, E., and Grisebach, H. 1981. Induction of phy-
    toalexin synthesis in soybean. Structure and reactions of naturally
    occurring and enzymatically prepared prenylated pterocarpans from
    elicitor-treated cotyledons and cell cultures of soybean. Z. Natur-
    forsch. 36c:234-241.

CHAPTER 32
Establishment of Nematode Germplasm Banks

R.M. Riedel, M. Alves de Lima, and *M. Martin*

The inability to establish plant-parasitic nematodes in pure culture has seriously retarded the study of these organisms as plant pathogens. Two important areas of phytopathological research, establishment of pathogenicity and crop loss assessment, are particularly difficult to pursue when large populations of microbiologically sterile test organisms are not available. Prior to 1961, nematologists interested in such questions were limited to working either with convenient organisms such as species of *Heterodera* or *Meloidogyne* that package inoculum in discrete, easily obtained, and sterilized units, or to developing laborious techniques, frequently involving hand sorting, to extract inoculum of less convenient nematodes, those more promiscuous in habits of egg laying, for field or greenhouse pot cultures.

Krusberg (1961) provided a breakthrough for those interested in some of the more important migratory parasites. Ample use has been made of this technique in a wide range of physiological, taxonomic, and pathogenicity studies on nematodes. Monoxenic callus or excised root cultures are excellent sources for quantities of sterile, debris-free nematodes for use in ultrastructural studies. In a thorough ultrastructural study of the body walls, vulval flap, intestine, esophagus, feeding apparatus, and spicules of *Pratylenchus penetrans*, monoxenic alfalfa callus cultures were used (Kisiel et al., 1972, 1974, 1976; Chen and Wen, 1972; Wen and Chen, 1976). Similarly Wergin and Orion (1981) and Orion et al. (1980) used *Meloidogyne incognita* from excised tomato root culture for scanning electron microscopy work.

Roman and Hirschmann (1969) cultured six species of *Pratylenchus* under standardized conditions on alfalfa callus to study inter- and intraspecific variations.   Tarte and Mai (1976) used *P. penetrans* from single female lines cultured on alfalfa callus to study intraspecific variation.   Roman and Triantophyllou (1969) and Hung and Jenkins (1969) used species of *Pratylenchus* from tissue culture to study gametogenesis, oogenesis, and embryology as well as reproductive system morphology and anatomy.

Tissue cultured nematodes have been used to study various aspects of biology.   Thistlethwayte (1969) used *P. penetrans* from tissue culture to study duration of life cycle in a range of incubation temperatures. Adamo et al. (1976) studied migratory behavior of *Aphelenchoides besseyi* produced on *Alternaria brassicae*.   Olowe and Corbett (1976) studied vertical migration and histopathology of diseases caused by *P. brachyurus* and *P. zeae* from aseptic excised corn roots.   *Pratylenchus penetrans* from alfalfa callus was used in studies of invasive behavior (Townshend, 1978).   Mitsui et al. (1975) used alfalfa callus culture to study temperature effects on fecundity of seven species of *Pratylenchus*. Eriksson (1972) used monoxenic culture to make reciprocal crosses of races of *Ditylenchus dipsaci*.   Perry et al. (1980) used sterile culture to study mating between *P. penetrans* and *P. fallax*.

Monoxenic cultures of nematodes are excellent sources of single species populations for physiological work, providing, for example, supplies of *Ditylenchus dipsaci* for studies of cell wall degrading enzymes (Riedel and Mai, 1971a,b) or *P.penetrans*, *D. dipsaci*, and *Aphelenchoides ritzemabosi* for respiration studies (Bhatt and Rohde, 1970).   Monoxenic cultures of nematodes have been used in studies on mechanisms of action of nematicides (Abawi and Mai, 1978).   Mass rearing of nematodes in tissue culture is a technique used extensively in industry to supply nematodes for nematicide screening programs.

The use of monoxenically cultured nematodes in important areas of agricultural research has not been fully exploited.   For example, in selection of resistant breeding material, monoxenically cultured nematodes have been used on a small number of forage legumes (Bingefors and Bingefors, 1976; Eriksson, 1980; Faulkner et al., 1974) and onions (Bergquist and Riedel, 1972).   In crop loss assessment studies use of cultured nematodes is restricted to a few greenhouses and growth chamber tests.   Field work with nematodes from monoxenic culture is almost untried (Martin et al., 1982).

To some extent this underutilization results from the limited number of species of plant-parasitic nematodes maintained in culture.   In some part it results from the lack of information concerning methods and materials required for implementation of the technique.   The object of this chapter is a compilation of such information necessary for the establishment and use of a collection of monoxenically cultured plant-parasitic nematodes.   The techniques included here are those that are in general use in a number of laboratories and are known to perform satisfactorily in a number of different conditions.   All these methods have undergone minor variations in the hands of different workers.   No attempt has been made here to include all these variations.

If by assembling these techniques here more people are induced to culture nematodes and more research on practical methods for culture of additional nematode species is stimulated, the chapter may in some way also serve to eliminate what has become a major obstacle to the study of plant parasitic nematodes.

ESTABLISHMENT OF CULTURES

Tissues Used to Culture Nematodes

A list of 41 species of plant parasitic nematodes cultured on plant tissues was included in a recent review (Krusberg and Babineau, 1977). Plant tissue used as substrate was drawn from 39 species of angiosperms and included tuber, pith, excised roots, seedling, and callus tissues; not callus by the definition of Street (1977), but rather a friable mass of plant cells arising from excised plant tissues or whole seedlings treated with plant growth promoters, usually 2,4-D (Krusberg and Babineau, 1977).

Nematodes are most commonly and successfully propagated on alfalfa callus. Four species of *Aphelenchoides*; one species each of *Aphelenchus*, *Bursaphelenchus*, *Hoplolaimus*, *Radopholus*, and *Telotylenchus*; three of *Ditylenchus*; eight of *Pratylenchus*; and three of *Tylenchorhynchus* have been propagated on alfalfa callus tissue (Krusberg and Babineau, 1977). Six additional species of *Pratylenchus*, *P. scribneri* (Roman and Hirschmann, 1977); *P. pinguicaudatus* (Corbett, 1969); *P. crenatus* and *P. loosi* (Mitsui et al., 1975); *P. convallariae* (Mitsui, 1977); and *P. agilis* (Krusberg, personal communication) have more recently been successfully cultured on alfalfa callus. *Tylenchorhynchus vulgaris* has been cultured on both callus and excised root tissue of corn (Upadhyay and Swarup, 1974). Researchers reported almost two times greater reproduction on corn callus as on corn roots. *Bursaphelenchus xylophilus* has been cultured on pine callus tissues (Tamura and Mamiya, 1979).

Although alfalfa callus seems to be the most suitable substrate for most of the culturable migratory ecto- and endoparasites, this may only reflect the fact that few nematode species have been cultured, and of those nematodes cultured, *Pratylenchus* species (14 cultured on alfalfa callus), which have a polyphagous nature, predominate. Yet even within the genus *Pratylenchus* there is some evidence that alfalfa callus is not the best substrate for all species. Mitsui et al. (1975), in comparing propagation ratios of eight species of *Pratylenchus*, found considerably lower populations of *P. crenatus* and *P. convallariae* on alfalfa callus. Work done in our laboratory comparing reproduction of four species of *Pratylenchus* (Table 1) has confirmed this evidence for *P. crenatus*, which showed significantly lower reproduction. Since *P. crenatus* has a mostly graminaceous host range (Loof, 1978) and *P. convallariae* develops almost exclusively in *Convallaria* rhizomes (Loof, 1978), it is not surprising that alfalfa was not a good host. To expand on this, there is no doubt that if reproductive rates of more species of the genera of the endo- and ectomigratory parasites already cultured are compared, alfalfa callus probably will not remain the best host substrate.

In contrast to the ecto- and endomigratory parasites, sedentary endoparasites such as *Heterodera*, *Meloidogyne*, and *Nacobbus* have

Table 1. Reproduction at 25 C of Four *Pratylenchus* Species on Alfalfa Callus Cultured on Riedel et al. (1973) Medium with 1 g/l Yeast Extract

	INITIAL INOCULUM	DAYS AFTER INOCULATION				
		20	40	60	80	100
P. scribneri	59[1] a[2]	20[3] b	331 a	4828 a	7236 a	9868 a
P. penetrans	39 c	25 a	186 b	716 b	1484 a	576 b
P. brachyurus	39 c	11 b	56 c	670 b	784 b	748 b
P. crenatus	52 b	13 b	37 c	92 c	156 c	256 c

[1]Values are the mean number of nematodes in ten 0.2 ml inoculation aliquots.

[2]Values in the columns followed by the same letter are not significantly different according to Duncan's multiple range test at P = 0.01.

[3]Values in columns 2-6 are the mean number of nematodes in 10 cultures, where each culture consists of callus from three alfalfa seedlings.

been cultured almost exclusively on excised roots and seedlings (Krusberg and Babineau, 1977). Tanda et al. (1980) reported that okra callus supported *Meloidogyne incognita.* However, the nematode did not reproduce well, and its life cycle was retarded. They concluded that this substrate was not practical for maintenance of *M. incognita.* Most recently Lauritis et al. (1981) reported successful excised root culture of *Heterodera glycinia* on soybean, while Rebois and Lauritis (1981) reported good reproduction of *Rotylenchulus reniformis* on tomato roots.

Fungal feeders such as species of *Aphelenchoides* and *Aphelenchus avenae* may be cultured on a wide range of fungi on axenic culture (Katznelson and Henderson, 1964; Pillai and Taylor, 1968; Shafer et al., 1981; Vanfleteren, 1978). Few stylet-bearing nematodes have been reared successfully in axenic culture. Myers (1967, 1968, 1971) and Buecher et al. (1970) cultured *Aphelenchoides sacchari* on a medium of relatively defined character. *Bursaphelenchus xylophilus* develops useful populations on *Caenorhabditis* medium containing heme (Dropkin, personal communication). Andreeva (1974) and Tarakanov (1974, 1975) reported successful axenic culture of potato rot nematode, *Ditylenchus destructor.* Components used by these researchers in their support media are not known. While axenic culture of stylet-bearing nematodes is the ultimate goal for plant nematologists interested in this area, the methods existing are not generally useful in practice.

Seed Selection and Treatment

Since many plant parasitic nematodes cultured to date can be maintained on alfalfa callus, the methods detailed here are specific for this host. Protocols described here for alfalfa can be easily adapted to other host plant species. Schematic representations of these methods are presented in Tables 2, 3, and 4. As a general reference for techniques of obtaining other kinds of tissue under aseptic conditions see Yeoman and Macleod (1977).

SELECTION. Selection of high-quality seeds is important for success in subsequent steps in establishment of nematode cultures (Viglierchio et al., 1973). Seeds must have high germination rates and be as free as possible of contaminating internal bacteria and fungi. In the United States alfalfa produced in drier areas of Western states is more likely to meet these criteria than are Eastern-grown seeds. Faulkner et al. (1974) suggested seed quality is improved by grading for size. They use seed retained on 16 mesh (1.19 mm pore size) screen.

TREATMENT. Pretreatment of seeds with heat is usually good insurance for success of future steps. Dry seeds of Ranger variety alfalfa can be rid of internal bacteria by a 10 min soak in hot water at 61 C. Faulkner et al. (1974) treat DuPuit, Vernal, or Team varieties in hot water for 1 min at 90 C. In general, time and temperature combinations are adjusted for contaminant-variety combination on the basis of experience. Seeds must be quickly cooled after treatment. Subsequent sterilization procedures frequently involve strong mineral acids. Hot-water-treated seeds must be thoroughly dried before subjecting them to acid treatments if serious damage to them is to be avoided.

To produce sterile seedlings from heat-treated seeds, surface sterilization of seeds prior to germination is necessary. The process varies from the simple to the complex, depending upon seed coat qualities and worker experience.

Eriksson (1972) surface sterilized with concentrated sulfuric acid for 10-20 min followed by a rinse of sterile, distilled water. We prefer a more complex procedure consisting of a 15 min bath in concentrated sulfuric acid (36 N), followed by three rinses in sterile, distilled water, a 15 min soak in 1:1000 $HgCl_2$ in 30% ethanol, and three final rinses in sterile, distilled water. Sulfuric acid in these procedures scarifies the seed while sterilizing them and improves germination. When sulfuric acid is used, care must be taken to drain the seeds of the majority of acid before rinsing and to make the volume of rinse water sufficiently large so that seeds are not overheated in the exothermic reaction of acid and water. Mercury vapors can be evolved from alcoholic solutions of $HgCl_2$. This solution should be used in good ventilation only.

Many workers prefer to surface sterilize with antibiotic solutions. Characteristic of these procedures is that used by Faulkner et al. (1974). They recommend that wet seeds immediately after heat treatment be soaked for 30 min in a solution of oxytetracycline hydroxide: streptomycin sulfate:captan (250 mg each in 200 ml water) mixed with ampicillin trihydrate:furazolidone solution (250 mg:100 mg in 100 ml water). This is followed by a bath in 1:500 $HgCl_2$ for 2 min and a final rinse in sterile distilled water.

We prefer not to use antibiotic solutions, because they tend to be bacteriostatic rather than bacteriocidal; they are rather expensive; and they are time consuming to prepare, especially since they generally must be prepared and used fresh each time seeds are sterilized.

CALLUS PRODUCTION. Surface-sterilized seeds are germinated to produce the plant organs from which callus will be produced. Seeds can be germinated on water agar (Eriksson, 1972) or other media that will aid in detection of contaminating microorganisms. Callusing media lacking growth factors are particularly suitable for this purpose, since these media are present in the environment in which contaminants are most likely to be active during the life of the cultures. Potato dextrose agar or nutrient agar are used for this purpose as well. When seeds are germinated in conditions in which contaminants may not be apparent, dips of antibiotic solutions can help prevent inclusion of contaminants in culture.

Alfalfa seed generally produces useful amounts of hypocotyl tissue after 7 days at room temperature (ca., 21 C) when subjected to natural photoperiods. When incubated for shorter periods or at high temperature (>25 C) less tissue is developed which will callus. Incubation periods shorter than 7 days may "mask" slow-growing contaminants, while longer incubation periods allow for extensive root development, making subsequent transfer operations difficult. For a thorough study on how to deal with persistant bacterial contamination in alfalfa and clover seed lots, see Viglierchio et al. (1973).

Callusing Media

For 20 years researchers at Uppsala, Sweden, have used a highly modified Knop's solution (Table 2) to culture several races of *Ditylenchus dipsaci, D. destructor, Aphelenchoides ritzemabosi,* and *A. fragariae* on alfalfa and red clover callus (Eriksson, 1972; Bingefors and Bingefors, 1976). They regularly inoculate each culture tube with 1000-3000 *D. dipsaci* and within 8 weeks obtain 14,000-40,000 nematodes.

Krusberg (1961) modified Hildebrandt et al. Medium (1946) (Table 2) for use in callusing alfalfa seedlings on which *D. dipsaci, A. ritzemabosi, Pratylenchus zeae, and P. penetrans* were successfully cultured. Reproduction of *Hoplolaimus coronatus* and *Tylenchorhynchus capitatus* was slow on alfalfa callus on this culture medium (Krusberg, 1961). In the intervening 20 years many researchers have reported using Krusberg medium (Lownsbery et al., 1967; Bhatt and Rohde, 1970; Sontirat and Chapman, 1970; Thistlethwayte, 1970; Riedel and Mai, 1971a,b; Viglierchio et al., 1973; Mitsui, 1977).

Media containing White's macro- and micronutrient solution (WH) (Table 2) have been widely used to grow excised roots which may be used for the culture of some plant parasitic nematodes (Perry et al., 1980; Mountain, 1955; Upadhyay and Swarup, 1972). Faulkner et al. (1974) used a modified WH with the addition of 2,4-D and NAA to callus alfalfa for mass rearing of *D. dipsaci.*

Because a major drawback to rearing nematodes in tissue culture was the complexity of available media, a simpler medium (Riedel et al., 1972) was developed for the culture of *P. penetrans* and *D. dipsaci* on alfalfa callus (Table 2). When first reported, this 4-component medium contained 5 g/l of yeast extract. Yeast extract concentrations of 1-2 g/l were shown later to improve reproduction of *P. penetrans* on this medium (Riedel et al., 1972). Reproduction of *Pratylenchus penetrans* on callus tissues grown on this medium was not further improved at lower concentrations of yeast extract (Table 3). Under closely standardized culture conditions the reproduction of *P. penetrans* on the simplified agar compares favorably to that found on more complex medium. With the use of alfalfa callused on Krusberg's medium at 23 C, Riedel and Foster (1970) reported an average of 36,000 *P. penetrans*/18 mm culture tube after 8 weeks. In our laboratory, using alfalfa callused on simplified medium at 20 C, we produce an average of 35,000 *P. penetrans*/25 mm culture tube, (range = 22,000 to 56,000) after 8 weeks. *Pratylenchus penetrans, P. crenatus, P. scribneri, P. agilis, P. brachyurus, Aphelenchoides fragariae, A. ritzemabosi,* and *Ditylenchus dipsaci* also reproduce well on this medium in our laboratory. *Bursaphelenchus xylophilus* has been maintained on it also (Tamura and Mamiya, 1979).

Table 2.   Components of Four Nutrient Media Used To Culture Plant Tissues on Which Nematodes Feed

COMPONENT	WHITE[1] (mg/l)	KRUSBERG[2] (mg/l)	BINGEFORS[3] (mg/l)	RIEDEL[4] (mg/l)
Na_2SO_4	200	800	–	–
$Ca(NO_3)\cdot 4H_2O$	300	400	500	–
$MgSO_4\cdot 7H_2O$	750	180	125	–
KNO_3	80	80	125	–
KCl	65	65	–	–
KH_2PO_4	–	–	125	–
$NaH_2PO_4\cdot 2H_2O$	19	33	–	–
$MnSO_4\cdot 4H_2O$	20	4.5	–	–
$MnCl_2\cdot 4H_2O$	–	–	0.360	–
$ZnSO_4\cdot 7H_2O$	12	6	0.044	–
H_3BO_3	6	0.375	1	–
KI	3	3	–	–
$FeSO_4\cdot 7H_2O$	–	–	3.3	–
$Fe_2(C_4H_4O_6)\cdot 3H_2O$	2.5	40	–	–
$CuSO_4\cdot 5H_2O$	0.4	–	0.016	–
MoO_3	0.004	–	–	–
$Na_2MoO_4\cdot 2H_2O$	–	–	0.025	–
Yeast extract	–	–	–	1000
Glycine	3	3	3	–
Thiamine-HCl	0.1	0.1	0.1	–
Nicotinic acid	0.5	–	0.5	–
Pyridoxine	0.1	–	0.1	–
Calcium panthotenate	–	2.5	–	–
NAA	0.1	0.1	–	–
2,4-D	2	2	2	2
CW (ml/l)	–	150	140	–
Sucrose (g/l)	10	10	11	10
Agar (g/l)	20	20	20	10

[1] White, 1963
[2] Krusberg, 1961
[3] Eriksson, 1972; Bingefors and Bingefors, 1976
[4] Riedel et al., 1973

Some researchers studying the influence of plant growth hormones on reproduction of plant parasitic nematodes in culture found that there was a positive correlation between callus growth and reproduction of *Aphelenchoides ritzemabosi* (Webster and Lowe, 1966; Dolliver et al., 1962) and *Ditylenchus dipsaci* (Viglierchio et al., 1973). On the other hand, Webster (1967) reported that the greatest number of *A. ritzema-*

bosi was not produced by the treatments that gave the most plant tissue. Krusberg and Blickenstaff (1964), while studying reproduction of *D. dipsaci, Pratylenchus penetrans,* and *P. zeae* on alfalfa tissue cultures, concluded that medium supporting best callus tissue growth was not necessarily best for nematode reproduction. Work done in our laboratory with *P. penetrans* supports the latter findings. Cultures of four different plant tissues, alfalfa seedlings, pea roots, carrot discs, and potato tuber discs growing for 7 weeks on simplified medium and on a modified MS (Murashige and Skoog, 1962) medium showed growth on MS medium two to four times heavier than similar cultures on simplified medium (Table 4). Nematode reproduction, however, was consistently better on the simplified medium.

Table 3. Reproduction of *Pratylenchus penetrans* on Alfalfa Callus Cultured on Riedel's et al. (1973) Medium with Three Levels of Yeast Extract

DAYS AFTER INOCULATION	GRAMS YEAST EXTRACT/LITER		
	0.0	0.5	1.0
20	56[1] b[2]	138 a	130 a
40	74 b	220 a	354 a
60	153 b	673 a	1097 a
80	490 b	1339 a	2816 a

[1] Values are the mean number of nematodes in 10 cultures, where each culture consists of callus from three alfalfa seedlings.

[2] Values within the same horizontal line followed by the same letters are not significantly different according to Duncan's multiple range test at P = 0.01.

Incubation Conditions

Temperature is a very important aspect of incubation environment, and it can be manipulated to accomplish various ends. Lownsbery et al. (1967) reported that *Pratylenchus vulnus* cultured on alfalfa callus increased more rapidly at 25 C than at 20 C and did not increase at 30 or 35 C. High populations could be maintained longer at 15 or 10 C than at 5, 20, or 25 C. Faulkner et al. (1974) reported optimal temperature of 20-25 C for population increase of *Ditylenchus dipsaci* on alfalfa callus. Mitsui et al. (1975) studied the effect of temperature on the propagation of seven species of *Pratylenchus.* For this study they established seven isolates of *P. penetrans* in monoxenic alfalfa callus, five of *P. loosi,* and one each of *P. crenatus, P. coffeae,* and *P. zeae.* Highest populations of *P. penetrans, P. fallax,* three isolates of *P. vulnus, P. loosi,* and *P. crenatus* developed at 25 C. A fourth isolate of *P. vulnus* reproduced optimally at 25-30 C. Optimum temperatures for reproduction of *P. coffeae* and *P. zeae* was 25-32 and 29-34 C, respectively.

Table 4. Reproduction of *Pratylenchus penetrans* on Four Crop Plants Cultured on Three Different Media[1]

	ALFALFA			CARROT			PEA			POTATO		
	WA	MS	R	WA	R	MS	WA	R	MS	WA	R	MS
[2]	61 x	559 z	283 y	139 x	278 y	533 z	87 x	228 y	382 z	191 x	323 x	833 y
[3]	170 i	514 h	6841 a	454 g	2865 b	510 f	123 y	2790 b	1211 c	539 f	1080 d	691 e

[1] WA = water agar, R = Riedel et al. Medium, and MS = Murashige and Skoog modified medium. Values followed by the same letter are not significantly different according to Duncan's multiple range test at $P = 0.01$.

[2] Values are the mean fresh weight of callus tissues in mg.

[3] Values are the mean number of nematodes in 10 cultures.

While in most cases larger populations are achieved more rapidly at higher temperatures, such cultures, generally, will decline rapidly (Faulkner et al., 1974; Lownsbery et al., 1967). For purposes of maintenance and routine subculture, cooler incubation temperatures prevent rapid buildup and depletion of medium and extend the period of time between subcultures. On a routine basis we incubate our cultures of *Pratylenchus, Ditylenchus,* and *Aphelenchoides* at 20 C. This temperature is suitable for both maintenance and good reproduction.

Olowe and Corbett (1976) compared the effect of temperature (5, 10, 15, 20, 25, 30, and 35 C) on generation time and reproduction rates of *Pratylenchus brachyurus* and *P. zeae* on excised maize roots. Generation time differed for the two species. At 15 C the generation time of *P. brachyurus* was 14 weeks and that of *P. zeae*, 12 weeks. *P. zeae* developed faster than *P. brachyurus* at all temperatures. Both developed fastest at 30 and 35 C, *P. brachyurus* completing one generation in 4 weeks and *P. zeae* in 3 weeks. At 5 or 10 C a generation was not completed within 14 weeks for either species. At 5, 10, and 15 C reproduction rates did not increase, the number remaining close to the inoculum added initially. For both species, final populations after 90 days of culture were highest at 30 C.

Subculturing

Faulkner et al. (1974) inoculated new callus cultures in 120 ml French square widemouthed bottles with 3 ml of a suspension containing 20,000–30,000 nematodes. Nematodes were dispensed with a Cornwall continuous pipetting syringe. We use 0.2 ml of nematode suspension per 30 ml French square dispensed with a Finnpipette (Ky Finnipipette, Helsinki, Finland). An aseptic nematode suspension can be prepared by adding sterile water to infested callus in the original culture containers or by extracting nematodes in Tiner Traps (Tiner, 1961) or sterilized modified Baermann Funnels. For research purposes when numbers of nematodes in inoculum must be known, counts of representative aliquots can be made.

Alternatively, new callus cultures can be started with bits of infected callus tissue. Bingefors adds 1000–3000 nematodes per culture in 25 x 150 mm tubes using this technique (Bingefors and Bingefors, 1976). Subcultures are routinely infested in our laboratory by this method also (Martin et al., 1982). Because populations in infected tissue inoculum vary with age of culture and species, nematode numbers in inoculum are roughly adjusted by the size of callus tissue bits transferred to new cultures. For example, larger pieces of callus inoculum must be transferred from cultures of a given age of *Pratylenchus crenatus* instead of equal age cultures of *P. scribneri* because of the differing reproduction rates of the two species on alfalfa callus (see Table 1).

Length of time between subculture transfers is species specific. At 20 C, *Aphelenchoides* must be subcultured at 2 months or cultures rapidly decline. *Pratylenchus* is subcultured at 3 month intervals at 20 C as a standard practice, but can easily go over 1 year between subcultures. *Pratylenchus* tubes are used routinely at 3 months for inoculum in field tests.

Extraction To Produce Aseptic Nematodes

Faulkner et al. (1974) extract aseptic nematodes from culture with the aid of antibiotic solutions. Contents of culture bottles are emptied onto Kimwipe (type 900 L) tissue on a 20 cm diameter sieve (pore size, 0.5 mm) and rested in a 30 cm diameter, unsterilized plastic wash basin. Tapwater is added to cover plant tissue and is decanted twice daily for 48 hr. All extracted nematodes are collected in a 1 liter Erlenmeyer flask resting in a slanted position. After nematodes have settled for at least 2 hours, the water is siphoned to concentrate the nematodes into about 100 ml. Afterward, 100 ml of 1:500 $HgCl_2$ in aqueous solution is added, and the mixture is agitated for 15 min. Following this, 700 ml of sterile distilled water is added. After settling for 2 hr, 800 ml of the supernatant is siphoned. This rinsing procedure is repeated, leaving nematodes again concentrated in about 100 ml. An equal volume of antibiotic stock solutions is added to the nematode suspension which serves as inoculum. Populations are adjusted by adjusting the volume of this suspension with half strength antibiotic solution.

Bingefors and Bingefors (1976) and Eriksson (1972) extracted nematode 6-7 weeks after inoculation of cultures by pouring water into the culture tubes that were covered with filter paper (Ederol 261). Covered tubes were inverted in a watch glass with water, and left for 20-24 hr. Tubes were shaken periodically during extraction. Nematodes pass through the filter paper and are collected in the water under the test tube. If extraction proceeds for more than 24 hr, 10-15% more nematodes are obtained.

We customarily collect aseptic nematodes using Baermann funnels made from 23 cm diameter pie pans. When aseptic inoculum is desirable, the pie pans are steam sterilized inside paper bags at 120 C for 15 min. As a further precaution against contamination, nematodes can be extracted in sterile distilled water, to which Merthiolate solution (Eli Lilly Co.) (8 ml/200 ml water) has been added. Nematodes are generally extracted from the cultures for 24 hr. The pie-pan assemblies are stored in a laminar flow transfer chamber during the extraction period.

Containers

Many kinds of containers can be adapted for nematodes cultured in callus tissue culture. Each container has advantages and disadvantages. In our laboratory, 25 x 150 mm tubes with plastic caps are used for mass rearing of all *Pratylenchus*, *Ditylenchus*, and *Aphelenchoides* species as well as for routine culture maintenance. Tubes are space efficient and easily manipulated throughout the culture process, provide fairly good visibility, and are easily emptied of their contents. These tubes generally contain 8-12 alfalfa seedlings on 14 ml of slanted culture medium. Many more seedlings could be used in these containers, however.

For research on tissue culture methods of *Pratylenchus*, 30 ml French square bottles have been used. They have the advantage of easy handling and storage, but the disadvantage of poor visibility, if close observation of the cultures is necessary.

Plastic petri dishes have been used when close periodic observations of the cultures are necessary. However, petri dishes become contaminated more frequently than tubes or bottles, even when sealed with parafilm (American Can Co., Dixie/Marathon, Greenwich, CT). Petri dishes, however, are readily available, and easy to manipulate and store.

Current research (authors' unpublished data) comparing reproduction in cultures sealed with PVC films in place of caps indicate no differences in population development when PVC film is used. This technique has been used successfully in plant cell culture work (Sondahl, personal communication). Increased gas exchange fostered by the use of PVC film does not appear to enhance culture efficiency, however.

Nematode Sterilization

Starting callus cultures from single, field-collected nematodes necessitates surface sterilization of specimen nematodes. A number of different methods have been used for nematode sterilization (Zuckerman, 1971).

Jones (1980) grouped sterilization methods according to nematode size. For large, mainly migratory nematodes such as species of *Xiphinema*, *Aphelenchoides*, and *Ditylenchus* he recommends nematodes be passed singly through different solutions using microneedles or mounted eye-lashes to handle specimens. For small nematodes, sedentary juvenile and eggs, washing and sedimentation in capped, conical, glass centrifuge tubes are recommended.

Faulkner and Darling (1961) surface sterilized nematodes by allowing nematodes to "walk" on sterile petri dishes containing acidified potato dextrose agar. After 4-5 days of incubation on this medium, gravid females were transferred through three baths of sterile, distilled water and then placed in culture.

Bingefors and Bingefors (1976) sterilized nematodes extracted from young greenhouse plants that had been inoculated previously with field-collected nematodes. The nematodes were axenized with 0.5% hibitane diacetate (bis (p-chlorophenyl diguanido) hexane diacetate) for 15-20 min and then rinsed in sterile distilled water. The transfer was accomplished by placing the nematodes, mostly 20-30 per tube, in a drop of sterile water suspended from the tip of a sterile needle fitted to a hypodermic syringe held in the required position by a specially adapted retort-stand. A culture tube with callus tissue was brought up over the needle and the drop of water with nematodes placed on the top of the callus piece by depressing the piston of the syringe. These transfers were carried out in a sterile transfer box equipped with a dissecting microscope (Eriksson, 1972).

In our laboratory *Aphelenchoides* and *Ditylenchus* are sterilized in large numbers by repeated centrifugation and rinsing in Merthiolate (Eli

Lilly Co.) (8 ml/200 ml sterile water). Soil nematodes are extracted by a modified Baermann funnel technique and singly sterilized by passing through aqueous Merthiolate solution in sterilized PBI dishes closed with 25 mm² cover slips. Four changes of solutions at 1 hr intervals are used. We have had good success passing single specimens through three baths of sterile, distilled water, one bath of 0.1% streptomycin sulfate and three final rinses of sterile, distilled water (Mountain, 1955). Nematodes should remain in the antibiotic for at least 15 min.

USE IN FIELD PLOTS

Nematodes can be reared in sufficient numbers and with sufficient ease in monoxenic culture to permit their use in field tests. Illustrative of this is work in microplots to evaluate the effects of *Pratylenchus* on the development of Verticillium Wilt in potatoes (Martin et al., 1982). The techniques would as easily fit test protocols for studies of crop loss and intrageneric competition of species on crop plants in field situations.

The primary problems surrounding the use of monoxenically cultured nematodes in field situations involved techniques to extract nematodes from cultures in a viable condition and to infest quantities of soil with precalculated reproducible populations of nematodes. These problems can be avoided by not extracting nematodes from cultures, instead adding cultures directly to soil as described below and in the Protocol.

Nematode inoculum is prepared by emptying the agar and callus tissue contents of each of 16, 25 x 150 mm tubes of cultures into a Waring blender. Each tube is rinsed with 1 ml of tapwater to ensure complete collection of nematode aggregations from tube walls. In blending the tube contents, the blender switch is flipped on and off as rapidly as possible. The contents are stirred by hand and blended once more by flipping the blender switch. This results in a semihomogenous mixture of callus and media and 0.5-1 cm segments of uncallused alfalfa seedlings. Complete homogenization would result in the destruction of too many nematodes. The blending steps are repeated for each batch of 16 tubes.

Blended contents of 64 culture tubes are added directly to 10 liters of fumigated soil and mixed well by hand. This results in 10 liters of highly concentrated, nematode-infested soil. The moisture content of the soil should be slightly less than potting consistency when this is added. Soil too wet will become sticky and be difficult to use. Soil that is too dry when homogenized material is added will decrease nematode survival. Soil texture is also an important influence on quality of inoculum produced by these techniques. The most uniform mixes result when nematodes are placed into fine organic or light-textured sandy mineral soils. Concentrated nematode infested soil is held in a plastic bag 12-24 hr at room temperature to permit redistribution of nematodes from callus bits to soil.

After incubation, volumes of concentrated infested soil, calculated to contain a desired population of nematodes, are added to 20 liter batches of fumigated soil in metal baskets and mixed by hand. It is further mixed for 10 revolutions in a 28 L capacity twin shell soil blender (Patterson-Kelly Co., Stroudsberg, PA). After emptying the soil from the blender to a bushel basket, two 100 ml samples are taken, from which populations in the mixed soil are determined by 24 hr incubation in modified Baermann funnels. Table 5 contains results from application of this technique in three soil types. Initial population levels in the microplot soil may be varied by altering the volume of concentrated infested soil added. With this technique 1500 25 x 150 mm culture tubes contained in a single 10 ft^3 incubator will provide inoculum for 1500 30 x 30 cm microplots infested with field levels of nematodes.

Table 5.   Initial Population of *Pratylenchus penetrans* in Rifle Peat, Kibbie Fine Sandy Loam, and Wooster Silt Loam Microplot Soil Using Callus Culture Inoculum

| TREATMENTS | NEMATODE #[1]/100 cm^3 MIXED MICROPLOT SOIL | | |
	Rifle Peat	Sandy Loam	Silt Loam
1979			
high	260[2] a[3]	233 a	185 a
medium high	118 b	120 b	45 b
medium	75 bc	61 bc	30 b
low	42 c	12 c	10 b
1980			
high 1	151 a	151 a	150 a
high 2	147 a	143 a	89 b
high 3	140 a	125 a	78 b
medium 1	56 b	41 b	32 c
medium 2	52 b	39 b	30 c
medium 3	44 bc	35 b	29 c
low 1	18 c	16 b	11 c
low 2	15 c	14 b	9 c
low 3	14 c	13 b	7 c
1981			
high 1	386 a	282 a	
high 2	364 a	260 a	150 a
high 3	334 a	228 a	
medium 1	153 b	100 b	
medium 2	133 bc	98 bc	24 b
medium 3	126 bc	91 bc	
low 1	50 c	21 bc	
low 2	47 c	15 bc	8 e
low 3	43 c	15 c	

[1]Nematodes extracted from soil using modified Baermann funnels in water for 24 hr at room temperature.

[2]Values in 1979 and 1981 are the means of eight samples, values in 1980 are the means of 15 samples

[3]In 1 year values in one soil type, followed by the same letters are not significantly different using Duncan's multiple range test at P = 0.05.

In general, using callus inoculum directly gives uniform controlled population levels for a particular test year across soil type. However, populations always tend to be lower in the silt loam soil (Table 5). This is partially due to greater but unavoidable nematode attrition in the heavier silt loam during mixing in the twin shell soil blender, and partially due to a practice of making concentrated nematode infested soil out of silt loam soil for the silt loam microplots. Silt loam soil makes a very sticky concentrated infested soil which is not optimal for nematode survival. However, a recent test (Table 6) has shown that if concentrated infested organic soil is used to add to the 20 liter batches of silt loam soil, low populations on heavier soils can be avoided. As a general rule we use organic soil to make all our highly concentrated nematode-infested soil mixes, and we use aliquots of this to infest Riflepeat, sandy loam, and silt loam soils.

Table 6.   Initial Populations of *Pratylenchus penetrans* in Rifle Peat and Wooster Silt Loam Microplot Soil Using Concentrated Nematode-Infested Soil Made from Organic Soil[1]

TREATMENTS	NEMATODE #/100 cm³ MIXED MICROPLOT SOIL
High Rifle peat[2]	259 a
Low Rifle peat	59 b
High silt loam	263 a
Low silt loam	55 b

[1] Nematodes extracted from soil using modified Baermann funnels in water for 24 hr at room temperature. Values followed by the same letters are not significantly different using Duncan's multiple range test at $P = 0.05$.

[2] Values are the means of eight samples.

The use of callus cultures of nematodes for field use requires standardization at each step in the culture process. For example, the nematode levels in these tests (Table 5) varied from year to year with the number of alfalfa seedlings used per culture tube. Culturing in 1979 was not standardized; anywhere from 6 to 12 seedlings per tube were used. In 1980, by contrast, 8-10 seedlings were used, and in 1981 an average of 12 seedlings were used. Populations vary accordingly

Culture ages must be standardized also. For example, in Table 7, cultures of *Pratylenchus scribneri* are 12 weeks old for the first plot, 13 for the second and 14-15 for the third. Population differences in these plots correlated to decline of populations of nematodes.

FAULKNER ET AL. (1974) PROTOCOL

1. High quality alfalfa seeds (cv. DuPuits, Vernal, or Team) should be selected, and only those seeds retained by a 16 mesh screen (pore size, 1.19 mm) used. Scarify the seeds.

2. Antibiotic stock solution is prepared as follows:
   (a) Dissolve 250 mg ampicillin trihydrate and 100 mg furazolidone in 100 ml sterile distilled water (solution A).
   (b) Dissolve oxytetracycline hydrochloride, streptomycin sulfate and captan, 250 mg each, in 200 ml sterile distilled water (solution B).
   (c) Mix solutions A and B, and bring volume to 500 ml.
   (d) This stock solution should be made fresh, because it loses effectiveness when stored for more than 2 days, even at 4 C.
3. Heat-treat seeds by placing ca. 10 g of seed in a 120 ml sterile French bottle filled with 90 C tapwater and shaking for 1 min.
4. Decant the hot water and treat seeds for 30 min with 50-60 ml of half-strength antibiotic solution prepared as above.
5. Decant antibiotic solution and treat seeds for 2 min with 1:500 $HgCl_2$.
6. Rinse seeds three times with sterile distilled water.
7. Incubate bottles for 24-36 hr at room temperature with caps loose and bottles laying flat. Rotate bottles from side to side periodically to hold moisture level near optimum for germination.
8. When radicles are 1-2 mm in length rinse seedlings in half-strength antibiotic solution.
9. Aseptically transfer about 200 seedlings to each culture bottle containing modified White's nutrient medium.
10. Incubate bottles at room temperature for 7 days and then inoculate with nematodes.

Table 7. Initial Population of *Pratylenchus scribneri* in Rifle Peat, Kibbie Fine Sandy Loam, and Wooster Silt Loam Microplot Soil Using Callus Culture Inoculum

| TREATMENTS | NEMATODE #[1]/100 cm^3 MIXED MICROPLOT SOIL | | |
	Rifle Peat	Sandy Loam	Silt Loam
High	282 a[2]	124 a	0 a
Medium	141 b	54 b	6 a
Low	54 c	13 c	3 a

[1] Nematodes extracted from soil using modified Baermann funnels in water for 24 hr at room temperature. Values in the same column followed by the same letters are not significantly different using Duncan's multiple range test at P = 0.05.

[2] Values are the means of eight samples

ERIKSSON (1980) PROTOCOL

1. Treat alfalfa seeds with concentrated sulfuric acid for 10-20 min.
2. Rinse in sterile distilled water.
3. Transfer seeds to tubes filled with 1% water agar.
4. After germination, transfer seedlings to tubes with nutrient medium.

5.   After callus development (from all parts of the seedlings), pieces of callus are transferred to new tubes filled with nutrient medium.

RIEDEL ET AL. (1973) PROTOCOL

1.   Alfalfa seeds (cv. Ranger) should be stored in a dessicator at room temperature.
2.   (a)   Heat treat seeds by soaking alfalfa seeds in 60 C tapwater for 10 min.   Cheesecloth, loosely tied to allow for good hot water circulation, is used for soaking.
     (b)   Immediately after heat treatment, spread seeds in a thin layer for rapid cooling.
     (c)   Allow seeds to dry at least 10 days before proceeding.
3.   The following treatments should be performed aseptically in a laminar airflow cabinet:
     (a)   For surface sterilization of seeds take eight 250 ml beakers and put a 9 cm diameter petri dish bottom on each for a lid. Put each separately in a brown sack and sterilize.
     (b)   Place a 10 ml volume of heat-treated alfalfa seeds in a small basket inside one sterile beaker with lid.
     (c)   Pour 75 ml concentrated sulfuric acid over the seeds in the basket and wait for 15 min.
     (d)   While waiting, fill the other three beakers each with 75 ml sterile distilled water for rinsing.
     (e)   After 15 min, rinse the basket of seeds in each of three sterile distilled water baths.
     (f)   Next, immerse the basket of seeds in a fourth sterile covered beaker containing 75 ml of 1:1000 mercuric chloride in 30% ethanol and wait 15 min (1 g reagent grade $HgCl_2$ dissolved in 1 liter 30% ethanol).
     (g)   Repeat previous two steps.
     (h)   To bioassay for contamination, plate ca. 0.5 ml seeds on petri dishes containing 20 ml of a rich medium.   (We use either 10 g sucrose, 1 g yeast extract and 10 g agar per liter, or potato dextrose agar.)
4.   Incubate plates at room temperature under regular day-night light conditions for 7 days.
5.   After 7 days, transfer only seedlings from completely clean plates to culture containers with callusing medium.   Nematodes are added 7 days after this transfer.

PROTOCOL FOR USING NEMATODE MONOXENIC CALLUS CULTURE
DIRECTLY FOR SOIL INFESTATION

1.   Shake agar and alfalfa callus contents of 16 three-month-old monoxenic culture tubes of *Pratylenchus* into Waring blender.
2.   Wash each tube with 1 ml tapwater to collect all nematodes.
3.   To mix, flip blender switch on and off immediately.
4.   Mix briefly by hand, and repeat previous step.

5.   Empty contents into beaker.
6.   Repeat steps 1-5 three times.
7.   Mix blended contents of 64 culture tubes with 10 liters of fumigated organic or sandy soil to make 10 liters of concentrated nematode infested soil.
8.   Incubate 12-24 hr at room temperature in a plastic bag.
9.   Add volumes of concentrated nematode infested soil to 20 liter batches of fumigated soil in metal bushel baskets. (We add 100, 300, and 700 ml to get field populations ranging from 15 to 250 *Pratylenchus*/100 cm^3 soil.)
10. Mix well by hand followed by 10 turns in a twin shell soil blender.
11. Two 100 ml samples are taken immediately after mixing each 20 liter batch; initial nematode populations in mixed soil are determined by a 24 hr extraction in modified Baermann funnels in water.
12. Infested soil may be used immediately or covered and stored 12-36 hr in a cool place.

FUTURE PROSPECTS

The labor intensive nature of culture production is often cited as a serious drawback to use of the technique.   Certainly proper maintenance of the cultures requires careful attention to detail.   This frequently makes the work onerous.   Perhaps because of this, culturing plant parasitic nematodes seems more time consuming than it actually is.   Viewed from another angle, however, the large populations of several species that can be housed in a few small incubators would require large greenhouse facilities and the extensive labor support required for greenhouse maintenance.   Add to this the energy consumed by greenhouses in temperate climates and the cost and labor requirements of monoxenic culture are not disproportionately larger than alternative culture techniques.   Nonetheless, further labor savings would be beneficial.   Use of simplified media and propagation techniques would be an easy way to arrive at each savings.
Labor savings inherent in improved and simplified storage have been touched upon in a recent review (Krusberg and Babineau, 1977).   The absence of a central collection of monoxenic cultures of plant-parasitic nematodes is another aspect of storage which deserves mention.   At the present time, cultures are widely scattered among private and public institutions.   This and the lack of catalogs to these collections inhibits the use of the material.   Lack of a central facility for culture maintenance fosters overlap of efforts in the field and improves the chance of losing important, irreplaceable material.   A central repository for nematode cultures would permit systematic studies of culture conditions on nematode reproduction.
Presently eight genera of plant-parasitic nematodes are or have been grown in monoxenic culture.   Important genera such as *Meloidogyne* and *Heterodera* are not among these, a fact recognized as a major drawback to the technique (Krusberg and Babineau, 1977; Eriksson,

1980). Further, only limited effort has been made to maintain in culture races or geographic isolated species of any genera. While many species could undoubtedly be added to cultures with present techniques if support for such work were available, more inclusive additions must await more complete understanding of nematode requirements.

The effect of physical conditions in culture on nematode reproduction has hardly been studied. Temperature, pH, viscosity, and osmotic pressure of substrates are known to affect behavior of nematodes in culture (Mitsui et al., 1975; Myers, 1971).

More species will be brought into culture when host tissue requirements are better understood. The ability to tailor tissue reaction to species requirements would not seem to be outside the possible, considering the state of the art of tissue culture today. What is lacking is the knowledge necessary to predict nematode requirments. Studies in this area would pay handsome dividends.

Monoxenic culture of nematodes on callus tissue represents a temporary solution to the problem of production of usable amounts of nematode inoculum for various types of research. Axenic culture of these animals in defined medium is the obvious best solution to the problem. Presently, axenization of large numbers of species appears to be far in the future. Since monoxenic cultures are likely therefore to be more than temporary expedients, research to improve the technique would be of long-term use to the science.

KEY REFERENCES

Bingefors, S. and Bingefors, S. 1976. Rearing stem nematode inoculum for plant breeding purposes. Swed. J. Agric. Res. 6:13-17.

Faulkner, L.R., Bower, D.B., Evans, D.W., and Elgin, J.H., Jr. 1974. Mass culturing of *Ditylenchus dipsaci* to yield large quantities of inoculum. J. Nematol. 6:126-129.

Jones, M.G.K. 1980. The interaction of plant parasitic nematodes with excised root and tissue cultures. In: Tissue Culture Methods for Plant Pathologists (D.S. Ingram and J.P. Helgeson, eds.) pp. 161-166. Blackwell Scientific, Oxford.

Krusberg, L.R. and Babineau, D.E. 1977. Application of plant tissue culture to plant nematology. In: Plant Cell and Tissue Culture (W.R. Sharp, P.O. Larsen, E.F. Paddock, and V. Raghavan, eds.) pp. 401-419. Ohio State Univ. Press, Columbus.

Martin, M.J., Riedel, R.M., and Rowe, R.C. 1982. *Verticillium dahliae* and *Pratylenchus penetrans*: Interactions in the early dying complex in potato in Ohio. Phytopathology 72:640-644.

Zuckerman, B.M. 1971. Gnotobiology. In: Plant Parasitic Nematodes, Volume II (B.M. Zuckerman, W.F. Mai, and R.A. Rohde, eds.) pp. 159-184. Academic Press, New York.

REFERENCES

Abawi, G.S. and Mai, W.F. 1978. Root diseases of fruit trees in New York State. X. Toxicity of oxamyl to *Pratylenchus penetrans*. Plant Dis. Rep. 62:637-641.

Adamo, J.A., Madamba, C.P., and Chen, T.A. 1976. Vertical migration of rice white tip nematode *Aphelenchoides besseyi*. J. Nematol. 8: 146-152.

Andreeva, G.N. 1974. Effect of culture medium on the viability and development of *Ditylenchus destructor* in axenic conditions (translation). Byull. Vses. Inst. Gel'mintol. 14:73-76.

Bhatt, B.D. and Rohde, R.A. 1970. The influence of environmental factors on the respiration of plant parasitic nematodes. J. Nematol. 2: 277-285.

Bergquist, R.R. and Riedel, R.M. 1972. Screening onion (*Allium cepa*) in a controlled environment for resistance to *Ditylenchus dipsaci*. Plant Dis. Rep. 57:603-605.

Buecher, E.J., Hansen, E.L., and Myers, R.F. 1970. Continuous axenic culture of *Aphelenchoides* sp. J. Nematol. 2:189-190.

Chen, T.A. and Wen, G.Y. 1972. Ultrastructure of the feeding apparatus of *Pratylenchus penetrans*. J. Nematol. 4:155-161.

Corbett, D.C.M. 1969. *Pratylenchus pinguicaudatus* n. sp. (Pratylenchinae: Nematoda) with a key to the genus *Pratylenchus*. Nematologica 15:550-556.

Dolliver, J.S., Hildebrandt, A.C., and Riker, A.J. 1962. Studies of reproduction of *Aphelenchoides ritzemabosi* Schwartz on plant tissues in culture. Nematologica 7:294-300.

Eriksson, K.B. 1965. Crossing experiments with races of *Ditylenchus dipsaci* on callus tissue cultures. Nematologica 11:244-248.

_____ 1972. Studies on *Ditylenchus dipsaci* (Kuehn) with reference to plant resistance. Ph.D. Thesis, Univ. Uppsala, Sweden.

_____ 1980. Nematode inoculum propagated in monoxenic plant tissue cultures - nematode bank. Paper presented at the European and Mediterranean Plant Protection Conference on Breakthroughs in Resistance Breeding, Sualoev (Sweden) June 20-22, 1978. EPPO Bull. 10: 371-378.

Faulkner, L.R. and Darling, H.M. 1961. Pathological histology, hosts, and culture of the potato rot nematode. Phytopathology 51:778-786.

Hildebrandt, A.C., Riker, H.J, and Duggar, B.M. 1946. The influence of the composition of the medium on growth "in vitro" of excised tobacco and sunflower tissue cultures. Am. J. Bot. 53:591-597.

Hung, C.-L. and Jenkins, W.R. 1969. Oogenesis and embryology of two plant parasitic nematodes, *Pratylenchus penetrans* and *P. zeae*. J. Nematol. 1:352-356.

Katznelson, H. and Henderson, V.E. 1964. Studies on the relationships between nematodes and other soil microorganisms. II. Interactions of *Aphelenchoides parietinus* (Bastian, 1865) Steiner 1932 with actinomycetes, bacteria and fungi. Can. J. Microbiol. 10:37-41.

Kisiel, M., Himmelhoch, S., and Zuckerman, B.M. 1972. Fine structure of the body wall and vulval area of *Pratylenchus penetrans*. Nematologica 18:234-238.

Kisiel, M., Himmelhoch, S., Castillo, J.M., and Zuckerman, B.M. 1974. Fine structure of the intestine of *Pratylenchus penetrans*. Nematologica 20:262-264.

Kisiel, M.J., Himmelhoch, S., and Zuckerman, B.M. 1976. Fine structure of the esophagus of *Pratylenchus penetrans*. J. Nematol. 8:218-228.

Krusberg, L.R. 1961. Studies on the culturing and parasitism of plant-parasitic nematodes, in particular *Ditylenchus dipsaci* and *Aphelenchoides ritzemabosi* on alfalfa tissues. Nematologica 6:181-200.

_____ and Blickenstaff, M.L. 1964. Influence of plant growth regulating substances on reproduction of *Ditylenchus dipsaci*, *Pratylenchus penetrans* and *Pratylenchus zeae* on alfalfa tissue cultures. Nematologica 10:145-150.

Lauritis, J.A., Rebois, R.V., and Endo, B.Y. 1981. Monoxenic culture of *Heterodera glycines* propagated on root cultures of susceptible soybean. J. Nematol. 13:447 (abstr.)

Loof, P.A.A. 1978. The genus *Pratylenchus*. Filipjev, 1936 (Nematoda: Pratylenchida): A review of its anatomy, morphology, distribution, systematics and identification. Vaxtskyddsrapporter, jordbruk 5, Sveriges Lantbruksuniversitet, Uppsala, Sweden.

Lownsbery, B.F., Huang, C.S., and Johnson, R.N. 1967. Tissue culture and maintenance of the root-lesion nematode, *Pratylenchus vulnus*. Nematologica 13:390-394.

Mitsui, Y. 1977. Comparison of different components of three media on the propagation of *Pratylenchus* spp. cultured on alfalfa callus tissues. Jpn. J. Nematol. 7:28-32.

_____, Yokozawa, R., and Ichinohe, M. 1975. Effect of temperature and pH on the propagation of *Pratylenchus* culturing with alfalfa callus tissues. Jpn. J. Nematol. 5:48-55.

Mountain, W.B. 1955. A method of culturing plant parasitic nematodes under sterile conditions. Proc. Helminthol. Soc. Wash. 22:49-52.

Murashige, T. and Skoog, F. 1962. A revised medium for rapid growth and bioassays with tobacco tissue cultures. Physiol. Plant. 15:473-497.

Myers, R.F. 1967. Axenic cultivation of plant parasitic nematodes. Nematologica 13:323.

_____ 1968. Nutrient medium for plant parasitic nematodes. 1. Axenic cultivation of *Aphelenchoides* sp. Exp. Parasitol. 23:96-103.

_____ 1971. Studies on nutrient media for plant parasitic nematodes. IV. Physical aspects of culturing *Aphelenchoides rutgersi*. Exp. Parasitol. 30:174-180.

Olowe, T. and Corbett, D.C.M. 1976. Aspects of the biology of *Pratylenchus brachyurus* and *P. zeae*. Nematologica 22:202-211.

Orion, D., Wergin, W.P., and Endo, B.Y. 1980. Inhibition of syncytia formation and root-knot nematode development on cultures of excised tomato roots. J. Nematol. 12:196-203.

Perry, R.N., Plowright, R.A., and Webb, R.M. 1980. Mating between *Pratylenchus penetrans* and *P. fallax* in sterile culture. Nematologica 26:125-129.

Pillai, J.K. and Taylor, D.P. 1968. Influence of fungi on host preference, host suitability, and morphometrics on five mycophagous nematodes. Nematologica 13:529–540.

Rebois, R.V. and Lauritis, J.A. 1981. Propagation of *Rotylenchus reniformis*, Linford and Oliveira, reniform nematode, on monoxenic excised
root tissue culture. J. Nematol. 13:457–458 (abstr.).

Riedel, R.M. and Foster, J.G. 1970. Monoxenic culture of *Ditylenchus
dipsaci* and *Pratylenchus penetrans* with modified Krusberg's and
White's media. Plant Dis. Rep. 54:251–254.

Riedel, R.M. and Mai, W.F. 1971a. Pectinases in aqueous extracts of
Ditylenchus dipsaci. J. Nematol. 3:28–38.

_____ 1971b. A comparison of pectinase from *Ditylenchus dipsaci* and
Allium cepa callus tissue. J. Nematol. 3:174–178.

Riedel, R.M., Foster, J.G., and Mai, W.F. 1972. Monoxenic culture of
Pratylenchus penetrans on a simplified medium. J. Nematol. 4:232–
233.

_____ 1973. A simplified medium for monoxenic culture of *Pratylenchus penetrans* and *Ditylenchus dipsaci*. J. Nematol. 5:71–72.

Roman, J. and Hirschmann, H. 1969. Morphology and morphometrics of
six species of *Pratylenchus*. J. Nematol. 1:363–386.

Roman, J. and Triantaphyllou, A.C. 1969. Gametogenesis and reproduction of seven species of *Pratylenchus*. J. Nematol. 1:357–362.

Shafer, S.R., Rhodes, L.H., and Riedel, R.M. 1981. In vitro parasitism
of endomycorrhizal fungi of ericaceous plants by the mycophagous
nematode *Aphelenchoides bicaudatus*. Mycologia 73:141–149.

Sontirat, S. and Chapman, R.A. 1970. Penetration of alfalfa roots by
different stages of *Pratylenchus penetrans* (Cobb.). J. Nematol. 2:
270–271.

Street, H.E. 1977. Introduction. In: Plant Tissue and Cell Culture
(H.E. Street, ed.) pp. 1–10. Univ. California Press, Berkeley and Los
Angeles.

Tamura, H. and Mamiya, Y. 1979. Reproduction of *Bursaphelenchus lig-
nicolus* on pine callus tissues. Nematologica 25:149–151.

Tanda, A.S, Atwal, A.S., and Bajaj, Y.P.S. 1980. Reproduction and
maintenance of root-knot nematode (*Meloidogyne incognita*) in excised
roots and callus cultures of okra (*Abelmoschus esculentus*). Indian J.
Exp. Biol. 18:1340–1342.

Tarakanov, V.I. 1974. Cultivation of the nematode *Ditylenchus destructor* in media of animal and plant origin (translation). Helminthologia
15:521–529.

_____ 1975. Axenic cultivation of *Ditylenchus destructor* and *Aphelenchus avenae* (translation). Materi. Nauchnykh Konferentsii Vsesoy-
uznogo Obshchestva Gel'mintol. 27:152–161.

Tarte, R. and Mai, W.F. 1976. Morphological variation in *Pratylenchus
penetrans*. J. Nematol. 8:185–195.

Thistlethwayte, B. 1969. Hatch of eggs and reproduction of *Pratylenchus penetrans* (Nematoda: Tylenchida) Ph.D. Thesis, Cornell University.

_____ 1970. Reproduction of *Pratylenchus penetrans* (Nematoda: Tylenchida). J. Nematol. 2:101–105.

Tiner, J.D. 1961. Cultures of the plant parasitic nematode genus *Pratylenchus* on sterile excised roots. II. A trap for collection of axenic nematodes and quantitative initiation of experiments. Exp. Parasitol. 11:231–240.

Townshend, J.L. 1978. Infectivity of *Pratylenchus penetrans* on alfalfa. J. Nematol. 10:318–323.

Upadhyay, K.D. and Swarup, G. 1972. Culturing, host range and factors affecting multiplication of *Tylenchorhyncus vulgaris* on maize. Indian J. Nematol. 2:139–145.

Vanfleteren, J.R. 1978. Axenic culture of free-living, plant parasitic, and insect-parasitic nematodes. Annu. Rev. Phytopathol. 16:131–157.

Viglierchio, D.R., Siddiqui, I.A., and Croll, N.A. 1973. Culturing and population studies of *Ditylenchus dipsaci* under monoxenic conditions. Hilgardia 42:177–213.

Webster, J.M. 1967. The influence of plant growth substances and their inhibitors on the host-parasite relationships of *Aphelenchoides ritzemabosi* in culture. Nematologica 13:256–262.

_____ and Lowe, D. 1966. The effect of the synthetic plant growth substance, 2,4-dichlorophenoxyacetic acid, on the host-parasite relationships of some plant-parasitic nematodes in monoxenic callus culture. Parasitology 56:313–322.

Wen, G.Y. and Chen, T.A. 1976. Ultrastructure of the spicules of *Pratylenchus penetrans*. J. Nematol. 8:69–74.

Wergin, W.P. and Orion, D. 1981. Scanning electron microscope study of the root-knot nematode (*Meloidogyne incognita*) on tomato root. J. Nematol. 13:358–366.

White, P.R. 1963. The Cultivation of Animal and Plant Cells. Ronald Press, New York.

Yeoman, M.M. and Macleod, A.J. 1977. Tissue (callus) cultures-techniques. In: Plant Tissue and Cell Culture (H.E. Street, ed.) pp. 31–59. Univ. California Press, Berkeley and Los Angeles.

CHAPTER 33
Breeding for Nematode Resistance

H. Medina-Filho and *S.D. Tanksley*

Since the pioneer work of Smithies (1955) and Hunter and Markert (1957) with the basic electrophoretic technique and staining procedure, many improvements have been made (Brewer, 1970; Gordon, 1975). Such improvements have stimulated a great deal of research on isozymes, revealing the presence of multimolecular forms of enzymes in nearly all organisms extensively investigated (Markert, 1975).

Plant geneticists readily acknowledged this new method for assaying genetic variation and, indeed, the literature of the past 15 years is replete with examples of use of gel electrophoresis for answering basic questions of genetics and population biology. Practical applications of isozymes as marker genes for breeding purposes is another area of growing interest (Pierce and Brewbaker, 1973; Torres et al., 1978a,b; Tanksley and Rick, 1980; Tanksley et al., 1981a).

As the methods of somatic hybridization in plants progress, electrophoresis is becoming an increasingly useful technique as a criterion to characterize hybrid plants generated by protoplast fusion (Evans et al., 1981). One of the advantages of zymogram-based screening is the unique opportunity to distinguish somatic hybrid plantlets of phenotypically similar parents or when morphological seedling markers are not available.

Recently, a great deal of effort has been concentrated in attempts to take advantage of the tremendous array of genetic variation found in progenies of tissue-culture-derived plants. Electrophoresis might also turn into an additional tool for surveying genetic variation generated through in vitro manipulation of plant cells and tissues. Indeed, it seems a new variant of locus 1 of alcohol dehydrogenase (Adh-1) has been detected in R_3 progenies of mesophyll derived plants of tomato (Medina-Filho et al., unpublished).

The present chapter was undertaken to critically review the studies on the feasibility of using electrophoresis in selecting for nematode resistance in tomato, as grounded in the initial information provided by Rick and Fobes (1974) that resistance was associated with a variant allele of acid phosphatase.

The investigations reported here were oriented toward a basic understanding of the potentials and limitations of isoenzyme analysis with respect to banding pattern consistency and predictability of nematode resistance, so as to establish the feasibility of extensive use of this technique in the future. For reasons that we now understand, many previous attempts to use this technique were hindered by problems of methodology, and in other cases the efforts were abandoned because resistant and susceptible cultivars failed to show any differences in the banding pattern.

THE NEMATODE PROBLEM

Root-knot nematodes (*Meloidogyne* spp.) are important plant parasites. Among the 37 recognized species of *Meloidogyne*, 7 are known to attack tomatoes. *M. incognita*, *M. javanica*, *M. arenaria*, and *M. hapla* are economically important. The first three species are widespread between 35 S and 35 N latitudes whereas *M. hapla* is restricted to temperate region species, frequently occuring north of 35 latitude (Taylor and Sasser, 1978). In California, *M. incognita* and *M. javanica* are often found in the Sacramento and San Joaquin Valleys, where most of the processed tomato crop is grown (Siddiqui et al., 1973).

Heavily infested plants have a very shallow and knotted root system. Normal development of feeder roots is impaired, and distribution of hormones, and translocation of minerals and photosynthates is altered (Wang and Bergeson, 1974). As a consequence, top growth is reduced, yields are low, and plants wilt during the hot periods of the day, even though soil water is plentiful. Symptoms of mineral deficiency are common, and often additional fertilizer is applied further increasing the costs of producing a poor crop.

The extent of economic loss due to nematode infestation is difficult to evaluate. It necessarily has to take into account the severity of attack and the cash value of the crop for each particular region. However, the limited data available represent compelling evidence of the seriousness of the problem. In Hawaii and Mexico, decreases in yield as much as 75% have been reported (MacFarlane et al., 1946; Palacios and Moss, 1972). Losses of 50-85% have been estimated for southern Italy, Malta, and North Carolina (Lamberti and Cirulli, 1970; Lamberti, 1971; Lamberti et al., 1976; Baker et al., 1976).

In addition to the losses caused by the direct effect of nematode infestation, predisposition or even breakdown of resistance to other root diseases is common (Wang and Bergeson, 1974). For example, nematode-susceptible lines resistant to bacterial wilt (*Pseudomonas solanacearum*) do not survive well if, in addition to nematodes, conditions for bacterial disease are present (Gilbert et al., 1974).

Chemical control is feasible, but it is costly and represents only a temporary solution by decreasing the soil inoculum. Recently, serious restrictions on the use of widely applied nematicides have been imposed by the US government. One nematicide, DBCP, has been reported to cause cancer in laboratory animals and sterility in man. Thus it has been a well-documented concept that breeding for genetic resistance is a wise approach to cope with plant parasitic nematodes, not only because of the financial aspect of the crop but also by avoiding possible ecological repercussions. In tomato more than 65 nematode-resistant cultivars are known (Fassuliotis, 1979) and undoubtedly many more are unknown. On a worldwide basis, many regions would still benefit from nematode-resistant cultivars. For instance, in California, a major tomato-growing and processing region, there is no acceptable processing open-pollinated cultivar with nematode resistance.

EARLY HISTORY OF NEMATODE RESISTANCE IN TOMATOES

D. M. Bailey at Tennessee Agricultural Experimental Station did a comprehensive survey of *Lycopersicon* species for nematode resistance. Of the *L. esculentum* lines (varieties) tested, 95 commercial cultivars and 420 USDA accessions failed to show any resistance. The same was also true for *L. glandulosum*, *L. hirsutum*, and *L. pimpinellifolium*. Resistance was found in *L. peruvianum*; among 25 introductions tested, 11 possessed a high level of resistance (Bailey, 1941).

The difficult cross between *L. esculentum* cv. Michigan State Forcing and the self-incompatible *L. peruvianum* var. dentatum (PI128657) was first achieved in 1944. By using the wild species as male parent to overcome the unilateral incompatibility barrier and aseptically culturing the immature embryos to surmount the hybrid inviability, P. G. Smith of the University of California at Davis was able to develop mature interspecific hybrids (Smith, 1944).

Cuttings of a single F_1 plant of that cross (P.G. Smith, personal communication) were secured by V. M. Watts in Arkansas, who then obtained the first two backcrosses to *L. esculentum* again using *L. esculentum* as female parent (Watts, 1947). Progenies of this second backcross were then sent to the Hawaiian Experimental Station (HES) where W. A. Frazier obtained additional backcrosses to the cultivated species (Frazier and Dennett, 1949).

The program of incorporation of nematode resistance was continued in Hawaii by Gilbert and associates and also in California by Smith using the material developed by Frazier. Further backcrosses and pedigree selections were carried out separately, culminating in the release of the first resistant cultivar VFN 8 (VFN 36-8) in California and of Anahu and other HES lines in Hawaii. Accordingly, the resistant cultivars now available are derived from one of these two original releases.

Although there exist some conflicting references on the inheritance of resistance derived from these programs (for review, see Sidhu and Webster, 1981), resistance is determined by a major dominant gene located at position 35 cM on chromosome 6. Since the original screening tests were performed with *Meloidogyne incognita* the symbol Mi was proposed for this gene (Gilbert and MacGuire, 1955; Gilbert, 1958). Subsequently, the Gilbert lines were investigated by Barham and Winstead (1957) who demonstrated that, in addition to *M. incognita*, the Mi gene also confers resistance to the prevalent species *M. javanica*, *M. acrita*, and *M. arenaria*.

SCREENING FOR NEMATODE RESISTANCE

The traditional method of evaluating nematode resistance involves direct observation of root galling developed by young plants grown in heavily infested soil.

Seeds are either directly sown into infested soil or germinated in sterile soil, with young plants later transplanted. To permit an accurate classification, both resistant and susceptible controls should be grown with the plants to be tested. When the susceptible control consistently shows a large number of well-developed galls and the resistant material is free from infestation (usually 4-8 weeks after exposure to the parasites) all the plants are pulled out, and the roots are washed and graded for the severity of symptoms. Homozygous resistant plants are usually devoid of galls but may occasionally have a few very small galls. Heterozygous resistant plants more frequently display the latter behavior (Barham and Winstead, 1957; Laterrot, 1973). The control of temperature is important, since the resistance conditioned by Mi is not effective at 28 C or above, and both resistant and susceptible plants are equally affected (Holtzmann, 1965; Dropkin, 1976).

To improve the efficiency of screening, several modifications on the methodology of screening for nematode resistance have been suggested (Fassuliotis and Corley, 1967; Dropkin et al., 1967; Hussey and Barker, 1973). However, a very important advance was proposed by Rick and Fobes (1974). Studying isozymes of tomatoes with starch gel electrophoresis, they found that VFN 8 as well as five other nematode-resistant cultivars carried a variant for locus 1 of acid phosphatase. This allele, Aps-1^1, was previously known to occur only in *L. peruvianum*. All other cultivars of *L. esculentum* carried the Aps-1$^+$ allele. An F_2 population from a cross between a resistant Aps-1^1 cultivar and a susceptible Aps-1$^+$ stock segregated 16 +/+ : 19 +/1 : 10 1/1 for Aps-1 with only the plants of genotype +/+ being susceptible to nematodes. A relationship between Mi and Aps-1 was suggested, resulting from either a tight linkage of the genes or pleiotropic effects of a single gene.

A number of cultivars and breeding lines from around the world were studied by Medina-Filho and Stevens (1980). Horticultural characteristics and nematode resistance were classified for genotype of locus 1 of acid phosphatase (Aps-1). Susceptible lines were consistently homozygous for Aps-1$^+$. Resistant open-pollinated cultivars were Aps-1$^{+/+}$ or Aps-1$^{1/1}$, whereas hybrid cultivars were Aps-1$^{+/+}$ or Aps-1$^{+/1}$ (Fig. 1). In all cases in which the pedigree of resistant open-pollinated cultivars Aps-1$^{+/+}$ could be checked, it was verified that they originated directly or indirectly from sources of the Hawaiian Agricultural Experiment

Aps-1 GENOTYPE

	+/+	+/1	1/1
SUSCEPTIBLE	YES	no	no
RESISTANT	YES	YES	YES
Open-pollinated	YES	no	YES
Hybrid	YES	YES	no

Figure 1. Banding pattern of homozygous and heterozygous genotypes for Aps-1 and their occurrence among tomato cultivars.

Station. The Aps-$1^{1/1}$ lines, on the other hand, were developed from materials of the breeding program of the University of California at Davis. A survey on the origin of the nematode resistant cultivars indicates they all originated from a single F_1 plant of the cross *L. esculentum* x *L. peruvianum*. Probably, a crossing over between Aps-1^1 and Mi occurred in early generations during the incorporation of nematode resistance into *L. esculentum*. This accounts for the Aps-$1^{+/+}$ resistant cultivars such as Anahu. VFN 8 and its derivatives maintained the original Aps-1^1 Mi association present in *L. peruvianum* (Fig. 2).

A linkage test for the markers yv coa c in addition to Aps-1 and Mi on chromosome 6L was performed in crosses of LA1178 with VFNT Cherry (Medina-Filho, 1980). In general, the relative map distances between yv, coa, and c conformed with previous data known for these loci. No recombinants were recovered for yv - Aps-1 - Mi. On the basis of 5591 gametes analyzed for recombination between yv and Aps-1, the estimate of maximum map distance between them is 0.082 cm. For the interval (yv, Aps-1) and Mi, the estimated maximum distance based on 513 plants is 0.894 cm indicating that the screening for resistance based on Aps-1^1 is remarkably reliable.

METHODOLOGY FOR ASSAYING APS-1

The fortuitous association of Mi with the electrophoretic variant Aps-1^1 has provided a new screening strategy in breeding tomatoes for nematode resistance and, in fact, has been successfully used by a few private plant-breeding organizations. However, the electrophoretic procedure for assaying acid phosphatase involves a number of variables that affect the reliability and quality of the results obtained. By experimenting with those variables it has been possible to improve the technique considerably allowing fast, accurate scoring of Aps-1.

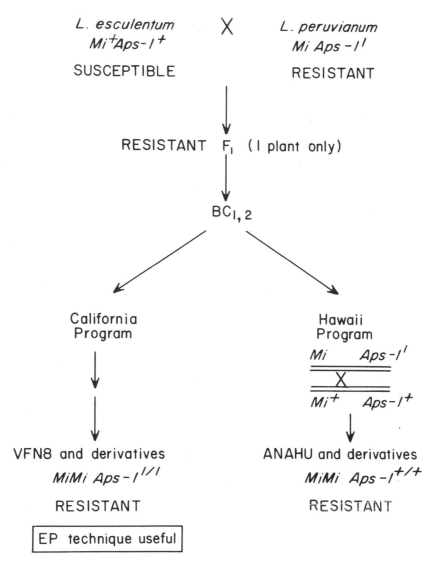

Figure 2. The development of nematode resistance in tomatoes. There was apparently a crossing over between Mi and Aps-1[1] in the Hawaii program that did not occur in the California program. The electrophoresis technique can be applied only to the California derived cultivars.

Electrophoresis Equipment

The apparatus designed by Tanksley (1979) has proven to be an excellent system for assaying isozymes with starch gels. To increase the number of individual samples per gel, a slight modification was made in the original apparatus by setting the wider part of the gels parallel to the buffer tanks.   The buffer tanks can be maintained with the same original dimensions, thus adding versatility to the apparatus that can still be used in the regular fashion when assaying for enzymes requiring longer runs for good band resolution.   The modified run is performed by substituting the original rectangular electrode sponges for T-shaped ones, with the wider part contacting the gel.   With such an arrangement, 35 wicks 3.2 x 5.0 mm can be inserted in a single gel without overcrowding (Fig. 3).   An increase in the number of samples can also be accomplished by using smaller wicks.   In this case the bands are narrower but still provide unambiguous readings.   Smaller wicks used with the above set up allow the assay of up to 40 individual plants in a single gel.

Gel and Electrode Buffer pH

A series of gel and electrode buffer pH values was tested in all combinations ranging from 7.0-9.0 at 0.4 intervals.   For the electrode buffer, a 0.3 M borate solution was adjusted to the desired pH with 4N NaOH.   Gel pH was adjusted by increasing the concentration of Tris or citrate as necessary, starting from a standard solution of 25 ml 0.0152 M Tris (hydroxymethyl) aminomethane and 25 ml 0.036 M citric acid in 1000 ml distilled water.   It is realized that this approach also changed the ionic strength of the buffer components.

No differences in the banding pattern and migration relative to the front were observed with the various borate (electrode buffer) pH values tested.   However, a striking effect was observed in resistance to the passage of current.   When the pH of buffer tanks was 7.4, it required 360 V to bring the amperage to 40 mA.   This figure was approximately 220, 210, and 190 V for pH 7.8, 8.2, and 8.5, respectively.   Attempts to speed up the run by increasing the voltage were unsuccessful with the lower pH, as it was impossible to raise the amperage.   In addition, with high voltage, too much heat was generated, which resulted in gels of unacceptable quality.   With a pH as high as 8.6, a cur-

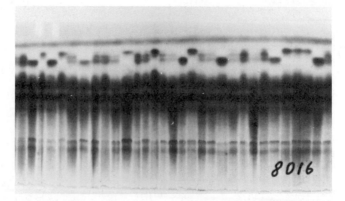

Figure 3.   Typical zymogram of F_2 population segregating for Aps-1 in a gel with 35 wicks.   From left to right: 1/1, +/1, +/+, 1/1, +/1...

rent of 65 mA can be applied with a voltage between 270 and 285. Under these conditions excessive heat is not generated, band resolution is excellent, and the running time reduced to 1.5-2.0 hr instead of 4-5 hr when borate solution of lower pH is used.

The relative migration, Rf = (band migration, mm/borate front migration, mm) x 100, and the resolution of Aps bands were strongly influenced by the concentration of the gel buffer. The different concentrations were obtained by changing the amount of Tris and monitored by the consequent change in the pH value of the gel buffer. At pH 7.0 the bands of Aps-1 migrate together with the front (Rf = 100). There was no resolution of this locus. However, the band(s) of locus Aps-2 that have a lower Rf were well resolved. Below the band(s) of Aps-2 the activity of acid phosphatase in these gels with pH 7.0 was very intense, but only a long, blurred strip was observed. As the pH was increased, the Rf of the bands decreased. As an example, at pH 8.6 the Rf of Aps-1$^+$ was 87. Furthermore, at high pH (8.6) there was good resolution of the bands at locus 1 and 2 plus two other bands of lower Rf values. In addition to the four bands mentioned above, a fifth faintly staining band was occasionally seen closer to the origin, with an Rf of 15. Although our concern here is with locus 1 because of its linkage to Mi, the resolution of other bands might be useful in other studies with *Lycopersicon* species. The pattern of gels at pH 9.0 was the same as at pH 8.6 except for a slight decrease in the activity of all bands.

Starch Concentration

Sigma potato starch hydrolyzed for electrophoresis was used exclusively. Starch concentrations from 10 to 14% were tried at intervals of 0.5%. No evident differences were found in gels in which the starch concentration was varied at intervals smaller than 0.5%.

No effects were observed in the sharpness of bands in gels of various starch concentrations. Differences, however, were found in the migration of bands relative to the front (Rf). As starch concentrations decreased, the Rf of the bands increased. As an example, Rf values for Aps-1$^+$ were 79.1, 80.0, 86.5, 100, and 100 for starch concentrations of 14, 13, 12, 11, and 10%, respectively. As seen from the Rf values, at starch concentrations lower than 12%, no resolution was possible for Aps-1, since the bands migrated with the front. However, it must be kept in mind that the optimum starch concentration is dependent on the particular batch of starch used. Thus the 12% routinely used in our work had to be decreased to 11.5% with a new batch of starch from the same company. Degree of hydrolysis in the commercial starch preparation seems to account for such variation.

Staining Solution

The effects of different substrates, dyes, and buffer pH were evaluated in gels containing a single paper wick 16 cm long imbibed with leaf extract of plants heterozygous for Aps-1. With the run completed, the bottom 2 mm slice was cut in as many vertical tracks as the test required. Each one was individually stained in germination boxes containing different assay solutions.

The following substrates were used: α-naphthyl acid phosphate, β-naphthyl acid phosphate, naphthol-AS-MX phosphoric acid, and naphthol AS-BI phosphoric acid. For each substrate three different dyes were tried: Fast Black K Salt, o-dianisidine tetrazotized (Diazo Blue), and Fast Garnet. Solutions were prepared in 0.5 M acetate buffer at pH 5.65. For all substrates and dyes, a concentration of 1 mg/ml was used. For the best combination of the above, the effect of pH of the staining solution was first evaluated in a wide range from 3.0 to 7.5 at 0.5 intervals and optimized in further tests by varying the pH from 5.0 to 6.0 at 0.1 intervals.

The several acid phosphatase isozymes present in tomato leaves seem to react differently with different substrates and dyes tested. The bands of locus 1 for example can be seen in gels reacted with α- and β-naphthyl acid phosphate but not with Sigma AS-MX. The substrate β-naphthyl acid phosphate was the best for assaying Aps-1. Black K Salt and o-dianisidine are satisfactory, but the latter dye gave slightly better results. Nevertheless, the use of o-dianisidine should be avoided, since it is a putative carcinogen.

With β-naphthyl acid phosphate and Black K Salt, the activity of acid phosphatase is detected in gels stained at pH from 3.5 to 7.5. Resolution for Aps-1 occurs at pH 4.0 or higher. The gel background, however, darkens as the pH increases, resulting in difficult scoring of genotypes when the pH is 6.0 or above. Furthermore at pH 6.0 or above the time required for development of bands is greatly increased. The best pH for the staining solution was found to be 5.5.

Slice Position

The standard gel 6 mm thick used in this study allows several horizontal slices to be taken. With glass rods of different thicknesses as slicing guides, it was possible to obtain four 1 mm and one 2 mm slices. Initial tests were performed by observing the resolution of bands in the several slices of two gels containing a single wick 10 cm long imbibed with extract of heterozygous plants. These observations were further tested in regular gels used for genotyping progenies.

The resolution quality of bands was different when several slices were assayed. The sharpest bands occur in the very bottom slice (2 mm thick) and bands become progressively less sharp in the top slices. The bands appear to be stained in a V shape with the vertex toward the glass mold. Increased sharpness of bands in the lower layers of the gel could also be seen in gels that were not horizontally but vertically sliced.

Time and Current Required for Electrophoresis

The rate of enzyme migration, and thus the running time, is proportional to the current applied to the gel. The possibility of decreasing the running time was tested in gels run at progressively higher currents. Gels were run until the borate front reached a distance of 7-8 cm from the origin. After workable current was established, a series of gels was studied in which the borate front was interrupted at distances progressively closer to the origin.

As explained before, the current applied to the gel can be increased when borate pH is high (8.6), resulting in faster migrations. The insertion of the crude extract into the gel can be performed very quickly at 40 mA, taking only 15 min instead of 40 min, as is required for gels with lower pH values. After removal of the paper wicks, the current can be adjusted as high as 65 mA, while maintaining the voltage between 270 and 285.

Power shutdown when the borate front reached 8 cm produced good separation of bands. Nevertheless it was possible to get good resolution of bands and unambiguous scoring of Aps-1 when the borate front had migrated only 5.5 cm. This occurs in less than 2 hr with the procedures described above. Reduction of the running time is convenient, since it allows one person to prepare four gels, crush 140 samples, perform the electrophoresis, and score the results within 8 hr.

Developmental Stages and Plant Tissues

A survey was conducted to determine suitable developmental stages of plants as well as possible alternative tissues for assaying Aps-1. Developmental stages ranging from dry seeds to mature plants were studied.

DRY SEEDS UP TO 2-WEEK-OLD SEEDLINGS. Seeds were germinated on moist blotter paper in germination boxes incubated at 24 C with 12 hr of fluorescent light (approximately 70 μE m^{-2} sec^{-2}). Electrophoresis was performed with crude extracts obtained by crushing the appropriate plant tissues with plastic rods after adding approximately 20 μl of extractant buffer described later. Homogenates were obtained from dry and sprouting whole seeds with a 5-10 mm radicle and also from the radicle alone. For seedlings at the stage of fully developed cotyledons, samples of their roots, basal stem, cotyledons alone, and cotyledons including the apical meristem were assayed.

Dry and germinating seeds as well as their correspondent basal stem and cotyledons do not show activity of Aps-1. Samples of cotyledons including the apical meristem showed good activity and can be used in presowing screening as indicated in the section on composite samples.

PLANTS 3-13 WEEKS OLD. To have a continuous series of plants of different ages to be evaluated at the same time, sowings were made weekly for 10 consecutive weeks. Plants were germinated and grown in metal flats filled with a mixture of 1:1 standard potting soil and sand and kept in a greenhouse at 26.6 C day and 22 C night.

The plants assayed included genotypes +/+, 1/1, and +/1 for Aps-1 as well as determinate (sp) and indeterminate (sp⁺) plant habits as follows: 77B175-1 (sp; 1/1), UCX 97-3 (sp; +/+), Short Red Cherry (sp ; 1/1), IAC 3946 (sp⁺; +/+), and F₁ SRC x LA1178 (sp⁺; +/1). For each cultivar leaflets of three plants of each age were assayed individually. For the 3-week-old plants (first true leaf pair) in addition to leaflets, apical meristem, cotyledons, and roots were also assayed.

With reference to the tests conducted with leaflets of plants from 3 to 13 weeks old, no major differences were observed in the intensity and resolution of the bands for all genotypes. For the 3-week-old plants there was intense activity of Aps-1 in the roots, basal stem, and apical meristem. Cotyledons of plants at this stage showed much lower activity and in some plants no Aps-1 bands were detected.

MATURE PLANTS. Leaves from plants of genotype +/+, +/1, and 1/1 transplanted to the field were also subjected to electrophoresis. Samples were taken periodically at intervals of approximately 30 days during the growing season up to harvest time. At midseason, electrophoresis was performed also for tissues of shoots, leaf petioles, flower peduncles, petals, sepals, immature and ripe anthers, pollen, immature fruits (2.0 cm diameter), fruits at breaker stage, and ripe fruits. For assaying pollen samples the technique used by Tanksley et al. (1981b) was adopted. Pollen of five dessicated anthers was mixed with fine sand, and 1% glutathione in 0.1 M Tris buffer (pH 7.5) and ground with glass rods inside plastic micro test tubes.

Plants grown in the field could be assayed throughout the season using leaf samples. The resolution of bands is slightly inferior when compared to young plants grown in the greenhouse but still allows unambiguous screening. It was observed in two consecutive seasons that occasionally the + band stains much lighter than the 1 band and the heterodimer. In extreme cases heterozygous plants showed only the 1 and the heterodimeric band, and the + band was almost imperceivable.

Despite this problem, classification of plants under such conditions is not precluded, since the homozygous 1/1 plants always produce a dark staining band with retarded migration, and the heterozygotes produce two distinct bands contrasting with an indistinguishable or very faint + band of the homozygous +/+ genotypes. With reference to the other tissues of adult plants, very low activity of Aps-1 was observed in the shoots, flower peduncle, sepals, and immature fruits and anthers. Virtually nothing could be detected in leaf petioles, petals, mature anthers, fruits at breaker stage, and ripe fruits. Very intense activity was observed in samples of pollen extracted from mature anthers. Pollen samples of heterozygous plants show only the bands for alleles + and 1 and lack the usual heterodimeric band seen in sporophytic tissues.

Absence of the intermediate heterodimeric band in pollen extracts of heterozygous plants is an indication of postmeiotic expression of the Aps-1 locus. Hence this locus not only functions in the sporophyte, but also it seems to be transcribed and translated during the gametophytic phase. This is not surprising, and indeed, a recent survey has revealed an extensive overlap of sporophytic and gametophytic gene expression in the tomato (Tanksley et al., 1981b).

Sample Storage

Electrophoresis was performed with extracts of fresh leaf tissue and compared with extracts obtained previously and kept frozen, as well as extracts from intact leaves kept frozen, and crushed just before analysis. The length of freeze storage ranged from 1 day to 1 year for crushed samples and up to 9 months for intact leaves. Frozen samples were kept in tightly closed micro test tubes stored at -20 C.

Aps-1 activity was quite stable under the freeze storage periods tested. Reasonable activity was observed in the crushed samples stored up to 9 months. The bands tend to become slightly more diffuse when compared with fresh samples. This effect occurs even after overnight freeze storage and was observed in both crushed and intact leaflets, although less pronounced in the latter. Unambiguous genotyping of plants can be done with frozen samples. For the reason explained above, it is preferable to store intact leaflets rather than the extract.

Extraction

One small leaflet piece (approximately 1.5 x 2.5 mm) taken from 3-4 week-old seedlings was placed in a plastic crushing board, two to three drops of extractant solution added, and the sample thoroughly crushed with plastic rods. The resultant juice was then absorbed into paper wicks (Beckman), placed in wick holders, and refrigerated until insertion into the gel. Various concentrations of solutions of the following chemicals were tested as extractants: Polyethylene Glycol (MW, 6000), Polyvinylpyrrolidone (MW, 36,000), Dowex-50, Amberlite, Glutathione (reduced form), β-mercaptoethanol, Tween-20, and Triton X-100. A buffer solution of 0.1 M Tris pH 7.5 was used as diluent. Pure extracts and extracts obtained with the buffer solution alone were used as controls.

Leaf extracts obtained without the addition of extractant produced satisfactory results. Among several chemical solutions tested, better results were observed with a mixture of Triton X-100 and polyvinylpyrrolidone (PVP) at concentrations of 2 and 5%, respectively, diluted in 0.1 M Tris buffer pH 7.5. Compared with the control, when the above extractant was used the bands were sharper with reduced activity between them thus increasing the resolution of Aps-1 and other loci. This effect is most pronounced for resolution of locus Aps-3 in leaf samples of young plants. Although not essential, the addition of extractant facilitates the crushing operation and allows the use of very small samples such as a single leaflet of three-week-old seedlings. Once prepared, the extractant solution keeps several months in the refrigerator.

Composite Samples

For reasons described later, single wicks imbibed with a mixture of extracts from plants of different genotypes were analyzed. With a cork borer, leaflet discs of 8 mm diameter were taken from plants of genotypes +/+ and +/1. Composite samples were prepared by crushing together discs of +/+ and +/1 genotypes in the proportion of 1:1 and 1:10, respectively. Samples of the single genotypes alone were used as controls. Additional tests were also extended to the mixture of genotypes including the genotype 1/1.

In samples composed of a mixture of +/+ and +/1 genotypes, in addition to the + band, the presence of the band of allele 1 and the heterodimer can be detected even at proportion of 7 +/+ to 1 +/1. In such highly diluted samples, the band 1 and the heterodimer are not actually seen, but the region in the gel correspondent to them shows a light smear easily distinguishable from a totally clear spot of the samples containing only +/+ genotype. In samples with a mixture of the same genotypes in proportion of 2:1, the difference from the control sample is striking. The same is true for mixtures of genotypes +/+ and 1/1. In this case, two bands (+/+ and 1/1) can be seen in the gel.

If a breeding program for nematode resistance is conducted by pedigree selection or by single seed descent, the best approach is to classify individual F_2 plants and advance only the homozygous 1/1 lines. However, situations do occur in which classification is delayed to F_3 or later generations. In those situations the use of composite samples might be a very useful procedure. To distinguish homozygous from segregating lines, six plants are usually assayed. By using composite samples this can be accomplished by running two composite samples of three plants each. Lines in which the two samples show only the band correspondent to Aps-1$^{1/1}$ are selected. The probability that a heterozygous or a homozygous +/+ plant in the population will not be included among six plants of a progeny of a heterozygous plant is equal to $(1/4)^6$ or 0.0002.

Six plants can be reliably used, since the locus Aps-1 segregates in monogenic fashion as indicated by the following two sets of data:

(1) A 1:1 segregation of backcross populations. These populations consisting of 881 plants comprise crosses involving six widely different genetic backgrounds (Table 1).

(2) Segregation into 1:2:1 ratio of selfed progenies of BC_2 plants (Fig. 3). The data of Table 2 show that a significant departure from 1:2:1 was observed in one of the families studied due to a deficient number of homozygous 1/1 individuals. It is noteworthy that even though the segregation data for the other families fit well the 1:2:1 ratio, they also yielded a reduced number of 1/1 genotypes as evidenced by the significant "pooled" χ^2. This trend is further supported by the heterogeneity χ^2 which was not significant (Table 2). The underlying cause of this deficiency is not known. One possibility is that homozygous 1/1 genotypes had a slower germination rate or were less vigorous and were preferentially rogued out during the thinning operation. Whatever the reason, it remains that composite sample is a reliable method for screening families. The shortage of 1/1 individuals

Table 1. Backcross Segregation for Aps-1[a]

PROGENY	+/+	+/1	χ^2 (1 df)
BC1 (USDA77B178-1 x UCX97-3) x UCX97-3	54	46	0.64 ns
BC2	55	55	0.00 ns
BC1 (USDA77B178-1 x UCX99m-1-2-3) x UCX99m-1-2-3	39	46	0.58 ns
BC2	33	27	0.60 ns
BC1 (USDA77B178-1 x LAC3946) x LAC3946	3	6	1.00 ns
BC1 (Short Red Cherry x LA1178) x LA1178	239	278	2.94 ns
Total	423	458	1.39 ns

[a]Statistical analysis for segregation: Sum = 6 d.f., χ^2 = 5.76 ns; Pooled = 1 d.f., χ^2 = 1.39 ns; Heterogeneity = 5 d.f., χ^2 = 4.37 ns. Unadjusted χ^2 as calculated against ratio of 1:1. Crosses listed comprise the reciprocals.

Table 2. Unadjusted χ^2 Segregation against 1:2:1 Ratio of Selfed Progenies of BC2 Segregating for Aps-1[a]

PROGENY	+/+	+/1	1/1	χ^2 (2 df)
80S A1	27	54	21	1.059 ns
80S A3	23	60	19	3.490 ns
80S A5	35	50	17	6.392 *
80S A6	31	57	18	3.792 ns
Total	116	221	75	10.345 **

[a]Statistical analysis for segregation: Sum = 8 d.f., χ^2 = 14.734 ns; Pooled = 2 d.f., χ^2 = 10.345**; Heterogeneity = 6 d.f., χ^2 = 4.389 ns. Progenies 80S A1 and A3 are backcrosses to UCX97-3, while 80S A5 and A6 are backcrosses to UCX99m-1-2-3 as listed in Table 1.

can actually improve the efficiency of composite screening by decreasing the probability that in a segregant progeny only the homozygous 1/1 plants will be sampled for electrophoresis.

With the four-gel apparatus routinely used in this study, 140 composite samples (70 families) can be screened in 1 day. These classifications can be done with 2-week-old seedlings growing in germination boxes. Such screening can be performed in the lab with minimal effort and at convenient time during winter or early spring. On the basis of this initial screening, only the homozygous 1/1 lines are then sown and subjected to field evaluations.

To minimize the number of gels to be run, an alternative approach would be to run initially only one composite sample of three plants per family. The families which are either homozygous +/+ or segregating are discarded. A second survey of another three plants with one composite sample is then performed only for those families which in the first survey had a band correspondent to the Aps-1$^{1/1}$ genotype. In this second run the families that again show the same banding pattern are homozygous 1/1 and therefore are selected to be sown for field evaluations. With this alternative approach, the number of gels to be run is minimal, yet the screening reliability is kept at the same level, since the selected families were chosen on the basis of the same number of plants.

PROTOCOL FOR ASSAYING ACID PHOSPHATASE

STOCK SOLUTION	CHEMICALS	FINAL VOLUME	REMARKS
(A) Electrode buffer	Boric acid, 334 g	18 l distilled H_2O	Good for several months at room temperature
(B) Gel buffer	Tris (hydroxymethyl) amino methane, 92 g	500 ml distilled H_2O	Good for several months under refrigeration
(C) Gel buffer	Citric acid·H_2O, 7.35 g	500 ml distilled H_2O	Same as B
(D) Staining buffer	Sodium acetate, 90 g	500 ml distilled H_2O pH 5.5 with concentrated acetic acid	Same as B
(E) Substrate	β-naphthyl acid phosphate 1.0 g	50 ml acetone; 50 ml distilled H_2O	Add substrate to the solution. Same as B. Light sensitive
(F) Sample extractant	Triton X-100 6.0 ml; polyvinylpyrrolidone, 17.5 g	300 ml 0.1 M Tris pH 7.5	Same as B
(G) Fixative	1 l glycerol 10 ml Thimerosal (Merthiolate)	Add 1 l distilled H_2O	Can be filtered and reused indefinitely. Same as A.

CURRENT. Insert and run 15 min at 40 mA. Take out wicks and adjust the current to 65 mA. Run until borate front reaches 6.5 cm from the origin (approximately 2 hr).

STAINING SOLUTION. 0.2 Black K Salt, 175 ml distilled H_2O, 20 ml of solution D. Add 5 ml of Solution E and stir for a few seconds.

Incubate the gels immediately in the dark for 2-4 hr. After bands are developed, pour off the solution, wash twice with H_2O, and add fixative solution (G).

To make four gels:

Gel Buffer. 50 ml of Solution B; 50 ml of Solution C; 900 ml distilled H_2O. The pH should be 8.6 ± 0.1.

Starch. 120 g of hydrolyzed starch for electrophoresis (12%).

Electrode Buffer. 1800 ml of Solution A. Adjust to pH 8.6 with 4N NaOH. Fill each tank with 200 ml and use the remaining 200 ml to rinse the sponges before placing them in the tanks.

Some Useful Hints

1. After crushing each sample put the corresponding paper wicks in appropriate holders. After samples of one gel are completed, close the holder and keep it refrigerated until the time of insertion into the gel.
2. To avoid exchange of juice among samples, do not saturate wicks with extract, and place them 1 mm above the glass mold.
3. Gels can be made on the eve of the run and allowed to sit at room temperature. Cool the gels in the refrigerator for 20 min before inserting samples. Keep the gels cool during operations of cutting the origin, insertion, and removal of wicks. These are done outside the refrigerator by placing them on top of cool pads made of wet newspaper sealed in plastic bags.
4. The electrode buffer can have the pH adjusted and can be put into the tanks inside the refrigerator on the eve of the run.
5. Before starting the run, warm up the power supply for at least 20 min. When gels are taken out for removal of wicks, leave the power supplies on the standby position.

CONCLUDING REMARKS

Studies on the linkage relationship of Aps-1 and Mi provided information on the reliability of the electrophoresis technique in breeding tomatoes for nematode resistance (Medina-Filho, 1980). Major details of the procedure itself have been analyzed, indicating that consistent results can be easily obtained.

There are several advantages of the electrophoretic technique over the traditional method which is based on direct infestation with nematodes:

Effort and Time Saving

No progeny test is necessary to distinguish homozygous from heterozygous resistant plants, because these two types have distinct banding

patterns.   In cases of direct screening with nematodes, repeated progeny tests may be required because of the dominance of Mi.   In addition, the same plants can be genotyped for Aps-1 and subjected to
field selections for horticultural characteristics and yield performance,
an unfeasible procedure with the traditional method.

Extended Period of Screening

For very extensive programs a number of alternative tissues and procedures can be used for screening, ranging from early classification of
families in germination boxes, screening of seedlings before transplanting, screening of plants growing in the field throughout the season, and
still two later alternatives for classifying individual plants even after
harvest by storing frozen leaf samples or dry anthers.

Dependability

The classification of Aps-1 can be reliably performed on 3-week-old
plants growing in sterilized soil.   This contrasts with the traditional
method that might take 10 weeks or more and is subject to nematode
escapes and loss of plants due to contamination with damping-off fungi.
Sometimes, even with the best care, very little or no infestation occurs
in the susceptible controls and the test must be repeated.

Economics

Electrophoresis equipment and chemicals used for assaying Aps are
inexpensive.   Also, a single person can perform the necessary operations and evaluate 140 individual plants or 70 or more families within
an 8 hr period.

Can Stimulate Cooperative Work

Because of the possibility of screening dry anthers or very young
seedlings, single plant selection for yield performance and horticultural
characteristics can be done in one area, and a few anthers or seeds of
selected plants saved and mailed to a lab in another region or country
to be genotyped for Aps-1.   The results can be available in less time
than when the traditional method is used at the same location as the
selection work.   A lab already equipped and routinely involved with
electrophoresis work can genotype large numbers of plants, an easy
task that can be performed without jeopardizing other work.   The possibility of such effective cooperative work can stimulate programs
aimed at incorporation of nematode resistance by investigators not
familiar with the methodology.
Considering the ease of the electrophoretic technique and its high
reliability, in addition to the aforementioned advantages, it is conceiv-

able that in the near future nematode resistance will be incorporated into all major tomato cultivars adapted to areas where root-knot nematodes are already a problem or constitute a potential crop loss risk.

ACKNOWLEDGMENTS

H.P. Medina-Filho was partly supported by CNPq. Research partly supported by The California Tomato Research Institute and Empresa Brasileira de Pesquisa Agropecuaria. The authors wish to acknowledge the invaluable assistance of M.A. Stevens, M.C.C. Medina, D. Hunt, W.R. Sharp, D.A. Evans, and J.E. Bravo.

KEY REFERENCES

Fassuliotis, G. 1979. Plant breeding for root-knot nematode resistance. In: Root-Knot Nematodes (*Meloidogyne* species): Systematics, Biology and Control (F. Lamberti and C.E. Taylor, eds.) pp. 425-453. Academic Press, New York.

Markert, C.L. (ed.) 1975. Isozymes, Vols. I, II, and III. Academic Press, New York.

Medina-Filho, H.P. and Stevens, M.A. 1980. Tomato breeding for nematode resistance: Survey of resistant varieties for horticultural characteristics and genotype of acid phosphatase. Acta Hortic. 100: 383-393.

REFERENCES

Bailey, D.M. 1941. The seedling test method for root-knot nematode resistance. Proc. Am. Soc. Hortic. Sci. 38:573-575.

Barham, W.S. and Winstead, N.N. 1957. Inheritance of resistance to root-knot nematodes. Rep. Tomato Genet. Coop. 7:3.

Baker, K.R., Shoemaker, P.B., and Nelson, L.A. 1976. Relationships of initial population densities of *Meloidogyne incognita* and *M. hapla* to yield of tomato. J. Nematol. 8:232-239.

Brewer, G.J. 1970. An Introduction to Isozyme Techniques. Academic Press, New York.

Dropkin, V.H. 1976. Plant response to root-knot nematodes at the cellular level. In: Proceedings Research Planning Conference on Root-Knot Nematodes, *Meloidogyne* spp. pp. 43-56. North Carolina State Univ., Raleigh.

_____, Davis, D.W., and Webb, R.E. 1967. Resistance of tomato to *Meloidogyne incognita* acrita and *M. hapla* (root-knot nematodes) as determined by a new technique. Proc. Am. Soc. Hortic. Sci. 90:316-323.

Evans, D.A., Flick, C.E., and Jensen, R.A. 1981. Disease resistance: Incorporation into sexually incompatible somatic hybrids of the genus *Nicotiana*. Science 213:907-909.

Fassuliotis, G. and Corley, E.C., Jr. 1967.   Use of seed growth pouches
   for root-knot nematode resistant tests. Plant Dis. Rep. 51:482-486.
Frazier, W.A. and Dennett, R.K. 1949.   Isolation of *Lycopersicon esculentum* type tomato lines essentially homozygous resistant to root-
   knot. Proc. Am. Soc. Hortic. Sci. 54:225-236.
Gilbert, J.C. 1958.   Some linkage studies with the Mi gene for resistance to root-knot nematodes. Rep. Tomato Genet. Coop. 8:15-17.
_____ and MacGuire, D.C. 1955.   One major gene for resistance to
   severe galling from *Meloidogyne incognita*.   Rep. Tomato Genet.
   Coop. 5:15.
_____, Tanaka, J.S., and Takeda, K.Y. 1974.   'KEWALO' tomato.
   Hortic. Sci. 9:481-482.
Gordon, A.H. 1975.   Electrophoresis of proteins in polyacrylamide and
   starch gels. North-Holland/Elsevier, Amsterdam, New York.
Holtzmann, O.V. 1965.   Effect of soil temperature on resistance of
   tomato to root-knot nematode (*Meloidogyne incognita*).   Phytopathology 55:990-992.
Hunter, R.L. and Markert, C.L. 1957.   Histochemical demonstration of
   enzymes separated by zone electrophoresis in starch gels.   Science
   125:1294-1295.
Hussey, R.S. and Barker, K.R. 1973.   A comparison of methods of collecting inocula of *Meloidogyne* spp. including a new technique. Plant
   Dis. Rep. 57:1025-1028.
Lamberti, F. 1971.   Prove di lotta chimica contro i nematodi galligeni
   del pomodoro in Puglia.   Phyliatr. Phytophram. Circum-medit 3:140-
   143.
_____ and Cirulli, M. 1970.   Prove prelimirari di lotta a nematodi del
   gen. *Meloidogyne* su pomodoro con due nematocidi sperimentali.   Ital.
   Agric. 170:721-723.
_____, Dandria, D., Vovlas, N., and Aquilina, J. 1976.   Prove di lotta
   contro i nematodi galligeni del pomodoro da mensa in serra a Malta.
   Coltore Protette 5:27-30.
Laterrot, H. 1973.   Selection de varietes de tomate resistantes aux
   Meloidogyne. OEPP/EPPO Bull. 3:89-92.
MacFarlane, J.S., Hartzler, E., and Frazier, W.A. 1946.   Breeding toma-
   toes for nematode resistance and for high vitamin C content in Ha-
   waii. Proc. Am. Soc. Hortic. Sci. 47:262-270.
Medina-Filho, H.P. 1980.   Linkage of Aps-1, Mi and other markers on
   chromosome 6. Rep. Tomato Genet. Coop. 30:26-28.
Palacios, A.A. and Sosa Moss, C. 1972.   Resistencia genetica de algun-
   as variedades de tomate (*Lycopersicon esculentum*) al ataque de Meloidogyne spp. Agrociencia 9:119-125.
Pierce, L.C. and Brewbaker, J.C. 1973.   Applications of isozyme analysis in horticultural science. HortScience 8:17-22.
Rick, C.M. and Fobes, J. 1974.   Association of an allozyme with nematode resistance. Rep. Tomato Genet. Coop. 24:25.
Siddiqui, I.A., Sher, S.A., and French, A.M. 1973.   Distribution of plant
   parasitic nematodes in California. Department Food and Agriculture,
   Division Plant Industry, Sacramento, California.
Sidhu, G.S. and Webster, J.M. 1981.   The genetics of plant nematode
   parasitic systems. Bot. Rev. 47:387-419.

Smith, P.G. 1944. Embryo culture of a tomato species hybrid. Proc. Am. Soc. Hortic. Sci. 44:413–416.

Smithies, O. 1955. Zone electrophoresis in starch gels: Group variation in the serum proteins of normal human adults. Biochem. J. 61: 629–641.

Tanksley, S.D. 1979. An efficient and economical design for starch gel electrophoresis. Rep. Tomato Genet. Coop. 29:37–38.

_____ and Rick, C.M. 1980. Isozymic gene linkage map of the tomato: Applications in genetics and breeding. Theor. Appl. Genet. 57:161–170.

_____, Medina-Filho, H.P., and Rick, C.M. 1981a. The effect of isozyme selection on metric characters in an interspecific backcross of tomato: Basis of an early screening procedure. Theor. Appl. Genet. 60:291–296.

_____, Zamir, D., and Rick, C.M. 1981b. Evidence for extensive overlap of sporophytic and gametophytic gene expression in *Lycopersicon esculentum*. Science 213:453–455.

Taylor, A.L. and Sasser, J.N. 1978. Biology, identification and control of root-knot nematodes. Department Plant Pathology, North Carolina State Univ., Raleigh.

Torres, A.M., Diedenhofen, U., Bergh, B.O., and Knight, R.J. 1978a. Enzyme polymorphisms as genetic markers in the avocado. Am. J. Bot. 65:134–139.

Torres, A.M., Soost, R.K., and Diedenhofen, U. 1978b. Leaf isozymes as genetic markers in citrus. Am. J. Bot. 65:869–881.

Wang, E.L.H. and Bergeson, G.B. 1974. Biochemical changes in root exudate and xylem sap of tomato plants infected with *Meloidogyne incognita*. J. Nematol. 6:194–202.

Watts, V.M. 1947. The use of *Lycopersicon peruvianum* as a source of nematode resistance in tomatoes. Proc. Am. Soc. Hortic. Sci. 49: 233–234.

CHAPTER 34

Mycorrhizae

L.H. Rhodes

Vesicular-arbuscular (VA) mycorrhizae are fungus-root symbioses that occur in the vast majority of vascular land plants, including nearly all major crop plants. In the VA mycorrhizal association, fungi aid roots in the uptake of inorganic nutrients while obtaining essential organic nutrients and growing space from the plant host. When plants are deprived of this beneficial relationship they often grow poorly and die. Perhaps the most dramatic example of what can happen when the symbiotic fungi are eliminated has been the stunting and death of citrus seedlings following soil fumigation with methyl bromide. Although this problem was initially attributed to several other causes, including toxicity from residual bromine, it was ultimately found to be caused by lack of mycorrhiza formation (Kleinschmidt and Gerdemann, 1972). It is now widely accepted that many other practices can reduce VA fungus populations or diminish the effectiveness of mycorrhizae (Rhodes, 1981). Likewise, mycorrhizal fungus populations in some soils may be naturally low or dominated by ineffective species. For all of these situations it may be possible in the future to introduce mycorrhizal fungi into the soil to achieve maximum plant productivity.

With the realization of the importance of mycorrhizae for plant growth has come an interest in the production of mycorrhizal inoculum suitable for use in various crop production systems. Perhaps the principal limitation to inoculum production and use in agriculture has been the fact that VA fungi are obligate symbionts and thus cannot be produced in axenic culture. The most common means of producing mycorrhizal inoculum has been to grow the symbionts in association with a plant host in pot cultures. This technique was first used by Mosse (1953), who produced a pure (i.e., single species) inoculum of *Endogone* species (*Glomus mosseae*) on the roots of potted strawberry plants growing in sterilized soil in the greenhouse. Nearly 30 years later the technology for production of fungal symbiont cultures has not progressed far beyond this point. Although techniques for production of mycorrhizae in gnotobiotic cultures in sterilized media have been developed, utilization of these techniques has been almost nil. Surface-sterilized spores have been used in some research, but virtually no experiments have been done with aseptically produced inoculum.

It is surprising that two such rapidly expanding areas of research as those involving plant tissue culture and mycorrhiza have not found common ground, particularly since the feasibility of producing mycorrhizal root organ cultures has been clearly demonstrated (Mosse and Hepper, 1975). The purpose of this chapter, therefore, is to present some basic information about VA fungi and to summarize the information on the establishment and growth of these fungi under aseptic conditions. It is hoped that such information can provide a base from which the use of gnotobiotic culture systems can be refined and expanded.

VESICULAR-ARBUSCULAR FUNGI

Mychorrhizal Relationships

The fungi that form VA mycorrhizae grow within the root cortex and at the same time develop extensive mycelial networks in the soil. In the root cortex, hyphae grow both inter- and intracellularly and often produce extensively branched haustorium-like structures called arbuscules, and sac-like bodies called vesicles. There is no obvious change in gross morphology of the root, and thus the mycorrhizal symbiosis is usually overlooked in macroscopic examination of roots. However, a typical mycorrhizal plant probably has 50% or more of its feeder root length colonized by VA fungi. When one considers this figure with the fact that over 90% of all vascular land plant species normally form VA mycorrhizae, it is easy to see why Gerdemann (1968) has said that "the amount of plant tissue infected by this group must exceed that infected by any other group of fungi."

The principal benefit of the VA mycorrhizal symbiosis for the higher plant is an increased supply of phosphorus, which is taken up by hyphae outside the root, translocated to internal fungal structures, and ultimately released into cortical cells of the root. Details of this process have recently been reviewed (Rhodes and Gerdemann, 1980; Smith, 1980). As the phosphorus level of the soil or growth medium is increased, mycorrhizal infection is reduced. This apparently results from physiological changes accompanying high concentrations of phosphorus in plant tissue, rather than a direct effect of phosphorus on growth of the mycorrhizal fungus (Sanders, 1975; Ratnayake et al., 1978).

Taxonomy and Ecology

Vesicular-arbuscular mycorrhizal fungi belong to the genera *Gigaspora*, *Acaulospora*, *Glomus*, *Sclerocystis*, and *Entrophospora*, which are

included in the family Endogonaceae, order Mucorales, class Zygomycetes. Over 80 species within these genera have been described (Trappe,
1982). Although mycorrhizal relationships have not been defined for
several species, it is likely that most, if not all, form VA mycorrhizae.

Vesicular-arbuscular fungi are widely distributed in soils throughout
the world. In fact, it is virtually impossible to find a natural or cultivated soil in which VA fungi are not present. Although the fungi are
obligate symbionts, very little host specificity exists within the group.
It is safe to say that most, if not all, species of VA fungi have the
capacity to form mycorrhizae with several thousand higher plant species. For example, a small fraction of the economically important
plants that form mycorrhizae with *Glomus mosseae* includes corn, soybean, wheat, rice, cotton, alfalfa, clover, tobacco, potato, tomato, pepper, lettuce, onion, apple, pear, peach, cherry, grape, strawberry, petunia, geranium, chrysanthemum, maple, and redwood. Likewise, any one
of these plants is probably capable of forming mycorrhiza with any of
the 80 or more known species of VA fungi.

Reproduction

Although VA fungi have apparent affinities with the Zygomycetes,
none of them has been known to produce zygospores. Reproduction
occurs by the formation of thick-walled, asexually formed resting spores
which are referred to as chlamydospores in the genera *Glomus* and
Sclerocystis, and as azygospores in the genera *Gigaspora* and *Acaulospora*. The spores are among the largest fungal spores known, and in
fact, those produced by certain species of *Gigaspora* regularly exceed
400 µm in diameter and are probably the largest of all fungal spores.
All spores are single-celled, at least when initially formed, and contain
a considerable quantity of lipid in the form of large droplets.

Little is known of the nuclear condition of spores or of nuclear behavior in life cycles of VA fungi. Mosse (1970a) indicated that in the
"honey colored sessile" spore type (*Acaulospora laevis*) the pregermina-
tion spore and germ tubes produced by it are multinucleate. Several
nuclei are present in the interior portion of the spore as well as in
the peripheral (germination) compartments. Azygospores of *Gigaspora
margarita* are also multinucleate (Sward, 1981). Karyogamy or meiosis
has not been described for any VA fungi.

Spore Germination

Although VA fungi do not undergo a complete life cycle in the absence of a host, the spores will germinate when placed in a moist environment and hyphae will make varying amounts of growth on agar.

The mode of spore germination varies between genera and even between species within a genus. In most species of *Glomus*, germ tubes
emerge through the broken end of the subtending hypha. *Glomus palli-
dus* (Hall, 1977) and *Glomus albidus* (Walker and Rhodes, 1981) are exceptions to this general type of germination in *Glomus*. In these spe-

cies germ tubes emerge directly through the chlamydospore wall. *Gigaspora* and *Acaulospora* species also have germ tubes that emerge directly through the spore wall. In some *Gigaspora* and *Acaulospora* species a unique type of cytoplasmic compartmentalization occurs immediately prior to germination (Mosse, 1970a; Old et al., 1973; Hall, 1977). A compartment of cytoplasm, appearing crescent-shaped in optical section, is formed between the inner and outer spore walls, near the subtending hypha. The cytoplasm within this compartment is further divided by the deposition of several radially oriented walls. This results in the formation of several germination compartments, from which germ tubes emerge through the outer spore wall. This type of compartmentalization does not occur in *Gigaspora margarita* (Sward, 1981) and probably not in some other *Gigaspora* species. The modes of germination of spores of *Sclerocystis* spp. and *Entrophospora infrequens*, the only species in this genus, have not been described.

Spores stored in moist soil at 4-10 C can remain viable for several years, and when taken from refrigerated pot cultures, surface disinfested, and placed on agar, usually begin to germinate within a few days. However, Daniels and Duff (1978) found that three of four isolates of *Glomus mosseae* did not readily germinate on agar. When these same isolates were buried in a pasteurized soil mix in which a *Coleus* plant was growing, germination was 35-40%, suggesting that soil or root diffusates may have been necessary to break spore dormancy. Mosse (1959) found that a soil dialysate stimulated germination. This stimulatory effect was not obtained with dialysate from sterilized soil. Similar results were obtained by Daniels and Graham (1976); however, when similar or greater amounts of dialysate from chloropicrin-treated soil were also added to the agar, spore germination was completely inhibited. They concluded that excessive nutrients released during soil sterilization were responsible for inhibition of germination. This conclusion was supported by the fact that both potato dextrose agar and nutrient broth agar were inhibitory to spore germination at standard concentrations, but inhibition diminished as the level of nutrients in the medium was decreased. Similarly, Mosse and Hepper (1975) reported that germination of *Endogone* (*Glomus*) *mosseae* was 95% after 14 days on water agar, but only 12% or less when spores were incubated on White's medium (White, 1963), which contains relatively high nutrient levels.

The effect of pH on germination varies with fungal species used and probably also with ecotypes within the species. Green et al. (1976) found that over a range of pH values from 4 to 8 *Gigaspora coralloidea* germinated optimally at pH 5, whereas *Glomus mosseae* and *Gigaspora heterogama* germinated optimally at pH 6 and 7, respectively. At pH 9 *Glomus mosseae* germination was still greater than 40%, whereas no germination occurred in either *Gigaspora* species. At pH 4 the converse was true, with *G. coralloidea* germination above 40% and *Glomus mosseae* germination 0%.

In contrast to the dramatic effects of high P concentrations on root infection by VA fungi, effects of P on spore germination have been minimal or absent (Mosse and Phillips, 1971; Daniels and Trappe, 1980; Koske, 1981). This supports the hypothesis that inhibitory effects of P result from physiological changes in the plant rather than a direct effect of P on the fungus.

The ions Zn^{2+}, Mn^{2+}, and Cu^{2+} have been shown to be inhibitory to spore germination, even when present in minute amounts. Hepper and Smith (1976) found that Zn^{2+} and Mn^{2+} added to water agar at concentrations below 1 μg/g inhibited germination of one or more isolates of *Endogone* (*Glomus*) *mosseae*. Preparations of some commercial agars with relatively high concentrations of these ions were inhibitory to spore germination (Hepper, 1979). Further studies substantiated the toxicity of $MnSO_4$, $ZnSO_4$, or $CuSO_4$ on germination of *Glomus caledonius* spores on water agar (Hepper, 1979). The effects of heavy metal ions are not permanent, however. Spores that failed to germinate on heavy metal-amended agar for 28 days germinated normally when transferred to water agar (Hepper, 1979).

Hirrel (1981) found that germination of *Gigaspora margarita* was reduced when spores were exposed to various Na^+ and Cl^- salt solutions. Germination rate appeared to be affected more by Cl^- than by Na^+. Sodium chloride, KCl, $CaCl_2$, $NaNO_3$, and Na_2SO_4 at 2.14×10^{-1} M, reduced germination to 20, 0, 13, 25, and 47%, respectively, as compared to 100% germination in the water control. Ionic concentrations of Na^+ and Cl^- at or below 8.6×10^{-2} M had no measureable effect on germination. Ungerminated spores transferred to 0.01 glucose after 12 days exposure to salt solutions recovered germinability at various rates. Sodium chloride and KCl pretreatments result in slower recovery rates and lower maximum percentage of germination.

Cycloheximide greatly reduced germination of two isolates of *Glomus caledonius* at 0.7 μg/ml and completely inhibited germination at 1.4 μg/ml (Hepper, 1979). This suggests that protein synthesis is required for germination. In this respect, spore germination of *G. caledonius* is more characteristic of saprophytic fungi than of obligate parasites (Hepper, 1979).

Hyphal Growth in Culture

Germ tubes of VA fungi grow along the agar surface or into the agar and initially are straight, thin-walled, and infrequently branched. Under the microscope they appear hyaline to somewhat yellowish. The total length of hyphae produced by one spore or surface-sterilized root segment usually does not exceed more than a few millimeters. As growth continues, the walls of older hyphae thicken. Thin-walled branches arise at various points along the main hyphae. These branches often become vacuolated, septate, and collapsed shortly after their formation. If the medium is suitable for relatively extensive hyphal growth, new spores begin to appear within a few days (Mosse, 1959; Hepper, 1979). These spores apparently never reach maturity, and attempts to subculture from axenic cultures with or without such spores have consistently failed.

A rapid, bidirectional cytoplasmic streaming in germ tubes produced in agar culture is easily observed with the light microscope (Rhodes and Gerdemann, 1980). Cytoplasmic streaming is probably important for germ tube growth, since continued hyphal elongation seems to be dependent on translocation of a compound or compounds from spores to hyphal tips.   Very little attention has been given to this phenomenon, which is unusual considering the role that cytoplasmic streaming must play in the translocation of nutrients to plant roots.   The cytoplasmic streaming inhibitor, cytochalasin B, stopped cytoplasmic streaming and consequently P translocation to roots through external hyphae of *G. mosseae* on an agar medium (Cooper and Tinker, 1981).   Cytochalasin B may also prove useful in studies on the relationship between streaming and continued germ tube growth.

A phenomenon known as wound healing occurs in germ tubes of *Gigaspora* species (Gerdemann, 1955b).   This phenomenon has apparently not been observed in other genera of VA fungi.   It was not observed in *Glomus macrocarpus* under conditions where numerous sites of wound healing were seen in *Gigaspora margarita* (Shafer et al., 1981).   If a localized region of the hypha dies, regrowth occurs from each living hyphal end adjacent to the dead hyphal segment.   The two growing points are attracted to each other and in a matter of a few hours grow together and anastomose.   This kind of wound healing reestablish-es a bridge of living cytoplasm from the original growing tip of the germ tube to the parent spore.   Although the significance of this phe-nomenon has not been completely determined, it nevertheless indicates the importance of a cytoplasmic connection between the germ tube growing point and the germinated spore.   It may be inferred that a specific stored nutrient or growth factor must be continuously trans-located from the spore to the growing hyphal tip, at least until root infection takes place.

Another feature of growth in agar culture which has only been re-ported for *Gigaspora* spp. is hyphal geotropism (Watrud et al., 1978a).   Germ tubes of *Gigaspora margarita* on water agar grew upward (i.e., were negatively geotropic) no matter what the orientation of the germinating spore.   Hyphal branches from these germ tubes invariably grew downward.   Possible effects of oxygen and light were ruled out on the basis of several variations in the germination experiments.   Similar results were obtained with *G. gigantea* in attempts to establish mycor-rhizal root organ cultures (J.W. Gerdemann, personal communication).   Azygospores of *G. gigantea* were placed in close proximity to excised tomato roots.   Aerial germ tubes were produced from spores on the agar surface.   Additional hyphae often penetrated the agar, with no apparent attraction to living roots.   This type of growth pattern is not easily explained, particularly in light of the ultimate necessity for the fungus to make root contact.

Chemotropism does not appear to play a significant role in the growth of hyphae prior to infection.   Hyphae are not attracted to roots (Mosse, 1959; Mosse and Hepper, 1975; Hepper and Mosse, 1975).   Contact of roots by hyphae occurs more or less by chance, with the first roots contacted almost never serving as infection sites (Hepper and Mosse, 1975).   Although root contact by hyphae may induce more profuse hyphal growth (Mosse, 1959), hyphae may actually grow over several roots with no apparent stimulus provided by root contact (Mosse and Hepper, 1975).

The only evidence for chemotropism in VA fungi is the attraction which occurs between hyphae prior to anastomosis.   This is particularly evident in the wound-healing response of *Gigaspora* spp. (Gerdemann, 1955b). The hypha originating next to the distal end of the wounded hyphal segment grows closely appressed to the wounded (dead) segment. The hypha regenerated near the proximal end of the wound does not show this close attraction to the wounded segment, but grows away from it, in more or less the same direction as the original germ tube. When the two regenerated hyphae are in close proximity, the growing tips bend toward each other, and growth continues until the tips anastomose. This pattern suggests two chemotropic responses, namely, the attraction of one hypha to the wounded segment and the attraction of both wound-healing hyphae toward the opposite, actively growing hyphal tip. The possibility that the former response is thigmotropic rather than chemotropic cannot be excluded.

GNOTOBIOTIC CULTURES

Surface Decontamination of Spores

Establishment of aseptic cultures is dependent on removal of adherent bacteria and fungi from the surfaces of VA fungus spores. When inoculum is produced in soil it may take 2-3 months to establish an aseptic culture. An abundant microflora usually develops on spore surfaces. Pits, ridges, and other types of spore ornamentation, as well as cracking and flaking of the outer spore wall, often make surface sterilization difficult. Mosse (1959, 1962) has taken advantage of the protective nature of the peridium of a sporocarpic species (*Glomus mosseae*) to obtain relatively uncontaminated spores. Chlamydospores dissected from the sporocarps are relatively free of surface contaminants. However, this procedure can only be used for isolates of this species with easily separable peridia. Other isolates of *G. mosseae* often have tightly adhering peridia (Daniels and Duff, 1978) from which spores are not easily dissected. Another source of relatively contaminant-free spores are those of *Glomus epigaeus* present in the interior of large epigeous sporocarps (Daniels and Thorpe, 1979).

Fortunately, the large, thick-walled spores of VA fungi are relatively resistant to some common sterilant solutions. The two most frequently used sterilants have been 2% chloramine T plus 200 ppm streptomycin (Mosse, 1959; Hepper and Mosse, 1980) and 0.5% sodium hypochlorite (10% commercial bleach) (Gerdemann, 1955a). In the former method the length of time in the sterilant solution is approximately 20 min. In the latter method the time of sterilization is usually 1-3 min. In both methods the sterilant solution is drained off and spores are rinsed through one or more changes of sterile distilled water. Tommerup and Kidby (1980) examined viability of VA fungi and associated microorganisms following treatment with oxidizing agents, antibiotics, moist heat, ultrasonic radiation, and ultraviolet radiation. Chlorine from chloramine-T or sodium hypochlorite was effective in eliminating contaminating surface microorganisms without reducing viability of VA fungi; however, chlorine concentrations in commercial preparations of sodium hypochlorite were found to be highly variable. The method ultimately selected for routine use in spore decontamination was: immersion in 5% chloramine-T (0.42% chlorine, w/v) at 30 C for 20-40 min, followed by rinsing and incubation on media containing 100 μg/ml chloramphenicol. It should be noted however, that incubation of spores on antibiotic-amended media may only temporarily suppress bacteria and thus mask the potential for contamination when germinated spores are transferred to more complete media. This is particularly true when subsequent incubation periods are to be longer than a few days. In some cases it may be advisable to plate treated spores on a complete medium such as nutrient agar or potato dextrose agar for a period of 1-2 days, to ensure the absence of surface contaminants (e.g., Ross and Harper, 1970). Clean spores may then be aseptically transferred to water agar or other medium suitable for spore germination.

The fungi *Glomus caledonius* and *Acaulospora laevis* were found to be extremely resistant to UV radiation, with germination and hyphal growth being unaffected by exposure to 5×10^8 ergs/cm^2 (Tommerup and Kidby, 1980). Although it was judged impractical for routine laboratory procedures, UV radiation deserves further attention, considering its potential usefulness in selective disinfestation of VA fungus spores.

Attempts at Axenic Culture

There have probably been a great many more attempts to establish VA fungi in axenic culture than have actually been reported in the literature, since negative results are invariably obtained. Barret (1947, 1961) reported the isolation and culture of a *Rhizophagus* spp. (≡ *Glomus* (Gerdemann and Trappe, 1974)) using a hemp seed baiting technique. Although his procedures are straightforward, his experiments, which include synthesis of mycorrhizae with one of the fungal cultures, have never been repeated.

It has normally been assumed that cessation of hyphal growth in culture is not a function of depletion of food reserves in spores. Mosse (1959) determined that if a germ tube that had stopped growing was severed from the spore, the spore was still capable of regermination and could support the same amount of germ tube growth a second and even third time as that which had occurred previously. This suggested that self-inhibition of hyphal growth could be occurring.

This hypothesis has been investigated by experiments incorporating activated charcoal into an agar medium on which *Gigaspora margarita* spores had germinated (Watrud et al., 1978b). The extent of hyphal growth was significantly greater in charcoal agar as opposed to water agar plates. In further experiments placement of charcoal agar wells in center or peripheral positions in the plate indicated that the growth stimulation occurred when hyphae were in close proximity to the charcoal. Finally, pretreatment of spores or water with charcoal did not stimulate hyphal growth. Although possible nutritive effects of added charcoal cannot be positively excluded, these experiments nevertheless strongly support the hypothesis that self-inhibitory compounds are produced by growing germ tubes of *G. margarita* and can be absorbed by activated charcoal present in the media. Such results are similar to those obtained by Day and Anagnostakis (1971) for *Ustilago maydis*, an organism previously considered an obligate parasite.

The limited growth of hyphae from germinated spores may be enhanced or inhibited by addition of several substances to the culture medium. Dialysate from sonically disintegrated roots stimulated hyphal growth of an *Endogone* sp. (*Glomus mosseae*) on agar (Mosse, 1959). Similar results were obtained with dialysate from disintegrated *Chlorella* cultures but not with disintegrated shoot tissue. Tartaric acid as well as some other organic acids stimulated hyphal growth, but maltose, sucrose, and glucose were inhibitory (Mosse, 1959). In contrast, Hepper (1979) found no inhibitory effect of a number of sugars and sugar alcohols on growth of *Glomus caledonius* when these were present at 0.1% in a medium with pieces of boiled lima bean seed. The lack of inhibition may have been attributable to differences between *G. mosseae* and *G. caledonius*; however, the boiled lima bean seed was shown to be stimulatory to hyphal growth and may have overcome any inhibitory effect of the sugars. In addition to boiled seed fragments, Hepper (1979) found that peptone, yeast extract, and thiamine promoted growth of *G. caledonius*. Several metabolic inhibitors, including cycloheximide, actinomycin D, proflavine hemisulphate, 5-fluorouracil, and ethidium bromide, reduced hyphal growth.

Monoxenic or "Two-Member" Culture

Cultures of aseptically synthesized mycorrhizae fall into two categories: whole plant cultures and excised root (root organ) cultures. Both types of cultures can be considered gnotobiotic or monoxenic systems (Dougherty, 1960) and are referred to in this chapter as monoxenic or "two-member" cultures, the latter term having been used frequently for mycorrhizal systems. Approximately 12 studies have involved the aseptic synthesis of VA fungus-host plant culture (Table 1). The majority of cultures have been combinations involving clovers (*Trifolium* spp.) and *Glomus* spp., particularly *G. mosseae*.

The establishment of pure two-member cultures was first reported by Mosse (1962). Her results indicated, however, that appressorium formation and root penetration were much more likely to occur if a *Pseudomonas* species was present in the culture. Cell-free or autoclaved extracts from cultures in which *Pseudomonas* was present, pectinase, or EDTA also facilitated the establishment of mycorrhizal infections in some cases, but the most consistent of effects on the proportion of successful inoculations were provided by the presence of *Pseudomonas* or by chance bacterial contamination. It was therefore suggested that three organisms, i.e., fungus, plant, and bacterium, might be necessary for synthesis of mycorrhizae. The medium used in these experiments was originally designed for legume nodule culture (Jensen, 1942) and while low in nitrogen, had a relatively high phosphorus content. It was later suggested (Mosse and Phillips, 1971) that the difficulties encountered in establishing pure two-member cultures were attributable to the high P levels of the medium. Presumably, the bacterial contami-

Table 1. VA Mycorrhizal Fungus/Host Plant Cultures Synthesized and Maintained under Aseptic Conditions

PLANT/FUNGUS	TYPE OF CULTURE	MEDIUM	PHOSPHATE SOURCE	REFERENCE
Trifolium parviflorum/Glomus mosseae	Whole plant	Agar (Jensen, 1942)	$CaHPO_4$ + $KHPO_4$	Mosse, 1962
Glycine max/Glomus sp.	Whole plant	Autoclaved soil	Native soil P	Ross & Harper, 1970
Trifolium parviflorum/Glomus mosseae	Whole plant	Agar	$CaHPO_4$; Ca–, Na–, and Fe-phytate; phytin; DNA; inositol; lecithin; glucose-6-phosphate; charcoal	Mosse & Phillips, 1971
Trifolium pratense; T. parviflorum; Lycopersicon esculentum/Glomus mosseae	Excised root (root organ)	Agar (modified White's medium)	KH_2PO_4; $CaHPO_4$; Ca-phytate	Mosse & Hepper, 1975 Hepper & Mosse, 1975, 1980
Trifolium repens/Glomus mosseae	Whole plant	Soil agar/water agar (divided plates)	Native soil P	Pearson & Tinker, 1975
Trifolium repens/Glomus mosseae	Whole plant	Soil agar/water agar (divided plates)	Native soil P	Cooper & Tinker, 1981
Trifolium parviflorum; T. repens/Glomus mosseae; G. caledonius	Whole plant	Agar; nutrient-soaked paper; liquid	KH_2PO_4; bone meal	Hepper, 1981
Bouteloua gracilis/Glomus fasciculatus	Whole plant	Agar	Ca-phytate; NaH_2PO_4 NaH_2PO_4 + Ca-phytate	Allen et al., 1979, 1981
Trifolium repens/Gigispora margarita	Whole plant	Sand/nutrient solution	Hydroxyapatite	St. John et al., 1981
Trifolium parviflorum/Glomus caledonius	Whole plant	Circulating nutrient solution	KH_2PO_4	Macdonald, 1981

933

nants absorbed P from the medium, allowing infection to occur. It is now widely accepted that organisms other than a VA fungus and appropriate plant host are unnecessary for establishment of mycorrhizal infection in agar culture.

With the exception of minor differences in the salts used to provide inorganic nutrients, the components of various media used in whole plant-mycorrhiza culture experiments have been very similar (St. John et al., 1981; Hepper, 1981; Allen et al., 1979, 1981; Macdonald, 1981). The medium must provide inorganic nutrients for plant growth, with organic nutrients being obtained by the fungus primarily or entirely from the plant symbiont.

Of particular significance has been the source and concentration of phosphate in the medium. It is well known that high soil P concentrations result in low levels of mycorrhizal infection. This inhibitory effect of P is also evident in aseptic cultures. Mosse and Phillips (1971) found that a number of organic P sources or $CaHPO_4$ were adequate for development of VA infection in agar media, provided that P concentration was not excessive. Ca-phytate and DNA, in particular, resulted in increased intensity of infection and external mycelium growth.

Allen et al. (1981) incorporated various P sources into a basic medium so that the final concentration of P was 1.4 mM in all media. Mycorrhizal infection of *Bouteloua gracilis* was 75% when Ca-phytate was the sole P source, and less than one-third of this value when NaH_2PO_4 or NaH_2PO_4 + Ca-phytate was used. When NaH_2PO_4 + inositol or NaH_2PO_4 + Ca + inositol was incorporated into the basic medium, no mycorrhizal infection occurred. These results would tend to indicate the importance of P source; however, P concentrations given (1.4 mM) reflect total P added to the medium rather than soluble P available for uptake by roots or mycorrhizal fungi. Thus the high level of infection which occurred in the Ca-phytate medium may simply reflect the insolubility of this form of organic phosphate and resultant lower P concentration in the medium. This is substantiated by the considerably lower concentrations of P in leaves of plants in the medium containing Ca-phytate as the sole P source (Allen et al., 1981).

Although concentration rather than source of P may be a more important factor affecting level of mycorrhizal infection in aseptic culture, the use of organic P sources such as Ca-phytate may be an ideal way to provide a balanced, slowly released supply of P to the mycorrhizal cultures. Other slowly soluble forms of P such as hydroxyapatite (St. John et al., 1981), and bonemeal (Hepper, 1981) have also proven useful.

A modification of White's (1963) medium has been used for the establishment of root organ cultures (Mosse and Hepper, 1975; Hepper and Mosse, 1980). This medium contains essential inorganic nutrients (including KI) plus glycine, thiamine HCl, nicotinic acid, pyridoxine, and sucrose. Mosse and Hepper (1975) used divided plates to separate a complete medium (modified White's medium), which was satisfactory for maintenance of root organ cultures, and a low nutrient medium, which had previously been shown to be conducive to mycorrhizal infection. This design was found to be satisfactory, with good infections developing in the low nutrient medium. However, infections were no more extensive than in a 1:1 mixture of the two media.

Because of difficulties encountered in obtaining germinated spores in complete agar media that would otherwise support growth of whole plants or excised roots, pregerminated spores have often been used to synthesize mycorrhizae in culture (Hepper, 1981; Mosse and Hepper, 1975; Macdonald, 1981).

Application of Gnotobiotic Culture Techniques

It has been suggested that definitive work on taxonomy or physiology of VA fungi must await axenic culture of these organisms. However, the taxonomy of many of the largest and economically most important fungal groups, including the rusts, powdery mildews, downy mildews and the majority of basidiomycetes, is based on the characteristic morphology of these fungi in association with their hosts or in their natural habitats. Considerable advances have been made in the taxonomy of VA fungi during the past 20 years, and it seems extremely unlikely that axenic culture will substantially alter current taxonomic criteria. Likewise, physiological studies have elucidated many of the mechanisms whereby mycorrhizal fungi increase the supply of phosphate and other nutrients for the plant symbiont and obtain and utilize organic compounds. These studies, as well, will not be negated by research on two-member systems. Researchers have also contended that commercial utilization of VA fungi will not be feasible until it can be assured that mycorrhizal inoculum is free of contaminating microorganisms, particularly plant pathogens. Considering the widespread use of nonsterile propagation media, mulches, and organic amendments used by many commercial horticulturists, it is unlikely that the addition of microbiologically impure inoculum would be viewed by growers as a serious hazard, if the potential for benefit from inoculation could be adequately demonstrated.

The preceding disclaimers notwithstanding, there are several reasons for pursuing expansion of gnotobiotic culture techniques. These include: (1) improvements in the standardization and purity of inocula used in mycorrhizal research; (2) elimination of fungal hyperparasites; and (3) potential improvement in symbiotic effectiveness of VA fungi through genetic manipulation of selected isolates.

There can be no question as to the need for improvements in inoculum production for research purposes. Although there is nothing inherently inferior with the pot culture system, in practice, single species pot cultures often turn into mixed bags of two or more species as a result of incomplete soil sterilization, of soil splash from overhead watering, or perhaps occasionally from dust or insect movement of spores. If research is conducted without careful examination of inoculum before each experiment, erroneous results are certain to occur.

The second disadvantage of the use of pot cultures is that almost all such cultures are initiated with several spores, usually hand-picked from a field population. Although these spores are morphologically similar, they could have considerable genetic diversity. Since single-propagule aseptic cultures are feasible, it would seem preferable that inoculum be produced in this manner. The use of well-characterized single spore isolates free from contaminant microorganisms could provide a more accurate picture of the potential use of VA fungi in various agricultural systems.

Finally, the use of pot cultures eliminates or greatly reduces the exchange of research materials across international boundaries. Such exchange is vital to studies comparing symbiotic effectiveness of various isolates and would be facilitated by widespread adoption of aseptic culture techniques.

Glomus and *Gigaspora* species are frequently parasitized by other fungi (Daniels and Menge, 1980; Ross and Ruttencutter, 1977; Schenck and Nicolson, 1977) and by mycophagous nematodes (Shafer et al., 1981). An *Endogone* species (*Acaulospora laevis*) was found to harbor bacteria-like organisms, but the pathogenicity of these organisms was not determined (Mosse, 1970b). Pot cultures of *Gigaspora* species tend to lose viability with time, which may be accounted for by hyperparasitism. Production of inocula under aseptic conditions could eliminate hyperparasites and allow for increased longevity of mycorrhizal inoculum.

The use of two-member cultures could also provide a system for screening various species and isolates of VA fungi for susceptibility to hyperparasitism. Although hyperparasitism has not been considered in experiments comparing the symbiotic effectiveness of different species of VA fungi, there is undoubtedly considerable variation in their susceptibility to these pathogens. Screening for disease resistance could become as important for improvement of fungal symbionts as it currently is for crop plants.

The anastomoses between germ tubes of different parent spores in *Glomus* species (Mosse, 1959, 1962) provide avenues for movement of nuclei, and although such nuclear migration has not been reported, it almost certainly occurs. Thus a form of parasexualism probably exists in VA fungi, and utilization of this phenomenon could allow for mating of selected isolates. Such work necessitates growth of the fungi in a relatively clear medium where the point of anastomosis can be easily observed. The use of root organ cultures could be particularly advantageous for this type of fungal hybridization. By bringing together two root pieces infected by two isolates of, e.g., *G. mosseae*, spores produced in external hyphae where anastomoses have clearly occurred could be used to inoculate new plants.

In addition to the potential for breeding improved isolates of VA fungi, the property of hyphal anastomosis could be useful for taxonomic purposes. For example, a widely used classification of *Rhizoctonia* is based on anastomosis with known isolates from four anastomosing groups (Parmeter, 1969). Mosse (1961) first suggested the use of this technique for VA fungi; however, there does not appear to have been any use of the method to date.

Currently, mycorrhizal fungi can be cultured only on the cortical tissue of roots. These roots may be separated from the whole plant, as in root organ cultures, but as yet no one has successfully subdivided or modified this unit, i.e., the root cortex, as a food base. The fact that VA fungi have such a high degree of specificity for the root cortex, yet readily cross the boundaries between orders and even classes of plants as long as this requirement is met, points out the importance of this particular type of parenchymatous tissue for their growth and development. Lesion nematodes (*Pratylenchus* spp.) are also obligate parasites of higher plants, whose nutritional base is the root cortex, yet those organisms are now produced in high populations under aseptic conditions on a routine basis using modified callus cultures (Chapter 32). Likewise, Beardmore and Pegg (1981) have recently synthesized orchid mycorrhizae in undifferentiated masses of orchid tissue maintained on agar media. Thus in the future it may be possible to use callus or modified callus cultures to produce VA fungus inoculum.

Only a few of the multitude of possible host-fungus combinations have been used in syntheses of two-member cultures. Consequently, much remains to be learned about the growth and development of various species in monoxenic culture. Certainly the enormous potential for use of VA mycorrhizal fungi to increase crop productivity should lend impetus to increased research activity in the gnotobiotic culturing of these organisms.

KEY REFERENCES

Gerdemann, J.W. 1968. Vesicular-arbuscular mycorrhiza and plant growth. Annu. Rev. Phytopathol. 6:397-418.

Hepper, C.M. 1979. Germination and growth of *Glomus caledonius* spores: The effects of inhibitors and nutrients. Soil Biol. Biochem. 11:269-277.

Mosse, B. 1953. Fructifications associated with mycorrhizal strawberry roots. Nature 171:974.

Rhodes, L.H. and Gerdemann, J.W. 1980. Nutrient translocation in vesicular-arbuscular mycorrhizae. In: Cellular Interactions in Symbiosis and Parasitism (C.B. Cook, P.W. Pappas, and E.D. Rudolph, eds.) pp. 173-195. Ohio State Univ. Press, Columbus.

REFERENCES

Allen, M.F., Moore, T.S., Christensen, M., and Stanton, N. 1979. Growth of vesicular-arbuscular-mycorrhizal and nonmycorrhizal *Bouteloua gracilis*. New Phytol. 87:687-694.

Allen, M.F., Sexton, J.C., Moore, T.S., Jr., and Christensen, M. 1981. Influence of phosphate source on vesicular-arbuscular mycorrhizae of *Bouteloua gracilis*. New Phytol. 87:687-694.

Barrett, J.T. 1947. Observations on the root endophyte, *Rhizophagus*, in culture. Phytopathology 37:359-360 (Abstract).

_____ 1961. Isolation, culture, and host relation of the phycomyce-toid vesicular-arbuscular mycorrhizal endophyte *Rhizophagus*. In: Recent Advances in Botany, Vol. 2, Univ. Toronto Press, Toronto.

Beardmore, J. and Pegg, G.F. 1981. A technique for the establishment of mycorrhizal infection in orchid tissue grown in aseptic culture. New Phytol. 87:527-535.

Cooper, K.M. and Tinker, P.B. 1981. Translocation and transfer of nutrients in vesicular-arbuscular mycorrhizas. IV. Effect of environmental variables on movement of phosphorus. New Phytol. 88:327-339.

Daniels, B.A. and Duff, D.M. 1978. Variation in germination and spore morphology among four isolates of *Glomus mosseae*. Mycologia 70: 1261-1267.

Daniels, B.A. and Graham, S.O. 1976. Effects of nutrition and soil extracts on germination of *Glomus mosseae* spores. Mycologia 68: 108-116.

Daniels, B.A. and Menge, J.A. 1980. Hyperparasitization of vesicular-arbuscular mycorrhizal fungi. Phytopathology 70:584-588.

Daniels, B.A. and Trappe, J.M. 1979. *Glomus epigaeus* sp. nov., a useful fungus for vesicular-arbuscular mycorrhizal research. Can. J. Bot. 57:539-542.

_____ 1980. Factors affecting spore germination of vesicular-arbuscular mycorrhizal fungus, *Glomus epigaeus*. Mycologia 72:457-471.

Day, P.R. and Anagnostakis, S.L. 1971. Corn smut dikaryon in culture. Nature 231:19-20.

Dougherty, E.C. 1960. Cultivation of aschelminths, especially rhabditoid nematodes. In: Nematology (J.N. Sasser and W.R. Jenkins, eds.) pp. 297-318. Univ. North Carolina Press, Chapel Hill.

Gerdemann, J.W. 1955a. Relation of a large soil-borne spore to phycomycetous mycorrhizal infections. Mycologia 47:619-632.

_____ 1955b. Wound-healing of hyphae in a phycomycetous mycorrhizal fungus. Mycologia 47:916-918.

_____ and Trappe, J.M. 1974. The Endogonaceae in the Pacific Northwest. Mycologia Memoir 5:1-75.

Green, N.E., Graham, S.O., and Schenck, N.C. 1976. The influence of pH on the germination of vesicular-arbuscular mycorrhizal spores. Mycologia 68:929-934.

Hall, I.R. 1977. Species and mycorrhizal infections of New Zealand Endogonaceae. Trans. Br. Mycol. Soc. 68:341-356.

Hepper, C.M. 1981. Techniques for studying the infection of plants by vesicular-arbuscular mycorrhizal fungi under axenic conditions. New Phytol. 88:641-647.

_____ and Mosse, B. 1975. Techniques used to study the interaction between *Endogone* and plant roots. In: Endomycorrhizas (F.E. Sanders, B. Mosse, and P.B. Tinker, eds.) pp. 65-75. Academic Press, New York.

_____ and Mosse, B. 1980. Vesicular-arbuscular mycorrhiza in root organ cultures. In: Tissue Culture Methods for Plant Phytologists (D.S. Ingram and J.P. Helgeson, eds.) pp. 167-171. Blackwell Sci. Pub., Oxford.

_____ and Smith, G.A. 1976. Observations on the germination of Endogone spores. Trans. Br. Mycol. Soc. 66:189-194.

Hirrel, M.C. 1981. The effect of sodium and chloride salts on the germination of *Gigaspora margarita*. Mycologia 73:610-617.

Jensen, H.L. 1942. Nitrogen fixation in leguminous plants. II. Is symbiotic nitrogen fixation influenced by Azotobacter? Proc. Linn. Soc. NSW 67:205–212.

Kleinschmidt, G.D. and Gerdemann, J.W. 1972. Stunting of citrus seedlings in fumigated nursery soils related to the absence of endomycorrhizae. Phytopathology 62:1447–1453.

Koske, R.E. 1981. *Gigaspora gigantea*: Observations on spore germination of a VA-mycorrhizal fungus. Mycologia 73:288–300.

Macdonald, R.M. 1981. Routine production of axenic vesicular-arbuscular mycorrhizas. New Phytol. 89:87–93.

Mosse, B. 1959. The regular germination or resting spores and some observations on the growth requirements of an *Endogone* sp. causing vesicular-arbuscular mycorrhiza. Trans. Br. Mycol. Soc. 42:273–286.

_____ 1961. Experimental techniques for obtaining a pure inoculum of of an *Endogone* sp., and some observations on the vesicular-arbuscular infections caused by it and by other fungi. In: Recent Advances in Botany, Vol. 2, pp. 1728–1732. Univ. Toronto Press, Toronto.

_____ 1962. The establishment of vesicular-arbuscular mycorrhiza under aseptic conditions. J. Gen. Microbiol. 27:509–520.

_____ 1970a. Honey-coloured, sessile *Endogone* spores. I. Life history. Arch. Mikrobiol. 70:167–175.

_____ 1970b. Honey-coloured, sessile *Endogone* spores. II. Changes in fine structure during spore development. Arch Mikrobiol. 74:129–145.

_____ and Hepper, C. 1975. Vesicular-arbuscular mycorrhizal infections in root organ cultures. Physiol. Plant Pathol. 5:215–223.

_____ and Phillips, J.M. 1971. The influence of phosphate and other nutrients on the development of vesicular-arbuscular mycorrhiza in culture. J. Gen. Microbiol. 59:157–166.

Olds, K.M., Nicolson, T.H., and Redhead, J.F. 1973. A new species of mycorrhizal *Endogone* from Nigeria with a distinctive spore wall. New Phytol. 72:817–823.

Parmeter, J.R. 1969. Anastomosis grouping among isolates of *Thanatephorus cucmeris*. Phytopathology 59:1270–1278.

Pearson, V. and Tinker, P.B. 1975. Measurement of phosphorus fluxes in the external hyphae of endomycorrhizas. In: Endomycorrhizas (F.E. Sanders, B. Mosse, and J.B. Tinker, eds.) pp. 277–287. Academic Press, London.

Ratnayake, M., Leonard, R.T., and Menge, J.A. 1978. Root exudation in relation to supply of phosphorus and its possible relevance to mycorrhizal formation. New Phytol. 81:543–552.

Rhodes, L.H. 1981. The use of mycorrhizae in crop production systems. Outlook Agric. 10:275–281.

Ross, J.P. and Harper, J.A. 1970. Effect of *Endogone* mycorrhiza on soybean yields. Phytopathology 70:1552–1556.

Ross, J.P. and Ruttencutter, R. 1977. Population dynamics of two vesicular-arbuscular endomycorrhizal fungi and the role of hyperparasitic fungi. Phytopathology 67:490–496.

Sanders, F.E. 1975. The effect of foliar-applied phosphate on the mycorrhizal infection of onion roots. In: Endomycorrhizas (F.E. Sanders, B. Mosse, and P.B. Tinker, eds.) pp. 261–276. Academic Press, London.

Schenck, N.C. and Nicolson, T.H. 1977. A zoosporic fungus occurring on species of *Gigaspora margarita* and other vesicular-arbuscular mycorrhizal fungi. Mycologia 69:1049-1053.

Shafer, S.R., Rhodes, L.H., and Riedel, R.M. 1981. Feeding of a mycophagous nematode, *Aphelenchoides bicaudatus*, on vesicular-arbuscular mycorrhizal fungi. Fifth North American Conference on Mycorrhizae, Programs and Abstracts, p. 67. Universite Leval, Leval, Quebec.

Smith, S.E. 1980. Mycorrhizas of autotrophic higher plants. Biol. Rev. 55:475-510.

St. John, T.V., Hays, R.I., and Reid, C.P.P. 1981. A new method for producing pure vesicular-arbuscular mycorrhiza-host cultures without specialized media. New Phytol. 89:81-86.

Sward, R.J. 1981. The structure of spores of *Gigaspora margarita*. II. Changes accompanying germination. New Phytol. 88:661-666.

Tommerup, I.C. and Kidby, D.K. 1980. Production of aseptic spores of vesicular-arbuscular endophytes and their viability after chemical and physical stress. App. Environ. Microbiol. 39:1111-1119.

Trappe, J.M. 1982. Synoptic keys to the genera and species of Zygomycetous (vesicular-arbuscular) mycorrhizal fungi. Phytopathology 72:1102-1108.

Walker, C. and Rhodes, L.H. 1981. *Glomus albidus*: A new species in the Endogonaceae. Mycotaxon 12:509-514.

Watrud, L.S., Heithaus, J.J., III, and Jaworski, E.G. 1978a. Geotropism in the endomycorrhizal fungus *Gigaspora margarita*. Mycologia 70:449-452.

_____ 1978b. Evidence for production of inhibitor by the vesicular-arbuscular-mycorrhizal fungus *Gigaspora margarita*. Mycologia 70:821-828.

White, P.R. 1963. Cultivation of Animal and Plant Cells, 2nd ed. Ronald Press, New York.

CHAPTER 35
Evaluation of Coffee Rust

M. Alves de Lima and *P.O. Larsen*

Cultures of *Hemileia vastatrix* Berk. & Br. on coffee leaf disks have been used to study some of the characteristics of host-pathogen interaction (Costa et al., 1978; Eskes, 1977; Narasimhswamy et al., 1961; Nutman and Roberts, 1963). The same resistance reaction to *H. vastatrix* observed on field-grown coffee plants is also expressed on leaf disks, allowing studies on the selection of resistant germplasm to be conducted under laboratory conditions (Costa et al., 1978). Tissue culture technology provides for the establishment of an in vitro coffee leaf disk-rust fungus association which is free of microbe contaminants. This in vitro host-parasite association permits maintenance of isolated races of *H. vastatrix* as well as providing an environment to make accurate physiological studies pertaining to *H. vastatrix*-plant host interaction.

Obtaining aseptic inoculum has been considered a limiting factor to the establishment of in vitro host-obligate parasite systems (Brian, 1967; Ingram, 1977; Ingram and Tommerup, 1973; Yarwood, 1956). Uredospores free from surface contaminants have usually been obtained by surface sterilizing infected plant parts at an early stage of uredosorum development, shortly before spores are freed from uredosori (Bushnell, 1968; Coffey et al., 1970; Coffey and Shaw, 1969; Coffey and Allen, 1973; Jones, 1972; Kuhl et al., 1971; Maheshwary et al., 1967; Singleton and Young, 1968; Turel, 1969; Turel and Ledingham, 1957; Williams et al., 1966). This chapter discusses a procedure that can be used to maintain pure cultures of *Hemileia vastatrix* on coffee leaf explants maintained in vitro.

Aseptic *H. vastatrix* uredospores can be obtained only from mature lesions on leaf explants excised from surface sterilized coffee leaves. A mature lesion is defined as a lesion that has developed beyond the yellow fleck stage with recognizable uredospores present in the center of the lesion. This procedure differs from those previously used to obtain aseptic uredospores of rust fungal species such as *Puccinia* spp. (Bushnell, 1968; Coffey and Shaw, 1969; Coffey and Allen, 1973; Kuhl et al., 1971; Maheshwari et al., 1967; Singleton and Young, 1968; Williams et al., 1966), *Melampsora lini* (Coffey et al., 1970; Turel, 1969; Turel and Ledingham, 1957) and *Uromyces dianthi* (Jones, 1972) in that they include surface sterilization of leaves having lesions at the yellow fleck stage. Morphological studies of *H. vastatrix* uredosori (Gopal-

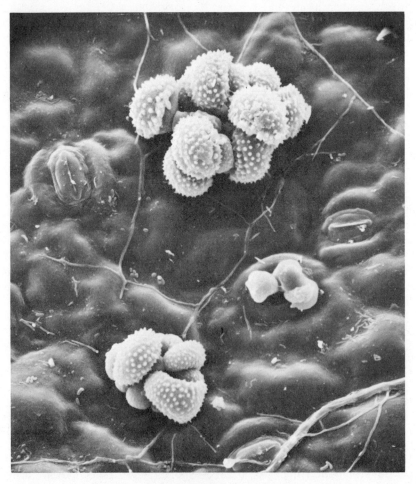

Figure 1.   Scanning electron micrograph of *Hemileia vastatrix* Berk. & Br. uredospore on *Coffea arabica* L. leaf (800x).   (a) Uredosorus with young immature spores protruding at the stomate opening; (b) Uredosorus with mature spores packed at the stomate opening.

krishna, 1951) have shown that horizontal hyphae extend immediately beneath the epidermis forming clusters or fascicles in the substomatal spaces. These clusters or fascicles pass through the stomatal opening and produce spores, either directly at the tips of hyphae or on specialized basal cells. Sporulation occurs only outside the host, without rupturing the guard cells or the adjacent epidermis. Uredia in *Puccinia*, *Uromyces*, and *Melampsora* species are erumpent (Cummins, 1959). Therefore, when lesions caused by those fungi are sterilized at the yellow fleck stage, sporogenous cells are still protected by the epidermis. In the case of *H. vastatrix* one may speculate that when sterilization treatment is conducted at the yellow fleck stage the hyphal tips or the specialized basal cells are directly exposed to the sterilizing solution and are likely to be damaged, thereby stopping the sporulation process. On the other hand, when the sorus is mature, the cluster of uredospores formed (Fig. 1) apparently acts as a barrier to the sterilizing agent in protecting the sporogenous cells.

PROTOCOLS

Leaves with mature coffee rust lesions from greenhouse-grown coffee plants are surface sterilized with sodium hypochlorite from which leaf explants containing lesions are excised. The explants are bioassayed for contamination by plating on potato dextrose agar (PDA) and then transferred to French square bottles containing water agar medium. When lesions resume sporulation, uredospore samples are collected and transferred to PDA plates to bioassay for contamination and viability (Fig. 2). Viable spores that prove to be aseptic are transferred to disinfected coffee leaf explants maintained in French square bottles on water agar medium. As new lesions develop from the inoculations, it is possible to maintain a continuous source of aseptic inoculum by monthly transfers of uredospores from mature lesions to sterile coffee leaf explants. The uredospores are routinely checked for contamination and viability at each transfer (Fig. 3).

A detailed stepwise description of the procedure followed to obtain pure cultures of *Hemileia vastatrix* on *Coffea arabica* leaf explants is provided below.

Coffee Plant Inoculation

1. Transfer uredospores with a camel's hair brush to the underside of the second and third youngest pairs of leaves of greenhouse-grown coffee plants.
2. Mist the plants with water using an atomizer, and seal the plants in plastic bags to maintain free water on the undersides of leaf surfaces.
3. Incubate the plants at 22-24 C for 72 hr in the dark.
4. Remove the bags and transfer plants to greenhouse benches until mature lesions develop, which occurs within 4-7 weeks.

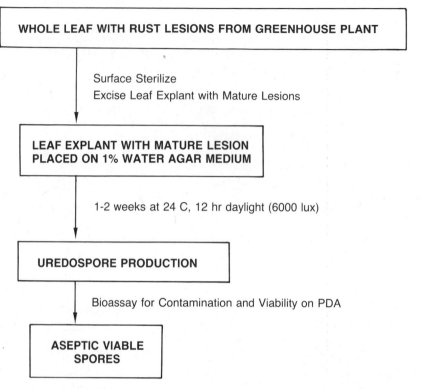

Figure 2.   Schematic representation of procedure followed to obtain aseptic uredospores of *Hemileia vastatrix* Berk. & Br. produced on *Coffea arabica* L. plants.

Surface Sterilization of Leaves with Mature Lesions

1.   Select leaves with mature lesions from inoculated plants.
2.   Wash the leaves in tapwater followed by double distilled water.
(The following steps should be performed under a laminar flow, filtered air hood.)
3.   Surface sterilize leaves in 1% sodium hypochlorite for 20 min (20% commercial bleach).
4.   Rinse the leaves three times in sterile double distilled water.
5.   Excise leaf explants with mature leasions using a sterile blade, leaving about 3 mm of surrounding healthy tissue.

Bioassay for Contamination on Leaf Explants

1.   Incubate excised leaf explants with mature lesions for 48 hr at 25 C in Petri dishes containing potato dextrose agar.
2.   Transfer uncontaminated leaf explants to 30 ml French square bottles containing 10 ml of 1% water agar medium.

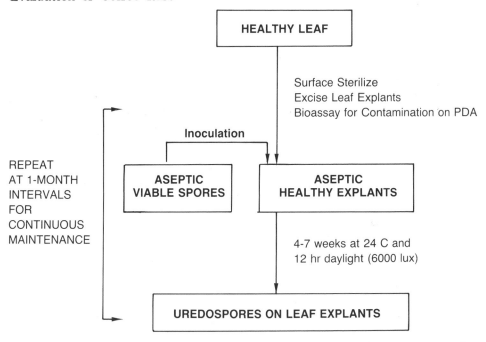

Figure 3. Schematic representation of the procedure followed to maintain *Hemileia vastatrix* Berk. & Br. pure cultures on *Coffea arabica* L. leaf explants.

Bioassay for Uredospore Contamination and Viability

1. When lesions on leaf explants transferred to water agar medium resume sporulation, collect a uredospore sample from each lesion with the tip of a sterile dissecting needle and transfer to potato dextrose agar plates.
2. Incubate transferred uredospore samples for 48 hr at 22-24 C in the dark.
3. Examine uredospores for contamination.
4. Spore germination is based on microscopic examination (100x magnification) of 100 spores from each sample. Spores are considered germinated when at least one of the five possible tubes exceeds 5 μm.

Leaf Explant Inoculation Under Aseptic Conditions

1. Surface sterilize healthy coffee leaves as above.
2. Excise leaf explants with a sterile blade, and bioassay for contamination as described above.
3. Transfer aseptic and viable uredospores from the lesion bioassayed with a sterile dissecting needle to healthy, sterile coffee leaf explants on water agar in French square bottles.

4. Incubate the inoculated explants at 22-24 C in the dark for 72 hr,
    then transfer them to conditions providing about 12 hr light (6000
    lux) daily.

Maintenance of Cultures

1. Uredospores from lesions produced on inoculated explants (above)
    are transferred to healthy sterile coffee leaf explants at 1 month
    intervals to provide a continuous supply of aseptic uredospores.
2. Uredospore samples are routinely bioassayed for contamination and
    viability as described above to ensure maintenance of monoxenic
    cultures.

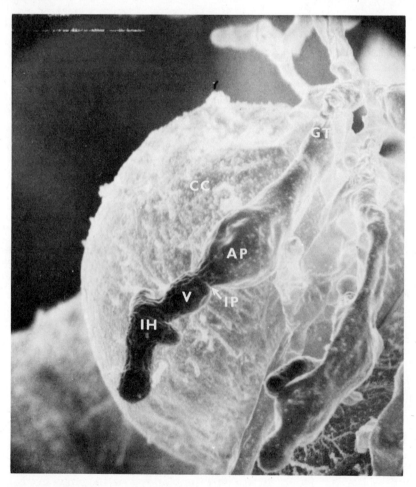

Figure 4.   Scanning electron micrograph of infection structures formed
by *Hemileia vastatrix* Berk. & Br. uredospore on an elongated paren-
chyma-like coffee cell (CC).   Note the appressorium (AP), infection peg
(IP) (arrow), vesicle (V), and germ tube (GT) with infection hypha (IH)
(2290x).

FUTURE PROSPECTS OF IN VITRO SELECTION OF COFFEE RUST-RESISTANT PLANTS

Disease-resistant plants have been regenerated from cells selected in culture for resistance to toxins produced by plant pathogens (Brettell and Ingram, 1979). Selection and regeneration of resistant cell lines using dual cultures of fungi and host callus tissue is another possibility to be considered. This has been proposed for *Coffea* breeding programs seeking resistance to *Hemileia vastatrix* (Sondahl and Sharp, 1977c). It was suggested that coffee plants regenerated from rust-resistant leaf explant cells from dual cultures (Sondahl and Sharp, 1977a,b) could provide a source for development of plants resistant to the coffee rust. Techniques to obtain dual cultures of obligately parasitic fungi, other than rust fungi with callus cultures, were recently reviewed (Buczacki, 1980; Ingram, 1980; Webb and Gay, 1980). Similar information is not available relating to dual cultures of rust fungi using callus cultures. Attempts to infect *Coffea arabica* callus tissue with *H. vastatrix* were not successful (Alves de Lima, 1980). Upon inoculation of callus, uredospores formed germ tubes, appressoria, infection pegs, vesicles, and infection hyphae on the callus surface but penetration did not subsequently occur (Fig. 4). Composition of the callus growth medium should be examined, since variation in combinations of growth hormones and some inorganic salts have been shown to influence rust infection of *Puccinia graminis* f. sp. *tritici* (Wang, 1959).

KEY REFERENCES

Brettell, R.I.S. and Ingram, D.A. 1979. Tissue culture in the production of novel disease-resistant crop plants. Biol. Rev. 54:329-345.

Ingram, D.S. 1977. Applications in plant pathology. In: Plant Tissue and Cell Culture (H.E. Street, ed.) pp. 463-500. Univ. California Press, Berkeley and Los Angeles.

Maheshwari, R., Hildebrandt, A.C., And Allen, P.J. 1967. Factors affecting the growth of rust fungi on host tissue cultures. Bot. Gaz. 128:153-159.

REFERENCES

Alves de Lima, M.M. 1980. Germination, appressorium formation and aseptic culture of *Hemileia vastatrix* on *Coffea arabica*. M.S. Thesis, Ohio State Univ., Columbus.

Brian, P.W. 1967. Obligate parasitism in fungi. Proc. R. Soc. London Ser. B 168:101-118.

Buczacki, S.T. 1980. Culture of *Plasmodiophora brassicae* in host callus tissue. In: Tissue Culture Methods for Plant Pathologists (D.S. Ingram and J.P. Helgeson, eds.) pp. 145-149. Blackwell Sci. Pub., Oxford.

Bushnell, W.R. 1968. In vitro development of an Australian isolate of *Puccinia graminis* f. sp. tritici. Phytopathology 58:526-527.

Coffey, M.D. and Allen, P.J. 1973.   Nutrition of *Melampsora lini* and *Puccinia helianthi*.   Trans. Br. Mycol. Soc. 60:245–260.

Coffey, M.D. and Shaw, M. 1969.   In vitro growth of gelatin suspensions of uredospores of *Puccinia graminis* f. sp. tritici.   Can. J. Bot. 47: 1291–1293.

Coffey, M.D., Bose, A., and Shaw, M. 1970.   In vitro culture of the flax rust, *Melampsora lini*.   Can. J. Bot. 48:773–776.

Costa, W.M., Eskes, A., and Ribeiro, I.J.A. 1978.   Avaliacao do nivel de resistencia do cafeeiro.   Bragantia 37:22–29.

Cummins, G.A. 1959.   Illustrated Genera of Rust Fungi.   Burgess Pub., Minneapolis.

Eskes, A.B. 1977.   Uso de discos de folhas para avaliar a resistencia do cafeeiro a *Hemileia vastatrix*:   Efeito da luminosidade e concentracao de inoculo.   V Congresso Brasileiro de Pesquisas Cafeeiras.   16–21 de outubro de 1977.   Guarapari, ES, Brasil.

Gopalkrishna, K.S. 1951.   Notes on the morphology of the genus *Hemileia*.   Mycologia 43:271–382.

Ingram, D.S. 1980.   The establishment of dual cultures of downy mildew fungi and their hosts.   In: Tissue Culture Methods for Plant Pathologists (D.S. Ingram and J.P. Helgeson, eds.) pp. 139–144.   Blackwell Sci. Pub., Oxford.

_____ and Tommerup, I.C. 1973.   The study of obligate parasites in vitro.   In: Fungal Pathogenicity and the Plant's Response (R.J.W. Byrde and C.V. Cutting, eds.) pp. 121–140.   Academic Press, London and New York.

Jones, D.R. 1972.   In vitro culture of carnation rust, *Uromyces dianthi*.   Trans. Br. Mycol. Soc. 58:29–36.

Kuhl, J.L., Maclean, D.J., Scott, K.J., and Williams, P.G. 1971.   The axenic culture of *Puccinia graminis* from uredospores: Experiments on nutrition and variation.   Can. J. Bot. 49:201–209.

Narasimhswamy, R.L., Nambiar, K.K.N., and Sreenivasan, M.S. 1961.   Report on the work of testing races of leaf disease fungus on coffee selections at Coffee Research Station Balehonnur.   Indian Coffee 25: 333–336.

Nutman, F.J. and Roberts, F.M. 1963.   Studies on the biology of *Hemileia vastatrix* Berk. & Br.   Trans. Br. Mycol. Soc. 46:27–48.

Singleton, L.L. and Young, H.C. 1968.   The in vitro culture of *Puccinia recondita* f. sp. tritici.   Phytopathology 58:1068 (Abstract).

Sondahl, M.R. and Sharp, W.R. 1977a.   High frequency induction of somatic embryos in cultured leaf explants of *Coffea arabica* L.   Z. Pflanzenphysiol. 81:395–408.

_____ 1977b.   Growth and embryogenesis in leaf tissues of *Coffea*.   Plant Physiol. 69:1.

_____ 1977c.   Research in *Coffea* spp. and applications of tissue culture methods.   In: Plant Cell and Tissue Culture (W.R. Sharp, P.O. Larsen, E.F. Paddock, and V. Raghavan, eds.) pp. 527–584.   Ohio State Univ. Press, Columbus.

Turel, F.L.M. 1969.   Saprophytic development of the flax rust *Melampsora lini*, race no. 3.   Can. J. Bot. 47:821–823.

_____ and Ledingham, G.A. 1957.   Production of aerial mycelium and uredospores by *Melampsora lini* (Pers.) Lev. on flax leaves in tissue culture.   Can. J. Microbiol. 3:813–819.

Wang, D. 1959. Effect of benzimidazole, dimethylbenzimidazole, glucose, and metal ions on the development of *Puccinia graminis tritici* on detached leaves of Khapli wheat. Can. J. Bot. 37:239–244.

Webb, K.J. and Gay, J.L. 1980. The recalcitrant powdery mildews— attempts to infect cultured tissues. In: Tissue Methods for Plant Pathologists (D.S. Ingram and J.P. Helgeson, eds.) pp. 151–159. Blackwell Sci. Pub., Oxford.

Williams, P.G., Scott, K.J., and Kuhl, J.L. 1966. Vegetative growth of *Puccinia graminis* sp. f. tritici in vitro. Phytopathology 56:1418–1419.

Yarwood, C.E. 1956. Obligate parasitism. Annu. Rev. Plant Physiol. 7: 115–152.

Species Index

Acacia koa, 29-30, 33
A. senegal, 730
Acaulospora, 927
A. laevis, 926, 931, 931
Acer pseudoplatanus, 396-397,
    400, 402, 403, 407, 791,
    796
Achromobacter, 839
Aegilops caudata x A. umbellu-
    lata, 232
Aegopodium podagraria, 329, 336
Aerva tomentosa, 251
Aesculus hippocastanum, 232
Agave, 52, 56-57
Agrobacterium tumefaciens,
    845
   Ti plasmid, 343, 502, 503,
      504, 507, 511, 516, 526,
      845
Agropyron repens, 232
Albizzia lebbek, 30, 89, 93
Alhagi camelorum, 29-30
Allium, 551
A. cepa, 52-53, 131, 259, 681,
    730, 733-734, 735, 738-
    742, 750

A. porrum, 53
A. sativum, 52-53, 94, 180, 559
Alnus, 837
A. glutinosa, 681
Aloe pretoriensis, 53
Alternaria brassicae, 881
Ammi major, 86
A. visnaga, 503, 504
Amoracia lapathifolia, 27-28
Anabaena azollae, 837
Anagallis arvensis, 40
Ananas cosmosus, 750, 757
A. sativus, 56
Andropogon gerardii, 46
Anemone, 232, 259
Anethum graveolens, 86
Angelica sylvestris, 730
Angiozanthos, 180, 207, 208
Anthemisia nobilis, 730, 732, 736
Anthurium andraeanum, 52, 56-
    57, 180
Antirrhinum majus, 92, 157
Aphelenchoides, 884, 890
A. besserji, 881
A. fragariae, 886
A. ritzemabosi, 881, 886-888

Subject Index